To Mother
With Love
Your Son
Al

DEVELOPMENTS IN MECHANICS

volume 6

PROCEEDINGS OF THE
12TH MIDWESTERN
MECHANICS CONFERENCE

A MIDWESTERN MECHANICS CONFERENCE PUBLICATION

DEVELOPMENTS IN MECHANICS

volume 6

PROCEEDINGS OF THE 12TH MIDWESTERN MECHANICS CONFERENCE HELD AT THE UNIVERSITY OF NOTRE DAME
AUGUST 16-18, 1971

EDITORIAL COMMITTEE

Conference Co-Chairmen

L. H. N. Lee
A. A. Szewczyk

Chairman, Papers Committee

K. T. Yang

Faculty Members, University of Notre Dame

T. Ariman
R. Betchov
R. M. Brach
N. C. Huang
C. K. K. Mak
V. W. Nee
S. T. McComas
N. D. Sylvester

UNIVERSITY OF NOTRE DAME PRESS
NOTRE DAME, INDIANA
1971

Printed in the U.S.A.

Library of Congress Catalog Card Number: 61-17719

Board of Directors

Midwestern Mechanics Conference, 1969-1971

Professor L. H. N. Lee, University of Notre Dame, Chairman
Professor A. A. Szewczyk, University of Notre Dame, Secretary
Professor H. J. Weiss, Iowa State University, Treasurer
Professor J. I. Abrams, University of Pittsburgh
Professor T. C. Woo, University of Pittsburgh
Professor J. E. Cermak, Colorado State University
Professor D. F. Young, Iowa State University

Past Midwestern Mechanics Conferences

MIDWEST CONFERENCES ON FLUID MECHANICS

No.	Year	Place	Publisher
1	1950	University of Illinois	Edwards Brothers Ann Arbor, Michigan
2	1951	Ohio State University	Bulletin 149 Engineering Experiment Station Ohio State University
3	1953	University of Minnesota	University of Minnesota Institute of Technology
4	1955	Purdue University	Purdue Engineering Experiment Station Res. Series 128
5	1957	University of Michigan	University of Michigan Press
6	1959	University of Texas	Engineering Institutes Extension Division University of Texas

MIDWESTERN CONFERENCES ON SOLID MECHANICS

1	1953	University of Illinois	Engineering Experiment Station
2	1955	Purdue University	Purdue Engineering Experiment Station Res. Series 129
3	1957	University of Michigan	University of Michigan Press
4	1959	University of Texas	Engineering Institutes University of Texas

MIDWESTERN MECHANICS CONFERENCES

7	1961	Michigan State University	1*	Plenum Press
8	1963	Case Institute of Technology	2*	Pergamon Press
9	1965	University of Wisconsin	3*	John Wiley & Sons
10	1967	Colorado State University	4*	Johnson Publishing Co.
11	1969	Iowa State University	5*	Iowa State University Press
12	1971	University of Notre Dame	6*	University of Notre Dame Press

*Developments in Mechanics, Volume No.

Reviewers

H.N. Abramson, D. Amas, W.F. Ames, L.W. Anderson, A.H.S. Ang, R. Aprahamian,. S.T. Ariaratnam, O.A. Arnas, A. Askar, H. Atassi, B. Avitzur, R.S. Ayre, C.D. Babcock, R.A. Bajura, C.S. Barton, R. Beckett, I.E. Beckwith, B. Bernstein, C.W. Bert, P.J. Blatz, H.H. Bleich, G.W. Bluman, D.B. Bogy, B.A. Boley, A.P. Boresi, W.A. Bradley, H. Brenner, F.P. Bretherton, B. Budiansky, D.R. Carver, T.K. Caughey, T. Cebeci, C.M. Chang, T.S. Chang, B.T. Chao, C.C. Chao, W.F. Chen, W.T. Chen, Y. Chen, S. Cheng, S.L. Chou, T.A. Clare, R.J. Clifton, R.W. Clough, J.D. Cole, W.J. Cook, S.C. Cowin, G.R. Cowper, W.O. Criminale, Jr., J.W. Dally, C. Dalton, R.T. Danes, I.M. Daniel, D.A. Danielson, J.M. Doyle, J.W. Deardorff, K.L. DeVries, R.J. Dohrmann, T.J. Dolan, S.B. Dong, E.H. Dowell, D.C. Drucker, J. Dunders, A.H. Emery, F. Erdogan, S. Eskinazi, F. Essenberg, J.R. Fitch, J.E. Fitzgerald, E.S. Folias, T.G. Foridis, V.G. Forsnes, H.O. Foster, R.L. Fox, J.C. Free, Y.C. Fung, I.S. Gartshore, J.M. Gere, J.E. Goldberg, J.R. Goodman, W.G. Gottenburg, R.A. Grot, D.T. Greenwood, M.E. Gurtin, N.B. Haaser, F.R. Hama, G.C. Hart, Z. Hashin, W.D. Hayes, G.A. Hegemier, G. Herrmann, J.C. Hill, E.A. Hirst, P.G. Hodge, Jr., D. Hoffman, E.J. Houg, Jr., T.C. Huang, T. Hughes, M.A. Hussain, C.T. Hsu, C.L. Hwang, E.F. Infante, W.D. Iwan, W. Jaunzemis, P.C. Jennings, F. John, M.W. Johnson, Jr., A. Kalnis, H. Kalsky, T.R. Kane, D. Karnopp, S. Kelsey, J.B. Keller, J. Kendall, M.J. Kenig, P.G. Kirmser, W.G. Knauss, J.K. Knowles, C. Kafadar, S.L. Koh, L. Kovasznay, R.H. Kraishnir, R.H. Lance, L. Landweher, H.L. Langhaar, J.P. LaSalle, A.W. Leissa, C.E. Leith, S.T. Li, G. Lianis, J.H. Lienhard, T.H. Lin, Y.K. Lin, N.C. Lind, J.L. Lumley, J.J. Marley, M.V. Markovin, C.W. Martin, J.B. Martin, E.F. Masur, G. Mattingly, W.C. Meecham, L. Meirovitch, R.N. Meroney, I.K. McIvor, J. Mikolwitz, R.E. Miller, M.V. Morkovin, J.B. Morton, F. Moses, T.J. Mueller, T. Mura, J.D. Murphy, P.M. Naghdi, J.F. Nash, W.A. Nash, S. Nemat-Nasser, F. Norwood, J.L. Nowinski, E.E. O'Brien, S.A. Orszag, R.H. Page, H.P. Pao, Y.C. Pao, Y.H. Pao, R. Parnes, B. Paul, H. Pih, A.C. Pipkin, E.A. Pipperger, T.H.H. Pian, A. Phillips, R.H. Plaut, R. Pletcher, P. Popov, S. Prager, A.G. Ratz, W.H. Reid, H. Reisman, E. Reissner, W.C. Reynolds, J.R. Rice, R.S. Rivlen, J.R. Robertson, L.C. Rogers, A.D. Russell, E.C. Russow, J.L. Sackman, A.E. Salama, H.J. Salane, J.L. Sanders, Jr., T. Sarpkaya, A. Schlack, Jr., L.A. Schmidt, Jr., W.C. Schnobrich, E.I. Suhubi, L.A. Scipio, E.E. Sechler, C.J. Scott, G.P. Sendeckyj, R.T. Shield, J.G. Simmonds, E. Skudrzyk, V. Smith, Jr., P.R. Smith, A.I. Solar, E.M. Sparrow, G.S. Springer, J.A. Stricklin, S.R. Swanson, J.L. Swedlow, A.L. Sweet, W.B. Swim, P.S. Symonds, R.I. Tanner, L.N. Tao, C.E. Taylor, T.C.T. Ting, D. Trifan, R.A. Toupin, M. VanDyke, F. Verhoff, F. Wan, W.E. Warren, R.I. Weiner, G. Wempner, R.K. Wen, O.E. Widera, J.F. Wilby, M.W. Wilcox, W.W. Willmarth, E.N. Wilson, W.O. Winer, W.J. Worley, C.E. Work, T.H. Wu, R. Wuki, W.H. Yang, J.T.P. Yao

Contents

Foreword

Developments in Mechanics, Volume 6, represents the Proceedings of the Twelfth Midwestern Mechanics Conference held at the University of Notre Dame, Notre Dame, Indiana, August 16-18, 1971. Publication of the proceedings is tangible evidence of the new and significant research contributions being made by the workers in all areas of applied mechanics and related fields.

Over one hundred and forty significant papers were submitted for the conference. The Editorial Committee worked toward the traditional goal of excellence in its final selection of papers as well as endeavoring to present a balanced program in a number of different areas of mechanics. Hence, many difficult decisions had to be made along the way to the final selection of papers to be presented at the Conference. We wish to express sincere thanks to all of the authors who submitted manuscripts, whether finally accepted or not.

It is also our pleasure to acknowledge the role of the invited lecturers whose contributions added greatly to the success of the Conference.

It has become a tradition of the Midwestern Mechanics Conference to have the publication of all papers available at the meeting. Without the help and cooperation of the authors, reviewers, editorial committee, publisher and secretaries it would have been an impossible task. We owe our thanks to the authors for their cooperation in preparing the final manuscripts. To our over 250 reviewers who accepted their task so generously and helped us make many difficult decisions we express special thanks. To Dr. K. T. Yang, Papers Committee Chairman, and the members of the Editorial Committee who labored so unselfishly we express our gratitude. We also acknowledge the efforts of Mrs. G. Curtis, Mrs. E. Friedman, Mrs. E. Levee and Miss K. Strantz and the University of Notre Dame Press for their efforts and counseling in the preparation of the final manuscripts.

Appreciation must also be given to the administrative officers of the University of Notre Dame, Dean Joseph C. Hogan and Associate Dean Edward W. Jerger, for their encouragement. To the staff of the Continuing Education Center for their cooperation and handling of the Conference, we express our appreciation.

Thanks should be given to a number of technical societies for their endorsement of and cooperation with the Conference: American Society of Civil Engineers, Engineering Mechanics Division; American Society of Mechanical Engineers, Applied Mechanics Division; Society of Engineering Science; Society of Experimental Stress Analysis; Society for Industrial and Applied Mathematics.

We also wish to acknowledge the financial support from the National Science Foundation and Air Force Office of Scientific Research.

L. H. N. Lee
A. A. Szewczyk
Co-Chairmen
UNIVERSITY OF NOTRE DAME
NOTRE DAME, INDIANA

PART I
GENERAL LECTURES

Developments in Mechanics, Vol. 6. Proceedings of the 12th Midwestern Mechanics Conference:

1

Metal Deformation Processing from the Viewpoint of Solid Mechanics

DANIEL C. DRUCKER
UNIVERSITY OF ILLINOIS, URBANA

ABSTRACT

Deformation processing design requires the selection of both the material and the deformation process to produce a given shape of part with given properties in specified regions. Although solid mechanics has much to offer the designer, it must be admitted that at present the solution is much more of an art than a science for the common alloys and is little of either for the esoteric metals. Some of the reasons for both the difficulties and the successes are explored.

The appropriateness of the perfectly plastic idealization is examined and reference is made to theorems for viscous behavior and to earlier work on metal cutting and Ekstein's paradox of the 1940s. The explanation of the paradox is repeated and the argument for perfectly plastic behavior restated in both mathematical and physical terms. Only rarely, however, even in high rate deformation processing would the conclusion that perfect plasticity represents physical reality for continuous metal cutting be transferable to metal working.

INTRODUCTION

The paper I should like to write would be specific and incisive in its use of fundamental concepts, addressed directly to the problems of practice. It would start with the need to produce a given part with given properties, such as ductility and strength, at specified points and in specified directions. Then it would tell you explicitly how to select both the material and the deformation process: rolling, drawing, spinning, extrusion, forging, . . . , to achieve the clearly stated objectives at a price you could afford to pay.

For reasons which will become more and more obvious as I proceed, and which have nothing to do with shortage of time and space, I shall disappoint you. Basic knowledge is so far from complete, that metal deformation processing is mainly an art, or perhaps black magic, at which some do well and some do poorly when asked to produce new shapes of new material. So we must retreat reluctantly from the grand goal of a combined systems approach to the design of the material and the design of the process with the aid of mechanics of solids. We ask instead what our understanding of mechanics can do to help us achieve an understanding of metal deformation processing.

Deforming metal, rather than removing it to achieve a given shape, enables us to do more than just save material (an objective which is not to be confused with saving money). The appropriate process can produce desired surface and subsurface properties through microstructural alignment of strong elements and of defects, and through the residual stress pattern induced. Unfortunately, an inappropriate process can produce severe additional flaws, and result in a part without adequate integrity prior to the application of load. Not nearly enough is known about how to characterize and how to handle even the moderately difficult to deform metals and alloys. Some conclusions after five years of study and discussion by an Ad Hoc Committee on Metalworking Processes and Equipment of the Materials Advisory Board, chaired by Professor W. Rostoker, are summarized in the Final Report. [1]

EXAMPLES OF DIFFICULTIES

The variety of problems encountered is fascinating. Rolling, Fig. 1 (from Alexander and Brewer [2]), can result in a long or loose edge, a cracked edge, a tight edge or loose center, or a herringbone pattern of inclined buckles instead of the desired flat shape. There is no need to call upon sophisticated theory or unusual production control to identify the cause and the method of correction for the loose edge or loose center condition. Longitudinal extension at the edge is increased or decreased by increasing or decreasing the pull-down force at the edge. The greater the force the looser the edge, the less the force the looser the center. Herringbone is something else. It might be caused by a periodic variation of roll force but just how is not clear and the corrective procedure is by no means obvious. It is, I believe, still one of the unsolved problems of rolling practice.

Wire drawing or extrusion, Fig. 1, can avoid surface or interior cracking and produce a surface of high quality but doesn't always. Designing the die of a complex forging to fill without folds, tears, or cracks, Fig. 2 (from Dieter [3]), is not a simple exercise in the absence of almost identical earlier experience.

IDEALIZATION AS VISCOUS OR PLASTIC

As a consequence of the myriad of difficulties, almost all books, papers, and reports start with assumptions which avoid the worst of the problems of the real world. They and I begin with the consideration of an infinitely ductile material to avoid temporarily the often over-riding limitations imposed by ductile or quasi-brittle fracture. Fortunately for the research worker in solid mechanics there are many fundamental questions of great interest quite apart from fracture. Also the picture of continuous ductile flow permits some estimate of the probability of fracture. Insight gained can help the practitioner to avoid fracture and other undesirable results.

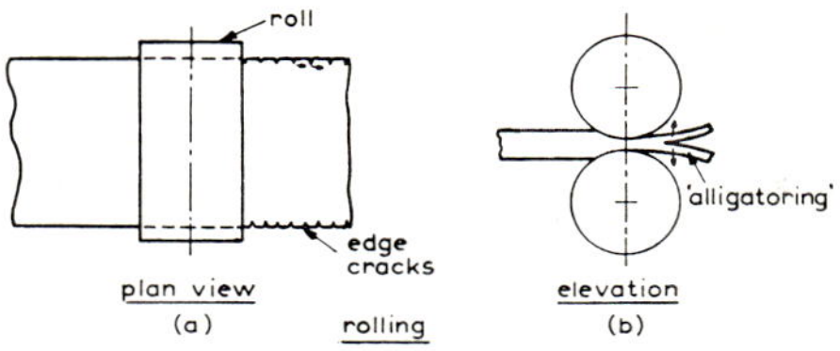

Defects in rolling:
(a) 'Edge cracking'
(b) 'Alligatoring'

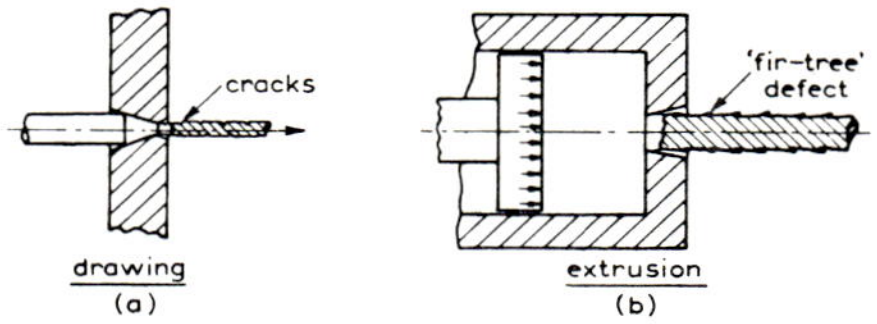

Surface defects in drawing and extrusion

Fig. 1 p 428 Alexander & Brewer [2]

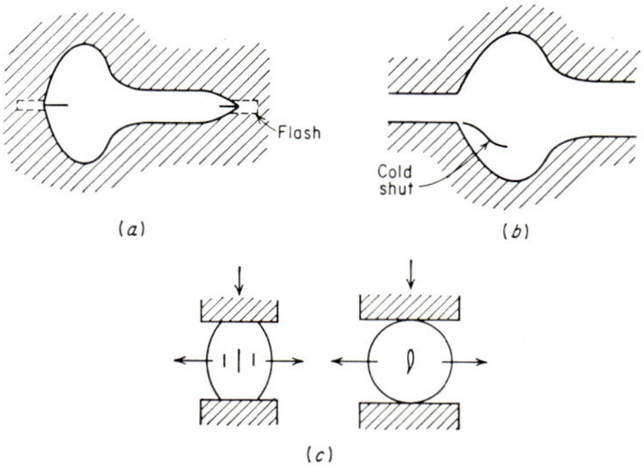

Typical forging defects. (*a*) Cracking at the flash; (*b*) cold shut or fold; (*c*) internal cracking due to secondary tensile stresses.

Fig. 2 From Dieter [3]

Hot and cold deformation processing then are treated together, although the hot metal is dominantly viscous while the cold is primarily elastic-plastic. Fig. 3 shows the sense of the close equivalence for the relevant ranges of interest. Certainly a time-dependent viscous material is a very different idealization from a time-independent elastic-plastic idealization. When the inelastic strain grows but the strain rate drops to a low value, the viscous material is at close to zero maximum shear stress while the plastic material is at a high shear stress. However, in most deformation processing, interest tends to be focussed on regions in which shear strain rates are high and, on some scale of dimension, span no more than a decade or two. The maximum shear stress τ_o^v at a typical shear strain rate $\dot{\gamma}_o$ then is reasonably close to the value of the maximum shear stress in most of the region of importance. Similarly, for a work-hardening material, the maximum shear stress τ_o^p , at a typcial high shear strain γ_o is representative of the shear stress at most points in the deforming regions. When force and stress calculations are made, both the strongly non-linear viscous behavior and the work-hardening behavior can be idealized to a fair first approximation as perfectly plastic with an appropriate yield stress τ_o . Such simplicity is gained at the sacrifice of much of the determinacy or uniqueness of strain and displacement.

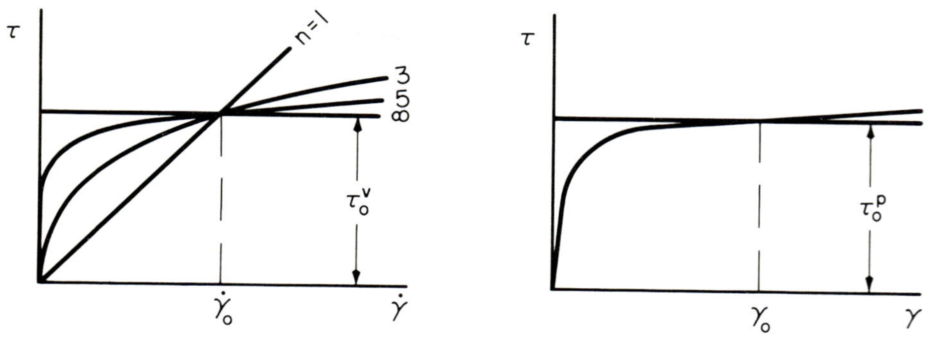

Fig. 3 Highly Non-linear Viscous Behavior Like Perfectly Plastic

AN ASIDE ON METAL CUTTING AND EKSTEIN'S PARADOX - THE VALIDITY OF THE PERFECTLY PLASTIC IDEALIZATION AT HIGH RATES OF STRAIN

Almost all metals do exhibit a gradual work-hardening in quasi-static simple shear from an initial yield stress to a considerably higher value at large strain. Annealed metals in particular may harden enormously with strain. Yet it turns out that at the high rates of strain in continuous chip, orthogonal, metal cutting, (the standard case examined by Piispannen [4] and by Ernst and Merchant [5], the shear stress on the shear plane is at or above the highest values observed in quasi-static testing; the perfectly plastic idealization is the real behavior to an excellent approximation [6].

The paradox was identified by H. Ekstein in a sharp and clear form in a 1945 Armour Research Foundation report to Shell Oil Company. As indicated by Fig. 4, the chip undergoes very large strain in a very short distance (close to zero on the scale of the drawing). A free-body sketch of the shear zone region and corresponding momentum balance leads to the rigorous result that $\tau_c - \tau_w$, the difference in average shear stress on planes parallel to the shear plane from the essentially undeformed workpiece to the tremendously hardened chip, is well below $2\rho v_w^2$ where ρ is the mass density of the material. At an extremely high cutting speed of 50 ft/sec, therefore, the average shear stress could increase by far less than 500 psi, an insignificant number in terms of the static work-hardening curve for structural metals and the very high shear stress on the shear plane in metal cutting.

$$F_c - F_w = \rho v_w b t \Delta v_F - R_T$$

or

$$\tau_c - \tau_w \approx 0 \quad (\rho v^2 \ll \tau_o)$$

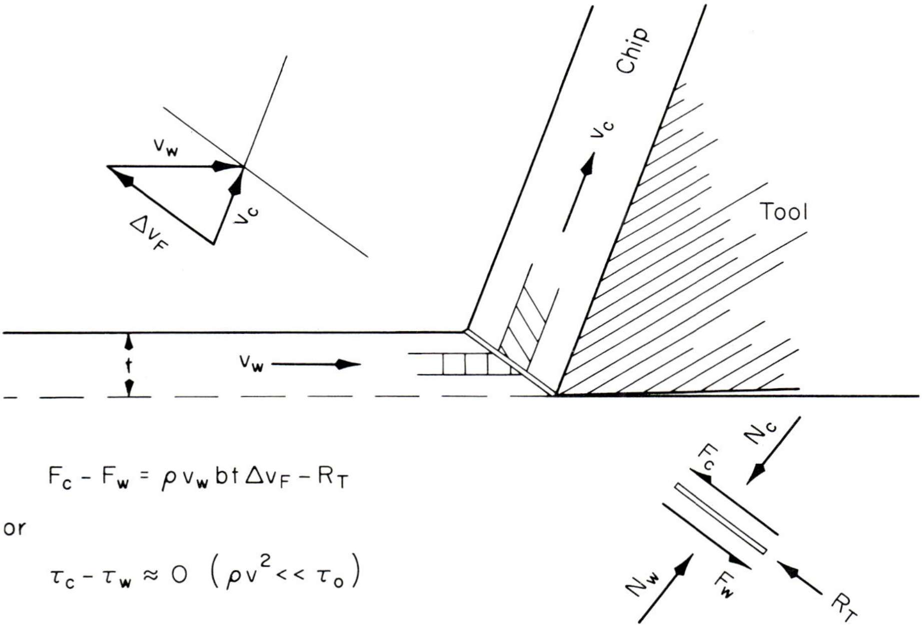

Fig. 4 Ekstein's Paradox (Enormous Shear Strain Occurs in the Traverse of a Thin Region of Constant Shear Force) Requires Perfect Plasticity

There is no escape from the conclusion that despite the enormous strains and corresponding hardening which takes place, the shear stress remains essentially constant throughout the zone of deformation. An alternate and fully equivalent statement is that the metal behaves as though it were perfectly plastic.

Dynamic or rate effects do provide the qualitative explanation for this paradox and do account as well for the elevation of the shear stress level far above the static values [6]. At the tremendous rates of strain (in excess of 10^5/sec), the process of generating and of moving dislocations in sufficient number over sufficient distances to produce shear strains of order unity throughout the volume of the chip is far more difficult than in a quasi-static deformation. Dislocations generated or initially present in the workpiece have no time to move enough to give visible plastic deformation before the metal enters the shear zone.

Metal cutting and metalworking are very different kinds of processes but some deformation processes do show local similarity to the pattern of deformation in orthogonal continuous chip metal cutting. However, in most instances the dimensions over which the deformation takes place are much larger, while the strains and strain rates are much smaller. Consequently a cam plastometer is a suitable instrument for the determination of flow strengths, and the metal's behavior is far closer to the quasi-static than to the high stress level, perfectly plastic, response in metal cutting. The argument for the use of perfect plasticity theory then rests more on convenience and mathematical approximation Fig. 3, than on physical grounds. One conclusion which does carry over, however, is that the inertia or momentum terms are almost always of negligible magnitude in comparision with the shear stresses associated with plastic deformation. Quasi-static solutions generally are satisfactory for the determination of stress fields.

PLASTIC LIMIT THEOREMS AND ALLIED EXTREMUM THEOREMS

In the sense described earlier it is permissible to employ either the limit theorems of perfect plasticity or the analogous extremum theorems of non-linear viscosity as one prefers in metalworking calculations.

The plastic limit theorems of Drucker, Greenberg, and Prager [7] state:

Lower Bound Theorem - If an equilibrium distribution of stress σ_{ij}^E exists which satisfies the boundary conditions on surface traction (T_i on A_t) and is everywhere below yield $F(\sigma_{ij}^E) < k_0^2$ the body will not collapse (continue to deform plastically at the given loads).

Upper Bound Theorem - The body will collapse (or the deformation process will take place) if there is any compatible pattern of plastic deformation $\dot{\varepsilon}_{ij}^p$, $\dot{u}_i^p$ ($\dot{u}_i^p = o$ on A_u) for which the rate at which the external forces do work equals or exceeds the rate of internal dissipation,

$$\int_{A_T} T_i \dot{u}_i^p \, dA + \int_V F_i \dot{u}_i^p \, dV \geq \int_V D(\dot{\varepsilon}_{ij}^p) \, dV = \int \sigma_{ij}^p \dot{\varepsilon}_{ij}^p \, dV$$

The corresponding upper and lower bounds for a non-linear viscous body [8], [9] are not quite as convenient but are simple enough for a power-term material, $\dot{\varepsilon}_{ij} = \partial\varphi / \partial\sigma_{ij}$, φ homogeneous of degree n in stress. As n approaches infinity, Fig. 3, the non-linear viscous theorems become indistinguishable from the limit theorems.

Either set of theorems, just like the extremum theorems of elasticity (minimum potential energy and minimum complementary energy) are global theorems in essence. They give total energy or energy dissipation, or deflection under a load, or a force applied to a collapsing or flowing body. The theorems are powerful and useful in one, two and three dimensional problems precisely because they are not much influenced by the crudity of the chosen local strain-rate pattern or by a moderate lack of homogeneity (say ± 20% from average) of material yield strength. Consequently, they permit remarkably accurate calculation of the forces needed or present in a deformation process. Consequently also, they do not permit close calculation of the strain-rate distribution in the process, the resulting deviations of final strains from the average and the residual stress pattern.

FRICTION AND LUBRICATION

Frictional boundary conditions are likely to be troublesome in all branches of mechanics of solids (elasticity, viscoelasticity, plasticity, . . .) whenever the forces taken by the frictional supporting surface cannot be computed from rigid body statics or dynamics alone. Coulomb friction leads to irreversibility and path dependence in an otherwise elastic and therefore path independent problem. The difficulty is typified by a pair of supports, Fig. 5, which can exert an undetermined set of distributed normal forces N, N^1 and a wide variety of associated frictional forces over and above the force distribution for any assumed idealized situation of zero initial force and zero initial clearance before load is applied. The distribution of surface tractions is not determined uniquely by the current value of the laod P. The situation is somewhat analogous to the indeterminacy produced by the possibility of initial stress in an conventional boundary value problem of elasticity or plasticity, but is made worse by the fact that, for Coulomb friction, an inequality $\tau \leq \mu \sigma_N$ governs instead of an equality. Correspondingly, although initial stresses are wiped out in the deforming regions of a perfectly plastic body, any frictional indeterminacy most often will remain. Except for the special conditions of zero friction or complete attachment, the usual extremum theorems of elasticity and the limit theorems of plasticity are not generally applicable to frictional systems. Theorems which are applicable may lead to bounds which are quite far apart [10].

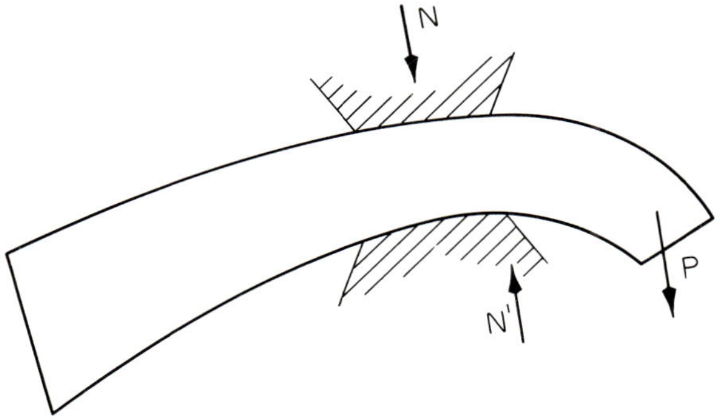

Fig. 5 Coulomb Friction Indeterminacy

As a consequence, the effect of adding or changing lubrication in drawing, extrusion, or rolling is not precisely predictable at best. This carries no implication that appreciable friction is undesirable. It is in fact essential for the success of some of the metalworking techniques.

SOLUTIONS FOR LOCAL REGIONS AND THEIR RELEVANCE

The serious limitation imposed by the frictional uncertainty and the restrictive global nature of the information obtained from the plastic limit theorems or other extremum theorems can be removed in part for two dimensional problems by obtaining more complete solutions to the problems based upon the appropriate idealizations of the material. Plane strain, slip line, fields are employed which take frictional slipping boundary conditions for the deforming region into account, as in Fig. (6), [11]. Completeness as well as uniqueness of solution, however, may be suspect in more complex situations with incomplete knowledge of boundary tractions. More important, however, the closeness of the idealization to the real material becomes far more crucial than for the computation of overall force.

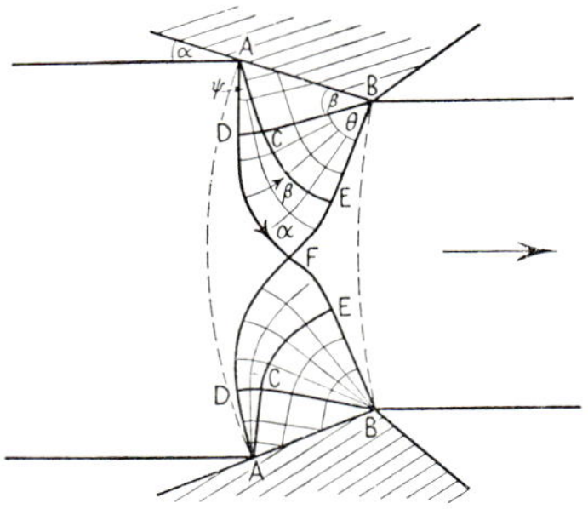

Slip-line field and plastic region for drawing through a rough wedge-shaped die giving a moderate reduction in thickness.

Fig. 6 Hill [11] p 173

Fig. 7 gives an over-simplified illustration of how sensitive the displacement field for the perfectly plastic idealization can be to a moderate inhomogeneity, whether pre-existing or resulting from deformation. Also, localization of strain along slip surfaces is typical in both the approximate upper bound computations of plastic limit theory and the solutions for plane strain problems. As in metal cutting, Fig. 4, when perfectly plastic material flows through these surfaces of discontinuity the strain is discontinuous but the displacements are continuous. When the flow is purely tangential, the displacement itself can be discontinuous. As discussed earlier, neither viscous behavior nor work-hardening would permit either the strong or the weaker form of discontinuity, because the mass-acceleration forces are too low to permit large changes of shear stress over short distances. Nevertheless, a discontinuity in strain or strain rate may be a fair approximation to the real answer.

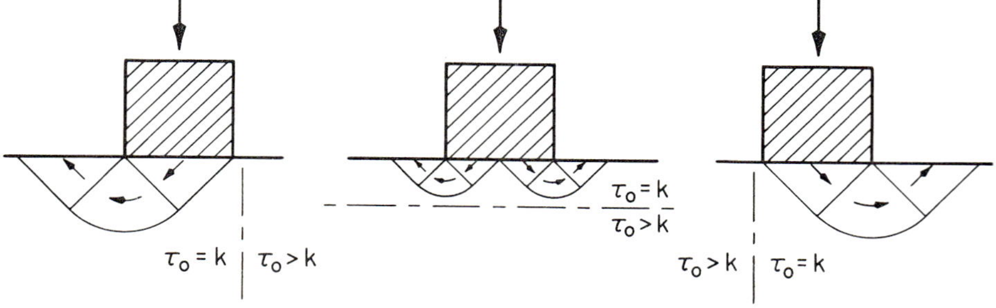

Fig. 7 A Variety of Modes of Deformation Frictionless Punch Inhomogeneous Half Space

Our choice of the technique of solution tends to bias our description of the problem. We speak, for example, of the undesirability of redundant work in a deformation process and look at the efficiency of the process. Die designs for an ideal process have been given by Richmond [12] and by Hill. Yet we do not really care about the extra work nearly so much as about the large local strains often associated with inhomogeneity of strain in the final product. The two are not unrelated because strain inhomogeneity in rolling, drawing, extruding, etc. implies redundant work, while zero redundant work requires homogeneity of strain in the process. However, minimization of the maximum strain would seem a more desirable goal than minimization of work when strain homogeneity is not possible. Of course, a meaningful evaluation of the situation demands far more than a look at strain alone. The state of stress under which the strain takes place is of critical importance. Inelastic strains in metals are shear strains, so that the presence of hydrostatic tension in regions of large strain is especially troublesome. Yet the maximum tensile stress itself also is significant because the cracking of inclusions and the opening of the cracks depends far more upon the uniaxial value than the mean or triaxial. All of these vitally important fine details are not likely to be given well by any of the drastic idealizations of the material which we use in order to obtain solutions and which in fact do represent reasonably well the gross common behavior of a wide variety of metals and alloys. Experiments or tests on model materials are subject to the same blurring of the details of the deformation.

There is, Fig. 8, a remarkable similarity in the gross features of the deformation patterns for the extrusion of water, oil, non-linear viscous fluids, hot metals, and ductile cold metals and alloys with and without lubrication. Extrusion under high hydrostatic pressure of a material which exhibits low ductility under normal conditions also leads to the same patterns. Pictures provided by Thomsen, Kobayashi, Yang [13] and their colleagues are very informative as are those given by a number of others including Johnson, Kudo, Mellor [14], [15] and their associates. Closer examination of the details of the deformation of a square grid shows up cusps or their absence, strong boundary layers of high shear, and opening of voids and cracks on the microscale.

Temperature increases due to the conversion of plastic and frictional work to heat can be significantly large and lead to strong localization of the flow field (adiabatic shear). Many of the macroscopic and microscopic properties of the material can have an important influence on the local picture, which in turn can drastically alter the properties.

The suggestion here is that model tests or theory can provide the appropriate input and boundary conditions to obtain the relevant, highly local, pieces of the solution in sufficient detail. Just how this can be done is an open question at present, extremely difficult but well worth exploring.

Yet, even in the absence of local solutions to provide this detailed quantitative information, our assumption of unlimited ductility does permit us to obtain insight into the problems of fracture control or avoidance during and after the processing. With reasonable ductility, control of the stress and strain pattern through thoughtful die design can avoid many of the serious problems. A rough indication of local stress and strain levels in a region permits some estimate of the likelihood of the cracking of inclusions, opening of voids, and coalescing of voids to produce fracture. Studies by Gurland and Plateau [16], Coffin and Rogers [17], and McClintock [18] are highly instructive. Brittle behavior in tension following large precompression strain also can be avoided or at least understood [19].

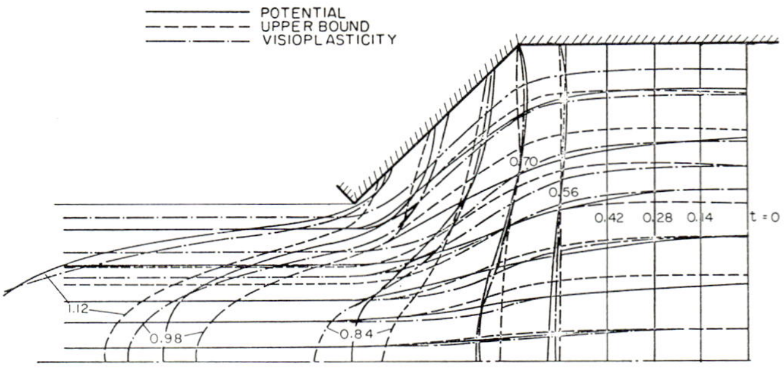

Comparison of the constant time contours

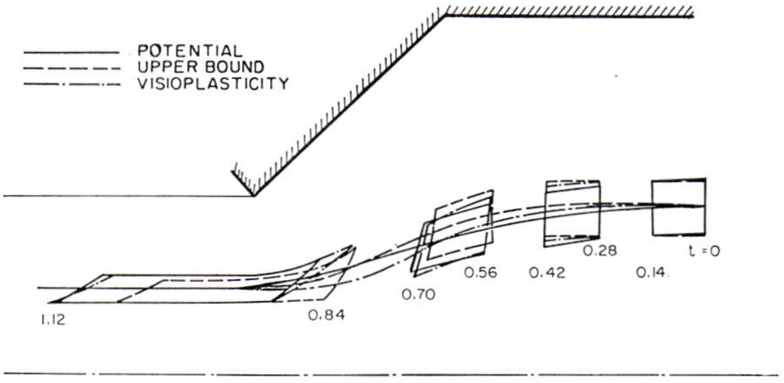

Comparison of the grid distortion

Fig. 8 Shabaik & Thomsen [24]

A more incisive understanding does require the development of both analytical and experimental techniques for the determination of the local stress and strain fields and their effect in a manner adequately sensitive to the real properties of the material undergoing processing. Along with this much more incisive look at the details from the viewpoint of mechanics of solids must go the experimental study of the ability of metals and alloys to undergo the required history of stress and strain and emerge with the properties desired in the part which is produced. Neither aspect is well-understood at present. Eventually, of course, the mechanics and the materials study must be combined to permit design of the process to produce the end product needed.

Superplasticity has been seized upon by many as the magic key to fracture-free deformation. However, superplasticity at any given temperature, as Backofen [20] has pointed out so clearly, is a manifestation of a relatively low stress range of linear viscous behavior or its close approximation and the consequent absence of necking in simple tension. The linearity of viscous behavior also permits the blowing of bubbles and the forming of intricate shapes without local thinning. In drawing, extrusion, and forging, the linearly viscous material in contrast to the non-linear viscous or the plastic will tend to have smoother strain gradients. However, as demonstrated in Fig. 8, the overall pattern for the constrained deformations of drawing and extruding is remarkably independent of this important characterization of the material and is less crucial to the success of the formation of the part than in tensile deformation. Also the strain rates at which superplastic behavior is observed tend to be rather too low for practical deformation processing.

SOME PROPERTIES OF HIGHLY DEFORMED METALS ON THE MICROSCALE AND MACROSCALE

It is well known that sheared edges or punched holes in structural steel plate are superb initiators of brittle fracture. The damaged material in the neighborhood of the sheared zone must be removed in any region of appreciable stress, or the properties of the steel must be restored through some appropriate thermomechanical treatment. Such traumatic straining of any metal or alloy is to be avoided whenever possible because of the microstructural damage produced, which along with severe work-hardening drastically reduces the effective ductility. Yet, the narrow shear zones characteristic of drawing, extruding, shear spinning, and forging do bear a remarkable resemblance to the region of a sheared edge. Cracking of brittle inclusions, decoherence of inclusions and matrix, opening of voids and cracks pose similar dangers in these processes. The one great difference, the essence of the success achieved, lies in the high compressive stresses induced by the dies or tools in the regions of high strain increments. A material, perfectly homogeneous on the microscale would have no tendency to separate under shear if the mean or hydrostatic compressive stress exceeded the yield stress in shear. The presence of spherical inclusions would increase the hydrostatic pressure needed by a small amount only, but polycrystalline metals with needle-like or plate-like inclusions can require much more. Difficult to deform metals need a high superposed pressure over and above that induced in the standard process because they have a very high yield stress in shear or they have a messy microstructure or both.

If a metal or alloy is ductile on the microscale, it will be ductile on the macroscale if there is only moderate localization of plastic deformation over distances comparable with the relevant microstructural dimensions. Both the localization in the solution at the continuum level which ignores the microstructure and the localization on the microscopic level are important, with the conventional mechanics solution providing the controlling environment for the microscale. Void coalescence [21] in a thin layer transverse to applied tensile stress results in a separation fracture of apparent negligible ductility, despite an infinitely ductile matrix. Therefore, the possibility of pre-existing cracks and voids must be allowed for in the thought process, along with those cracks and voids which form as a result of the deformation. Void coalescence in a fully ductile matrix can lead to separation or ductile fracture under far less severe conditions and at a much earlier stage when inclusions are not bonded to the matrix than when the bond is perfect. No macroscopic tensile stress need be applied across the surface of separation, however. Voids which are closely spaced compared with their diameter can be initiated and join up in a zone of simple shear which is locally distorted by inclusions and other inhomogeneities. The entire thermomechanical treatment prior to and during the metal deformation process must be considered very carefully if the bond strength between inclusion and matrix is not to be taken as close to zero.

Lateral compression orients defects as well as strengthening elements nicely for subsequent longitudinal tension. Rolled sheets or plates, drawn wires, extruded bars and tubes, and the surface regions of forgings properly made, do have this desirable fibrous or sheetlike texture. Their tensile properties in the perpendicular direction, however, are likely to be poor. Should this through-the-thickness tensile strength or ductility be needed, as it may for certain forms of attachments, the result may be disappointing or worse. Moderately large compressive pre-strain of the order of 10 or 20% is not especially deleterious in this respect but much larger strains perpendicular to the surface are common in metalworking. Alligatoring in rolling, Fig. 1, provides a good example of layer separation. As is known from the studies with Mylonas and our co-workers, for common structural steels, compressive prestrains of the order of 50 or 60% at room temperature and half that amount at the blue brittle temperature will lead to an extremely low ductility fracture in subsequent tension. When there is a macroscopic strain concentration, a quasi-brittle fracture will occur well down in the nominal elastic range. Spall resistance for shock waves traversing the thickness is likely to be low.

THE FREE SURFACE PROBLEM AND INCREMENTAL FORGING

The indeterminacy induced by frictional effects at boundaries occurs in forging as well as in drawing and extrusion. In addition the freedom of the metal to choose its deformed shape may conflict with the desire of the designer and manufacturer to produce a given shape. As mentioned previously and illustrated somewhat by Fig. 7, free boundary problems are always troublesome in mechanics. They are especially so far the perfectly plastic idealization with its inherent lack of uniqueness of strain and displacement and the consequent great sensitivity to small variations in properties from point to point of the material.

Incremental forging appears to be a promising technique to overcome the problems of properly filling a closed die forging and eliminating the difficult design and expensive construction of conventional dies for open or closed die forging. Incremental forging was recommend in the MAB report [1] as the analog of machining. Instead of being removed, metal would be pushed from region to region with a simple sturdy tool like a punch or roll. A primitive version of this technique is employed in the first stage of production which transforms a steel ingot into a rectangular, square, or circular billet. Experience and experimentation are relied upon to determine the transverse and longitudinal changes in dimension due to lateral pressing or forging. The elementary and typical free surface mechanics problem of a rectangular bar pressed between two equal and opposite platens on a portion of two of its surfaces with rather indeterminate friction conditions for the platen faces is far from simple. Wistreich at BISRA [22] developed useable empirical relations only after an extensive series of tests.

Despite the apparent analog to the process of metal removal by machining, a single tool or a single pair of tools is unlikely to have the capability of creating a given desired shape of three-dimensional character. The freedom of the metal to choose its path of deformation may well be incompatible in general with an arbitrarily chosen shape. In mathematical terms, no solution may exist. Empirically, Schey and his co-workers [23] have abandoned the attempt and gone over to a double pair of tools for a practical example of reasonable but not extreme complexity. It is not obvious in principle that this degree of constraint will be sufficient for a more general situation. Here then is another of the yet to be answered problems of mechanics of solids as applied to metal deformation processing. The solution for the degree of constraint needed to produce any desired change in shape obviously requires as a first step the ability to predict with moderate accuracy the change in shape of a three-dimensional body produced by a given simple tool or set of simple tools.

CONCLUDING REMARKS

The application of the mechanics of solids to metal deformation processing is direct and valuable in the prediction of global quantities such as energy dissipation and the magnitude of the forces required. Less direct but still useful is the insight given on preventive measures to be taken to avoid fracture in the process or insufficient ductility in the finished part. Realistic local solutions based on idealized overall flow patterns are needed, both with and without consideration of the microstructure, to make further significant progress. Free surface problems basic to forging must be solved if incremental forging is to become a practical art for complex shapes and if conventional forging is to move from an esoteric art to an applied science and standard engineering technique. The friction problem, so troublesome in all of mechanics of solids, also needs considerable attention. Consequently, despite the great progress which has been made and the many books and papers which have appeared in the last two decades, much remains to be done before a given process with a given and well-defined material can be analyzed correctly in adequate detail. The real problem is to minimize trial and error in the choice of a new material and the design of a process to produce a new part with specified properties. The achievement of this goal will remain a worthwhile challenge to analysts and experimentalists alike for a long time to come.

REFERENCES

[1] Final Report by the Materials Advisory Board Committee on Metalworking Processes and Equipment to the National Research Council, National Academy of Sciences/National Academy of Engineering, Washington, D. C., MAB-206-M(7), September 1968.

[2] J. M. Alexander and R. C. Brewer: Manufacturing Properties of Materials, van Nostrand, p. 428 (1963)

[3] G. E. Dieter, Jr.: Mechanical Metallurgy. McGraw Hill, p. 485 (1961)

[4] V. Piispannen: Lastunmuodostumisen Teoriaa. Teknillinen Aikakauslehti, v. 27, n. 9, pp. 315-322. (1937)

[5] H. Ernst and M. E. Merchant: Chip Formation, Friction, and Finish. Cincinnati Milling Machine Company, 1940, also Trans. ASM v 29, pp. 299-328 (1941)

[6] D. C. Drucker: An Analysis of the Mechanics of Metal Cutting. J. of Applied Physics v. 20, pp. 1013-1021 (1949)

[7] D. C. Drucker, H. J. Greenberg, and W. Prager: Extended Limit Design Theorems for Continuous Media. Quarterly of Applied Mathematics, v. 9, pp. 381-389 (Jan. 1952)

[8] N. J. Hoff: Approximate Analysis of Structures in the Presence of Moderately Large Creep Deformations. Quarterly of Applied Mathematics, v. 12, pp. 49-53 (1954)

[9] C. R. Calladine and D. C. Drucker: Nesting Surfaces of Constant Rate of Energy Dissipation in Creep. Quarterly of Applied Mathematics, v. 20, pp. 79-84 (1962)

[10] D. C. Drucker: Coulomb Friction, Plasticity, and Limit Loads. J. of Applied Mechanics, v. 21, Trans. ASME, v. 76, pp. 71-74 (1954)

[11] R. Hill: The Mathematical Theory of Plasticity. Oxford University Clarendon Press (1950)

[12] O. Richmond and M. L. Devenpeck: A Die Profile for Maximum Efficiency in Strip Drawing. Proceedings, Fourth U. S. National Congress of Applied Mechanics, v. 2, ASME, pp. 1053-1057 (1962)

[13] E. G. Thomsen, C. T. Yang, S. Kobayashi: Mechanics of Plastic Deformation in Metal Processing. Macmillan (1965)

[14] W. Johnson and H. Kudo: The Mechanics of Metal Extrusion. Manchester University Press (1962)

[15] W. Johnson and P. B. Mellor: Plasticity for Mechanical Engineers. D van Nostrand Co., (1962)

[16] J. Gurland and J. Plateau: The Mechanism of Ductile Rupture of Metals Containing Inclusions. Trans. Quarterly American Society for Metals, v. 56, pp. 442-454 (1963)

[17] L. F. Coffin, Jr. and H. C. Rogers: Influence of Pressure on the Structural Damage in Metal Forming Processes. ASM Trans. Quarterly, v. 60, pp. 672-686 (1967)

[18] F. A. McClintock: A Criterion for Ductile Fracture by the Growth of Holes. Journal of Applied Mechanics, ASME, v. 35, pp. 363-371 (1968)

[19] D. C. Drucker, C. Mylonas, and G. Lianis: Exhaustion of Ductility of E-Steel in Tension Following Compressive Prestrain. Welding Journal Research Supplement, v. 25, pp. 117s-120s (1960)

[20] W. A. Backofen, I. R. Turner, and D. H. Avery: Superplasticity in an Al-Zn Alloy. ASM Trans. Quarterly, v. 57, pp. 980-990 (1964)

[21] D. C. Drucker: The Continuum Theory of Plasticity on the Macroscale and Microscale. Marburg Lecture, Journal of Materials, ASTM, v. 1, pp. 873-910 (1966)

[22] J. G. Wistreich and A. Shutt: Theoretical Analysis of Bloom and Billet Forging. Journal Iron and Steel Institute, v. 193, pp. 163-176 (1959)

[23] P. H. Abramoqitz and J. A. Schey: Principles of Incremental Forging. Report No. 70-6 to Naval Air Systems Command from University of Illinois-Chicago Circle, June (1970)

[24] A. H. Shabaik and E. G. Thomsen: Comparison of Two Complete Solutions in an Axisymmetric Extrusion with Experiment. Journal of Engineering for Industry, Trans. ASME, Series B v. 91, pp. 543-548 (1969)

Developments in Mechanics, Vol. 6. Proceedings of the 12th Midwestern Mechanics Conference: 2

The Future of Applied Mechanics

HOWARD W. EMMONS
HARVARD UNIVERSITY

INTRODUCTION

The name Applied Mechanics suggests the study of Newton's Laws as related to the real world. Both phenomena of academic interest and the solution of practical problems are included, although the latter predominate as they should. Rigid body mechanics, the vibrating motions of flexible solids and the motion of liquids and gases are all included. Finally, we should note the various related questions which are legitimate inclusions; applied mathematics, thermodynamics, materials properties, heat transfer, etc.

Over the years many of the simpler problems in all these areas have been solved for most practical purposes and we have begun to consider more difficult questions. The buckling of complex structures, the deformation of plastic solids, the failure mechanics of solids, the flow of multicomponent fluids, the mechanics of composite materials, turbulent flow are the continuing problems which Applied Mechanics must help to solve. And yet if we look at a few issues of the applied mechanics journals, we still find many papers reporting on residual minor points of the time honored problems with relatively little effort put onto the larger problems which yet remain. In fact we need to reexamine our potential contribution to the engineering of the future if we are to continue to justify our salary. We have become so expert at and so enamored with taking a few laws, a few properties and some applied math and deriving often beautiful but complex solutions to special problems that we often forget to look at the real world and see which problems are now most important.

For example, consider a specialist in flat plate laminar boundary layer theory who has contributed much to our understanding of laminar boundary layer applications to civil, mechanical, and aeronautical engineering. There are many points of interest still to be clarified. These now fall mostly into two classes. The first is relatively unimportant minor points (as an alternate derivation of a well-known result or a solution with slightly altered boundary conditions), and the second are problems of great difficulty for which new inspiration or new techniques are required (as laminar flow separation). It is now time for such an expert to lend a hand with some of today's more urgent problems. All the old areas of Applied Mechanics similarly have their share of inflexible experts while what is needed at the present day is a reappraisal of the important problems and a reorientation of Applied Mechanics to their solution.

In most cases the urgent technical problems of engineering and the environment would greatly benefit from a major effort of the old methods of Applied Mechanics bringing in new laws of materials and new empirical facts as needed.

Like all practical problems, the newly recognized ones must be solved. Engineers will solve them one way or another and it is up to us in Applied Mechanics to so develop the basic understanding that rational solutions may be used to replace empirical ones when this is possible.

To illustrate what I mean, I am going to look at a part of the process of paper making. At every step of such an ancient process there are interesting and important questions which can be understood more completely only by applying the methods of applied mechanics first at a perhaps more empirical level than we currently prefer and then as our knowledge matures at an ever increasing rational level.

THE APPLIED MECHANICS OF PAPER MAKING

Paper is made by suspending cellulose fibers in water, pouring the mixture onto a screen, sucking off the water, and drying the resultant sheet on a hot drum. There is much applied mechanics in the production of the cellulose fibers. The trees must be cut and chipped, poured into piles, watered to control temperature rise, cooked with chemicals to remove lignin, washed and suspended in water for the paper machine or dried in bales for shipment. This part of the story is not for today. We begin with the dilute slurry ready for the paper machine. It has a composition of about .002 gms of dry fiber per gm of mixture. In a paper machine, this watery mixture is poured onto a wire (the fourdrinier wire) in a layer about an inch deep and 20 feet wide at a velocity of some 3000 ft/minute. After passing over a number of support rolls and over several vacuum boxes, the bulk of the water has been removed so that the fiber sheet attached to the fourdrinier wire screen (about 80 mesh) is ready for drying. The composition is now about .2 gm of dry fibers with .8 gms of water. The wet tangled fiber sheet is still too weak to support itself. By pressing a damp felt belt against the fibers they are picked up off the wire and transferred to a yankey drier, a steam heated drum some 20 feet in diameter. At the drum a pressure roll loads the felt and fiber mat at about 300 lb per lineal inch and thus transfers the fibers to the hot yankey and squeezes out more water to about .4 mass fraction fibers. The yankey and a series of other smaller hot drums complete the drying of the paper up to the .9 to .95 mass fraction fibers with which we are all familiar as ordinary paper.

The accomplishment of all these steps is the art of paper making. Now, let us look at some of the applied mechanics.

The original dilute fiber suspension consists of cellulose fibers suspended in so much water that the fibers can float freely about. I say "can" because in general they do not. The density of dry cellulose is about $\rho_s = 1.55$ gm/cm^3. However, in suspension they absorb so much water that they swell in diameter by a factor of about 2 with little change in length. The density of the swollen fiber in water is about $\rho_f = 1.15$ gm/cm^3, hence it will settle out if not continually stirred.

While being stirred, the fibers do not remain uniformly distributed in the water. Instead they form flocs. As one stirs the slurry the flocs continually interact by joining and separating again along a different parting line. Thus the fibers exchange partners frequently. This is most easily shown experimentally by pouring together into a beaker a slurry of white fibers and a slurry of colored fibers. The resultant flocs in the beaker are wholly flocs of mixed colors. What is the process of floc formation? Is it adhesion between fibers or merely mechanical entanglement? What is the size of the flocs? How do these behave during flow? How must the slurry be treated if a uniform non-floced slurry is desired? These are open questions of applied mechanics, the solid mechanics of floc formation and breakup and the fluid mechanics of the flow of floced slurries.

Thus the slurry poured on the fourdrinier wire is not so simple, but we will from here on ignore the flocs and assume uniform fiber distribution to exist in the original slurry. Of course as soon as the slurry is put onto the screen the water begins to run through while the longer fibers are brought by the water and fall on top of those already held by the screen. Thus, a "solid" mat of fibers builds up as the water is removed. At this stage we have some of the original slurry with a distribution of sizes of cellulose fibers on top of a growing mat of "solid" fiber structure containing all of the long fibers and many of the shorter pieces through which flows a liquid consisting of water and some of the small pieces of fiber. The fraction of fibers that appear in the paper is called the "retention". A statistical theory of retnetion has been attempted by Estridge [1], but much more work is required before a satisfactory quantitative prediction of the retention aspect of paper making is attained.

I have called the forming paper sheet a solid. It consists of cellulose fibers whose orientation is essentially in the plane of the sheet and has a direction in that plane which more or less favors the direction of motion of the wire, the "machine direction". There is much empirical knowledge about fiber orientation because it has an important effect on the "cross strength ratio" of the paper formed. The "cross strength ratio" is the tensile strength of the finished paper in the machine direction divided by the tensile strength in the sheet perpendicular to the machine direction. The theoretical prediction of this fiber orientation is still to be accomplished in a satisfactory way.

If we regard the fiber mat as a porous solid it has strength and deformation properties which in principle can be predicted from the fiber orientation statistics and the mechanical properties of the swollen fibers but which in practice must be measured. In the first place the elastic and plastic properties of the swollen wet fibers are very complex and only partially known [2]. Second the calculation of the properties of the bulk porous fiber mat from known properties of individual fibers has barely begun [3].

What are the bulk fiber mat properties? The most important for paper making are its plastic failure in compression and its very low tensile strength. The first controls the way it compacts during the removal of water, while the second probably controls the size of flocs. I say "probably" for the second because no theory nor experiment to prove the point is as yet available.

Figure 1 shows experimental compression properties of a cellulose fiber mat. This figure was taken from reference [4] but similar data have been obtained many times [5, 6]. The load is the force per area of porous fiber mat (not the stress in a fiber) and is applied to the mat by a screen which permits the water to flow through. The density coordinate is the easily measured mass of dry fibers per unit volume of mixture - the consistency on a volume basis. Figure 1 shows the consistency to which the fiber mat collapses under the load. If the load is removed there is a small recovery partly elastic, partly plastic. I know of no studies of these recovery and elastic properties.

At this point a short discussion of the water filled mat is appropriate. We visualize the mixture as a movable liquid filling the pores of a mat of swollen wet fibers which is relatively stationary. Not only the cellulose but the absorbed water and the "stationary" water inside of hollow fibers must be regarded as parts of the "solid fiber structure". The movable liquid, on the other hand, consists of the water free to move together with all the small bits of swollen fiber which move through the mat pores with the liquid.

The exact division between "solid" and "liquid" phase in the fiber mat is both experimentally and theoretically difficult. How should the water which moves very slowly in a very small pore be classified? How should we classify a small piece of fiber which is stuck on the mat most of the time but which occasionally breaks loose and moves a short distance with the liquid? A much more extensive study of retention as related to particle size distribution and mat density is needed. For the present we conceive of the mixture as consisting of only two phases, liquid (movable) and solid (stationary except for compaction motions).

If we hold the mat (with a screen) and pass liquid through it, the liquid will encounter a resistance to flow which transfers load from the pressure in the liquid to the solid structure and ultimately to the screen. Thus, a pressure drop coefficient, α, is defined.

$$\frac{d\rho_L}{dx} = \alpha U \qquad (1)$$

where ρ_L is the pressure in the liquid at position x, the mean of the pressure in all the pores at x;
U is the mean velocity of the liquid=volume flow/mixture cross section (not to be confused with the mean velocity of the liquid in the pores);
α is the pressure drop coefficient and is the reciprocal of the usual Darcy coefficient.

The data shown in Figure 1 does not quite give a straight line between x and c_p on a log scale. It is much more nearly straight than data for α obtained with inert fibers (fibers which absorb little or no water) shown by the dashed line in Figure 1. For the present discussion, it will be taken as straight and therefore of the form.

$$\alpha = \alpha_o c^\delta \qquad (2)$$

We also note in Figure 1 that the compression data is well represented by

$$p_p = \beta c^\gamma \qquad (3)$$

And finally and most importantly we can use $\gamma = \delta$.

The reasons behind these straight lines and the equality of γ and δ are still open applied mechanics questions.

To return to paper making, we visualize the water removal process as in Figure 2. Water being removed comes both from the original slurry above and from the compaction of the mat already formed. To calculate these phenomena we must write the continuity equations and the equations of motion for both the liquid and the fibers.

The one dimensional equations of motion are

for the structure:

$$\frac{\delta\rho_p}{\partial t} + \frac{\delta \rho_p v_p}{\partial x} = 0 \qquad (4)$$

$$\rho_p \frac{dv_p}{dt} = -\frac{\partial p_p}{\partial x} + \rho_p g + F_{wp} \qquad (5)$$

for the liquid:

$$\frac{\partial \rho_w}{\partial t} + \frac{\partial \rho_w v_w}{\partial x} = 0 \qquad (6)$$

$$\rho_L \frac{dv_w}{dt} = -\frac{\partial p_w}{\partial x} + \rho_w g - F_{wp} \quad (7)$$

where the notation is standard except that subscript, p - structure, and w - liquid, refers to structure and liquid properties per unit of mixture area or volume and

F_{wp} is the force by the liquid on the structure in the x direction.

The interphase force F_{wp} must consist of a buoyancy and a flow resistance term. However, these terms give the force on a volume of structure whose top and bottom surfaces are exposed to the liquid. If a volume element is considered inside of the structure, the cut structure surface between pores (at x and x + dx) are not exposed to the liquid forces and a correction is required. Thus, we get

$$F_{wp} = -\rho_L g X_p + \alpha U + \frac{\partial}{\partial x}(p_L A_p) \quad (8)$$

where; subscript L denotes the mean property of the liquid phase itself

X_p the structure volume fraction

A_p the structure section area fraction.

Now making use of the facts that for a suitably random structure

$$A_p = X_p = 1 - \frac{\rho_w}{\rho_L} \quad (9)$$

and

$$U = (1-X_p)(v_w - v_p) \quad (10)$$

the equations of motion become

$$\rho_p \frac{dv_p}{dt} = -\frac{d}{dx}(p-p_L) + (\rho-\rho_L) g + \alpha \left(1-\frac{\rho_w}{\rho_L}\right)(v_w - v_p) \quad (11)$$

and

$$\rho_w \frac{dv_w}{dt} = -\frac{\partial p_L}{\partial x} + \rho_L g - \alpha \left(1-\frac{\rho_w}{\rho_L}\right)(v_w - v_p) \quad (12)$$

where

$$(p-p_L) = \beta \rho_p^{\gamma}, \quad \alpha = \alpha_o \rho_p^{\delta} \quad (13)$$

Several things should be noted.

1. The measurable consistency of Equations 2 and 3 have been replaced (with the required change of numerical values of β, α_o) by the conceptually important but unmeasurable structure density ρ_p.
2. These equations apply only during the collapse phase, therefore the load-up phase of the fiber motion.
3. The equations of motion of solid and liquid which are symmetrical in the form of equations 5 and 7 are no longer symmetrical as equations 11 and 12 since in fact the liquid can move through the solid without a density change while the solid structure cannot move except by collapse.

Now with the basic equations in hand, we can solve various problems, check them against experimental results and then try to solve the paper making problem.

If we fill a tall cylinder with a fiber slurry and let it stand the fibers will settle. In general there will be three regions after settlement.

1. A top layer of liquid water
2. A bottom layer of compressed fibers
3. A middle layer of uncompressed initial slurry

Only the top layer is visible in the settled fibers and Figure 3 shows the comparison of the measured and predicted settlement results. The theory was developed without adjustable constants since all the empirical structure and flow constants were predetermined by compression and flow tests as in Figure 1.

To study the dynamic process of water removal from a slurry during growth of a paper fiber mat, we can reduce the equation of motion to a simpler form by using the fact that both the liquid density ρ_L and the intrinsic swollen fiber density ρ_f are constants (not the fiber mat density ρ_L and the intrinsic swollen fiber density ρ_f are constants (not the fiber mat desnity ρ_p) and hence there is a relation between the velocity of the liquid and the solid. This results in the equation (see reference 4 for details) for the solid mass flow rate by compaction of

$$\rho_p v_p = -\frac{\beta\gamma}{\alpha_o} \rho_p^{\gamma-\delta} \frac{\partial \rho_p}{\partial x} \tag{14}$$

The importance of the approximate relation $\gamma = \delta$ is now evident since it removes the last non-linearity from the equations since with equation 14 inserted into the continuity Equation 4 there results the linear diffusion equation

$$\frac{\partial \rho_p}{\partial t} = D \frac{\partial^2 \rho_p}{\partial x^2} \tag{15}$$

where $D = \frac{\beta\gamma}{\alpha_o}$

If we now consider an experiment in which water is drawn from a slurry through a screen at constant pressure drop, we can compute the whole process. For the simplest case of an infinite body of slurry the boundary moves as the half power of the time.

$$x = 2k\,(Dt)^{1/2} \tag{16}$$

and the density distribution is given by

$$\frac{\rho_p}{\rho_o} = \frac{k \operatorname{erfc}(\eta/2+k) + \operatorname{ierfc} k}{\operatorname{ierfc} k} \tag{17}$$

where k is dependent upon the initial and final density by

$$\frac{\rho_\infty - \rho_o}{\rho_\infty} = \sqrt{\pi}\; k e^{k^2} \operatorname{erfc} k \tag{18}$$

Again without adjustment of any constants, the theory and experiment agrees as in Figure 4. The experiments were, of course, performed with a finite layer of slurry and hence the long time results differ from the above theory. This region has also been predicted with about the same precision.

It is clear that the making of paper involves the flow of water out of the forming mat according to the diffusion equation as a first approximation. Thus, for rough practical estimates a few measurements of fiber properties can determine the required constants.

That this initial understanding is not the whole story is shown by the fact that in Figure 4 the 1/2 power dependence of slurry compression on time is only approximate. The best fitting straight line has a smaller slope averaging about .485 for a wide range of pulps from a long fiber pine to groundwood. This result has also been found by others [7] and is considered important, but is not yet used in paper machine design. The explanation of the effect is assumed to be the continual relocation of small pieces of fiber during the dynamic compression process not properly accounted for by the static tests from which the basic constants were determined. The theory needs further development.

Another fact needing a theoretical basis is shown by Figure 5 in which the compression constants of Equation 3 are plotted against each other for a wide range of pulps. There must be some general law of fiber mats behind this simple relation. The compression law is thus reduced to a single parameter equation.

$$\Delta p = 17.8\ (11.2c)^{\gamma} \quad \Delta p \text{ in gm force/cm}^2,\ c \text{ in gms dry fibers/ml of mixture} \qquad (19)$$

where γ is the only parameter; the numbers being universal constants.

We have perhaps looked long enough at this one step in the whole complex process of paper formation. A thorough review of progress through 1965 is given in a report of the Institute of Paper Chemistry [6]. You should not conclude that the above exhausts the applied mechanics of paper making. Far from it.

I will now quickly review some of the other outstanding problems.

1. The removal of water from the slurry. This may sound like what we have just been discussing. However, nothing has yet been said about how the pressure difference is applied to remove the water. Figure 6 shows a table roll (R) and a suction box (B) over which runs the fourdrinier wire carrying the slurry. Gravity removes some water but not very much. The diverging space between the roll supporting the fourdrinier wire and the wire itself exerts a major suction on the water. This problem on which Sir Geoffrey Taylor [8] did some pioneer work is moderately well understood. The suction box, of course applies a pressure difference directly.

2. The paper pickup. The problem of the mechanism by which a damp felt belt picks up the paper sheet from the wire is most interesting and is essentially untouched from an applied mechanics point of view.

3. The properties of the three phase mixture. Before reaching the pickup felt, "all" the loose water is sucked out of the fiber mat. In the process a meniscus separating liquid and gas is formed in the pores. Thus, the mat is additionally compressed and some of the liquid is left behind in small pores. Little work has been done on the mechanical properties of and dynamics of this mechanically very weak three phase mixture.

4. The pressure roll. In Figure 7 the weak wet paper, 20% by weight dry fiber, stuck to a felt is run under a roll a foot or so in diamter. The roll is heavily loaded to apply 300 lb/in to the paper and felt. Some water is squeezed out. Some is forced from the paper into the relatively drier felt. What is the applied mechanics of this process? Clearly, we need information about the compression properties of both the paper and the felt. Both materials are three phase mixtures and the properties of neither one are known. Here is an excellent applied mechanics problem for the future.

5. The drying of the paper sheet. All of the problems of mechanical properties of the paper mat are here enlarged by introducing the temperature variation that will drive out the water. Water evaporates and flows out of the pores. It evaporates and diffuses about and sometimes recondenses in cooler locations. The water may flow as liquid in small pores, thus producing the heat pipe effect. At present the whole drying process - other than some overall heat balances - are purely empirical and await the applied mechanics of the future to

clarify the important details and make the design process more rational.

CONCLUSIONS

This is perhaps sufficient discussion of paper making, even though that subject is by no means exhausted. I hope I have convinced you that there are interesting, important, and doable applied mechanics problems awaiting at every step of the paper making machine. And further, it is clear that what is true of paper making is equally true of every empirically developed manufacturing process.

You may ask - why are any of these things important - the processes work well, do they not? The answer is that present processes have some minor difficulties but do work remarkably well. The applied mechanics understanding might help a little with present processes but not much. Their real value is in the further development of the processes. Use of alternate materials, run the machine faster, use cheaper methods, etc. Every minor change is now a major evolutionary process. A quantitative understanding would permit an occasional giant step to something really new.

I do not believe the future of Applied Mechanics lies in the refinement of our solutions of the old problems in the old areas, but the acceptance of the challange to analyze and understand the urgent problems of today and tomorrow. I hope to see the applied mechanics papers of the future encompass a much wider range of problems at a high level of understanding just as the papers of the past have solved the urgent problems of a bygone day.

REFERENCES

[1] R. Esteridge: The Initial Retention of Fibers by Wire Grids. 16th Tappi Engineering Conference, October (1961).

[2] C. A. Jentzen: The Effect of Stress Applied During Drying on Some of the Properties of Individual Pulp Fibers. Tappi, Vol. 47, No. 7, pp. 412-418 (1964).

[3] R. Meyer and D. Wahren: On the Elastic Properties of Three-Dimensional Fiber Networks. Svensk Papperstidn, Vol. 67, No. 10, pp. 432-436. (1964)

[4] H. W. Emmons: The Continuum Properties of Fiber Suspensions. Tappi, Vol. 48, No. 12, pp. 679-687. (1965)

[5] H. D. Wilder: The Compression Creep Properties of Wet Pulp Mats. Tappi, Vol. 43, No. 8, p. 715. (1960)

[6] The Status of the Sheet - Forming Process - A Critical Review. Institute of Paper Chemistry, Appleton, Wisconsin, December 31. (1965)

[7] B. Wahlström, G. O'Blenes: The Drainage of Pulps at Paper-Making Rates and Consistencies Using a New Drainage Tester. Pulp and Paper Magazine of Canada, Vol. 63, pp. 405-417. (1962)

[8] G. I. Taylor: Drainage at a Table Roll. Pulp and Paper Magazine of Canada, Vol. 57, pp. 267-273. (1956)

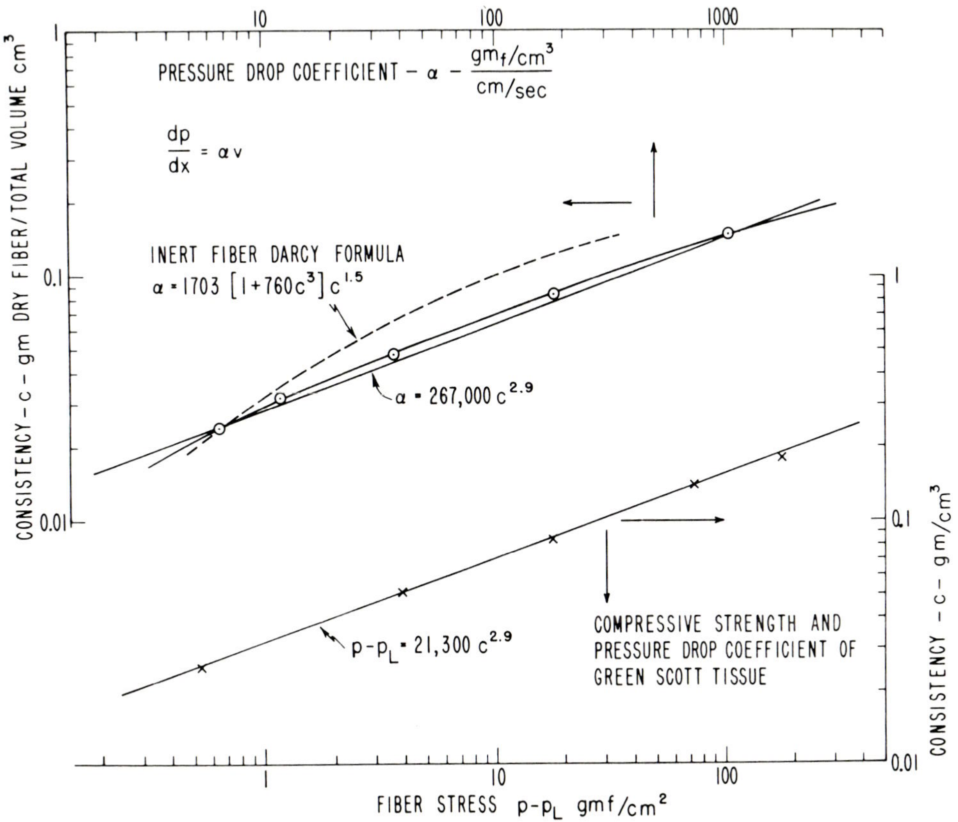

Figure 1. Compressive strength and pressure drop coefficients for a specific sample of green Scott tissue, (softwood bleached sulfite 39%, hardwood bleached Kraft 39%, mechanical fiber 22%).

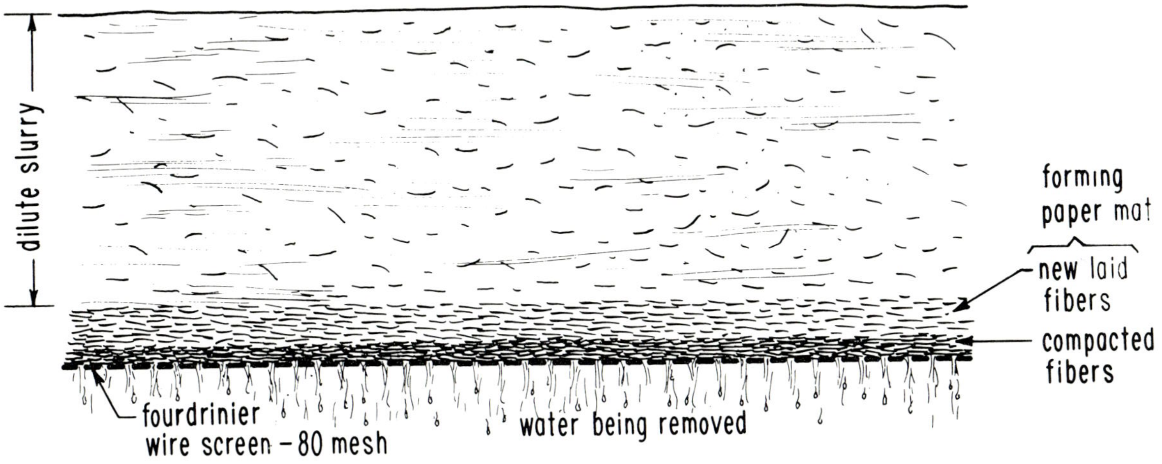

Figure 2. The removal of water from paper fibers on a fourdrinier wire.

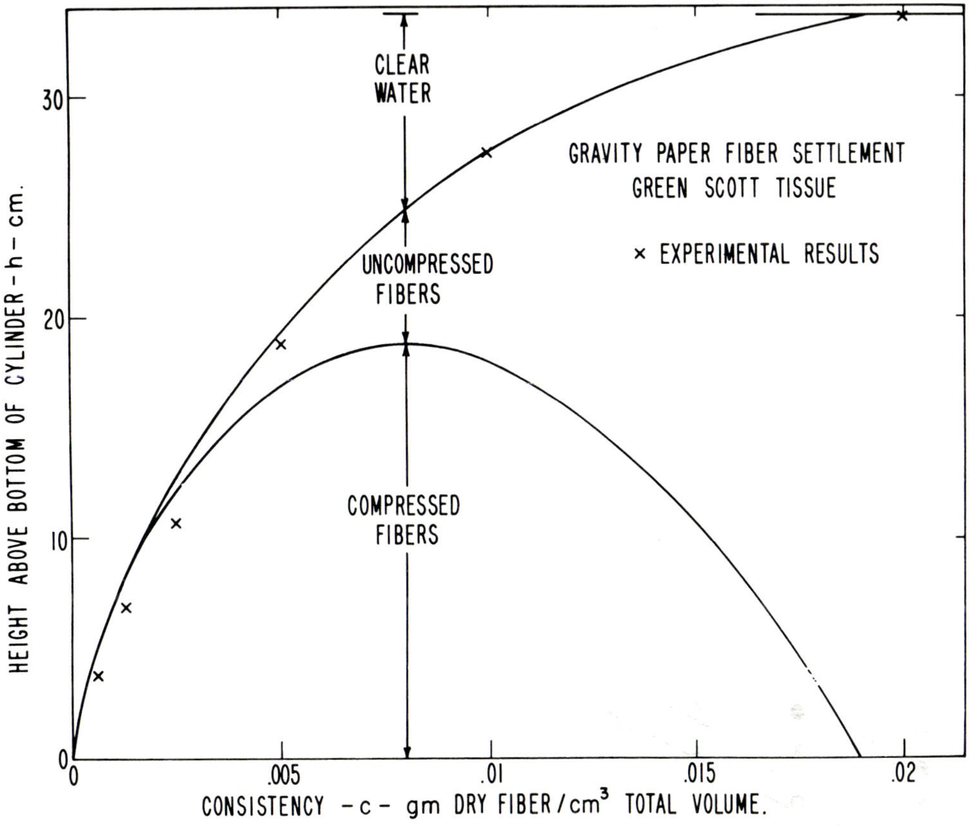

Figure 3. Comparison of measured and predicted thickness of fiber mat formed by gravity settlement in a liter cylinder. Dotted line shows effect of zero pressure density of $C^{o} = .002$.

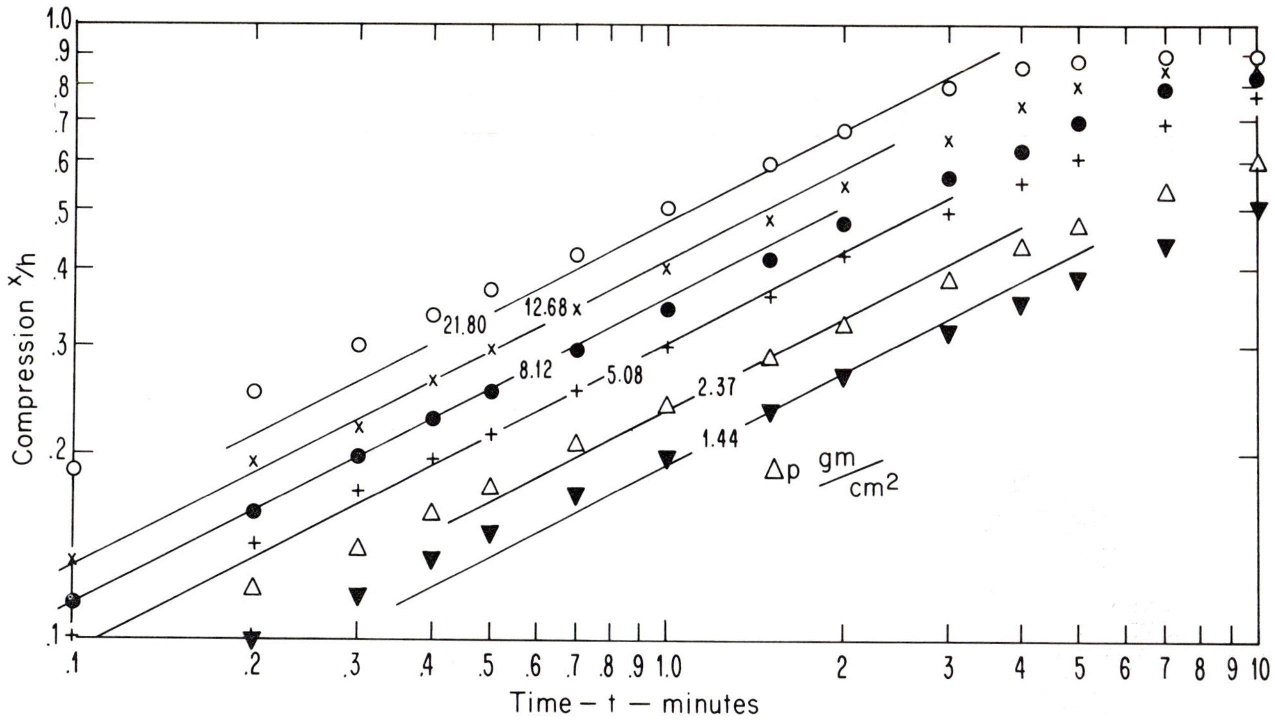

Figure 4. The constant pressure compaction of a paper fiber mat in water. Lines are by theory, points by experiment.

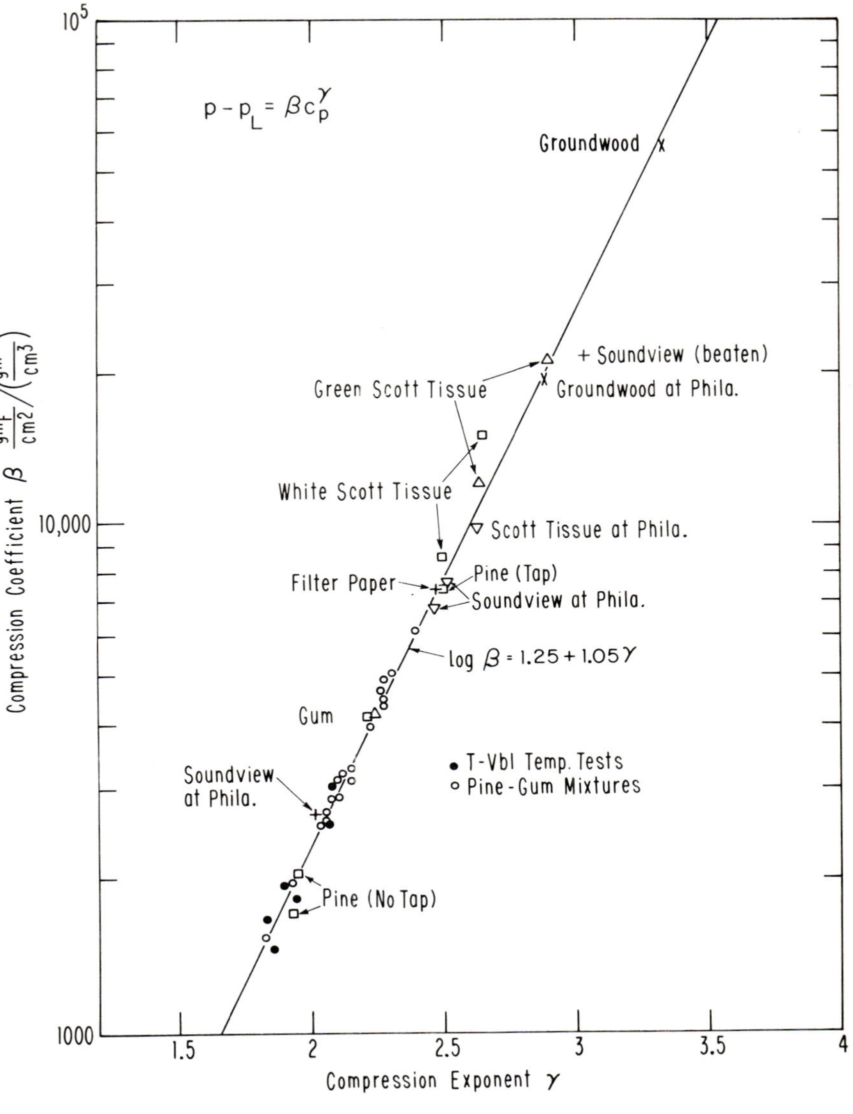

Figure 5. On empirical relationship between compression constants for fiber beds.

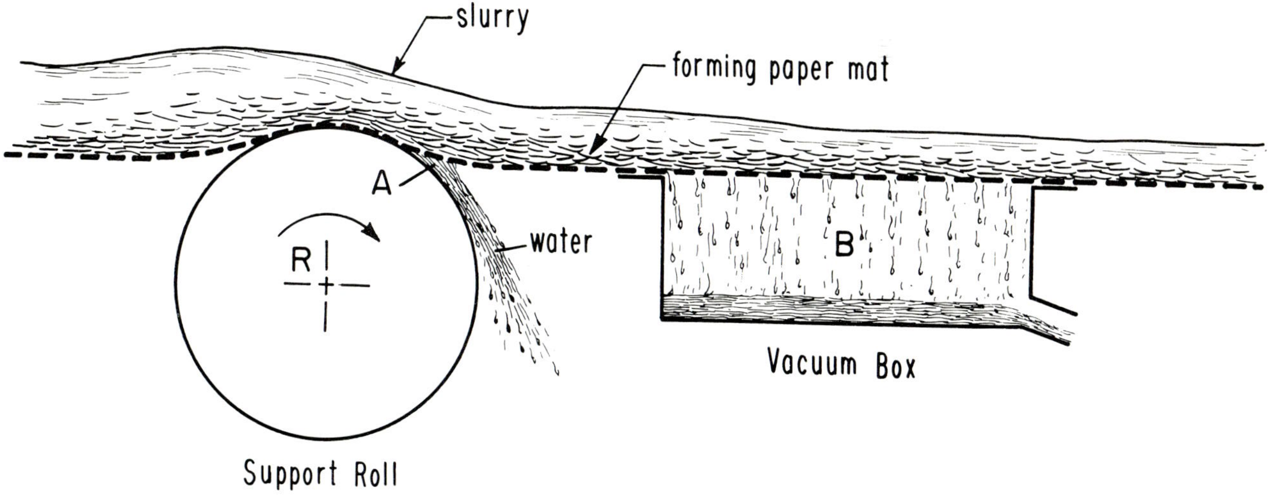

Figure 6. Two processes for water removal from a paper fiber slurry.

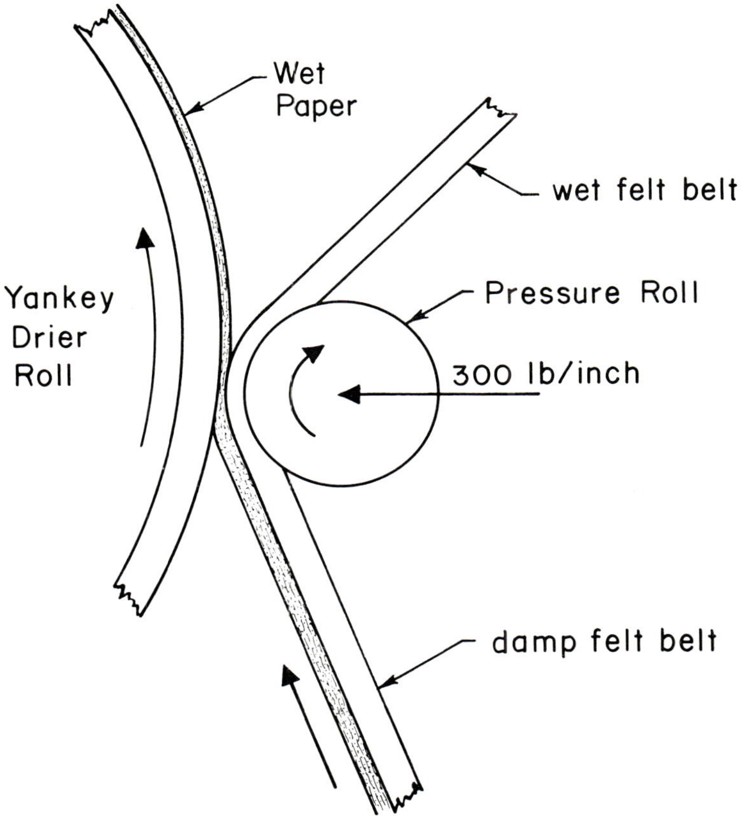

Figure 7. Paper transfer and water removal by a pressure roll.

Developments in Mechanics, Vol. 6. Proceedings of the 12th Midwestern Mechanics Conference: 3

Muscle Controlled Flow

Y. C. FUNG
UNIVERSITY OF CALIFORNIA, SAN DIEGO

ABSTRACT

Body fluids are generally contained in muscular organs. Thus we cannot understand blood circulation, lymph flow etc. without knowledge of the function of the muscles. In this article a basic formulation of the mechanics of the muscular organ is presented, and is illustrated by the problem of ureteral peristalsis.

INTRODUCTION

Our blood is pumped by the heart, and the heart is a muscle. Our ureters collect urine from the kidneys and peristaltically send it to the bladder, and both ureters and bladders are muscles. Arteries and veins contain muscles, which control the dimensions and elasticity of the vessels. Lymph nodes pump constantly; so do the intestines. It is clear that one cannot understand these biological fluid flows without understanding these muscles.

If we know how muscles behave under stress and strain, then we can wrap them around the fluids and formulate a biofluid mechanics exactly as we formulated the theory of aeroelasticity. Unfortunately, it is at the very first step that we get hung up. We run to the physiologists and demand a description of the mechanical properties of the muscles. We ask for "constitutive equations" for the muscles. (In physiology, the terminology is the length-tension-velocity-time relationship; here "length" refers to the total length of the muscle, "velocity" refers to the shortening velocity of the contractile element, and "time" refers to the time after stimulation). But we shall be disappointed. The physiologists can give us a tremendous amount of information, a library of literature, and mountains of data, but not a simple constitutive equation. For the skeletal muscles and the heart muscle, such equations are almost there; but for the smooth muscles of the blood vessels, ureter, intestine, etc., they are not in sight.

We cannot blame the physiologists for this lack of information. They are interested in other things. They are not interested in the details of fluid flow or stress and strain distributions. Not too many physiologists are equipped with the kind of mathematical tools required to handle the particulate flow of the blood encased in a vessel whose material is a complex composite, and suffers finite deformation most of the time. Therefore the responsibility falls on the bio-engineer who wants to clarify these basic problems.

The engineer's first reaction is probably to take a piece of muscle out of the body and perform a mechanical test. Ideally one should test single cells or single fibers; but this is in general difficult for smooth muscles because of the extreme shortness of their fibers - in the arterial wall, for example, they are only 0.02 mm long (the "average length" for mammalian tissue is given as 50 - 150 μ and the breadth 5 - 50 μ in central region). The next best is to test isolated segments of the tissues (a segment of an artery, a ureter, etc.). There are practical difficulties in doing this: in addition to the normal precautions of material testing, we must try to keep the tissues alive, provide them with the correct physiological environment, with controlled temperature, ph values, osmolarity, etc. There are always the questions of evaluating the effects of injury and deviations from physiological conditions. Since cells in an animal are kept alive by blood circulation, when circulation is interrupted we must provide other means to nourish them. The simplest approach is to use diffusion directly: by bubbling oxygen and CO_2 in a suitable solution (such as Kreb's). This requires rather small specimens to keep the path of diffusion short.

The alternative approach is to observe the intact animal. For example, although it is hard to get a nice, living specimen of heart muscle to be tested in a mechanical testing machine, it is quite possible to measure the geometric details of a beating heart by biplane X-ray cinematography. Similarly, although we do not have a good method to get a specimen of ureter without inflicting tremendous injury to the specimen, we can quite easily observe the peristalsis in a normal ureter through x-ray. Thus the question arises: can we deduce the mechanical properties of the tissues in the heart and in the ureter by physiological observations on intact animals?

Here we encounter an age-old dilemma in scientific research: an easy experiment is not easy to analyze; whereas the conditions for a simple theory are difficult to produce in the laboratory. In the world of the living the distance between an observable quantity and the basic laws and basic physical constants is usually tremendous. To understand a physiological phenomenon from the basic theoretical point of view requires extensive analysis. To extract the basic laws and basic physical constants from convenient observables requires Herculean effort.

It is perhaps the attitude one takes at this point that distinguishes a bio-engineer from a physiologist. Both aim at the clarification of physiological phenomena. But the engineer tends to make exhaustive analysis, to study the effects of the variation of the physical parameters and boundary conditions. Isn't this the hallmark of Applied Mechanics? On the other hand, the tradition in physiology is to create incisive experiments and make immediate deductions. In a traditional physiological paper the distance between hypotheses and conclusions is usually very short. Although modern physiology is a quantitative science, most of it has managed to be presented without mathematical formalism.

Whatever be the personal reason for an engineer to invest his time and energy to study biology and medicine, there is perhaps an intuitive conviction that his method of approach will bring new developments to the field. Only time will tell how fruitful the engineers' labor will be. But a beginning has to be made.

In this article the mathematical representation of the mechanical properties of the muscles will be discussed, then a basic scheme of analyzing living organs will be presented. Finally, using ureteral peristalsis as an example, we shall illustrate an application of the theory.

MATHEMATICAL DESCRIPTION OF THE MECHANICAL PROPERTIES OF MUSCLE.

(A) Hill's Equation

There are many kinds of muscles. Different muscles may behave quite differently mechanically. Therefore in speaking of muscles we should specify clearly which muscle. The bulk of knowledge about muscle mechanics exists today only for two muscles: the skeletal muscle of the frog, and the heart papillary muscle of the cat and the rabbit. The reason why these two muscles are specifically favored is because they meet the requirements on size, shape, and preparation without extensive injury. Other muscles have been studied too, of course, but not as extensively.

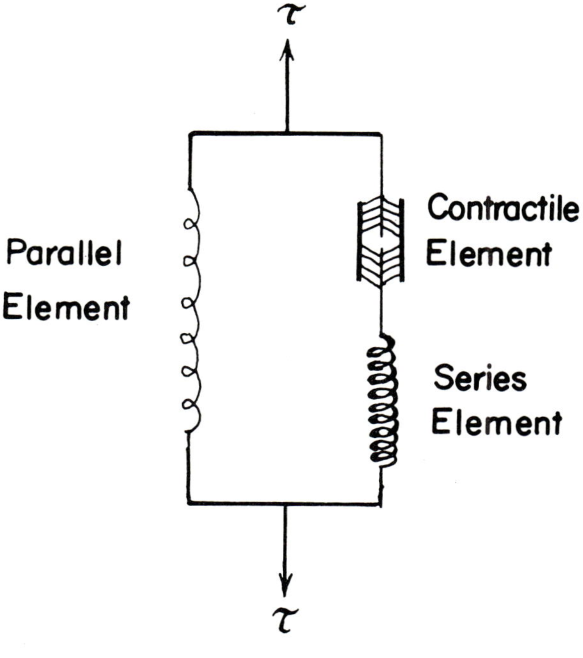

Fig. 1 Hill's functional model of the muscle

For the skeletal muscle many theories have been proposed, but Hill's ideas are most widely accepted. Hill distinguishes two states of the muscle. When it is relaxed and not stimulated, the muscle behaves passively like other polymeric material. When it is stimulated, it is capable to contract and generate tension. Hill presents the muscle in a functional diagram as shown in Figure 1. In the unstimulated condition, the contractile element is assumed to be stress free; the molecules that make up the contractile element can slide against each other freely. Mechanically, then, the stress-strain relationship of an unstimulated muscle can be represented by that of a "parallel element" as shown in the figure. Upon stimulation, the contractile element will shorten at a specific velocity, and tension will be generated. The function of shortening is represented by the sliding elements of the contractile element; the generation of tension is represented by the elasticity of the "series" element. The activity due to stimulation is regarded as the engagement of a ratchet mechanism between the sliding elements of the contractile element.

Since early 1950's A. F. Huxley, H. E. Huxley and their associates have studied the contractile element with electron microscope, x-ray diffraction, and chemical kinetics; and a "sliding elements" theory is established [16, 17]. Briefly the sliding elements consist of thick myosin-containing fibers and thin actin-containing fibers, arranged in a regular array which is shown schematically in Figure 2a. Repetition of these units as shown in Figure 2b makes up the striated appearance of these muscles. The units repeat themselves within the cell along the Z-lines. When muscle fibers are macerated they tend to break at the Z-line, hence the unit, or sarcomere, of the fiber is defined as the region enclosed between two successive Z-lines. As indicated in the figure, the bands formed by the myosin fibers are called A-bands, the remainder are called I-bands.

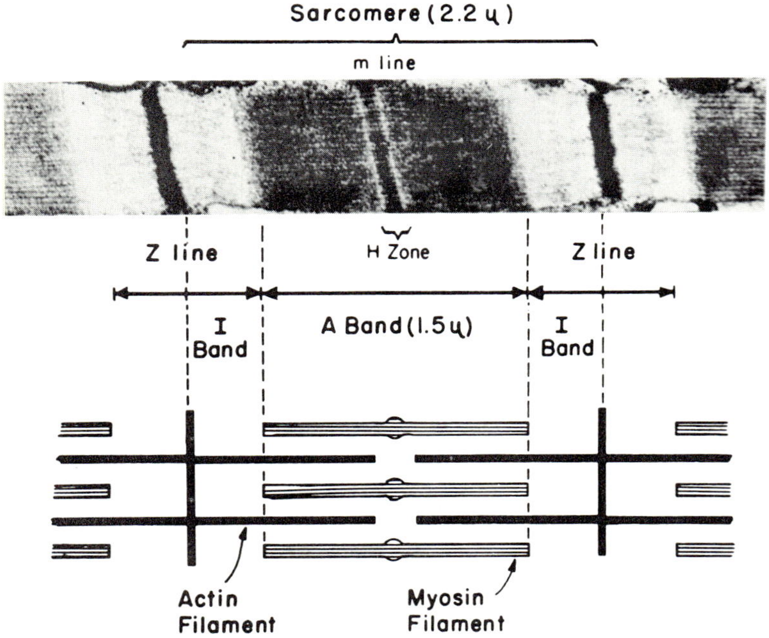

Fig. 2a The sliding elements of a striated muscle
Courtesy of Dr. James Covell, UCSD.

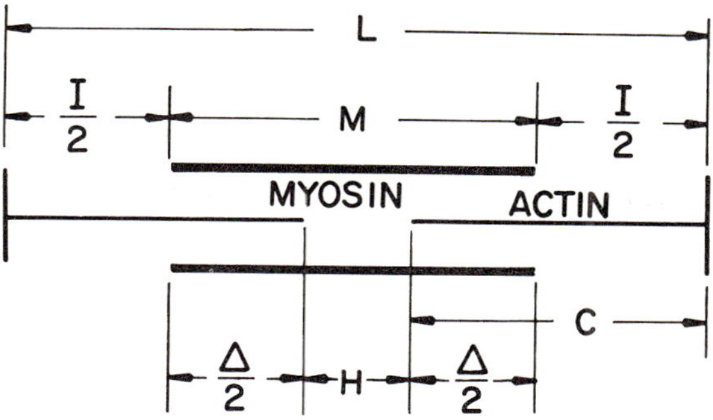

Fig. 2b Nomenclature of dimensions within a sarcomere.

Between the sliding elements are molecules that form "bridges" between actin and myosin. Upon stimulation these bridges are activated; the myosin and actin fibers are drawn toward each other and tension is generated.

If we accept Hill's model for the function of the skeletal model, we see that the mechanical properties of the parallel element can be determined by testing the muscle in the unstimulated condition. The elasticity of the series element can be measured by a sudden change of tension in a stimulated muscle.

If a skeletal muscle is stimulated by a steady train of electric pulses at a high frequency it can be maintained in a state of constant maximum tension and the muscle is said to be tetanized. If the tension in a tetanized muscle is suddenly reduced to a lower level, at first there will be a sudden change of length because of the shortening of the series elastic element, then a continuous shortening will follow. Since at a steady tension the length of the series elastic element is constant, the continued shortening is due entirely to that of the contractile element. The velocity of shortening of the contractile element can be measured. From the experimental results Hill [15] obtained his famous equation

$$(P + a)(v + b) = (P_0 + a)b = \text{constant} \qquad (1)$$

i. e.,

$$v = \frac{b(P_0 - P)}{a + P} \qquad (1a)$$

where

P = tension in the muscle
P_0 = maximum (tetanized) tension in the muscle
v = velocity of shortening of the contractile element
a, b = constants. For the frog sartorius muscle, $a/P_0 \doteq 0.25$, $b \doteq 0.34$ length/sec. at 0° c.

This equation was originally derived by Hill in 1939 [15] from his experimental results on the heat of liberation associated with shortening, (related to the constant a). Figure 3 illustrates Hill's velocity-tension relationship.

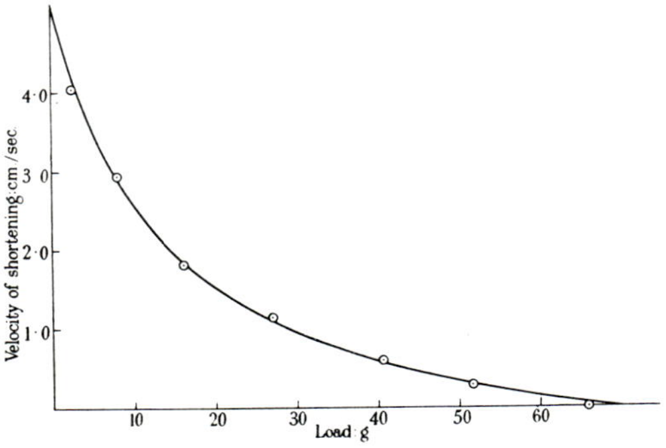

Relation between load (g. wt.) and speed of shortening (cm./sec.) in isotonic contraction. Each circle is the mean of two observations in a series and reverse. Muscle, 165 mg., 38 mm., in Ringer's solution at 0° C. Tetanus 11·4 shocks/sec.: observations at 4 min. intervals: 5·6 mm. shortening. The curve is calculated from the equation $(P+14\cdot35)(v+1\cdot03)=87\cdot6$. Hence $a = 14\cdot35$ g. wt. = 357 g./sq. cm., $a/P_0 = 0\cdot22$, $b = 1\cdot03$ cm./sec. = 0·27 length/sec.

Fig. 3 Hill's velocity-tension relationship. From Ref. 15.

Hill's equation is the cornerstone of the edifice of muscle mechanics. For the frog sartori the tension developed in the parallel element is negligible in comparison with P_0, and that element can be ignored.

(B) The Modified Hill's Equation

Heart muscle is also a striated muscle, microscopically similar to the skeletal muscle, and hence presumably the sliding element concept can be applied. The heart muscle differs from the skeletal muscle in two important aspects: (1) it cannot be tetanized, (2) the tension contributed by the parallel element cannot be ignored in comparison with that developed in the series element. Because of (1), the applicability of Hill's equation should be questioned. Because of (2), the parallel element should be considered in detail.

An unstimulated papillary muscle behaves very much like other soft tissues. See, for example, Figure 4. Analytically, if L denotes the length of the muscle, and P denotes the tension in the parallel element, then the loading curve (increasing load and increasing length) can be represented analytically in the form

$$P = (P^* + \hat{\beta})\, e^{\hat{\alpha}(L - L^*)} - \hat{\beta} \tag{2}$$

where $\hat{\alpha}$, $\hat{\beta}$ and P^*, L^* are physiological constants. P^* is the value of P when $L = L^*$.

The elastic behavior of the series element has been studied by Sonnenblick and others, [22] [5] [23]. The data of Parmley and Sonnenblick [19] can be represented by the following equation:

$$S = (S^* + \beta)e^{\alpha(\eta - \eta^*)} - \beta \tag{3}$$

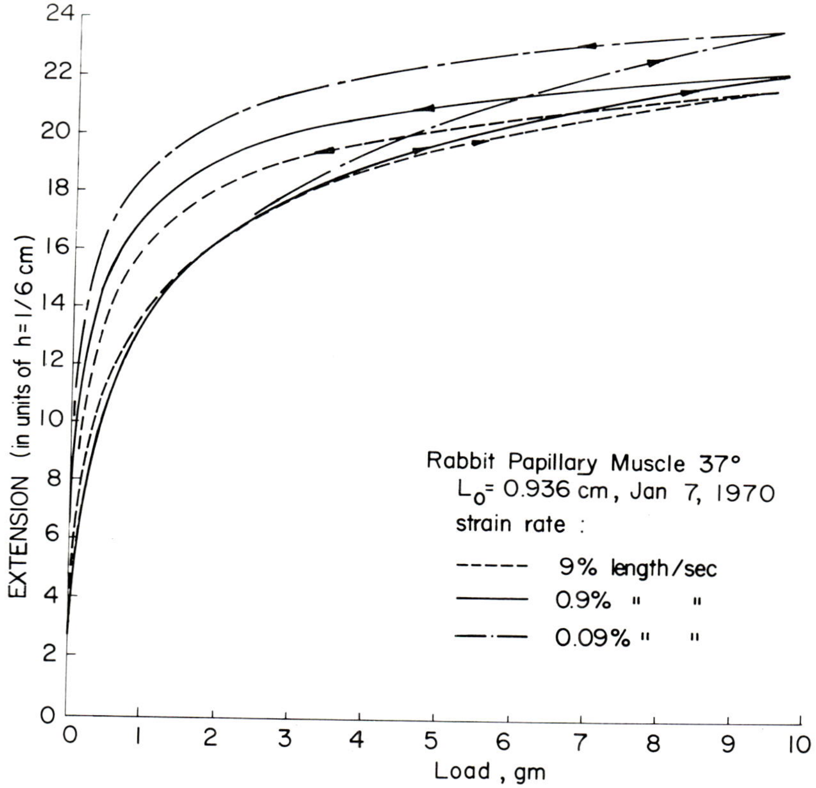

Figure 4. Stress-strain relationship of an unstimulated papillary muscle of the rabbit. The parallel element characteristics.

where S denotes the tension in the series element, η denotes the extension of the series elastic element, and α, β, S^*, η^* are physiological constants.

Figure 5 shows that the series element has an elasticity similar to but much stiffer than the parallel element; the exponent α being almost twice as large as $\hat{\alpha}$.

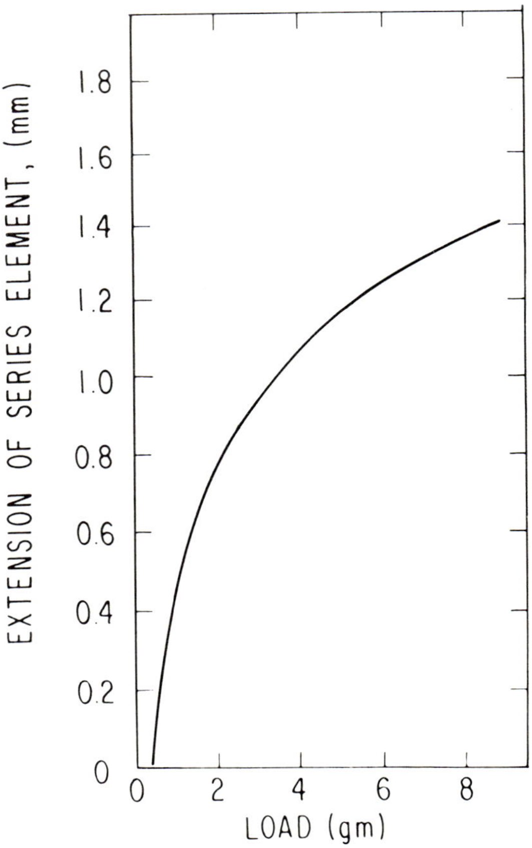

Fig. 5 The stress-strain relationship of the series element of the heart muscle of the cat. From Parmley and Sonnenblick, Ref. 19.

The total tension in a specimen of a stimulated muscle is the sum of the tensions in the series and parallel elements:

$$\tau = P + S. \tag{4}$$

The elastic stretch η and the muscle length L are related in the model shown in Figure 1. From Figure 2 it is seen that when the muscle is unstimulated its sarcomere length L is equal to the sum of the lengths of the actin fiber C and the myosin fiber M, minus the overlap Δ, i. e. $L_0 = (M + 2C) - \Delta$. When it is stimulated both the series and the parallel elements extend by an amount η. Hence

$$L = (M + 2C) - \Delta + \eta. \tag{5}$$

We shall call Δ the "insertion" of the actin and myosin fibers. Although we speak of these lengths in terms of a single sarcomere, it is clearly possible to define the length, insertion, and stretch for a whole muscle in the same fashion by a change of units, because a muscle cell consists of a repetition of sarcomeres, and the whole muscle is composed of muscle cells. In the same token the word tension used above may be converted to tensile stress after reduction of the data.

It remains to describe the contractile element. For a skeletal muscle Hill's equation (1a) provides the desired relation if we identify v with $d\Delta/dt$ and P with S. Extensive work by Sonnenblick and others has shown that in many ways Hill's equation offers a valid model for the contractile element of the heart muscle when P_0 is interpreted as the maximum tension reached in an isometric twitch. However, some workers, notably Brady, point out the many difficulties with Hill's equation. Currently there is a controversy about how the series elements should be arranged; (questioning whether the experimental data are better simulated by the so-called Maxwell model of Figure 1, or the Voigt model which differs from that shown in Figure 1 by placing the series element outside the square box next to τ). In my opinion these arguments are misdirected because if one does not assume that the elastic modulus of the series element is independent of the length of the muscle, then these two models are equivalent; and they are convertible into each other. The real issue is simply that Hill's equation must be modified to include time as a factor. The dependence of contractile velocity on time after stimulation must be brought in. In the original form of Eq. (1), time is absent, and the course of a single twitch cannot be described. For example, if Eq. (1) is applied to an isometric contraction, it can be shown that the maximum tension can be reached only as $t \to \infty$. Similarly, in an isotonic contraction Hill's equation would predict that the shortening velocity will monotonically decrease to zero only at $t = \infty$.

How to modify Hill's equation is the problem. In Ref. [12], the author proposed the following equation:

$$v = \frac{d\Delta}{dt} = \frac{b[S_0 f(t) - S]^n}{a + S} \tag{6}$$

in which v is the contractile element velocity, a, b are functions of the length L, S is the tension in the contractile element, S_0 is the peak tension arrived in an isometric contraction at length L, n is a positive exponent, t is the time after stimulation, and $f(t)$ is a function of time. Comparison with experimental results of Ross et al shows that n is about 0.5. Brady's plot of $(S_0 - S)/v$ against S for rabbits' papillary muscle, which sharply differs from Hill's equation, can be fitted by the equation above with $n = 0.5$. The active state described by Edman and Nilsson suggests that f(t) can be represented by a half-sine wave:

$$f(t) = \lambda \sin [\pi(t_0 + t)/(2t_m)], \tag{7}$$

where λ, t_0, t_m are constants, with λ to be chosen so that in an isometric contraction, $f(t) = 1$ at the instant of time when the tension S reaches the peak value S_0. Physically, t_0 defines the initiation of an active state at stimulation, t_m defines the half-time to peak activity. Equation (6) becomes ambiguous when $S_0 f(t) - S$ is negative; then it is replaced by

$$v = -\frac{b|S_0 f(t) - S|^n}{a + S} \tag{6a}$$

Figure 6 shows the relationships (6) and (6a) in a normalized form. Thus our modified equation resembles Hill's hyperbola. At any instant of time t the difference lies mainly in the neighborhood of the maximum tension S_0.

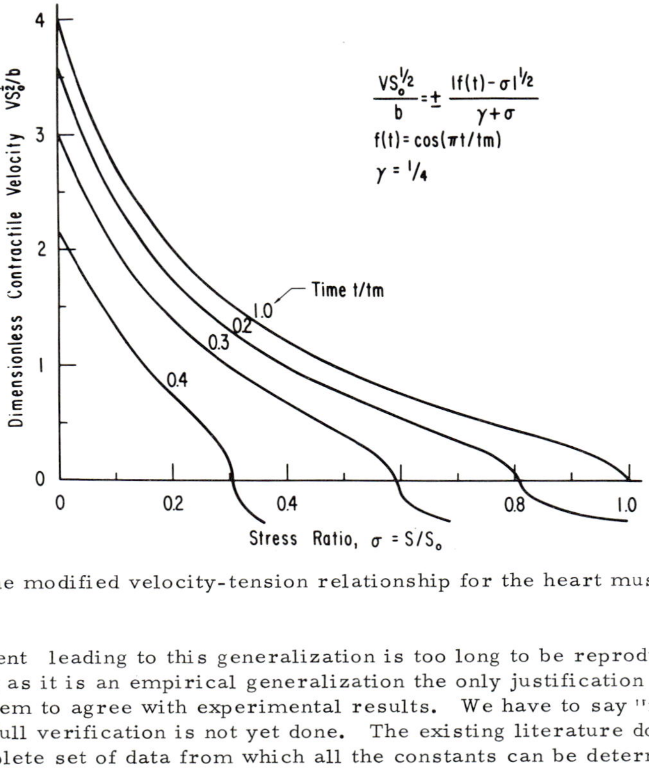

Fig. 6 The modified velocity-tension relationship for the heart muscle.

The argument leading to this generalization is too long to be reproduced here. But as long as it is an empirical generalization the only justification is that its features seem to agree with experimental results. We have to say "it seems" because a full verification is not yet done. The existing literature does not contain a complete set of data from which all the constants can be determined for a single muscle. In particular the details of relaxation after contraction are not known. Work is in progress along this direction.

A different proposition was made by Apter and Graessley [1] who made use of the fact that the generation of tension in a muscle requires the transfer of calcium ions. A viscoelastic model was assumed whose physical constants (elastic moduli and viscosity) as well as the initial length (with respect to which the strain is calculated) were made to depend upon an ionic current $n(t)$ whose time course is specified. Her scheme is also empirical, and in application it is considerably more complex than the procedure presented above.

We remark that Eqs. (2) to (6) describe the mechanical behavior of a muscle fiber in a general course of tension and stretch. If L is kept constant we have an isometric process. If τ is kept constant we have an isotonic process.

Before leaving the subject let us give some typical values for the cat papillary muscle. The functions $a(L)$ and $b(L)$ are nearly constants. When data were reduced with $n = 1$ Sonnenblick gave $a/S_0 = 0.45$. If $S = 0$, and $f(t) = 1$, Eq. (6) yields the maximum velocity

$$v_{max} = \frac{bS_0^{\ n}}{a} \tag{8}$$

which is of the order of 1 to 2 muscle lengths per sec. The maximum isometric tension S_0 is a function of length. Brutsaert and Sonnenblick [5] gave the result

$$S_0 = k_1 L + k_2 \tag{9}$$

with k_1 about 5 g/mm^2 / (muscle length) k_2 about $-3g/mm^2$. The elastic stress in the series elastic element is a function of η and L. The derivative $\partial S/\partial\eta$ is proportional to S:

$$\frac{\partial S}{\partial \eta} \doteq \alpha\ (S + \beta) \tag{10}$$

Parmley and Sonnenblick [19] show that α is of the order of 40 per muscle length, β is about $2g/mm^2$. The other derivative $\partial S/\partial L$ is usually treated as zero. For the parallel element, $\hat{\alpha}$ is of order 19, $\hat{\beta}$ is about zero. All these numbers are not accurately known, and are subjected to future revision.

COMPOSITION OF MUSCLE FIBERS INTO AN ORGAN

Each organ has its own particular structure. When muscles in the organ are stimulated in certain order, each fiber cannot contract freely as if it were isolated. Mutual interference must occur in order to maintain the whole as a continuum. A formalism to deal with this situation will be presented in this section.

We shall consider the heart as a typical example. The muscle structure of the left ventricle of the dog has been detailed by Streeter et al. [25]. See also [24]. Figure 7 shows the orderly variation of the fiber direction in the heart from the outside (epicardium) to the inside (endocardium). For the purpose of the following discussion, we assume that the structure is known. Let x_i $(i = 1, 2, 3)$ be a set of fixed rectangular cartesian coordinates. Let e_{ij} and E_{ij} be the Eulerian and Lagrangian strain tensors respectively, which describe the deformation of the heart from a reference state (preferably the diastolic state). Finite strains may be considered. If da_i is a differential element in the reference

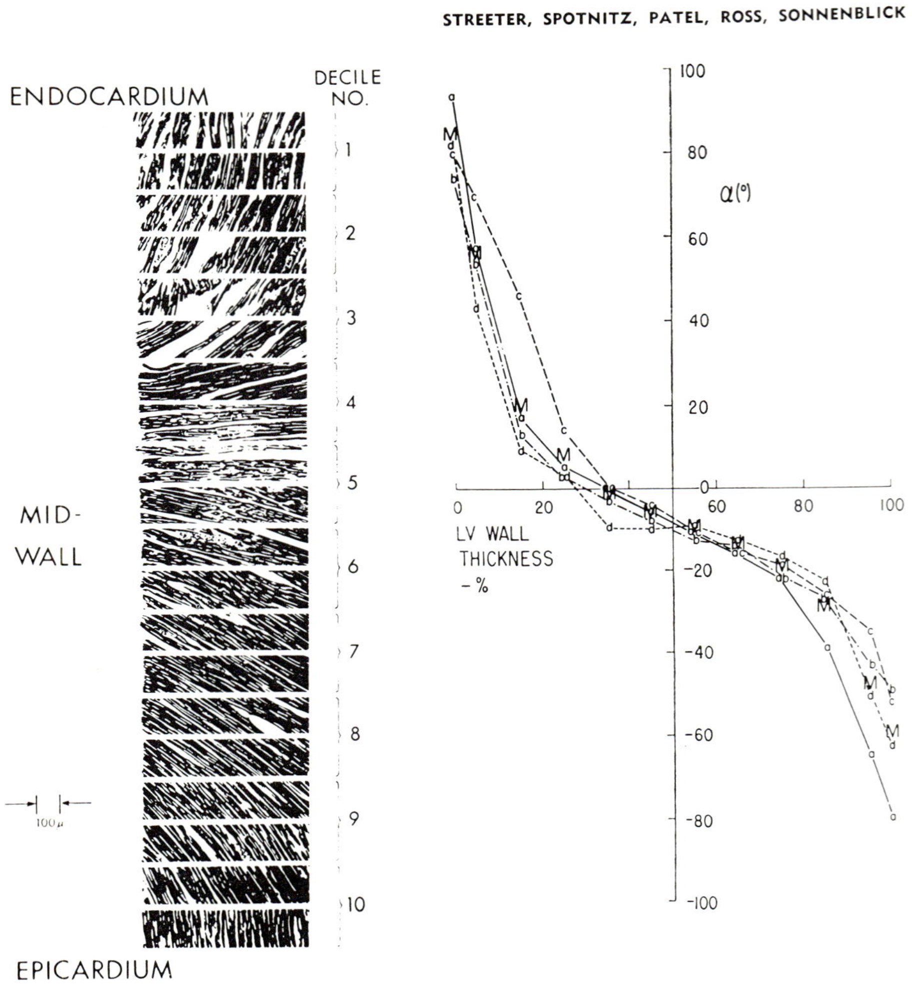

Fig. 7 The fiber orientation in the left ventricle of the dog.
Left-fibers as seen under the microscope.
Right-angle between fiber and equator.
From Streeter, Spotnitz, Patel and Sonnenblick, Ref. 25.

state, and dx_i is the corresponding element after deformation, and if ds_0 and ds represent the lengths of the elements da_i and dx_i respectively, then by definition (in the notations of Ref. [9]):

$$ds^2 - ds_0^2 = 2E_{ij}\, da_i\, da_j = 2e_{ij}\, dx_i\, dx_j \tag{11}$$

Hence the stretch ratio

$$\lambda = ds/ds_0 \tag{12}$$

is related to the direction cosines

$$n_i = da_i/ds_0, \quad \nu_i = dx_i/ds_0 \tag{13}$$

by the equations

$$\lambda^2 - 1 = 2E_{ij}\, n_i\, n_j \tag{14}$$

$$1 - \frac{1}{\lambda^2} = 2e_{ij}\, \nu_i \nu_j \tag{15}$$

Now if $\mathbf{da}_i$ is an element (say, a sarcomere) of a muscle fiber (so that ds_0 represents the sarcomere length in the reference state), then dx_i is the deformed element and ds represents the instantaneous sarcomere length which can be identified with the length L in Eq. (5). We notice that in our previous formulation of the muscle action we need to distinguish L from the "insertion" $-\Delta$ and the series elastic element extension η. The same must be true for the heart which is a composite of such fibers. Therefore, the strain tensor e_{ij} must be decomposable into the insertion tensor $-\Delta_{ij}$ and the series elastic strain tensor η_{ij}. But all the muscle fibers cannot act individually without interference. If Δ_{ij} and η_{ij} were computed as if each fiber can deform freely, then the strain $-\Delta_{ij} + \eta_{ij}$ in general will not satisfy the condition of compatibility. Therefore, a compatibility strain-tensor ξ_{ij} must be introduced so that

$$e_{ij} = -\Delta_{ij} + \eta_{ij} + \xi_{ij} \tag{16}$$

satisfies the compatibility equation

$$R_{ijk\ell} = 0 \tag{17}$$

If e_{ij} were infinitesimal, $R_{ijk\ell}$ reduces to

$$R_{ijk\ell} = e_{ij,k\ell} + e_{k\ell,ij} - e_{ik,j\ell} - e_{j\ell,ik} = 0 \tag{18}$$

When e_{ij} is finite, $R_{ijk\ell}$ is more complex, see, for example, Green and Zerna, [30]. It is not our concern at this time to worry about the exact form of

$R_{ijk\ell}$, but we note that ξ_{ij} is defined by way of Eq. (16).

Let a set of local cartesian coordinate axes y_1, y_2, y_3 be chosen so that y_1 lies in the direction of the muscle fiber. Then on account of the incompressibility of the muscle we may write:

$$(\bar{\eta}_{ij}) = \begin{pmatrix} \eta & 0 & 0 \\ 0 & -\mu\eta & 0 \\ 0 & 0 & -\mu\eta \end{pmatrix} \tag{19}$$

$$(\bar{\Delta}_{ij}) = \begin{pmatrix} \Delta & 0 & 0 \\ 0 & -\mu\Delta & 0 \\ 0 & 0 & -\mu\Delta \end{pmatrix} \tag{20}$$

where Δ and η have the same significance as in Eq. (5). The factor μ corresponds to the Poisson's ratio. For the compatibility strain tensor, we require that $\xi_{11} = 0$, so that the length-tension relationship for each muscle fiber is not disturbed. Hence,

$$(\bar{\xi}_{ij}) = \begin{pmatrix} 0 & \bar{\xi}_{12} & \bar{\xi}_{13} \\ \bar{\xi}_{21} & \bar{\xi}_{22} & \bar{\xi}_{23} \\ \bar{\xi}_{31} & \bar{\xi}_{32} & \bar{\xi}_{33} \end{pmatrix} \tag{21}$$

if the axes x_1, x_2, x_3 are related to y_1, y_2, y_3 by the equations

$$x_i = \beta_{ij} \, y_j \tag{22}$$

then

$$\eta_{ij} = \bar{\eta}_{k\ell}\beta_{ik}\beta_{j\ell}$$
$$\Delta_{ij} = \bar{\Delta}_{k\ell}\beta_{iik}\beta_{j\ell} \tag{23}$$

$$\xi_{ij} = \bar{\xi}_{ij}\beta_{ik}\beta_{j\ell} \tag{24}$$

Now turn our attention to the stress tensor, τ_{ij}, which can be decomposed into the sum of a stress tensor for the "parallel element" and one for the "series element", $\tau_{ij}^{(p)}$, $\tau_{ij}^{(s)}$ respectively:

$$\tau_{ij} = \tau_{ij}^{(s)} + \tau_{ij}^{(p)}. \tag{25}$$

Referred to the muscle axes y_1, y_2, y_3, we have

$$(\bar{\tau}_{ij}^{(s)}) = \begin{pmatrix} S & 0 & 0 \\ 0 & 0 & 0 \\ 0 & 0 & 0 \end{pmatrix} \qquad (26)$$

Hence

$$\tau_{ij}^{(s)} = \bar{\tau}_{k\ell}\, \beta_{ik}\beta_{j\ell} \qquad (27)$$

Finally, we can pose the general formulation as follows:

(1) $\tau_{ij}^{(p)}$ is related to e_{ij} by a generalization of Eq. (2), to be determined by methods already developed in the theory of finite elasticity.

(2) S is related to η by Eq. (3).

(3) The velocity of insertion of actin and myosin fibers, $d\Delta/dt$, is given by Eq. (6).

Then the stress tensor τ_{ij} is fully determined when the history of stimulation and constraints are prescribed. The structural details of the heart are prescribed by the tensor β_{ij}. The equation of motion is, as usual,

$$\rho \frac{DV_i}{Dt} = \tau_{ij,j} + \rho X_i \qquad (28)$$

where V_i is the velocity vector, DV_i/Dt is the acceleration, X_i is the body force per unit mass, and ρ is the density. Together with the condition of incompressibility, the system of field equations is now complete.

A key point in the formulation above is the introduction of the compatibility tensor ξ_{ij}, which makes individual muscle fibers compatible in a continuum.

The equations above provide a foundation for the mechanics of the heart. Boundary-value problems can then be formulated and solved, so that experimental evaluation can eventually be done. The details of the stress distribution in the heart during systole and diastole, the details of its movement, the field of flow, the dynamics of the valves, the influence of hypertension, and so on can then be investigated. It is axiomatic to all engineers that we must be able to calculate the stress and strain and dynamic responses accurately and in detail. It would be inconceivable to do advanced engineering if we could not do so. We feel that the same must be true for the heart specialist. We are not suggesting that every physician and surgeon should be able to handle tensor analysis or finite-elements computer programs; not anymore than we expect the aircraft managers and users to analyze the airplane design. But somewhere in the organization this capability must exist.

We would like very much to illustrate the theory with practical results, but the matter is very complex and not enough definitive results have been achieved.

For the purpose of illustration we shall turn to the problem of ureteral peristalsis, which is of a simpler geometry and slower motion, and greater simplification can be theoretically justified.

APPLICATION TO URETERAL PERISTALSIS

The hydrodynamics of the upper urinary tract, from the kidney to the bladder, attracted attention from the earliest days of biology and medicine, (it was discussed by Aristotle), but even today the hydrodynamics of the ureter is not fully understood. The present status of research can be seen from Ref. [3]. The ultrastructure of the ureter, the mechanical properties of the ureteral smooth muscle, etc., need further clarification.

Theoretical and experimental studies on peristalsis considered as a problem of a harmonic traveling wave on a channel or a cylinder, have been reported by Fung, Yih, Yin, Shapiro, Jaffrin, Weinberg, Barton, Raynor, Burns, Parkes, Hanin, Lykoudis, Zien, Ostrach, Li, and others. [6, 11, 14, 18, 21, 26, 28, 29]. The principal differences between these theories are concerned mainly with the effects of amplitude, wave length, and Reynolds number. Shapiro et al. solved the problem elegantly and simply under the assumption of very long waves. Some other works are concerned with the higher order effects.

In the limiting case where all theories apply, all the papers agree, and the experimental results of Yin and Fung [28] do confirm the theory where applicable. Shapiro and I did have different definitions for the term reflux, but if it helps the matter I am willing to accept Shaprio's definition, and rename what I used to call reflux the mean flow reversal, and limit its use only to waves of small amplitude.

One of the conclusions of these works is that peristalsis is a very ineffective means of propulsion if the tube does not contract down so as to almost occlude the flow at the minimum section. The real, healthy ureter does contract this way. How does the ureter do it cannot be understood without taking the muscles into account. But we could not find much data in the literature about the mechanical properties of the ureteral smooth muscle.

It is commonplace for a radiologist to measure the geometry of the moving urine bolus in the ureter. It is also possible to measure the pressure history in the ureter with pressure gages inserted on a catheter. We therefore ask whether it is possible to deduce some of the physiological constants of the ureter from these measurements. To test this idea, a theoretical study was made [13]. Subsequently, supplementary biological data about the ureter were collected [27]. We shall present an improved version below.

Before analyzing the problem, let us present some physiological data. The ureter collects urine at the upper end and sends it down in the form of successive boluses of fluid as shown in the x-ray photograph in Figure 8. On the average, the dimensions and flow characteristics of normal ureters are given by Boyarsky [3] as follows:

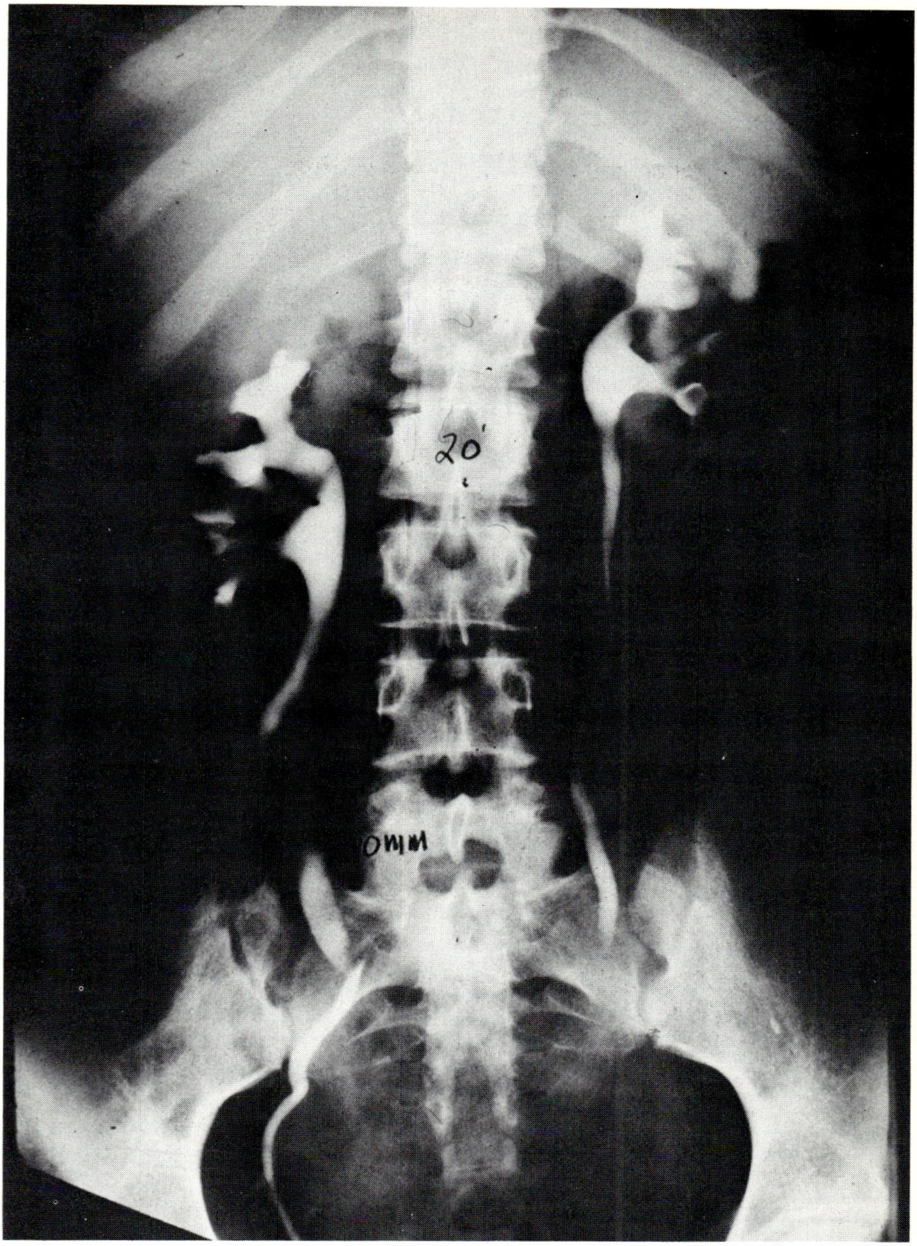

Fig. 8 X-ray photograph of ureteral peristalsis. Normal human. Diuretic condition. The funnel-shaped bodies next to the point marked "20" are in the kidneys. Two fluid boluses are visible in each ureter. Courtesy Drs. R. Gittes and Lee Talner, UCSD.

	Human	Dog
Length, cm	27-37	25 - 27
Frequency, number per min.	1 - 5	2 - 8
Volume flow, ml/min/ureter	0.5 - 3	0.1 - 6.0
ml/bolus	0.1 - 0.6	
Conduction rate, cm/sec	3	3
Back peristaltic pressure, mm Hg	10-40	10-50
Renal pelvic capacity, ml	4 - 8	1 - 2
Bolus: max. diameter, mm	1 - 4	5
length, mm	30 - 130	10 - 80

At the upper end where the ureter joins the kidney at the pelvis, there is no valve. At the lower end the ureter enters the bladder in a very delicate fold which forms a one-way valve. The intramural portion is about 1.25 cm long. If this valve - the so-called ureterovesicular juncture - is damaged or if the peristalsis is too weak urine will be accumulated in the ureter, which will then be dilated. In that pathological condition the kidney will be injured.

For the analysis of peristalsis in a normal ureter, we note that the fluid is Newtonian, the tissues are incompressible, and the Reynolds number of the flow is small (of the order of 1) so that the inertia force may be neglected. We then make the following assumptions:

(1) The fluid bolus is axisymmetric and is so slender that its diameter is much smaller than its length.

(2) The muscle behavior can be described by Eqs. (2) to (6).

Because of the slender geometry, the compatibility strain tensor discussed in the previous section has no effect on the average stress in the ureter wall. Hence when we focus our attention on the fluid and only ask for an approximate solution for the mean stress in the ureter, we may ignore the compatibility tensor. Furthermore, it is easy to show that the tangential shear stress due to fluid motion is negligible in comparison with the circumferential stress, and that the longitudinal tension in the ureter has a negligible effect on the bolus.

Let us take a set of polar coordinates (r, θ, x) with the x-axis coinciding with the axis of the ureter. See Figure 9. We denote the velocity components by v_r, v_θ, v_x, the normal stresses by $\sigma_r, \sigma_\theta, \sigma_x$, and the shear stresses by $\sigma_{r\theta}, \sigma_{rx}, \sigma_{x\theta}$. The inner wall of the ureter is located at $r = r_i\,(x,t)$; the outer at $r = r_0\,(x,t)$, where t denotes time. The equation of equilibrium of the tube wall yields the relation:

$$p_i r_i - p_0 r_0 = (r_0 - r_i) \cdot \langle T\rangle \equiv \int_{r_i}^{r_0} \sigma_\theta \, dr \qquad (29)$$

where p_i, p_0 are the pressures in the lumen and outside the ureter respectively, and $\langle T\rangle$ is the average tensile stress in the wall. For the fluid, the Stokes equations

$$-\frac{\partial p_i}{\partial x} + \mu \frac{1}{r}\frac{\partial}{\partial r}\left(r \frac{\partial v_x}{\partial r}\right) = 0, \qquad \frac{\partial p_i}{\partial r} = 0, \qquad (30a, 30b)$$

and the equation of continuity for an incompressible fluid,

$$\frac{1}{r}\frac{\partial}{\partial r}(rv_r) + \frac{1}{r}\frac{\partial v_\theta}{\partial \theta} + \frac{\partial v_x}{\partial x} = 0, \tag{31}$$

are subjected to the boundary conditions

$$v_x = v_r = 0 \quad \text{at} \quad r = r_i, \qquad v_r = \frac{\partial v_x}{\partial r} = 0 \quad \text{at} \quad r = 0. \tag{32}$$

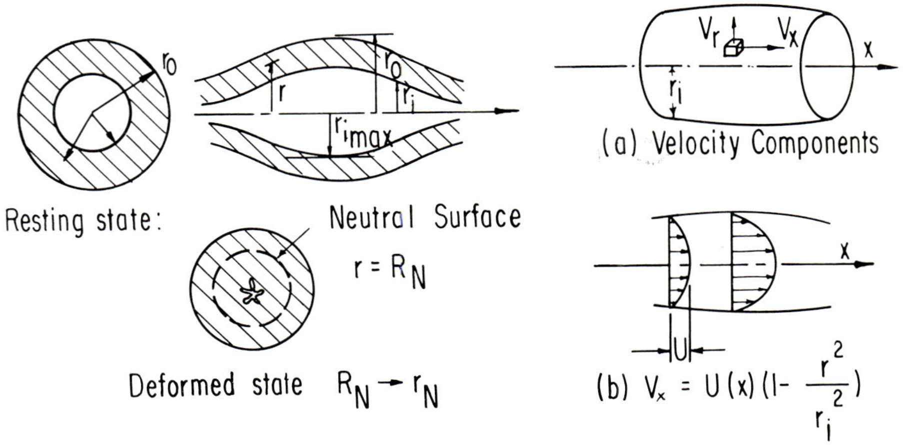

Fig. 9 Notations and coordinates.

Noting from (30b) that p is independent of r, we obtain from (30a) and (32) that

$$v_x(r, x, t) = U(x, t)\left(1 - \frac{r^2}{r_i^2}\right) . \tag{33}$$

On the other hand, (31) and (32) yields, for an axisymmetric flow, the relation

$$v_r\,(r, x, t) = -\frac{1}{r}\int_0^r r\,\frac{\partial v_x(r, x, t)}{\partial x}\,dr . \tag{34}$$

On the inner wall of the tube, the radial velocity $v_r\,(r_i, x, t)$ is precisely the velocity of the wall $\partial r_i(x, t)/\partial t$. On substituting (33) into (34), integrating, then setting $r = r_i$, and equating it with $\partial r_i\,(x, t)/\partial t$, we obtain

$$\frac{\partial r_i(x, t)}{\partial t} = -\frac{r_i}{4}\,\frac{\partial U(x, t)}{\partial x} . \tag{35}$$

In a steady peristaltic motion for which the whole pattern moves to the right at a constant velocity c, r_i and U are functions of the single variable $x - ct \equiv \zeta$, and Eq. (35) can be integrated to

$$\log \frac{r_i(\zeta)}{A} = \frac{1}{4c} U(\zeta), \tag{36}$$

with an integration constant A. Thus the velocity U is positive (agreeing with the direction of propagation of the peristaltic wave) when $r_i > A$; it is negative when $r_i < A$. Backward flow (U negative) occurs if the tube is open with radius $< A$, as a consequence of the conservation of mass. This backward flow can be stopped if the ureter is closed off both at the front and the rear when $r_i = A$ as shown in Figure 10; as indeed it does in reality.

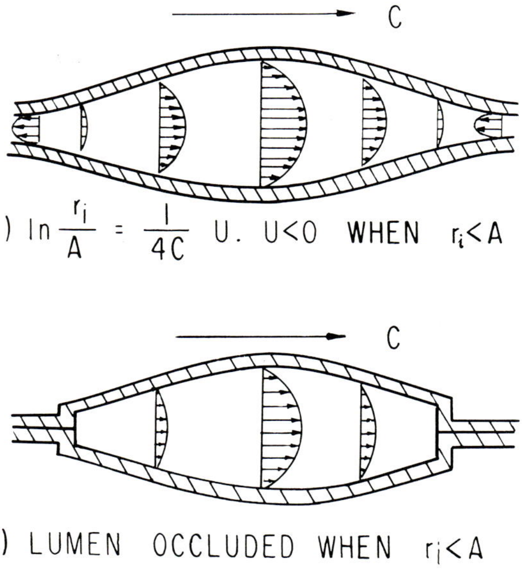

Fig. 10 Fluid velocity distribution as demanded by the equation of continuity.

Now we must consider the action of the muscle. A complete analysis of the ureter is complicated because the tensile stress σ_θ varies throughout the thickness of the wall. To simplify the analysis, note that according to Eq. (29) we need only the mean stress $\langle T \rangle$. Let a neutral surface with radius r_N be defined so that

$$\langle T \rangle = \sigma_\theta(r_N) . \tag{37}$$

Let the neutral surface be located at $r = R_N$ when the ureter is unstressed,* and at $r = r_N$ when the ureter is distended. Then we can take R_N as the unit of length for the muscle and write r_N/R_N and r_N^*/R_N in place of L and L^* in Eq. (2), and η/R_N and η^*/R_N in place of η and η^* in Eq. (3). Then, for the ureteral wall we have, from Eqs. (2) - (8),

$$P \equiv (P^* + \hat{\beta})e^{\hat{\alpha}\left(\frac{r_N}{R_N} - \lambda_1^*\right)} - \hat{\beta} \tag{38}$$

$$S \equiv (S^* + \beta)e^{\alpha\left(\frac{\eta}{R_N} - \frac{\eta^*}{R_N}\right)} - \beta \tag{39}$$

where P^*, S^*, $\hat{\alpha}$, a, $\hat{\beta}$, β are values appropriate for the ureter. Note that according to (38), (39)

$$\frac{\partial P}{\partial r_N} = \frac{\hat{\alpha}}{R_N} (P + \hat{\beta}) \tag{40a}$$

$$\frac{\partial S}{\partial \eta} = \frac{\alpha}{R_N} (S + \beta) . \tag{40b}$$

The rate of change with respect to time of the average tension $< T >$ (which will be written as T for simplicity), is, from (4)

$$\frac{dT}{dt} = \frac{\partial P}{\partial r_N} \frac{dr_N}{dt} + \frac{\partial S}{\partial \eta}\frac{d\eta}{dt} + \frac{\partial S}{\partial \Delta}\frac{d\Delta}{dt} \tag{41}$$

Since $d\eta/dt = dr_N/dt + d\Delta/dt$ according to Eq. (5),

$$\frac{dT}{dt} = \left(\frac{\partial P}{\partial r_N} + \frac{\partial S}{\partial \eta}\right) \frac{dr_N}{dt} + \left(\frac{\partial S}{\partial \eta} + \frac{\partial S}{\partial \Delta}\right) \frac{d\Delta}{dt} \tag{42}$$

Finally, $d\Delta/dt$, the rate of insertion of actin and myosin, is given by the modified Hill's equation (6), (7), and (8). Thus

---

*A definition of the "natural state" or "unstressed state" of the ureter is required. It is natural to consider a ureter as unstressed as it is grown in the body under a constant peritoneal (approximately atmospheric) pressure. Then the muscle tension $T = P + S$ is the stress above the peritoneal pressure p_0. With the internal pressure p_i measured as the gauge pressure with respect to p_0, then we should set $p_0 = 0$ in our equations.

$$\frac{dT}{dt} = \left[\frac{\hat{\alpha}}{R_N}(P + \hat{\beta}) + \frac{\alpha}{R_N}(S + \beta)\right]\frac{dr_N}{dt}$$

$$+ \left[\frac{\alpha}{R_N}(S + \beta) + \frac{\partial S}{\partial \Delta}\right]\frac{b(L)\,\mathrm{sgn}\,|S_0 f(t) - S|^n}{a(L) + S}, \tag{43}$$

In a peristaltic wave of constant speed both r_N and T as functions of space and time are functions of a single variable $\zeta = x - ct$. Hence, $d/dt = -c\,d/d\zeta$. Substituting this and $S = T - P$ into (43), we obtain the first basic equation

$$\frac{dT}{d\zeta} = L\frac{dr_N}{d\zeta} + M \tag{44}$$

where

$$L = \left[\alpha T - (\alpha - \hat{\alpha})\,P + \alpha\beta + \hat{\alpha}\hat{\beta}\right]\frac{1}{R_N} \tag{45}$$

$$M = -\frac{1}{c}\left[\alpha(T - P) + \alpha\beta + \frac{\partial S}{\partial \Delta}R_N\right]\frac{1}{R_N}\;\frac{b\,\mathrm{sgn}|S_0 f(t) - T + P|^n}{\alpha + T - P} \tag{46}$$

To derive a second equation, we combine (33), (36) and (30a), and introduce ζ to obtain

$$-\frac{dp_i}{d\zeta} = \frac{4\mu}{r_i^2}U = \frac{16c\mu}{r_i^2}\log\frac{r_i}{A}. \tag{47}$$

We then differentiate Eq. (29) with respect to ζ to obtain

$$(r_0 - r_i)\frac{dT}{d\zeta} + \left(\frac{\partial r_0}{\partial \zeta} - \frac{\partial r_i}{\partial \zeta}\right)T = r_i\frac{dp_i}{d\zeta} + p_i\frac{dr_i}{d\zeta} - p_0\frac{dr_0}{d\zeta} - r_0\frac{\partial p_0}{\partial \zeta}. \tag{48}$$

Substituting p_i and $dp_i/d\zeta$ from (29), (47) into (48), setting $p_0 = 0$, and expressing the radii r_i, r_0, and their derivatives in terms of r_N and $dr_N/d\zeta$ by the condition of incompressibility of the ureteral wall,

$$r_N^2 - r_i^2 = R_N^2 - R_i^2, \qquad r_0^2 - r_N^2 = R_0^2 - R_N^2 \tag{49}$$

where R_i, R_0, R_N are, respectively, the radii of the inner wall, outer wall, and neutral surface of the contracted ureter (hence $R_i = 0$), we obtain

$$\frac{dT}{d\zeta} = K \frac{dr_N}{d\zeta} + N, \tag{50}$$

where

$$K = T \frac{r_N}{r_i} \left(\frac{1}{r_0} + \frac{1}{r_i} \right) \tag{51}$$

$$N = - \frac{16 c \mu}{r_i (r_0 - r_i)} \log \frac{r_i}{A} \tag{52}$$

Equations (44) and (50), governing the segment of the ureter in which the muscle is active, can be written as

$$\frac{dr_N}{d\zeta} = \frac{N - M}{L - K}, \quad \frac{dT}{d\zeta} = M + L \left(\frac{N - M}{L - K} \right) \qquad \text{(Active)} \tag{53}$$

$$\frac{dT}{dr_N} = L + \frac{M(L - K)}{N - M} \qquad \text{(Active)} \tag{54}$$

These equations are of the type discussed in the theory of nonlinear vibrations of autonomous systems. They can be solved numerically without difficulty. The right hand sides of both equations do not vanish in the range of interest, hence there are no singular points.

An integration of (53) - (54) yields the tension and the radius of the bolus in the segment when the muscle is actively contracting. The corresponding equation in the passive segment can be obtained by setting $S = T - P = 0$, because then no tension exists in the contractile element. On substituting P for T in Eq. (50), noting that $dP/d\zeta = (dP/dr_N)(dr_N/d\zeta)$ and using Eq. (40a), we obtain,

$$\frac{d\zeta}{dr_N} = \frac{1}{N} \left[\frac{\hat{\alpha}}{R_N} (P + \hat{\beta}) - K \right] \qquad \text{(Passive)} \tag{55}$$

These segments are jointed together at a section which may be designated $\zeta = 0$, where $T = P$. This completes the basic theory.

To integrate these differential equations it is easiest to start from the point $\zeta = 0$ where the diameter of the neutral surface is the maximum. The integration constants are the values of r_N and T at $\zeta = 0$. If the bolus moves to the right, the segment $\zeta > 0$ corresponds to the passive state of the ureter, whereas $\zeta < 0$ corresponds to the active segment. The section at $\zeta = 0$ marks the beginning of muscle contraction. It can be verified a posteriori that $r = r_{N\,max}$ when $\zeta = 0$.

The constants P^*, λ^*, may be expressed in terms of $P(r_{N\,max})$:

$$p(r_N) = \left[P(r_{N\,max}) + \hat{\beta}\right] e^{\hat{\alpha}\left(\frac{r_N}{R_N} - \frac{r_{N\,max}}{R_N}\right)} - \hat{\beta} \qquad (56)$$

The derivative $\partial S/\partial \Delta$ is unknown and is presumed to be zero.

With the initial values $r = r_{N\,max}$, $T = T(r_{N\,max}) = P(r_{N\,max})$, Eq. (54) can be integrated numerically to obtain T as a function of r_N; then ζ can be obtained by integrating the first of Eq. (53):

$$\zeta(r_N) = - \int_{r_N}^{r_{N\,max}} \frac{(L - K)}{N - M} dr_N \qquad \text{(Active)} \qquad (57)$$

because $\zeta(r_{N\,max}) = 0$. For the passive segment, Eq. (55) can be integrated directly.

ILLUSTRATIVE EXAMPLE

The physiological problem of the ureter contains a large number of parameters which are unknown at the present time. To illustrate the theoretical procedure, as well as to gain insight to the physiological problem, an example is given. The results are shown in Figure 11; the assumed values of the parameters are listed in the caption. All pressures are gauged to the peritoneal cavity. The course of the active state of the muscle, specified by f(t) in Eq. (6), is not known for the ureter. In contrast to the heart muscle which required electric stimulation, the ureter can be excited mechanically by stretching. The relationship between the active state and stretching in smooth muscles is one of the most important and interesting problems being investigated. For the purpose of this illustration we assume that the muscle can be passively stretched to certain extent, then it becomes active; so that a course of activity $f(t) = \sin \pi(t + t_0)/t_m$ similar to that for the heart muscle is followed. Negative activity is considered unlikely so that f(t) is set to zero for $t > t_m - t_0$. Three values of the ratio $t_0/t_m = 0$, 1/4 and 1/2 are shown in the figure.

The constant A is indeterminate if mean flow reversal (or reflux) is permitted. But if such reversed flow is prevented then the fluid must move with the wave. At the maximum diameter the fluid velocity is parallel to the cylinder axis and the mean fluid speed must be equal to c. Hence the maximum speed $U = 2c$ and Eq. (36) yields $A = r_{i\,max}/\sqrt{e}$.

The upper drawing of Fig. 11 shows the calculated contours of the fluid bolus under the assumed constants. The values of $\hat{\alpha}$, $\hat{\beta}$ are the mean values for the dog recently measured in our laboratory. The point $\zeta = 0$ on the abscissa marks the beginning of the muscle contraction. The bolus moves to the right. To the right of the origin the ureter is passive, to the left the muscle is actively contracting. The middle curves refer to a normal ureter free of obstruction. The lower curve refers to the pressure experienced by a pressure gage inserted on the side of a catheter. The diameter of the catheter is assumed to be 1.58 mm (size 5F) so that the inner lumen of the ureter is limited to $r_i = 0.79$ mm. It is seen that with the assumed parameters, there is a rise in pressure due to isometric contraction of the muscle onto the catheter tip after the passage of the bolus. In this particular case the high pressure rise is an artifact caused by the

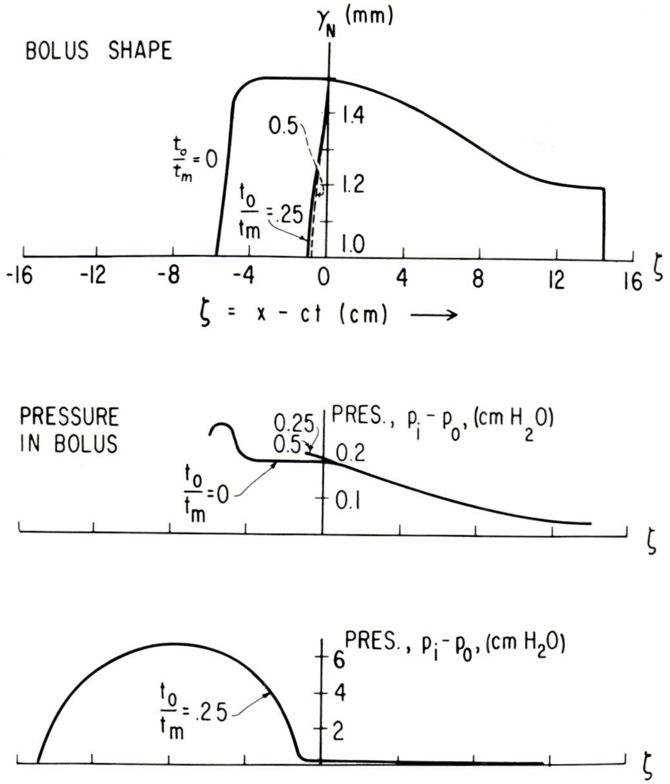

Figure 11.

A theoretical example. Upper curves: Shape of the neutral surface (upper curve) and the corresponding distribution of differential pressure across the ureteral wall. Parameters:

$\hat{\alpha}$ = 7.53 per muscle length,

α = 33.3 per muscle length,

$r_{N\ max}$ = 1.5 mm

R_i = 0

$T(0.15) = P(0.15)$ = 200 dynes/cm^2

c = 3 cm/sec.

$S_0(0.15) = 10^4$ dynes/cm^2.

b = 0.05

n = 1.

$$f(t) = \frac{\sin \pi (t + t_0)}{t_m}$$

$\hat{\beta}$ = 0.283 dynes/cm^2

β = 0 dynes/cm^2

R_N = 1 mm

R_0 = 2 mm

μ = 0.009 poise

v_{max} = 0.2 length/sec.

γ = 1/4

$A = r_{im}/\sqrt{e}$

t_m = 10 sec.

Lower curve: Course of pressure measured on the surface of a catheter of diameter 1.58 mm inserted in the ureter. Contraction becomes isometric when the inner lumen of the ureter is reduced to the catheter diameter.

catheter; the rise depends on the catheter size. The pressure peak is caused by the continued contraction of the muscle, which will subside only when the active state decreases to zero. One expects, in general a phase shift between the largest diameter of the fluid bolus and the peak muscle tension; hence the intra-lumen pressure is out of phase with the lumen diameter. This phase lag has been found by Barry, Absher and Boyarski [2] with catheter tip pressure measurements in the dog, see Fig. 12.

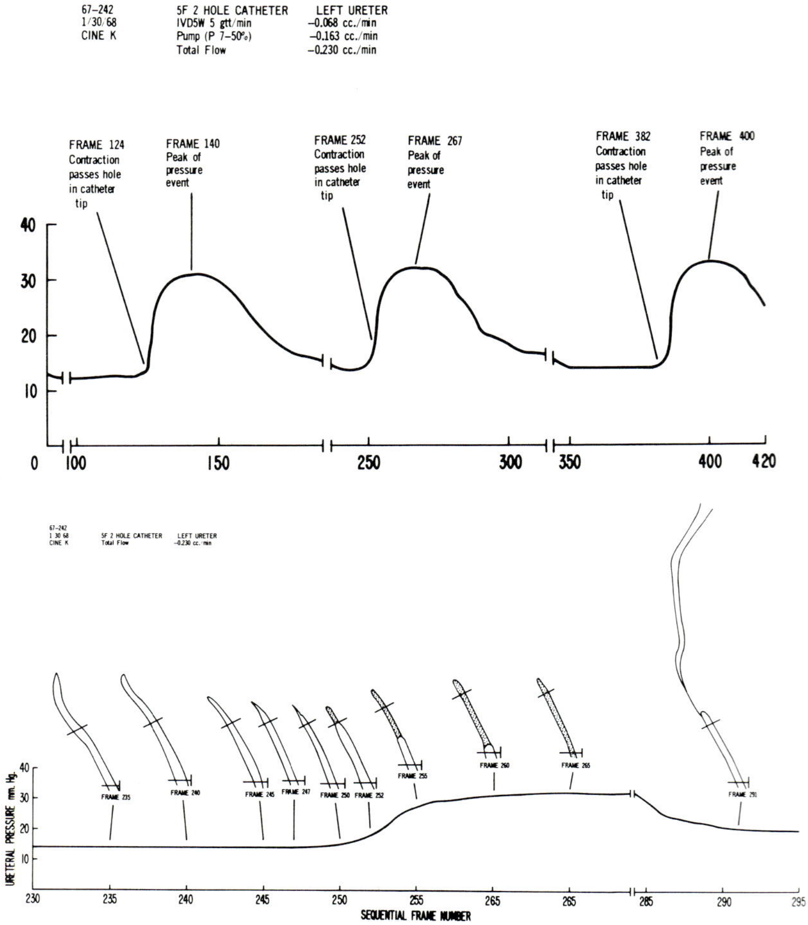

Fig. 12. Correlation of peristaltic wave with pressure measured by catheter tip pressure gages.

Upper curve: Pressure wave in ureter correlated with the location of fluid bolus relative to the catheter tip holes for pressure gages. Ordinate pressure in mm Hg. Abscissa. Frame number related to time sequence.

Lower curve: Details showing fluid bolus relative to the catheter. The inclined line marks the location of the pressure hole. Shaded areas indicate catheter. In frames 235 - 250 the catheter is enveloped in fluid bolus.

Courtesy Drs. Barry, Absher, and Boyarsky. From Ref. 2.

A remarkable feature is the small value of the ureteral wall tension required to accommodate and to move the urine. In Figure 11 the maximum wall stress $P(0.15) = T(0.15)$ is only 200 dynes/cm^2. The corresponding transmural pressure $p_i - p_0$, is also small. This small demand on muscle shows the ease with which a ureter containing a fluid can form a bolus. It explains why no special organ is needed at the renal pelvis to initiate a urine bolus. It also suggests that most of the power of the muscle is spent not to move the bolus but to assure a tight closure behind the bolus, and to force the fluid into the bladder.

A survey of the effects of varying the physical parameters on the bolus shape and pressure history was made. Increasing the wall stress $P(r_{N\,max})$ will make the bolus longer. Decreasing the contractile element velocity b, or the initial activity time t_0/t_m has the same effect. Higher t_0/t_m implies stronger initiation of muscle activity and therefore shorter length of bolus on the active side. Variation of the elastic constants $\alpha, \hat{\beta}, \alpha, \beta$ has only a minor effect on the bolus length. The maximum radius $r_{N\,max}$ and the maximum tension S_0 are the most important parameters influencing the peak pressure in the lumen after the passage of the fluid bolus. When $r_{N\,max}$ is large such high peak pressure does not arise. The initiation time t_0 also affects the peak pressure significantly when there is an obstruction.

To complete our understanding, further experimental data on muscle dynamics must be obtained. However, the theoretical analysis does provide a framework in which the effect of various physical parameters can be studied; thus it can help unravel pathological and comparative aspects of ureteral hydrodynamics.

ACKNOWLEDGEMENT

This work is supported by U. S. PHS NIH Grant HE 12494 and NSF Grant GK 10553.

BIBLIOGRAPHY

[1] J. T. Apter and W. W. Graessley: A physical model for muscular behavior. Biophys. J. 10, pp. 539-555 (1970).

[2] W. F. Barry, R. G. Absher and S. Boyarsky: Correlation of cineradiographic image and pressure tracing of ureteral activity. NRC Workshop on Hydrodynamics of the Upper Urinary Tract, Oct. 1969.

[3] S. Boyarski (Editor): Urodynamics: Hydrodynamics of the Ureter and Renal Pelvis. Academic Press, Inc. (In press).

[4] A. J. Brady: Time and displacement dependence of cardiac contractility: problems in defining the active state and force-velocity relations. Fedn. Proc. 24, pp. 1410-1420 (1965).

[5] D. L. Brutsaert and E. H. Sonnenblick: Force-velocity-length-time relations of the contractile elements in heart muscle of the cat. Circulation Res., 24, pp. 137-149 (1969).

[6] J. C. Burns and T. Parkes: Peristaltic motion. J. Fluid Mechanics, 29, pp. 731-743 (1967).

[7] K. A. P. Edman and E. Nilsson: The mechanical parameters of myocardial contraction studied at a constant length of the contractile element. Acta physiol. scand, 72, pp. 205-219 (1968).

[8] D. L. Fry, D. M. Griggs, and J. C. Greenfield: Myocardial mechanics: tension-velocity-length relationships of heart muscle. Circulation Res. 14, pp. 73-85 (1964).

[9] Y. C. Fung: Foundations of Solid Mechanics, Prentice-Hall, Englewood Cliffs, N. J. p. 91 (1965).

[10] Y. C. Fung: Elasticity of soft tissues in simple elongation. Am. J. Physiol, 213, pp. 1532-1544 (1967).

[11] Y. C. Fung and C. S. Yih: Peristaltic transport. Trans. ASME, J. Appl. Mechanics. Vol. 35, pp. 669-675, Dec. (1968).

[12] Y. C. Fung: Mathematical representation of the mechanical properties of the heart muscle. J. Biomechanics, Vol. 3, pp. 381-404. Pergamon Press. (1970).

[13] Y. C. Fung: Peristaltic pumping: a bioengineering model. NRC Workshop, Chicago, (In press). Oct. (1969) In: Urodynamics: Hydrodynamics of the Ureter and Renal Pelvis. Academic Press, Inc. (1971).

[14] M. Hanin: The flow through a channel due to transverally oscillating walls. Israel J. of Tech., 6, pp. 67-71 (1968).

[15] A. V. Hill: Heat of shortening and dynamic constants of muscle. Proc. Royal Soc. Lond., B126, pp. 136-195 (1939).

[16] A. F. Huxley: Muscle structure and theories of contraction. Progress in Biophysics and Biophysical Chemistry. (Edited by J. A. V. Butler and B. Katz), Vol. 7, pp. 255-318, Pergamon Press, Oxford (1957).

[17] H. E. Huxley: The Mechanism of Muscular Contraction. Science 164, pp. 1356-1366 (1969).

[18] C. H. Li: Peristaltic transport in circular cylindrical tubes. J. of Biomechanics, 3, pp. 513-523 (1970).

[19] W. W. Parmley and E. H. Sonnenblick: Series elasticity in heart muscle: its relation to contractile element velocity and proposed muscle models. Circulation Res., 20, pp. 112-123 (1967).

[20] J. Ross, Jr., E. H. Sonnenblick, J. W. Covell, G. A. Kaiser, and D. Spiro: The architecture of the heart in systole and diastole. Circulation Res., 21, pp. 409-421 (1967).

[21] A. H. Shaprio, M. Y. Jaffrin, and S. L. Weinberg: Peristaltic pumping with long wavelengths at low reynolds number. Fluid Mechanics Lab., Pub. No. 68:5, Dept. Mech. Engr. MIT (1968).

[22] E. H. Sonnenblick: Series elastic and contractile elements in heart muscle: changes in muscle length. Am. J. Physiol., 207, pp. 1330-1338, (1964).

[23] E. H. Sonnenblick: The mechanics of myocardial contraction. The Myocardial Cell Structure Function, and Modification by Cardiac Drugs. (Edited by S. A. Briller and H. L. Conn, Jr.) pp. 173-250. Heart Association of Southeastern Pennsylvania, Third International Symposium.

[24] E. H. Sonnenblick, J. Ross, Jr., J. W. Covell, H. M. Spotnitz, and D. Spiro: The ultrastructure of the heart in systole and diastole. Circulation Res., 21, pp. 423-431, October (1967).

[25] D. D. Streeter, Jr., H. M. Spotnitz, D. P. Patel, J. Ross, Jr., and E. H. Sonnenblick: Fiber orientation in the canine left ventricle during diastole and systole. Circulation Res., 24, pp. 339-347, March (1969).

[26] F. Yin and Y. C. Fung: Peristaltic waves in circular cylindrical tubes. Trans. ASME. J. Applied Mechanics, Paper No. 69-APMW-3, (1969).

[27] F. Yin and Y. C. Fung: Mechanical properties of isolated mammalian ureteral segments. I., Longitudinal tests. (To be published).

[28] F. Yin and Y. C. Fung: Comparison of theory and experiment in peristaltic transport. (To appear in J. Fluid Mechanics).

[29] T. F. Zien and S. Ostrach: Peristaltic pumping. (To appear in J. Biomechanics).

[30] A. E. Green and W. Zerna: Theoretical Elasticity, London, Oxford, Univ. Press, p. 62 (1954).

Developments in Mechanics, Vol. 6. Proceedings of the 12th Midwestern Mechanics Conference: 4

Some Comments on the Energy Method

J. L. LUMLEY
THE PENNSYLVANIA STATE UNIVERSITY

ABSTRACT

It is suggested that the large organized motions which are observed in inhomogeneous turbulent flows may be those the net energy of which as the largest growth rate (the smaller scale turbulence being replaced by a constitutive relation). This is similar to the classical energy method for stability analysis; a modern extension of this method (by Serrin, 1959) is described, and the implications of the method carefully evaluated. It is concluded that the method describes dynamically possible motions; the work of Petrov (1938), in which the contrary conclusion is reached, is examined. The motions are found to be necessarily nonlinear, and to evolve. The method is applied to longitudinal rolls in the wall region of boundary layer flow (found by Joseph, 1966, to be least stable in Couette flow, and observed in the wall region). The resulting equations are solved by asymptotic techniques; these make clear that unstable eddies experience viscosity only near the wall; that streamwise disturbances are produced from the mean gradient by vertical disturbances; that in the viscous region transverse disturbances decay, while still producing streamwise disturbances. As a check on the asymptotic analysis, it is also applied to rectlinear Couette flow, where it gives the classical value for the zero-growth-rate Reynolds number (Serrin, 1959; Joseph, 1966). The eigenvalue relation for the wall layer indicates very large growth rates (in sublayer variables) at observed Reynolds numbers, in agreement with observation. Predicted wavenumbers are larger than observed; a way is suggested by which the reduction in shear due to the presence of the eddies would reduce the wavenumber to that observed.

INTRODUCTION

It is now well known that inhomogeneous turbulent flows exhibit recurrent structures in the velocity field, which are referred to as "large eddies" (Townsend, 1956). The dynamical role which these motions play was discussed at length in Townsend (1956), and several guesses made as to the form of eddy required to explain the measurements in various flows. Suggestions have been made (Lumley, 1967, 1970b) as to how these motions might be identified objectively in correlation data. The dynamical equations obeyed by these eddies have been discussed (Townsend, 1956; Lumley, 1967). Both discussions suggest replacing the turbulent velocities of smaller scale by a constitutive relation; that is, treating the large eddy deterministically, as weather is treated meteorologically, while the smaller scale turbulence is treated statistically. Although this is justified in meteorology, it is probably not in flows in the laboratory: it is not hard to show that a spectral gap half a decade wide is required, for such a treatment to be rigorously justified (Lumley, 1970c). In laboratory flows, there is sometimes a gap of sorts, but it is neither wide enough nor deep enough. Nevertheless, there are even less propitious applications of this technique (to cases with no gap at all) which have been remarkably successful, such as the Heisenberg (1948) treatment of the spectrum. Certainly this approach does not violate physical laws (such as the second law of thermodynamics), it can be made to meet global requirements (such as dissipating the proper amount of energy) and can be endowed with certain known characteristics of the real situation (such as viscoelastic behavior; c. f. Lumley, 1970a); hence there is a certain justification in hoping for qualitatively correct results. The method, in addition has the virtue of simplicity.

The equations of motion for the large eddy are now reduced to the Navier-Stokes equations, if a simple eddy viscosity is assumed, or some more complicated equivalent, if a more realistic constitutive equation is adopted. Although this represents a considerable simplification, it still does not make possible straight forward calculation: the number of known exact (or even asymptotic) solutions to the Navier-Stokes equations is remarkably small. Ordinarily, for given boundary conditions, the variety of solutions that can be generated by various initial conditions is considerable; which one is the large eddy? Accordingly, another principle appears to be needed, to isolate the large eddy. It has been suggested (Lumley, 1967) that the mean velocity profile might be neutrally stable (in the small disturbance sense) to the large eddy, thus identifying the latter as the eigenfunction of the Orr-Sommerfeld (or equivalent) equation corresponding to the minimum value of Reynolds number. Unfortunately, direct calculation (Reynolds, Tiederman, 1967) indicates that turbulent boundary layers, at any rate, are stable to all such disturbances.

We are proposing here another principle, that the large eddy observed is that motion which can most efficiently extract energy from the mean motion and give up as little as possible to the (turbulent) dissipation. This is the motion which can therefore grow largest under given circumstances. Application of this principle gives a well defined and relatively simple eigenvalue problem, at least for the case of a Newtonian constitutive relation. In this case, it is formally equivalent to what is known classically as the energy method of stability analysis.

We will restrict ourselves to a Newtonian constitutive relation, since we have in mind applying this analysis to the wall region of a boundary layer or pipe flow. In such a flow, the maximum shear and minimum viscosity is in the sublayer, which is Newtonian; one may reasonably expect that the dynamics of such large eddies as exist may be dominated by Newtonian phenomena. In fact, we will find from the asymptotic analysis that the nature of the effective viscosity outside the sublayer has no influence on the energy balance of the eddies. This would certainly not be generally true; in attempting to calculate the form of large eddies in shear layers, for example, one might expect the character of the effective constitutive relation to be important, since the maximum shear region occurs where the turbulence is strong; the energy balance then will be between production and turbulent dissipation, and the effective constitutive relation is certainly not Newtonian (Lumley, 1970a).

THE ENERGY METHOD OF SERRIN

The classical energy method is associated with Orr (1907). This was limited to two-dimensional disturbances to the basic flow, and continuity was satisfied by the use of a stream function. In 1959, Serrin extended the method to three-dimensional disturbances, using Lagrange multipliers. Our derivation will be restricted to a Newtonian constitutive relation, and is a trivial generalization of Serrin's, using a slightly different method.

We are seeking the disturbance u_i to the basic flow U_i which, subject to the condition

$$u_{i,i} = 0 \tag{1}$$

has, at a fixed time t, for a fixed value and distribution of ν, an extremal value of

$$\sigma = (1/2E)dE/dt \tag{2}$$

where E is the global disturbance energy. That is, the disturbance which can most efficiently extract energy from the basic flow, losing as little as possible, will have the largest growth rate. That this extremism is truly a maximum can be seen more mathematically by considering that, for a fixed distribution of ν, the dissipation(which must always be nonnegative) may be made arbitrarily large by selecting u_i having sufficiently small characteristic length scales. The production is bounded (relative to the energy) above and below (by the negative of the least and greatest eigenvalues of the strain rate tensor of the basic flow). Due to the dissipation, however, the range of growth rates is bounded only above, and unbounded below.

Using the integral of the equations of motion, we have from (2)

$$\sigma = [-\int S_{ij}u_iu_jdV - \int u_{i,j}\nu 2s_{ij}dV] / \int u_iu_idV \tag{3}$$

where S_{ij} and s_{ij} are respectively the strain rate tensors of the basic flow and of the disturbance. Note that all transport terms vanish, since they are conservative. Applying the usual techniques of the calculus of variations, we obtain (indicating by σ^+ the maximum value)

$$\int\{\sigma^+u_i + S_{ij}u_j - 2(\nu s_{ij})_{,j}\}\delta u_i dV = 0 \tag{4}$$

Now, if the variation is over incompressible motions, the δu_i are not independent, but are related by

$$\delta u_{i,i} = 0 \tag{5}$$

We may specify δu_1 and δu_2 independently, for example, and then δu_3 is given. If we write

$$\sigma^+u_i + S_{ij}u_j - 2(\nu s_{ij})_{,j} = X_i \tag{6}$$

for convenience, then (4) becomes

$$\int X_i\delta u_i dV = \int (X_1\delta u_1 + X_2\delta u_2 + X_3\delta u_3)\, dV = 0 \tag{7}$$

If we set $X_3 = -\pi_{,3}/\rho$, where π is an arbitrary function, then integration by parts (using the condition that u_i and δu_i either vanish on the boundaries or satisfy a cyclic boundary condition) gives

$$\begin{aligned}\int X_i\delta u_i dV &= \int (X_1\delta u_1 + X_2\delta u_2 + \delta u_{3,3}\,\pi/\rho)\, dV \\ &= \int (X_1\delta u_1 + X_2\delta u_2 - [\delta u_{1,1} + \delta u_{2,2}]\,\pi/\rho)\, dV \\ &= \int [X_1 + \pi_{,1}/\rho)\delta u_1 + (X_2 + \pi_{,2}/\rho)\delta u_2]\, dV\end{aligned} \tag{8}$$

Since δu_1 and δu_2 are independent, we have

$$\sigma^+ u_i + S_{ij} u_j = - \pi_{,i}/\rho + 2(\nu s_{ij})_{,j} \quad u_{i,i} = 0 \tag{9}$$

This is the same as that derived in Serrin (1959) with the exception of the inclusion of a non-zero growth rate, and variable viscosity. The pressure π is that required to assure (1), and hence may be regarded as a Lagrange multiplier.

This has been applied by Serrin (1959) to Couette flow with uniform viscosity and $\sigma^+ = 0$. Two-dimensional disturbances of the same flow were, of course, calculated by Orr (1907); he found that the Reynolds number at which $\sigma^+ = 0$ (the critical Reynolds number) for such disturbances is 6.65, based on the shear velocity and the half width. Although Orr (1907) said "Analogy with other problems leads us to assume that disturbances in two dimensions will be less stable than those in three; ... " as pointed out by Joseph (1966), Serrin (1959) found that longitudinal rolls could survive to a lower Reynolds number, namely 4.54 (on the same basis); evidently such disturbances can extract energy more efficiently. Serrin (1959) did not show that this type of disturbance was the most efficient, i.e. - that it corresponded to the largest growth rate (=0) at a given Reynolds number, or to the lowest Reynolds number at a given growth rate. Joseph (1966) however, by considering the combined problem of Couette flow of a Boussinesq fluid heated from below, showed that, in fact, the longitudinal rolls are the most efficient.

Taylor (1960) suggested using this type of analysis for a purpose different from ours: namely, viewing a Couette flow as two viscous sublayers face to face, the thickness Reynolds number (defined as above) corresponding to a maximum growth rate of zero appears to give the thickness of the layer next to the wall in which no disturbance can survive without importing energy from above. He used the analysis of Lorentz (1907) who did not actually solve the extremum problem, but guessed at a solution. Using Serrin and Joseph's solution, this would suggest a viscous sublayer thickness of $y^+ = 4.54$, which is near the point at which the mean velocity profile bends away from the linear, although well below the point ($\sim$9) at which measured dissipation first exceeds production (Townsend, 1956).

Before applying (9) to a specific calculation, we must examine the implications of the method, so that we will know what to expect of it.

CRITIQUE OF THE ENERGY METHOD

The energy method, as applied to flows without buoyancy*, has been subject to extensive criticism. For example, Lin (1955) states: "Even here only the lower limit for the critical Reynolds number can be expected, because stability must be established for all disturbances in the present method, while in reality only those satisfying the hydrodynamic equations need to be considered. The inclusion of the spurious disturbances ..." (page 59) and "The energy equation thus gives critical Reynolds numbers that are too low," (page 61). Serrin (1959) states: "It is important to note that the energy method cannot provide accurate knowledge of the limits of stability, such as can be gained from the linearized perturbation theory ... The reason is that in the energy method one establishes stability relative to arbitrary disturbances, while in reality only those satisfying the hydrodynamical equations need be considered, "(page 4). Joseph (1966) states "The functions which are admitted for review need not be possible solutions of the conservation equations (momentum and energy). It is in this sense that one may speak of dynamically inadmissible disturbances," (page 181). And further: "Evidently ... the energy method reflects a sensitivity to spurious and dynamically inadmissible disturbances," (page 182). None of these allegations is supported. However, Monin and Yaglom (1971) state "Analyzing this fact [that the energy method gives lower critical Reynolds numbers than the linear theory] Petrov (1938) came to the conclusion that the values of ψ at which

*with $\sigma^+ = 0$, and uniform viscosity.

$F[\psi]$ takes a maximum, taking into account the time variation of all the functions, will apparently in no case generate a dynamically possible motion. Thus the energy method can never give an exact value of $Re_{cr\ min}$, but is only suitable for making preliminary, very rough estimates of this value."

The question of why the critical Reynolds numbers predicted by the energy method in flows without buoyancy are substantially lower than those predicted by the linear theory* is certainly an interesting one, although whether they are "too low" as stated by Lin (supra) and in what sense, remains to be seen. The explanations given by Lin, Serrin and Joseph are clearly not satisfactory, since functions are "admitted for review" at a fixed time only; any velocity field at a fixed time satisfies the hydrodynamical equations, which serve to determine its evolution, i.e. - the time derivative. The only substantial argument appears to be that of Petrov (1938).

Since the paper of Petrov (1938) appears never to have been translated, a translation of the relevant section (pages 20-24) is appended (some obvious misprints in the equations have been corrected). The remainder of the Petrov paper is a concise introduction to classical small disturbance stability theory and the energy method, and need not concern us here. Petrov, of course, is reasoning primarily about linearized disturbances having exponential behavior in time.

His argument may be summarized as follows: the small disturbance equations obtained from the Navier-Stokes equations, and the equations obtained from the energy method are not the same; both cannot be satisfied simultaneously (the difference is Petrov's equation (22)); hence, disturbances considered in the energy method do not satisfy the (small disturbance limit of the) Navier-Stokes equations.

This conclusion is correct when applied to the small disturbance limit of the Navier-Stokes equations (though misleading); it is not correct for the full equations. The difference lies in the interpretation of the time derivative. The energy method specifies a global growth rate; locally the growth rate may be greater or less, with the non-linear terms transferring energy from one region to another as required. These non-linear transport terms, being conservative, do not appear in the integrals, so that the equations of the energy method apply to large disturbances whether this is intended or not. The difference between the equations of the energy method and the Navier-Stokes equations serves to determine the local growth rate. Only in the linearized case, where no mechanism is present to transport energy from one level to another (in a parallel flow) is the growth rate required to be the same at all levels, and only then does inconsistency result. Hence, it is the assumption of linearity that is inconsistent. Let us examine the equations.

The condition (2) does not imply that $\partial u_i / \partial t = \sigma u_i$ at each point in the fluid at the instant in question. The growth rate of the disturbance may be larger or smaller than optimum locally, so long as (2) is satisfied. In fact, of course, u_i must also satisfy the Navier-Stokes equation

$$\dot{u}_i + U_{i,j} u_j + u_{i,j} U_j + u_{i,j} u_j = - p_{,i}/\rho + (\nu(u_{i,j} + u_{j,i}))_{,j} \tag{10}$$

and the local growth rate can be obtained by substracting (9) from (10)

$$\dot{u}_i - \sigma^{+} u_i = - \Omega_{ij} u_j - u_{i,j} U_j - u_{i,j} u_j - (p-\pi)_{,i}/\rho \tag{11}$$

* In Joseph (1966) it is shown that in several flows with buoyancy, the critical Reynolds numbers predicted by the energy method are close to, or the same as, those predicted by the linear theory.

where $U_{i,j} = S_{ij} + \Omega_{ij}$, Ω_{ij} being the antisymmetric part. As required by (2), multiplication of (11) by u_i and integration over the region gives

$$dE/dt = (d/dt)\int u_i u_i dV/2 = \int u_i \dot{u}_i \, dV = \sigma + \int u_i u_i dV \qquad (12)$$

If our explanation is correct, equation (11), when restricted to a linearized two-dimensional disturbance of a two-dimensional parallel flow, should give Petrov's equation (22). Let $U_i = (U(x_2), 0, 0)$,

$$S_{ij} = \begin{matrix} 0 & U'/2 & 0 \\ U'/2 & 0 & 0 \\ 0 & 0 & 0 \end{matrix}, \qquad \Omega_{ij} = \begin{matrix} 0 & U'/2 & 0 \\ -U'/2 & 0 & 0 \\ 0 & 0 & 0 \end{matrix} \qquad (13)$$

where $U' = U_{,2}$. Let us satisfy (1) by the use of a stream function $u_1 = \psi_{,2}$, $u_2 = -\psi_{,1}$. Eliminating the pressure between 1 and 2 components of the linearized (11) gives

$$(\partial/\partial t)\nabla^2\psi - \sigma^+\nabla^2\psi = U''\psi_{,1}/2 - U\nabla^2\psi_{,1} - U'\psi_{,21} \qquad (14)$$

Now, in the linear case (considering constant, uniform viscosity) the coefficients of (10), or of the equation for the stream function obtained from it, are independent of x_1, t, and hence the solution may be written as a sum of terms of the form

$$\psi = \phi \, e^{i\alpha(x_1 - ct)} \qquad c = c_r + ic_i \qquad (15)$$

where, from (2) and (15) we necessarily have

$$\sigma^+ = \alpha c_i \qquad (16)$$

Substituting this in (14) we obtain immediately Petrov's equation (22) -

$$(U - c_r)(\phi'' - \alpha^2\phi) - U''\phi/2 + U'\phi' = 0 \qquad (17)$$

That this is inconsistent is clear because, as Petrov points out, one parameter (the viscosity) is absent from (17); hence an eigen-solution of (17) cannot be consistent with an eigen-solution of the Orr-Sommerfeld equation (obtained from the linearized equation for the stream function obtained from (10), on substitution of (15)), since the latter contains the viscosity.

Hence, we must conclude that the extremal disturbance generated by the energy method is necessarily non-linear, entailing transport of disturbance energy, and necessarily has a local growth rate of the disturbance energy differing from the global growth rate, but averaging spacially to the latter. The disturbance does satisfy the hydrodynamical equations, which serve to determine the local growth rate. Since the class of disturbances considered is no smaller than, and may be larger than that considered in the linear theory (which necessarily have no transport, and the same growth rate everywhere) the Reynolds number obtained for the extremal value $\sigma^+ = 0$ is no larger than, and may be less than, that obtained from the linear theory.

As an aside, it may be noted that, since the (necessarily) non-linear disturbance generated by the energy method for $\sigma = \sigma^+$ (necessarily) has local growth rates different from σ^+ it will evolve in form; at an instant following (or preceding) the instant of analysis, it will have changed and may no longer be optimum. But, if the optimum disturbance evolved continuously through optimum states, equation (9) would have to be satisfied in some neighborhood. We would thus have two eigenvalue problems in this neighborhood: either (9) and (10), or equivalently (9) and (11). But we may use Petrov's argument that the sets (9) and (11) cannot be simultaneously satisfied, due to the absence of the viscosity in (11). Hence, (9) and (11) can at most be simultaneously valid at a point, and the optimum disturbance must evolve to one which is not optimum, and hence has a growth rate less than the extremal value σ^+. Hence, the disturbance energy has, at the instant of analysis, a global growth rate of $2\sigma^+$, and in some neighborhood of that time a smaller growth rate. At no time can the energy have a larger growth rate than the optimal one; if the optimal disturbance were not unique, the disturbance energy might at later times evolve to other optimum states, producing points of growth rate equal to $2\sigma^+$; by the preceding argument, however, these must be isolated, so that the disturbance energy has a growth rate almost everywhere less than $2\sigma^+$. Hence, the optimal disturbance energy is bounded from above by the exponential with growth rate $2\sigma^+$ (a result of Serrin, 1959, obtained in a different way), and has this growth rate at the instant of analysis; at other times, the growth rate is almost everywhere smaller. Thus, optimum disturbances obtained for $\sigma^+ = 0$ have a disturbance energy which is monotone decreasing almost everywhere.

We must note that we have not excluded the possibility that disturbances which have, at the instant of analysis, positive global growth rates, will ultimately decay. These would also be counted stable. For this reason also (and perhaps primarily) the critical Reynolds number given by the energy method for $\sigma^+ = 0$ is conservative: it assures that no disturbance can grow ultimately, and that at a higher value some disturbance can grow initially; it does not guarantee that there is a disturbance which can grow ultimately at a higher Reynolds number. Thus, one says that at a given Reynolds number the energy method determines stability, but cannot determine instability. In fact, Joseph has recently proved (1971) that longitudinal rolls in a shear flow, although they have an initially increasing disturbance energy, always decay ultimately.

We may speculate that, in flows with heat transfer, there is a smaller disparity between the linear and non-linear critical values because the buoyancy provides a mechanism for transport of disturbance energy normal to the flow even in the linearized case.

APPLICATION OF THE METHOD TO THE CONSTANT STRESS LAYER; THE ASYMPTOTIC SOLUTION

We are going to apply this method to the constant stress wall layer of a turbulent boundary layer or channel flow with zero pressure gradient. Rather than attempt to find the general type of disturbance which is most efficient, we will use the result of Serrin (1959) and Joseph (1966) for the Couette flow, presuming that here also longitudinal rolls will be most efficient; thus we will assume a disturbance independent of x_1 (the streamwise direction) and periodic cross-stream. Setting $U_i = (U(x_2), 0, 0)$,

$$u_1 = u\, e^{ikx_3}$$
$$u_2 = ik\psi\, e^{ikx_3}, \quad u_3 = -\psi'\, e^{ikx_3} \tag{18}$$

where u, and ψ are functions of x_2, and a prime denotes differentiation with respect to x_2, equation (9) becomes

$$\sigma^+ u + U' ik\psi/2 = (\nu u')' - k^2 \nu u$$
$$ik\sigma^+\psi + U' u/2 = -\pi'/\rho + 2\,(\nu ik\psi')' - ik\nu(\psi'' + k^2\psi) \tag{19}$$
$$-\sigma^+\psi' = -\pi ik/\rho - [\nu(\psi'' + k^2\psi)]' + 2\nu k^2\psi'$$

Eliminating the pressure π between the second and third of (19) results in (writing $\nabla^2 = D^2 - k^2$, $D = d/dy$)

$$\sigma^+ \nabla^2 \psi + U' iku/2 = (\nu \nabla^2 \nabla^2 + 2\nu' \nabla^2 D + \nu'' [D^2 + k^2]) \psi$$

$$\sigma^+ u + U' ik\psi/2 = (\nu \nabla^2 + \nu' D) u \tag{20}$$

In the constant stress layer of a turbulent boundary layer we may write $\nu = u_\tau{}^2/U'$, where u_τ is the shear velocity. Thus, in the viscous sublayer,

$$\nu = \nu_m \tag{21}$$

the molecular value, while in the logarithmic region we have

$$\nu \sim u_\tau \kappa y \tag{22}$$

where κ is von Karman's constant, ~ 0.4. We may use Reichardt's (1951) expression, which agrees with measurement reasonable well through the entire region (see Figure 1):

$$\nu = \nu_m [1 + \kappa (y^+ - R \tanh y^+/R)] \tag{23}$$

where R is the point where the extended wall profile meets the extended logarithmic profile, roughly 11.6 and $y^+ = yu_\tau/\nu_m$. We will find that our final answer is only weakly dependent on the particular form chosen for (23); it is only essential that the first two derivatives vanish at the wall. Using (23), it can be seen that ν' is zero in the sublayer, rises near R, and takes the constant value κ in the logarithmic region; ν'' is zero in the sublayer and in the logarithmic region, being non-zero (and positive) only near $y^+ = R$. From the relationship $\nu = u_\tau{}^2/U'$, we can evidently express the mean velocity gradient also in terms of the function (23). If we non-dimensionalize y by the value corresponding to $y^+ = R$, say y_* so that the dimensionless $y = y^+/R = y/y_*$, and with $\lambda = y_*^2 \sigma^+/\nu_m$, then the equations can be written as

$$\lambda \nabla^2 \psi + (R^2 ik/2f) u = f \nabla^2 \nabla^2 \psi + 2f' \nabla^2 D\psi + f'' (D^2 + k^2) \psi$$

$$\lambda u + (R^2 ik/2f) \psi = f \nabla^2 u + f' Du \tag{24}$$

with Reichardt's form for f

$$f = 1 + \kappa R(y - \tanh y) \tag{25}$$

We see (c.f. Figure 2) that the term in R^2 is, for $R = 11.6$ (the true value in the boundary layer) nearly 10^2 times the terms on the right-hand side near $y = 1$. As y increases, the terms do not become comparable until $y \sim 10^1$. This suggests an asymptotic solution (Cole, 1968) using R^2 as the large parameter (but holding the implicit R which appears in the definition of f fixed*). In order to make a sensible (non-trivial) problem, we must assume that $\lambda = 0(R^2)$ also; that is, that outside the (viscous) wall layer production and growth balance, while in the wall layer, viscous loss is important.

---

*It appears to be possible to construct an expansion allowing this R to vary also; it is more complicated, however, in that three distinguished limits are found, and the behavior of the solutions in our range of parameters is the same.

Sufficiently far outside (for $y > 10^1$) there must again be a region where viscosity is important, but there the solution is already so small that we may safely ignore the transition to this new region**. Thus, writing $\beta^2 = \lambda/R^2$, $R = \epsilon^{-1}$, we have

$$\beta^2 \nabla^2 \psi + (ik/2f)\, u = \epsilon^2 (f \nabla^2 \nabla^2 \psi + 2f' \nabla^2 D\psi + f''(D^2 + k^2)\psi)$$

$$\beta^2 u + (ik/2f)\, \psi = \epsilon^2 (f \nabla^2 u + f' u') \tag{26}$$

Now, for the inner problem (next to the wall) we can write $\eta = y/\epsilon$, and

$$\psi = \nu_0(\epsilon)\, (g_0(\eta) + \epsilon^2 g_1(\eta) + \cdots)$$

$$u = \nu_0(\epsilon)\, (h_0(\eta) + \epsilon^2 h_1(\eta) + \cdots) \tag{27}$$

The order of the boundary layer thickness was determined from the necessity of keeping the highest order derivative in the equations. The order of the second order terms in (27) was determined from the desire to obtain a g_1, h_1 different from g_0, h_0. If necessary for the matching (it will not be) another g_0 and h_0 of order between ϵ^2 and 1 might be added in (27). The equality of the orders of the leading terms arises from the need to retain ψ terms in the u equation, and u terms in the ψ equation. The equations are (after cross-substituting)

$$g_0'''' - \beta^2 g_0'' = 0; \quad g_1'''' - \beta^2 g_1'' = ikh_0/2 + 2k^2 g_0'' - \beta^2 k^2 g_0$$

$$h_0'' - \beta^2 h_0 = ikg_0/2; \quad h_1'' - \beta^2 h_1 = k^2 h_0 + ikg_1/2 \tag{28}$$

The boundary conditions at the wall, of course, are $\psi(0) = \psi'(0) = 0$; $u(0) = 0$. Applying these, the solutions which we need become

$$g_0 = D\,[1 - \beta\eta - e^{-\beta\eta}]$$

$$h_0 = D\,[e^{-\beta\eta} - 1 + \beta\eta + (\beta\eta/2)e^{-\beta\eta}]\, ik/2\beta^2$$

$$g_1 = B(e^{-\beta\eta} - 1) + \beta\eta[B + (9/4 + 4\beta^4)k^2 D/24\beta^6] - (1 - 4\beta^4)(\beta^2\eta^2 - \beta^3\eta^3/3)k^2 D/8\beta^6$$

$$-\beta\eta\, e^{-\beta\eta}(9/4 + 4\beta^4)k^2 D/24\beta^6 + \beta^2\eta^2\, e^{-\beta\eta} k^2 D/32\beta^6 \tag{29}$$

We have also discarded terms that are transcendentally large as $\eta \to \infty$, which would preclude matching. The constant B must be found in terms of D by the matching. D, of course, must remain arbitrary, since the equations are homogeneous. In addition, of course, we will have to determine ν_0 from the matching.

It is interesting to note in equation (28) that to zeroth order ψ (i. e. - the lateral motion) simply experiences a balance between decay and dissipation; the streamwise motion, however, experiences some production due to the lateral motion.

** That is, we will find that it is the matching to the solution between 1 and 10^1 that determines the eigenvalue relations to the order in which we are interested.

This agrees with the conclusions of Joseph (1971). From the equation for h_0 it is clear that the vortices are sweeping slow moving fluid up from the wall (and vice versa) producing disturbances in the streamwise motion.

We may now attack the outer problem. Since the amplitude of the solution is arbitrary, we may take it to be of order unity, writing

$$\psi = G_0(y) + \varepsilon^2 G_1(y) + \cdots$$
$$u = H_0(y) + \varepsilon^2 H_1(y) + \cdots \tag{30}$$

where, again, the order of the second term is fixed by requiring that G_1 be different from G_0 (a term of intermediate order similar to G_0 could be included, but will not be necessary). To first order we have

$$\beta^2 H_0 + (ik/2f)\, G_0 = 0$$
$$\beta^2 \nabla^2 G_0 + (ik/2f)\, H_0 = 0 \tag{31}$$

The first of these is the "mixing length" assumption, frequently made (see Bakewell, Lumley, 1967): that the perturbations in u are those produced from the mean velocity gradient by the lateral motion. Combining the two equations (31), we obtain

$$G_0'' - k^2 (1 - 1/4f^2\beta^4)\, G_0 = 0 \tag{32}$$

It is fairly simple to show that this solution cannot match with (29) unless there is a turning point - i. e. - unless $1 - 1/4f^2\beta^4$ has a zero somewhere. This is a simple kinematic requirement: if (32) has no turning point, the lateral velocity (outside the viscous layer) is monotone with distance from the surface, producing upward and downward streaming jets, rather than cells. These do not satisfy the requirement that the stream function and its derivative must vanish at infinity. If (32) has one turning point, we have one cell (see Figure 3). Two turning points correspond to two cells, each on top of the other, and so on. Clearly, more cells means a smaller length scale, and hence greater dissipation. Thus, the cases of two and more turning points correspond to a lower growth rate. We will consider only the case of a single turning point.

A simple form of the solution for (32) may be obtained (Cole, 1968) formally valid for large k (but probably satisfactory for k as small as 3) by the two variable expansion procedure. If we designate by y_c the point where

$$2\beta^2 f(y_c) = 1 \tag{33}$$

then for $y > y_c$ we can write

$$G_0 = [1 - 1/4f^2\beta^4]^{-1/4} e^{-kY} +, \quad Y_+ = \int_{y_c}^{y} [1 - 1/4f^2\beta^4]^{1/2} dy' \tag{34}$$

where we have discarded the positive exponential. Now, for $y < y_c$ the solution takes the form of trigonometric functions. The proper branch (i. e. - the continuation of (34) may be selected by constructing an inner expansion of (32) near the turning point, as is discussed in Cole (1968) (see Figure 4). The result is for $y < y_c$,

$$G_0 = 2\sqrt{2/3}\,(\sin\pi/12 + \cos\pi/12)\,[1/4\beta^4 f^2 - 1]^{-1/4}\sin(3\pi/4 - kY_-)$$
$$Y_- = \int_y^{y_c} [1/4\beta^4 f^2 - 1]^{1/2}\,dy' \tag{35}$$

For smaller k, the coefficient in (12) becomes a series in k^{-1}, the terms involving integrals and derivatives of $1/4\beta^4 f^2 - 1$. A little manipulation of the equations is sufficient to show, however, that, as written in (35), the coefficient has vanishing first and second derivatives at the wall, while the full series representation of the coefficient has vanishing first through third derivatives, due to the vanishing of the first two derivatives of f. (Substitution of a form such as (35) with arbitrary coefficient into (32) gives an equation for the coefficient, from which the derivatives at the wall may be determined; if the first derivative vanishes, the first three derivatives vanish. But a non-vanishing first derivative produces a term which could only be matched with (29) if f' or f'' had been assumed not to vanish at the wall.) Thus, although the proper form of the coefficient for finite k will change the shape of G_0 away from the wall somewhat, it will not affect the matching through third order; we will need only second order.

The second order equations for G_1 and H_1 are, of course,

$$\beta^2\nabla^2 G_1 + ikH_1/2f = f\nabla^2\nabla^2 G_0 + 2f'\nabla^2 DG_0 + f''(D^2 + k^2)\,G_0$$
$$\beta^2 H_1 + ikG_1/2f = f\nabla^2 H_0 + f'H_0' \tag{36}$$

and the right handsides may be evaluated in terms of G_0 and G_0' using equations (31) and (32). We are primarily interested in the behavior of the solutions near the wall, however. If account is taken of the fact that $G_0(0) = 0(\varepsilon), G_0'(0) = 0(1)$ (which we can only know from the first order matching; i.e. that the inviscid boundary condition is nearly satisfied, while the no slip condition is not) then the leading term (through linear terms in y) is

$$\beta^2\nabla^2 G_1 + ikH_1/2 \cong (k^2/4\beta^4)^2 G_0$$
$$\beta^2 H_1 + ikG_1/2 \cong (ik^3/8\beta^6)G_0 \tag{37}$$

or

$$G_1'' + k^2(1/4\beta^4 - 1)\,G_1 = (2k^4/16\beta^{10})\,G_0 \tag{38}$$

To linear terms, then, the particular solution will be

$$(G_1)_p = (2k^2/16\beta^{10})\,(1/4\beta^4 - 1)^{-1} G_0 \tag{39}$$

This result is the mathematical expression of the physical truth that near the wall, the variation of viscosity with y is not important.

The general solution will be exactly like (35) - we can thus write to the order of linear terms in y,

$$\psi = G_0 + \varepsilon^2\{C\,G_0 + (2k^2/16\beta^{10})\,(1/4\beta^4 - 1)^{-1} G_0\} + \cdots \tag{40}$$

where C is an unknown constant that must be determined from the matching.

We may now proceed to the matching, using an intermediate limit in which $y/\delta = y_\delta = O(1)$, $\delta(\epsilon)$ 0, $\delta/\epsilon \to \infty$, so that $\eta = \delta y_\delta/\epsilon \to \infty$. The inner solution becomes (we will do the matching only in ψ)

$$\psi \cong \nu_0 D(1-\beta\delta y_\delta/\epsilon)+\epsilon^2\nu_0\{-B+\beta(\delta y_\delta/\epsilon)\,[B+(9/4+4\beta^4)k^2D/24\beta^6]$$
$$-(1-4\beta^4)\,[\beta^2(\delta y_\delta/\epsilon)^2-\beta^3(\delta y_\delta/\epsilon)^3/3]k^2D/8\beta^6\}+\cdots \tag{41}$$

neglecting transcendentally small terms.

Writing

$$3\pi/4 - kY_- = \mu(\epsilon) + k(1/4\beta^4 - 1)^{1/2}\,\delta y_\delta + O(\delta^4) \tag{42}$$

near the wall, we can write (to third order)

$$G_0 = P\sin[\mu(\epsilon) + k(1/4\beta^4 - 1)^{1/2}\delta y_\delta + \cdots] \tag{43}$$

where P is a constant. Thus, through terms $O(\delta^3)$, $O(\mu^3)$,

$$G_0 = P\{\mu-\mu^3/6+\delta y_\delta k(1/4\beta^4-1)^{1/2}(1-\mu^2/2)-k^2(1/4\beta^4-1)(\mu-\mu^3/6)(\delta y_\delta)^2/2$$
$$-k^3(1/4\beta^4-1)^{3/2}(1-\mu^2/2)(\delta y_\delta)^3/6+\cdots\} \tag{44}$$

and, from (40), ψ is given by

$$\psi = (44)+P\epsilon^2\{C+(2k^2/16\beta^{10})(1/4\beta^4-1)^{-1}\}$$
$$\{\mu-\mu^3/6+\delta y_\delta k(1/4\beta^4-1)^{1/2}(1-\mu^2/2)+\cdots\} \tag{45}$$

where the second term is correct to $O(\delta)$.

Matching the term of $O(\delta)$ in (45) we find that we must have $\nu_0 = \epsilon$,

$$-\beta D = Pk(1/4\beta^4-1)^{1/2} \tag{46}$$

Writing

$$\mu(\epsilon) = \omega_1\epsilon + \omega_2\epsilon^2 + \omega_3\epsilon^3 + \cdots \tag{47}$$

we may match the term of $O(\epsilon)$ in (41), giving

$$D = P\omega_1 \tag{48}$$

Since there is no term of $O(\epsilon^2)$ in (41), $\omega_2 = 0$. Matching terms of $O(\epsilon^3)$, we have

$$-B = P\omega_3 - \omega_1^3 P/6 + P[C + (2k^2/16\beta^{10})(1/4\beta^4-1)^{-1}]\,\omega_1 \tag{49}$$

Terms of $0(\epsilon^2\delta)$ give

$$\beta\,[B + (9/4 + 4\beta^4)k^2 D/24\beta^6] = -\,Pk(1/4\beta^4 - 1)^{1/2}\,\omega_1{}^2/2$$

$$+\,P[C + (2k^2/16\beta^{10})\,(1/4\beta^4 - 1)^{-1}]k(1/4\beta^4 - 1)^{1/2} \qquad (50)$$

Terms of $0\,(\epsilon\delta^2)$ give

$$-(1 - 4\beta^4)\,\beta^2 k^2 D/8\beta^6 = -\,Pk^2(1/4\beta^4 - 1)\omega_1/2 \qquad (51)$$

and of $0\,(\delta^3)$

$$+\,(1-4\beta^4)\beta^3 k^2 D/24\beta^6 = -\,Pk^3(1/4\beta^4 - 1)^{3/2}/6 \qquad (52)$$

As a result of (46), (51) and (52) are satisfied exactly. Equations (49) and (50) represent two equations in three unknowns; fortunately two of the unknowns occur in the same combination, giving

$$\omega_3 = -\,(1/4\beta^4 - 1)^{1/2}\,(1/48\beta^4 + 1)\,k^3/2\beta^3 \qquad (53)$$

The other combination of constants need not concern us.

We can thus write, using (46), (47), (48) and (53), our eigenvalue conditions:

$$3\pi/4 - kY_{-0} = -(1/4\beta^4 - 1)^{1/2}\epsilon k/\beta - (1/4\beta^4 - 1)^{1/2}(1/48\beta^4 + 1)\epsilon^3 k^3/2\beta^3 + \cdots \qquad (54)$$

to $0(\epsilon^5)$, where Y_{-0} is given by

$$Y_{-0} = \int_0^{y_c} [1/4\beta^4 f^2 - 1]^{1/2}\,dy' \qquad (55)$$

Expression (54) gives the offset (see Figure 5) by which the outer (inviscid) solution fails to meet the inviscid boundary condition to leave room $0(\epsilon)$ for the viscous (eddy) boundary layer. Equation (54) has, in general, two positive roots for k (for fixed ϵ, β); we are interested in the critical condition when there is only one (double) root. This may be obtained by differentiating the expression (54) with respect to k, requiring that the derivative vanish also at the same value of k. If we indicate by k_+ this critical value, we obtain (noting that $\beta/\epsilon = \sqrt{\lambda}$)

$$(\sqrt{\lambda}/k_+)^3 = (4/3\pi)\,(1/4\beta^4 - 1)^{1/2}(1/48\beta^4 + 1)$$

$$Y_{-0}\sqrt{\lambda} = (1/4\beta^4 - 1)^{1/2} + (1/4\beta^4 - 1)^{1/6}(1/48\beta^4 + 1)^{1/3}(3/2^{1/3})(3\pi/8)^{2/3} \qquad (56)$$

If now a value of β is selected, Y_{-0} may be calculated, as well as $k_+/\sqrt{\lambda}$ and $\sqrt{\lambda}\ Y_{-0}$; from the latter then we have $\sqrt{\lambda}$, and from the former, k_+, while from $R = \sqrt{\lambda}/\beta$ we obtain R.

If Figure 6 we show a plot of λ and k_+ versus R (holding the R in f fixed at 11.6). It is worth noting that the form of the viscosity outside the sublayer did not influence the solution; the only property we used was the vanishing of the first two derivatives at the wall, a kinematic requirement in a zero pressure gradient. The form of the mean shear only influences expression (55).

COUETTE FLOW

The amount of algebra involved in arriving at (56) is such as to make one suspicious of the result in the absence of an independent check. Fortunately, one is available. The case of Couette flow, solved by Serrin (1959) and Joseph (1966) can be handled by the same technique. The inner solution (29) remains the same, as does the outer solution; we have only to set $f \equiv 1$; the fact that f varied affected the matching only through the value of Y_{-0}. To determine the value of Y_{-0}, we have only to determine the value of y_c, and here we must replace our previous condition (that ψ decay exponentially) by a requirement that ψ be symmetric about the mid-point of the channel. Thus y_c must be fixed so that $3\pi/4 - kY_-|_{1/2} = \pi/2$, or

$$3\pi/4 - k(1/4\beta^4 - 1)^{1/2}(y_c - 1) = \pi/2 \tag{57}$$

Substituting in (54), we have

$$\pi/2 - k(1/4\beta^4 - 1)^{1/2} = -(1/4\beta^4 - 1)^{1/2}\epsilon k/\beta - (1/4\beta^4 - 1)^{1/2}(1/48\beta^4 + 1)\epsilon^3 k^3/2\beta^3 + \cdots \tag{58}$$

We may thus go directly to (56), substituting $\pi/2$ for $3\pi/4$, and $(1/4\beta^4 - 1)^{1/2}$ for Y_{-0}; thus, we have

$$(\sqrt{\lambda}/k_+)^3 = (2/\pi)(1/4\beta^4 - 1)^{1/2}(1/48\beta^4 + 1)$$

$$(1/4\beta^4 - 1)^{1/2}\sqrt{\lambda} = (1/4\beta^4 - 1)^{1/2}$$

$$+ \ (1/4\beta^4 - 1)^{1/6}(1/48\beta^4 + 1)^{1/3}(3/2^{1/3})(\pi/4)^{2/3}$$

The same computations have been carried out using these equations; the results are also shown in Figure 6. It may be seen that the curve of λ vs. R extrapolates through the exact value of $R = 4.54$ as $\lambda \rightarrow 0$, while the curve of k_+ extrapolates to the exact value of 1.56 at the same value of R.

We may conclude from this that our estimate for λ is satisfactory for $\lambda \geq 20$, while the k estimate is somewhat less reliable, being good for perhaps $k \geq 5$. The bending away of both curves below these points must be attributed to progressive failure of the approximation.

DISCUSSION OF THE RESULTS

Examining Figure 6, we evidently have at $R = 11.6$ a growth rate of $\lambda \sim 31.5$ and a $k_+ \sim 3.7$ (the latter value being somewhat less reliable). The critical Reynolds number for this type of flow (corresponding to $\lambda = 0$) is evidently about 9.2. That this should be larger than that for the Couette flow can be justified by noting that in the Couette flow the top halves of the eddies are subjected to the same strain rate, and hence are self-sustaining; in the wall-layer flow, the shear drops rapidly, and the outer parts of the eddies must be sustained by the inner parts; hence the shear must be proportionately higher, or the dissipation lower, resulting in a higher critical Reynolds number.

The growth rates are quite large measured in sublayer variables; since sublayer time scales are themselves the shortest in a wall flow, the evolution of these eddies must be regarded as extremely rapid. In this connection, the observations of Kline and his group (e. g. - Kline, et. al. , 1967) are relevant: wall eddies of this general form were observed to "burst," i. e. - evolve very rapidly relative to other sublayer phenomena.

In other respects the dynamics of the eddies (from the asymptotic analysis) agree well with observations, in particular the "mixing length approximation" (eg. 31a), which has often been used (Townsend, 1956; Bakewell and Lumley, 1967; Payne and Lumley, 1967).

The single disappointing feature of these eddies is their size. Referred to sublayer variables, $k_+ = 3.7$ corresponds to a transverse wave length of roughly 20, while observed wave lengths (e.g. Bakewell and Lumley, 1967) are closer to 100. Our eddies have their apparent centers (the maxima of ψ) at about 10, while observed eddies seem to have their centers closer to 30. Evidently these eddies are too small by a factor between 3 and 5; can this be explained?

In our considerations, we have not taken into account the effect of the presence of the eddy on the undisturbed velocity profile; to an inviscid first order approximation the mean shear will be reduced (outside the viscous region at the wall) by an amount $k^2 \psi^2 / 2\beta^2$ ($1/2\beta^2 \sim 2$ when $R = 11.6$). This reduction in shear will favor the growth of an eddy with the same value of k_+, but a smaller k due to the greater thickness of the layer. That is, as the shear is reduced, all length scales will grow to keep R, k_+ roughly constant. Hence, the k we have found is evidently a starting value; evolution of the eddy will involve progressive decrease in the dimensional k, until the disturbance energy reaches its peak and begins to decay. Viewed another way, as the shear is reduced, the effective Reynolds number is being reduced* - the eddy still feels the same effect of viscosity at the wall, but has a smaller energy source; we may expect the eddy to move down the curve toward $\lambda = 0$, having at that point a $k_+ \sim 0.7$ (from rather uncertain extrapolation of the curve of Figure 6). This corresponds roughly to the value observed, in sublayer variables, of ~ 100.

ACKNOWLEDGEMENTS

Supported in part by the Garfield Thomas Water Tunnel of the Ordnance Research Laboratory.

This work grew out of a numerical analysis carried out by Elswick (1967) for the same flow with $\sigma = 0$; although the results of that analysis now appear to be somewhat questionable, and have not been used here, its seminal role in this investigation must be acknowledged.

I am particularly grateful to L. Katherine Jones, who carried out extensive preliminary computations in connection with this investigation, finally resulting in Figure 6.

BIBLIOGRAPHY

[1] H. P. Bakewell and J. L. Lumley: The Viscous Sublayer and Adjacent Wall Region in Turbulent Pipe Flow. Physics of Fluids, 10, 1880 (1967).

[2] J. D. Cole: Perturbation Methods in Applied Mathematics. Blaisdell Publ. Co., Waltham, Mass (1968).

[3] R. C. Elswick, Jr.: Investigation of a Theory for the Structure of the Viscous Sublayer in Wall Turbulence. M. S. Thesis in Aerospace Engineering, The Pennsylvania State University (1967).

[4] W. Heisenberg: Zur Statistischen Theorie der Turbulenz. Z. Phys., 124, 628 (1948).

---

* Note that the form of the velocity profile only affects Y_{-0}; a reduction in shear reduces Y_{-0} but keeps the relation (54) otherwise unchanged.

[5] D. D. Joseph: Nonlinear Stability of the Boussinesq Equations by the Method of Energy. Archive for Rational Mechanics and Analysis, 22, 163 (1966).

[6] D. D. Joseph and W. Hung: Contributions to the Nonlinear Theory of Stability of Viscous Flow in Pipes and Between Rotating Cylinders. Submitted for Publicatioin (1971)

[7] S. J. Kline, W. C. Reynolds, F. A. Schraub and P. W. Runstadler; The Structure of Turbulent Boundary Layers. J. Fluid Mech., 30, 741 (1967).

[8] C. C. Lin: The Theory of Hydrodynamic Stability. Cambridge, The University Press (1955).

[9] H. A. Lorentz: Ueber die Entstehung Turbulenter Fluessigkeits-Bewegungen und Ueber den Einfluss Dieser Bewegungen dei der Stroemung Durch Rohren. Abhandlung Ueber Theoretische Physik, Leipzig, 1, 43 (1907).

[10] J. L. Lumley: Structure of Inhomogeneous Turbulent Flows. Atmospheric Turbulence and Radio Wave Propagation, Nauka, Moscow, 166 (1967)

[11] J. L. Lumley: Toward a Turbulent Constitutive Relation. J. Fluid Mechanics, 41, 413 (1970a).

[12] J.L. Lumley: Stochastic Tools in Turbulence. Academic Press, New York (1970b).

[13] J. L. Lumley: Structure of Turbulent Shear Flows. A. I. Ch. E. Advanced Seminar: Turbulent Transport, Chicago (1970c).

[14] A. S. Monin and A. M. Yaglom: Statistical Fluid Mechanics (Trans.: Scripta Technica; J. Lumley, ed.), MIT Press, Cambridge, Mass (1971).

[15] W. McF. Orr: The Stability or Instability of Steady Motions of a Liquid. Part II: A Viscous Liquid. Proc. Roy. Irish Acad. (A) 28, 122 (1907).

[16] F. R. Payne and J. L. Lumley: Large Eddy Structure of the Turbulent Wake Behind a Circular Cylinder. Physics of Fluids Supplement, 10, S194 (1967).

[17] G. I. Petrov: O Rasprostraneniy Kolebaniy v Vyazkoy Zhidkosti i Vozniknoveniy Turbulent-Nosti. Vuypusk 345, Trudi Tsentral 'novo Aero-Gidrodynamicheskovo Unstituta, Moscow (1938).

[18] H. Reichardt: Vollstaendige Darstellung der Turbulenten Geschwindigkeitsverteilung in Glatten Leitungen. ZAMM 31, 208 (1951).

[19] W. C. Reynolds and W. G. Tiederman: Stability of Turbulent Channel Flow, with Application to Malkus' Theory. J. Fluid Mech., 27, 253 (1967).

[20] J. Serrin: On the Stability of Viscous Fluid Motions. Archive for Rational Mechanics and Analysis, 3, 1 (1959).

[21] G. I. Taylor: Scientific Papers II. The University Press, Cambridge, 27 (1960).

[22] A. A. Townsend: The Structure of Turbulent Shear Flow. The University Press, Cambridge (1956).

APPENDIX I

Excerpt from: G. I. Petrov, On the Growth of Oscillations in a Viscous Liquid and Transition to Turbulence. Publication Number 345, Transactions of the Central Aero-Hydrodynamic Institute, Moscow, 1938: pp. 20-24.

Lorentz, selecting an initial velocity distribution in the form of elliptical vortices, obtained:

$$R = U_{max} h/\nu = 72$$

From a more precise statement of the problem, Karman and Orr obtained an even lower value $Re \sim 42$, which obviously does not correspond to reality.

Below, we will attempt to give an explanation of this fact, and to show how this problem may be correctly formulated.

The linearized equation for the disturbance stream function, superimposed on a stationary plane parallel flow, has the form:

$$\Delta\Delta\psi - R[U\,\partial\Delta\psi/\partial x - U''\,\partial\psi/dx + \partial\Delta\psi/\partial\tau] = 0. \qquad (8)$$

As we have already shown, it is convenient to assume a solution of the form $f(x, y)e^{-i\beta t}$, i. e., in the form of an oscillation, the amplitude of which is a function of x and y. The distribution of the initial amplitude, or the function f, may be obtained from the equation:

$$\Delta\Delta f - R[U\,\partial\Delta f/\partial x - U''\,\partial f/\partial x - i\beta\Delta f] = 0 \qquad (9)$$

and corresponding boundary conditions. This equation is of the fourth order, and cannot be placed in self-adjoint form. The coefficients of the equation contain two characterizing parameters R and β, of which the first, in accordance with its physical interpretation, is always real, while the second is in general complex. Thus, the problem may be stated as: for a given value of R (which defines the basic flow) find a system of eigenvalues β and a corresponding system of functions $f(x, y)$, assuming that an arbitrary initial disturbance may be expanded in a series of these functions, or, in a different formulation, find a value of R which is associated in some way with a definite region of the values of β (for example, the minimum value of R, below which all given oscillations will decay, i. e., the imaginary part of β will be negative). This value of R is ordinarily used in calculating critical numbers, both as we have seen, in connection with the problem of stability of the flow of a viscous fluid and, as it is ordinarily formulated, the determination of the point of transition of a laminar flow to turbulence.

Neither of these formulations is, at first glance, obvious. The first, because it is not known with certainty whether the complete solution, expanded in a series of periodic oscillations of the form $f(x, y)\, e^{-i\beta t}$, will decay. And the second because the critical number R determines an initial growth diverging from the real velocity of the fluid, and this change in the regime determines a change in the mean velocity field.

Let us look at several general properties of our system. As we have already remarked, equation (9) cannot be placed in a self-adjoint form, and the eigenvalues of our system will in general be complex, in addition to the obvious difference from systems ordinarily considered in the theory of oscillations, that the characteristic parameters appear in the coefficients of the derivatives.

Let us examine the following expression:

$$\iint \bar{f}(x, y) L(f(x, y)) dxdy$$

the integral being taken over the region of the flow.

Here:

$$L(f) = \Delta\Delta f - R(U\,\partial\Delta f/\partial x - U''\,\partial f/\partial x - i\beta\Delta f),$$

i. e., the right-hand (sic) side of equation (9).

This expression can be transformed to the following form:

$$\iint \bar{f}(x, y) L(f(x, y) dxdy = \iint \Delta f \Delta f dxdy + R\iint (U\,\partial f/\partial x\,\Delta f + U''\,\bar{f}\,\partial f/\partial x + i\beta\bar{f}\,\Delta f) dxdy + \cdots \tag{20}$$

plus a contour integral.

If the function $f(x, y)$ satisfies equation (9), then expression (20) must equal zero, and we may determine R as the ratio of the two integral expressions, and seek the critical value as the minimum of this ratio, setting β_i equal to several values.

Since we have given boundary confitions, let us examine the variation of just the double integrals. The problem of the determination of the critical value may be set as an isoperimetric problem, i. e., to find the extremum of the functional $I_1 = \iint \Delta\bar{f}\Delta f dxdy$ under the condition

$$I_2 = \iint - [U\,\partial\bar{f}/\partial x\,\Delta f + U''\,\bar{f}\,\partial f/\partial x + i\beta\bar{f}\Delta f] dxdy = 1.$$

The Euler equation for our problem will be equation (9), and for $\bar{f}$ the equation of its adjoint:

$$\Delta\Delta\bar{f} - \lambda\,[U\,\partial\Delta\bar{f}/\partial x + 2U'\,\partial^2\bar{f}/\partial x\partial y - i\beta\Delta\bar{f}] = 0. \tag{21}$$

But, in order that our integrals have the same physical interpretation as in the work of Lorentz, Karman, et. al. which we described above, we must assume that $\bar{f}$ is the complex conjugate of a complex function f, i. e., that

$$f = f_r + if_i, \text{ to } \bar{f} = f_r - if_i.$$

All of our conclusions follow from the form of the ititial disturbance, i. e., by taking into account that the coefficients of our equation do not depend on x, taking $f = \phi e^{i\alpha x}$ and then $f = \bar{\phi}(y)e^{-i\alpha x}$ (where $\bar{\phi}(y)$ is the complex conjugate of $\phi(y)$).

Then the expression (2) takes the following form:

$$\int_0^1 (\bar{\phi}''\phi'' + 2\alpha^2\bar{\phi}'\phi' + \alpha^4\bar{\phi}\phi)\,dy + R\int[\alpha U\,1/2(\bar{\phi}\,\phi' - \phi\bar{\phi}') - \beta_i(\phi'\bar{\phi}' + \alpha\phi\bar{\phi})]\,dy$$
$$- i\alpha R \int_0^1 [(U - c_r)(\bar{\phi}'\phi' + \alpha^2\bar{\phi}\phi) + 1/2\,U''\bar{\phi}\phi]\,dy \tag{20a}$$

The integrals are taken over a region contained bwtween the walls, and over a wavelength of the given disturbance, so that the contour integral will vanish.

The real part of expression (20a) can be regarded as the change in energy in a strip of one wavelength of the disturbance, and may be obtained by examining the expression for dE/dt dx dy (see expression (6)), taking into account that the real part of the stream function is $\psi = \phi(y)e^{i(\alpha x-\beta t)}$ and integrating in x over one wave length.

The same functional can be obtained as an expression $\int_0^1 (\bar{\phi}L(\phi) + \phi\bar{L}(\bar{\phi}))dy$, where $L(\phi)$ is the left-hand side of equation (13), while $\bar{L}(\bar{\phi})$ is the corresponding expression for the conjugate function, and has the form:

$$\bar{L}(\bar{\phi}) = \bar{\phi}^{IV} - 2\alpha^2\bar{\phi}'' + \alpha^4\bar{\phi} + i\alpha R[(U - \bar{c})(\bar{\phi}'' - \alpha^2\bar{\phi}) - U''\bar{\phi}],$$

i. e. - the left-hand side of equation (13) with the complex parameters changed to their conjugates.

The imaginary part of expression (2) may be obtained as an expression

$$\int_0^1 (\bar{\phi}\, L(\phi) - \phi\bar{L}(\bar{\phi}))\, dy$$

Both of these expressions for functions, satisfying equation (13), must vanish. Then from the real part we may express R, or β for given R, as a ratio of functionals. The imaginary part satisfies a Zol'berg condition, from which it follows that for undisturbed velocity profiles not having an inflection point, c_r is always less than U_{max}, and we may obtain the phase velocity c_r or the wave number β_r as a ratio of functionals.

But in order that the disturbance be dynamically possible, i. e., satisfy equation (13), the real and imaginary parts must vanish simultaneously, and the problem of the critical value must be formulated as a problem of the extremum of $\int_0^1 (\bar{\phi}''\phi'' + 2\alpha^2\bar{\phi}'\phi' + \alpha^4\bar{\phi}\phi)dy$ under the condition

$$\int_0^1 \{1/2\, \alpha U(\bar{\phi}'\phi - \bar{\phi}\phi') + i\alpha\,|\,(U-c_r)(\phi'\bar{\phi}' + \alpha^2\phi\bar{\phi}) + 1/2\, U\, \phi'\bar{\phi}\,|\}dy = 1$$

(which includes the condition that the imaginary part vanish). The Euler equations for this problem are equation (13) for ϕ and its adjoint for $\bar{\phi}$, i. e.,

$$\bar{\phi}^{IV} - 2\alpha^2\bar{\phi}'' + \alpha^4\bar{\phi} - i\alpha R[(U - c)(\bar{\phi}'' - \alpha^2\bar{\phi}) + 2U'\bar{\phi}'] = 0.$$

But, if ϕ satisfies equation (13), then the function must satisfy the equation $\bar{L}(\bar{\phi}) = 0$. This is possible only if the following relation holds:

$$(U - c_r)(\bar{\phi}'' - \alpha^2\bar{\phi}) - 1/2\, U''\bar{\phi} + U'\bar{\phi}' = 0, \qquad (22)$$

i. e. - if equation (13) and its conjugate are anti-symmetric.

But the condition (22) very much reduces the class of functions which can give our solution. Since the solution of equation (22) depends only on two arbitrary parameters, the solution of the variational problem in such a formulation is apparently impossible. This explains the failure of all the attemps to resolve the problem of the so-called energy method, since apparently disturbances obtained from the minimum value of the ratio which determines R, are dynamically impossible.

Thus, the critical number R may be found as the smallest value of the ratio $I_1/R(I_2)$ (where $R(I_2)$ is the real part of the complex quantity I_2), in which are substituted solutions of equation (9), but these functions do not give an extremum of this ratio. It may also be said regarding the eigenvalues $\beta = \beta_r + i\beta_i$, that in the system under consideration they apparently do not possess an extremal character.

Translator's Note: Equation (13) is:

$$\phi'''' - 2\alpha^2\phi'' + \alpha^4\phi - i\alpha R[(\bar{U} - c)(\phi'' - \alpha^2\phi) - U''\phi] = 0$$

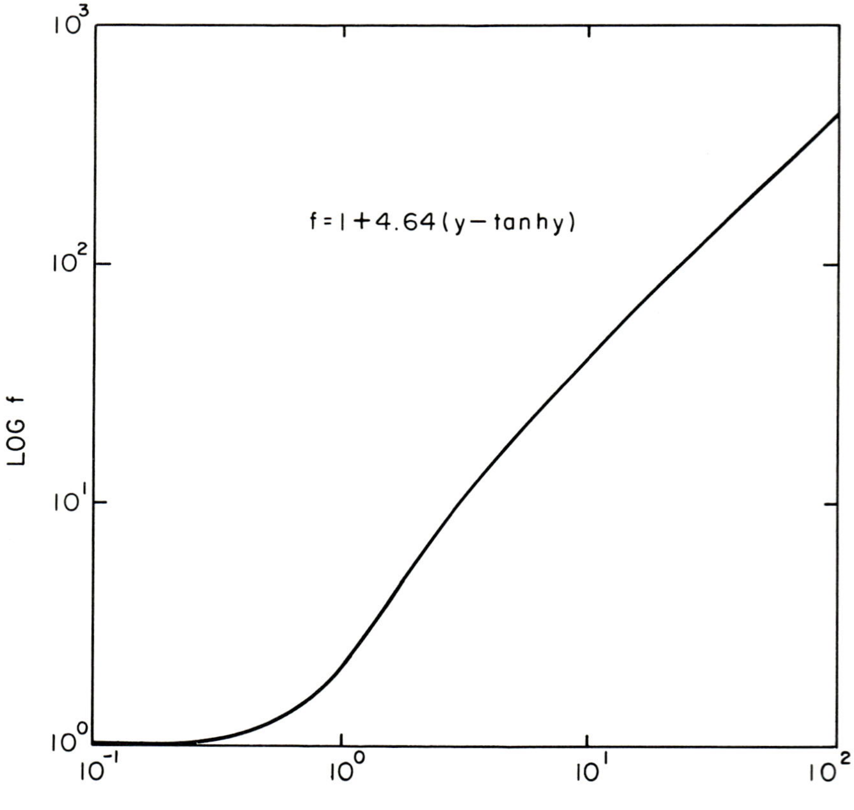

Figure 1 Reichardt's (1951) Viscosity Expression.

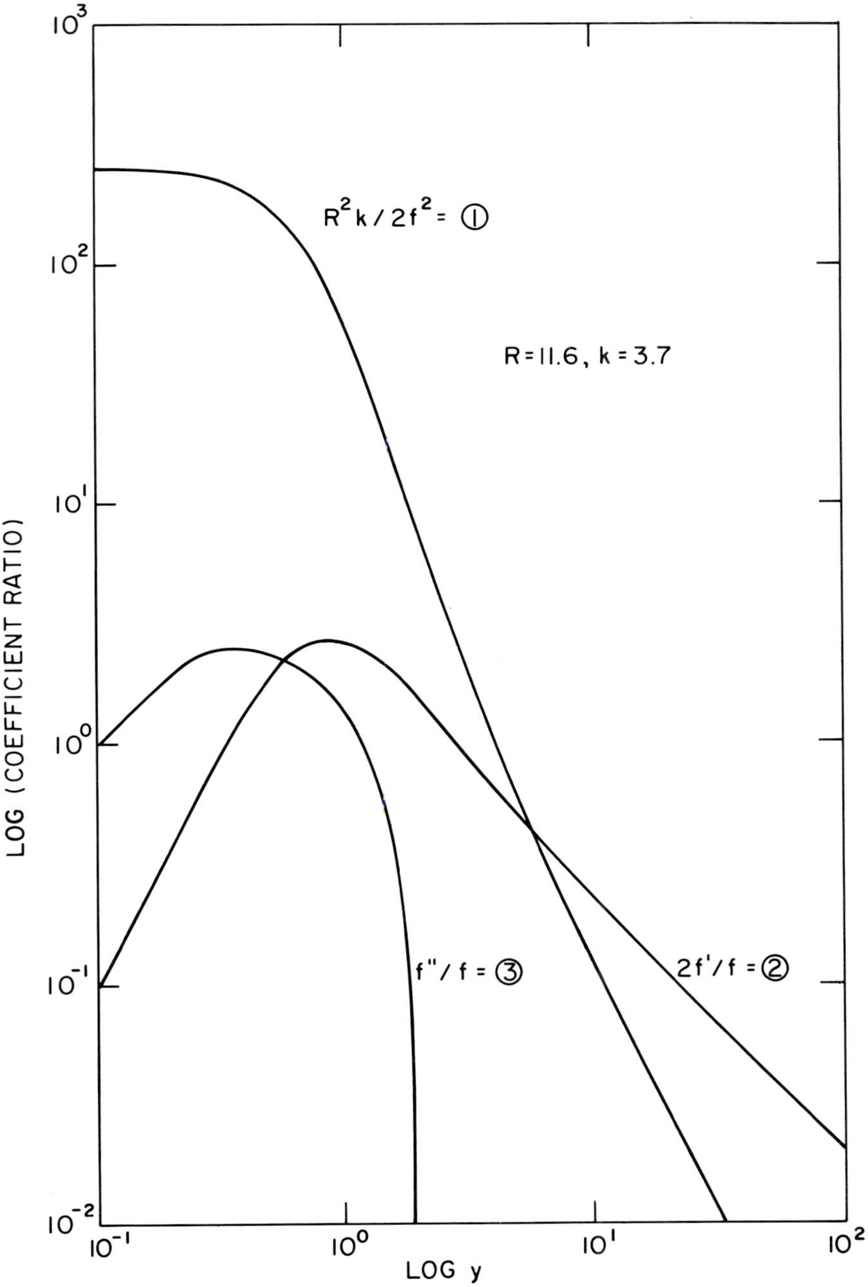

Figure 2. Ratio of the Coefficients of the Production Term 1, the f' Term 2, and the f'' Term 3 to f.

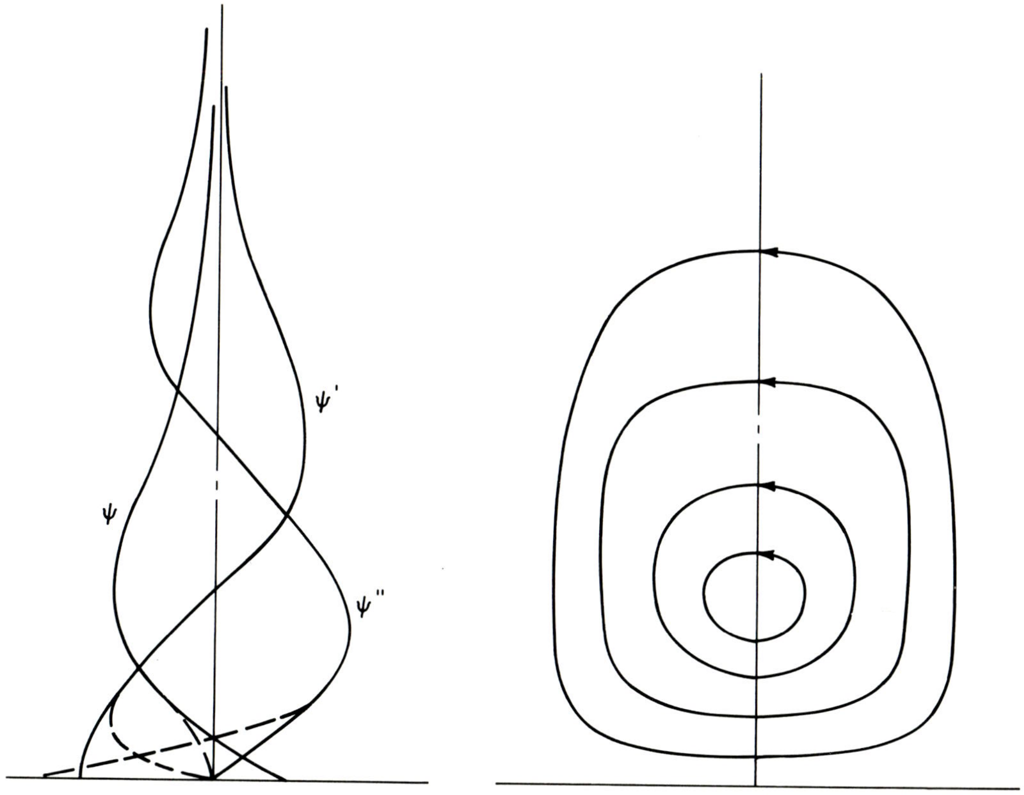

(a) ONE TURNING POINT

Figure 3 Structure of ψ and Corresponding Cells for One and Two Turning Points. Note Change due to Viscosity (dashed curves).

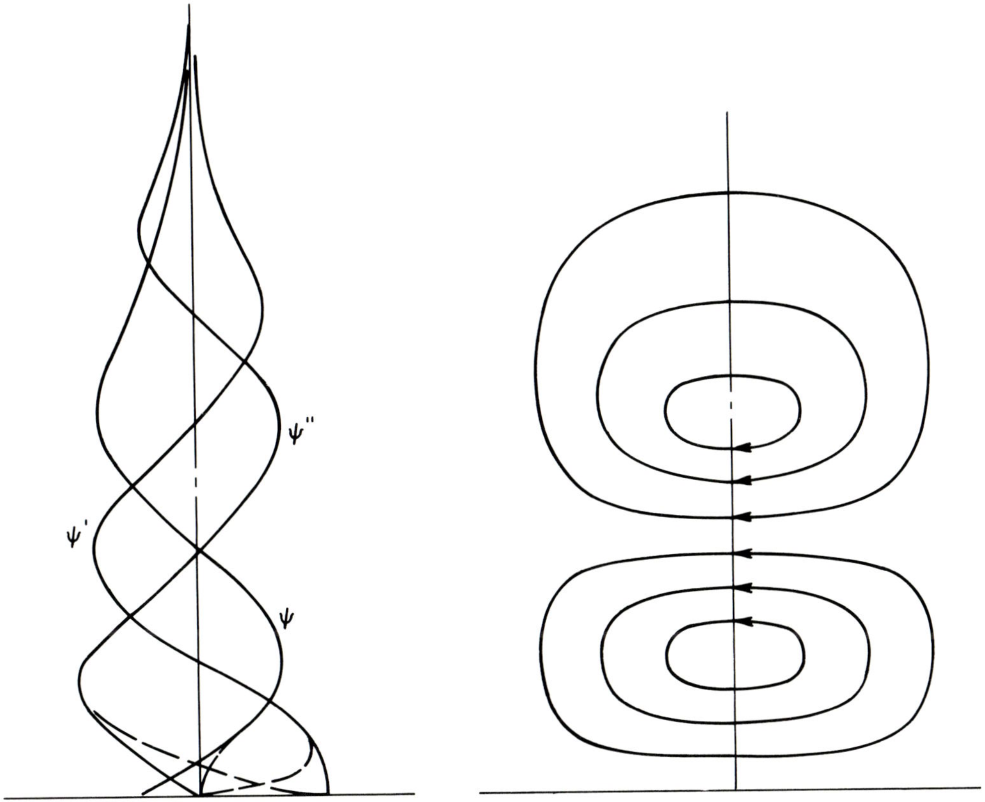

(b) TWO TURNING POINTS

Figure 3 Continued

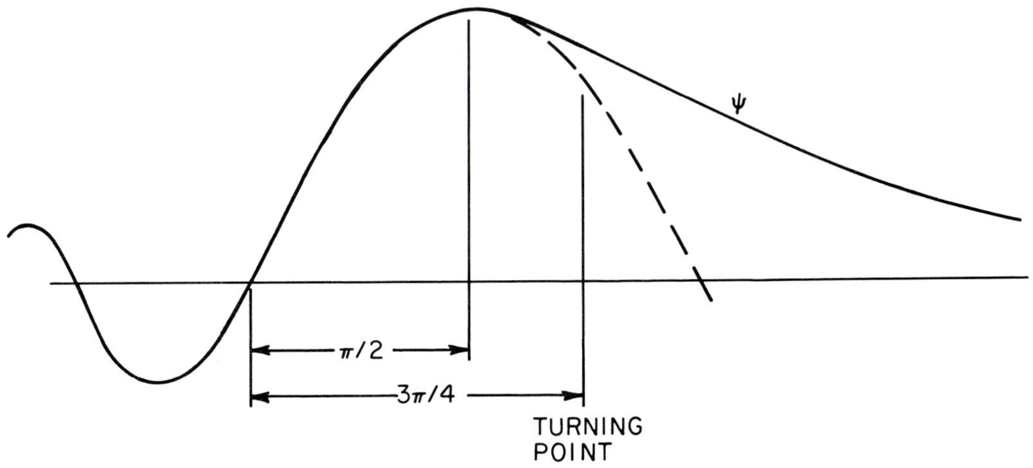

Figure 4 Form of the First Order Outer Solution, After Cole (1968). The Corresponding Solution for Couette Flow is Sketched (dashed).

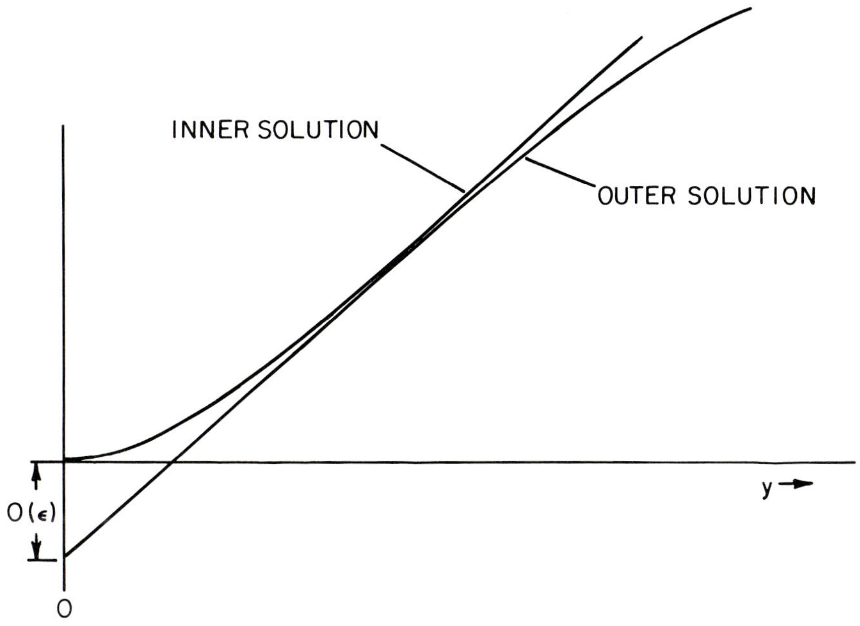

Figure 5 Illustration of the Offset Required for Matching.

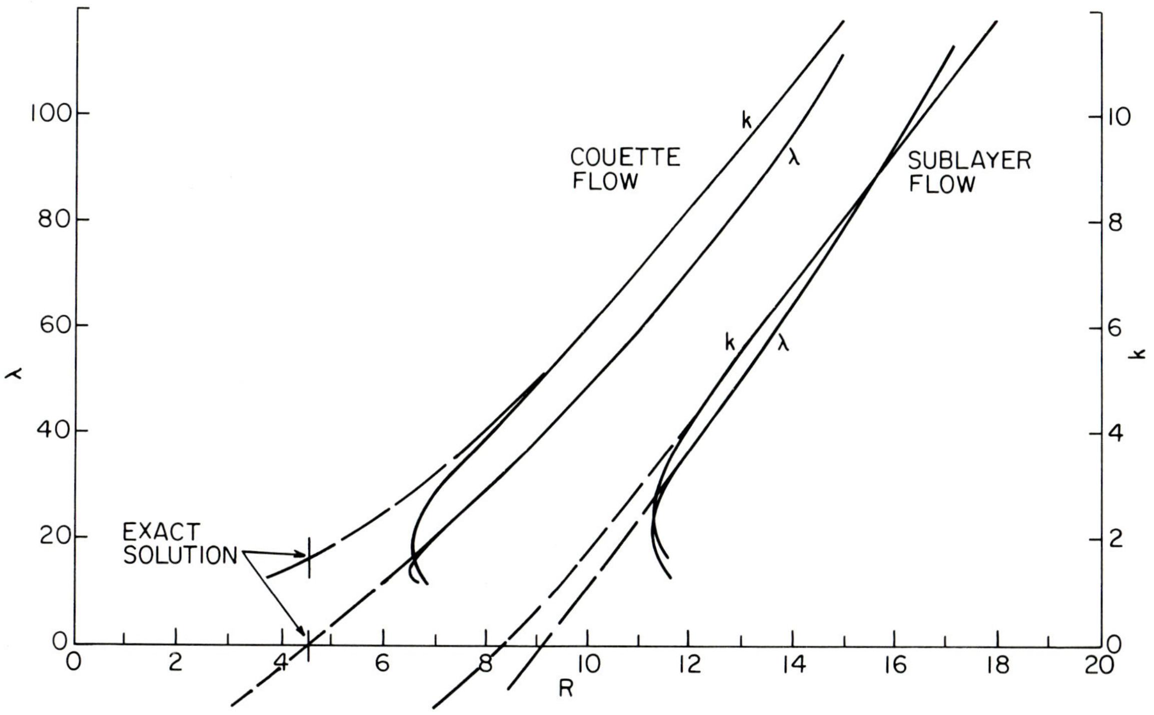

Figure 6 Plot of Growth Rate and Wave Number vs. Reynolds Number for Couette Flow and for the Boundary Layer.

Developments in Mechanics, Vol. 6. Proceedings of the 12th Midwestern Mechanics Conference: 5

Fluid Mechanics and Desalination

RONALD F. PROBSTEIN
MASSACHUSETTS INSTITUTE OF TECHNOLOGY

ABSTRACT

Two promising techniques for the desalination of brackish water, which has a salt content about a tenth that of sea water, are reverse osmosis and electrodialysis. Both of these processes employ membranes. One of the main limitations of reverse osmosis has been the relatively slow rates of product water flow through the membrane, while in electrodialysis a limitation has been excessive membrane deterioration under high rates of salt removal.

In reverse osmosis the saline water is separated from the fresh water by a membrane supported by a structure. The application of a large enough hydrostatic pressure to the saline water causes desalted water to pass through the membrane, which is relatively permeable to the water as opposed to the salt which remains behind. Electrodialysis makes use of the fact that salt in water exists in the form of positively and negatively charged ions. Channels through which the saline water passes are made up by alternating positive ion and then negative ion permeable membranes, many such channels consituting a cell. The passage of a direct current through the cell results in the depletion and enrichment of the salt in alternating channels, because the ions move toward the electrode with opposite charge.

In the present paper a review of a fluid mechanic analysis of porous reverse osmosis membranes is presented in which the membrane is considered to be composed of pores whose surfaces are electrically charged when the brackish water flows through. This phenomenon sets up an electric field which tends to retard the salt flow compared to the flow of water. It is shown both by theory and experiment that it may be possible to construct membranes of charged porous materials with much higher product flow rates though somewhat lower salt rejection rates than present cellulose acetate membranes, which reject salt by an activated diffusion mechanism.

In an electrodialysis cell the total current through a cell pair, which is proportional to the amount of salt removed, at first increases linearly with the applied voltage and then at higher voltages levels off as the salt concentration decreases, thereby causing an increase in the electrical resistance of the fluid. This leveling off or "polarization" phenomena, which results in membrane deterioration depends on the flow conditions. It is shown that polarization does not take place until a higher salt removal rate with increasing flow velocity and/or fluid turbulence level. The polarization behavior is described both theoretically and experimentally for laminar, intermediate and turbulent flow conditions, with and without the presence of membrane fouling material in the water. It is indicated how the results obtained may be applied to operate systems at higher salt removal rates without excessive polarization.

Developments in Mechanics, Vol. 6. Proceedings of the 12th Midwestern Mechanics Conference:

6

Nonlinear Models of Normal and Abnormal Heart Rhythms

R. M. ROSENBERG and C. H. CHAO
UNIVERSITY OF CALIFORNIA, BERKELEY

ABSTRACT

A mathematical model of electrical heart activity and of the electrocardiograph is described. The electrocardiograms of normal and abnormal heart rhythms of this model are compared with those obtained clinically; the agreement between them is satisfying.

INTRODUCTION

It is a startling fact that the human heart pumps 75 gallons of blood in one hour, or nearly 18 million barrels in 70 years, and it does this with a pump having an average capacity of only 18 fluid ounces [1]. The heart consists of two halves (right and left), and each has two chambers, the atrium and the ventricle. In each half, a valve permits flow from the atrium to the ventricle, but not in the opposite direction. The pumping is accomplished by a periodic contraction (systole) and subsequent relaxation (diastole) of the heart chambers. The abrupt closings of the valves produce a thumping noise that can be heard at the chest wall, and leaking valves produce a jet noise called a heart "murmur."

The heart performs the periodic contractions and relaxations; i.e., it beats, because it is a muscle--a very complex one to be sure, but nevertheless a muscle--, and muscles have the ability to shorten under suitable excitation. The muscle fibers surround the heart in a spiral arrangement; when they shorten they produce a volume reduction of the chambers.

The excitation which produces the muscle contraction in the heart is electrical. A muscle fiber may be considered as a chain of cylindrical cells, arranged end to end. When a potential difference is applied to the extremities of such a fiber, it shortens; when the potential is removed, the fiber regains its original length again.

Certain conditions must be satisfied for the heart to function properly; among them are these: (a) all components of the heart must "beat" with the same frequency; (b) there must be a suitable phase shift between the action of these components; (c) the wave forms of the component motions must be "correct." It is easy to understand therefore, that the heart action is governed by an intricate electrical control system which activates each component of the heart at the proper instant of time, and in the correct manner. This control system is not external to, but is located in, the heart itself.

The electrical activity of the heart produces currents throughout the heart which give rise to an electric vector field; that field is conveniently represented by a dipole distribution over the heart surface. The measurement of the resultant of the dipole distribution is achieved by means of a recording galvanometer called an electrocardiograph which measures components of the resultant dipole directions which are determined by the choice of points of application of the electrocardiograph terminals. The recordings of the cardiograph are called electrocardiograms (ECG's), and a modern ECG consists of twelve different recordings, as described in more detail later on.

The purpose of this paper is to describe a mathematical model of the heart and of an electrocardiograph which can compute the ECG of the heart model. More precisely, we propose to construct a system of nonlinear oscillators and a coupling between them so that this system reproduces the control system of the heart. In addition, we construct a model of the electrocardiograph which produces six of the twelve traces of the ECG. There is no difficulty in producing the six others as well, but this was not considered essential for the purpose of this paper.

The muscle action which is excited by the electrical heart activity is ignored in our model because we are interested here only in modelling the control system of the heart. Nevertheless, it is probably possible to generalize our model so as to include the effects of infarctation.

It is a simple matter to introduce certain arrhythmias (and also mechanical pacemakers) into our heart model. If the corresponding ECG's of our model resemble those obtained from patients suffering from these arrhythmias, there is good evidence to believe that this model is adequate. In that case, it is anticipated that the model ECG's will be characterized by their Fourier coefficients and are, then, available for matching with the coefficients of clinical ECG's.

2. THE MACROSCOPIC DESCRIPTION

The heart is made up of two characteristically different kinds of muscle tissue. One of these consists of lumps and/or bundles of muscle fiber which can and do spontaneously generate periodic electrical impulses. They are called the pacemakers, and their periods are, in general, unequal. The remaining tissue does the contractive work and, in electrical terms, it transmits, but does not create, electrical impulses.

In Fig. 2, the pacemakers are shown in black, the remaining tissue in gray [1]. The sinoatrial node (S-A node) generates approximately seventy-two impulses per minute. It has the largest frequency of all pacemakers. Its impulse is transmitted in the directions shown by the arrows in Fig. 2. When the impulse arrives at the atrioventricular node (A-V node) that node fires (with some delay) at the same frequency as the arriving impulse, <u>not</u> at its own (lesser) frequency. The two branches of pacemaker tissue of Fig. 2 which leave the V-A node and lie along the ventricle walls are the so-called bundles of His. Conceptually, one may divide them into pieces of finite length, each with a lesser frequency than its neighbor nearer the V-A node. Nevertheless, all fire at the frequency of the S-A node signal as transmitted by the V-A node, rather than at their own still lesser frequencies. Therefore, the S-A node controls the frequency of the beat of all parts of the normal heart.

The numbers in Fig. 2 give the time in seconds, elapsed since the firing of the S-A node when that impulse arrives at the point where each number is printed. For instance, the S-A node impulse arrives at the left atrium approximately .06 seconds after firing, but it requires nearly .2 seconds before that signal arrives at the top of the ventricle walls.

It is not certain whether the A-V node is actually a pacemaker or not. For instance, Hoffman and Cranefield thought so in 1960 [2], but changed their minds in 1964 [3]. However, it is generally believed that "... the rate of the heart depends on the rate of discharge [impulse] of its most rapidly firing automatic cells"[3], and Lister et al [4] stated in 1967 "That the fastest pacemaker of the heart will activate the entire heart in an axiom...".

3. THE MICROSCOPIC DESCRIPTION

The cells of the heart may be grouped in two sets: they are either "automatic cells," or they are not [3].

Consider an automatic cell. In the state of (electrical) rest, the interior of the cell has a negative potential difference of approximately -90 mv relative to the cell exterior; the cell is said to be "polarized." In an automatic cell there occurs now a slow potential increase up to a threshold value of approximately -60 mv, at which time the increase in potential becomes very rapid and rises to approximately +10 mv. This phase of cell activity is called "depolarization." The potential now decreases slowly (this is usually referred to as "repolarization") until it reaches its minimum of -90 mv again. This process of de- and repolarization is a self-excited oscillation whose mechanism it not completely understood. It is known, however, that the increase in "transmembrane action potential" is produced by a penetration into the cell interior of positively charged sodium ions and, that the permeability of the cell membrane to these ions is relatively small at large potential differences but increases greatly when this difference reaches, or is less than, the threshold value of -60 mv.

A non-automatic cell does not develop spontaneous oscillations of its transmembrane potential. However, "when a cell is excited (... by the arrival of a propagated action potential), there is a rapid depolarization... A gradual change of potential back toward the resting state then follows, terminated by a more rapid repolarization." [3]* The voltage-time curves of an automatic cell and of a non-automatic one, shown in Fig. 3, are taken from Hoffman and Cranefield [3]. The curve labelled S A is that of an automatic cell, the one labelled A is that of a non-automatic cell.

It is evident that the functional relation between automatic and non-automatic cells is very similar to that between pacemaker and non-pacemaker muscles of the heart. It is for this reason that one model may be used for both, the macroscopic and the microscopic description of electrical heart activity.

---

*The term "action potential" is juxtaposed with "resting state potential." It refers to that portion of the voltage-time curve where the voltage is not constant.

4. THE ELECTROCARDIOGRAPH

The electrocardiograph is simply a recording galvanometer which measures the potential difference between two points of the body. It was invented in 1903 by Einthoven [5]. From Einthoven's time until the middle of the 1940's, the electrocardiogram remained a useful but empirical clinical tool; the gross relations between the ECG and electrical heart activity were being discovered, but little was known about electrical activity of the heart cells in generating and conducting electrical impulses.

Consider Fig. 4 which shows an element of heart volume of area dA and thickness δ, charged with a charge q. The resulting dipole is well known to be $\underline{v} = q\,\delta\,\hat{n}$, where $\hat{n}$ is the outward unit normal. Then the potential relative to infinity at a point P, due to this dipole is

$$dV_p = \frac{1}{4\pi\varepsilon_o} \frac{\underline{v}\cdot\underline{r}}{r^3}\, dA = \frac{1}{4\pi\varepsilon_o} \frac{v\cos\theta}{r^2}\, dA$$

where ε_o is the dielectric constant of the body, $\underline{r}$ is the position vector of the centroid of dA relative to P, and θ is the angle subtended by $\underline{v}$ and $\underline{r}$. Thus, if we measure the voltage at a point P_j, produced by n elements of equal area ΔA having centroid positions $\underline{r}_{ij}$ relative to P_j, each with dipole strength $v_i(t)$, we find

$$V_j = K \sum_{i=1}^{n} v_i(t)\cos\theta_{ij}\, r_{ij}^{-2}$$

where K is a constant, and v_i is the time dependent potential of the i'th element. The potential difference between two points P_j and P_k is, then,

$$V_{kj} = V_k - V_j = K \sum_{i=1}^{n} v_i(\cos\theta_{ik}\, r_{ik}^{-2} - \cos\theta_{ij}\, r_{ij}^{-2}). \qquad (1)$$

Equation (1) is an analytical expression for the potential difference between two points, as measured by the electrocardiograph. It is a well-known result which follows directly from elementary theory.

5. THE VAN DER POL MODEL

In 1928, van der Pol and van der Mark published a famous paper [6] divided into two parts. The first contains a second-order nonlinear differential equation with a parameter, now known as the van der Pol equation; the second describes a heart model.

The van der Pol equation is shown to have as solution a self-excited oscillation when the parameter is positive. This oscillation resembles simple harmonic motion when the parameter is small, but it has a very different character when the parameter is large. The latter type of motion was called "relaxation oscillation" by the authors.

The heart modelled in the second part is considered to consist of the S-A node (called sinus in the paper), one atrium and one ventricle. The simulation of this model was accomplished by three coupled relaxation oscillators and a retardation system; the latter was included to produce the time delay between the atrium and ventricle responses.

The authors also present an ECG produced by the heart simulator, and this ECG shows all but one of the features of a normal electrocardiogram. The missing feature is the so-called T-wave. The authors explain that it was impossible for them to reproduce the T-wave because of "...the uncertainty (as far as we are aware) of what causes the T tops in the electrocardiogram." The basic cause for the T-wave was indeed not known in 1928. It is now known to be produced by the rapid repolarization of the non-automatic ventricle fibers [7]. For this reason, it is difficult to understand how van der Pol and his co-author failed to find a T-wave in their ECG because "repolarization" is an essential feature of their relaxation oscillators; in fact, it may point to a deficiency in their electrocardiograph model.

The great importance of the paper of van der Pol and van der Mark is that theirs was the first attempt to model human heart activity on a simulator. The fact that their model could produce ECG's greatly resembling clinical ones, including certain arrhytmias, was a scientific triumph. Since the publication of van der Pol's paper, none have appeared (to our knowledge) which model the entire heart-electrocardiograph complex.

6. ELEMENTS AND SUBSYSTEMS OF THE HEART MODEL

The characteristic feature of electrical heart activity is that it is produced by oscillators which have distinct frequencies when uncoupled, but they all oscillate at the frequency of their fastest member when coupled together.

We shall construct such a system of oscillators from elements which obey second-order, nonlinear differential equations with discontinuous right-hand sides. Consider the oscillator shown in Fig. 5. It consists of a see-saw having a scoop S at one extremity, a counterweight W at the other, and above the scoop there is a fluid reservoir R, which discharges fluid at the mass flow rate q through a discharge pipe, as shown. In addition, there is a trigger mechanism T whose function will be explained below. The system parameters are so adjusted that the see-saw has the angular position $-x_o$ when the scoop is empty. However, as the scoop is being filled from the reservoir, there will be a time beyond which the scoop and fluid "outweigh" the counterweight. The see-saw would now move in the positive (down) direction if it were not held in position by the trigger mechanism. Let it require some additional fluid mass M before the trigger mechanism releases the scoop and permits the see-saw to move in the positive direction. The purpose of the trigger mechanism is to increase the unbalance at the onset of the motion so that, when the see-saw is released, it will move more rapidly from $-x_o$ to x_1 than it would have done without the trigger. Perhaps with the exception of the trigger mechanism, the see-saw described here is a well-known oscillator. However, rather than satisfying a van der Pol equation, its motion may be regarded as satisfying a differential equation with discontinuous right-hand side.

Let the see-saw have the length 2ℓ with the fulcrum at the middle; let the mass of the empty scoop be m_1 and that of the counterweight m_2. Then, with M already defined as the additional mass required to release the trigger, the equation of motion of the see-saw is, under some simplifying assumptions,

$$\ddot{x} = f_i\ (x,t), \begin{cases} i = 1 & \text{for } \dot{x} > 0, \\ i = 2 & \text{for } \dot{x} < 0, \end{cases}$$

where

$$f_1 = \frac{g}{\ell} \frac{m_1+q(t-t_o)+M-m_2}{m_1+q(t-t_o)+M+m_2} \cos x,$$

$$f_2 = \frac{g}{\ell} \frac{m_1-m_2}{m_1+m_2} \cos x. \qquad (2)$$

In these equations it was assumed that the fluid empties out of the scoop in zero time, and that the fluid flow into the scoop during the return motion $\dot{x}<0$ may be ignored. The latter assumption may be removed without causing an increase in the theoretical complexity, but it does increase the computational labor. The time history of the angular deflection of the mechanism looks somewhat as shown in Fig. 6. The see-saw oscillator described here is the basic element from which the proposed heart model is constructed.

First we show how a system of oscillators, each with a different frequency, may be arranged in such a way that the fastest entrains all others. Consider n see-saw oscillators of the type described. For simpliticity, let them have identical mechanical parameters. Let them be arranged vertically beneath each other in the manner and numerical order shown in Fig. 6, and let the i'th oscillator be filled from the reservoir at the rate q_i such that

$$q_1 < q_2 < \ldots < q_n.$$

Then, if w_i is the frequency of the i'th oscillator, one also has for the uncoupled frequencies

$$w_1 < w_2 < \ldots < w_n,$$

because the frequency of each of the otherwise identical oscillators varies at the rate of mass flow into each scoop. Stated differently, the oscillators are arranged vertically beneath each other in descending order of their uncoupled frequencies.

The coupling consists in the vertical arrangement of the oscillators. If the external flow into all scoops is started at the same time, the top scoop will trip first, and it will empty its discharge into the second. The second will now tip because the fluid contribution from above is sufficient to overbalance the counterweight and overcome the trigger mechanism. A time delay between the responses of adjacent oscillators is produced by increasing the vertical distance between them. Decoupling is achieved by deflecting the spill from the scoop above. For instance, the element D between the first and second oscillator of Fig. 7 is a deflecting vane which leaves the coupling intact when the vane is in the vertical position, but breaks it when in the slanted position (shown by the dotted line in Fig. 7). A partial rotation of the vane permits some, but not all of the spill from above to reach the scoop of the next-lower oscillator; thus it produces partial decoupling. In the model it is supposed that there is a deflecting vane between all neighboring oscillators.

Next we show how a system of pacemaker and non-pacemaker muscle fibers can be modelled by a proper arrangement of oscillators like that shown in Fig. 5. Each discharge pipe in Fig. 7 is supplied with a control valve V which regulates the mass flow out of that pipe. Let a number of these be shut off, but let the flow q_1 remain undisturbed. In that case, the closing of some valves will have no effect of the oscillation of each oscillator, or on the time delay between their motions; the system performs in all details as it did before. However, the see-saws without external fluid supply are no longer self-excited oscillators; they transmit a signal, but they cannot oscillate without that signal.

The see-saw is also useful for modelling either automatic or non-automatic cells or "pacemaker cells" and "latent pacemaker cells" [2]. Every possible type of cell can be modelled by controlling suitably the fluid supply between that resulting from spill and that from the external source. Even cells which are, or have become, completely inert in that they neither produce nor transmit signals can be modelled by a see-saw whose scoop volume is less than that required for releasing the trigger. It is this latter feature that could be utilized for modelling damaged muscle tissue such as results from infarctation.

7. THE MODEL

For purposes of analysis, the heart is divided into eight components listed in Table I

TABLE I

No	Component	Notation
1	S-A Node	S-A
2	Left Atrium	A_L
3	Right Atrium	A_R
4	A-V Node	A-V
5	Left Ventricle, Upper Half	V_{L1}
6	Left Ventricle, Lower Half	V_{L2}
7	Right Ventricle, Upper Half	V_{R1}
8	Right Ventricle, Lower Half	V_{R2}

The components are shown in a schematic arrangement in Fig. 8. The manner in which they are coupled together has already been described, and the arrangement of see-saws used to model the electrical interaction and activity of the heart is shown in Fig. 9. (It should, of course, be realized that we do not propose to construct a mechanical arrangement like that of Fig. 9; that illustration is merely a pictorial representation of a coupled system of differential equations.)

During normal heart acitivity, only the valves (a) and (d) in Fig. 9 are open, and the flow rate at (a) is approximately twice that at (d). If, however, the A-V node is not regarded as a pacemaker, but the bundle of His is, then the valve at (d) would be closed, but those at (e) and (f) would be open instead.

To construct the model of the electrocardiograph we note from Section 4 that the potential at a point P is

$$v_p = \frac{1}{4\pi\varepsilon_o} \int_A \frac{v(t)\cos\theta}{r^2}\, dA$$

$$= \frac{1}{4\pi\varepsilon_o} \int_\Omega v(t)\, d\Omega \qquad (3)$$

where the symbols have the meaning given in Section 4, and

$$d\Omega = r^{-2} \cos\theta \, dA \tag{4}$$

is the solid angle subtended by dA as seen from P.

We utilize (3) in the form

$$V_j = \frac{1}{4\pi\varepsilon_o} \sum_{i=1}^{8} \int_{\Omega_i} v_i(t) \, d\Omega \tag{5}$$

where the index j identifies the point P_j at which the potential is measured, and the index i refers to the numbers in the first column of Table I.

The eight areas over which this integration is carried out are shown in Fig. 8. Actually, the areas of the two nodes are so small compared to the others that the solid angles subtended by them are considered vanishingly small. In consequence, the contributions of the nodes to the potential at P_j will be ignored.

If the dipole distribution of the remaining components is represented by the resultant dipole of each, acting at the (electrical) centroid of each area, one has for each

$$\int_{\Omega_i} v_i(t) d\,\Omega_i = F_i(t)\,(\cos\theta_{ij}) r_{ij}^{-2}, \quad (i = 2,3,5,6,7,8),$$

where $F_i(t)$ is the strength of the resultant dipole of the i'th area, and r_{ij}, is the length of the position vector of its point of application relative to P_j. Then, the potential difference between two points P_j and P_k is

$$V_{kj} = V_k - V_j = K \sum_{i=2}^{8} F_i(t) \left[(\cos\theta_{ik}) r_{ik}^{-2} - (\cos\theta_{ij}) r_{ij}^{-2}\right], \quad (i \neq 4) \tag{6}$$

where $F_i(t)$ bears a simple relation to the solution of (2). In fact, let $x = x(t)$ be the solution to (2); then, remembering that there are eight equations like (2) for the eight oscillators, one has

$$F_i(t) = x_{io} + x_i(t).$$

Except for smoothing, this is the mathematical model of the electrocardiograph.

Every electrical circuit, and every moving mechanical element having mass is, necessarily, a bandpass filter. In other words, if its response is regarded as a periodic function of time, terms of a frequency larger than some limiting value cannot be present in its Fourier series representation. This filtering process is not contained in (6). Therefore, if we represent the smoothing operation by an operator S, the electrocardiogram is modelled by

$$\tilde{V}_{kj} = S(V_{kj}) \tag{7}$$

where V_{kj} is given in (6).

Bellman et al [8] have posed the inverse problem: given V_{kj} in (6), what are the $F_i(t)$? They have answered this question by demonstrating that their identification problem may be viewed as a nonlinear multipoint boundary value problem which they solve by quasilinearization and high-speed computer techniques.

8. THE NORMAL ELECTROCARDIOGRAM

A normal Lead II electrocardiogram [9] is shown in Fig. 10. The maxima and minima are labelled from left to right P,Q,R,S,T and U and are usually referred to as the P-wave, Q-wave, etc.; the Q,R and S-waves are frequently referred to as the QRS-complex.

From a knowledge of the occurrences of the action potentials of the components of the heart, it is simple to determine that the P-wave is produced by the atria, the QRS-complex by ventricle depolarization, the T-wave by repolarization of the right and left ventricle, and the U-wave by that of the left ventricle. The relation between the electrical heart activity and the appearance of the ECG is illustrated in Fig. 11.

The twelve-point ECG consists of twelve different traces, obtained by placing the leads of the electrocardiograph at twelve different points on the body. Six of these are shown in Fig. 12 [9]. The arrows indicate the direction in which the recorded components of the resultant dipole of the heart are taken. Leads I, II and III were suggested by Einthoven; leads aVR, aVL and aVF were devised by Wilson [10]. Six more traces are obtained by placing the leads in a standardized way on the rib cage.

The traces of the normal ECG differ from person to person depending on the orientation of the heart within the body, and also on the orientation of the mean electrical axis; these two orientations are largely independent of each other. We have chosen the case in which the mean electrical axis points approximately in the direction of the arrow of Lead II in Fig. 12. For that case the six traces of a normal ECG are shown in Fig. 13 taken from [9].

9. MODEL PERFORMANCE

In our model the action potential of each of the components of the heart is given by one dipole produced by that component. Now, the quantities θ_{ij}, θ_{ik}, r_{ij} and r_{ik} in (6) are fixed for any chosen points P_j and P_k on the body, and for any one component i. Therefore, (6) may be written as

$$V_{kj} = \sum_i C^i_{kj} F_i(t) \tag{7}$$

where the quantities C^i_{kj} depend on the position vectors of the i'th component relative to P_j and P_k, and on the direction of the dipole of the i'th component. A schematic representation of the heart components and the direction of their dipoles is shown in Fig. 14. The arrows indicate the approximate directions of the dipoles; these may be understood by comparing these directions with the arrows in Fig. 2 giving the signal flow directions. The constants used in computing the ECG's of our model are given in Table II. An understanding of the signs and magnitudes of these constants may be had by comparing the directions of the arrows in Fig. 14 with those of Fig. 12.

TABLE II

Component	LI	LII	LIII	aVR	aVL	aVF
A_R	.038	.04	.002	-.038	0	.014
A_L	.038	.04	.002	-.038	0	.014
V_{R1}	-.23	-.45	-.24	.23	-.11	-.24
V_{R2}	-.30	-.60	-.33	.30	-.15	-.33
V_{L1}	.50	1.0	.55	-.50	.25	.55
V_{L2}	.05	.10	.06	-.05	.025	.06

A plot of Lead II of our model before and after smoothing is shown in Fig. 15. The latter should be compared with Fig. 10 showing a clinical (Lead II) ECG. The six leads of our heart model after smoothing are shown in Fig. 16 (a), (b), (c), (d), (e) and (f). It is seen that, on the whole, the ECG's produced by our heart model agree very well with those of Fig. 13.

There is a distinct difference in the rise of the T-wave between the clinical result of Fig. 10 and that of Fig. 15, made by our model. This portion of the ECG is produced by the repolarization phase of the ventricles. The explanation for the discrepancy is that, in our model, the waveform of the ventricle action potential does not have the same flat top and abrupt repolarization feature as that of the actual heart. This slight defect could be remedied by replacing the see-saws of the ventricles by other oscillators having the required waveform, and being coupled into the system in exactly the same way as those used here.

While it is, of course, possible to introduce many arrhytmias into our model, only one will be illustrated, namely, A-V block. In our model, this is produced by turning the vane above the oscillator representing the A-V node of Fig. 9. Inasmuch as our model was so constructed that the uncoupled frequency of the V-A node is precisely half of that of the S-A node, one would expect a curve in which alternate QRS-complexes and T-waves are missing. This is, indeed, the case as shown in Fig. 17. For comparison, a clinical ECG of a patient suffering from A-V block is shown in Fig. 18 [9]. It is seen that our model reproduces the anomaly very satisfactorily.

REFERENCES

[1] Wiggers, C. J., The Heart, Scientific American, Vol. 196, No. 5, p. 75, (May 1957).

[2] Hoffman, B. F. and Cranefield, P. F., Electrophysiology of the Heart, McGraw-Hill Book Co. (1960).

[3] Hoffman, B. F. and Cranefield, P. F., The Physiological Basis of Cardiac Arrhythmias, American Journal of Medicine, vol. 37, pp. 670 (November 1964).

[4] Lister, J. W., Abner, J. D., Stein, E., Grunwald, R. and Robinson, G., The Dominant Pacemaker of the Human Heart, Circulation, Vol. 35, p. 22, (January 1957).

[5] Einthoven, W., Ein neues Galvanometer, Annalen der Physik, Vol. 12, p. 1059 (1903).

[6] Van der Pol, B. and van der Mark, J., The Heartbeat considered as a Relaxation Oscillation, and an Electrical Model of the Heart, Philosophical Magazine, Vol. 6, p. 486 (1928).

[7] Hoffman, B. F., Cranefield, P. F. and Wallace, A. G., Physiological Basis of Cardiac Arrhythmias, American Heart Association-39th Scientific Session, New York, p. 103 (October 1966).

[8] Bellman, R., Collier, C., Kagiwada, H., Kalaba, R. and Selvester, R., Estimation of Heart Parameters Using Skin Potential, Communications of the ACM, Vol. 7, No. 11, p.   (1964).

[9] Arbeit, S. R., Rubin, I. L. and Gross, H., Differential Diagnosis of the Electrocardiogram, F. A. Davis Company, Philadelphia, (1960).

[10] Wilson, F. N., MacLeod, A. G. and Barker, P. S., The Potential Variations Produced by the Heart Beat at the Apeces of Einthoven's Triangle, American Heart Journal, Vol. 7, p. 207, (1931).

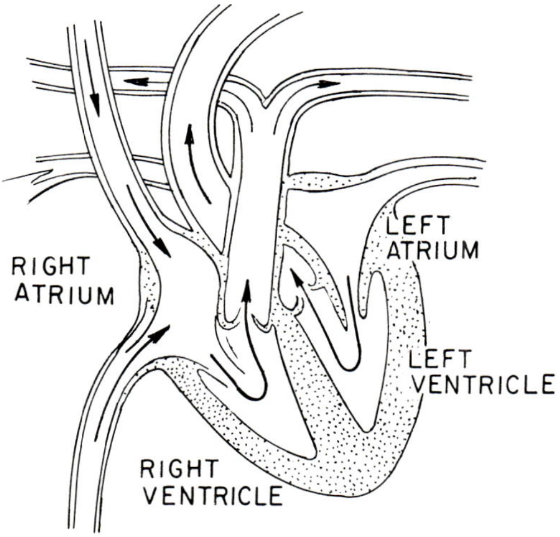

Fig. 1 The heart

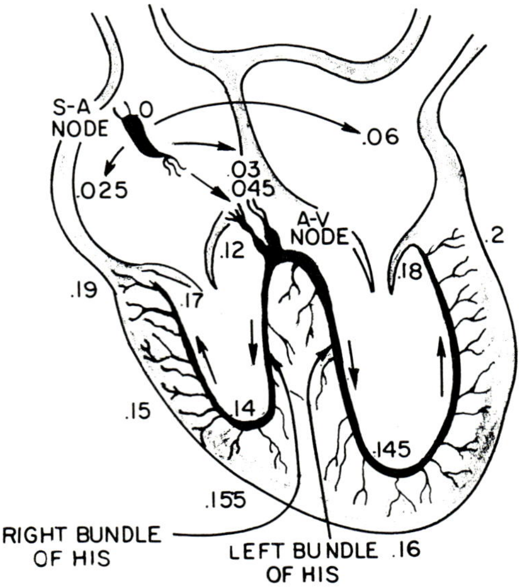

Fig. 2 The pacemakers (in black)

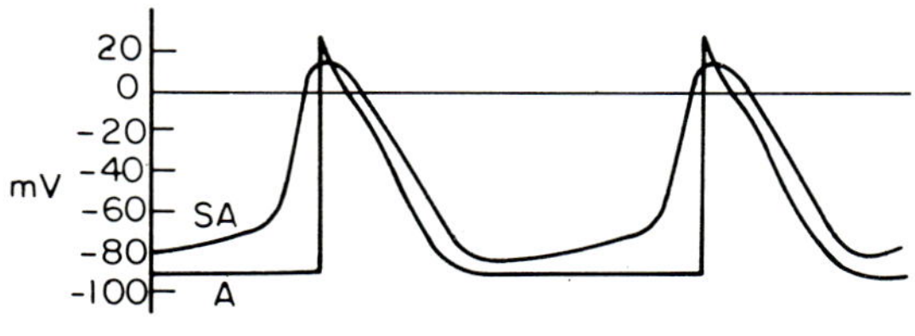

Fig. 3 Action potentials of heart cells

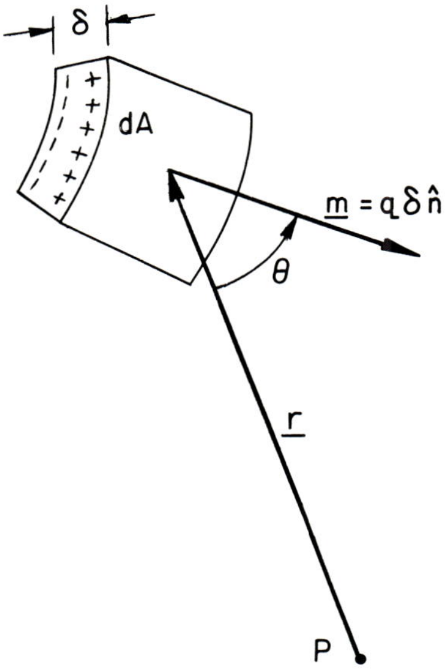

Fig. 4 Dipole of volume element

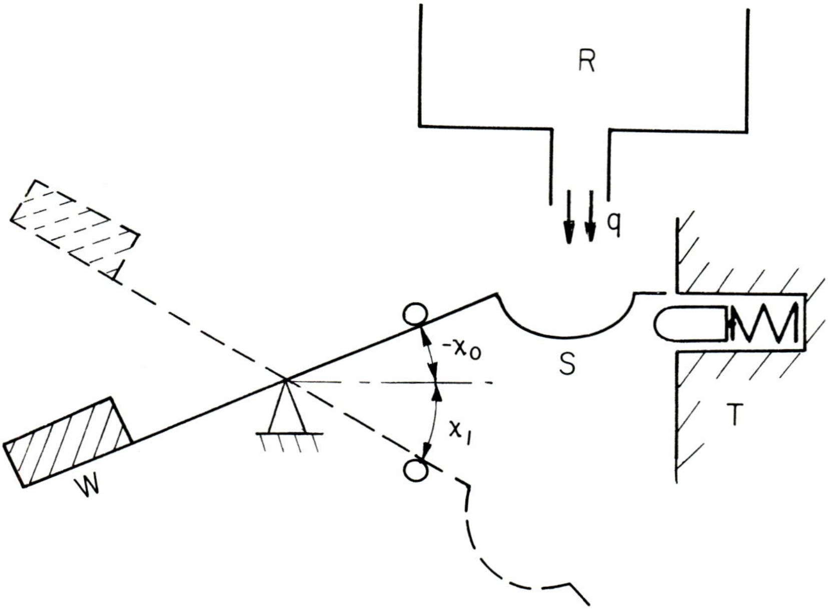

Fig. 5 See-saw, a component of the heart model

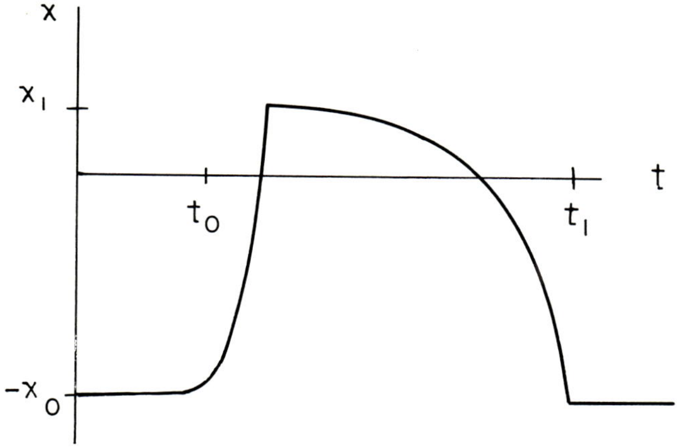

Fig. 6 Angular position of see-saw as function of time

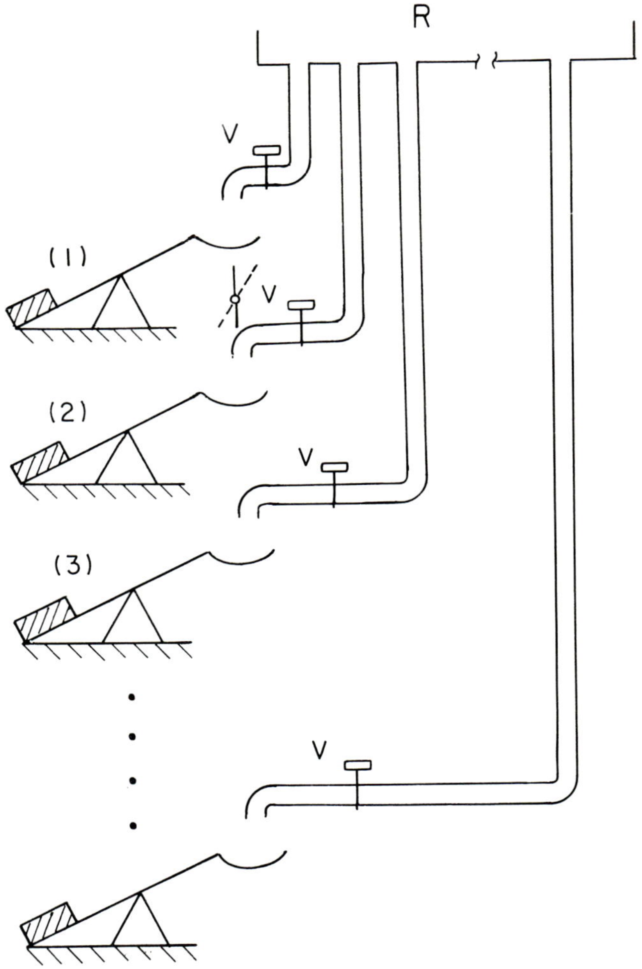

Fig. 7 Coupling between see-saws

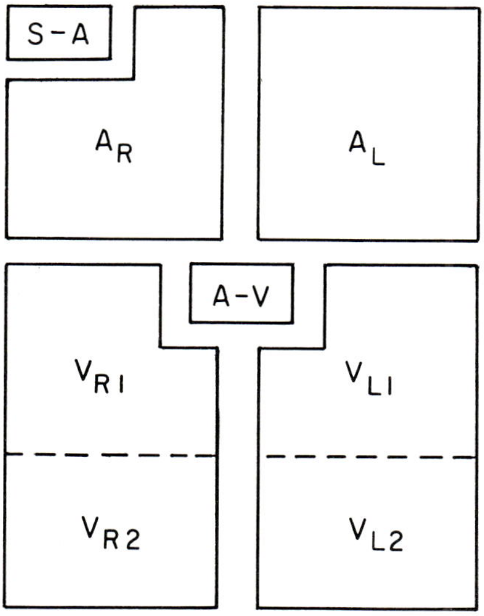

Fig. 8 Schematic representation of the heart

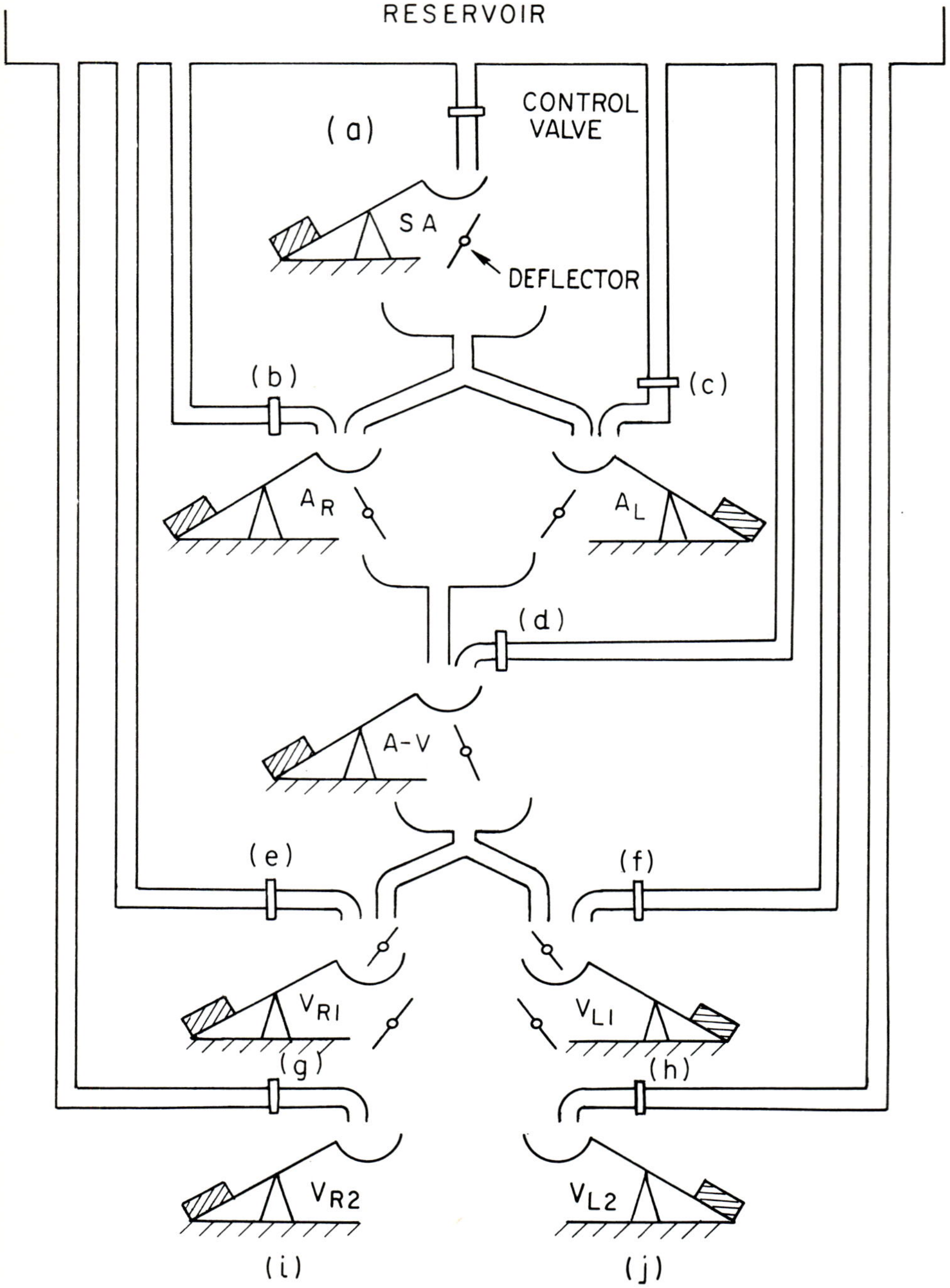

Fig. 9 The heart model

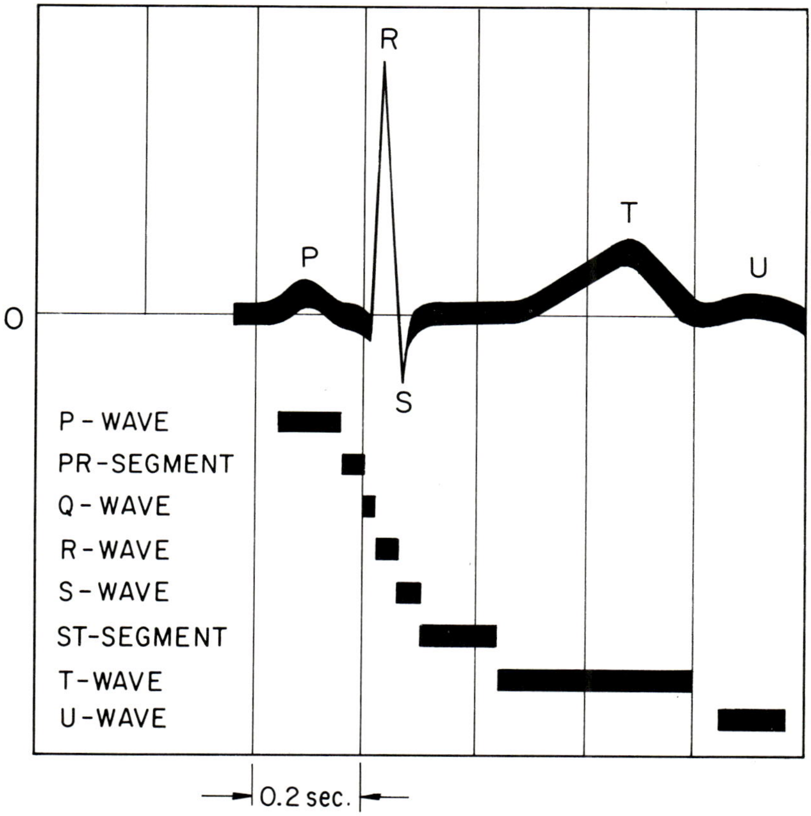

Fig. 10 Normal (Lead II) electrocardiograph

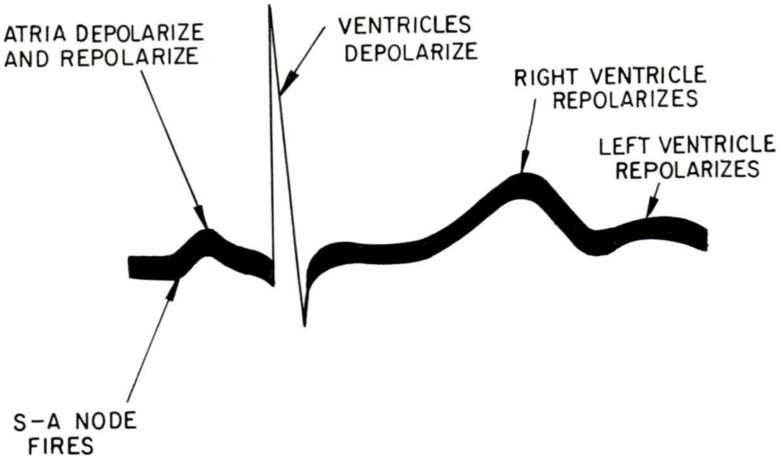

Fig. 11 ECG and heart action

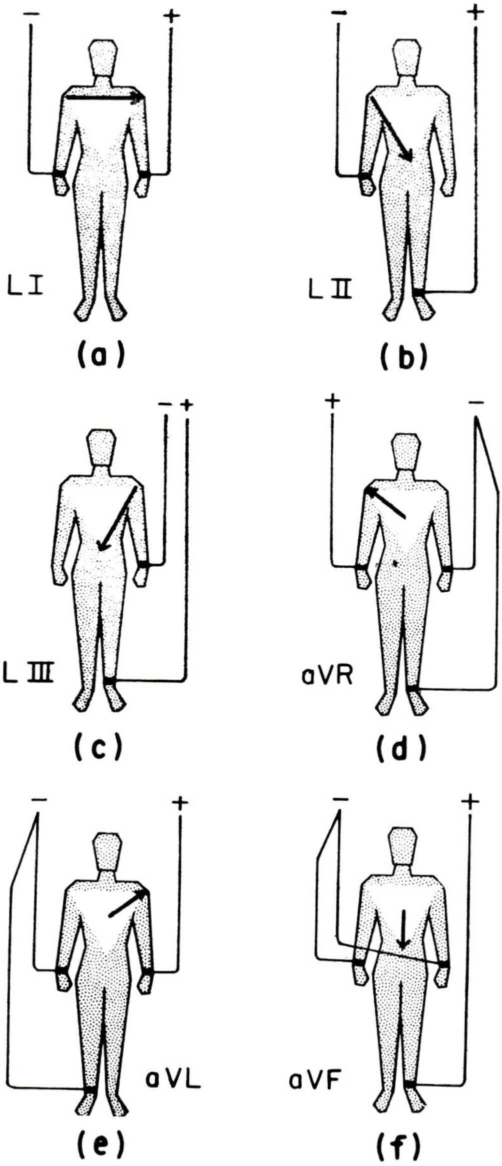

Fig. 12 Placement of electrocardiograph terminals for the six standard leads

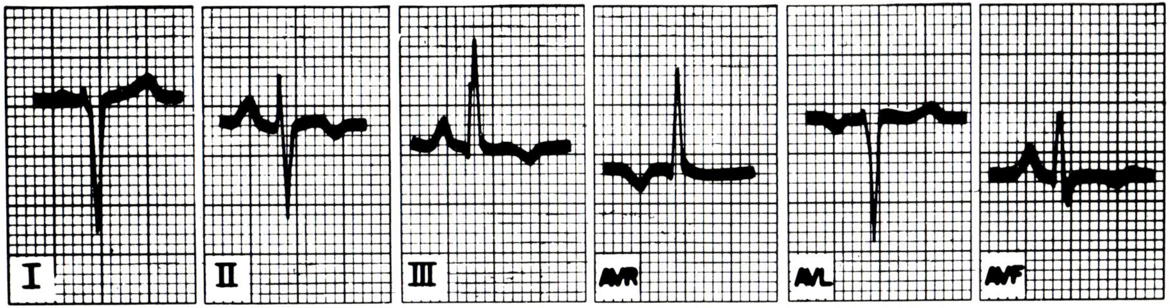

Fig. 13 Normal clinical ECG's for standard leads

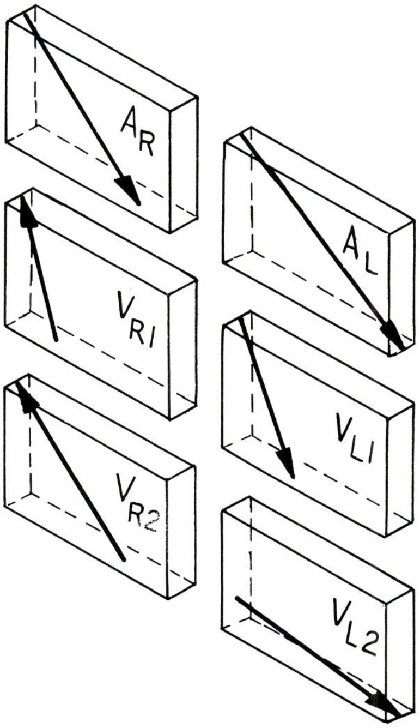

Fig. 14 Schematic representation of dipoles of heart components

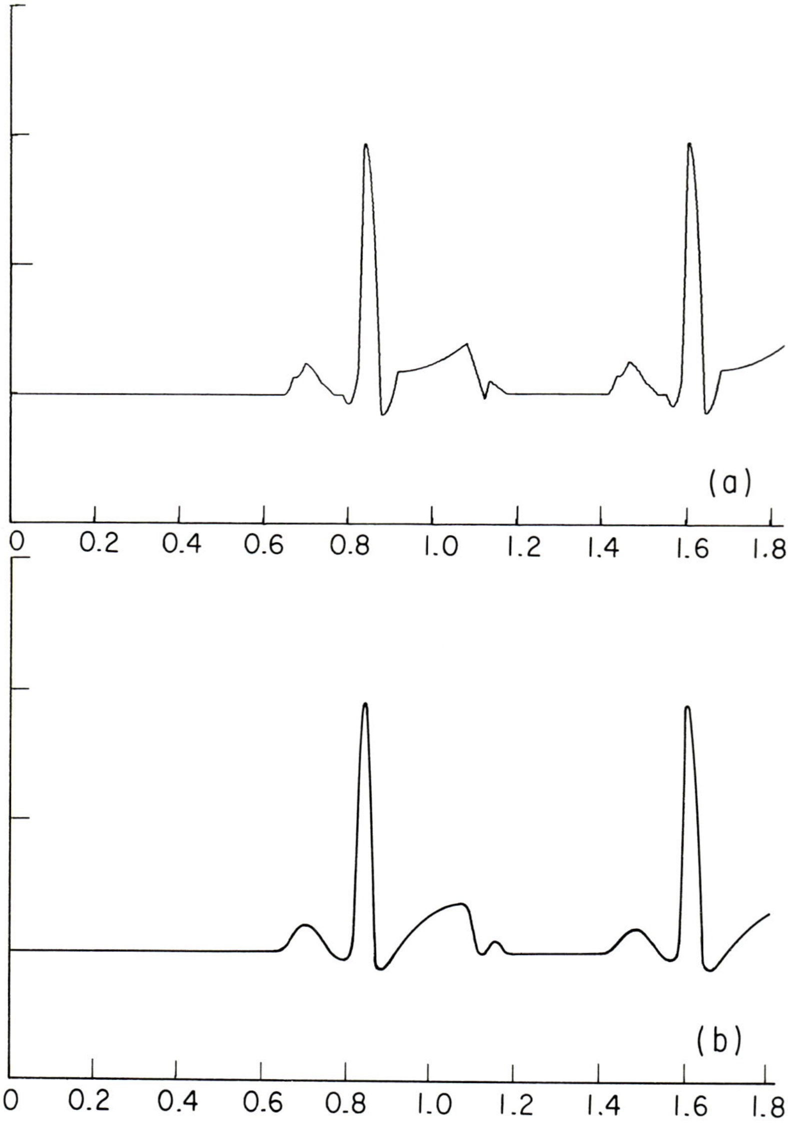

Fig. 15 ECG (Lead II) of heart model before and after smoothing

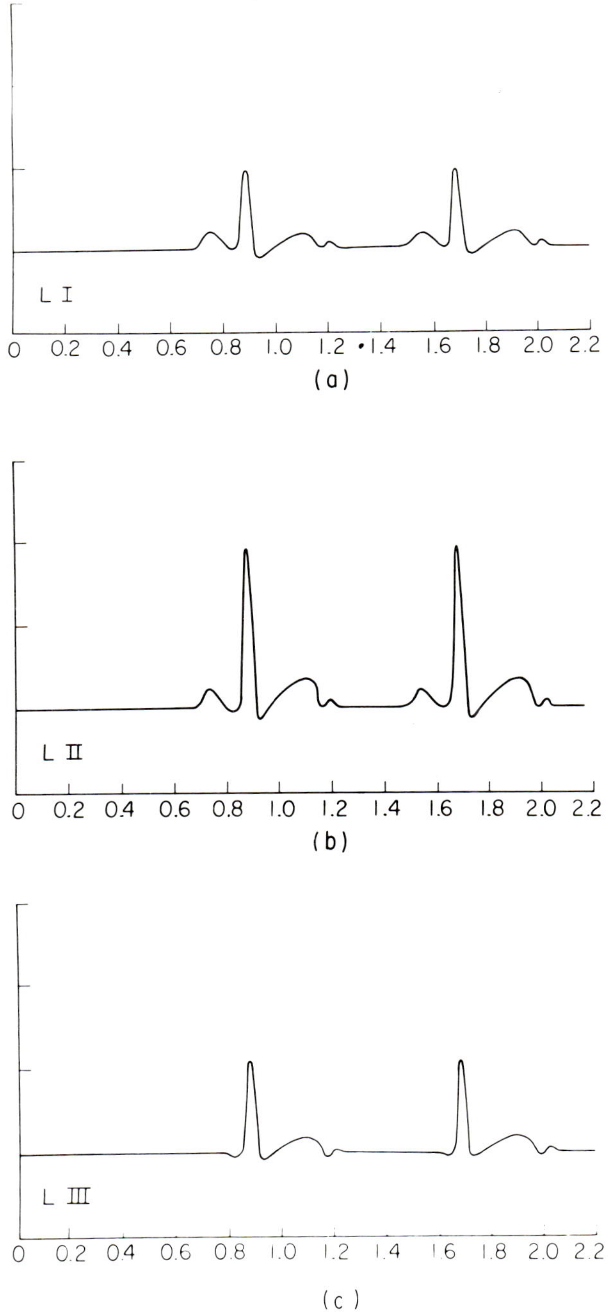

Fig. 16 ECG's of standard leads of heart model

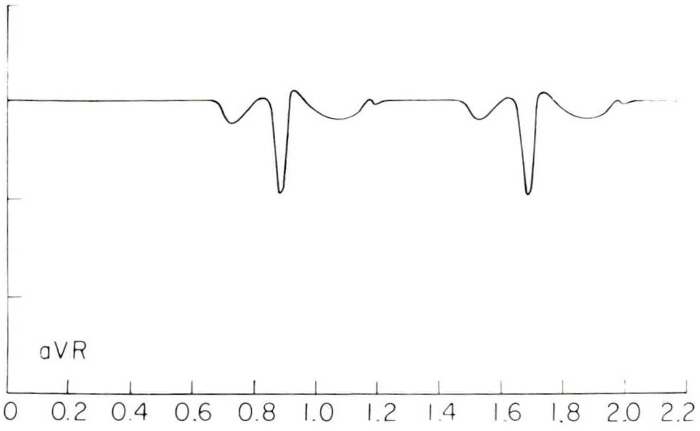

(d)

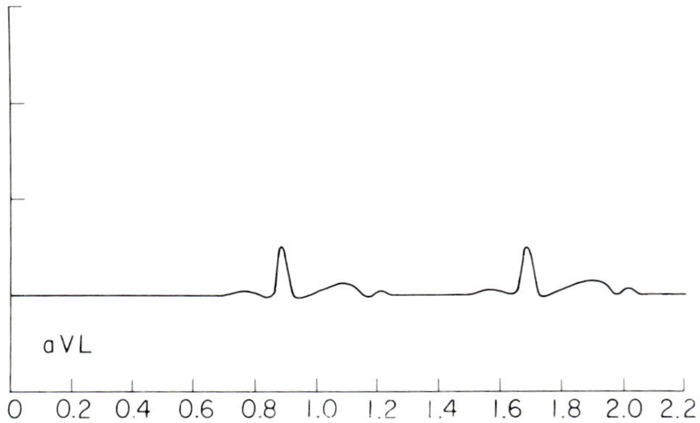

(e)

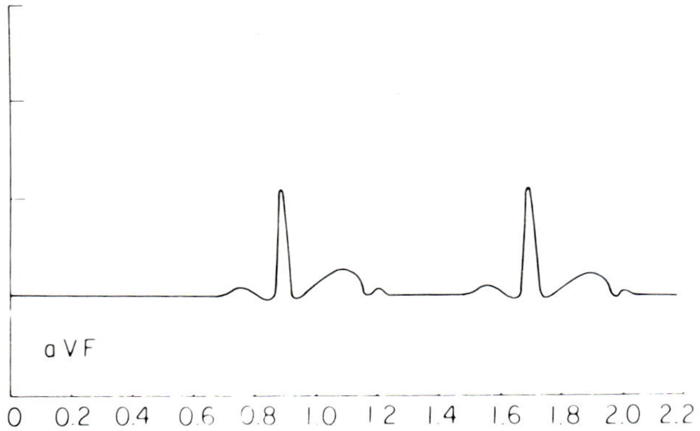

(f)

Fig. 16 con't. ECG's of standard leads of heart model

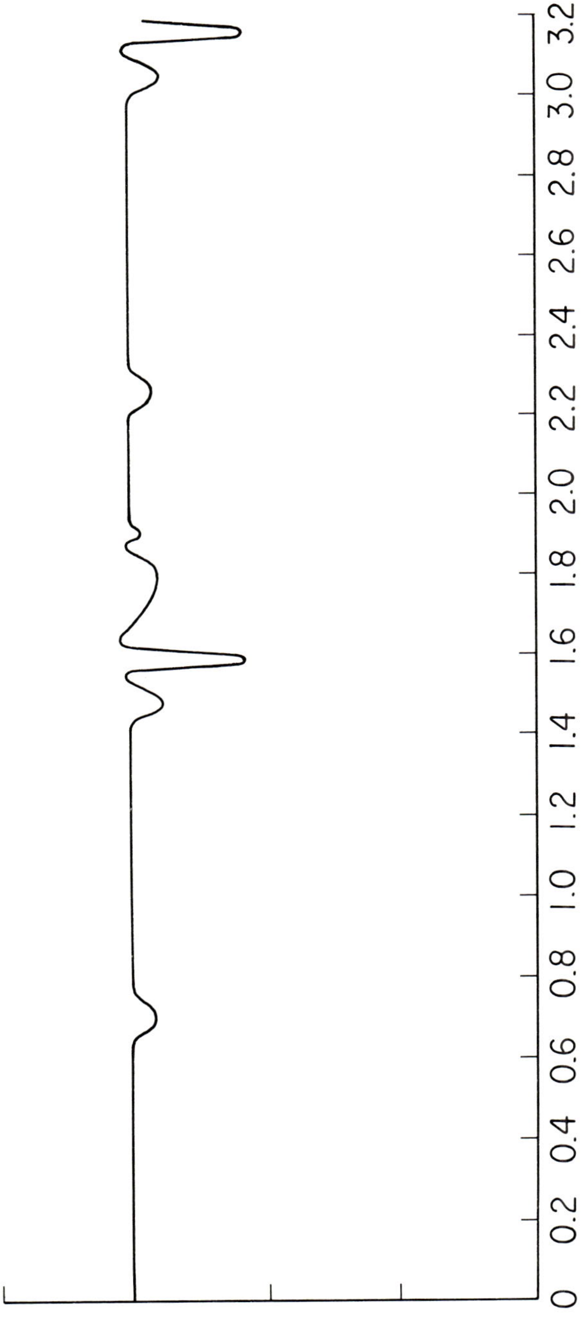

Fig. 17 Model ECG of complete heart block (Lead aVR)

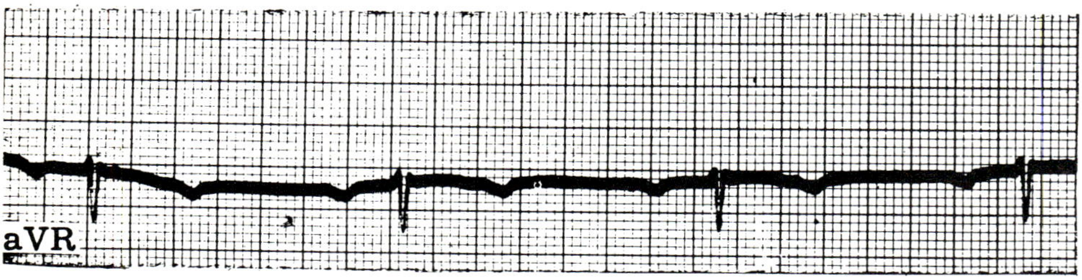

Fig. 18 Clinical ECG of complete heart block (Lead aVR)

PART II
FLUID MECHANICS

EXTERNAL FLOW

Developments in Mechanics, Vol. 6. Proceedings of the 12th Midwestern Mechanics Conference: 7

Slow Free-molecule Flow Past a Sphere

CHIA-HAN LIU and MAURICE L. RASMUSSEN
UNIVERSITY OF OKLAHOMA

ABSTRACT

Steady, slow free-molecule flow past a sphere is studied by means of asymptotic expansions for small molecular-speed ratio. The molecular distribution function is obtained by the method of characteristics from the collisionless kinetic equation and boundary conditions specified at infinity and on the sphere surface. The flow at infinity is in equilibrium, and the surface boundary conditions are prescribed in terms of an accommodation coefficient and wall temperature. Explicit expressions for the number-density and velocity fields are obtained, appropriate for slow flow. The higher-order terms of the expansions elicit an embryonic development of a wake behind the sphere. Comparison of velocity profiles and streamline shapes are made with incompressible Stokes' flow and incompressible potential flow, the continuum counterparts of the slow-flow problem.

INTRODUCTION

When the flow of a gas past a solid body is characterized by a mean free path of the constituent molecules that is much larger than the characteristic length of the body, the conventional methods of continuum fluid mechanics are inapplicable, and the problem must be attacked by means of the kinetic theory. When such flows are extremely rarefied, the collisions of the molecules in a region near a solid body can be completely ignored, and the flow is called free-molecule flow. The molecules are regarded to move in straight-line paths interrupted only by collisions with the solid body. The first to draw attention to the aerodynamics aspects of free-molecule flow were Zahm [1] and Tsien [2,3]. Schaaf and Chambre [4] and Patterson [5,6] have presented surveys of the various features of rarefied gas dynamics together with many references.

Because of applications to flight in the upper atmosphere, the aspect of free-molecule flow commanding the most attention has been the computation of moments at the surfaces of solid bodies, leading to aerodynamic forces and heat transfer.

Aerodynamic characteristics of various bodies in free-molecule flow are well known [7,8,9]. On the other hand, determination of free-molecule flow fields about finite solid bodies has received little attention, in part because surface conditions can be found without explicit computation of the flow field. In many practical situations, a free-molecule flow field, or the molecular distribution function at any point of the flow field, is not as important as the conditions on a solid surface. Nevertheless, if the free-molecule solution is viewed as the first approximation in a systematic procedure that begins by ignoring all molecular collisions in Boltzmann's equation, then the free-molecule distribution function itself becomes more important [10,11]. Some features of the corrections to free-molecule flow are discussed by Willis [10].

There are a number of problems in which the free-molecule flow field is of practical as well as theoretical interest. Problems of rarefied plasma flows require a knowledge of the number density field (ignoring magnetic effects) in order to obtain self-consistent solutions for the electric field. Free-molecule solutions can be used as approximations for hyperthermal ion flow fields [12,13]. Flows of dusty gases in which the dust particles are exceedingly small (compared to the mean free path) are also of practical interest. Settling of dust in the atmosphere or flow of particulate matter in vacuum systems are some examples. In these problems the relative flow past the dust particles can be regarded as slow. Aspects of such flows are discussed by Soo [14]. Aside from practical considerations, there is also a basic scientific interest in studying free-molecule flow fields as part of the spectrum of rarefied gas dynamics behavior. Establishing various mathematical techniques associated with free-molecule flow is also important.

Investigation of free-molecule flow fields has been devoted mainly to spheres. In the context of solutions of Boltzmann's equation, Szymanski [15] and Goldberg [16] discussed flow fields past spheres for very slow flow. Their calculations were concerned primarily with the density field. Karamcheti and Sentman [17] examined collisionless flows past spheres in the presence of potential fields. In this reference and a later one [18] they obtained density fields for specular reflection by numerical evaluation of the number-density quadratures. Al'pert et al. [12] and Prager and Rasmussen [13], among others, have used free-molecule solutions as first approximations to ion density fields for rarefied plasma flows past spheres. Although Al'pert et al. wrote down the general quadrature expressions for the number density, they calculated explicit field results only for high-speed flow in which the thermal motions of the molecules were neglected. Generally, velocity fields, streamlines, and other gas dynamic variables have not been explicitly calculated. The essential features of free-molecule flow past a cone were laid out by Elliott and Rasmussen [19].

The object of this paper is to determine the general structure of steady slow free-molecule flow fields past spheres. For this situation, explicit expressions for the density and velocity fields, as well as other variables, can be obtained in terms of expansions for small molecular-speed ratio $S \equiv (mU_\infty^2/2kT_\infty)^{\frac{1}{2}}$. This expansion is the counterpart of the Janzen-Rayleigh expansion for small Mach number for potential flow past a sphere [20]. The particles emitted from the sphere surface are described in terms of an accommodation coefficient such that part of the molecules are reflected diffusely and the remainder specularly. Streamlines are obtained and compared with the continuum counterparts of Stokes' flow and potential flow. Higher-order results show some of the structure of the wake development behind the sphere. Because analytical results are obtained, various mathematical features of free-molecule flow are portrayed.

FORMULATION OF THE PROBLEM

Consider the flow of a neutral rarefied gas past a sphere of radius a. Spherical coordinates r,θ,ψ are introduced as shown in Fig. 1. The spherical velocity components are given by $\xi_r,\xi_\theta,\xi_\psi$. The freestream velocity is denoted by $\vec{U}_\infty$ and is in the direction of the z-axis. Let n_∞ and T_∞ denote the freestream number density and temperature. Let T_w denote the temperature of the sphere surface.

The description of free-molecule flow starts with the distribution function $f(\vec{r},\vec{\xi},t)$, defined such that $f(\vec{r},\vec{\xi},t)\, d^3r d^3\xi$ represents the probable number of molecules lying in the range between $\vec{r}$ and $\vec{r} + d\vec{r}$ and having velocities in the range between $\vec{\xi}$ and $\vec{\xi} + d\vec{\xi}$. In the absence of collisions, the kinetic equation that governs f is

$$\frac{\partial f}{\partial t} + \vec{\xi} \cdot \frac{\partial f}{\partial \vec{r}} + \frac{\vec{F}}{m} \cdot \frac{\partial f}{\partial \vec{\xi}} = 0 \qquad (1)$$

where $\vec{F}$ is the external force acting on the particles, which have mass m.

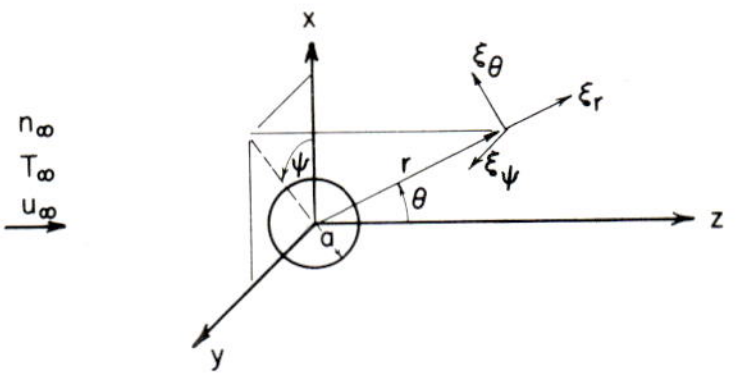

Fig. 1. Coordinate System for a Sphere.

The general solution to this equation can be found by means of the theory of characteristics. The characteristic equations for (1) are

$$\frac{df}{dt} = 0 \;, \quad \frac{d\vec{r}}{dt} = \vec{\xi} \;, \quad \frac{d\vec{\xi}}{dt} = \frac{\vec{F}}{m} \qquad (2,3,4)$$

Eq. (2) states that f is a constant on the characteristic curves. Eqs. (3) and (4) are simply the equations of motion of a particle of mass m with the force F acting on it. In this analysis, we shall assume that there are no external forces, that is $\vec{F} = 0$. In this case the integrals of (3) and (4) are

$$\vec{\xi} = \vec{\xi}_o = \text{const} \;, \quad \vec{r} = \vec{r}_o + \vec{\xi} t \qquad (5,6)$$

Here $\vec{\xi}_o$ and $\vec{r}_o$ are the constants of integration evaluated at t = 0. The particle trajectory is a straight line in configuration space. Because a trajectory is determined by $\vec{r}_o$ and $\vec{\xi}_o$, it follows that f is an arbitrary function of $\vec{r}_o$ and $\vec{\xi}_o$. Thus the general solution is

$$f(\vec{r},\vec{\xi},t) = G(\vec{r}_o,\vec{\xi}_o) = G(\vec{r} - \vec{\xi}t,\vec{\xi}) \qquad (7)$$

where G is an arbitrary function.

If r,θ,ψ denote the scalar representation of $\vec{r}$ in spherical coordinates and r_o,θ_o,ψ_o the corresponding scalars for $\vec{r}_o$, then vector analysis of (6) shows that

$$r_o^2 = (r - \xi_r t)^2 + (\xi_\theta t)^2 + (\xi_\psi t)^2 \qquad (8)$$

$$\cos\theta_o = \frac{1}{r_o}\left[(r - \xi_r t)\cos\theta + (\xi_\theta t)\sin\theta\right] \qquad (9)$$

$$\tan\psi_o = \frac{(r - \xi_r t)\sin\psi\sin\theta - (\xi_\theta t)\sin\psi\cos\theta - (\xi_\psi t)\cos\psi}{(r - \xi_r t)\cos\psi\sin\theta - (\xi_\theta t)\cos\psi\cos\theta + (\xi_\psi t)\sin\psi} \qquad (10)$$

It is further convenient, in many problems, to represent the scalar components of $\vec{\xi} = \vec{\xi}_o$ in terms of energy E and angular momentum $\vec{L} = m\,\vec{r} \times \vec{\xi}$. We find

$$E = \tfrac{1}{2} m \xi^2 = \text{const} \;, \quad L^2 = m^2 r^2 (\xi_\theta^2 + \xi_\psi^2) = \text{const} \qquad (11,12)$$

$$L_z \equiv \vec{L} \cdot \hat{e}_z = m\, r\, \xi_\psi \sin\theta = \text{const} \qquad (13)$$

If the flow is steady, then the time t can be treated as a parameter, solved for in (8), and substituted into (9) and (10). If the flow is axisymmetric, then ψ will not appear in the problem and (10) drops out. These are the simplifications we shall now make use of for steady flow past a sphere.

To construct a particular solution for steady flow past a sphere, consider an arbitrary point $\vec{r}$ in the flow field. Different particle trajectories passing through this point are determined by different values of $\vec{\xi}$. If $\vec{\xi}$ is such that its corresponding trajectory traced backwards from the point $\vec{r}$ does not intersect the sphere, then the distribution function f has the same value as that at infinity. If $\vec{\xi}$ is such that this trajectory intersects the sphere, the distribution function at $\vec{r}$ has the same value as that on the surface of the sphere at the point of intersection.

For a fixed value of $\vec{r}$, let τ denote that part of velocity space for which the particles come from infinity, and let σ denote that part of velocity space for which the particles come from the sphere. Then the distribution function can be considered to be the sum of two distributions,

$$f = f_\tau + f_\sigma \tag{14}$$

where f_τ is determined by the boundary conditions at infinity and f_σ is determined by the boundary conditions at the sphere surface. Here it is understood that f_τ is zero when $\vec{\xi}$ is in σ and f_σ is zero when $\vec{\xi}$ is in τ. The division of velocity space into τ and σ regions for the sphere problem is shown in Fig. 2. The inside of semi-infinite cone of semi-vertex angle $\gamma_o = \sin^{-1}(a/r)$ is denoted by σ and the remainder of velocity space is denoted by τ.

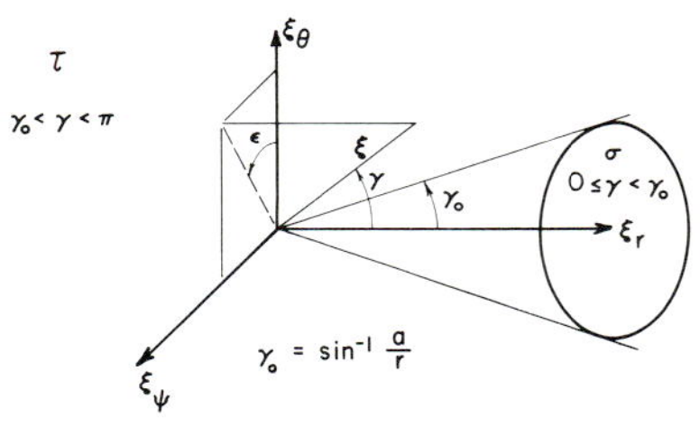

Fig. 2. Velocity Space Associated with a Sphere.

We now proceed to determine the distribution function for special boundary conditions. The surface boundary conditions must satisfy the condition that there is no mass flow across the sphere surface. This is given by

$$\int_{\xi_r < 0} \xi_r f_\tau d^3\xi + \int_{\xi_r > 0} \xi_r f_s d^3\xi = 0 \text{ on } r = a \tag{15}$$

The above moments $\int \xi_r f \, d^3 \xi$ are the mean mass fluxes impinging on and leaving the sphere surface. The function f_s represents the prescribed surface distribution function.

SOLUTIONS FOR THE DISTRIBUTION FUNCTION

INFINITY CONDITION

At infinity we assume that the distribution function is a displaced Maxwellian, that is, the flow is in thermodynamic equilibrium:

$$f_\infty(\vec{r},\vec{\xi}) = A \exp\left[- B (\vec{\xi} - \vec{U}_\infty)^2\right], \quad A \equiv n_\infty \left(\frac{m}{2\pi k T_\infty}\right)^{3/2}, \quad B \equiv \frac{m}{2kT_\infty} \tag{16a}$$

where k is the Boltzmann constant. In spherical coordinates (16a) becomes

$$f_\infty(\vec{r},\vec{\xi}) = A \exp\left[- B (\xi^2 - 2U_\infty\xi_r \cos\theta + 2U_\infty\xi_\theta \sin\theta + U_\infty^2)\right] \tag{16b}$$

Because f is a constant on straight-line trajectories and (16) satisfies the steady kinetic equation (with F = 0), we deduce that (16) is the distribution function for the particles that come from infinity, that is,

$$f_\tau(\vec{r},\vec{\xi}) = A \exp\left[- B (\vec{\xi} - \vec{U}_\infty)^2\right], \vec{\xi} \text{ in } \tau \tag{17}$$

SPECULAR REFLECTION

For specular reflection it is assumed that the normal component of molecular velocity is reversed as it undergoes reflection from a solid surface. Thus we have

$$f_{s_{sp}}(\vec{r},\vec{\xi}) = f_\tau(\vec{r},\vec{\xi} - 2\vec{\xi}\cdot\hat{n}\,\hat{n}), \quad r = a \tag{18}$$

This expression satisfies (15). Since $\hat{n} = \hat{e}_r$, we deduce that

$$f_{s_{sp}}(\vec{r},\vec{\xi}) = A \exp\left[-B\,(\xi^2 + 2U_\infty\xi_r \cos\theta + 2U_\infty\xi_\theta \sin\theta + U_\infty^2\right], \quad r = a \tag{19}$$

This has the same form as (16b) except that the algebraic sign on ξ_r has been reversed. From the theory of characteristics, we now deduce that

$$f_{\sigma_s}(\vec{r},\vec{\xi}) = f_{s_{sp}}(\vec{r}_o(\vec{r},\vec{\xi}), \vec{\xi}_o(\vec{r},\vec{\xi}))$$

$$= A \exp\left[-B(\xi^2 + 2U_\infty\xi_{r_o} \cos\theta_o + 2U_\infty\xi_{\theta_o} \sin\theta_o + U_\infty^2)\right] \tag{20}$$

Here we have already accounted for $\xi = \xi_o$. We can obtain an expression for $\cos\theta_o$ by solving for t in (8) and substituting into (9). Setting $r_o = a$, $y \equiv r/a$, and $\xi_\rho^2 \equiv \xi_\theta^2 + \xi_\psi^2$, we obtain

$$\cos\theta_o = \left[y\,\xi_\rho^2/\xi^2 + (\xi_r/\xi)\,\{1 - y^2(\xi_\rho^2/\xi^2)\}^{\frac{1}{2}}\right]\cos\theta$$

$$+ (\xi_\theta/\xi)\left[y\,\xi_r/\xi - \{1 - y^2(\xi_\rho^2/\xi^2)\}^{\frac{1}{2}}\right]\sin\theta \tag{21}$$

From (11) and (12), we get

$$\xi_{r_o} = \xi_r\left\{y^2 - (y^2 - 1)(\xi^2/\xi_r^2)\right\}^{\frac{1}{2}} = \xi\left\{1 - y^2(\xi_\rho^2/\xi^2)\right\}^{\frac{1}{2}} \tag{22}$$

From (12),(13), and (21), we find

$$\xi_{\theta_o}\sin\theta_o = y\,\xi_\theta \sin\theta\left\{(\xi_\rho^2 \sin^2\theta_o - \xi_\psi^2 \sin^2\theta)/\xi_\theta^2 \sin^2\theta\right\}^{\frac{1}{2}}$$

$$= \left[\xi_\rho^2/\xi\,(- y^2\,\xi_r/\xi + y\left\{1 - y^2(\xi_\rho^2/\xi^2)\right\}^{\frac{1}{2}})\right]\cos\theta$$

$$+ y\,\xi_\theta\left[y\,\xi_\rho^2/\xi^2 + \xi_r/\xi\,\{1 - y^2(\xi_\rho^2/\xi^2)\}^{\frac{1}{2}}\right]\sin\theta \tag{23}$$

The combination $\xi\cos\Theta \equiv \xi_{r_o}\cos\theta_o + \xi_{\theta_o}\sin\theta_o$ appearing in (20) now takes the form

$$\xi\cos\Theta = \left[\xi_r\,\{1 - 2y^2(\xi_\rho^2/\xi^2)\} + 2y(\xi_\rho^2/\xi)\{1 - y^2(\xi_\rho^2/\xi^2)\}^{\frac{1}{2}}\right]\cos\theta$$

$$+ \xi_\theta\left[- 1 + 2y^2(\xi_\rho^2/\xi^2) + 2y\,(\xi_r/\xi)\{1 - y^2(\xi_\rho^2/\xi^2)\}^{\frac{1}{2}}\right]\sin\theta \tag{24}$$

When this is substituted into (20), the distribution function f_{σ_s} for specular reflection is thus obtained:

$$f_{\sigma_s}(\vec{r},\vec{\xi}) = A \exp\left[- B(\xi^2 + 2U_\infty \xi \cos \Theta + U_\infty^2)\right], \quad \xi \text{ in } \sigma \tag{25}$$

Although obtained by slightly different means, this expression is the same as that of Karamcheti and Sentman [17,18], who obtained explicit results only for specular reflection. For specular reflection, the wall temperature T_W does not enter into the problem, and there is no heat transfer.

DIFFUSE REFLECTION

For diffuse reflection, we assume that the particles are emitted from the sphere surface with a Gaussian distribution of the form

$$f_{s_d}(\vec{r},\vec{\xi}) = \bar{n}_d(\cos \theta)\, A\, T^{3/2} \exp\left[- B\, T\, \xi^2\right] \tag{26}$$

where A and B are defined the same as for (16) and $T \equiv T_\infty / T_W$ is the ratio of the temperature at infinity and the temperature T_W of the particles emitted from the wall of the sphere. The function $\bar{n}_d$ $(\cos \theta)$ must be selected to satisfy the condition (15). Substituting (26) and (17) into (15), evaluating the integrals, and solving for $\bar{n}_d$ gives

$$\bar{n}_d(\cos \theta) = T^{\frac{1}{2}}\left[e^{-S^2 \cos^2 \theta} - \sqrt{\pi}\, S \cos \theta \operatorname{erfc}(S \cos \theta)\right] \tag{27}$$

where $S \equiv \sqrt{B}\, U_\infty$ is the molecular-speed ratio and erfc (x) is the complimentary error function with argument x. We shall assume that T_W, and thus T, is a constant.

The solution for $f_{\sigma_d}(\vec{r},\vec{\xi})$ is now obtained from f_{s_d} by replacing $\cos \theta$ by $\cos \theta_o$, which is given by (21), and writing

$$f_{\sigma_d}(\vec{r},\vec{\xi}) = A\, T^2 \left[e^{-S^2 \cos^2 \theta_o} - \sqrt{\pi}\, S \cos \theta_o \operatorname{erfc}(S \cos \theta_o)\right] e^{-B\, T\, \xi^2}, \quad \xi \text{ in } \sigma \tag{28}$$

The subscript d denotes diffuse reflection.

ACCOMMODATION COEFFICIENT

The materials of most solid surfaces are found to reflect particles neither completely specularly nor completely diffusely (see Schaaf and Chambre [4]). A common engineering approximation is to assume that part of the particles are emitted diffusely and the remainder specularly. In this case we write

$$f_s = \alpha f_{s_d} + (1 - \alpha) f_{s_{sp}} \tag{29}$$

where α is the accommodation coefficient. The value of α lies in the range $0 \le \alpha \le 1$, but for a large number of physical situations α is about 0.9. When α is unity the surface is said to be perfectly accommodating. Expression (29) satisfies the condition (15) since f_{s_d} and $f_{s_{sp}}$, as given by (26) and (19), do separately. For the surface distribution function given in terms of the accommodation coefficient, the distribution function for the emitted particles, f_σ, is found to be

$$f_\sigma = \alpha f_{\sigma_d} + (1 - \alpha) f_{\sigma_s} \tag{30}$$

where f_{σ_d} and f_{σ_s} are given by (28) and (25).

MOMENTS

If $\Phi(\vec{\xi})$ is any molecular property, then $\int \Phi(\xi)\, f\, d^3\xi$ over all velocity space is called a moment of the distribution function. The average, or expected value, of the molecular property $\Phi(\xi)$, denoted by $< \Phi >$, is defined as

$$n < \Phi > \equiv \int \Phi f d^3\xi = \int_{\tau} \Phi f_{\tau} d^3\xi + \alpha \int_{\sigma} \Phi f_{\sigma_d} d^3\xi + (1 - \alpha) \int_{\sigma} \Phi f_{\sigma_s} d^3\xi$$

$$= n_{\tau} < \Phi >_{\tau} + \alpha\, n_d < \Phi >_{\sigma_d} + (1 - \alpha)\, n_{sp} < \Phi >_{\sigma_s} \tag{31}$$

Here $n(\vec{r})$ is the number density, obtained from (31) when $\Phi = 1$, that is,

$$n(\vec{r}) = \int f d^3\xi = \int_{\tau} f_{\tau}\, d^3\xi + \alpha \int_{\sigma} f_{\sigma_d}\, d^3\xi + (1 - \alpha) \int_{\sigma} f_{\sigma_s}\, d^3\xi$$

$$= n_{\tau}(\vec{r}) + \alpha\, n_d(\vec{r}) + (1 - \alpha) n_{sp}\, (\vec{r}) \tag{32}$$

where n_{τ}, n_d, and n_{sp} are identified with their respective distribution functions.

Because the distribution function has been determined, any expected value or moment can be obtained in principle by appropriate integration over the τ and σ regions of velocity space shown in Fig. 2. In practice, however, the general analytic evaluation of these integrals has not been possible. Thus it is of interest to evaluate some of these moments in certain limiting situations. When the distribution function is expanded in a power series for small molecular-speed ratio, S, the integrals involved in the expressions for the moments can be evaluated and a resulting power series in S obtained. We shall now investigate this slow-flow approximation.

SLOW-FLOW APPROXIMATION

DISTRIBUTION FUNCTIONS

Let us now treat $S \equiv \sqrt{B}\, U_{\infty}$ as a small parameter and expand f_{τ}, f_{σ_s}, and f_{σ_d} in power series for small S. From (17), (25), and (28) we obtain

$$f_{\tau} = A\, e^{-B\,\xi^2} \Big[1 + 2S\sqrt{B}\,(\xi_r \cos\theta - \xi_{\theta} \sin\theta) + S^2 \left\{ 2B(\xi_r \cos\theta - \xi_{\theta}\sin\theta)^2 - 1 \right\} + 0(S^3) \Big] \tag{33}$$

$$f_{\sigma_s} = A\, e^{-B\,\xi^2} \Big[1 - 2S\sqrt{B}\,\xi \cos\Theta + S^2 \left\{ 2B\,\xi^2 \cos^2\Theta - 1 \right\} + 0(S^3) \Big] \tag{34}$$

$$f_{\sigma_d} = A\, T^2\, e^{-B\,T\,\xi^2} \Big[1 - \sqrt{\pi}\, S \cos\theta_o + S^2 \cos^2\theta_o + 0(S^4) \Big] \tag{35}$$

It is convenient to carry out the integrations over velocity space in spherical coordinates, shown in Fig. 2. We have

$$\xi_r = \xi \cos\gamma, \quad \xi_{\theta} = \xi \sin\gamma \cos\epsilon, \quad \xi_{\psi} = \xi \sin\gamma \sin\epsilon,$$

$$d^3\xi = \xi^2 \sin\gamma\, d\xi\, d\gamma\, d\epsilon \tag{36}$$

If $\gamma_o \equiv \sin^{-1}\frac{a}{r}$, the τ region corresponds to $\gamma_o \le \gamma \le \pi$, and the σ region (inside the cone) corresponds to $0 \le \gamma \le \gamma_o$. The other variables for both regions integrate over the ranges $0 \le \xi \le \infty$ and $0 \le \epsilon \le 2\pi$.

NUMBER DENSITY

With the above expansions for the distribution functions, we now obtain to second order, with $x \equiv a/r$ and $\cos\gamma_o \equiv (1 - x^2)^{\frac{1}{2}}$,

$$n_\tau(\vec{r}) = \frac{n_\infty}{2}\left[1 + \cos\gamma_o - S\frac{2}{\sqrt{\pi}}x^2\cos\theta + \tfrac{1}{2}S^2 x^2\cos\gamma_o\,(1 - 3\cos^2\theta)\right] \tag{37}$$

$$n_{sp}(\vec{r}) = \frac{n_\infty}{2}\left[1 - \cos\gamma_o - S\frac{4}{\sqrt{\pi}}I_1\cos\theta + \tfrac{1}{2}S^2(1 - \cos\gamma_o - 3I_2)(1 - 3\cos^2\theta)\right] \tag{38}$$

$$n_d(\vec{r}) = \frac{n_\infty}{2}T^{\frac{1}{2}}\left[1 - \cos\gamma_o - S\sqrt{\pi}\,K_1\cos\theta + \tfrac{1}{2}S^2\left(1 - \cos\gamma_o - K_2 + (3K_2 - 1 + \cos\gamma_o)\cos^2\theta\right)\right] \tag{39}$$

where $I_1 \equiv \frac{1}{4x^2}\left[3x - x^3 - \left(\frac{3}{2} - x^2 - \tfrac{1}{2}x^4\right)\ln\frac{1 + x}{1 - x}\right]$

$$I_2 \equiv \frac{4}{x^4}\left[-\frac{8}{21} + \frac{2}{5}x^2 + \frac{1}{12}x^4 - \frac{2}{105}x^7 + \frac{7}{12}\cos^3\gamma_o - \frac{7}{30}\cos^5\gamma_o + \frac{13}{420}\cos^7\gamma_o\right]$$

$$K_1 \equiv \frac{1}{3x}\left[2 + x^3 - 3\cos\gamma_o + \cos^3\gamma_o\right]$$

$$K_2 \equiv \frac{1}{x^2}\left[\frac{2}{5} + \frac{1}{3}x^2 + \frac{4}{15}x^5 - \cos\gamma_o + \frac{2}{3}\cos^3\gamma_o - \frac{1}{15}\cos^5\gamma_o\right]$$

Expressions (37), (38), and (39) substituted into (32) give the number-density field correct to and including order S^2. These expressions will be discussed later.

VELOCITY

For the axisymmetric flow the azimuthal component of velocity V_ψ is zero. Thus the components of interest are V_r and V_θ. The radial component V_r is obtained from (31) by substituting $\Phi = \xi_r$. We obtain

$$n V_r = n_\tau < \xi_r >_\tau + \alpha\, n_d < \xi_r >_{\sigma_d} + (1 - \alpha)n_{sp} < \xi_r >_{\sigma_s} \tag{40}$$

With the expanded distribution functions (33-35) we obtain to second order

$$n_\tau < \xi_r >_\tau = \frac{n_\infty}{2\sqrt{\pi B}}\left[-x^2 + S\sqrt{\pi}\,(1 + \cos^3\gamma_o)\cos\theta + S^2\left\{x^2 - x^4 - (4x^2 - 3x^4)\cos^2\theta\right\}\right] \tag{41}$$

$$n_{sp} < \xi_r >_{\sigma_s} = \frac{n_\infty}{2\sqrt{\pi B}}\left[x^2 - S\,3\sqrt{\pi}\,I_3\cos\theta + S^2\left\{x^2 - 4I_4 + 4\,(3I_4 - \tfrac{1}{2}x^2)\cos^2\theta\right\}\right] \tag{42}$$

$$n_d < \xi_r >_{\sigma_d} = \frac{n_\infty}{2\sqrt{\pi B}}\left[x^2 - S\,2\sqrt{\pi}\,K_3\cos\theta + S^2\left\{\tfrac{1}{2}x^2 - K_4 + (3K_4 - \tfrac{1}{2}x^2)\cos^2\theta\right\}\right] \tag{43}$$

where $I_3 \equiv \frac{2}{x^2}\left[-\frac{2}{15} + \frac{1}{6}x^2 + \frac{2}{15}x^5 + \frac{1}{3}\left(1 - \frac{1}{2}x^2\right)\cos^3\gamma_o - \frac{1}{5}\cos^5\gamma_o\right]$

$$I_4 \equiv \frac{1}{x^4}\left[-\frac{5}{16}x + \frac{19}{48}x^3 - \frac{1}{48}x^5 + \frac{1}{6}x^6 - \frac{1}{16}x^7 + \frac{1}{12}x^8 + \left(\frac{5}{32} - \frac{1}{4}x^2 + \frac{1}{16}x^4 + \frac{1}{32}x^8\right)\ln\frac{1 + x}{1 - x}\right]$$

$$K_3 \equiv \frac{1}{8x}\left[x + x^3 + 2x^4 - \tfrac{1}{2}(1 - x^2)2 \ln \frac{1 + x}{1 - x}\right]$$

$$K_4 \equiv \frac{1}{x^2}\left[\frac{1}{8} x - \frac{1}{12} x^3 + \frac{1}{4} x^4 + \frac{1}{8} x^5 + \frac{1}{12} x^6 - \frac{1}{16}(1 - x^2 - x^4 + x^6) \ln \frac{1 + x}{1 - x}\right]$$

When these expressions are combined, the leading terms in each of (41-43) cancel, and we have

$$nV_r = \frac{n_\infty U_\infty}{2}\left[\cos\theta\left\{1 + \cos^3\gamma_o - 3(1 - \alpha) I_3 - 2\alpha K_3\right\}\right.$$

$$\left. + \frac{S}{\sqrt{\pi}}(1 - 3\cos^2\theta)\left\{\frac{4 - \alpha}{2} x^2 - x^4 - 4(1 - \alpha) I_4 - \alpha K_4\right\}\right] \qquad (44)$$

For the V_θ component of velocity we have

$$nV_\theta = n_\tau < \xi_\theta >_\tau + \alpha n_d < \xi_\theta >_{\sigma_d} + (1 - \alpha) n_{sp} < \xi_\theta >_{\sigma_s} \qquad (45)$$

To order S^2, we obtain

$$n_\tau < \xi_\theta >_\tau = -\frac{n_\infty U_\infty}{2}\left[\sin\theta\left\{1 + \frac{3}{2}\cos\gamma_o - \frac{1}{2}\cos^3\gamma_o\right\} - \frac{S}{\sqrt{\pi}} x^4 \sin 2\theta\right] \qquad (46)$$

$$n_{sp} < \xi_\theta >_{\sigma_s} = -\frac{n_\infty U_\infty}{2}\left[\frac{3}{2} I_5 \sin\theta - \frac{4S}{\sqrt{\pi}} I_6 \sin 2\theta\right] \qquad (47)$$

$$n_d < \xi_\theta >_{\sigma_d} = -\frac{n_\infty U_\infty}{2}\left[K_5 \sin\theta - \frac{S}{\sqrt{\pi}} K_6 \sin 2\theta\right], \text{ where} \qquad (48)$$

$$I_5 \equiv \frac{2}{x^2}\left[\frac{8}{15} - \frac{1}{3} x^2 + \frac{2}{15} x^5 - (1 - \tfrac{1}{2} x^2)\cos\gamma_o + \frac{2}{3}(1 - \frac{1}{4} x^2)\cos^3\gamma_o - \frac{1}{5}\cos^5\gamma_o\right]$$

$$I_6 \equiv \frac{1}{x^4}\left[\frac{15}{16} x - \frac{11}{16} x^3 - \frac{1}{48} x^5 - \frac{1}{16} x^7 + \frac{1}{12} x^8\right.$$

$$\left. - \left(\frac{15}{32} - \frac{1}{2} x^2 + \frac{1}{16} x^4 - \frac{1}{32} x^8\right) \ln \frac{1 + x}{1 - x}\right]$$

$$K_5 \equiv \frac{1}{4x}\left[-\frac{3}{2} x + \frac{1}{2} x^3 + x^4 + (1 - x^2)\left(\frac{3}{4} + \frac{1}{4} x^2\right) \ln \frac{1 + x}{1 - x}\right]$$

$$K_6 \equiv \frac{1}{4x^2}\left[-x + \frac{1}{6} x^3 + \frac{1}{2} x^5 + \frac{1}{3} x^6 + \frac{1}{2}(1 - \frac{1}{2} x^2 - \frac{1}{2} x^6) \ln \frac{1 + x}{1 - x}\right]$$

When the different parts are combined, we obtain

$$nV_\theta = -\frac{n_\infty U_\infty}{2}\left[\sin\theta\left\{1 + \frac{3}{2}\cos\gamma_o - \frac{1}{2}\cos^3\gamma_o + \frac{3}{2}(1 - \alpha) I_5 + \alpha K_5\right\}\right.$$

$$\left. - \frac{S}{\sqrt{\pi}}\left\{x^4 + 4(1 - \alpha) I_6 + \alpha K_6\right\} \sin 2\theta\right] \qquad (49)$$

STREAM FUNCTION

It is of some interest to examine the shape of the streamlines for free molecule flow. The first moment of the kinetic equation (1) for steady flow yields the familiar continuity equation of fluid mechanics:

$$\text{div}(n \vec{V}) = 0 \qquad (50)$$

We can satisfy this equation identically by means of a stream function Ψ defined as

$$nV_r = \frac{1}{r^2 \sin\theta}\frac{\partial\Psi}{\partial\theta}, \quad nV_\theta = \frac{-1}{r \sin\theta}\frac{\partial\Psi}{\partial r} \tag{51}$$

By means of (44) and (49), we can integrate these equations for Ψ and obtain to order S

$$\frac{\Psi}{n_\infty U_\infty a^2} = \frac{1}{4}\frac{r^2}{a^2}\sin^2\theta\left[\left\{1 + \cos^3\gamma_o - 3(1-\alpha)I_3 - 2\alpha K_3\right\} - \frac{2S}{\sqrt{\pi}}\cos\theta\left\{\frac{4-\alpha}{2}\frac{a^2}{r^2} - \frac{a^4}{r^4} - 4(1-\alpha)I_4 - \alpha K_4\right\}\right] \tag{52}$$

The value of Ψ is constant along a streamline. By means of (52) it is possible to plot streamlines for slow free-molecule flow past a sphere.

DISCUSSION OF RESULTS

The free-molecule density field for slow flow is given by (32) together with the individual contributions (37,38,39). The first-order term, proportional to S, depends on $\cos\theta$, whereas the second-order term, proportional to S^2, depends on $\cos^2\theta$. First- and second-order constant-density curves for specular reflection ($\alpha = 0$) are shown in Fig. 3 for S = 0.1. The constant-density curve $n/n_\infty = 1$, which intersects the sphere at $\theta = \pi/2$, begins to curve downstream in the second approximation, showing the initial development of a wake. The constant-density curves for the second approximation move closer to the sphere on the upstream side and farther from the sphere on the downstream side. Similar results for purely diffuse reflection ($\alpha = 1$) with T = 1 are shown in Fig. 4. Comparison of the two figures shows that the flow is more dense upstream and less dense downstream of the sphere for the diffuse reflection than for the specular. Szymanski [15] obtained expressions for the density and velocity fields (for $\alpha = 1$ and $T \equiv T_\infty/T_W = 1$) to one less order in S than the present analysis. His results thus do not contain the asymmetry that is characteristic of the wake development.

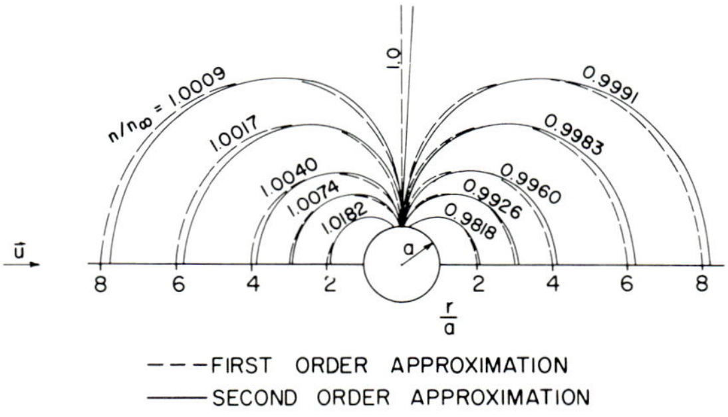

Fig. 3 Number-Density Field for S = 0.1 and $\alpha = 0$.

It is interesting to compare the free-molecule velocity field with two well-known counterparts from continuum mechanics. These are the results for viscous incompressible Stokes' flow and inviscid incompressible potential flow. For Stokes' flow, the stream function is

$$\frac{\psi}{n_\infty U_\infty a^2} = \frac{1}{2}\frac{r^2}{a^2}\left(1 - \frac{3}{2}\frac{a}{r} + \frac{1}{2}\frac{a^3}{r^3}\right)\sin^2\theta \tag{53}$$

and for potential flow the stream function is

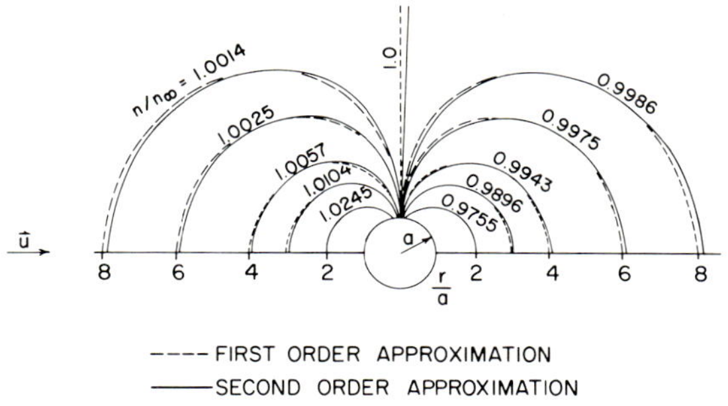

Fig. 4 Number-Density Field for $S = 0.1$, $\alpha = 1$, and $T \equiv T_\infty / T_w = 1$.

$$\frac{\psi}{n_\infty U_\infty a^2} = \frac{1}{2} \frac{r^2}{a^2} \left(1 - \frac{a^3}{r^3}\right) \sin^2 \theta \tag{54}$$

The velocity components are obtained from (51) (with $n = n_\infty$ for incompressible flow).

The place that these formulas occupy in rarefied gas dynamics can be established on the basis of the Knudsen number $K_n \equiv \lambda/a$, the ratio of the mean free path to the characteristic length of the body. For free-molecule flow $K_n \to \infty$, whereas in the continuum limit $K_n \to 0$. The Knudsen number can also be written so that it is proportional to the ratio of Mach number and Reynolds number [4], $K_n \propto M/Re$. The molecular speed ratio is proportional to the Mach number. On this basis, one can make the following comparison:

Slow free-molecule flow:	$K_n \to \infty$, $M \to 0$, $Re \to 0$
Incompressible Stokes' flow:	$K_n \to 0$, $M \to 0$, $Re \to 0$
Incompressible potential flow:	$K_n \to 0$, $M \to 0$, $Re \to \infty$

Fig. 5 compares the V_θ components of velocity for the different cases at the shoulder of the sphere ($\theta = \pi/2$). The zeroth- and first-order approximations are the same in Eq. (49) since $\sin 2\theta$ vanishes when $\theta = \pi/2$. The Stokes' flow has zero velocity at the sphere surface whereas the free molecule flow shows slip at the surface. The specular reflection ($\alpha = 0$) shows greater slip ($V_\theta / U_\infty = -1$) than the diffuse reflection ($V_\theta / U_\infty = -\frac{1}{2}$). Further, there is a slight overshoot for specular reflection. The potential flow

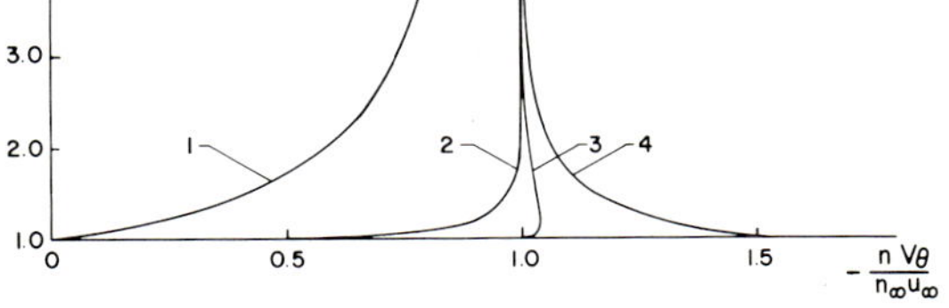

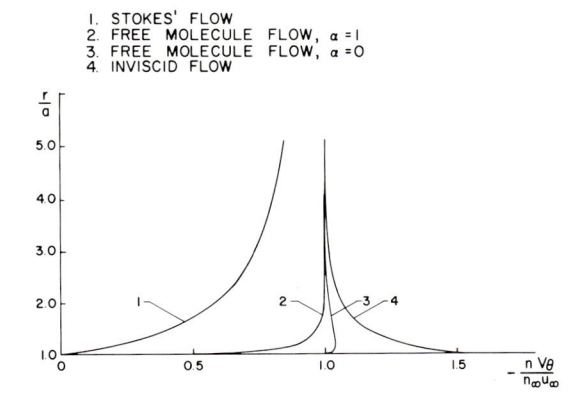

Fig. 5 Comparison of Velocity Profiles at $\theta = \pi/2$.

also shows a slip at the surface ($V_\theta/U_\infty = -3/2$), but there is a viscous boundary layer that strictly should be accounted for. The normal derivative of the velocity is infinite for the free-molecule flow cases. This does not mean that the stress is infinite, however, because stress is not reckoned on the basis of a linear stress-strain relationship as for the limiting continuum flows.

Some typical streamlines for free-molecule flow are shown in Fig. 6 for $\alpha = 0$ and in Fig. 7 for $\alpha = T = 1$. In order to clearly illustrate first-order effects in (52), a value of $S = 0.4$ was used. It can be seen that the higher-order streamlines lie closer to the sphere on the windward side and farther away on the leeward side. For the same value of Ψ, a streamline is deflected less for specular reflection ($\alpha = 0$) than for the diffuse reflection ($\alpha = 1$).

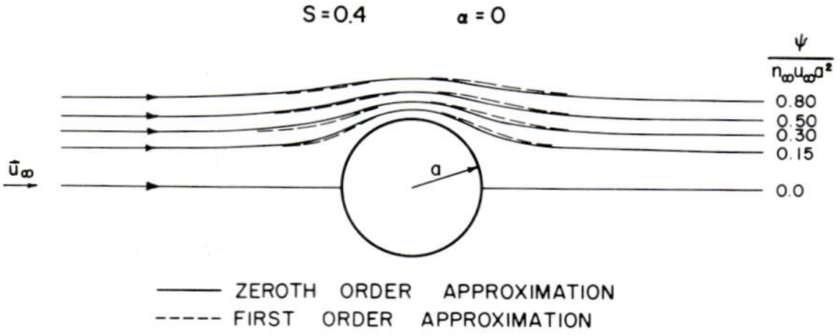

Fig. 6 Streamlines for S = 0.4 and Specular Reflection.

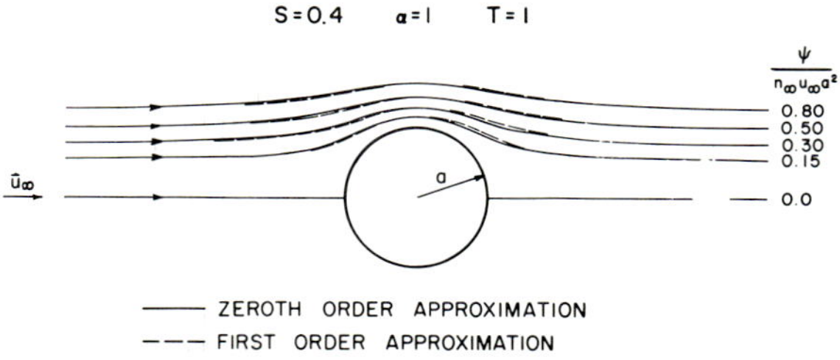

Fig. 7 Streamlines for S = 0.4 and Diffuse Reflection.

The streamlines for free-molecule flow ($\alpha = 0$ and $\alpha = 1$), Stokes' flow, and potential flow are shown in Fig. 8 for $\psi/n_\infty U_\infty a^2 = 0.8$. The streamline for Stokes' flow shows the most deflection as the flow passes the sphere; this occurs, of course, because the no-slip condition requires a greater area to pass the same amount of flow past a sphere. By the same token, the potential-flow streamline undergoes the least deflection as it passes the sphere.

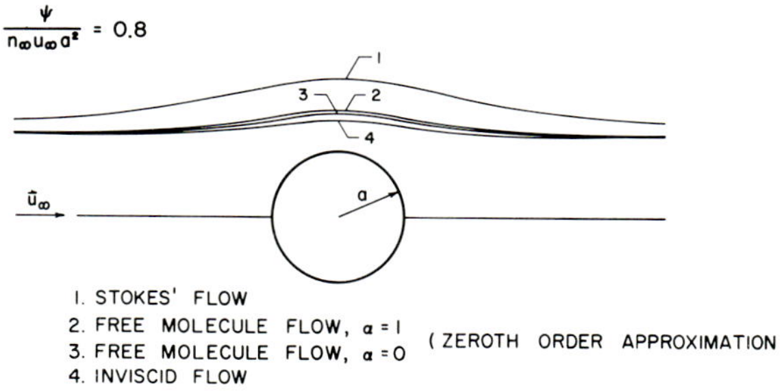

Fig. 8 Comparison of Streamlines for Slow Flow Past a Sphere.

CONCLUDING REMARKS

Slow free-molecule flow past a sphere has been examined by means of an expansion for small molecular speed ratio. This flow is akin to Stokes' flow and potential flow past a sphere, which are valid when the Knudsen number tends to zero. The free-molecule flow is valid when the Knudsen number tends to infinity. The three problems thus constitute limiting situations for the rarefied gas dynamic spectrum of slow flow past a sphere. Among other applications, slow free-molecule flow would be applicable to the relative motions of small particulate matter (dust particles, for instance) in a sufficiently rarefied medium, such as occurs at high altitudes. A worthwhile extension of this work would be to examine other moments of the distribution function, such as the temperature, stress, and heat-flux fields.

ACKNOWLEDGMENT

This work was supported by NASA Grant NGL 37-003-026.

REFERENCES

[1] A.F. Zahm, "Superaerodynamics," Journal of the Franklin Institute, Vol. 217, 1934, pp. 153-166.

[2] H. Tsien, "Superaerodynamics, Mechanics of Rarefied Gases," Journal of the Aeronautical Sciences, Vol. 13, No. 12, Dec. 1946, pp. 653-664.

[3] H. Tsien, "Wind Tunnel Testing Problems in Superaerodynamics," Journal of the Aeronautical Sciences, Vol. 15, No. 10, Oct. 1948, pp. 573-580.

[4] S.A. Schaaf, P.L. Chambre, "Flow of Rarefied Gases," High Speed Aerodynamics and Jet Propulsion, Vol. 3, edited by H.W. Emmons, Princeton University Press, Princeton, N.J., 1958, Sec. H, pp. 687-739.

[5] G.N. Patterson, "A State-of-the-Art Survey of Some Aspects of the Mechanics of Rarefied Gases and Plasmas," UTIAS Review 18, 1964, University of Toronto, Institute for Aerospace Studies, Toronto, Canada.

[6] G.N. Patterson, "A Synthetic View of the Mechanics of Rarefied Gases," AIAA Journal, Vol. 3, No. 4, April 1965, pp. 577-590.

[7] M. Heineman, "Theory of Drag in Highly Rarefied Gases," Communications on Pure and Applied Mathematics, Vol. 1, No. 3, Sept. 1948, pp. 259-273.

[8] H. Ashley, "Applications of the Theory of Free-Molecule Flow to Aeronautics," Journal of the Aeronautical Sciences, Vol. 16, No. 2, Feb. 1949, pp. 95-104.

[9] J.R. Stalder and V.J. Zurick, "Theoretical Aerodynamic Characteristics of Bodies in a Free-Molecule-Flow Field," TN 2423, July 1951, NACA.

[10] D.R. Willis, " A Study of Some Nearly Free-Molecule Flow Problems," Rept. 440, 1958, Princeton University, Aeronautical Engineering Lab., Princeton, N.J.

[11] J.B. Keller, "On the Solution of the Boltzmann Equation for Rarefied Gases," Communications on Pure and Applied Mathematics, Vol. 1, No. 3, Sept. 1948, pp. 275-285.

[12] Ya. L. Al'pert, A.V. Gurevich, and L.P. Pitaevskii, Space Physics with Artificial Satellites, Consultants Bureau, New York, 1965.

[13] D.J. Prager and M.L. Rasmussen, "The Flow of a Rarefied Plasma Past a Sphere," SUDAAR 299, Jan. 1967, Stanford University, Department of Aeronautics and Astronautics, Stanford, Calif.

[14] S.L. Soo, Fluid Dynamics and Multiphase Systems, Blaisdell Publishing Co., Waltham, Mass., First Edition, 1967, pp. 1-85.

[15] Z. Szymanski, "Some Flow Problems of Rarefied Gases," Arch. Mech. Stos. (Warsaw), Vol. 8, No. 4, 1956 and Vol. 9, No. 1, 1957.

[16] R. Goldberg, "Flow of a Rarefied Perfect Gas Past a Fixed Spherical Obstacle," Ph.D. thesis, 1954, New York University.

[17] K. Karamcheti and L.H. Sentman, "Some Aspects of the Problems of Solid Surfaces in Kinetic Theory," SUDAAR 236, May 1965, Stanford Univ., Dept. of Aeronautics and Astronautics, Stanford, Calif.

[18] L.H. Sentman and K. Karamcheti, "Rarefied Flow Past a Sphere," AIAA Journal, Vol. 7, No. 1, January 1969, pp. 161-163.

[19] J.P. Elliott and M.L. Rasmussen, "Free-Molecule Flow Past a Cone," AIAA Journal, Vol. 7, No. 1, January 1969, pp. 78-86.

[20] M. Van Dyke, Perturbation Methods in Fluid Mechanics, Academic Press, New York, 1964, pp. 4,15,16.

Developments in Mechanics, Vol. 6. Proceedings of the 12th Midwestern Mechanics Conference: 8

Laminar Plane Wall Jet

R. S. REDDY GORLA and D. R. JENG
THE UNIVERSITY OF TOLEDO

ABSTRACT

An analysis is made for the laminar, incompressible plane wall jet based on the integral method. Numerical results for both the potential core region and fully developed region are presented for slot Reynolds number ranging from 200 to 900. Several parameters characterizing the nature of flow, namely potential core length, boundary layer thickness, maximum velocity, similarity exponents, virtual origin length, exterior momentum flux and skin friction coefficient are obtained and compared with the available experimental data.

Potential core length, constans of proportionality for the decay of maximum velocity and the growth of the boundary layer thickness, virtual origin length and exterior momentum flux are correlated and expressed explicitly in terms of the slot Reynolds number.

INTRODUCTION

The flow pattern that results when a jet of fluid is discharged tangentially over a plane wall into surroundings of similar composition is called a plane wall jet. Such a flow is produced at the lower section of a canal having different water levels in its two sections, when a sluice separating the two sections is slightly raised. Plane wall jet flows are used in many practical applications, for example, in window deicing, tempering of glass by heated air blown over the surface, on blowing flaps to give additional thrust and lift on aircraft wings, etc.

Let's consider a fluid emerging through a nozzle with slot height s made of convergent parallel walls. For this case, it has been shown in Reference [1] that the velocity distribution across the slot will be mostly square-headed. The velocity deviates markedly from the maximum velocity u_o only at the walls and will be uniform across almost the whole cross section of the slot. (According to Prandtl-Tietjens [2], in the case of a tube, a convergence of 1mm in radius over a length of 1m, while Reynolds number is 1000, produces a substantial amount of flattening in the velocity distribution.) Since there has been no reliable measurement of the velocity profiles either at the slot exit or in its immediate neighborhood in the case of two-dimensional wall jet, the assumption of a uniform velocity distribution at the slot exit seems to be justified. The flow development is illustrated in Figure 1. The flow region is divided into two parts, namely, potential core region (OP) and fully developed region (downstream of the point P). Owing to viscous friction, the boundary layer will be formed on solid wall and the viscous layer will be developed some distance away from the wall. Its width will increase in the downstream direction. In the potential core region, the behaviour of the flow close to the solid wall is similar to the boundary layer development in the case of a uniform flow over a flat plate and the flow mechanism at far away distances from the solid wall is similar to that of a free boundary layer flow. In between these two limits, there exists a region of constant velocity u_o. As the streamwise distance from the slot end increases, the two viscous layers gradually merge into each other and finally the potential core disappears completely, whereby the velocity profile is transformed asymptotically into a similarity form. The distance from the slot end to the point where the potential core disappears is defined as potential core length and is denoted by ℓ. As we proceed further downstream of the potential core region, the local maximum velocity starts decaying. The zone downstream of the point where the maximum local velocity starts decaying is called fully developed region. The part of the wall jet which is found in between the plane surface and the locus of points of maximum velocity is known as inner layer and the remaining portion is called outer layer. Here the inner layer is similar to the boundary layer flow over a flat plate at zero incidence and the outer layer is similar to one-half of a plane free jet.

A numerical comparison of the theoretical and experimental normalized velocity profile of the plane wall jet with the theoretical velocity profiles of the plane free jet by Bickley [3]* in the outer layer and with the Blasius velocity profile in the inner layer is carried out by Bajura and Szewczyk [4]. They concluded that the velocity distribution in the laminar plane wall jet can be closely approximated, both theoretically and experimentally, by a suitably scaled combination of Blasius and plane free jet profiles.

The laminar plane wall jet was first studied theoretically by Tetervin [5]. With a hypothesis that the maximum velocity and boundary layer thickness vary as the downstream distance raised to a power, namely, the similarity exponents, he obtained a velocity profile. Tetervin showed that the boundary layer thickness will increase as 3/4 power of the downstream distance and the maximum velocity decays as 1/2 power of downstream distance. However, the above similarity rule is only valid when the downstream distance is measured not from the slot end but from a rather fictitious point behind the slot, termed virtual origin. This means that the flow is assumed to be issued from this point. In his analysis, Tetervin concluded that the dimensionless shear stress at the wall is independent of the viscosity which is physically an unacceptable conclusion. In 1956, Glauert [6] analyzed both laminar plane and laminar radial wall jets. In his analysis, he has shown that the flux of exterior momentum F is an integral invariant and the similarity solution for a dimensionless stream function is independent of F. He also indicated that F may be determined experimentally at any downstream station, either in fully developed region or in potential core region. Since F is proportional to a reference velocity cubed and a reference length squared, the selection of any two parameters determines the third.** Although the numerical values of the dimensionless stream function and its derivatives are presented in Glauert's paper, none of these parameters have been determined theoretically or experimentally by him. Therefore the local velocity and shear stress values are also not

*Numbers in brackets designate References at the end of the paper.

**The authors are indebted to the reviewer, who pointed out this fact.

evaluated. His predictions for boundary layer growth and maximum velocity are in agreement with Tetervin, but the wall shear stress is shown to depend on viscosity of the fluid. In 1958, Riley [7] investigated the effects of compressibility on a radial laminar wall jet. More recently, Bajura and Szewczyk [4] studied the boundary layer growth and decay of maximum velocity for laminar plane wall jet experimentally. They also conducted some experiments related to the stability of plane laminar wall jet.

All the previous analytical methods are valid for fully developed region of the wall jet and no analysis is existing in literature to predict the potential core lengths, virtual origin and F. The present investigation aims at predicting these and other flow characteristics of the plane wall jet. The analysis is made for both potential core and fully developed regions by using integral method. Results are compared with the experimental data of [4].

ANALYSIS

Consider a laminar plane wall jet emerging from a long, thin slot parallel with an infinite plane. The basic conservation laws for an incompressible, steady, laminar flow in a zone of no body forces and pressure gradient are:

Continuity:
$$\frac{\partial u}{\partial x} + \frac{\partial v}{\partial y} = 0 \qquad (1)$$

Momentum:
$$u\frac{\partial u}{\partial x} + v\frac{\partial u}{\partial y} = \nu\frac{\partial^2 u}{\partial y^2} \qquad (2)$$

Where x and y denote distances along and normal to the wall, x being measured from the slot exit, u and v the corresponding velocity components, and ν the kinematic viscosity.
The boundary conditions are:

$$u = v = o \quad \text{at} \quad y = o, \quad u \to o \quad \text{as} \quad y \to \infty \qquad (3)$$

The momentum and kinetic energy integral equations can be readily derived from (1), (2) and (3) and may be written as:

Momentum Integral Equation:
$$\frac{d}{dx}\int_0^\infty u^2 dy = -\frac{\tau_w}{\rho} \qquad (4)$$

Kinetic Energy Integral Equation:
$$\frac{d}{dx}\int_0^\infty u^3 dy = -2\nu\int_0^\infty \left(\frac{\partial u}{\partial y}\right)^2 dy \qquad (5)$$

It is now required to find suitable velocity profiles for both the potential core region and fully developed region in order to seek a solution by the integral technique. To this end, we introduce a dimensionless variable $\eta = \frac{y}{\delta}$ where δ is a boundary layer width which is an unknown function of x and is to be determined by solving (4) and (5). Since the streamwise velocity vanishes both at the plane surface and at very large distances from it, a normalized similarity velocity profile of the form

$$u^* = \frac{u}{u_m} = \frac{\eta}{\eta_m}\exp(\eta_m^c)\exp(-\eta^c) \qquad 0 \le \eta \le \infty \qquad (6)$$

is chosen for the fully developed region, where c is a constant and η_m, which is a function of c, is the value of η at which it gives u a maximum. This normalized velocity profile satisfies the boundary conditions at $\eta = 0$ and $\eta \to \infty$ and gives rise to a maximum velocity at a finite value of η. The unknown constant c appearing in (6) was determined as follows. The stream function of Glauert [6] for a laminar plane wall jet was recalculated by us using the electronic computer (since only the graphical presentation of the stream function was given in [6]) and the figure of the normalized velocity profile u* plotted against the dimensionless distance $\frac{\eta}{\eta_m}$ is obtained. Rewriting the equation (6) in terms of $\frac{\eta}{\eta_m}$ variable, the constant c was adjusted so that the resulting normalized profile coincides with that of Glauert as shown in Figure 2. It was determined that c = 2.20 satisfies the above requirement and the corresponding value of η_m is found to be 0.6988. The advantage of matching the equation (6) in $\frac{\eta}{\eta_m}$ coordinate is that we may determine the constant c for the dimensionless velocity profile without knowing the boundary layer width δ(x). The velocity profile in the potential core region is assumed to have a zone of constant velocity in between the inner and outer layers. Hence, for potential core region, we have $u_m = u_o$ and the normalized velocity profile is given by:

$$\frac{u}{u_o} = \begin{cases} A\,\eta \exp(-\eta^c) & \text{for } o \le \eta \le \eta_m \\ 1 & \eta_m \le \eta \le \eta_e \\ A\,(\eta-\eta_d)\exp[-(\eta-\eta_d)^c] & \eta_e \le \eta \le \infty \end{cases} \tag{7a}$$

$$\text{with } A = \eta_m^{-1} \exp(\eta_m^{\,c}) = 2.2545 \tag{7b}$$

$$\text{and } \eta_d = \eta_e - \eta_m$$

where η_e is the dimensionless distance to the inner edge of the free boundary layer flow.

Introducing the dimensionless variables

$$x^* = \frac{x}{s}, \quad y^* = \frac{y}{s}, \quad \delta^* = \frac{\delta}{s}, \quad y_d^* = \frac{y_d}{s} \text{ and } u^+ = \frac{u_m}{u_o}$$

into (4) and (5) and substituting equations (6) and (7) respectively for fully developed and potential core regions, we have:

For potential core region:

Momentum:

$$Re_s\ \delta^* \left[0.3544\frac{d\delta^*}{dx^*} + 0.4436\frac{dy_d^*}{dx^*}\right] + 1 = 0 \tag{8}$$

Kinetic energy:

$$Re_s\ \delta^* \left[0.2935\frac{d\delta^*}{dx^*} + \frac{dy_d^*}{dx^*}\right] + 5.2569 = 0 \tag{9}$$

For fully developed region:

Momentum:

$$0.3544\ Re_s\ \delta^* \left[u^+\frac{d\delta^*}{dx^*} + 2\delta^*\frac{du^+}{dx^*}\right] + 1 = 0 \tag{10}$$

Kinetic energy:

$$0.1302\ Re_s\ \delta^* \left[u^+\frac{d\delta^*}{dx^*} + 3\delta^*\frac{du^+}{dx^*}\right] + 1.0342 = 0 \tag{11}$$

with the initial conditions

$$\delta^* = 0,\ y_d^+ = 1, \text{ at } x^* = 0 \tag{12}$$

$$u^+ = 1 \qquad \text{at } x^* = \ell^* \tag{13}$$

where Re_s is the slot Reynolds number defined by $\frac{u_o s}{\nu}$.

The above governing equations were solved simultaneously by using modified Euler's method of numerical integration for slot Reynolds number ranging from 200 to 900. Potential core region was first solved numerically on a computer for $x^* = 0$ to $x^* = \ell^*$ at which y_d^* vanishes. Then using this results of δ^* at $x^* = \ell^*$ and (13) as initial conditions, (10) and (11) were solved simultaneously. The results of δ^* versus x^* are plotted in Figure 3 for several slot Reynolds numbers indicated. These data will be used later in the calculation of parameters characterizing the nature of a plane wall jet.

RESULTS AND DISCUSSIONS

POTENTIAL CORE LENGTH

The dimensionless potential core length is corresponding to the value of x^* at which y_d^* vanishes in the solution of (8) and (9). It is found to vary with slot Reynolds number. The numerical results of the potential core length for the slot Reynolds number ranging from 200 to 900 are obtained and satisfactorily correlated by a formula

$$\ell^* = 0.001558\,(Re_s + 119)^{1.273} \tag{14}$$

Above equation was obtained by employing a method of least squares to make the best fitting of the curve to the numerical results.

At the present time, neither the experimental results of the velocity distributions in the core region nor the direct measurements of the core length have been reported in the literature. This was because of the problems posed by instrumentation. Accuracy of flow measurement in a station where wire size is comparable in magnitude to the source size is bound to be poor. In Reference [4], effort has been made toward the measurement of the velocity profiles at a station 0.015 inches downstream from the wall jet slot. It was reported that the position of the wire of the anemometer above the wall could not be determined and the velocity profiles are constructed by linear interpolation of the data. The exit profiles in the experiment are predicted approximately in parabolic form. If we approximate the velocity profile in the potential core region by equation (7), at least a qualitative comparison with the experimental results can be made. It is, however, anticipated that there will be some discrepency between the present results and experimental results due to the above assumption.

In Figure 4, the numerical results of the present analysis are compared with those obtained by extrapolating the data in Reference [4]. The extrapolation is done as follows. The experimental data for u^+ versus $x^* + x_o^*$ is plotted on log-log coordinates for a given slot Reynolds number. The experimental points seem to lie on a straight line, indicating that $u^+ \sim (x^* + x_o^*)^r$ where r is a constant. The intersection of this straight line with horizontal line corresponding to $u^+ = 1$ determines the values of $\ell^* + x_o^*$ and ℓ^* may be readily reduced from known value of x_o^*. Agreement between the theory and the experimental data is very good for the cases of $Re_s = 530$ and 750. The data for $Re_s = 300$, 400 and 650 fall somewhat below the analytical curves.

BOUNDARY LAYER THICKNESS, MAXIMUM VELOCITY, THEIR SIMILARITY EXPONENTS AND VIRTUAL ORIGIN LENGTH

Two important characteristics of the laminar plane wall jet are the decay of maximum velocity and the growth of the boundary layer thickness as the downstream distance increases. The boundary layer thickness in this paper will be arbitrarily defined as the distance from the wall to the point of the half maximum velocity in the outer layer. Value of $\eta_{\frac{m}{2}} = 1.294$ equivalent to the locus of the dimensionless boundary layer thickness may be determined by solving (6) with $u^* = \frac{1}{2}$. The boundary layer thickness is given by

$$y_{\frac{m}{2}} = 1.294\ \delta(x) \tag{15}$$

By substituting the known value of $\delta(x)$ for the fully developed region into (15), the boundary layer thickness for several Reynolds numbers is calculated. They are shown graphically in Figure 5 together with the experimental data from Reference [4]. All the experimental data were obtained for air at 80°F. Agreement between the present theoretical value and the experimental data is good considering the approximate method used and the approximation of the slot velocity profile in the analysis. The theoretical curve tends to fall somewhat below the experimental one except for the case of $Re_s = 300$.

The maximum velocity $u_m(x)$ can also be obtained by solving equations (10) and (11). The data is plotted against downstream distance measured from slot end for various values of exit Reynolds number in Figure 6. Comparison is also made with the experimental data and the agreement is very good.

To determine the similarity exponents and virtual origin, the correlations of the maximum velocity and the boundary layer thickness to the numerical data shall be made. To this end, correlations for the dimensionless maximum velocity and boundary layer thickness are assumed to be of the form

$$u^+ = K_1\ (x^* + x_o^*)^r \tag{16}$$

and

$$y^*_{\frac{m}{2}} = K_2\ (x^* + x_o^*)^t \tag{17}$$

The least squares method of analysis was applied to determine the three parameters, namely x_o^*, r and K_1 or x_o^*, t and K_2 for the best fitting of the equations (16) and (17) to its analytical data. Two similarity exponents are found to be $r = -\frac{1}{2}$ and $t = 3/4$. They are in excellent agreement with the theoretical results of [5] and [6] and in close agreement with the experimental results given in [4].

The results of the virtual origin can be correlated in dimensionless form as

$$x_o^* = 0.0074\ (Re_s - 99)^{1.3168} \tag{18}$$

for the range of slot Reynolds numbers investigated. Equation (18) indicates that the length of the virtual origin increases monotomically as the slot Reynolds number increases, similar to the results for the two dimensional free jet by Andrade [8], and Sato and Sakao [9]. A comparison between (18) and experimental results is made in Figure 7. Present results predict slightly lower values. The two constants of proportionality K_1 and K_2 for describing the decay of dimensionless maximum velocity and the growth of dimensionless boundary layer thickness in the downstream direction are found to be

$$K_1 = 0.107\ (28 + Re_s)^{0.618} \tag{19}$$

and

$$K_2 = 264.725\ (247.5 + Re_s)^{-1.225} \tag{20}$$

which depend solely on slot Reynolds number. Equations (19) and (20) are compared with experimental results by plotting in suitable coordinates in Figures 8 and 9.

EXTERIOR MOMENTUM FLUX

In the analysis of other boundary layer problems, the condition such as external free stream for flow over a flat plate or the constancy of momentum flux for a free jet is required to determine the similarity exponents for its similarity solutions. However, neither external free stream nor constancy of momentum flux is present in the case of wall jet problem. Glauert [6] has shown that the non-trivial solutions to the governing momentum equation of the wall jet exist under the condition if the flux of exterior momentum flux F defined by

$$F = \int_0^\infty u \left[\int_y^\infty u^2 dy \right] dy = \text{constant} \qquad (21)$$

is conserved. The value of F is indeterminate in his analysis. More recently, Bajura and Szewczyk [4] have found F experimentally for slot Reynold numbers ranging from 250 to 800. Nevertheless, the expression for F may be readily deduced from the results we obtained previously. Rewriting the equation (21) in the dimensionless form, substituting equation (6) in it and evaluating the double integral, one obtains

$$F = 587.409 \frac{\nu^3}{S} \frac{Re_s^3 (Re_s + 28.0)^{1.855}}{(Re_s + 247.5)^{2.450}} I \qquad (22)$$

where

$$I = 0.0351$$

A derivation of (22) is given in the Appendix which also gives analytical expression of I.

Equation (22) may be compared with the experimental results. Several values of F determined from [4] for the twelve slot Reynolds numbers are listed in Table I. Included also are the results computed from (22). It is seen that the agreement is very satisfactory.

TABLE I

COMPARISON OF RESULTS FOR $F^{(a)}$

	Re_s	250	300	350	400	450	500
F $\frac{ft^5}{sec^3}$	Determined From Ref. [4]	0.00568	0.0105	0.0174	0.0266	0.0389	0.0539
	From (22)	0.00571	0.0106	0.0177	0.0273	0.0398	0.0554

	Re_s	550	600	650	700	750	800
F $\frac{ft^5}{sec^3}$	Determined From Ref. [4]	0.0728	0.0963	0.123	0.155	0.192	0.236
	From (22)	0.0744	0.0970	0.124	0.154	0.189	0.229

(a) Values of F were calculated for air at 80°F.

SKIN FRICTION COEFFICIENT

The skin friction coefficient is given by

$$C_f = \frac{\tau_w}{\frac{1}{2}\rho u_o^2} = \frac{4.509}{Re_s \, \delta^*} \qquad \text{for } 0 < x^* \leq \ell^*$$

$$C_f = \frac{4.509}{Re_s} \frac{u^+}{\delta^*} \qquad \text{for } l^* \leq x^* \tag{23}$$

Employing the results for the variation of δ^* and u^+ with x^* the numerical values of C_f are evaluated and presented in Figure 10 for various slot Reynolds numbers. C_f increases with decreasing slot Reynolds number at any location downstream of the slot end.

No correlation could be made for the skin friction coefficient in the potential core region, however, in the fully developed region, (23) may be correlated as

$$C_f = K_3 \; (x^* + x_o^*)^{-5/4} \tag{24a}$$

with

$$K_3 = 2.358.10^{-3} \; \frac{(28 + Re_s)^{0.618}(247.5 + Re_s)^{1.225}}{Re_s} \tag{24b}$$

No experimental data for C_f are available at the present time for comparison, however, (24) may directly be compared with that of Glauert [6]. Using our notation, the analytical results of Glauert may be written as

$$C_f = K_4 \; (x^* + x_o^*)^{-5/4} \tag{25a}$$

with

$$K_4 = \frac{2}{9} \frac{1}{Re_s^2} \left(\frac{125 \; s^3 F^3}{8\nu^9} \right)^{\frac{1}{4}} \tag{25b}$$

The numerical results for C_f from our analysis can be compared with those of Glauert if the experimental value for F is substituted in (25b), for any slot Reynolds number. The same can be achieved by comparing K_3 in (24b) and K_4 in (25b). Using the data of Table I for the value of F, we can compare K_3 and K_4 for any slot Reynolds number. As an example, we let $Re_s = 500$ and calculate K_3 and K_4 for air at 80°F with s = 27 mills. It is found that $K_3 = 0.752$ and $K_4 = 0.638$ for this case. So, the skin friction values calculated by us exceed those evaluated by using Glauert's equation, by about 18%.

ACKNOWLEDGMENTS

One of the authors (R.S.R.G.) is grateful to Dr. W. R. Miller, Chairman of the Mechanical Engineering Department and Dr. O. Zmeskal, Dean of the College of Engineering at The University of Toledo for their financial arrangement to make this investigation possible.

The authors wish to thank Mr. Muhammed Al-Salihi of the Mechanical Engineering Department for his help in making the final drawings.

APPENDIX

Derivation of equation (22): By transforming (21) in the dimensionless variables and substituting (6) into it, we obtain

$$F = \delta^2 \; u_m^3 \; A^3 \; I \tag{26}$$

where

$$I = \int_0^\infty \eta \; \exp\,(-\eta^{2.2}) \left[\int_\eta^\infty \eta^2 \; \exp\,(-2\eta^{2.2}) d\eta \right] d\eta$$

Evaluating the inner integral of I in incomplete gamma functions expanding it in a series, followed by evaluating the outer integral in gamma functions and inserting the expressions for δ and u_m and the numerical value of A, we obtain (22) with

$$I = \frac{\Gamma\left(\frac{15}{11}\right)\Gamma\left(\frac{10}{11}\right)}{4.84 \times 2^{\frac{3}{2\cdot 2}}} - \frac{1}{4.84 \times 2^{\frac{5}{2\cdot 2}}} \sum_{n=0}^{\infty} \left(\frac{2}{3}\right)^{\frac{25}{11}+n} \frac{\Gamma\left(\frac{25}{11}+n\right)}{\Gamma\left(\frac{26}{11}+n\right)} \tag{27}$$

The numerical value of I was calculated using n = 20 in (27).

NOMENCLATURE

A	=	a constant defined by equation (7b)
C	=	constant defined in (7a)
C_f	=	skin friction coefficient = $\frac{\tau_w}{\frac{1}{2}\rho u_o^2}$
F	=	exterior momentum flux defined by (21)
K_1	=	proportionality constant defined in (16)
K_2	=	proportionality constant defined in (17)
K_3	=	proportionality constant defined in (24a)
K_4	=	proportionality constant defined in (25a)
L	=	potential core length
L^*	=	dimensionless core length $\frac{L}{s}$
r	=	similarity exponent defined in (16)
Re_s	=	slot Reynolds number $\frac{u_o s}{\nu}$
s	=	slot height
t	=	similarity exponent defined in (17)
u	=	velocity component in x direction
u_o	=	maximum slot exit velocity
u_m	=	local maximum velocity
u^*	=	dimensionless velocity $\frac{u}{u_m}$
u^+	=	ratio of local maximum velocity to slot exit velocity $\frac{u_m}{u_o}$
v	=	velocity component in y direction
x	=	distance along streamwise direction measured from slot exit
x^*	=	dimensionless distance $\frac{x}{s}$
x_o	=	distance from slot exit to virtual origin
x_o^*	=	dimensionless virtual origin $\frac{x_o}{s}$
y	=	distance normal to the plane wall
y^*	=	dimensionless distance $\frac{y}{s}$
δ	=	characteristic length which is a function x
δ^*	=	ratio of δ to s, $\frac{\delta}{s}$
η	=	dimensionless distance defined by $\frac{y}{\delta}$
η_m	=	value of η at which $\frac{du}{d\eta} = 0$
ν	=	kinematic viscosity of the fluid
ρ	=	density of the fluid
τ_w	=	wall shear stress

SUBSCRIPTS

e: conditions at the edge of free boundary layer flow in the potential core region

m: conditions at the maximum velocity position

$\frac{m}{2}$: conditions at the half maximum velocity position in the outer layer

REFERENCES

1. Tollmien, Handbook Experimental Physics, Vol. 4, Pt. 1, pp. 257 (Leipzig, 1931).
2. Prandtl-Tietjens, Applied Hydro-and Aeromechanics, P. 54 (London, 1934).
3. W. Bickley, "The Plane Jet", Philosophical Magazine, Ser. 7, Vol. 23, 1939, pp. 727.
4. R.A. Bajura and A.A. Szewczyk, "An Experimental Investigation of a Laminar Two-Dimensional Plane Wall Jet", Technical Report No. 67-17, University of Notre Dame, 1967.
5. N. Tetervin, "Laminar Flow of a Slightly Viscous Incompressible Fluid that Issues from a Slit and Passes Over a Flat Plate", NACA TN 1644, 1948.
6. M.B. Glauert, "The Wall Jet", Journal of Fluid Mechanics, Vol. 1, 1956, pp. 625-643.
7. N. Riley, "Effects of Compressibility on a Laminar Wall Jet", Journal of Fluid Mechanics, Vol. 4, 1958, pp. 615-628.
8. E.N. Andrade, "The Velocity Distribution in a Liquid-into-Liquid Jet, Part 2: The Plane Jet", Proceedings of Physical Society of London, Vol. 51, 1939, pp. 784-793.
9. H. Sato and F. Sakao, "An Experimental Investigation of the Instability of a Two-Dimensional Jet at Low Reynolds Numbers", Journal of Fluid Mechanics, Vol. 20, 1964, pp. 337-352.

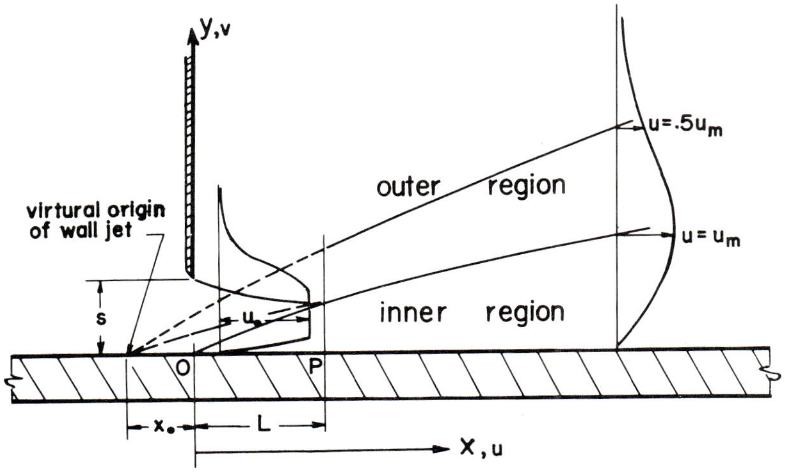

Figure 1. Co-ordinate System and Flow Development of a Laminar Plane Wall Jet.

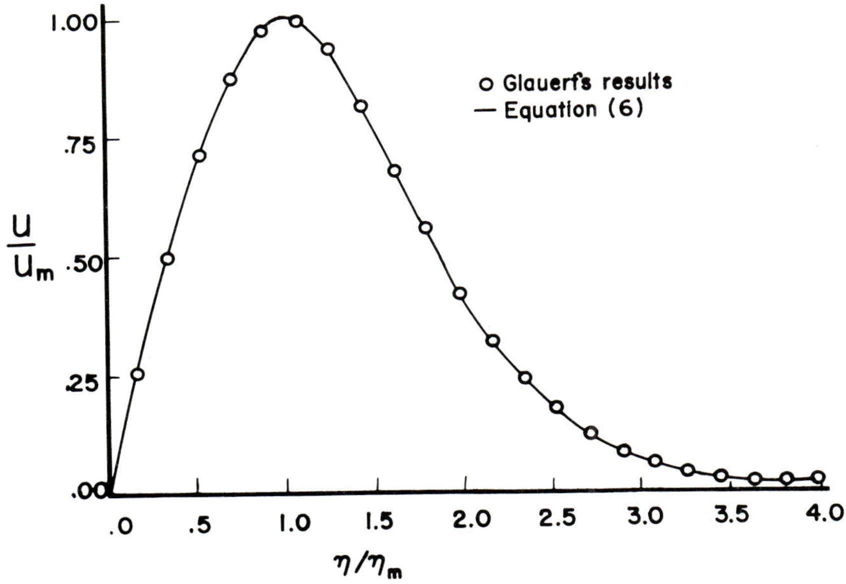

Figure 2. Comparison of Equation (6) with Glauert's Results in η/η_m coordinate.

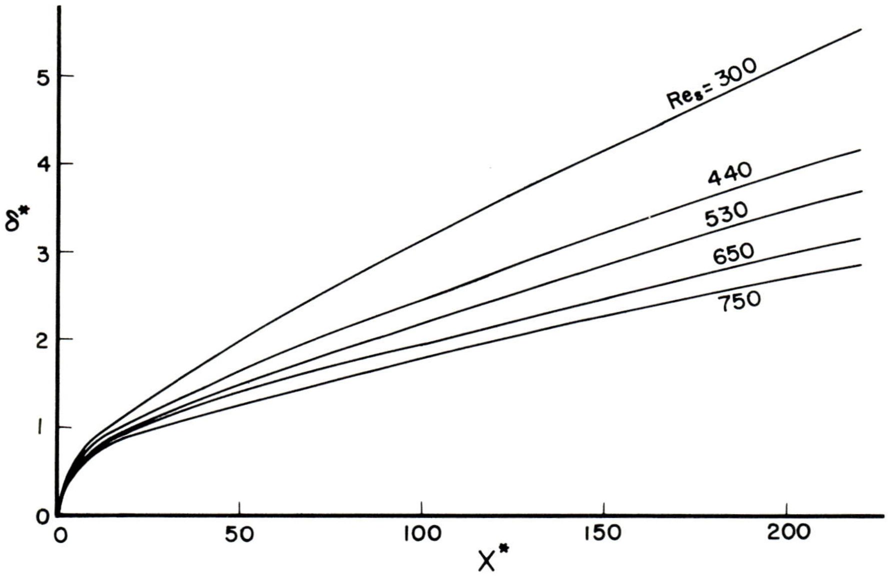

Figure 3. Variation of the non-dimensional Variable δ^* with Dimensionless Downstream Distance from the Slot End.

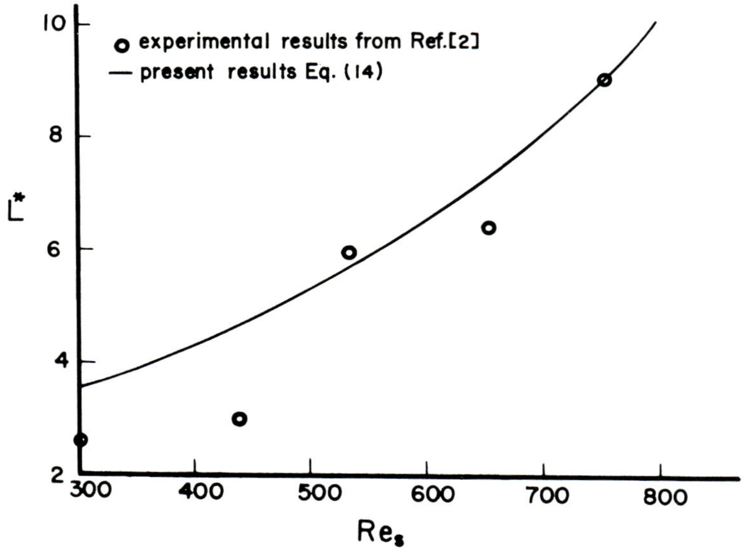

Figure 4. Variation of Dimensionless Potential Core Lengths with Slot Reynolds Number.

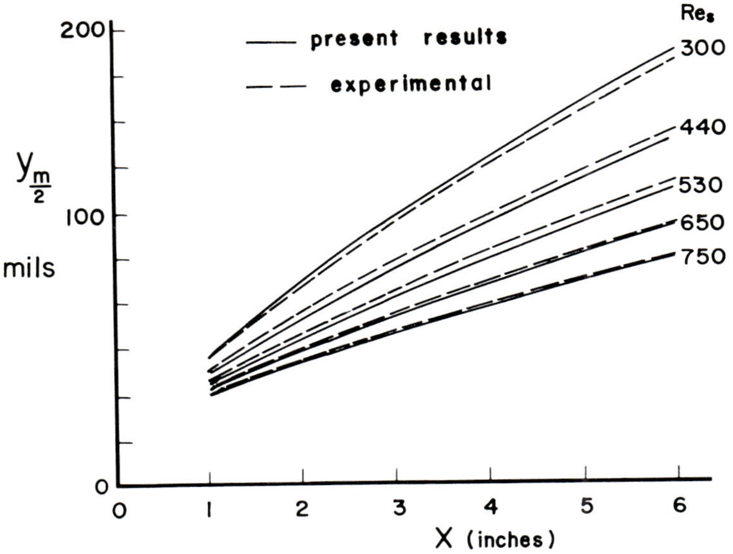

Figure 5. Variation of Boundary Layer Thickness with Downstream Distance from the Slot End.

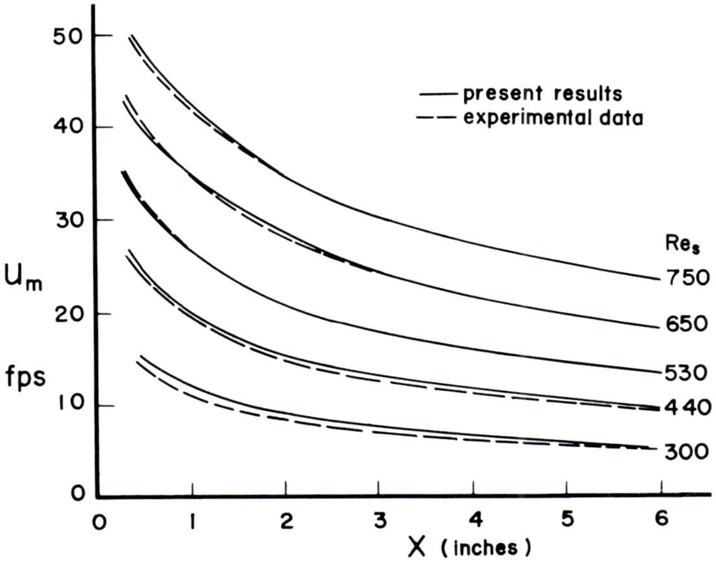

Figure 6. Maximum Velocity Decay with Downstream Distance from the Slot End.

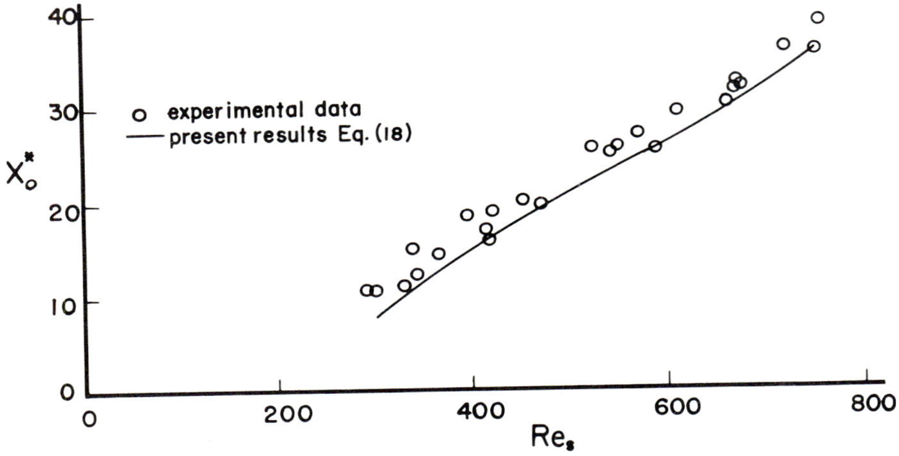

Figure 7. Variation of Dimensionless Virtual Origin Length with Slot Reynolds Number.

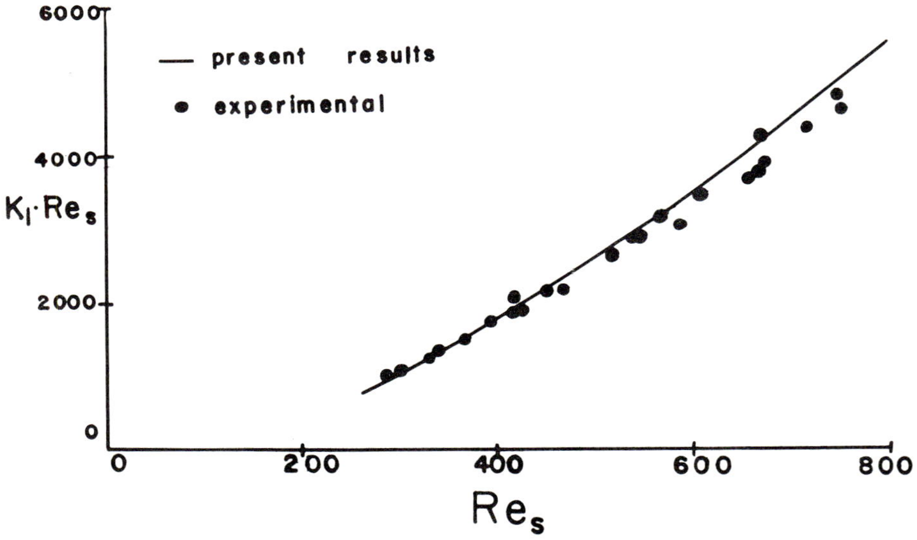

Figure 8. Variation of the Proportionality Constant K_1 Re_s to Describe Maximum Velocity Decay with Re_s.

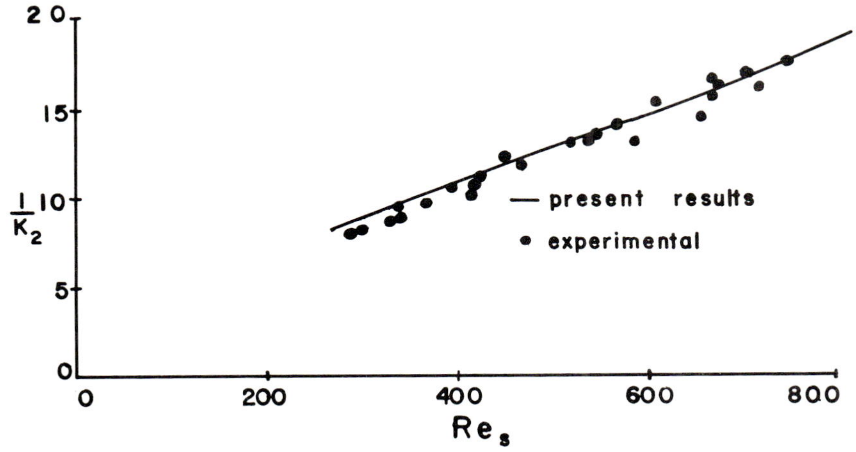

Figure 9. Variation of the Proportionality Constant $\frac{1}{K_2}$ to Describe Boundary Layer Growth with Re_s.

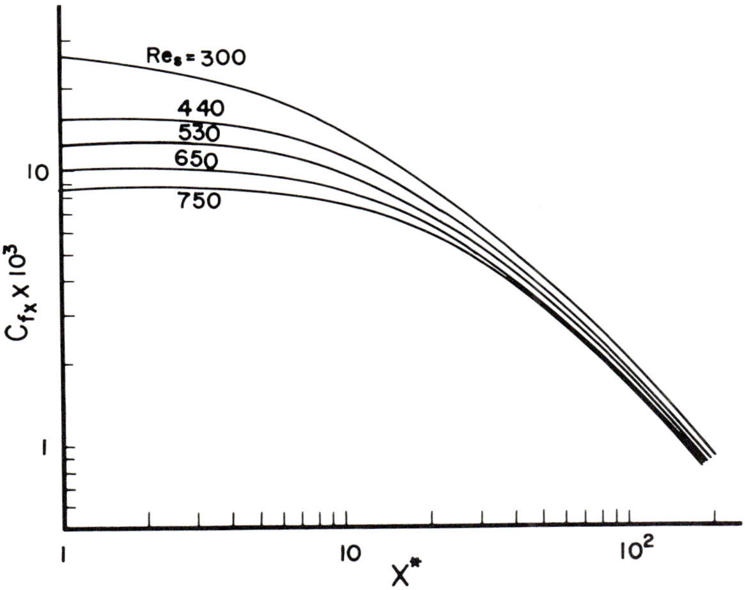

Figure 10. Variation of Skin Friction Coefficient with Dimensionless Downstream Distance.

Developments in Mechanics, Vol. 6. Proceedings of the 12th Midwestern Mechanics Conference:

9

Boundary-layer Solutions for Viscoelastic Liquids

JOHN PEDDIESON, JR.
TENNESSEE TECHNOLOGICAL UNIVERSITY

ABSTRACT

The second-order constitutive equation, due to Coleman and Noll, is used as a model to test a method of solution for viscoelastic boundary-layer flows. Several flows are considered, and solutions are obtained which are correct to the same order of approximation as the constitutive equation. The qualitative trends of the results are noted and compared, where possible, with those of previous investigators. The significance of the findings of this paper to more realistic models of viscoelastic behavior is discussed.

INTRODUCTION

Interest in boundary-layer flows of viscoelastic liquids has been high during the past decade because of the occurrence of such flows in industrial processes. In order to determine the effects of elasticity, several investigators have performed boundary-layer calculations using various idealized constitutive equations of the differential type. The purpose of the present paper is to suggest a consistent method for solving the boundary-layer equations for a class of viscoelastic liquids and to use the method to solve several boundary-layer problems. For simplicity attention is restricted to the second-order liquid of Coleman and Noll [1]. The point of view adopted here is that while the second-order model is not quantitatively accurate at the high rates of shear occurring in boundary layers, it can serve as a simple prototype for more accurate models which have similar, but more complicated, constitutive equations. This is similar to the way in which the Oseen equations serve as a prototype for the Navier-Stokes equations in Newtonian fluid mechanics.

GOVERNING EQUATIONS

Davis [2] has given the most comprehensive formulation of the boundary-layer theory for a second-order liquid. Referred to the coordinate system to be used in the present paper his equations expressing conservation of mass and conservation of linear momentum are respectively

$$u,_s + (j/s)u + v,_n = 0$$

$$uu,_s + vu,_n = U_eU_e' + u,_{nn} - \alpha_1\ [uu,_{snn} + u,_s u,_{nn} + vu,_{nnn} - u,_n u,_{sn}$$
$$-(j/s)\ (2uu,_{nn} + u,_n^2)] - \alpha_2\ (j/s)\ (2uu,_{nn} + u,_n^2) \quad (1)$$

In the above equations a comma denotes partial differentiation with respect to the following subscript, a prime denotes total differentiation with respect to s, and all quantities are dimensionless. The coordinate parallel to the main-flow direction is s, and the velocity in that direction is u(s,n). The normal coordinate is n, and the velocity in that direction is v(s,n). The inviscid surface speed is $U_e(s)$. When j = 0 the flow is two-dimensional and the coordinate system is a plane cartesian one parallel to the plane of the motion. When j = 1 the flow is axisymmetric with n being the axis of symmetry and s being a radial coordinate measured perpendicular to the axis of symmetry. The dimensionless parameters α_1 and α_2 are defined as follows:

$$\alpha_1 = -\mu_1\ U/\eta_0 L,\ \alpha_2 = \mu_2\ U/\eta_0 L$$

where U is a characteristic velocity and L is a characteristic length. The material constants of the second-order liquid are η_0, μ_1, and μ_2 (for a Newtonian liquid $\mu_1 = \mu_2 = 0$). Thermodynamic considerations require that $\eta_0 > 0$ and $\mu_1 < 0$ but place no restriction on the sign of μ_2 (see [3]). All lengths in (1) are made dimensionless by L and all velocities by U. In addition the normal coordinate and normal velocity are stretched by the usual factor of $R^{-1/2}$ where $R = \rho UL/\eta_0$ is the Reynolds number.

STAGNATION-POINT FLOW

Consider the plane or axisymmetric stagnation-point flow against a flat plate of a second-order liquid. The stagnation point is located on the plate at the origin of the coordinate system. The s axis is tangent to the plate and the n axis is normal to it. For this flow

$$U_e(s) = s \quad (2)$$

As in the Newtonian case a similarity solution is possible. Thus defining new independent and dependent variables by the relationships

$$s = \xi,\quad n = \eta$$

$$u(s,n) = \xi F(\eta),\quad v(s,n) = (1 + j)\ V(\eta) \quad (3)$$

and substituting (2) and (3) into (1) leads to the ordinary differential equations

$$V' + F = 0$$

$$F'' - (1 + j)\ VF' - F^2 + 1 + \alpha_1\ [(1 + j)\ (F'^2 - VF''') - 2(1 - j)\ FF'']$$
$$-\alpha_2 j\ (F'^2 + 2FF'') = 0 \quad (4)$$

where a prime denotes differentiation with respect to η and the parameters α_1 and α_2 are defined as follows:

$$\alpha_1 = -\mu_1 U_a/\eta_0 a, \quad \alpha_2 = \mu_2 U_a/\eta_0 a$$

In the above equations U_a is the dimensional inviscid surface speed at a dimensional distance $s^* = a$ from the stagnation point. The appropriate boundary conditions are

$$F(0) = V(0) = 0, \quad F(\eta) \sim 1 \quad \text{as } \eta \to \infty \tag{5}$$

Before attempting the solution of the boundary-value problem posed by (4) and (5) a brief discussion of the second-order model is in order.

Previous experience with the second-order model (see, for instance, the simple and clear discussion of the so-called CDM paradox in [4]) indicates that it is valid only for slightly viscoelastic liquids. In this interpretation the second-order constitutive equation is seen as resulting from a two-term asymptotic expansion of a very general constitutive equation (that of the simple fluid [1]) for small elastic effects. A three-term asymptotic expansion results in the constitutive equation for the third-order fluid (see [5]), a four-term expansion in the fourth-order fluid, etc. In all cases the first term of the expansion is the constitutive equation for a Newtonian fluid. To be consistent with the order of accuracy inherent in the second-order constitutive equation it is clear that the solution to (4) should also have the form of a two-term asymptotic expansion for small elastic effects. Thus the solution

$$F(\eta;\alpha_1,\alpha_2) \sim F_0(\eta) + \alpha_1 F_{11}(\eta) + \alpha_2 F_{12}(\eta) + \ldots$$

$$V(\eta;\alpha_1,\alpha_2) \sim V_0(\eta) + \alpha_1 V_{11}(\eta) + \alpha_2 V_{12}(\eta) + \ldots \tag{6}$$

is accurate to the same order as the constitutive equation. It should be noted that the asymptotic expansion cannot be continued until the third-order terms are included in the constitutive equation.

Substituting (6) into (4) and (5) leads to the problems

$\underline{0}$: $V_0' + F_0 = 0$ and $F_0'' - (1 + j)\, V_0 F_0' - F_0^2 + 1 = 0$

$$F_0(0) = V_0(0) = 0, \quad F_0(\eta) \sim 1 \quad \text{as } \eta \to \infty \tag{7}$$

$\underline{11}$: $V_{11}' + F_{11} = 0$

$$F_{11}'' - (1 + j)\,(F_0' V_{11} + V_0 F_{11}') - 2F_0 F_{11} + (1 + j)\,(F_0'^2 - V_0 F_0''')$$

$$- 2(1 - j)\, F_0 F_0'' = 0$$

$$F_{11}(0) = V_{11}(0) = 0, \quad F_{11}(\eta) \sim 0 \quad \text{as } \eta \to \infty \tag{8}$$

$\underline{12}$: $V_{12}' + F_{12} = 0$

$$F_{12}'' - (1 + j)\,(F_0' V_{12} + V_0 F_{12}') - 2F_0 F_{12} - j(F_0'^2 + 2F_0 F_0'') = 0$$

$$F_{12}(0) = V_{12}(0) = 0, \quad F_{12}(\eta) \sim 0 \quad \text{as } \eta \to \infty \tag{9}$$

Numerical integration of the above problems is easily accomplished by any of a variety of methods. In this paper a finite-difference method was used. The functions F_0 and F_{11} are plotted in Figure 1 for plane flow ($F_{12} = 0$ for $j = 0$) and the functions F_0, F_{11}, and F_{12} are plotted in Figure 2 for axisymmetric flow. The following results are found for the skin friction function and the displacement thickness function respectively.

$$F'(0) \sim \begin{cases} 1.233 + 1.140\alpha_1 + \ldots & j = 0 \\ 1.312 + 1.423\alpha_1 + 0.030\alpha_2 + \ldots & j = 1 \end{cases} \tag{10}$$

$$\int_0^{\infty}(1 - F)d\eta \sim \begin{cases} 0.648 - 0.826\alpha_1 + \ldots & j = 0 \\ 0.570 - 0.681\alpha_1 - 0.273\alpha_2 + \ldots & j = 1 \end{cases} \tag{11}$$

The results for plane flow are in agreement with the solutions previously given by Beard and Walters [6] and Denn [7] and verify that the numerical method used in this paper is suitable. The results for axisymmetric flow appear to be new. It is seen that the effect of elasticity is to increase the skin friction and decrease the displacement thickness.

Rajeswari and Rathna [8], Davies [9], and Davis [2] have previously solved the complete equations (4). The former two used integral methods while the latter employed numerical integration. Provided that the solution is unique (the work of Frater [10] raises some questions about this) there is nothing wrong with this procedure except that it is wasteful. Davis [2], for instance, solves (4) for several values of α_1 and α_2. Since no significance can be attached to any deviation from a linear dependence on α_1 and α_2, this being a third-order effect, all numerical integrations beyond those required to establish two points on the straight lines are unnecessary. Davis's [2] results for skin friction are, incidentally, in agreement with (10a) and (10b) for sufficiently small values of α_1 and α_2.

FLAT PLATE FLOW

The complicated nature of the constitutive equations for various viscoelastic liquids has greatly inhibited the solution of problems dealing with realistic geometries. On the one hand, the complete partial differential equations are extremely difficult to integrate numerically; on the other, very few similarity solutions exist. It is proposed here to combine the idea of local similarity (see, for instance, [11]) with the perturbation procedure described in the previous section to obtain simple approximate solutions for nonsimilar flows. In this section the method is tested on the problem of uniform flow past a flat plate for which some previous results exist. The origin of the coordinate system is taken to be at the leading edge of the plate with s measured tangent to the plate and n measured normal to it. For this flow

$$U_e(s) = 1 \tag{12}$$

New independent and dependent variables (the Blasius variables) are defined by the following transformations:

$$s = \xi, \quad n = (2\xi)^{1/2}\eta$$

$$u(s,n) = F(\eta), \quad v(s,\eta) = (2\xi)^{-1/2} \; [V(\eta) + \eta F(\eta)] \tag{13}$$

Substituting (12) and (13) into (1) and setting j = 0 for plane flow leads to

$$V' + F = 0$$

$$F'' - VF' + \alpha_{1x} \; (2FF'' - F'^2 - VF''') = 0 \tag{14}$$

where a prime denotes differentiation with respect to η and

$$\alpha_{1x} = -\mu_1 U_\infty / 2\eta_0 x^*$$

In the above equation U_∞ is the dimensional free-stream velocity and x^* is the dimensional distance from the leading edge. In the local similarity approximation it is assumed that the dependence of F and V on the tangential coordinate enters only through the parameter α_{1x}. This will be approximately true when α_{1x} is a slowly varying function of the tangential coordinate. For this reason and also because the effect of elasticity must be small to make the second-order model reasonable it is clear that the local similarity approximation is only valid far down stream on the plate.

As in the previous section an asymptotic expansion is sought in the form

$$F(\eta;\alpha_{1x}) \sim F_0(\eta) + \alpha_{1x}F_1(\eta) + \cdots$$

$$V(\eta;\alpha_{1x}) \sim V_0(\eta) + \alpha_{1x}V_1(\eta) + \cdots \quad (15)$$

Substituting (15) into (14) and into the appropriate boundary conditions

$$F(0) = V(0) = 0, \quad F(\eta) \sim 1 \text{ as } \eta \to \infty \quad (16)$$

yields the problems

$$\underline{0}: V_0' + F_0 = 0 \quad \text{and} \quad F_0'' - V_0F_0' = 0$$

$$F_0(0) = V_0(0) = 0, \quad F_0(\eta) \sim 1 \text{ as } \eta \to \infty \quad (17)$$

$$\underline{1}: V_1' + F_1 = 0$$

$$F_1'' - (V_0F_1' + F_0'V_1) - V_0F_0''' - F_0'^2 + 2F_0F_0'' = 0$$

$$F_1(0) = V_1(0) = 0, \quad F_1(\eta) \sim 0 \text{ as } \eta \to \infty \quad (18)$$

Numerical integration yields the functions F_0 and F_1 shown in Figure 3 and the following results for the skin friction function and displacement thickness function respectively.

$$F'(0) \sim 0.470 - 0.546\alpha_{1x} + \cdots$$

$$\int_0^\infty (1 - F)\,d\eta \sim 1.217 + 0.869\alpha_{1x} + \cdots \quad (19)$$

Thus the present results indicate that far down stream on a flat plate the effect of elasticity is to decrease skin friction and increase displacement thickness. These results are in agreement with the trends found by White [12] who used an integral technique and by Davis [2] who used an asymptotic expansion for large ξ. (Denn [7] and Lockett [5] also attempted solutions similar to that of Davis [2]. Both of these authors, however, arrived at incorrect results because they neglected to enforce the condition of exponential decay of vorticity.) Thus it appears that in variables for which a similarity solution exists in the Newtonian case the local similarity method yields qualitatively correct predictions. In the next two sections the method is applied to problems for which no previous results have been given.

PLANE SINK FLOW

Consider the plane flow toward a sink located at the intersection of two flat plates. The origin of coordinates is at the sink, the coordinate s is measured tangent to one of the plates, and the coordinate n is normal to it. The appropriate inviscid surface speed is

$$U_e(s) = -s^{-1} \quad (20)$$

New variables are defined by the transformations

$$s = \xi, \quad n = \xi\eta$$

$$u(s,n) = -\xi^{-1} F(\eta), \quad v(s,n) = -\xi^{-1} \eta F(\eta) \tag{21}$$

Equations (21c) and (21d) satisfy (1a) identically. Substituting (20) and (21) into (1b) (with j = 0) results in

$$F'' - F^2 + 1 + \alpha_{1x} (F'^2 - 2FF'') = 0 \tag{22}$$

The appropriate boundary conditions are

$$F(0) = 0, \quad F(\eta) \sim 1 \text{ as } \eta \to \infty \tag{23}$$

The parameter α_{1x} in (22) is defined as follows:

$$\alpha_{1x} = -2\mu_1 aU_a/\eta_0 x^{*2}$$

where U_a is the dimensional surface speed at the dimensional distance $x^* = a$ from the sink.

As in the previous section the solution is assumed to have the form

$$F(\eta;\alpha_{1x}) \sim F_0(\eta) + \alpha_{1x} F_1(\eta) + \ldots \tag{24}$$

Substituting (24) into (22) and (23) results in the problems

$$\underline{0} : F_0'' - F_0^2 + 1 = 0$$

$$F_0(0) = 0, \quad F_0(\eta) \sim 1 \text{ as } \eta \to \infty \tag{25}$$

$$\underline{1} : F_1'' - 2F_0F_1 + F_0'^2 - 2F_0F_0'' = 0$$

$$F_1(0) = 0, \quad F_1(\eta) \sim 0 \text{ as } \eta \to \infty \tag{26}$$

The functions F_0 and F_1 are plotted in Figure 4. For the skin friction function and the displacement thickness function one obtains respectively

$$F'(0) \sim 1.155 + 0.867\alpha_{1x} + \ldots$$

$$\int_0^\infty (1 - F)d\eta \sim 0.779 - 0.665\alpha_{1x} + \ldots \tag{27}$$

In this flow the effect of elasticity is to increase the skin friction and reduce the displacement thickness. For reasons similar to those discussed in the last section the local similarity approximation is valid only for large ξ.

EXPONENTIAL PRESSURE GRADIENT FLOW

In this section the flow past an infinite flat plate due to an exponential pressure gradient is considered. The origin is located on the plate at the point where the inviscid surface speed is unity. The s axis is tangent to the plate and the n axis is normal to it. For this flow

$$U_e(s) = \exp(2\xi) \tag{28}$$

In the Newtonian case a similarity solution exists in terms of the new variables defined by the following equations.

$$s = \xi, \quad n = \exp(-\xi)\eta \tag{29}$$

$$u(s,n) = \exp(2\xi)\, F(\eta), \quad v(s,n) = \exp(\xi)\, [V(\eta) - \eta F(\eta)] \tag{30}$$

Thus these variables form the basis of the local similarity analysis. Substituting (28), (29), and (30) into (1) yields

$$V' + F = 0$$

$$F'' - VF' - 2F^2 + 2 + \alpha_{1x}(3F'^2 - 6FF'' - VF''') = 0 \tag{31}$$

The appropriate boundary conditions are

$$F(0) = V(0) = 0, \quad F(\eta) \sim 1 \text{ as } \eta \to \infty \tag{32}$$

In (31b) the dimensionless parameter α_{1x} is given by

$$\alpha_{1x} = -(\rho\mu_1 U_0^2/\eta_0^2)\exp(2\rho U_0 x^*/\eta_0) \tag{33}$$

where U_0 is the dimensional inviscid surface speed at $x^* = 0$ (x^* being the dimensional tangential coordinate). It is apparent from the form of (33) that the perturbation solution will be valid for $-\xi >> 1$.

The solution has the now familiar form

$$F(\eta;\alpha_{1x}) \sim F_0(\eta) + \alpha_{1x}F_1(\eta) + \ldots \tag{34}$$

Substituting (34) into (31) and (32) leads to the problems

$\underline{0}$: $V_0' + F_0 = 0$ and $F_0'' - V_0F_0' - 2F_0^2 + 2 = 0$

$$F_0(0) = V_0(0) = 0, \; F_0(\eta) \sim 1 \text{ as } \eta \to \infty \tag{35}$$

$\underline{1}$: $V_1' + F_1 = 0$

$$F_1'' - (V_0F_1' + V_1F_0') - 4F_0F_1 + 3F_0'^2 - 6F_0F_0'' - V_0F_0''' = 0$$

$$F_1(0) = V_1(0) = 0, \; F_1(\eta) \sim 0 \text{ as } \eta \to \infty \tag{36}$$

The functions F_0 and F_1 are obtained by numerical integration and shown on figure 5. The numerical results for the skin friction function and the displacement thickness function respectively are as follows

$$F'(0) \sim 1.688 + 4.548\alpha_{1x} + \ldots$$

$$\int_0^\infty (1 - F)d\eta \sim 0.498 - 1.598\alpha_{1x} + \ldots \tag{37}$$

From (37) it is apparent that the effect of elasticity is to increase the skin friction and reduce the displacement thickness.

It is interesting to note that the results for skin friction and displacement thickness obtained in the previous sections do not exhibit a consistent pattern. For flat plate flow the skin friction is reduced and the displacement thickness increased by elasticity. For the other flows considered the skin friction is increased and the displacement thickness is reduced. This is further confirmation of an observation made by several previous authors; namely, that the effect of elasticity is dependent on the flow conditions.

ADDITIONAL COMMENTS

Consider an Oseen linearization of equation (22) about unity. The result is

$$(1 - 2\alpha_{1x})F'' - F + 1 = 0 \tag{38}$$

which, subject to (23), has the solution

$$F = 1 - \exp [-\sqrt{1/(1 - 2\alpha_{1x})}\eta] \tag{39}$$

A two-term expansion of this result is

$$F \sim 1 - \exp (-\eta) + \alpha_{1x}\eta \exp (-\eta) + \cdots \tag{40}$$

This is equivalent to the form of solution used in this paper. Frater [10] points out, in the context of a similar problem, that while (39) predicts $F \leq 1$ for all η, (40) predicts $F > 1$ for sufficiently large η. From this one must conclude that any velocity overshoot predicted by the present method has no physical significance. A small amount of overshoot should be ignored. A large amount of overshoot should be interpreted as a signal that the value of α_{1x} has exceeded the range of values for which a two-term expansion is adequate. This conclusion is further supported by numerical integration of (22). It is again found that while direct solution of (22) predicts $F \leq 1$ for all η, the perturbation solution as found previously predicts $F > 1$ for sufficiently large η.

Frater [10] suggests that (40) be replaced by the expression

$$F \sim 1 - \exp [- (1 + \alpha_{1x})\eta] + \cdots \tag{41}$$

which is equivalent to (39) for $\alpha_{1x} << 1$. This expression is superior to (40) in that it preserves the no-overshoot property of the exact solution. It is difficult to see, however, how this method could be applied if the exact solution cannot be found in closed form. It is interesting to note that both (40) and (41) lead to the identical results

$$F'(0) \sim 1 + \alpha_{1x} + \cdots$$

$$\int_0^\infty (1 - F)d\eta \sim 1 - \alpha_{1x} + \cdots$$

which lends confidence to the results of this paper.

Consider the dissipation function for the linearized problem. It is

$$G = \exp [- 2\sqrt{1/(1 - 2\alpha_{1x})}\eta] \tag{42}$$

which has the two-term expansion

$$G \sim \exp (-2\eta) - \alpha_{1x} 2\eta \exp (-2\eta) + \cdots \tag{43}$$

It is easily seen that for all values of α_{1x} for which a solution exists ($\alpha_{1x} < 1/2$) (42) predicts $G \geq 0$ for all η while (43) predicts $G < 0$ for sufficiently large η. This appearance of negative dissipation (which occurs in all the examples calculated in this paper) must be regarded as a result of the perturbation process and as having no physical significance. Direct numerical integration of (22) confirms that the exact solution predicts non-negative dissipation for all η.

CONCLUDING REMARKS

The reader undoubtedly realizes that the function F_0 in each of the problems solved represents the solution for a Newtonian fluid. These classical results have not been individually referenced because they are all contained in any of the standard works on boundary-layer theory such as the book by Schlichting [13].

In the present paper a perturbation method has been combined with the concept of local similarity and used to obtain approximate solutions to several boundary-layer problems for the second-order viscoelastic liquid. This mathematical method should find direct application when dealing with more complicated and realistic models of the second-order type such as the power-law models used in [7] and [12]. It also appears that the qualitative trends for skin friction and displacement thickness are in agreement with the predictions of the power-law models (for sink flow see [7], for flat plate flow see [12]). In [14] it is shown that for a particular power-law model a similarity solution is possible in the case of exponential pressure gradient flow. The appropriate ordinary differential equation is obtained but not solved. It would be of interest to carry out this solution in order to compare the results with those of the present work.

The present results also represent a necessary preliminary to finding solutions for the more realistic third-order model. Lockett [5] has attempted to do this for flat plate flow but, as mentioned previously, his results are incorrect. Thus it is seen that, although unrealistic in itself, the second-order model has its uses in the prediction of viscoelastic boundary-layer flows.

REFERENCES

[1] B. D. Coleman and W. Noll: "An Approximation Theorem for Functionals with Applications in Continuum Mechanics," Archive for Rational Mechanics and Analysis, Vol. 6, No. 5, pp. 335-370 (1960).

[2] R. T. Davis: "Boundary-Layer Theory for Viscoelastic Liquids," presented at the 10th Midwestern Mechanics Conference (1967).

[3] H. Markovitz and B. D. Coleman: "Incompressible Second-Order Fluids," Advances in Applied Mechanics, Vol. 8, Academic Press, New York, pp. 69-101 (1964).

[4] A. D. D. Craik: "A Note on the Static Stability of an Elastico-Viscous Fluid," Journal of Fluid Mechanics, Vol. 33, No. 1, pp. 33-38 (1968).

[5] F. J. Lockett: "Boundary-Layer Flow of a Viscoelastic Fluid," National Physical Laboratory Report No. N69-21342 (January, 1969).

[6] D. W. Beard and K. Walters: "Elastico-Viscous Boundary-Layer Flows I. Two-Dimensional Flow Near a Stagnation Point," Proceedings of the Cambridge Philosophical Society, Vol. 60, No. 3, pp. 667-674 (1964).

[7] M. M. Denn: "Boundary Layer Flows of a Class of Elastic Fluids," Chemical Engineering Science, Vol. 22, No. 3, pp. 395-404 (1967).

[8] G. K. Rajeswari and S. L. Rathna: "Flow of a Particular Class of Non-Newtonian Visco-Elastic and Visco-Inelastic Fluids Near a Stagnation Point," Journal of Applied Mathematics and Physics (ZAMP), Vol. 13, No. 1, pp. 43-57 (1962).

[9] M. H. Davies: "A Note on Elastico-Viscous Boundary-Layer Flows," Journal of Applied Mathematics and Physics (ZAMP), Vol. 17, No. 1, pp. 189-191 (1966).

[10] K. R. Frater: "On the Solution of Some Boundary-Value Problems Arising in Elastico-Viscous Fluid Mechanics," Journal of Applied Mathematics and Physics (ZAMP), Vol. 21, No. 1, pp. 134-137 (1970).

[11] M. J. Werle: "Solutions of the Second-Order Boundary-Layer Equations for Laminar Incompressible Flow," U. S. Naval Ordinance Laboratory Report No. NOL TR 68-19 (January, 1968).

[12] J. L. White: "Application of Integral Momentum Methods to Viscoelastic Fluids: Flow About Submerged Objects," American Institute of Chemical Engineers Journal, Vol. 12, No. 5, pp. 1019-1022 (1966).

[13] H. Schlichting: Boundary Layer Theory, McGraw-Hill, New York, Fourth Edition (1960).

[14] J. L. White and A. B. Metzner: "Constitutive Equations for Viscoelastic Fluids with Application to Rapid External Flows," American Institute of Chemical Engineers Journal, Vol. 11, No. 2, pp. 324-330 (1965).

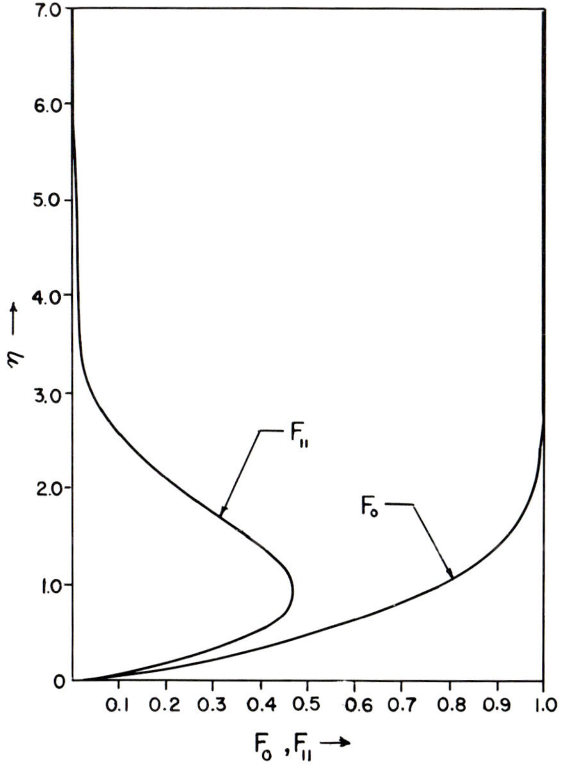

Figure 1
Plane Stagnation-Point Velocity Profile Functions

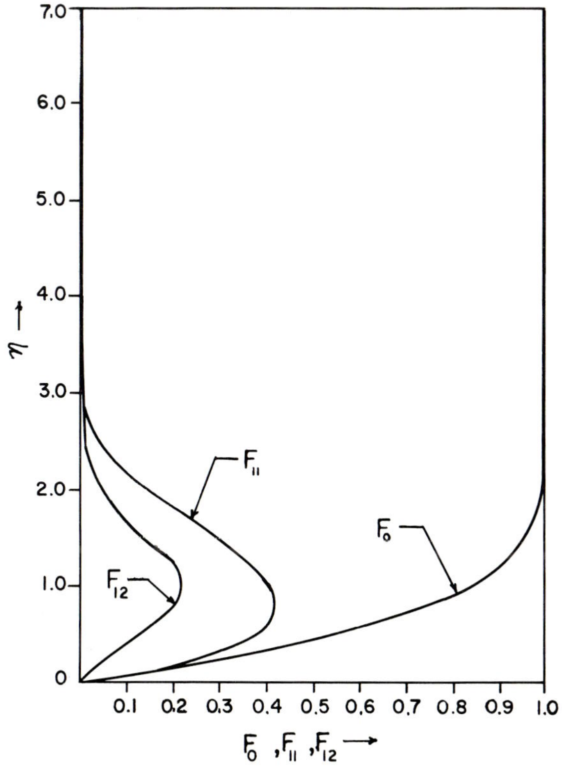

Figure 2
Axisymmetric Stagnation-Point Velocity Profile Functions

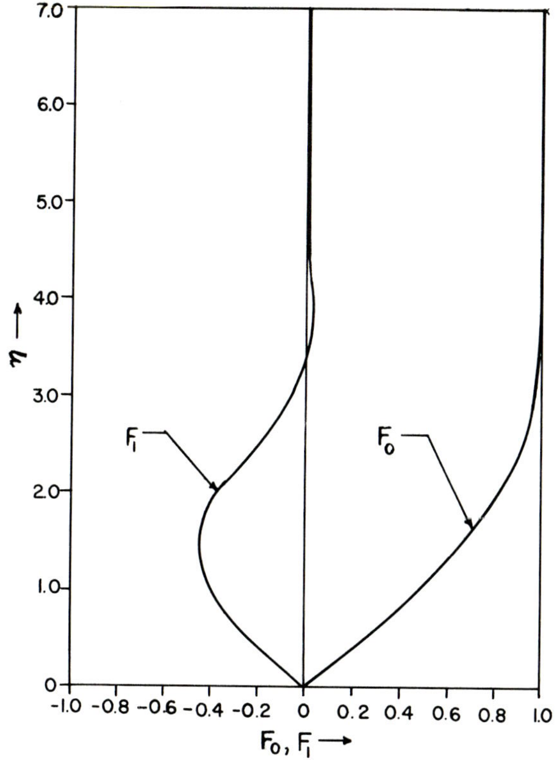

Figure 3
Flat Plate Velocity Profile Functions

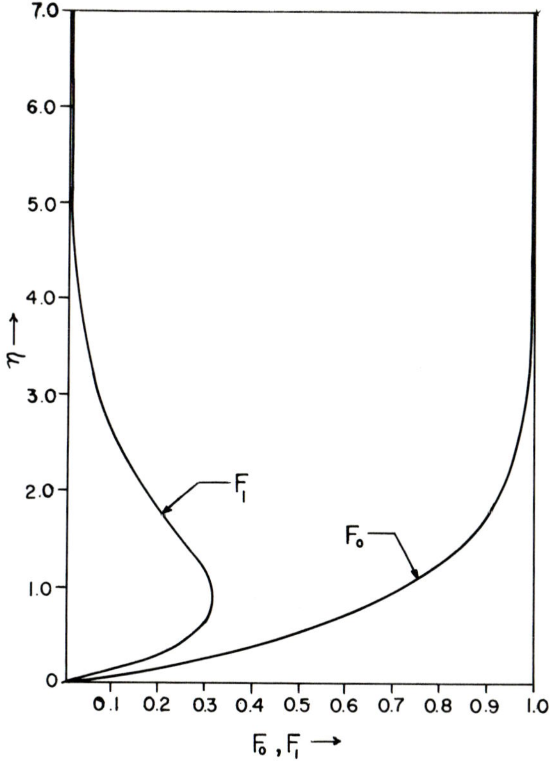

Figure 4
Sink Velocity Profile Functions

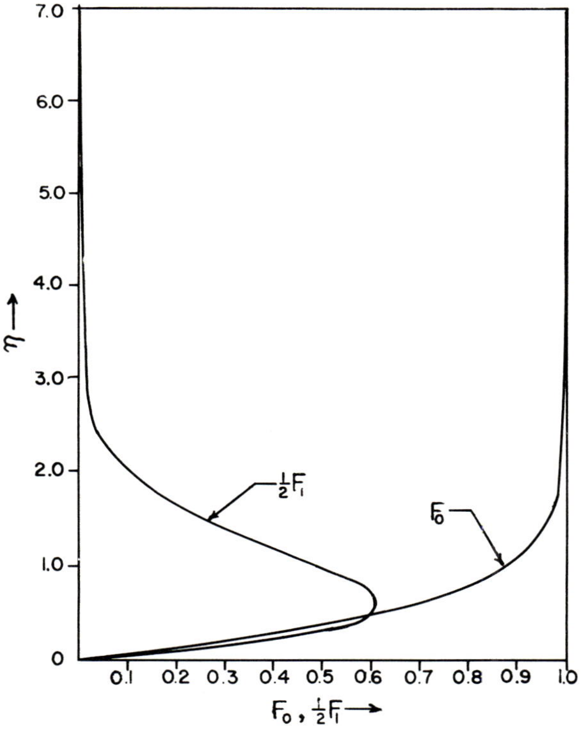

Figure 5

Exponential Pressure Gradient Velocity Profile Functions

Developments in Mechanics, Vol. 6. Proceedings of the 12th Midwestern Mechanics Conference: 10

On Certain Exact Solutions of the Boundary Layer Equations

JAMES L. S. CHEN
UNIVERSITY OF PITTSBURGH

ABSTRACT

Some exact solutions of the unsteady state boundary layer equations which are applicable to both laminar flow systems in the presence of a free surface and the transfer of heat or mass in fluids having vanishingly small Prandtl or Schmidt number are presented. A simple and more generally applicable solution approach is suggested. Sample problem has been worked out and the advantages and disadvantages of the method have also been pointed out.

INTRODUCTION

This paper is concerned with some exact solutions of the boundary layer equations which arise in laminar flows at large Reynolds number in the presence of a free surface, and in the transfer of heat or mass from a body placed in a flowing fluid having vanishingly small Prandtl or Schmidt number. It happens in the first case that the governing equations can be linearized, and the simplified equations possess pronounced similarities to those in the second case. These linear partial differential equations involving three independent variables for the non-steady two-dimensional or non-steady axisymmetrical case are solved by a transformation whereby the number of independent variables is reduced by one. This reduction, indeed, leads to the familiar form of heat conduction equation for which a solution is found without considering any boundary condition. While such an approach has not fully been exploited in literature, its advantages and disadvantages are discussed briefly.

A free boundary differs from a rigid boundary in one main aspect: The former, which may be typified as a gas-liquid interface, requires a non-zero jump in velocity derivatives and generates a finite vorticity, whereas the latter requires a non-zero jump in velocity at the boundary of a region of irrotational flow and hence the vorticity is initially infinite at the rigid boundary. Consequently, the velocity variation across the boundary layer at a free boundary is small compared with that in the case of a solid boundary. This suggests that the equation of motion for the boundary layer at a free boundary may be linearized in the departure of the velocity from the irrotational solution. There exist a number of important problems of fluid flow involving a free boundary, for example, a gas bubble rising through a viscous liquid. Steady-state solutions of this problem were first investigated by Levich [1] and later by Moore [2,3] and by Chao [4].

In the second case of heat or mass flow to fluids with vanishingly small Prandtl or Schmidt numbers, one actually deals with thermal or concentration boundary layer in a potential flow near a material boundary. In this connection, Boussinesque [5] should be credited for his first introducing the concept of the thermal boundary layer in analyzing the steady heat transfer rate from a fluid sphere. Since then a considerable number of investigations along this line have been reported, including a recent paper by Soliman and Chambré [6]. In studying the thermal response behavior of laminar boundary layers in wedge flow, Chen and Chao [7] also examined the steady temperature field in small Prandtl number fluids by considering the limiting case of zero Prandtl number. Their agreement with known data for Prandtl number equal to 0.01 is quite good. In what follows, both the flow and the temperature or mass concentration may be regarded as non-steady, and thus, it represents a further generalization of Boussinesque's theory.

THE BOUNDARY-LAYER EQUATIONS

Under the assumption of incompressible flow with constant properties and negligible dissipation, the fundamental equations for non-steady boundary layers are:

$$\frac{\partial(r^i u)}{\partial x} + \frac{\partial(r^i v)}{\partial y} = 0 \qquad (2.1)$$

$$\frac{\partial u}{\partial t} + u\frac{\partial u}{\partial x} + v\frac{\partial u}{\partial y} = -\frac{1}{\rho}\frac{\partial p}{\partial x} + \nu\frac{\partial^2 u}{\partial y^2} \qquad (2.2)$$

$$\frac{\partial T}{\partial t} + u\frac{\partial T}{\partial x} + v\frac{\partial T}{\partial y} = \alpha\frac{\partial^2 T}{\partial y^2} \qquad (2.3)$$

$$\frac{\partial c}{\partial t} + u\frac{\partial c}{\partial x} + v\frac{\partial c}{\partial y} = D\frac{\partial^2 c}{\partial y^2} \qquad (2.4)$$

where t is the time variable, (x,y) are the space coordinates, x being measured from the front stagnation point along the contour for two-dimensional case and along the meridian for axisymmetrical case, and y being measured normal to the material boundary; (u,v) the velocity components in these directions, p the pressure, T the temperature, c the solute concentration; and ρ, ν, α, D are density, kinematic viscosity, thermal diffusivity, and mass diffusivity, respectively. In (2.1), i = 0, 1 for the respective two-dimensional and axisymmetrical cases, and r(x) denotes the distance from a surface element to the axis of symmetry of the body. Since the pressure in the boundary layer is virtually that of the main stream, then it satisfies the approximate equation

$$-\frac{1}{\rho}\frac{\partial p}{\partial x} = \frac{\partial U}{\partial t} + U\frac{\partial U}{\partial x} \tag{2.5}$$

where U(x,t) is the x-component of velocity just outside the boundary layer.

From the physical argument presented in the foregoing section, one may write

$$u = U + u' \quad ; \qquad |u'| << |u| \tag{2.6a}$$

$$v = V + v' \quad ; \qquad |v'| << |v| \tag{2.6b}$$

in which V is the y-component of velocity in the main stream, and the perturbed velocity components are designated by primes. One may, thus, assert that $U \sim O(1)$, $V \sim O(\delta)$, where δ is the boundary layer thickness; $u' \sim O(\delta\Delta\omega)$, where $\Delta\omega$ is the finite jump in vorticity across the boundary layer; and $v' \sim O(\delta^2\Delta\omega)$. Consequently, the continuity equation (2.1) gives

$$v \doteq V = -\frac{y}{r^i}\frac{\partial(r^i U)}{\partial x} \tag{2.7}$$

and (2.2), (2.5) and (2.6) then yield, with consistent approximation,

$$\frac{\partial u'}{\partial t} + u'\frac{\partial U}{\partial x} + U\frac{\partial u'}{\partial x} - \frac{y}{r^i}\frac{\partial(r^i U)}{\partial x}\frac{\partial u'}{\partial y} = \nu\frac{\partial^2 u'}{\partial y^2} \tag{2.8}$$

Let us now seek solutions in which U(x,t) is of the form

$$U(x,t) = t^n W(x) \tag{2.9}$$

so that (2.8) becomes

$$\frac{\partial f}{\partial t} + t^n W\frac{\partial f}{\partial x} - t^n G y\frac{\partial f}{\partial y} = \nu\frac{\partial^2 f}{\partial y^2} \tag{2.10}$$

where

$$f = W u' \tag{2.11}$$

$$G = \frac{dW}{dx} + \frac{W}{r^i}\frac{dr^i}{dx} \tag{2.12}$$

For the thermal and the concentration boundary layers in a potential flow, Equations (2.3) and (2.4), respectively, reduce to, after using (2.7), (2.9) and (2.12)

$$\frac{\partial T}{\partial t} + t^n W\frac{\partial T}{\partial x} - t^n G y\frac{\partial T}{\partial y} = \alpha\frac{\partial^2 T}{\partial y^2} \tag{2.13}$$

$$\frac{\partial c}{\partial t} + t^n W \frac{\partial c}{\partial x} - t^n G y \frac{\partial c}{\partial y} = D \frac{\partial^2 c}{\partial y^2} \quad (2.14)$$

The similarities of the three equations (2.10), (2.13) and (2.14), are obvious. In fact, one may compactly describe the transfer of momentum, heat or mass by a single equation as follows:

$$\frac{\partial F}{\partial t} + t^n W \frac{\partial F}{\partial x} - t^n G y \frac{\partial F}{\partial y} = A \frac{\partial^2 F}{\partial y^2} \quad (2.15)$$

where F = f,T,c and A = ν, α, D, according to (2.10), (2.13) and (2.14), respectively.

With appropriate boundary conditions, equation (2.15) can be solved in a variety of ways, for example, the standard methods of integral transform might be used. However, since there are no particular boundary conditions to be satisfied we may attempt a solution approach as follows.

THE SOLUTION APPROACH

To reduce the number of independent variables by one, we introduce:

$$\eta = y r^i(x) W(x) \; ; \qquad \tau = \tau(x, t) \quad (3.1)$$

so that (2.15) transforms to the form of heat conduction equation

$$\frac{\partial F}{\partial \tau} = A \frac{\partial^2 F}{\partial \eta^2} \quad (3.2)$$

with τ satisfying the first-order partial differential equation

$$\frac{\partial \tau}{\partial t} + t^n W \frac{\partial \tau}{\partial x} = (r^i)^2 W^2 \quad (3.3)$$

which may be solved by the method of characteristics [8]. The subsidiary differential system of (3.3) is

$$\frac{dt}{1} = \frac{dx}{t^n W} = \frac{d\tau}{(r^i)^2 W^2} \quad (3.4)$$

The solutions of this set obviously are

$$Y_1(x, t, \tau) = \beta - (n+1)^{-1} t^{n+1} = C_1 \quad (3.5)$$

$$Y_2(x, t, \tau) = \tau - (n+1)^{-m} \int^x (r^i)^2 W (\beta - C_1)^{-m} dx = C_2 \quad (3.6)$$

where C_1 and C_2 are arbitrary constants, m = n/(n + 1), and

$$\beta = \int^x W^{-1} dx \quad (3.7)$$

on the condition that the integrals in (3.6) and (3.7) exist. Hence the general solution of (3.3) assumes the form

$$Y_2 = g(Y_1) \quad (3.8)$$

whence, upon substituting (3.5) and (3.6), there results

$$\tau = (n+1)^{-m} \int^{x} (r^{i})^{2} W(\beta - C_{1})^{-m} dx + g[\beta - (n+1)^{-1} t^{n+1}] \tag{3.9}$$

where g is an arbitrary function of its argument. In view of the physical ground that the diffusing distance is, following (3.2), of the order $(A\tau)^{\frac{1}{2}}$, the initial condition

$$\lim_{t \to 0} (\tau) = 0 \tag{3.10}$$

may be used in the determination of g, so that

$$g(\beta) = -(n+1)^{-m} \int^{X(\beta)} [r^{i}(z)]^{2} W(z) [\beta(z) - \beta(x)]^{-m} dz \tag{3.11}$$

in which

$$x = X(\beta) \tag{3.12}$$

has been inserted under the assumption that (3.7) can be inverted and there is a one-to-one correspondence between x and β. Therefore at time t, one has

$$\tau = (n+1)^{-m} \int_{X(\beta - \frac{t^{n+1}}{n+1})}^{x} [r^{i}(z)]^{2} W(z) \left[\frac{t^{n+1}}{n+1} + \beta(z) - \beta(x)\right]^{-m} dz \tag{3.13}$$

Returning to (3.2), we notice that since this equation is exactly the same as that governing the one-dimensional temperature in a solid of thermal diffusivity A, a number of mathematical results already reported (see, e.q., Carslaw and Jaeger, [9]) can be adopted for the present purpose. One of the most useful of these results is the solution describing the temperature distribution in the infinite region $-\infty < \eta < \infty$, viz.

$$F(\tau;\eta) = \frac{1}{2}(\pi A \tau)^{-\frac{1}{2}} \int_{-\infty}^{\infty} \exp\left[\frac{-(\eta - z)^{2}}{4A\tau}\right] [\phi_{1}(z)\mu(z) + \phi_{2}(z)\mu(-z)] dz \tag{3.14}$$

where $\phi_1(z)$ and $\phi_2(z)$ are arbitrary functions, and $\mu(z)$ is the unit step function, assuming that this integral is convergent. The physical meaning of the two arbitrary functions may be realized by taking the limit of $\tau \to 0$ in (3.14) so that $\phi_1(\eta)\mu(\eta)$ and $\phi_2(\eta)\mu(-\eta)$ are the initial temperature distributions for $\eta > 0$ and $\eta < 0$ respectively.

If one assumes that these unknown functions ϕ_1 and ϕ_2 are analytic and can be expanded in power series, i.e.

$$\phi_1(\eta) = \sum_{k=0}^{\infty} \phi_{1k} \eta^{k} \; ; \quad \phi_2(\eta) = \sum_{k=0}^{\infty} \phi_{2k} \eta^{k} \tag{3.15}$$

then, upon substituting into (3.14), one has

$$F(\xi;\eta) = \pi^{-\frac{1}{2}} \int_{-\frac{\eta}{\xi}}^{\infty} \exp(-z^{2}) \sum_{k=0}^{\infty} \phi_{1k} (\eta + z\xi)^{k} dz + \pi^{-\frac{1}{2}} \int_{-\infty}^{-\frac{\eta}{\xi}} \exp(-z^{2}) \sum_{k=0}^{\infty} \phi_{2k} (\eta + z\xi)^{k} dz \tag{3.16}$$

This equation can be expressed in terms of the k-th repeated integral of the complementary error function, viz.

$$F(\xi;\eta)=\frac{1}{2}\sum_{k=0}^{\infty} k!\,\xi^k\left[\phi_{1k}\, i^k\mathrm{erfc}\left(-\frac{\eta}{\xi}\right)+(-1)^k\phi_{2k}\, i^k\mathrm{erfc}\left(\frac{\eta}{\xi}\right)\right] \tag{3.17}$$

where $\xi = 2(A\tau)^{1/2}$.

The essential condition for this formal solution to be valid is that the initial and boundary conditions on the original problem must be expressible in terms of the new variables according to (3.1), and then they can be expanded as power series. When (3.16) or (3.17) is used in each of the boundary conditions, the resultant equations are first repeatedly differentiated with respect to ξ and the results then evaluated at $\xi = 0$ (this is possible because of the uniform convergence in ξ of the integrals in (3.16) and (3.17)). There results a system of simultaneous algebraic equations which may be solved successively for the coefficients of the ϕ functions.

Let us now consider the general case when the initial condition assumes the form

$$F(0;\eta)=h(\eta)=\sum_{k=0}^{\infty} h_k\,\eta^k \qquad 0<\eta<\infty \tag{3.18}$$

By regarding (3.18) as the limiting case of (3.17) when $\tau\to 0$ one obtains

$$\phi_{1k}=h_k \tag{3.19}$$

whereby the boundary condition when $y\to\infty$ should also be satisfied. The function $\phi_2(\eta)$ may be regarded in this case as a fictitious initial distribution of $F(\tau;\eta)$ in the region $\eta<0$. Its coefficients ϕ_{2k} can be determined by using the boundary condition at $y = 0$ in such a manner as described above.

ILLUSTRATION OF THE THEORY

It remains to show how the present theory can be applied. As an example, let us consider the problem of the steady motion of a spherical gas bubble in a liquid having small viscosity, for which the results have already been found by Moore [3]. Denoting the free-stream velocity by U_∞, the radius of the bubble by R, and putting $\theta = x/R$ one has in this case

$$r=R\sin\theta\;;\quad U(\theta,t)=W(\theta)=\frac{3}{2}U_\infty\sin\theta \tag{4.1}$$

so that the governing equation (2.8) becomes

$$\frac{3}{2}\frac{\partial}{\partial\theta}(u'\sin\theta)-3y\cos\theta\frac{\partial u'}{\partial y}=\frac{\nu R}{U_\infty}\frac{\partial^2 u'}{\partial y^2} \tag{4.2}$$

with the boundary conditions [3]

$$\lim_{\theta\to 0}(u')=0 \qquad \text{for all } y \tag{4.3}$$

$$\frac{\partial u'}{\partial y} = 3\frac{U_\infty}{R}\sin\theta \qquad \text{for} \qquad y=0, \quad 0 \le \theta \le \pi \tag{4.4}$$

$$\lim_{y\to\infty}(u') = 0 \qquad \text{for} \qquad 0 \le \theta \le \pi \tag{4.5}$$

Introducing the new dependent and independent variables according to (2.11) and (3.1) with (3.13), respectively, one has

$$f = \frac{3}{2}U_\infty \sin\theta \, u' \tag{4.6}$$

$$\eta = \frac{3}{2}U_\infty R y \sin^2\theta \tag{4.7a}$$

and

$$\tau = \int_0^x r^2 w\, dx = \frac{3}{2}R^3 U_\infty\left(\frac{2}{3} + \frac{1}{3}\cos^3\theta - \cos\theta\right) \tag{4.7b}$$

where the lower limit of the integral in (4.7b) is zero since X = 0 as $t\to\infty$ for the steady-state case. The solution to this problem is of the form (3.17) which must now satisfy the boundary conditions on $f(\tau, \eta)$, viz.

$$\lim_{\tau\to 0}(f) = 0 \tag{4.8}$$

$$\frac{\partial f}{\partial \eta} = 3U_\infty R^{-2} \qquad \text{when} \qquad \eta = 0 \tag{4.9}$$

$$\lim_{\eta\to\infty}(f) = 0 \tag{4.10}$$

It is apparent that conditions (4.8) and (4.10) require

$$\phi_{1k} = 0 \qquad \text{for all } k \tag{4.11}$$

Substituting (3.17) into the boundary condition (4.9) gives

$$3U_\infty R^{-2} = \frac{1}{2}\sum_{k=0}^{\infty} k!\,\xi^{k-1}\left[-\frac{(-1)^k \phi_{2k}}{2^{k-1}\Gamma\left(\frac{k-1}{2}+1\right)}\right] \tag{4.12}$$

As $\xi \to 0$ this yields

$$\phi_{20} = 0, \quad \phi_{21} = 6U_\infty R^{-2}$$

The k-th derivatives of (4.12) evaluated at $\xi = 0$ give

$$\phi_{2k} = 0 \qquad \text{for} \qquad k \ge 2$$

Therefore the solution is

$$f(\tau, \eta) = -3U_\infty R^{-2}\xi \, i\,\mathrm{erfc}\left(\frac{\eta}{\xi}\right) \tag{4.13}$$

Returning to the physical plane (θ , y), we have, for the perturbed velocity,

$$\frac{u'}{U_\infty} = -4\left(\frac{3}{Re}\right)^{\frac{1}{2}} \lambda^{\frac{1}{2}} \sin\theta \; ierfc\left[\frac{(3Re)^{\frac{1}{2}}}{4} \frac{y}{R} \lambda^{-\frac{1}{2}}\right] \tag{4.14}$$

where

$$\lambda(\theta) = \csc^4\theta \left(\frac{2}{3} + \frac{1}{3}\cos^3\theta - \cos\theta\right)$$

and

$$Re = \frac{2RU_\infty}{\nu}$$

is the Reynolds number of the flow. Equation (4.14) can be easily shown to be identical to equation (2.32) of Reference [3].

CONCLUDING REMARKS

We have presented some exact solutions of the unsteady boundary-layer equations arising in a number of important engineering problems. The example described in Section 4 serves to illustrate the development of the theory. Solutions to other particular problems may be readily deduced from the present framework. For example, one might study the boundary layer growth at the interface of a translating fluid sphere at large Reynolds number; to the author's knowledge, solution for this problem has not been previously reported in the literature. If both the external and internal velocity fields are of interest, the respective governing equations are exactly analogous and the two transition relations at the interface, continuity of velocity and continuity of stress, must be used.

There are two advantages to employing the solution approach described herein. The first is that it may be possible to cover a wider range of problems in view of the reduction of the number of independent variables by one without continuing to obtain an ordinary differential equation as is usually done in similarity analysis. Second, it is rather simple to apply. The determination of the unknown ϕ functions by using the boundary conditions only involves simple algebra and differentiation. It is also noted that when a single similarity solution might not exist for certain cases, such combinations of simple solutions as is given by (3.17) may yield desired results. Furthermore, the method is particularly advantageous to problems involving two or more media in which the quantity (velocity, temperature or concentration) at the interface(s) is unknown a priori.

On the other hand, the main limitation of the method is that no steps have been taken to insure the boundary conditions are expressible in terms of the new variables. In spite of this disadvantage it is worthwhile to explore further the possibility of employing such an approach in solving the boundary layer equations while little has been done in this aspect.

ACKNOWLEDGEMENTS

This work was supported by a Faculty Development Grant from

the Department of Mechanical Engineering, and the School of Engineering, University of Pittsburgh, for which the author expresses his appreciation. Thanks also go to Mrs. Phyllisjo Salinski for the typing of the manuscript.

BIBLIOGRAPHY

[1] Levich, V. G., Physicochemical Hydrodynamics, Prentice-Hall, N. J., 1962, pp. 436-448.

[2] Moore, D. W., "The Rise of a Gas Bubble in a Viscous Liquid," Journal of Fluid Mechanics, Vol. 6, 1959, pp. 113-130.

[3] Moore, D. W., "The Boundary Layer on a Spherical Gas Bubble," Journal of Fluid Mechanics, Vol. 16, 1963, pp. 161-176.

[4] Chao, B. T., "Motion of Spherical Gas Bubbles in a Viscous Liquid at Large Reynolds Numbers," Physics of Fluids, Vol. 5, No. 1, 1962, pp. 69-79.

[5] Boussinesque, M. J., "Calcul du pouvoir refroidissant des courant fluids," Journal of Pure and Applied Mathematics, Vol. 70, 1905, p. 285.

[6] Soliman, M. and Chambré, P. L., "On the Time-Dependent Convective Heat Transfer in Fluids with Vanishing Prandtl Number," International Journal of Heat and Mass Transfer, Vol. 12, 1969, pp. 1221-1230.

[7] Chen, J. L. S. and Chao, B. T., "Thermal Response Behavior of Laminar Boundary Layers in Wedge Flow," International Journal of Heat and Mass Transfer, Vol. 13, 1970, pp. 1101-1114; also presented at the Sixth U.S. National Congress of Applied Mechanics, Harvard University, Cambridge, Mass., June 15-19, 1970.

[8] Courant, R. and Hilbert D., Methods of Mathematical Physics, Vol. II, John Wiley and Sons, N.Y., 1962, p. 62.

[9] Carslaw, H. S. and Jaeger, J. C., Conduction of Heat in Solids, Oxford University Press, London, Second Edition, 1959.

STABILITY OF FLUIDS

Developments in Mechanics, Vol. 6. Proceedings of the 12th Midwestern Mechanics Conference: 11

The Hydrodynamic Stability of Two Axisymmetric Annular Flows

JOHN M. GERSTING, JR*
INDIANA UNIVERSITY-PURDUE UNIVERSITY AT INDIANAPOLIS

DANIEL F. JANKOWSKI
ARIZONA STATE UNIVERSITY

ABSTRACT

The linear hydrodynamic stability of two axisymmetric annular flows is investigated by the application of numerical techniques. The primary flows, which represent exact solutions of the Navier-Stokes system, are caused by a constant axial pressure gradient and an axial translation of the outer wall of the annulus or a cross-flow through porous annulus walls. The new solution method employs a selective application of the Gram-Schmidt process to control parasitic error during the numerical integration. Results presented in the form of neutral stability curves show that the outer wall translation or the cross-flow may produce a destabilizing effect. Several previously solved stability problems both in planar and annular geometries are obtained as special cases. Selected eigenfunctions are included.

INTRODUCTION

Orr-Sommerfeld problems of linear hydrodynamic stability theory are of interest since results obtained from them are generally taken as a starting point for the complicated transition process. Classically, plane primary flows have received the most attention, but recently primary flows involving circular geometry (which seem somewhat better suited to experimental verification) have been widely studied. The most obvious flow in this class is Hagen-Poiseuille ("pipe") flow for which only stable behavior is predicted for axially symmetric disturbances. (The recent paper by Graebel [1] reviews earlier work and establishes unstable behavior for non-symmetric disturbances.) On the other hand, axial flow in an annulus exhibits instability even for axially symmetric disturbances [2]. In the present paper, the stability problems for two modifications of the axial flow in an annulus are studied; in particular, the modifications due to an axial translation

*Formerly at Arizona State University, Tempe, Arizona.

of the outer wall of the annulus and the presence of a "cross-flow" through porous annulus walls are considered. The resulting primary flows, which represent exact solutions of the Navier-Stokes system, are referred to as Couette-Poiseuille annular flow and porous annular flow, respectively.

The stability characteristics for the Couette-Poiseuille annular flow should be of interest in considering problems concerning the extraction of a core sample from a drill casing or the application of a coating of insulation on an electrical wire while the stability characteristics of the porous annular flow would be useful in bio-medical applications, such as the design of systems for oxygenation or dialysis of body fluids.

The primary flows and disturbance equations are discussed first. A discussion of the numerical solution method, which had not previously been employed in Orr-Sommerfeld problems, follows. Detailed results and discussion close the paper.

GOVERNING EQUATIONS

THE PRIMARY FLOWS. - Consider the steady fully-developed flow of an incompressible Newtonian fluid in the region between two concentric circular cylinders. Figure 1 depicts the geometry and coordinate system being used, along with other pertinent symbols, such as U_e the cross-flow efflux velocity for the porous annular flow, and U_s the speed of the outer cylinder for the Couette-Poiseuille annular flow. A constant axial pressure gradient is present and the "no-slip" condition applies at all surfaces.

Consider first the primary flow for the flow in a porous annulus with a constant volume efflux cross-flow perpendicular to the axis of the annulus. In this case the velocity field is axially symmetric and

$$\bar{v}_r = \bar{v}_r(\bar{r}),\ \bar{v}_\theta = 0,\ \bar{v}_z = \bar{v}_z(\bar{r})$$

where $\bar{\ }$ indicates a dimensional quantity. The governing differential system for this flow consists of the Navier-Stokes equations and the continuity equation. Here the latter becomes

$$\frac{d}{d\bar{r}}(\bar{r}\,\bar{v}_r) = 0$$

so that

$$\bar{v}_r = \frac{U_e\, b}{\bar{r}} \tag{1}$$

The axial equation ($\bar{z}$ equation) of the Navier-Stokes system becomes for this case

$$\rho\left(\bar{v}_r \frac{d\bar{v}}{d\bar{r}}\right) = -\frac{\Delta\bar{p}}{\Delta\bar{z}} + \mu\left(\frac{d^2\bar{v}}{d\bar{r}^2} + \frac{1}{\bar{r}}\frac{d\bar{v}}{d\bar{r}}\right) \tag{2}$$

where $\Delta\bar{p}/\Delta\bar{z}$, μ and ρ are the pressure gradient, viscosity and density, respectively. The boundary conditions needed to complete the formulation are

$$\bar{v}_z(a) = \bar{v}_z(b) = 0 \tag{3}$$

where a and b are the inner and outer radii of the annulus.

The solution to Eq. 2, considering Eq. 1 and Eq. 3, is (for $\sigma \neq 2$)

$$\bar{v}_{zPA}(\bar{r}) = -\frac{1}{\mu}\frac{\Delta\bar{p}}{\Delta\bar{z}}\frac{a^2}{2(2-\sigma)}\left(1-\left(\frac{\bar{r}}{a}\right)^2 - \left[\frac{1-k^2}{1-k^\sigma}\right]\left[1-\left(\frac{\bar{r}}{a}\right)^\sigma\right]\right)$$

The subscript "PA" denotes porous annular flow; the two non-dimensional parameters appearing in this equation are σ, the cross-flow Reynolds number, and k the radius ratio,

$$\sigma = \frac{\rho U_e b}{\mu} \quad , \qquad k = \frac{b}{a}$$

A particularly powerful procedure for non-dimensionalization of annular stability problems suggested by Mott and Joseph [2] is employed here. The non-dimensionalization is effected by introducing the following characteristic velocity and length

$$U_c = \bar{v}_{zPAmax} \quad , \qquad \ell_c = \left(\frac{k-1}{2}\right) a$$

and a new independent variable

$$\eta = \frac{\bar{r}}{a} = \frac{k-1}{2}\, r \tag{5}$$

Then Eq. 4 becomes (for $\sigma \neq 2$)

$$W_{PA}(\eta) \equiv \frac{\bar{v}_{zPA}}{\bar{v}_{zPAmax}} = \frac{1 - \eta^2 - \left[\frac{1-k^2}{1-k^\sigma}\right](1-\eta^\sigma)}{1 - \eta_m^2 - \left[\frac{1-k^2}{1-k^\sigma}\right](1-\eta_m^2)} \tag{6}$$

with

$$\eta_m = \left(\frac{2}{\sigma}\left[\frac{1-k^\sigma}{1-k^2}\right]\right)^{\frac{1}{\sigma-2}}$$

The primary velocity field for Couette-Poiseuille annular flow consists of the axial flow in an annulus due to a constant translational motion of the outer cylinder and a constant axial pressure gradient. The form of the velocity field is

$$\bar{v}_r = 0, \quad \bar{v}_\theta = 0, \quad \bar{v}_z = \bar{v}_z(\bar{r})$$

and the governing differential system for the flow reduces to

$$\frac{1}{\bar{r}} \frac{d}{d\bar{r}} \left(r \frac{d\bar{v}_z}{d\bar{r}} \right) = \frac{1}{\mu} \frac{\Delta \bar{p}}{\Delta \bar{z}}$$

together with the boundary and conditions

$$\bar{v}_z(a) = 0 \quad , \qquad \bar{v}_z(b) = U_s$$

where U_s is the speed of the outer cylinder relative to the inner cylinder. The solution to this boundary-value problem is

$$\bar{v}_{zCPA}(\eta) = -\frac{1}{\mu} \frac{\Delta \bar{p}}{\Delta \bar{z}} \frac{a^2}{4} \left[1-\eta^2 + \frac{k^2-1}{\ell n\, k}\, \ell n\, \eta^2 \right] + U_s \frac{\ell n\, \eta}{\ell n\, k}$$

$$= \bar{v}_{zA}(\eta) + U_s \frac{\ell n\, \eta}{\ell n\, k} \tag{7}$$

where a partial non-dimensionalization has been carried out by using Eq. 5. The subscript "CPA" denotes Couette-Poiseuille annular flow. The first term in the above expression $\bar{v}_{zA}(\eta)$ is the velocity field for a fixed-walled annulus as given by Mott and Joseph [2]. To complete the non-dimensionalization and to formalize it, the characteristic velocity and length used are

$$U_c = \bar{v}_{zAmax} + \frac{U_s}{2} \quad , \quad \ell_c = \frac{(k-1)a}{2}$$

Based on these, Eq. 7 becomes

$$W_{CPA}(\eta) = \frac{1}{1+\beta} W_A + \frac{2\beta}{1+\beta} \frac{\ell n \, \eta^2}{\ell n \, k^2} \tag{8}$$

where

$$W_A(\eta) = \frac{1 - \eta^2 + \eta_m^2 \, \ell n \, \eta^2}{1 - \eta_m^2 + \eta_m^2 \, \ell n \, \eta_m^2}$$

and

$$\eta_m^2 = \frac{k^2-1}{\ell n \, k^2}$$

the parameter β is defined as

$$\beta = \frac{U_s}{2\bar{v}_{zAmax}}$$

Finally, the power of the non-dimensionalization may be seen by examining Eq. 5 for specific values of $\bar{r}$, namely $\bar{r} = a$ and $\bar{r} = b$. The result is

$$r_{\substack{\text{inner}\\ \text{cylinder}}} = \frac{2}{k-1} \equiv r_I \qquad r_{\substack{\text{outer}\\ \text{cylinder}}} = \frac{2k}{k-1} \equiv r_0$$

which shows that the choice of the radius ratio k completely specifies the configuration of the annulus, e.g., for k = 2.0, $r_I = 2$ and $r_0 = 4$.

The general annular geometry is reduced by this non-dimensionalization to the consideration of annuli with a fixed spacing of 2 units between the cylinders and an inner cylinder radius specified by the choice of k. (The fixed spacing is of great value in the numerical integration procedure.) In addition, the form of the non-diemnsionalization allows the execution of the limiting process $\lim_{a\to\infty}$ for $b - a = 2d$ and $\bar{r} = a + y$ (where 2d is the plate spacing and y is the Cartesian independent variable) on Eq. 4 and the dimensional form of Eq. 7, to recover the two limiting primary flows, plane porous wall flow [3], and plane Couette-Poiseuille flow [4], [5]. The annular flow [2] is recovered if $\sigma \to 0$ or $\beta \to 0$ in the appropriate equation, and if this result and the limiting process mentioned above are combined the classical plane Poiseuille flow is recovered.

THE DISTURBANCE EQUATIONS

The goal of the analysis undertaken here is to determine under what conditions an infinitesimal axisymmetric disturbance superimposed on either of the primary flows just presented will grow or decay. To continue the analysis perturbed velocity and pressure fields are considered. These fields are obtained by superimposing on the primary flow velocity components axisymmetric time-varying components of a disturbance velocity field, and on the primary flow pressure field a disturbance pressure field. The Navier-Stokes equations together with the continuity equation (non-dimensionalized in a manner consistent with the primary flow) and appropriate boundary conditions form the governing differential system for the flow. Substitution of the perturbation fields into the governing differential system produces, in keeping with restrictions of the linear analysis, a set of three linear partial differential equations.

Examination of these equations indicates that use of a stream function, Ψ, is permissible and that an appropriate form for a solution is

$$\Psi(r,z,t) = \phi(r)\, e^{i\alpha(z-ct)}, \qquad p(r,z,t) = q(r)\, e^{i\alpha(z-ct)}$$

where ϕ and q are amplitude functions, α is the wave number and $c = c_r + ic_i$ in which c_r is the wave speed and c_i determines the growth ($c_i>0$) or decay ($c_i<0$) of the disturbance. Neutral stability occurs when c_i is zero.

Use of the expressions for Ψ and p and elimination of q from the equations results in the following equation for the porous annular flow

$$\phi^{iv} - \frac{2}{r}\phi''' + \frac{3}{r^2}\phi'' - \frac{3}{r^3}\phi' + \alpha^2(-2\phi'' + 2\frac{\phi'}{r} + \alpha^2\phi)$$

$$+ i\alpha R\{(c-W_{PA})(\phi'' - \frac{\phi'}{r} - \alpha^2\phi) + (W''_{PA} + \frac{W'_{PA}}{r})\phi\}$$

$$- \frac{\sigma}{r}(\phi'''-\alpha^2\phi') + \frac{\sigma}{r^2}(3\phi''' - \frac{3\phi'}{r} - 2\alpha^2\phi) = 0 \qquad (9)$$

with boundary conditions

$$\phi = \phi' = 0 \text{ at } r = \begin{cases} r_I = \dfrac{2}{k-1} & \text{inner cylinder} \\ r_0 = \dfrac{2k}{k-1} & \text{outer cylinder} \end{cases}$$

where

$$R = \frac{\rho U_c \ell_c}{\mu} = \frac{\rho \bar{v}_{zPAmax}\,(k-1)a}{2\,\mu}$$

is the Reynolds number and ' indicates differentiation with respect to r. Equation 9 is the equation which governs the behavior of the disturbance (the disturbance equation) imposed on the flow in the porous annulus.

The disturbance equation for the flow in the Couette-Poiseuille annulus is obtained by the same procedure previously outlined and the resulting equation is the same as Eq. 9 when the terms involving σ are deleted, and W_{PA} is replaced by W_{CPA}. It is noted that due to the difference in the characteristic velocity for the two flows, the Reynolds numbers will also differ.

SOLUTION METHOD

SELECTION OF A METHOD - For convenience the governing differential system for the two problems being considered will be written as

$$\phi^{iv} + L_t(\phi, \phi', \phi'', \phi''', W_r; \gamma_t, \alpha, R, c) = 0 \qquad (10)$$

with boundary conditions

$$\phi(r_I) = \phi'(r_I) = \phi(r_0) = \phi'(r_0) = 0$$

where the subscript t denotes the problem of interest. Let t = 1 for the porous annular flow problem, thus $W_1 \equiv W_{PA}$, $\gamma_1 = \sigma$ and L_1 is obtained from Eq. 9. For the Couette-Poiseuille annular flow t = 2 and $W_2 \equiv W_{CPA}$, $\gamma_2 \equiv \beta$ and L_2 is obtained from Eq. 9 after making the aforementioned changes. Equation 10 represents an eigenvalue problem involving a fourth-order ordinary differential equation with complex variable coefficients.

The eigenvalue problems posed in Eq. 10 cannot be "solved" in the strict definition of the word, since no exact solutions are known. Without exact solutions, approximate methods must be examined. Approximate methods do not yield solutions but approximations to solutions, hopefully "good" approximations. Henceforth, the word "solution" is taken to mean an approximation to the "exact" solution.

In searching for an approximate method of solution it is noted that the equations contain a parameter R which on the basis of past experience can be expected to be large. This fact suggests application of the methods of asymptotic analysis. Only recently through the work of Graebel [6] has a formal procedure for applying asymptotic theories to problems of this complexity (the problems have, in general, two distinct critical layers) become available. As yet results of the application of this procedure have not appeared in the literature. For this reason, and due to the increasing power of numerical methods, asymptotic analysis of these problems was set aside.

The numerical procedures available to solve eigenvalue problems of the type posed by Eq. 10 fall into two broad classifications: algebraic and differential. Algebraic techniques are those methods whereby the original differential eigenvalue problem is approximated by an algebraic eigenvalue problem. Differential techniques basically involve direct integration of the differential system using, say, a Runge-Kutta procedure. A variety of these numerical procedures have been applied successfully at one time or another to various problems in hydrodynamic stability. Reference [7] contains a detailed comparison and evaluation of all of the successfully employed numerical methods as applied to the plane Poiseuille flow stability problem. The algebraic finite difference method and a particular differential method, here called the method of near-orthonormalized integration, were found to be the best currently available numerical solution methods. The method of near-orthonormalized integration, however, proved to be less problem dependent and was therefore chosen as the solution method for the two problems posed by Eq. 10.

THE METHOD OF NEAR-ORTHONORMALIZED INTEGRATION - Differential techniques in general are based on initial value procedures for solving boundary value problems related to the eigenvalue problems given by Eq. 10. The eigenvalue problem to be examined (t=1 or t=2) is converted into a boundary value problem by specifying values for each of the parameters α, R, γ_t, and c. Note that none of the parameters is singled out as an eigenvalue; any two parameters will suffice as an "eigenvalue."

Equation 10 is then reduced to a system of first order equations; let $\phi = y_1$; then

$$y_1' = y_2, \quad y_2' = y_3, \quad y_3' = y_4, \quad y_4' = -L_t$$

with boundary conditions

$$y_1(r_I) = y_2(r_I) = y_1(r_O) = y_2(r_O) = 0 \tag{11}$$

In matrix form these become

$$\underset{\sim}{y}' = \underset{\sim}{A}\,\underset{\sim}{y}$$

$$\underset{\sim}{B}\,\underset{\sim}{y}\,(r_I) = \underset{\sim}{0} \qquad \underset{\sim}{D}\,\underset{\sim}{y}\,(r_O) = \underset{\sim}{0} \tag{12}$$

where $\underset{\sim}{y}$ is a 4 x 1 matrix, and the matrices $\underset{\sim}{A}$, $\underset{\sim}{B}$ and $\underset{\sim}{D}$ are identified by comparing Eq. 11 and Eq. 12. A special form for the solution of the boundary value problem (Eq. 12) is constructed. Let

$$\underset{\sim}{y}(r) = \beta_1 \underset{\sim}{y}^{(1)}(r) + \beta_2 \underset{\sim}{y}^{(2)}(r)$$

$$= \begin{bmatrix} y_1^{(1)} & y_1^{(2)} \\ y_2^{(1)} & y_2^{(2)} \\ y_3^{(1)} & y_3^{(2)} \\ y_4^{(1)} & y_4^{(2)} \end{bmatrix} \begin{bmatrix} \beta_1 \\ \beta_2 \end{bmatrix}$$

$$= \underset{\sim}{Y}(r)\, \underset{\sim}{\beta} \qquad (13)$$

where $\underset{\sim}{\beta}$ is a matrix of (complex) constants. To insure that $\underset{\sim}{y}(r)$ in Eq. 13 is a solution to Eq. 12 the matrices $\underset{\sim}{Y}$ and $\underset{\sim}{\beta}$ will be chosen as follows: The matrix $\underset{\sim}{Y}$ must satisfy the differential equation

$$\underset{\sim}{Y}'(r) = \underset{\sim}{A}(r)\, \underset{\sim}{Y}(r) \qquad (14)$$

subject to the initial conditions

$$\underset{\sim}{Y}(r_I) = \underset{\sim}{U}(r_I)$$

where $\underset{\sim}{U}(r_I)$ is a 4 x 2 matrix whose columns are linearly independent satisfying the condition

$$\underset{\sim}{B}\, \underset{\sim}{U}(r_I) = \underset{\sim}{0}$$

The constant matrix $\underset{\sim}{\beta}$ is then chosen so as to satisfy the second boundary condition in Eq. 12, that is,

$$\begin{bmatrix} y_1^{(1)}(r_0) & y_1^{(2)}(r_0) \\ y_2^{(1)}(r_0) & y_2^{(2)}(r_0) \end{bmatrix} \begin{bmatrix} \beta_1 \\ \beta_2 \end{bmatrix} = 0$$

For a non-trivial solution to this set of equations to exist

$$\begin{vmatrix} y_1^{(1)}(r_0) & y_1^{(2)}(r_0) \\ y_2^{(1)}(r_0) & y_2^{(2)}(r_0) \end{vmatrix} = 0 \qquad (15)$$

Equation 15 then provides a criterion by which the eigenvalue may be selected. An initial estimate of the eigenvalue is refined using an iterative procedure such as Muller's method [8] until Eq. 15 is satisfied within some tolerance. Equation 15, since the quantities are complex, provides two conditions, one for each eigenvalue. The initial value problems are usually separated into real and imaginary parts to avoid complex arithmetic in the integrations. Care must be exercised so that Eq. 15 is separated consistently with the initial value problems so as to yield two real conditions.

The aforementioned procedure would seem to be sufficient to solve the problem; unfortunately, for the problems being considered here, it is computationally impossible as it stands. There is a subtle difficulty which arises in these problems when they are solved by numerical integration. The difficulty is caused by the fact that the four general solutions to the fourth order differential equations represented by Eq. 10 grow at vastly different rates. Even though the boundary conditions exclude the rapidly growing solution in a mathematically exact way, the procedure outlined above when executed on a digital computer using a finite number of significant figures cannot be freed of the influence of these unwanted solutions. The rapidly growing portion of the general solution dominates the solutions to the initial value problems posed by Eq. 14; this domination is often called parasitic error. The result of parasitic error in a procedure of this type is that the columns in Eq. 15 become identical, thereby destroying the eigenvalue selection criterion.

Conte [9] suggested a way of controlling the parasitic error which develops during the integration procedure. His procedure, which had not previously been employed in Orr-Sommerfeld problems, is a generalization of the method of "fully-orthonormalized integration" used by Wazzan et al. [10]. The integration is started with the base solutions $y^{(i)}$ orthonormal. During the integration from the first mesh point to the next mesh point the round-off noise produced by the finite arithmetic of the computer introduces parasitic error into the base solutions. Since the existence of parasitic error causes the solutions to tend toward linear dependence, exact orthonormality of the solutions has generally been destroyed at this point and the "angle" between the solution vectors changes from 90°. It may be possible to integrate over several subintervals before the solution vectors become computationally dependent (when the "angle" is zero); if they do become dependent the solution is lost. Conte suggests that before this point is reached the integration be interrupted and the Gram-Schmidt process be used to replace the current solution set by a set of vectors which are orthonormal. The integration is then continued using the newly computed orthonormal vectors as the initial conditions until either the end of the interval is reached or another orthonormalization is required. Unlike the method of "fully-orthonormalized integration," which requires orthonormalization at each mesh point, this method thus requires orthonormalization only at selected mesh points, hence the name "near-orthonormalized integration." The result of this process is a set of linearly independent vectors at the end point of the interval which may be used to meaningfully compute the value of the determinant in Eq. 15.

The equations required to perform the orthonormalization may be stated in general as follows: given the N x K matrix of solution vectors $\underset{\sim}{Y}(r_i)$ at the ith mesh point, which are linearly independent, they may be orthonormalized by post-multiplication by the K x K matrix $P^{(i)}$ to obtain an orthonormal set in the form of the N x K matrix denoted $\underset{\sim}{Z}(r_i)$, i.e.,

$$\underset{\sim}{Z}(r_i) = \underset{\sim}{Y}(R_i)\, \underset{\sim}{P}^{(i)}$$

or

$$\begin{bmatrix} z_1^{(1)} & z_1^{(2)} \cdots & z_1^{(K)} \\ z_2^{(1)} & z_2^{(2)} \cdots & \cdot \\ \cdot & & \cdot \\ \cdot & & \cdot \\ \cdot & & \cdot \\ z_N^{(1)} & \cdots & z_N^{(K)} \end{bmatrix} = \begin{bmatrix} y_1^{(1)} & y_1^{(2)} \cdots & y_1^{(K)} \\ y_2^{(1)} & y_2^{(2)} \cdots & \cdot \\ \cdot & & \cdot \\ \cdot & & \cdot \\ \cdot & & \cdot \\ y_N^{(1)} & \cdots & y_N^{(K)} \end{bmatrix} \begin{bmatrix} P_{11} & P_{12} \cdots & P_{1K} \\ 0 & P_{22} & \cdot\ \cdot \\ \cdot & & \cdot \\ \cdot & & \cdot \\ \cdot & & \cdot \\ 0 & \cdot\ \cdot & P_{KK} \end{bmatrix}$$

where P_{ij} are the elements of $\underset{\sim}{P}^{(i)}$ and are given by

$$P_{ij} = \begin{cases} -\sum_{\ell=i}^{j-k} (z^{\ell}, y^{j}) \dfrac{P_{i\ell}}{w_{jj}} & i < j \\ 0 & i > j \\ \dfrac{1}{w_{jj}} & i = j \end{cases}$$

where $w_{jj} = ||t^j|| = (t^j, t^j)^{1/2}$, the parentheses denoting the N-tuple inner product. Here

$$t^j = y^j - \sum_{i=1}^{j-1} (y^j, z^i) z^i$$

Conte discusses how a solution of the form of $\underset{\sim}{Y}(r)$ in Eq. 13 may be formed without reintegration provided certain information is retained at each mesh point.

The criterion used for determining whether orthonormalization is required at a particular mesh point involves determination of the "angle" between the vectors $y^{(1)}, y^{(2)}, \ldots, y^{(K)}$ taken two at a time. Orthonormalization is carried out whenever any of these "angles" becomes too small. This criterion may be expressed as follows:

$$\min_{(i,j)} \cos^{-1} \left| \frac{(y^{(i)}, y^{(j)})}{||y^{(i)}|| \quad ||y^{(j)}||} \right| < \Omega$$

The value of Ω may range from 0° to 90°. If Ω = 90° orthonormalization will be performed at each mesh point (this is fully-orthonormalized integration, c.f. Wazzan et al. [10]);if Ω = 0° no orthonormalizations will be performed. Conte discusses procedures for choosing an appropriate value for the criterion. In the work presented here Ω = 45° was employed.

RESULTS AND CONCLUSIONS

RESULTS - The stability problems for the porous annular flow and Couette-Poiseuille annular flow as posed by Eq. 10 were solved using the method of near-orthonormalized integration. The numerical computations* were carried out on the CDC 3400 at Arizona State University using single precision FORTRAN IV programs. The results of these computations are presented in Figs. 2 and 3 in the form of neutral stability curves, curves for which $c_i = 0$. It is to be noted that the method of near-orthonormalized integration permitted the neutral stability curves to be obtained directly by setting $c_i = 0$ and using α and c_r or R and c_r as the "eigenvalues." Muller's method was used for the initial refinement of eigenvalue estimates while differential correction (a scheme based upon the derivative of a function of two variables) was used for the final fine adjustments in the results.

Figures 2 and 3 present results for three radius ratios k = 1.01, k = 2.00 and k = 3.33 for various values of the parameters σ and β. The following information may be obtained from these curves. First, it must be added that $c_i > 0$ "inside" of each curve, that is, to the right of the neutral stability curve and $c_i < 0$ to the left of the neutral curve. Thus for a flow situation with a given Reynolds number, R, and known parameters σ or β, these curves indicate whether a disturbance (infinitesimal disturbance) with a given wave number, α, will grow or decay; some will and some will not. A real disturbance, however, consists of a spectrum of wave numbers. Through the linear character of the analysis used here a superposition may be carried out to approximate a real (small) disturbance. Then the curves in Figs. 2 and 3 give the minimum Reynolds number at which a disturbance will grow. This is known as R_c, the critical Reynolds number; for example, for k = 1.01, σ = 0 (or β = 0) R_c = 5800. This is simply plane Poiseuille flow.

Next consider the interesting special cases. For k = 1.01 excellent agreement is obtained between the data presented and the planar problems, plane Poiseuille flow (σ = 0, or β = 0), plane porous wall flow [3] ($\sigma \neq 0$), and plane Couette-Poiseuille flow [4], [5] ($\beta \neq 0$). Also for all three radius ratios with σ = 0, or β = 0, these results compare very well with the work of Mott and Joseph [2] for the solid-fixed walled annulus.

*The detailed programs used are contained in [11]

Now consider Fig. 2 in detail. As the cross-flow Reynolds number, σ, is changed from zero the neutral curves shift away from the $\sigma = 0$ curves (the solid walled annulus). For $k = 1.01$ no destabilization is noted; that is, no decrease in the critical Reynolds number occurs as σ is changed. As expected the problem is "symmetric" in σ in that the curves for $\sigma = -20.2$, -60.6 and -121.2 fall exactly on the curves shown for positive σ. This, however, is not the case where $k = 2.00$ or 3.33. For $k = 2.00$, increasing σ negatively causes no destabilization, but for positive σ the curves shift to the right significantly, implying a destabilization. For $k = 3.33$ the destabilization is also present and even, to some extent, for negative values of σ. The effect of the cross-flow is presented in another form in Fig. 4. There the critical Reynolds numbers obtained from Fig. 2 are plotted against σ/b. Again the decrease in R_c is noticeable for $k = 2.00$ and $k = 3.33$ as σ increases. However, for $k = 2.00$ the destabilization is only temporary. If σ is increased enough the flow is again stabilized. This conclusion is reached by examining the dashed curve. This curve is not strictly a plot of critical Reynolds numbers but a plot of the Reynolds numbers on the neutral curves for $\alpha = .95$ which is approximately the critical α for these curves. No prediction of trends for large σ for $k = 3.33$ could be made from the data.

The curves for the Couette-Poiseuille annulus in Fig. 3, although seemingly more complicated than the previous curves, are of a similar nature. Here only a slight destabilization is present for $k = 2.00$. Again $k = 3.33$ shows a distinct destabilizing trend. The plot of critical Reynolds numbers for this flow in Fig. 5 confirms the trends.

To complete the results of this analysis some selected eigenfunctions* are presented in Fig. 6.

CONCLUSIONS

In both stability analyses, for appropriate values of the pertinent parameter (σ or β), the flow is stabilized relative to the flow in a solid-walled annulus; that is, the critical Reynolds number is increased. There are, however, cases where destabilization occurs. The most pronounced cases are small cross-flows for the porous annular flow with $k = 2.00$. Possible explanations for these phenomena follow.

It is known that the presence of viscosity plays an important role in the generation of instability in problems of the type being considered here. When $\sigma = 0$ or $\beta = 0$, the solid-walled annulus results in which the mechanism for vorticity transport is viscous diffusion from the solid walls. However, in the porous annular flow in which a cross-flow is present a second mechanism also exists in the form of the convection of vorticity.

In the two limiting cases of very large cross-flow, either positive or negative, the flow seems to be completely stabilized by the cross-flow. This most probably is due to the domination of the convective mechanism whereby the vorticity is confined to a narrow region near one of the walls, leaving the majority of the flow as an essentially "linear" or "Couette-like" profile which appears to be inherently stable.

For intermediate values of the cross-flow the situation is considerably more complicated. Here, there is evidently an interplay between the diffusion mechanism and the convection mechanism. The existence of this interplay may be seen in the imaginary parts of the eigenfunctions where the amplitude decreases but the oscillations become more pronounced. These shape changes near the critical layers (layers where $c_r - W_t = 0$) indicate modification of the action of viscosity within the critical layers. Further quantification of these mechanisms has been elusive.

*Additional eigenfunctions are contained in [12].

The stability characteristics in Couette-Poiseuille annular flow must be influenced solely by viscous action within the critical layers since no cross-flow vorticity convection mechanism is present. Investigation of plane Couette-Poiseuille flow by Potter [4] indicates that the increase in stability of this flow is attributable to the movement of the critical layer near the moving wall toward and eventually past the wall. Thus as β increases, the plane Couette-Poiseuille flow problem changes from a two critical layer problem to a one critical layer problem.

Although for the cases presented in Fig. 3 the single critical layer configuration has not been reached (it should occur for $\beta = 0.10$), examination of the location of the critical layers shows that the critical layer near the outer cylinder does in fact approach the outer cylinder as β is increased positively from zero. For $k = 2.00$, $\beta = 0.5$, the critical layer and the wall layer begin to coalesce. (Increasing β negatively moves the outer critical layer slightly inward relative to the $\beta = 0$ case.) Throughout, the critical layer near the inner cylinder remains essentially unchanged.

Changes occurring in the critical layer regions can be noted by observing the modifications in the amplitude and change in size of the oscillations of the imaginary parts of the eigenfunction in Fig. 6. The conclusion then is that the translation of the cylinder produces a primarily stabilizing influence through interaction with the critical layers, primarily the one near the outer cylinder.

ACKNOWLEDGEMENTS

This research was supported in part by the University Grants Committee and the Graduate College of Arizona State University.

BIBLIOGRAPHY

1. Graebel, W. P., "The Stability of Pipe Flow. Part 1. Asymptotic Analysis for Small Wave-Numbers," Journal of Fluid Mechanics, Vol. 43, 1970, pp. 279-290.

2. Mott, J. E. and Joseph, D. D., "Stability of Parallel Flow Between Concentric Cylinders," Physics of Fluids, Vol. II, 1968, pp. 2065-2073.

3. Sheppard, D. M., "The Hydrodynamic Stability of the Flow Between Parallel Porous Walls," unpublished Doctoral Dissertation, Arizona State University, 1969.

4. Potter, M. C., "Stability of Plane Couette-Poiseuille Flow," Journal of Fluid Mechanics, Vol. 24, 1966, pp. 609-619.

5. Vargo, J. J., "The Hydrodynamic Stability of Plane Couette-Poiseuille Flow," unpublished M.S.E. Thesis, Arizona State University, 1969.

6. Graebel, W. P., "On Determination of Characteristic Equations for the Stability of Parallel Flows," Journal of Fluid Mechanics, Vol. 24, 1966, pp. 497-508.

7. Gersting, J. M., Jr., and Jankowski, D. F., "Numerical Methods for Orr-Sommerfeld Problems," International Journal for Numerical Methods in Engineering, to appear.

8. Muller, D. E., "A Method for Solving Algebraic Equations Using an Automatic Computer," Mathematics of Computation, Vol. 10, 1956, pp. 208-215.

9. Conte, S. D., "The Numerical Solution of Linear Boundary Value Problems," SIAM Review, Vol. 8, 1966, pp. 309-321.

10. Wazzan, A. R., Okamura, T. T., Smith, A. M. O., "Stability of Laminar Boundary Layers at Separation," Physics of Fluids, Vol. 10, 1967, pp. 2540-2545.

11. Gersting, J. M., Jr., "Computer Programs for the Analysis of Certain Problems in Hydrodynamic Stability," Arizona State University Engineering Report, 1970.

12. Gersting, J. M., Jr., "The Hydrodynamic Stability of Two Axisymmetric Annular Flows," unpublished Doctoral Dissertation, Arizona State University, 1970.

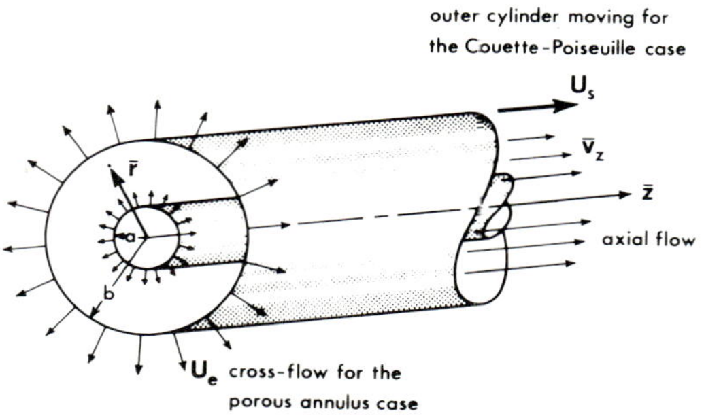

Fig. 1. Definition sketch.

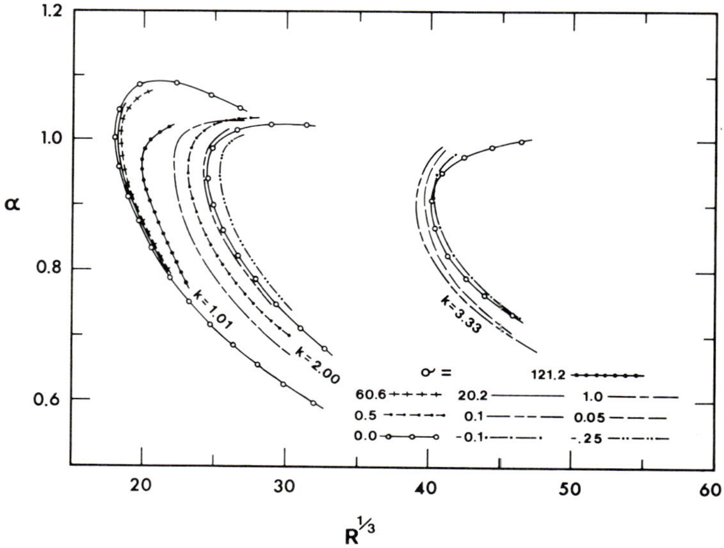

Fig. 2. Neutral stability curves for porous annular flow.

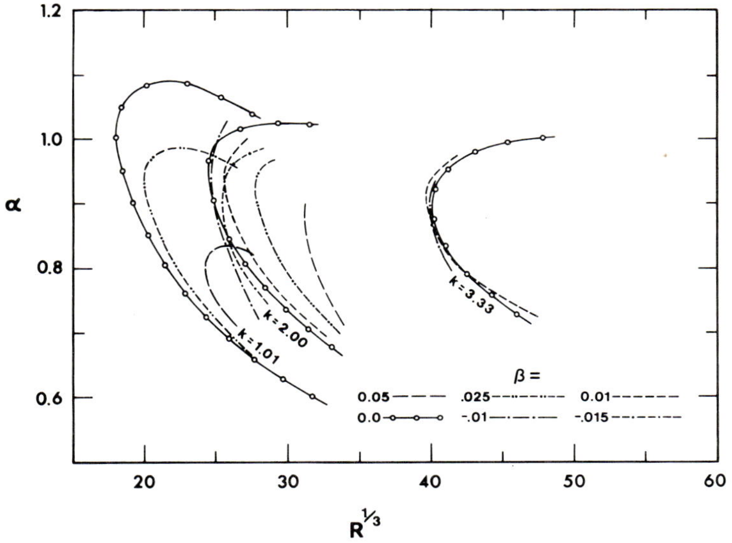

Fig. 3. Neutral stability curves for Couette-Poiseuille annular flow.

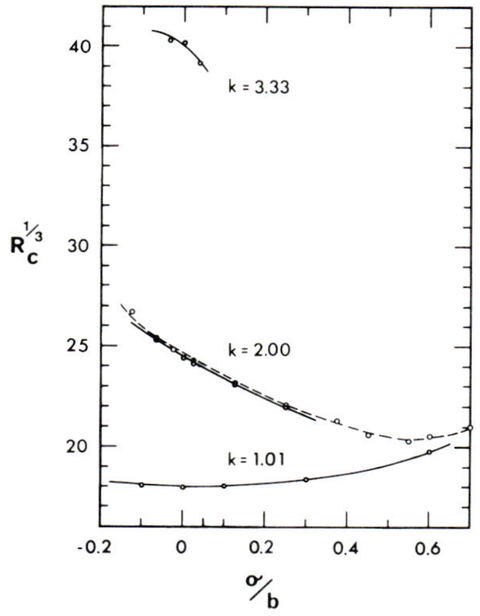

Fig. 4. Critical Reynolds numbers for porous annular flow. The dashed line represents Reynolds numbers on the neutral curve for α = 0.95.

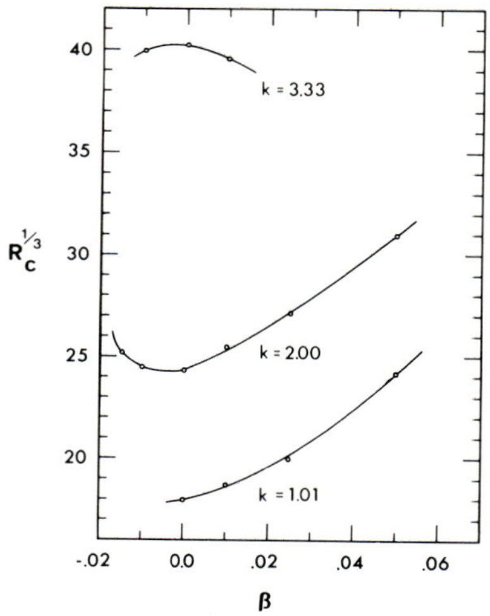

Fig. 5. Critical Reynolds numbers for Couette-Poiseuille annular flow.

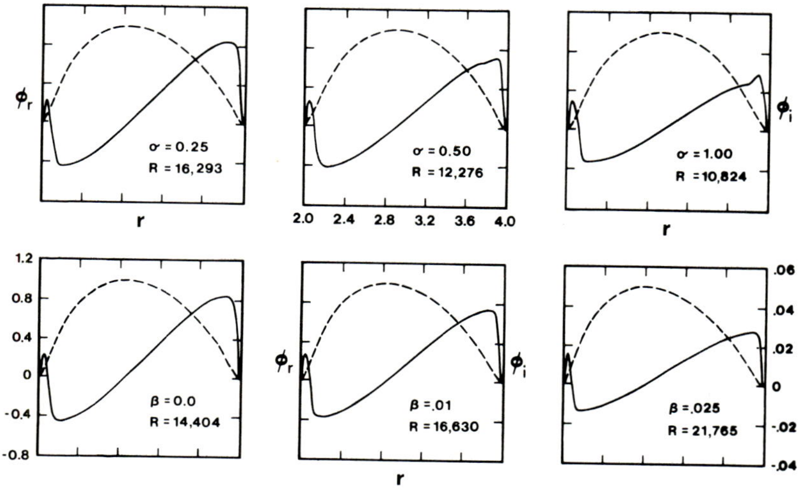

Fig. 6. Selected eigenfunctions. The dashed line represents the real part of the eigenfunction; in each case, $k = 2.0$ and $\alpha = 0.95$.

Developments in Mechanics, Vol. 6. Proceedings of the 12th Midwestern Mechanics Conference: 12

Compressible Boundary Layer Stability Equations Without the Parallel Flow Assumption

LOUIS I. BOEHMAN
UNIVERSITY OF DAYTON RESEARCH INSTITUTE

ABSTRACT

In order to assess the reasons why the supersonic boundary layer stability calculations of Brown do not compare favorably with the experimental observations of Kendall or with the calculations of Mack, an analysis was made of the small disturbance stability equations developed by Brown. His equations, which contain the effect of boundary layer growth in the form of a mean flow velocity (v) normal to a flat plate, are compared to equations presented in this paper which contain all of the terms representing the growth of boundary layer thickness and which include the v terms as well as the longitudinal derivatives of mean flow quantities. These derivatives are evaluated under the assumption that the mean boundary layer flow is similar and it is shown that these terms are of the same order as the v terms. Further, it is shown that the addition of these terms has a large effect on some of the coefficients of the stability equations, but does not significantly change the continuity equation. It is concluded that only complete stability calculations with the equations presented herein can demonstrate whether or not the relaxation of the parallel flow assumption is necessary at high Mach Numbers and whether or not the inclusion of only the v terms in the manner of Brown is valid.

INTRODUCTION

Recent trends in the study of the stability of compressible, laminar boundary layer flows have centered on the following two approaches: (1) improving the accuracy of the small disturbance equations by inclusion of more terms in the equations, and (2) through the use of digital computers to obtain more accurate solutions to the small disturbance equations. The work of W. B. Brown [1, 2] has actually included both of these trends. Evidence was presented by Dunn [3] and Cheng [4], that if all the terms in the original Lees-Lin [5] small disturbance equations were to be used in a numerical solution, then the terms involving the velocity of the mean boundary layer flow perpendicular to the flat plate (v-component) should be included also. Brown added the appropriate v terms to the Lees-Lin equations and found that at Mach 5, the augmented Lees-Lin equations considerably overestimated the critical Reynolds Number (R_{crit}). He then used Dunn and Lin's [6] approximate method of allowing for the three-dimensional aspect of the disturbance velocities and was able to get an improved estimate of R_{crit}, but a poor estimate of the upper branch of the neutral stability curve at Mach 5. Finally, by using a set of stability equations including all three momentum equations and also the v-component terms, he was able to predict a neutral stability curve for Mach 5 which agreed very well with the experimental data of Demetriades [7].

L. M. Mack [8, 9, 10] has extensively developed the linearized theory of compressible boundary layer stability through the latter approach. He has shown that higher order unstable modes exist in supersonic boundary layer flows and his analysis also indicated that above M=3 viscosity has only a stabilizing effect with the consequence that inviscid instability is dominant above M=3. Both Mack and Brown have shown that in compressible flow, unlike in incompressible flow, three-dimensional disturbances are more unstable than two-dimensional disturbances.

The above comments relative to the work of Mack and Brown are not, however, altogether self-consistent as has been noted by Morkovin [11]. Morkovin argues that Brown's neutral stability curve for M=5 cannot validly be compared to the experimental results obtained by Demetriades because of the uncertain nature of the disturbances present in those experiments. Morkovin notes that Kendall's [12] experiments at M=4.5 are the only existing source of data which can be used to compare with supersonic linear stability predictions and that Mack's predictions are much better than Brown's when compared to Kendall's data. In [11], Morkovin argues that the inclusion of the v terms (v, v', v'') without including the x-derivatives of mean flow quantities may be the reason why Brown's results do not agree with Kendall's data since the equation of continuity of the mean flow is not satisfied when only the v terms are included.

In this paper, it is shown that the x-derivatives of mean flow quantities can easily be included into the linear stability equations. It is shown that at M=5, these terms are of the same order as the v terms and thus the stability characteristics of supersonic boundary layers may be sensitive to the inclusion of these terms. The complete linear stability equations for a two-dimensional supersonic boundary layer with three-dimensional disturbances are presented with all the v terms and x-derivatives of mean flow quantities included. That is, the principal terms which represent the growth of the boundary layer thickness are included in the analysis. Some additional reasons why Brown's results do not compare favorably with Kendall's data are also presented.

DERIVATION OF AUGMENTED STABILITY EQUATIONS

As has been noted in the Introduction, Brown has shown that the stability characteristics of laminar boundary layers in supersonic flow are very sensitive to the inclusion of the v terms. However, Morkovin [11] argues that the inclusion of the v terms without also including the x^*-derivatives of the mean flow quantities is less likely to be correct than the quasi-parallel theory for the two reasons given in the Introduction. These two considerations, when taken together, suggest that the x^*-derivatives of the mean flow quantities if included in a linear stability analysis should counteract the effect of the v terms. In [4], Cheng concluded, however, that the effect of the v-component of velocity is more critical than the x^*-derivative terms, although it appears that his method of introducing the mean flow quantities into the equations in the form of Taylor series expansions about the critical layer may not have been adequate to support his assessment of the relative importance of these terms. No calculations were given in [4] to support Cheng's conclusion. Therefore, in order to provide a firm basis for establishing the relative importance of these terms, the complete linearized small disturbance equations for three-dimensional disturbances in a two-dimensional boundary layer are developed.

The continuity equation for three-dimensional small disturbances in a two-dimensional boundary layer flow is [3, 4]

$$\frac{\partial \rho^{(1)}}{\partial t} + \bar{\rho}\left(\frac{\partial u_1^{(1)}}{\partial x^*} + \frac{\partial u_2^{(1)}}{\partial y^*} + \frac{\partial u_3^{(1)}}{\partial z^*}\right) + \left(\frac{\partial w}{\partial x^*} + \frac{\partial v}{\partial y^*}\right)\rho^{(1)} + \frac{\partial \bar{\rho}}{\partial x^*}u_1^{(1)} + \frac{\partial \bar{\rho}}{\partial y^*}u_2^{(1)}$$

$$+ w\frac{\partial \rho^{(1)}}{\partial x^*} + v\frac{\partial \rho^{(1)}}{\partial y^*} = 0 \qquad (1)$$

where the superscript (1) denotes fluctuating quantities or alternately the perturbation order if one considers any quantity to be represented by the perturbation series*

$$u_i = \sum_0^\infty u_i^{(\nu)}\epsilon^\nu \qquad (2)$$

and where all dependent variables have been non-dimensionalized by their free stream values. Usually terms such as $\frac{\partial w}{\partial x^*}$, $\frac{\partial v}{\partial y^*}$, $\frac{\partial \bar{\rho}}{\partial x^*}$, and v are ignored in linear stability analyses, although as mentioned earlier, Brown [1, 2] did include the v terms. Following the usual practice of considering the mean flow to be self-similar, w, v, $\bar{\rho}$, and $\bar{T}$ are given as functions of the Blasius similarity variable $y = \frac{y^*}{x^*}R$ and v is therefore determined from the relation [1]

$$v(y) = \frac{1}{2R}\left[wy - \bar{T}\int_0^y \frac{w}{\bar{T}}\,dy\right] \qquad (3)$$

A systematic method of determining both x^* and y^* derivatives of mean flow quantities is to consider a coordinate transformation from the (x, y, z) plane to the (x^*, y^*, z^*) plane where $x=x^*/\ell$, $y=y^*/\ell$, $z=z^*/\ell$ and where ℓ is defined as x^*/R. The Jacobian of the transformation is

$$J = \begin{vmatrix} \frac{\partial x}{\partial x^*} & \frac{\partial y}{\partial x^*} & \frac{\partial z}{\partial x^*} \\ \frac{\partial x}{\partial y^*} & \frac{\partial y}{\partial y^*} & \frac{\partial z}{\partial y^*} \\ \frac{\partial x}{\partial x^*} & \frac{\partial y}{\partial z^*} & \frac{\partial z}{\partial z^*} \end{vmatrix} = \begin{vmatrix} \frac{R}{2x^*} & -\frac{y}{2x^*} & -\frac{z}{2x^*} \\ 0 & \frac{R}{x^*} & 0 \\ 0 & 0 & \frac{R}{x^*} \end{vmatrix} \qquad (4)$$

---

* For instance, if u_1 is to be considered, the perturbation series would be

$$u_1 = w + u_1^{(1)}\epsilon + u_1^{(2)}\epsilon^2 + \ldots.$$

and therefore

$$\frac{\partial}{\partial x^*} = \frac{R}{2x^*}\frac{\partial}{\partial x} - \frac{y}{2x^*}\frac{\partial}{\partial y} - \frac{z}{2x^*}\frac{\partial}{\partial z} \tag{5}$$

$$\frac{\partial}{\partial y^*} = \frac{R}{x^*}\frac{\partial}{\partial y} \tag{6}$$

$$\frac{\partial}{\partial z^*} = \frac{R}{x^*}\frac{\partial}{\partial z} \tag{7}$$

For any mean flow quantity, $\frac{\partial}{\partial x} = \frac{\partial}{\partial z} = 0$ and therefore the x* gradient of mean flow quantities is simply given by

$$\frac{\partial}{\partial x^*} = -\frac{y}{2x^*}\frac{\partial}{\partial y} \tag{8}$$

The disturbance forms assumed are those of Dunn [3] as used by both Mack and Brown where any fluctuating quantity, Q, is expressed as

$$Q(x, y, z, t) = q(y)e^{i(\alpha_1 x + \alpha_2 z - \alpha_1 ct)} \tag{9}$$

which in the (x*, y*, z*, t*) plane becomes

$$Q(x^*, y^*, z^*, t^*) = q(\frac{y^*}{x^*} R)\, e^{i(\alpha_1^* x^* + \alpha_2^* z^* - \alpha_1^* c^* t^*)} \tag{10}$$

where $\alpha_1 = \alpha_1^* \ell$, $\alpha_2 = \alpha_2^* \ell$, and $c = c^*/U_\infty^*$. Following the practice of previous investigators, α_1^*, α_2^*, and c^* will be assumed to be constant although, as has been noted in [13], the amplification function c_i^* (or $\alpha_{1_i}^*$ if spatial amplification is being considered) is strongly dependent on R and that some error may occur in the calculated values of the amplification function. The usual practice of considering the amplitude of a fluctuating quantity to be expressed as a function of y only is of course necessary in order to reduce the problem to one of solving a system of ordinary linear differential equations* although as has been noted in [13] the consequence of this assumption is that some of the amplification of small disturbances is carried by the amplitude functions as well as by the amplification function with the result that a calculated neutral stability curve based solely on c_i^* (or α_{1i}^*) = 0 does not agree with an experimentally determined one based on $\partial u^{(1)}/\partial x^* = 0$.

In order to facilitate the process of obtaining the derivatives of fluctuating quantities required in Equation 1, Equation 9 is rewritten in the form

$$Q(x^*, y, z^*, t^*) = q(y)e^{i(\alpha_1^* x^* + \alpha_2^* z^* - \alpha_1^* c^* t^*)} \tag{11}$$

with Equations 6 and 8 used to obtain the x* and y* derivatives of q(y). The x* derivative of a fluctuating quantity Q is then, with ( )' denoting a derivative with respect to y,

* Often the term "parallel flow assumption" refers to forcing the small disturbance equations to become a system of ordinary linear differential equations. It is clear however that the flow in the boundary layer can be assumed to be locally parallel and yet not obtain ordinary differential equations if the amplitude functions are taken to be dependent on R as well as on y. In this paper, the term "parallel flow assumption" refers to considering the flow in the boundary layer to be locally parallel.

$$\frac{\partial Q}{\partial x^*} = [i\alpha_1^* q(y) - q'(y)\frac{y}{2x^*}]\, e^{i(\alpha_1^* x^* + \alpha_3^* z^* - \alpha_1^* c^* t^*)} \quad (12)$$

The term within square brackets can be written as

$$\frac{1}{\ell}[i\alpha_1 q(y) - q'(y)\frac{y}{2R}] \quad (13)$$

In what follows, the second term in Equation 12, which represents the x^*-derivative of a fluctuation amplitude, will be ignored since in most cases of interest $\alpha_1 R>>1$ and q (y) is usually of the same order as $q'(y)y$. It should be noted, however, that in what follows, it would be a simple matter to include the x^*-derivatives of the fluctuation amplitudes if it became necessary to relax the assumption that $\alpha_1 R>>1$.

When the fluctuation quantities as given by Equation 11 are substituted into Equation 1, the following equation is obtained

$$[i\alpha_1(w-c) + (v' - \frac{w'y}{2R})]r + \bar{\rho}(i\alpha_1 f + i\alpha_3 h + \alpha_1\phi') - \frac{\bar{\rho}'y}{2R} f + \alpha_1 \bar{T}'\phi + v r' = 0 \quad (14)$$

Following the procedure of Brown [1, 2], the equation of state for a perfect gas is used to replace r and r' in the continuity equation

$$r = \frac{\pi}{\bar{T}} - \frac{\theta}{\bar{T}^2} \quad (15)$$

$$r' = \frac{\pi'\bar{T} - \pi\bar{T}'}{\bar{T}^2} - \frac{(\theta'\bar{T}^2 - 2\theta\bar{T}\bar{T}')}{\bar{T}^4} \quad (16)$$

and

$$\bar{\rho} = \frac{1}{\bar{T}} \quad (17)$$

since the mean pressure in the boundary layer is assumed to be constant. Upon substitution of Equations 15, 16, and 17 into Equation 14, the continuity equation becomes

$$-\phi' - \frac{vM^2}{\alpha_1}(\frac{\pi'}{M^2}) = [i + \frac{\bar{T}'y}{2\bar{T}\alpha_1 R}]f - \frac{\bar{T}'}{\bar{T}}\phi + \frac{\pi}{M^2}\{\frac{M^2}{\alpha_1}[(w-c) + (v' - \frac{w'y}{2R}) - v\frac{\bar{T}'}{\bar{T}}]\}$$
$$- \theta\{\frac{1}{\alpha_1\bar{T}}[i\alpha_1(w-c) + (v' - \frac{w'y}{2R}) - 2v\frac{\bar{T}'}{\bar{T}}]\} - \frac{v}{\alpha_1\bar{T}}\theta' + i\frac{\alpha_3}{\alpha_1}h \quad (18)$$

When Equation 18 is compared to Brown's continuity equation, (Equation 5 in [2]), one observes that the only difference is that v' in Brown's equation is replaced by $v' - \frac{w'y}{2R}$. This quantity can be evaluated by taking the derivative of Equation 3 with respect to y with the result that

$$(v'R - \frac{w'y}{2}) = -\frac{\bar{T}'}{2}\int_0^y \frac{w}{\bar{T}}\,dy \quad (19)$$

In Figure 1, a comparison of v'R and Equation 19 is presented for flow of air over an insulated flat plate at M=5. The curves shown in Figure 1 were obtained by solving the compressible flow laminar boundary layer equations using the procedure described by Mack in Section II of Reference 9. The shapes of the two curves are very nearly the same and except for very near the plate, they have the same order of magnitude. Thus, one may tentatively conclude that the inclusion of the x^*-derivatives of mean flow quantities to Brown's equations should not change his results significantly, at least as far as satisfying the continuity equation is concerned. However, inspection of Equation 18 indicates that at the critical layer where $w=c_r$, the influence of the term $\frac{w'y}{2R}$ on neutral stability curves may be large since at this point in the boundary layer, the coefficients of π and θ are entirely dependent on terms which are not included in parallel flow stability analyses.

In Figure 2, the quantity $-v\overline{T}'/\overline{T}$ is shown as a function of y. Comparison of Figure 1 and 2 indicates that $-v\overline{T}'/\overline{T}$ is positive throughout the boundary layer and always greater than $Rv'-w'y/2$, but of the same order at this term. Thus, at the critical layer, the coefficients of π and θ do not change sign or are otherwise strongly affected by the presence of the $w'y/2R$ term for typical values of R.

The effect of including the x^*-derivatives of mean flow quantities on the three momentum equations and the energy equation can be assessed in the same manner as for the continuity equation. When the procedures used to develop Equation 18 are also applied to the complete linearized small disturbance forms of these equations for three-dimensional disturbances in a two-dimensional mean boundary layer, a ninth-order system of linear first order differential equations is obtained. As noted by Brown [1, 2], the inclusion of only the v terms produces a ninth-order system. The system is set up as follows:

$$a_{ij} Z'_j = b_{ij} Z_j \qquad (20)$$

where

$Z_1 = f,\ Z_2 = f' = Z_1',\ Z_3 = \phi,\ Z_4 = \pi/M^2,\ Z_5 = \theta,\ Z_6 = \theta' = Z_5',\ Z_7 = h,$

$Z_8 = h' = Z_7',\ Z_9 = \phi' = Z_3'$

and where the row index, i, in Equation 20, is set up in the following order:

Row Index	Equation
1	$Z_1' = Z_2$
2	First momentum (x-direction)
3	Continuity
4	Second momentum (y-direction)
5	$Z_5' = Z_6$
6	Energy
7	$Z_7' = Z_8$
8	Third momentum (z-direction)
9	$Z_3' = Z_9$

Brown [1, 2] reduced his ninth order system to an eighth order system by taking the derivative of the continuity equation with respect to y and then using the resulting equation to eliminate ϕ'' from the second momentum equation. Brown's second momentum equation with this ϕ'' replacement contained the term $(v/\alpha_1)\pi''$ which he showed to be negligible and which he dropped from his analysis in order to generate his eighth order system. Inspection of Equation 18 shows that the inclusion of the x^*-derivatives of mean flow quantities does not affect the coefficient of π' in that equation so that when Brown's aforementioned procedure is applied, an eighth order system is obtained.* The a_{ij} and b_{ij} matrices of the eighth order system are given in Appendix 1.

A comparison of Brown's a_{ij} and b_{ij} matrices with those presented in Appendix 1 indicates that some of the elements in the b_{ij} matrices are now much more complex, to the point where only complete stability calculations can conclusively demonstrate the sensitivity of these calculations to the inclusion of the x^*-derivative terms although there is some reason to suspect that their effect may be large. Consider for instance the element, b_{64}, where the term $(v-wy/2R)$ appears. In Figure 1, $(Rv-wy/2)$ is plotted as a function of y and when compared to v, we see that $(Rv-wy/2)$ is negative while Rv is, of course, positive and that the absolute value of $(Rv-wy/2)$ is somewhat less than that of Rv. The term $(v-wy/2R)$ appears in a number of the other b_{ij} elements -- b_{24}, b_{25}, b_{44}, b_{45}, b_{65} -- so that its influence may be significant. This term represents the mean flow convective operator $(w\frac{\partial}{\partial x^*} + v\frac{\partial}{\partial y^*})$ and arises because, following the procedure of Brown, the mean flow momentum and energy equations are not used to eliminate the mean flow convective terms from the small disturbance equations. Figure 3, for instance, shows that the presence of this term causes b_{64} to be always positive, whereas Brown's form of b_{64} undergoes a reversal in sign. Therefore, it is reasonable to expect that the x^*-derivatives of mean flow quantities may be important, but as mentioned earlier, only a complete solution to the stability equations can provide conclusive proof.

A RE-EXAMINATION OF BROWN'S ANALYSIS

Even if in the future, it is shown that the x^*-derivative terms are unimportant, there are some other reasons why Brown's results do not agree with Kendall's data. One of these reasons is based on some differences that exist between Brown's analysis as presented in [2] with that presented in [1]. As a preliminary to this part of the paper, the following short review of the differences that exist in the terminology of viscosity coefficients is necessary.

The stress tensor for a fluid is usually written as

$$\sigma_{ij} = -p\,\delta_{ij} + 2\mu\dot{\epsilon}_{ij} + \lambda\dot{\epsilon}_{k,k}\,\delta_{ij}$$

whereas, Lees-Lin [5] write it as

$$\sigma_{ij} = -p\,\delta_{ij} + 2\mu\dot{\epsilon}_{ij} + 2/3(\mu_2-\mu_1)\,\dot{\epsilon}_{k,k}\,\delta_{ij}$$

and the bulk viscosity coefficient is usually written as**

$$\xi = \lambda + 2/3\,\mu\ .$$

* In [10] Mack presents curves of π versus y which show that π'' is negligible for the first mode but can become large for some of the higher modes. Therefore it would be desirable to solve the full ninth order system in order to determine under what conditions $v/\alpha_1\pi''$ is really negligible. The ninth order system however contains a singularity at the wall where v=0 so the solution of the ninth order system is not a trivial extension of the solution of the eighth order system.

** A wide variety of symbols are currently in usage for the bulk viscosity coefficient, i.e., ; μ_2, K, μ_B, ξ.

For a Stokesian fluid (and in incompressible fluids and monatomic gases), $\lambda = -2/3\mu$ and $\xi = 0$. Thus, $\mu_2 = 3/2\xi$ and $\mu = \mu_1$. The term "second coefficient of viscosity" usually is applied to λ, although the terminology used by Lees and Lin would suggest the same name for their μ_2. Apparently, because of this confusion in terminology, Brown in [1] took $\mu_2 = -2/3\mu_1$ and proceeded to develop his stability equations. In [2], Brown presents the same equations with $\mu_2 = 0$ although the results appear to be identical to those given in [1]. When Brown became aware of the fact that $\mu_2 \neq \lambda$, he took $\mu_2 = 0$ and redid his calculations for M=5 near the critical Reynolds number with no appreciable change being observed in the results [14]. This result is somewhat surprising insofar as a value of $\mu_2 = -2/3\mu_1$ means that $\xi = -4/9\mu_1$ and as is well known, the Second Law of Thermodynamics requires that $\xi \geq 0$. Therefore, it is somewhat unexpected that Brown's results did not change appreciably when he changed μ_2 from $-2/3\mu_1$ to 0, although it is possible that near the critical Reynolds number, the effect of μ_2 may be less than at Reynolds numbers considerably greater than at $R_{crit.}$ since Mack [8] has shown that the rate of amplification of unstable disturbances grows very rapidly when $R >> R_{crit.}$ It appears from Reference [8], that Mack has used a value of $\mu_2 = 0.8\mu$ in his calculations. It is not known if he has performed any studies to test the sensitivity of stability characteristics to μ_2.

One also notes that although Brown did include the v terms in his analysis, his eighth order system of equations is not complete. The terms missing in his system of equations correspond to a_{83}, b_{27}, b_{47}, b_{48}, b_{81}, b_{83}, and b_{85} of this paper and it should be noted that these terms, while small, can be of the same order as the v terms. Also, his method of solving for the characteristic values of his C^*_{ij} matrix ($C^*_{ij} = (a_{ij})^{-1}b_{ij}$ evaluated at $y = \delta$) which describe the solution to the disturbance equations outside the boundary layer may not be sufficiently accurate to yield correct results. By a direct expansion of the eighth order characteristic determinant, Brown forms the coefficients of the characteristic polynomial and then solves for the eight complex roots of this polynomial by an unspecified numerical technique. As noted in the text book by Conte [15], this direct expansion technique is not recommended for solving higher order characteristic determinants because of the loss of significant figures in the numerical computations. The need for a very accurate technique is crucial since the author has observed that the $(C^*_{ij} - \lambda\delta_{ij})$ determinant for the complete system of equations with the v terms included yields repeated characteristic values. For instance, a calculation of the characteristic values of the $(C^*_{ij} - \lambda\delta_{ij})$ determinant using the a_{ij} and b_{ij} given in Appendix 1, yields the following results: for $\alpha_1 = 0.0266$, $\alpha_3 = 1.43\alpha_1$, $R = 1160$, $C_r = 0.8315$, $C_i = 0.0$, $M = 5$, $\sigma = 0.7$, $\gamma = 0.7$, $C_v = 1.0$ and $\mu(T)$ from [8], the $\lambda_j (j = 1, 2, \ldots, 8)$ are given in the following Table:

A Set of Characteristic Values for the C^*_{ij} Matrix

j	λ_r	λ_i
1	-0.13914058189585	-0.78803761670985
2	-0.10471924126878	-0.81116790429344
3	-0.10471924126864	-0.81116790429337
4	$-0.40658936476968 \times 10^{-1}$	$0.59803706152014 \times 10^{-3}$
5	$0.40659622829522 \times 10^{-1}$	$0.60031517979635 \times 10^{-3}$
6	$0.44787045664176 \times 10^{1}$	0.78806958129969
7	$0.63039782412686 \times 10^{1}$	0.81116790429337
8	$0.63039782412688 \times 10^{1}$	0.81116790429279

These values of λ_j were calculated on an RCA Spectra 70/46 using the double precision complex arithmetic feature of that machine with the ALLMAT Arbitrary Matrix Eigensystem Solver algorithm [16] being used to generate the λ_j. It should be noted that the C_{ij}^* matrix is unaffected by the x^*-derivatives of mean flow quantities. Therefore, it is not the addition of these terms which is responsible for the appearance of the repeated roots.

Most elementary techniques for solving for the characteristic functions corresponding to the characteristic values do not generate a system of linearly independent characteristic functions when repeated characteristic values exist. In particular, it appears that the method used by Brown falls into this category. Thus, there is some reason to question the accuracy of Brown's numerical techniques.

CONCLUSIONS

The main conclusion of this paper is that Brown's calculations should be redone in order to conclusively demonstrate the relative importance of including the v-terms and the x^*-derivative terms in supersonic laminar boundary layer stability calculations. Their is also a need to improve Brown's numerical techniques for calculating the solutions to the small disturbance equations in the region of flow outside the boundary layer.

ACKNOWLEDGEMENTS

This work was supported in part by Grant GY-3517 from the National Science Foundation and by the University of Dayton Research Institute. Support in the form of computer time was provided by the Air Force Flight Dynamics Laboratory, Wright-Patterson Air Force Base, Ohio, and by the Office of Computing Activities, University of Dayton.

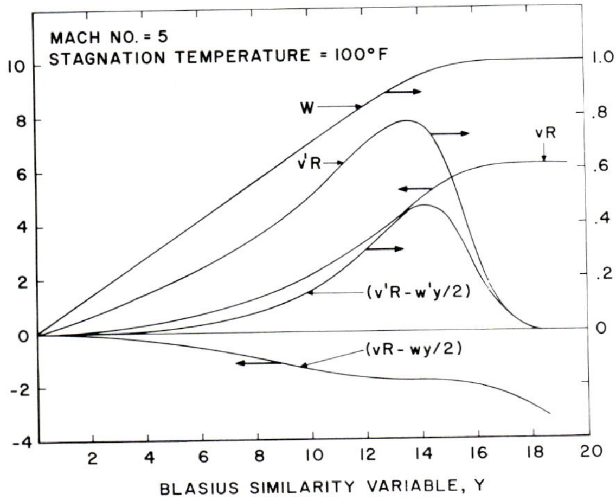

Fig. 1. Mean Flow Quantities in Insulated Flat Plate Boundary Layer.

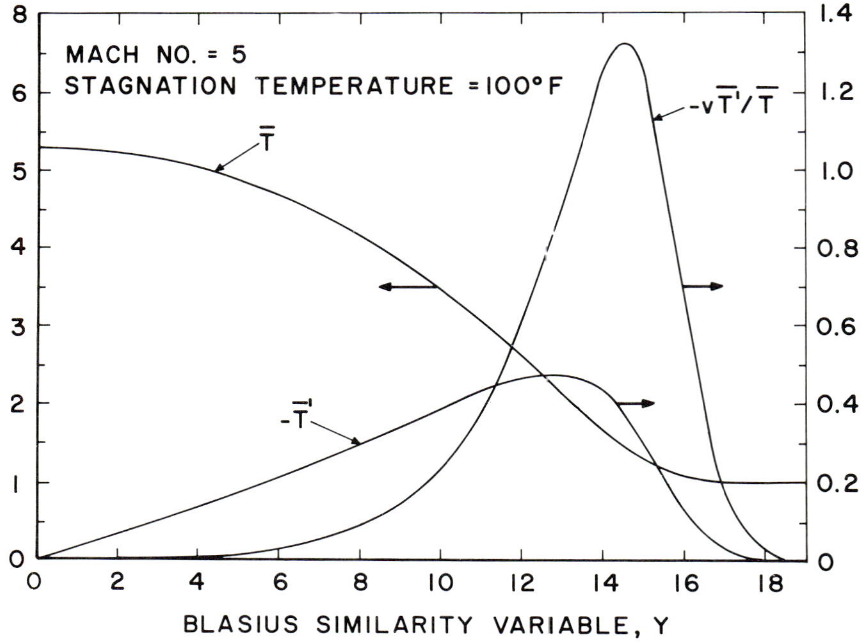

Fig. 2. $\overline{T}$, $\overline{T}'$, and $-v\overline{T}'/\overline{T}$ in Boundary Layer - Insulated Flat Plate.

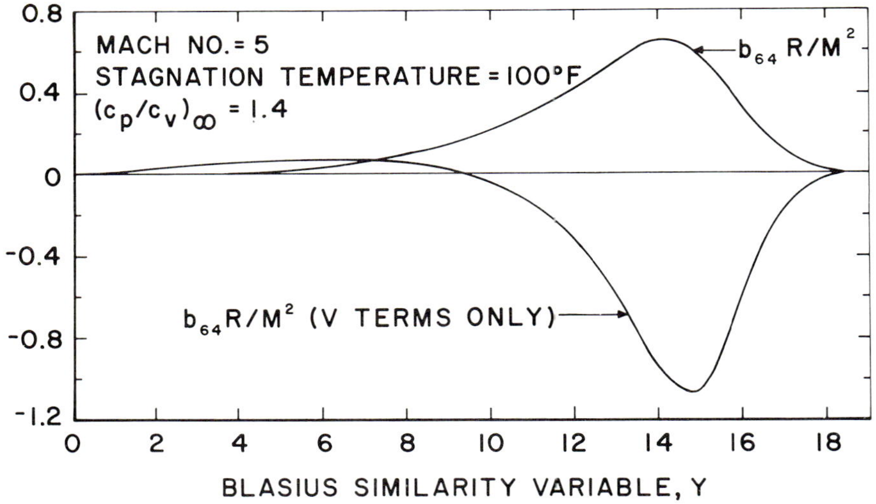

Fig. 3. Effect of x^*-derivatives on Matrix Element b_{64}.

NOMENCLATURE

Symbol	Definition	Characteristic Measure
c	- phase velocity of disturbance = c_r+ic_i	U^*_∞
c_v	- specific heat at constant volume	$c^*_{v_\infty}$
f	- velocity disturbance amplitude in x^*direction	U^*_∞
h	- velocity disturbance amplitude in z^*direction	U^*_∞
i	- unit imaginary number $\sqrt{-1}$ or coordinate index	
k	- thermal conductivity	k^*_∞
ℓ	- characteristic length = x^*/R	
M	- Mach number	
P	- pressure	P^*_∞
R_x	- Reynolds number=$\rho^*_\infty U^*_\infty x^*/\mu^*_\infty$	
R	- $\sqrt{R_x}$	
r	- amplitude of density fluctuation	ρ^*_∞
T	- temperature = T^*/T^*_∞	T^*_∞
t	- time	ℓ/U^*_∞
U^*_∞	- mean free stream flow	
u_j	- instantaneous velocity component in Cartesian coordinate system	U^*_∞
v	- mean flow component of velocity normal to wall (y^*-direction)	U^*_∞
w	- mean flow component of velocity parallel to wall (x^*-direction)	U^*_∞
x^*	- dimensional distance parallel to wall in direction of free stream flow	
x	- non-dimensional distance = x^*/ℓ	ℓ
y^*	- dimensional distance perpendicular to wall	
y	- Blasius similarity variable = y^*R/x^*	
z^*	- dimensional distance parallel to wall in direction transverse to free stream flow	
z	- non-dimensional distance = z^*/ℓ	
α_1	- disturbance wave number in x^* direction	ℓ^{-1}
α_3	- disturbance wave number in z^* direction	ℓ^{-1}
β	- angular frequency	U^*_∞/ℓ
γ	- specific heat ratio = $c^*_{p\infty}/c^*_{v\infty}$	
δ_{ij}	- Kronecker delta	
ϵ	- perturbation parameter	
$\dot{\epsilon}_{ij}$	- strain rate tensor	
θ	- amplitude of temperature fluctuation	T^*_∞
λ	- dilatational coefficient of viscosity, = $-2/3\mu$ for a Stokesian fluid	μ^*_∞
μ, μ_1	- first viscosity coefficient	μ^*_∞
μ_2	- Lees-Lin definition of bulk viscosity coefficient = $3/2\xi$	
ξ	- coefficient of bulk viscosity, =0 for a Stokesian fluid	μ^*_∞
π	- amplitude of pressure fluctuation	P^*_∞
ρ	- fluid density	ρ^*_∞
σ	- Prandtl number	
σ_{ij}	- stress tensor	
$\alpha_1\phi$	- velocity disturbance amplitude in y^* direction	U^*_∞

Superscripts	Definition
$(\ )'$	- derivative with respect to y
$(\ )^*$	- dimensional quantity
$(\ )^{(i)}$	- perturbation index
$(\bar{\ })$	- mean flow property

Subscripts	Definition
∞	- free stream value
r, i	- real and imaginary parts, respectively

APPENDIX 1

ELEMENTS OF THE a_{ij} and b_{ij} MATRICES

$a_{11} = 1 \qquad a_{22} = \mu/R \qquad a_{23} = \frac{i\alpha_1^2}{R}(\frac{1}{3}\mu + \xi) + \frac{\alpha_1 \bar{T}' y}{2R^2}(\frac{2}{3}\frac{d\mu}{d\bar{\bar{T}}} - \frac{d\xi}{d\bar{\bar{T}}})$

$a_{33} = -1 \qquad a_{34} = \frac{-vM^2}{\alpha_1} \qquad a_{43} = \frac{\alpha_1}{R}(\ell n\,\bar{T})'(\frac{4}{3}\mu + \xi) + \frac{\alpha_1 \bar{T}'}{R}(\frac{4}{3}\frac{d\mu}{d\bar{\bar{T}}} + \frac{d\xi}{d\bar{\bar{T}}}) - \frac{\alpha_1 v}{\bar{\bar{T}}}$

$a_{44} = \frac{-1}{\gamma} - \frac{M^2}{R}(\frac{4}{3}\mu + \xi)[-v(\ell n\bar{T})' + 2v' - \frac{w'y}{2R} + i\alpha_1(w-c)] \qquad a_{46} = \frac{v}{R\bar{\bar{T}}}(\frac{4}{3}\mu + \xi)$

$a_{55} = 1 \qquad a_{63} = -\alpha_1(\gamma - 1) + \frac{\alpha_1}{R}\gamma(\gamma-1)M^2[2(\xi + \frac{4}{3}\mu)v' + 2(\frac{2}{3}\mu - \xi)\frac{w'y}{2R}]$

$a_{66} = \frac{\gamma k}{\sigma R} \qquad a_{77} = 1 \qquad a_{83} = \frac{i\alpha_1\alpha_3}{R}(\frac{1}{3}\mu + \xi) \qquad a_{88} = \mu/R \qquad b_{12} = 1$

$b_{21} = \frac{i\alpha_1}{\bar{\bar{T}}}(w-c) + \frac{1}{R}[\mu(\frac{4}{3}\alpha_1^2 + \alpha_3^2) + \xi\alpha_1^2] + \left[\frac{\bar{T}'y}{2R^2}(\frac{4}{3}\frac{d\mu}{d\bar{\bar{T}}} + \frac{d\xi}{d\bar{\bar{T}}})\right] i\alpha_1 - \frac{w'y}{2R\bar{\bar{T}}}$

$b_{22} = \frac{v}{\bar{\bar{T}}} - \frac{\bar{T}'}{R}\frac{d\mu}{d\bar{\bar{T}}} \qquad b_{23} = \frac{\alpha_1 w'}{\bar{\bar{T}}} - i\alpha_1^2\frac{\bar{T}'}{R}\frac{d\mu}{d\bar{\bar{T}}} \qquad b_{24} = \frac{M^2 w'}{\bar{\bar{T}}}(v - \frac{wy}{2R}) + \frac{i\alpha_1}{\gamma}$

$b_{25} = -\frac{w'}{\bar{\bar{T}}^2}(v - \frac{wy}{2R}) - \frac{w''}{R}\frac{d\mu}{d\bar{\bar{T}}} - \frac{d\mu}{d\bar{\bar{T}}}[\frac{w'y}{R^3} + \frac{w''y^2}{3R^3} - \frac{1}{6}(\frac{v'}{R^2} + \frac{v''y}{R^2})]$

$\quad - \frac{d\xi}{d\bar{\bar{T}}}[\frac{3}{4}\frac{w'y}{R^3} + \frac{1}{4}\frac{w''y^2}{R^3} - \frac{v'}{2R^2} - \frac{v''y}{2R^2}] - \frac{d^2\mu}{d\bar{\bar{T}}^2}[\frac{1}{3}\frac{\bar{T}'y}{R^2}(v' + \frac{w'y}{R})$

$\quad + (w' - \frac{v'y}{2R})\frac{\bar{T}'}{R}] + \frac{d^2\xi}{d\bar{\bar{T}}^2}[\frac{\bar{T}'}{2R}\frac{y}{R}(v' - \frac{w'y}{2R})] + \frac{i\alpha_1}{R}[\frac{2}{3}\frac{d\mu}{d\bar{\bar{T}}}(v' + \frac{w'y}{R}) - \frac{d\xi}{d\bar{\bar{T}}}(v' - \frac{w'y}{2R})]$

$b_{26} = -(w' - \frac{v'y}{2R})\frac{1}{R}\frac{d\mu}{d\bar{\bar{T}}} \qquad b_{27} = \frac{\alpha_1\alpha_3}{R}(\frac{1}{3}\mu + \xi) - \frac{i\alpha_3\bar{T}'y}{2R^2}(\frac{2}{3}\frac{d\mu}{d\bar{\bar{T}}} - \frac{d\xi}{d\bar{\bar{T}}})$

$b_{31} = i + \frac{y}{2\alpha_1 R}(\ell n\,\bar{T})' \qquad b_{33} = -(\ell n\,\bar{T})'$

$b_{34} = \frac{M^2}{\alpha_1}[-v(\ell n\,\bar{T})' + i\alpha_1(w - c) + v' - \frac{w'y}{2R}]$

$b_{35} = \frac{1}{\alpha_1\bar{\bar{T}}}[2v(\ell n\bar{T})' - i\alpha_1(w-c) - v' + \frac{w'y}{2R}] \qquad b_{36} = -\frac{v}{\alpha_1\bar{\bar{T}}} \qquad b_{37} = i\frac{\alpha_3}{\alpha_1}$

$b_{41} = \frac{v'y}{2R\bar{\bar{T}}} + i\alpha_1\frac{\bar{T}'}{R}(\frac{2}{3}\frac{d\mu}{d\bar{\bar{T}}} - \frac{d\xi}{d\bar{\bar{T}}}) + (\frac{4}{3}\mu + \xi)[\frac{1}{2R^2}(y(\ell n\bar{T})'' + (\ell n\,\bar{T})')]$

$b_{42} = \frac{i\alpha_1\mu}{R} + \frac{\bar{T}'y}{2R^2}[\frac{1}{\bar{\bar{T}}}(\frac{4}{3}\mu + \xi) + \frac{d\mu}{d\bar{\bar{T}}}]$

$b_{43} = \frac{i\alpha_1^2}{\bar{\bar{T}}}(w-c) + \frac{\alpha_1 v'}{\bar{\bar{T}}} + \frac{\mu}{R}(\alpha_1^3 + \alpha_1\alpha_3^2) + i\alpha_1^2[\frac{\bar{T}'y}{2R^2}\frac{d\mu}{d\bar{\bar{T}}}] - \frac{\alpha_1}{R}(\frac{4}{3}\mu + \xi)(\ell n\bar{T})''$

$b_{44} = \frac{M^2}{\bar{\bar{T}}}v'(v - \frac{wy}{2R}) + \frac{M^2}{R}(\frac{4}{3}\mu + \xi)[-v(\ell n\,\bar{T})'' - v'(\ell n\bar{T})' + i\alpha_1 w' + v'' - \frac{w''y}{2R} - \frac{w'}{2R}]$

$b_{45} = -\frac{v'}{\bar{\bar{T}}^2}(v - \frac{wy}{2R}) - \frac{d\mu}{d\bar{\bar{T}}}[\frac{4}{3}\frac{v''}{R} + \frac{3}{4}\frac{y}{R^3}v' + \frac{1}{4}\frac{y^2}{R^3}v'' - \frac{1}{6}(\frac{w'}{R^2} + \frac{w''y}{R^2})] - \frac{d\xi}{d\bar{\bar{T}}}[\frac{v''}{R} - \frac{w'}{2R^2} - \frac{w''y}{2R^2}]$

$\quad + \frac{d^2\mu}{d\bar{\bar{T}}^2}[-(\frac{4}{3}\frac{v'}{R} + \frac{w'y}{3R^2})\bar{T}' + (\frac{w'}{R} - \frac{v'y}{2R^2})\frac{\bar{T}'y}{2R}] - \frac{d^2\xi}{d\bar{\bar{T}}^2}[\frac{v'}{R} - \frac{w'y}{2R^2}]\bar{T}' + (\frac{4}{3}\mu + \xi)\frac{1}{R\bar{\bar{T}}}[3v'(\ell n\bar{T})'$

$\quad + 2v(\ell n\bar{T})'' - 2v(\ell n\bar{T})'^2 - v'' + \frac{w'}{2R} + \frac{w''y}{2R} - \frac{w'y}{2R}(\ell n\bar{T})' + i\alpha_1[i\ell n\bar{T})'(w-c) - w']] - \frac{i\alpha_1}{R}(w' - \frac{v'y}{2R})\frac{d\mu}{d\bar{\bar{T}}}$

$$b_{46} = \frac{d\mu}{d\overline{T}}\left(-\frac{4}{3}\frac{v'}{R} - \frac{1}{3}\frac{w'y}{R^2}\right) - \frac{d\xi}{d\overline{T}}\left(\frac{v'}{R} - \frac{w'y}{2R^2}\right) + \left(\frac{4}{3}\mu + \xi\right)\frac{1}{R\overline{T}}\left[3v(\ell n\overline{T})' - 2v' + \frac{w'y}{2R} - i\alpha_1(w-c)\right]$$

$$b_{47} = \frac{i\alpha_3}{R}\overline{T}'\left(\frac{2}{3}\frac{d\mu}{d\overline{T}} - \frac{d\xi}{d\overline{T}}\right) \qquad b_{48} = \frac{i\alpha_3\mu}{R} \qquad b_{56} = 1$$

$$b_{61} = -\frac{C_v y}{2R}(\ell n\overline{T})' + i\alpha_1(\gamma-1) + \gamma(\gamma-1)\frac{M^2}{R}\left[2\left(\xi + \frac{4}{3}\mu\right)\frac{w'y}{2R} + \left(\frac{4}{3}\mu - 2\xi\right)v'\right]i\alpha_1$$

$$b_{63} = \alpha_1 C_v(\ell n\overline{T})' - 2i\alpha_1^2\left(w' - \frac{v'y}{2R}\right)\frac{\mu}{R}\gamma(\gamma-1)M^2 \qquad b_{62} = -2\left(w' - \frac{v'y}{2R}\right)\frac{\mu}{R}\gamma(\gamma-1)M^2$$

$$b_{64} = (\gamma-1)\left(v' - \frac{w'y}{2R}\right)M^2 + C_v M^2\left(v - \frac{wy}{2R}\right)(\ell n\overline{T})'$$

$$b_{65} = \frac{C_v}{\overline{T}} i\alpha_1(w-c) + \frac{\gamma}{\sigma R}\left[k(\alpha_1^2 + \alpha_3^2 + i\alpha_1\frac{dk}{d\overline{T}}\frac{\overline{T}'y}{R}\right] - \frac{\overline{T}''\gamma}{\sigma R}\frac{dk}{d\overline{T}} - \frac{\gamma}{\sigma R}(\overline{T}')^2\frac{d^2k}{d\overline{T}^2}$$

$$- \frac{\gamma(\gamma-1)M^2}{R}\left[\frac{d\mu}{d\overline{T}}\left\{\frac{4}{3}\left[(v')^2 + \left(\frac{w'y}{2R}\right)^2\right] + (w')^2 + \left(\frac{v'y}{2R}\right)^2 - \frac{w'v'y}{3R}\right\} + \left(v' - \frac{w'y}{2R}\right)\frac{d\xi}{d\overline{T}}\right]$$

$$- \frac{C_v}{\overline{T}}(\ell n\overline{T})'\left(v - \frac{wy}{2R}\right)$$

$$b_{66} = \frac{C_v v}{\overline{T}} - \frac{2\overline{T}'\gamma}{\sigma R}\frac{dk}{d\overline{T}} \qquad b_{67} = i\alpha_3(\gamma-1)\left[1 + 2\left(\frac{2}{3}\mu - \xi\right)\left(v' - \frac{w'y}{2R}\right)\frac{\gamma M^2}{R}\right.$$

$$b_{78} = 1 \qquad b_{81} = \frac{\alpha_1\alpha_3}{R}\left(\frac{1}{3}\mu + \xi\right) + i\alpha_3\frac{d\mu}{d\overline{T}}\frac{\overline{T}'y}{2R^2} \qquad b_{83} = -\frac{i\alpha_1\alpha_3}{R}\overline{T}'\frac{d\mu}{d\overline{T}}$$

$$b_{84} = \frac{i\alpha_3}{\gamma} \qquad b_{85} = \frac{i\alpha_3}{R}\left(v' - \frac{w'y}{2R}\right)\left(\frac{2}{3}\frac{d\mu}{d\overline{T}} - \frac{d\xi}{d\overline{T}}\right)$$

$$b_{87} = \frac{i\alpha_1}{\overline{T}}(w-c) + \frac{\mu}{R}(\alpha_1^2 + \alpha_3^2) + \left(\frac{1}{3}\mu + \xi\right)\frac{\alpha_3^2}{R} + i\alpha_1\frac{\overline{T}'y}{2R^2}\frac{d\mu}{d\overline{T}} \qquad b_{88} = \frac{v}{\overline{T}} - \frac{\overline{T}'}{R}\frac{d\mu}{d\overline{T}}$$

REFERENCES

[1] Raetz, G. S. and Brown, W. B., "Theoretical Investigations of Boundary Layer Stability, Parts 2 and 3," AFFDL Technical Report No. AFFDL-TR-64-184, Air Force Flight Dynamics Laboratory, Wright-Patterson Air Force Base, Ohio, September, 1966.

[2] Brown, W. B., "Stability of Compressible Boundary Layers," AIAA J., Vol. 5, No. 10, 1967, pp. 1753-1759.

[3] Dunn, D. W., "On the Stability of the Laminar Boundary Layer in a Compressible Fluid," Ph. D. Thesis, Massachusetts Institute of Technology, 1953.

[4] Cheng, Sin-I, "On the Stability of Laminar Boundary Layer Flow," Quarterly of Applied Mathematics, Vol. XI, No. 3, 1953, pp. 346-350.

[5] Lees, L. and Lin, C. C., "Investigation of the Stability of the Laminar Boundary Layer in a Compressible Fluid," NACA TN-1115, National Advisory Committee for Aeronautics, Washington, D. C., September, 1946.

[6] Dunn, D. W. and Lin, C. C., "On the Stability of the Laminar Boundary Layer in a Compressible Fluid," Journal of Aeronautical Sciences, Vol. 22, No. 7, 1955, pp. 455-477.

[7] Demetriades, A., "An Experimental Investigation of the Stability of the Hypersonic Laminar Boundary Layer," Memorandum No. 3, Hypersonic Research Project, Guggenheim Aeronautical Laboratory, California Institute of Technology, May 15, 1958.

[8] Mack, L. M., "Computation of the Stability of the Laminar Compressible Boundary Layer," in Methods of Computational Physics: Volume 4, Academic Press, New York, 1965, pp. 247-299.

[9] Mack, L. M., "The Stability of the Compressible Laminar Boundary Layer According to a Direct Numerical Solution," in Recent Developments in Boundary Layer Research, AGARDograph 97, part I, 1965, pp. 329-362.

[10] Mack, L. M., "Boundary-Layer Stability Theory," AIAA Professional Study Series, High Speed Boundary Layer Stability and Transition. Published by the Jet Propulsion Laboratory, California Institute of Technology, May, 1969.

[11] Morkovin, M. V., "Critical Evaluation of Transition from Laminar to Turbulent Shear Layers With Emphasis on Hypersonically Travelling Bodies," AFFDL Technical Report No. AFFDL-TR-68-149, Air Force Flight Dynamics Laboratory, Wright-Patterson Air Force Base, Ohio, March, 1969.

[12] Kendall, J. M., Jr., "Boundary-Layer Stability Experiments," in Jet Propulsion Laboratory Space Programs Summary No. 37-39, pp. 147-149, California Institute of Technology, 1966.

[13] Ross, J.A., Barnes, F.H., Burns, J.G., and Ross, M.A.S., "The Flat Plate Boundary Layer. Part 3. Comparison of Theory With Experiment," Journal of Fluid Mechanics, Vol. 43, Part 4, 1970, pp. 819-832.

[14] Brown, W. B., Private Communication.

[15] Conte, S.D., Elementary Numerical Analysis, McGraw-Hill, New York, 1965, p. 188.

[16] Funderlic, R.E., and Rinzel, J., "ALLMAT, A FORTRAN IV Arbitrary Matrix Eigensystem Solver," SHARE Program Library, August 20, 1968.

Developments in Mechanics, Vol. 6. Proceedings of the 12th Midwestern Mechanics Conference: 13

A Boundary Vorticity Method for Finite Amplitude Convection in Plane Poiseuille Flow

G. J. HWANG and K. C. CHENG
UNIVERSITY OF ALBERTA

ABSTRACT

Finite amplitude convection with longitudinal vortex rolls for hydrodynamically and thermally fully developed laminar flow between horizontal flat plates where the lower plate is heated isothermally and the upper plate is cooled isothermally is approached by a boundary vorticity method using a line iterative relaxation technique. The governing equations with Boussinesq approximation are solved numerically for a range of Rayleigh number $Ra^* \leq Ra \leq 3 \times 10^5$, where Ra^* = 1707.8 is a critical value for the onset of longitudinal vortex rolls with infinitesimal disturbance. An algorithm for the boundary vorticity method is presented which represents a new approach to the numerical solution of the vorticity transport equation in fluid mechanics, and is believed to have a wide applicability in many related problems. The numerical results for Pr = 0.7 are compared with theoretical and experimental data available in the literature, and show that Stuart's integral method leads to appreciable error at high Rayleigh number.

INTRODUCTION

Consider a steady fully developed laminar forced convection between two horizontal flat plates where the lower plate is heated isothermally and the upper plate is cooled isothermally. When the vertical temperature gradient of the fluid layer exceeds a critical value, longitudinal vortex rolls appear in the channel. This thermal instability problem with plane Poiseuille flow superimposed is known to be identical mathematically to the two-dimensional Bénard convection problem. Mori and Uchida [1] studied the problem theoretically in post-critical regime with Rayleigh number ranging from the critical value up to nine times the critical value by adapting Stuart's approximate energy method for

* The work reported here is based on part of the Ph.D. thesis[4] of G.J. Hwang.

nonlinear mechanics of hydrodynamic stability [2], and compared the results with experimental data. It is noted that Stuart's method represents an approximation to a perturbation theory about the critical characteristic number and cannot be expected to be valid for a wide range of Rayleigh number above the critical value. For example, for the analogous problem of rotating-cylinder flow, it is estimated that for a Taylor number up to ten times the critical value the largest error is about seven per cent for the torque. In view of the limitation of Stuart's energy-balance method, a numerical solution appears to be the only practical approach for the accurate solution of the present finite amplitude convection problem.

The purpose of this paper is to show that finite amplitude convection problem can be approached by a recently developed boundary vorticity method [3,4] using the present problem as an example. In addition to presenting flow and heat transfer results, with comparison made against the available data in the literature and pointing out the discrepancy, the detailed algorithm for the boundary vorticity method will be given.

For the solution of steady two-dimensional Navier-Stokes equations one is often faced with the solution of the vorticity transport equation [5] which is a quasi-linear partial differential equation with a biharmonic operator. The analytical solution of the fourth-order quasi-linear elliptic partial differential equation is, in general, very difficult; and a finite-difference method is most commonly used in recent years. In order to achieve a faster convergence rate, a vorticity may be introduced enabling one to reduce the vorticity transport equation into a second-order quasi-linear elliptic partial differential equation and a Poisson's equation. The new numerical method of determining the vorticity at the boundary exactly represents a new approach in a finite-difference formulation of the two-dimensional Navier-Stokes equations, and proves to be computationally very stable with a considerable saving in computing time as compared with the conventional methods.

Recently, Ogura and Yagihashi [6] carried out a numerical solution of the present problem using unsteady method for boundary vorticity. It should be pointed out that the present post-critical convection with plane Poiseuille flow superimposed is mathematically analogous to the two-dimensional Bénard convection problem solved numerically, for example, by Plows [7]. The main feature of the present paper is the utilization of a rather simple linear relationship between stream function and vorticity at boundaries which represents an alternative approach to the solution of finite amplitude convection problem. The method enables one to solve for the vorticities and stream functions along a line simultaneously and is different from the conventional unsteady method [6] using an explicit forward integration technique.

THEORETICAL ANALYSIS

In order to facilitate the analysis for a steady fully developed laminar forced convection in the post-critical regime between two infinite horizontal plates where the lower plate is heated isothermally (T_1) and the upper plate is cooled isothermally (T_2), a Boussinesq approximation will be used. In addition, the wave number a for the post-critical regime will be taken to be the one predicted by the linear stability theory [8] with the wave number a = 3.116 for the critical Rayleigh number Ra* = 1707.8.

The existence and the steadiness of the finite amplitude limit cycle in the post-critical regime for the present problem were confirmed by the experiments reported in [9].

The coordinate system for the present problem is shown in Fig.1 where the region OABC represents a domain for a single vortex roll. In order to investigate the flow structure and temperature field in the post-critical regime, perturbation quantities are superimposed on the basic flow quantities as

$$U = U_b + u', \quad V = v', \quad W = w', \quad T = T_b + \theta', \text{ and } P = P_b + p' \tag{1}$$

where basic flow quantities U_b, T_b and P_b satisfy the well known equations for plane Poiseuille flow. The solutions for the unperturbed state are:

$$U_b = 4U_0(z-z^2) = U_0\,\phi_u/2, \quad T_b = T_1 - z(T_1 - T_2) = T_1 - z\Delta T \tag{2}$$

where $\phi_u = 8(z - z^2)$, $\Delta T = T_1 - T_2$, $z = z'/h$, h is the distance between two plates, and U_0 is the maximum velocity in the unperturbed state.

Applying the Boussinesq approximation and equation (1) to the Navier-Stokes equations and the energy equation, and subtracting the equations for plane Poiseuille flow from the respective Navier-Stokes equations and the energy equation, the following equations for the perturbation quantities are obtained after eliminating pressure terms between y'- and z'-direction momentum equations by cross-differentiation.

Axial momentum equation

$$v'\frac{\partial u'}{\partial y'} + w'\frac{\partial u'}{\partial z'} + w'\frac{\partial U_b}{\partial z'} = \nu\,\nabla_1^2 u'; \quad \text{where } \nabla_1^2 = \partial^2/\partial y'^2 + \partial^2/\partial z'^2 \tag{3}$$

Vorticity transport equation for secondary flow

$$v'\frac{\partial \xi'}{\partial y'} + w'\frac{\partial \xi'}{\partial z'} = \nu\,\nabla_1^2\,\xi' - \beta g\frac{\partial \theta'}{\partial y'} \tag{4}$$

where the vorticity is defined by $\xi' = \nabla_1^2\psi'$ (5)

and the stream function ψ' is defined by $v' = \dfrac{\partial\psi'}{\partial z'}, \quad w' = -\dfrac{\partial\psi'}{\partial y'}$ (6)

Energy equation $v'\dfrac{\partial\theta'}{\partial y'} + w'\dfrac{\partial T_b}{\partial z'} + w'\dfrac{\partial\theta'}{\partial z'} = \kappa\,\nabla_1^2\,\theta'$ (7)

Note that the vorticity is introduced here to apply the boundary vorticity method. For the purpose of simplification and convenience, the equations (3) to (7) are reduced to non-dimensional forms by using the following transformations:

$(y',z') = (y,z)$; $(u',v',w') = (\text{Re}\ u, v, w)\nu/h$; $\theta' = \Delta T\ \theta$; $\xi' = \xi\nu/h^2$; $\psi' = \nu\psi$;
the parameters $\text{Re} = U_0\ h/(2\nu)$; $\text{Gr} = g\ \beta\ \Delta T\ h^3/\nu^2$; and $\text{Ra} = \text{PrGr}$.

The resulting equations are:

$$v\frac{\partial u}{\partial y} + w\frac{\partial u}{\partial z} + w\frac{\partial\phi_u}{\partial z} = \nabla^2 u, \quad \text{where } \nabla^2 = \partial^2/\partial y^2 + \partial^2/\partial z^2 \tag{8}$$

$$v\frac{\partial\xi}{\partial y} + w\frac{\partial\xi}{\partial z} = \nabla^2\xi - \frac{\text{Ra}}{\text{Pr}}\frac{\partial\theta}{\partial y} \tag{9}$$

$$\xi = \nabla^2\psi \tag{10}$$

$$v = \frac{\partial\psi}{\partial z}, \qquad w = -\frac{\partial\psi}{\partial y} \tag{11}$$

$$v\frac{\partial\theta}{\partial y} + w\frac{\partial\theta}{\partial z} - w = \frac{1}{\text{Pr}}\nabla^2\theta \tag{12}$$

The associated boundary conditions are:

$u = \psi = \partial\psi/\partial z = \theta = 0$ at top and bottom plates (13)

$\partial u/\partial y = \psi = \xi = \partial\theta/\partial y = 0$ along vertical lines $y=0$, $0 < z < 1$ and $y = \pi/a$, $0 < z < 1$

At this point it is noted that the no-slip boundary conditions $\partial\psi/\partial z = 0$ at the top and bottom plates cannot be applied directly, and furthermore the boundary vorticity is not defined. For the present problem two parameters Pr and Ra appear in the governing equations. The governing equations are seen to be identical to those for the two-dimensional Bénard convection problem solved, for example, by Plows [7]. An alternative formulation of the governing equations in the forms described in references [3,6] is also possible.

FINITE-DIFFERENCE FORMULATION AND A BOUNDARY VORTICITY METHOD

In view of the complexity of the governing equations (8) to (12) and the fact that appreciable error is observed with Stuart's approximate method, a combination of boundary vorticity method and line iterative method will be employed for the numerical solution of the present rather complicated problem. As stated earlier, one of the objectives of this paper is to present an algorithm for the boundary vorticity method. By using a three-point central-difference approximation and a dummy variable f for dependent variables ξ, ψ, u and θ, a general finite-difference equation can be obtained as follows, (see Fig.1):

$$\sigma \left(v^{o}_{j,k} \frac{f_{j+1,k} - f_{j-1,k}}{2\,\Delta y} + w^{o}_{j,k} \frac{f_{j,k+1} - f_{j,k-1}}{2\,\Delta z}\right)$$

$$= \frac{f_{j+1,k} - 2\,f_{j,k} + f_{j-1,k}}{(\Delta y)^2} + \frac{f_{j,k+1} - 2\,f_{j,k} + f_{j,k-1}}{(\Delta z)^2} + G_{j,k} \qquad (14)$$

where $\sigma = 1,\ G_{j,k} = -\,w^{o}_{j,k}\left(\frac{\partial \phi_u}{\partial z}\right)_{j,k}$ for $f_{j,k} = u_{j,k}$

$\sigma = 1,\ G_{j,k} = -\,\frac{Ra}{Pr}\left(\frac{\partial \theta}{\partial y}\right)_{j,k}$ for $f_{j,k} = \xi_{j,k}$

$\sigma = 0,\ G_{j,k} = -\,\xi_{j,k}$ for $f_{j,k} = \psi_{j,k}$

$\sigma = Pr,\ G_{j,k} = Pr\ w^{o}_{j,k}$ for $f_{j,k} = \theta_{j,k}$

$j,k = 2,3\ldots M$ for $f_{j,k} = \psi_{j,k}$ or $\xi_{j,k}$

$j = 1,2,\ldots M+1,\quad K = 2,3,\ldots M$ for $f_{j,k} = u_{j,k}$ or $\theta_{j,k}$

The superscript o stands for the value obtained at the previous iteration step.

To illustrate the computational procedure for boundary vorticity method, consideration will be given to the vorticity transport equation (9) and the Poisson's equation (10) for stream function. A set of finite-difference equations for $\xi_{j,k}$ and $\psi_{j,k}$ along the line j = 2 can be put into a matrix form as,

$$\begin{bmatrix} B_2 & C_2 & & & & & & \\ A_3 & B_3 & C_3 & & & & & \\ & A_4 & B_4 & C_4 & & 0 & & \\ & & \cdot & \cdot & \cdot & & & \\ & & & \cdot & \cdot & \cdot & & \\ & & & & A_{M-2} & B_{M-2} & C_{M-2} & \\ & 0 & & & & A_{M-1} & B_{M-1} & C_{M-1} \\ & & & & & & A_M & B_M \end{bmatrix} \begin{bmatrix} \xi_{2,2} \\ \xi_{2,3} \\ \xi_{2,4} \\ \cdot \\ \cdot \\ \xi_{2,M-2} \\ \xi_{2,M-1} \\ \xi_{2,M} \end{bmatrix} = \begin{bmatrix} H_2 - A_2\xi_{2,1} \\ H_3 \\ H_4 \\ \cdot \\ \cdot \\ H_{M-2} \\ H_{M-1} \\ H_M - C_M\xi_{2,M+1} \end{bmatrix} \tag{15}$$

$$\begin{bmatrix} B_1' & (A_1'+C_1') & & & & & & \\ A_2' & B_2' & C_2' & & & 0 & & \\ & A_3' & B_3' & C_3' & & & & \\ & & \cdot & \cdot & \cdot & & & \\ & & & \cdot & \cdot & \cdot & & \\ & & & & A_{M-1}' & B_{M-1}' & C_{M-1} & \\ & 0 & & & & A_M' & B_M' & C_M' \\ & & & & & & (A_{M+1}'+C_{M+1}') & B_{M+1}' \end{bmatrix} \begin{bmatrix} \psi_{2,1} \\ \psi_{2,2} \\ \psi_{2,3} \\ \cdot \\ \cdot \\ \psi_{2,M-1} \\ \psi_{2,M} \\ \psi_{2,M+1} \end{bmatrix} = \begin{bmatrix} H_1' \\ H_2' \\ H_3' \\ \cdot \\ \cdot \\ H_{M-1}' \\ H_M' \\ H_{M+1}' \end{bmatrix} \tag{16}$$

where,

$$A_k = 1 - \frac{\Delta z}{2} \left(\frac{\partial\psi}{\partial y}\right)^o_{2,k} ; \quad B_k = B_k' = -2\left[1 + \left(\frac{\Delta z}{\Delta y}\right)^2\right]$$

$$C_k = 1 + \frac{\Delta z}{2} \left(\frac{\partial\psi}{\partial y}\right)^o_{2,k} ; \quad H_k = -\left(\frac{\Delta z}{\Delta y}\right)^2 \left[1 - \frac{\Delta y}{2} \left(\frac{\partial\psi}{\partial y}\right)^o_{2,k}\right] \xi_{3,k} + (\Delta z)^2 \frac{Ra}{Pr}\left(\frac{\partial\theta}{\partial y}\right)_{2,k}$$

$$A_k' = C_k' = 1; \quad H_k' = -\left(\frac{\Delta z}{\Delta y}\right)^2 \psi_{3,k} + (\Delta z)^2 \xi_{2,k}$$

It is noted that the unknown values for $\xi_{2,1}$ and $\xi_{2,M+1}$ on the right hand side of equations (15) and (16) are the unknown boundary conditions to be determined. Two methods are available for the numerical determination of boundary vorticity.

Method (1): Referring to Fig.1, the equations and boundary conditions under consideration are

Equation	On boundaries OA and BC	On lines of symmetry AB and OC	
(9)	$\xi = ?$	$\xi = 0$	(17)
(10)	$\partial\psi/\partial z = 0$	$\psi = 0$	

It is noted that the vorticity ξ along the boundaries OA and BC must be determined such that $\psi = 0$ along these lines. In principle, the following linear relationship between vorticity and stream function at boundaries may be obtained after eliminating 2(M-1) unknowns from 2M equations for equations (15) and (16).

$$\begin{bmatrix} \psi_{2,1} \\ \psi_{2,M+1} \end{bmatrix} = \begin{bmatrix} a_{11} & a_{12} \\ a_{21} & a_{22} \end{bmatrix} \begin{bmatrix} \xi_{2,1} \\ \xi_{2,M+1} \end{bmatrix} + \begin{bmatrix} k_1 \\ k_2 \end{bmatrix} \qquad (18)$$

The discovery of the above linear relation is a key to the success of boundary vorticity method. Boundary vorticities $\xi_{2,1}$ and $\xi_{2,M+1}$ can now be obtained by observing that $\psi_{2,1} = \psi_{2,M+1} = 0$ on boundaries. The procedure of finding elements a_{11}, a_{12}, a_{21}, a_{22}, k_1 and k_2 is shown schematically below:

Assign $\xi_{2,1} = c_1$ and $\xi_{2,M+1} = c_2$

↓

Apply Gaussian elimination method to equations (15) and (16) and solve for ξ and ψ

↓

Store values for $\psi_{2,1} = c_1'$ and $\psi_{2,M+1} = c_2'$

Similarly, by assigning further $\xi_{2,1} = c_3$, $\xi_{2,M+1} = c_4$ and $\xi_{2,1} = c_5$, $\xi_{2,M+1} = c_6$, one obtains the corresponding values $\psi_{2,1} = c_3'$, $\psi_{2,M+1} = c_4'$ and $\psi_{2,1} = c_5'$, $\psi_{2,M+1} = c_6'$, respectively. Using the values c_i and c_i' (i=1,2,...6) and equation (18), one obtains the following six linear algebraic equations for $a_{i,j}$ and k_i (i,j=1,2).

$$\begin{bmatrix} c_1 & c_2 & 0 & 0 & 1 & 0 \\ 0 & 0 & c_1 & c_2 & 0 & 1 \\ c_3 & c_4 & 0 & 0 & 1 & 0 \\ 0 & 0 & c_3 & c_4 & 0 & 1 \\ c_5 & c_6 & 0 & 0 & 1 & 0 \\ 0 & 0 & c_5 & c_6 & 0 & 1 \end{bmatrix} \begin{bmatrix} a_{11} \\ a_{12} \\ a_{21} \\ a_{22} \\ k_1 \\ k_2 \end{bmatrix} = \begin{bmatrix} c_1' \\ c_2' \\ c_3' \\ c_4' \\ c_5' \\ c_6' \end{bmatrix} \qquad (19)$$

The elements $a_{i,j}$ and k_i (i,j=1,2) can now be found provided that the matrix is not singular. Consequently $\xi_{2,1}$ and $\xi_{2,M+1}$ in equation (18) can be determined.

Method (2): Alternatively the problem may be formulated as,

Equation	On boundaries OA and BC	On lines of symmetry AB and OC
(9)	$\xi = ?$	$\xi = 0$
(10)	$\psi = 0$	$\psi = 0$

In contrast to method (1), the values of ξ on boundaries OA and BC are to be determined such that $\partial\psi/\partial z = 0$ on these lines. Detailed procedure for this method is given in [4] and for simplicity will be omitted here.

With the numerical method of determining boundary vorticity established, the conventional line iterative relaxation method [3] may be used for the numerical solution of a set of governing equations. The general iterative procedure is given elsewhere [3]. It suffices to mention that the number of inner iteration for ξ, ψ and θ is set to be one. The line iterative method is repeated until the following prescribed error is satisfied.

$$\varepsilon = \sum_{i,j}|f_{i,j}^{(n)} - f_{i,j}^{(n-1)}| \; / \; \sum_{i,j}|f_{i,j}^{(n)}| < 10^{-5} \qquad (20)$$

The method of unsteady state solution for the determination of boundary vorticity used by Samuels and Churchill [10] is also applied to the present problem in an attempt to compare with the boundary vorticity method. The table listed after the References shows the numerical results from these two methods for Pr = 0.7 and Ra = 8010.52 = 4.69Ra* with M = 10 and 20. It can be stated that the method of unsteady state solution requires a longer computing time, a larger number of outer iterations, and one additional storage step than the present numerical method. For the unsteady state method using explicit forward integration, the time increment is restricted by the mesh size because of numerical stability and furthermore it requires a complete solution for stream function at each time step. Thus, it is believed that except for a problem where the transient solution is of interest, the present method is more economical from the viewpoint of computing time and storage space. The effect of grid size on the convergence of flow and heat transfer results is also studied, and the details are given in [4]. All the numerical results for flow and heat transfer are obtained by using mesh size M = 24, 28 or 40, depending on the order of magnitude of Rayleigh number.

RESULTS AND DISCUSSION

The problem under consideration is mathematically analogous to the classical steady laminar Bénard convection with two-dimensional rolls. Quantitative measurements for temperature and velocity fields for the two-dimensional Bénard convection are not available in the post-critical regime. For this reason Mori and Uchida [1] superimposed fully developed main flow to the steady laminar Bénard convection between two infinite horizontal plates in order to carry out detailed quantitative measurements. Mori and Uchida's theoretical and experimental results will be compared with the present numerical results. Before proceeding to the presentation of flow and heat transfer results it is emphasized that all the results reported here are based on Pr = 0.7 and wave number a=3.116. In this connection, some secondary flow patterns in the post-critical regime for the present problem obtained by flow visualization technique are reported in [9].

The axial velocity and temperature distributions along three vertical lines from the present numerical solution are compared with the results of Mori and Uchida [1] in Figs.2,3, respectively, for the case of Pr = 0.7, Ra = 8010.52 = 4.69(Ra*). For the velocity profiles shown in Fig.2, it is seen that the experimental data generally lie between the present numerical prediction and Mori and Uchida's analytical result. In particular, one notes that the analytical method apparently predicts a wrong trend for the velocity profile along the vertical line 2 near the central region (z = 0.5) exhibiting a saddle shape velocity profile. This phenomenon contradicts both experimental data and numerical solution, and cannot be explained physically.

As shown in Fig.3, the temperature profile along the vertical line 3 from the numerical solution agrees very well with the experimental data as compared with the analytical result. In particular, along the line 1 and 3, the analytical result predicts positive vertical temperature gradient in the central region in contrast to the negative one demonstrated by both the experimental data and the numerical solution. For the temperature profile along line 2, both the numerical solution and the analytical solution exhibit positive vertical temperature gradient near the central region with the point of inflection located at $z = 0.5$. In contrast, the experimental data do not appear to show this trend.

It is instructive to examine the distributions of constant axial velocity lines, isothermals, constant vorticity lines and secondary flow streamlines for $Pr = 0.7$ at two different values of Rayleigh number, namely $Ra = 8010.52$ and 4×10^4. Fig.4 shows the constant axial velocity lines. One notes that without secondary motion the constant axial velocity lines are straight and horizontal; but with the secondary motion the constant velccity lines are distorted. It is clearly seen in this figure that the saddle shape velocity profile shown in Fig.2 along the line 2 based on approximate analytical prediction [1] is unreasonable. It is also noted that at higher value of Rayleigh number $Ra = 4 \times 10^4$, the value of the maximum axial velocity is lower than that at $Ra = 8010.52$

Fig.5(a) and (b) show the distribution of isothermals for $Pr = 0.7$ at $Ra = 8010.52$ and 4×10^4 respectively, based on the dimensicnless temperature difference $(T-T_1)/(T_1-T_2)$. Without the secondary motion, the isothermals are horizontal straight lines; but with the secondary flow effect the isothermals are distorted. As the Rayleigh number increases, so does the distortion of the isothermals. By looking at this figure one can clearly distinguish the hot and cold regions by noting the isothermal line for $(T-T_1)/(T_1-T_2) = -0.5$. One can also clearly see that plumes of warm (cool) current impinge strongly upon the cool (warm) plate indicating the strong effect due to ronlinear advection caused by the buoyancy forces.

The distributions of constant vorticity lines for the case of $Pr = 0.7$ at $Ra = 8010.52$ and 4×10^4 are shown in Fig.6(a) and (b) respectively. One knows that the solid boundary represents the source of vorticity, and the vorticity is generated at the top and bottom plates (with negative sign) and dissipated into the central region as a sink of the vorticity (with positive sign). At $Ra = 4 \times 10^4$, one sees that there are two sinks with $\xi = 520$ in the central region for the two strong sources of vorticity located near $y = 0.65$, $z = 0$ and $y = 0.35$, $z = 1.0$ on the rigid boundary. This is in contrast to the one sink and two sources for $Ra = 8010.52$. One is particularly impressed with the drastic change of pattern for the vorticity as the value of the Rayleigh number increases from 8010.52 to 4×10^4.

Fig.7 illustrates the streamlines for the case of $Pr = 0.7$ at $Ra = 8010.52$ and 4×10^4. For the present problem one can see that the two diagonal lines are lines of symmetry by substituting $(u, -v, -w, -\theta, \psi, \xi, -y$ and $-z)$ for $(u, v, w, \theta, \psi, \xi, y$ and $z)$ into equations (8) to (12). Fig.7 shows that as the value of Ra increases, the streamlines are seen to be compressed in the direction of the line joining points $y = 0$, $z = 0$ and $y = 1.008$, $z = 1.0$ representing slight distortion into an oblong shape. It is interesting to note that the distributions of the stream function are much more smooth than those of the vorticity due to the fact that stream function is a twice-integrated version of vorticity.

For the finite amplitude thermal convection problem under consideration, such overall quantities as resistance coefficient and Nusselt number are usually of interest in design. The product of friction factor and Reynolds number, fRe, is defined in several different ways as follows:

$$(fRe)_1 = \frac{\overline{\left|4 + \frac{1}{2}(\partial u/\partial z)_{z=0}\right|}}{\overline{[4(z-z^2) + \frac{u}{2}]^2}} \text{ , at lower plate}$$

$$(fRe)_2 = (fRe)_1 \text{ , at upper plate} \tag{21}$$

$$(fRe)_3 = [(fRe)_1 + (fRe)_2] / 2 = 4 / \overline{[4(z-z^2) + \frac{u}{2}]^2} \text{, from overall force balance}$$

where $f_1 = 2\mu\left|\frac{\partial \bar{U}}{\partial z'}\right|_{z'=0} / \rho\bar{U}^2$, $\quad f_2 = 2\mu\left|\frac{\partial \bar{U}}{\partial z'}\right|_{z'=h} / \rho\bar{U}^2$

$f_3 = (\partial P/\partial x')\, h/\rho\bar{U}^2$ and $Re = U_0\, h/(2\nu)$

The Nusselt number is defined as,

$$(Nu)_1 = q_1 h/\Delta T\, k = 1 - \overline{(\partial\theta/\partial z)}_{z=0}$$
$$(Nu)_2 = q_2 h/\Delta T\, k = -(Nu)_1, \qquad (Nu)_3 = q_3 h/\Delta T\, k = 0 \tag{22}$$

where q_1 and q_2 are the amount of heat conducted from the lower and upper plates respectively, to the fluid per unit time per unit area, and q_3 is the net energy increase in the fluid per unit time per unit cross-sectional area.

Fig.8 shows the flow result in the form of the ratio $(fRe)_3/(fRe)_0$ versus Ra for the case of Pr = 0.7. One notes that $(fRe)_0 = 9.0$ for the case without free convection effect. With the effect of secondary motion, $(fRe)_3/(fRe)_0$ reaches a value of 2.6 at Ra = 10^5. Unfortunately no other data are available for a direct comparison with the present result. A numerical experiment is also made to evaluate the wave number effect. At Ra = 10^5, the difference of about ten per cent and twenty-five per cent for the flow result is observed as the wave number is varied from 3.116 to 3.5 and 4.0 respectively.

Fig.9 shows the heat transfer result from the present investigation with comparison made against the results of Mori and Uchida [1], Plows [7], Fromm [11] and Silveston [12].

Plows [7] studied two-dimensional steady laminar Bénard convection and considered the effect of Prandtl number for Rayleigh number ranging from 2000 to 20000. However, only the curve for Pr = 0.5 ~ 1 is taken for comparison here. It is not surprising that the result of the present numerical solution and Plows' numerical solution coincide exactly since the governing equations and boundary conditions are identical for the present problem and the two-dimensional Bénard convection problem. The agreement of the two results serves to confirm the accuracy of the present numerical technique.

Mori and Uchida's analytical solution based on Stuart's method starts deviation at Ra = 3000 from the two numerical solutions and considerable difference is observed at higher Rayleigh number. At Ra/Ra* = 9.0, Mori and Uchida's result is seen to have about twenty-five per cent error as compared with the numerical result. Mori and Uchida's experimental data are also seen to lie below the two numerical solutions.

A comparison with Fromm's numerical solution for Bénard problem [11] is also possible. Since his graphical result cannot be read accurately, only one datum point at Ra = 10^5 is plotted in Fig.9 for comparison, and an excellent agreement is again observed.

Although the present problem is not exactly identical to the Bénard problem except for the case of two-dimensional rolls, nevertheless the heat transfer results for the two problems are known to be similar, and a comparison of numerical solution with experimental data [12] for Bénard convection is also made in Fig.9. A difference of twenty per cent, based on the numerical solution is observed at Ra = 3.5 x 10^4 for the experimental data.

A numerical experiment is also made to study the effect of variable wave number. As shown in Fig.9, only a few per cent difference is found for a rather wide range of wave number from a = 3.116 to 5.0 at a high Rayleigh number of Ra = 10^5. For smaller variation of wave number and at lower Rayleigh numbers, the difference is found to be even smaller. Consequently, one can conclude that the present assumption on wave number is a reasonable approximation for the heat transfer result.

CONCLUDING REMARKS

1. The boundary vorticity method has been shown to be quite effective for the numerical solution of the present rather complicated problem. With respect to the determination of the boundary vorticity, the method of unsteady state solution employed in [10] requires a longer computation time than the present technique, at least for the present problem.

2. The heat transfer result from the present numerical study agrees exactly with the numerical result reported in [7] for the Bénard convection problem. This agreement serves to confirm the accuracy of the present numerical technique. Mori and Uchida's approximate analytical result [1] starts deviation from the present numerical result at Ra = 3000, and about twenty-five per cent error is observed at Ra = 9.0 Ra*. Thus, one can conclude that Stuart's method is valid only for Rayleigh number ranging up to say Ra = 3.0 Ra* with a maximum error of about ten per cent. In addition to the present problem, the boundary vorticity method is believed to be equally effective for the numerical solution of a host of other similar finite amplitude convection problems in the post-critical regime with Taylor-Goertler vortices.

3. The assumption that the wave number in post-critical regime does not deviate from the critical value determined by the linear stability analysis is found to be a fairly good approximation for the heat transfer result; but a rather poor approximation for the flow result. A question regarding the physical process which controls the mode of finite-amplitude roll cells has been studied quite extensively for the Bénard-Rayleigh convection problem. In contrast, for the present problem it is suggested that the wave number in the post-critical regime should be determined from a study in the thermal entrance region instead of the fully developed region.

From physical reasoning one has no reason to expect that the size and structure of the vortex roll should be determined in the fully developed region alone. It is believed that a study in the thermal entrance region might shed some light regarding the variation of wave number in the post-critical regime.

4. A remark regarding the possibility of extending the present numerical method to unsteady state finite amplitude convection problems may be of interest. By using the boundary vorticity method, the unsteady state problems can be approached by employing one of the implicit methods. This is in contrast to the conventional unsteady method which is restricted to the explicit forward integration technique.

5. A comparison between the boundary vorticity method and the usual straightforward method of obtaining boundary vorticity using Taylor series such as $\xi_{j,1} = 2\psi_{j,2}/(\Delta z)^2$ is of considerable practical interest. The approximation of the boundary vorticity is known to have a significant effect on the stability of numerical solution. In this connection, a numerical experiment with the related problem [13] reveals that the boundary vorticity method has definite advantage

in terms of computational stability, particularly in the high parameter regime.

6. The role of secondary flow on Nusselt number for the present problem is similar to that of cellular motion in the Bénard-Rayleigh convection problem. Due to the secondary flow, the pressure-drop parameter (fRe) also increases with Rayleigh number.

NOMENCLATURE

a = wave number, $2\pi h/\lambda$, f = friction factor, Gr = Grashof number, $g\beta\Delta Th^3/\nu^2$, g = gravitational acceleration, h = distance between two parallel plates, $\bar{h}$ = average heat transfer coefficient, k = thermal conductivity, M = number of divisions in both y'- and z'-directions, Nu = Nusselt number, P = pressure, p' = disturbance pressure, Pr = Prandtl number, ν/κ, Ra = Rayleigh number, $PrGr$, Re = Reynolds number, $U_0h/2\nu$, T = temperature, U,V,W = velocity components in x', y' and z' directions, u,v,w = dimensionless velocity disturbances in x', y' and z' directions, u',v',w' = velocity disturbances in x', y' and z' directions, x,y,z = dimensionless Cartesian coordinates, x',y',z' = Cartesian coordinates, ∇^2 = dimensionless Laplacian operator, $\partial^2/\partial y^2 + \partial^2/\partial z^2$, ∇_1^2 = Laplacian operator, $\partial^2/\partial y'^2 + \partial^2/\partial z'^2$, $|\ |$ = absolute value, $[\ ]$ = matrix, β = coefficient of thermal expansion, ΔT = temperature difference between two plates, T_1-T_2, θ = dimensionless temperature disturbance, θ' = temperature disturbance, κ = thermal diffusivity, λ = wave length of vortex rolls, ν = kinematic viscosity, ξ = dimensionless vorticity function, ξ' = vorticity function, ρ = density, ϕ_u = basic velocity profile function, ψ = dimensionless stream function, ψ' = stream function.

Superscript

$^-$ = mean value, $*$ = critical value, o = value at previous iteration step.

Subscript

b = boundary point or a basic quantity in unperturbed state,
j,k = space subscripts of a grid point,
0 = condition for pure forced convection or maximum quantity,
1 = value at lower plate, 2 = value at upper plate,
3 = value obtained from overall balance.

ACKNOWLEDGEMENTS

This work was supported by the National Research Council of Canada. The first author (G.J.H.) is grateful to the National Research Council of Canada for a post-graduate scholarship. The authors wish to thank Mrs. Evelyn Buchanan for typing the manuscript.

REFERENCES

[1] Mori, Y. and Uchida, Y., "Forced Convective Heat Transfer Between Horizontal Flat Plates," International Journal of Heat and Mass Transfer, vol.9, 1966, pp.803-817.

[2] Stuart, J.T., "On the Non-linear Mechanics of Hydrodynamic Stability," Journal of Fluid Mechanics, vol.4, 1958, pp.1-21.

[3] Hwang, G.J. and Cheng, K.C., "Boundary Vorticity Method for Convective Heat Transfer with Secondary Flow-Application to the Combined Free and Forced Laminar Convection in Horizontal Tubes," Heat Transfer 1970 (Proceedings, Fourth International Heat Transfer Conference, Paris-Versailles), Elsevier Publishing Company, Amsterdam, vol.4, NC3.5.

[4] Hwang, G.J., "Thermal Instability and Finite Amplitude Convection with Secondary Flow," Ph.D. Thesis, University of Alberta, Edmonton, Alberta, Canada, 1970.

[5] Schlichting, H., Boundary-Layer Theory, McGraw-Hill, New York, Sixth Edition, 1968, pp.68-70.
[6] Ogura, Y. and Yagihashi, A., "A Numerical Study of Convection Rolls in a Flow Between Horizontal Parallel Plates," Journal of the Meteorological Society of Japan, vol.47, 1969, pp.205-217.
[7] Plows, W.H., "Some Numerical Results for Two-Dimensional Steady Laminar Bénard Convection," The Physics of Fluids, vol.11, 1968, pp.1593-1599.
[8] Nakayama, W., Hwang, G.J. and Cheng, K.C., "Thermal Instability in Plane Poiseuille Flow," Journal of Heat Transfer, Trans. ASME, Series C, vol.92, 1970, pp.61-68.
[9] Akiyama, M., Hwang, G.J. and Cheng, K.C., "Experiments on the Onset of Longitudinal Vortices in Laminar Forced Convection Between Horizontal Plates," To be published in Journal of Heat Transfer, 1971.
[10] Samuels, M.R. and Churchill, S.W., "Stability of a Fluid in a Rectangular Region Heated from Below," A.I.Ch.E. Journal, vol.13, 1967, pp.77-85.
[11] Fromm, J.E., "Numerical Solution of the Nonlinear Equations for a Heated Fluid Layer," The Physics of Fluids, vol.8, 1965, pp.1757-1774.
[12] Silveston, P.L., "Wärmedurchgang in Waagerechten Flüssigkeitsschichten," Part 1, Forsch. Ing. Wes., vol.24, 1958, pp.29-32 and 59-60.
[13] Akiyama, M. and Cheng, K.C., "Boundary Vorticity Method for Laminar Forced Convection Heat Transfer in Curved Pipes," To be published in International Journal of Heat and Mass Transfer.

A comparison of boundary vorticity method with unsteady method at Ra/Ra* = 4.69

	Unsteady Method (1)		Boundary Vorticity Method(2)	
M x M	10 x 10	20 x 20	10 x 10	20 x 20
Computing time for each outer iteration (sec.)	0.35 ~ 0.4	2.3	0.2	0.82
Total number of outer iterations	191	764 (estimated)	126	356
Dimensions	3		2	
Number of iteration steps stored	2		1	
Solution obtained	unsteady and steady		steady	

(1) A point successive over-relaxation method is employed to obtain stream function at each time step.
(2) A relaxation factor of unity is used.

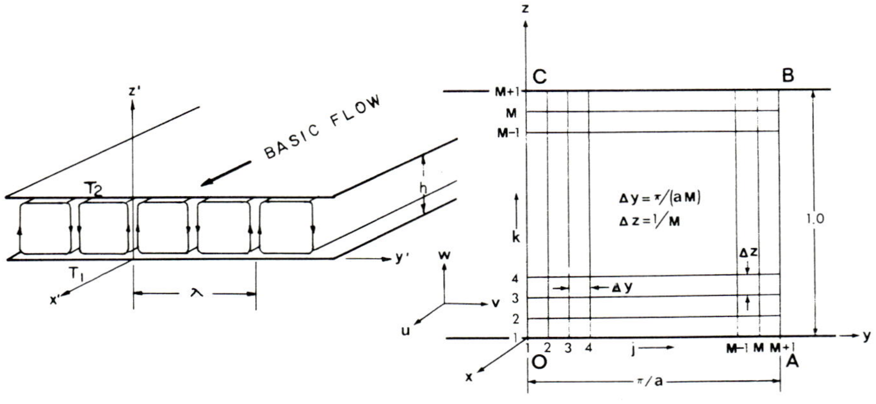

Fig.1 Coordinate System and Numerical Grid

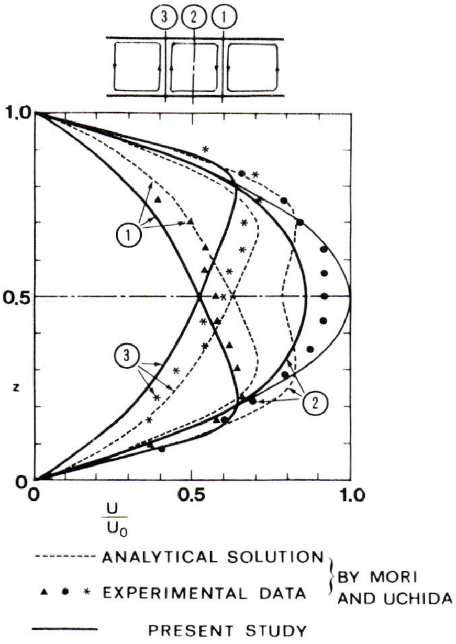

Fig.2 Comparison of Velocity Profiles for Pr = 0.7 and Ra = 8010.52

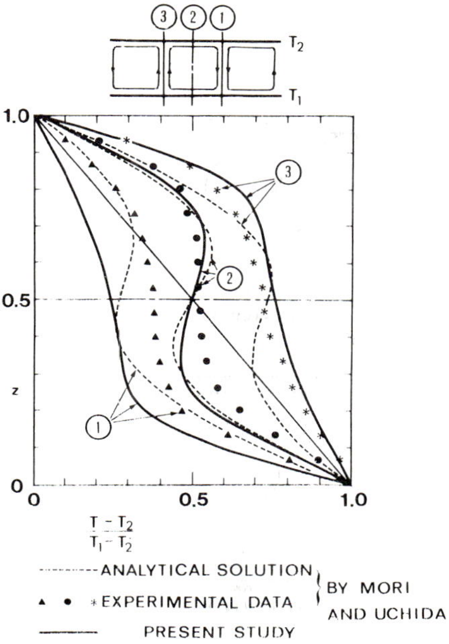

Fig.3 Comparison of Temperature Profiles for Pr = 0.7 and Ra = 8010.52

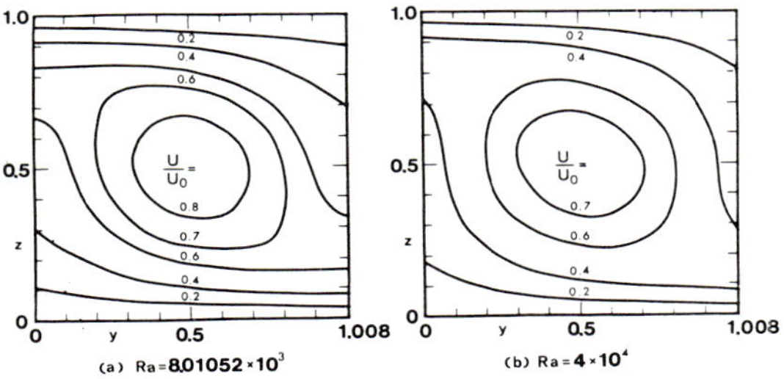

Fig.4 Lines of Constant Axial Velocity for Pr = 0.7

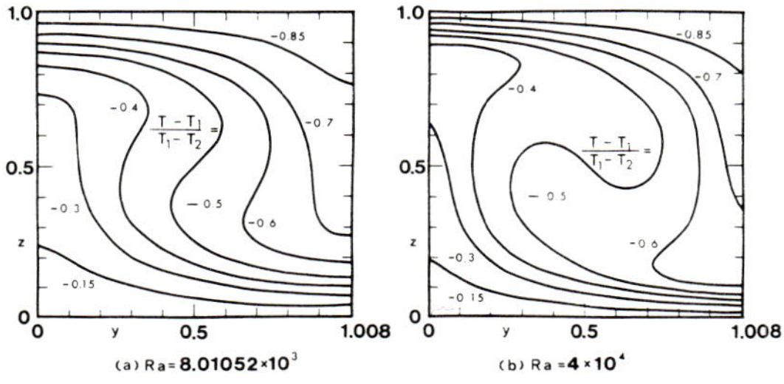

Fig.5 Isotherms for Pr = 0.7

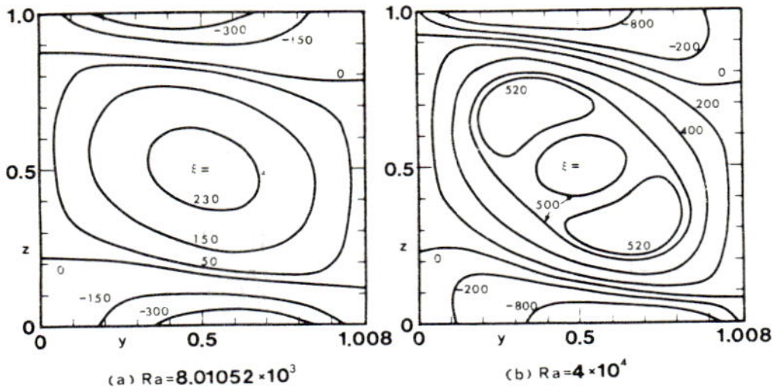

Fig.6 Lines of Constant Vorticity for Pr = 0.7

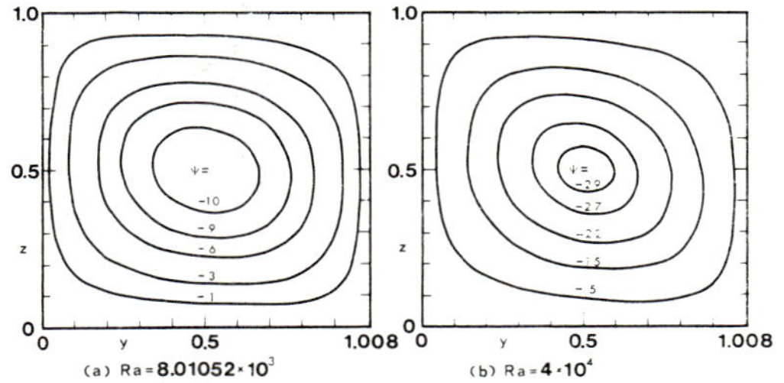

Fig.7 Secondary Flow Streamlines for Pr = 0.7

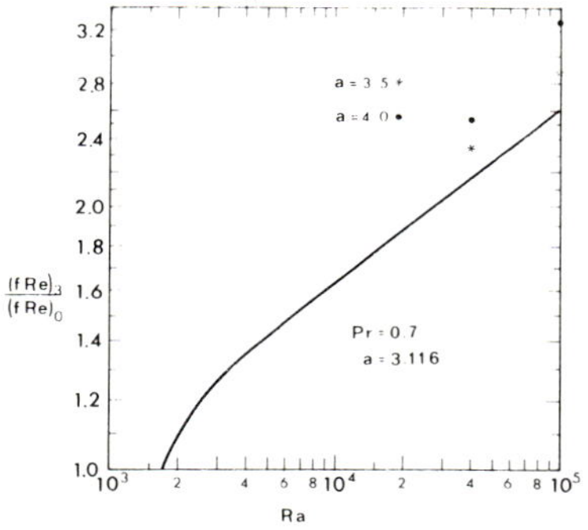

Fig.8 Flow Result for Pr = 0.7

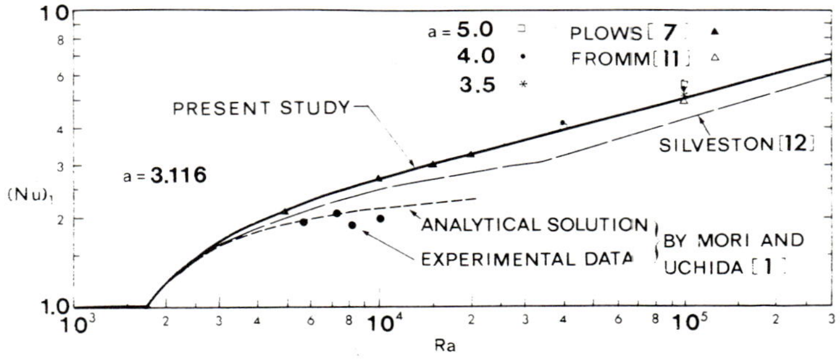

Fig.9 Heat Transfer Result for Pr = 0.7

Developments in Mechanics, Vol. 6. Proceedings of the 12th Midwestern Mechanics Conference: 14

Shape Stability of a Spherical Gas Bubble with Mass Transfer

WEN-JEI YANG
THE UNIVERSITY OF MICHIGAN
JENN-WUU OU
GONZAGA UNIVERSITY

ABSTRACT

The general problem of shape stability of an initially spherical gas bubble in a miscible liquid with chemical reaction is formulated. Utilizing the small perturbation theory the conditions for the stability or instability of the interface are obtained for the case where the liquid inertia effects are minor and mass transfer between the bubble and the liquid plays a dominant role in the dynamics. It is disclosed that interface distortions of small amplitudes undergo a time-varying oscillation with small time-varying damping factors. The amplitudes grow in magnitude when the bubble is situated in an oversaturated solution, and diminish when the bubble is placed in an undersaturated solution. In other words, a growing spherical bubble is stable, whereas a collapsing one is always unstable. When the bubble is situated in a saturated liquid these distortion amplitudes undergo a self-sustaining oscillation due to the action of surface tension force. The effect of chemical reaction on the stability of the spherical interface is of secondary importance. According to the analogy between heat and mass transfer, the stability conditions may also be applied to the case where heat transfer is the controlling mechanism of the bubble growth or collapse.

INTRODUCTION

The general problem of the dynamics of a vapor or gas cavity in a liquid is quite complex due to a coupling between the momentum equation and the equation for the temperature and/or concentration field which govern the physical phenomenon. For convenience in the investigation of the mechanism controlling collapse and growth rates, the mode of bubble collapse (and analogously growth) may be classified into three categories, (i) where the collapse is controlled by liquid inertia, (ii) where heat and/or mass transfer from the bubble to the liquid is the controlling mechanism, and (iii) the intermediate case where both effects are of comparable importance. Florschuetz and Chao [1] have identified a dimensionless parameter which characterizes the mode of collapse. Analytical solutions can be found for the first and second categories following the simplification of the governing differential equations. On the other hand, the analytical solution for the general case is formidable and only numerical solution may be obtained with the aid of a computing machine.

The stability problem of a spherical shape of a vapor bubble may be established by considering whether interface distortions of small amplitude grow or diminish. The stability criterion depends on the mechanism which controls collapse and growth rates. Some stability problems of a spherical bubble have been studied [2-10] under the conditions that heat or mass transfer effects are minor but the liquid inertia plays a dominant role. They include the work of Plesset [2] for the stability of the interface between two immiscible incompressible fluids in radial motion, Plesset and Mitchell [3] and Naude and Ellis [4] for the shape of a stationary bubble in a quiescent liquid with uniform system pressure, Hartunian and Sears [5] for the instability of a translating bubble in a liquid with uniform pressure, Walters and Davidson [6] for the deformation of bubble shape caused by the translating motion and the pressure gradient in the surrounding liquid, Ivay, et al. [7] for photographic study of cavitation bubbles in a liquid flowing through a venturi and Yeh, et al. [8-10] for the velocity, size and deformation of a bubble moving in the flow field around a point source or sink [8,9] and in a quiescent liquid subject to a sudden pressure change [8,10]. In all the analyses, the two fluids are considered immiscible and the gas inside the bubble undergoes a polytropic process during growth or collapse.

A gas bubble in a liquid-gas solution will grow or shrink by mass diffusion according as the solution is oversaturated or undersaturated. With the neglect of the translational motion of the bubble, Epstein and Plesset [11] have obtained approximate solutions for both the rate of solution by diffusion of a gas bubble in an undersaturated liquid-gas solution and the rate of growth of a bubble in an oversaturated liquid-gas solution. The effect of surface tension on the diffusion process and consequently bubble dynamics is found to be entirely negligible except near complete collapse.

In this paper the stability of the spherical shape of a gas bubble in a miscible liquid when the growth or collapse is controlled by mass transfer between the bubble and the liquid. The temperature at the bubble surface is assumed to take the saturation value corresponding to the external pressure and hence the pressure difference becomes essentially zero. The liquid inertia may thus be neglected. Consideration is given to chemical reaction in the solution. The situation is equivalent to the injection of an oxygen bubble into blood or a carbondioxide bubble into plasma. During the small times that follow, chemical reaction in the solution can be approximated as of the first order [12].

ANALYSIS

Consider at the initial time $t = 0$ a spherical gas bubble of radius R_o expands or collapses from rest in a liquid-gas solution in which the concentration of dissolved gas is uniform and equal to C_∞. The solution is maintained at constant temperature T_∞ and pressure P_∞. Chemical reaction takes place between the liquid and the dissolved gas at a rate proportional to the concentration of the dissolved gas C. At $t = 0$, the bubble has a distortion of small amplitude a_o with the initial velocity amplitude for distortion v_o. It is the purpose of the study to examine the stability of spherical interface; that is whether a distortion of the interface of small amplitude grows or diminishes.

GENERAL FORMULATION

For convenience in analysis, the center of the stationary gas bubble is taken as the origin of a spherical polar coordinate system. Then, the dissolved gas concentration at a location in the solution at a distance from the origin is governed by the equation

$$\partial C/\partial t + u.\nabla C = D\nabla^2 C - kC \tag{1}$$

where t is the time, u is the velocity produced in the medium by bubble growth or collapse, D is the mass diffusivity and k is the reaction rate constant. The appropriate initial and boundary conditions are

$$C(r,o) = C_\infty;\ C(\infty,t) = C_\infty;\ C(R,t) = C_s \tag{2}$$

in which R(t) is the radius of the spherical bubble and C_s denotes the dissolved gas concentration for a saturated solution at the system temperature and pressure. The mass balance at the bubble surface requires

$$dR/dt = (D/\rho_g)(\partial C/\partial r)_{r=R} \tag{3}$$

wherein ρ_g is the gas density.

The stability of the spherical interface will be established by considering whether a distortion of the interface of small amplitude grows or diminishes. Therefore, consider a distortion of the interface from the surface of a sphere or radius R to a surface with radius vector of magnitude r_s, where

$$r_s(t,\theta) = R(t) + \sum_{n=1}^{\infty} a_n(t)P_n(\cos\theta) \tag{4}$$

a_n are the time varying coefficient to be determined, $P_n(\cos\theta)$ are the Legendre polynomials and θ is the angle measured from a reference line. The stability analysis given here will be limited to disturbances of the spherical interface of small amplitude. This means

$$|a_n(t)| << R(t)$$

and that terms of order higher than the first in a_n and da_n/dt are negligible.

Now, the velocity potential is determined which corresponds to a disturbance which decreases away from the interface in both the inward and outward directions. With the aid of this velocity potential, the Bernoulli integral is used to evaluate the pressure on either side of the interface. Next, one writes for balance on a differential element of bubble surface. In this equation, the terms independent of P_n give the equation of motion for the unperturbed interface, while the terms proportional to P_n give the differential equation for a_n:

$$R\frac{d^2R}{dt^2} + \frac{3}{2}\left(\frac{dR}{dt}\right)^2 = \frac{P_g - P_\infty - 2\sigma/R}{\rho - \rho_g} \tag{5}$$

$$\frac{d^2a_n}{dt^2} + \frac{3}{R}\frac{dR}{dt}\frac{da_n}{dt} - Aa_n = 0 \tag{6}$$

with

$$A = \frac{[n(n-1)\rho-(n+1)(n+2)\rho_g]d^2R/dt^2-(n-1)n(n+2)\sigma/R^2}{[n\rho + (n+1)\rho_g]R} \qquad (7)$$

where ρ is the solution density, P_g is the gas pressure inside the bubble and σ is the surface tension. Details of this derivation are presented in reference [2].

The appropriate initial conditions of Eqs. (5) and (6) are

$$R(0) = R_o, \; dR(0)/dt = 0 \qquad (8)$$

$$a_n(0) = A_o, \; da_n(0)/dt = v_o \qquad (9)$$

SOLUTION OF THE STABILITY PROBLEM

The solution of the stability problem will be obtained for the case where the liquid inertia may be neglected and mass transfer effects play a dominant role. The effect of surface tension on the diffusion process is also neglected. Taking into account that the gas density is negligible in comparison with the liquid density, the function A of Eq. (7) can then be reduced to

$$A = \frac{(n-1)}{R}\frac{d^2R}{dt^2} \qquad (10)$$

Equation (1), with the neglect of the $u \cdot \nabla C$ term, can be solved by the series expansion method. Let

$$C^*(\eta,\tau) = C_o(\eta) + \tau^{1/2}C_1(\eta) + \tau C_2(\eta) + \tau^{3/2}C_3(\eta) + \ldots \qquad (11)$$

$$C^*(\eta,\tau) = C(r_1t)/C_s, \; \eta = y/\tau^{1/2}, \; y = r/R_o-1, \; \tau = \frac{Dt}{R_o^2}, \; K = \frac{kR_o^2}{D} \qquad (12)$$

By substituting Eq. (11) into Eqs. (1) and (2) followed by the separation of terms according to the powers of τ, it can be shown that the functions C_n must satisfy the following equations

$$C_o'' + \frac{\eta}{2}C_o' = 0$$

$$C_1'' + \frac{\eta}{2}C_1' - \frac{1}{2}C_1 = -2C_o'$$

$$C_2'' + \frac{\eta}{2}C_2' - C_2 = -2(C_1'-C_o'\eta) + KC_o \qquad (13)$$

$$C_3'' + \frac{\eta}{2}C_3' - \frac{3}{2}C_3 = -2(C_2'-C_1'\eta + C_o'\eta^2) + KC_1$$

$$C_4'' + \frac{\eta}{2}C_4' - 2C_4 = -2(C_3'-C_2'\eta + C_1'\eta^2-C_o'\eta^3) + KC_2$$

....

subject to the boundary conditions

$$\eta = 0: \quad C_o = 1, \; C_n = 0 \text{ for } n \geq 1$$
$$\eta = \infty: \quad C_o = \begin{cases} C_\infty/C_s & \text{for } K = 0 \\ 0 & \text{for } K \neq 0, \; C_n = 0 \text{ for } n \geq 1 \end{cases} \tag{14}$$

The substitution of Eq. (11) into Eq. (3) followed the integration of the resulting expression produces

$$R^*(\tau) = 1 + (C_s/\rho_g)[2\tau^{1/2}C_o'(0) + \tau C_1'(0) + (2/3)\tau^{3/2}C_2'(0) + (1/2)\tau^2 C_3'(0) + (2/5)\tau^{5/2}C_4'(0) + \ldots] \tag{15}$$

where $R^*(\tau) = R(\tau)/R_o$. The initial bubble radius R_o is a characteristic length for the system. The exact solution of the C_o equation subject to the appropriate boundary conditions is found to be

$$C_o(\eta) = (C_\infty/C_s - 1)\mathrm{erf}(\eta/2) + 1 \tag{16}$$

Therefore,

$$C_o'(0) = (C_\infty/C_s - 1)/\pi^{1/2} \tag{16}$$

RESULTS AND DISCUSSION

Equations (13) for C_n with $n \geq 1$ are numerically integrated with the aid of an IBM 360 digital computer. One representative result for the case of $kR_o^2/D = 50$ and $C_\infty/C_s = 0$, coresponding to a carbon dioxide bubble of 2 mm diameter in degassed water, gives $C_o'(0) = 0.5642$, $C_1'(0) = -0.9872$, $C_2'(0) = -27.4585$, $C_3'(0) = 1.8492$, ... The first two values, $C_o'(0)$ and $C_1'(0)$, are universal, i.e., for all values of kR_o^2/D.

The derivatives $dR^*/d\tau$ and $d^2R^*/d\tau^2$ can be readily obtained from Eq. (15) as

$$\frac{dR^*}{d\tau} = \left(\frac{C_s}{\rho_g}\right)\left[\frac{C_\infty/C_s - 1}{(\pi\tau)^{1/2}} + C_1'(0) + \tau^{1/2}C_2'(0) + \tau C_3'(0) + \ldots\right] \tag{17}$$

and

$$\frac{d^2R^*}{d\tau^2} = \left(\frac{C_s}{\rho_g}\right)\left[-\frac{C_\infty/C_s - 1}{2\tau(\tau\pi)^{1/2}} + \frac{1}{2\tau^{1/2}}C_2'(0) + C_3'(0) + \ldots\right] \tag{18}$$

respectively. Equation (6) may be rewritten in dimensionless form as

$$\frac{d^2a_n^*}{d\tau^2} + \frac{3}{R^*}\left(\frac{dR^*}{d\tau}\right)\frac{da_n^*}{d\tau} - \frac{n-1}{R^*}\frac{d^2R^*}{d\tau^2}a_n^* = 0 \tag{19}$$

wherein $a_n^* = a_n/a_o$. Equation (19) (as well as Eq. (6)) is the characteristic equation which determines the stability of a small-amplitude distortion of a spherical interface. By means of transformation [2]

$$a_n^* = b_n^*/(R^*)^{3/2} \tag{20}$$

Eq. (19) can be reduced to

$$\frac{d^2 b_n^*}{d\tau^2} - F(\tau) b_n^* = 0 \tag{21}$$

where

$$F(\tau) = \frac{3}{4}\left(\frac{dR^*/d\tau}{R^*}\right)^2 + \left(n+\frac{1}{2}\right)\frac{1}{R^*}\frac{d^2R^*}{d\tau^2} \tag{22}$$

Judging from the nature of Eq. (21), the deformation represented by b_n^* is stable if $F(\tau)$ is negative, but it becomes unstable when $F(\tau)$ is positive. Furthermore, Eq. (20) implies that the distortion amplitudes a_n^* are diminishing when R^* is increasing and growing when R^* is decreasing.

For small times, for which the solutions are valid, R^* and their derivatives may be approximated as

$$R^* \sim 1 + \frac{2C_s}{\rho_g}\left(\frac{C_\infty}{C_s} - 1\right)\left(\frac{\tau}{\pi}\right)^{1/2} \tag{23}$$

$$\frac{dR^*}{d\tau} \sim \frac{C_s}{\rho_g}\left(\frac{C_\infty}{C_s} - 1\right)/(\pi\tau)^{1/2} \tag{24}$$

$$\frac{d^2R^*}{d\tau^2} \sim -\frac{C_s}{\rho_g}\left(\frac{C_\infty}{C_s} - 1\right)/[2\tau(\pi\tau)^{1/2}] \tag{25}$$

Since $da_n^*/d\tau$ is small and $dR^*/d\tau$ is small compared to $d^2R^*/d\tau^2$, the second term of Eq. (19) is very small, indicating that distortions oscillate with time-varying amplitude and very small damping factor.

When $C_\infty/C_s<1$ or in an undersaturated solution, $d^2R^*/d\tau^2$ is always positive. Hence $F(\tau)>0$ and one has instability. Whereas, when $C_\infty/C_s>1$ or in an oversaturated solution, the function $F(\tau)$ becomes

$$F(\tau) = \frac{3}{4}\left(\frac{dR^*/d\tau}{R^*}\right)^2 - \frac{|d^2R^*/d\tau^2|}{R^*}\left(n + \frac{1}{2}\right)$$

It can be shown that $(3/4)\left(\frac{dR^*/d\tau}{R^*}\right)^2 < \frac{|d^2R^*/d\tau^2|}{R^*}\left(n + \frac{1}{2}\right)$ is always true. Hence, $F(\tau) < 0$ for $C_\infty/C_s > 1$ and one has stability. Since a gas bubble in a solution will grow or shrink according as the solution is oversaturated or undersaturated as described in Eq. (23), it can be stated that a growing spherical bubble is always stable, while a collapsing one is unstable. In a saturated solution, $C_\infty/C_s = 1$, a bubble will remain its original size, $R^* = 1$. When a distortion is applied, the bubble may restore to its initial shape through the action of surface tension force. Under such circumstances, the quatity A in Eq. (7) reduces to

$$A = (n - 1)(n + 1)(n + 2)\sigma/\rho R^3$$

and the stability Eq. (6) becomes

$$\frac{d^2 a_n^*}{d\tau^2} + B^2 a_n^* = 0$$

where

$$B^2 = (n - 1)(n + 1)(n + 2)\sigma^*, \quad \sigma^* = \sigma R_o/\rho D^2.$$

The solution of this equation subject to the appropriate initial conditions $a_n^*(0) = 1$ and $da_n^*(0)/d\tau = v_o/a_o$ is

$$a_n^* = \left(\frac{v_o}{a_o}\right)\frac{1}{B} \sin B\tau + \cos B\tau$$

This means, the bubble undergoes self-sustaining oscillation with both the amplitude and frequency depending upon the surface tension.

The effect of chemical reaction of the first order on the growth and stability of an initially spherical bubble can be revealed through an examination of Eq. (13). The term involving K first appears in the third equation for C_2, indicating that its effects on R(t) and a_n are indeed secondary during small times. This is true not only for the first order case but also for the zeroth and nth order cases as can be shown by the series expansion.

CONCLUSION

The stability of an initially spherical gas bubble in a liquid is investigated for the case where mass transfer between the bubble and the liquid is the controlling mechanism. In an oversaturated solution in which the bubble tends to expand the distortion amplitudes grow, while in an undersaturated solution in which the bubble tends to shrink, the distortion amplitudes diminish. However, in a saturated solution, the interface distortions undergo a self-sustaining harmonic oscillation due to the action of surface tension force. The effect of chemical reaction on the shape stability is of secondary importance. The stability conditions may be applied to the case where the growth or collapse of the bubble is controlled by heat transfer between the bubble and the liquid through the analogy between heat and mass transfer.

NOMENCLATURE

A	defined by Eq. (7) or (10).
a_o	initial amplitude of distortion of bubble surface.
$a_n(t)$	time-vary coefficients as defined in Eq. (4).
a_n^*	a_n/a_o.
b_n^*	$a_n^*(R^*)^{3/2}$.

$C(r,t)$	concentration of the gas dissolved in the liquid; C_s, saturated value; C_∞, at infinity.
$C_n(\eta)$	universal concentration-distribution function.
$C^*(\eta,\tau)$	$C(r,t)/C_s$.
D	mass diffusivity of the gas dissolved in the liquid.
$F(\tau)$	function defined by Eq. (22).
K	kR_o^2/D.
k	reaction rate constant.
n	integer.
P	pressure; P_g, of the gas inside the bubble; P_∞, of the liquid at infinity.
P_n	Legendre polynomials
$R(t)$	bubble radius; R_o, initial value.
$R^*(\tau)$	R/R_o.
r	radial distance measured from the bubble center.
r_s	radius vector of distorted bubble surface.
T_∞	liquid or system temperature.
t	time.
u	velocity produced in the solution by bubble growth or shrinkage.
v_o	initial velocity amplitude for distortion.
y	$r/R_o - 1$.
η	$y/\tau^{1/2}$.
θ	angle measured from a reference line.
ρ	liquid density; ρ_g, gas density.
σ	surface tension.
τ	Dt/R_o^2.

Superscripts:

$'$, $''$	first and second derivatives with respect to η.

Subscripts:

o	initial value.
s	at bubble surface.
∞	at infinity.

BIBLIOGRAPHY

[1] L. W. Florschuetz, and B. T. Chao: On the Mechanics of Vapor Bubble Collapse. Journal of Heat Transfer, Transactions ASME, Series C, Vol. 87, No. 2, p. 209-220 (1965).

[2] M. S. Plesset: On the Stability of Fluid Flows with Spherical Symmetry. Journal of Applied Physics, Vol. 25, No. 1, p. 96-98 (1954).

[3] M. S. Plesset, and T. P. Mitchell: On the Stability of the Spherical Shape of a Vapor Cavity in a Liquid. Quarterly Applied Mathematics, Vol. 13, No. 4, p. 419-430 (1956).

[4] C. F. Naude, and A. T. Ellis: On the Mechanism of Cavitation Damage by Nonspherical Cavities Collapsing in Contact with a Solid Boundaries. Journal of Basic Engineering, Transactions ASME, Series D, Vol. 83, p. 648 (1961).

[5] R. A. Hartunian, and W. R. Sears: On the Instability of Small Gas Bubbles Moving Uniformly in Various Liquids. Journal of Fluid Mechanics, Vol. 3, p. 27 (1957).

[6] J. K. Walters, and J. F. Davidson: The Initial Motion of a Gas Bubble Formed in an Inviscid Liquid. Part 1 The Two-Dimensional Bubble. Journal of Fluid Mechanics, Vol. 12, p. 408 (1962); Part 2 The Three-Dimensional Bubble. Journal of Fluid Mechanics, Vol. 17, p. 321 (1963).

[7] R. D. Ivany, F. G. Hammitt, and T. M. Mitchell: Cavitation Bubble Collapse Observations in a Venture. Journal of Basic Engineering, Transactions ASME, Series D, Vol. 88, p. 649 (1966).

[8] H. C. Yeh: The Dynamics of Gas Bubbles Moving in Liquids with Pressure Gradient. Ph.D. Thesis, The University of Michigan (1967).

[9] H. C. Yeh, and W. J. Yang: Dynamics of Bubbles Moving in Liquids with Pressure Gradient. Journal of Applied Physics, Vol. 39, No. 7, p. 3156-3165 (1968).

[10] H. C. Yeh, and W. J. Yang: Dynamics of a Gas Bubble Moving in an Inviscid Liquid Subjected to a Sudden Pressure Change. Journal of Applied Physics, Vol. 40, No. 4, p. 1763-1768 (1969).

[11] P. S. Epstein, and M. S. Plesset: On the Stability of Gas Bubbles in Liquid-Gas Solutions. Journal Chemical Physics, Vol. 18, No. 11, p. 1505-1509 (1950).

[12] W. J. Yang: Dynamics of Gas Bubbles in Whole Blood and Plasma. Journal of Biomechanics, Vol. 3, (1970).

INTERNAL FLOW

Developments in Mechanics, Vol. 6. Proceedings of the 12th Midwestern Mechanics Conference:

15

A Unified Method For Unsteady Flow in Polygonal Ducts

JAMES C. M. YU and C. H. CHEN
AUBURN UNIVERSITY

ABSTRACT

The problem of an unsteady laminar flow of a viscous incompressible fluid in a duct with an arbitrary cross section due to a time-dependent axial pressure gradient is transformed, by a special type of variation, to a variational problem of a functional which is defined in an arbitrary region. This region is mapped by a holomorphic function onto a unit circular region to facilitate the choice of the coordinate functions. The stationary value problem in the circular region is then solved by the Rayleigh-Ritz method. As an illustration, the velocity profiles, friction factors, and the rates of energy dissipation factors in regular polygonal ducts are evaluated. The velocity profile in a square duct is compared with that of the exact solution and is in good agreement.

INTRODUCTION

The unsteady laminar flow of a viscous, incompressible fluid in a circular pipe was investigated by Mithal [1]. A similar problem in a square duct was studied by Drake [2] under a time-dependent periodic pressure, and by Fan and Choa [3] under an arbitrary pressure gradient. Jeng [4] calculated the velocity profile of the problem in [3] by the method of point-matching.

The classical variational method was applied by Sparrow and Siegal [5] to the study of a laminar flow in a square duct under constant pressure gradient. An extended variational method has been applied by Hays [6] to Poiseuille flow with temperature dependent viscosity. The extended variational method differs from the classical one in the fact that the same physical quantity is represented by two symbols, one subject to variation and the other not. This paper presents another aspect of the application of the extended variational method. In this approach, the problem of an unsteady laminar flow of a viscous incompressible fluid in cylindrical ducts with arbitrary cross sections will be solved.

The basic idea of this new approach is to transform the differential equation into an extended variational problem of a functional which is defined in an arbitrary region. This region is then conformally mapped onto a unit circular region to facilitate the choice of the coordinate functions [7]. The variational problem can then be easily solved by the Rayleigh-Ritz method.

GENERAL FORMULATION

For an unsteady laminar flow of a viscous, incompressible fluid in a cylindrical duct of an arbitrary cross section R+C, the axial velocity v = v(x,y,t) due to an impulsive pressure gradient satisfies the following equations:

$$\frac{\partial v}{\partial t} - \nu \nabla^2 v = \frac{1}{\rho}\frac{\partial p}{\partial z} = \delta(t) \qquad \text{in } R \tag{1a}$$

$$v = 0 \quad \text{at } t = 0 \quad \text{in } R+C \tag{1b}$$
$$v = 0 \quad \text{on } C \text{ for any } t \tag{1c}$$

From the definition of $\delta(t)$ it was shown that Eq.(1) is equivalent to the following [3]:

$$\frac{\partial v}{\partial t} - \nu \nabla^2 v = 0 \qquad \text{in } R \tag{2a}$$

$$v = 1 \quad \text{at } t = 0 \quad \text{in } R+C \tag{2b}$$
$$v = 0 \quad \text{on } C \text{ for any } t \tag{2c}$$

In the variational problem the governing equation is the Euler equation of a functional with a stationary value. It should be obevious that there are mathematical advantages in using the homogeneous form of Eq.(2a) rather than Eq.(1a).

Multiplying Eq.(2a) by the variation δv and integrating over R and from t = 0 to t = ∞ yield

$$\int_0^\infty \left\{ \iint_R \frac{\partial v}{\partial t}\,\delta v\, dxdy + \delta \iint_R \frac{\nu}{2}\left[\left(\frac{\partial v}{\partial x}\right)^2 + \left(\frac{\partial v}{\partial y}\right)^2\right] dxdy \right\} dt = 0 \tag{3}$$

In Eq.(3) it is observed that if $\frac{\partial v}{\partial t}$ is replaced by $\frac{\partial v^o}{\partial t}$ to symbolize that it is not subject to any variation, then the following variational principle is obtained:

$$\delta F[v] = \delta \int_0^\infty L[v]\, dt = 0 \tag{4}$$

where

$$L[v] = \iint_R \left\{ v \frac{\partial v^o}{\partial t} + \frac{\nu}{2}\left[\left(\frac{\partial v}{\partial x}\right)^2 + \left(\frac{\partial v}{\partial y}\right)^2\right] \right\} dxdy \tag{5}$$

The method in which the same physical quantity is represented by two different symbols where one is subject to variation and the other not has been used in [6,8,9]. Since both symbols v and v^o represent the same physical quantity, v^o must be set equal to v after the variational process. The condition to set $v^o = v$ after the variational process is called a subsidiary condition. The application of this technique to nonlinear heat conduction problems of polygonal plates was made by the first author [10].

On taking the variation of Eq.(4) and using the subsidiary condition one obtains Eq.(2a) as the Euler equation. This completes the proof of the equivalence between Eq.(2a) and Eq.(4). The variational problem in Eq.(4) will be solved by the Rayleigh-Ritz method.

If the velocity is expressed as follows:

$$v = \sum_{i=1}^{\infty} A_i \ \phi_i(x,y) \ \exp(-\alpha_i t) \tag{6}$$

with $\phi_i = 0$ on C and $\sum_{i=1}^{\infty} A_i\phi_i(x,y) = 1$ to satisfy Eqs.(2b) and (2c), then the coefficients α_i can be determined by Eq.(4). The coordinate functions $\phi_i(x,y)$ for a region R bounded by a finite number of smooth arcs can be constructed by the formulation given in [11]. However, the formulation in [11] is very complicated in manipulations for a many-sided region; and furthermore, the variational process must be performed anew for each different cross section. It is one of the purposes of this paper to provide a new approach such that the manipulations can be executed once and for all.

If the region R+C can be obtained by mapping a unit circular region $R_\zeta + C_\zeta$ by a holomorphic function

$$w = f(\zeta) \quad \text{with} \quad f'(\zeta) \neq 0 \text{ in } R_\zeta \tag{7}$$

then, through some manipulations, Eq.(4) is reduced to the following:

$$\delta F[v] = \delta \int_0^{\infty} L[v] \, d\tau = 0 \qquad \text{in } R_\zeta \tag{8}$$

where

$$L[v] = \int_0^{2\pi}\int_0^1 \left(\frac{1}{a_c^2} \left\| f'(\zeta) \right\|^2 v \frac{\partial v^o}{\partial \tau} + 2 \left\| v_\zeta \right\|^2 \right) r dr d\theta \tag{9}$$

The initial and boundary conditions in R_ζ are reduced to

$$v^o(r,\theta,0) = v(r,\theta,0) = 1 \tag{10a}$$
$$v^o(1,\theta,\tau) = v(1,\theta,\tau) = 0 \tag{10b}$$

By neglecting the change of the velocity with respect to the polar angle θ in R_ζ, the N-term approximation of the velocity can be expressed as follows:

$$v = \sum_{m=1}^{N} A_m \, J_0(\beta_m r) \, \exp(-\alpha_m \tau) \tag{11a}$$

$$v^o = \sum_{m=1}^{N} A_m \, J_0(\beta_m r) \, \exp(-\alpha_m^o \tau) \tag{11b}$$

The initial and the boundary conditions will be automatically satisfied if

$$A_m = 2 \, [\beta_m \, J_1(\beta_m)]^{-1} \tag{12}$$

$$J_0(\beta_m) = 0 \tag{13}$$

The roots β_m of the Bessel function in Eq.(13) will be arranged in an ascending order $\beta_1 < \beta_2 < \ldots < \beta_m < \ldots$

Due to the holomorphic property of the napping function, its derivative can be expanded in a polynomial with a preassigned degree of accuracy. Thus,

$$f'(\zeta) = a_c \Gamma \sum_{p=0}^{M} b_p (\zeta^\Delta)^p \tag{14}$$

where a_c, Γ, b_p, and Δ are given constants. Substitution of Eqs. (11) and (14) into Eq.(9) and integration with respect to time yield the following expression:

$$F(\alpha_1 \ldots \alpha_N) = -\Gamma^2 \sum_{m,n=1}^{N} \sum_{p,q=0}^{M} A_m A_n b_p b_q \frac{B_{mnpq} \alpha_m^o}{\alpha_m^o + \alpha_n}$$

$$+ 2 \sum_{m,n=1}^{N} A_m A_n \frac{C_{mn}}{\alpha_m + \alpha_n} \tag{15}$$

where $$C_{mn} = \int_0^{2\pi} \int_0^1 P_m \overline{P_n} \, r dr d\theta \tag{16}$$

$$B_{mnpq} = \int_0^{2\pi}\int_0^1 J_0(\beta_m r)\ J_0(\beta_n r)\ (\zeta^\Delta)^p\,\overline{(\zeta^\Delta)^q}\ r\,dr\,d\theta \tag{17}$$

$$P_m = \frac{\partial}{\partial\zeta}[J_0(\beta_m r)] \quad \text{and} \quad \zeta = r\ e^{i\theta} \tag{18}$$

By substitution of Eq.(15) into Eq.(8) which is now equivalent to

$$\frac{\partial}{\partial\alpha_j}[F(\alpha_1\ldots\alpha_N)] = 0 \tag{19}$$

and by use of the following identities:

$$B_{mjpq} = 0 \quad \text{for } p \neq q \tag{20}$$

$$B_{mj00} = \pi\delta_{mj}\ [J_1(\beta_m)]^2 \tag{21}$$

$$B_{mjpp} = 2\pi\ D_{mjp} \tag{22}$$

$$D_{mjp} = \int_0^1 r^{(2\Delta p+1)}\ J_0(\beta_m r)\ J_0(\beta_j r)\ dr \tag{23}$$

$$C_{mj} = 0 \quad \text{for } m \neq j \tag{24}$$

$$C_{jj} = \frac{\pi}{4}\ [\beta_j\ J_1(\beta_j)]^2 \tag{25}$$

tedious but straightfoward calculations yield the following system of N equations for the N unknowns $\alpha_1\ldots\alpha_N$:

$$R_j = \Gamma^2 \sum_{m=1}^{N}\sum_{p=1}^{M} A_m(b_p)^2\ D_{mjp}\ \frac{\alpha_m(\alpha_j)^2}{(\alpha_m+\alpha_j)^2}$$

$$+ \frac{A_j}{8}[J_1(\beta_j)]^2\ [\Gamma^2\alpha_j-(\beta_j)^2] = 0, \quad j=1,2\ldots N \tag{26}$$

To satisfy the subsidiary condition, α_j^o in Eq.(26) have been set equal to α_j after the differentiation process in Eq.(19) was performed. The solution of Eq.(26) for α_j can be obtained by the procedures established in [10] or any other numerical technique. In particular, when $\Gamma = 1$, $b_j = 0$ for any j, Eq.(26) yield the expected solution $\alpha_j = (\beta_j)^2$ for a circular region.

To employ the method in [10], the derivatives of R_j with respect to the parameters α_k are needed. Through certain rearrangements, they are

$$\frac{\partial R_j}{\partial \alpha_k} = \frac{\Gamma^2}{8} A_j [J_1(\beta_j)]^2 \delta_{jk} + 2\Gamma^2 \sum_{m=1}^{N} \sum_{p=1}^{M} A_m (b_p)^2 D_{mpj} \frac{\alpha_j (\alpha_m)^2}{(\alpha_j + \alpha_m)^3} \delta_{jk}$$

$$+ \Gamma^2 \sum_{p=1}^{M} A_k (b_p)^2 D_{kjp} \frac{\alpha_j (\alpha_j - \alpha_k)}{(\alpha_j + \alpha_k)^3}, \qquad j,k=1,2...N. \tag{27}$$

With the values of A_m, β_m, and α_m determined by Eqs.(12), (13), and (26), respectively, the velocity distribution due to an impulsive pressure gradient is given by Eq.(11). This velocity distribution can be considered as a generating solution, since the velocity due to an arbitrary time-dependent pressure gradient can be obtained by the convolution integral [12].

If v_s is the velocity profile due to an arbitrary pressure gradient P(t), then

$$v_s = \int_0^t P(s)\, v(t-s)\, ds \tag{28}$$

To be specific, let

$$P(s) = k_o u(s) = 0 \quad \text{for } s<0$$
$$= k_o \quad \text{for } s\geq 0 \tag{29}$$

Substitution of Eqs.(11a) and (29) into Eq.(28) yields

$$v_s = \frac{1}{4} \left(\frac{4}{\nu} k_o a_c^2 \right) \sum_{m=1}^{N} \frac{A_m}{\alpha_m} J_0(\beta_m r)[1 - \exp(-\alpha_m \tau)] \tag{30}$$

From Eq.(30), the expressions for $\langle v_s \rangle_\infty$, D_f, $R_e f_r$, and Λ_∞^E can be evaluated.

$$\langle v_s \rangle_\infty = 4 \left(\frac{4}{\nu} k_o a_c^2 \right) \sum_{m=1}^{N} [\alpha_m (\beta_m)^2]^{-1} \tag{31}$$

$$D_f = -\mu \int_0^{2\pi} \left[\left(\frac{\partial v_s}{\partial r} \right)_{\substack{r=1 \\ \tau=\infty}} \right] d\theta = \pi\mu \left(\frac{4}{\nu} k_o a_c^2 \right) \sum_{m=1}^{N} (\alpha_m)^{-1} \tag{32}$$

$$R_e f_r = \frac{R_e}{2\pi} \frac{4D_f}{(\rho/2)(<v_s>_\infty)^2} = \frac{8 \sum_{m=1}^{N} (\alpha_m)^{-1}}{\sum_{m=1}^{N} [\alpha_m(\beta_m)^2]^{-1}} \tag{33}$$

The rate of energy dissipation factor per unit length along the duct is defined as follows [13]:

$$\Lambda^E = \frac{2\mu}{(\rho/2)(<v_s>_\infty)^2} \iint_R \left[\left(\frac{\partial v_s}{\partial x}\right)^2 + \left(\frac{\partial v_s}{\partial y}\right)^2\right] dxdy \tag{34}$$

Equation (34) can be transformed into the following form:

$$\Lambda^E = \frac{8\mu}{(\rho/2)(<v_s>_\infty)^2} \int_0^{2\pi}\int_0^1 \left\|\frac{\partial v_s}{\partial \zeta}\right\|^2 rdrd\theta \tag{35}$$

Substitution of Eq.(30) with $\tau = \infty$ into Eq.(35) yields

$$\Lambda_\infty^E = \frac{\pi}{Q} \sum_{m=1}^{N} (\alpha_m)^{-2} \tag{36}$$

where

$$Q = \sum_{m,n=1}^{N} [\alpha_m \alpha_n (\beta_m \beta_n)^2]^{-1} \tag{37}$$

It should be pointed out that Eq.(12) has been used in Eqs.(31-37) to simplify the expressions.

NUMERICAL RESULTS

For regular polygonal ducts, the derivatives of the mapping functions are

$$f'(\zeta) = a_c \Gamma \, (1 - \zeta^\Delta)^{-2/\Delta} \tag{38}$$

where Δ is the number of the sides of the polygon, a_c is the apothem, and Γ represents the mapping coefficients given in Table I [14]. Comparison of Eq.(38) with Eq.(14) yields

$$b_0 = 1 \quad \text{and}$$
$$b_n = \frac{2}{n!\Delta^n} (2+\Delta)(2+2\Delta)\ldots[2+(n-1)\Delta] \tag{39}$$

TABLE I

MAPPING COEFFICIENTS Γ FOR REGULAR POLYGONS

Shape	Γ
Triangle	1.135
Square	1.079
Pentagon	1.052
Hexagon	1.038
Heptagon	1.028
Octagon	1.022

With the known information Eq.(26) has been solved for $\alpha_1 \ldots \alpha_{10}$ with sixteen terms retained in the derivatives of the mapping functions, i.e., N = 10 and M = 15. The results are listed in Table II. Substitution of the ten coefficients $\alpha_1 \ldots \alpha_{10}$ back into Eq.(26) yields ten residues $R_1 \ldots R_{10}$ which indicate the degree of the unsatisfaction of the equations. The root mean square errors $\hat{R}$ are listed at the bottom of Table II. It is seen that the root mean square errors for each case can be considered neglegible when it is compared with even the smallest coefficient α_1. In Table III are listed the numerical results of $\langle v_s \rangle_\infty$, D_f, $R_e f_r$, and Λ_∞^E. It is interesting to observe that Λ_∞^E is a constant within three decimals. It can thus be concluded that, with the degree of accuracy, the rate of energy dissipation for all regular polygons is proportional to its steady average kinetic energy.

In Fig. 1 the velocity profiles of a square duct for different values of τ are compared with the exact solutions [3] with excellent agreement for $\tau < 0.3$. Even for $\tau = \infty$, the profile may be considered accurate enough for most applications.

In Fig. 2 are shown the velocity profiles for different shapes of regular polygons at different values of τ. It is noticed in Fig. 2 that the influence of the shape of the duct on the velocity profile is pronounced for increase in τ as it is expected.

CONCLUSION

The problem of an unsteady laminar flow of a viscous incompressible fluid in ducts with arbitrary cross sections under time-dependent pressure gradient was solved. Without reiterating the many advantages of the variational method, the procedure developed in this paper provides the following significant features: unification of all cross sections of ducts and economy in time for numerical results

The method in this paper is an approximation in nature. However, with the development of the high speed computer, the degree of accuracy can be increased at one's desire with almost no limitation in applications. Even for a ten-term approximation, the accuracy of the velocity profile compared with the exact one is excellent. Also seldom, if ever, does a duct maintain its designed shape after installation. Therefore, this paper provides a method of estimation of the flow field based on the distorted cross section. It is approximate but might be more realistic than the exact solution based on the designed geometry.

ACKNOWLEDGEMENT

The authors wish to acknowledge the assistance of Professor W. A. Shaw in editing the paper. The computations were performed on computer IBM 360 at Auburn University Computer Center.

REFERENCES

[1] M. L. Mithal: Unsteady Flow of a Viscous Homogeneous Incompressible Fluid in a Circular Pipe of Uniform Cross Section. Bulletin of the Calcutta Mathematical Society, Vol. 52, No. 3, pp. 147-154 (1960).

[2] D. G. Drake: On the Flow in a Channel due to a Periodic Pressure Gradient. Quarterly Journal of Mechanics and Applied Mathematics, Vol. XVIII, pt. 1, pp. 1-10 (1965).

[3] C. Fan and B-T Chao: Unsteady, Laminar, Incompressible Flow through Rectangular Ducts. Zeitschreff für angewandte Mathematik und Physik, Vol. 16, pp. 351-360 (1965).

[4] D. R. Jeng: Calculation of Unsteady Flow in Ducts of Arbitrary Shape by the Point-Matching Method. Journal of Applied Mechanics, Transaction of the ASME, Vol. 34, pp. 764-766(1967).

[5] E. M. Sparrow and R. Siegel: A Variational Method for Fully Developed Laminar Heat Transfer in Ducts. Journal of Heat Transfer, Transaction of the ASME, Vol. 81, pp. 157-167 (1959).

[6] D. F. Hays: An Extended Variational Method Applied to Poiseuille Flow: Temperature Dependent Viscosity. International Journal of Heat Mass Transfer, Vol. 9, pp. 165-170 (1966).

[7] J. C. M. Yu: Application of Conformal Transformation to the Variational Method: Buckling Loads of Polygonal Plates. Presented at the Fifth Southeastern Conference on Theoretical and Applied Mechanics, Raleigh/Durham, April 16-17, 1970, and accepted for publication in the Developments in Theoretical and Applied Mechanics, Vol. 5 (in Print).

[8] P. Rosen: Use of Restricted Variational Principles for Solution of Differential Equations. Journal of Applied Physics, Vol. 25, pp. 336-338 (1954).

[9] LL. G. Chambers: A Variational Principle for the Conduction of Heat. Quarterly Journal of Mechanics and Applied Mathematics, Vol. IX, pt. 2, pp. 234-235 (1956).

[10] J. C. M. Yu: Application of Conformal Mapping and Variational Method to the Study of Heat Conduction in Polygonal Plates with Temperature-Dependent Conductivity. International Journal of Heat Mass Transfer, (1971, in print).

[11] M. Yoshiko and T. Kawai: On the Method of Application of Energy Principles to Problems of Elastic Plates. Proceedings of the Eleventh International Congress of Applied Mechanics, pp. 461-468 (1964).

[12] A. Bronwell: Advanced Mathematics in Physics and Engineering. McGraw-Hill, New York, First Edition, p. 432 (1953).

[13] R. B. Bird, W. E. Stewart and E. N. Lightfoot: Transport Phenomena. John Wiley & Sons, New York, p. 91 (1960).

[14] P. A. Shahady, R. Passarelli, and P. A. Laura: Application of Complex-Variable Theory to the Determination of the Fundamental Frequency of Vibrating Plates. The Journal of the Acoustical Society of America, Vol. 42, pp. 806-809 (1967).

TABLE II

COEFFICIENTS α_j FOR REGULAR POLYGONS

α_j	$\Delta=3$	$\Delta=4$	$\Delta=5$	$\Delta=6$	$\Delta=7$	$\Delta=8$
α_1	4.4377	4.9514	5.2194	5.3647	5.4710	5.5360
α_2	23.492	26.118	27.510	28.270	28.828	29.170
α_3	58.668	64.517	67.756	69.551	70.890	71.714
α_4	107.07	119.19	125.47	128.95	131.50	133.07
α_5	175.12	192.25	201.84	207.16	211.12	213.55
α_6	249.01	278.69	294.10	302.49	308.62	312.38
α_7	359.01	390.48	408.71	418.89	426.60	431.34
α_8	442.60	501.61	531.73	547.98	559.63	566.76
α_9	622.09	663.56	690.02	705.63	717.78	725.33
α_{10}	680.19	785.63	837.71	865.23	884.51	896.22
$\hat{R}\times10^7$	5.58	5.02	3.31	2.67	5.44	1.66

TABLE III

PARAMETERS $<v_s>_\infty/K$, $D_f/(\mu K)$, $R_e f_r$,

AND Λ_∞^E FOR REGULAR POLYGONS

PAR.	$\Delta=3$	$\Delta=4$	$\Delta=5$	$\Delta=6$	$\Delta=7$	$\Delta=8$
$<v_s>_\infty/K$	0.1628	0.1460	0.1385	0.1347	0.1321	0.1306
$D_f/(\mu K)$	0.9806	0.8805	0.8356	0.8131	0.7974	0.7880
$R_e f_r$	61.341	61.440	61.464	61.471	61.474	61.475
Λ_∞^E	100.54	100.54	100.54	100.54	100.54	100.54

LIST OF SYMBOLS

A_m parameters chosen to satisfy the initial conditions;
a_c characteristic dimension of a duct;
$B_{mjpq}, D_{mjp}, C_{mj}$ sets of constants defined in Eqs.(16),(17), and (23);

D_f friction force per unit axial length in R_ζ;

$F[v]$, $L[v]$	functionals of any function v;
$F(\alpha_1 \ldots \alpha_N)$	function of coefficients $\alpha_1 \ldots \alpha_N$;
f_r	friction factor;
$f'(\zeta)$	derivative of $f(\zeta)$ with respect to ζ;
$J_0()$, $J_1()$	first kind Bessel functions of zero and first orders;
$K=(4k_o a_c^2)/\nu$	a constant parameter;
k_o	a constant pressure gradient;
$\underline{P}(t)$	an arbitrary pressure gradient as function of t;
$\overline{Q}$	complex conjugate of any quantity Q;
$\langle Q \rangle_\infty$	cross sectional average of any quantity Q with $\tau=\infty$;
$\Vert Q \Vert=(Q\overline{Q})^{1/2}$	modulus of any quantity Q;
$R_e=(2\rho/\mu)\langle v_s \rangle_\infty$	Reynold number;
R+C	region R with a boundary C in the w-plane;
$R_\zeta+C_\zeta$	region R_ζ with a boundary C_ζ in the ζ-plane;
R_j, $\hat{R}$	residues and the root mean square error;
(r,θ)	polar coordinates in the ζ-plane;
t, $\tau=\nu t/a_c^2$	time and dimensionless time;
$u(s)$	unit step function;
v, v^o	velocities subject and not subject to variation;
v_s	velocity due to the pressure gradient P(t);
v_ζ	derivative of v with respect to ζ;
w, ζ	complex variables;
$w=f(\zeta)$	conformal mapping function;
(x,y)	rectangular coordinates in the w-plane;
α_i	parameters chosen to satisfy Eq.(26);
Γ, b_i	mapping coefficients;
β_m	roots of Bessel function of zero order;
Δ	number of sides of a regular polygon;
δq	variation of any quantity q;
$\delta(t)$	Dirac delta function;
Λ^E	rate of energy dissipation factor;
μ, ν	dynamic and kinematic viscosities;
ρ	fluid density;
ϕ_i	coordinate functions
δ_{km}	Kronecker symbol;
∇^2	two dimensional Laplace operator;
$\frac{\partial p}{\partial z}$	axial pressure gradient.

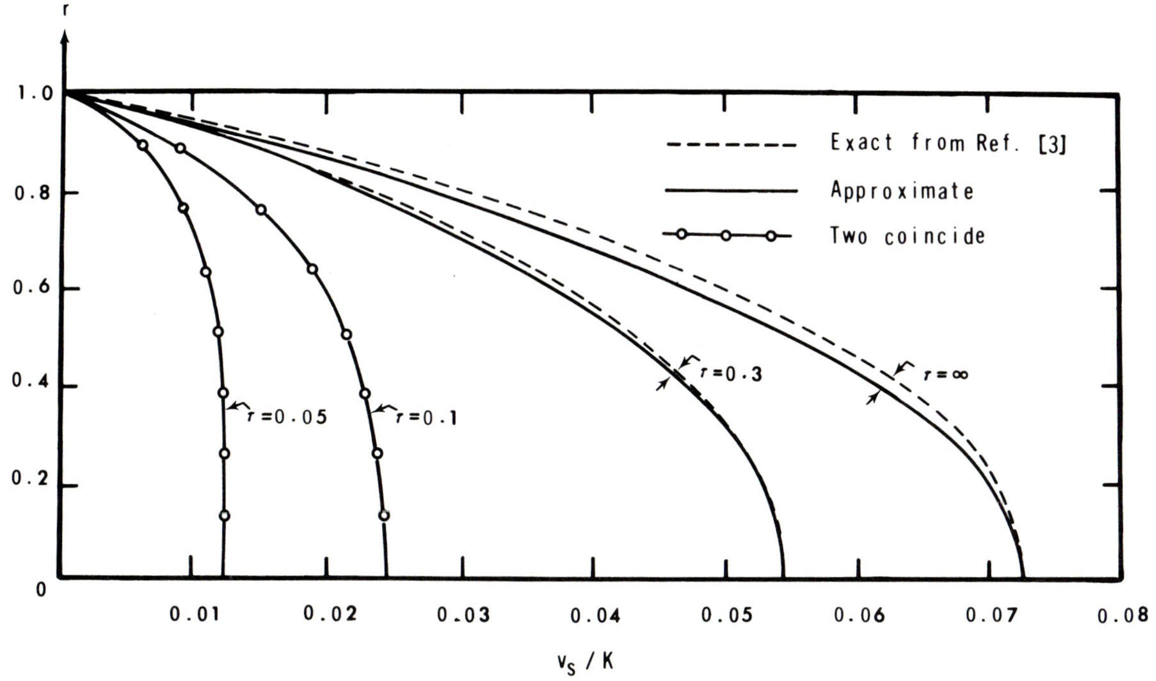

Fig. 1. Comparison of velocity profiles for a square duct.

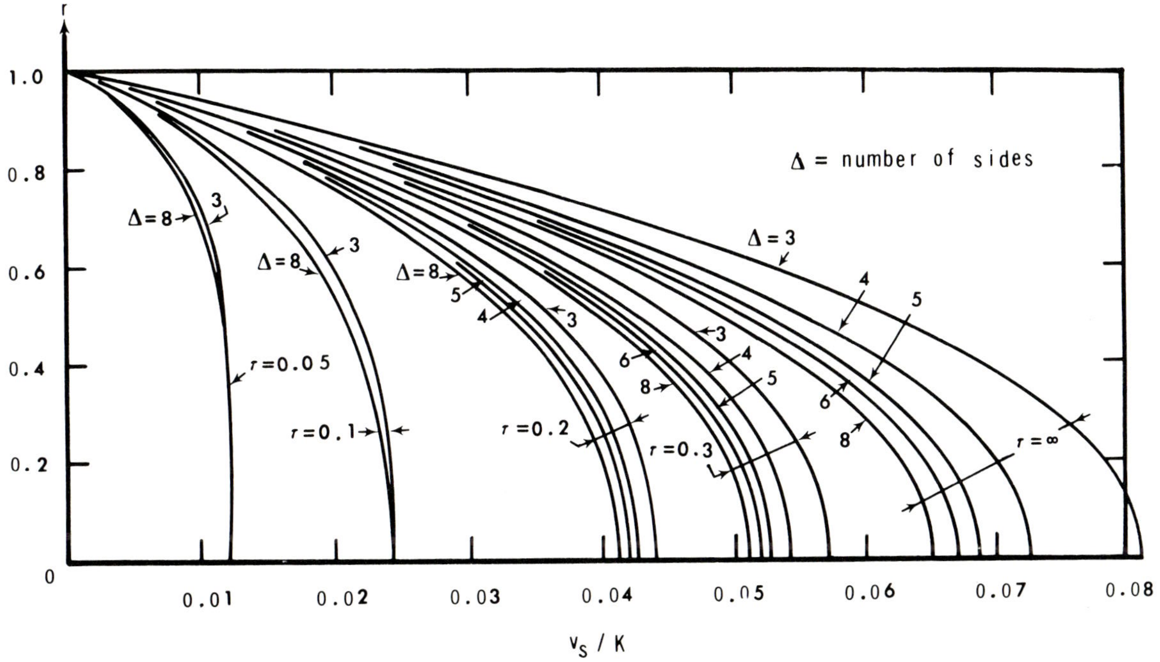

Fig. 2. Dimensionless velocity profiles for regular polygons.

Developments in Mechanics, Vol. 6. Proceedings of the 12th Midwestern Mechanics Conference: 16

On Laminar Flow in Wavy Channels

JOSEPH C. F. CHOW, KUNISIHA SODA and CRAIG DEAN
UNIVERSITY OF ILLINOIS AT CHICAGO CIRCLE

ABSTRACT

An analytical solution is presented for the flow of an incompressible viscous fluid in a symmetric channel with sinusoidal wall variation. The explicit forms for the stream function and the pressure are obtained using a perturbation technique for small wall roughness up to the second order. Possible occurrence of separation and the formation of the vortices in the region of the separation in the channel are examined as well as the pressure drop along the channel.

INTRODUCTION. - In this paper we shall consider a viscous fluid of constant density and viscosity flowing through a symmetric sinusoidal channel with constant volume rate of flow. Previous work on this subject has been limited to the situation where the flow field can be assumed to be the Stokes flow [1] and where the ratio of the amplitude, a, of the wall variation to the mean width, d, of the channel is considered to be very small.[2] Here we shall remove these two restrictions and obtain the closed form solution for the case where the channel width is small compared to the wavelength, λ, of the wall for moderate Reynolds number, and the channel constriction, a/d, up to the second order.

One of the practical applications to this problem is the blood flowing through a membrane oxygenator with irregular wall surface which can be simulated as a wavy wall.[3,4] In a rational design of a membrane oxygenator, it is necessary to know the velocity distribution, pressure drop, and strength of vorticity inside the oxygenator. These are, respectively, related to the efficiency of the oxygenator due to convective mixing, allowable pressure drop across the cardiac assisted oxygenator, and the possible damage to the red cells due to existence of high shear stress inside the channel

In the following, the effects of the various pertinent parameters upon the flow field, contours of the constant streamlines and vorticities, and the pressure drop along the channel are given, and the possible occurrence of separation and the formation of vortices in the channel are examined. In conclusion, comparison is offered for straight parallel flow with similar boundary conditions and the sensitivity of the geometric parameters upon the flow field is discussed.

FORMULATION OF THE PROBLEM. - Consider the steady laminar flow of an incompressible Newtonian fluid in a horizontal symmetric sinusoidal channel. In terms of stream function ψ, the equation governing the motion of fluid in Cartesian coordinates is

$$\psi_{,y} \nabla^2\psi_{,x} - \psi_{,x} \nabla^2\psi_{,y} = \nu\nabla^4\psi \tag{1}$$

where x, y, ∇^2 and ν are, respectively, the longitudinal axis along the channel, the axis normal to the channel, Laplacian operator, and kinematic viscosity of the fluid. The subscripts after the commas denote partial differentiation.

The boundary conditions are the nonslip condition on the wall, the symmetric condition, and the constant volume flux along the channel. They are, respectively,

$$\psi_{,x} = \psi_{,y} = 0 \qquad \text{at } y = \pm y_w \tag{2}$$

$$\psi_{,x} = 0 \qquad \text{at } y = 0 \tag{3}$$

$$\int_{-y_w}^{+y_w} \psi_{,y}\, dy = Q \qquad \text{for all } x \tag{4}$$

where Q is the volume flow rate per unit depth of the channel. The wall of the channel is given by

$$y_w = \pm\, (d + a \sin\frac{2\pi x}{\lambda}) \tag{5}$$

In order to properly assess the effects of various parameters upon the flow field, let us introduce the following transformations

$$x' = \frac{x}{\lambda}, \quad y' = \frac{y}{d}, \quad \psi' = \frac{\psi}{u_o d}, \quad Q' = \frac{Q}{u_o d} \tag{6}$$

where u_o is the average velocity of the fluid at the cross-sectional area 2d. In terms of nondimensional variables, Eqs. (1), (2), (3) and (4) become, respectively,

$$Re\delta(\psi'_{,y'} \nabla'^2\psi'_{,x'} - \psi'_{,x'} \nabla'^2\psi'_{,y'}) = \nabla'^4\psi' \tag{7}$$

$$\psi'_{,x'} = \psi'_{,y'} = 0 \qquad \text{at } y' = \pm\,\eta \tag{8}$$

$$\psi'_{,x'} = 0 \qquad \text{at } y' = 0 \tag{9}$$

$$\int_{-\eta}^{\eta} \psi'_{,y'}\, dy' = 2 \qquad \text{for all } x' \tag{10}$$

where

$$\nabla'^2 = \delta^2 \frac{\partial^2}{\partial x'^2} + \frac{\partial^2}{\partial y'^2}, \quad \delta = \frac{d}{\lambda}, \quad Re = \frac{u_o d}{\nu}$$

$$\eta = 1 + \varepsilon \sin\beta x', \quad \varepsilon = \frac{a}{d}, \quad \beta = 2\pi \tag{11}$$

The pressure gradients along the x- and y-axis can be readily obtained from the momentum equation and are, respectively, in nondimensional form in terms of the stream function:

$$p'_{,x'} = \frac{1}{Re}(\delta^2\psi'_{,x'x'y'} + \psi'_{,y'y'y'}) + \delta(-\psi'_{,y'}\psi'_{,y'x'} + \psi'_{,x'}\psi'_{,y'y'}) \tag{12}$$

$$p'_{,y'} = \frac{-\delta^2}{Re}(\delta^2\psi'_{,x'x'x'} + \psi'_{,x'y'y'}) + \delta^3(\psi'_{,y'}\psi'_{,x'x'} - \psi'_{,x'}\psi'_{,x'y'}) \tag{13}$$

where
$$p' = \frac{p}{\rho u_o^2}\frac{d}{\lambda} \tag{14}$$

p and ρ are, respectively, the pressure and density of the fluid.

The significant nondimensional parameters entering the problems are: 1) two geometric parameters, namely, δ, a measure of wall roughness and ε, a measure of the channel constriction, and 2) one flow parameter, Re, a measure of relative importance of the inertia force to the viscous force. It follows that the dependent variables $\underline{u}$, the fluid velocity vector and p are related to the above parameters in the following ways:

$$\underline{u}' = \frac{\underline{u}}{u_o} = \underline{f}_1(x',y';Re,\varepsilon,\delta) \tag{15}$$

$$\nabla p' = \frac{\nabla p}{\rho u_o^2} = f_2(x',y';Re,\varepsilon,\delta) \tag{16}$$

where ∇p is the pressure difference between two points in the flow field.

Our objective here is to obtain the explicit relationships between the dependent variables and their pertinent parameters and to show quantitatively the effects of those parameters upon the flow field.

METHOD OF SOLUTION. - We shall obtain the solution of the stream function and the pressure gradient by expanding them in series in terms of the geometric parameter δ and seek the asymtotic solution of ψ, $p_{,x}$, and $p_{,y}$ in the limit of $\delta \to 0$.

$$\psi(x,y;Re,\varepsilon,\delta) \simeq \sum_{n=0}^{N} \delta^n\psi_n(x,y;Re,\varepsilon) \tag{17}$$

$$p_{,x}(x,y;Re,\varepsilon,\delta) \simeq \sum_{n=0}^{N} \delta^n p_{,x}(x,y;Re,\varepsilon) \tag{18}$$

$$p_{,y}(x,y;Re,\varepsilon,\delta) \simeq \sum_{n=0}^{N} \delta^n p_{,y}(x,y;Re,\varepsilon) \tag{19}$$

For convenience, we have deleted the prime from nondimensional quantities.

Upon substitution of the above relations into Eqs. (7) through (13) and subsequent collection of terms with equal power of δ yield the following sets of perturbed equations up to the second order.

Zeroth order:

$$\psi_{0,yyyy} = 0 \tag{20}$$

$$p_{0,x} = 1/Re\ \psi_{0,yyy} \tag{21}$$

$$p_{0,y} = 0 \tag{22}$$

$$\psi_{0,x} = \psi_{0,y} = 0 \qquad \text{at } y = \pm\eta \tag{23}$$

$$\psi_{,x} = 0 \qquad \text{at } y = 0 \tag{24}$$

First order:

$$\psi_{0,y}\psi_{0,xyy} - \psi_{0,x}\psi_{0,yyy} = \frac{1}{Re}\psi_{1,yyyy} \quad (25)$$

$$P_{1,x} = 1/Re\ \psi_{0,xxyy} + \psi_{0,x}\psi_{0,yy} - \psi_{0,y}\psi_{0,yx} \quad (26)$$

$$P_{1,y} = 0 \quad (27)$$

$$\psi_{1,x} = \psi_{1,y} = 0 \qquad \text{at } y = \pm\eta \quad (28)$$

$$\psi_{1,x} = 0 \qquad \text{at } y = 0 \quad (29)$$

Second order:

$$-\frac{2}{Re}\psi_{0,xxyy} + \psi_{0,y}\psi_{1.xyy} + \psi_{1,y}\psi_{0,xyy} - \psi_{0,x}\psi_{1,yyy} - \psi_{1,x}\psi_{0,yyy} = \frac{1}{Re}\psi_{2,yyyy} \quad (30)$$

$$P_{2,x} = \frac{1}{Re}(\psi_{0,xxy} + \psi_{2,yyy}) + \psi_{0,x}\psi_{1,yy} + \psi_{0,x}\psi_{1,yy} + \psi_{1,x}\psi_{0,yy} - \psi_{0,y}\psi_{1,yx} - \psi_{1,y}\psi_{0,yx} \quad (31)$$

$$P,_{2y} = -\frac{1}{Re}\psi_{0,xyy} \quad (32)$$

$$\psi_{2,x} = \psi_{2,y} = 0 \qquad \text{at } y = \pm\eta \quad (33)$$

$$\psi_{2,x} = 0 \qquad \text{at } y = 0 \quad (34)$$

In addition, the flow rate across each cross-sectional area is equal to a constant; this gives from Eq. (10)

$$\psi_0 = 1, \quad \psi_1 = \psi_2 = 0 \qquad \text{at } y = \pm\eta \quad (35)$$

STREAM FUNCTION AND PRESSURE. - Here, we shall obtain the solutions for the stream function ψ and pressure p up to the second order satisfying their appropriate boundary conditions. For clarity, we shall give the expressions ψ and p without going through the detailed integrations of the governing equations. The analytical solutions of ψ and p for different orders are given below.

Zeroth order:

$$\psi_0 = C(Y^3 - 3Y) \quad (36)$$

$$p_0 = \frac{6C}{\beta Re}\left[\frac{\eta,_x}{2\beta(1-\varepsilon^2)\eta^2}\left(1 + \frac{3\eta}{1-\varepsilon^2}\right) - \frac{\varepsilon(4-\varepsilon^2)}{2(1-\varepsilon^2)^2} + \frac{2+\varepsilon^2}{(1-\varepsilon^2)^{5/2}}\left\{\tan^{-1}\left(\frac{\tan\pi x+\varepsilon}{\sqrt{1-\varepsilon^2}}\right) - \tan^{-1}\left(\frac{\varepsilon}{\sqrt{1-\varepsilon^2}}\right)\right\}\right] + p_0(0) \quad (37)$$

First order:

$$\psi_1 = C^2 Re\eta,_x f(Y) \quad (38)$$

$$p_1 = \frac{108C^2}{35}\left(\frac{\eta^2-1}{\eta^2}\right) + p_1(0) \quad (39)$$

Second order:

$$\psi_2 = -\frac{3C}{10}(4\eta_{,x}^2 - \eta\eta_{,xx})\,[(Y^2-1)^2Y] + Re^2C^3\eta\eta_{,xx}F(Y)$$

$$- Re^2C^3(\eta_{,x})^2G(Y) \qquad (40)$$

$$p_2 = \left(\frac{\eta_{,x}}{\eta^2} - \beta\varepsilon\right)\left[\frac{9C}{Re}\left(Y^2 - \frac{7}{15}\right) + ReC^3\left(\frac{1248}{13475}\right)\right]$$

$$- 2J(x)\left[\frac{18C}{Re}\left(Y^2 - \frac{3}{10}\right) + ReC^3\left(\frac{1248}{13475}\right)\right] + p_2(0) \qquad (41)$$

where

$$Y = \frac{Y}{\eta}, \quad C = -\frac{1}{2}$$

$$f(Y) = -\frac{3}{70}Y^7 + \frac{3}{10}Y^5 - \frac{33}{70}Y^3 + \frac{3}{14}Y$$

$$F(Y) = -\frac{1}{1540}\left(Y^{11} - 11Y^9 + \frac{1518}{35}Y^7 - \frac{462}{5}Y^5 + \frac{3279}{35}Y^3 - \frac{1213}{35}Y\right)$$

$$G(Y) = -\frac{1}{550}\left(Y^{11} - \frac{165}{14}Y^9 + \frac{2244}{49}Y^7 - \frac{627}{7}Y^5 + \frac{4111}{49}Y^3 - \frac{2875}{98}Y\right)$$

$$J(x) = \int_0^x \frac{\eta_{,xx}}{\eta^2}dx = \frac{-\beta\varepsilon^2}{1+\varepsilon^2}\left[\beta x - \frac{2}{\sqrt{1-\varepsilon^2}}\arctan\frac{\tan\pi x+\varepsilon}{\sqrt{1-\varepsilon^2}}\right]_0^x$$

The expression for the pressure is obtained by integrating

$$dp = p_{,x}\,dx + p_{,y}\,dy \qquad (42)$$

from x = 0 to x = x and y = 0 to y = y with p(0) corresponding to the pressure at x = 0 and y = 0.

SEPARATION POINT. - Here, we shall examine the possibility of the separation inside the channel. The region of the separation is given by

$$\frac{\partial u_s}{\partial n} \leq 0 \qquad \text{at } y = \eta \qquad (43)$$

where u_s is the tangential velocity along the contour of the channel wall and n is the axis normal to the channel wall. The equal sign in Eq.(66) corresponds to the separation point. In Fig. 1, the normalized wall contour is

$$y = Y\eta \qquad 0 \leq Y \leq 1 \qquad (44)$$

The velocity u_s is

$$u_s = u\cos\theta + v\sin\theta, \quad \tan\theta = Y\beta\varepsilon\cos\beta x \qquad (45)$$

where u and v are the velocity components along the x and y axes, respectively. Differentiating u_s with respect to n, we obtain

$$u_{s,n} = u_{s,x}\sin\theta - u_{s,y}\cos\theta = \sin\theta\cos\theta[u_{,x} + v_{,x}\tan\theta + (v-u\tan\theta)\theta_{,x}]$$

$$- \cos^2\theta[u_{,y} + v_{,y}\tan\theta + (v-u\tan\theta)\theta_{,y}] \qquad (46)$$

Eq. (46) becomes at the channel wall

$$u_{s,n} = -6C\left(\frac{1 + \eta_{,x}^2}{\eta^2}\right)K(x) \qquad (47)$$

where

$$K(x) = 1 - \frac{8\varepsilon\delta^2\pi^2}{5}\sin\beta x - \frac{32\varepsilon^2\delta^2\pi^2}{5}(\cos^2\beta x + \frac{\sin^2\beta x}{4}) - \frac{8Re\delta\varepsilon\pi}{35}\cos\beta x$$

$$- \frac{(Re\delta\varepsilon\pi)^2}{3}\left[\frac{1264}{13475}\cos^2\beta x + \frac{128\sin\beta x}{2695}(\sin\beta x + \frac{1}{\varepsilon})\right] \quad (48)$$

DISCUSSION OF THE RESULTS. - We shall present the solution in graphical form in Figs. 2 through 8. The representative flow patterns for ψ_0, ψ_1, and ψ_2 are shown in Figs. 2, 3, and 4, respectively, for $\delta = 0.1$, $\varepsilon = 0.2$ and $Re = 12.5$. The zeroth order solution corresponds to the flow with vanishing wall slope and reduces to the flow between parallel plates for $\varepsilon = 0$. As expected, the streamlines are relatively straight in the center of the channel and closely conform to the shape of the channel near the wall. The first order solution induces clockwise and counterclockwise rotational motion, respectively, in the convergent and divergent parts of the channel as shown in Fig. 3. Thus, it is possible to have a separated region in the divergent section of the channel provided that the parameter $\delta\varepsilon Re$ is larger than 2.755 (see Eq. (47)).

Fig. 4 shows that the second order solution reinforces the first order solution in the divergent part of the channel, except near the throat; thus, the separation would occur at a lower Reynolds number for given δ and ε as compared to the solution up to the first order (see Fig. 6).

The point of separation can be obtained from Eq. (47) by finding the combination of the parameters δ, ε and Re which makes $(\partial u_s/\partial n)|_w \leq 0$. For example, the following combination of the parameters ($\delta = 0.1$, $\varepsilon = 0.2$, $Re = 41.15$), ($\delta = 0.1$, $\varepsilon = 0.4$, $Re = 21.35$), and ($\delta = 0.2$, $\varepsilon = 0.2$, $Re = 18.94$) causes the separation to occur at $x/\lambda = 0.098$, 0.055, and 0.098, respectively. It is found that the separation point occurs at the vicinity of $x = 0$ where the slope of the channel is the steepest.

Fig. 6 shows the line of flow separation, laminar vortices in the region of separation, and the reattachment point of the line of separation for $\delta = 0.1$, $\varepsilon = 0.4$ and $Re = 21.34$. By increasing either Re or ε, the separation point would move down toward the throat in the divergent part of the channel with subsequent enlargement of the region of separation. The question is how stable is the flow field after the separation point reaches the throat. It is quite likely that the flow field will be unstable at a much lower Reynolds number compared to the flow in the straight channel. It may be desirable to conduct an experiment in order to verify the theory and to determine the transition condition.

The contours of the constant vorticity are shown in Fig. 5 for $\delta = 0.1$, $\varepsilon = 0.2$, and $Re = 12.5$. The shear stress on the wall is linearly proportional to the strength of the vorticity at the corresponding point.

Fig. 7 shows the pressure drop along the channel with respect to the channel constriction ε and the roughness ratio δ for $Re = 12.5$. The pressure drop increases at a greater rate as the constriction gets larger and approaches infinity as ε approaches 1, the latter condition corresponding to the complete blockage of the channel. The increase of δ also increases the pressure drop as expected. The effect of Reynolds number on the pressure drop is shown in Fig. 8 for $\delta = 0.1$ and $\varepsilon = 0.2$. It shows that the increase in pressure drop is relatively small compared to the similar flow in a parallel channel for $\delta = 0.1$ and $\varepsilon = 0.2$.

ACKNOWLEDGMENTS. - The authors wish to thank the partial support given by the University of Illinois Chicago Circle Research Board for carrying out the project.

REFERENCES. - [1] Langlois, W. E., Slow Viscous Flow. MacMillan, New York, pp. 174-184 (1964).

[2] Belinfante, D. C., "On Viscous Flow in a Pipe with Constrictions," Proc. Camb. Phil. Soc., 58, pp. 405-416 (1962).

[3] Peirce II, E. C., Extracorporeal Circulation for Open-Heart Surgery. Charles C. Thomas, Publisher, Springfield, Illinois (1969).

[4] Kolobow, T., Zapol, W., and Sigman, R. L., "Design considerations and long-terms in vivo studies with disposable spiral membrane lung," Presented at International Symposium on Blood Oxygenation held at the Univ. of Cincinatti, Cincinatti, Ohio, Dec. 1-3 (1969).

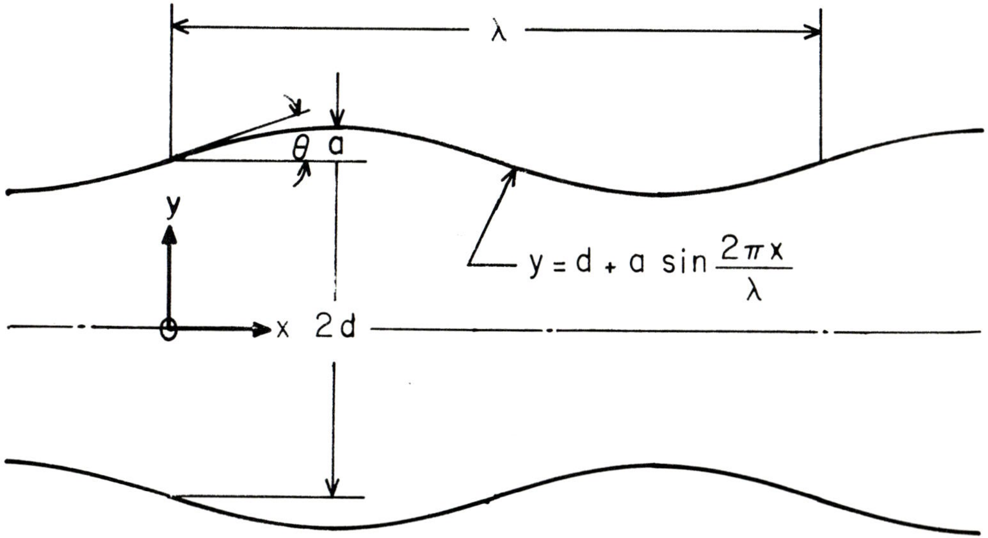

Fig. 1. Sinusoidal wavy channel.

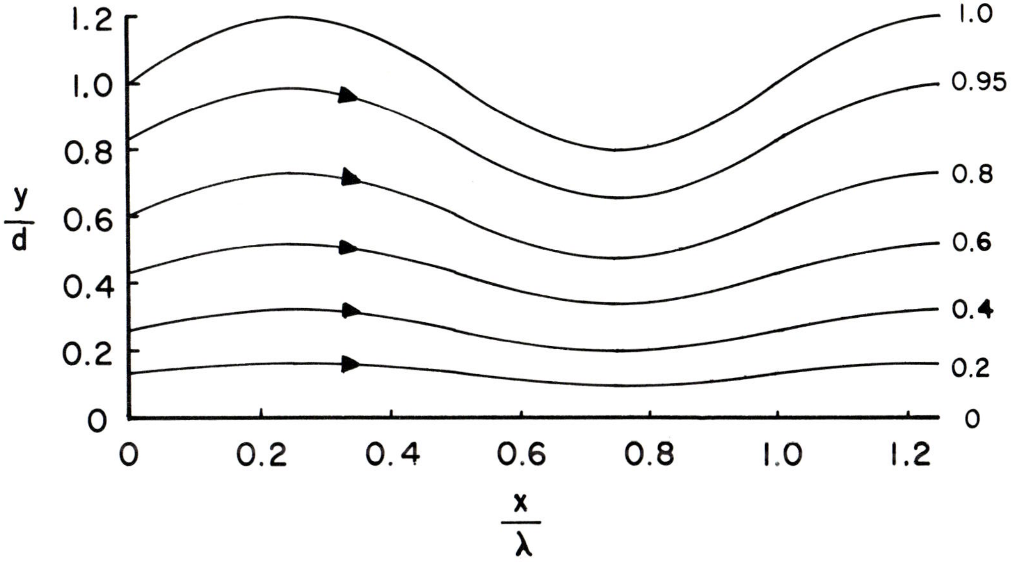

Fig. 2. Zeroth order streamlines with $\delta = 0.1$, $\varepsilon = 0.2$ and Re = 12.5.

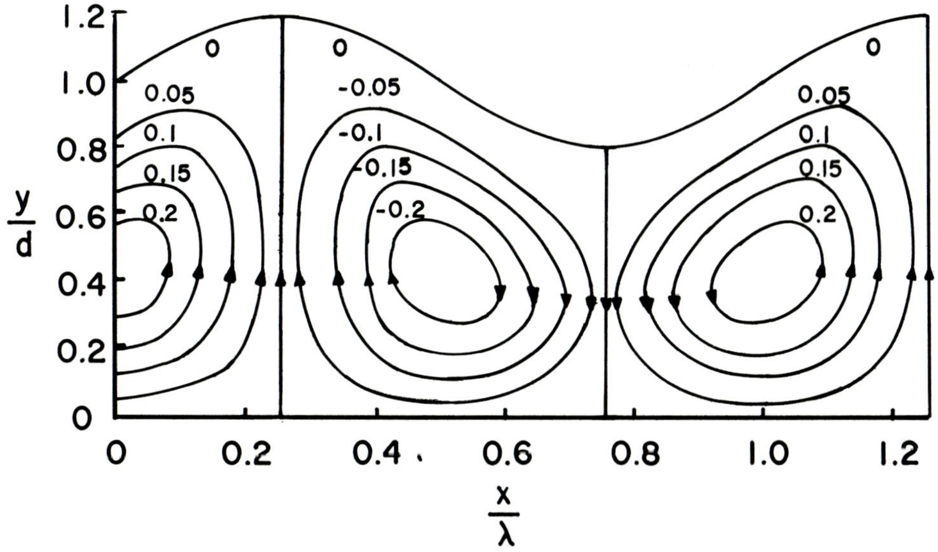

Fig. 3. First order streamlines with $\delta = 0.1$, $\varepsilon = 0.2$ and Re = 12.5.

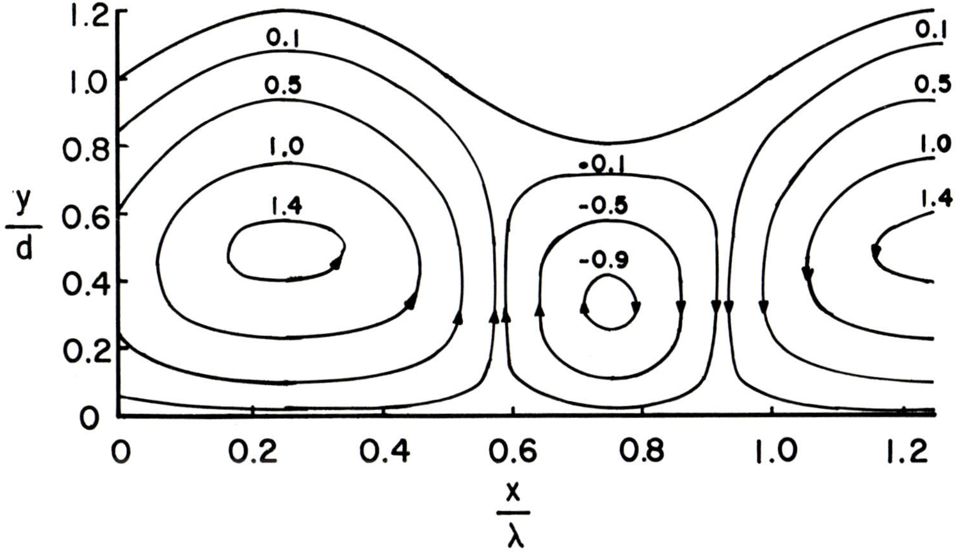

Fig. 4. Second order streamlines with $\delta = 0.1$, $\varepsilon = 0.2$ and $Re = 12.5$.

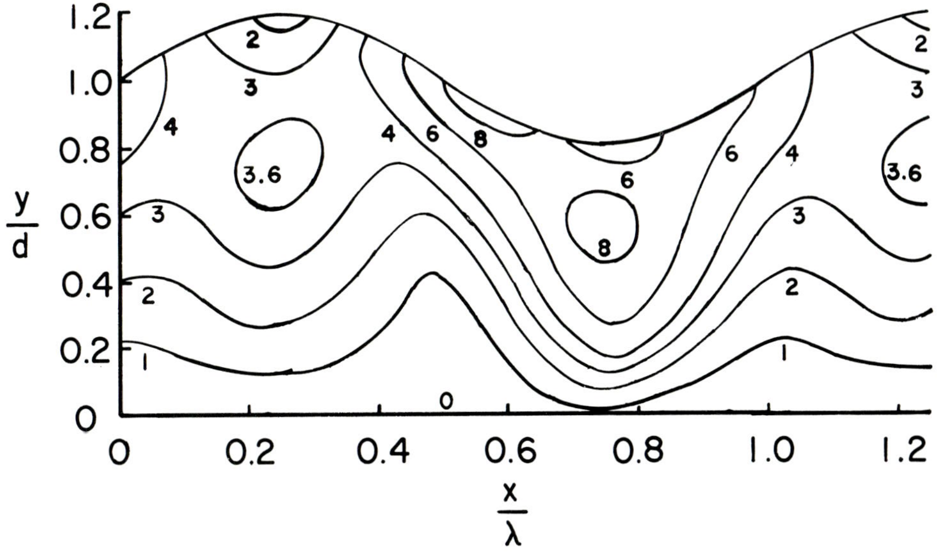

Fig. 5. Vorticity contours up to second order for $\delta = 0.1$, $\varepsilon = 0.2$ and Re = 12.5.

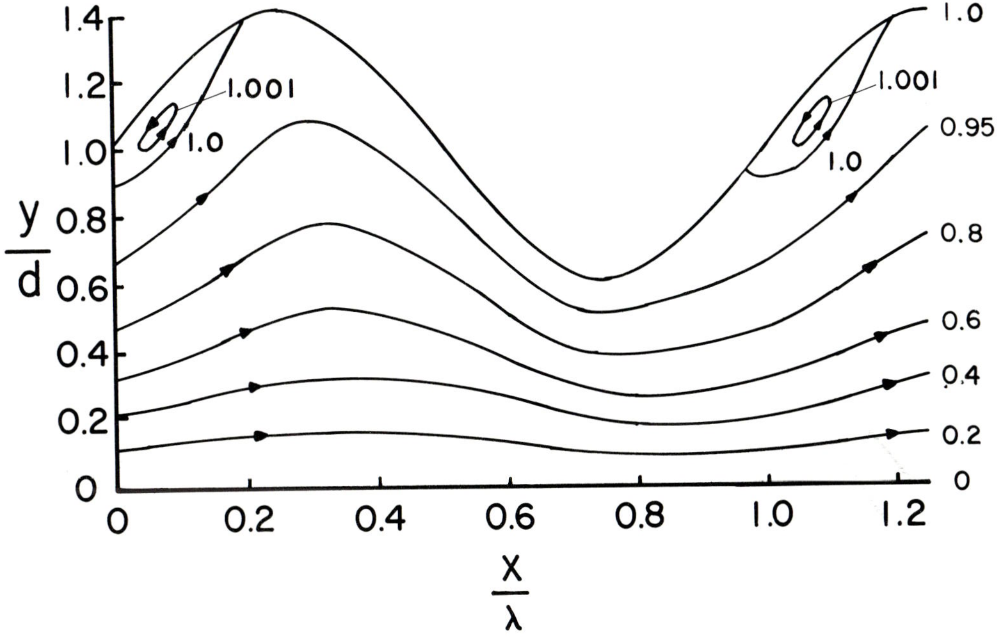

Fig. 6. Streamlines up to second order for $\delta = 0.1$, $\varepsilon = 0.4$ and $Re = 25$.

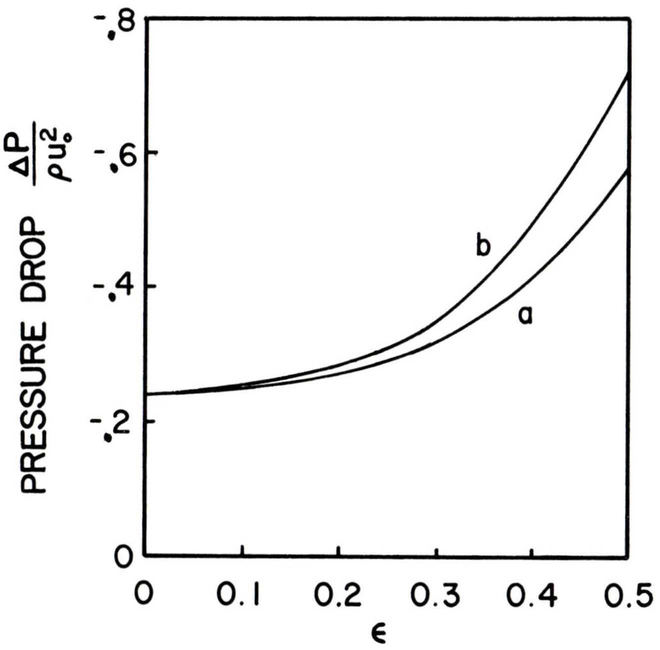

Fig. 7. Pressure drop per wavelength versus channel constriction for $\delta = 0.1$ (a) and $\delta = 0.3$ (b) with Re = 12.5.

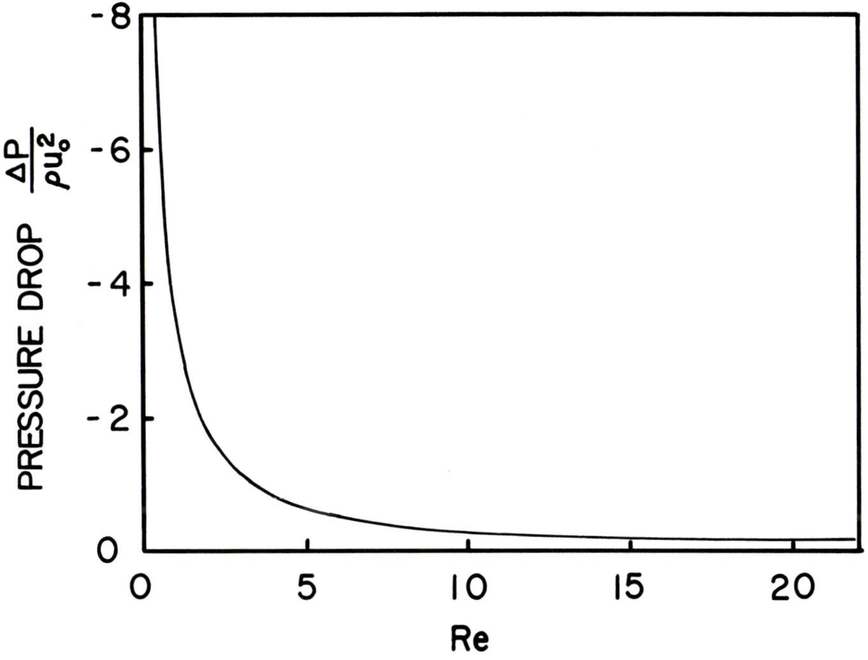

Fig. 8. Pressure drop per wavelength versus Reynolds number for δ = 0.1 and ε = 0.2.

Developments in Mechanics, Vol. 6. Proceedings of the 12th Midwestern Mechanics Conference: 17

Secondary Flow Effects upon Heat Transfer in Asymmetrically Roughened Ducts

DAVID P. GUTMAN and HOWARD N. McMANUS, JR.
CORNELL UNIVERSITY

ABSTRACT

Heat transfer characteristics to a single-phase simulation approximating the gas core of horizontal-annular two-phase flow were experimentally investigated.

Temperature distributions within the flow and around the wall were measured and used to calculate local heat transfer coefficients. Temperature fields and heat transfer rates showed an asymmetric pattern which in turn were shown to be compatible with the basically double spiral isovel pattern known to be present in this system.

It is seen that the secondary flow patterns played an important role in determining the distribution of local heat transfer coefficients as well as shear stresses.

INTRODUCTION

Two-phase flow has long concerned researchers because of its occurrence in engineering devices, e.g., pipelines, heat exchangers, rocket nozzles, reactors. Early experimental and analytical efforts were directed to acquisition of design data (correlations) and qualitative descriptions of the flow [1, 2, 3, 4]. These gross approaches gradually gave way to studies more concerned with obtaining a basic understanding of the flow mechanisms and phase interactions involved [5, 6]. In 1965 the idea of circumferential (secondary) gas core velocities was advanced [7, 8] to explain how in a horizontal annular flow, i.e., a central gas core with a liquid wall film, a continuous and essentially constant film could be maintained while liquid was being drained to the bottom by gravity.

The effect of a secondary flow was explored analytically by Pletcher and McManus. While the analysis predicted that such a flow would effect the film thickness, no experimental evidence for the presence of such a flow existed. However, argument could be advanced that the highly asymmetrical wave height (roughness) known to exist on the film, i.e., high waves at the bottom, small waves at the top, would create a shear situation which would give rise to a secondary flow. The existence of such a flow was confirmed by the experimental work presented in Ref. [9].

This work employed a single phase simulation of the film roughness and found that a vigorous secondary flow developed in the gas with a significant effect upon axial flow distribution. The magnitude of these flows was found to be of the order required to effect film dimensions [7].

A review of two-phase heat transfer work [10] discloses that it has dealt mainly with averaged quantities and in general has been concerned with prediction and control of the so-called burn-out point. One experimental study [11] has concerned itself with local heat transfer characteristics. This has provided knowledge of the temperature distribution and heat transfer coefficients to be expected.

The work being presently reported on can be viewed as ambivalent in that it can be treated as an isolated but interesting forced convection problem or viewed as a facet of the complex two-phase flow field. The point of view entertained in reporting is the latter since the motivation for this work and that of Ref. [9] stemmed from a long-term interest in annular two-phase flow and the associated burn-out problem. The results presented can be expected to be meaningful and useful in appropriate single phase situations. The present work is directed toward a better understanding of the effect of this secondary flow upon the heat transfer problem both from a qualitative and quantitative point of view.

APPARATUS

The experiments were carried out in a straight aluminum duct of 4.375-inch outside diameter and a total assembled length of 262 inches. This included 29 sections of ordinary duct, a test section, and two insulating spacers. Figure 1 contains a schematic drawing of the entire system including instrumentation, and details of the design and assembly can be found in [12].

The inner surface of the duct was machined to approximate the configuration of the liquid film ordinarily observed in horizontal-annular two-phase flow. The simulation consisted of a centered bore of 3.875-inch diameter into which was cut a straight sided vee screw thread of 1/16-inch depth, displaced 1/16 inch below the centerline of the duct. This resulted in a smooth upper surface and a thread of 1/8-inch depth at the bottom of the duct. The axial waviness of the interface was approximated by the 1/10-inch pitch of the thread. When comparison of this inner surface is made to the actual gas-liquid interface in a typical two-phase system, it is found that the groove depth corresponds well with the wave amplitude but the thread pitch is decidedly shorter than the wavelength.

Air entered the system through a filter box made of angle iron and fitted on each of five sides with two 20" x 20" x 1" standard air conditioner filters. The sixth side, which faced the duct entrance, was covered with a plywood sheet into which a flow nozzle was fitted. This nozzle had a smooth inner surface which reduced to a size just equal to the duct's inner diameter. The air was drawn into the filter box by a centrifugal blower connected to the exit end of the duct by a 48-inch long diffuser and a 12-inch diameter flow straightener containing egg crate type sections. The blower exhausted into the room, and a sliding plate over the exhaust duct provided a control on the air flow rate.

The first 134 inches of duct were unheated and served as an isothermal section in which the flow pattern became fully developed. The following 99 inches, which were thermally insulated from the rest of the duct, were wrapped with a heating strip and provided a section of uniform heat flux to the core flow. The heat was generated from an A.C. power supply capable of delivering 5000 watts at 250 volts, and could be regulated to within 1/2 volt.

The measuring station was 1/2 inch in length, and was located just after the heated section. It was machined with a smooth inner bore which coincided with the thread peaks of the other duct sections. The exterior cross section of this station was square and was machined with grooves to allow the seating of an instrument holder through which the thermocouple or anemometer probe was secured. It was possible to insert the probe through a series of parallel holes spaced from 0.3 to 0.5 inches apart. Each hole had its own individually machined plug which, when screwed fully into the hole, matched the smooth inner wall. The depth of the probe was controlled by a micrometer screw on the instrument holder.

EXPERIMENTAL PROCEDURE

Each data set consisted of a complete test section temperature profile, an upstream temperature profile along the vertical diameter, wall temperature readings, a flow rate measurement, and measurement of the electrical power used by the heater. The test section temperature distribution consisted of readings taken at 45 probe locations spaced over the cross-sectional area on one side of the vertical diameter. All of the readings within the flow field were taken with a single chromel-alumel 30-gauge thermocouple so as to eliminate possible errors arising from nonuniform thermocouple production. The thermocouple was mounted inside a tube of about 1/10-inch diameter so that a length of only 1/8 inch, including the bead, was exposed to the flow. Temperature measurements of the inside surface of the duct were made on eleven radial lines spaced over the perimeter of one side of the duct. Both flow field and wall temperatures were recorded on the same side of the vertical diameter. Thermocouples were positioned for these measurements by drilling small holes (a number 53 drill was used) in the plane perpendicular to the duct axis and at such an angle that the radial lines were intersected at only the inner duct surface so that essentially no material was removed from the path of the radial lines as they traversed the duct wall. The thermocouples were cemented into these holes so that their beads were flush with the inner surface of the duct. Each of the thermocouples was soldered to a silver-tipped selector switch which allowed readings of the entire set in rapid succession.

Temperature readings were made with a Vidar 260 voltage-to-frequency converter in conjunction with a Hewlett Packard 521C Industrial Electronic Counter equipped with a crystal controlled electronic time base. The converter was capable of producing pulses linearly proportional to input voltage to within 0.1%, including time drift effects up to 24 hours, ambient temperature change of 5°C., and line voltage fluctuations of 10%, while the crystal controlled counter had an accuracy of 0.01%. By utilizing the 10 millivolt full-scale range, 5-digit readings were obtained from the thermocouple, and under steady state running conditions readings were repeatable to within 10 counts or 0.001 millivolts. Upon completion of the flow field temperature data set from the test section, the thermocouple probe was removed from its holder and transferred to an upstream station to measure the temperature of the air before entering the heated section. Before each set of readings was taken, the instrumentation was adjusted against a standard cell. The reference temperature for the thermocouple circuit was a frequently replenished water-ice bath contained within a standard thermos vacuum bottle.

The velocity of the incoming air was measured by a Kiel type total pressure probe. The probe was located very near the entrance to the first section of the duct. A static pressure tap located at the same axial position was used to measure the dynamic pressure by differencing the two on a Meriam 10-inch micromanometer. Inasmuch as the velocity profile at this station was found to be very flat, after several checks only a centerline reading was recorded.

The power level of each run was determined by two methods. The first method was to read it directly from a wattmeter accurate to 1/4% and attached to the power supply. The second method was to record the output voltage of the power supply and use this in conjunction with a measurement of the heating element resistance taken at the time of assembly of the duct with a Wheatstone Bridge from which the power was computed.

Because one of the goals of this program was to correlate the heat transfer results obtained with measurements of secondary velocities in [9], axial velocity profiles at the test section were taken for comparison with those in this reference. This validity check was made by a Disa Model 55A01 Constant Temperature Anemometer. After careful calibration of the hot wire against a micromanometer and probe system, the test section was traversed using the same probe holder as for the thermocouple. All possible leaks around the stem were eliminated with modeling clay, and readings spaced 0.2 inches along the vertical diameter were taken.

EXPERIMENTAL RESULTS

Results were obtained under six different operating conditions. These consisted of two power levels nominally 3850 Btu/hr and 7700 Btu/hr as well as three mass flow rates characterized by centerline velocities of 83, 143, and 166 ft/sec. In addition, runs at 116 ft/sec were used for axial velocity measurements.

Figure 2 compares the normalized axial velocity profile at 116 ft/sec taken along the vertical diameter at the test section with similar data at 67.31 and 192.6 ft/sec from Ref. [9]. All runs shown in this figure were under isothermal conditions. However, a profile taken also at 116 ft/sec centerline velocity and at a heating value of 3800 Btu/hr proved to be identical to that presented in Fig. 2.

Heat balance calculations were made by comparing the electrical power input to the energy gained by the fluid while passing through the heated section of the duct. The power dissipated by the heating tape was determined by the wattmeter and also calculated from the source voltage and resistance across the heater. The largest absolute difference between the energy input and gain was found to be 7.0%. Because of inevitable heat losses through the outer insulating jacket the input power was expected to be slightly higher than that captured by the fluid, i.e., a slightly negative heat balance. This trend was displayed by the wattmeter readings, but even after adjusting the resistance measurement for the increased temperature, the voltmeter usually gave a slightly positive, up to 6%, balance. Hence, either a larger resistance than accounted for existed between the voltmeter terminals at the time of measurement, or the voltmeter was not properly calibrated, or the tape was not purely resistive when carrying current. It was decided that the direct wattmeter measurement was the more reliable of the two in that it would correct for any nonresistive effect.

Temperature measurements within the flow field at the test section were combined on polar plots showing lines of constant temperature. All of the runs resulted in similar patterns, and typical plots are presented on Figs. 3 and 4. In addition, the temperature of the inner duct surface as recorded by the buried thermocouples is presented by comparing the inner wall temperature and the fluid temperature near the wall on Figs. 5 and 6 for the same runs as on Figs. 3 and 4 respectively.

DISCUSSION

The convective phenomena being considered here is a complex one and subject to several superposed effects. There is a basic axial flow upon which is imposed a secondary flow. The result of the secondary flow is a marked distortion of the velocity field with a consequent effect upon the wall shear distribution. The wall shear effect coupled with the wall roughness causes a highly asymmetric turbulence field to exist. All of the previously mentioned flow parameters inevitably effect the heat transfer situation individually and in an interrelated manner. Because of this interrelation the discussion, while treating these effects individually, does of necessity consider the interaction.

Axial Velocity Profiles

When the axial velocity profiles of the test section were compared to profiles of single-phase flow in pipes of various but uniform roughness [13], basic differences were seen. In all cases of uniformly roughened tubes, the velocity profile could be expressed by a power law of the form

$$\frac{U}{U_{cl}} = \left(\frac{y}{R}\right)^{1/n}$$

For the Reynolds number range used in this study, Ref. [13] indicated that for a smooth-walled tube, n would have a value within the bounds of 7.0 to 7.4. Roughening of the tube has the effect of lowering the value of n. The amount of this change depends on the roughness size as compared with the tube diameter. The most obvious difference between the velocity profiles of uniformly roughened tubes and those of Fig. 2 was that the point of maximum velocity was displaced from the centerline to a point closer to the bottom of the duct, i.e., towards the rougher wall. The consequences of this displacement as evidence of the existence of a secondary velocity field are discussed in [9].

By normalizing the velocities with the maximum velocity, and normalizing the radial coordinate with the distance from the maximum velocity point to the wall, a somewhat more standardized profile was obtained. The profiles adjusted in this manner were then compared with the rough pipe velocity distributions of Ref. [13]. It was found that the points near the wall of the upper half profile, $\theta = 0°$, compared well with a curve made for a Reynolds number of 10^6 and a roughness factor, R/k_s, of about 30. Points below r/R of 0.7 were found to fall below those of the curve cited. The points of the lower half profile exhibited a shape similar to the standard curves but at a lower roughness factor, i.e., greater roughness size, than any of the curves included in this reference. It can be seen in Fig. 2 that the lower portion of the profile continuously curves, while in the upper half there is a long relatively flat section thus causing the deviation from the normal power law curves below $r/R = 0.7$. It is concluded that the shift in the point of maximum velocity as well as the deviation from the shape of the power law velocity profile were both direct effects of the presence of a secondary velocity field to be discussed in a subsequent section.

When the normalized velocity and position coordinates were plotted on a log-log graph, although an exact straight line did not result, the curve was gentle enough to be approximated by straight lines. By comparing the different slopes obtained from the upper and lower profiles, the degree of change between the maximum and minimum roughness was assessed. The slope of the top-half profile was equivalent to $n = 5.1$ while the lower-half profile had a value of $n = 2.9$. The curves of Ref. [13] covered the range of n from 4 to 5 with roughness factors from 30.6 and greater. The calculated values indicate equivalent sand roughness sizes greater than the depths of the threads. This was expected since the shape and distribution of roughness elements can be as important to the friction characteristics as is the actual depth of roughness. Experiments on tubes artificially roughened with thread-type grooves [14] have confirmed this, and it has been reported that for an equivalent friction factor with this configuration the sand roughness needed is generally greater than the depth of the groove.

Temperature and Secondary Flow Fields

The isotherms presented on Figs. 3 and 4 are oriented to show the temperature pattern when looking downstream. The figures exhibit the four distinguishing features found in all data sets.

1. Fluid near the center of the duct was coolest.

2. The point of minimum temperature was below the centerline of the duct.

3. The maximum fluid temperature near the wall was at the top of the duct.

4. The fluid near the wall decreased in temperature to a minimum point located between $\theta = 90^\circ$ and $\theta = 120^\circ$, beyond which the temperature increased until $\theta = 180^\circ$. However, the temperature of the fluid at the bottom of the duct was always lower than that at the top.

At the same time, the temperatures recorded by the wall thermocouples, Figs. 5 and 6, exhibited a somewhat different pattern. It was observed that for each run, the temperature at the top of the duct was highest and decreased continuously to a minimum at the bottom of the duct. This circumferential variation of wall temperature for any one run was between 5°F and 30°F. When comparing the wall temperatures in Figs. 5 and 6 to results of Ref. [11], in which an actual two-phase system was employed, it is seen that in both cases a similar decreasing pattern existed from top to bottom, and also the magnitudes of the temperature variations compared favorably.

By combining the data of the present work and that of [9] it is possible to consider the effects of secondary flow magnitude and direction upon the temperature field. In general, the secondary flow pattern was found to be downward near the vertical diameter and upward along the side walls. A careful examination of data from [9] shows that only in the upper half of the duct did the flow near the wall adhere to a circumferential direction. In the lower half of the duct the flow direction is nearly radially outward as far from the center as $r/R = 0.65$. Then, at approximately $r/R = 0.80$, it was turned to a mostly circumferential direction with a slight inward component. At the same time the magnitude of the circumferential vectors near the bottom of the duct were quite small when compared to the others mentioned. The tangential velocity reached a relative maximum of 7% of the local axial velocity just above the horizontal diameter. The pattern inferred by this data (see Fig. 7 taken from [9]) is that flow from the center of the duct was directed outwards towards the wall over the entire lower half of the duct. When the fluid at the bottom neared the wall, it was heated and turned to a circumferential direction where it combined with cooler fluid from the central region which reached the wall at a higher position. This fluid then was all turned to the new circumferential direction and resulted in a higher velocity tangent to the wall, as well as a mixture temperature lower than the smaller amount of fluid heated at a lower position. At approximately $\theta = 90^\circ$ the process of turning and mixing fluid from the center was completed and the velocity in the upper half was more closely circumferential at greater distances from the wall. Thus, in the upper half where no new fluid was introduced directly from the central section, the temperature increased continuously until the top of the duct was reached. It can be seen that the isovel patterns of [9] and the isotherm maps of this study have a high degree of compatability.

Shear Stresses and Turbulence Fields

A review of the literature pertaining to heat transfer in roughened tubes [15] reveals that for turbulent horizontal flow the friction factor, and hence the shear stress in uniformly roughened tubes, increases with the roughness size. This leads to the assumption that under these conditions the heat transfer coefficient will also increase with wall roughness size. However, in view of the present situation, in which the effective wall roughness varies with angular position, this direct correspondance breaks down and thus additional influences must be considered.

The secondary flow field was found to be influential in determining the variation of wall stress with circumferential position. The major component of the shear stress was due to the axial velocities. In the range of Reynolds numbers under consideration the turbulent stress will dominate the laminar stress by a factor of about one hundred. Turbulent axial shear stress was calculated from data of Ref. [9] by using the Law of the Wall for rough ducts. The equation for such velocity profiles is:

$$\frac{U}{u_*} = 2.5 \ln \frac{y}{k_s} + B$$

where the constant B is a function of the size and type of roughness present, and k_s is the size of the equivalent roughness. The term u_* is the friction velocity and is equal to $\sqrt{\tau_w/\rho}$. It was assumed that points near the wall at any location were controlled by the roughness at that spot. Due to the continuously varying roughness in the circumferential direction points farther toward the center of the flow could be considered to be influenced by a wider range of the roughness spectrum. Thus, τ_w was calculated by finding u_* along different radial lines. For each set of calculations the velocity and location of each of four points, all closer to the wall than $y/R = 0.20$, were measured from graphical representations of the axial velocity profile along that radial line. Then u_* could be calculated for each pair of points by:

$$u_* = \frac{U_j - U_{j-1}}{2.5 \ln \left(\frac{y_j}{y_{j-1}}\right)}$$

It should be noted that both k_s and B drop out of the final calculation. This is because of the assumption that within such close range of the wall, the roughness factors are constant for all four points in each set. The fact that k_s and B change for different sets is of no consequence since only points of the same set are included in any one calculation. It may seem desirable to base the estimation on only the two points closest to the wall, thus reducing the influence of the surrounding roughness size. However, it is important to consider the order of the error possible with this calculational technique. The first point was so close to the wall that an error in position of 1/2 of the finest division on the measuring scale used could result in an error of 24% in the τ_w calculated from just the first two points. On the other hand, the same error made when calculating τ_w from the third and fourth points would have resulted in an uncertainty in τ_w of only 2.5%. Therefore, in order to reduce calculational error and still keep a strong influence from the local roughness elements, the results on Fig. 8 come from an average of values of τ_w found from successive pairings of the first four points.

As previously mentioned, roughness shape and distribution can effect the friction factor of the flow as much as depth. However, in a situation such as the present one, where the number of ridges per inch is constant and the shape of the groove is similar regardless of the angular position, while the thread depth changes, it can be worthwhile to compare shear stress variation with the depth variation. Figure 8 shows that the shear stress curve resembles a cosine curve with the center somewhat elongated. On the other hand, the thread depth variation has the shape of an exact cosine curve. Therefore, the shear stress curve is seen to be influenced by the roughness depth in its general trend but is noticeably distorted when compared in detail.

This distortion can be examined by again considering the secondary velocity field. Over the entire lower half of the duct, form $\theta = 90°$ to $\theta = 180°$, fluid was being supplied to the wall from the fast moving center core. Throughout this region there was high mixing as the fluid was being turned towards the top and was thus attaining a circumferential velocity. This high activity seems to have evened out the effect of the varying roughness and resulted in the flattened portion of the shear stress curve. Simultaneously, in the upper portion of the duct, as the flow followed the wall more closely, the response to roughness variation was more precise; and the shear stress curve conformed more closely to the cosine shape.

The cross shear component resulting from the circumferential velocities were presented in Ref. [9]. It was found that the percent cross shear was zero at both the top and bottom of the duct and reached a maximum of 7% at the horizontal position, $\theta = 90°$. Therefore, when adding this to the axial stress component the effect is only minor, but the trend would be to accentuate the flatness of the center portion while increasing the steepness of the ends.

The lower curve of Fig. 8 depicts the variation of turbulence intensity at $r/R = .85$. This data is taken from Fig. 12 of Ref. [9]. It is readily observed that the variation is reasonably cosine in form and quite compatible with the shear distribution. The relatively high intensity in the lower position of the duct confirms the previously mentioned high mixing and further explains the flatness of the shear distribution.

Heat Transfer Coefficients

The heat transfer coefficients presented in Figs. 9 and 10 are local values that have been calculated from the convection relation $\dot{q} = hA\Delta T$. In using this relation it was assumed that the heat flow through the tube wall was purely radial and the ΔT for meaningful local values was $T_w - T_{fw}$. Of these assumptions the poorer is probably that of radial conduction. The wall temperature data of Figs. 5 and 6 suggest that at least in the topmost part of the tube circumferential conduction may be significant since the temperature gradient is rather steep. In the lower portion of the tube, $\theta > 40°$, circumferential conduction will be of little importance since the gradient is slight and constant and the conduction path is being reduced by the increasing thread depth.

If it is assumed that the observed wall gradient is representative of the situation across the conduction path, then an order of magnitude calculation shows that for the tube top the conduction path is capable of ducting all heat generated in this sector to the adjacent portion of the tube. Hence, the first assumption causes the convection coefficient for this region to be much too great. For practical purposes this region, $350° \leq \theta \leq 10°$, is one of very low heat convection. The values of those coefficients at 20° and 40° also would be effected by the circumferential conduction but to a lesser degree. Their values would be increased slightly.

It is usual in a convection problem to use that area where heat transfers from one phase to another. In this system the actual area varies as a cosine function along the circumference. If the actual area is used to compute a heat transfer coefficient, then the lower curves B of Figs. 9 and 10 result. However, if an area based on a cylindrical surface coincident with the thread peaks is employed, then curves A are obtained. The comparison of these curves A and B suggests, not surprisingly, that the slant area of the threads is being used ineffectively insofar as heat transfer is concerned. A more basic interpretation is that there is little flow of fluid into the grooves and consequently the fluid in the grooves is of long residence and indeed presents a barrier to heat transfer. It is well to note that in either case as the tube bottom is approached the coefficient of heat transfer increases. In the variable area case the improved convection process overcomes the depressing effect of the increasing area. The mechanism responsible for this is the increased turbulence in the lower portion of the duct and to a lesser extent the downward directed central flow.

When curves A of Figs. 9 and 10 are normalized and plotted along with the shear and turbulence data on Fig. 11, it is seen that the form of the heat transfer curves differ markedly from the shear and turbulence. In a more conventional case greater similarity of shear and heat transfer could be expected. The most striking difference is the reversed curvature of the heat transfer relations. The secondary flow is evidently of sufficient effect to suppress the heat transfer in the upper portion of the tube when it is directed away from the wall and enhance it at the bottom when directed toward the wall. In both regions the secondary flow distorts the temperature and turbulence fields with a drastic effect upon the heat transfer.

When comparing the values of heat transfer coefficient to those for which the actual two-phase system was employed [11], it was found that the variational trends discussed were supported. However, the analogy is not complete due to the fact that the values of the heat transfer coefficient calculated for the simulation were one order of magnitude lower. This is not surprising in light of other studies, [16] and [17], which dealt with single phase systems including cases of artificial roughness. Both of these references reported heat transfer values consistent with those presented in this study, and the importance of the variational trends is considered greater than the actual numerical values involved.

From these results it is seen that both shear stress and heat transfer coefficients do not increase directly with increased roughness. What this implies is that in addition to this mechanism, the influence of the secondary velocity field is strong enough to cause very noticeable changes. The fact that fluid is mixed from the central flow region only in the lower half of the duct, and that it flows nearly tangent to the wall before returning to the center from the top of the duct, is just as important as the roughness factor in determining the transport coefficients.

Burn-Out Considerations

Since it was earlier indicated that part of the impetus for this work came from an interest in the burn-out problem of two-phase flow, a brief consideration of the work in such a context is appropriate. The finding in this work that a region of very low heat transfer exists at the top of the tube is important. While the duct used here was aluminum, which has a high conductivity, in a practical system the material is frequently a high alloy steel, e.g., stainless, with a quite low conductivity. If significant amounts of heat must be transferred circumferentially, it could only be done if there were a very steep temperature gradient, hence a tube hot spot.

The presence of a liquid layer on the wall would undoubtedly result in a higher transfer coefficient from wall to liquid. However, it is reasonable that the transfer coefficient from liquid to gas would be very low as in this study. Vaporization of liquid would take place at an accelerated pace and, if flow conditions were not such as to replenish this liquid, a dry spot would form. The low gas heat transfer in this region with the failure of the film to protect would lead to progressive failure, i.e., burn-out.

While the above has inferred that circumferential heat transfer is necessary to precipitate the burn-out situation, it is well to note that the existence of the secondary flow will cause the conditions for the hot spot to exist. Viewing the present work from the context of two-phase flow offers a fundamental insight into the mechanism of the burn-out problem.

CONCLUSIONS

From the work presented and the related discussion several conclusions can be drawn:

1. In a flow passage of asymmetrical roughness a secondary flow will develop which will markedly alter the isotherms.

2. In such a system heat transfer coefficients will vary with circumferential position with a region of very low transfer occurring at the top of the tube.

3. The heat transfer distribution differs markedly from the shear stress distribution indicating a significant effect of the secondary flow.

4. A definite similarity between the wall temperature in a two-phase system and those of this single-phase system exist.

5. A more detailed understanding of the coupling between metal, liquid, and gas in the burn-out problem is possible in light of this work.

NOMENCLATURE

A	Area across which heat is transferred (ft^2)
B	Constant in Law of The Wall
h	Heat transfer coefficient (Btu/hr ft^2 °F)
k_s	Equivalent sand grain roughness size
n	Exponent in Power Law velocity profile
$\dot{q}$	Heat transfer rate (Btu/hr)
R	Inside radius of duct (ft)
r	Radial distance from center of duct (ft)
T	Temperature (°F)
ΔT	Temperature difference (°F)
U	Axial velocity component (ft/sec)
u_*	Friction velocity (ft/sec)
y	Distance from duct wall (ft)
θ	Angular position
ρ	Density (lbm/ft^3)
τ	Shear stress (lbf/ft^2)

Subscripts

cl	Along centerline of duct
fw	Flow property extrapolated to duct wall
o	Value at $\theta = 0°$
w	Property of the wall

REFERENCES

1. Alves, G. E., "Co-current Liquid Gas Flow in a Pipe-Line Contractor", Chem. Eng. Prog., 50, 1954, p. 449-456.

2. Ambrose, T. W., "Literature Survey of Flow Patterns Associated with Two-Phase Flow", USAEC HW-52927, General Electric Co., Hanford Labs, October, 1957.

3. Krasiakova, L. I., "Some Characteristics of the Flow of a Two-Phase Mixture in a Horizontal Pipe", AERE Trans. 695, January, 1957.

4. Martinelli, R. C. and Boetler, M. R., et al, "Isothermal Pressure Drop for Two-Phase Two-Component Flow in a Horizontal Pipe", Trans. of ASME, 66, 1944, pp. 139-151.

5. Chang, F. W. and Dukler, A. E., "The Influence of a Wavy Moving Interface on Pressure Drop for Flow in Conduits", Int. J. Heat and Mass Trans., 7, 1964, pp. 1395-1404.

6. Laird, A. D. K., "Stability of a Gas Core in a Tube as Related to Vertical Annular Gas-Liquid Flow", Trans. of ASME, 76, 1954, pp. 1005-1010.

7. Pletcher, R. H. and McManus, H. N., Jr., "The Fluid Dynamics of Three Dimensional Liquid Films with Free Surface Shear: A Finite Difference Approach", Dev. in Mech.", 3, 1965, pp. 305-318.

8. Russel, T. W. F. and Lamb, D. E., "Flow Mechanism of Two-Phase Annular Flow", Can. J. of Chem. Eng., October, 1965, pp. 237-245.

9. Darling, R. S. and McManus, H. N., Jr., "Flow Patterns in Circular Ducts with Circumferential Variation of Roughness: a Two-Phase Flow Analog", Dev. in Mech., 5, 1969, pp. 153-170.

10. Silvestri, Mario, "Fluid Mechanics and Heat Transfer of Two-Phase Annular Dispersed Flow", Adv. in Heat Trans., 1, 1964, pp. 355-446.

11. Pletcher, R. H. and McManus, H. N., Jr., "Heat Transfer and Pressure Drop in Horizontal Annular Two-Phase, Two-Component Flow", Int. J. Heat and Mass Trans., 11, 1968, pp. 1087-1104.

12. Gutman, D. P., "An Experimental Study of Heat Transfer in a Simulation of Horizontal Annular Two-Phase Flow", unpublished M.S. Thesis, Cornell Univ., 1970.

13. Schlichting, H., Boundary Layer Theory, 6th Edition, McGraw Hill Book Co., 1968, pp. 544-589.

14. Streeter, V. L., "Frictional Resistance in Artificially Roughened Pipes", Proc. Am. Soc. of Civil Eng., 61, 1935, pp. 163-186.

15. Kolar, V., "Heat Transfer in Turbulent Flow of Fluids through Smooth and Rough Tubes", Int. J. Heat and Mass Trans., 8, 1965, pp. 639-653.

16. Furber, B. N. and Cox, D. N., "Heat Transfer and Pressure Drop Measurements in Channels with Whitworth Thread Form Roughness", J. Mech. Eng. Sci., 9, 1967, pp. 339-350.

17. Wilson, N. W. and Medwell, J. O., "An Analysis of Heat Transfer for Fully Developed Turbulent Flow in Concentric Annuli", ASME paper 67-WA/HT-4, 1967.

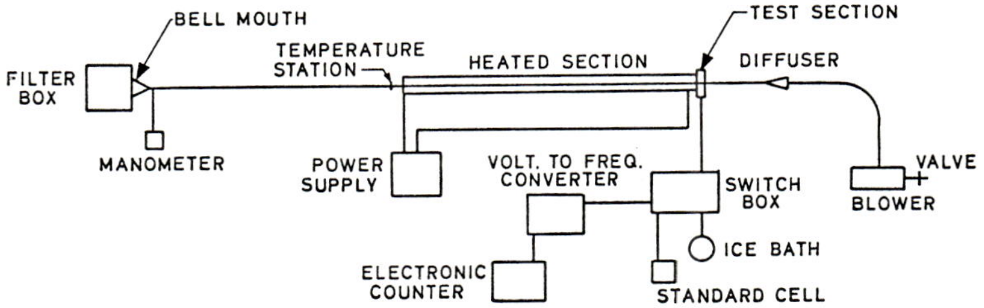

Fig. 1. Experimental Apparatus

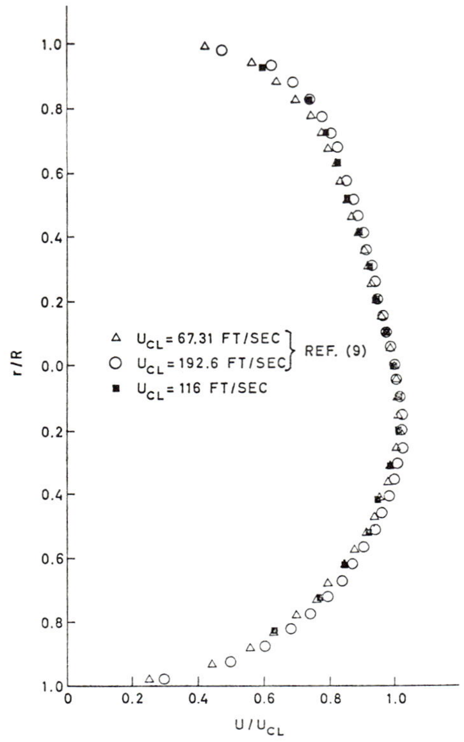

Fig. 2. Normalized Axial Velocity Profile

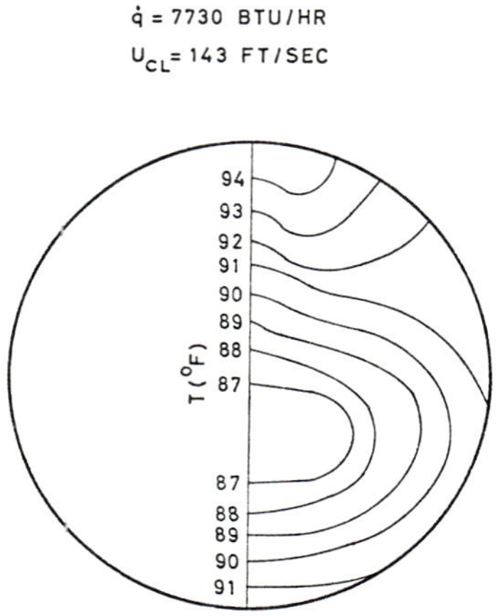

Fig. 3. Isotherm Map Run 15

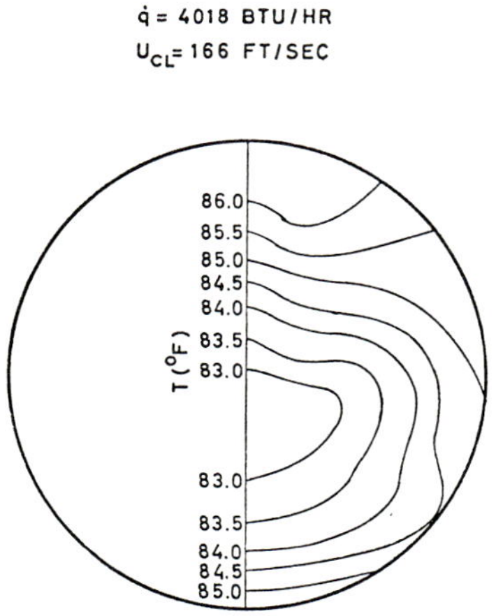

Fig. 4. Isotherm Map Run 20

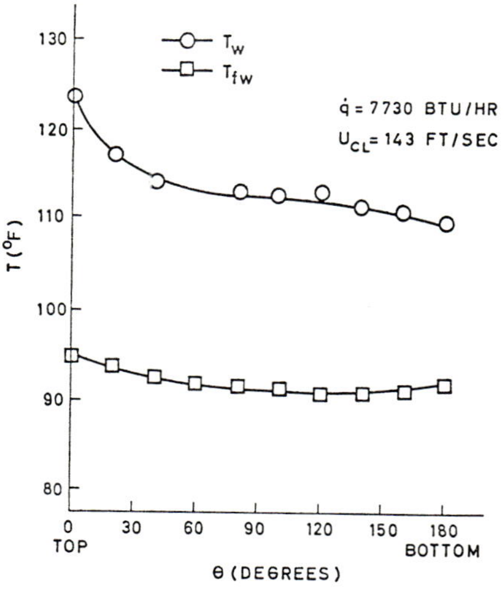

Fig. 5. Wall Temperatures Run 15

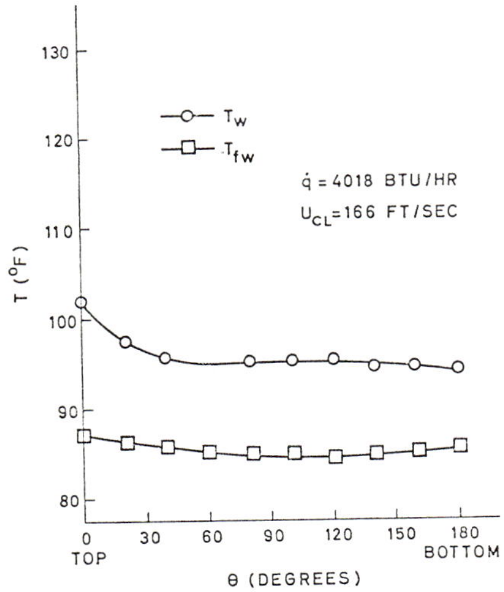

Fig. 6. Wall Temperatures Run 20

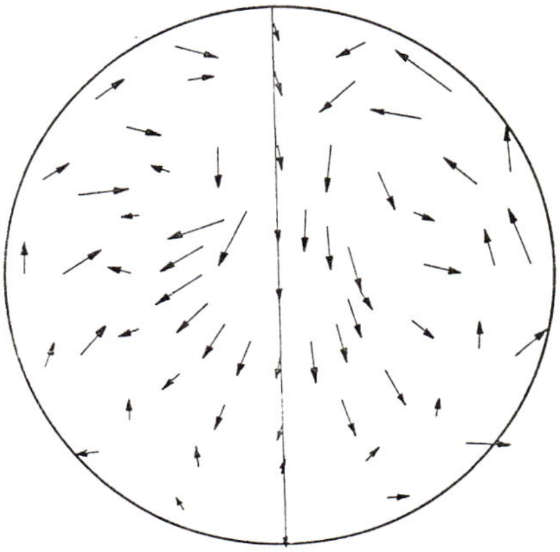

Fig. 7. Secondary Velocity Field

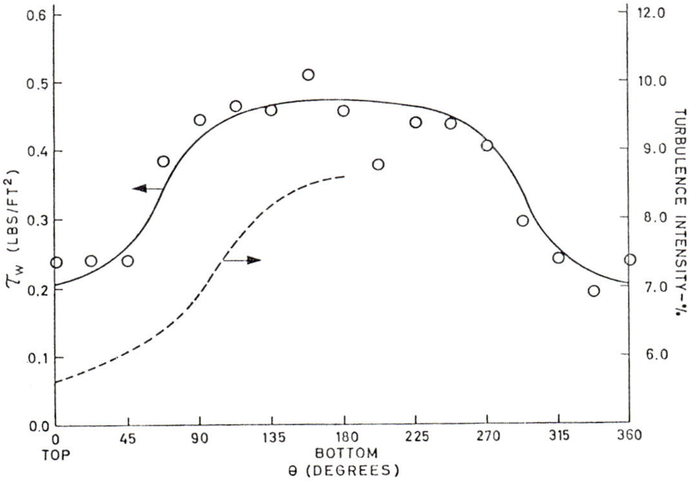

Fig. 8. Turbulent Axial Wall Shear Stress and Turbulence Intensity

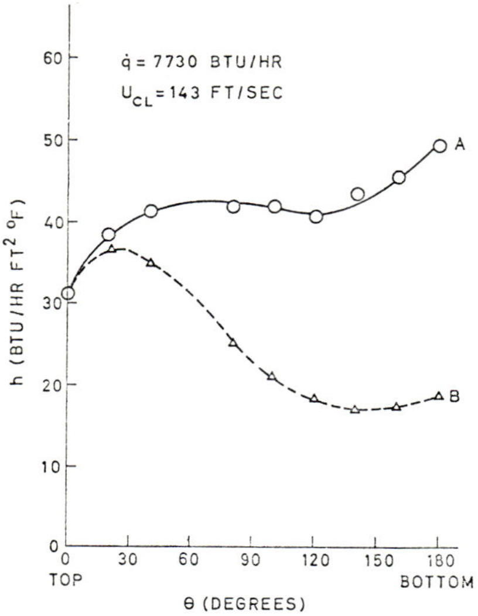

Fig. 9. Heat Transfer Coefficients Run 15

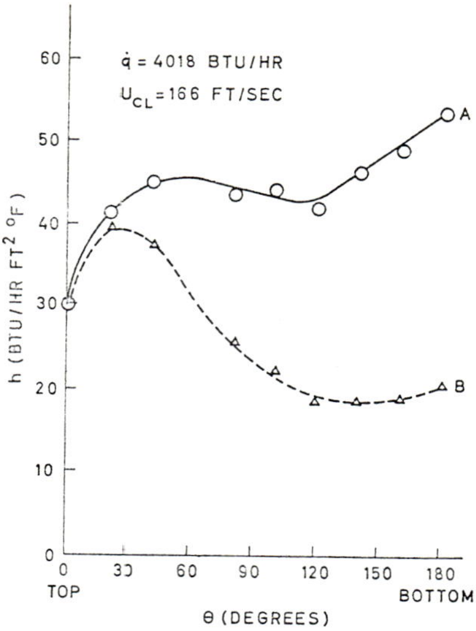

Fig. 10. Heat Transfer Coefficients Run 20

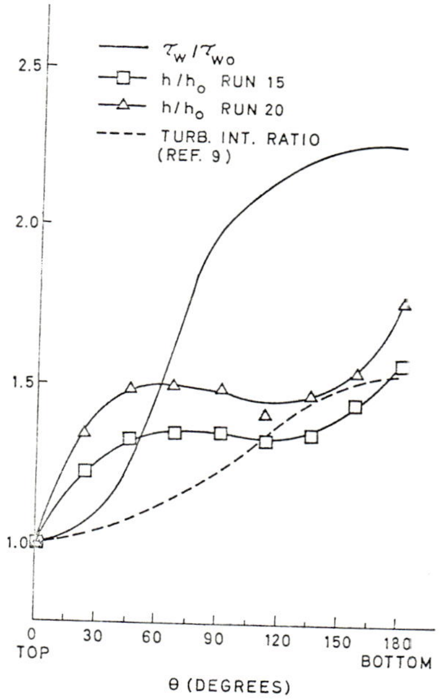

Fig. 11. Normalized Stresses, Heat Transfer Coefficients, and Turbulence Intensities

Developments in Mechanics, Vol. 6. Proceedings of the 12th Midwestern Mechanics Conference: 18

Flow through a Semicircular Pipe by Three-dimensional Flow Birefringence Method

W. J. McAFEE and H. PIH
THE UNIVERSITY OF TENNESSEE

ABSTRACT

The recently developed scattered light method for three-dimensional flow birefringence has been applied to investigate the problem of flow through a pipe with semicircular cross section. The flow rate and temperature were chosen so that the fluid remained in the Newtonian region. The aqueous colloidal suspensions of Milling Yellow, an organic dye, was used as the fluid in this investigation. A special scattered light polariscope designed for three-dimensional flow birefringence was utilized in this research.

The rheological properties of the fluid used were determined over a temperature range to insure that the flow remains in the Newtonian region. The relative birefringent fringe orders in two orthogonal planes were determined as functions of the distance along the light beam from chosen reference lines. Spacial derivatives were obtained on points of a square grid system on the cross section of the semicircular pipe. The shear strain rates were calculated using previously developed flow-optic relationships and calibration information. For convenience of comparison with the existing theoretical solution, the shear strain rates were transformed to a polar coordinate grid system. The velocity profile was obtained by numerical integration of the shear strain rate field. The experimentally determined velocity and shear strain rates were compared with theoretically calculated values along three different radii. Excellent agreement was obtained. This investigation provided further evidence of the validity of the developed flow-optic relations. It also indicated the applicability of the method to the general three-dimensional flow problem. It is felt that the scattered light flow birefringent method will provide a new powerful tool for three-dimensional fluid flow analysis.

INTRODUCTION

The application of flow birefringence to fluid dynamics has been limited to two-dimensional problems so far. Recently the scattered light method which has been used in photoelasticity was extended to flow birefringence for investigating three-dimensional fluid flow problems. The basic flow-optic relations were developed for three-dimensional scattered light birefringence [1].* In order to show that the three-dimensional scattered light birefringent method is not limited to flow problems with highly symmetrical cross sections such as those used in verifying the basic relations and to provide further supporting evidence for the validity of the flow-optic relationships, the flow through a pipe of semicircular cross section was analyzed using this method. This problem was chosen to represent a general cross section for it is asymmetric. An analytical solution of this problem is available. This paper presents the analysis of this flow problem and the comparison of the experimental results with theoretically calculated values along three different radial lines.

EQUIPMENT AND PROCEDURE

The Special Scattered Light Polariscope

The special scattered light polariscope designed and built for three-dimensional flow birefringence was utilized for this analysis. The detailed construction of this polariscope was reported elsewhere [2]. Only a brief description will be given here. Figure 1 shows a schematic of this polariscope and Fig. 2 is a photograph of the actual instrument. The polariscope consists of three major sections, the light source and related optics, the flow channel receptacle and immersion tank, and the observation optical system. The light source used is a 10 mw He-Ne gas laser which provides a small diameter polarized, collimated, monochromatic light beam of high intensity. A prism is used to direct the horizontal light beam to a vertical direction for convenient observations. A quarter wave plate is inserted following the prism to produce circular polarized light when needed. Provision is made to tilt the light beam for oblique incident observations. A Babinet-Soleil compensator can be inserted below the quarter wave plate for determining fractional fringe orders. The flow channel receptacle assembly consists of two reservoirs, one at each end, which serve as buffers to minimize any short term fluctuation in the flow. The test channel is connected to the reservoirs by flanges. Different channels with the desired geometry can be interchanged for different studies. At the intersections of the flow channel with the immersion tank two flexible bellows were provided for sealing the immersion fluid while allowing relative motions between the test channel and the tank. The arrangement is necessary for measurement in different areas in the cross section. The flow channel receptacle assembly is mounted on a cross-feed slide table to provide the movement required. The immersion fluid is used to approximately match the indices of refraction in order to make observations through curved walls and in oblique directions. The immersion tank is a circular cylinder made of plexiglass. The light beam normally passes vertically along the axis of the tank. Since observations are made along the direction normal to the light path the observation optical bench is mounted in the radial direction of the tank on a rotary indexing table. For a narrow light beam (1-2 mm.), the lens effect at the outer surface of the tank is negligible. The lens system and the recording camera or photoelectric cell used to take readings can be mounted on the optical bench on either side of the tank. The photoelectric cell is mounted on a vertical indexing slide for scanning the height of the test channel.

The Flow Channel

The flow channel used in this investigation was made of transparent plexiglass circular pipe. Figure 3 shows a general view of the circular pipe assembly, two flow reducers, and the semicircular insert used in the experiment. One flow reducer was used at each end of the circular pipe to provide a smooth flow transition from the square reservoirs to the circular pipe. The large end flanges were

*Number in brackets refers to bibliography

designed to mate with the reservoirs as shown in Fig. 2. The smaller inner flanges were joined to the flexible bellows on the immersion tank to serve as closures. The test section of the channel was made as a separate insert. Both the inner and outer surfaces of a plexiglass tube were machined and polished to eliminate the slight variation of thickness and the small eccentricity of the as-received pipe. The final dimensions of the circular insert were 12 in. long, 1.506 in. ID, and 1.630 in. OD. A semicircular insert was machined from transparent plexiglass to fit the inside diameter of this circular tube closely, thus yielding a semicircular flow cross section. The ends of this semicircular insert (Figure 3) were tapered toward the tube wall to make the flow transition smoother.

The Fluid Circulation System

The fluid circulation system used in this investigation is illustrated in the schematic shown in Fig. 4. A constant temperature bath with both heating and cooling capabilities was used to maintain the fluid temperature within ±0.1°C. To aid in the regulation of the fluid temperature the room temperature was controlled. A by-pass was also provided to reduce the heat generated due to pumping at low flow rates since the pump used had only a single speed. The fluid temperature during the test was recorded continuously with a strip-chart recorder using a thermocouple inserted in the inlet reservoir. The fluid flow rate was controlled by a bronze gate valve. The flow rate was measured using a glass rotameter with a mercury loaded plexiglas bob. The rotameter reading was calibrated versus flow rate and temperature.

The birefringent fluid used in this investigation was the aqueous colloidal suspensions of a commercial organic dye, Milling Yellow NGS, with a dye concentration of 1.84% by weight. This type of suspension has been used for two-dimensional flow birefringence studies by other investigators [3].

Rheological Measurement

Colloidal suspensions of Milling Yellow display both Newtonian and non-Newtonian characteristics. Hirsch [4] observed three discreet regions of viscosity behavior of this fluid. It is Newtonian up to shear strain rates of 3-10 $sec.^{-1}$. The transition value depends on dye concentration and temperature. The next region is non-Newtonian and extends up to shear strain rates of about 500 $sec.^{-1}$. Above this shear strain rate level the fluid behaves essentially Newtonian again. The Newtonian region at the lower shear strain rates was the one used in this study. The rheological measurements were performed to yield the viscosity and the shear strain rate limits for the fluid to remain Newtonian.

A falling head capillary viscometer was used to determine the rheological data of the fluid. The use of this instrument has been discussed in the literature [4, 5]. The method of analysis of the data is due to Maron, Krieger, and Sisko [6].

Flow Birefringent Measurements

The flow channel was positioned in the polariscope such that the flat side of the semicircle was parallel to the vertical laser beam (y-direction referring to Cartesian Coordinates). The flow was in the x-direction. The channel was moved horizontally in the z-direction by indexing the slide table in 0.050 in. increments. At each observation station the optical bench was rotated to align with the direction of maximum contrast of the interference fringes. The birefringence patterns were either photographed for later measurement or directly scanned with the photoelectric search unit mounted on the vertical sliding table. A typical fringe pattern is shown in Fig. 5. The optical bench was then rotated 45°, and data was again taken. This rotation was necessary to observe fringes that might be missed due to the rotation effect of the optic axes of birefringence. The photometer was used only in areas where fringe orders were determined by the aid of the compensator. In areas where both photometric and photographic methods were used to take the data the results agreed with each other very well. Flow rates were adjusted such that the maximum shear strain rates were within the Newtonian region of the fluid for all tests. The fluid temperature for this test was 24.80°C and the flow rate was 0.140 in.^{3}/sec.

Next, the flow channel was rotated 90° so that the flat side of the semicircle was normal to the light beam. The same flow rate-temperature combination was repeated, and data was again taken in a similar fashion.

ANALYSIS OF FLOW BIREFRINGENCE DATA

Data Reduction

For the data that was recorded photographically, each negative was placed on a back lighted tracing table, and a traveling microscope was used to scan the image to determine the relative fringe order versus distance. The spacial gradiants of the fringe order, or the slopes of the plots of fringe order versus distance along the light path (dn/ds), were determined at the nodal points of the square grid as shown in Fig. 6. The method used in determining (dn/ds) was a combination of plotting by hand and a curve fitting routine on an electronic computer. The hand plotting was used where data showed more scatter and curve fitting failed to give regular derivatives of the curve. The values of (dn/ds) in both the xy- and xz-planes were determined for the three-dimensional analysis of the problem.

Data Analysis

The three-dimensional flow-optic relations [1] used to determine the shear strain rates from the experimentally determined (dn/ds) are represented by the following equations:

$$\left(\frac{dn}{ds}\right)^2_{xz} = \left(C'_1\right)^2 \varepsilon_{xz}^2 + \left(C'_2\right)^2 \varepsilon_{xy}^2 , \qquad (1)$$

$$\left(\frac{dn}{ds}\right)^2_{xy} = \left(C'_1\right)^2 \varepsilon_{xy}^2 + \left(C'_2\right)^2 \varepsilon_{xz}^2 , \qquad (2)$$

where $(dn/dx)_{xz}$ and $(dn/ds)_{xy}$ are the spatial derivatives of the fringe orders when the light beam is in y- and z-direction respectively, ε_{xz} and ε_{xy} are the shear strain rate tensor components, and C'_1 and C'_2 are the flow-optic constants at the test temperature for the fluid used (9.91 and 2.40 respectively for this test). Solving equations (1) and (2) for the shear strain rates yeilds:

$$\varepsilon_{xz} = \left\{\frac{1}{\left(C'_1\right)^4 - \left(C'_2\right)^4}\left[\left(C'_1\right)^2\left(\frac{dn}{ds}\right)^2_{xz} - \left(C'_2\right)^2\left(\frac{dn}{dx}\right)^2_{xy}\right]\right\}^{1/2} , \qquad (3)$$

and

$$\varepsilon_{xy} = \left\{\frac{1}{\left(C'_1\right) - \left(C'_2\right)}\left[\left(C'_1\right)^2\left(\frac{dn}{dx}\right)^2_{xy} - \left(C'_2\right)^2\left(\frac{dn}{ds}\right)_{xz}\right]\right\}^{1/2} . \qquad (4)$$

These equations allow the calculation of the shear strain rates which are used to solve for the velocity field from the following equation:

$$\nabla^2 u - 2\frac{\partial^2 u}{\partial x^2} = 2\left[\frac{\partial \varepsilon_{xy}}{\partial y} + \frac{\partial \varepsilon_{xz}}{\partial z}\right] , \qquad (5)$$

where ∇^2 is the Laplacian operator and u is the velocity field of the flow. The velocity field is determined by a numerical integration of equation (5) on a digital computer.

In the problem presented in this paper it is more convenient to use polar coordinates. The shear strain rates were transformed into polar coordinate components, ε_{rx} and $\varepsilon_{\theta x}$, where (r, θ, x) are the coordinates. The velocity field can then be determined from the numerical integration of the following equation:

$$\nabla^2 u - 2\frac{\partial^2 u}{\partial x^2} = 2\left(\frac{1}{r}\frac{\partial \varepsilon_{\theta x}}{\partial \theta} + \frac{\varepsilon_{rx}}{r} + \frac{\partial \varepsilon_{rx}}{\partial r}\right) . \qquad (6)$$

RESULTS

The rheological properties of the Milling Yellow suspensions depend on the temperature. The variation of viscosity with temperature of the fluid used is given in Fig. 7. The maximum shear strain rate for the beginning of non-Newtonian behavior of the fluid versus temperature is also shown in Fig. 7.

The analytical solution of the problem of laminar flow through a semicircular pipe was first given by Greenhill [7] in the form of an infinite series. The series solution was evaluated with the aid of a digital computer. The shear strain rate tensor components and velocity profile along three radii, $\theta = 0°$, 40°, and 70°, were computed and compared with the experimentally determined values. Figure 8 shows the comparison for the $\theta = 0°$ radial line. This is the only line of symmetry in this cross section. Only one shear strain rate exists at points on this line. The comparison of the two shear strain rates and velocity profile for $\theta = 40°$ radial line is shown in Fig. 9. Similar comparison for the $\theta = 70°$ line is presented in Fig. 10.

DISCUSSION AND CONCLUSIONS

It can be observed in Figures 8, 9, and 10 that with the exception of a few points the experimentally determined shear strain rates and velocity profiles agree very well with the theoretically calculated values. Due to the reduced area of the cross section, the volumetric flow rates used for the semicircular channel were very low (0.073 - 0.140 in.^{3}/sec.). These low flow rates were difficult to maintain very accurately with the flow and control system used. It is felt that this was the major contribution to the error of the data. The experimental values correspond to a higher flow rate than was used for the theoretical calculations, but the deviation is within the probable error of the calibration for the rotameter at these low rates.

The results for the semicircular pipe problem are significant for two reasons. First, the flow cross section was asymmetric, thus, features of symmetry which had been used in earlier investigation [1] were not present. Within the limits previously stated, this case represents a general three-dimensional flow problem. Second, the method of taking data required that the flow system be shut down when the semicircular section was rotated 90°. This required reproducing both the temperature and the flow rate of the fluid in the system, a technique that would be necessary for the evaluation of any complex flow geometry. Some error was probably introduced by a slight mismatch of these parameters. The experimental results represent the practical limits of the method. The kind of agreement with theoretical results shows that the flow-optic relations and the calibration data for the fluid are applicable to a general three-dimensional flow problem. It is felt that the scattered light flow birefringent method used here will provide a new powerful tool for three-dimensional fluid flow analysis.

BIBLIOGRAPHY

[1] W. J. McAfee and H. Pih: Scattered Light Flow-Optic Relations in Three-Dimensional Flow Birefringence. Submitted to ASME Applied Mechanics Division to be published.

[2] W. J. McAfee and H. Pih: A Scattered-Light Polariscope for Three-Dimensional Birefringent Flow Studies. The Review of Scientific Instrument, Vol. 42, No. 2, p. 221-223 (1971).

[3] F. N. Peebles, H. J. Garber, and S. H. Jury: Preliminary Studies of Flow Birefringence Utilizing a Double Refracting Liquid. Proceedings of the Third Midwestern Conference on Fluid Mechanics, Minneapolis, The University of Minnesota Press, p. 441-454 (1953).

[4] A. E. Hirsch: The Flow of Non-Newtonian Fluid in a Diverging Duct, Experimental. Unpublished Ph.D. Dissertation, The University of Tennessee, (1964).

[5] E. Tejeira: Numerical and Experimental Investigation of Two-Dimensional Laminar Flow with Non-Regular Boundaries. Unpublished Ph.D. dissertation, The University of Tennessee, (1966).

[6] S. H. Maron, I. M. Krieger, and A. W. Sisko: A Capillary Viscometer with Continuously Varying Pressure Head. Journal of Applied Physics, Vol. 25, No. 3, p. 971-976 (1954).

[7] A. G. Greenhill: On the Flow of Viscous Liquid in a Pipe or Channel. Proceedings of London Mathematical Society, Vol. 13, p. 43-47 (1881).

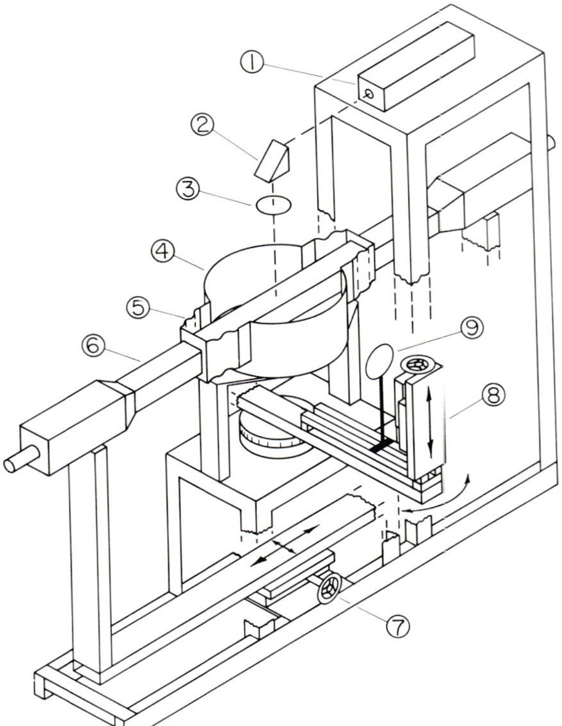

Figure 1. Schematic of the Special Scattered Light Polariscope. (1)--Laser, (2)--prism, (3)--quarter wave plate, (4)--immersion tank, (5)--flexible bellows, (6)--flow channel, (7)--cross feed slide table, (8)--observation optical bench with photomultiplier search unit, and (9)--lens.

Figure 2. Overall View of the Special Scattered Light Polariscope.

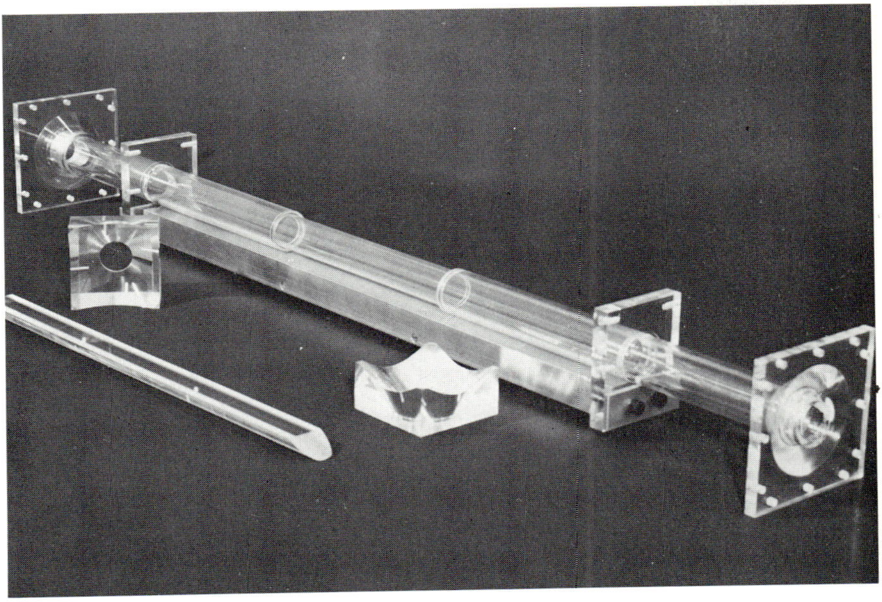

Figure 3. The Circular Flow Channel Assembly, the Flow Reducers and the Semicircular Insert.

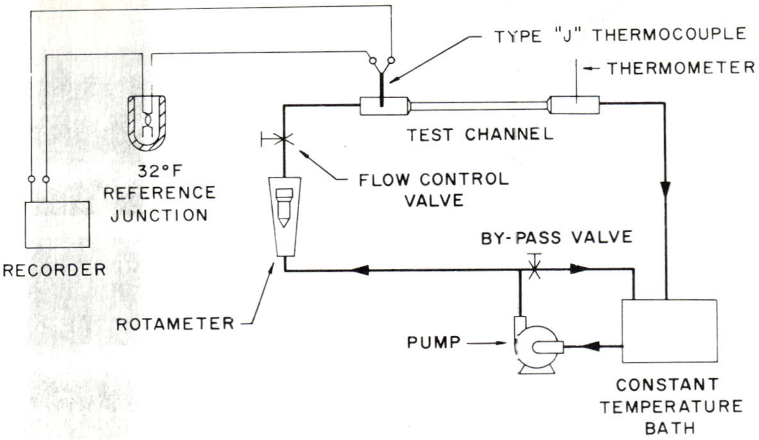

Figure 4. Schematic of the Fluid Circulation System.

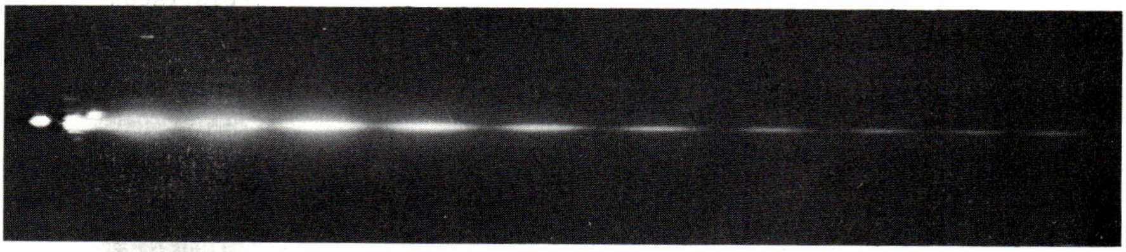

Figure 5. Typical Fringe Pattern for Semicircular Channel (Light Beam in Z-Direction)

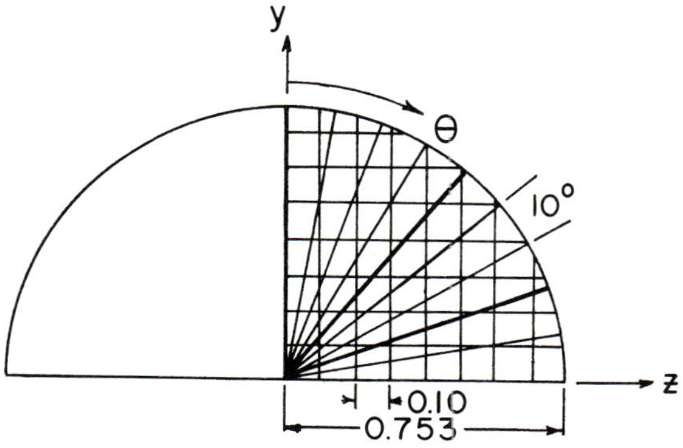

Figure 6. The Grid System in the Semicircular Cross Section.

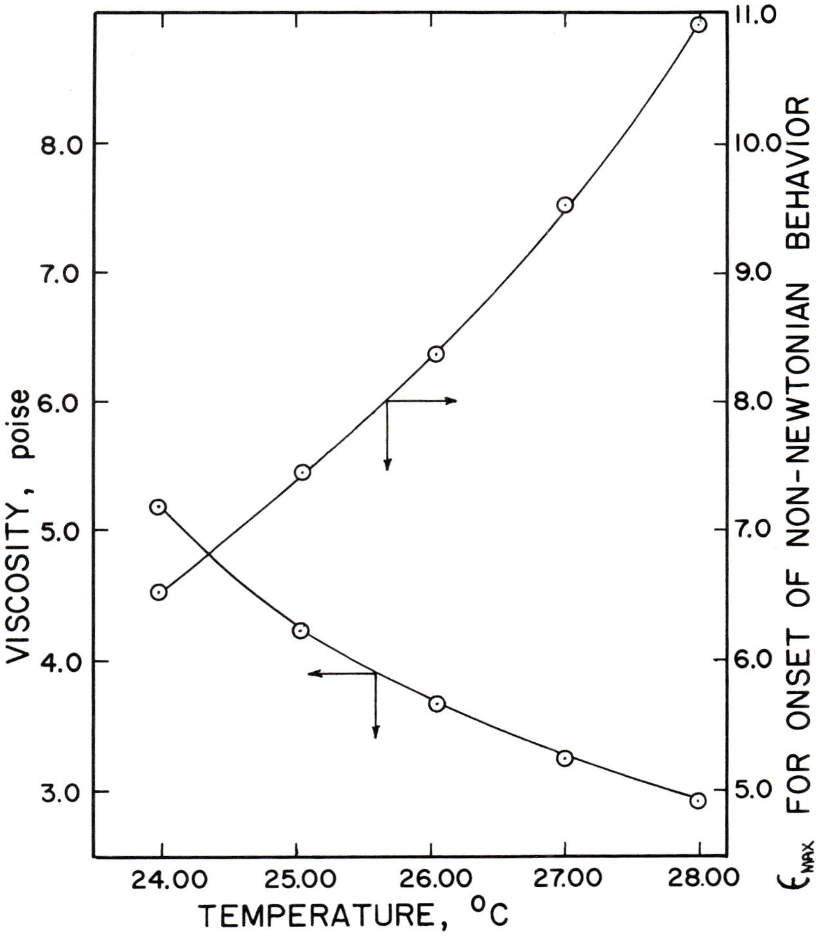

Figure 7. Variation of Rheological Properties with Temperature.

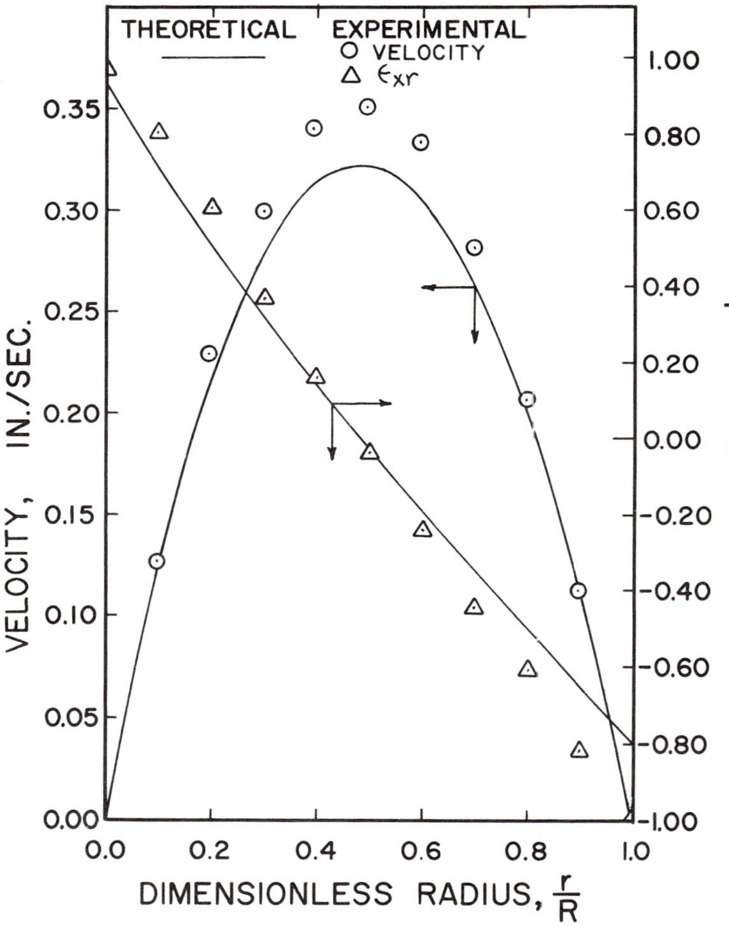

Figure 8. Comparison of Experimental Shear Strain Rates and Velocity with Theoretical Values for θ = 0° Radial Line.

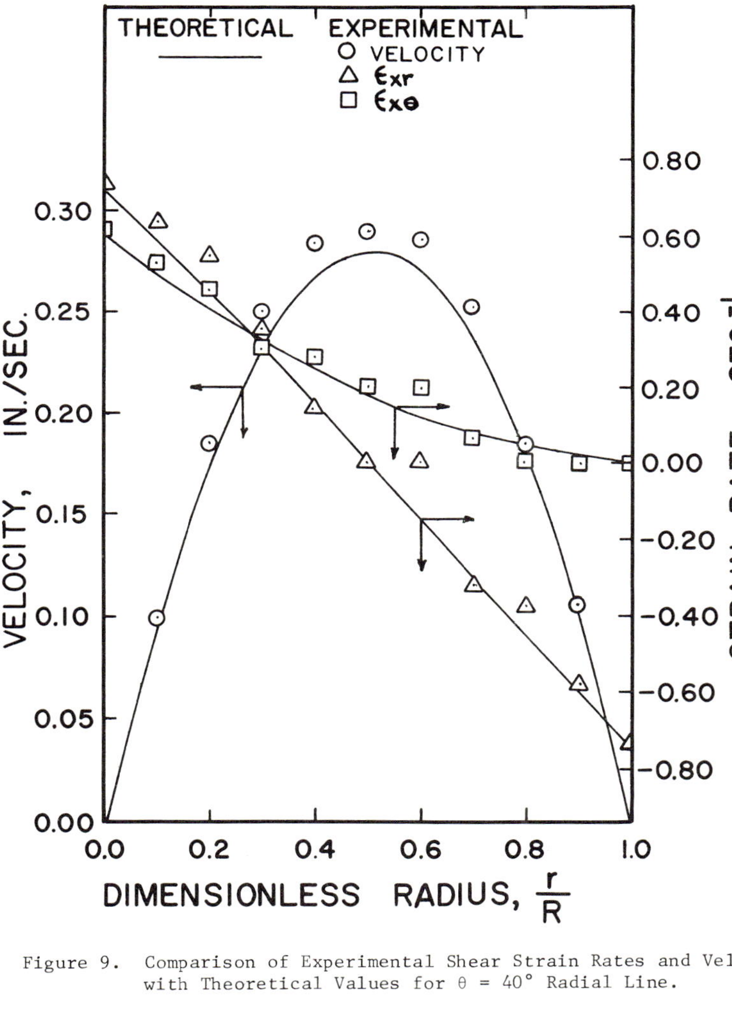

Figure 9. Comparison of Experimental Shear Strain Rates and Velocity with Theoretical Values for θ = 40° Radial Line.

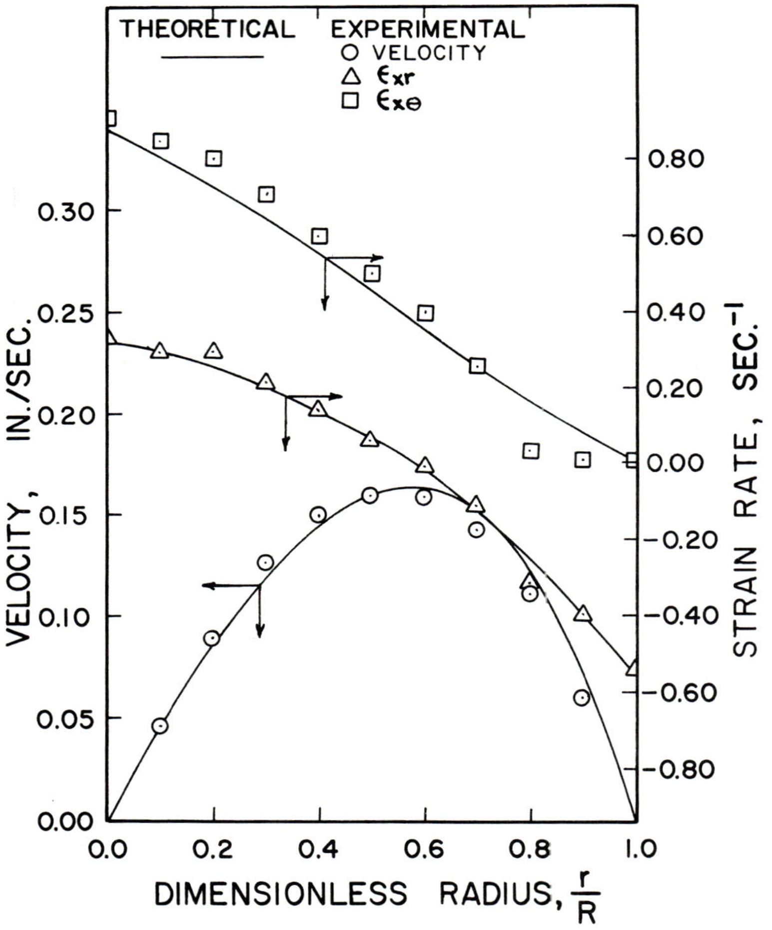

Figure 10. Comparison of Experimental Shear Strain Rates and Velocity with Theoretical Values for $\theta = 70°$ Radial Line.

TURBULENCE

Developments in Mechanics, Vol. 6. Proceedings of the 12th Midwestern Mechanics Conference: 19

Creation of a Pseudo-turbulent Velocity Field

GOODARZ AHMADI
PAHLAVI UNIVERSITY
VICTOR W. GOLDSCHMIDT
PURDUE UNIVERSITY and IMPERIAL COLLEGE

ABSTRACT

In attempting to define the motion of a particle suspended in a turbulent field a description of the fluid velocity field itself is first necessary. A random superposition of vortices is not sufficient unless the Navier-Stokes equation is satisfied.

In this work a turbulent field is numerically simulated. It is forced to satisfy the Navier-Stokes equation and stationarity. In order to reach stationarity energy is re-injected into the field at the same rate at which viscous effects dissipate it. The mode of energy injection influences the final state; simulating the manner in which turbulent energy is absorbed from the mean flow.

For simplicity the field treated is incompressible, homogeneous isotropic and two-dimensional. Two cases are treated. One in which the energy input is normally distributed over the wave number space and a second where energy is introduced only through the low wave number region.

INTRODUCTION

In earlier studies [1] a simulation of a turbulent field by the superposition of randomly distributed vortices was described. "Energy spectra" and "correlations" were obtained and the motion of neutrally buoyant particles was tracked kinematically.

The simulated field, although appearing like turbulence, had two basic limitations. First it was limited to be a two-dimensional field, and second it did not satisfy the Navier-Stokes equations. The extension to three-dimensions was suggested by [2], but is not considered at this time. Instead, the field will be forced to satisfy both the Navier-Stokes and energy conservation equations. Stationary conditions will be sought by re-injecting into the turbulent field an amount of energy equivalent to that dissipated by viscous effects. The assumption of two-dimensionality may be somewhat restrictive.

Such a field would lack the vortex stretching mechanism for energy transfer and may be quite unlike three-dimensional turbulence [3]. In [4] the created flow is used to solve the dynamic equations of motion of an individual particle with definite size and specific gravity, greatly extending the applicability of [1] and showing agreement with measured values. The field to be created will be limited, for simplicity, to be incompressible, homogeneous and isotropic and for computational expediency to be two-dimensional.

DESCRIPTION

An incompressible turbulent velocity field is in essence a random vector field whose time development is governed by the Navier-Stokes equations.

$$\frac{\partial \underline{u}_f}{\partial t} + \underline{u}_f \cdot \nabla \underline{u}_f = - \frac{1}{\rho_f} \nabla P + \frac{1}{Re_L} \nabla^2 \underline{u}_f \tag{1}$$

$$\nabla \cdot \underline{u}_f = 0 \tag{2}$$

(In the above all velocities and lengths are made dimensionless by a characteristic length and velocity scales L and U_o.) The purpose is to generate in a digital computer a continuous turbulent field. A method is suggested by Giorgini [5] for obtaining such a field. There the simplest case of one-dimensional turbulence (i.e., Burgers' equation), is considered; we will now extend that approach to a two-dimensional flow. (The procedure, although independently done [6], is somewhat similar to that followed by Lilly [7]).

Following [5] periodic boundary conditions are assumed over the length $2\pi L$, where L is assumed large when compared with the turbulent macroscales. The velocity vector may then be expanded as

$$\underline{u}_f(\underline{x},t) = \sum_{\underline{K}} \underline{u}(K,t) e^{i\underline{K}\cdot\underline{x}} \tag{3}$$

where $\underline{K}$ is a dimensionless wave vector (dimensionless in terms of L). Substituting Eq.(3) in (1) gives the Navier-Stokes equation in wave vector space, which for a two-dimensional field takes the following form [6]:

$$\frac{\partial u}{\partial t}(K_x,K_y,t) = - \frac{K_x^2 + K_y^2}{Re_L} u(K_x,K_y,t)$$

$$- i\{K_x \left[1 - \frac{K_x^2}{K_x^2 + K_y^2}\right] \sum_{K'_x} \sum_{K'_y} u(K'_x,K'_y,t)\, u(K_x - K'_x, K_y - K'_y, t)$$

$$+ K_y \left[1 - \frac{2K_x^2}{K_x^2 + K_y^2}\right] \sum_{K'_x} \sum_{K'_y} u(K'_x, K'_y, t)\, v(K_x - K'_x, K_y - K'_y, t)$$

$$+ K_y \left[\frac{-K_x K_y}{K_x^2 + K_y^2}\right] \sum_{K'_x} \sum_{K'_y} v(K'_x, K'_y, t)\, v(K_x - K'_x, K_y - K'_y, t)\}$$

$$\frac{\partial}{\partial t} v(K_x,K_y,t) = - \left[\frac{K_x^2 + K_y^2}{Re_L}\right] v(K_x,K_y t)$$

$$- i\{K_x \left[- \frac{K_x K_y}{K_x^2 + K_y^2}\right] \sum_{K'_x} \sum_{K'_y} u(K'_x, K'_y, t)\, u(K_x - K'_x, K_y - K'_y, t)$$

$$+ K_x \left[1 - \frac{2K_y^2}{K_x^2 + K_y^2} \right] \sum_{K_x'} \sum_{K_y'} u(K_x', K_y', t)\, v\,(K_x - K_x', K_y - K_y', t)$$

$$+ K_y \left[1 - \frac{K_y^2}{K_x^2 + K_y^2} \right] \sum_{K_x'} \sum_{K_y'} v(K_x', K_y', t)\, v\,(K_x - K_x', K_y - K_y', t) \qquad (4)$$

Similarly the continuity equation (2) in wave vector space becomes

$$u(K_x, K_y, t) \cdot K_x + v(K_x, K_y, t) \cdot K_y = 0 \,. \qquad (5)$$

Equations (4) and (5) are valid for any K_x, K_y. They imply that the rate of decrease of energy amplitude at a wave vector $\underline{K}$ is proportional to the square of $\underline{K}$ and is inversely proportional to the Reynolds number.

The nonlinear summation terms represent the interaction among the modes. Physically, it means that in order to know the time variation of a particular mode, the interaction of this mode with all other modes must be known.

The purpose is now to construct a solution to equations (4) and (5). For such, an integration will have to be performed.

NUMERICAL INTEGRATION

The set of equations (4) and (5) is an infinite set of equations corresponding to infinite possible values of K_x and K_y. In order to find a numerical solution two approximations are unavoidable.

1) The set must be truncated at some wave vector $\underline{N}$ in order to make the set finite.

2) The equations must be integrated by choosing a finite time increment.

Before attempting a solution it must be recognized whether or not the solution to the truncated equation is physically realizable. It is well known that the turbulent energy spectrum decays to zero very rapidly at large wave numbers, accompanied by energy transfers down the spectrum and dissipation at large wave numbers (cascade process). Therefore, the cut-off wave vector $\underline{N}$ with components $N_x=N_y=N$ may be selected large enough so that the energy transfer through it is negligibly small; hopefully permitting a physically realizable solution. The error resulting from this truncation is related to ε , where

$$\varepsilon = \frac{\text{energy corresponding to wave vectors} > \underline{N}}{\text{energy of wave vectors} \le \underline{N}} \,.$$

Using Benton's [8] solution to Burgers' equation, Giorgini [5] studied the wave number cut-off, $\underline{N}$, versus Reynolds number for different ε . Arbitrarily choosing $\varepsilon = 0.0\overline{1}$, hence discarding 1 percent of the total dissipation, Giorgini [5] found that for a one-dimensional model $N \le 1.1\ Re_L$. Using the same criteria for our two-dimensional case, and arbitrarily expecting the turbulent Reynolds number to be about 10, then

$$N_x = N_y = \pm 10 \qquad (6)$$

The infinite set of equations is limited then to a summation on 21 values of N_x and N_y, or 441 simultaneous differential equations which can be hopefully solved numerically. The value of 10 for the wave number cut-off is in agreement with that chosen by Bray and Batchelor (See [9]).

It is possible to extend this method to a three-dimensional space. However, the number of differential equations increases rapidly. For instance, with the same criteria as above, the number of differential equations for the three-dimensional case would become

$$(2N + 1)^3 = 9261$$

which presently can not be integrated within a reasonable computer time.

The selection of time increment depends on the accuracy required, and for the present case was selected as 0.002τ where the time scale

$$\tau = \frac{L}{U_o} .$$

A Runge-Kutta method was used to integrate the system of Equations (4). The details of the integration procedure as well as computer programs are given in [6].

ENERGY BALANCE

In the described model the energy will decay with time due to viscous dissipation. In order to simulate a stationary turbulent field some mechanism must be considered to retrieve the energy into the flow. We would like this mechanism to be such that the expected value of the energy remains constant. We hope that with this addition of energy, which compensates the loss through viscosity, the flow would be forced to be stationary after a given time T. Of course, this conjecture must be checked by at least studying the stationarity of the correlation and energy spectrum.

The mechanism which is used in the present study is a "random mixer in wave vector space" which contributes a specific amount of energy to each mode at every time step. It does this in such a way that the expected value of the total energy remains constant.

The energy contribution mode must satisfy the following conditions:
1) Reality of the velocity field
2) Continuity equation

The first condition implies that the variation in the conjugate modes $\underline{u}(\underline{K},t)$ and $\underline{u}(-\underline{K},t)$ must be the same. The second condition imposes a restriction on the pulsation of energy to the u and v components; only one of these contributions can be chosen independently.

The random mixer acts in wave vector space and varies the Fourier amplitudes $u(\underline{K})$ and $v(\underline{K})$ by an amount Δu and Δv respectively. The real and imaginary parts of Δu and Δv are normal random variables chosen in such a way that the expected value of the energy remains unchanged once stationarity is reached.

If the difference between the total initial energy and that at time t is given by ΔE, and as from continuity

$$\Delta v = -\frac{K_x}{K_y} \Delta u \tag{7}$$

then the variance [6]

$$\sigma_{\Delta u}^2 (K_x,K_y) = \frac{K_y^2}{K^2} \frac{\Delta E}{2(N^2 - 1)} . \tag{8}$$

Let $\Delta e(K_x,K_y)$ be the energy received by the mode (K_x,K_y) from the random mixer at time t, then

$$\Delta e(K_x,K_y) = |\Delta u(K_x,K_y)|^2 + |\Delta v(K_x,K_y)|^2 . \tag{9}$$

Combining Equations (7) and (9) one obtains

$$\Delta e(K_x,K_y) = \frac{K^2}{K_y^2} |\Delta u(K_x,K_y)|^2 . \tag{10}$$

It is interesting to note that the expected value of $\Delta e(K_x,Ky)$, using Equations (10) and (8) becomes

$$< \Delta e(K_x,K_y)> = \frac{\Delta E}{N^2 - 1} . \tag{11}$$

The expected value of the total energy then becomes

$$< \sum_{K_x} \sum_{K_y} \Delta e(K_x,K_y) > = \sum_{K_x} \sum_{K_y} \frac{\Delta E}{N^2 - 1} = \Delta E \tag{12}$$

a quantity unchanged with time once stationary conditions are reached.

There is one arbitrariness remaining in the random mixer. Even though the variances are fixed the distribution of either Δu and Δv is not. This means that the wave number distribution through which the energy is re-injected into the flow is arbitrary. In essence it is such a distribution which in real flows relates the turbulent dissipating region to the mean shear flow from which the energy comes. To simulate this the random mixer would have to be accordingly weighted in the low wave number region. Two cases are treated herein. In the first a normal distribution (limited to truncated wave space) is selected. (This is similar to the distribution selected for the "forcing function" of reference [7]). In the second case the random mixing is accomplished solely through random pulsations on the wave number region $|K| \leq 1$. (See Reference [6]). This second case might better simulate the manner in which turbulence gains energy from the mean flow as it concentrates in the low wave number region.

RESULTS AND CONCLUSIONS

At time t = 0 a random velocity field is arbitrarily selected whose only requirement is to satisfy the continuity equation (7). The time development of the field is governed by the forced Navier-Stokes equation generalized as

$$\frac{\partial u}{\partial t}(K_x,K_y,t) = \mu_u(u,v,) + \phi_u(K_x,K_y,t)$$

$$\frac{\partial v}{\partial t}(K_x,K_y,t) = \mu_v(u,v) + \phi_v(K_x,K_y,t) \qquad (13)$$

where μ_u and μ_v are nonlinear operators on the right hand sides of Equation (4); ϕ_u and ϕ_v are random forcing functions due to the random mixer in wave vector space.

The set of equations (7) was integrated for 300 steps until the stationarity of the field was confirmed by looking at the two-dimensional energy spectrum. As would be expected the results became insensitive to the initial field, although dependent on the random mixer. The statistical properties of the field were obtained assuming ergodicity in order to equate time with ensemble averaging.

It was also assumed that the field was isotropic and that any rotation of the velocity field in the wave vector space around the origin became also a realization of the ensemble.

In taking time averages seventy time steps were used. The two-dimensional energy spectrum was computed using the following equation: [6],

$$E(|\underline{K}|) = \sum_{|\underline{K}|-1<|\underline{K}|<|K|+1} \langle\{u^2(K_x', K_y') + v^2(K_x', K_y')\}\rangle . \qquad (14)$$

The Eulerian correlations and scales may be related to the energy spectrum. For a three-dimensional field such a relationship is well documented, for instance, in Chapter III of [10]. In our special case of a two-dimensional field, as shown in the appendix,

$$f(r) = \frac{1}{\overline{u^2}} \sum_{K=0}^{\infty} E(K) \frac{J_1(Kr)}{Kr} \qquad (15)$$

and

$$g(r) = \frac{1}{\overline{u^2}} \sum_{K=0}^{\infty} E(K) \left[J_o(Kr) - \frac{J_1(Kr)}{Kr} \right] \qquad (16)$$

The turbulence microscales are given by

$$\lambda_f = \frac{4}{\frac{1}{\overline{u^2}} \sum_{K=0}^{\infty} K^2 E(K)} \qquad (17)$$

and

$$\lambda_g = \frac{\lambda_f}{\sqrt{3}} \tag{18}$$

The macroscales are defined as

$$\Lambda_f = \int_0^\infty f(r')dr = \frac{1}{\overline{u^2}} \sum_{K=0}^{\infty} \frac{E(K)}{K} \tag{19}$$

and

$$\Lambda_{g_1} = \int_0^\infty g(r')dr' = 0 \tag{20}$$

$$\Lambda_{g_2} = \int_0^r \int_0^{r_1} g(r')dr' \tag{21}$$

where r_1 is such that $g(r') \geq 0$ for $r' \leq r_1$ and $g(r') < 0$ for $r' > r_1$.

RANDOM MIXER NORMALLY DISTRIBUTED

The first concern is to make certain that the computation scheme is carried along sufficiently in time. The easiest criterion is to compute the energy spectrum and correlations for different times after the field appears to be stationary. Figures 1 through 3 show the energy spectrum for time steps 0, 299, and 300. The stationarity is apparent. In addition Figures 4 through 9 show the corresponding space correlations computed with Equations (15) and (16).

The corresponding scales given by Equations (17) to (21) were computed in the apparently stationary range as:

longitudinal microscale:	λ_f = 0.3010 L
lateral microscale:	λ_g = 0.1738 L
longitudinal macroscale:	Λ_f = 0.4722 L
lateral macroscale:	Λ_{g_1} = 0
	Λ_{g_2} = 0.2352 L

The mean square of the velocity is

$$\overline{u^2} = 1.33\ U_o^2$$

where L and U_o are the length and velocity scales.

The integration of the Navier-Stokes equation was performed in the wave vector space and t. Therefore, the Fourier amplitudes of the velocity field are known at any time. The velocity field can be obtained by the inversion of Equation (3).

$$\underline{u}(x,y,t) = \sum_{K_x} \sum_{K_y} \underline{u}(K_x,K_y,t)e^{i(K_x \cdot x + K_y \cdot y)}$$

The summation may be simplified due to the reality of the velocity field:

$$\underline{u}(x,y,t) = \sum_{K_x=-N}^{N} \left[2 \sum_{K_y=1}^{N} \text{Re}\{\underline{u}(K_x,K_y,t)e^{i(K_x x + K_y y)}\}\right] \tag{22}$$

where Re{ } means the real part of the expression inside the bracket.

RANDOM MIXER THROUGH RANDOM PULSATIONS IN $K \leq 1$.

The same procedure was repeated for the second case. Details of the computer programming are in [6] except for the nature of the random mixer. The computed correlations were comparable to those of Figures 4 - 9 except that the corresponding scales were found to be:

$$\lambda_f = 0.7138\ L$$
$$\lambda_g = 0.4121\ L$$
$$\Lambda_f = 1.7124\ L$$
$$\Lambda_{g_2} = 0.92\ L; \quad \Lambda_{g_1} = 0$$

The intensities for this field, where again U_o is an arbitrary scale, were computed as

$$\overline{u^2} = \overline{v^2} = 1.536\ U_o^2$$

With the random mixer acting predominantly through the large scale region the expected relative orders of magnitude of the micro and macroscales result. The difference between the micro and macroscales was not as pronounced for the first case where the energy re-injection was simulated to occur normally distributed through all wave numbers. It is the second case which is probably more realistic. In effect it suggests an interesting method to infer aspects of the energy transfer mechanism.

ACKNOWLEDGMENTS

The authors are jointly indebted to Professor Aldo Giorgini for his criticisms and suggestions. The work was in part sponsored by the National Science Foundation, Grant GK2729. The second author acknowledges the Department of Aeronautics of The Imperial College of Science and Technology, London, and The Freeman Fund for support during the final stages of the work.

APPENDIX: KINEMATICS OF HOMOGENEOUS ISOTROPIC TWO DIMENSIONAL TURBULENCE

Consider a homogeneous turbulent flow field with the restriction that the component of the velocity in a particular direction (for simplicity this direction is assumed to be the z-direction) is zero at all times. Furthermore, it is assumed that the velocity field in the plane normal to the z-direction is isotropic.

Similar to the three-dimensional case there are only two independent scalars in the components of the two point velocity correlation tensor R_{ij}. From [9] the correlation tensor is

$$R_{ij} = \overline{u^2}\ \{[f(r) - g(r)]\ \frac{r_i\ r_j}{r^2} + g(r)\ \delta_{ij}\} \tag{A1}$$

Where in the present case i, and j can only be 1 or 2 due to the two-dimensionality of the field. Incompressibility implies

$$\frac{\partial R_{ij}}{\partial r_i} = \frac{\partial R_{ij}}{\partial r_j} = 0 \tag{A2}$$

Combining the above, and recognizing that now $\delta_{ii} = 2$, then

$$g(r) = f(r) + r\ \frac{\partial f}{\partial r}\ (rf)\ , \tag{A3}$$

slightly different from the corresponding three-dimensional case.

The two point velocity correlation is related to the power spectrum tensor E_{ij} in the following form [9]

$$R_{ij}(\underline{r}) = \sum_{\underline{K}} E_{ij}(K)e^{i\underline{K}\cdot\underline{r}}\ . \tag{A4}$$

Contracting the above and combining with

$$R_{ii} = \overline{u^2}\ [f(r) + g(r)] = \frac{\overline{u^2}}{r}\ \frac{\partial}{\partial r}\ (r^2 f) \tag{A5}$$

yields

$$\frac{\overline{u^2}}{r}\ \frac{\partial}{\partial r}\ (r^2 f) = \sum_{\underline{K}} E_{ii}(K)e^{i\underline{K}\cdot\underline{r}} \tag{A6}$$

The double summation on the right hand side of Equation (A6) in polar coordinates becomes

$$\frac{\overline{u^2}}{r}\frac{\partial}{\partial r}(r^2 f) = \sum_{K=0}^{\infty}\int_0^{2\pi} E_{ii}(K)\,\frac{KL}{2\pi}\,e^{iKr\cos\theta}\,d\theta \qquad (A7)$$

Where $\frac{K \cdot L \cdot d\theta}{2\pi}$ is the number of modes on an infinitesimal area with radius K. (L is the length scale).

Recalling the integral representation of the Bessel function equation (A7) becomes

$$\frac{\overline{u^2}}{r}\frac{\partial}{\partial r}(r^2 f) = \sum_{K=0}^{\infty} LK\,E_{ii}\,J_o(Kr)\,. \qquad (A8)$$

Multiplying both sides by r and integrating

$$\overline{u^2} f(r) = \sum_{K=0}^{\infty} LK\,E_{ii}\,\frac{J_1(Kr)}{Kr} \qquad (A9)$$

The two-dimensional energy spectrum can be defined as the total energy on the wave number K, or

$$E(K) = LK\,E_{ii} \qquad (A10)$$

Thus, the longitudinal correlation may be expressed in terms of the two-dimensional energy spectrum as

$$f(r) = \frac{1}{\overline{u^2}}\sum_{K=0}^{\infty} E(K)\,\frac{J_1(Kr)}{Kr}\,. \qquad (A11)$$

The lateral correlation g(r) can be easily obtained in terms of E(K). Substituting Equation (A11) in Equation (A3) then

$$g(r) = \frac{1}{\overline{u^2}}\sum_{K=0}^{\infty} E(K)\left[J_o(K_r) - \frac{J_1(Kr)}{Kr}\right]. \qquad (A12)$$

In order to compute the microscales of turbulence f(r) and g(r) are expanded near the origin. The power series expansion for the Bessel function of order n is

$$J_n(x) = \sum_{S=0}^{\infty}\frac{(-1)^S}{S!(n+S)!}\left(\frac{x}{2}\right)^{n+2S}\,. \qquad (A13)$$

Thus, f(r) for small r becomes

$$f(r) = 1 - \frac{1}{16}\sum_{K=0}^{\infty} K^2\frac{E(K)}{\overline{u^2}}\,r^2 + \cdots\cdot$$

and, the longitudinal microscale λ_f can be obtained as

$$\lambda_f = \frac{1}{\left[\frac{1}{16\overline{u^2}}\sum_{K=0}^{\infty} K^2 E(K)\right]^{\frac{1}{2}}}\,. \qquad (A14)$$

Using a power expansion of f(r) in (A3) yields

$$\lambda_g = \frac{\lambda_f}{\sqrt{3}} \qquad (A15)$$

Thus, the lateral microscale can also be obtained in terms of E(K) as

$$\lambda_g = \frac{1}{\left[\frac{3}{16\overline{u^2}}\sum_{K=0}^{\infty} K^2 E(K)\right]^{\frac{1}{2}}} \qquad (A16)$$

With the conventional definition for the longitudinal macroscale, and using Equation (A11) one obtains

$$\Lambda_f = \frac{1}{\overline{u^2}} \sum_{K=0}^{\infty} \frac{E(K)}{K} \qquad (A17)$$

Similarily the lateral macroscale of turbulence defined as

$$\Lambda_{g_1} = \int_0^{\infty} g(r)dr \qquad (A18)$$

simply becomes

$$\Lambda_{g_1} = 0 \qquad (A19)$$

Therefore the kinematic properties of a two-dimensional homogeneous isotropic turbulence are similar to those of the three-dimensional case differing within numerical factors and the interchange of Bessel functions with sine and cosine functions.

NOMENCLATURE

E	two-dimensional energy spectrum
E_{ij}	energy spectrum tensor
$f(r)$	longitudinal correlation
$g(r)$	lateral correlation
$\underline{K}$	wave vector
K	wave number equal to $\lvert K \rvert$
K_x	first component of wave vector
K_y	second component of wave vector
L	a length scale
N	wave vector cut off
N_x	first component of the wave vector cut off
N_y	second component of the wave vector cut off
p	pressure
r	radial distance
r_i	components of r
Re_L	Reynolds number based on U_o and L
R_{ij}	correlation tensor
u	first component of the velocity vector
$\underline{u}(\underline{K},t)$	velocity vector in wave vector space
$u(K_x,K_y)$	first component of the velocity amplitude
$\underline{u}_f(\underline{x},t)$	velocity vector in real space
U_o	a velocity scale
v	second component of the velocity vector
$v(K_x,K_y)$	second component of the velocity amplitude
x	coordinate axis
x_i	i-component of the coordinate
λ_f	longitudinal microscale
λ_g	lateral microscale
Λ_f	longitudinal macroscale
Λ_g	lateral macroscale

BIBLIOGRAPHICAL REFERENCES

[1] Ahmadi,G. and Goldschmidt,V.W., "Kinematic Computer Simulation of the Turbulent Dispersion of Neutrally Buoyant Particles", Developments in Mechanics, Vol.5, Proceedings of the 11th Midwestern Mechanics Conference, pp.201-214.(1969).

[2] Base,T.E. and Davies,P.O.A.L., "Computer Studies of Vortex Models to Represent Turbulent Fluid Flow", Aeronautical Research Council,Noise Research Committee A.R.C. 29072, No.519,(1967).

[3] Batchelor, G.K., The Theory of Homogeneous Turbulence , Cambridge University Press, Section 8.4, (1960).

[4] Ahmadi,G. and Goldschmidt,V.W., "Dynamic Simulation of the Turbulent Diffusion of Small Particles", Hydrotransport I, BHRA, Cranfield, Bedford, England, September 1-4, (1970).

[5] Giorgini,A., "A Numerical Experiment on a Turbulent Model", National Center for Atmospheric Research, pre-publication Review Number 422.

[6] Ahmadi,G., "Analytical Prediction of Turbulent Diffusion of Finite Size Particles", Ph.D. Thesis, School of Mechanical Engineering, Purdue University; also available as Technical Report FMTR-70-3, March (1970).

[7] Lilly,D.K., "Numerical Simulation of Two-Dimensional Turbulence", High Speed Computing in Fluid Dynamics, The Physics of Fluids Supplement II, Volume 12, Number 12, pp.240-249, December (1969).

[8] Benton,E.R., "Some New Exact,Viscous,Nonsteady Solutions of Burgers' Equation", Physics of Fluids, 9,5,1247-1248, (1966).

[9] Batchelor,G.K., "Computation of the Energy Spectrum in Homogeneous Two-Dimensional Turbulence", High-Speed Computing in Fluid Dynamics, The Physics of Fluids Supplement II, Volume 12, Number 12, pp.233-239, December (1969).

[10] Hinze,J.O.,Turbulence, An Introduction to its Mechanism and Theory, McGraw Hill Book Co., Inc.,(1959).

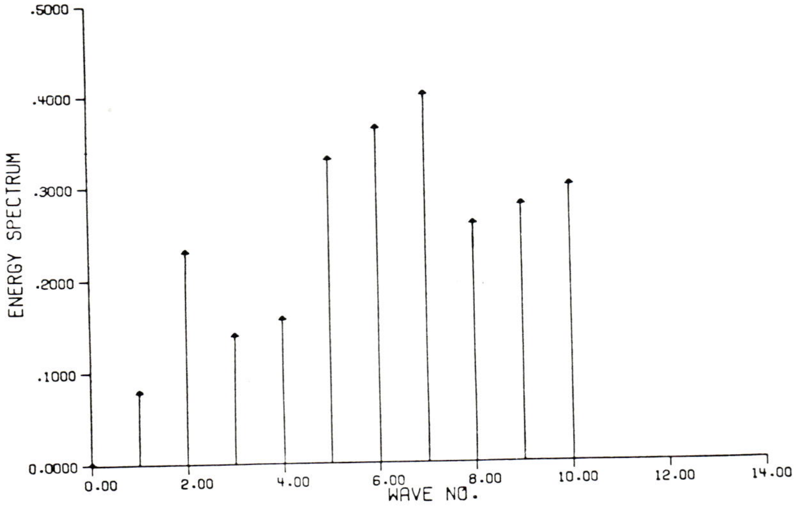

1. Energy Spectrum at Initial Step

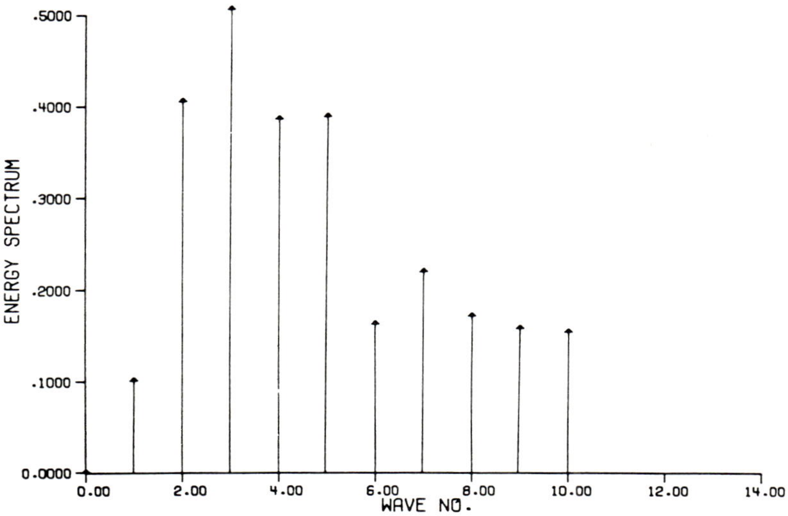

2. Energy Spectrum at Time Step 299

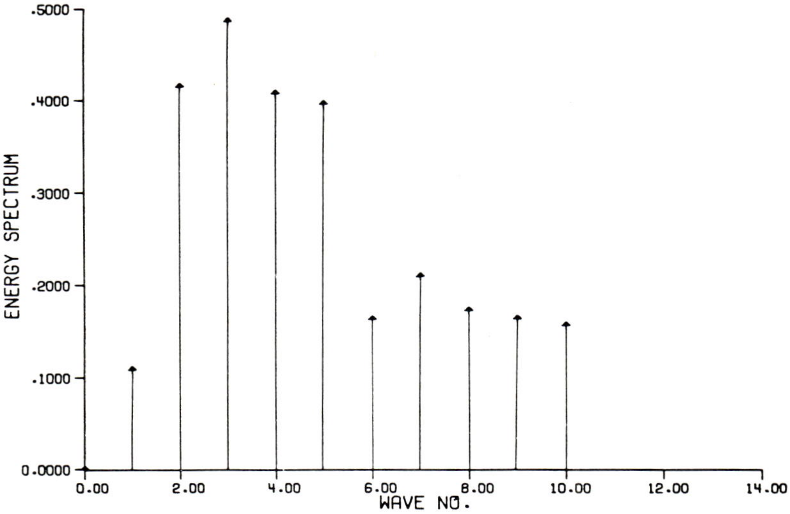

3. Energy Spectrum at Time Step 300

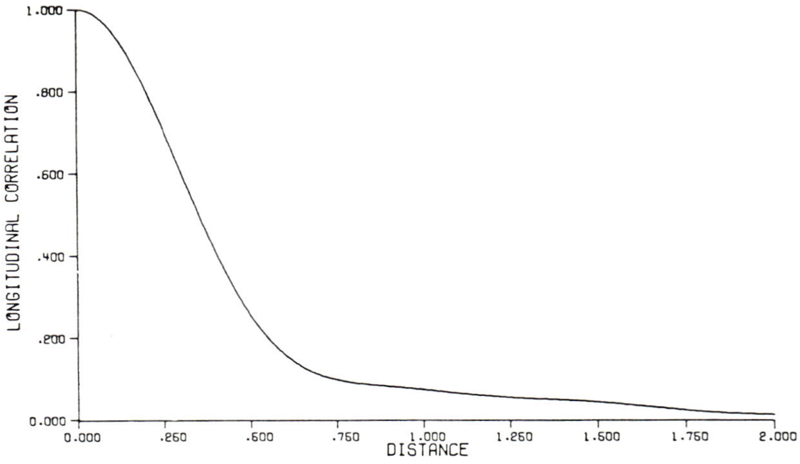

4. Longitudinal Correlation at Initial Step

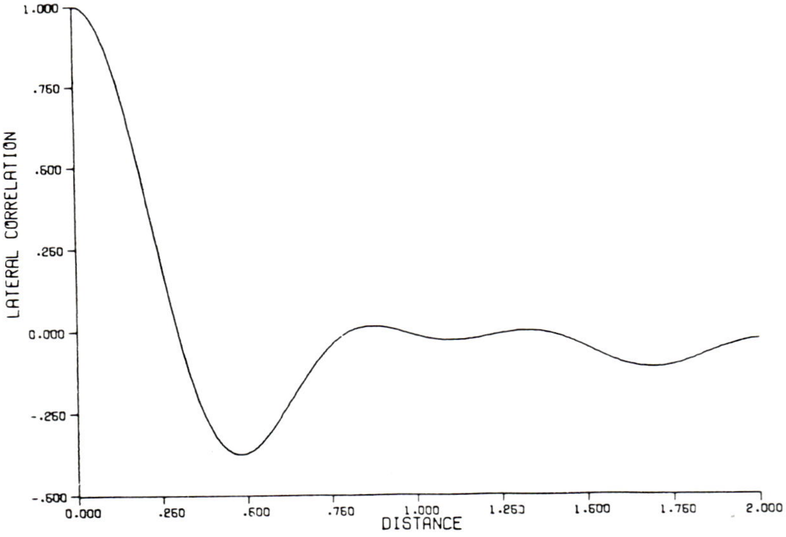

5. Lateral Correlation at Initial Step

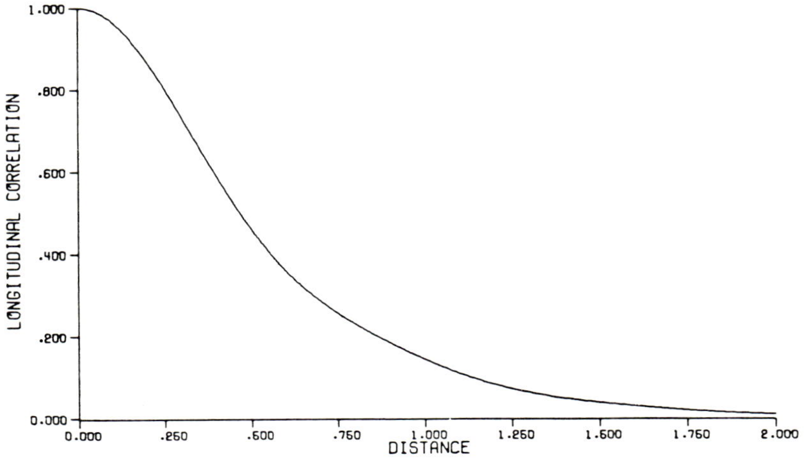

6. Longitudinal Correlation at Time Step 299

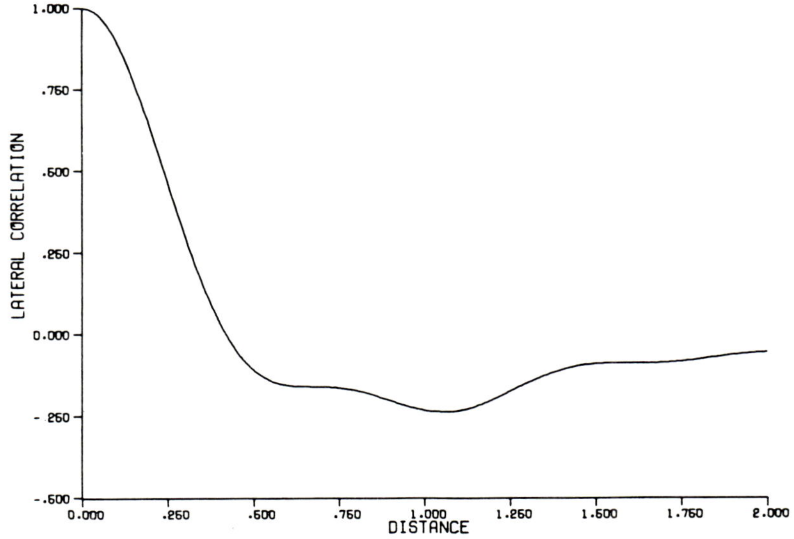

7. Lateral Correlation at Time Step 299

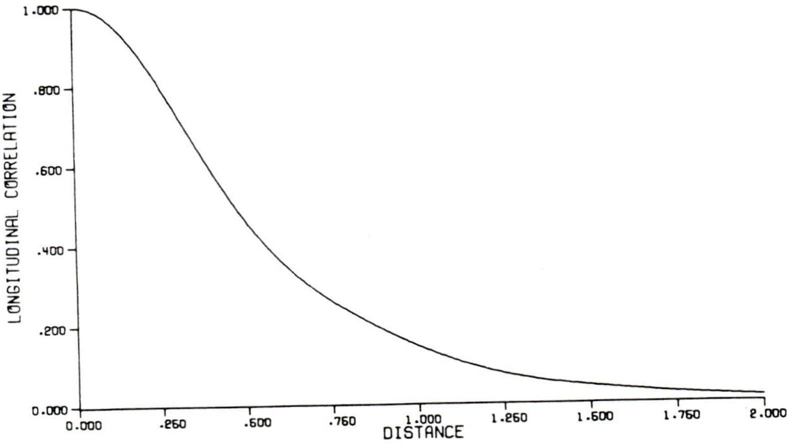

8. Longitudinal Correlation at Time Step 300

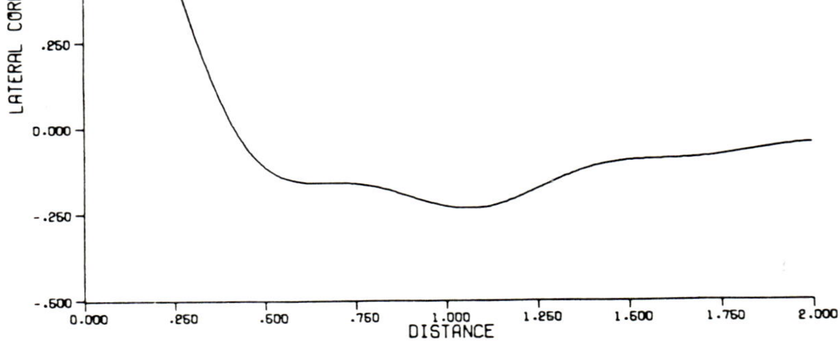

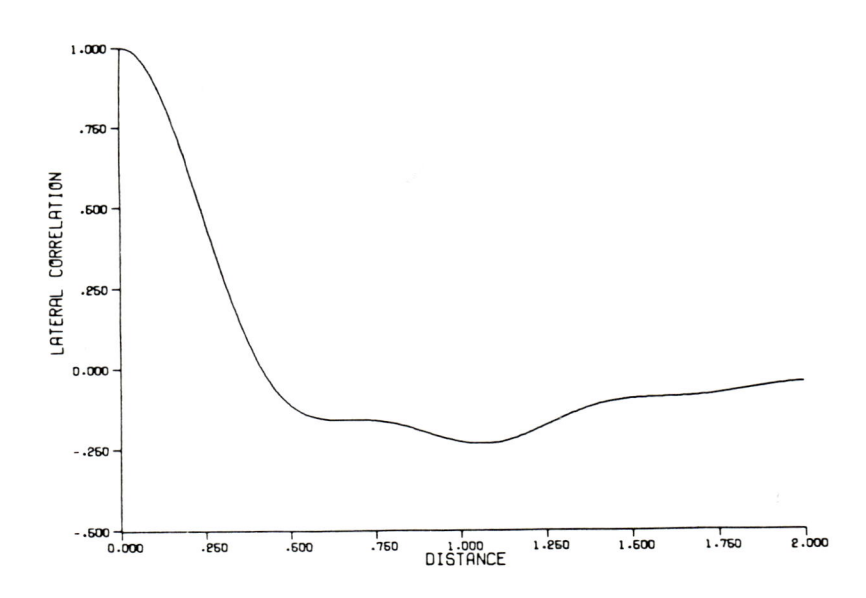

9. Lateral Correlation at Time Step 300

Developments in Mechanics, Vol. 6. Proceedings of the 12th Midwestern Mechanics Conference: 20

Influence of Surface Roughness and Mass Transfer on Boundary Layer and Friction Coefficient

EDWARD LUMSDAINE and HARVEY W. WEN
SOUTH DAKOTA STATE UNIVERSITY
FRANCIS K. KING
INTERNATIONAL BUSINESS MACHINE CORP.

ABSTRACT

This paper deals with the influence of surface roughness and mass transfer on the boundary layer and friction coefficient for turbulent flow over a flat plate. The roughness is introduced as a vortex generation factor in the sublayer following the work of van Driest, but the van Driest constant is modified using Rotta's data to account for suction and injection. The analysis first considers the region near the wall, with the effect of a pressure gradient being neglected and with the assumption of a constant injection or suction velocity. The resulting equation for the velocity profile is in the form of a first order ordinary differential equation which is easily solved by numerical integration and gives the boundary layer characteristics near the wall for varying degrees of surface roughness. With the finite difference method developed recently by Pletcher calculations are made which include mass transfer and roughness. Agreement between the present friction coefficient values for smooth plates and the experimental data by Mickley and Davis is quite good using the modified van Driest constant. Friction coefficients for plates with varying degrees of roughness and mass transfer conditions have been calculated and are presented graphically. The solution has been applied to partially rough plates; for no surface mass transfer it agrees quite well with data published in Schlichting's text.

INTRODUCTION

Papers dealing with turbulent boundary layers under the influence of surface mass transfer are numerous. The approach to the theoretical solution varies, but in general the semi-empirical theories for shear flow with suction and injection are based on the momentum equation coupled with Prandtl's mixing-length theory. A review of the existing theories on the turbulent boundary layer with suction or injection is given by Stevenson [1], and experimental results on flow with mass transfer have also been reported [1-7]. Recent years have seen substantial experimentation on flow over impervious rough materials, but relatively few have included mass transfer. To the authors' knowledge both theory and experimentation are completely lacking for flow over porous materials with mass transfer which includes the effect of roughness. In any application, the effect of roughness on the sublayer is always present and can be of equal magnitude as the effect of injection or suction. The materials used as porous linings are usually quite rough, as for example those used in transpiration-cooled turbine blades. In more recent applications, turbofan engines are being lined with acoustic materials in the inlet and bypass ducts [8]. These panels are rough and porous, and the acoustic pressure drives the air near the wall into the cavities, thus creating a fluctuating velocity at the surface.

The present investigation was undertaken in order to arrive at some idea of the wall-adjoining layer when the fluid is simultaneously affected by mass transfer and roughness and to obtain dimensionless profiles and friction coefficients which could be used under various roughness and surface mass transfer conditions. The first part of the analysis deals with the simpler problem of the influence of roughness and surface mass transfer on the boundary layer near the wall. This analysis follows the method used by Rotta [9] but includes a factor for roughness derived by van Driest [10]. Roughness is assumed to act as a vortex generator decreasing the effect of damping at the surface of the plate. Then, Pletcher's recently developed method of a stable explicit finite difference formulation [11] is used to compare with the solution obtained by the simpler analysis (the damping function used by Pletcher is modified and roughness is added to the mixing-length equation). The finite difference method can also be applied to problems with pressure gradients as well as variable surface mass transfer.

ANALYSIS

In the presence of a pressure gradient and with variable surface mass transfer the equation of steady motion for flow over a flat plate and the equation of continuity reduce to

$$u \frac{\partial u}{\partial x} + v \frac{\partial u}{\partial y} + \frac{1}{\rho} \frac{dP}{dx} = \frac{1}{\rho} \frac{\partial \tau}{\partial y} \tag{1}$$

$$\frac{\partial u}{\partial x} + \frac{\partial v}{\partial y} = 0 \tag{2}$$

where the appropriate boundary conditions are

$$\text{for } y = 0, \quad x > 0 : \quad u = 0, \quad v = V_W(x) \tag{3}$$

$$y \to \infty, \quad \text{for all } x : \quad u = U_\infty(x), \quad v = 0 \tag{4}$$

$$\text{for all } y, \quad x = 0 : \quad u = U_\infty \tag{5}$$

Usually, a certain value for the thickness σ_s of the viscous sublayer is assumed in the boundary layer; accordingly, for $y < \sigma_s$ there should be no turbulent motion of the fluid at all. Experiments have shown conclusively that this expectation of no turbulent motion in the sublayer is not true. Rather, they suggest that turbulent motion is damped out by the wall but still exists. Stokes [12] showed that for an infinite plate undergoing simple harmonic oscillations parallel to the plate in the infinite fluid, the amplitude of oscillating motion of the fluid diminishes from the wall as a function of the factor $\exp(-y/A)$, where A is a constant depending upon the frequency of oscillation of the plate and the kinematic viscosity of the fluid. The factor $1 - \exp(-y/A)$ was later introduced by van Driest [10] in the turbulent shear stress term and applied to the case in which the plate is fixed with the fluid oscillating relatively to the plate (i.e. turbulent fluid flow).

Solution Near the Wall

For the special case of zero pressure gradient and constant distribution of injection or suction along the plate, the derivatives of flow parameters with respect to x are so small that they can be neglected in the range very close to the wall, or

$$\frac{\partial}{\partial x} = \frac{d}{dx} = 0 \qquad (6)$$

The continuity equation (2) is then reduced to the form

$$\frac{dv}{dy} = 0 \qquad (7)$$

After integration of equation (7) and substitution in the boundary conditions (3), the result is

$$v = V_w = C_1 U_\infty \qquad (8)$$

where V_w is an arbitrary constant which is positive for injection and negative for suction. With the substitution of $v = V_w$ and $\partial/\partial x = 0$ into equation (1), the equation of motion becomes

$$V_w \frac{du}{dy} = \frac{1}{\rho}\frac{\partial \tau}{\partial y} \qquad (9)$$

By integrating equation (9) and using boundary conditions (3) one obtains

$$V_w u = \frac{1}{\rho}(\tau - \tau_w) \qquad (10)$$

where τ_w is the shear stress at the wall and is assumed constant along the plate. By rearranging equation (10), one gets

$$1 + V_{w\star} u_\star = \frac{\tau}{\tau_w} \qquad (11)$$

In an alternate form, this equation is

$$\frac{\tau}{\tau_w} = 1 + B\phi$$

where $B = V_{w\star} U_{\infty\star}$, $\phi = \frac{u_\star}{U_{\infty\star}}$

which is used by several authors (equation (2), [13]) and agrees well with experimental results for values of $0 \leq \phi \leq 0.5$ [13]. This is consistent with the assumption in this section of restricting the analysis to the region near the wall.

Prandtl's mixing-length theory gives the relation

$$-\rho\overline{u'v'} = \rho\chi^2 y^2 \left(\frac{du}{dy}\right)^2 \qquad (12)$$

(where χ is a universal constant and is approximately equal to 0.41.) since the total shear stress is defined as

$$\tau = \mu\frac{du}{dy} - \rho\overline{u'v'} \qquad (13)$$

For a rough wall van Driest suggests the formula

$$1 - \exp(-y/A) + \exp(-yD/AE) \qquad (14)$$

The last term in expression (14) is used by van Driest to counterbalance the vortex damping at the surface, thus considering roughness as a vortex generation factor. To include injection the roughness damping factor can be written in the form

$$RF = 1 - \exp\left(\frac{-y_* \sqrt{1+V_{w*}u_*}}{A_*}\right) + \exp\left(\frac{-y_* D_* \sqrt{1+V_{w*}u_*}}{A_* E_*}\right) \quad (15)$$

where, for $V_{w*} = 0$, this equation reduces to that given by van Driest for no mass transfer. Introducing equations (15) and (12) into (13) and dividing by τ_w one obtains

$$\frac{\tau}{\tau_w} = \frac{du_*}{dy_*} + \chi^2 y_*^2 \left[1 - \exp\left(\frac{-y_*\sqrt{1+V_{w*}u_*}}{A_*}\right) + \exp\left(\frac{-y_* D_* \sqrt{1+V_{w*}u_*}}{A_* E_*}\right)\right] \left(\frac{du_*}{dy_*}\right)^2 \quad (16)$$

When equations (11) and (16) are equated and simplified, the result is

$$\frac{du_*}{dy_*} = \frac{2(1 + V_{w*}u_*)}{1 + \sqrt{(1+V_{w*}u_*)}\sqrt{1+4\chi^2 y_*^2\left[1-\exp\left(\frac{-y_*\sqrt{1+V_{w*}u_*}}{A_*}\right)+\exp\left(\frac{-y_* D_*\sqrt{1+V_{w*}u_*}}{A_* E_*}\right)\right]^2}} \quad (17)$$

As the Reynolds' stress $\rho\overline{u'v'}$ is usually written as τ_t (turbulent shear stress), and since $\tau_\ell = \mu\, du/dy$ for the laminar shear stress, the total shear is $\tau = \tau_\ell + \tau_t$. With the eddy viscosity ε defined as $\tau_t = \varepsilon\, du/dy$, then

$$\tau_t = \tau - \tau_w\left(\frac{du_*}{dy_*}\right) \quad (18)$$

or

$$\frac{\tau_t}{\tau_w} = \frac{\tau}{\tau_w} - \left(\frac{du_*}{dy_*}\right) \quad (19)$$

By the relation from equation (11), (19) finally reduces to

$$\frac{\tau_t}{\tau_w} = (1 + V_{w*}u_*) - \left(\frac{du_*}{dy_*}\right) \quad (20)$$

Since $\tau = (\mu + \varepsilon)du/dy$, then by non-dimensionalizing and rearranging, one obtains

$$\frac{\varepsilon}{\mu} = \frac{\tau/\tau_w}{du_*/dy_*} - 1 \quad (21)$$

The substitution of equation (11) into (21) yields

$$\frac{\varepsilon}{\mu} = \frac{1 + V_{w*}u_*}{du_*/dy_*} - 1 \quad (22)$$

which is the relation between viscosity and eddy viscosity. By multiplying the numerator and denominator of the left-hand term of equation (22) by du/dy, one finds that

$$\frac{\tau_t}{\tau_\ell} = \frac{1 + V_{w*}u_*}{du_*/dy_*} - 1 \quad (23)$$

Equation (23) provides a relationship between turbulent shear stress and laminar shear stress throughout the boundary layer.

From the conservation of energy, Rotta [9] derived a relation for the sublayer:

$$\int_0^{\sigma_s} \tau_\ell \, \tau_t dy - \int_0^{\sigma_s} \rho \phi dy = 0 \tag{24}$$

or

$$\int_0^{\sigma_s} \tau_\ell \, \tau_t dy - \Phi = 0 \tag{25}$$

or

$$\int_0^{\sigma_{s\star}} \tau_\ell \, \tau_t dy_\star = \left(\frac{\tau_w}{\rho}\right)^{3/2} \int_0^{\sigma_{s\star}} \frac{d^2}{dy_\star^2} \left(\frac{u_\star^2}{2}\right) dy \tag{26}$$

In equation (25), the first term is the production of energy and the second term is the total energy dissipation in the sublayer. The energy production can be written as

$$\tau_\ell \tau_t = \mu \frac{du}{dy} \tau_t \quad \text{or} \quad \tau_\ell \tau_t = \frac{du_\star}{dy_\star} \frac{\tau_t}{\tau_w} \tag{27}$$

Equations (24) and (27) provide an understanding of how the energy dissipation acts in the sublayer.

There are two unknowns $D_\star$ and $A_\star$ which should be determined before solving equations (16) or (17). Driest [10] suggests that $D_\star = 60$ is a good approximation. From Rotta's data [9], $A_\star$ can be represented approximately by

$$A_\star = 26 - 25 \, V_{w\star} \quad \text{(for the injection case)} \tag{28}$$

and

$$A_\star = 26 - 200 \, V_{w\star} \quad \text{(for the suction case)} \tag{29}$$

Figure 1 shows a comparison between the present approximation using the experimental data of Rotta [9] and Cebeci's approximation using Simpson's data [6,14]. It can be seen that for suction the present linear approximation, the exponential approximation of Cebeci, Simpson's data and Rotta's data are all very close: however, for blowing the present method utilizing Rotta's data extends to quite large values, whereas Simpson's experimental data is limited to small values. For large values of blowing, the difference between the two approximations is quite significant.

The boundary conditions used to solve equation (17) are

$$y_\star = 0 \qquad u_\star = 0 \qquad \frac{du_\star}{dy_\star} = 1 \tag{30}$$

and

$$y_\star \to \infty \quad : \quad \frac{du_\star}{dy_\star} = 0 \tag{31}$$

The integration constant of equation (17) is taken care of by numerical integration procedures (for the computer program, see pp. 89-91, Reference 15).

Figure 2 shows the characteristic of the velocity profile for increasing roughness and several mass transfer rates. For values of $E_\star$ less than 60 the plate is considered partially rough, and for values larger than 60 ($E_\star > D_\star$) the plate is considered fully rough. It is seen that the roughness has a very significant effect on the shape of the profile. For very small values of $y_\star$ in the sublayer it is seen that mass transfer and roughness have very little effect on the value of $u_\star$. This is consistent with the fact that very near the wall the viscosity effect of the fluid dominates. The influence of roughness is quite pronounced outside this sublayer region.

This is brought out more clearly in Figure 3. For values of $y_\star$ less than two and for $E_\star = 0$ (smooth plate), the value of turbulent shear is zero for all values of injection and suction. However, for a rough plate with $E_\star = 60$ and for $y_\star$ less than two, the value of τ_t/τ_ℓ is less than one. Thus the turbulent shear is still less than the laminar shear although the viscous shear no longer dominates completely as in the case of a smooth plate. It is interesting to note that even under large values of injection there is a region in which laminar shear completely dominates, whereas under the action of roughness, which is likewise assumed to create turbulence in the region near the wall, the turbulent shear is no longer zero near the wall but is of equal value in magnitude to the laminar shear.

Figure 4 is a graph of equation (20) for the case of zero roughness and for a roughness of $E_\star = 60$. Roughness decreases the ratio of turbulent to wall shear for injection but increases it for suction. The numerical values for equation (27) are given in Figure 5. The comparison between the lines for rough and smooth plates shows that injection increases the energy production because it adds energy into the sublayer. Roughness decreases energy production because more energy dissipates to the wall, increasing the value of shear stress at the wall. In view of the conservation of energy, with equal shear stress at the wall, the thickness of the sublayer is reduced by injection or roughness but increased with suction. However, the total boundary layer thickness is increased by injection or roughness and decreased by suction.

Finite Difference Solution

To compare the results of the preceeding analysis and to provide a method where a pressure gradient and variable surface mass transfer can be included, the finite difference method developed by Pletcher [11] is used. From the mixing-length theory the turbulent shear can be written in the form

$$-\rho\overline{u'v'} = \rho\ell^2 \left|\frac{\partial u}{\partial y}\right| \frac{\partial u}{\partial y} \qquad (32)$$

Thus the momentum equation becomes

$$u_\star \frac{\partial u_\star}{\partial x_\star} + v_\star \frac{\partial u_\star}{\partial y_\star} = U_{\infty\star}(x) \frac{\partial U_{\infty\star}}{\partial x_\star} + \frac{\partial}{\partial y_\star}\left[\left(1 + \ell_\star^2 \left|\frac{\partial u_\star}{\partial y_\star}\right|\right)\frac{\partial u_\star}{\partial y_\star}\right] \qquad (33)$$

and the continuity equation is

$$\frac{\partial u_\star}{\partial x_\star} + \frac{\partial v_\star}{\partial y_\star} = 0 \qquad (34)$$

with boundary conditions

$$\left.\begin{array}{lll} \text{for } x_\star \geq 0,\ y_\star = 0 & : & u_\star = 0, \quad v_\star = V_{w\star}(x_\star) \\ \text{for all } x_\star\ \ y_\star \to \infty & : & u_\star = U_{\infty\star}(x_\star), \quad v_\star = 0 \end{array}\right\} \qquad (35)$$

Using the mixing-length distribution given by Pletcher (modified slightly) together with the introduction of the roughness factor $\exp(-y_\star D_\star/A_\star E_\star)$ the result is

$$\ell/\delta = 0.41\left[1 - \exp(-y_\star/A_\star) + \exp(-y_\star D_\star/A_\star E_\star)\right](y/\delta), \qquad y/\delta < 0.1 \qquad (36a)$$

$$\ell/\delta = 0.41[1 - \exp(-y_*/A_*) + \exp(-y_*D_*/A_*E_*)](y/\delta) - 1.53506(y/\delta - 0.1)^2$$
$$+ 2.75625(y/\delta - 0.1)^3 - 1.88425(y/\delta - 0.1)^4 \qquad 0.1 \leq y/\delta < 0.6 \qquad (36b)$$

$$\ell/\delta = 0.089 \qquad 0.6 \leq y/\delta \qquad (36c)$$

With these modifications to include roughness, Pletcher's finite difference method [11] is followed to obtain the solution to equations (33) and (34) which are then programmed to give numerical results shown in the following figures. In the presence of a pressure gradient, A_* as given by equations (28) and (29) should be further modified.

Figure 6 gives the friction coefficient curves for zero roughness and for $U_\infty E/\nu = 500$ (the definition of E here is the same as for k_s in Schlichting). The values obtained by the present method are slightly lower than those given by the Prandtl-Schlichting formula. However, this fact is consistent with the results of other theoretical methods and with experimental data (see [16] for a smooth plate).

Figure 7 is a comparison of the dimensionless velocity profile using equation (17) and the finite difference method for zero mass transfer. For small values of y_* (near the wall) the agreement is excellent. The agreement is quite good even for moderate values of y_*, thus indicates an upper limit for the accuracy and range of equation (17). Figure 8 compares the two methods for several degrees of roughness but no mass injection. A variation in surface mass transfer or roughness has a substantial influence on the velocity in the outer region. Near the wall this disturbance is damped out; however, away from the wall it promotes turbulence and thus changes the value of the velocity profile. Figure 9 is a comparison between the present finite difference solution and the solution presented by Smith and Cebeci [16]. The agreement is excellent.

Figures 10, 11 are graphs of friction coefficients for several values of roughness, injection and suction. For the case of the smooth plate with injection, comparisons with the Mickley and Davis data [2] show close agreement. The modification of the van Driest constant is slight for small values of mass transfer (see Figure 1): the value of $A_* = 26$ used by Pletcher to calculate the friction coefficients deviates only slightly from the present result and is thus not plotted. Figure 12 shows a comparison between the present calculation and data from several sources, for $V_w/U_\infty = 0.003$. Although the present method gives higher coefficients than those obtained by using $A_* = 26$, they are still lower than those given by the law-of-the-wall equation proposed by Stevenson. The corrections made by Stevenson to account for the pressure gradient in Mickley and Davis's data are also plotted. Since the present value of A_* was determined from flat-plate experimental data, no attempt has been made to make a correction for A_* in the presence of a pressure gradient. Figure 13 is a graph of friction coefficients for several values of roughness, with $V_w/U_\infty = 0.003$.

Limited experimental tests conducted at South Dakota State University with a small wind tunnel and porous panels used in aircraft linings showed that for suction the agreement between the present results and experimental data is quite good but for blowing the agreement is rather poor (see [15] for additional information). Because of the difficulty in taking accurate measurements near the wall in the case of mass injection, the discussion and results are too lengthy and inconclusive to be included here.

CONCLUSION

Results have been presented to show the influence of roughness on the boundary layer for flow over a flat plate with mass transfer. Friction coefficients and velocity profiles as well as sublayer parameters have been calculated for various degrees of roughness and mass transfer conditions. Comparisons of the present results with experimental data and previous analyses show close agreement.

ACKNOWLEDGMENT

The authors wish to thank South Dakota State University for supporting this work under Research Grant No. 5393 awarded to the senior author.

REFERENCES

1 Stevenson, T.N., "A Law of the Wall for Turbulent Boundary Layers With Suction or Injection," The College of Aeronautics, Cranfield, Co.A. Report No. 166, July 1963.

2 Mickley, H.S., Davis, R.S., "Momentum Transfer for Flow Over a Flat Plate With Blowing," NACA Technical Note 4017, 1957.

3 Rosenbaum, H., Margolis, D.P., "Pressure Fluctuations Beneath an Incompressible Turbulent Boundary Layer with Mass Addition," The Physics of Fluids, Volume 10, No. 6, June 1967, pp. 1231-1235.

4 Black, T.J., Sarnecki, A.J., "The Turbulent Boundary Layer with Suction or Injection," Aeronautical Research Council Report 20,501, London, 1958.

5 Favre, A., Dumas, R., Verollet, E., Coantic, M., "Turbulent Boundary Layer on Porous Wall with Blowing," Jo.Mec. 5, 1958, pp. 3-28.

6 Simpson, R.L., "The Turbulent Boundary Layer on a Porous Plate: An Experimental Study of the Fluid Dynamics with Injection and Suction," Ph.D. Thesis, Thermosciences Division, Mechanical Engineering Department, Stanford University, 1967.

7 Simpson, R.L., Moffet, R.J., Kays, W.M., "The Turbulent Boundary Layer on a Porous Plate: Experimental Skin Friction with Variable Injection and Suction," International Journal of Heat and Mass Transfer, Vol. 12, 1969, pp. 771-791.

8 "Progress of NASA Research Relating to Noise Alleviation of Large Subsonic Jet Aircraft," NASA SP-189, Conference held at Langley Research Center, Hampton, Virginia, Oct. 8-10, 1968.

9 Rotta, J.C., "On the Velocity Distribution with Turbulent Flow in the Vicinity of Porous Walls," Deutsche Luft- und Raumfahrt, 1966.

10 Van Driest, E.R., "On Turbulent Flow Near a Wall," Journal of Aeronautical Science, Vol. 23, 1956, pp. 1007-1011.

11 Pletcher, R.H., "On a Finite-Difference Solution for the Constant-Property Turbulent Boundary Layer," AIAA Journal, Vol. 7, No. 2, February 1968, pp. 305-311.

12 Stokes, G.G., "On the Effect of Internal Friction of Fluids on the Motion of Pendulums," Transactions Cambridge Philosophical Society, Vol. 9, 1951.

13 Wooldridge, C.E., Muzzy, R.J., "Boundary-Layer Turbulence Measurements with Mass Addition and Combustion," AIAA Journal, Vol. 4, No. 11, November 1966, pp. 2009-2016.

14 Cebeci, T., "Behavior of Turbulent Flow near a Porous Wall with Pressure Gradient," AIAA Journal, Vol. 8, No. 12, December 1970, pp. 2152-2156.

15 King, F.K., "Flow over a Rough Porous Flat Plate with Surface Mass Transfer," unpublished M.S. Thesis, South Dakota State University, 1968.

16 Smith, A.M.O., Cebeci, T., "Numerical Solution of the Turbulent Boundary Layer Equations," Report No. PAC 33735, McDonnell Douglas, May 29, 1967.

NOMENCLATURE

Symbol		Units
A	= van Driest constant	ft
C_f	= local friction coefficient	
D	= disturbance constant	ft
D_*	= dimensionless disturbance constant $\sqrt{\tau_w/\rho}\ D/\nu$	
E	= average roughness height	ft
E_*	= dimensionless average roughness height $\sqrt{\tau_w/\rho}\ E/\nu$	
ℓ	= mixing length	ft
R_x	= Reynolds number $U_\infty x/\nu$	
u	= velocity component parallel to wall	ft/sec
U_∞	= free stream velocity parallel to wall	ft/sec
u_*	= dimensionless velocity $u/\sqrt{\tau_w/\rho}$ (shear velocity)	
$U_{\infty}*$	= dimensionless free stream velocity $U_\infty/\sqrt{\tau_w/\rho}$	
u', v'	= fluctuation velocities of u and v	ft/sec
v	= velocity component perpendicular to wall	ft/sec
v_*	= dimensionless velocity $v/\sqrt{\tau_w/\rho}$	
V_w	= perpendicular velocity component at the wall	ft/sec
V_{w*}	= dimensionless velocity $V_w/\sqrt{\tau_w/\rho}$	
x	= coordinates along the wall	ft
x_*	= dimensionless distance $\sqrt{\tau_w/\rho}\ x/\nu$	
y	= coordinates perpendicular to the wall	ft
y_*	= dimensionless distance $\sqrt{\tau_w/\rho}\ y/\nu$	
Greek Letters		
β	= pressure gradient parameter $(\delta/\tau_w)dP/dx$	
δ	= displacement thickness $\int_0^\infty [1 - (u/U_\infty)]dy$	ft
δ_s	= thickness of the sublayer	ft
δ_{s*}	= dimensionless sublayer thickness $\delta_s/\sqrt{\tau_w/\rho}$	
θ	= momentum thickness $\int_0^\infty [(u/U_\infty)[1 - (u/U_\infty)]]dy$	ft
ε	= eddy viscosity	lb_f-sec/ft^2
χ	= universal constant from the mixing-length theory	
μ	= dynamic viscosity	lb_f-sec/ft^2
ν	= kinematic viscosity μ/ρ	ft^2/sec
ρ	= density	lb_m/ft^2
τ	= total shear stress	lb_f/ft^2
τ_t	= turbulent shear stress (Reynolds' stress)	lb_f/ft^2
τ_ℓ	= laminar shear stress	lb_f/ft^2
τ_w	= shear stress at the wall = $\mu(\partial u/\partial y)_{wall}$	lb_f/ft^2
ϕ	= energy dissipation	lb_f^2/ft^4
Φ	= total energy dissipation in the sublayer	lb_f^2/ft^3

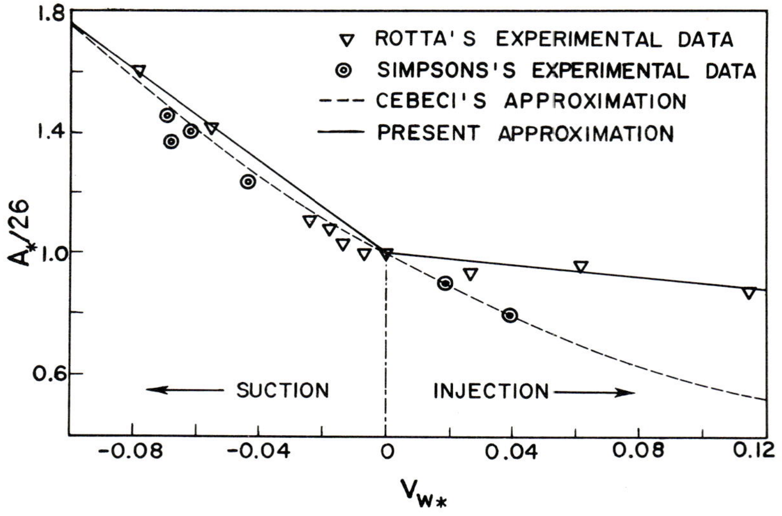

Fig. 1. Comparison of the damping constant for a porous flat plate.

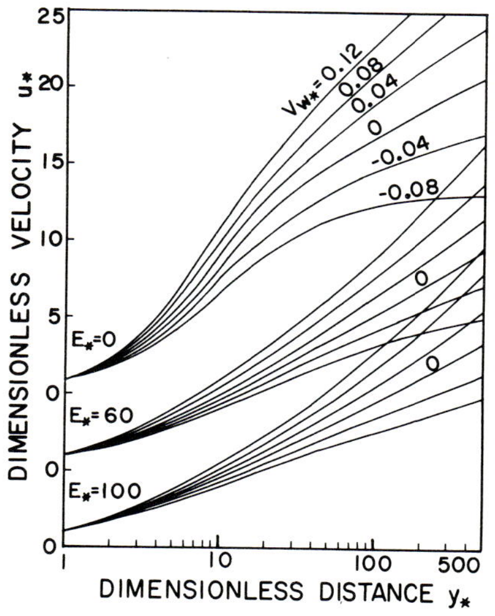

Fig. 2. Theoretical velocity distribution at various $V_{w\star}$ and $E_\star = 0, 60, 100$.

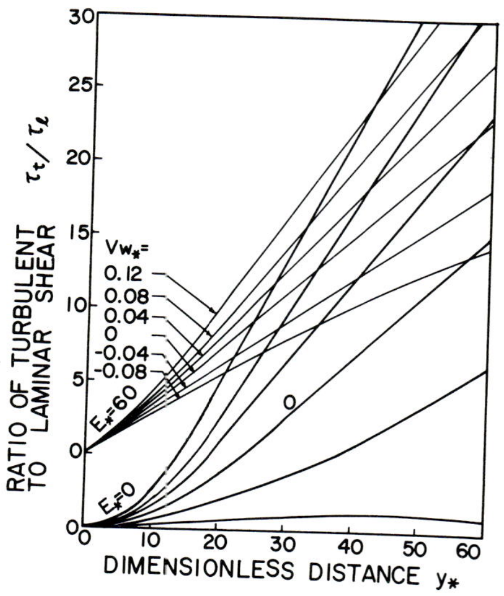

Fig. 3. τ_t/τ_ℓ versus $y_\star$ at $E_\star = 0$, 60 and various $V_{w\star}$.

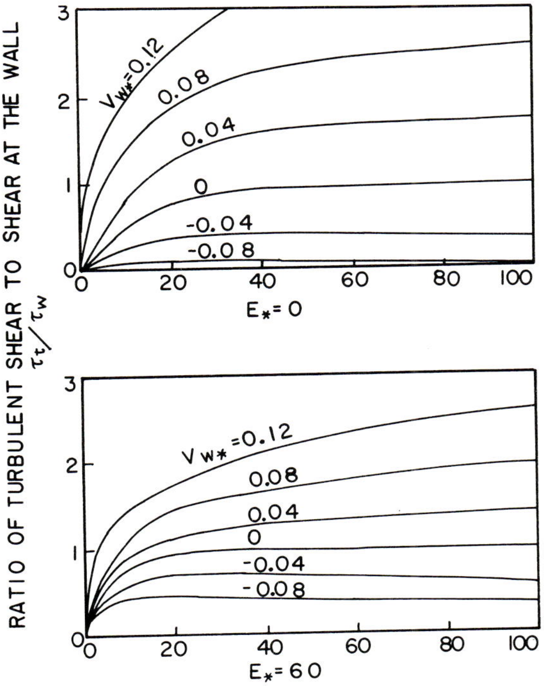

Fig. 4. τ_t/τ_w versus $E_\star = 0$, 60 for various $V_{w\star}$.

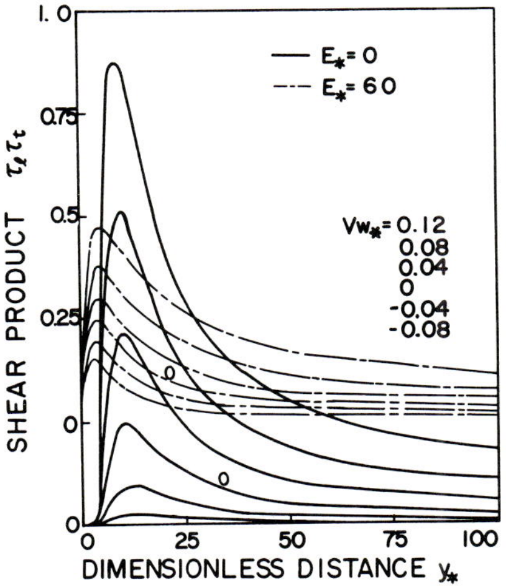

Fig. 5. $\tau_\ell \tau_t$ versus $y_\star$ at $E_\star = 0$, 60 for various $V_{w\star}$.

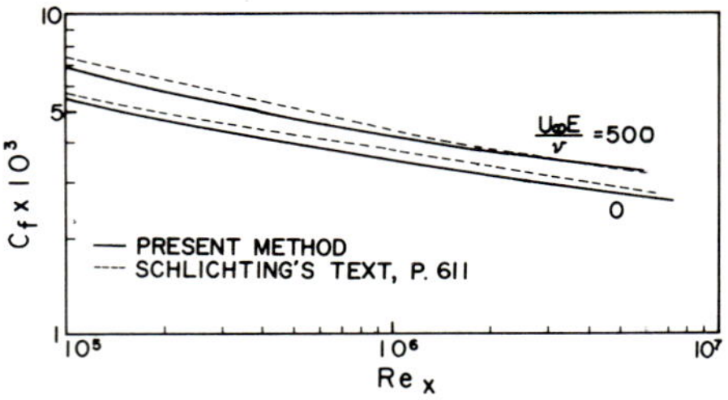

Fig. 6. Friction coefficient C_f versus Reynolds number.

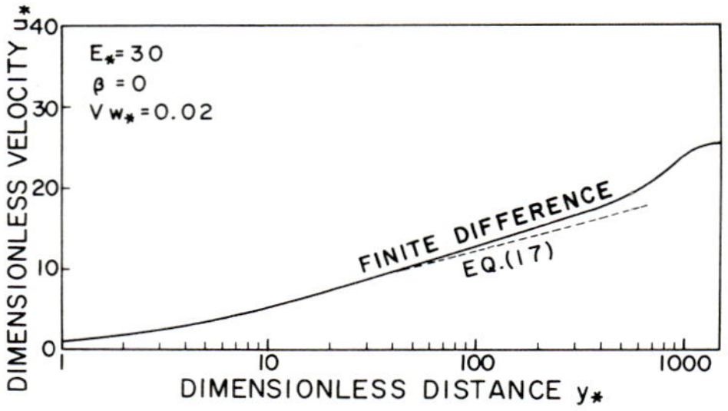

Fig. 7. Comparison of velocity profiles.

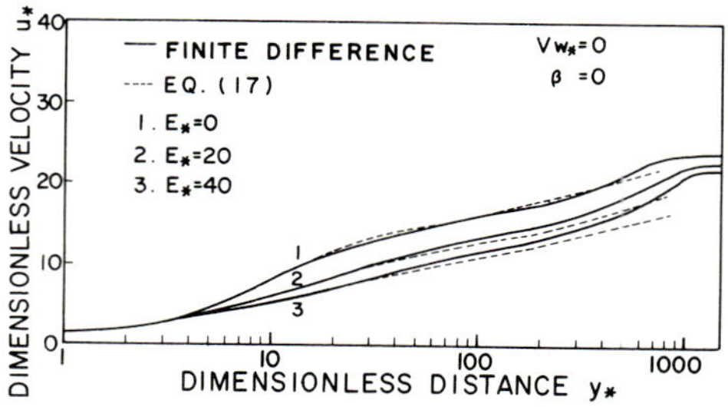

Fig. 8. Comparison of velocity profiles for various degrees of roughness.

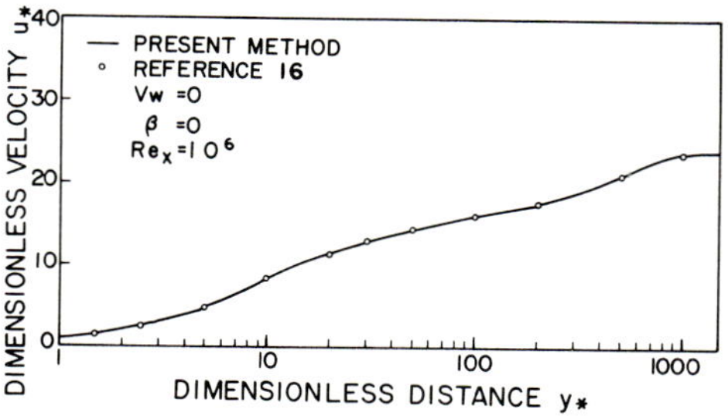

Fig. 9. Comparison between present finite difference solution and Smith's data.

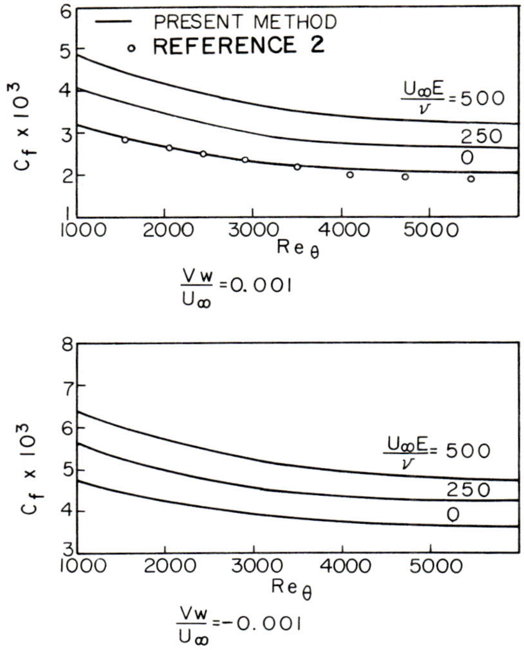

Fig. 10. Friction coefficients.

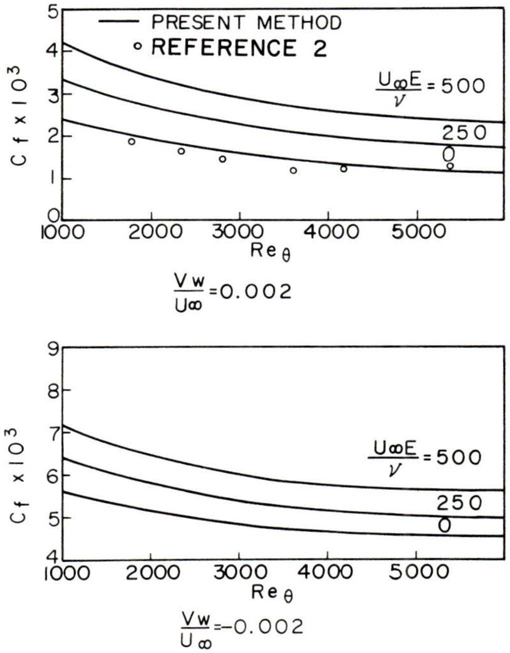

Fig. 11. Friction coefficients.

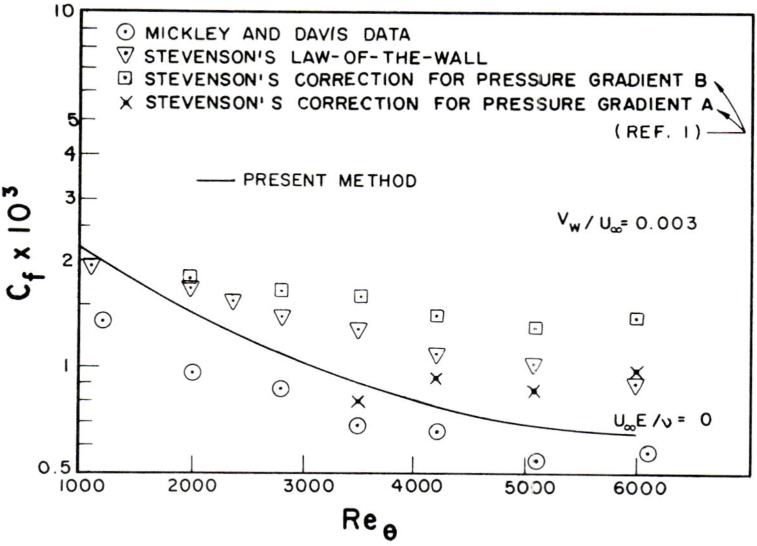

Fig. 12. Comparison of various friction coefficients with present method.

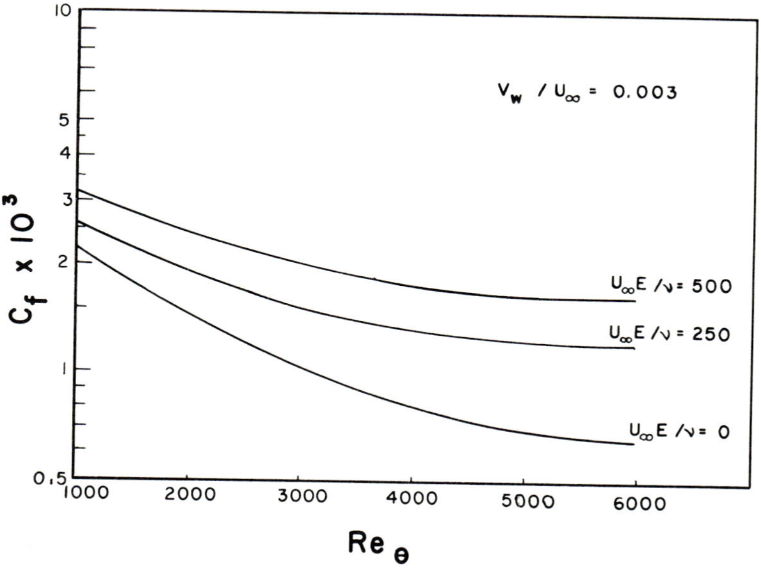

Fig. 13. Friction coefficient for several degrees of roughness.

Developments in Mechanics, Vol. 6. Proceedings of the 12th Midwestern Mechanics Conference: 21

Recompression of a Two-dimensional Supersonic Turbulent Free Shear Layer

WEN L. CHOW
UNIVERSITY OF ILLINOIS

ABSTRACT

A flow model has been devised to study the turbulent recompression process associated with a two-dimensional supersonic free shear layer. The flow field was divided into two subregions along the dividing streamline. The external supersonic free stream guides and interacts with the upper viscous layer. A velocity profile of third-degree polynomial was assumed for the upper viscous layer and the pressure difference across this layer was estimated from the normal momentum relationship. The lower viscous layer consisted of a forward flow with linear velocity profile and a back flow with a cosine profile. The difference in pressure across this layer was also accounted for. It was also pointed out that this analysis is equally applicable for cases with and without the downstream bounding wall. Conservation principles were subsequently applied to these regions and a system of ordinary differential and algebraic equations was obtained. In conjunction with the flow conditions prevailing at the end of the constant pressure jet mixing region, the system of equations may be integrated and solved numerically. For a given flow problem, the correct value of base pressure and the location along the wake boundary where recompression starts were established through iterations until the conditions at the rear stagnation point were satisfied. This procedure of calculations fully illustrated the typical elliptic behavior of all separated flow problems. Calculations for isoenergetic flow cases with thin initial boundary layers have been carried out and the results showed good agreement with the experimental data. It was also found from the results of calculations that the eddy diffusivity remained to be in the same order of magnitude as that of the constant pressure mixing region. With the suggested estimation for the eddy diffusivity within this recompression region, the turbulent normal stresses were also found to be one to two orders of magnitude smaller than the important flow quantities of such problems. These findings fully supported and justified the method of analysis suggested for this recompression process.

INTRODUCTION

For a flow passing a body with blunt trailing edge, the flow always separates away from the body ahead or at the base, creating a wake behind the body. The pressure within the wake, usually termed as the base pressure, is usually much lower than that of the free stream thus accounting for the large drag suffered by the body. This phenomenon exists irrespective of whether the problem is in the low speed or high speed flow regime. Since the advent of high speed flight, the base drag was recognized to be a serious problem, and a considerable amount of research has been directed into this area within the last two decades.

For a supersonic flow past a back step (see Fig. 1), it is recognized that the flow separates at the corner and the main flow will expand from the initial pressure to the lower base pressure. The initial boundary layer flow which is more or less guided by the free stream will also follow this expansion process. Along the early part of the wake boundary, the pressure is reasonably uniform, and for flows with large Reynolds number, a constant pressure turbulent jet mixing process occurs. Nearing the end of the wake, the main flow has to realign itself to the horizontal flow direction, initiating thereby a compressive process. As a result of this recompression, part of the fluid entrained within the viscous layer is turned back to form the recirculatory wake flow, while the rest will proceed downstream. Originally dealing with a simplified model for this flow problem, Korst [1] suggested an "escape criterion" associated with this recompression process. It specifies that the "dividing streamline" which separates the jet fluid from the wake fluid should assume such a mechanical energy level at the end of the jet mixing region that when it stagnates at the rear stagnation point through an isentropic (although irreversible-diabatic) process, its pressure is equal to the static pressure impressed behind the shock at the end of the wake. Employing this escape criterion would yield a unique base pressure solution for the problem. Experimental data have indicated, however, that the pressure at the rear stagnation point is much lower than what is impressed behind the shock. Nash [2], suggested empirical correlations of the two pressure levels. Page [3], Carriere and Sirieix [4], also developed empirical schemes individually to correlate characteristics associated with this recompression process.

Lees and his associates [5,6] considered this type of problem on an entirely different approach. Following the original idea suggested by Crocco and Lees [7], they treated the attached and separated viscous layers under one single framework, and showed that a critical point existed at the end of the wake (downstream of the rear stagnation point) which is somewhat similar to the nozzle throat. The correct base pressure associated with this type of flow assures a smooth flow passing through this critical point. Calculations for cases of turbulent flows have also been performed [8]. Extension to include the normal pressure gradient appeared recently [9]. Similar ideas were applied to laminar flow with axial-symmetric configurations by performing detailed numerical calculations for the outer flow [10]. However, the recompression flow process, particularly for turbulent flow, was never properly studied in detail. Other studies of wake flow problems are based on the method of integral relations [11], or complete numerical solutions of the Navier-Stokes equation [12]. These calculations require a considerable amount of time even with high speed computers and the turbulent flow cases have not been studied.

The present investigation was intended to consider this turbulent recompression process associated with a two-dimensional supersonic flow past a back step. A flow model which is equally applicable to laminar flows is presented. The establishment of the initial conditions for this recompression process is subsequently discussed. Numerical calculation procedures which illustrate the typical elliptic behavior of all separated flow problems are described. Finally, the results of calculations for flow cases with thin initial boundary layers and their comparison with the experimental data are presented and discussed.

Theoretical Flow Model

The present study of this recompression process is exclusively based on an integral approach. Referring to Fig. 2, the flow model guiding the present analysis can be described and discussed as follows:

1. The recompression region is split into two parts along the dividing streamline. The fluid above the dividing streamline will eventually proceed downstream, while the fluid below will be turned back to form the recirculatory wake flow as a result of recompression. The upper viscous layer interacts with the external inviscid stream; the latter guides and receives the influence of the former by following itself a Prandtl-Meyer compression relationship. This interaction is described by the fact that the transverse velocity component at the edge of the viscous layer induces an increase of pressure in the free stream which, in turn, influences the flow properties within the viscous layer, including this transverse velocity component at the edge of the viscous layer.

Integrating the continuity equation,

$$\frac{\partial(\rho u)}{\partial x} + \frac{\partial(\rho v)}{\partial y} = 0 \tag{1}$$

across the upper viscous layer, one obtains for the streamline angle at the edge of the viscous layer (see Fig. 3a)

$$\tan\beta_e = \frac{v_e}{u_e} = \frac{d}{dx}\left[\delta_a\int_0^1\left(1 - \frac{\rho}{\rho_e}\phi\right)d\zeta\right] - \frac{\delta_a\int_0^1 \frac{\rho}{\rho_e}\phi\, d\zeta}{(1 - c_e^2)^{1/(\gamma-1)}c_e}\frac{d}{dx}\left[(1 - c_e^2)^{1/(\gamma-1)}c_e\right] \tag{2}$$

where

$$\zeta = \frac{y}{\delta_a}, \quad \phi = \frac{u}{u_e}, \quad \text{and } c_e = \frac{V_e}{V_{max}} \doteq \frac{u_e}{V_{max}}$$

Since the free stream follows the Prandtl-Meyer relationship, the streamline angle β_e is related to the Prandtl-Meyer function ω by

$$\beta_e = \beta_\infty + \omega(c_\infty) - \omega(c_e) \tag{3}$$

where β_∞ is the difference of streamline angles between the dividing streamline and the free stream within the upstream constant pressure jet mixing region.

It is recognized, however, that the pressure difference across this upper viscous layer is not negligible. This difference can be calculated from

$$\frac{p_d}{p_e} = 1 + \frac{2\gamma}{\gamma - 1}\left\{\frac{c_e^2}{1 - c_e^2}\left(\tan^2\beta_e - \tan\beta_e\frac{d\delta_a}{dx}\right) + \frac{1}{(1 - c_e^2)^{\gamma/(\gamma-1)}} \cdot \frac{d}{dx}\left[(1 - c_e^2)^{1/(\gamma-1)}c_e^2\left(\delta_a\int_0^1\frac{\rho}{\rho_e}\phi^2\tan\beta\, d\zeta - \frac{\varepsilon}{u_e}\int_{\phi_d}^1\frac{\rho}{\rho_e}d\phi\right)\right]\right\} \tag{4}$$

with $\tan\beta = \tan\beta_e(2\zeta - \zeta^2)$, which is obtained by integrating the normal momentum equation,

$$\frac{\partial(\rho uv)}{\partial x} + \frac{\partial(\rho v^2)}{\partial y} = -\frac{\partial p}{\partial y} + \frac{\partial\tau_{xy}}{\partial x} \tag{5}$$

across the layer δ_a. The shear stress τ_{xy} in Eq. (5) has been evaluated by

$$\tau_{xy} = \rho\varepsilon\frac{\partial u}{\partial y}$$

where ε is the average eddy diffusivity across this layer and is, thus, a function of x only.†

2. For this upper viscous layer, a velocity profile of third-degree polynomials is assumed; namely,

$$\phi = \phi_d + \left.\frac{\partial\phi}{\partial\zeta}\right|_d \zeta + \left[3(1-\phi_d) - 2\left.\frac{\partial\phi}{\partial\zeta}\right|_d\right]\zeta^2 + \left[\left.\frac{\partial\phi}{\partial\zeta}\right|_d - 2(1-\phi_d)\right]\zeta^3 \tag{6}$$

which obviously satisfies conditions of

$$\phi = \phi_d\ ,\quad \frac{\partial\phi}{\partial\zeta} = \left.\frac{\partial\phi}{\partial\zeta}\right|_d \quad \text{at } \zeta = 0\ ,\quad \phi = 1\ ,\quad \frac{\partial\phi}{\partial\zeta} = 0 \text{ at } \zeta = 1$$

In view of the fact that, at the rear stagnation point ($\phi_d = 0$), the shear stress at the dividing streamline should vanish, a simple correlation between the slope parameter $(\partial\phi/\partial\zeta)_d$ and ϕ_d would be that they shall be linearly proportional to each other; the constant of proportionality being determined from the initial condition prevailing at the end of the mixing region.

3. The lower viscous layer consists of a forward flow characterized by the dividing streamline velocity and a back flow characterized by a maximum back flow velocity. A linear velocity profile is assumed for the forward flow and a cosine profile for the back flow (see Fig. 3b). It was also recognized that the pressure difference across the lower viscous layer would influence the rate of recompression. For simplification purposes, it was assumed that the forward flow has the constant pressure p_d of the dividing streamline, while the back flow has the constant wall pressure p_w. The difference in pressure across this layer may be calculated from

$$\frac{p_w}{p_d} = 1 + \frac{1}{p_d}\left(\frac{dx}{dx_w}\sin\theta_\infty\tau_d - \frac{d}{dx_w}\int_0^{\delta_b}\rho u^2\sin\theta_\infty\,dy\right) \tag{7}$$

which is the momentum relationship†† normal to x_w-direction for the small region as shown in Fig. 4a. If one further assumes that the dividing streamline and the line of zero-longitudinal velocity component follow straight line trajectories, Eq. (7) may be written for isoenergetic flows as

$$\frac{p_w}{p_d} = 1 + \frac{2\gamma}{\gamma-1}\frac{\sin\theta_\infty}{(p_d/p_e)(p_e/p_{o\infty})}\left\{\frac{\cos\theta_d}{\cos(\theta_\infty-\theta_d)}\frac{\tau_d}{\rho_\infty u_\infty^2}(1-c_\infty^2)^{1/(\gamma-1)}c_\infty^2 - \frac{d}{dx_w}\left[\frac{p_d}{p_e}\frac{p_e}{p_{o\infty}}\delta_b\left(\frac{1}{2c_d}\ln\frac{1+c_d}{1-c_d} - 1\right)\right]\right\} \tag{8}$$

---

†The expression for ε within such a recompression region is given later.

††Note that the contribution from the lateral shear stresses is small as the eddy diffusivity within the wake is expected to be at a lower level.

It should be mentioned that, for cases with the presence of the lower wall, one is compelled to consider the wall boundary layer of the back flow. It can be shown for the isoenergetic flow that, if the wall boundary layer has also a cosine flow profile, the total back flow mass and momentum fluxes would be the same as long as the back flow height and the maximum velocity are the same. The only difference between cases of with and without the wall is the wall shear stress, which is very small and can be neglected. Thus, flow cases of reattachment onto a solid wall are also included in the present formulation.

The geometry of the wake gives also relations such as

$$\frac{\delta_b}{\delta_{b\infty}} = \frac{\ell_{wk} - (\ell_m + x)}{\ell_{wk} - \ell_m} \tag{9}$$

$$\frac{\delta_b}{h_b} = \frac{\sin\theta_d}{\sin(\theta_\infty - \theta_d)} \tag{10}$$

where ℓ_{wk} is the length of the wake boundary. Application of the continuity principle across the lower viscous layer would produce a relationship which correlates the flow properties by (for isoenergetic flows)

$$\frac{\delta_b}{h_b} = \frac{4}{\pi}\frac{p_w}{p_d}\frac{c_d}{\sqrt{1 - c_b^2}}\tan^{-1}\frac{c_b}{\sqrt{1 - c_b^2}} \Bigg/ \left[-\ell n\,(1 - c_d^2)\right] \tag{11}$$

It should be noted that fluxes associated with the transverse velocity component (normal to the indicated velocity profiles) would not contribute significantly in the foregoing considerations because they are small in their order of magnitude; in addition, they tend to cancel each other.

4. Integral momentum principle is applied to both parts of the viscous layer (see Fig. 4) and two differential equations are obtained,

$$\frac{d}{dx}\left[\rho_e u_e^2 \delta_a \int_0^1 \frac{\rho}{\rho_e}\frac{u}{u_e}\left(1 - \frac{u}{u_e}\right) d\zeta\right] - \rho_e u_e \delta_a \int_0^1 \frac{\rho}{\rho_e}\frac{u}{u_e}\, d\zeta\, \frac{du_e}{dx}$$
$$+ p_e \frac{d\delta_a}{dx} - \frac{d}{dx}\left(p_e \delta_a \int_0^1 \frac{p}{p_e}\, d\zeta\right) = \tau_d \tag{12}$$

$$\frac{d}{dx}\left[p_d \delta_b + p_w h_b \cos\theta_\infty\right] + p_w \sin\theta_\infty \frac{dx_w}{dx} + \frac{d}{dx}\left[\rho_d u_d^2 \delta_b \int_0^1 \frac{\rho}{\rho_d}\left(\frac{u}{u_d}\right)^2 d\zeta\right.$$
$$\left. + \rho_b u_b^2 h_b \int_0^1 \frac{\rho}{\rho_b}\left(\frac{u}{u_b}\right)^2 d\zeta \cos\theta_\infty\right] = \tau_d \tag{13}$$

Upon introducing

$$\frac{p}{p_e} = \frac{p_d}{p_e} - \frac{3}{2}\left(\frac{p_d}{p_e} - 1\right)\zeta + \frac{1}{2}\left(\frac{p_d}{p_e} - 1\right)\zeta^3$$

into Eq. (12) for the pressure variation across the upper layer, and the assumed profiles for the lower viscous layer into Eq. (13) and normalizing, Eqs. (12) and (13) become, for isoenergetic flows,

$$\frac{d}{dx}\left[(1 - c_e^2)^{1/(\gamma-1)} c_e^2 \delta_a \int_0^1 \frac{\rho}{\rho_e} \phi(1 - \phi)\, d\zeta\right] + (1 - c_e^2)^{1/(\gamma-1)} c_e \delta_a$$

$$\cdot \int_0^1 \left(1 - \frac{\rho}{\rho_e}\phi\right) d\zeta \frac{dc_e}{dx} - \frac{3}{16}\frac{\gamma - 1}{\gamma}\frac{d}{dx}\left[(1 - c_e^2)^{\gamma/(\gamma-1)} \delta_a \left(\frac{p_d}{p_e} - 1\right)\right]$$

$$= \frac{\tau_d}{\rho_\infty u_\infty^2}(1 - c_\infty^2)^{1/(\gamma-1)} c_\infty^2 \qquad (12a)$$

and

$$\frac{\gamma - 1}{2\gamma}\frac{d}{dx}\left[\frac{p_d}{p_e}\frac{p_e}{p_{o\infty}}\left(\delta_b + \frac{p_w}{p_d} h_b \cos\theta_\infty\right)\right] + \frac{\gamma - 1}{2\gamma}\sin\theta_\infty \frac{\cos(\theta_\infty - \theta_d)}{\cos\theta_d}$$

$$\cdot \frac{p_w}{p_d}\frac{p_d}{p_e}\frac{p_e}{p_{o\infty}} + \frac{d}{dx}\left\{\frac{p_d}{p_e}\frac{p_e}{p_{o\infty}}\left[\delta_b\left(\frac{1}{2c_d}\ell n\frac{1 + c_d}{1 - c_d} - 1\right) + \frac{p_w}{p_d} h_b \cos\theta_\infty\right.\right.$$

$$\left.\left.\cdot \left(\frac{1}{\sqrt{1 - c_b^2}} - 1\right)\right]\right\} = \frac{\tau_d}{\rho_\infty u_\infty^2}(1 - c_\infty^2)^{1/(\gamma-1)} c_\infty^2 \qquad (13a)$$

The shear stress τ_d which appears at the right sides of these two equations is evaluated from an eddy diffusivity formulation for this recompression process which assumes the same form as that of the preceding constant pressure mixing region and the equivalent σ value is assumed to vary inversely as the upper viscous layer thickness; namely

$$\varepsilon = \frac{1}{4\sigma^2}(\ell_m + x)\, u_e \quad \text{with} \quad \frac{\sigma}{\sigma_\infty} = \frac{\delta_{a\infty}}{\delta_a}$$

Thus, the shear stress term appearing in both Eqs. (12a) and (13a) can be evaluated from

$$\frac{\tau_d}{\rho_\infty u_\infty^2}(1 - c_\infty^2)^{1/\gamma-1} c_\infty^2 = \frac{\varepsilon}{u_e \delta_a}\frac{\rho}{\rho_e}\bigg|_d \frac{\partial\phi}{\partial\zeta}\bigg|_d (1 - c_e^2)^{1/\gamma-1} c_e^2 \qquad (14)$$

5. The initial conditions are provided for from the upstream constant pressure jet mixing process. In order to achieve a smooth joining between the mixing and recompression regions, a constant pressure turbulent jet mixing analysis was performed by adopting a velocity profile compatible with that for the recompression study.

Fully Developed Constant Pressure Turbulent Jet Mixing

Upon employing the momentum principle

$$\frac{d}{dx}\int_{-\delta_b}^{0} \rho u^2\, dy = \tau_d = \frac{d}{dx}\int_0^{\delta_a} \rho u(u_\infty - u)\, dy$$

and the stipulation that the velocity profile slope at the dividing streamline matches with that of an error function profile, *i.e.*,

$$\frac{d\phi}{d\eta}\bigg|_d = \frac{1}{\sqrt{\pi}}$$

It may be shown that, for such an isoenergetic fully developed jet mixing region, all flow quantities are functions of a dimensionless homogeneous coordinate $\eta[\eta = \sigma_\infty(y/x)]$ only and following relations are obtained

$$\frac{c_\infty^3}{4\pi} = (1 - c_{d\infty}^2)\left(\frac{1}{2}\ln\frac{1 + c_{d\infty}}{1 - c_{d\infty}} - c_{d\infty}\right) \tag{15}$$

$$\left.\frac{d\phi}{d\zeta}\right|_d = \frac{\eta_a}{\sqrt{\pi}} \tag{16}$$

$$\eta_b = \phi_{d\infty}\sqrt{\pi} \tag{17}$$

and

$$\frac{1}{4\sqrt{\pi}}\,\frac{1 - c_\infty^2}{1 - c_{d\infty}^2} = \eta_a \int_0^1 \frac{\rho}{\rho_\infty}\,\phi(1 - \phi)\,d\zeta \tag{18}$$

where

$$\phi = \phi_{d\infty} + \left.\frac{d\phi}{d\zeta}\right|_{d\infty}\zeta + \left[3(1 - \phi_{d\infty}) - 2\left.\frac{d\phi}{d\zeta}\right|_{d\infty}\right]\zeta^2 + \left[\left.\frac{d\phi}{d\zeta}\right|_{d\infty} - 2(1 - \phi_{d\infty})\right]\zeta^3 \quad \left(\text{for } \zeta > 0,\ \phi = \frac{u}{u_\infty},\ \zeta = \frac{y}{\delta_{a\infty}}\right)$$

and

$$\phi = (1 + \zeta) \quad \left(\text{for } -1 < \zeta < 0,\ \zeta = \frac{y}{\delta_{b\infty}},\ \phi = \frac{u}{u_{d\infty}}\right)$$

have been used as the velocity profile. Values of $\eta_a = \sigma_\infty \delta_{a\infty}/\ell_m$ and $\eta_b = \sigma_\infty \delta_{b\infty}/\ell_m$ are solved from these relations.

Developing Flows

For all practical flow cases, the flow at the end of the mixing region is never fully developed and the initial boundary layer has predominant influence on the recompression process. For thin initial boundary layers,† it was shown by Hill and Page, Carriere and Sirieix, and Korst and Chow that the effect of its presence may be accounted for through the origin shift concept [13] and the equivalent bleed concept [4,14]. Thus, the initial flow properties of the recompression region can be obtained from the corrections of the fully developed flow through

$$\delta_{a\infty} = \eta_a(\ell_m + x_o)/\sigma_\infty \tag{19}$$

$$\delta_{b\infty} = \eta_b(\ell_m + x_o)/\sigma_\infty \tag{20}$$

where

$$x_o = \frac{\sigma\theta_2}{\left.\int_{-\eta_b}^{\eta_a} \frac{\rho}{\rho_\infty}\frac{u}{u_\infty}\left(1 - \frac{u}{u_\infty}\right)d\eta\right|_{fd}} \tag{21}††$$

is the origin shift, θ_2 being the momentum thickness of the profile at the beginning of the mixing region. The dividing streamline velocity ($c_{d_{nfd}}$) for the non-fully developed flow is found from

$$[-\ln(1 - c_{d_{nfd}}^2)] = \frac{2c_{d\infty}^2}{(1 - c_\infty^2)\phi_{d\infty}\eta_b}\left[\left.\int_{-\eta_b}^{\eta_a} \frac{\rho}{\rho_\infty}\frac{u}{u_\infty}\left(1 - \frac{u}{u_\infty}\right)d\eta\right|_{fd} - \frac{\sigma_\infty\theta_2}{\ell_m}\right] \tag{22}$$

The streamline angle β_∞ appearing in Eqs. (3) can be approximately estimated by

---

†The effect of lip shock is also disregarded. Thus, the present calculation is also restricted to low supersonic free stream Mach numbers.

††Henceforth, fd denotes fully developed, while nfd denotes non-fully developed flows.

$$\tan \beta_\infty = \frac{\eta_a}{\sigma_\infty}\int_0^1 \left(1 - \frac{\rho}{\rho_\infty}\phi\right) d\zeta\Big|_{nfd}$$

and the relationship concerning the spread rate parameter

$$\sigma_\infty = 12 + 2.76M_\infty$$

has been tacitly employed [13].

It should be emphasized that these correction techniques are valid only for thin approaching initial boundary layers.

CALCULATION PROCEDURES

Specializing for isoenergetic flows, one may select for given initial conditions a pair of values for the base pressure ratio p_b/p_1 and the length ℓ_m† along the wake where recompression starts. The momentum thickness of the viscous layer at the beginning of jet mixing may be calculated by White's formula [15]

$$\frac{\theta_2}{\theta_1} = \left(\frac{c_1}{c_\infty}\right)^{3.2}\left(\frac{1 - c_1^2}{1 - c_\infty^2}\right)^{1.4} \qquad (24)$$

or the streamtube expansion method and the initial conditions required for recompression study may be established according to the previously described scheme. One may start to integrate the system of differential equations through a step-by-step procedure. At each location along the course of recompression, the free stream Crocco number c_e and the dimensionless dividing streamline velocity ϕ_d may be selected and iterated upon until the system of equations is satisfied. It may be noted that, for each pair of values of c_e and ϕ_d, δ_a may be found from Eq. (2), p_d/p_e from Eq. (4), δ_b from Eq. (9), p_w/p_d from Eq. (8), h_b from Eq. (10), and c_b from Eq. (11). Upon substituting all this information into Eqs. (12a) and (13a), two residues are usually obtained. Values of c_e and ϕ_d at this location should be iterated upon until these residues vanish.

These calculations can be continued until the rear stagnation point is reached. At this location, ϕ_d is set to zero and the wall pressure becomes the stagnation pressures of both the dividing streamline and the representative back flow. From the normal momentum relationship given by Eq. (8), the correct free stream Crocco number at the stagnation point can be established when the wall pressure is obtained from the intersection (and averaging) of previously established curves for p_{od} and p_{ob}. Again, two residues are usually obtained from Eqs. (12a) and (13a) at the rear stagnation point.

The initially selected values of the base pressure ratio p_b/p_1 and the location ℓ_m where recompression starts have to be iterated upon until the residues of the system of equations at the rear stagnation point are reduced to zero. The correct flow field is thus established up to the rear stagnation point.

It is worthwhile to point out that this scheme of calculations and iterations exhibits the typical elliptic behavior of all separated flow problems. The fact that the value of the base pressure ratio is uniquely determined according to the conditions associated with the rear stagnation point is well evidenced by the method of calculation of this recompression process up to the rear stagnation point. In addition, the correct flow pattern established at the point of reattachment serves also as the initial condition for the downstream flow field where additional recompression and flow rehabilitation occur.

---

†Note that, for numerical calculations, all lengths have been normalized by the step height H.

RESULTS OF CALCULATIONS†

Figure 5 shows the pressure distribution on the wall for the flow case of $M_1 = 2.0$, $\theta_1/H = 0.014$. The base pressure ratio p_b/p_1 determined from these calculations is 0.362. It can be seen that the calculated pressure distribution up to the point of reattachment showed excellent agreement with the experimental data which were obtained in the blow down facilities at the University of Illinois. Figure 6 shows also results indicating variations of p_{od}, p_{ob}, ϕ_b, and ϕ_b ($\phi_b = u_b/u_e$). It is particularly interesting to see that the dividing streamline is being energized continuously throughout this part of the recompression process.

Figure 7 shows another calculation for $M_1 = 2.25$, $\theta_1/H = 0.01$. Again, the base pressure agreed well with the experimental data.

Figure 8 shows the variation of base pressure with respect to the initial momentum thickness of the boundary layer at $M_1 = 2.0$. It also shows that the theoretical results presented by Alber and Lees [8] are too high, especially for cases with thin initial boundary layers. Pertinent experimental data from elsewhere are also included in the same figure.

Figure 9 presents results of calculations for $M_1 = 1.56$ as functions of the initial momentum thickness. Results for one set of flow conditions of $M_1 = 3.0$, $\theta_1/H = 0.003$ have also been obtained. The base pressure of $p_b/p_1 = 0.133$ agrees very well with the experimental data compiled by Reda and Page [6] and reproduced in Fig. 10.

It should be pointed out that, in all these calculations, the influence of the pressure variations across the upper viscous layer to the density was not accounted for and the density was estimated as if the viscous layer had the constant free stream pressure; *i.e.*,

$$\rho/\rho_e = (1 - c_e^2)/(1 - c_e^2\phi^2)$$

Also, in estimating the pressure difference across the layer, the shear stress term has been ignored in Eq. (8).

DISCUSSION AND CONCLUSIONS

In comparing theoretical calculations with the experimental data, it appears that the method suggested here for the reattachment process produced reasonable results. In view of the lack of knowledge in the eddy diffusivity within the recompression region and the inadequate data concerning the spread rate of the preceding constant pressure mixing process, it is felt that the analysis suggested here can only be considered as a workable scheme. In addition, there is considerable room for improvement to the present analysis. Particularly for cases of thick initial boundary layers, the upstream jet mixing as well as the recompression region may well have been imbedded within a rotational flow field and a "local freestream" concept would seem to be useful under these situations.

Nevertheless, it is interesting to point out that, under the present formulation of the eddy diffusivity, the normal turbulent stresses have been found to be one to two orders of magnitude smaller than the important flow quantities. In addition, the eddy diffusivity was found to remain in the same order of magnitude as that of the preceding constant pressure mixing region. These findings fully support and justify the method of analysis suggested for such a recompression process.

Further calculations toward downstream direction are possible by adopting the

---

†All calculations were carried out on a digital computer IBM 7094, Department of Physics, University of Illinois at Urbana-Champaign.

same system of equations. Referring to Fig. 11 where the initial conditions for downstream calculations can be estimated from the information obtained at the rear stagnation point, it may be anticipated that the characteristic feature of the flow is the realignment of the external inviscid stream into the original horizontal flow direction and it is believed that any shear stress acting along the centerline may be again ignored. From Eqs. (2) and (12a), it may be observed that relaxation of the pressure difference across the viscous layer provides the main balancing factor, at least in the early part of the flow rehabilitation. The final equilibrium state is reached only when the main flow is in the horizontal direction.

Finally, it is worthwhile to point out that the suggested flow model may be employed to study flow problems in many other flow regimes. Upon combining with the conformal mapping technique, it is hoped that many of the incompressible flow problems may be studied.

ACKNOWEDGMENTS

This work is partially supported by NASA through Research Grant NGL 14-005-140 entitled "Fluid Dynamic and Heat Transfer Problems Associated with Modern Propulsive Systems." This problem was initiated during the summer of 1969 while the author was a summer consultant with RTF, ARO, Inc., Tullahoma, Tennessee. Acknowledgment is due to Mr. F. Minger for his interest and encouragement, to Messers C. E. Peters, R. C. Bauer, and C. E. Willbanks for their interesting discussions. The author is also indebted to Professor R. A. White and Mr. P. Gerhart for providing the experimental data obtained at the University of Illinois.

REFERENCES

1. Korst, H. H., "A Theory for Base Pressures in Transonic and Supersonic Flow," Journal of Applied Mechanics, Vol. 23, 1956, pp. 593-600.

2. Nash, J. F., "An Analysis of Two-Dimensional Turbulent Base Flow Including the Effect of the Approaching Boundary Layer," NPL Aero Report No. 1036, 1962.

3. Page, R. H., Hill, W. G., Jr., and Kessler, T. J., "Reattachment of Two-Dimensional Supersonic Turbulent Flows," ASME Paper 67-F-20, presented at the Fluids Engineering Conference, Chicago, Illinois, May 8-11, 1967.

4. Carriere, P., and Sirieix, M., "Résultats Récents dans l'Etude des Problèmes de Mélange et de Recollement," XI Congres Internal de Mecaniquee, Munich, T. P. 165 (1964) ONERA, September 1964.

5. Lees, L., and Reeves, B. L., "Supersonic Separated and Reattaching Laminar Flows: 1. General Theory and Application to Adiabatic Boundary Layer/Shock Wave Interactions," AIAA Journal, Vol. 2, 1964, pp. 1907-1920.

6. Reeves, B. L., and Lees, L., "Theory of the Laminar Near Wake of Blunt Bodies in Hypersonic Flow," AIAA Journal, No. 3, 1965, pp. 2061-2074.

7. Crocco, L., and Lees, L., "A Mixing Theory for the Interaction between Dissipative Flows and Nearly Isentropic Streams," Journal of Aerospace Science, Vol. 19, 1952, pp. 649-676.

8. Alber, I. E., and Lees, L., "Integral Theory for Supersonic Turbulent Base Flows," AIAA Journal, Vol. 6, 1968, pp. 1343-1351.

9. Shammoth, S. J., and McDonald, H., "A New Solution of the Near Wake Recompression Problem," paper presented at AIAA Eighth Aerospace Sciences Meeting, December 1969.

10. Ohrenberger, J. T., and Baum, E., "A Theoretical Model of the Near Wake of a Slender Body in Supersonic Flow," AIAA paper No. 70-792, presented in AIAA Third Fluid and Plasma Dynamics Conference, Los Angeles, Calif., June 1970.

11. Nielsen, J. N., Lynes, L. K., and Goodwin, F. K., "Calculation of Laminar Separation with Free Interaction by the Method of Integral Relations," Report AFFDL-TR-65-107, Part II, VIDYA Division, ITEK Corporation, January 1966.

12. Roache, P. J., and Mueller, T. J., "Numerical Solutions of Laminar Separated Flows," AIAA Journal, Vol. 8, 1970, pp. 530-538.

13. Hill, W. G., Jr., and Page, R. H., "Initial Development of Turbulent, Compressible Free Shear Layers," Journal of Basic Engineering, Vol. 91, 1969, pp. 67-73.

14. Korst, H. H., and Chow, W. L., "Non-Isoenergetic Turbulent Jet Mixing between Two Compressible Streams at Constant Pressure," NASA CR-419, 1965.

15. White, R. A., "Effect of Sudden Expansions or Compressions on the Turbulent Boundary Layer," AIAA Journal, Vol. 4, 1966, pp. 2232-2234.

16. Reda, D. C., and Page, R. H., "Supersonic Turbulent Flow Reattachment Downstream of a Two-Dimensional Backstep," Report AFOSR 69-1592 TR or RU-TR 125-MAE-F, Department of Mechanical and Aerospace Engineering, Rutgers University, New Brunswick, New Jersey, May 1969.

NOMENCLATURE

c	V/V_{max}, Crocco number
c_b	Crocco number for back flow of maximum velocity
H	step height
h_b	height for back flow
ℓ_m	length of the constant pressure region along the wake boundary, measured from separation corner
ℓ_{wk}	$H/\sin\theta_\infty$, length of the wake boundary measured from the separation corner to the point of reattachment
M	Mach number
p	pressure
u	x-velocity component
V	magnitude of velocity
v	y-velocity component
x	coordinate in main flow direction
y	coordinate normal to x
ρ	density
β	streamline angle
δ	thickness of viscous layer
ϕ	u/u_e, dimensionless velocity
ζ	y/δ_a or y/δ_b
$\omega(c)$	$\sqrt{(\gamma+1)/(\gamma-1)}\,\tan^{-1}\sqrt{[(c^2-(\gamma-1)/(\gamma+1)]/[1-c^2)}-\tan^{-1}\sqrt{\frac{[(\gamma+1)/(\gamma-1)\,c^2-1]}{1-c^2}}$, Prandtl-Meyer function
γ	ratio of specific heat
τ	shear stress
ε	eddy diffusivity
θ	angle or momentum thickness
η	$\sigma y/x$
σ_∞	similar parameters for constant pressure jet mixing regions

Subscripts

a	viscous layer above the dividing streamline
b	viscous layer below the dividing streamline or back flow
d	dividing streamline
e	external inviscid stream
w	wall or centerline state
0	stagnation state
1	approaching flow state
2	flow state after the Prandtl-Meyer expansion
∞	station at the beginning of recompression

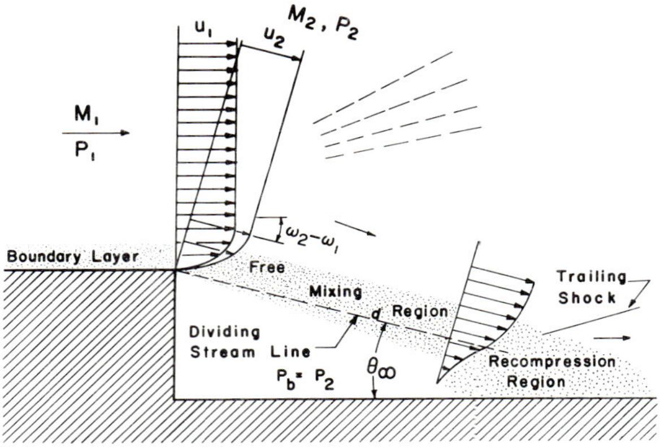

Figure 1 Two-Dimensional Supersonic Flow Past a Back Step

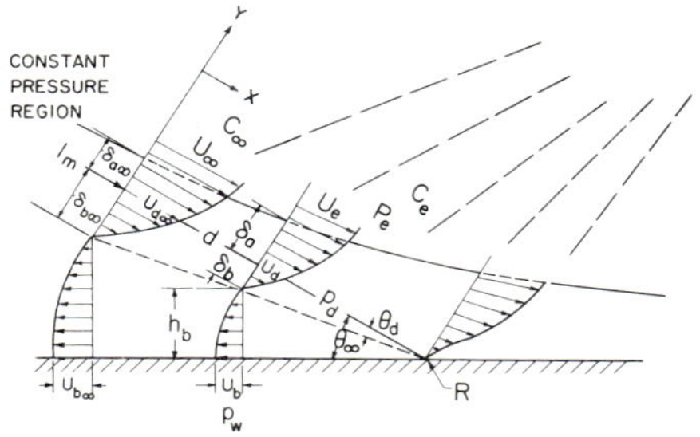

Figure 2 The Flow Model for Recompression Study

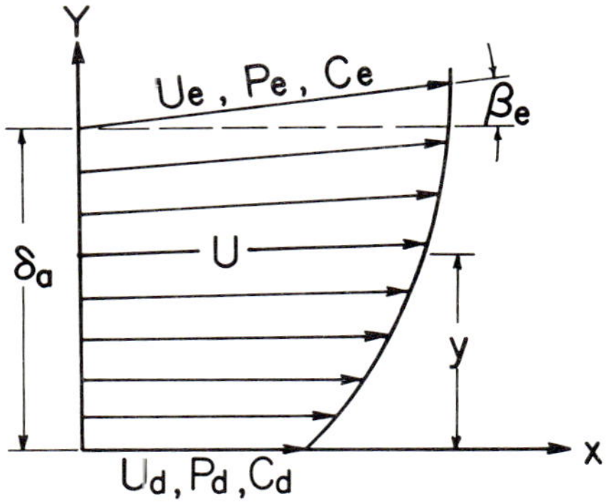

Figure 3a The Upper Viscous Layer

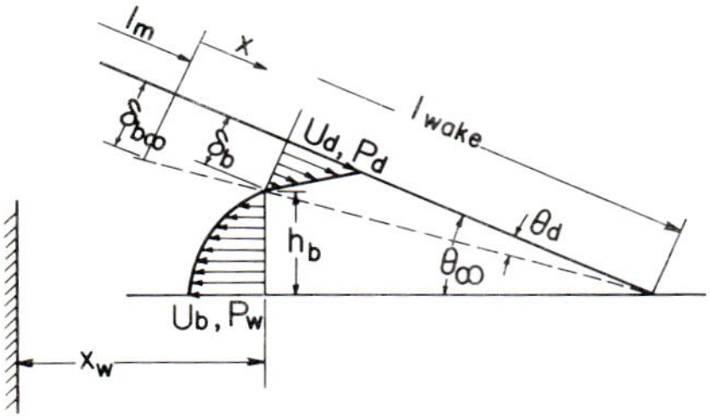

Figure 3b The Lower Viscous Layer

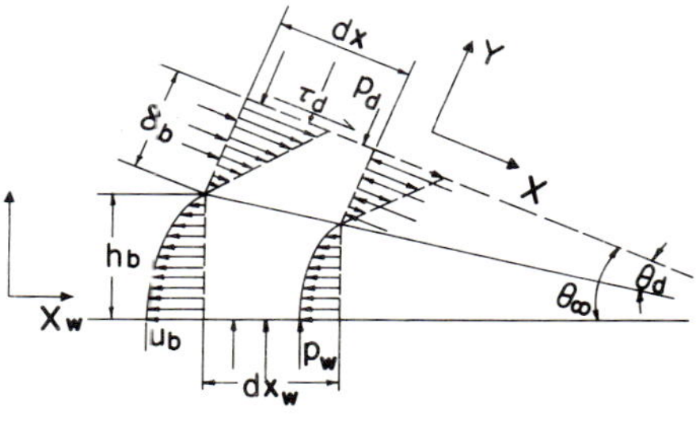

Figure 4a Elementary Control Volume for Lower Viscous Flow Region

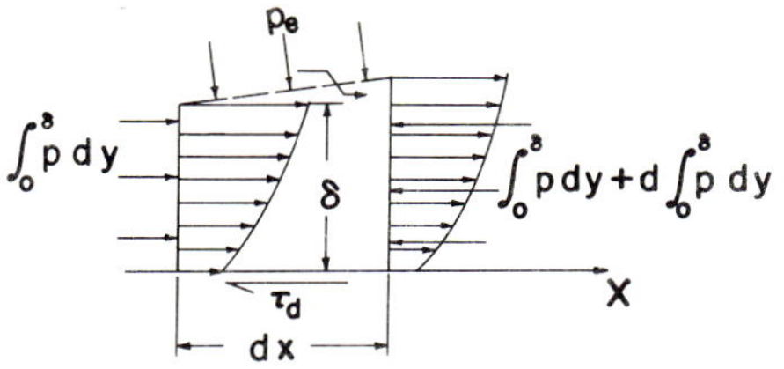

Figure 4b Elementary Control Volume for Upper Viscous Layer

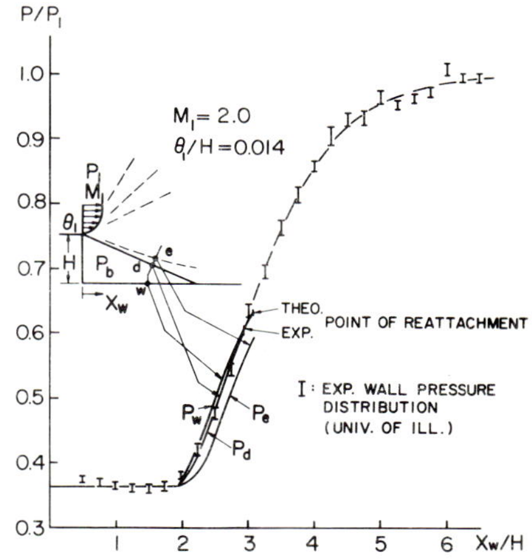

Figure 5 Theoretical and Experimental Wall Pressure Distribution for $M_1 = 2$, $\theta_1/H = 0.01$

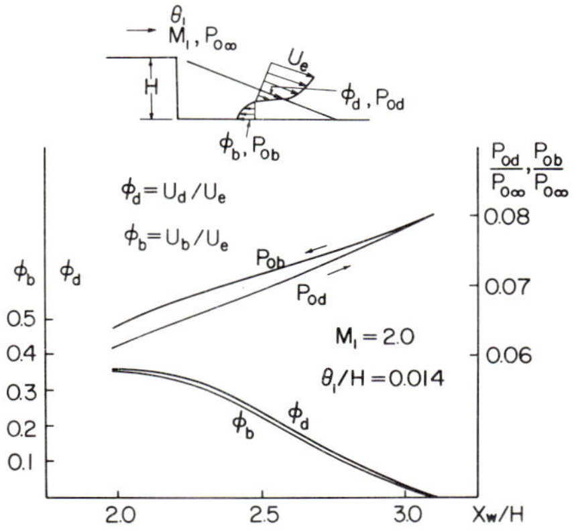

Figure 6 Variation of p_{od}, p_{ob}, ϕ_d, and ϕ_b

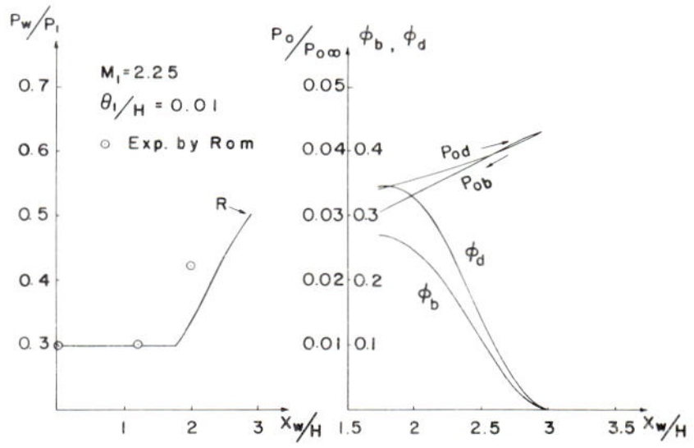

Figure 7 Results for $M = 2.25, \theta_1/H = 0.01$

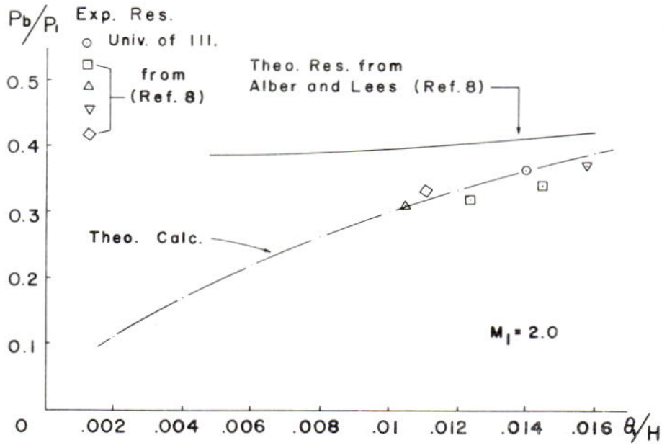

Figure 8 Influence of Initial Momentum Thickness on the Base Pressure Ratio (M = 2.0)

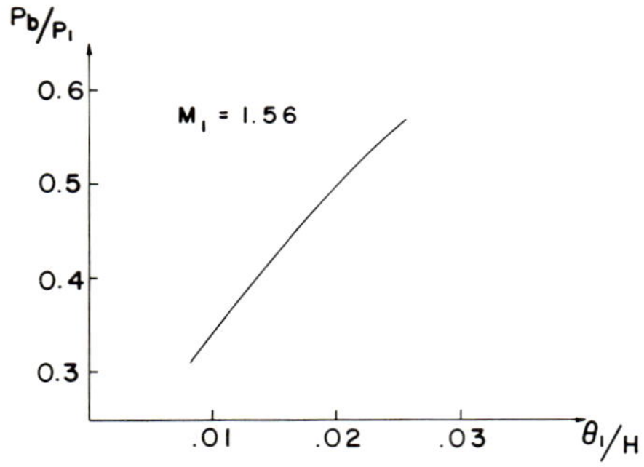

Figure 9 Influence of Initial Momentum Thickness on the Base Pressure ratio (M = 1.56)

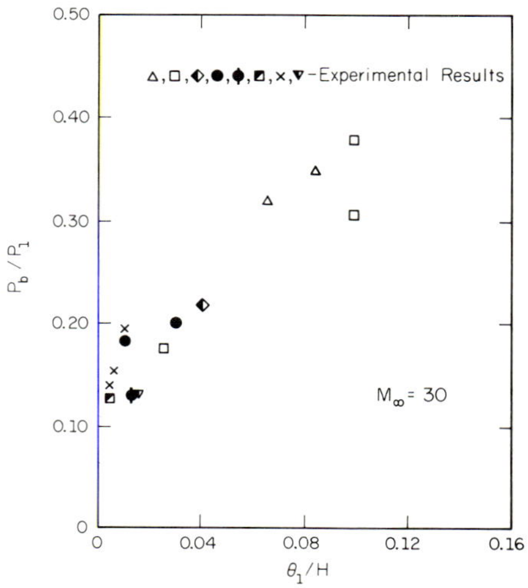

Figure 10 Experimental Base Pressure Ratio for M = 3.0 (Adapted from Ref. 16)

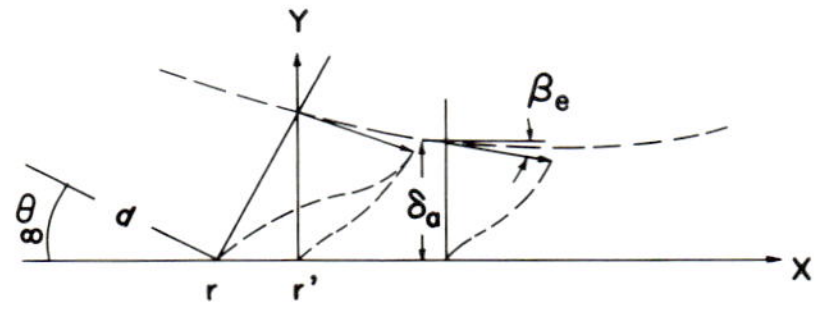

Figure 11 Downstream Flow Field

Developments in Mechanics, Vol. 6. Proceedings of the 12th Midwestern Mechanics Conference: 22

The Turbulent Near-wake of Infinite Axisymmetric Bodies at Mach Four

W. R. SIELING
BELL TELEPHONE LABORATORIES

C. E. G. PRZIREMBEL and R H. PAGE
RUTGERS UNIVERSITY

ABSTRACT

An experimental investigation of the axisymmetric turbulent near-wakes generated by two base configurations (blunt and hemispherical) in a free stream with a nominal Mach number of four and high Reynolds number is reported. A special annular wind tunnel, which completely eliminated model support and forebody disturbances, was used to obtain detailed pressure measurements in all regions of the near-wakes, as well as on the surface of the models. Various flow visualization techniques were exploited to verify the location of the separation and rear stagnation points, and to determine qualitatively the overall flow patterns. Comparison of the data for the two base geometries indicated a noticeable similarity between the two reattachment processes.

INTRODUCTION

The rear geometry of a body in supersonic flight usually causes the attached boundary layer and the external supersonic flow to separate at or near the base. The resulting separation zone, usually denoted as the near-wake, is dominated by complex viscous-inviscid flow interactions. Detailed knowledge of the fluid mechanics of this region is essential in the determination of convective base heat transfer rates and base drag. Furthermore, the near-wake flow properties are the initial conditions for the far wake, understanding of which is necessary for identification of re-entry vehicles by radar discrimination techniques.

A comprehensive investigation of the separated flow configuration must include the study of the attached approaching boundary layer, the separation process, the free shear layer, the reattachment or realignment process, the recirculating flow, and the adjacent flow. These flow components for the hemispherical base model are depicted in the schematic shown in Fig. 1.

Experimental investigations of the near-wake of axisymmetric bodies may be subdivided into various categories depending upon approaching flow conditions, body configuration, and means of support. The approaching flow may be laminar, transitional, or turbulent in nature. The bodies investigated take a variety of shapes and may be supported by numerous methods such as wires, struts, stings or magnetic devices.

The body configurations which have been most widely used in studies of this type are cones, cone-cylinders, and ogive-cylinders. These are convenient body shapes, because they may be supported in conventional wind tunnels and may also be fired in ballistic and free-flight tests. An extensive literature review of axisymmetric blunt base models in turbulent supersonic flow has been compiled by Przirembel (1). It should be noted that most studies of the near-wake region have been limited to the measurement of base pressure rather than measurements of the entire flow field.

There is no hemispherical base data by other authors. Investigations of the near-wake of fully circular surfaces have been limited to spheres, cylinders, and wedge-cylinders.

Support structures invariably create disturbances in the flow field. These effects have been well documented by numerous investigators, and have been reviewed recently in references (1-3). On the other hand, ballistic and free-flight tests have the disadvantages of low data yield per run, a relatively short test time, high cost, and probable misalignment.

The two testing configurations which eliminate support interference are annular nozzles and magnetic suspension of bodies. However, the present state of the art is such that the use of magnetic suspension systems is restricted to investigations involving very small models under laminar flow conditions. Thus, an annular type tunnel is currently the most suitable experimental facility for turbulent near-wake studies. Also, this configuration generally eliminates any possible upstream influences on the main flow (such as that due to the leading edge of a cone), and therefore is the most advantageous arrangement to isolate the base region for detailed investigation.

Annular type tunnels have previously been utilized by Donaldson (4), Badrinarayanan (5), Reid and Hastings (6,7), Korst (8), Zakkay and Fox (9), and more recently by Sirieix and Delery (10) and Hong (11). Detailed measurements in the near-wake were made only in the investigations reported in the last two references. It should be noted that some of these tunnels (4,5) were fabricated by inserting upstream stings in conventional nozzles (two-dimensional or axisymmetric) and cannot be expected to produce a uniform flow.

The purpose of this paper is to present the results of a comprehensive experimental investigation of supersonic axisymmetric turbulent near-wake flow fields behind blunt and hemispherical bases in the absence of model support interference. The present data provide the details for each of the flow components necessary for the quantitative evaluation of presently available analytical models and for the formulation of improved solutions to this problem.

APPARATUS

All tests were carried out in the Rutgers Axisymmetric Near-Wake Tunnel (RANT) in the Department of Mechanical and Aerospace Engineering at Rutgers University. It is a blow-down facility (average run time - 20 seconds) which uses dry air (less than 5 ppm of water vapor) as the working fluid and exhausts to the atmosphere. RANT is shown schematically in Fig. 2. A detailed description of RANT has been reported by Przirembel (12).

The annular nozzle was designed by the method of characteristics and produces a uniform flow field at a nominal Mach number of 4. The centerbody and test models are integral parts of the nozzle configuration and have constant outside diameters of 3.00 inches. This design essentially eliminates support interference effects. The centerbody is hollow, thus allowing pressure and temperature leads to pass from the test models to measuring instruments located outside the tunnel.

The test section is equipped with rectangular windows to allow for optical observations. It was designed to avoid separation from the windows or detrimental shock formation caused by the nozzle test section interface.

EXPERIMENTAL PROCEDURE

A brief description of the present measurements and associated instrumentation is presented in this section. More details about the experimental procedure may be found in reference (13).

The stagnation pressure and temperature-time histories during each test run were recorded by means of a pressure transducer and a thermocouple, respectively. A constant stagnation pressure ($\pm$1/2 psi) was maintained during each test. Curves of stagnation pressure versus time of run and curves demonstrating repeatability are presented in reference (12). It was found that during steady state the temperature dropped approximately one degree per second. Thus, considering the stagnation temperature level of approximately 520° R, it can be assumed that the stagnation temperature remained constant during each run.

All pressure transducer signals were recorded on X-Y plotters. Each individual transducer-plotter system was calibrated with a Texas Instruments precision pressure gage with a resultant accuracy of 0.5% full scale.

Each test model had three or four shoulder pressure taps which were equally spaced on the circumference. These were utilized to measure the approaching free stream static pressure and to check alignment of the centerbody, thus insuring uniform approaching flow conditions.

Stagnation and static pressures on the centerline were measured by means of a standard subsonic pitot-static probe, 1/8 inch in diameter, which was introduced from the center of the base. Its axial location was adjustable from outside the tunnel. Due to the large velocity gradients which exist on the centerline, the conventional interpretation of the probe data could not be used. The static pressure measured at a specific location of the static holes, X, was taken to be the static pressure at X, not as the static pressure at the pitot orifice location. The presence of this small centerline probe did not noticeably affect the base pressure.

The approaching boundary layer, the shear layer, and the recirculation region were investigated with specially designed pitot probes. The outer and inner diameters of the probe tips were 0.042 inches and 0.022 inches, respectively. They were introduced into the region of interest through the base wall at a point 1/2 inch from the centerline of the body. By rotating the probe, investigation of flow properties over a one inch continuous range of radii with a single probe was effected. The axial location of the probes was fixed by the use of collars. On the basis of color Schlieren photographs, the shear layer probe was aligned with the mean flow direction (approximately 16 degrees from the axis). Although exact alignment is not guaranteed at all probe positions, the errors are found to be negligible (13).

Conventional and color Schlieren photographs were obtained of the near-wake flow fields. High speed motion pictures (5 K to 7 K frames per second) were also taken. For additional flow visualization, talc particles were injected into the base region for some of the high speed films. The latter provided information about the tunnel starting process and about the shape of path lines in the near-wake region.

EXPERIMENTAL RESULTS

All experiments were carried out at values of stagnation pressure of 182.5 psia (blunt base), or 152.5 psia (hemispherical base). The stagnation temperature in both cases was 520° R ± 20°. The corresponding free-stream Mach numbers and Reynolds numbers were (3.91, 1.82×10^7 per foot) and (3.88, 1.56×10^7 per foot), respectively.

From the measurements in the attached boundary layer, distributions of Mach number and velocity were calculated by standard methods. In light of previous data on axisymmetric turbulent boundary layers (14), the assumption that the static pressure is constant normal to the surface was used. Using the boundary layer thickness and the boundary layer displacement thickness, as determined from the experimental data, in conjunction with tabulated results (15), the other boundary layer properties were calculated and are presented in the Table below. It can be concluded that the turbulent boundary layer is fully developed.

BOUNDARY LAYER PROPERTIES

P_o (psia)	δ	δ^*	θ	n
182.5	0.338"	0.138"	0.0168"	6.9
152.5	0.344"	0.140"	0.0171"	7.3

BLUNT BASE NEAR-WAKE

The base wall pressure distribution, Fig. 3, is found to be constant across the base except in the immediate vicinity of the corner. Schlieren, water injection, and oil film visualization techniques clearly indicated that the separation point is located precisely at the corner. In a similar investigation at a free stream Mach number of 2.42, Hong (11) observed an apparent separation line on the base approximately 0.15 base radii below the corner. The ratio of the approaching boundary layer thickness to the base diameter in Hong's facility was substantially larger than in the present investigation. Since viscous effects dominate the separation process, it may be speculated that the difference in separation line location can be attributed to the relative difference in approaching boundary layer thicknesses. More detailed measurements of base flows with variable approaching boundary layer parameters are needed to clarify this problem.

Presented in Fig. 4 are the centerline static and pitot pressures. The data show clearly that the rear stagnation point is located at X/D = 0.84. These pressure data, in conjunction with the assumption of isoenergetic flow, were used to calculate the centerline Mach number distribution, which is compared with data by Sirieix and Delery (10) for M = 4.0 in Fig. 5. For the RANT data, the peak recirculation velocity was calculated at X/D = 0.5 to be approximately equal to one-third the free stream velocity.

Pitot pressure measurements were made in the shear layer at stations X/D = 0.360, 0.587, and 0.833. These are presented in Fig. 6. From these profiles, and from Schlieren observations, there is no evidence of the existence of a separation, or "lip" shock.

As previously mentioned, high speed motion pictures (16) were taken to observe the flow pattern during start-up and steady state. A tri-color Schlieren system (17) was utilized for most of these studies. During one phase of this investigation talc particles were introduced into the base region through two small holes located close to the corner. The particles averaged about 2×10^{-5} inches in diameter. Although the resolution of the optical system was such that individual particles could not be traced, groups of particles could be followed using special projection equipment. The resulting path lines are shown in Fig. 7. It was observed that all particles are contained within an hour glass shaped region outlined by the dashed line.

HEMISPHERICAL BASE NEAR-WAKE

The surface pressure distribution on the hemispherical base is presented in Fig. 8. Also shown in the figure is Hama's laminar, two-dimensional 6° half-angle, wedge cylinder data (18), and two-dimensional and axisymmetric (method of characteristics) predictions. The surface pressure in the separated flow region is reasonably uniform. The sharp change in pressure at $\theta = 66°$ located the separation point. This is verified by Schlieren photography, and oil film and water injection visualization techniques. Shown in Fig. 9 is a compilation of experimental data and theoretical results relating Re/M to location of minimum surface pressure, θ_s, for cylinders, wedge-cylinders and the present configuration. The Mach number for the approaching free stream varied between 2.6 and 15.4; the attached boundary layers were laminar, transitional, or turbulent. Extrapolation of these data beyond the indicated values is not warranted at this time, due to the limited information available on the separation process.

The centerline static pressure and Mach number distributions are given in Figs. 10 and 11. The centerline pressure in the recompression region increases to a value which is greater than the approaching free stream value. The rear stagnation point is located at $X_h/D = 0.81$. A peak recirculation velocity, equal in magnitude to 1/3 of the free stream value, exists at $X_h/D = 0.29$.

Pitot pressure measurements were made in the shear layer at stations $X_h/D = -0.215$, -0.033, 0.333, and 0.667. The results of this investigation are presented on the pressure "map," Fig. 12. Those points marked "high and low" are the relative maxima and minima of the pitot curves in the vicinity of the separation shock. There is excellent agreement between the Schlieren data and the "high" point data. The Schlieren data also indicate a slight curvature of the separation shock near the surface of the hemisphere. The theoretical prediction of the shock location calculated by the rotational method of characteristics is included for comparison. The dotted lines are estimated contours of constant Mach number calculated using the following assumptions: (1) behind the shock wave, the static pressure is constant and equal to either the surface or centerline static pressure at the particular axial location, and (2) ahead of the shock wave, the total pressure is equal to the tunnel stagnation pressure. The small recompression wavelets which are located between $r = 0.6$ and 1.1 inches, $X_h = 2.0$ inches, were also observed in the color Schlieren photographs.

DISCUSSION

Some rather illuminating correlations and comparisons of the near-wake properties of the two configurations are presented in this section.

The lowest pressure recorded on the surface of the hemisphere before separation is identically equal to the pressure on the blunt base.

A comparison of the centerline recirculation velocities shows that the peak velocity occurs closer to the body in the hemispherical case. This is not surprising as this shape naturally offers less resistance to the recirculating flow. The peak values of velocity, however, are nearly identical in magnitude.

The centerline static pressure distributions at first seem quite different. However, upon choosing an appropriate reference for axial location, a remarkable similarity between the two flow situations is seen to exist. A natural reference point is the point of separation, $Xs = 0$; whence $Xs = 0$ when $X = 0$ and $X_h = -0.9$. Both distributions are plotted as a function of Xs/D in Fig. 13. It is seen that the curves are similar in most respects except for a slight axial displacement. This leads to the following conclusions: (1) starting at $X_s/D = 0.65$ the free realignment process is the same in both cases and (2) because of this fact, the respective shear layers must be fully developed, i.e., independent of the history of the flow. Of course, the upstream flow pattern, $X_s/D < 0.65$, is very dependent on the geometry of the base.

To check the above conclusions, the pitot pressure profiles for both cases have been plotted at X_S/D of approximately 0.60 in Fig. 14. These pitot pressures are normalized with respect to the free-stream pitot pressures at their respective axial locations. As seen, the inner edges of the shear layers, r = 0.6 to 1.3 are indeed similar. A complete estimation as to the scope of this phenomena is dependent on further investigations with other base configurations.

SUMMARY OF RESULTS

The turbulent near-wakes of both blunt and hemispherical based axisymmetric models have been experimentally investigated. The facility used for these tests completely eliminated any support interference effects.

The major results of the blunt-base near-wake investigation are: (1) all flow visualization studies indicated that the separation point was exactly on the corner, (2) pitot pressure measurements and optical studies provided no evidence of a "lip" shock, and (3) the centerline static pressure in the recirculation region was approximately constant, and equal to the base pressure from the base to X/D = 0.6.

The principal findings of the hemispherical-base near-wake study are: (1) pressure measurements and various optical techniques accurately located the point of separation, and (2) pitot pressure and optical measurements indicated the separation shock to be very nearly a straight line.

NOMENCLATURE

D	diameter of body (in)
M	Mach number
n	velocity profile power law coefficient
P	pressure (lb_f/in^2)
r	radius measured from centerline (in)
Re	Reynolds number per foot
S	location of separation point
X	distance from blunt base along centerline (in)
X_h	distance from hemispherical base along centerline (in)
X_s	distance from separation point along centerline (in)
δ	boundary layer thickness (in)
δ^*	boundary layer displacement thickness (in)
θ	boundary layer momentum thickness (in)
θ	angular position on hemispherical base (deg)

SUBSCRIPTS

b base

F.S. free-stream conditions at specified X

h hemispherical near-wake data

s separation

∞ free-stream conditions

ACKNOWLEDGEMENTS

This research was sponsored by the Air Force Office of Scientific Research, Office of Aerospace Research, United States Air Force, under contract F44620-68-C-0018.

BIBLIOGRAPHY

[1] Przirembel, C.E.G., "The Turbulent Near-Wake of an Axisymmetric Body at Supersonic Speeds," Ph.D. Thesis, Mechanical and Aerospace Engineering Department, Rutgers University, June 1967.

[2] Sieling, W.R., "The Effect of Sting Diameter and Length on Base Pressure at M = 3.88," The Aeronautical Quarterly, Vol. 19, 1968, pp. 368-374.

[3] Sieling, W.R. and Page, R.H., "A Re-examination of Sting Interference Effects," AIAA Paper No. 70-585, AIAA Fifth Aerodynamic Testing Conference, May 1970.

[4] Donaldson, I.S., "The Effect of Sting Supports on the Base Pressure of a Blunt-Based Body in a Supersonic Stream," The Aeronautical Quarterly, Vol. 6, 1955, pp. 221-224.

[5] Badrinarayanan, M.A., "An Experimental Investigation of Base Flows at Supersonic Speeds," Journal of the Royal Aeronautical Society, Vol. 65, No. 7, 1961, pp. 475-482.

[6] Reid, J. and Hastings, R.C., "Experiments on the Axisymmetric Flow Over Afterbodies and Bases at M = 2.0," Aero No. 2628, Royal Aircraft Establishment, Farnborough, England, October 1959.

[7] Reid, J. and Hastings, R.C., "The Effect of a Central Jet on the Base Pressure of a Cylindrical Afterbody in a Supersonic Stream," Aero No. 2621, Royal Aircraft Establishment, Farnborough, England, December 1959.

[8] Korst, H.H., Chow, W.L., and Zumwalt, G.W., "Research on Transonic and Supersonic Flow of a Real Fluid at Abrupt Increases in Cross Section," ME TR 392-5, Department of Mechanical and Industrial Engineering, University of Illinois, Urbana, Illinois, December 1959.

[9] Zakkay, V. and Fox, H., "Experimental and Analytical Consideration of Turbulent Heterogeneous Mixing in the Wake," Report AA-66-54, School of Engineering and Science, New York University, New York, New York, April 1965.

[10] Sirieix, M. and Delery, J., "Analyse Experimentale du Proche Sillage d'un Corps Elance Libre de Tout Support Lateral," AGARD Proceedings No. 19, Fluid Physics of Hypersonic Wakes, May 1967.

[11] Hong, Y.S., "Experimental Study of Axially Symmetric Base Flow with Turbulent Initial Boundary Layer at M = 2.42," AIAA Paper No. 70-796, AIAA Third Fluid and Plasma Dynamics Conference, July 1970.

[12] Przirembel, C.E.G., "Construction and Operation of RANT," Technical Report No. 110-ME-F, Department of Mechanical and Aerospace Engineering, Rutgers University, New Brunswick, N.J., November 1966.

[13] Sieling, W.R., Przirembel, C.E.G., and Page, R.H., "Axisymmetric Turbulent Near-Wake Studies at Mach Four: Blunt and Hemispherical Bases," RU-TR 122-MAE-F, OSR No. 68-2465, Department of Mechanical and Aerospace Engineering, Rutgers University, New Brunswick, N.J., November 1968.

[14] Clutter, D.W. and Kaups, K., "Wind-Tunnel Investigation of Turbulent Boundary Layers on Axially Symmetric Bodies at Supersonic Speeds," Report No. LB31425, Douglas Aircraft Division, Long Beach, California, February 1964.

[15] Tucker, M., "Approximate Calculation of Turbulent Boundary Layer Development in Compressible Flow," NACA TN 2337, National Advisory Committee on Aeronautics, April 1951.

[16] Sieling, W.R., "High Speed Motion Pictures of the Near-Wakes of Axisymmetric Bodies," Proceedings - Ninth International Congress on High-Speed Photography, August 1970.

[17] Kessler, T.J. and Hill, W.G., Jr., "Schlieren Analysis Goes to Color," Astronautics and Aeronautics, Vol. 4, No. 1, 1966, p. 38.

[18] Hama, F.R., "Experimental Studies on the Lip Shock," AIAA Journal, Vol. 6, No. 2, 1968, pp. 212-219.

[19] McCarthy, J.F., Jr. and Kubota, T., "A Study of Wakes Behind a Circular Cylinder at M = 5.7," AIAA Journal, Vol. 2, No. 4, 1964, pp. 629-636.

[20] Tewfik, O.K. and Giedt, W.H., "Heat Transfer, Recovery Factor and Pressure Distributions Around a Cylinder Normal to a Supersonic Rarefied Air Stream: Part 1, Experimental Data," TR HE-150-162, University of California, Berkeley, California, January 1959.

[21] Baum, E., "An Interaction Model of a Supersonic Laminar Boundary Layer on Sharp and Rounded Backward Facing Steps," AIAA Journal, Vol. 6, No. 3, 1968, pp. 440-447.

[22] Gregorek, G.M. and Korkan, K.D., "An Experimental Observation of the Mach- and Reynolds Number Independence of Cylinders in Hypersonic Flow," AIAA Journal, Vol. 1, No. 1, 1963, pp. 210-211.

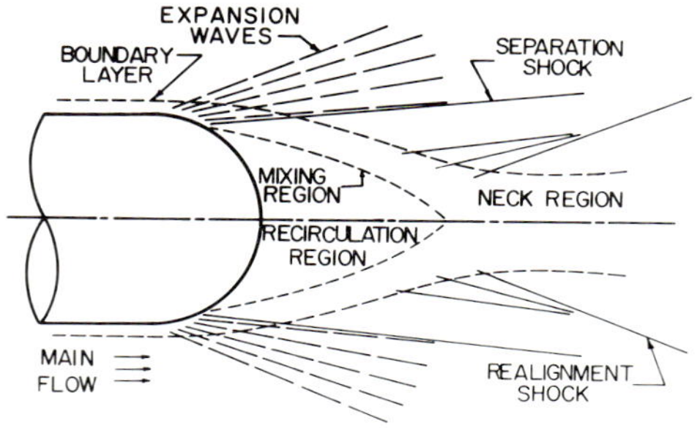

Fig. 1 Hemispherical Base - Flow Schematic

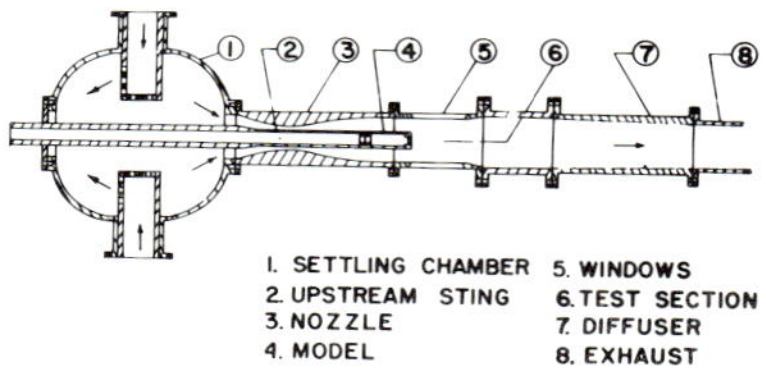

Fig. 2 Schematic of RANT

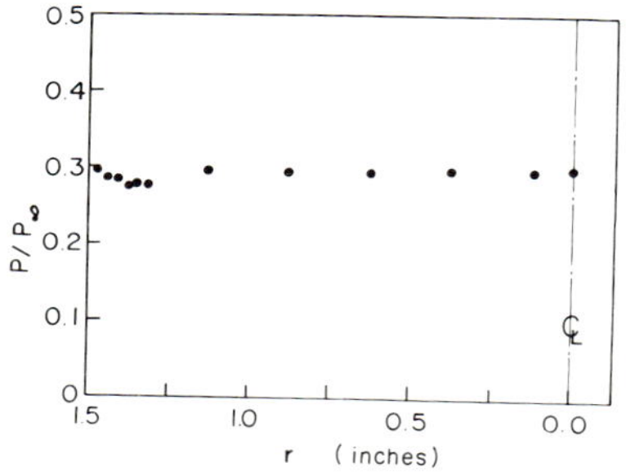

Fig. 3 Base Pressure Distribution - Blunt

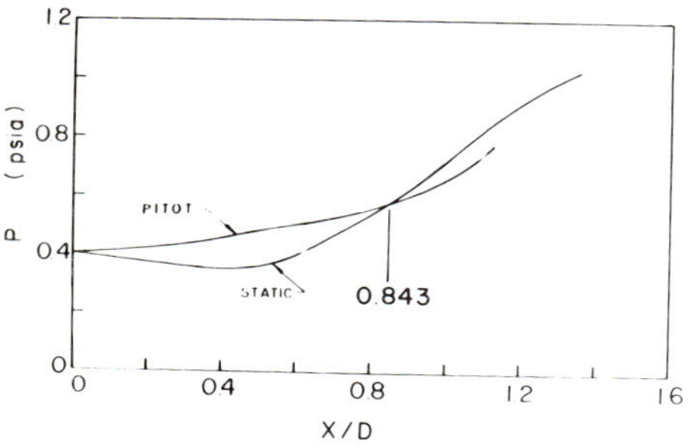

Fig. 4 Centerline Pressures - Blunt

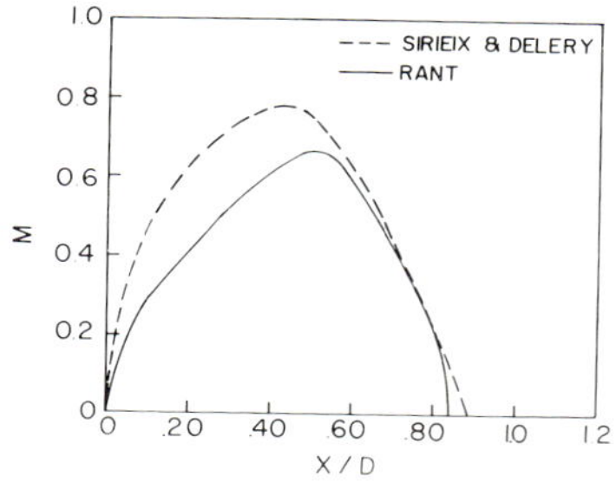

Fig. 5 Centerline Mach Number - Blunt

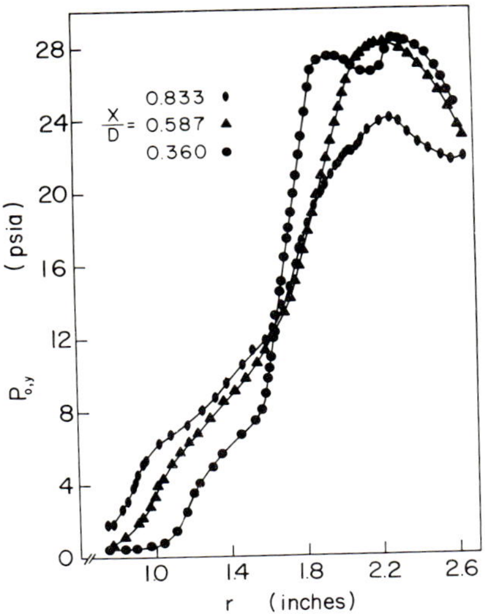

Fig. 6 Shear Layer Pitot Pressure - Blunt

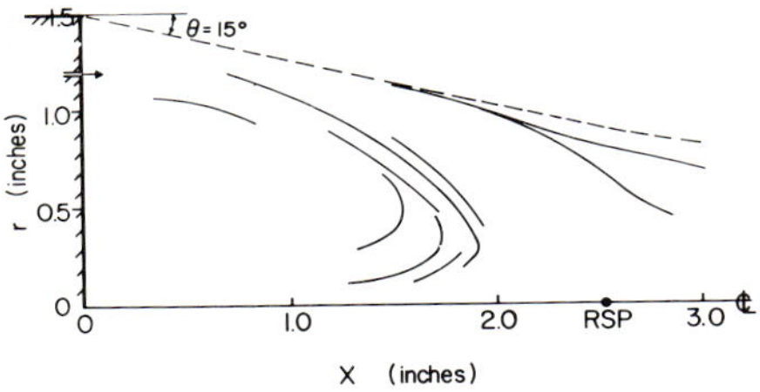

Fig. 7 Talc Particle Tracing - Blunt

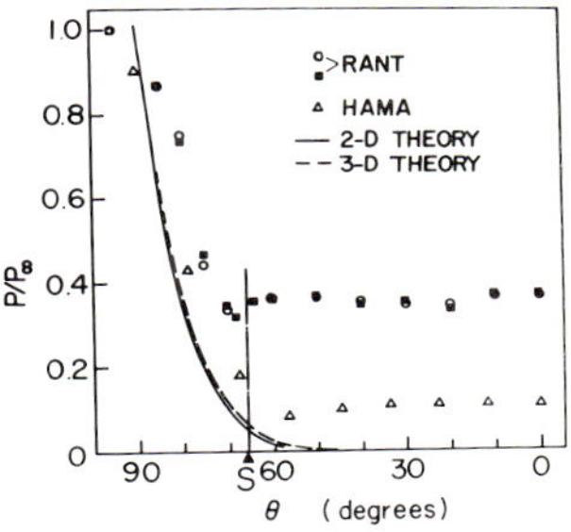

Fig. 8 Surface Pressure - Hemisphere

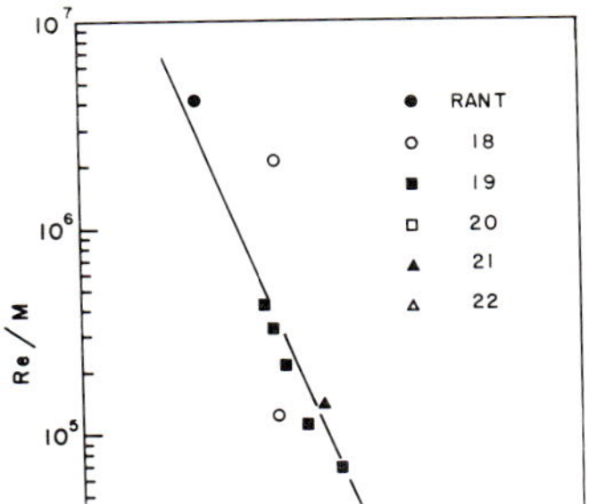

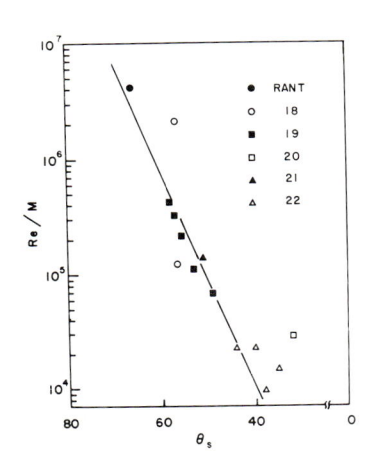

Fig. 9 Curved Afterbody Data

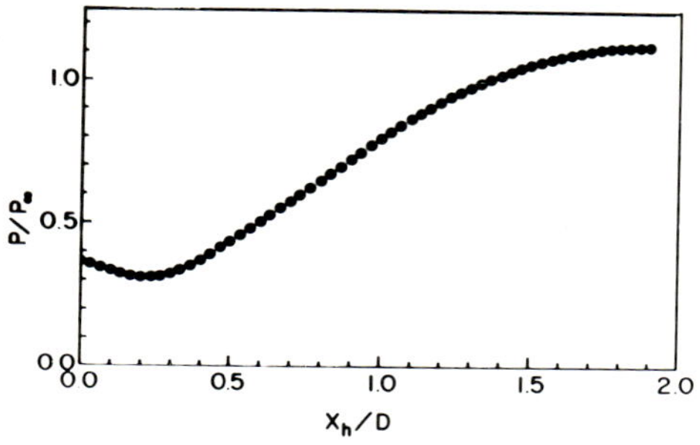

Fig. 10 Centerline Static Pressure - Hemisphere

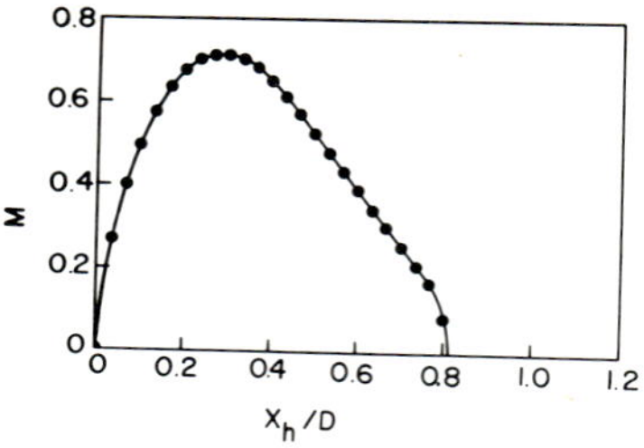

Fig. 11 Centerline Mach Number - Hemisphere

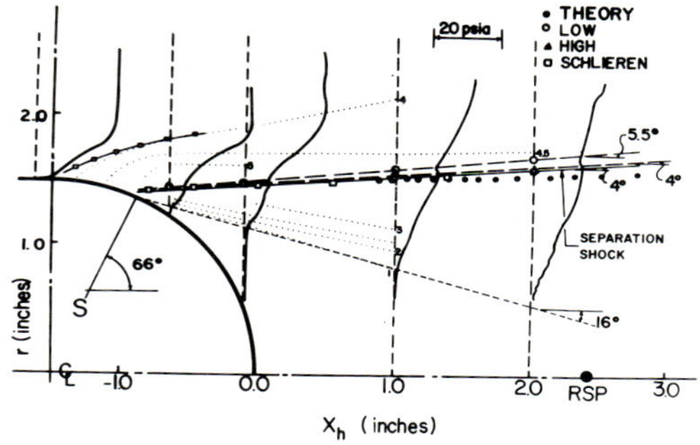

Fig. 12 Flow Field Map - Hemisphere

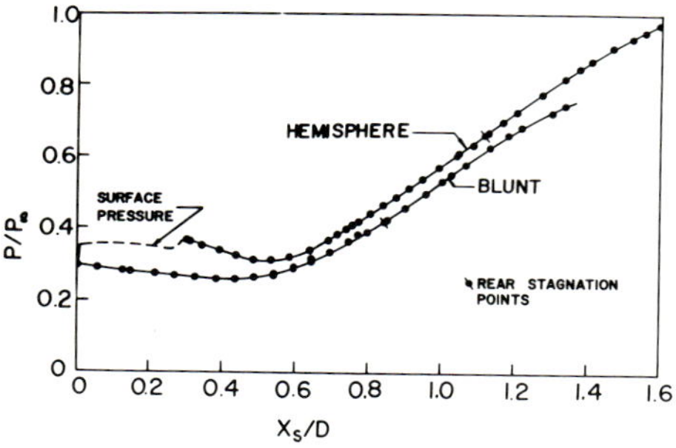

Fig. 13 Centerline Static Pressure - Composite

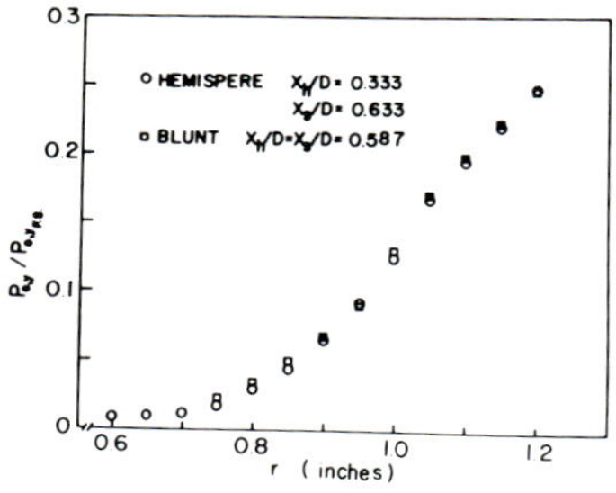

Fig. 14 Shear Layer Pitot Pressure - Composite

PART III
SOLID MECHANICS

PLASTICITY

Developments in Mechanics, Vol. 6. Proceedings of the 12th Midwestern Mechanics Conference: 23

Three Dislocation Concepts and Micromorphic Mechanics

W. D. CLAUS, JR.
FABRIC RESEARCH LABORATORIES, INC.

A. C. ERINGEN
PRINCETON UNIVERSITY

ABSTRACT

The ultimate objective of continuum dislocation theory is to provide a framework for the development of a consistent and useful macroscopic plasticity theory. Until that goal is reached, fragmentary results which illuminate specific facets of the general problem will be the rule. This paper contains three such separate results. First, the initial stress-couple stress problem is formulated using incompatible micropolar elasticity; continuous distributions of dislocations and disclinations are treated. Second, dislocation rate effects are investigated by including the dislocation rate tensor in the construction of constitutive relations. The relationship between this rate approach and a recent Russian plasticity theory is discussed. Finally, motivated by an interpretation of micromorphic kinematics, a slip motion is defined which is shown to lead to classical plasticity and the plastic strain tensor.

THE INITIAL STRESS-COUPLE STRESS PROBLEM

The initial stress problem is the original problem of continuum dislocation theory. For reviews see [1,2,3,4]. Lattice defects cause internal stresses in solids, and, in particular, a continuous distribution of dislocations causes internal stresses, "internal" as distinguished from those stresses arising from external loads on the body. Given the static distribution of dislocations, the stresses are to be determined. The initial dislocations existing in a body may be viewed as having been created by prior plastic deformation. If, in addition to displacements, microrotations are considered, a second type of lattice defect called a disclination [5] or rotational dislocation is included in the initial stress theory. In fact in the following treatment [6] the continuous distribution of disclinations is shown to cause couple stresses.

In the initial stress theory, the specified distribution of dislocations appears as a source term in the classical elasticity compatibility equations. Thus the phrase "incompatibile elasticity" is used to refer to the initial stress problem. In the following sections, the initial stress-couple stress problem is formulated by developing an incompatible micropolar elasticity. First the basic equations of micropolar elasticity are summarized and then continuous distributions of dislocations and disclinations are included in an incompatible micropolar elasticity. For background on static dislocation theories the reader is referred to the above referenced reviews and for a more detailed discussion of micropolar elasticity to [7].

Compatible Micropolar Elasticity

Micropolar elasticity is a special case of the micromorphic theory. For our purposes here, we begin with a short summary of the basic equations. The balance laws are

$$t_{k\ell,k} + \rho f_\ell = \rho \ddot{u}_\ell \qquad \text{(momentum)} \tag{1}$$

$$m_{k\ell,k} + e_{\ell mn}\, t_{mn} + \rho \ell_\ell = \rho j \ddot{\phi}_\ell \qquad \text{(moment of momentum)} \tag{2}$$

and the linear constitutive equations are explicitly

$$t_{k\ell} = \lambda \varepsilon_{mm}\delta_{k\ell} + (\mu+\kappa)\, \varepsilon_{k\ell} + \mu\varepsilon_{\ell k} \tag{3}$$

$$m_{k\ell} = \alpha\, \kappa_{mm}\delta_{k\ell} + \beta\kappa_{\ell k} + \gamma\kappa_{k\ell} \tag{4}$$

where

$$\varepsilon_{k\ell} = u_{\ell,k} - e_{k\ell m}\, \phi_m, \qquad \kappa_{k\ell} = \phi_{\ell,k} \tag{5}$$

In the above equations, $m_{k\ell}$ is the couple stress tensor, ℓ the body couple and ϕ the microrotation vector. A set of boundary conditions for micropolar elasticity are

$$t_{k\ell}n_k = T_\ell, \qquad m_{k\ell}n_k = M_\ell \qquad \text{on } S_t \tag{6}$$

$$u_k = U_k, \qquad \phi_k = \Phi_k \qquad \text{on } S_u = S-S_t$$

where T_ℓ and M_ℓ are prescribed surface tractions and moments on a part of the boundary S_t, and U_k and Φ_k are prescribed displacements and microrotations on the remaining part of the surface S_u.

The compatibility equations of micropolar elasticity are obtained from Eq. 5 by eliminating $\underset{\sim}{u}$ and $\underset{\sim}{\phi}$. They are

$$e_{kmn}\, \varepsilon_{n\ell,m} - \kappa_{\ell k} + \kappa_{mm}\delta_{k\ell} = 0 \tag{7}$$

$$e_{kmn}\kappa_{n\ell,m} = 0$$

Notice that these are first order partial differential equations in the strains ε and κ as contrasted with the second order equations of classical elasticity.

Our purpose in the next section is to extend the micropolar elasticity to include dislocation and disclinations.

Incompatible Micropolar Elasticity

The total micropolar strains, now denoted by $\varepsilon^T_{k\ell}$ and $\kappa^T_{k\ell}$, are considered to be the sum of elastic and plastic parts, i.e.,

$$\varepsilon^T_{k\ell} = u_{\ell,k} - e_{k\ell m}\phi_m = \varepsilon^E_{k\ell} + \varepsilon^P_{k\ell} \tag{8}$$

$$\kappa^T_{k\ell} = \phi_{\ell,k} = \kappa^E_{k\ell} + \kappa^P_{k\ell} \tag{9}$$

Compatibility equations on the strains are obtained by differentiating Eq. 8 and Eq. 9 and eliminating u_k and ϕ_k. Thus

$$e_{kmn}\varepsilon^E_{n\ell,m} - \kappa^E_{\ell k} + \kappa^E_{mm}\delta_{k\ell} = \Pi_{k\ell} \tag{10}$$

$$e_{kmn}\kappa^E_{n\ell,m} = \theta_k \tag{11}$$

where

$$\Pi_{k\ell} = -e_{kmn}\,\varepsilon^P_{n\ell,m} + \kappa^P_{\ell k} - \kappa^P_{mm}\delta_{k\ell} \tag{12}$$

$$\theta_{k\ell} = -e_{kmn}\,\kappa^P_{n\ell,m} \tag{13}$$

We call Π the micropolar <u>dislocation density</u> and θ the micropolar <u>disclination density</u> for reasons which will be discussed below. Eq. 12 shows that ε^P and κ^P are the sources of micropolar dislocations; Eq. 13 shows that disclinations arise from the gradients of $\underset{\sim}{\kappa}^P$.

By differentiating Eq. 12 and Eq. 13 we obtain

$$\Pi_{k\ell,k} + e_{\ell mn}\,\theta_{mn} = 0 \tag{14}$$

$$\theta_{k\ell,k} = 0 \tag{15}$$

which are identities which must be satisfied by $\underset{\sim}{\Pi}$ and $\underset{\sim}{\theta}$. Also differentiation of Eq. 10 with respect to a free index and using Eq. 11 yields

$$e_{kmp}e_{\ell nq}\varepsilon^E_{mn,pq} + e_{k\ell m}\,\kappa^E_{rr,m} = e_{\ell mn}\Pi_{kn,m} + \theta_{\ell k} \tag{16}$$

The symmetric part of this equation is

$$e_{kmp}\,e_{\ell nq}\,\varepsilon^E_{(mn),pq} = -e_{(k|mn|}\Pi_{\ell)m,n} + \theta_{(k\ell)} \tag{17}$$

When $\theta_{(k\ell)}$ vanishes, Eq. 17 is formally the same (with a minus sign) as Kroner's incompatibility equations [8].

The skew-symmetric part of Eq. 16 gives the incompatibility equations for the skew part of the micropolar strain tensor

$$e_{mnp}\,\varepsilon^E_{[np],km} + 2\kappa^E_{mm,k} = \Pi_{mm,k} \tag{18}$$

Eq. 18 appears to be new; it reflects the fact that the micropolar theory is based on non-symmetric strain and stress tensors.

The static, initial stress-couple stress problem can now be formulated using the incompatible micropolar theory presented above. Here, one considers the sources of internal stress (the dislocation density $\underset{\sim}{\Pi}$ and the disclination density $\underset{\sim}{\theta}$) to be given throughout a body, and the problem is to determine the initial stresses and couple stresses. The basic system of equations to be solved is

$$t_{k\ell,k} = 0 \tag{19}$$

in V

$$m_{k\ell,k} + e_{\ell mn}\,t_{mn} = 0 \tag{20}$$

$$t_{k\ell}n_k = 0, \; m_{k\ell}n_k = 0 \quad \text{on } S \tag{21}$$

with $\underset{\sim}{\Pi}(\underset{\sim}{x})$ and $\underset{\sim}{\theta}(\underset{\sim}{x})$ that appear in the incompatibility equations

$$e_{kmn}\, \varepsilon^E_{n\ell,m} - \kappa^E_{\ell k} + \kappa^E_{mm}\delta_{k\ell} = \Pi_{k\ell}(\underset{\sim}{x}) \tag{22}$$

in V

$$e_{kmn}\, \kappa^E_{n\ell,m} = \theta_{k\ell}(\underset{\sim}{x}) \tag{23}$$

to be specified throughout the body. The constitutive equations to be used in the isotropic case are

$$t_{k\ell} = \lambda\, \varepsilon^E_{mm}\, \delta_{k\ell} + (\mu+\kappa)\varepsilon^E_{k\ell} + \mu\, \varepsilon^E_{\ell k} \tag{24}$$

$$m_{k\ell} = \alpha\, \kappa^E_{mm}\, \delta_{k\ell} + \beta\, \kappa^E_{\ell k} + \gamma\, \kappa^E_{k\ell} \tag{25}$$

Recall that $\underset{\sim}{\Pi}(\underset{\sim}{x})$ and $\underset{\sim}{\theta}(\underset{\sim}{x})$ must be specified subject to Eq. 14 and Eq. 15.

Other work on continuum theories of disclinations include deWit [9], Anthony et al [10]. Anthony et al present a short discussion of the definitions of the dislocation and disclination tensors only; they have no balance laws, incompatibility equations, or constitutive relations. On the other hand, deWit covers all of the above points using a symmetric strain tensor and comments that his work could be extended to non-symmetric strains by using the Cosserat deformation tensors.

DISLOCATION RATE EFFECTS

Using micronorphic mechanics, a dynamical model of dislocations moving elastically about their equilibrium configuration was developed in [6] and [11]. The applicability of that model to dislocation dispersion of elastic waves was qualitatively demonstrated in [6] and [12]. In this section, dislocation rate effects are studied within the micromorphic model; it is believed that this approach offers a potential continuum method of attack on the dislocation damping problem [13]. For a more detailed discussion of the material presented here see [6].

The elaborate basic equations of the dynamical model for reversible elastic motions are in [6,11] and will not be unnecessarily reproduced here. The kinematics and balance laws are used here in unchanged form; additional constitutive equations [6] are constructed below. We consider the free energy ψ to be of the form

$$\psi = \psi(\underset{\sim}{e}, \underset{\sim}{\varepsilon}, \underset{\sim}{a}, \dot{\underset{\sim}{a}}, \theta) \tag{26}$$

where the strain tensors are related to the displacement vector u_k and the microdeformation tensor $\phi_{k\ell}$ by

$$\begin{aligned} e_{k\ell} &= \frac{1}{2}(u_{k,\ell} + u_{\ell,k}) \\ \varepsilon_{k\ell} &= u_{\ell,k} + \phi_{k\ell} \end{aligned} \tag{27}$$

$\underset{\sim}{a}$ is a third order tensor related to the dislocation density $\underset{\sim}{\alpha}$ by

$$\begin{aligned} a_{k\ell m} &= -\frac{1}{2}\, e_{\ell mn}\, \alpha_{kn} \\ \alpha_{k\ell} &= -\, e_{\ell mn}\, \phi_{kn,m} \end{aligned} \tag{28}$$

and θ is the temperature. Also, the stress $t_{k\ell}$, the volume stress average $s_{k\ell}$, the first stress moments $\lambda_{k\ell m}$ and the entropy η are all assumed to be functions of the variables listed in Eq. 26. A superposed dot denotes time differentiation.

The linear spatial form of the micromorphic energy equation is

$$\rho\dot{\varepsilon} = [2t_{(k\ell)}-s_{k\ell}]\,\dot{e}_{k\ell} + (s_{k\ell}-t_{k\ell})\,\dot{\varepsilon}_{\ell k} - \lambda_{k\ell m}\dot{\gamma}_{\ell mk} + q_{k,k} + \rho h \qquad (29)$$

q_k is the heat flux vector and h is the heat source per unit mass. The free energy is related to the internal energy density ε by

$$\psi = \varepsilon - \theta\eta \qquad (30)$$

For construction of the constitutive relations, the second law of thermodynamics is needed in the form

$$\rho\dot{\eta} - \left(\frac{q_k}{\theta}\right)_{,k} - \frac{\rho h}{\theta} \geq 0 \qquad (31)$$

Combining Eq. 29 and Eq. 30 with Eq. 31 results in the second law in the form

$$-\rho(\dot{\psi}+\dot{\theta}\eta) + [2t_{(k\ell)} - s_{k\ell}]\,\dot{e}_{k\ell} + (s_{k\ell}-t_{k\ell})\,\dot{\varepsilon}_{\ell k}$$
$$-\lambda_{k\ell m}\,\dot{\gamma}_{\ell mk} + \frac{q_k\theta_{,k}}{\theta} \geq 0 \qquad (32)$$

Substituting the constitutive assumptions Eq. 26 into Eq. 32 yields

$$[2t_{(k\ell)} - s_{k\ell} - \rho\frac{\partial\psi}{\partial e_{k\ell}}]\,\dot{e}_{k\ell} + [s_{k\ell} - t_{k\ell} - \rho\frac{\partial\psi}{\partial\varepsilon_{\ell k}}]\,\dot{\varepsilon}_{\ell k}$$
$$- [\lambda_{k\ell m} + \rho\frac{\partial\psi}{\partial a_{\ell mk}}]\,\dot{a}_{\ell mk} - \lambda_{(k|\ell|m)}\,\dot{\gamma}_{\ell(mk)} \qquad (33)$$
$$- \rho\frac{\partial\psi}{\partial\dot{a}_{\ell mk}}\,\ddot{a}_{\ell mk} + \frac{q_k\theta_{,k}}{\theta} \geq 0$$

Since $\dot{e}_{k\ell}$, $\dot{\varepsilon}_{\ell k}$, $\dot{\gamma}_{\ell(mk)}$, $\ddot{a}_{\ell mk}$ and $\theta_{,k}$ can all be varied independently and the coefficients of these terms are not functions of them, it follows from Eq. 33 that

$$2t_{(k\ell)} - s_{k\ell} = \rho\frac{\partial\psi}{\partial e_{k\ell}}$$
$$s_{k\ell} - t_{k\ell} = \rho\frac{\partial\psi}{\partial\varepsilon_{\ell k}} \qquad (34)$$
$$\lambda_{(k|\ell|m)} = 0,\quad q_k = 0,\quad \frac{\partial\psi}{\partial\dot{a}_{\ell mk}} = 0$$

Thus Eq. 33 reduces to

$$-\left(\lambda_{k\ell m} + \rho\frac{\partial\psi}{\partial a_{\ell mk}}\right)\dot{a}_{\ell mk} \geq 0 \qquad (35)$$

Since ψ is independent of $\dot{\underset{\sim}{a}}$ we introduce ${}_E\underset{\sim}{\lambda}$ the recoverable part of $\underset{\sim}{\lambda}$ by

$${}_E\lambda_{k\ell m} = -\rho\frac{\partial\psi}{\partial a_{\ell mk}} \qquad (36)$$

and call ${}_D\tilde{\lambda}$ the dissipative part of $\tilde{\lambda}$ defined by

$$ {}_D\lambda_{k\ell m} = \lambda_{k\ell m} - {}_E\lambda_{k\ell m} \tag{37} $$

Thus Eq. 35 reduces to

$$ -{}_D\lambda_{k\ell m}\, \dot{a}_{\ell mk} \geq 0 \tag{38} $$

and the constitutive relation for the tensor lambda can be written as

$$ \lambda_{k\ell m} = {}_E\lambda_{k\ell m}(\tilde{a}) + {}_D\lambda_{k\ell m}(\tilde{a},\dot{\tilde{a}}) \tag{39} $$

For later comparison with Russian work Eq. 39 is written in terms of the dislocation density $\tilde{\alpha}$ instead of $\tilde{a}$ by using Eq. 28

$$ \lambda_{k\ell m} = e_{krm} \frac{\partial \Sigma}{\partial \alpha_{\ell r}} + {}_D\lambda_{k\ell m}(\tilde{\alpha},\dot{\tilde{\alpha}}) \tag{40} $$

and where

$$ \Sigma = \rho\psi \tag{41} $$

Since ${}_D\lambda_{k\ell m} = -{}_D\lambda_{m\ell k}$ we introduce the second order tensor ${}_D\lambda_{k\ell}$ by putting

$$ {}_D\lambda_{k\ell} = \frac{1}{2} e_{kmn}\, {}_D\lambda_{k\ell m}, \quad {}_D\lambda_{k\ell m} = e_{kmn}\, {}_D\lambda_{n\ell} \tag{42} $$

Using Eq. 40 and Eq. 42 the micromorphic moment of momentum balance law can be written as

$$ e_{krm}\left(\frac{\partial \Sigma}{\partial \alpha_{\ell r}} - {}_D\lambda_{r\ell}\right)_{,k} + t_{m\ell} - s_{m\ell} + \rho \ell_{\ell m} = \rho \dot{\sigma}_{\ell m} \tag{43} $$

Russian School

The recent work of Berdichevskii and Sedov [14] is closely related to micromorphic mechanics in that they also introduce nine new generalized displacement degrees of freedom in addition to the classical three. These authors construct general balance equations from a variational prinicple, and then they show how the classical elastic, plastic, and fluid constitutive relations can be constructed within their framework. In their final section they choose appropriate variational variables and obtain a dislocation-plasticity theory which utilizes the concepts of dislocation density, plastic yield function, strain-rate, and entropy production. For comparison purposes, their theory is summarized as follows:

$$ p_{k\ell,k} + F_\ell = \rho \dot{v}_\ell \tag{44} $$

$$ \frac{1}{2} e_{\ell mn}(R_{mk} + \Sigma_{mk})_{,n} + p_{k\ell} - Q_{k\ell} = 0 \tag{45} $$

$$ \rho\theta\dot{\eta} = -\theta\left(\frac{q_k}{\theta}\right)_{,k} + \sigma_1(\nabla\theta) + \sigma_2\,(\dot{\tilde{\varepsilon}}'',\dot{\tilde{S}}) \tag{46} $$

No names are given to the new quantities appearing in these balance laws, but they can be identified from the constitutive relations:

$$ p_{k\ell} = \frac{\partial U}{\partial \varepsilon'_{k\ell}}, \quad \Sigma_{k\ell} = \frac{\partial U}{\partial S_{k\ell}} \quad \text{(reversible part)} \tag{47} $$

$$ Q_{k\ell} = Q_{\ell k} = \frac{\partial \sigma_2}{\partial \dot{\varepsilon}''_{k\ell}}, \quad R_{k\ell} = \frac{\partial \sigma_2}{\partial \dot{S}_{k\ell}} \quad \text{(irreversible part)} \tag{48} $$

$$ q_k = -\frac{\partial \sigma_1}{\partial \theta_{,k}} $$

Also a yield function is postulated in the form

$$f(Q_{k\ell}, R_{k\ell}) = 0 \tag{49}$$

In the above equations, ε' and ε'' are the elastic and plastic strains, $S_{k\ell}$ the dislocation density, and a superposed dot denotes time rate.

In Eq. 45 the contribution to the kinetic energy in the Lagrangian from the dislocations has been neglected; hence, there is no inertia term on the right-hand side. The micromorphic moment of momentum equation, Eq. 43, is similar to Eq. 45 which Berdichevskii and Sedov refer to as the "internal parameters" equation. They interpret the skew part of Eq. 45 as an angular momentum equation. No interpretation is given to the symmetric part. Their use of the entropy rate equation in constructing constitutive equations is quite different from the current practice.

SLIP MOTIONS AND PLASTICITY

An approach to a theory of plastic flow is presented in this section [6]; the definition of slip motion introduced here is motivated by the technique used to obtain the microdeformation tensor in micromorphic mechanics.

The classical motion without slip of a material particle $\underset{\sim}{X}$ is

$$x_k = x_k(\underset{\sim}{X}, t) \tag{50}$$

Since the classical motion is smooth, the motion $\underset{\sim}{x}'$ of a neighboring particle $\underset{\sim}{X}'$ is given approximately by

$$x'_k = x_k(\underset{\sim}{X}, t) + x_{k,K} dX_K + \cdots \tag{51}$$

The micromorphic kinematics is here interpreted as the slipped motion of a particle $\underset{\sim}{X}'$ near to $\underset{\sim}{X}$ due to plastic deformation.

The slipped motion is

$$x'_k = x_k(\underset{\sim}{X}, t) + \chi_{kK}(\underset{\sim}{X}, t) dX_K + \cdots \tag{52}$$

We take Δx_{kK} as a measure of the extent of slip

$$\Delta x_{kK} = x_{k,K} - \chi_{kK} \tag{53}$$

The difference between the actual slipped motion and the motion of $\underset{\sim}{X}'$ if the deformation were smooth is measured by Δx_{kK}. If $\Delta x_{kK} = 0$, the motion is not slipped and classical.

The slip concept introduced above leads us to consider a solid which possesses a stress potential of the form

$$\Sigma = \Sigma(x_{k,K}, \Delta x_{kK}) \tag{54}$$

If Δx_{kK} is zero, Eq. 54 describes an elastic solid; with Δx_{kK} nonzero, Eq. 54 describes a plastic solid. The principle of objectivity requires that Σ be a function of $\underset{\sim}{C}$ and $\underset{\sim}{A}$ where

$$C_{KL} = x_{k,K} x_{k,L}; \quad A_{KL} = x_{k,K} \Delta x_{kL} \tag{55}$$

Instead of the deformation tensors $\underset{\sim}{C}$ and $\underset{\sim}{A}$, we introduce the strain tensors

$$E_{KL} = \frac{1}{2}(C_{KL} - \delta_{KL}); \quad E'_{KL} = \frac{1}{2}(A_{KL} - \delta_{KL}) \tag{56}$$

Σ may also be considered a function of $\underset{\sim}{E}$ and $\underset{\sim}{E}'$. We next identify the plastic strain $\underset{\sim}{E}''$ as

$$2E''_{KL} = x_{k,K}\chi_{kL} \tag{57}$$

so that we have, for the linear theory

$$E_{KL} = E'_{KL} + E''_{KL} \tag{58}$$

The micromorphic energy equation can be written as [6]

$$\rho_o\dot{\varepsilon} = 2T_{KL}\dot{E}_{KL} + 2S_{KL}\dot{E}''_{KL} + \Lambda_{KLM}\dot{\Gamma}_{KLM} \tag{59}$$

where T_{KL}, S_{KL} and Λ_{KLM} are related to $t_{k\ell}$, $s_{k\ell}$ and $\lambda_{k\ell m}$ by

$$\begin{aligned} T_{KL} &= \frac{j}{2}[2t_{(k\ell)} - s_{k\ell}]X_{K,k}X_{L,\ell} \\ S_{KL} &= j(t_{k\ell} - s_{k,\ell})X_{K,\ell}\chi_{Lk} + j\,\lambda_{k\ell m}X_{K,\ell}\chi_{Lm,k} \\ \Lambda_{KLM} &= j\,\lambda_{k\ell m}X_{K,\ell}\chi_{Lm}X_{M,k} \\ j &= \det(x_{k,K}) \\ \Gamma_{KLM} &= x_{k,K}\chi_{kL,M} \end{aligned} \tag{60}$$

$X_{K,k}$ and χ_{Kk} are the inverse deformation gradient and inverse micro-deformation tensors.

Heat conduction is neglected in writing Eq. 59. The constitutive assumptions of this model are

$$\begin{aligned} \psi &= \psi(E'_{(KL)}, E''_{KL}, \theta) \\ \eta &= \eta(E'_{(KL)}, E''_{KL}, \theta) \end{aligned} \tag{61}$$

where $E'_{[KL]}$ is not included since from Eq. 58

$$E'_{[KL]} = -E''_{[KL]} \tag{62}$$

The stress rates $\dot{T}_{KL}$, $\dot{S}_{KL}$, and $\dot{\Lambda}_{KLM}$, assumed to be functions of

$$E'_{(KL)}, \quad E''_{KL}, \theta, \quad \dot{E}'_{(KL)}, \quad \dot{E}''_{KL}, \quad \dot{\theta} \tag{63}$$

are homogeneous of degree one in the rate variables in order to render the stress rates independent of the time scale used to calculate them.

We wish to consider the special case of Eq. 61 and Eq. 63 where moment effects are negligible. We assume that the constitutive coefficients for $\underset{\sim}{\Lambda}$ are negligible as well as body moments and spin moments. Hence $\underset{\sim}{\Lambda} \cong 0$ and from the moment of momentum equation $(\lambda_{k\ell m,k} + t_{m\ell} - s_{m\ell} = o)$

$$t_{m\ell} - s_{m\ell} = o \tag{64}$$

From Eq. 60

$$\begin{aligned} T_{KL} &= \frac{j}{2}t_{k\ell}X_{K,k}X_{L,\ell} \\ S_{KL} &= 0, \ \Lambda_{KLM} = o \end{aligned} \tag{65}$$

The special forms adopted for $\dot{T}_{KL}$ and $\dot{S}_{KL}$ are that $\dot{\tilde{T}}$ is integrable in the time giving a relation of the form

$$T_{KL} = T_{KL}(E'_{(MN)}, E''_{MN}, \theta) \tag{66}$$

which is invertible for $E'_{(MN)}$ as a function of $\tilde{T}$, $\tilde{E}''$ and θ. From $\dot{\tilde{S}} = 0$ we extract a relation of the form

$$\dot{E}''_{KL} = A_{KLMN}\, \dot{T}_{MN} + B_{KL}\, \dot{\theta} \tag{67}$$

when $f = \kappa(\kappa \neq 0)$ and $\frac{\partial f}{\partial T_{KL}} \dot{T}_{KL} + \frac{\partial f}{\partial \theta} \dot{\theta} \geq 0$

where $f = f(\tilde{T}, \tilde{E}'', \theta) = \kappa$ is the yield function and

$$\dot{\kappa} = H_{KL}(\tilde{T}, \tilde{E}'', \theta)\, E''_{KL} \tag{68}$$

During the neutral loading and unloading

$$\dot{E}''_{KL} = 0$$

when

$$\frac{\partial f}{\partial T_{KL}} \dot{T}_{KL} + \frac{\partial f}{\partial \theta} \dot{\theta} < 0 \text{ on } f = \kappa,\ \dot{\kappa} = 0 \tag{69}$$

or when

$$f < \kappa, \qquad \dot{\kappa} = 0.$$

Recently Green and Naghdi [15] introduced a non-symmetric plastic strain tensor and showed how their earlier non-polar plasticity theory [16] could be viewed in the context of multipolar mechanics. In the section above we have demonstrated that their plasticity model can also be viewed as a constrained micromorphic model. The "constraints" are the definition of the plastic strain tensor and the omission of moment effects.

An explicit comparison of the above approach with that of Green and Naghi [15] is given in Table I.

TABLE I

Green and Naghdi [15]		Micromorphic	
$x_{i,A}$	x_{iB}	$x_{k,K}$	χ_{kK}
S'_{KA}, $\bar{S}_{AB}$,	S_{KAB}	$2T_{KL}$, S_{KL},	Λ_{KLM}
e_{KA},	e''_{AB}	E_{KL},	Ψ_{KL}
e_{ABK}		Γ_{KLM}	

ACKNOWLEDGMENTS

Part of this work was done while the first author was a Research Fellow of the Textile Research Institute, Princeton, New Jersey. The second author wishes to thank the Office of Naval Research for partial support.

REFERENCES

[1] Kazuo Kondo: On the Analytical and Physical Foundations of the Theory of Dislocations and Yielding by the Differential Geometry of Continua. Int. J. Engng. Sci., Vol. 12, p.219-251 (1964).

[2] B. A. Bilby: Continuous Distributions of Dislocations, in Progress in Solid Mechanics, Vol. I. Edited by I. N. Sneddon and R. Hill, Amsterdam: North Holland Publishing Co., 1960.

[3] E. Kroner: Dislocation Field Theory, in Theory of Crystal Defects. Edited by Boris Gruber, New York: Academic Press, 1966.

[4] Roland deWit: The Continuum Theory of Stationary Dislocations, in Solid State Physics, Supplement 10, p.249-292. Edited by F. Seitz and D. Turnbull, New York: Academic Press, 1960.

[5] F. R. N. Nabarro: Theory of Crystal Dislocations. London: Oxford University Press, 1967.

[6] W. D. Claus, Jr.: On Dislocations, Plasticity and Micromorphic Mechanics, a dissertation submitted to the Department of Aerospace and Mechanical Sciences, Princeton University, October 1969.

[7] A. C. Eringen: Theory of Micropolar Elasticity, in Fracture: An Advanced Treatise. Vol. II. Edited by H. Liebowitz, New York: Academic Press, 1968.

[8] E. Kroner: Plastizitat und Versetzungen, in Arnold Sommerfield, Mechanik der Deformierbaren Medien, 5th Edition. Leipzig: Geest and Portig K.-G., 1964.

[9] Roland deWit: Linear Theory of Static Disclinations, to appear in Proc. Conf. on Fundamental Aspects of Dislocation Theory. NBS. Gaithersburg, Md. April 21-25, 1969.

[10] K. Anthony, U. Essmann, A. Seeger, and H. Traüble: Disclinations and the Cosserat-Continuum with Incompatible Rotations, in Mechanics of Generalized Continua. p.355-358. Edited by E. Kröner, Berlin: Springer-Verlag, 1968.

[11] A. C. Eringen and W. D. Claus, Jr.: A Micromorphic Approach to Dislocation Theory and Its Relation to Several Existing Theories, to appear in Proc. Conf. on Fundamental Aspect of Dislocation Theory. NBS. Gaithersburg, Md. April 21-25, 1969.

[12] W. D. Claus, Jr. and A. C. Eringen: Dislocation Dispersion of Elastic Waves, to appear in the Int. J. Engng. Sci.

[13] A. V. Granato and A. C. Lücke: The Vibrating String Model of Dislocation Damping, in Physical Acoustics, Vol. 4A, p.225-276, Edited by W. P. Mason. New York: Academic Press, 1966.

[14] V. L. Berdichevskii and L. I. Sedov: Dynamic Theory of Continuously Distributed Dislocations. Its Relation to Plastic Theory, PPM, Vol. 31, p.989-1006 (1967).

[15] A. E. Green and P. M. Nagdhi: Plasticity Theory and Multipolar Continuum Mechanics, Mathematika, Vol. 12, p.21-26 (1965).

[16] A. E. Green and P. M. Nagdhi: A General Theory of Elastic-Plastic Continuum, Arch.Rat.Mech.Anal., Vol.18,p.251-281 (1965).

Developments in Mechanics, Vol. 6. Proceedings of the 12th Midwestern Mechanics Conference: 24

A Method for Micro-hardness Analysis of an Elastoplastic Material

JULIAN J. WU and FREDERICK F. LING
RENNSSELAER POLYTECHNIC INSTITUTE

ABSTRACT

A method is presented to correlate the characteristics of an elastoplastic, work-hardening material with measurable quantities from micro-hardness experiments.

First, the problem of an elastic plate pressed on by a rigid indenter of a given profile is considered. Using Hankel transforms, the relation of surface parameters is expressed in terms of a singular integral equation. The non-singular part of the kernel of this singular integral equation is closely approximated by a polynomial of even order. Then, for a given load, the radius of contact circle, the depth of indentation and the pressure distribution are obtained by solving a set of linear, simultaneous algebraic equations. Stress distributions within the plate are next computed by convolution.

For the post-yield behavior of the plate, the formality of simultaneous equations as in the case of the elasticity solution is assumed, except that the constant coefficients which characterize material properties are determined by successive approximations using the finite-element method. A numerical example is given to illustrate the method.

INTRODUCTION

In simplest words, hardness means resistance to deformation, and in the selection of materials for application, hardness is a singularly important index. Since hardness measurements involve mainly a process of deformation, how are the measurable parameters related to basic properties such as Young's modulus and yield stress? For indentation hardness, the first important correlation was due to Hertz [1,2].

Though much is now known about contact problems and hardness [3,4,5], there has been little work done in three-dimensional contact problems of elastoplastic material other than Ishlinsky's ideal plastic indentation soluting using an artificial stress-strain relation of Karman and Haar.

This paper presents a method by means of which an analytic correlation between the measurable quantities from a micro-hardness test and the properties of an elastoplastic material can be made.

In Section 2, the solution of an elastic plate under a rigid indenter of a general profile is considered. The formulation of the problem is similar to that of Sneddon [6]. It is shown that the solution can be obtained by approximating the non-singular part of the kernel function by an even-ordered polynomial. Assuming that the analytic form of solution for elastic state is valid for the state of incipient plasticity, the finite-element method is used to deduce the effect of each small increment of load. Thus a step-by-step procedure is developed for the post-yield state of stress. This is described in Section 3. Finally in Section 4, a numerical example is worked out to show how this method is applied.

SOLUTIONS FOR AN ELASTIC PLATE

Formulation of the Problem

Consider an infinite plate of finite thickness H resting on a rigid foundation and pressed on by a rigid indenter with load N as shown in Fig.1. Reference axes (r,z) are as shown. Love's stress function χ for elastic, axisymmetric problems is defined by Eq.(1).

$$\nabla^4\chi = \left(\frac{\partial^2}{\partial r^2} + \frac{1}{r}\frac{\partial}{\partial r} + \frac{\partial^2}{\partial z^2}\right)^2 \chi = 0, \tag{1}$$

with stress and displacement components expressible in terms of χ:

$$\begin{aligned}
\sigma_{rr} &= \frac{\partial}{\partial z}\left(\nu\nabla^2 - \frac{\partial^2}{\partial r^2}\right)\chi \\
\sigma_{\theta\theta} &= \frac{\partial}{\partial z}\left(\nu\nabla^2 - \frac{1}{r}\frac{\partial}{\partial r}\right)\chi \\
\sigma_{zz} &= \frac{\partial}{\partial z}\left[(2-\nu)\nabla^2 - \frac{\partial^2}{\partial z^2}\right]\chi \\
\sigma_{rz} &= \frac{\partial}{\partial r}\left[(1-\nu)\nabla^2 - \frac{\partial^2}{\partial z^2}\right]\chi \\
\sigma_{r\theta} &= \sigma_{z\theta} = 0,
\end{aligned} \tag{2}$$

and

$$\begin{aligned}
u &= -\frac{1+\nu}{E}\frac{\partial^2\chi}{\partial r\partial z} \\
v &= \frac{1+\nu}{E}\left[2(1-\nu)\nabla^2 - \frac{\partial^2}{\partial z^2}\right]\chi.
\end{aligned} \tag{3}$$

Material constants ν and E are Poisson's ratio and Young's modulus, respectively. The displacement component in r-direction is denoted by u, and in z-direction by v.

The boundary conditions between the plate and the rigid base (at z = 0) are given by Eq.(4) if the interface is lubricated and by Eq.(5) if the interface is bonded:

$$\sigma_{rz}\Big|_{z=0} = 0, \qquad v\Big|_{z=0} = 0 \tag{4}$$

$$u\Big|_{z=0} = 0, \qquad v\Big|_{z=0} = 0 \tag{5}$$

Although Eq.(4) is used in this paper, the approach applies equally well to boundary conditions described by Eq.(5).

On the top of the plate, the conditions are that: (1) there is no shearing traction inside and outside of the contact area; (2) the vertical displacement within the contact area is prescribed by the shape of the indenter; and (3) the normal traction is zero outside the contact area. Hence,

$$\sigma_{rz}\big|_{z=H} = 0, \quad v\big|_{z=H,0\leq r\leq A} = -D + \varphi(r) \text{ and } \sigma_{zz}\big|_{z=H,r>A} = 0 \;. \tag{6}$$

The function $\varphi(r)$ describes the profile of the indenter, which is given. Let $\varphi(r) = 0$, so that D represents the depth of indentation, an unknown quantity. Note the radius of contact area A is unknown. Also unknown is the normal pressure distribution f(r) within the contact area. Of course,

$$2\pi \int_0^A \rho f(\rho)\, d\rho = N \tag{7}$$

is the total load applied.

For computation, it is convenient to introduce dimensionless variables wherein all lengths are normalized by A and stresses by the shear modulus $G = E/2(1+\nu)$. Subsequent reference to Eqs.(1) through (7) will be the dimensionless form. Also, no new symbols will be introduced except $\lambda = H/A$.

Formal Solution by Integral Transforms

Apply the Hankel transform,

$$\bar{f}(p) = \int_0^\infty r f(r) J_0(pr)\, dr \tag{9}$$

whose inverse is

$$f(r) = \int_0^\infty p\bar{f}(p) J_0(pr)\, dp, \tag{10}$$

to relevant parts of Eqs.(1) through (8) with super bar denoting the transformed quantities. Thus, result the differential equation

$$\left(\frac{\partial^2}{\partial z^2} - p^2\right)\bar{X} = 0, \tag{11}$$

and boundary conditions

$$\left.\begin{aligned} &\nu \frac{\partial^2}{\partial z^2} X + (1-\nu)p^2 \bar{X} = 0 \\ &(1-2\nu)\frac{\partial^2}{\partial z^2}\bar{X} - 2(1-\nu)p^2\bar{X} = 0, \end{aligned}\right\} (z = 0) \tag{12}$$

$$\left.\begin{aligned} &\nu \frac{\partial^2}{\partial z^2}\bar{X} + (1-\nu)p^2\bar{X} \\ &(1-\nu)\frac{\partial^3}{\partial z^3}\bar{X} - (2-\nu)p^2\frac{\partial\bar{X}}{\partial z} = g(p), \end{aligned}\right\} (z = \lambda) \tag{13}$$

where

$$g(p) = \int_0^1 r f(r) J_0(pr)\, dr. \tag{14}$$

Equation (11) with boundary conditions (12) and (13) possesses the solution

$$\bar{\chi} = A \sinh pz + Bz \cosh pz, \tag{15}$$

where

$$pA = \frac{-2(2\nu \sinh p\lambda + p\lambda \cosh p\lambda)g(p)}{p^2(\sinh 2p\lambda + 2p\lambda)}, \quad B = \frac{(2 \sinh p\lambda)g(p)}{p^2(\sinh 2p\lambda + 2p\lambda} . \tag{16}$$

Using the inversion theorem and assuming f(r) is such that the order of integration can be interchanged, χ can be written as

$$\chi = 2\pi \int_0^1 \rho f(\rho)\mathcal{K}(r,\rho)\, d\rho, \tag{17}$$

where

$$\mathcal{K}(r,\rho) = \frac{1}{2\pi} \int_0^\infty J_o(pr)J_o(p\rho) \cdot$$

$$\cdot \left\{\frac{-2(2\nu \sinh p\lambda + p\lambda \cosh p\lambda)\sinh pz + 2(\sinh p\lambda)pz \cosh pz}{p^2(\sinh 2p\lambda + 2p\lambda}\right\} dp. \tag{18}$$

The fundamental solution of Love's stress function for the plate, i.e., solution for a ring load, is $\mathcal{K}(r,\rho)$. In special cases when $\lambda \to \infty$, $\mathcal{K}(r,\rho)$ becomes the fundamental solution of a half-space body due to a ring load.

With χ known, the stress and displacement components can be obtained from Eqs.(2) and (3). They are given in the Appendix of Ref.[10]. In particular,

$$v|_{z=\lambda} = 2\pi \int_0^1 \rho f(\rho)K(r,\rho)\, d\rho, \tag{19}$$

where

$$K(r,\rho) = \frac{1-\nu}{2\pi} \int_0^\infty J_o(pr)J_o(p\rho) \frac{\cosh 2p\lambda - 1}{\sinh 2p\lambda + 2p\lambda}\, dp. \tag{20}$$

In order that the second of the boundary condition (6) is satisfied:

$$-D + \varphi(r) = 2\pi \int_0^1 \rho f(\rho)K(r,\rho)\, d\rho \qquad (0 \le r \le 1). \tag{21}$$

Also, from Eq.(7),

$$2\pi \int_0^1 \rho f(\rho)\, d\rho = N. \tag{7a}$$

Equations (21) and (7a) are used to solve for the unknown function f(r) and the quantity D.

Solution by an Approximate Kernel Function

In the formal solution of the previous section, the key is seen to be the dual integral equations [Eqs.(21) and (7a)]. In general, $\varphi(r)$ of Eq.(21) can be expressed by an even-ordered polynomial, i.e.,

$$-D + \varphi(r) = \sum_{i=0}^{n} \varphi_i r^{2i}, \tag{22}$$

where φ_i, $i = 1, 2, \ldots, n$ and the value of n are all given as the profile of the indenter. But

$$\varphi_o = -D \tag{23}$$

is to be solved.

The solution presented here is based on the facts that (1) the kernel function $K(r,\rho)$ can be separated into two parts: a singular part and a non-singular part; and (2) the solution for the same equations corresponding to a half-space body is known.

Using the identity,

$$\frac{\cosh x - 1}{\sinh x + x} = 1 - \left[\frac{2e^{-x}}{1 + e^{-x}} + \frac{x}{\sinh x + x} - \frac{2xe^{-x}}{(1 + e^{-x})(\sinh x + x)}\right], \tag{24}$$

in Eq.(20), one can write

$$K(r,\rho) = K_o(r,\rho) - K_1(r,\rho) \tag{25}$$

with

$$K_o(r,\rho) = \frac{1 - \nu}{2\pi} \int_o^1 J_o(pr)J_o(p\rho)\, dp, \tag{26}$$

and

$$K_1(r,\rho) = \frac{1 - \nu}{2\pi} \int_o^1 J_o(pr)J_o(p\rho) \cdot$$

$$\cdot \left[\frac{2e^{-2p\lambda}}{1 + e^{-2p\lambda}} + \frac{2p\lambda}{\sinh 2p\lambda + 2p\lambda} - \frac{4p\lambda e^{-2p\lambda}}{(1 + e^{-2p\lambda})(\sinh 2p\lambda + 2p\lambda)}\right] dp. \tag{27}$$

The fundamental solution for a half-space body due to a ring load is $K_o(r,\rho)$, which is singular. The contribution due to the thickness of the plate is $K_1(r,\rho)$ and is non-singular. Also, it is observed that

$$2\pi \int_o^1 \rho(1 - \rho^2)^{i-\frac{1}{2}} K_o(r,\rho)\, d\rho = \sum_{j=0}^{i} A_j^i\, r^{2j},$$

where A_j^i's are known constants dependent on values of i and j only. This observation emanates from the fact that the Legendre polynomials are eigenfunctions of the kernel $K_o(r,\rho)$ functions [7].

Equations (21) and (25) lead to

$$\varphi_o + \varphi(r) = \sum_{i=0}^{n} \varphi_i r^{2i} = 2\pi \int_o^1 \rho f(\rho)K_o(r,\rho)\, d\rho - 2\pi \int_o^1 \rho f(\rho)K_1(r,\rho)\, d\rho. \tag{28}$$

Since $K_1(r,\rho)$ is a well-behaved function and, in view of the axisymmetry, it can be approximated by an even-ordered polynomial symmetric with respect to r and ρ. Thus the (2n)th ordered polynomial approximation for $K_1(r,\rho)$ can be written as

$$K_1(r,\rho) = \sum_{m=0}^{n} \sum_{i=0}^{m'} C_i^m\, r^{2i} \rho^{2i} (r^{2m-4i} + \rho^{2m-4i}) \tag{29}$$

where $m' = m/2$ for m to be even, and $m' = (m - 1)/2$ for m to be odd. Assume the form of $f(\rho)$ to be

$$f(\rho) = (1 - \rho^2)^{\frac{1}{2}} \sum_{i=0}^{n} f_i (1 - \rho^2)^i. \tag{30}$$

No generality will be lost if the values of n in Eqs.(28), (29), and (30) are taken to be the same. One may take the largest of them all, for example. Using Eqs.(29) and (30) in integration, one obtains

$$2\pi \int_0^1 \rho f(\rho) K_1(r,\rho)\, d\rho = \sum_{i=0}^{n} \sum_{j=0}^{n} b_{ij} f_j r^{2i}, \tag{31}$$

where b_{ij} are constants depending on i, j and can be calculated from given C_k^m's. Now, using Eqs.(31) and (30) in Eq.(28) and identifying the terms with the same power of r, one obtains, after some rearrangement,

$$\varphi_i = \sum_{j=1}^{n} B_{ij} f_j \qquad (i = 1, 2, \ldots, n) \tag{32}$$

and

$$N = \sum_{j=1}^{n} a_j f_j. \tag{33}$$

When the f_j's and A have been obtained, the depth of indentation φ_o can be calculated, i.e.,

$$\varphi_o = \sum_{j=1}^{n} B_{oj} f_j. \tag{34}$$

In order to see how some of the coefficients look like and what are the actual operations involved, a case for n = 2 is worked out here. For this case,

$$v|_{z=\lambda} = \varphi_o + \varphi(r) = \varphi_o + \varphi_1 r^2 + \varphi_2 r^4,$$

$$K_1(r,\rho) = C_{00} + C_{01}(r^2 + \rho^2) + C_{02}(r^4 + \rho^4) + C_{12} r^2 \rho^2$$

and

$$f(\rho) = f_1(1 - \rho^2)^{\frac{1}{2}} + f_2(1 - \rho^2)^{\frac{3}{2}}.$$

It is noted that

$$2\pi \int_0^1 \rho f(\rho) K_o(r,\rho)\, d\rho = f_1(A_{10} + A_{11} r^2) + f_2(A_{20} + A_{21} r^2 + A_{22} r^4),$$

where

$$A_{10} = (1 - \nu)\pi/4, \quad A_{11} = -(1 - \nu)\pi/8$$

$$A_{20} = 9(1 - \nu)\pi/128, \quad A_{21} = -69(1 - \nu)\pi/224, \quad A_{22} = 663(1 - \nu)\pi/2240.$$

Also

$$2\pi \int_0^1 \rho f(\rho) K_1(r,\rho)\, d\rho$$

$$= f_1\left[(C_{00}a_{01} + C_{01}a_{11} + C_{02}a_{12}) + (C_{01}a_{01} + C_{12}a_{11})r^2 + C_{02}a_{01}r^4\right]$$

$$+ f_2\left[(C_{00}a_{02} + C_{01}a_{12} + C_{02}a_{22}) + (C_{01}a_{02} + C_{12}a_{12})r^2 + C_{02}a_{02}r^4\right],$$

where

$$a_{ij} = 2\pi \int_0^1 \rho^{2i+1}(1 - \rho^2)^{j - \frac{1}{2}}\, d\rho .$$

Applying the above to Eq.(21) and identifying terms of equal powers in r, one obtains

$$\varphi_o = B_{01}f_1 + B_{02}f_2 \tag{35}$$

$$\varphi_1 = B_{11}f_1 + B_{12}f_2, \quad \varphi_2 = B_{21}f_1 + B_{22}f_2 \tag{36}$$

where

$$B_{01} = A_{10} - (C_{00}a_{01} + C_{01}a_{11} + C_{02}a_{21}), \quad B_{02} = A_{20} - (C_{00}a_{02} + C_{01}a_{12} + C_{02}a_{22})$$

$$B_{11} = A_{11} - (C_{01}a_{01} + C_{12}a_{11}), \quad B_{12} = A_{21} - (C_{01}a_{02} + C_{12}a_{12})$$

$$B_{21} = -C_{02}a_{01}, \quad B_{22} = A_{22} - C_{02}a_{02}.$$

Equation (7a) gives

$$N = a_1f_1 + a_2f_2, \tag{37}$$

where

$$a_1 = 2\pi/3, \quad a_2 = 4\pi/15.$$

Next, Eqs.(35), (36), and (37) are used to solve for the unknowns f_1, f_2, and A. In actual dimensions, these equations become

$$\tilde{\varphi}_1 = \frac{1}{A}(B_{11}f_1 + B_{12}f_2), \quad \tilde{\varphi}_2 = \frac{1}{A^3}(B_{21}f_1 + B_{22}f_2)$$

and

$$\tilde{N} = GA^2(a_1f_1 + a_2f_2)$$

where the super ~ indicates quantities in their actual dimensions. This process can apparently be carried out with the same ease for any value of n.

ELASTOPLASTIC SOLUTION USING THE FINITE ELEMENT METHOD

In the region of plasticity near the incipient yield, the stress-strain relation is assumed to be of the incremental work-hardening type. In addition to this nonlinearity, the surface of the stress boundary conditions also changes with the total load applied in case of the indentation hardness analysis. A step-by-step procedure with the assistance of a finite element computer program is adapted for this part of the problem.

The finite element method has become increasingly important to the solution of practical problems of continuum mechanics [8]. Only the following points are to be noted here:

(1) The use of the finite element method in contact problems is hindered by the unknown boundary conditions. This is overcome here by a step-by-step iteration procedure.

(2) The stress-strain relation in the state of plasticity can be quite general. Various work-hardening rules as well as unloading can be incorporated with some programming considerations.

The procedure for solutions is now described. In the previous section, it was shown that the elastic solution of a thick plate under a rigid indenter can be reduced to the solution of a set of algebraic simultaneous equations. When two terms were used, for example, two constants f_1 and f_2 were to be found from the equations

$$\tilde{\varphi}_1 = \frac{1}{A} (B_{11}f_1 + B_{12}f_2) \tag{38}$$

$$\tilde{\varphi}_2 \quad \frac{1}{A^3} (B_{21}f_1 + B_{22}f_2), \tag{39}$$

where A is assumed to be given instead of N. We can then obtain N and $\tilde{\varphi}_o$ from the equations

$$N = GA^2(a_1f_1 + a_2f_2), \tag{40}$$

and

$$\tilde{\varphi}_o = A(B_{01}f_1 + B_{02}f_2). \tag{41}$$

All quantities shown above have been defined in the previous section. It is to be emphasized in particular that B_{ij}'s are constants which are dependent on the plate parameter $\lambda = \frac{H}{A}$ and the material properties of the plate.

It will be assumed that, in states of incipient plastic yielding, the solution can be expressed in the same form as those of Eqs.(38) through (41), and that B_{ij}'s will depend also on the stress state in the plate.

Equations (38) and (39) can be rewritten in the following form:

$$\tilde{\varphi}_1 = \frac{1}{A} B_{11}f_1(1 + r_1) \tag{42}$$

$$\tilde{\varphi}_2 = \frac{1}{A^3} B_{22}f_2(r_2 + 1), \tag{43}$$

where

$$r_1 = \frac{B_{12}f_2}{B_{11}f_1} \quad \text{and} \quad r_2 = \frac{B_{21}f_1}{B_{22}f_2} .$$

If the ratios r_1 and r_2 are constant, the values of f_1, f_2, and B_{ij}'s can be obtained by the following procedure of successive approximation:

(1) Select an arbitrary set of f_1', f_2', and B_{ij}'s.

(2) Apply on the contact area of radius A, a pressure distribution

$$f(\rho) = f_1'(1 - \rho^2)^{\frac{1}{2}} + f_2'(1 - \rho^2)^{\frac{3}{2}}$$

as part of the boundary condition in using the finite element program. The results from the finite element program give the values of $\tilde{\varphi}_1'$ and $\tilde{\varphi}_2'$, hence

$$k_1 = \frac{\tilde{\varphi}'_1}{\tilde{\varphi}_1}, \quad k_2 = \frac{\tilde{\varphi}'_2}{\tilde{\varphi}_2}.$$

(3) If either k_1 or k_2 is not unity, make the following adjustments

$$f''_1 = \frac{1}{\sqrt{k_1}} f'_1, \quad f''_2 = \frac{1}{\sqrt{k_2}} f'_2$$

$$B''_{11} = \frac{1}{\sqrt{k_1}} B'_{11}, \quad B''_{21} = \frac{\sqrt{k_1}}{\sqrt{k_2}} B'_{21}$$

$$B''_{12} = \frac{\sqrt{k_2}}{\sqrt{k_1}} B'_{12}, \quad B''_{22} = \frac{1}{\sqrt{k_2}} B'_{22}.$$

Use this new set of f''_1, f''_2, and B''_{ij}'s and return to Step (1). If both k_1 and k_2 are unity, the assumed (or current) values of f_1, f_2, and B_{ij}'s are the correct solution. Since $\tilde{\varphi}_1$, $\tilde{\varphi}_2$ are directly proportional to f_1, f_2 respectively for a small increment of load, as can be seen from Eqs.(42) and (43), this process is self-adjusting due to Step (3) above. The requirement that r_1, r_2 must be constant is met for the case of perturbed solutions of incipient plasticity where the elastic solution just prior to this point is already known.

A NUMERICAL EXAMPLE

Statement of Problem

Consider a steel plate of 5 mil thick pressed on by a rigid indenter having a parabolic profile

$$\tilde{v} = \tilde{\varphi}_o + \tilde{\varphi}_1 \tilde{r}^2, \tag{44}$$

where $\tilde{\varphi}_1$ is given and $\tilde{\varphi}_o$ is yet to be found.

The problem here is to determine the relation between a given load $\tilde{N}$ and the radius of contact circle A, the depth of indentation $D = -\tilde{\varphi}_o$ and the pressure distribution within the contact area $f(\rho)$.

Stress-Strain Relation

The elastic properties are given by Young's modulus E and Poisson's ratio ν which are taken to be 26×10^6 psi and 0.3 respectively. In the region of plasticity, the material is assumed to behave according to a linear work-hardening rule of Mises yield condition and Prandtl-Reuss flow equation,

$$f(\sigma_{ij}) = J'_2 = \frac{1}{2} \sigma'_{ij}\sigma'_{ij} = \frac{1}{3} \bar{\sigma}^2 \tag{45}$$

$$d\epsilon_{ij} = \frac{d\sigma}{3K} \delta_{ij} + \frac{1}{2G} d\sigma'_{ij} + H\sigma'_{ij}\sigma'_{k\ell} d\sigma'_{k\ell} \tag{46}$$

where σ'_{ij}, $\bar{\sigma}$ are deviatoric stress components and uniaxial yield stress in tension; J'_2 is the second deviatoric stress invariant; ϵ_{ij} denotes strain components; K, G are the bulk and shear moduli, respectively, and σ is the hydrostatic component of the stress tensor. Also

$$H = \frac{3}{4J_2'}\left(\frac{1}{E_t} - \frac{1}{E}\right), \tag{47}$$

where E, E_t are Young's modulus and tangent modulus, respectively.

The inverse relation of Eq.(46) can be shown [9] to be

$$d\sigma_{ij} = 2G\left[d\varepsilon_{ij} - \sigma'_{ij}\frac{\sigma'_{k\ell}d\varepsilon_{k\ell}}{S} + \frac{\nu}{1-2\nu}d\varepsilon_{kk}\delta_{ij}\right], \tag{48}$$

where

$$S = 2J_2'\left[1 + \frac{1}{3G\left(\frac{1}{E_t} - \frac{1}{E}\right)}\right].$$

In matrix notation and for the axisymmetric state of stress, Eq.(48) becomes

$$\{d\sigma\} = [D]^P\{d\varepsilon\}, \tag{49}$$

where

$$\{d\sigma\} = \begin{bmatrix} d\sigma_{rr} \\ d\sigma_{\theta\theta} \\ d\sigma_{zz} \\ d\sigma_{rz} \end{bmatrix} \qquad \{d\varepsilon\} = \begin{bmatrix} d\varepsilon_{rr} \\ d\varepsilon_{\theta\theta} \\ d\varepsilon_{zz} \\ d\varepsilon_{rz} \end{bmatrix} \tag{50, 51}$$

$$[D]^P = [D]^e - [D]^S, \tag{52}$$

with

$$[D]^e = \frac{2G(1-\nu)}{1-2\nu}\begin{bmatrix} 1 & & & \text{SYM.} \\ \frac{\nu}{1-\nu} & & & \\ \frac{\nu}{1-\nu} & \frac{\nu}{1-\nu} & 1 & \\ 0 & 0 & 0 & \frac{1-2\nu}{2(1-\nu)} \end{bmatrix} \tag{53}$$

$$[D]^S = \frac{2G}{S}\begin{bmatrix} \sigma'^{2}_{rr} & & & \text{SYM.} \\ \sigma'_{rr}\sigma'_{\theta\theta} & \sigma'^{2}_{\theta\theta} & & \\ \sigma'_{rr}\sigma'_{zz} & \sigma'_{\theta\theta}\sigma'_{zz} & \sigma'^{2}_{zz} & \\ \sigma'_{rr}\sigma'_{rz} & \sigma'_{\theta\theta}\sigma_{rz} & \sigma'_{zz}\sigma'_{rz} & \sigma'^{2}_{rz} \end{bmatrix} \tag{54}$$

Approximation for the Function $K_1(r,\rho,\lambda)$

The elastic solution to the problem is based on the fact that $K_1(r,\rho)$ of Eq.(27) can be approximated closely by a power series of r and ρ. In this example, it is assumed that

$$K_1(r,\rho) = \frac{1-\nu}{2\pi}\left(C_0 + C_1(r^2 + \rho^2) + C_2 r^2\rho^2\right). \tag{55}$$

Since $K_1(r,\rho)$ given in Eq.(27) is non-singular, its value at any particular point can be evaluated accurately by numerical integration. The coefficients of Eq.(55) are evaluated by matching the polynomial with the values of K_1 actually calculated from Eq.(27) for various points within the range of interest. The number of terms required depends on the accuracy desired and the value of $\lambda = H/A$, which characterizes the relative thickness of the plate. The approximating $K_1(r,\rho)$ of Eq.(55) and the original $K_1(r,\rho)$ of Eq.(27) can then be compared point by point over the entire region of interest: $0 \le r \le 1$, $0 \le \rho \le 1$ in the r-ρ plane. The table given below shows this result for a total of $6 \times 6 = 36$ points.

TABLE - ACCURACY EVALUATION OF $K_1(r,\rho)$ APPROXIMATIONS

λ	Average K_1	Standard Deviation	Maximum Error	C_o	C_1	C_2
2	0.54	0.0008	0.0015	0.5838	-0.0457	0.0119
1.5	0.70	0.003	0.005	0.7784	-0.1022	0.0408
1.0	1.00	0.15	0.025	1.1675	-0.2963	0.1905
0.6	1.60	0.10	0.16	1.9436	-0.9312	0.8537
0.4	2.00	0.30	0.55	2.869	-1.857	1.908

It is clear from the table that

(1) for $\lambda > 2$, the contribution of $K_1(r,\rho)$ is mainly due to a constant term which affects only the depth of indentation but not the pressure distribution;

(2) with rather simple expression of $K_1(r,\rho)$ in Eq.(55), reasonably accurate solutions can be obtained for the value of λ in the neighborhood of unity;

(3) for smaller λ, good approximation can be expected by taking more terms in the polynomial and the solutions can be obtained in the same simple fashion.

Elastic Solutions

Equation (55) is very similar to the sample calculation in Section 3 and the solution may be put in the form

$$D = -\tilde{\varphi}_o = 2A^2 \tilde{\varphi}_1 \alpha \tag{56}$$

$$\tilde{N} = \frac{16 GA^3 \tilde{\varphi}_1}{3(1 - \nu)} \beta \tag{57}$$

where

$$\alpha = \frac{15\pi - 4(5C_0 + 2C_1)}{15\pi + 8(5C_1 + 2C_2)} , \quad \beta = \frac{15\pi}{15\pi + 8(5C_1 + 2C_2)} . \tag{58}$$

When α, $\beta = 1$, Eqs.(56) and (57) become the solution of the half-space. Thus α, β indicate the effect due to the thickness of the plate. Figure 2 shows the values of α and β as functions of $\lambda = \frac{H}{A}$. Other results of calculation are shown in Figs.3 and 4. In Fig.3, the depth of indentation D, the radius of the contact area A and the maximum pressure under the indenter are plotted against the total load for $\tilde{\varphi}_1 = 1$, i.e., the radius of curvature at the asperity has a value $R = \frac{1}{2\tilde{\varphi}_1} = \frac{1}{2}$". Figure 4 shows the contours of J_2', the second invariant of the deviatoric stress tensor. It is obtained through numerical integration of the stress expressions given in [10] and with the pressure distribution function

$$f(\rho) = f_1(1 - \rho^2)^{\frac{1}{2}}$$

where

$$f_1 = A\tilde{\varphi}_1 / B_{11}.$$

Elastoplastic Solutions

A perturbed solution for incipient plastic stress state is now sought for this problem. Following the method described in Section 3, the actual steps of calculation are as follows:

(1) Let the elastic solution be known just prior to incipient plasticity from the previous sub-section. That is, for a given A, the values of f_1, $\tilde{N}$, B_{11}, and B_{01} can all be calculated. An increase in A (correspondingly an increase in $\tilde{N}$) will initiate a plastic state of stress. Thus

(2) Apply the finite element analysis using the pressure distribution function

$$f(\rho) = f_1(1 - \rho^2)^{\frac{1}{2}}$$

with a contact circle of radius $A' = A + \Delta A$. This calculation will lead to the values of $\tilde{\varphi}'_o$, $\tilde{\varphi}'_1$ and $k_1 = \tilde{\varphi}'_1/\tilde{\varphi}_1$. In the process of calculating, if any element is found to have become plastic, its stiffness matrix is to be adjusted for the next cycle of iteration.

(3) If k_1 obtained in Step (2) is not unity, make adjustment

$$f'_1 = f_1/\sqrt{k_1}$$

and return to Step (2). If k_1 is unity, the set of A', $\tilde{\varphi}'_o$, f_1, and $\tilde{N}$ constitutes the solution of onset plasticity sought.

Results of the calculation are shown in Figs.5, 6, and 7 for D, A, and $\tilde{f}_{max}$ vs. $\tilde{N}$, respectively. These curves are for $E = 26 \times 10^6$ psi, $\nu = 0.3$. For plastic work-hardening solutions, they are given for $E_t = E/10$, $E_t = 0$, $\bar{\sigma} = 3.0 \times 10^6$ psi and $\bar{\sigma} = 6.0 \times 10^6$ psi as shown by the short branch curves.

CONCLUSIONS

The following conclusions can be drawn from this investigation:

1) The solution of an infinite plate of any thickness pressed upon by an indenter of arbitrary shape with axial symmetry can be obtained by approximating a non-singular part of a kernel function by a polynomial. This approximation can be made as close as one wishes to the original function by taking enough terms in the polynomial. In general, the thinner is the plate compared with the radius of contact area, the more terms are required for the approximating kernel.

2) By so doing, the elastic solutions become that of a set of simultaneous algebraic equations. The coefficients of these equations depend on material constants and also on an important parameter $\lambda = \frac{H}{A}$ where H is the thickness of plate and A, the radius of contact area.

3) Assuming that the form of solutions obtained is valid in the plasticity state as well as in the elasticity state, a scheme of iteration process can be established with the help of the finite element method to extend solutions into the incipient plastic yielding. The use of an iteration is essential in connection with the finite element method to overcome the difficulty of changing boundary.

It is clear that more accurate results can be obtained by taking more terms in the approximate kernel functions, taking more terms in the contact pressure distribution function, improving the accuracy of specified displacement boundary conditions at the edges of the plate for the finite element program, and using finer grid scheme, especially within the contact area and in the region of expected plasticity.

ACKNOWLEDGEMENT

The authors wish to acknowledge the generous fellowship support provided by the Gillete Safety Razor Company for J.J. Wu.

REFERENCES

[1] Hertz, H.: On the Contact of Elastic Solids. Miscellaneous Papers, London, p.148 (1896).

[2] Hertz, H.: On the Contact of Rigid Elastic Solids and on Hardness, ibid., p.163.

[3] Timoshenko, S. and J.N. Goodier: Theory of Elasticity. McGraw-Hill, p.362 (1951).

[4] Tabor, D.: The Hardness of Metals. Oxford, p.40 (1951).

[5] Davis, R.M.: The Determination of Yield Stresses Using a Steel Ball. Proc. Roy. Soc. A, Vol. 197, p.416 (1949).

[6] Sneddon, I.N.: Fourier Transforms. McGraw-Hill, p.469 (1951).

[7] Tu, Y. and D.C. Gazis: The Contact Problem of a Plate Pressed Between Two Spheres. J. Applied Mech., Vol. 31, No. 4, p.659 (1964).

[8] Zienkiewicz, O.C. and Y.K. Cheung: The Finite Element Method in Structural and Continuum Mechanics, McGraw-Hill (1967).

[9] Yamada, Y., N. Yoshimura, and T. Sakurai: Plastic Stress-Strain Matrix and its Application for the Solution of Elastic-Plastic Problems by the Finite Element Method. Int. J. Mech. Sci., Vol. 10, p.343 (1968).

[10] Wu, J.J.: Ph.D. Thesis, Mechanics Division, Rensselaer Polytechnic Institute, Troy, N.Y. (1970).

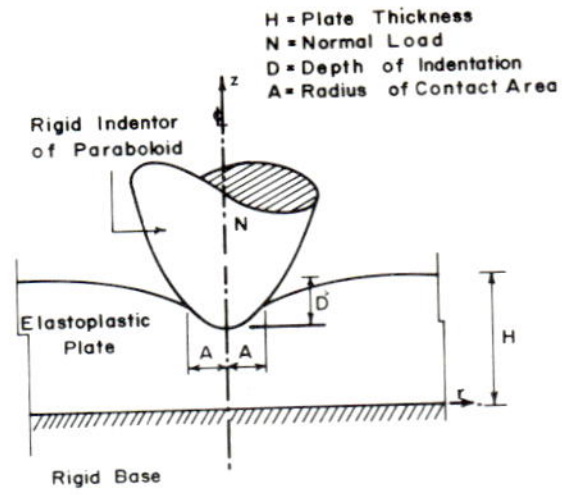

Figure 1. Configuration of the Problem

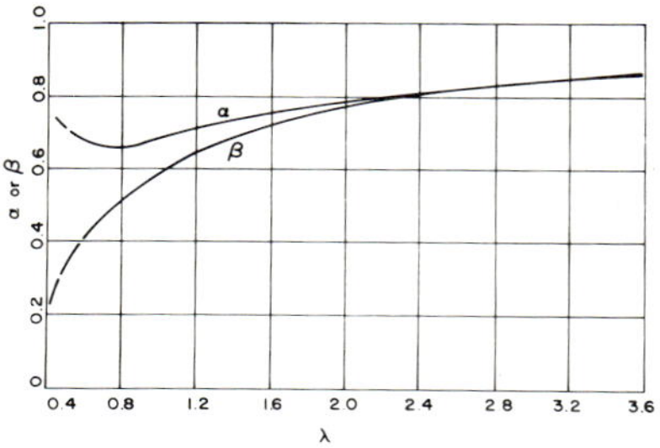

Figure 2. α and β vs. λ

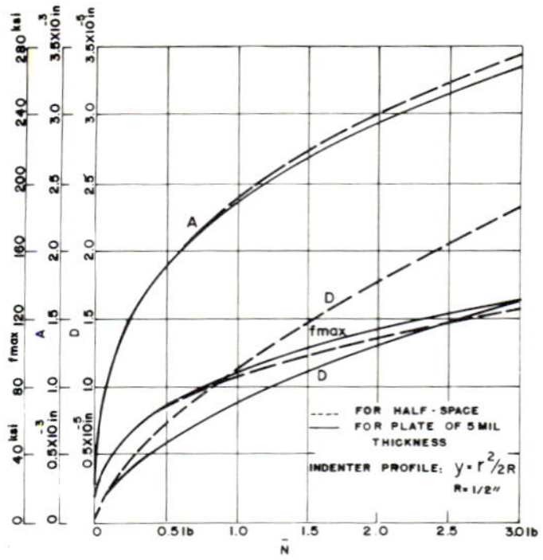

Figure 3.
Relations of Surface Parameters

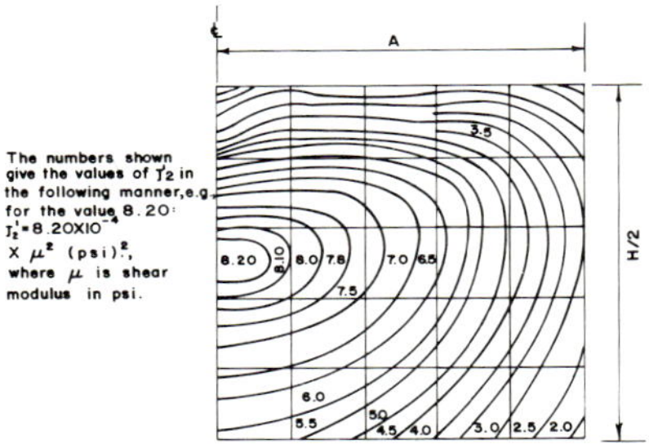

Figure 4. Contours for J_2' within a Plate, (λ = 2)

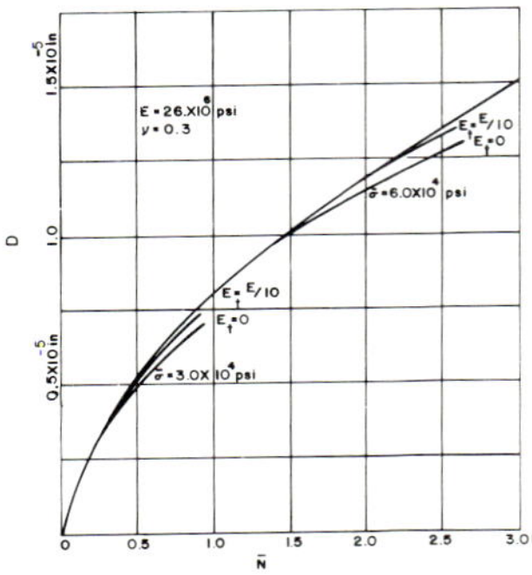

Figure 5.
D vs. N - Elastoplastic Solution

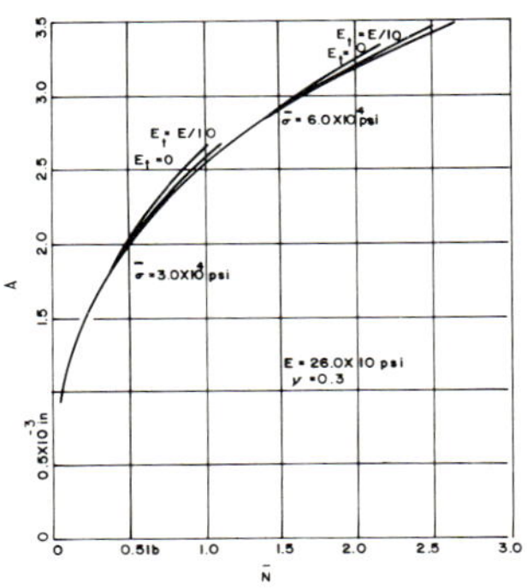

Figure 6. A vs. N - Elastoplastic Solution

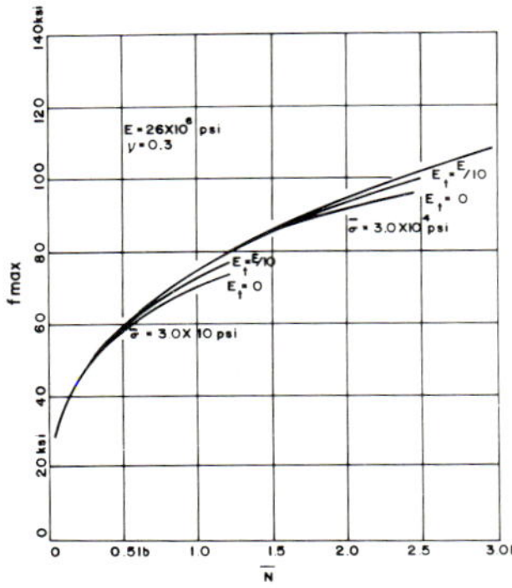

Figure 7.
f_{max} vs. N - Elastoplastic Solution

Developments in Mechanics, Vol. 6. Proceedings of the 12th Midwestern Mechanics Conference: 25

A Perturbation Method for Compressible Elastic-plastic Materials: An Axisymmetric Problem

MARTIN A. EISENBERG and GENE W. HEMP
UNIVERSITY OF FLORIDA

ABSTRACT

A method of solution of elastic-plastic boundary value problems for compressible materials is described. A perturbation technique is employed. The solution to the corresponding problem for an incompressible material is used as the generating solution since the assumption of incompressibility often leads to considerable simplification in the governing differential equations thereby permitting complete solutions to be obtained. The perturbation parameter is chosen to be proportional to the ratio of the shear modulus to the bulk modulus. This technique is illustrated by treating an axially symmetric boundary value problem - the elastic plastic deformation of a thick-walled cylinder obeying the Mises yield criterion and Prandtl-Reuss flow law.

INTRODUCTION

Exact solutions to all but the simplest boundary value problems in the theory of elastic-perfectly plastic solids are difficult to obtain. It is usually necessary either to neglect elastic compressibility, to neglect elastic strains completely, to use deformation theories, to approximate yield functions and/or to employ non-associative flow laws. In many cases numerical techniques are necessary, even within the context of the above approximations.

Another alternative is to attempt asymptotic solutions by expanding the dependent variables in a power series in terms of a small parameter. Such perturbation techniques [1-9] have been employed primarily within the context of rigid-perfectly plastic theory. Hodge [1] has treated an incompressible problem by employing the rigid-perfectly plastic solution as a generating solution; Ivlev [2] and Spencer [3] have considered rigid plastic problems in which the perturbation was applied to the boundary conditions; Kuznetsov [4] and Spencer [5] have treated rigid materials with small non-homogeneity; Onat [6] obtained solutions for incipient plastic flow in a rigid material by employing time as a small parameter; Spencer [7] handled dynamic problems in rigid plastic materials by employing the quasi-static solution as a generating solution and also considered the effect of small body forces [8] by perturbation techniques. Richmond and Morrison [9] have employed a technique similar to that employed by Spencer for the solution of rigid plastic problems, by using exact characteristics instead of the characteristics of the generating system.

In the present work, a perturbation technique is employed for elastic-perfectly plastic problems. The incompressible solution is used as the generating solution since the assumption of incompressibility often leads to considerable simplification in the governing differential equations thereby permitting complete solutions to be obtained. The perturbation parameter is chosen to be proportional to the ratio of the shear modulus to the bulk modulus. This technique should prove fruitful in many problems. The technique is illustrated by treating the elastic plastic deformation of a thick-walled cylinder by means of the Mises yield criterion and Prandtl-Reuss flow law. The results are compared with solutions obtained by other techniques [10,11].

THICK-WALLED CYLINDER PROBLEM

Given a thick-walled cylindrical tube consisting of an elastic-perfectly plastic material, and given that the tube is subject to internal pressure and edge loading so as to induce a state of plane strain, there are several possible approaches to the solution for the internal stress distribution and the displacements of the tube. It remains to specify the yield criterion, the flow law, flow or deformation theory and numerical or analytic solution representation. Certainly, the simplest solution is obtained by assuming the material to be incompressible. Under this assumption the solution obtained from the flow and deformation theories and from the Mises and Tresca yield conditions are identical and can be exhibited in closed form [12]. However, the incompressibility assumption is not tenable when the total strains are of the order of magnitude of the elastic components of strain.

Beliaev and Sinitskii [13] have obtained numerical solutions and Allen and Sopwith [14] obtained a closed form, albeit extremely complicated, solution using the Hencky deformation theory. However, the use of deformation theory is objectionable in principle since conditions of proportional loading are not achieved. Solutions based on flow theory were presented by Hill, Lee and Tupper [15], Hodge and White [10] and Koiter [11]. The latter two solutions employ the Mises and Tresca yield conditions, respectively, and their associated flow laws, whereas Hill et al obtain their solution based on the less satisfactory non-associative Tresca, Prandtl-Reuss assumption.

Thus, of all the existing compressible solutions, only Hodge and White's and Koiter's are internally consistent and free from objection on theoretical grounds.*

*For a more complete survey of work on this and similar problems see Hill [16], Steele [17], and Nadai [18].

The Hodge and White formulation leads to a system of nonlinear-partial differential equations which are then solved numerically using a finite difference method, whereas Koiter obtains simple closed form solutions based upon the Tresca criterion. The solution exhibited below is obtained by applying a perturbation technique to the Hodge and White equations, thereby reducing the problem to a sequence of linear problems for which the equations of the Mth perturbation may depend on the solutions of the lower order perturbation equations.

GOVERNING EQUATIONS

Consider a circular cylindrical tube of internal radius a and external radius b and let r, θ, z denote a system of cylindrical coordinates where the z-axis coincides with the geometrical axis of the tube. Let the tube be initially stress free and let a monotonically increasing pressure p = p(t) be applied in a quasistatic manner to the inner surface. If the ends of the tube are restrained against axial motion, then the only non-zero stress components are $\sigma_z(r,t)$, $\sigma_r(r,t)$ and $\sigma_\theta(r,t)$. The corresponding non-zero strain components are $\varepsilon_r = \partial u/\partial r$ and $\varepsilon_\theta = u/r$, where u = u(r,t) is the radial displacement. In the following, spatial coordinates shall be non-dimensionalized by dividing by the outer radius b, all variables with the dimensions of stress shall be divided by k, the yield stress in pure shear, and all displacement related variables shall be multiplied by G/kb, where G is the elastic shear modulus.

The only non-trivial equilibrium equation is

$$\frac{\partial \sigma_r}{\partial r} + \frac{\sigma_r - \sigma_\theta}{r} = 0 \; . \tag{1}$$

Also, the compatibility equation

$$\frac{\partial e}{\partial r} - \frac{\partial \phi}{\partial r} = \frac{2\phi}{r} \tag{2}$$

must be satisfied, where the kinematic variables e and ϕ are defined as

$$e = \frac{1}{3}(\varepsilon_r + \varepsilon_\theta) \; , \tag{3}$$

$$\phi = \frac{1}{3}(\varepsilon_r - \varepsilon_\theta) \; . \tag{4}$$

The strains may be divided into elastic and plastic portions, ε'_{KL} and ε''_{KL}, and the ε'_{KL} are related to the stress state by Hooke's law

$$s_{KL} = 2\, e'_{KL} \tag{5}$$

$$\varepsilon\sigma_{KK} = \varepsilon'_{KK} \tag{6}$$

where s_{KL} and e_{KL} represent the deviatoric stress and strain tensors and $\varepsilon = G/3K$, i.e., the shear modulus divided by three times the bulk modulus. Since $0<\varepsilon<1$, $\varepsilon = 0$ for an incompressible material, and ε is much less than unity for most materials, it is a natural choice for a perturbation parameter.

When the pressure p is sufficiently small and there is no plastic flow, equations (1,2,5,6) may be integrated subject to the boundary conditions

$$\sigma_r(a/b) = -p, \ \sigma_r(1) = 0 \ . \tag{7}$$

This is the classical Lame solution for thick-walled cylinders and is valid as long as the Mises yield condition

$$J_2 = \frac{1}{2} s_{LK} s_{KL} = 1 \tag{8}$$

is not satisfied anywhere in the domain $a/b \leq r \leq 1$.

It can be demonstrated that J_2 is a maximum on the inner wall. Therefore, as the pressure increases, plastic flow spreads outward from the inner wall. The elastic-plastic interface, Γ, is a cylindrical surface of non-dimensional radius ρ, $a/b \leq \rho \leq 1$. The solution in the elastic region ($\rho \leq r \leq 1$) is now obtained by replacing equation (7a) with the plasticity condition

$$J_2(\rho) = 1 \ . \tag{9}$$

The solution in the plastic region ($a/b \leq r \leq \rho$) is obtained by employing equations (1-6) and the associated flow law (Prandtl-Reuss)

$$\frac{\partial \varepsilon''_{KL}}{\partial \rho} = \lambda \frac{\partial J_2}{\partial \sigma_{KL}} = \lambda s_{KL} \tag{10}$$

for the rate of change of the plastic strains. If one eliminates the positive parameter λ by combining equations (10) and expresses s_z and s_θ in terms of s_r by means of equation (8), then the governing differential equations may be written in form

$$\frac{\partial e}{\partial r} - \frac{\partial \phi}{\partial r} = \frac{2\phi}{r} \tag{11}$$

$$\frac{\partial s_r}{\partial r} + \frac{1}{\varepsilon}\frac{\partial e}{\partial r} = \frac{-3s_r + \sqrt{4-3s_r^2}}{2r} \ , \tag{12}$$

$$\frac{\partial s_r}{\partial r} - \frac{1}{4}\left(4 - 3s_r^2 - 3s_r\sqrt{4-3s_r^2}\right)\frac{\partial e}{\partial \rho} -$$

$$- \frac{3}{4}\left(4 - 3s_r^2 + s_r\sqrt{4-3s_r^2}\right)\frac{\partial \phi}{\partial \rho} = 0 . \tag{13}$$

Equations (11-13) constitute a system of first order hyperbolic quasi-linear differential equations which must be solved in the triangular domain $a/b \leq r \leq \rho$, $a/b \leq \rho \leq 1$ subject to continuity conditions in J_2, σ_r and u along the non-characteristics line $r = \rho$. Hodge and White [10] obtain numerical solutions by employing finite difference approximations and satisfying the indicated continuity conditions with the elastic region solution which may be represented in non-dimensional form as

$$u = \frac{3P}{2}\left[\frac{r\varepsilon}{1+\varepsilon} + \frac{1}{3r}\right] \ , \tag{14}$$

$$\sigma_r = P\left(1 - \frac{1}{r^2}\right), \ \sigma_\theta = P\left(1 + \frac{1}{r^2}\right), \ \sigma_z = P\,\frac{1-2\varepsilon}{1+\varepsilon} = 2\nu P \tag{15}$$

where ν is Poisson's ratio and

$$P = \left[3\left[\frac{\varepsilon}{1+\varepsilon}\right]^{2} - \frac{1}{\rho^{4}} \right]^{-1/2} . \quad (16)$$

PERTURBATION EQUATIONS

Let the solution of Equations (11-13) be written in the form

$$e(r,\rho;\varepsilon) = e_o(r,\rho) + \varepsilon e_1(r,\rho) + \varepsilon^2 e_2(r,\rho) + \ldots$$

$$\phi(r,\rho;\varepsilon) = \phi_o(r,\rho) + \varepsilon\phi_1(r,\rho) + \varepsilon^2\phi_2(r,\rho) + \ldots$$

$$s_r(r,\rho;\varepsilon) = s_{r_o}(r,\rho) + \varepsilon s_{r_1}(r,\rho) + \varepsilon^2 s_{r_2}(r,\rho) + \ldots \quad (17)$$

Also, let the displacements, strains and stresses derivable from these variables be expanded in similar fashion as needed. In the limit as $\varepsilon \to 0$ the following incompressible solution [12] is obtained

$$e_o = 0, \quad s_{r_o} = -1, \quad \phi_o = -\frac{1}{3}\left(\frac{\rho}{r}\right)^2 . \quad (18)$$

Expressions (17) and (18) may be substituted into Equations (11-13) and the resultant terms grouped according to the power of the perturbation term. By successively passing to the limit of small ε we can conclude that the resulting differential equations must be satisfied independently in each power of the small parameter ε. The perturbation equations are linear first-order differential equations with variable coefficients and non-homogeneous terms which are functions of the lower order perturbation solutions, and therefore known functions of r and ρ. The Nth order perturbation system is of the form

$$\frac{\partial e_N}{\partial r} = f_N(r,\rho,e_M,\phi_M,s_{r_M}) \quad (19)$$

$$\frac{\partial e_N}{\partial r} - \frac{\partial \phi_N}{\partial r} - 2\,\frac{\phi_N}{r} = 0 \quad (20)$$

$$\frac{\partial s_{r_N}}{\partial \rho} - \frac{\partial e_N}{\partial \rho} + \frac{2\rho}{r^2}\, s_{r_N} = g_N(r,\rho,e_M,\phi_M,s_{r_M}) \quad (21)$$

where M<N, N = 1,2,..., and the lowest order equations are defined by

$$f_1 = \frac{2}{r}, \qquad f_2 = -\frac{\partial s_{r_1}}{\partial r},$$

$$f_3 = -\frac{\partial s_{r_2}}{\partial r} - \frac{3 s_{r_1}^2}{r}, \qquad g_1 = 0 .$$

$$g_2 = 3 s_{r_1} \frac{\partial}{\partial \rho}(e_1 + \phi_1) - \frac{3\rho}{r^2} s_{r_1}^2 . \quad (22)$$

It should also be noted that the governing differential equations are in characteristic form and that the characteristics of the generating system equations are the same as the exact characteristics, i.e., r = constant and ρ = constant. The proposed method is, for this problem, analogous to both Spencer's [3,5,7,8] and Richmond and Morrison's [9] methods for rigid-plastic theory.

SOLUTION OF PERTURBATION EQUATIONS

The perturbation equations may now be solved in the same triangular domain as the original equations by satisfying continuity conditions across Γ. To accomplish this we expand the exact elastic solution (Equations (14-16)) in power series in ε. The interface conditions are then satisfied order-by-order for each perturbation.

The stresses and displacements are determined from e, ϕ, and s_r by the relations

$$u = \frac{3}{2} r(e-\phi) \ , \tag{23}$$

$$\sigma_{ij} = s_{ij} + \delta_{ij} \, e/\varepsilon \ . \tag{24}$$

It should be noted that the Nth order contribution to the stresses must be computed from the Nth order stress deviation and the $(N+1)^{th}$ order dilation contribution.

In order to obtain the first two complete perturbation solutions we then require the solution for e_1, e_2, e_3, ϕ_1, ϕ_2, s_{r1}, and s_{r2}. Equations (19-21) may be solved for the above functions and Equations (23) and (24) used to obtain the following second order perturbation solutions for the displacement and stress fields in the plastically deforming region

$$\begin{aligned} u(r,\rho) = {} & \frac{\rho^2}{2r} + \varepsilon\frac{3r}{2} [\rho^2-1 - \ln(x) + x] \\ & -\varepsilon^2 \frac{3r}{2} \{[r^2 - e^{-x}[Ei(x) - Ei(1)] + x[1+\frac{\rho^2}{2} + \frac{\rho^4}{2} + \\ & +Ei(1)[E_1(1) - E_1(x)] - \mathcal{E}_1(x,-1,1)] -1 - \frac{r^2}{2}]\} \end{aligned} \tag{25}$$

$$\begin{aligned} \sigma_r(r,\rho) = {} & \rho^2 - \ln(x) - 1 + \varepsilon^2 \frac{3}{2} \{- \frac{r^4}{2} + \frac{3}{2} \rho^4 - \rho^6 - \mathcal{J}(x) + \\ & + Ei^2(1)[E_1(2) - E_1(2x)] - 2\rho^2 \mathcal{E}_2(x,-1,1) + \\ & + 2\rho^2 Ei(1)[e^{-1} - \frac{e^{-x}}{x} + E_1(x) - E_1(1)] + \\ & + \mathcal{E}_1(x,-2,1)[Ei(x) - 2Ei(1)]\} \end{aligned} \tag{26}$$

$$\sigma_\theta(r,\rho) = \rho^2 - \ln(x) + 1 + \varepsilon^2 \{\tfrac{3}{2} (- \tfrac{r^4}{2} + \tfrac{3}{2} \rho^4 - \rho^6$$

$$- \mathcal{F}(x) + Ei^2(1)[E_1(2) - E_1(2x)] -$$

$$- 2\rho^2 \mathcal{E}_2(x,-1,1) + 2\rho^2 Ei(1)[e^{-1} - \frac{e^{-x}}{x} +$$

$$+ E_1(x) - E_1(1)] + \mathcal{E}_1(x,-2,1)[Ei(x) - 2Ei(1)])$$

$$- 3(r^2 - e^{-x}[Ei(x) - Ei(1)])^2\} \tag{27}$$

$$\sigma_Z(r,\rho) = \rho^2 - \ln(x) - 3\varepsilon[r^2 - e^{-x}[Ei(x) - Ei(1)] +$$

$$+ 3\varepsilon^2 \{2 + 3r^2 - \tfrac{3}{4} r^4 + \tfrac{3}{4} \rho^4 - \tfrac{1}{2} \rho^6 - 2(1+r^2)e^{1-x}$$

$$- \tfrac{1}{2} \mathcal{F}(x) + \tfrac{1}{2} Ei^2(1) [E_1(2) - E_1(2x)] - \rho^2 \mathcal{E}_2(x,-1,1)$$

$$+ \rho^2 Ei(1) [e^{-1} - \frac{e^{-x}}{x} + E_1(x) - E_1(1)] +$$

$$+ \tfrac{1}{2} \mathcal{E}_1(x,-2,1) [Ei(x) - 2Ei(1)] - \tfrac{1}{2} e^{-2x}[Ei(x)-Ei(1)]^2$$

$$- e^{-x}[2x + 1][Ei(x) - Ei(1)]\} \tag{28}$$

In the above

$$x = (\rho/r)^2 \tag{29}$$

$$Ei(x) = \int_{-\infty}^{x} \frac{e^x}{x} dx \tag{30}$$

$$E_1(x) = \int_{x}^{\infty} \frac{e^{-x}}{x} dx \tag{31}$$

$$\mathcal{E}_N(x,\alpha,\beta) = \int_{1}^{x} \frac{e^{\alpha x}}{x^N} Ei(\beta x) dx \tag{32}$$

$$\mathcal{F}(x) = \int_{1}^{x} \frac{e^x}{x} \quad {}_1(x,-2,1) dx \tag{33}$$

$Ei(x)$ is the exponential integral and $E_1(x) = -Ei(-x)$ is the exponential integral of the first kind. Both functions are tabulated [19]. The functions $\mathcal{E}_n(x,\alpha,\beta)$ and $\mathcal{F}(x)$ are not known to be available in the literature. These functions can be represented by expanding the integrands in Taylor series and integrating term by term. They are monotonically increasing functions of x for the range of values of the parameters of interest in this paper.

DISCUSSION

In order to compare our solution with the results of Hodge and White's numerical integration of the nonlinear equations, with Koiter's solution for the Tresca material and with the incompressible solution, the displacements and stresses were evaluated for a tube with an inner to outer diameter ratio of 1/2, for k/G = 0.003, and a Poisson's ratio of 0.3. The perturbation parameter ε then has the value 0.154.

The radial and tangential stresses given by all solutions are essentially the same. This result is to be expected, since Equations (26) and (27) show that the first order contribution to σ_r and σ_θ vanishes identically.

The axial stresses for the compressible cases differ markedly from the incompressible solution. The perturbation solutions correspond closely with the Hodge and White numerical solution, while Koiter's solution shows a more linear variation with ρ than the Mises solutions.

Figure 1 shows the displacement solutions of the cylinder walls. It can be seen that the incompressible or zero order solution differs markedly from the compressible solutions. The first order perturbation solution overshoots the exact solution somewhat and the second order solution is essentially the same as the exact solution.

As a result of these observations we may conclude that the perturbation method proposed does indeed lead to solutions which closely approximate the numerical solutions to the exact nonlinear equations and that no more than two perturbations need be considered, while one would suffice for most purposes. The analytic solution which results is certainly more convenient for studying the effects of the relevant parameters than having to solve anew each problem by finite difference methods. In the particular problem considered, it should be noted that for extremely thick-walled cylinders the arguments of the exponential integral and its related integrals $\mathcal{E}_N$ and $\mathcal{F}$ can become large and their evaluation may be simplified considerably by using appropriate asymptotic forms [19].

This perturbation approach has recently been applied to the elastic-plastic wedge problem [20] and should also permit analytic solutions to be developed for other problems involving compressible elastic-perfectly plastic flow of materials obeying the Mises yield criterion. The use of this criterion in solving boundary value problems in plasticity may then become in some cases a more attractive alternate to the Tresca formulations which require a priori assumptions of the ordering of the principal stresses, and which conform less closely with experimental data for metal yielding than does the Mises criterion.

ACKNOWLEDGMENT

The authors wish to acknowledge the assistance of Messrs. E. V. Browell and R. Edelstein for their contributions to exploratory studies in this area and to the development of a computer code to reconstruct the Hodge and White solution. The authors also wish to thank the National Science Foundation for their support through its Undergraduate Research Participation Program for portions of this investigation.

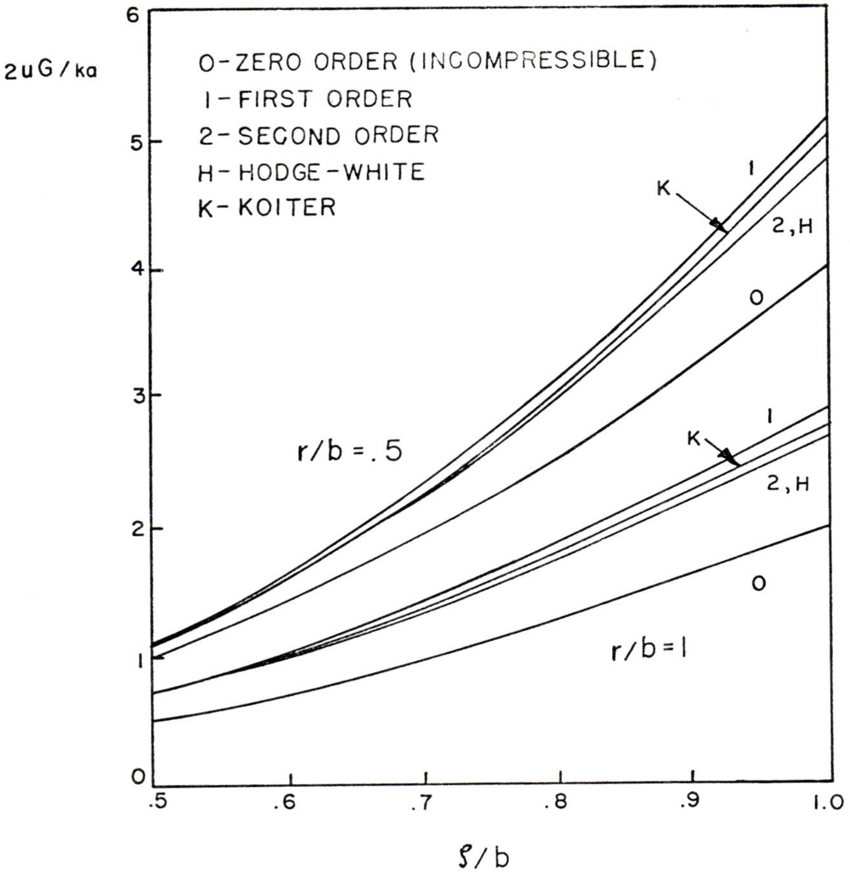

Fig. 1. Radial Displacement Solution

BIBLIOGRAPHY

[1] P. G. Hodge, Jr.: On the Plastic Strains in Slabs with Cutouts. J. Appl. Mech., Vol. 20, p. 183-188 (1953).

[2] D. D. Ivlev: Approximate Computation by the Small Parameter Method for Plane Elasto-Plastic Problems in the Theory of Ideal Plasticity. Vestnik Moskovskogo Univ., Ser. Mat., Mekh., Vol. 12, p. 17-26 (1957).

[3] A. J. M. Spencer: Perturbation Method in Plasticity-II. Plane Strain of Slight Irregular Bodies. J. Mechanics and Physics of Solids, Vol. 10, p. 17-26 (1962).

[4] A. I. Kuznetsov: The Problem of Torsion and Plane Strain of Nonhomogeneous Plastic Solids. Archiwum Mechaniki Stosowanej, Vol. 10, p. 447-462 (1958).

[5] A. J. M. Spencer: Perturbation Methods in Plasticity-I. Plane Strain of Nonhomogeneous Plastic Solids. J. Mechanics and Physics of Solids, Vol. 9, p. 279-288 (1961).

[6] E. T. Onat: The Effects of Nonhomogeneity Caused by Strain Hardening on the Small Deformations of a Rigid-Plastic Solid, Nonhomogeneity in Elasticity and Plasticity, Pergamon, New York, p. 171-180 (1959).

[7] A. J. M. Spencer: The Dynamic Plane Deformation of an Ideal Plastic-Rigid Solid. J. Mechanics and Physics of Solids, Vol. 8, p. 262-279 (1960).

[8] A. J. M. Spencer: Perturbation Methods in Plasticity -III. Plane Strain of Ideal Solids and Plastic Solids with Body Forces. J. Mechanics and Physics of Solids, Vol. 10, p. 165-177 (1962).

[9] O. Richmond and H. L. Morrison: Application of a Perturbation Technique Based on the Method of Characteristics. J. Appl. Mech., Vol. 35, p. 117-122 (1968).

[10] P. G. Hodge, Jr. and G. N. White, Jr.: A Quantitative Comparison of Flow and Deformation Theories of Plasticity. J. Appl. Mech., Vol. 17, p. 180-184 (1950).

[11] W. T. Koiter: On Partially Plastic Thick-Walled Tubes. C. B. Biezeno Anniversary Volume of Applied Mechanics, Haarlem, p. 232-251 (1953).

[12] W. Prager and P. G. Hodge, Jr.: Theory of Perfectly Plastic Solids, Wiley, New York (1951).

[13] N. M. Beliaev and A. K. Sinitskii: Stresses and Strains in Thick-Walled Cylinders in the Elastic-Plastic State. Izvestia Ak. Nauk S.S.S.R. No. 2, 3-54 (1938), Brown University translation for David Taylor Model Basin RMB-29 (1947).

[14] D. N. de G. Allen and D. G. Sopwith: The Stresses and Strains in a Partly Plastic Thick Tube under Internal Pressure and End Load. Proceedings Royal Society, 205A, p. 69-83 (1951).

[15] R. Hill, E. H. Lee, and S. J. Tupper: The Theory of Combined Plastic and Elastic Deformation with Particular Reference to a Thick Tube under Internal Pressure, Proceedings Royal Society, 191A, p. 278-303 (1947).

[16] R. Hill: Mathematical Theory of Plasticity. Oxford, London (1950).

[17] M. C. Steele: Partially Plastic Thick-Walled Cylinders. J. Appl. Mech., Vol. 19, p. 133-140 (1952).

[18] A. Nadai: Plasticity. McGraw-Hill, New York (1931).

[19] M. Abramowitz and I. A. Stegun: Handbook of Mathematical Functions, National Bureau of Standards Applied Mathematics Series-55, Washington, D. C (1964).

[20] M. A. Eisenberg and W. R. Cribb: Perturbation Solution for the Compressible Elastic-Perfectly Plastic Wedge. To be presented at the Third Canadian Congress of Applied Mechanics, University of Calgary, Calgary 44, Alberta, Canada, 17-21 May (1971).

Developments in Mechanics, Vol. 6. Proceedings of the 12th Midwestern Mechanics Conference: 26

Upper Bound Solutions to Symmetrical Extrusion Problems through Curved Dies

K. T. CHANG and J. C. CHOI
KOREA INSTITUTE OF SCIENCE AND TECHNOLOGY

ABSTRACT

Velocity fields of plain-strain and axially symmetrical extrusion problems through curved dies are presented for an incompressible material. These velocity fields are also applicable to wedge shaped dies in plain-strain and conical dies in axially symmetrical extrusion. Thus, in principle, upper bound solutions for plain-strain and axially symmetrical extrusion problems through arbitrarily shaped dies are obtained. As an illustration, solutions of axially symetrical extrusion processes through parabolic dies have been developed.

INTRODUCTION

Upper-bound technique, based on limit theorems [1, 2] has been used extensively to calculate an overestimate for the average ram pressure of plane-strain and axially symmetrical problems. Plain-strain extrusion problems through square and wedge shaped dies and axially symmetrical extrusion problems through cylindrical square and conical converging dies have been extensively investigated [3, 4, 5, 8, 9, 10] .
Plain-strain extrusion through curved dies, i.e., circular, cosine, elliptic, and hyperbolic types, have also been investigated [6, 7] . It seems, however, that no upper-bound solutions have previously been proposed for plain-strain and axially symmetrical extrusion through other types of curved dies, which is the concern of the present paper.

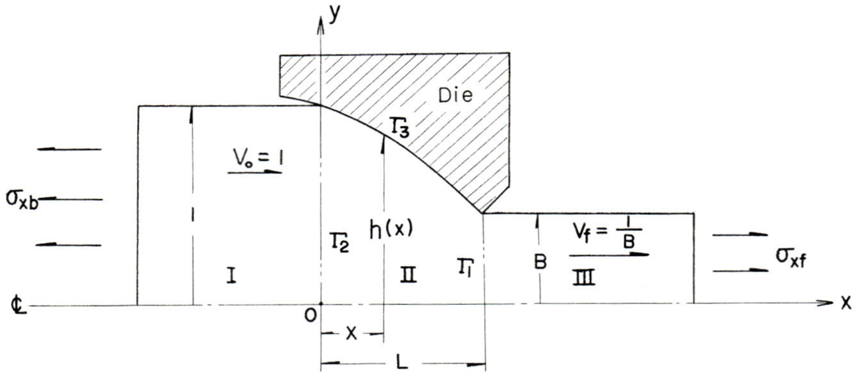

Fig. 1. Plain-strain extrusion through curved die

Velocity Field

The equation of the curved die profile is given by

$$y = h(x)$$
$$\text{and } h(0) = 1,\ h(L) = B. \qquad \ldots (1)$$

The strip is divided into three zones as shown in Figure 1. In zone I, the undeformed region, a rigid body motion in the x-direction is described by $V_o (=1)$.
In zone III, the already deformed strip, no further deformation occurs; the rigid body motion is described by $V_f (= \frac{1}{B})$, parallel to the x-direction. Assuming no change in width and because of volume constancy,

$$V_f = \frac{1}{B} \qquad \ldots (2)$$

Zone II is bounded by the two curved planes with an initial gap 2 and final gap 2B and the vertical surfaces Γ_1 and Γ_2. In zone II, it is assumed that the velocity fields are described in a Cartesian coordinate system.

$$U_x = \frac{1}{h(x)}$$
$$U_y = y \cdot \frac{h'(x)}{h^2(x)} \qquad \ldots (3)$$
$$U_z = 0$$

For a velocity field to be kinematically admissible, volume constancy must be satisfied. The material crossing boundaries Γ_1 and Γ_2 is identical whether computed for the velocity field coming into the surface or going out of it.
For Γ_2, the velocity normal to the surface is $U_x\big|_{x=0} = 1$ for both sides.
For Γ_1, the velocity normal to the surface is $U_x\big|_{x=L} = \frac{1}{B}$
Parallel to these surfaces, velocity discontinuities exist:

$$\text{Along } \Gamma_2 \quad \Delta V_2 = y \cdot |h'(0)|$$
$$\text{Along } \Gamma_1 \quad \Delta V_1 = \frac{y}{B^2} \cdot |h'(L)| \qquad \ldots \quad (4)$$

Along the die surfaces, the resultant velocity is parallel to it, namely

$$\frac{U_y}{U_x}\bigg|_{y=h(x)} = h'(x) \qquad \ldots \quad (5)$$

Thus, the normal velocity to the die surface is zero, and the tangential velocity is equal to the resultant velocity, namely

$$\text{Along } \Gamma_3 \quad \Delta V_3 = \frac{1}{h(x)} \sqrt{1 + h'^2(x)} \qquad \ldots \quad (6)$$

Deformation occurs only in zone II.
Strain rates derived from equation (3) are:

$$\dot{\epsilon}_{xx} = -\frac{1}{h^2(x)} \cdot h'(x)$$
$$\dot{\epsilon}_{yy} = \frac{1}{h^2(x)} \cdot h'(x) \qquad \ldots \quad (7)$$
$$\dot{\epsilon}_{zz} = \dot{\epsilon}_{xz} = \dot{\epsilon}_{yz} = 0$$
$$\dot{\epsilon}_{xy} = \frac{y}{2h^3(x)} \left[h(x) \cdot h''(x) - 2h'^2(x)\right]$$

Checking the volume constancy,

$$\dot{\epsilon}_{xx} + \dot{\epsilon}_{yy} + \dot{\epsilon}_{zz} = 0 \qquad \ldots \quad (8)$$

The proposed velocity field of equation (3) is admissible, since it satisfies the condition of incompressibility and continuity and it also satisfies the velocity boundaries.

Upper Bound

Avitzur [8] formulated the upper-bound theorem. "If surface of velocity discontinuities are to be included, it reads: Theorem 2. Among all kinematically admissible strain rate fields the actual one minimizes the expression".

$$J^* = \frac{2}{\sqrt{3}} \sigma_o \int_V \sqrt{\frac{1}{2} \dot{\epsilon}_{ij} \cdot \dot{\epsilon}_{ij}} \; dV + \int_{S\Gamma} \tau \cdot \Delta v \, ds - \int_{St} T_i \, v_i \, ds \quad \ldots (9)$$

The individual terms of equation (9) are now computed, for a unit width of the strip.
The power dissipated along the traction prescribed boundary St is

$$\dot{E}_t = + \int_{St} T_i \, V_i \, ds = 2 \cdot \sigma_{xf} \quad \ldots (10)$$

The internal power of deformation is computed over zone II alone:

$$\dot{E}_i = \frac{2}{\sqrt{3}} \sigma_o \int_V \sqrt{\frac{1}{2} \dot{\epsilon}_{ij} \cdot \dot{\epsilon}_{ij}} \; dV$$

$$= \left| \frac{4}{\sqrt{3}} \sigma_o \int_0^L dx \int_0^{h(x)} dy \cdot \frac{1}{h^2(x)} \cdot \sqrt{h'^2(x) + \frac{y^2}{4h^2(x)} F_1^2(x)} \right| \cdot \quad \ldots (11)$$

where

$$F_1(x) = h(x) \cdot h''(x) - 2h'^2(x)$$

Shear losses over surfaces of velocity discontinuities Γ_1 and Γ_2 are

$$\dot{E}_{S2} = \int_{S\Gamma_2} \tau_2 \cdot \Delta V_2 ds = \frac{\sigma_o}{3} \left| h'(0) \right| \quad \ldots (12)$$

and

$$\dot{E}_{S1} = \int_{S\Gamma_1} \tau_1 \cdot \Delta V_1 \, ds = \frac{\sigma_o}{\sqrt{3}} \left| h'(L) \right| \quad \ldots (13)$$

Power consumption over the surface Γ_3 becomes.

$$\dot{E}_{S3} = \int_{ST_3} \tau_3 \cdot \Delta V_3 \, ds$$

. . . (14)

$$= \left| \frac{2}{\sqrt{3}} m \sigma_o \int_0^L \frac{1}{h(x)} \left[1 + h'^2(x) \right] dx \right|$$

where m is shear factor ($0 \leq m \leq 1$).
Setting the applied powers equal to the upper-bound on energies, one obtains:

$$J^* = -2 \sigma_{xb} \quad \text{. . . (15)}$$

Substituting equations (10), (11), (12), (13), (14), and (15) into equation (9), one obtains:

$$-\frac{\sigma_{xb}}{\sigma_o} = \frac{\dot{E}_i}{2\sigma_o} + \frac{1}{2\sqrt{3}} \left[|h'(0)| + |h'(L)| \right] + \frac{\dot{E}_{S3}}{2\sigma_o} - \frac{\sigma_{xf}}{\sigma_o} \quad \text{. . . (16)}$$

As an example, if die is cosine curve given by

$$h(x) = \frac{1}{2}(1 + B) + \frac{1}{2}(1 - B)\cos(\pi x/L) \quad \text{. . . (17)}$$

We obtain velocity field given by

$$U_x = h^{-1}(x)$$

. . . (18)

$$U_y = -\frac{1}{2}(1 - B)\, y \cdot h^{-2}(x) \cdot \frac{\pi}{L} \sin(\pi x/L)$$

Velocity field of equation (18) is the same as equation (8) of Ref. [6], and the usefulness of velocity field equation (3) is clear in the case of curved dies.

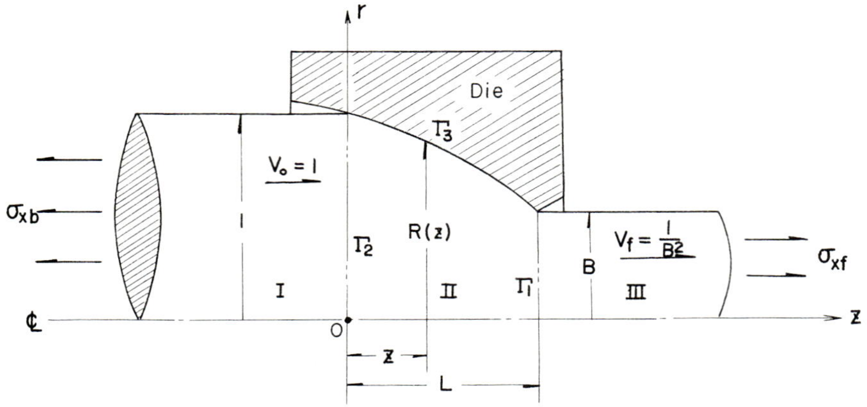

Fig. 2. Axially symmetrical extrusion through curved die

Velocity Field

The equation of the curved die profile is given by

$$r = R(z)$$

$$\text{and} \quad R(0) = 1, \quad R(L) = B. \qquad \ldots \quad (19)$$

The round bar is divided into three zones (see, Fig. 2). In zone I, the undeformed region, a rigid body motion in the Z-direction is described by $V_0(=1)$. In zone III, the already deformed bar, no further deformation occurs; the rigid body motion is described by V_f, parallel to Z-direction. Because of volume constancy,

$$V_f = \frac{1}{B^2} \qquad \ldots \quad (20)$$

Zone II is bounded by the one curved surfaces with an initial diameter 2 and final diameter 2B and the vertical surfaces Γ_1 and Γ_2 .
In zone II, it is assumed that the velocity fields are described in a cylindrical coordinate system,

$$U_r = \frac{r}{R^3(z)} R'(z)$$

$$U_z = \frac{1}{R^2(z)} \qquad \ldots (21)$$

$$U_\theta = 0$$

For Γ_2 , the velocity normal to the surface is $U_z|_{z=0} = 1$ for both sides. For Γ_1 , the velocity normal to the surface is $U_z|_{z=L} = \frac{1}{B^2}$.

Parallel to these surfaces, a velocity discontinuity exist of

$$\text{Along } \Gamma_2 \quad \Delta V_2 = r \cdot |R'(0)|$$

$$\ldots (22)$$

$$\text{Along } \Gamma_1 \quad \Delta V_1 = \frac{r}{B^3} \cdot |R'(L)|$$

Along the die surfaces, the resultant velocity is parallel to it, namely

$$\left. \frac{U_r}{U_z} \right|_{r=R(z)} = R'(z) \qquad \ldots (23)$$

Thus, the normal velocity to the die surface is zero, and the tangential velocity is equal to the resultant velocity, namely

$$\text{Along } \Gamma_3 \quad \Delta V_3 = \frac{1}{R^2(z)} \sqrt{1 + R'^2(z)} \qquad \ldots (24)$$

Deformation occurs only in zone II. In a cylindrical coordinate system, the strain rates derived from equation (21) are:

$$\dot{\epsilon}_{rr} = \frac{1}{R^3(z)} R'(z)$$

$$\dot{\epsilon}_{\theta\theta} = \frac{1}{R^3(z)} R'(z)$$

$$\dot{\epsilon}_{zz} = -\frac{2}{R^3(z)} R'(z) \qquad \ldots (25)$$

$$\dot{\epsilon}_{r\theta} = \dot{\epsilon}_{\theta z} = 0$$

$$\dot{\epsilon}_{rz} = \frac{r}{2R^4(z)} \left[R''(z) R(z) - 3 R'^2(z) \right]$$

Checking the volume constancy,

$$\dot{\epsilon}_{rr} + \dot{\epsilon}_{\theta\theta} + \dot{\epsilon}_{zz} = 0 \qquad \ldots (26)$$

The proposed velocity field of equation (21) is admissible, since it satisfies the condition of incompressibility and continuity and it also satisfies the velocity boundaries.

Upper Bound

The individual terms of equation (9) are now computed. The power dissipated along the traction prescribed boundary St is

$$\dot{E}_t = +\int_{St} T_i\, V_i\, ds = \pi\, \sigma_{xf} \qquad \ldots (27)$$

The internal power of deformation is computed over zone II alone:

$$\dot{E}_i = \frac{2\sigma_o}{\sqrt{3}} \int_V \sqrt{\frac{1}{2}\, \dot{\epsilon}_{ij} \cdot \dot{\epsilon}_{ij}}\; dV \qquad \ldots (28)$$

$$= \left| \frac{4}{\sqrt{3}}\, \pi\, \sigma_o \int_0^L dz \int_0^{R(z)} dr \cdot \frac{r}{R^3(z)}\, x \sqrt{3R'^2(z) + \frac{r^2}{4R(z)^2} F_2^2(z)} \right|$$

where

$$F_2(z) = R''(z)\, R(z) - 3R'^2(z)$$

Shear losses over surfaces of velocity discontinuities Γ_1 and Γ_2 are

$$\dot{E}_{S1} = \int_{S\Gamma_1} \tau_1 \cdot \Delta V_1 \cdot ds = \frac{2\pi\sigma_o}{3\sqrt{3}} \left| R'(L) \right| \qquad \ldots (29)$$

$$\dot{E}_{S2} = \int_{S\Gamma_2} \tau_2 \cdot \Delta V_2 \cdot ds = \frac{2\pi\sigma_o}{3\sqrt{3}} \left| R'(0) \right| \qquad \ldots (30)$$

Power Consumption over surface Γ_3 becomes

$$\dot{E}_{S3} = \int_{S\Gamma} \tau_3 \cdot \Delta V_3 dS$$

$$= \left| \frac{2}{\sqrt{3}} \pi m \sigma_o \int_0^L \frac{1}{R(z)} \left[1 + R'^2(z)\right] dz \right| \qquad \ldots (31)$$

Setting the applied powers equal to the upper bound on energies, one obtains;

$$J^* = -\pi \sigma_{xb} \qquad \ldots (32)$$

Substituting equations (27), (28), (29), (30), (31) and (33) into equation (9), one obtains;

$$-\frac{\sigma_{xb}}{\sigma_o} = \frac{\dot{E}_i}{\pi \sigma_o} + \frac{2}{3\sqrt{3}} \left[\left| R'(0) \right| + \left| R'(L) \right| \right] + \frac{\dot{E}_{S3}}{\pi \sigma_o} - \frac{\sigma_{xf}}{\sigma_o} \qquad \ldots (33)$$

The integration involved in equations (16) and (33) must be done numerically and the solutions are clear in principle.

AXIALLY SYMMETRICAL EXTRUSION THROUGH PARABOLIC DIES

The equation of the parabolic die profile is given by (See Fig. 2)

$$R(z) = \sqrt{az + 1} \qquad \ldots (34)$$

where

$$a = (B^2 - 1)/L$$

Substituting equation (34) into equation (21), we obtain the velocity fields in a cylindrical coordinate system.

$$U_r = \frac{ar}{2} (az + 1)^{-2}$$

$$U_z = (az + 1)^{-1} \qquad \ldots (35)$$

$$U_\theta = 0$$

In a cylindrical coordinate system, the strain rates derived from equation (25) are:

$$\dot{\epsilon}_{rr} = \frac{a}{2}(az+1)^{-2}$$

$$\dot{\epsilon}_{\theta\theta} = \frac{a}{2}(az+1)^{-2}$$

$$\dot{\epsilon}_{zz} = -a(az+1)^{-2} \qquad \ldots (36)$$

$$\dot{\epsilon}_{r\theta} = \dot{\epsilon}_{\theta z} = 0$$

$$\dot{\epsilon}_{rz} = -\frac{a^2 r}{2}(az+1)^{-3}$$

Velocity discontinuites exist of:

Along Γ_1 $\quad \Delta V_1 = \frac{|a|\,r}{2B^4}$

Along Γ_2 $\quad \Delta V_2 = \frac{|a|\,r}{2}$ $\qquad \ldots (37)$

Along Γ_3 $\quad \Delta V_3 = (az+1)^{-1}\sqrt{1+\frac{a^2}{4}(az+1)^{-1}}$

The individual terms of equation (33) are now computed:

$$\dot{E}_i = \frac{2}{\sqrt{3}}\sigma_0\int_V \sqrt{\frac{1}{2}\dot{\epsilon}_{ij}\cdot\dot{\epsilon}_{ij}}\;dV$$

$$= \left|\frac{2}{\sqrt{3}}\sigma_0\int_0^L dz\int_0^{R(z)} 2\pi r\cdot dr\cdot\frac{1}{2}(az+1)^{-2}a\cdot\left\{3+r^2\cdot a^2\cdot(az+1)^{-2}\right\}^{1/2}\right| \qquad \ldots (38)$$

$$= \left|\frac{4\sigma_0\pi}{a^2}\left[\frac{a^2}{2}\ln\frac{\left(B+\sqrt{B^2+\frac{a^2}{3}}\right)}{\left(1+\sqrt{1+\frac{a^2}{3}}\right)} + \frac{3}{2}\left\{B\left(B^2+\frac{a^2}{3}\right)^{1/2}-\left(1+\frac{a^2}{3}\right)^{1/2}\right\} + \frac{1}{2}(1-B^2)+\left\{\left(1+\frac{a^2}{3}\right)^{3/2}-\frac{\left(B^2+\frac{a^2}{3}\right)^{3/2}}{B}\right\}\right]\right|$$

Since for long die length $|a| \ll 1$, equation (38) reduces to

$$\dot{E}_i = 2\sigma_0 \pi \ln \frac{1}{B} \quad . . . (39)$$

Which is the ideal deformation energy.

$$\dot{E}_{S1} = \int_{S\Gamma_1} \tau_1 \cdot \Delta V_1 \cdot ds$$

$$= \left| \frac{\sigma_0}{\sqrt{3}} \int_0^B \frac{|a| \cdot r}{2B^4} \cdot 2\pi r \cdot dr \right| = \frac{\pi \sigma_0 |a|}{3\sqrt{3}\, B} \quad . . . (40)$$

$$\dot{E}_{S2} = \int_{S\Gamma_2} \tau_2 \cdot \Delta V_2 \cdot ds$$

$$= \left| \frac{\sigma_0}{\sqrt{3}} \int_0^1 \frac{|a| \cdot r}{2} \cdot 2\pi r \cdot dr \right| = \frac{\pi \sigma_0 |a|}{3\sqrt{3}} \quad . . . (41)$$

$$\dot{E}_{S3} = \int_{S\Gamma_3} \tau_3 \cdot \Delta V_3 \cdot ds$$

$$= \left| \frac{m\sigma_0}{\sqrt{3}} \int_0^L 2\pi \cdot (az+1)^{1/2} \cdot (az+1)^{-1} \sqrt{1 + \frac{a^2}{4}(az+1)^{-1}} \cdot dz \right|$$

$$= \frac{2\pi m \sigma_0}{\sqrt{3}\,|a|} \left| 2\left[\left(B^2 + \frac{a^2}{4}\right)^{1/2} - \left(1 + \frac{a^2}{4}\right)^{1/2} \right] + \frac{|a|}{2} \ln \frac{\left(\sqrt{B^2 + \frac{a^2}{4}} - \frac{|a|}{2}\right)\left(\sqrt{1 + \frac{a^2}{4}} + \frac{|a|}{2}\right)}{\left(\sqrt{B^2 + \frac{a^2}{4}} + \frac{|a|}{2}\right)\left(\sqrt{1 + \frac{a^2}{4}} - \frac{|a|}{2}\right)} \right| \quad . . . (42)$$

Substituting equations (38) - (42) into equation (33), one obtains:

$$- \frac{\sigma_{xb}}{\sigma_0} = \frac{4}{a^2} \left| \frac{a^2}{2} \ln \frac{\left(B + \sqrt{B^2 + \frac{a^2}{3}}\right)}{\left(1 + \sqrt{1 + \frac{a^2}{3}}\right)} + \frac{3}{2}\left[B\left(B^2 + \frac{a^2}{3}\right)^{1/2} - \left(1 + \frac{a^2}{3}\right)^{1/2} \right] + \frac{1}{2}(1 - B^2) + \left[\left(1 + \frac{a^2}{3}\right)^{3/2} - \frac{\left(B^2 + \frac{a^2}{3}\right)^{3/2}}{B} \right] \right| + \frac{|a|}{3\sqrt{3}}\left(1 + \frac{1}{B}\right)$$

$$+ \frac{2m}{\sqrt{3}\,|a|} \left| 2\left[\left(B^2 + \frac{a^2}{4}\right)^{1/2} - \left(1 + \frac{a^2}{4}\right)^{1/2} \right] + \frac{|a|}{2} \ln \frac{\left(\sqrt{B^2 + \frac{a^2}{4}} - \frac{|a|}{2}\right)\left(\sqrt{1 + \frac{a^2}{4}} + \frac{|a|}{2}\right)}{\left(\sqrt{B^2 + \frac{a^2}{4}} + \frac{|a|}{2}\right)\left(\sqrt{1 + \frac{a^2}{4}} - \frac{|a|}{2}\right)} \right| \quad . . (43)$$

DISCUSSION

Radial flow velocity field proposed by Avitzur [8] can also be used to obtain upper bound solutions to axisymmetric extrusion through parabolic dies as shown in Fig. 3.

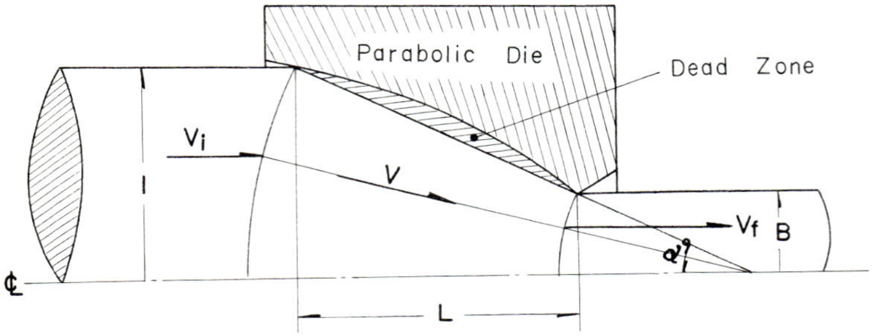

Fig. 3. Radial flow velocity field proposed by Avitzur

In this case, equation (30) of Ref. [8] reduces to:

$$-\frac{\sigma_{xb}}{\sigma_o} = 2f(\alpha)\ln\frac{1}{B} + \frac{2}{\sqrt{3}}\left[\frac{\alpha}{\sin^2\alpha} - \cot\alpha\left(1 - m\cdot\ln\frac{1}{B}\right)\right] \quad ...\ (44)$$

Where α is equal to $\tan^{-1}\frac{(1-B)}{L}$, and assumed to be smaller than α_1 defined by equation (6) of Ref. [11]. Substituting 1 for m in equation (44), one obtains upper-bound solutions to axisymmetric extrusion through parabolic dies.
To check the usefulness of the present solutions, the results of equation (43) were compared with that of equation (44) as shown in the following Table 1.

Table 1.

Process Variables		Solution of Eq. 43			Solution of Eq. 44		
		m=0.01	0.1	1	m=0.01	0.1	1
B=1/2, L=1/2,	=45°	2.6956	2.7447	3.2362	2.0701	2.1420	2.8522
L=1,	=26.5°	2.0922	2.2088	3.3743	1.7827	1.9266	3.3669
L=2,	=14°	1.8130	2.1019	4.9909	1.6093	1.8973	4.7758
B=3/4, L=1/2,	=26.5°	1.0421	1.1120	1.8011	0.9535	1.0137	1.6170
L=1,	=14°	1.1117	1.2281	2.3923	0.7737	0.9024	2.0770
L=2,	=7°	1.5638	1.8040	4.2056	0.6021	0.8358	3.4045

To smooth parabolic dies, equation (43) gives lower upper-bound than equation (44), but, to rough parabolic dies, equation (44) is lower.
Under the same frictional condition along die surfaces and the same die projected length, conical dies gives lower upper-bound than parabolic dies.

CONCLUSIONS

1. Velocity fields and upper-bound solution of plain-strain and axially symmetrical extrusion through curved die surfaces of arbitrary functions are obtained through equations (3), (16), (21) and (33).
2. By setting the shear factor, m, equal to one, the upper bound solutions of square-cornered dies are also obtained through equations (16) and (33).
3. Analytical expressions for upper bound solutions of axially symmetrical extrusion through parabolic dies were obtained through equation (43), and it yields lower upper bound than Avitzur's solution for comparatively wide range of process variables.

NOMENCLATURE

B = half thickness of the strip at die exit in plain-strain and radius of the round bar at die exit in axially symmetrical extrusion
$\dot{E}_i$ = the internal power of deformation
$\dot{E}_{SK}$ = the power due to shear or friction along surface of velocity discontinuity where k=1, 2 or 3
$\dot{E}_t$ = the power dissipated along the traction prescribed boundary
$h(x)$ = half thickness of the strip at a generic section
J^* = the applied power
L = Axial (projected) length of the curved die
m = friction or shear factor
O = origin of the coordinate system
$R(z)$ = radius of the wire at a generic section
r, θ, z = coordinates of a cylindrical coordinate system
x, y, z = coordinates of a Cartesian coordinate system
U_r, U_θ, U_z = components of the velocity field in cylindrical coordinate system
U_x, U_y, U_z = components of the velocity field in Cartesian coordinate system
V_f = the exit velocity of the strip and wire
V_0 = the entrance velocity of the strip and wire (=1)
$\Delta V_1, \Delta V_2, \Delta V_3$ = velocity difference
S_t = surface of integration where the surface traction is prescribed
S_Γ = surface of integration where the velocity discontinuities is involved
T = the prescribed applied surface traction
V = volume
σ_o = yield limit at uniaxial load
σ_{xb} = stress exerted on the strip and wire at the entrance of the die

σ_{xf} = stress exerted on the strip and wire at the exit of the die

$\dot{\epsilon}_{ij}$ = components of strain rate

τ = shear stress

τ_1, τ_2, τ_3 = shear stress along the velocity discontinuity

$\Gamma_1, \Gamma_2, \Gamma_3$ = surfaces of velocity discontinuities

REFERENCES

[1] R. Hill, "On the State of Stress in a Plastic-Rigid Body at the Yield Point," Philosophical Magazine, Vol. 42, 1951, p. 868.

[2] D. C. Drucker, W. Prager, and H. J. Greenberg, "Extended Limit Design Theorems for Continuous Media," Quarterly of Applied Mechanics, Vol. 9, 1952, p. 381.

[3] W. Johnson, and P. B. Mellor, " Plasticity for Mechanical Engineers," D. Van Nostrand press, England, 1962.

[4] E. G. Thomsen, C. T. Yang, and S. Kobayashi, " Mechanics of Plastic Deformation in Metal Processing," Macmillan Company, New York, 1965.

[5] B. Avitzur, J. Fueyo, and H. Thompson, " Analysis of Plastic Flow Through Inclined Planes in Plain-Strain Extrusion," Journal of Engineering for Industry, TRANS, ASME, Series, B, Vol. 89, 1967, pp. 503-513.

[6] P.C.T. Chen, " Upper Bound Solutions to Plane-Strain Extrusion Problems," ASME Paper No. 69-WA/Prod-13.

[7] W. Johnson, " Upper Bound Loads for Extrusion Through Circular Dies," Applied Science Research, Vol. 7A, 1958, pp. 437-448.

[8] B. Avitzur, " Analysis of Wire Drawing and Extrusion Through Conical Dies of Large Cone Angle," Journal of Engineering for Industry, TRANS, ASME, Series B, Vol. 86, No. 4, Nov. 1964, pp. 305-316.

[9] B. Avitzur, " Analysis of Metal Extrusion," Journal of Engineering for Industry, TRANS, ASME, Series B, Vol. 87, No. 1, Feb., 1965, pp. 57-70.

[10] D. Z. Zimerman, and B. Avitzur, " Metal Flow Through Conical Converging Dies-A Lower Upper Bound Approach Using Generalized Boundaries of Plastic Zone," ASME, Paper No. 69-WA/prod-19.

[11] B. Avitzur, " Hydrostatic Extrusions," Journal of Engineering for Industry, TRANS, ASME, Series B, Nov. 1965, pp. 487-494.

Developments in Mechanics, Vol. 6. Proceedings of the 12th Midwestern Mechanics Conference: 27

Determining Hill's Anisotropic Yield Parameters

ROBERT W. BUND
GENERAL MOTORS INSTITUTE
LAWRENCE E. MALVERN
UNIVERSITY OF FLORIDA

ABSTRACT

Assuming the plane stress case, this paper presents the results of an experimental study to determine the initial parameters in Hill's anisotropic yield function and plastic potential function. The material investigated was commercially - produced AISI 1006 sheet steel, usually called aluminum-killed steel. A single-shear test coupon was used to determine the initial value of the anisotropic shear parameter N.

Tensile test coupons were elongated on an Instron to determine the initial yield stress as a function of orientation; then some of these same tensile coupons were further elongated to determine the strain ratio R as a function of orientation. This data permitted the initial anisotropic tensile parameters F, G, and H to be determined by both the direct and the strain-ratio methods. A comparison of these two methods is included.

INTRODUCTION

The anisotropy initially present in rolled sheet metal is generally believed to control the onset of such phenomena as earing during the cupping operation, when a sheet metal blank is plastically transformed into a cup-shaped part using a die in a mechanical press.

Two methods of determining the initial anisotropy are compared in this paper. Both methods are based on determining the anisotropic parameters in R. Hill's orthotropic yield function, assuming a plane stress state. The first method (direct method) utilizes the initial yield stress from tensile and shear coupons. The second (indirect method) utilizes strain ratios after yielding and hence determines the parameter ratios of the plastic potential function, which is assumed to coincide with the yield function initially. Experimental evidence in the literature will be cited, which indicates that the strain ratios in the tensile coupons remain essentially constant during the test, so that finite strains can be used to get the strain ratios with greater accuracy.

R. Hill has suggested that a generalization of the Mises Yield equation could be used as the yield criterion for some anisotropic materials. See [1], Chapter 12. This generalization requires that four parameters, for the plane stress case, be determined by experimental studies of the sheet material used in the plastic-working operation.

Hill's anisotropic yield equation can be specialized for the plane stress case to

$$2f(\sigma_{ij}) = (G + H)\sigma_x^2 - 2H\sigma_x\sigma_y + (H + F)\sigma_y^2 + 2N\tau_{xy}^2 = 1 \qquad (1)$$

Equation 1 assumes orthotropic symmetry, with three mutually-orthogonal planes of symmetry in the metal. The intersection of these planes of symmetry defines the principal anisotropic axes. We take the x-direction as the direction of rolling, the y-direction as the transverse direction in the plane of the sheet metal, and the z-direction as the thickness direction (See Figure 1). Corresponding to these directions, X_0 represents the value of σ_x at initial yield for uniaxial tension in the x-direction, while Y_0 and Z_0 represent the initial tensile yield stresses in the y-direction and the z-direction, respectively. When a specimen yields in pure shear, the initial yield value of τ_{xy} is called T_0.

There are four parameters (F, G, H, and N) in the plane-stress, anisotropic yield condition, Equation 1, to be evaluated from experimental data. The parameter N was independently determined by experimentally finding the shear yield stress, T_0. Two methods were used to find the remaining three parameters. One of these two methods can be called the direct method, whereby the parameters F, G, and H were determined by experimentally finding the tensile yield stress as a function of orientation with respect to the direction of rolling. The second method investigated is called the indirect method or strain-ratio method; this method used the ratio of the width strain to the thickness strain to evaluate the three initial anisotropic parameters F, G, and H.

The purpose of this investigation was to compare and evaluate the direct and the indirect methods of determining the initial anisotropic yield parameters. The material studied was commercially-produced AISI 1006 sheet steel, often called aluminum-killed steel.

THEORY FOR THE DIRECT METHOD

If the anisotropic yield function, Equation 1, is specialized for the case of a tensile test coupon oriented at angle α to the rolling or x-direction, the transformation equations

$$\sigma_x = \sigma \cos^2 \alpha \qquad \sigma_y = \sigma \sin^2 \alpha \qquad \tau_{xy} = \sigma \sin \alpha \cos \alpha \qquad (2)$$

reduce it to

$$F \sin^2 \alpha + G \cos^2 \alpha + H + (2N - F - G - 4H)\sin^2 \alpha \cos^2 \alpha = \frac{1}{\sigma_\alpha^2} \qquad (3)$$

The four parameters F, G, H, and N can not be evaluated by experimentally determining the yield stress for tensile coupons in four directions because the four equations are not independent. Hence, the initial value of N was independently determined by specializing Equation 1 for the pure shear case to get

$$N = \frac{1}{2T_0^2} \qquad (4)$$

After the value for N was determined, then only three equations were needed to solve for the remaining three parameters F, G, and H.

COMPUTATION OF THE SHEAR PARAMETER

The value of T_0, the initial yield stress in shear, to be used in Equation 4 to determine the initial shear parameter N was determined by pulling twelve single-shear test coupons, Figure 2, on an Instron Tensile Testing Instrument, type TT-C. The shear coupon was based on the work of Yen [2], published in 1960, and particularly the discussion by Bradley [3], which included a study of the geometry parameters of the coupon and their effect on the elastic shear stress distribution in the stressed zone. The average shear yield stress for these twelve samples was computed to be 16,923 psi with a standard deviation of 533 psi. With this average yield stress, Equation 4 gave

$$N = \frac{1}{2T_0^2} = \frac{1}{2(16{,}923)^2} = 1.745 \times 10^{-9} \text{ inches}^4/\text{ pound}^2 \tag{5}$$

Since the load for initial yielding in shear, the thickness, and the shear length were each measured to three significant figures, the accuracy of the shear parameter N is also limited to three significant figures.

COMPUTATION OF THE TENSILE PARAMETERS USING THE DIRECT METHOD

A study of the initial tensile yield stress as a function of orientation was required when using the direct method to determine the tensile anisotropic parameters F, G, and H. Twelve tensile coupons at each of nine orientations were pulled on the Instron. Each coupon was approximately 7 inches long, and 0.532 inches wide. The yield stress was defined, as indicated in Figure 3, as the point on the stress strain diagram where the transition zone meets the plastic zone. The results of this testing program are shown in Table I.

An examination of these results of the tensile yield strength study clearly shows the problem associated with the direct method of computing the tensile parameters: the tensile yield strength changes only a small amount with a change in orientation. Hill, [1] page 321, has suggested another measure of anisotropy, the strain-ratio method, which will be discussed later in this paper.

It is theoretically possible to compute F, G, and H by choosing any three orientations of tensile coupons and their corresponding initial yield stresses, as given in Table I, along with the value of N from Equation 5. However, because of the experimental scatter, F, G, and H were determined in this investigation from the data for all nine orientations using the least squares method. See, for example, Wylie [4]. Equation 3 was rewritten in the form:

$$\frac{1}{\sigma_\alpha^2} = F(\sin\alpha)^4 + G(\cos\alpha)^4 + H(\cos 2\alpha)^2 + N\frac{(\sin 2\alpha)^2}{2} \tag{6}$$

The least squares method consisted of setting up a difference equation by subtracting the left member of Equation 6 from the right member. This difference was called δ_α.

$$\delta_\alpha = F(\sin\alpha)^4 + G(\cos\alpha)^4 + H(\cos 2\alpha)^2 - \frac{1}{\sigma_\alpha^2} + \frac{N(\sin 2\alpha)^2}{2} \tag{7}$$

Each pair (α, σ_α), from Table I, gave a difference equation. To get the "best" values of F, G, and H, an error function "E" was defined

$$E = \sum_\alpha (\delta_\alpha)^2 \quad \text{or}$$

$$E = \sum_\alpha \left[F(\sin\alpha)^4 + G(\cos\alpha)^4 + H(\cos 2\alpha)^2 - \frac{1}{\sigma_\alpha^2} + \frac{N(\sin 2\alpha)^2}{2}\right]^2 \tag{8}$$

The error function was minimized in the usual way by setting equal to zero the partial derivatives of the function, with respect to the three variables F, G, and H.

$$\frac{\partial E}{\partial F} = 0 \qquad \frac{\partial E}{\partial G} = 0 \qquad \frac{\partial E}{\partial H} = 0 \tag{9}$$

This resulted in three equations to solve for the three unknowns F, G, and H, using the computed value of $N = 1.745 \times 10^{-9}$ as given in Equation 5.

These simultaneous linear equations were solved for F, G, and H using the standard library BASIC language computer program "SIMEQN" on the General Electric 265 Time-Sharing Computer. The "best" values were

$$\begin{aligned} F &= 6.94 \times 10^{-10} \text{ inches}^4/\text{ pound}^2 \\ G &= 7.60 \times 10^{-10} \text{ inches}^4/\text{ pound}^2 \\ H &= 5.96 \times 10^{-10} \text{ inches}^4/\text{ pound}^2 \end{aligned} \tag{10}$$

As was the case with "N," the accuracy of the "best-fit" tensile parameters is not more than three significant figures. As the discussion of Figure 4 in the appraisal section indicates, the accuracy was actually less than this.

COMPUTATION OF THE TENSILE PARAMETERS USING THE INDIRECT METHOD

The indirect method of computing the anisotropic parameters requires a consideration of the plastic potential theory as proposed by Mises [5,6], and as generalized by Hill [1,7] for the anisotropic case. A rigid, work-hardening material is assumed; hence the plastic strain increment is the total strain increment. Plastic potential theory assumes that the stress, strain-increment relation is derivable from the plastic potential function $f(\sigma_{ij})$ by the relationship

$$d\varepsilon_{ij} = d\lambda \frac{\partial f}{\partial \sigma_{ij}} \tag{11}$$

where $f(\sigma_{ij})$ for the general case is defined by Hill [1] as follows:

$$2f(\sigma_{ij}) = F(\sigma_y - \sigma_z)^2 + G(\sigma_z - \sigma_x)^2 + H(\sigma_x - \sigma_y)^2 + 2L\tau_{yz}^2 + 2M\tau_{zx}^2 + 2N\tau_{xy}^2 \tag{12}$$

It should be noted that Equation 12 can be specialized to Equation 1 for the plane stress case where

$$\sigma_z = \tau_{zx} = \tau_{zy} = 0 \tag{13}$$

Equations 11 and 12, give the following relationships:

$$\begin{aligned} d\varepsilon_x &= d\lambda[(G + H)\sigma_x - H\sigma_y - G\sigma_z] \\ d\varepsilon_y &= d\lambda[(H + F)\sigma_y - F\sigma_z - H\sigma_x] \\ d\varepsilon_z &= d\lambda[(F + G)\sigma_z - G\sigma_x - F\sigma_y] \\ d\varepsilon_{xy} &= d\lambda[N\tau_{xy}] \end{aligned} \tag{14}$$

It should be noted that $d\varepsilon_x + d\varepsilon_y + d\varepsilon_z = 0$, which is consistent with the assumption of a rigid, work-hardening material, if plastic deformation is assumed to produce no volume change.

For a tensile test coupon oriented in the x-direction, the direction of rolling, $\sigma_y = \sigma_z = \tau_{xy} = 0$, and Equations 14 reduce to:

$$d\varepsilon_x = d\lambda(G + H)\sigma_x, \qquad d\varepsilon_y = -d\lambda(H)\sigma_x, \qquad d\varepsilon_z = -d\lambda(G)\sigma_x \tag{15}$$

The ratio of width to thickness strain increments for this test coupon at $\alpha = 0°$ then can be written:

$$R_0 = \frac{d\varepsilon_w}{d\varepsilon_t} = \frac{d\varepsilon_y}{d\varepsilon_z} = \frac{H}{G} \tag{16}$$

Similarly, a tensile test coupon oriented in the y-direction ($\sigma_x = \sigma_z = \tau_{xy} = 0$ and $\alpha = 90°$) gives the following ratio for width to thickness strain increments:

$$R_{90} = \frac{d\varepsilon_w}{d\varepsilon_t} = \frac{d\varepsilon_x}{d\varepsilon_z} = \frac{H}{F} \tag{17}$$

For the more general case of a tensile test coupon oriented at the arbitrary angle "α" to the direction of rolling, the width strain increment must be found from the strain transformation equation:

$$d\varepsilon_w = d\varepsilon_x \sin^2\alpha + d\varepsilon_y \cos^2\alpha - 2d\varepsilon_{xy} \sin\alpha \cos\alpha \tag{18}$$

This gives a ratio of width to thickness strain increment as follows:

$$R_\alpha = \frac{d\varepsilon_w}{d\varepsilon_t} = \frac{d\varepsilon_x \sin^2\alpha + d\varepsilon_y \cos^2\alpha - 2d\varepsilon_{xy} \sin\alpha \cos\alpha}{d\varepsilon_z} \tag{19}$$

Equations 14 can be transformed to refer to a tensile coupon oriented at "α" to the direction of rolling by using stress transformation Equations 2 to get the following stress strain increment relations:

$$\begin{aligned} d\varepsilon_x &= d\lambda[(G + H)\sigma\cos^2\alpha - H\sigma\sin^2\alpha] \\ d\varepsilon_y &= d\lambda[(H + F)\sigma\sin^2\alpha - H\sigma\cos^2\alpha] \\ d\varepsilon_z &= d\lambda[-G\sigma\cos^2\alpha - F\sigma\sin^2\alpha] \\ d\varepsilon_{xy} &= d\lambda[N\sigma\sin\alpha\cos\alpha] \end{aligned} \tag{20}$$

whence the general strain ratio Equation 19 becomes

$$R_\alpha = \frac{H + (2N - F - G - 4H)\sin^2\alpha\cos^2\alpha}{F\sin^2\alpha + G\cos^2\alpha} = \frac{d\varepsilon_w}{d\varepsilon_t} \tag{21}$$

This reduces to Equations 16 and 17 for the special cases of $\alpha = 0°$ and $\alpha = 90°$.

Tests by several investigators, including Bramley and Mellor [8,9], Atkinson [10], and Lankford, Snyder and Bauscher [11] indicate that, for low carbon steel, the width to thickness strain-ratios do not vary as the material strain hardens during the tensile test. This fact permits the strain ratios to be computed at larger values of strain to get more accurate results and permits finite strains to be used in place of strain increments in Equation 21.

$$R = \frac{\text{Width Strain}}{\text{Thickness Strain}} = \frac{\ln\left[\frac{w_0}{w}\right]}{\ln\left[\frac{t_0}{t}\right]} = \frac{\ln\left[\frac{w_0}{w}\right]}{\ln\left[\frac{\ell w}{\ell_0 w_0}\right]} \tag{22}$$

In Equation 22 "w," "ℓ," and "t" represent the current width, gage length, and thickness respectively; similarly "w_0," "ℓ_0," and "t_0" represent the initial measurements of these quantities. Atkinson [10,12] recommends that the strain ratio should be measured just prior to necking, at a logarithmic length strain of about 0.20 for low carbon steel. Following Atkinson's recommendation, experimental strain-ratio data were obtained for the same tensile coupons used in the first six series reported in Table I. Since each series had one coupon at each of nine orientations, a total of fifty-four coupons were included in this strain-ratio study. The results of this study are given in Table II.

It is possible to use Equation 21 in conjunction with any three orientations and the associated strain-ratio average values from Table II to evaluate the three tensile parameters F, G, and H using the computed value of N from Equation 5. This was done by Bund and is reported on page 42 of [13]. In the present study, the least squares method was employed to utilize all nine average values of strain ratio and their corresponding α-values.

Equation 21 was used to define a difference function δ_α by transposing R_α. The error function to be minimized was

$$E = \sum_{\alpha} \left[\frac{H + (2N - F - G - 4H) \sin^2 \alpha \cos^2 \alpha}{F \sin^2 \alpha + G \cos^2 \alpha} - R_\alpha \right]^2 \qquad (23)$$

While it is possible to minimize this function by using Equations 9, this approach would require solving three non-linear equations for the three unknowns. Although this can be done using iterative procedures, an alternative approach seemed more desirable. A computer program was written in the Basic Fortran IV language which permitted non-negative functions to be minimized on a specified interval. The function was initially evaluated at a hundred randomly-picked points on the interval to determine at which of these points a minimum functional value existed. A second, smaller interval was next defined by the program in the neighborhood of this "minimum" point; then the program reiterated by evaluating the function at a second group of a hundred, randomly-picked points on this smaller interval. For most reasonable functions, a satisfactory minimum can be found after ten to fifteen iterations.

This Basic Fortran IV program was used on an IBM 360 model 40 computer to determine the following "best-fit" values for the tensile anisotropic parameters:

$$\begin{aligned} F &= 5.2898 \times 10^{-10} \text{ inches}^4/\text{pound}^2 \\ G &= 5.7036 \times 10^{-10} \text{ inches}^4/\text{pound}^2 \\ H &= 8.2869 \times 10^{-10} \text{ inches}^4/\text{pound}^2 \end{aligned} \qquad (24)$$

Again, the accuracy of the results given by Equations 24 is at most three significant figures.

AN APPRAISAL OF THE COMPUTED ANISOTROPIC PARAMETERS

There are many ways to judge the reasonableness of the computed anisotropic parameters. Several of these will be discussed in this section. As a first check of the direct method, the values of N, F, G, and H from Equations 5 and 10 were used in conjunction with Equation 3 to calculate yield stress versus orientation. This result is plotted in Figure 4 along with the nine average values of the experimental data from Table I. From Figure 4, it is evident that the experimental data required a lot of "smoothing out" during the least squares procedure.

For a second appraisal of the direct method, the anisotropic parameters computed by the strain-ratio method, Equations 24 were inserted into Equation 3.

The resulting yield stress versus orientation output is also plotted in Figure 4 to provide a convenient comparison with the results of the direct method. It is obvious that the strain-ratio method indicates a greater degree of planar anisotropy than does the direct method.

The anisotropic parameters, Equations 24, computed by the strain-ratio method were used in conjunction with Equation 21 to provide output of strain ratio versus orientation. These results are plotted in Figure 5 along with the nine average strain-ratio values from Table II.

It is evident that the experimental data is more nearly represented by the "best-fit" curve in Figure 5 than is true for the results presented in Figure 4. This fact suggests that more credence should be given to the strain-ratio results than to the results based on the direct method.

As a final check, strain ratio versus orientation curves were compared using the two sets of anisotropic parameters. The results are shown in Figure 5. It should be noted that the strain-ratio curve from the direct method is always less than unity, while the strain-ratio curve based on the measured strain ratios is always greater than unity.

CONCLUSTIONS AND RECOMMENDATIONS

The following conclusions are based on the experimental procedures and results:

1. The yield stress data as presented in Table I and in Figure 4 indicate the difficulty of obtaining sufficiently accurate data. This may be due to the difficulty of defining the yield stress as well as the difficulty of measuring the small variations being studied.

2. Figures 4 and 5 indicate that an anomaly exists between the anisotropic parameters determined by the direct and the indirect method.

3. The strain-ratio data is an indication of the material state in the plastic regime, but possibly not in the transition zone (between the elastic and plastic regime).

The following recommendations are based on the study:

1. If the plastic-working operation being studied involves large plastic deformation, the indirect (strain-ratio) method should be used to determine the anisotropic parameters.

2. If the plastic-working operation being studied involves small deformations, it might be better to use the direct method to determine the anisotropic parameters.

3. A strain-ratio study should be made which extends back into the transition zone. This might provide information to help resolve the anomaly as indicated by the results shown in Figures 4 and 5.

BIBLIOGRAPHY

[1] Hill, R., The Mathematical Theory of Plasticity, Oxford University Press, London, 1950, pp. 317-340.

[2] Yen, C. S., "Stress Distribution in Single-Shear Sheet Specimens," ASTM Special Technical Publication Number 289, Symposium on Shear and Torsion Testing, Sixty-Third Annual Meeting Papers, American Society for Testing Materials, 1960, pp. 15-20.

[3] Bradley, W. A., "Discussion on Stress Distribution in Single-Shear Sheet," ASTM Special Technical Publication Number 289, Symposium on Shear and Torsion Testing, Sixty-Third Annual Meeting Papers, American Society for Testing Materials, 1960, pp. 21-25.

[4] Wylie, C. R., Jr., Advanced Engineering Mathematics, McGraw-Hill Book Company, Inc., New York, Second Edition, 1960, pp. 175-193.

[5] Mises, R. Von, "Mechanik der Plastischen Formänderung von Kristallen," Zeitschrift Für Angewandte Mathematick und Mechanik, Volume 8, Number 3, June 1928, pp. 161-185.

[6] Mises, R. von, "Mechanik der festen Körper im plastisch-deformablen Zustand," Göttingen Nachrichten, 1913, pp. 582-592.

[7] Hill, R., "A Theory of the Yielding and Plastic Flow of Anisotropic Metals," Royal Society of London, Proceedings A, Volume 193, 1948, pp. 281-297.

[8] Bramley, A. N., and Mellor, P. B., "Some Strain-Rate and Anisotropy Effects in the Stretch-Forming of Steel Sheet," International Journal of Machine Tool Design and Research, Volume 5, 1965, pp. 43-55.

[9] Bramley, A. N., and Mellor, P. B., "Plastic Flow in Stabilized Sheet Steel," International Journal of Mechanical Sciences, Volume 8, February 1966, pp. 101-114.

[10] Atkinson, M., and Maclean, I. M., "The Measurement of Normal Plastic Anisotropy in Sheet Metal," Sheet Metal Industries, Volume 42, Number 456, April 1965, pp. 290-298.

[11] Lankford, W. T., Snyder, S. C., and Bauscher, J. A., "New Criteria for Predicting the Press Performance of Deep Drawing Sheets," Transactions of the American Society for Metals, Volume 42, 1950, pp. 1197-1232.

[12] Atkinson, M., "Assessing Normal Anisotropic Plasticity of Sheet Metals," Sheet Metal Industries, Volume 44, Number 479, March 1967, pp. 167-178.

[13] Bund, R. W., "Earing in Cupping Experiments Related to Anisotropic Plasticity Theory," Ph.D. Dissertation, Michigan State University, East Lansing, Michigan, 1969.

ACKNOWLEDGEMENT

* The results presented in this paper are based upon a thesis submitted by R. W. Bund to Michigan State University in partial fulfillment of the requirements for the degree of Doctor of Philosophy.

Table I Summary Results of the Tensile Yield Stress Study

SERIES NO.	DEGREES TO ROLL DIRECTION 0°	30°	40°	42.5°	45°	47.5°	50°	60°	90°
1	27,709	27,033	28,331	28,597	28,544	28,973	28,548	28,627	27,729
2	27,463	27,506	28,469	28,412	28,812	28,250	28,551	28,597	27,948
3	27,492	27,728	28,276	28,626	28,973	28,736	28,816	28,842	27,869
4	27,282	27,333	28,548	28,519	29,105	28,358	28,360	28,548	27,703
5	28,030	27,175	28,067	28,418	28,702	28,310	28,633	28,471	27,729
6	27,199	27,766	28,111	28,471	28,809	28,218	28,571	28,571	27,456
7	27,272	27,313	28,329	28,417	28,755	28,540	28,610	28,555	27,710
8	27,165	26,981	28,386	28,494	28,648	28,380	28,439	28,494	27,708
9	27,327	27,423	28,122	———	28,970	28,732	28,493	28,355	27,784
10	26,983	27,456	27,729	28,380	28,755	28,916	28,742	28,409	27,784
11	27,381	27,222	27,784	28,364	28,678	28,418	28,734	28,301	27,456
12	27,489	27,732	28,230	28,418	28,702	28,310	28,386	28,355	27,566
Average	27,399	27,389	28,198	28,465	28,788	28,512	28,573	28,510	27,703
Standard Deviation	261.27	253.28	240.674	82.07	151.9	251.9	136.7	142.1	142.2

Table II Summary Results of the Strain-Ratio Study

	DEGREES TO ROLL DIRECTION								
Series No.	0°	30°	40°	42.5°	45°	47.5°	50°	60°	90°
1	1.421	1.196	1.140	1.060	1.137	1.097	1.033	1.262	1.613
2	1.444	1.178	1.150	1.117	1.122	1.089	1.078	1.261	1.592
3	1.433	1.152	1.061	1.107	1.034	1.067	1.012	1.163	1.541
4	1.509	1.187	1.112	1.141	1.154	1.029	1.064	1.241	1.545
5	1.448	1.180	1.119	1.123	1.070	1.053	1.031	1.254	1.527
6	1.403	1.212	1.117	1.147	1.113	1.034	1.032	1.255	1.592
Average	1.443	1.184	1.116	1.116	1.105	1.061	1.042	1.239	1.568
Standard Deviation	.036	.020	.031	.031	.045	.028	.024	.038	.035

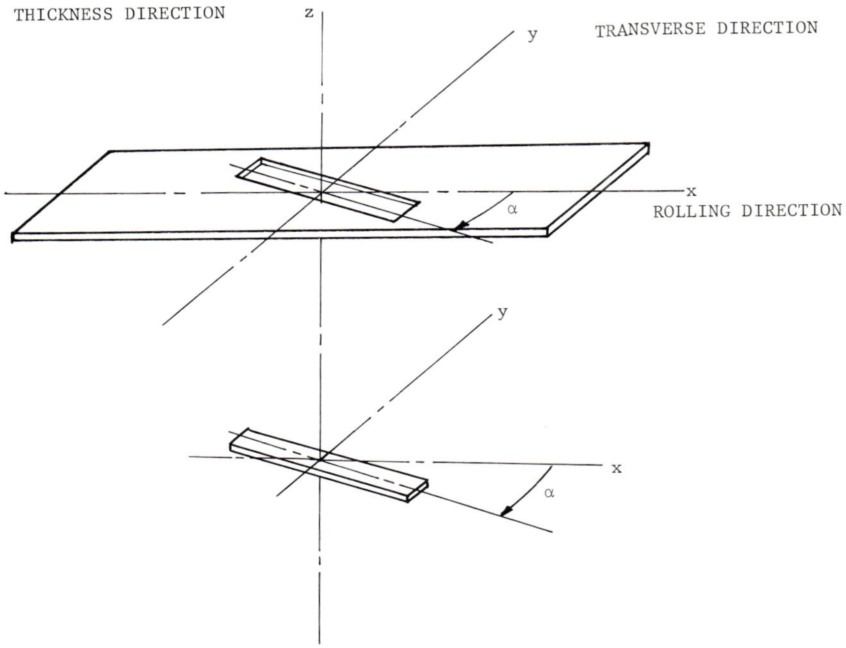

Figure 1. Notation Used for Orientation of Tensile Coupons with Respect to the Principal Anisotropic Axes

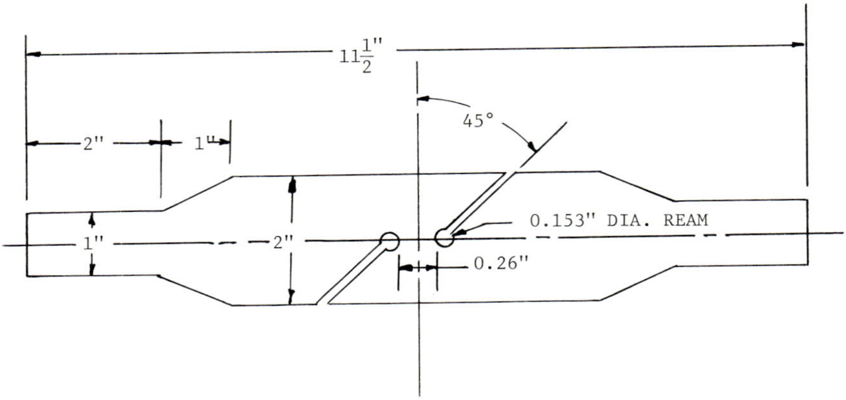

Figure 2. Half-Size Layout of the Single-Shear Test Coupon

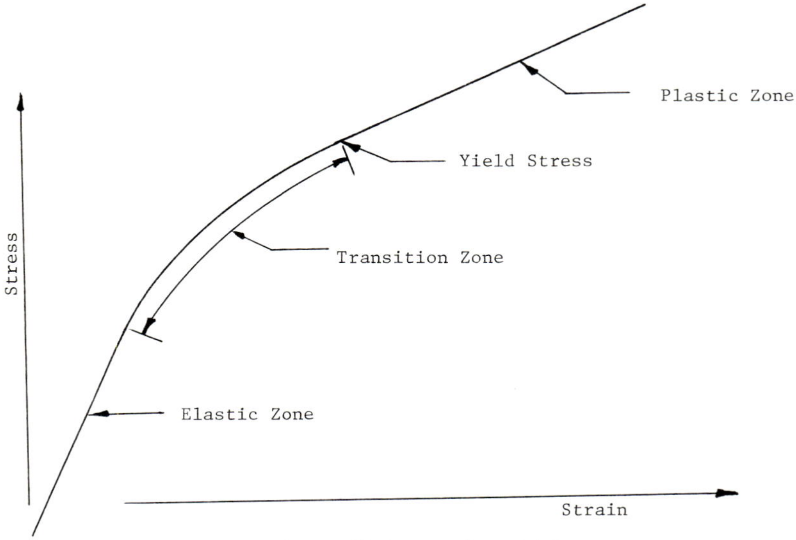

Figure 3. Definition of Yield Stress

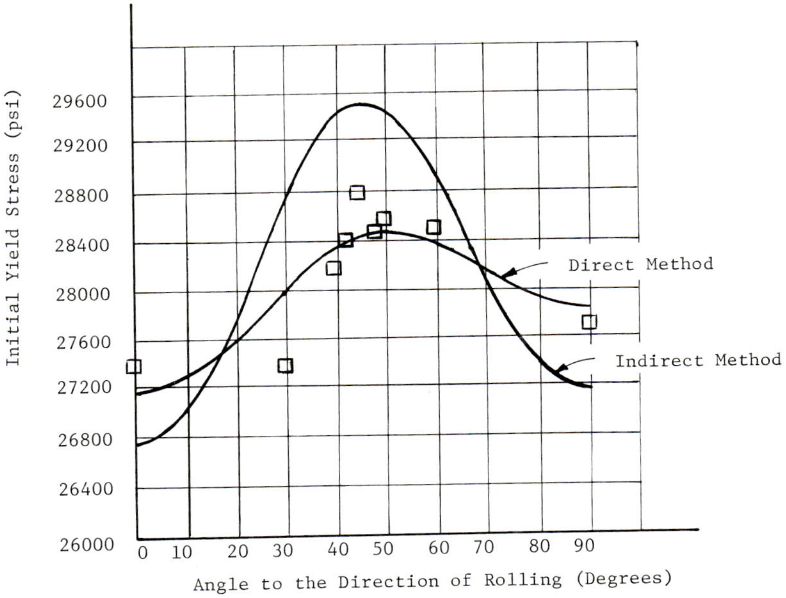

Figure 4. Initial Yield Stress Versus Orientation Curves, Based on the Direct and the Indirect Methods.

□ Experimental Yield Stress Data from Table I

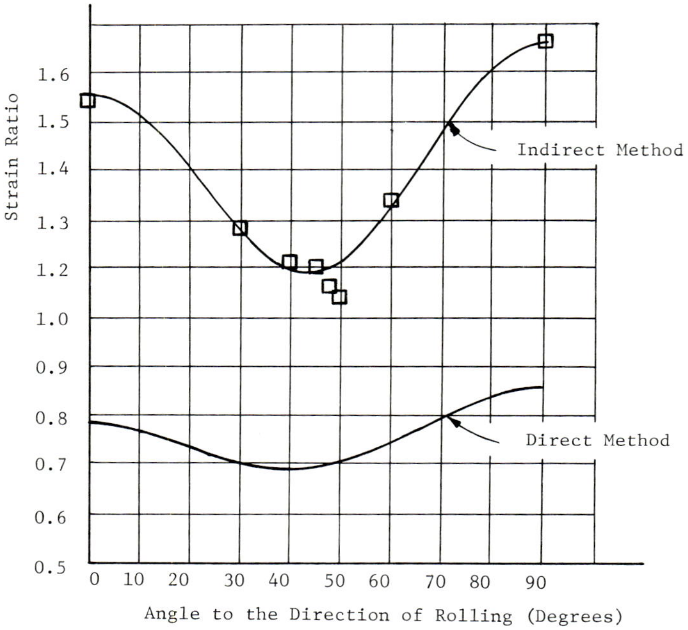

Figure 5. Strain-Ratio Versus Orientation Curves, Based on the Direct and the Indirect Methods.

□ Experimental Strain Ratio Data from Table II

Developments in Mechanics, Vol. 6. Proceedings of the 12th Midwestern Mechanics Conference: 28

Steady-state Response of Hysteretic Systems with Multifold Yielding Distribution

I. C. JONG and Y. S. CHEN
UNIVERSITY OF ARKANSAS

ABSTRACT

This study introduces the concept and the formulation of hysteretic models with a multifold yielding distribution. The steady-state response of such models to sinusoidal excitations is investigated. It is found that in hysteretic systems with a threefold or higher order of yielding distribution there may exist the phenomenon of discontinuous jumps in amplitude as either the frequency or the level of the sinusoidal excitation varies smoothly and continuously. Furthermore, it is shown that the phenomenon of unbounded amplitude resonance under a certain finite level of the sinusoidal excitation may be suppressed by increasing the order of yielding distribution in the hysteretic system.

INTRODUCTION

The phenomenon of yielding and associated hysteresis has been observed in many physical systems carrying cyclic loads or withstanding dynamic disturbances. The cause of this phenomenon may be due to the occurrence of slipping of the system material on certain crystallographic planes and the presence of Coulomb friction associated with the slipping.

In a given system if the yield limit (or slip level) is the same for all material elements, the action of yielding of the system will be accomplished in an abrupt manner. In other words, such a system will exhibit a concentrated (or sudden) yielding. The hysteresis loop associated with this kind of system is generally composed of straight line segments of two different slopes (i.e., bilinear).

Hysteresis loops of real systems are, however, rarely bilinear. Thus, it stands to reason that yield limits for the material elements of most physical systems are generally not all the same. This concept was mentioned by Timoshenko [1] some four decades ago but appears to have received significant attention from investigators only in recent years.

Based on the aforementioned concept, a straightforward distributed-element model for hysteresis was lately formulated and studied by Iwan [2]. He presented the analytical results of the steady-state response of the model and compared them with the experimental results from an actual structure. The basic elements used for distribution in his model [2] are the so-called Jenkin's elements. Each Jenkin's element consists of a linear spring in series with a Coulomb (or slip) damper which can withstand forces that are smaller than its yield limit. The linear spring in each basic element is assumed to have the same stiffness, while the yield limit for the Coulomb damper in the basic element is allowed to vary from one basic element to the other. Hence it is the yield limit of the Coulomb damper, rather than the stiffness of the linear spring, that makes a basic element different from the others in the model. Consequently, the distribution of basic elements with various yield limits in the model can be viewed as the distribution (or the spreading out) of the action of yielding of the model. Furthermore, noting that the yielding action of a Jenkin's element is of an undistributed (or a concentrated) type, the hysteretic model formulated by Iwan in [2] may be classified as a hysteretic model with a onefold (or single) yielding distribution.

The work presented in [2] can be, in a way, regarded as an extension of a previous study by Caughey [3] who accomplished the analytical investigation of the sinusoidal excitation of a one-degree-of-freedom system with bilinear hysteresis. The studies and results of hysteretic models with a concentrated yielding [3] and hysteretic models with a onefold yielding distribution [2] are of basic significance and quite interesting.

As a logical extension of [2,3], the present study introduces the concept and the formulation of hysteretic models with a multifold yielding distribution. A hysteretic model with a twofold (or double) yielding distribution is defined as a distributed-element hysteretic model whose basic elements for distribution are hysteretic models with a onefold yielding distribution. Similarly, a hysteretic model with an n-fold yielding distribution is defined as a distributed-element hysteretic model whose basic elements for distribution are hysteretic models with an (n-1)-fold yielding distribution. As an application, such models are herein employed for a further investigation of the steady-state response of one-degree-of-freedom hysteretic systems to sinusoidal excitations.

The purpose of this investigation is to show that there may exist in hysteretic systems with a threefold or higher order of yielding distribution the phenomenon of discontinuous jumps in amplitude as either the frequency or the level of the driving force is varied smoothly and continuously. It is to be pointed out that such a phenomenon, characteristic of certain nonlinear systems [4], has not, heretofore, been found to exist in hysteretic systems whose hysteresis loops do not pinch together at the origin [2,3]. Furthermore, the results of this investigation show that the phenomenon of unbounded amplitude resonance under a certain finite level of the sinusoidal excitation may be suppressed by increasing the order of yielding distribution in the hysteretic system.

THE MODEL

We know that a standard linear solid in viscoelasticity is converted into a general bilinear hysteretic model if the viscous damper (i.e., the dashpot) in the standard linear solid is replaced with a Coulomb (or slip) damper. Since there are two non-degenerate representations [5] of the standard linear solid, there also exist two non-degenerate representations of the general bilinear hysteretic model. One of these two representations is a model which consists of a linear spring and a Jenkin's element in parallel as illustrated in Fig. 1(a). The other representation is a model which consists of a linear spring in series with an assemblage composed of a linear spring and a Coulomb damper in parallel (not shown).

In Fig. 1(a), we let the stiffness of the linear spring in the Jenkin's element be μ, the stiffness of the linear spring in parallel with the Jenkin's element be 1-μ, and the yield limit (or slip level) for the Coulomb damper in the Jenkin's element be μa^*.

As a first step, let the general bilinear hysteretic models, illustrated in Fig. 1(a), be the basic elements for distribution in the formulation of the hysteretic model with a onefold yielding distribution. This approach is followed and is schematically illustrated in Fig. 1(b). In every basic element as illustrated in Fig. 1(b), it is to be noted that the stiffness of the linear spring in the Jenkin's element takes now the value of μ/N, and the stiffness of the linear spring in parallel with the Jenkin's element takes now the value of (1-μ)/N, where N is the total number of basic elements distributed. The yield limit for the Coulomb damper in the i-th basic element in Fig. 1(b) is $\mu a_i/N$ where a_i is allowed to vary from basic element to basic element.

Let the basic elements in Fig. 1(b) be arranged in the order of increasing yield limit, and the highest yield limit for an element be designated as a^*/N. Then, it is clear that when the value of N is large and tends to infinity, the hysteretic model with a onefold yielding distribution is obtained.

Employing the foregoing approach, a hysteretic model with a twofold yielding distribution may be formulated using hysteretic models with a onefold yielding distribution as the basic elements for distribution. This is done and is schematically illustrated in Fig. 1(c).

Hysteretic models with a twofold yielding distribution as illustrated in Fig. 1(c) are, in turn, used as the basic elements for distribution in the formulation of the hysteretic model with a threefold yielding distribution as illustrated in Fig. 1(d). Any hysteretic model with a higher order of yielding distribution can, of course, be formulated using a similar approach.

The force-displacement relationship for a hysteretic model with an n-fold yielding distribution as described above is shown in Fig. 2(a). In this figure, the function $Y(a)$ is used to define the number or the fraction (or percentage) of the basic elements whose ultimate yield limit is a, x is the displacement, μ is the hysteretic parameter, t is the time variable, $F = F[x,\mu,t,Y(a)]$ is a functional denoting the hysteretic restoring force, and A is the amplitude of the displacement. Although $Y(a)$ may be specified in many different ways, it will be assumed to be a band-limited step function as shown in Fig. 2(b), where a^* is equivalent to the smallest displacement beyond which all of the basic elements for distribution will yield and the total area of the distribution diagram is $\eta(1/\eta) = 1$.

From Fig. 2(a), it is obvious that the hysteretic parameter μ controls the degree of hysteresis damping to be experienced by the system. As a matter of fact, we have

$$\lim_{\mu \to 0} F[x,\mu,t,Y(a)] = x \qquad (1)$$

It can be shown that the forward-loading part and the reversed-loading part of the hysteresis loop for F versus x are smooth curves without sharp corners if $0 < \eta \leqslant a^*$ in Fig. 2(b). Moreover, the bilinear hysteresis loop can be obtained in the special case in which $Y(a) = \delta(a-a^*)$, a Dirac delta function.

STEADY-STATE EQUATIONS

The hysteretic model formulated in the preceding section will now be used for the analysis of the steady-state response of a one-degree-of-freedom hysteretic system acted on by a sinusoidal force P cosωt. The mass m of the system is mounted on one end of a hysteretic spring. The other end of the hysteretic spring is connected to a rigid support. The restoring force developed in the hysteretic spring is $k\,F[x,\mu,t,Y(a)]$ where k is a constant and $F[x,\mu,t,Y(a)]$ is defined in Fig. 2. Thus, the equation of motion for the system may be written as

$$m\ddot{x} + k\ F(x,\mu,t,Y) = P\ \cos\omega t \tag{2}$$

where

$$Y = Y(a) = \frac{1}{\eta}\,[H(a-a^*+\eta) - H(a-a^*)] \tag{3}$$

and $H(a)$ is the Heavyside's unit step function.

For convenience, let the following variables be introduced:

$$\omega_o^2 = k/m \tag{4}$$

$$\tau = \omega_o t \tag{5}$$

$$f = P/(\mu a^* k) \tag{6}$$

$$\beta = \omega/\omega_o \tag{7}$$

$$X = x/a^* \tag{8}$$

$$R = A/a^* \tag{9}$$

Then, following the method of equivalent linearization [6], Eq. 2 may be written in the form

$$X'' + \lambda X' + \kappa X + \varepsilon(X,X',\tau) = \mu f\ \cos\beta\tau \tag{10}$$

where a prime is used to indicate a differentiation with respect to the dimensionless time variable τ, and

$$\varepsilon(X,X',\tau) = F(X,\mu,\tau,Y) - \lambda X' - \kappa X \tag{11}$$

which is termed the equation deficiency [6]. In this method, we first minimize the equation deficiency with respect to the parameters λ and κ and then neglect it in Eq. 10 to obtain an equivalent linearized equation of motion

$$X'' + \lambda X' + \kappa X = \mu f\ \cos\beta\tau \tag{12}$$

whose steady-state solution is known to be of the form

$$X = R\ \cos\theta \tag{13}$$

$$\theta = \beta\tau - \psi \tag{14}$$

In the present study, let the mean squared value of the equation deficiency be minimized with respect to λ and κ; i.e.,

$$\frac{\partial}{\partial\lambda}\ \overline{[\varepsilon(X,X',\tau)]^2} = 0 \tag{15}$$

$$\frac{\partial}{\partial\kappa}\ \overline{[\varepsilon(X,X',\tau)]^2} = 0 \tag{16}$$

where the bar superscript denotes a time average.

The equivalent linearized parameters λ and κ may now be evaluated using Eqs. 11, 13, 14 and conditions 15 and 16. This yields

$$\lambda = -\frac{S(R)}{\beta R} \tag{17}$$

$$\kappa = \frac{C(R)}{R} \tag{18}$$

where

$$S(R) = \frac{1}{\pi} \int_0^{2\pi} F(R\cos\theta, \mu, \tau, Y) \sin\theta \, d\theta \tag{19}$$

$$C(R) = \frac{1}{\pi} \int_0^{2\pi} F(R\cos\theta, \mu, \tau, Y) \cos\theta \, d\theta \tag{20}$$

Substituting Eqs. 13 and 14 into Eq. 12 and equating the coefficients of $\sin\theta$ and those of $\cos\theta$ on both sides, we obtain the steady-state response equations

$$\lambda\beta R = \mu f \sin\psi \tag{21}$$

$$-\beta^2 R + \kappa R = \mu f \cos\psi \tag{22}$$

Using Eqs. 17 and 18 and eliminating ψ in Eqs. 21 and 22, we arrive at the steady-state response equation

$$\beta^2 = \frac{1}{R}\left\{ C(R) \pm \sqrt{(\mu f)^2 - [S(R)]^2} \right\} \tag{23}$$

which relates the frequency to the amplitude. Moreover, using Eqs. 17, 18, and 23, we obtain from Eqs. 21 and 22, the phase shift

$$\psi = \pm \tan^{-1}\left[S(R) \Big/ \sqrt{(\mu f)^2 - [S(R)]^2} \right] \tag{24}$$

Since the maximum amplitude will occur at the point where two roots of β^2 coalesce, it is readily seen from Eq. 23 that the amplitude resonance occurs at the point where

$$f = \frac{|S(R)|}{\mu} \tag{25}$$

$$\beta^2 = \frac{C(R)}{R} \tag{26}$$

Meanwhile, substitution of Eq. 25 into Eq. 24 yields $\psi = \pm \pi/2$. Thus, the phase resonance and the amplitude resonance always occur at the same frequency in the present system.

EVALUATION OF S(R) AND C(R)

In steady-state response, the hysteresis loop is symmetric about the origin. The integrals associated with S(R) and C(R) as defined in Eqs. 19 and 20 may be taken as twice of the same integrals evaluated over one half cycle of oscillation. This has been done for: (1) the case of a hysteretic model with undistributed yielding (i.e., bilinear hysteresis) in [3] (and also in [7]), (2) the case of a hysteretic model with a onefold yielding distribution in [2] (and also in [8]). Results for these two cases have been published and need not be derived again. With appropriate adjustments in notation, they will be, in keeping with Fig. 2, taken from [7] and [8] and listed below. Results for the cases of hysteretic models with a twofold and a threefold yielding distribution are derived in this section.

For convenience, the following notations will be used:

$$\bar{a} = a/a^* \tag{27}$$

$$\gamma = \eta/a^* \tag{28}$$

$$\chi = \cos^{-1}\left[1 - \frac{2(1-\gamma)}{R}\right] \tag{29}$$

$$\zeta = \cos^{-1}\left(1-\frac{2}{R}\right) \tag{30}$$

$$\zeta^* = \cos^{-1}\left(1-\frac{2\bar{a}}{R}\right) \tag{31}$$

(I) Hysteretic Model with Undistributed Yielding

With appropriate adjustments in notation, results for the present case are [7]:

<u>For $0 < R \leqslant 1$</u>

$$S(R) = 0 \tag{32}$$

$$C(R) = R \tag{33}$$

<u>For $R > 1$</u>

$$S(R) = -\frac{\mu R}{\pi}\sin^2\zeta \tag{34}$$

$$C(R) = R - \frac{\mu R}{\pi}(\pi - \zeta + \sin\zeta\cos\zeta) \tag{35}$$

(II) Hysteretic Model with a Onefold Yielding Distribution

With appropriate adjustments in notation, results for the present case are [8]:

<u>For $0 < R \leqslant 1 - \gamma$</u>

$$S(R) = 0 \tag{36}$$

$$C(R) = R \tag{37}$$

<u>For $1 - \gamma < R \leqslant 1$</u>

$$S(R) = -\frac{\mu R^2}{6\pi\gamma}\left[(3-\cos^2\chi)\cos\chi + 2\right] \tag{38}$$

$$C(R) = R - \frac{\mu R^2}{6\pi\gamma}\left[3(\pi-\chi)\cos\chi + (2+\cos^2\chi)\sin\chi\right] \tag{39}$$

<u>For $R > 1$</u>

$$S(R) = -\frac{\mu R^2}{6\pi\gamma}\left[(3-\cos^2\chi)\cos\chi - (3-\cos^2\zeta)\cos\zeta\right] \tag{40}$$

$$C(R) = R - \frac{\mu R^2}{6\pi\gamma}\left[3(\pi-\chi)\cos\chi + (2+\cos^2\chi)\sin\chi - 3(\pi-\zeta)\cos\zeta - (2+\cos^2\zeta)\sin\zeta\right] \tag{41}$$

(III) Hysteretic Model with a Twofold Yielding Distribution

In the present model, the basic elements for distribution are the hysteretic models with a onefold yielding distribution. For simplicity, these basic elements are assumed to have a full yielding distribution (i.e., $\gamma = 1$) in themselves. In terms of dimensionless variables, the distribution function Y as defined in Eq. 3 may be written in the form

$$Y = Y(\bar{a}) = \frac{1}{\gamma}\left[H(\bar{a}-1+\gamma) - H(\bar{a}-1)\right] \tag{42}$$

Thus, making use of the assumed nature of the individual basic elements of the present model and using Eqs. 38 through 41 (where γ is set equal to 1) as well as Fig. 2, we find that Eqs. 19 and 20 as applied to the present model are equivalent to the following two equations:

$$S(R) = -\frac{\mu R^2}{6\pi}\int_0^R [2-(3-\cos^2\zeta^*)\cos\zeta^*]\,Y(\bar{a})\,d\bar{a} - \frac{2\mu R^2}{3\pi}\int_R^\infty Y(\bar{a})\,d\bar{a} \tag{43}$$

$$C(R) = \int_0^R \left\{R - \frac{\mu R^2}{6\pi}[3\pi - 3(\pi-\zeta^*)\cos\zeta^* - (2+\cos^2\zeta^*)\sin\zeta^*]\right\} Y(\bar{a})\,d\bar{a}$$
$$+ (R - \frac{1}{2}\mu R^2)\int_R^\infty Y(\bar{a})\,d\bar{a} \tag{44}$$

Substituting Eqs. 31 and 42 into Eqs. 43 and 44, we obtain:

For $0 < R \leqslant 1 - \gamma$

$$S(R) = -\frac{2\mu R^2}{3\pi} \tag{45}$$

$$C(R) = R - \frac{1}{2}\mu R^2 \tag{46}$$

For $1 - \gamma < R \leqslant 1$

$$S(R) = -\frac{\mu R^2}{384\pi\gamma}[128(1+\gamma) - R(109+20\cos 2\chi-\cos 4\chi)] \tag{47}$$

$$C(R) = R - \frac{\mu R^2}{384\pi\gamma}\left\{192\pi\gamma - R[36(\pi-\chi) + 24(\pi-\chi)\cos 2\chi + 28\sin 2\chi + \sin 4\chi]\right\} \tag{48}$$

For $R > 1$

$$S(R) = -\frac{\mu R^2}{384\pi\gamma}[128\gamma - R(20\cos 2\chi-\cos 4\chi-20\cos 2\zeta+\cos 4\zeta)] \tag{49}$$

$$C(R) = R - \frac{\mu R^2}{384\pi\gamma}\left\{192\pi\gamma - R[-36\chi + 24(\pi-\chi)\cos 2\chi + 28\sin 2\chi + \sin 4\chi + 36\zeta\right.$$
$$\left. - 24(\pi-\zeta)\cos 2\zeta - 28\sin 2\zeta - \sin 4\zeta]\right\} \tag{50}$$

(IV) Hysteretic Model with a Threefold Yielding Distribution

The basic elements for distribution in the present model are the hysteretic models with a twofold yielding distribution. For simplicity, these basic elements are also assumed to have a full yielding distribution (i.e., $\gamma = 1$) in themselves. Thus, following a similar approach used in the preceding case, we find that Eqs. 19 and 20 as applied to the present model are equivalent to the following two equations:

$$S(R) = -\frac{\mu R^2}{384\pi}\int_0^R [128 - R(19-20\cos 2\zeta^*+\cos 4\zeta^*)]\,Y(\bar{a})\,d\bar{a}$$
$$-\frac{\mu R^2}{3\pi}(2-R)\int_R^\infty Y(\bar{a})\,d\bar{a} \tag{51}$$

$$C(R) = \int_0^R \left\{R - \frac{\mu R^2}{384\pi}[192\pi - R(24\pi+36\zeta^*-24\pi\cos 2\zeta^*+24\zeta^*\cos 2\zeta^*-28\sin 2\zeta^*\right.$$
$$\left. - \sin 4\zeta^*)]\right\} Y(\bar{a})\,d\bar{a} + [R - \frac{\mu R^2}{32}(16-5R)]\int_R^\infty Y(\bar{a})\,d\bar{a} \tag{52}$$

Substituting Eqs. 31 and 42 into Eqs. 51 and 52, we obtain:

For $0 < R \leqslant 1 - \gamma$

$$S(R) = -\frac{\mu R^2}{3\pi}(2-R) \quad (53)$$

$$C(R) = R - \frac{\mu R^2}{32}(16-5R) \quad (54)$$

For $1 - \gamma < R \leqslant 1$

$$S(R) = -\frac{\mu R^2}{480\pi\gamma}[160\gamma(2-R) - 80R(1+\cos\chi) + R^2(64+55\cos\chi+10\cos^3\chi-\cos^5\chi)] \quad (55)$$

$$C(R) = R - \frac{\mu R^2}{2880\pi\gamma}\Big\{90\pi\gamma(16-5R) + R^2[15(\pi-\chi)(3+4\cos^2\chi)\cos\chi + (105-95\sin^2\chi+6\sin^4\chi)\sin\chi]\Big\} \quad (56)$$

For $R > 1$

$$S(R) = -\frac{\mu R^2}{480\pi\gamma}[160\gamma(2-R) - 80R\cos\chi + R^2(55\cos\chi+10\cos^3\chi-\cos^5\chi) + 80R\cos\zeta - R^2(55\cos\zeta+10\cos^3\zeta-\cos^5\zeta)] \quad (57)$$

$$C(R) = R - \frac{\mu R^2}{2880\pi\gamma}\Big\{90\pi\gamma(16-5R) + R^2[15(\pi-\chi)(3+4\cos^2\chi)\cos\chi + (105-95\sin^2\chi+6\sin^4\chi)\sin\chi] - R^2[15(\pi-\zeta)(3+4\cos^2\zeta)\cos\zeta + (105-95\sin^2\zeta+6\sin^4\zeta)\sin\zeta]\Big\} \quad (58)$$

(V) Hysteretic Models with a Higher Order Than Threefold Yielding Distribution

The functions S(R) and C(R) associated with hysteretic models with a higher order than threefold yielding distribution can be evaluated in a similar manner. Because of limited space, no more of them will be presented in this study.

AMPLITUDE-FREQUENCY CURVES

Knowing the functions S(R) and C(R), we can now readily investigate the relation between the steady-state response amplitude and the frequency of the external sinusoidal forcing function using Eqs. 23, 25, and 26. Since results in the preceding section reveal that S(R) is a quantity of the order of μ and C(R) is equal to R plus another quantity of the order of μ, we may, for convenience in the study, introduce the frequency parameter

$$\Omega = \frac{\beta^2 - 1}{\mu} \quad (59)$$

Results of the steady-state response of the hysteretic system with onefold, twofold, and threefold yielding distributions are plotted in Figs. 3, 4, and 5 respectively, for f = 0.2 and several values of γ.

In the hysteretic system with a onefold yielding distribution and f = 0.2, it is shown in Fig. 3 that when the width γ of the distribution diagram is increased from zero to unity, the resonance peak first decreases and then increases and in the meanwhile moves to the left. It is also shown that all curves here are single-valued. The curve corresponding to γ = 0.00 is, in fact, the steady-state response curve of a bilinear hysteretic system.

The general feature of the steady-state response of the hysteretic system with a twofold yielding distribution is in several respects different from the preceding one. As illustrated in Fig. 4, when the width γ is increased from zero to unity, the resonance peak will move upward and at the same time move to the right. Moreover, the resonance peak is seen to become sharper and sharper as γ varies from zero to unity. It is to be recognized that the curve corresponding to $\gamma = 0.00$ in Fig. 4 is the same as the curve corresponding to $\gamma = 1.00$ in Fig. 3. All curves in Fig. 4 are also single-valued.

Quite different from the above results and the results of earlier investigators [2,3], the steady-state amplitude-frequency curves of the hysteretic system with a threefold yielding distribution are found to be multi-valued as shown in Fig. 5. This feature has a strong resemblance to the general feature of the steady-state response of a mass mounted on a cubic soft spring and experiencing some viscous damping [4]. It is seen from Fig. 5 that the feature of multi-valuedness becomes more pronounced when the parameter γ varies from zero to unity. The resonance peak is found to decrease and move to the left when γ is increased from zero to unity. Again, we note that the curve corresponding to $\gamma = 0.00$ in Fig. 5 is the same as the curve corresponding to $\gamma = 1.00$ in Fig. 4, and is singled-valued.

The features of the steady-state response of hysteretic systems with a higher order of yielding distribution can be more complicated. The foregoing results are presented to show some characteristic features that exist in hysteretic systems with a multifold yielding distribution.

EXHIBITION OF JUMP PENOMENON

It is found that two types of jump phenomenon may exist in the steady-state response of hysteretic systems with a threefold yielding distribution and those with a higher order of yielding distribution.

The first one is the phenomenon of discontinuous jumps in amplitude R as the forcing frequency parameter Ω varies. This phenomenon is depicted in Fig. 6 for various values of the forcing level f of the hysteretic system with a threefold full yielding distribution (i.e., $\gamma = 1$). For the curve corresponding to $f = 2.5$ in Fig. 6, starting at P_1, let the frequency be increased gradually. The operating point will first travel through P_2 to P_3. Then, on reaching P_3, the amplitude R will jump suddenly to P_6 and continue to move along to P_7 with increase in Ω. From P_7 suppose the frequency is decreased gradually. The operating point will first travel through P_6 and P_5 to P_4. Then, on reaching P_4, the amplitude R will drop abruptly to P_2 and continue to move along to P_1 with decrease in Ω. Hence, there is a range of frequencies, located between the dashed lines P_2P_4 and P_3P_6 of this figure, for which either of two different values of R may exist. Which value does exist depends upon the past history of the operation of the system. The portion P_3P_4 of the response curve is unstable, and operation along P_3P_4 cannot be realized in practice. Similar comments apply to the curves corresponding to $f = 1.0$ and $f = 0.2$ in Fig. 6.

By the way, it is seen in Fig. 6 that the resonance peak moves to a lower frequency as the level f of the driving force is increased. Thus, the system exhibits a "soft" type resonance as in [2,3].

The other type of jump phenomenon is the discontinuous jumps in amplitude R as the forcing level f varies. Such a jump phenomenon is indicated in Fig. 7 for $\Omega = -1$ and $\gamma = 1$ of the hysteretic system with a threefold yielding distribution. It is clearly seen in this figure that, with the forcing frequency held constant ($\Omega = -1$), discontinuous jumps in R will occur if f is varied smoothly and continuously. Moreover, we similarly note in this figure that there is a range of values of the forcing level f for which three values of R are possible. The upper and the lower of these values are stable, while the middle value is unstable and cannot be realized in practice.

The phenomenon of discontinuous jumps is also noted to exist in other hysteretic systems with a higher order of yielding distribution. However, such a phenomenon has not been found to exist in the steady-state response of hysteretic systems whose hysteresis loops do not pinch together at the origin [2,3].

POSSIBILITY OF UNBOUNDED RESONANCE

The possibility of having an unbounded resonance in amplitude R under a certain finite forcing level f may be investigated using Eq. 25 and the appropriate S(R) for the hysteretic model under consideration.

Substituting Eqs. 30 and 34 into Eq. 25 and then solving for R, we get

$$R = \frac{4}{4 - \pi f} \tag{60}$$

Thus, when the forcing level takes the value of

$$f = \frac{4}{\pi} \tag{61}$$

the hysteretic system with undistributed yielding will exhibit an unbounded resonance in amplitude.

Substituting Eqs. 29, 30, and 40 into Eq. 25 and then solving for R, we get

$$R = \frac{4(3-3\gamma+\gamma^2)}{3[2(2-\gamma) - \pi f]} \tag{62}$$

Thus, whenever the forcing level is given by

$$f = \frac{2}{\pi}(2-\gamma) \tag{63}$$

the hysteretic system with a onefold yielding distribution will also exhibit an unbounded resonance. It is to be noted that Eq. 63 becomes Eq. 61 as expected if we set $\gamma = 0$ in Eq. 63.

Next, let us substitute Eqs. 29, 30, and 49 into Eq. 25 and then solve for R. This yields

$$R = \frac{4 - 6\gamma + 4\gamma^2 - \gamma^3}{2(3-3\gamma+\gamma^2) - 3\pi f} \tag{64}$$

From Eq. 64, we note that when

$$f = \frac{2}{3\pi}(3-3\gamma+\gamma^2) \tag{65}$$

the hysteretic system with a twofold yielding distribution will have an unbounded resonance. If we set $\gamma = 1$ in Eq. 63 and $\gamma = 0$ in Eq. 65, we see that these two equations give identical values for f as expected.

Similarly, let us now substitute Eqs. 29, 30, and 57 into Eq. 25. This gives

$$f = \frac{1}{30\pi}\left[5\gamma R^2 + 5(4-6\gamma+4\gamma^2-\gamma^3) - \frac{2}{R}\cdot(5-10\gamma+10\gamma^2-5\gamma^3+\gamma^4)\right] \tag{66}$$

It is clear from Eq. 66 that f approaches infinity as R approaches infinity. For $1 \geqslant \gamma > 0$, it is seen that no solution of R equal to infinity and f equal to a finite value is possible for Eq. 66. Thus, the present hysteretic system with a threefold yielding distribution cannot exhibit an unbounded resonance under a certain finite forcing level.

The foregoing analyses demonstrate that the phenomenon of unbounded resonance in amplitude under a certain finite forcing level may be suppressed by increasing the order of yielding distribution in the hysteretic system.

SUMMARY AND CONCLUDING REMARKS

The present study introduces the concept and the formulation of hysteretic models with a multifold yielding distribution. Specifically, a hysteretic model with an n-fold yielding distribution is defined as a distributed-element hysteretic model whose basic elements for distribution are hysteretic models with an (n-1)-fold yielding distribution. Such models are then employed for the further investigation of the steady-state response of one-degree-of-freedom hysteretic systems.

Results of the investigation point out that:

(1) In hysteretic systems with a threefold or higher order of yielding distribution, there may exist the phenomenon of discontinuous jumps in amplitude as either the frequency or the level of the driving force is varied smoothly and continuously.

(2) The phenomenon of unbounded amplitude resonance under a certain finite level of the driving force may be suppressed by increasing the order of yielding distribution in the hysteretic system.

The features of the foregoing results are, of course, consequences of the yielding and hysteretic mechanisms proposed for the hysteretic model. Since the present hysteretic model possesses many features of real physical systems, the results of this study may provide additional useful insight into the possible types of the steady-state response of physical systems exhibiting yielding and hysteresis effects.

ACKNOWLEDGEMENT

This work was supported by the National Science Foundation under Grant GK-13533.

REFERENCES

[1] Timoshenko, S.P., Strength of Materials, Part 2, D.Van Nostrand, New York, First Edition, 1930, pp. 679-680.

[2] Iwan, W.D., "A distributed-Element Model for Hysteresis and Its Steady-State Dynamic Response," Journal of Applied Mechanics, Vol. 33, No. 4, Trans. ASME, Vol. 88, Series E. Dec. 1966, pp. 893-900.

[3] Caughey, T.K., "Sinusoidal Excitation of a System with Bilinear Hysteresis," Journal of Applied Mechanics, Vol. 27, No. 4, Trans. ASME, Vol. 82, Series E, Dec. 1960, pp. 640-643.

[4] Cunningham, W.J., Introduction to Nonlinear Analysis, McGraw-Hill, New York, 1958.

[5] Bland, D.R., The Theory of Linear Viscoelasticity, Pergamon Press, New York, 1960, pp. 1-5.

[6] Caughey, T.K., "Random Excitation of a System with Bilinear Hysteresis," Journal of Applied Mechanics, Vol. 27, No. 4, Trans. ASME, Vol. 82, Series E, Dec. 1960, pp. 649-652.

[7] Jong, I.C., "On Stability of a Circulatory System with Bilinear Hysteresis," Journal of Applied Mechanics, Vol. 36, No. 1, Trans. ASME, Vol. 91, Series E, Mar. 1969, pp. 76-82.

[8] Jong, I.C., "On Stability of a Circulatory System with Weak Distributed Yielding Effects," Journal of Applied Mechanics, Vol. 37, No. 3, Trans. ASME, Vol. 92, Series E, Sept. 1970, pp. 870-873.

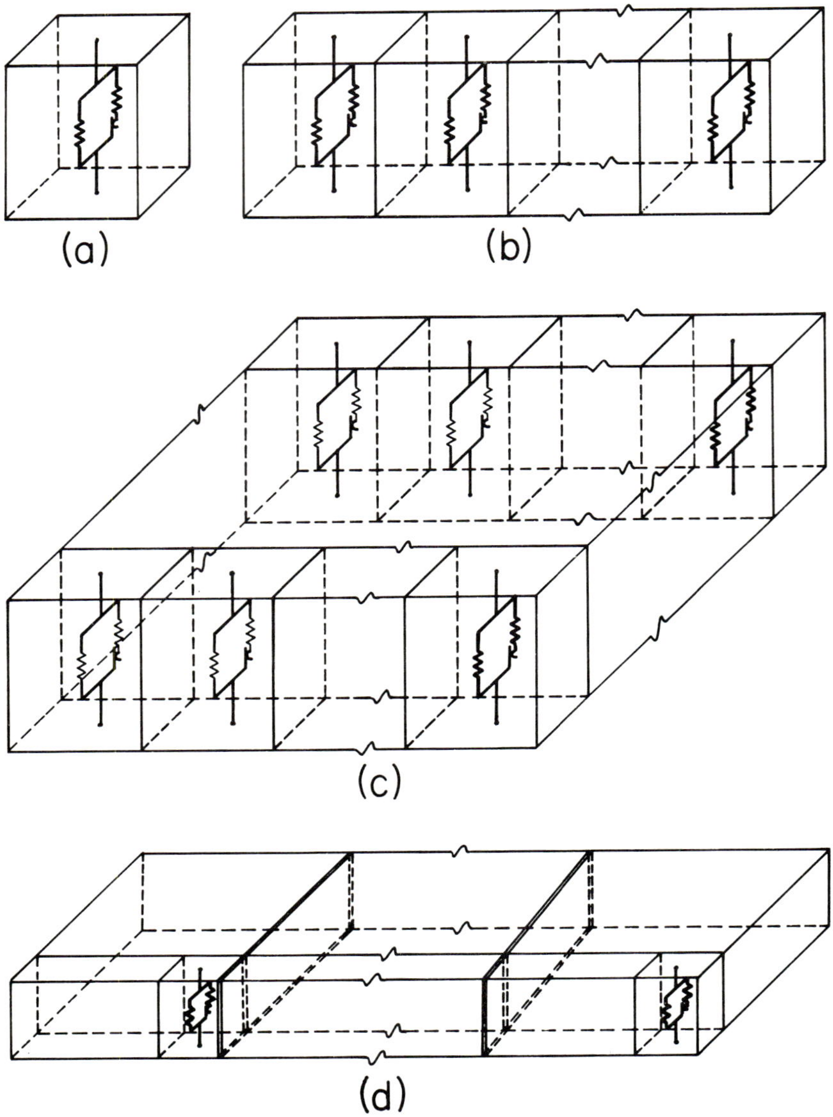

Fig. 1. Formulation of hysteretic models with a multifold yielding distribution.

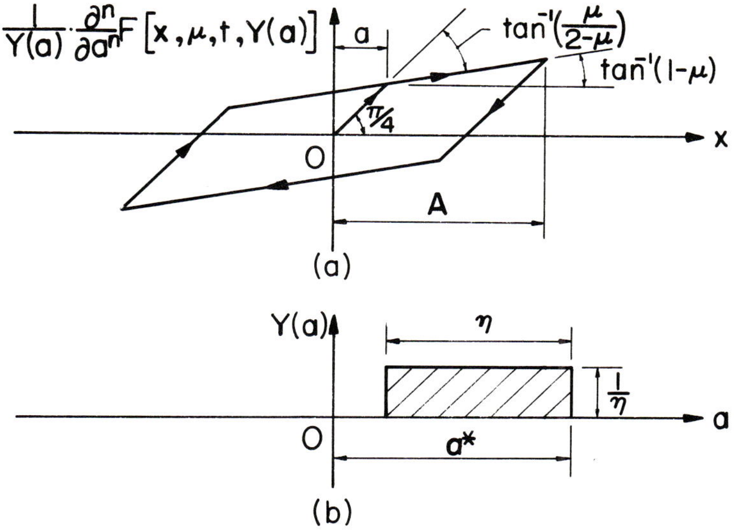

Fig. 2. Characteristics of the hysteretic model with an n-fold yielding distribution: (a) force-displacement relationship, (b) distribution function $Y(a)$.

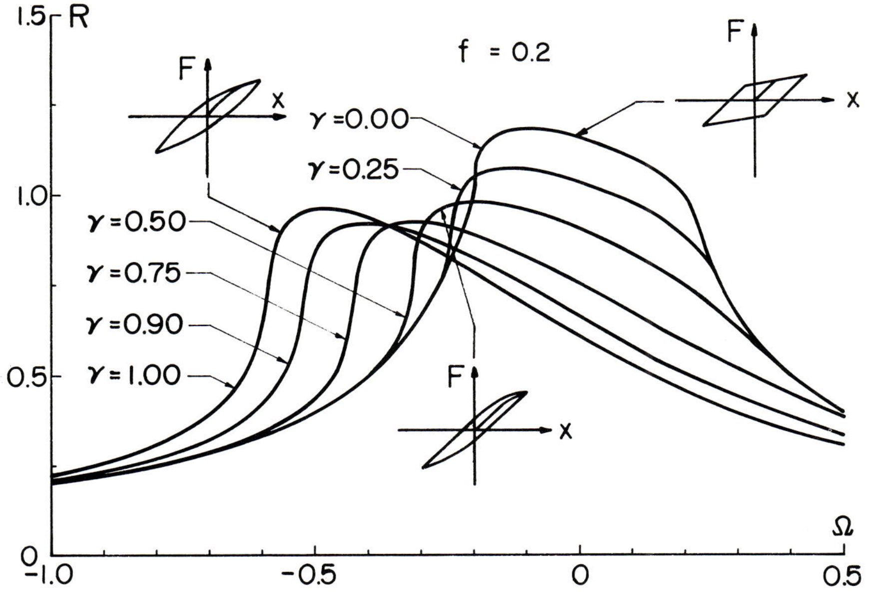

Fig. 3. Family of amplitude-frequency curves, onefold yielding distribution.

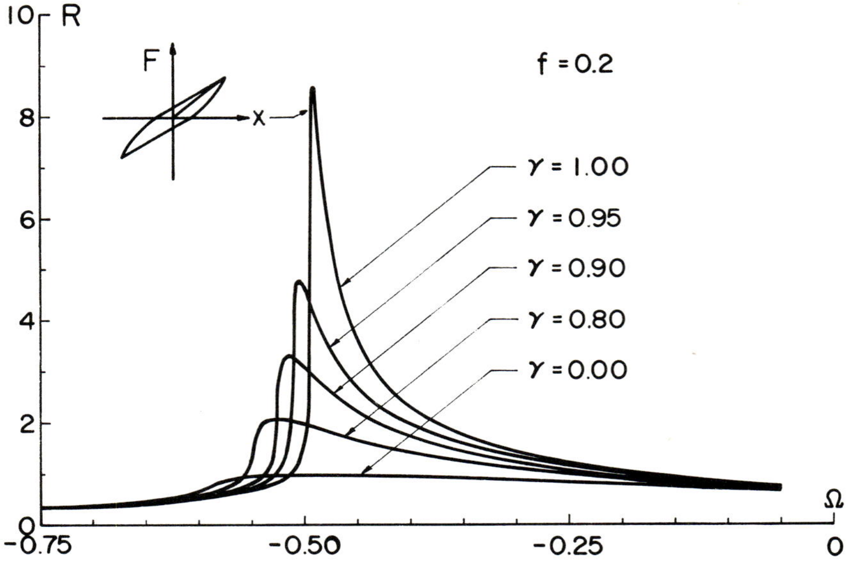

Fig. 4. Family of amplitude-frequency curves, twofold yielding distribution.

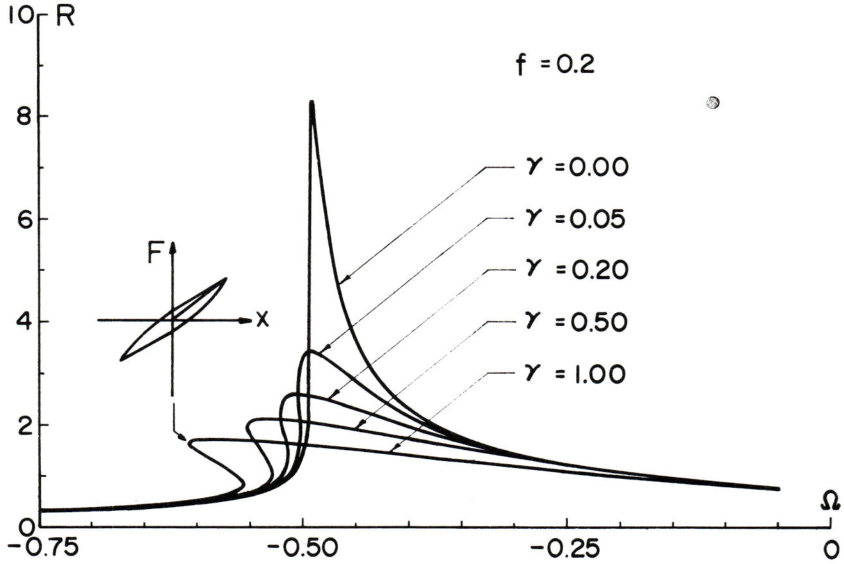

Fig. 5. Family of amplitude-frequency curves, threefold yielding distribution.

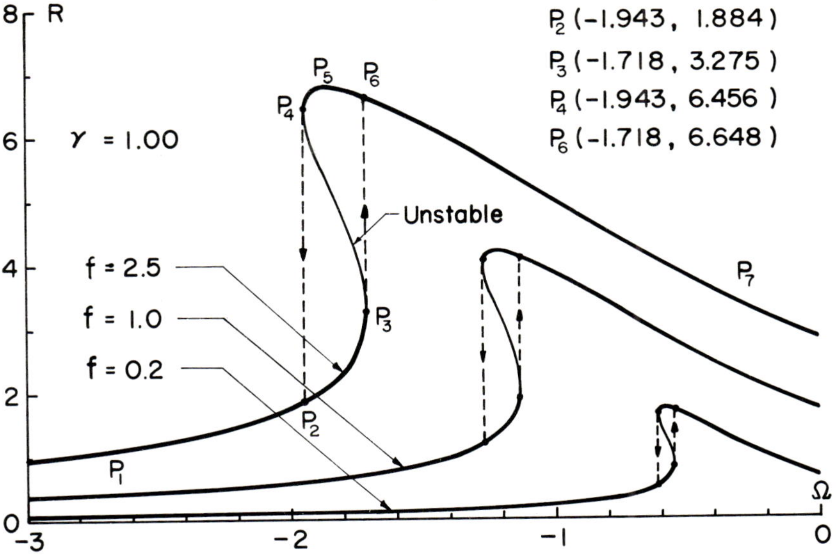

Fig. 6. Discontinuous jumps in amplitude as forcing frequency varies, threefold yielding distribution.

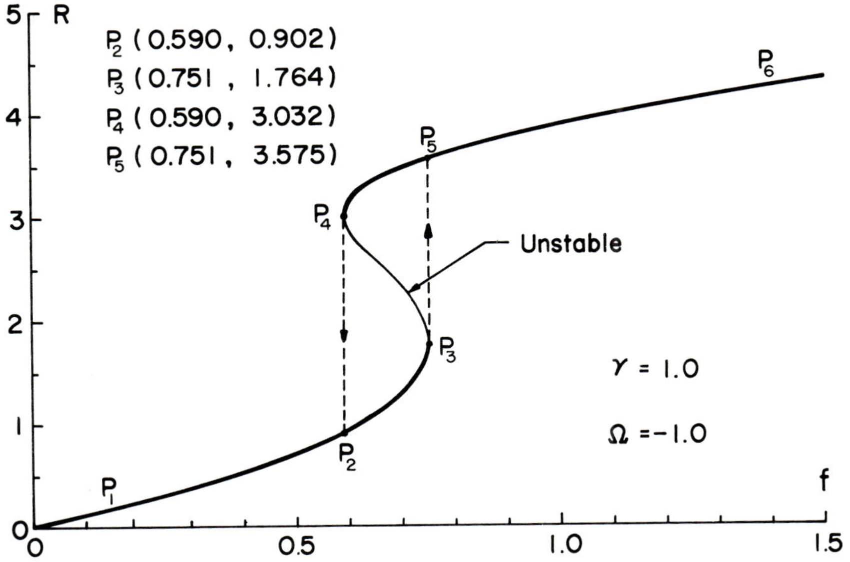

Fig. 7. Discontinuous jumps in amplitude as forcing level varies, threefold yielding distribution.

VISCOELASTICITY

Developments in Mechanics, Vol. 6. Proceedings of the 12th Midwestern Mechanics Conference 29

Transient and Residual Thermal Stresses in a Viscoelastic Cylinder

T. C. WOO
UNIVERSITY OF PITTSBURGH
J. W. JONES
WESTINGHOUSE CORPORATION
T. C. T. TING
UNIVERSITY OF ILLINOIS AT CHICAGO CIRCLE

ABSTRACT

A viscoelastic circular cylinder undergoing temperature changes is solved with a method based on superposition principle. The cylinder has no forces on the lateral surface and is in the plane strain condition. The viscoelastic property prescribed is linear but otherwise general, and its temperature dependence follows thermo-rheologically simple law. Transient and residual stresses are given using the numerical data for the commercial soda-lime glass at elevated temperatures.

INTRODUCTION

Viscoelastic stress analysis has been largely stimulated by the requirement of a more critical design of plastics and polymer components. The development started with the representation of viscoelastic characteristics [1] and was recently directed toward analytical and numerical techniques of solutions to various engineering problems, with dynamic effects and temperature influence included [2]. The present status of the linear viscoelastic stress analysis has been summarized by Lee [3]. Aside from applications in the high-polymer field, uses have also been found for numerous other cases in which strain rates and thermal effect become predominant. Investigations of bio-mechanical systems, sub-grade foundation analysis, and thermal treatment of glass are such pertinent examples. This paper presents a numerical solution, based on superposition, to a circular cylinder with general linear viscoelastic properties undergoing temperature changes while the lateral surface remains traction free. Solution by the usual method of elimination of variables would result in an enormous set of linear equations, for which both computer storage and solution time would be prohibitive. The numerical data presented here are those of glass at elevated temperatures and the actual results will therefore indicate the thermal stresses in a glass rod under the hypothetical heat treatments.

BASIC FORMULATION

We consider a solid viscoelastic cylinder in the state of plane strain with no forces on the curved surface. Initially it is at a uniform temperature T_o and free of stress. The heat transfer is characterized by Newtonian cooling at the lateral surface and conduction radially in the cylinder. The dilatational response for this material is considered to be elastic. The deviatoric behavior is given by the relaxation function which follows the thermorheologically simple law for its temperature dependence [4]. The object of this analysis is to predict the residual as well as transient stresses in the cylinder from its known viscoelastic properties.

In terms of cylindrical coordinates the problem is formulated as follows:

$$\epsilon_{zr} = \epsilon_{z\theta} = \epsilon_{r\theta} = \epsilon_z = 0 \qquad \text{for all } t,$$

$$\sigma_r(a,t) = 0 \qquad \text{for all } t,$$

where "a" is the radius of the cylinder.

The basic equations with usual notations are given as:

$$\epsilon_r = \frac{\partial u}{\partial r} \tag{1}$$

$$\epsilon_\theta = \frac{u}{r} \tag{2}$$

$$\frac{\partial \sigma_r}{\partial r} + \frac{\sigma_r - \sigma_\theta}{r} = 0 \tag{3}$$

$$\sigma_r + \sigma_\theta + \sigma_z = 3k[\epsilon_r + \epsilon_\theta + \epsilon_z - 3\alpha \tilde{T}(r,t)] \tag{4}$$

$$\sigma_\theta - \sigma_r = 2\int_0^\xi G(\xi - \xi')\frac{\partial}{\partial \xi'}(\epsilon_\theta - \epsilon_r)\,d\xi' \tag{5}$$

$$\sigma_\theta - \sigma_z = 2\int_0^\xi G(\xi - \xi')\frac{\partial}{\partial \xi'}(\epsilon_\theta - \epsilon_z)\,d\xi' \tag{6}$$

where k is the constant bulk modulus, $\tilde{T}$ the difference between the current and initial temperatures, $G(\xi)$ the relaxation function in shear, and the variable ξ gives the reduced time scale in adjusting the viscoelastic properties to the pertinent temperature [5].

$$\xi(r,t) = \int_0^t \varphi[T(r,t')]\,dt' = \int_0^t 10^{A[T(r,t') - T_B]}\,dt' \tag{7}$$

in which A is a linear shift constant and T_B is the base temperature from which relaxation function shifts on the logarithmic time scale.

Equations (1) and (2) imply the following compatibility condition:

$$\frac{\partial}{\partial r}(r\epsilon_\theta) = \epsilon_r \tag{8}$$

Equation (3) can be rewritten as:

$$\frac{\partial}{\partial r}(r\sigma_r) - \sigma_\theta = 0 \tag{9}$$

Equations (4) and (6) have the following forms:

$$\sigma_r = k(\epsilon_r + \epsilon_\theta - 3\alpha\tilde{T}) + \frac{2}{3}\int_0^\xi G(\xi-\xi')\frac{\partial}{\partial\xi'}(2\epsilon_r - \epsilon_\theta)\,d\xi' \tag{10}$$

$$\sigma_\theta = k(\epsilon_r + \epsilon_\theta - 3\alpha\tilde{T}) + \frac{2}{3}\int_0^\xi G(\xi-\xi')\frac{\partial}{\partial\xi'}(2\epsilon_\theta - \epsilon_r)\,d\xi' \tag{11}$$

$$\sigma_z = k(\epsilon_r + \epsilon_\theta - 3\alpha\tilde{T}) - \frac{2}{3}\int_0^\xi G(\xi-\xi')\frac{\partial}{\partial\xi'}(\epsilon_r + \epsilon_\theta)\,d\xi' \tag{12}$$

We integrate equations (8) and (9) which yield:

$$\epsilon_\theta(r,t) = \frac{1}{r}\int_0^r \epsilon_r(r',t)\,dr' \tag{13}$$

$$\sigma_r(r,t) = \frac{1}{r}\int_0^r \sigma_\theta(r',t)\,dr' \tag{14}$$

As mentioned in the previous section, it is difficult to obtain a solution by the usual method of elimination of variables unless we make a drastic assumption to simplify the material properties and their temperature dependence. To maintain the viscoelastic law as general as prescribed above, and yet obtain the solution, we resort to the method based on the superposition principle as given in the next section.

Before we discuss the method of solution in detail, we will outline the approach as follows:

For each fixed time $t-\Delta t$ up to which the solution is assumed to have been obtained, if we assume a value for $\epsilon_r(0,t)$, then from equations (13), (10) and (11) we can determine $\epsilon_\theta(0,t)$, $\sigma_r(0,t)$ and $\sigma_\theta(0,t)$. Now by using equations (13), (10), (11) and (14) we can determine $\epsilon_r(\Delta r,t)$, $\epsilon_\theta(\Delta r,t)$, $\sigma_r(\Delta r,t)$ and $\sigma_\theta(\Delta r,t)$. Repeating this process until $r = a$ we will get $\sigma_r(a,t)$, which is in general not zero as required. In view of the results obtained in Appendix A, we can immediately determine the correct stresses and strains which will satisfy the condition $\sigma_r(a,t) = 0$. The procedures of the solution are given in the sequel.

METHOD OF SOLUTION

We assume that:

$$\sigma_r(a,t) = P(t) \tag{15}$$

where P(t) is given and equals to zero in this particular case. We integrate equation (8) from $r-\Delta r$ to r and make approximations to obtain:

$$\epsilon_\theta(r,t) \doteq \frac{\Delta r}{2r}\epsilon_r(r,t) + Q_1 \tag{16}$$

where $Q_1 = \left[\left(1-\frac{\Delta r}{r}\right)\epsilon_\theta(r-\Delta r,t) + \frac{\Delta r}{2r}\epsilon_r(r-\Delta r,t)\right]$

Equation (10) can be approximated by:

$$\sigma_r(r,t) \doteq C_1\epsilon_r(r,t) + C_2\epsilon_\theta(r,t) + C_3 \tag{17}$$

where:

$$C_1 = k + \frac{2}{3}\left\{G(0) + G[\xi(r,t) - \xi(r,t-\Delta t)]\right\}$$

$$C_2 = k - \frac{1}{3}\left\{G(0) + G[\xi(r,t) - \xi(r,t-\Delta t)]\right\}$$

$$C_3 = \frac{2}{3}\int_0^{t-\Delta t} G[\xi(r,t)-\xi(r,t')]\frac{\partial}{\partial t'}[2\epsilon_r(r,t')-\epsilon_\theta(r,t')]dt'$$
$$-\frac{1}{3}\{G(0)+G[\xi(r,t)-\xi(r,t-\Delta t)]\}[2\epsilon_r(r,t-\Delta t)-\epsilon_\theta(r,t-\Delta t)] - 3\alpha k\tilde{T}$$

Similarly equation (11) is given by:

$$\sigma_\theta(r,t) \doteq \sigma_r(r,t) + C_4[\epsilon_\theta(r,t)-\epsilon_r(r,t)] + C_5 \quad (18)$$

where

$$C_4 = G(0) + G[\xi(r,t)-\xi(r,t-\Delta t)]$$

and

$$C_5 = 2\int_0^{t-\Delta t} G[\xi(r,t)-\xi(r,t')]\frac{\partial}{\partial t'}[\epsilon_\theta(r,t')-\epsilon_r(r,t')]dt'$$
$$-\{G(0)+G[\xi(r,t)-\xi(r,t-\Delta t)]\}[\epsilon_\theta(r,t-\Delta t)-\epsilon_r(r,t-\Delta t)]$$

Equation (9) is integrated from $r-\Delta r$ to r as

$$\sigma_r(r,t) \doteq \frac{\Delta r}{2r}\sigma_\theta(r,t) + Q_2 \quad (19)$$

where $Q_2 = (1-\frac{\Delta r}{r})\sigma_r(r-\Delta r,t) + \frac{\Delta r}{2r}\sigma_\theta(r-\Delta r,t)$

Equations (16) to (19) then can be used to solve for ϵ_r, ϵ_θ, σ_θ, and σ_r. Note that C_1, C_2, C_3, C_4, and C_5 are known up to and including position r but previous to time t and Q_1 and Q_2 are known at time t but just preceding the position r. The equation for $\epsilon_r(r,t)$ can be obtained from the above as:

$$\epsilon_r(r,t) = \frac{(Q_2-C_2Q_1-C_3)+[C_3+C_5+C_1Q_1]\frac{\Delta r}{2r}}{C_1[1-(\frac{\Delta r}{2r})^2]} \quad (20)$$

By assuming a value $\epsilon_r^*(0,t)$ we can, from equations (16) to (20), calculate all the strains and stresses, denoted by $\epsilon_r^*(r,t)$, $\epsilon_\theta^*(r,t)$, $\sigma_r^*(r,t)$ and $\sigma_\theta^*(r,t)$, say. In a similar manner we assume a value $\epsilon_r'(0,t)$ and solve for the strains and stresses corresponding to the same problem except with $\alpha = 0$ (cf. Appendix A), denoted by $\epsilon_r'(r,t)$, $\epsilon_\theta'(r,t)$, $\sigma_r'(r,t)$ and $\sigma_\theta'(r,t)$. The correct solution is then obtained by determining β from the following equation:

$$\sigma_r(a,t) = \sigma_r^*(a,t) + \beta\sigma_r'(a,t) = 0$$

i.e. $\beta = -\sigma_r^*(a,t)/\sigma_r'(a,t)$. (21)

And the solution is given by:

$$\left.\begin{aligned}
\epsilon_r(r,t) &= \epsilon_r^*(r,t) + \beta\epsilon_r'(r,t)\\
\epsilon_\theta(r,t) &= \epsilon_\theta^*(r,t) + \beta\epsilon_\theta'(r,t)\\
\sigma_r(r,t) &= \sigma_r^*(r,t) + \beta\sigma_r'(r,t)\\
\sigma_\theta(r,t) &= \sigma_\theta^*(r,t) + \beta\sigma_\theta'(r,t)
\end{aligned}\right\} \quad (22)$$

To obtain the residual stresses we have to calculate until the whole cylinder cools down to the ambient temperature. The amount of storage and the computing time would both increase exponentially with the steps of calculations. It is imperative to make approximations when the viscoelastic effect ceases to be significant and all the later computation can be lumped together using elastic theory. This type of treatment has been used in [5] and discussed in [6] in conjunction with a heuristic argument regarding residual stress determination in glass. A more general result has been obtained and will be given in the Appendix B. For the present problem we simply have:

$$\sigma_r(r,\infty) = \sigma_r(r,t^*) - \sigma_r^e(r,t^*)$$

$$\sigma_\theta(r,\infty) = \sigma_\theta(r,t^*) - \sigma_\theta^e(r,t^*)$$

where t* represents the "hardening time" beyond which the material is essentially elastic, and superscript "e" signifies elastic stresses under identical conditions.

The numerical data used in this problem are taken from the published viscoelastic properties of commercial soda-lime glass at elevated temperatures [7]. The reasons for choosing glass as the material for illustration are: the wide range on the logarithmic time scale of the viscoelastic functions at different temperatures (about twenty decades for a change of 500° C.); the usefulness of the knowledge on such transient and residual stresses; and the possibilities of using photoelasticity directly on the prototype rather than a model for any verification after the cylinder is cooled down completely.

The calculations here require the relaxation modulus function in shear G(t), while the data available in [7] are that in flexure, E(t). They are related through an integral equation [8]:

$$E(t) = 2G(t)[1+\nu(0)] + 2\int_0^t \dot{\nu}(t-t')G(t')\,dt' \tag{23}$$

where $\nu(t)$ is the viscoelastic counterpart of Poisson's ratio, and dot denotes derivative with respective to its argument. Since we assume a purely elastic dilatational response, $\nu(t)$ takes the form:

$$\nu(t) = \frac{1}{2}\left[1 - \frac{E(t)}{3k}\right] \tag{24}$$

and, with E(t) given in Figure 1, the function G(t) can be obtained from the following integral equation by numerical iteration and is shown in Figure 2:

$$E(t) = \left[3 - \frac{E(0)}{3k}\right]G(t) - \frac{1}{3k}\int_0^t \dot{E}(t-t')G(t')\,dt' \tag{25}$$

The temperature problem is solved with Newtonian cooling on the lateral surface and uniform initial temperature T_o while the ambient temperature is T_a. The solution is given by the following series [9]:

$$T(r,t) = T_a + \sum_{j=1}^{\infty} A_j J_o(\lambda_j r) e^{-\lambda_j^2 \kappa t} \tag{26}$$

where κ is the thermal diffusivity and A_j are given by:

$$A_j = \frac{2\lambda_j a (T_o - T_a) J_1(\lambda_j a)}{(\lambda_j^2 a^2 + h^2)[J_o(\lambda_j a)]^2}$$

The constant h is the surface heat transfer coefficient and λ_j are given as roots of the equation:

$$h' J_o(\lambda a) + \lambda a J_o'(\lambda a) = 0$$

where h' is the Biot's number. The series (26) is summed on the computer and then used for the calculation of $\xi(r,t)$ by equation (7).

The numerical values used in the calculations are given as:

$\alpha = 92. \times 10^{-7}$, $A = 0.0386$, $T_o = 560$, $\kappa = 0.00333$, $T_a = 30$

all in c.g.s. units. The material constants are given as $k = E(o) = 9.6 \times 10^6$ psi. The following cases have been calculated from the method outlined:

Case	1	2	3	4	5
Radius of cylinder (cm)	0.635	0.635	0.635	0.3175	0.3175
Heat transfer Coefficient h (Cal./cm^2 - ^{o}C - sec.)	0.01	0.0075	0.005	0.005	0.002

The results are shown in Figures 3 to 7. They represent different degrees of tempering with heat transfer rates varying from 0.01 to 0.002 Cal./cm^2 - ^{o}C - sec.

DISCUSSIONS

The figures in which the transient and residual tangential stresses are given resemble those for an infinite slab under symmetric Newtonian cooling on both sides [5]. Initial response is invariably low compression throughout except near the surface where there is high tension. This is due to the constraint between the layers which hinders the surface layers from shrinking on a sudden chill. Later in time the viscoelastic effect, thermal contraction and the continuity among the layers interact in a rather complex way and intuitive reasoning becomes somewhat difficult. The residual stresses, however, are found to be of very realistic order of magnitudes and serve to predict the toughening of glass rods very well.

The motivation for studing the thermo-viscoelastic stresses in a cylinder is twofold. On a laboratory scale it is easier to fabricate test specimen and experimental measurement is more readily accomplished using rods. Also, useful applications for the cylindrical shapes are known. For instance, the cooling rate of glass rods used in laser devices can be determined so that the residual stresses are reduced to the lowest practical value.

The problem has been treated for the plane strain condition. The stress σ_z has not been calculated since it is of minor interest. It can, however, easily be evaluated from the algebraic equation (4) and would then give the constraint forces at the ends necessary to maintain the condition $\epsilon_z = 0$ at all times. There are situations in which given forces N(t), say, act in the axial direction and the stretch is uniform in z-direction. In this case, ϵ_z is a function of t only and

$$\int_0^a \sigma_z(r,t)\, 2\pi r\, dr = N(t) \qquad (27)$$

When N(t) is indentically zero, we will have the state of generalized plane strain. The numerical procedure can be readily modified by assuming an $\epsilon_z^*(t)$ in addition to $\epsilon_r^*(o,t)$. It would be analogous to the present one.

BIBLIOGRAPHY

[1] Alfrey, T., Mechanical Behavior of High Polymers, Interscience, New York, 1948.

[2] Lee, E. H., "Viscoelastic Stress Analysis", Proc. First Symposium on Naval Structural Mechanics, Pergamon Press, New York, 1960.

[3] Lee, E. H., "Some Recent Developments in Linear Viscoelastic Stress Analysis", Proc. 11th International Congress of Applied Mechanics, Springer-Verlag, Berlin, 1964.

[4] Morland, L. W. and Lee, E. H., "Stress Analysis for Linear Viscoelastic Materials with Temperature Variation", Trans. Soc. Rheology IV, 1960, pp. 233-263.

[5] Lee, E. H., Rogers, T. G. and Woo, T. C., "Residual Stresses in a Glass Plate Cooled Symmetrically from Both Surfaces", J. Amer. Ceram. Soc., Vol. 48, No. 9, 1965, pp. 480-487.

[6] Woo, T. C., "Thermal Stress Problems in Glass", J. Appl. Phys., Vol. 39, No. 4, 1968, pp. 2082-2087.

[7] Van Zee, A. F. and Noritake, H. M., "Measurement of Stress-Optical Coefficient and Rate of Stress Release in Commercial Soda-Lime Glasses", J. Amer. Ceram. Soc., Vol. 41, No. 5, 1958, pp. 164-175.

[8] Stuart, H. A. (editor), Die Physik der Hochpolymeren, Vol. II, Academic Press, New York, 1957.

[9] Rohsenow, W. M. and Choi, H., Heat, Mass, and Momentum Transfer, Prentice-Hall, New Jersey, 1961.

APPENDIX A - SUPERPOSITION PRINCIPLE

Basic equations for thermo-rheologically simple viscoelasticity are:

$$\epsilon_{ij} = \frac{1}{2}(u_{i,j} + u_{j,i}) \tag{A1}$$

$$\sigma_{ij,j} = 0 \tag{A2}$$

$$s_{ij}(x,\xi) = 2\int_0^{\xi} G(\xi-\xi')\frac{\partial}{\partial\xi'}(e_{ij})\,d\xi' \tag{A3}$$

$$\sigma(x,\xi) = 3\int_0^{\xi} k(\xi-\xi')\frac{\partial}{\partial\xi'}[\epsilon(x,\xi') - 3\alpha\tilde{T}(x,\xi')]\,d\xi' \tag{A4}$$

where

$$\sigma = \sigma_{11} + \sigma_{22} + \sigma_{33}, \qquad \epsilon = \epsilon_{11} + \epsilon_{22} + \epsilon_{33},$$

$$s_{ij} = \sigma_{ij} - \frac{1}{3}\delta_{ij}\sigma, \qquad e_{ij} = \epsilon_{ij} - \frac{1}{3}\delta_{ij}\epsilon,$$

and

$$\xi(x,t) = \int_0^t \varphi[T(x,t')]\,dt'$$

with usual Cartesian tensor notations. Boundary conditions are:

$$u_i = 0 \quad \text{on boundary } \Sigma_u \tag{A5}$$

$$\sigma_{ij} n_j = P_i \quad \text{on boundary } \Sigma_P \tag{A6}$$

A viscoelastic solution which satisfies (A1) to (A6) will be denoted by $[S_v]$. Let us consider another viscoelastic solution which satisfies (A1) to (A5) and

$$\sigma_{ij} n_j = P_i^* \quad \text{on} \quad \Sigma_P \tag{A6*}$$

and denote this solution by $[S_v^*]$.

In addition to $[S_v]$ and $[S_v^*]$, we introduce the following three solutions. In place of (A4) and (A6) we write:

$$\sigma(x,\xi) = 3\int_0^{\xi} k(\xi-\xi')\frac{\partial}{\partial \xi'}[\epsilon(x,\xi')]d\xi' \tag{A7}$$

$$\sigma_{ij} n_j = P_i' \quad \text{on} \quad \Sigma_P \tag{A6'}$$

$$\sigma_{ij} n_j = P_i'' \quad \text{on} \quad \Sigma_P \tag{A6''}$$

$$\sigma_{ij} n_j = P_i''' \quad \text{on} \quad \Sigma_P \tag{A6'''}$$

Now we will denote by $[S_v']$ a solution which satisfies equations (A1), (A2), (A3), (A7), (A5) and (A6'); by $[S_v'']$ a solution which satisfies the same equations as above except with (A6') replaced by (A6''); by $[S_v''']$ a solution with (A6') replaced by (A6''').

With the definitions it can readily be shown that:

$$S_v = S_v^* + \beta_1 S_v' + \beta_2 S_v'' + \beta_3 S_v''' \tag{A8}$$

with β_i determined by:

$$P_i = P_i^* + \beta_1 P_i' + \beta_2 P_i'' + \beta_3 P_i''' \tag{A9}$$

It should be noted that (A9) gives three equations when i = 1, 2, 3 and S_v's can represent displacements, strains, or stresses.

APPENDIX B - RESIDUAL STRESS CALCULATIONS

The basic equations are given by equations (A1) to (A6) in Appendix A. For $t \geq t^*$ where t^* is the hardening time, all the equations remain unchanged, except for (A3) and (A4) which can be approximated by:

$$s_{ij}(x,t) \doteq \int_0^{t^*} 2G[\xi(x,t^*)-\xi(x,t')]\frac{\partial}{\partial t'}[e_{ij}(x,t')]dt' + 2G(0)[e_{ij}(x,t)-e_{ij}(x,t^*)] \tag{A3'}$$

$$\sigma(x,t) \doteq \int_0^{t^*} 3k[\xi(x,t^*)-\xi(x,t')]\frac{\partial}{\partial t'}[\epsilon(x,t')-3\alpha\tilde{T}(x,t')]dt' + 3k(0)\{\epsilon(x,t)-\epsilon(x,t^*)-3\alpha[\tilde{T}(x,t)-\tilde{T}(x,t^*)]\} \tag{A4'}$$

Hence, if $S_v(x,t)$ is the viscoelastic solution for $t \geq t^*$ we may write:

$$S_v(x,t) = S_v(x,t^*) + \hat{S}(x,t,t^*)$$

where S_v represents solutions which satisfy (A1) to (A6) and $\hat{S}(x,t,t^*)$ should satisfy the following equations:

$$\left.\begin{aligned} \hat{\epsilon}_{ij} &= \tfrac{1}{2}(\hat{u}_{i,j} + \hat{u}_{j,i}) \\ \hat{\sigma}_{ij,j} &= 0 \end{aligned}\right\}$$

$$\left.\begin{aligned}
\hat{s}_{ij} &= 2G(0)[e_{ij}(x,t) - e_{ij}(x,t^*)] \\
\hat{\sigma} &= 3k(0)\{\epsilon(x,t) - \epsilon(x,t^*) - 3\alpha[\tilde{T}(x,t) - \tilde{T}(x,t^*)]\} \\
\hat{u}_i &= 0 \quad \text{on } \Sigma_u \\
\hat{\sigma}_{ij} n_j &= 0 \quad \text{on } \Sigma_p
\end{aligned}\right\} \tag{A10}$$

But the set of equations (A10) is satisfied by:

$S_e(x,t) - S_e(x,t^*)$

where $S_e(x,t)$ is the elastic solution of (A1) to (A6) with $G(\xi)$ and $k(\xi)$ taken to be constant values. Hence,

$$S_v(x,t) - S_e(x,t) \doteq S_v(x,t^*) - S_e(x,t^*), \qquad t \geq t^* \tag{A11}$$

In particular, when $t \to \infty$, we have $S_e(x,\infty) \to 0$, if the initial and final remperatures are both uniform. We then obtain the general result for evaluating residual values:

$$S_v(x,\infty) \doteq S_v(x,t^*) - S_e(x,t^*) \tag{A12}$$

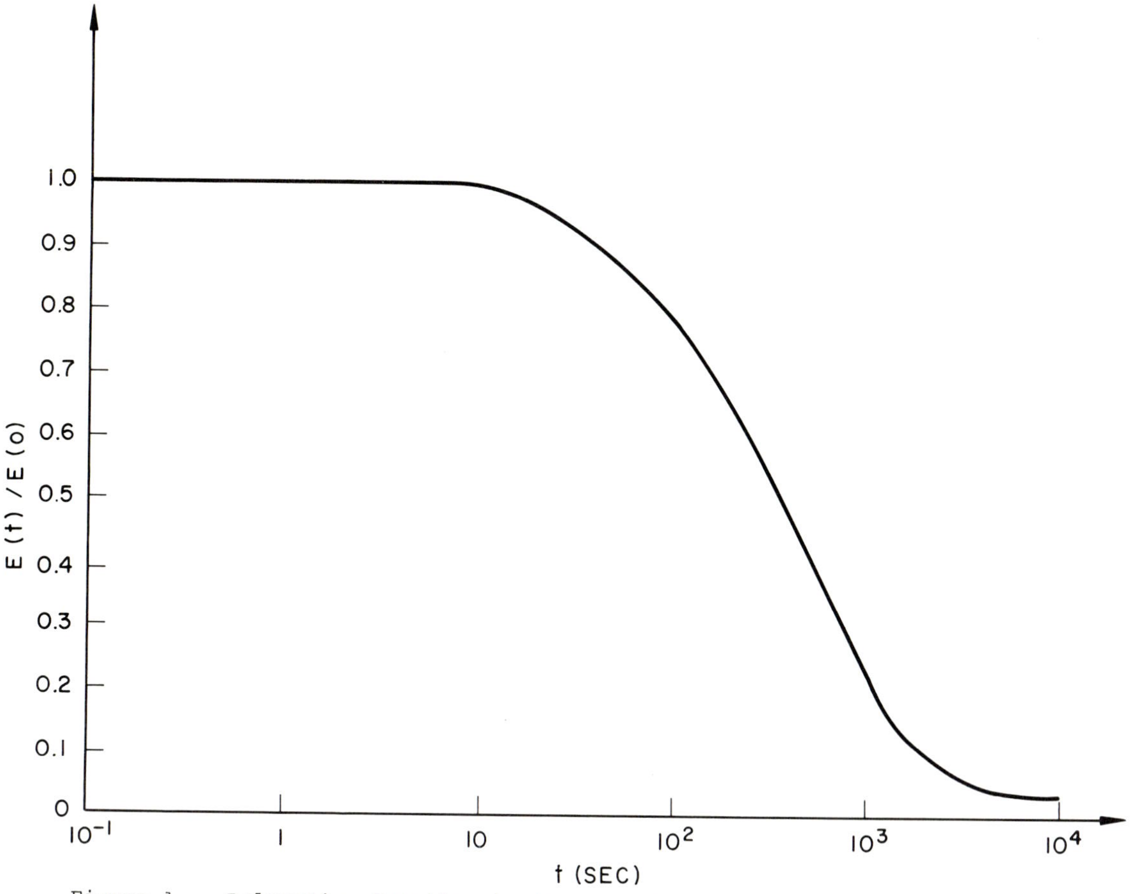

Figure 1 Relaxation Function in Flexure of Glass

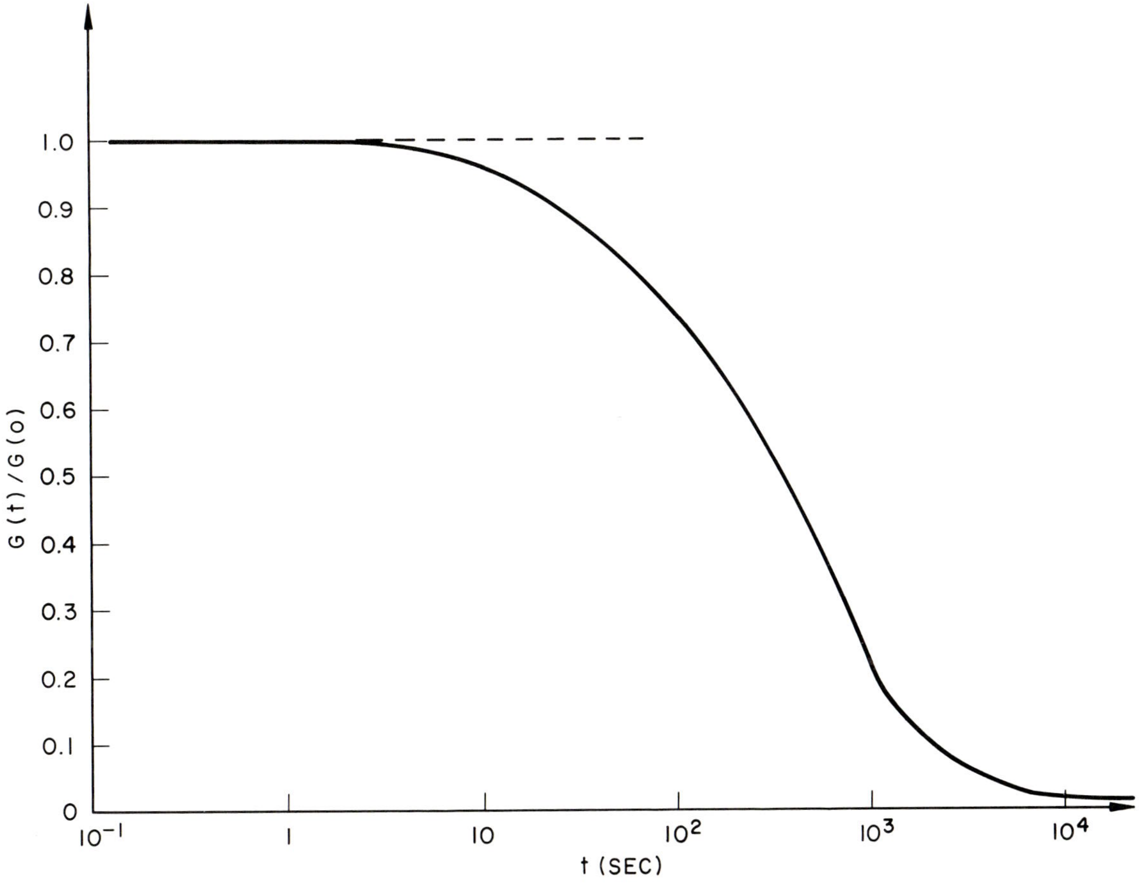

Figure 2 Relaxation Function in Torsion of Glass

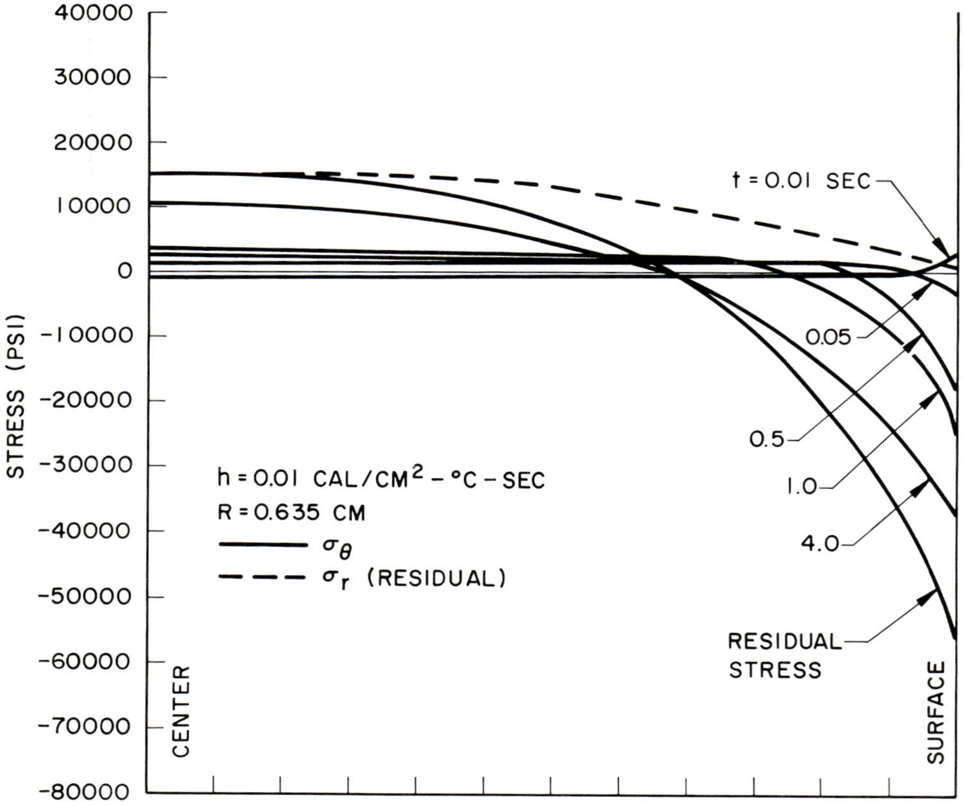

Figure 3 Stress for Cylinder of 0.635 cm. Radius with h = 0.01

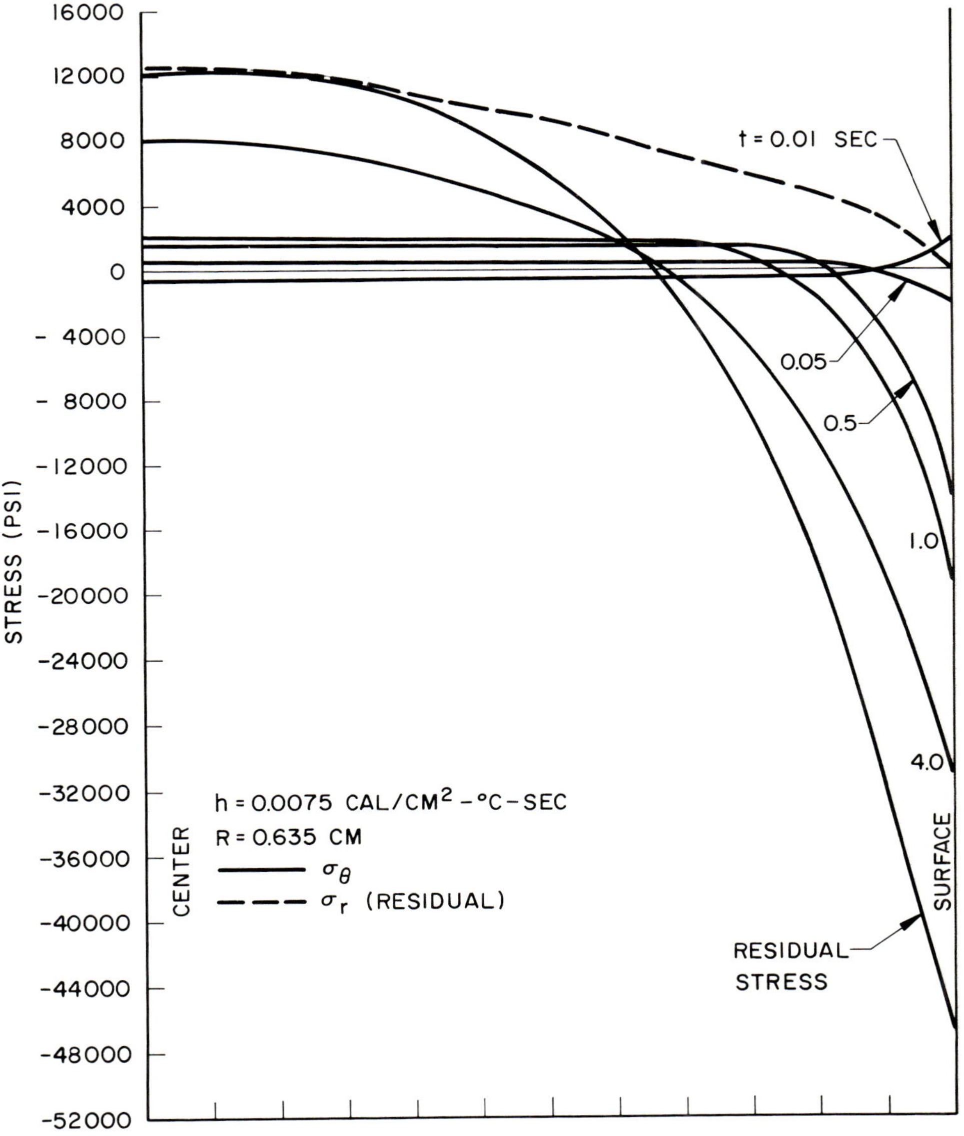

Figure 4 Stresses for Cylinder of 0.635 cm. Radius with h =.0075

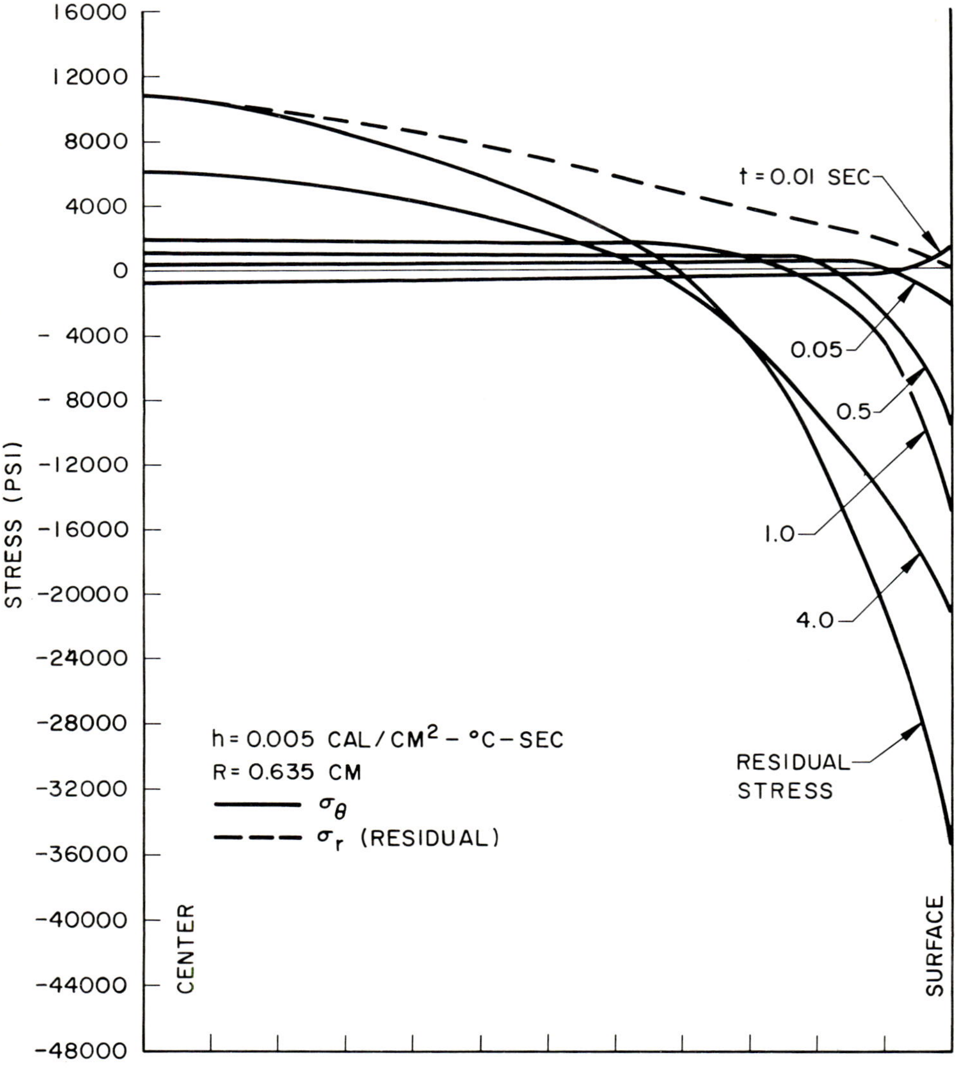

Figure 5 Stresses for Cylinder of 0.635 cm. Radius with h = .005

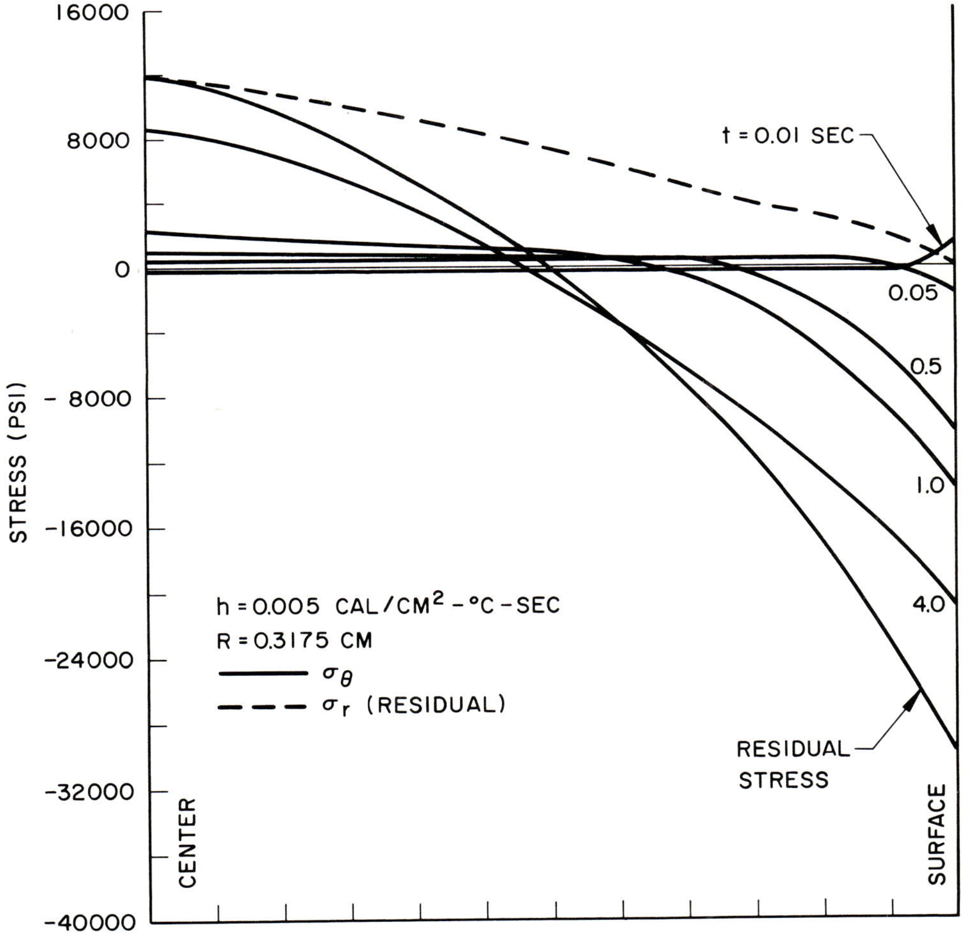

Figure 6 Stresses for Cylinder of 0.3175 cm. Radius with h = 0.005

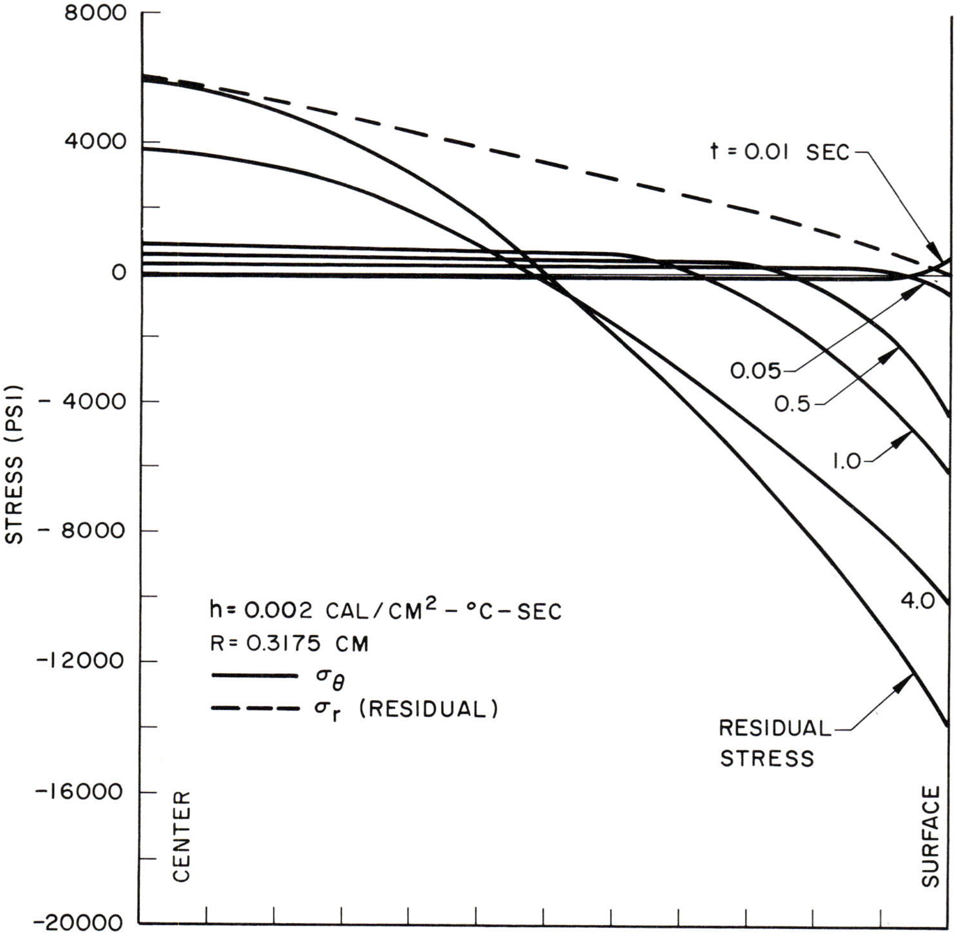

Figure 7 Stresses for Cylinder of 0.3175 cm. Radius with h = 0.002

Developments in Mechanics, Vol. 6. Proceedings of the 12th Midwestern Mechanics Conference: 30

Stress Concentration around Holes in a Class of Rheological Materials Displaying a Poroelastic Structure

J. L. NOWINSKI
UNIVERSITY OF DELAWARE

ABSTRACT

Two-phase poroelastic rheological material is taken as the object of the analysis, the solid phase being treated as a linear isotropic perfectly elastic solid and the fluid substances filling the pores as a Newtonian viscous fluid. Using Biot equations, derived in his consolidation theory, and assuming a two-dimensional state of deformation governing equations involving a stress function and the fluid excess pressure are derived. Upon using the Laplace transformation the resulting general equations are solved by a procedure extending the complex variable method of Muskhelishvili. An illustrative example involving stress concentration around a cylindrical hole under a uniform pressure in an infinite body subject to a uniaxial tension is solved in detail.

INTRODUCTION

Although the earliest problem of stress concentration around holes in solids has been solved by Kirsch at the end of the past century, the last two decades have witnessed a revival of interest of this subject mostly due to an intensive study of the brittle fracture in which all stress concentrators play a dominate role.

A number of results involving both two- and three- dimensional problems of stress concentration was presented on the eve of the second World War by Neuber in his known book [1], 1937. After the war, Savin in 1951, in his monograph [2], collected many new solutions to two-dimensional problems obtained by means of the complex variable technique promoted in the book by Muskhelishvili [3]. Recently, a second edition of Savin's book [2], 1968 was published containing, however, so much new material that it has to be considered a separate publication [4]. In fact, in this book Savin offered a collection of almost all known solutions to stress concentration problems in two dimensions in the elastic, plastic, nonlinear, thermoelastic, brittle and dynamic cases, as well as for the Casserat medium and the shells with holes. The book includes also a short chapter on the relatively less explored stress concentration in viscoelastic bodies, based on the linear theory of viscoelastic continua.

It is well known that the time factor plays a more or less substantial role in deformation and stresses generated in rheological materials such as metals in elevated temperatures, high polymers, wood, concrete and even bones, see e.g. [10]. With elapsing time the associated transient phenomena of stress relaxation or creep originate, and the mechanical processes involved cannot be considered as stationary. Consequently, the elastic constitutive equations which characterize an instantaneous onset of final stationary mechanical states have to be substituted by relations that reflect the time dependence of the properties of the materials.

In the present study we investigate the stress concentration problem in a particular class of rheological materials called poroelastic and studied by Biot on the occasion of his consolidation theory, see e.g. [6]. As shown in a number of reports, see e.g. [5], [7], Biot's theory can successfully simulate the behavior of osseous tissue acknowledged by experimenters [10], [11] and practitioners [12]. The model so adopted involves a number of simplifications. First, despite the complexity of the bone structure the osseous material is treated as a statistically homogeneous medium. Secondly, it is assumed that the porosity of the material is uniform and that the solid skeleton is linearly and perfectly elastic. Third, the liquid phase filling the pores is treated as a Newtonian viscous fluid governed by Darcy's law. Finally, it is postulated that the stresses on the bulk material are smoothly divided between the solid phase, as the stress σ_{ij}, and the liquid phase as the pressure σ, both per unit area of the bulk material.

It is probable that in the bone tissue the cavities act as some kind of stress concentrators. There prevails, however, an opinion that the discontinuities in the bone are generally arranged in such a fashion as to create only minimum stress concentrations. Whatever the role of pores, one of the problems of practical importance is also connected with the effect of holes that are large as compared with pores*. Such holes are produced intentionally during surgical interventions in order to insert various fixing devices. In view of our assumptions the material surrounding the holes is treated approximately as a homogeneous continuum. In this paper we first give a brief derivation of the governing equations for the two-dimensional deformation of poroelastic media in terms of the stress function F and the fluid pressure σ. The time variable is eliminated by means of a Laplace transformation, and the solution to the problem reduced to a disclosure of one harmonic, one biharmonic and two Helmholtz type functions. Application of a generalized Muskhelishvili method of complex variables enables one to represent the general solution in terms of two analytic functions only, determined by means of one of the known techniques.

An illustrative example is solved in more detail involving stress concentration around a cylindrical hole, subjected to a uniform

*At the present time, the stress concentration around drilled holes is estimated by the orthopaedists globally by means of experimental coefficients.

pressure, in an infinite body under uniaxial tension. While the viscoelastic properties of the adopted model seem to agree with the data obtained from experiments on bones (see e.g. the treatise [10] and the text [11]), it belongs to future investigations to decide whether such an approach fully describes the actual behavior of bones in vivo.

1. GENERAL EQUATIONS

In Biot's theory of the poroelastic materials the material is treated as a statistically homogeneous medium composed of an elastic skeleton with tiny interconnected pores filled with a viscous fluid governed by Darcy's law. In the two-dimensional case, analyzed in the present paper, and with σ_{ij} denoting the stress acting on the solid phase and σ the fluid pressure (both per unit area of the bulk material), the constitutive equations of the medium in cylindrical coordinates r, θ, z are

$$\begin{aligned} \sigma_{rr} &= 2Ne_{rr} + Ae + Q\varepsilon , \\ \sigma_{\theta\theta} &= 2Ne_{\theta\theta} + Ae + Q\varepsilon , \\ \sigma_{zz} &= Ae + Q\varepsilon , \\ \sigma_{r\theta} &= 2Ne_{r\theta} , \\ \sigma &= Qe + R\varepsilon . \end{aligned} \tag{1}$$

Here e and ε denote the dilatation of the solid and the fluid phase, respectively, e_{ij} is the strain tensor, the $r\theta$ - plane is the plane of deformation and A, N, Q and R are material constants. For definiteness it is assumed that the two-dimensional state of deformation is that of plane strain. Treating the problem as a quasi-static, that is disregarding the influence of the inertia terms, we satisfy the balance equations

$$\begin{aligned} &\frac{\partial(\sigma_{rr} + \sigma)}{\partial r} + \frac{1}{r}\frac{\partial \sigma_{r\theta}}{\partial \theta} + \frac{\sigma_{rr} - \sigma_{\theta\theta}}{r} = 0, \\ &\frac{1}{r}\frac{\partial(\sigma_{\theta\theta} + \sigma)}{\partial \sigma} + \frac{\partial \sigma_{r\theta}}{\partial r} + \frac{2\sigma_{r\theta}}{r} = 0, \end{aligned} \tag{2}$$

upon introducing the stress function $F(r, \theta; t)$ defined by

$$\begin{aligned} &\sigma_{rr} + \sigma = \frac{1}{r}\frac{\partial F}{\partial r} + \frac{1}{r^2}\frac{\partial^2 F}{\partial \theta^2} , \quad \sigma_{\theta\theta} + \sigma = \frac{\partial^2 F}{\partial r^2} \\ &\sigma_{r\theta} = \frac{1}{r^2}\frac{\partial F}{\partial \theta} - \frac{1}{r}\frac{\partial^2 F}{\partial r \partial \theta} . \end{aligned} \tag{3}$$

Biot has shown ([6], Equation (42)) that the fulfillment of the compatibility equation leads to the following relation between the stress function and the fluid pressure

$$\kappa \nabla^4 F = \nabla^2 \sigma, \tag{4}$$

where $\kappa = [(A + 2N)R - Q^2]/2N(Q + R)$.

On the other hand, Darcy's law in combination with the strain-displacement relations furnishes the governing equations in the form

$$K \nabla^6 F = b \frac{\partial}{\partial t} \nabla^4 F, \qquad K \nabla^4 \sigma = b \frac{\partial}{\partial t} \nabla^2 \sigma, \tag{5}$$

where $K = [(A + 2N) R - Q^2]/[A + R + 2 (N + Q)]$ and b is the permeability coefficient. Henceforth we map the problem into the Laplace subspace by means of the one-sided transformation

$$f^*(r,\theta;\xi) = \int_0^\infty f(r,\theta;t)e^{-\xi t}dt, \tag{6}$$

where the asterisk denotes the transform. It is assumed here that the derivative $\partial f(r,\theta;t)/\partial t$ as a function of t remains bounded at infinity.

With this in mind, the $F^*(r,\theta;\xi)$ and $\sigma^*(r,\theta;\xi)$ functions may be represented as sums of two functions, in view of the commutativity of the operators $\partial/\partial t$ and ∇^2 and ∇^4,

$$F^* = F_1^* + F_2^*, \qquad \sigma^* = \sigma_1^* + \sigma_2^*, \tag{7}$$

where F_1^* is a biharmonic function satisfying the equation

$$\nabla^4 F_1^* = 0, \tag{8}$$

σ_1^* is a harmonic function satisfying the equation

$$\nabla^2 \sigma_1^* = 0, \tag{9}$$

and F_2^* and σ_2^* are solutions of the Helmholtz equations

$$r_0^2 \nabla^2 F_2^* - F_2^* = 0, \qquad r_0^2 \nabla^2 \sigma_2^* - \sigma_2^* = 0. \tag{10}$$

Clearly, $\nabla^2 = \partial^2/\partial r^2 + \partial/r\partial r + \partial^2/r^2\partial\theta^2$ and $r_0^2 \equiv K/b\xi$. Let us denote the solutions of the equations (10) by $\Phi_0^*(r,\theta,\xi)$ and $\Sigma^*(r,\theta;\xi)$, respectively. Since σ_1^* is a harmonic and F_1^* a biharmonic function then*

$$\sigma^* = 2\mathrm{Re}\, s^*(z) + \Sigma^*, \tag{11}$$

where $s^*(z)$ is an analytic function of the complex variable $z = re^{i\theta}$, and

$$F^* = \mathrm{Re}[\bar{z}\phi^*(z) + \chi^*(z)] + \Phi_0^*, \tag{12}$$

*In what follows all starred functions with an explicit argument z should be understood as depending on the parameter ξ (omitted for brevity, e.g. $\phi^*(z) \equiv \phi^*(z,\xi)$.

with ϕ^* and χ^* being Goursat potentials holomorphic in simply connected regions, see e.g. [3].

An easy computation using the equation (4) furnishes the following connection between Φ_0^* and Σ^*,

$$\Phi_0^* = \frac{r_o^2}{\kappa} \Sigma^* . \tag{13}$$

In complex variable representation

$$\sigma_{rr}^* + \sigma^* = 2 \left[\frac{\partial^2 F^*}{\partial z \partial \bar{z}} - \mathrm{Re}\left(\frac{z}{\bar{z}} \frac{\partial^2 F^*}{\partial z^2}\right)\right] ,$$

$$\sigma_{\theta\theta}^* + \sigma^* = 2 \left[\frac{\partial^2 F^*}{\partial z \partial \bar{z}} + \mathrm{Re}\left(\frac{z}{\bar{z}} \frac{\partial^2 F^*}{\partial z^2}\right)\right] , \tag{14}$$

$$\sigma_{r\theta}^* = 2 \ \mathrm{Im} \left(\frac{z}{\bar{z}} \frac{\partial^2 F^*}{\partial z^2}\right),$$

where, clearly, $\bar{z}$ is the complex conjugate of z.

Upon using (14) a slight manipulation yields the basic relations

$$\sigma_{rr}^* + \sigma_{\theta\theta}^* = 2\mathrm{Re}[2 \ \Phi^*(z) - 2s^*(z)] + \frac{1-2\kappa}{\kappa} \Sigma^*,$$

$$\sigma_{\theta\theta}^* - \sigma_{rr}^* + 2i\sigma_{r\theta}^* = 2 \ H^*(z,\bar{z}) , \tag{15}$$

$$\sigma_{rr}^* - i\sigma_{r\theta}^* = \mathrm{Re}[2\Phi^*(z) - 2s^*(z)] - H^* (z,\bar{z}) + \frac{1-2\kappa}{2\kappa} \Sigma^* ,$$

with

$$H^*(z,\bar{z}) = \frac{z}{\bar{z}} [\bar{z} \ \Phi^{*\prime}(z) + \Psi^*(z) + \frac{2r_0^2}{\kappa} \frac{\partial^2 \Sigma^*}{\partial z^2}] \tag{16}$$

and $\Phi^*(z) = \phi^{*\prime}(z)$, $\Psi^*(z) = \chi^{*\prime\prime}(z)$.

A longer computation enables us to represent the transforms of the complex displacements u^* and v^* (in the x- and y direction, respectively) in terms of the Goursat potentials as follows,

$$2N(u^* + i \ v^*) = (3-2T)\phi^*(z) - z\overline{\phi^{*\prime}(z)} - \overline{\psi^*(z)} + t(T-1)S^*(z,\bar{z})$$
$$\frac{2r_0^2}{\kappa} \frac{\partial \Sigma^*}{\partial \bar{z}} \tag{17}$$

with the notation

$$T = \frac{AR-Q^2}{(N+A)R-Q^2} , \quad t = \frac{Q+R}{R} , \quad S^* = \int \sigma_1^* \, dx + i \int \sigma_1^* dy ,$$
$$\sigma_1^* = 2\mathrm{Re} \ s^*(z) . \tag{18}$$

An inspection of the equations (15) and (17) shows that these equations generalize the well known elastic equations of the Muskhelishvili theory. In fact, upon disregarding the asterisks (as immaterial in the statical theory) and making the functions $s(z)$ and $\Sigma(z)$ vanish (as for a one phase elastic material), equations (15) transform into equations (4) and (5) in Section 39 of [3]. Similarly, by taking the constants N and A to coincide with Lamé constants μ and λ, respectively, suppressing the constants Q and R and proceeding as before we easily reduce (17) to the equation (1) in Section 32 of [3].

This completes the formulation of the two-dimensional theory of the class of rheological media under investigation, in a complex variable setting.

It is well known that the Kolosoff-Muskhelishvili method of solving two-dimensional elastostatic problems, based on a conversion of the solution of the boundary-value problems to the solution of certain functional equations in complex domain, has proved to be highly effective*, in that it uses the entire apparatus of the powerful theory of analytic functions.

In fact, a solution of any two-dimensional problem whether related to a body of a simple or multiple connectivity, and subjected to arbitrary traction or displacement type boundary conditions, requires no more than a determination of two functions of a complex variable. This can be done, for instance, by using the expansions in complex Fourier series, by means of Cauchy integral method or by a reduction to problems of linear relationship.

In what follows we show the effectiveness of the Muskhelishvili complex variable approach generalized so as to encompass the poroelastic materials. For definiteness, we employ the Fourier expansion method of solution.

Let us consider an infinite plane with removed finite portions bounded by simple closed contours, that is an infinite plane with holes. In particular, let us analyze in more detail a plane with a single circular hole. However, the generalized method applied is effective in any other two-dimensional case, exactly in the same manner as the classical method of Muskhelishvili.

We first circumscribe, from the origin of coordinates as the center, a circumference such that all hole contours remain inside the circumference. Next we assume that the stress acting on the solid phase and the fluid pressure remain bounded in the entire domain of definition including an infinitely distant point $r = \infty$.

Under these conditions, the potential part of the fluid pressure in (11) may be represented by the main part of the Laurent's expansion

$$s^*(z) = \sum_{n=0}^{\infty} \frac{d_n}{z^n}, \tag{19}$$

where the coefficients d_n are, in general, complex numbers depending on the Laplace parameter ξ. Similarly the requirement of boundedness of the functions $\Phi^*(z)$ and $\Psi^*(z)$ appearing in (15)

*Prior to the work of Kolosoff and Muskhelishvili the complex variable approach in two-dimensional elasticity was propounded by L. Filon in Philos. Trans., Vol. 201, p. 69 (1903).

suggests the form (compare [3], Section 56)

$$\Phi^*(z) = \sum_{k=0}^{\infty} \frac{a_k}{z^k} , \quad \Psi^*(z) = \sum_{k=0}^{\infty} \frac{b_k}{z^k} \tag{20}$$

with functions $\phi^*(z)$, $\chi^*(z)$ and F_1^* derived from Φ^* and Ψ^* by integration.

There remains now to find the representation of the Helmholtz part of the functions F^* and σ^* depending actually on a single function Σ^*. To this end note that a general solution of the Helmholtz equation has the form [8]

$$\Sigma^*(r,\theta,\xi) = g_o K_o(\frac{r}{r_o}) + \sum_{n=1}^{\infty} (g_n \cos n\theta + h_n \sin n\theta) K_n(\frac{r}{r_o})$$
$$+ k_o I_o(\frac{r}{r_o}) + \sum_{n=1}^{\infty} (k_n \cos n\theta + \ell_n \sin n\theta) I_n(\frac{r}{r_o}), \tag{21}$$

where K_n and I_n are the modified Bessel functions of the n-th order. Apparently, in the general problem considered here the modified functions of the first kind should be disregarded as unbounded at $r = \infty$.

2. ILLUSTRATIVE EXAMPLE

As an illustrative example, let us consider an infinite plane with a single circular hole of radius a . Assume that at time $t = +0$, the surface of the hole is suddenly subjected to a uniform hydrostatic pressure q_o acting upon the solid phase of the bulk material. Assume also that the surface $r = a$ is unsealed, and that the fluid is suddenly subjected to an excess pressure σ_o . At infinity, the liquid pressure, say, vanishes and the plane is under the action of the suddenly applied (at $t = +0$) uniform horizontal tractions such that at $x = \infty$, Figure 1,

$$\sigma_{xx} = -p_o , \quad \sigma_{yy} = \sigma_{xy} = 0. \tag{22}$$

It is assumed that at all times $t \geqslant +0$ the forces q_o , σ_o and p_o acting on the body are maintained constant. The boundary conditions just described may be cast into a symbolic form as follows,

$$\sigma^*(a,\theta;\xi) = \frac{\sigma_o}{\xi} ,$$
$$\sigma^*_{rr}(a,\theta;\xi) = -\frac{q_o}{\xi} , \tag{23}$$
$$\sigma^*_{r\theta}(a,\theta;\xi) = 0 ,$$

$$\sigma^*(\infty,\theta;\xi) = 0 ,$$
$$\sigma^*_{rr}(\infty,\theta;\xi) = -\frac{p_o}{2\xi}(1 + \cos 2\theta) ,$$
$$\sigma^*_{\theta\theta}(\infty,\theta;\xi) = -\frac{p_o}{2\xi}(1 - \cos 2\theta) , \tag{24}$$
$$\sigma^*_{r\theta}(\infty,\theta;\xi) = \frac{p_o}{2\xi}\sin 2\theta ,$$

at all times $t \geqslant +0$. Clearly, the foregoing conditions should be satisfied identically for all values of the angle θ. With (11) (19) and (21) in mind, the boundary conditions $(23)_1$ and $(24)_1$ give

$$d'_o = 0 \tag{25.1}$$

and

$$g_o = \frac{\sigma_o}{K_{oa\xi}}, \tag{25.2}$$

$$g_n = h_n = 0, \quad d'_n = 0 \quad \text{for} \quad n = 1,2,\ldots,$$

with the notation* $d_n = d'_n + id''_n$ and $K_{oa} \equiv K_o(\frac{a}{r_o})$.

The imaginary parts of the parameters d_n remain undetermined, but as seen from (11) their values are immaterial in the present case and may be set equal to zero. This gives finally

$$\sigma^* = \frac{\sigma_o}{K_{oa}\xi} K_{or}, \tag{26}$$

and similarly

$$\Sigma^* = \frac{\sigma_o}{K_{oa}\xi} K_{or}. \tag{27}$$

Since

$$\frac{\partial^2}{\partial z^2} = \frac{1}{4}\left[\frac{\partial^2}{\partial r^2} - \frac{1}{r}\frac{\partial}{\partial r} - \frac{1}{r^2}\frac{\partial^2}{\partial \theta^2} - \frac{2}{r} i\left(\frac{\partial^2}{\partial r \partial \theta} - \frac{1}{r}\frac{\partial}{\partial \theta}\right)\right]e^{-2i\theta}, \tag{28.1}$$

then

$$\frac{\partial^2 \Sigma^*}{\partial z^2} = \frac{\sigma_o}{4\ K_{oa}\ \xi r_o}\left(\frac{1}{r_o} K_{or} + \frac{2}{r} K_{1r}\right)e^{-2i\theta}, \tag{28.2}$$

and for $(14)_3$ we find

$$\sigma^*_{r\theta} = \sum_{k=0}^{\infty} \frac{1}{r^k}\left[k(a'_k \sin k\theta - a''_k \cos k\theta)\right.$$
$$\left. - (b'_k \sin (k-2)\theta - b''_k \cos (k-2)\theta)\right]. \tag{29}$$

Fulfillment of the boundary conditions $(24)_3$ and $(23)_3$ then yields, respectively,

$$b'_o = \frac{p_o}{2\xi}, \quad b''_o = 0 \quad \text{or} \quad b_o = \frac{p_o}{2\xi},$$

and

*Similarly we use the symbols $K_{or} \equiv K_o(\frac{r}{r_o})$, $K'_{oa} \equiv \left.\frac{\partial K_o(r/r_o)}{\partial r}\right|_{r=a}$, etc.

$$b''_2 = 0, \quad a'_1 + b'_1 - b'_3/a^2 = 0, \quad a''_1 - b''_1 - b''_3/a^2 = 0,$$

$$2a'_2 - b'_4/a^2 = -\frac{p_o}{2\xi} a^2, \quad 2a''_2 - b''_4/a^2 = 0, \tag{30}$$

$$na'_n - b'_{n+2}/a^2 = 0, \quad na''_n - b''_{n+2}/a^2 = 0,$$

$$\text{for } n > 3.$$

A longer but trivial calculation leads to the following explicit equations for the remaining stress components,

$$\sigma^*_{rr} = -\frac{\sigma_o}{\kappa K_{oa}\xi}(K_{or} + \frac{r_o}{r} K_{1r}) + \sum_{k=0}^{\infty} \frac{1}{r^k}[(2+k)\alpha_K - \beta_k], \tag{31}$$

$$\sigma^*_{\theta\theta} = \frac{\sigma_o}{\kappa K_{oa}\xi}[(1-\kappa)K_{or} + \frac{r_o}{r} K_{1r}] + \sum_{k=0}^{\infty} \frac{1}{r^k}[(2-k)\alpha_k + \beta_k], \tag{32}$$

where

$$\alpha_k = a'_k \cos k\,\theta + a''_k \sin k\,\theta\,,$$
$$\beta_k = b'_k \cos (k-2)\theta + b''_k \sin (k-2)\theta\,. \tag{33}$$

Upon satisfying the remaining three boundary conditions one gets the following results

$$a'_0 = -\frac{p_o}{4\xi}, \quad b'_0 = \frac{p_o}{2\xi},$$

$$b'_2 = a^2\left\{-\frac{\sigma_o}{\kappa K_{oa}\xi}(K_{oa} + \frac{r_o}{a}) K_{1a}) - \frac{p_o}{2\xi} + \frac{q_o}{\xi}\right\}, \tag{34}$$

$$3a'_1 - b'_1 - b'_3/a^2 = 0, \qquad 3a''_1 + b''_1 - b''_3/a^2 = 0,$$

$$4a'_2 - b'_4/a^2 = \frac{p_o}{2\xi} a^2, \qquad 4a''_2 - b''_4/a^2 = 0, \tag{35}$$

$$(n+2)a'_n - b'_{n+2}/a^2 = 0, \qquad (n+2)a''_n - b''_{n+2}/a^2 = 0$$

$$\text{for } n > 3.$$

An inspection of (30) and (35) shows that: (1°) by (30) and $(35)_{1,2}$ is $a_1 = b_1 = b_3 = 0$; (2°) by $(30)_{4,5}$ and $(35)_{3,4}$ is $a''_2 = b''_4 = 0$ and

$$a'_2 = \frac{p_o a^2}{2\xi},$$
$$b'_4 = \frac{3p_o a^4}{2\xi}; \tag{36}$$

(3°), by the last two equations of (30) and (35) all

$$a_n = 0 \,, \qquad b_{n+2} = 0 \qquad \text{for} \geqslant 3. \tag{37}$$

Summarizing all a_n and b_n except for the real parts of a_o, a_2, b_o, b_2 and b_4 appear to be equal to zero.

Bearing this in mind the final equations for the stress components acting on the solid phase and for the fluid pressure become,

$$\sigma^*_{rr} = - \frac{p_o}{2\xi} (1 - \frac{a^2}{r^2}) - \frac{p_o}{2\xi} (1-4 \frac{a^2}{r^2} + 3 \frac{a^4}{r^4}) \cos 2\theta - \frac{q_o}{\xi} \frac{a^2}{r^2}$$

$$- \frac{\sigma_o}{\kappa K_{oa} \xi} [K_{or} + \frac{r_o}{r} K_{1r} - (K_{oa} + \frac{r_o}{a} K_{1a}) \frac{a^2}{r^2}] \,, \tag{38.1}$$

$$\sigma^*_{\theta\theta} = - \frac{p_o}{2\xi} (1+\frac{a^2}{r^2}) - \frac{p_o}{2\xi} (1 + 3\frac{a^4}{r^4})\cos 2\theta + \frac{q_o}{\xi} \frac{a^2}{r^2} + \tag{38.2}$$

$$+ \frac{\sigma_o}{\kappa K_{oa} \xi} \left\{ (1-\kappa) K_{or} + \frac{r_o}{r} K_{1r} - [(1-\kappa) K_{oa} + \frac{r_o}{a} K_{1a}] \frac{a^2}{r^2} \right\}.$$

$$\sigma^*_{r\theta} = \frac{p_o}{2\xi} (1 + 2\frac{a^2}{r^2} - 3\frac{a^4}{r^4}) \sin 2\theta \,, \tag{38.3}$$

$$\sigma^* = \frac{\sigma_o}{K_{oa} \xi} K_{or} \,. \tag{38.4}$$

These equations generalize the following three known results.

(a) Assuming independence of time, that is taking $\xi = 1$, and posing $p_o = \sigma_o = 0$ one obtains the classical solution to the Lamé's problem for a tube under internal pressure with the outer radius tending to infinity, that is for a cylindrical hole in an infinite body.

(b) Proceeding as before but taking $\sigma_o = q_o = 0$ one obtains the classical solution to the problem of an uniaxial uniform stretch of a plate with a circular hole [3], Section 56a, Equations (4).

(c) Taking $q_o = p_o = 0$ and returning to the unspecified coefficients one obtains the equations which can be derived, after certain manipulations, from the solution given by Jana for an axially symmetric case, with the aid of a displacement function, see [9], his Equations (3.5) in combination with (4.11) for $\Psi \equiv 0$. It is seen that even in the absence of the external load acting on the solid phase, stresses arise in this phase, due to the excess fluid pressure applied on the boundary of the body. On the other hand, from the linearity of the phenomena under discussion clearly exhibited in the linear dependence of stress components on the individual load categories, there follows the following viscoelastic-elastic analogy based on the principle of superposition of effects:

In the viscoelastic case the ξ multiplied Laplace transforms of the stress components acting on the solid phase are equal to the sums of the corresponding components of: (α) the stresses generated by the given external load if acting on an elastic body identical with the given viscoelastic body, and (β) the stresses generated in the solid phase by the excess fluid pressure if acting alone on the boundary of the given viscoelastic body. Clearly, the latter are found once and for all for the given body and the preassigned load.

3. STRESS COMPONENTS AND FLUID PRESSURE

The equations for the stress components (38) are obtained in a transformed form, so that their retransformation to the physical space is needed. However, because of the presence of the modified Bessel functions in (38) the evaluation of the inverse transforms, say, by means of contour integrations is clumsy and would rather obscure than clarify the discussion. Instead, we choose a simpler program adequate for our purposes. We first multiply both sides of equations (38) by the parameter ξ and note that all terms except those with the factor σ_o become independent of ξ (and consequently of time, implicit in r_o). We call these terms $\sigma_{ij\ \mathrm{elast}}$, denote the remaining terms with the factor σ_o by $\sigma_{ij\ \mathrm{fluid}}$ and convert equations (38) into the following form,

$$\xi\sigma^*_{rr} = \sigma_{rr\ \mathrm{elast}} + \sigma^*_{rr\ \mathrm{fluid}}\ ,$$

$$\xi\sigma^*_{\theta\theta} = \sigma_{\theta\theta\ \mathrm{elast}} + \sigma^*_{\theta\theta\ \mathrm{fluid}}\ , \tag{39}$$

$$\sigma_{r\theta} = \sigma_{r\theta\ \mathrm{elast}}.$$

The last equation shows that the shear stress is not influenced by the fluid pressure.

We first discuss the azimuthal stress component $\sigma^*_{\theta\theta}$ as the one that is responsible for the stress concentration at the surface of the hole.

We use two limit theorems of the operational calculus expressed symbolically by

$$f(t = +0) = \lim_{\xi\to\infty} \xi\, f^*(\xi)\ ,$$
$$f(t = +\infty) = \lim_{\xi\to 0} \xi\, f^*(\xi)\ , \tag{40}$$

and giving the values of a function from the value of its transform in two instances: at $t = 0$, that is in our case directly after the application of the load, and secondly after the elapse of an infinitely long time. In the first case, the Macdonald functions $K_n(x)$ may be approximated by their asymptotic expression $e^{-x}(\pi/2x)^{\frac{1}{2}}$. In the second case, the asymptotic approximations $K_o(x) \approx -\ln x$ and $K_1(x) \approx 1/x$ are permissible. Confining ourselves to the circle $r = a$ where the stress concentration reaches its peak we find after passing to the appropriate limits

$$\sigma_{\theta\theta}\ (a,\theta;\ t=0) = \sigma_o + \sigma_{\theta\theta}(a,\theta)_{\mathrm{elast}}\ ,$$
$$\sigma_{\theta\theta}\ (a,\theta;\ t=\infty) = \sigma_o \frac{1-\kappa}{\kappa} + \sigma_{\theta\theta}(a,\theta)_{\mathrm{elast}}\ . \tag{41}$$

A similar computation leads to the following values of the radial stress component,

$$\sigma_{rr}(r,\theta;\ t=0) = \begin{Bmatrix} 0 & \text{for } r = a \\ \sigma_o \dfrac{a^2}{r^2} & \text{for } r \neq a \end{Bmatrix} + \sigma_{rr\ \mathrm{elast}}, \tag{42}$$

$$\sigma_{rr}(r,\theta;\ t=\infty) = \sigma_o\left(\frac{a^2}{r^2} - 1\right) + \sigma_{rr\ \mathrm{elast}};$$

the last equation retains the validity in the vicinity of the hole only.

The equations for the fluid pressure now become

$$\sigma(r,\theta;\ t=0) = \begin{cases} 0 & \text{for } r \neq a\ , \\ \sigma_o & \text{for } r = a\ . \end{cases} \tag{43}$$

$$\sigma(r,\theta;\ t=\infty) = \sigma_o \quad \text{in the neighborhood of} \quad r = a.$$

It is of interest to make an estimate of the variation of the azimuthal component $(41)_2$ with time. To this purpose we have to evaluate the coefficient κ made up of the material coefficients. For the osseous tissue the coefficients remain for the most part unknown. It is possible, however, to make a rough assessment of the value of κ by using an analysis of the poroelastic coefficients similar to that given in [9] and indicated in more detail in [5]. The analysis furnishes the value $\kappa \approx 0.3$, so that the equations for the azimuthal stress at the surface of the hole (where the existing tensions may originate a crack departing from any tiny fissure) become

$$\begin{aligned} \sigma_{\theta\theta}(a,\theta;\ t=0) &= \sigma_o + \sigma_{\theta\theta\ \mathrm{elast}}, \\ \sigma_{\theta\theta}(a,\theta;\ t=\infty) &= \tfrac{7}{3}\,\sigma_o + \sigma_{\theta\theta\ \mathrm{elast}}. \end{aligned} \tag{44}$$

Clearly the first terms represent here the stress increments to the classical expression provided by the assumed rheology of the material. It is also seen that, as the time goes by, the azimuthal stress at the hole intensifies, in view of the time independence of the classical components. On the other hand, the increase of the σ_o term is certainly bounded.

The variation in time of any deformation characteristic, such as for instance the dilatation, can now be trivially found. Thus from (1) we find at $r = a$.

$$e = \frac{R}{2[(A+N)R-Q^2]}\left(\sigma_{rr} + \sigma_{\theta\theta} - 2\frac{Q}{R}\,\sigma\right), \tag{45}$$

or, in view of (41) - (43) explicitly at $\theta = \frac{\pi}{2}$, say,

$$e = \frac{R}{(A+2N)R-Q^2}\cdot\begin{cases} -2\frac{Q}{R}\,\sigma_o + \sigma_o & \text{at } t = 0 \\ -2\frac{Q}{R}\,\sigma_o + \frac{7}{3}\,\sigma_o & \text{at } t = \infty \end{cases} + (\sigma_{rr} + \sigma_{\theta\theta})_{\mathrm{elast}} \tag{46}$$

Since Q/R is found to be positive (see [5]), it is seen that under a sudden application of load (in the present case of the tension p_o and the pressures q_o and σ_o) an instantaneous deformation originates which, with elapsing time, increases asymptotically to a final bounded value. Such a behavior of the body under the action of a constant load represents what is known as the creep associated with the standard rheological (three parameter) model composed of a Kelvin-Voigt element in series with an elastic spring. It is remarkable that such a model of osseous tissue was proposed by Sedlin in a fundamental treatise [10] based on comprehensive experimental observations. In fact, according to Sedlin, under load well below the fracture load the actual behavior of bones is qualitatively described by the standard viscoelastic model. This opinion seems to be shared by practitioners. For instance, Frankel, orthopaedic surgeon and an authority in orthopaedic biomechanics, in the text coauthored by Burstein, [12], p. 112, notes that "the results of tests conducted . . . on bovine bones. . . (show) excellent qualitative agreement, up to the point of yield, between the (standard) model behavior. . . and the experimentally observed behavior".

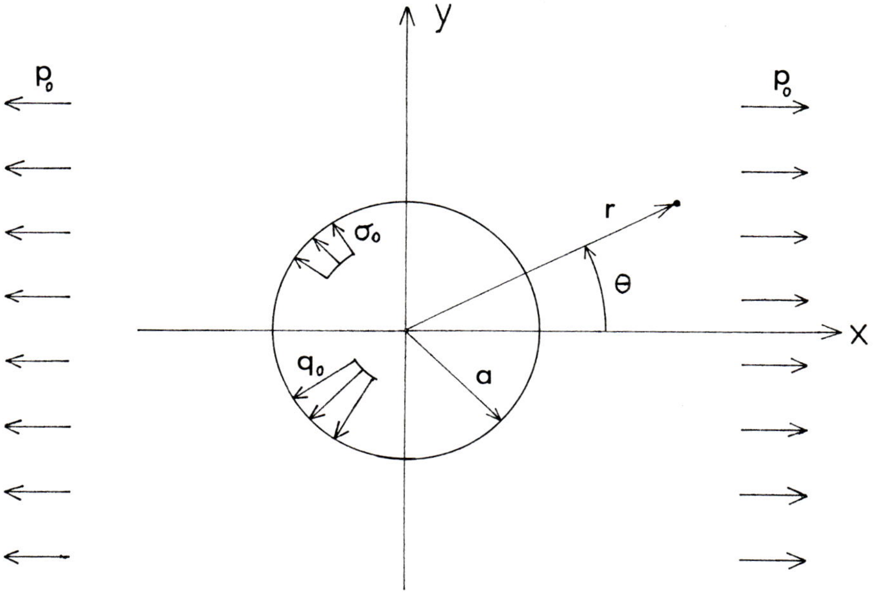

Fig. 1 Geometry and load

NOMENCLATURE

r, θ, z	= cylindrical coordinates
t	= time
σ_{ij}, e_{ij}	= stress and strain tensors
σ	= fluid pressure
e, ε	= dilatations of solid and fluid
A,N,Q,R.	= material constants, $P = A + 2N$
κ	= $(PR - Q^2)/2N(Q + R)$
K	= $(PR - Q^2)/(A + 2N + R + 2Q)$
b	= permeability coefficient
r_o^2	$K/b\xi$
f^*	= Laplace transform
ξ	= transform parameter
∇^2	= $\partial^2/\partial r^2 + \partial/r\partial r + \partial^2/r^2\partial\theta^2$
$F(r,\theta;t)$	= stress function
Φ_o^* and Σ^*	= solutions of Equations (10)
ϕ^* and χ^*	= Goursat potentials
s^*	= the potential part of fluid pressure
$\bar{z}$	= conjugate of complex variable $z = re^{i\theta}$
Re and Im	= real and imaginary parts

a_k, b_k, d_k = coefficients of Laurent's expansions
α_k, β_k = defined by Equations (33)
I_o, K_o = modified Bessel functions
$K_{or} \equiv K_o(\frac{r}{r_o})$, $K'_{oa} = \partial K_o(r/r_o)/\partial r$ at $r = a$
q_o, σ_o = hole pressure on solid and liquid phase
p_o = tension on solid phase at infinity

REFERENCES

[1] H. Neuber: Kerbspannungslehre, Berlin (1937) (translated as Theory of Notch Stresses, Edwards (1946)).

[2] G. N. Savin: Stress Concentration Around Holes (in Russian), GITTL, Moscow (1951) (translated by Pergamon Press, (1961).

[3] N. I. Mushelishvili: Some Basic Problems of the Mathematical Theory of Elasticity (in Russian), 5th edition, Nauka, Moscow (1966) (translation of the 3rd edition by P. Noordhoff, (1963)).

[4] G. N. Savin: Stress Concentration Around Holes (in Russian), Naukova Dumka, Kiev, (1968).

[5] J. L. Nowinski: Stress Concentration Around a Cylindrical Cavity in a Bone Treated as a Poroelastic Body, Forthcoming Acta Mechanica.

[6] M. A. Biot: General Solutions of the Equations of Elasticity and Consolidation for a Porous Material, J. Appl. Mech., Vol. 23, p. 91-96, (1956).

[7] J. L. Nowinski: Cylindrical Bones as Anisotropic Poroelastic Members Subjected to Hydrostatic Pressure, Forthcoming in Proc. Fifth Southeastern Conf. Theoret., Appl. Mech., Raleigh-Durham, (1970).

[8] G. N. Watson: A Treatise on the Theory of Bessel Functions, Cambridge (1922).

[9] M. A. Biot: The Elastic Coefficients of the Theory of Consolidation, J. Appl. Mech., Vol. 24, p. 594-601, (1957).

[10] E. D. Sedlin: A Rheological Model of Cortical Bone, Acta Orthop. Scandinav., Suppl. 83, Munksgaadr, Copenhagen, p. 1-77, (1965).

[11] H. Kraus: On the Mechanical Properties and Behavior of Human Compact Bone, Adv. Biom. Eng. Med. Phys., Vol. 2, (1968).

[12] V. H. Frankel and A. H. Burnstein: Orthopeadic Biomechanics, Lea and Febiger, Philadelphia, (1970).

Developments in Mechanics, Vol. 6. Proceedings of the 12th Midwestern Mechanics Conference: 31

Constitutive Equations for Mixtures of Newtonian Fluids and Viscoelastic Solids

FARHAD TABADDOR
B. F. GOODRICH COMPANY
ROBERT Wm. LITTLE
MICHIGAN STATE UNIVERSITY

ABSTRACT

Using a general thermodynamic theory of interacting media, the constitutive equations for interacting continuum composed of a linear viscoelastic solid and a Newtonian fluid are developed. The case of incompressible fluid is also discussed.

INTRODUCTION

A general theory for heterogeneous media has been first proposed by Truesdell and Toupin [1]. The equations of mass, momentum and energy balance were postulated for each component of the mixture, some alternative formalisms were also proposed by several authors. Green and Naghdi [2], on the other hand, proposed a single energy and entropy production inequality for the whole mixture. The basic equations, governing the mixture, are then derived from the postulated equations by a systematic application of invariance requirements.

Based on the latter theory, the constitutive equations for a mixture of elastic solids and Newtonian fluids were derived by Green and Steel [3] and also by Crochet and Naghdi [4]. The constitutive equations for a binary mixture of elastic solids and incompressible Newtonian fluids were derived by the authors [5]. Some methods of solutions to the governing equations of such binary mixtures, for steady-state case, were also presented and classical theories of flow through undeformable porous media [6] and deformable porous media [7] also discussed in the light of continuum approach.

In this paper the constitutive equations for binary mixtures of Newtonian fluids and viscoelastic solids will be given. These equations will be derived by following the theory of Green and Naghdi together with a thermodynamic theory of irreversible systems [8]. Such constitutive equations are applicable to a great variety of engineering problems such as flow of oil or water in petroleum reserviors, seepage problems and soil consolidation just to mention a few. This problem has also been considered by Biot [9].

BASIC EQUATIONS

The basic equations of a binary mixture are given in the following for self-consistency. The motion of a mixture of two components, S_1 and S_2, is referred to fixed Cartesian coordinates with material coordinates designated by X and Y respectively. The position of particles at time t is given by

$$x_i = x_i(X_j,t) \qquad y_i = y_i(Y_j,t) \tag{1}$$

These particles are considered to occupy the same position at time t so that

$$y_i = x_i \tag{2}$$

The velocity and acceleration vectors of S_1 are given by u_i and a_i and those of S_2 by v_i and g_i. The densities at time t are ρ_1 and ρ_2 and the rate of deformation tensors are defined to be:

$$d_{ij} = 1/2(u_{i,j} + u_{j,i}) \quad f_{ij} = 1/2(v_{i,j} + v_{j,i}) \tag{3}$$

where a comma denotes partial differentiation with respect to x_i or y_i.

The summation convention is being used in the above equations and throughout the paper unless stated otherwise.

The vorticity tensors and other mixture variables are defined as follows:

$$\Gamma_{ij} = 1/2(u_{i,j}-u_{j,i}) \quad \Lambda_{ij} = 1/2(v_{i,j}-v_{j,i}) \quad \rho = \rho_1 + \rho_2$$

$$\rho\overline{v}_i = \rho_1 u_i + \rho_2 v_i \quad \frac{D}{Dt} = \frac{\partial}{\partial t} + \overline{v}_m \frac{\partial}{\partial x_m} \quad \rho\frac{D}{Dt} = \rho_1 \frac{{}^{(1)}D}{Dt} + \rho\frac{{}^{(2)}D}{Dt} \tag{4}$$

where the numerical superscript on the material time derivative refers to the component in question presented here for reference only. The continuity equations for a binary mixture in the absence of chemical reactions are as follows

$$\frac{{}^{(1)}D\rho_1}{Dt} + \rho_1 u_{k,k} = 0 \qquad \frac{{}^{(2)}D\rho_2}{Dt} + \rho_2 v_{k,k} = 0 \tag{5}$$

The equations of motion for the mixture are

$$(\sigma_{ki}+\tau_{ki})_{,k}+\rho_1F_i+\rho_2G_i = \rho_1a_i+\rho_2g_i \tag{6}$$

where σ_{ki} and τ_{ki} are partial stresses of the solid and the fluid respectively and F_i and G_i are the body forces per unit mass of each continuum. The partial stresses satisfy the following symmetry relation

$$\sigma_{ki} + \tau_{ki} = \sigma_{ik} + \tau_{ik} \tag{7}$$

The diffusive force π_i is given by

$$\pi_i = 1/2(\sigma_{ki}-\tau_{ki})_{,k} + 1/2\rho_1(F_i-a_i)-1/2\rho_2(G_i-g_i) \tag{8}$$

The entropy production inequality, under isothermal conditions, may be written [3]

$$-\rho\frac{DA}{Dt} + \pi_i(u_i-v_i) + 1/2(\sigma_{ki}+\sigma_{ik})d_{ik} + 1/2(\tau_{ki}+\tau_{ik})f_{ik}$$

$$+ 1/2(\sigma_{ki}-\sigma_{ik}) \quad (\Gamma_{ik}-\Lambda_{ik}) \geq 0 \tag{9}$$

where $A = U - TS$ is the Helmholtz free energy, U is the internal energy per unit mass, S is the entropy per unit mass and T is the temperature.

We finally define the notion of hidden coordinates of irreversible thermodynamic which will be used in conjunction with the above basic results.

The thermodynamic system is assumed to have n degrees of freedom defined by n state variables $\xi_1 \ldots . \xi_n$. These independent state variables are alternatively called generalized coordinates. These coordinates are divided into two groups of hidden and observed ones. The system is assumed to be under the action of n generalized forces Q_j in such a manner that $Q_i d\xi_i$ represents an incremental amount of work done on the system. The hidden variables are those whose corresponding conjugate forces are zero and are of interest only to the extent of their influence upon our observed variables. As an example, in a body under external loading, strain components are considered as observed variables and stress components as their conjugate external force, while the effect of "molecular configuration", interstitial atoms, dislocation, grain boundaries, etc., on stress-strain law can be accounted by hidden variables. At this stage we are not concerned about the explicit form of ξ_i. However, they are assumed to be functionals of the history of the observed variables.

DEVELOPMENT OF THE CONSTITUTIVE EQUATIONS

The following assumptions will be used in this paper

a) - Both the viscoelastic solid and the fluid are initially at rest under zero initial stresses.

b) - The continuum is initially homogeneous and undergoes an isothermal deformation.

c) - The displacements of the solid, the change in density and velocity as well as their space and time derivatives, remain small during the motion so that higher order terms may be neglected.

We now postulate the following constitutive equations in accordance with the equipresence principle.

$$A = A(e_{ij},\ \rho_2,\ \xi_\ell,\ f_{ij},\ d_{ij},\ u_i - v_i) \tag{10}$$

$$S = S(e_{ij},\ \rho_2,\ \xi_\ell\ \ f_{ij},\ d_{ij},\ u_i - v_i) \tag{11}$$

$$\frac{1}{2}(\sigma_{ki} + \sigma_{ik}) = A_{ik} + A_{ikj}(u_j - v_j) + A_{ikrs} f_{rs} + \bar{A}_{ikrs} d_{rs} \tag{12}$$

$$\frac{1}{2}(\tau_{ki} + \tau_{ik}) = B_{ik} + B_{ikj}(u_j - v_j) + B_{ikrs} f_{rs} + \bar{B}_{ikrs} d_{rs} \tag{13}$$

$$\frac{1}{2}(\sigma_{ki} - \sigma_{ik}) = -\frac{1}{2}(\tau_{ki} - \tau_{ik}) = D_{ki} + D_{kij}(u_j - v_j) + D_{kirs} f_{rs} + \bar{D}_{ikrs} d_{rs} \tag{14}$$

$$\pi_i = a_i + a_{ij}(u_j - v_j) + a_{irs} f_{rs} + \bar{a}_{irs} d_{rs} \tag{15}$$

where A_{ik}...depend on e_{ij}, ρ_2, ξ_ℓ. The dependence of f_{ij}, d_{ij}, $u_i - v_i$ can be omitted from (eq. 10) and (eq. 11) by thermodynamic consideration as shown by Crochet and Naghdi [4].

$$A = A(e_{ij},\ \rho_2,\ \xi_\ell) \tag{16}$$

$$S = S(e_{ij},\ \rho_2\ \ \xi_\ell) \tag{17}$$

The entropy production inequality yields

$$\frac{D\gamma}{Dt} = -\rho\frac{DA}{Dt} + \pi_i(u_i - v_i) + \frac{1}{2}(\sigma_{ki} + \sigma_{ik})d_{ik} + \frac{1}{2}(\tau_{ki} + \tau_{ik})f_{ik}$$

$$+ \frac{1}{2}(\sigma_{ki} - \sigma_{ik})\ \ (\Gamma_{ik} - \Lambda_{ik}) \geqslant 0 \tag{18}$$

where $D\gamma/Dt$ is the rate of irreversible entropy production.

Differentiating (eq. 16) gives,

$$\rho \frac{DA}{Dt} = \rho \frac{\partial A}{\partial \rho_2} \frac{D\rho_2}{Dt} + \rho \frac{\partial A}{\partial e_{rs}} \frac{De_{rs}}{Dt} + \rho \frac{\partial A}{\partial \xi_\ell} \frac{D\xi_\ell}{Dt} \tag{19}$$

In view of (eq. 9), (eq. 19) becomes

$$\begin{aligned}\rho \frac{DA}{Dt} = & -\rho\rho_2 \frac{\partial A}{\partial \rho_2} f_{kk} + \frac{1}{2} \frac{\partial x_i}{\partial X_r} \frac{\partial x_j}{\partial X_s} \left(\frac{\partial A}{\partial e_{rs}} + \frac{\partial A}{\partial e_{sr}} \right) d_{ij} \\ & + (u_k - v_k) \left[\rho_1 \frac{\partial A}{\partial \rho_2} \frac{\partial \rho_2}{\partial y_k} - \frac{1}{2} \rho_2 \left(\frac{\partial A}{\partial e_{rs}} + \frac{\partial A}{\partial e_{sr}} \right) \frac{\partial e_{rs}}{\partial x_k} \right] \\ & + \rho \frac{\partial A}{\partial \xi_\ell} \frac{D\xi_\ell}{Dt}\end{aligned} \tag{20}$$

With the help of (eq. 20) and (eqs. 12 to 15), (eq. 18) becomes

$$\begin{aligned}\frac{D\gamma}{Dt} = & \left[A_{ik} - \frac{1}{2}\rho \frac{\partial x_i}{\partial X_r} \frac{\partial x_k}{\partial X_s} \left(\frac{\partial A}{\partial e_{rs}} + \frac{\partial A}{\partial e_{sr}} \right) \right] d_{ik} + \left(B_{ik} + \rho\rho_2 \frac{\partial A}{\partial \rho_2} \delta_{ik} \right) f_{ik} \\ & + \left[a_i - \rho_1 \frac{\partial A}{\partial \rho_2} \frac{\partial \rho_2}{\partial y_i} + \frac{1}{2} \rho_2 \left(\frac{\partial A}{\partial e_{rs}} + \frac{\partial A}{\partial e_{sr}} \right) \frac{\partial e_{rs}}{\partial x_i} \right] (u_i - v_i) \\ & + (B_{rsi} + a_{rsi}) f_{rs} (u_i - v_i) + (A_{rsi} + \bar{a}_{rsi}) d_{rs} (u_i - v_i) \\ & + A_{ikrs} d_{ik} f_{rs} + a_{ij} (u_i - v_i)(u_j - v_j) + B_{ikrs} f_{ik} f_{rs} \\ & + D_{ki} (\Gamma_{ik} - \Lambda_{ik}) + D_{kij} (\Gamma_{ik} - \Lambda_{ik})(u_i - v_i) + D_{kirs} f_{rs} (\Gamma_{ik} - \Lambda_{ik}) \\ & - \rho \frac{\partial A}{\partial \xi_\ell} \frac{D\xi_\ell}{Dt} + \bar{A}_{ikrs} d_{rs} d_{ik} + \bar{B}_{ikrs} d_{rs} f_{ik} + \bar{D}_{kirs} (\Gamma_{ik} - \Lambda_{ik}) \geq 0\end{aligned} \tag{21}$$

the summation over ℓ carries from $\ell = 1, \ldots\ldots, n$, where n is the number of independent thermodynamic hidden coordinates.

Applying the same argument as in [3] to the above inequality and incorporating the results into the constitutive equations, we obtain

$$\sigma_{ki} = \sigma_{ik} = \rho \frac{\partial x_i}{\partial X_r} \frac{\partial x_k}{\partial X_s} \frac{\partial A}{\partial e_{rs}} + A_{ikrs} f_{rs} + \bar{A}_{ikrs} d_{rs} \tag{22}$$

$$\tau_{ki} = \tau_{ik} = -\rho\rho_2 \frac{\partial A}{\partial \rho_2} \delta_{ik} + B_{ikrs} f_{rs} + \bar{B}_{ikrs} d_{rs} \tag{23}$$

$$\pi_i = \rho_1 \frac{\partial A}{\partial \rho_2} \frac{\partial \rho_2}{\partial y_i} - \frac{1}{2} \rho_2 \frac{\partial A}{\partial e_{rs}} \frac{\partial e_{rs}}{\partial x_i} + a(u_i - v_i) \qquad (24)$$

It is found that

$$B_{ikj} = 0 \quad A_{ikj} = 0 \quad a_{ikj} = 0 \quad \overline{a}_{ikj} = 0 \qquad (25)$$

for isotropic mixtures only, however, to keep the equations in a simpler form and yet preserve the generality to anisotropic media, the dependence of partial stresses on $(u_i - v_i)$ is neglected in the above analysis.

In view of the constitutive equations proposed and worked out for the rate of entropy production by Biot [9], Schapery [10], Valanis [11] and the others, we postulate the following constitutive equations for $\frac{D\gamma}{Dt}$.

$$\frac{D}{Dt} = f(\dot{\xi}_\ell,\ d_{ij},\ f_{ij},\ u_i - v_i,\ e_{ij},\ \rho,\ \xi_\ell) \geq 0 \qquad (26)$$

$$\frac{D\xi}{Dt} = \dot{\xi} \qquad (27)$$

where $\dot{\xi}_\ell$ is the material time derivative of ℓth hidden coordinate. If ξ_ℓ is expressed in terms of material coordinates, the above derivative reduces to partial derivative with respect to time.

The rate of entropy production is required to be zero in equilibrium state.

$$f(0,0,0,0,e_{ij},\ \rho,\ \xi_\ell) = 0 \qquad (28)$$

Therefore the appearance of e_{ij}, ρ, ξ_ℓ in the above constitutive equation is implicit and consequently (eq. 26) can be written as

$$\frac{D}{Dt} = f(\dot{\xi}_\ell,\ d_{ij},\ f_{ij},\ u_i - v_i) \geq 0 \qquad (29)$$

where the implicit dependence on e_{ij}, ρ, ξ_ℓ is understood. Assuming required smoothness of the function f, it can be expanded in Taylor series. Neglecting terms higher than the second and eliminating terms less than two on account of the assumption of zero initial stresses, yields,

$$\frac{D\gamma}{Dt} = (b_{i\alpha}\dot{\xi}_\alpha\dot{\xi}_\alpha + b_{ij\alpha}d_{ij}\dot{\xi}_\alpha + \overline{b}_{ij\alpha}f_{ij}\dot{\xi}_\alpha + b_{i\alpha}(u_i - v_i)\dot{\xi}_\alpha)$$

$$+ \text{ terms not including } \dot{\xi}. \qquad (30)$$

The rate of irreversible entropy generation must be always non-zero. This would then lead to the following additional relations.

$$b_{ijk} = 0 \qquad b_{ijk} = 0 \qquad b_{ik} = 0 \tag{31}$$

it is further deduced that $b_{i\alpha}\dot{\xi}_i\dot{\xi}_\alpha$ has to be a positive definite. The n thermodynamic coordinates ξ_α can be transformed to their canonical form, where the new independent coordinates constitute an orthogonal system. Considering the above result and comparing (eq. 30) with (eq. 21) leads to:

$$b\,\dot{\xi} + \rho \frac{\partial A}{\partial \xi_\alpha} = 0 \qquad \text{no summation } \alpha = 1,\ldots.n \tag{32}$$

Expanding A in Taylor series under the same conditions as

$$\overline{\rho}A = \frac{1}{2} C_{ijk\ell}e_{ij}e_{k\ell} + C_{ij\alpha}e_{ij}\xi_\alpha + \Sigma\,\frac{1}{2}\; C_\alpha\xi_\alpha\xi_\alpha + - b\eta^2 +$$

$$C_{ij}e_{ij}\eta + \overline{C}_\alpha\xi_\alpha\eta \tag{33}$$

where $\eta = \rho_2 - \overline{\rho}_2$.

In view of (eq. 33), (eq. 32) becomes

$$\frac{\rho}{\overline{\rho}}\,(C_{ij\alpha}e_{ij} + C_\alpha\xi_\alpha + C_\alpha\eta) + b_\alpha\dot{\xi}_\alpha = 0 \tag{34}$$

We write the above equation in the following form

$$\dot{\xi}_\alpha + \frac{C_\alpha}{b_\alpha}\,\xi_\alpha = -\,Q(t) \tag{35}$$

where

$$Q(t) = \frac{1}{b_\alpha}\left(C_{ij\alpha}e_{ij} + \overline{C}_\alpha\eta\right)$$

$\rho/\overline{\rho}$ is considered to be unity in view of the second order effect of neglected terms. The solution of equation (eq. 35) is

$$\xi_\alpha = -\int_{-\infty}^{t} Q(\tau)e^{-\frac{C_\alpha}{b_\alpha}(t-\tau)}\,d\tau$$

or

$$\xi_\alpha = -\left[\frac{b_\alpha}{C_\alpha}\;Q(t) - \int_{-\infty}^{t} e^{-\frac{C_\alpha}{b_\alpha}(t-\tau)}\,\frac{\partial Q(t)}{\partial t}\right]d\tau \tag{36}$$

Substituting for A from (eq. 33) into (eqs. 22-24) and retaining linear terms we obtain:

$$\sigma_{ik} = C_{ik\ell m}e_{\ell m} + \Sigma\, a_{ik\alpha}\xi_\alpha + a_{ik}\eta + A_{ikrs}f_{rs} + \overline{A}_{kirs}d_{rs} \tag{37}$$

$$\tau_{ik} = -\,\overline{\rho}_2(b\eta + C_{\ell m}e_{\ell m} + \Sigma\,\overline{C}_\alpha q_\alpha)\,\delta_{ik} + B_{ikrs}f_{rs} + \overline{B}_{ikrs}d_{rs} \tag{38}$$

$$\pi_i = a(u_i - v_i) \tag{39}$$

Substituting for ξ_α from (eq. 36) into constitutive equations for partial stresses yields:

$$\sigma_{ik} = \int_{-\infty}^{t} \overset{(1)}{G}_{ik\ell m}(t-\tau)\,\frac{\partial e_{\ell m}(\tau)}{\partial\tau}\,d\tau + \int_{-\infty}^{t} \overset{(2)}{G}_{ik\ell m}(t-\tau)\,\frac{\partial f_{\ell m}(\tau)}{\partial\tau}\,d\tau$$

$$+ \int_{-\infty}^{t} \overset{(3)}{G}_{ik\ell m}(t-\tau)\,\frac{\partial d_{\ell m}(\tau)}{\partial\tau}\,d\tau + \int_{-\infty}^{t} \overset{(4)}{G}_{ik}(t-\tau)\,\frac{\partial\eta(\tau)}{\partial\tau}\,d\tau \tag{40}$$

$$\tau_{ik} = \left\{\int_{-\infty}^{t} \overset{(1)}{F}_{\ell m}(t-\tau)\,\frac{\partial e_{\ell m}}{\partial\tau}\,d\tau + \int_{-\infty}^{t} \overset{(4)}{F}(t-\tau)\,\frac{\partial\eta(\tau)}{\partial\tau}\,d\tau\right\}\delta_{ik}$$

$$+ \int_{-\infty}^{t} \overset{(2)}{F}_{ik\ell m}(t-\tau)\,\frac{\partial f_{\ell m}}{\partial\tau}\,d\tau + \int_{-\infty}^{t} \overset{(3)}{F}_{ik\ell m}(t-\tau)\,\frac{\alpha d_{\ell m}}{\partial\tau}\,d\tau \tag{41}$$

where

$$\overset{(1)}{G}_{ik\ell m} = C_{ik\ell m} - \Sigma\frac{a_{ik}\,C_{\ell m\alpha}}{C_\alpha}\left(1 - e^{-\frac{C_\alpha}{b_\alpha}(t-\tau)}\right) \tag{42}$$

$$\overset{(2)}{G}_{ik\ell m} = A_{ik\ell m} \tag{43}$$

$$\overset{(3)}{G}_{ik\ell m} = \overline{A}_{ik\ell m} \tag{44}$$

$$\overset{(4)}{G}_{ik} = a_{ik} - \Sigma\frac{a_{ik\alpha}\overline{C}_\alpha}{C_\alpha}\left(1 - e^{-\frac{C_\alpha}{b_\alpha}(t-\tau)}\right) \tag{45}$$

and

$$\overset{(1)}{F}_{\ell m} = -\overline{\rho}_2 C_{\ell m} + \overline{\rho}_2\Sigma\frac{\overline{C}_\alpha C_{\ell m\alpha}}{C_\alpha}\left(1 - e^{-\frac{C_\alpha}{b_\alpha}(t-\tau)}\right) \tag{46}$$

$$\overset{(2)}{F}_{ik\ell m} = B_{ik\ell m} \tag{47}$$

$$\overset{(3)}{F}_{ik\ell m} = \overline{B}_{ik\ell m} \tag{48}$$

$$F^{(4)} = -\bar{\rho} b + \bar{\rho}_2 \, \Sigma \, \frac{\bar{C}_\alpha^{\,2}}{C_\alpha} \left(1 - e^{-\frac{C_\alpha}{b_\alpha}(t - \tau)} \right) \qquad (49)$$

Where $G^{(1)}_{ik\ell m}$......are relaxation functions. The above constitutive equations hold for the most general anisotropic case. The interaction was assumed to render the constitute equations for partial stresses a function of all state variables. However some of the relaxation functions might be either a constant or zero at all times. This remains to be determined through experiments and we do not discuss it any further.

INCOMPRESSIBLE MEDIA

Consider the case where the fluid is incompressible. The incompressibility condition was derived to be [5].

$$f_{kk} + a_1 d_{kk} = 0 \qquad (50)$$

where $a_1 = (R-P_o)/P_o$, P_o denotes the initial porosity of the visco-elastic solid and R is a constant expressing the ratio of pores compressibility to the total compressibility.

Similar to that of reference [5] if the quantity

$$\bar{p}\,[f_{kk} + a_1 d_{kk}] = \bar{p}\,[f_{ik} + a_1 d_{ik}]\,\delta_{ik} \qquad (51)$$

where $\bar{p}$ is an unknown Scalar function, is added to the entropy production inequality. The resulting equation may be treated as a variational problem. The resulting constitutive equations are

$$\sigma^{(I)}_{ik} = \sigma_{ik} + a_1 \bar{p}\,\delta_{ik} \qquad (52)$$

$$\tau^{(I)}_{ik} = \tau_{ik} + \bar{p}\,\delta_{ik} \qquad (53)$$

where σ_{ik} and τ_{ik} are partial stresses for the case where fluid is incompressible. σ_{ik} and τ_{ik} are to be substituted from (eqs. 37 and 38). It is seen that $\bar{p}$ introduces a new unknown, however, the incompressibility condition provides the additional equation. It is further seen that η can be related to solid dilatation e_{mm} through the integration of (eq. 50) and use (eq. 5). This would yield

$$\eta = a_1 \bar{p}_2 e_{mm} \qquad (54)$$

It is convenient to introduce a new unknown Scalar function p given by

$$p = \bar{p} - \rho_2 (b\eta + C_{\ell m} e_{\ell m} + \bar{C}_\alpha \xi_\alpha) \qquad (55)$$

The final constitutive equations are

$$\sigma_{ik}^{(I)} = \int_{-\infty}^{t} H_{iklm}(t-\tau)\frac{\partial e_{lm}}{\partial \tau}d\tau + a_1 p\delta_{ik}$$

$$+ A_{iklm} f_{lm} + \bar{A}_{iklm} d_{lm} \qquad (56)$$

$$\tau_{ik}^{(I)} = p\delta_{ik} + B_{iklm} f_{lm} + \bar{B}_{iklm} d_{lm} \qquad (57)$$

H_{iklm} is a new relaxation function.

Similar to the case of classical elastic solid and Newtonian fluid, the Scalar p may be defined as thermodynamic pressure. It should be pointed out that the effect of thermodynamic hidden coordinates are contained in the expression for p, as may physically be expected.

The thermodynamic pressure p, may also be defined for compressible case

$$p = -\bar{\rho}\,(b\eta + C_{lm} e_{lm} + \bar{C}_{\alpha}\,\xi_{\alpha}) \qquad (58)$$

In analogy to fluid mechanics, the above equation may be considered as the equation of state. It should be remembered that ξ_{α} is a known functional of η and e_{lm} through (eq. 36). The above equations together with basic equations of section two constitute a consistent mathematical system describing the binary mixture of viscoelastic solids and Newtonian fluids.

REFERENCES

[1] Truesdell, C. and Toupin, R.A., The Classical Field Theories, Handluch der physik, Vol. III/I.

[2] Green, A.E. and Naghdi, P.M.,: A dynamic theory of interacting continua. Int. Journal, Engng. Sci. Vol. 3, 1965, pp. 231-241.

[3] Green, A.E., and Steel, T.R.,: Constitutive equations for interacting continua. Int. J. Engng. Sci. Vol. 4, 1966, pp. 483-500.

[4] Crochet, M.J. and Naghdi, P.M.,: On constitutive equations for flow of fluid through an elastic solid. Int. Engng. Sci. Vol. 4, 1966, pp. 383-401.

[5] Tabaddor, F. and Little, R.W.,: Interacting continuous medium of an elastic solid and an incompressible Newtonian fluid. Inter. Journal of solids and structures, (in press).

[6] Biot, M.A.,: General theory of three-dimensional consolidation. Journal of applied physics, Vol. 12, 1964, p. 155.

[7] Musket, N.: The flow of homogeneous fluids through porous media. New York: McGraw-Hill Book Co., 1967.

[8] Biot, M.A.,: Linear thermodynamics and mechanics of solids. Proc. 3rd., U.S. National Congress of applied mechanics, ASME, 1958, pp. 1-18.

[9] Biot, M.A.,: Theory of stress-strain relations in anistropic viscoelasticity and relaxation phenomena. Journal, appl. phys., Vol. 25, No. 11, 1954, pp. 1385-1391.

[10] Shapery, R.A.,: A theory of non-linear thermoviscoelastic based on irreversible thermodynamics. Proc. Fifth, U.S. Nat. Congress of Appl. Mech., New York: Am. Soc. Mech. Engrs.,1966.

[11] Valanis, K.C.,: Thermodynamics of large viscoelastic deformation. J.Methematics and physics, Vol. XLV, No. 2, 1966.

WAVE PROPAGATION

Developments in Mechanics, Vol. 6. Proceedings of the 12th Midwestern Mechanics Conference: 32

Response of a Penny-shaped Crack to Impact Waves[1]

G. T. EMBLEY[2] and G. C. Sih[3]
LEHIGH UNIVERSITY

ABSTRACT

The transient response of a penny-shaped crack subjected to time-dependent loadings is investigated using the method of integral transforms. Treated in detail is the problem of a load applied indicially on the crack surface. By means of superposition of appropriate solutions with the crack absent, the present investigation also yields the solutions to (1) the problem of diffraction of a normally incident tension wave by a penny-shaped crack, and (2) the sudden appearance of a penny-shaped crack in a uniform tensile stress field. The stress wave pattern produced by the impact load is examined and the results show that the intensity of the local stress field reaches a peak very quickly and then decreases in magnitude oscillating about the static value. The time at which the local stresses peak also corresponds to the maximum crack opening displacement, a useful quantity for estimating the crack extension force.

---

[1]This research work was performed under Contract N00014-68-A-514 with the Office of Naval Research, Department of Defense.

[2]Assistant Professor of Mechanics, Lehigh University, Bethlehem, Pennsylvania 18015.

[3]Professor of Mechanics, Lehigh University, Bethlehem, Pennsylvania 18015.

INTRODUCTION

The response of a solid medium to the sudden application of loads has been a subject of interest to many investigators in the past. However, only a few studies have been concerned with the redistribution of stresses around geometric discontinuities such as crack-like imperfections under impact. One question which may be raised in this connection is whether or not the medium weakened by flaws or cracks behaves in any critical manner as the load is applied indicially in comparison with static loading.

Certainly this is of importance to the study of earthquake phenomena and in the analysis of structural failure under dynamic loading. The problem considered here has a direct bearing on the case where the disturbance due to an explosive loading is reflected at a free surface thus causing a tensile wave to pass across the plane of a penny-shaped crack.

Major contributions to the area of dynamic crack analysis have been reviewed in "Some Elastodynamic Problems of Cracks" [1]. Some other works that are pertinent to this study are due to Flitman [2], Kostrov [3], Eshelby [4], and Achenbach [5,6]. Flitman considered the sudden appearance of a crack in an infinite elastic media. He obtained far-field stress and displacement solutions for a number of geometric configurations, this being of importance in geophysics. However, his approach did not yield information about the stress field near the crack tip, the important region when considering structural failure. Kostrov, Eshelby, and Achenbach have obtained solutions for propagation of longitudinal shear cracks under static and transient longitudinal shear loading. Their methods are based on mathematical techniques which, until now, are only applicable to the case of anti-plane strain. Ravera and Sih [7] have used a different method to study the character of the dynamic stress field near a line crack for the case of anti-plane strain. Their approach was to reduce the problem to a standard integral equation in the Laplace transform variable which was inverted by application of the Cagniard technique as rephrased and simplified by deHoop. An asymptotic stress solution was given for large values of time. This work was later extended to the more difficult but analogous problem in the plane theory of elasticity by Sih and Ravera [8]. More recently, Sih and Embley [9] proposed a three-dimensional crack model consisting of a penny-shaped flaw engulfed by transient torsional waves.

This paper is concerned with a new technique which will produce a complete near-field solution, valid for small and large values of time. The technique involves an isolation of the singular stress solution in the Laplace transformed domain. Time inversion is performed only on the near-field solution which provides the necessary and sufficient information for determining the energy release rate of the entire crack system as Sih [10] has pointed out in a recent paper. Results are presented for the problem of a penny-shaped crack subjected to a suddenly applied tensile load and are compared with previously published results for a line crack.

STATEMENT OF THE PROBLEM

In an impact problem the medium under consideration is initially at rest. Either a stress or displacement is prescribed at a boundary for $t \geq 0$. The resulting motion is determined by solving the initial-boundary value problem consisting of the partial-differential equations of motion and the initial and boundary conditions.

The geometry of the present problem is that of a penny-shaped crack in an infinite homogeneous isotropic elastic solid. The crack is of radius a and lies in the plane z=0 in a cylindrical coordinate system (r,θ,z) as shown in Fig. 1. The applied load is represented in the form of a tension wave normally incident to the plane of the crack. The displacement expression which represents the wave with no crack is

$$u_r = u_\theta = 0, \quad u_z = \frac{\sigma_0}{\lambda+2\mu}(z-c_1 t)H(t-\frac{z}{c_1}) \tag{1}$$

where σ_0 is a constant with the dimension of stress, λ and μ are Lamé's constants, $c_1 = \sqrt{(\lambda+\mu)/\rho}$ is the dilatational wave velocity with ρ being the mass density, and $H(t)$ is the Heaviside step function. The corresponding non-zero stress component is,

$$\sigma_{zz} = \sigma_0 H(t-\frac{z}{c_1}) \tag{2}$$

The solution to the problem in which the wave passes across the plane of the crack at t=0 is given by the superposition of the stress and displacement field of Eqs. (1) and (2) and the stresses and displacements due to the application of uniform pressure, $\sigma_0 H(t)$, to the surface of the penny-shaped crack in an initially unstressed medium. It should be noted that, in the region through which the wave front has passed, the resultant stress solution is identical to that due to the sudden appearance of a crack in a medium under uniform tensile stress σ_0.

The boundary conditions for the problem of normal compression on the crack surface are tractions $\sigma_0 H(t)k$ and $-\sigma_0 H(t)k$ applied to the upper and lower crack surfaces respectively and vanishing displacement at infinity (the unit vector k is directed along the positive z-axis). The initial conditions are taken to be zero. Considerable simplification is possible if the symmetry properties of the problem are taken into account. The condition of axial symmetry requires that u_θ be zero throughout the body and that u_r and u_z be independent of θ. That is

$$u_\theta = 0, \quad u_r = u_r(r,z,t), \quad u_z = u_z(r,z,t) \tag{3}$$

In turn, this form of the displacement components dictates that $\sigma_{r\theta}$ and $\sigma_{z\theta}$ be zero everywhere.

The symmetry condition across the plane z=0 allows the reformulation of the problem for the upper half-space z>0 where the stress field due to the presence of the crack is obtained by specifying appropriate boundary conditions on the plane of symmetry and at infinity. These are

$$\sigma_{rz} = 0 \text{ for all } r \text{ and } z=0, \tag{4}$$

$$\sigma_{zz} = -\sigma_0 H(t) \text{ for } r<a \text{ and } z=0, \tag{5}$$

$$u_z = 0 \text{ for } r>a \text{ and } z=0, \tag{6}$$

and

$$u_z \to 0 \text{ and } u_r \to 0 \text{ as } \sqrt{r^2+z^2} \to \infty \tag{7}$$

For the sake of generality, integral expressions will be obtained in the following section for the stress and displacement fields for the class of axisymmetric half-space problems where $\sigma_{rz} = 0$ on the plane z=0 and where displacements vanish at infinity.

STRESS AND DISPLACEMENT COMPONENTS IN AXISYMMETRIC HALF-SPACE PROBLEMS

For axial symmetry the displacement components in Eq. (3) may be expressed in terms of two scalar potentials, $\phi(r,z,t)$ and $\psi(r,z,t)$.

$$u_r = \frac{\partial \phi}{\partial r} - \frac{\partial \psi}{\partial z} \tag{8}$$

and

$$u_z = \frac{\partial \phi}{\partial z} + \frac{1}{r}\frac{\partial}{\partial r}(r\psi) \tag{9}$$

Navier's equations are satisfied if

$$\frac{1}{r}\frac{\partial}{\partial r}\left(r\frac{\partial \phi}{\partial r}\right) + \frac{\partial^2 \phi}{\partial z^2} = \frac{1}{c_1^2}\frac{\partial^2 \phi}{\partial t^2} \tag{10}$$

$$\frac{1}{r}\frac{\partial}{\partial r}\left(r\frac{\partial \psi}{\partial r}\right) - \frac{\psi}{r^2} + \frac{\partial^2 \psi}{\partial z^2} = \frac{1}{c_2^2}\frac{\partial^2 \psi}{\partial t^2} \tag{11}$$

where the dilatational and transverse wave velocities are respectively

$$c_1 = [(\lambda+2\mu)/\rho]^{1/2} \tag{12}$$

$$c_2 = (\mu/\rho)^{1/2} \tag{13}$$

Recalling that the initial conditions are zero, the Laplace transform may be applied to Eqs. (10) and (11) with the results

$$\frac{\partial^2 \phi^*}{\partial r^2} + \frac{1}{r}\frac{\partial \phi^*}{\partial r} + \frac{\partial^2 \phi^*}{\partial z^2} = \frac{p^2}{c_1^2}\phi^* \tag{14}$$

and

$$\frac{\partial^2 \psi^*}{\partial r^2} + \frac{1}{r}\frac{\partial \psi^*}{\partial r} - \frac{\psi^*}{r^2} + \frac{\partial^2 \psi^*}{\partial z^2} = \frac{p^2}{c_2^2}\psi^* \tag{15}$$

where the Laplace transform pair[11] is defined by the equations

$$f^*(p) = \int_0^\infty f(t)\exp(-pt)dt \tag{16}$$

$$f(t) = \frac{1}{2\pi i}\int_{\gamma-i\infty}^{\gamma+i\infty} f^*(p)\exp(pt)dp \tag{17}$$

By application of Hankel transforms [12] of the first and second order, it is easily shown that a solution satisfying the governing differential equations (14) and (15) and the boundary conditions of Eqs. (4) and (7) may be written

$$\phi^* = \frac{\sigma_o c_2^2}{\mu p^3}\int_0^\infty \frac{A^*(s,p)s\left(s^2+\frac{p^2}{2c_2^2}\right)}{\gamma_1}\exp(-\gamma_1 z)J_o(sr)ds \tag{18}$$

$$\psi^* = \frac{\sigma_o c_2^2}{\mu p^3}\int_0^\infty A^*(s,p)s^2\exp(-\gamma_2 z)J_1(sr)ds \tag{19}$$

where

$$\gamma_1 = \left(s^2+\frac{p^2}{c_1^2}\right)^{1/2} \tag{20}$$

$$\gamma_2 = \left(s^2+\frac{p^2}{c_2^2}\right)^{1/2} \tag{21}$$

and where, in order that displacement vanishes at infinity, the s-plane is cut so that $\gamma_1 \geq 0$ and $\gamma_2 \geq 0$ for $0 \leq s < \infty$. $J_o(sr)$ and $J_1(sr)$ are Bessel functions of zeroth and first order respectively.

The corresponding displacement and stress expressions are

$$u_z^* = -\frac{\sigma_o c_2^2}{\mu p^3} \int_0^\infty A^*(s,p)\left[s\left(s^2+\frac{p^2}{2c_2^2}\right)\exp(-\gamma_1 z) - s^3 \exp(-\gamma_2 z)\right] J_o(sr)\,ds \tag{22}$$

$$u_r^* = -\frac{\sigma_o c_2^2}{\mu p^3} \int_0^\infty A^*(s,p)\left[\frac{s^2\left(s^2+\frac{p^2}{2c_2^2}\right)}{\gamma_1} \exp(-\gamma_1 z) - s^2\gamma_2 \exp(-\gamma_2 z)\right] J_1(sr)\,ds \tag{23}$$

$$\theta^* = \sigma_{rr}^* + \sigma_{zz}^* + \sigma_{\theta\theta}^*$$

$$= \frac{\sigma_o}{p}(3-4c_o^2) \int_0^\infty A^*(s,p)\,\frac{\left(s^2+\frac{p^2}{2c_2^2}\right)^2}{\gamma_1} \exp(-\gamma_1 z) J_o(sr)\,s\,ds \tag{24}$$

$$\sigma_{zz}^* = \frac{2\sigma_o c_2^2}{p^3} \int_0^\infty A^*(s,p)\left[\frac{\left(s^2+\frac{p^2}{2c_2^2}\right)^2}{\gamma_1} \exp(-\gamma_1 z) - s^2\gamma_2 \exp(-\gamma_2 z)\right] J_o(sr)\,s\,ds \tag{25}$$

$$\sigma_{rr}^* - \sigma_{\theta\theta}^* = \frac{2\sigma_o c_2^2}{p^3} \int_0^\infty A^*(s,p)\left[\frac{\left(s^2+\frac{p^2}{2c_2^2}\right)^2}{\gamma_1} \exp(-\gamma_1 z) - s^2\gamma_2 \exp(-\gamma_2 z)\right] J_2(sr)\,s\,ds \tag{26}$$

$$\sigma_{rz}^* = \frac{2\sigma_o c_2^2}{p^3} \int_0^\infty A^*(s,p)\,s^2\left(s^2+\frac{p^2}{2c_2^2}\right)\left[\exp(-\gamma_1 z) - \exp(-\gamma_2 z)\right] J_1(sr)\,ds \tag{27}$$

FORMULATION AND SOLUTION OF DUAL INTEGRAL EQUATIONS

In order to determine a stress and displacement solution for the half-space problem that is also the solution for the crack problem under consideration, the function $A^*(s,p)$ must be evaluated so that the mixed boundary conditions specified by Eqs. (5) and (6) are met. The Laplace transforms of these expressions are

$$\sigma_{zz}^* = -\frac{\sigma_o}{p} \quad \text{for } r<a \text{ and } z=0 \tag{28}$$

$$u_z^* = 0 \quad \text{for } r>a \text{ and } z=0 \tag{29}$$

Substitution of Eqs. (25) and (22) into (28) and (29) yields the following set of simultaneous integral equations

$$\int_0^\infty A^*(s,p)t(sk)J_0(sr)s^2ds = -1, \quad r<a, \tag{30}$$

$$\int_0^\infty A^*(s,p)J_0(sr)sds = 0, \quad r>a. \tag{31}$$

where

$$t(sk) = \frac{2(sk)^2}{\left(1+\frac{c_o^2}{(sk)^2}\right)^{1/2}}\left[\left(1+\frac{1}{2(sk)^2}\right)^2\right.$$

$$\left. - \left(1+\frac{1}{(sk)^2}\right)^{1/2}\left(1+\frac{c_o^2}{(sk)^2}\right)^{1/2}\right], \tag{32}$$

$$k = \frac{c_2}{p} \tag{33}$$

$$c_o^2 = c_2^2/c_1^2 \tag{34}$$

In order that the simultaneous integral equations be in a form suitable for solution by known means the two equations must have the same physical dimensions. This may be accomplished by multiplying Eq. (30) by r and integrating from zero to r. The result is

$$\int_0^\infty A^*(s,p)t(sk)J_1(sr)sds = -\frac{r}{2}, \quad r<a \tag{35}$$

In order to accomplish the solution of the set of dual integral equations, (35) and (31), the form of the expression for u_z for r<a and z=0 must be constructed to render the expected asymptotic behavior at the crack border. That is, let

$$u_z^* = \frac{\sigma_o}{2\mu p}\int_r^a \frac{U^*(\tau,p)\tau d\tau}{(\tau^2-r^2)^{1/2}} \quad r<a, \; z=0 \tag{36}$$

where $U^*(\tau,p)$ is assumed to be continuous in τ on the closed interval [0,a].

It should be noted that Laplace inversion and integration by parts of Eq. (36) leads to the expression

$$u_z = \frac{\sigma_o}{2\mu p}\left[U(a,t)\sqrt{a^2-r^2} - \int_r^a \frac{\partial U(\tau,t)}{\partial\tau}\sqrt{\tau^2-r^2}\,d\tau\right] \quad r<a, \; z=0 \tag{37}$$

where the leading term implies that, near the edge of the crack, the displacement is proportional to the square root of the distance from the edge. Referring to Eqs. (22) and (31) the function $U^*(\tau,p)$ may be related to $A^*(s,p)$ by the equation

$$\int_0^\infty A^*(s,p)J_0(sr)sds = \begin{cases} -\int_r^a \frac{U^*(\tau,p)\tau d\tau}{(\tau^2-r^2)^{1/2}} & r<a \\ 0 & r>a \end{cases} \tag{38}$$

The left side of Eq. (38) is the Hankel transform of the zeroth order of $A^*(s,p)$. Thus, the Hankel inversion theorem may be applied with the result being an expression for $A^*(s,p)$ that identically satisfies Eq. (31).

$$A^*(s,p) = -\int_0^a [J_0(sr)r \int_r^a \frac{U^*(\tau,p)\tau d\tau}{(\tau^2-r^2)^{1/2}}]dr \tag{39}$$

If the identity [13]

$$\int_0^\tau \frac{J_0(rs)rdr}{(\tau^2-r^2)^{1/2}} = \frac{\sin(s\tau)}{s}, \tag{40}$$

is introduced, then reversal of the order of integration in Eq. (39) leads to

$$A^*(s,p) = -\frac{1}{s}\int_0^a \sin(s\tau)U^*(\tau,p)\tau d\tau \tag{41}$$

Substitution of Eq. (41) into Eq. (35) will result in a singular integral equation whose solution is $U^*(\tau,p)$. It will be the purpose of subsequent analysis to reduce this equation to a Fredholm integral equation suitable for numerical analysis. To this end, a function $f(ks)$ of order $(ks)^{-2}$ for large values of the argument is defined by the relation

$$f(ks) = b_0 - t(ks) \tag{42}$$

where

$$b_0 = 1 - c_0^2 \tag{43}$$

Then, combining Eqs. (41) and (42) with Eq. (35) the following form of Abel's integral equation [14] is obtained.

$$\int_0^r \frac{U^*(\tau,p)\tau^2 d\tau}{(r^2-\tau^2)^{1/2}} = \frac{r}{b_0}\{\frac{r}{2} + \int_0^a U^*(\tau,p)\tau[\int_0^\infty f(ks)J_1(sr)\sin(s\tau)ds]d\tau\}, \quad r<a, \tag{44}$$

where use has been made of the identity

$$\int_0^\infty \sin(s\tau)J_1(sr)ds = \begin{cases} 0 & 0<r<\tau . \\ \dfrac{\tau}{r(r^2-\tau^2)^{1/2}}, & 0\le\tau<r . \end{cases} \tag{45}$$

Inversion of Eq. (44) yields

$$b_0 \frac{\pi}{2} \tau U^*(\tau,p) = \tau + \int_0^\tau \frac{r}{(\tau^2-r^2)^{1/2}} \{\int_0^a [U^*(\zeta,p)\zeta \int_0^\infty f(ks)J_0(sr) \cdot \sin(s\zeta)ds]d\zeta\}dr \quad 0<\tau\le a \tag{46}$$

Application of Eq. (40) results in the equation

$$b_o \frac{\pi}{2} \tau U^*(\tau,p) = \tau + \int_0^a [U^*(\zeta,p)\zeta \int_0^\infty f(ks)\sin(s\tau)\sin(s\zeta)ds]d\zeta \qquad 0<\tau\leq a \tag{47}$$

Non-dimensionalize Eq. (47) by the transformation of variables,

$$\xi = \tau/a, \quad \eta = \zeta/a, \quad \alpha = sa, \tag{48}$$

and define

$$\kappa = k/a \tag{49}$$

$$\Lambda^*(\xi,\kappa) = \frac{b_o\pi\xi}{2} U^*(\xi a,p) \tag{50}$$

Then, Eq. (47) may be written in the form of a standard Fredholm integral equation of the second kind,

$$\Lambda^*(\xi,\kappa) - \int_0^1 \Lambda^*(\eta,\kappa)K(\xi,\eta)d\eta = \xi, \quad 0<\xi\leq 1 \tag{51}$$

whose kernel, being symmetric in ξ and η, is

$$K(\xi,\eta) = \frac{2}{\pi b_o} \int_0^\infty f(\alpha\kappa)\sin(\alpha\xi)\sin(\alpha\eta)d\alpha \tag{52}$$

Convergence in the numerical computations is facilitated by introduction of the function

$$g(\alpha\kappa) = f(\alpha\kappa) + \frac{D}{(\alpha\kappa)^2+E^2} \tag{53}$$

where, in order that $g(\alpha\kappa)$ be of order $(\alpha\kappa)^{-6}$ for large $(\alpha\kappa)$, D and E^2 are chosen as

$$D = \frac{1}{4} [3c_o^4-4c_o^2+3] \tag{54}$$

$$E^2 = \frac{8D}{[5c_o^6-6c_o^4+2c_o^2+1]} \tag{55}$$

Since

$$\int_0^\infty \frac{D}{(\alpha\kappa)^2+E^2} \sin(\alpha\xi)\sin(\alpha\eta)d\alpha = \frac{\pi}{2}\frac{D}{E\kappa} \exp(-E\eta/\kappa)\sinh(E\xi/\kappa), \qquad \eta>\xi \tag{56}$$

the expression for the kernel in the Fredholm integral equation may be written

$$K(\xi,\eta) = \frac{1}{b_o} \{- \frac{D}{E\kappa} \exp(-E\eta/\kappa)\sinh(E\xi/\kappa) + \frac{2}{\pi} \int_0^\infty g(\alpha\kappa)\sin(\alpha\xi)\sin(\alpha\eta)d\alpha\}, \quad \eta>\xi \tag{57}$$

Recalling Eqs. (41) and (50) the function $A^*(s,p)$ may be written in terms of the solution to the Fredholm integral equation as

$$A^*(s,p) = -\frac{2}{\pi}\frac{a^2}{b_o s}\int_0^1 \sin(sa\xi)\Lambda^*(\xi,\kappa)d\xi \tag{58}$$

TIME DEPENDENT STRESS FIELD

The dynamic stress field may be obtained by determining inverse Laplace transforms of the expressions given by Eqs. (24) through (27). The near field solution, i.e., the singular stresses near the crack edge, will be obtained here. Since the integrands in Eqs. (24) through (27) are bounded for all s any divergence of the integrals along the crack border must be due to behavior as $s\to\infty$. Hence the integrands are expanded asymptotically and those terms dominant for large s retained. If Eq. (58) is integrated by parts, it can be seen that, for the portion of the solution that is singular at the crack border, $A^*(s,p)$ may be written

$$A^*(s,p) = \frac{2a}{\pi b_o s^2}\cos(sa)\Lambda^*(1,\kappa) + \ldots \tag{59}$$

If the above equation is substituted into the equations for the stress field and if the lowest order terms in expansions of the integrands for large s are retained the resulting expressions are

$$\sigma^*_{rr} = \frac{2}{\pi}\sigma_o a\,\frac{\Lambda^*(1,\kappa)}{p}\int_0^\infty (1-sz)\cos(sa)\exp(-sz)J_o(sr)ds + \ldots \tag{60}$$

$$\sigma^*_{\theta\theta} = \frac{2}{\pi}\sigma_o a(1-2c_o^2)\,\frac{\Lambda^*(1,\kappa)}{c_o^2 p}\int_0^\infty \cos(sa)\exp(-sz)J_o(sr)ds + \ldots \tag{61}$$

$$\sigma^*_{zr} = \frac{2}{\pi}\sigma_o a\,\frac{\Lambda^*(1,\kappa)}{p}\int_0^\infty (sz)\cos(sa)\exp(-sz)J_1(sr)ds + \ldots \tag{62}$$

and

$$\sigma^*_{zz} = \frac{2}{\pi}\sigma_o a\,\frac{\Lambda^*(1,\kappa)}{p}\int_0^\infty (1+sz)\cos(sa)\exp(-sz)J_o(sr)ds + \ldots \tag{63}$$

Upon application of the Laplace inversion theorem and well known Bessel integral identities [13], the following results are obtained:

$$\sigma_{rr} = \frac{2}{\pi}M(T)\sigma_o\frac{a}{\sqrt{r_1 r_2}}\{\cos[\tfrac{1}{2}(\theta_1-\theta_2)] - \frac{az}{r_1 r_2}\sin[\tfrac{3}{2}(\theta_1-\theta_2)]\} + \ldots \tag{64}$$

$$\sigma_{\theta\theta} = \frac{4}{\pi}M(T)\nu\sigma_o\frac{a}{\sqrt{r_1 r_2}}\cos[\tfrac{1}{2}(\theta_1-\theta_2] + \ldots \tag{65}$$

$$\sigma_{zr} = \frac{2}{\pi}M(T)\sigma_o\frac{a}{\sqrt{r_1 r_2}}\{\frac{rz}{r_1 r_2}\cos[\tfrac{3}{2}(\theta_1-\theta_2)]\} + \ldots \tag{66}$$

and

$$\sigma_{zz} = \frac{2}{\pi}M(T)\sigma_o\frac{a}{\sqrt{r_1 r_2}}\{\cos[\tfrac{1}{2}(\theta_1-\theta_2)] + \frac{az}{r_1 r_2}\sin[\tfrac{3}{2}(\theta_1-\theta_2)] + \ldots \tag{67}$$

where r_1, r_2, r, θ_1, and θ_2 are shown in Fig. 1 and

$$M(T) = \frac{1}{2\pi i} \int_{\gamma-i\infty}^{\gamma+i\infty} \frac{\Lambda^*(1,\kappa)}{p} \exp(pt)dp \tag{68}$$

in which $T = c_2t/a$ is a dimensionless time variable. The procedure used for numerical evaluation of M(T) is due to Papoulis [15] and Miller and Guy [16] and is summarized in [9].

In a manner analogous to the static theory of linear fracture mechanics the local stress field at the edge of the crack may be obtained by letting $\theta_2 \to 0$ and $r_2 \to 2a$. Then, in terms of r_1 and θ_1, the stresses are

$$\sigma_{rr} = \frac{2}{\pi} M(T)\sigma_o \frac{\sqrt{a}}{\sqrt{2r_1}} \cos(\frac{\theta_1}{2}) \{1-\sin(\frac{\theta_1}{2})\sin(\frac{3\theta_1}{2})\} + \ldots \tag{69}$$

$$\sigma_{\theta\theta} = \frac{4\nu}{\pi} M(T)\sigma_o \frac{\sqrt{a}}{\sqrt{2r_1}} \cos(\frac{\theta_1}{2}) + \ldots \tag{70}$$

$$\sigma_{zr} = \frac{2}{\pi} M(T)\sigma_o \frac{\sqrt{a}}{\sqrt{2r_1}} \cos(\frac{\theta_1}{2})\sin(\frac{\theta_1}{2})\cos(\frac{3\theta_1}{2}) + \ldots \tag{71}$$

$$\sigma_{zz} = \frac{2}{\pi} M(T)\sigma_o \frac{\sqrt{a}}{\sqrt{2r_1}} \cos(\frac{\theta_1}{2}) \{1+\sin(\frac{\theta_1}{2})\sin(\frac{3\theta_1}{2})\} + \ldots \tag{72}$$

Following the definition used by Sih [10] a dynamic stress-intensity factor, which represents the strength of the stress singularity at the crack edge, may be defined here as

$$k_1(T) = \frac{2}{\pi} \sigma_o\sqrt{a}\, M(T) \tag{73}$$

which, in the limit as $T\to\infty$, reduces to the static value of

$$(k_1)_s = \frac{2}{\pi} \sigma_o\sqrt{a} \tag{74}$$

Thus, M(T) represents the ratio of the dynamic to the static stress-intensity factor.

The solution to the Fredholm integral equation for $\Lambda^*(\xi,\kappa)$ depends on the material parameter c_o where, from Eq. (34), it may be shown that

$$c_o^2 = \frac{1-2\nu}{2(1-\nu)} \tag{75}$$

Values of $\Lambda^*(1,\kappa)$ corresponding to various Poisson's ratio ν in the range $.10 \leq \nu < .35$ have been plotted as a function of κ in Fig. 2. The differences in the curves for the common structural materials Aluminum (ν=.34), Steel (ν=.29), and Glass (ν=.25), are insignificant; however, a noticeable change is observed when the difference in ν is sufficiently large, between ν=.10 and ν=.29 for example.

Once $\Lambda^*(1,\kappa)$ is known, M(T) can be obtained from Eq. (68). The curves representing the ratio $k_1/(k_1)_s$ as a function of c_2t/a for ν=.10 and ν=.29 are shown in Fig. 3. For steel, the maximum value of k_1 is observed to be 23% greater than the corresponding static value and occurs at $c_2t/a \approx 1.25$. It can also be seen that the smaller value of ν corresponds to a smaller peak value in the stress-intensity factor.

As was pointed out in [3] for the torsional case, the near field solution obtained here is valid as long as $c_2t>r_1$. Thus, the local region in which the stress-intensity factor is defined must be kept arbitrarily small in order that the solution be good for all time t (except near t=0).

CRACK SURFACE DISPLACEMENTS

Examination of the displacements of the crack surface as a function of time provides additional information regarding the nature of transient response of a solid interrupted by a crack. Making use of Eqs. (36) and (50), the expression for u_z^* on the surface of the crack is found to be

$$u_z^* = \frac{\sigma_0 a}{\mu p b_0 \pi} \int_{r/a}^{1} \frac{\Lambda^*(\xi,\kappa)d\xi}{(\xi^2-r^2/a^2)^{1/2}}, \quad r<a, \quad z=0 \tag{76}$$

The above integral may be evaluated numerically for varying κ and r/a. The displacement can then be calculated as a function of T by means of the numerical inversion process described in [9]. The position of the crack surface for values of T in the range $0<T\leq 3.0$ is shown in Fig. 4.

For small values of T there is a circular region of radius $(a-c_1t)$ on the crack surface that the wave emanating from the crack border has not reached. The uniform displacement within this region is equal to $c_1\sigma_0 t/\lambda+2\mu$, the displacement of the surface of a semi-infinite half space upon which uniform pressure $\sigma_0 H(t)$ has been applied. Thus for example, the normalized displacement for T=0.25 is equal to 0.3 everywhere within a radius of 0.54a (since $c_1t=.46$) and only begins to be affected by the crack edge for $r>0.54a$.

As c_2t/a increases further, the effect of the crack edge becomes dominant. In a manner analogous to the behavior of the stress-intensity factor the crack surface displacement reaches a maximum value greater than that found for the static case and subsequently oscillates around the static curve with decreasing amplitude. The displacement reaches a maximum at approximately the same time as maximum k_1 occurs.

CLOSURE

The most important result contained here is that the nature of the transient stress field near the crack border allows the definition of a dynamic stress-intensity factor, k_1, which may be compared to the static value. Although the actual expressions that have been obtained have been for the problem of normal compression applied to the crack surface, the addition of the displacement and stress components given in Eqs. (1) and (2) yields the solution to the problem of diffraction of a normally incident tension wave by the penny-shaped crack. It is apparent that these additional terms will not affect the singularity of the stress field or the difference between the displacements of the upper and lower crack surfaces.

The principle of superposition also allows the solution corresponding to more complicated inputs. Thus, a square wave of duration Δt may be generated by the addition to the original tension wave of a compression wave of the same magnitude which passes across the plane of the crack at a time Δt later than the tension wave. Referring again to Fig. 3 it can be seen that if Δt is greater than $.7a/c_2$ the dynamic stress-intensity factor will reach a value greater than the static k_1.

Figure 5 allows the comparison of k_1 for normal compression on the penny-shaped crack with k_1 for the same loading on the line crack under conditions of plane strain. Although dynamic magnification of the static stress-intensity factor seems to be the same for both cases, the curve for the penny-shaped crack reaches a peak value at $T \approx 1.5$, in approximately half the time required for the value of k_1 associated with the through-the-thickness line crack to reach a maximum. This difference may be attributed to the nature of the disturbances emanating from the crack border in the two cases. These stress waves were compared for the case of mode 3 loading in [3]. For the case of the straight edged tunnel crack model in normal compression treated by Sih and Ravera [2] the disturbance initially takes the form of two outgoing cylindrical waves of radius c_1t which radiate from the crack edges. The wave fronts do not interact until $c_1t=a$. As should be expected the value of k_1 reaches a maximum after interference at $c_2t \approx 3a$ in this case.

In the axisymmetric problem considered here the crack geometry produces a different stress wave pattern. The initial disturbance is toroidal in shape and interference, or interaction of stress waves, occurs much more rapidly due to the effect of curvature of the crack border. This is reflected in the fact that k_1 reaches a peak value much sooner than in the plane case.

The disturbed and undisturbed wave fronts for the problem of the tension wave normally incident on the penny-shaped crack are shown in Fig. 6. The wave front of the original tension wave is shown by the straight solid line, the diffracted wave is contained with the circle of radius c_1t, and the front of the wave reflected by the crack surface is the dotted line. The unshaded regions 1 and 2 are stress free. On region 1 the displacement is zero everywhere. In region 2, which is stress free due to superposition of the reflected and incident waves, the particles, including those on the lower crack surface, are moving at a constant velocity $2c_1\sigma_0/\lambda+2\mu$ in the negative z-direction. For steel, this velocity when normalized as in Fig. 4 is equal to $2.4\ c_2t/a$. Thus, in the associated problem of normal compression on the crack, the portion of each crack surface which the diffracted wave has not yet reached must be moving with a velocity of $1.2\ c_2t/a$. For example, for $c_2t/a = .25$, the portion of the crack surface within a distance of .54a of the origin must have a uniform displacement of 0.3. This is confirmed by the displacement curve in Fig. 4.

REFERENCES

[1] G. C. Sih, "Some Elastodynamic Problems of Cracks", Int. J. Fracture Mech. 4, 51-68, 1968.

[2] L. M. Flitman, "Waves Caused by a Sudden Crack in a Continuous Elastic Media", J. Appl. Math. Mech. (PMM) 27, 938-953, 1963.

[3] B. V. Kostrov, "Unsteady Propagation of Longitudinal Shear Cracks", J. Appl. Math. Mech. (PMM) 30, 1241-1248, 1966.

[4] J. D. Eshelby, "The Elastic Field of a Crack Extending Non-Uniformly under General Anti-Plane Loading", J. Mech. Phys. Solids

17, 177-199, 1969.

[5] J. D. Achenbach, "Crack Propagation Generated by a Horizontally Polarized Shear Wave", J. Mech. Phys. Solids 18, 245-259, 1970.

[6] J. D. Achenbach, "Brittle and Ductile Extension of a Finite Crack by a Horizontally Polarized Shear Wave", Int. J. Engng. Sci. 8, 947-966, 1970.

[7] R. J. Ravera and G. C. Sih, "Transient Analysis of Stress Waves Around Cracks under Antiplane Strain", J. Acoust. Soc. of Amer. 47, 875-881, 1970.

[8] G. C. Sih and R. J. Ravera, "Transient Response of a Finite Crack in Plane Elasticity", (to be published).

[9] G. C. Sih and G. T. Embley, "Sudden Twisting of a Penny-Shaped Crack", J. Appl. Mech. (to be published).

[10] G. C. Sih, "Dynamic Aspects of Crack Propagation", Inelastic Behavior of Solids, p. 607, McGraw-Hill, New York, 1970.

[11] G. Doetsch, Guide to Applications of Laplace Transforms, D. Van Nostrand, 1961.

[12] I. N. Sneddon, Fourier Transforms, McGraw-Hill, 1951.

[13] G. N. Watson, A Treatise on the Theory of Bessel Functions, 2nd. Ed., The MacMillan Co., New York, 1948.

[14] E. T. Whittaker and G. N. Watson, A Course of Modern Analysis, 4th Ed., Cambridge University Press, 1927.

[15] A. Papoulis, "A New Method of Inversion of the Laplace Transform", Quart. Appl. Math. 14, 405-414, 1957.

[16] M. K. Miller and W. T. Guy, "Numerical Inversion of the Laplace Transform by use of Jacobi Polynomials", SIAM J. Numer. Anal. 3, 624-635, 1966.

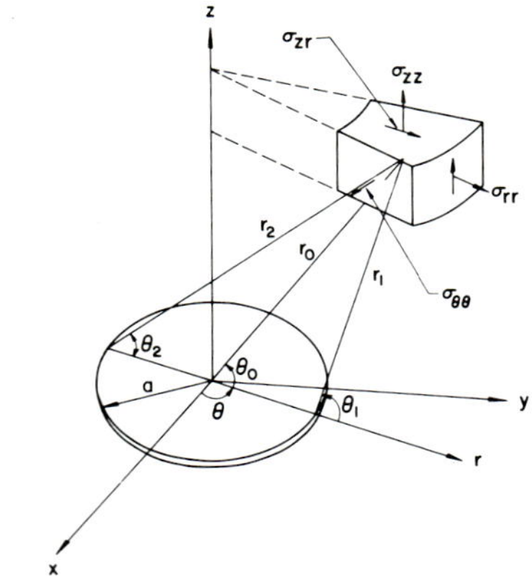

Figure 1. Stress Components and Geometry Referred to a Penny-Shaped Crack.

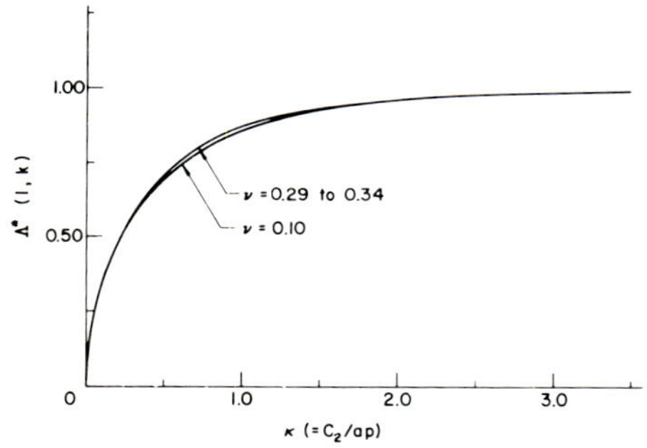

Figure 2. Solutions to the Fredholm Integral Equation.

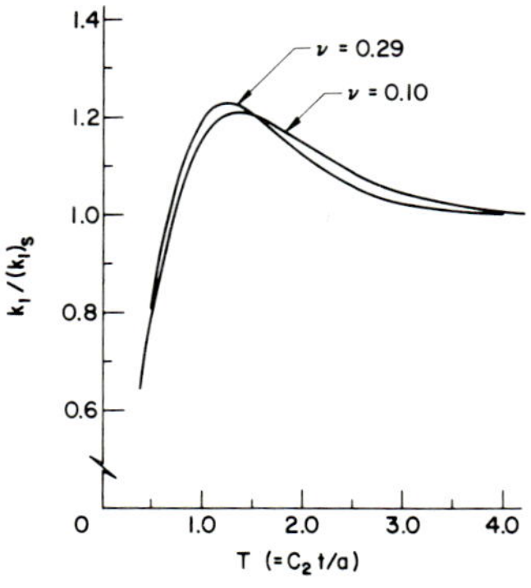

Figure 3. Poisson's Ratio Effect on Stress-Intensity Factor for Case of Normal Compression on a Penny-Shaped Crack.

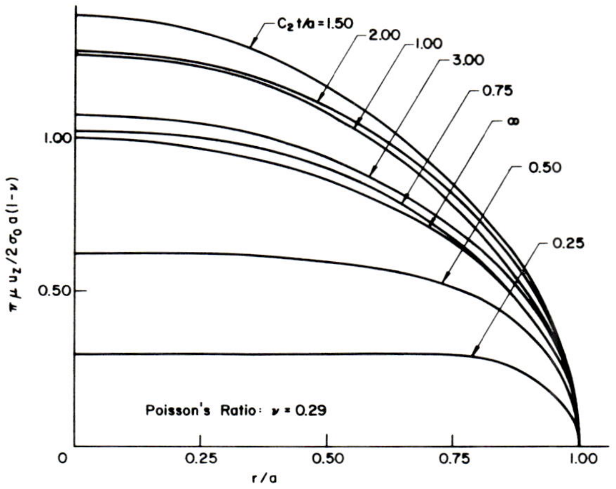

Figure 4. Crack Surface Displacements as a Function of Time.

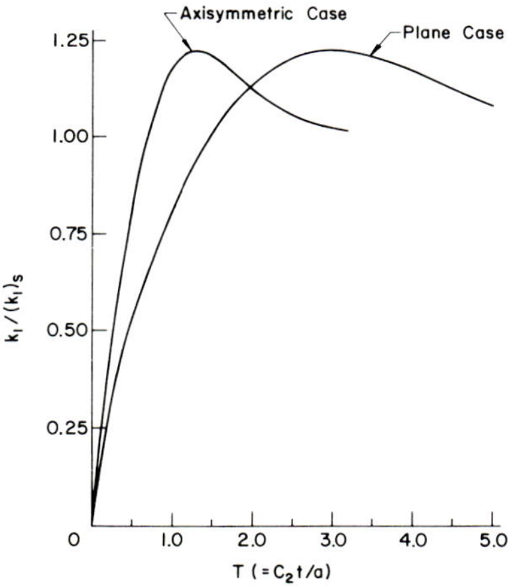

Figure 5. Comparison of Dynamic Stress-Intensity Factors for Normal Compression on the Penny-Shaped Crack of Radius a and the Line Crack of Radius 2a.

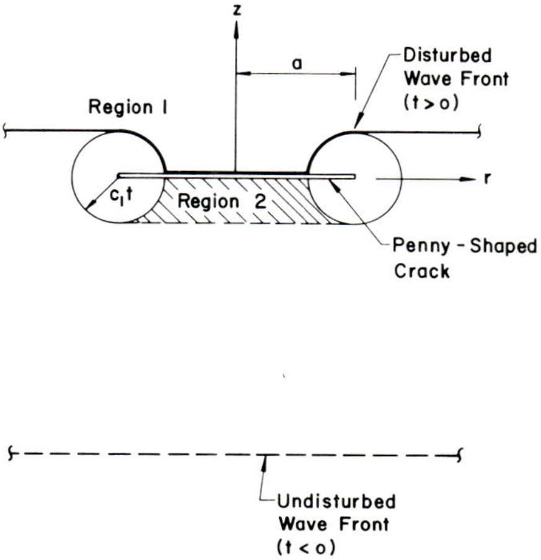

Figure 6. Undisturbed and Disturbed Wave Configurations.

Developments in Mechanics, Vol. 6. Proceedings of the 12th Midwestern Mechanics Conference: 33

The Reflection of Rayleigh Waves from a Crack Tip

L. B. FREUND
BROWN UNIVERSITY

ABSTRACT

The problem studied is the reflection of Rayleigh waves, which propagate on the faces of an open crack in an elastic solid, from the crack tip. A semi-infinite plane crack in an unbounded isotropic elastic solid is considered. The incident surface waves are time-harmonic and are prescribed in such a way that a state of plane strain exists. The problem is solved by integral transform methods. Upon reflection, the incident surface waves are found to give rise to cylindrical dilatational and shear waves, along with their associated head waves, diverging from the crack tip. In addition, reflected surface waves are generated on the faces of the crack, which is the contribution of primary interest here. In particular, numerical values of the relative amplitudes and phases of the reflected surface waves on each face of the crack are presented for several combinations of incident waves. From these results, the rate at which surface wave energy is lost due to mode conversion to body wave modes can be determined.

INTRODUCTION

It has been known for some time that elastic surface waves may be guided over significant distances by forming waveguides on surfaces of elastic solids in various ways. It now appears that a slit (an open crack of finite width and indefinite length) running through the interior of an elastic solid can also act as a waveguide for surface waves. In order to analyze the latter problem, some understanding of the interaction of surface waves propagating on the faces of the crack (or slit) with the edges of the crack must be developed. As a first step in this direction, it is the purpose of this paper to discuss the normal incidence of elastic surface waves on the edge of a half-plane crack. The results may be of interest in other areas, for example, ultrasonic flaw detection.

A homogeneous, isotropic, elastic solid is assumed to contain a semi-infinite open crack occupying a half-plane. The excitation is in the form of time-harmonic Rayleigh surface waves on one or both faces of the crack, propagating toward the tip of the crack. The reflected surface waves on the faces of the crack are sought under the conditions of plane strain. This situation may arise if surface waves are generated on the faces of the crack by transducers, or it may develop from the following diffraction process. Suppose a plane, harmonic wave is incident on the edge of a long planar crack. The plane wave is diffracted by the edge of the crack, and the diffracted field includes cylindrical dilatational and shear waves, along with their associated head waves, diverging from the crack edge [1]. The amplitude of these waves decays quite rapidly with distance from the edge. Also included in the diffracted field are surface waves on the faces of the crack, which propagate without attenuation down the crack. The only significant elements of the diffracted field to reach the far end of the crack are, therefore, the surface waves on the faces of the crack. It is the arrival of these surface waves at the far edge of the crack which is the subject under consideration here. In particular, the portion of the energy of these surface waves which is reflected from the edge as surface waves and the portion reconverted to body wave modes is determined.

The reflection of a Rayleigh wave by a rigid, smooth obstacle on the surface of a half-space, which is a special case of the problem considered here, was considered by Fredrick and Knopoff [2]. The solution to the corresponding three-dimensional problem was given by Freund [3]. In both cases, the problem was solved by integral transform methods and the Wiener-Hopf technique [4]. In [5], a method is presented for obtaining the dynamic stress concentration factor for wave motion in the vicinity of a finite crack. Several references to papers dealing with waves in a body containing a crack are given in [5].

FORMULATION

Consider an unbounded homogeneous elastic solid containing a semi-infinite crack. Cartesian coordinates $x_i = (x, y, z)$ are prescribed in the body in such a way that the crack occupies the half-plane $z = 0$, $x < 0$; see Figure 1. The faces of the crack are taken to be free of traction, that is, the crack is open. The displacement field due to the incident surface waves in this problem is uniform in the y-direction. The problem is therefore two-dimensional, and the plane strain formulation is employed. Most expressions are written in the index notation of Cartesian tensor analysis, and the range of the indices is usually 1, 3.

In the absence of body forces, the equation governing the non-zero components of the displacement vector $U_i(x, z, t)$ is

$$C_{ijk\ell}\, U_{k,j\ell} - \rho\, \ddot{U}_i = 0 \ , \ i = 1, 3 \tag{1}$$

where ρ is mass density. For the isotropic material being considered, $C_{ijk\ell}$ can be expressed in terms of the Lamé constants Λ and μ as

$$C_{ijk\ell} = \Lambda\, \delta_{ij}\delta_{k\ell} + \mu(\delta_{ik}\delta_{j\ell} + \delta_{i\ell}\delta_{jk}) \quad . \tag{2}$$

The boundary conditions which must be satisfied for the faces of the crack to be traction-free are

$$\Sigma_{33}(x, 0^{\pm}, t) = 0 \ , \ \Sigma_{13}(x, 0^{\pm}, t) = 0 \ , \quad x < 0 \tag{3}$$

where the notation $0^{\pm}$ means that the indicated stress components must vanish as $z = 0$ is approached through positive or negative values of z. The stress matrix is related to the displacement by the constitutive relation

$$\Sigma_{ij} = C_{ijk\ell} U_{k,\ell} \quad .$$

The excitation is in the form of surface waves on one face or on both faces of the crack. A steady-state situation is assumed to exist with surface waves, harmonic in time with circular frequency ω, propagating on the faces of the crack from $x = -\infty$ toward the tip of the crack. Three particular cases are considered: (*i*) surface wave on one face of the crack, incident on the crack tip, (*ii*) incident surface waves on each face, the waves having normal displacements which are of the same amplitude and are in phase, and (*iii*) incident surface waves on each face, the waves having normal displacements which are of the same amplitude but are π radians out of phase. Any other situation may be viewed as a linear combination of the above three cases. (In fact, (*ii*) and (*iii*) above may be combined to yield case (*i*). Case (*i*) is a sufficiently important special case, however, to deserve explicit consideration).

The three cases are now cast into mathematical form, making use of the notation for discontinuities,

$$\Delta U_i(x, t) \equiv U_i(x, 0^+, t) - U_i(x, 0^-, t) \quad . \tag{5}$$

The discontinuity in the incident disturbance is written as

$$\Delta U_j^{inc.} = A_j \, e^{i(\omega t - \gamma x)} \quad , \quad x < 0 \quad , \tag{6}$$

where γ is the wavenumber of Rayleigh surface waves. The various cases are then considered by assigning appropriate values to A_j :

$$(i) \qquad A_1 = i\Gamma \quad , \quad A_3 = 1 \tag{7a}$$

$$(ii) \qquad A_1 = 2 \quad , \quad A_3 = 0 \tag{7b}$$

$$(iii) \qquad A_1 = 0 \quad , \quad A_3 = 2 \tag{7c}$$

where

$$\Gamma \equiv (2\gamma^2 - \kappa_b^2)/2\gamma(\gamma^2 - \kappa_b^2)^{1/2}$$

and κ_b is the wavenumber for shear waves. All values of A_j are normalized with respect to the magnitude of the z-component of displacement on $z = 0^+$.

It is assumed that the scattered field has the same harmonic time-dependence as the incident waves, that is,

$$U_i(x, z, t) = u_i(x, z)\, e^{i\omega t} \quad , \tag{8a}$$

$$\Delta U_i(x, t) = \Delta u_i(x)\, e^{i\omega t} \quad , \tag{8b}$$

$$\Sigma_{ij}(x, z, t) = \sigma_{ij}(x, z)\, e^{i\omega t} \quad . \tag{8c}$$

The amplitudes σ_{ij} , u_i and Δu_i are determined by means of a modified form of a displacement representation theorem due to de Hoop [6]. In Section 5 of [6] the plane strain problem of an elastic solid bounded by a simple closed curve in the x, z - plane is considered. Time-dependent boundary data are specified in such a way that the elasto-dynamic problem is well-posed. It is then shown that the displacement at any interior point and at any time can be represented as a sum of two line integrals over the bounding curve (plus a body force term). The Laplace transform on time of the representation theorem is also presented.

A displacement representation theorem of use in solving the problem at hand may be easily obtained in two ways. Since the representation of [6] is valid for any time-dependence, the simple harmonic time-dependence may be made explicit and some integrals over time may be evaluated to yield a representation for $u_i(x, z)$. The representation may also be derived by substituting $i\omega$ for the Laplace transform parameter in the transformed representation theorem of [6]. In either case, the result is (neglecting body force terms)

$$\begin{aligned} u_i(x) = & \int_C C_{jk\ell m}\, \Gamma_{ij}(x, \xi)\, u_{\ell,m}(\xi) n_k(\xi)\, d\xi \\ & + \frac{\partial}{\partial x_m} \int_C C_{jk\ell m}\, \Gamma_{i\ell}(x, \xi)\, u_j(\xi)\, n_k(\xi)\, d\xi \end{aligned} \tag{9}$$

where C is the boundary curve, $x = (x, z)$ is any point in the body, $\xi = (\xi, \zeta)$ is a point on C , n is the outward unit normal to C , and

$$\begin{aligned} \Gamma_{ij}(x, \xi) = \frac{i}{4\rho\omega^2} \Big\{ & \frac{\partial^2}{\partial x_i \partial x_j} [H_o^{(2)}(\kappa_a r) - H_o^{(2)}(\kappa_b r)] \\ & - \delta_{ij}\, \kappa_b^2\, H_o^{(2)}(\kappa_b r) \Big\} \end{aligned} \tag{10}$$

$$r = [(\xi - x)^2 + (\zeta - z)^2]^{1/2} \tag{11}$$

In (10), $H_o^{(2)}$ is the Hankel function of the second kind and κ_a is the wavenumber of dilatational waves in the elastic solid.

For the problem being considered here, the curve C is made up of the straight segment coinciding with the side of the crack $z = 0^-$, a circle of infinitesimal radius around the crack tip, the straight segment coinciding with the face of the crack $z = 0^+$, and a circle of infinitely large radius centered at the crack tip, as shown in Figure 1. In view of the boundary conditions (3), the representation (9) reduces to

$$u_i(x, z) = -\frac{\partial}{\partial x_m} \int_{-\infty}^{0} C_{j3\ell m}\, \Gamma_{i\ell}(x, z, \xi, 0)\, \Delta u_j(\xi)\, d\xi \quad . \tag{12}$$

As is indicated by the form of (12), the contributions from each of the circular segments of C are vanishingly small. Some of the steps in going from (9) to (12) are outlined in [6]. The representation (12) is now employed to obtain a solution to the posed problem by Laplace transform methods and the Wiener-Hopf technique.

REFLECTED SURFACE WAVES ON $z = 0^{\pm}$

The general scheme for obtaining the reflected surface waves on the faces of the crack, due to the incident surface waves, is the following. A bilateral Laplace transform is defined over the spatial variable x. This transform is then applied to the representation formula (12), to the stress-strain relation (4) and to the boundary conditions (3). The transformed representation of displacements is substituted into the transformed stress-strain relation, and the boundary conditions are then imposed. The result of these steps is a pair of equations relating functions of the complex transform parameter which, with proper interpretation, are of the standard Wiener-Hopf type [4]. Once these equations are solved, the transformed displacements may be inverted.

The bilateral Laplace transform of a function is denoted by a bar over the function and is defined by

$$\bar{\phi}\,(\lambda, z) = \int_{-\infty}^{\infty} e^{-\lambda x}\, \phi\,(x, z)\, dx \quad . \tag{13}$$

The anticipated form of the diffracted fields suggests that all quantities are dominated by a term like $(x)^{-1/2} \exp(-i\kappa_a x)$ as $x \to +\infty$ and like $\exp(i\gamma x)$ as $x \to -\infty$, which implies that all transforms converge only on the imaginary axis. To apply the Wiener-Hopf method, however, a strip of convergence is required. The usual means of achieving this strip is to assume the material to be slightly dissipative. It is well-known from the linear theory of viscoelasticity that this is accomplished simply by replacing the constants Λ and μ in (2) by their corresponding complex moduli, which in general are functions of frequency. All wavenumbers are then complex, with negative imaginary parts, which in turn implies that all field variables are dominated by a decaying exponential as $|x|$ becomes large. With this result, the existence of the requisite strip of convergence of (13) can be established. The Wiener-Hopf method is then applied, after which the dissipation is assumed to vanish [4, 7]. This procedure is well-established in the literature and the details will not be discussed here. A main result of this scheme which must be mentioned is that the inversion path for the transform (13) is the imaginary axis, approached from the right in the lower half-plane and from the left in the upper half-plane, as shown in Figure 2. The details of how the appropriate inversion path is determined are well-known, and may be found in [4], for example.

The transform is first applied to (12), making use of the familiar result for transforming a convolution integral. The transform of the Hankel function is [8]

$$\int_{-\infty}^{\infty} e^{-\lambda x} H_o^{(2)}[\kappa_a(x^2 + z^2)^{1/2}]\, dx = -\frac{2}{\alpha} e^{i\alpha|z|} \quad , \tag{14}$$

where $\alpha = (\kappa_a^2 + \lambda^2)^{1/2}$ and $Im(\alpha) \geq 0$. Making use of (14), the transform of (12) is

$$\bar{u}_i(\lambda, z) = M_{ij}(\lambda, z)\, \overline{\Delta u_j}(\lambda) \quad , \tag{15}$$

where

$$M_{11} = \pm[-\, 2\lambda^2 \exp(i\alpha|z|) + (\kappa_b^2 + 2\lambda^2) \exp(i\beta|z|)]/2\kappa_b^2 \tag{16a}$$

$$M_{12} = i\lambda[-(\kappa_b^2 + 2\lambda^2) \exp(i\alpha|z|) + 2\alpha\beta \exp(i\beta|z|)]/2\alpha\kappa_b^2 \tag{16b}$$

$$M_{21} = i\lambda[2\alpha\beta \exp(i\alpha|z|) - (\kappa_b^2 + 2\lambda^2) \exp(i\beta|z|)]/2\beta\kappa_b^2 \tag{16c}$$

$$M_{22} = \pm[(\kappa_b^2 + 2\lambda^2) \exp(i\alpha|z|) - 2\lambda^2 \exp(i\beta|z|)]/2\kappa_b^2 \tag{16d}$$

$$\beta = (\kappa_b^2 + \lambda^2)^{1/2} \quad , \quad Im(\beta) \geq 0 \quad , \tag{17}$$

$$\overline{\Delta u_j}(\lambda) = \int_{-\infty}^{0} \Delta u_j(x)\, e^{-\lambda x}\, dx \quad . \tag{18}$$

In (16) the upper signs apply for $z > 0$ and the lower signs for $z < 0$. The result of applying the transform to the stress-strain relation is

$$\frac{1}{\mu} \bar{\sigma}_{13}(\lambda, z) = \frac{\partial \bar{u}_1}{\partial x_3}(\lambda, z) + \lambda u_3(\lambda, z) \tag{19a}$$

$$\frac{1}{\mu} \bar{\sigma}_{33}(\lambda, z) = \lambda\left(\frac{\kappa_b^2}{\kappa_a^2} - 2\right) \bar{u}_1(\lambda, z) + \frac{\kappa_b^2}{\kappa_a^2} \frac{\partial \bar{u}_3}{\partial x_3}(\lambda, z) \quad . \tag{19b}$$

Finally, the transformed boundary conditions are

$$\bar{\sigma}_{i3}(\lambda, 0^{\pm}) = R_i(\lambda) \tag{20}$$

where R_i is an unknown function which is analytic on, and to the right of, the inversion contour in the λ-plane. Implicit in (20) is the condition that the traction is continuous across $z = 0$ for $x > 0$.

The Wiener-Hopf equations are obtained by substituting (15) and (20) into (19), and the result is

$$\frac{1}{\mu} R_1(\lambda) = \frac{i\, \overline{\Delta u_1}(\lambda)\, d(\lambda)}{2\kappa_b^2\, \beta} \quad , \tag{21a}$$

$$\frac{1}{\mu} R_3(\lambda) = \frac{i\, \overline{\Delta u_3}(\lambda)\, d(\lambda)}{2\kappa_b^2\, \alpha} \tag{21b}$$

where $d(\lambda)$ is the Rayleigh wave function

$$d(\lambda) \equiv (2\lambda^2 + \kappa_b^2)^2 - 4\lambda^2\alpha\beta \quad . \tag{22}$$

The equation $d(\lambda) = 0$ has two roots in the cut λ-plane (for all realistic material parameters) located at $\lambda = \pm i\gamma$. It is helpful at this point to write $\overline{\Delta u_j}$ as a sum of two terms

$$\overline{\Delta u_j}(\lambda) = - A_j/(\lambda + i\gamma) + L_j(\lambda) \quad , \tag{23}$$

which separates out the discontinuity in displacement due to the incident waves. The function L_j is the transform of the discontinuity in displacement across $z = 0$ due to outgoing waves only and this function is, therefore, analytic on, and to the left of, the inversion path in the λ-plane [7].

The functions α and β may be factored into products of sectionally analytic functions $\alpha_+\alpha_-$ and $\beta_+\beta_-$ where

$$\alpha_\pm = (\lambda \pm i\kappa_a)^{1/2} \quad , \quad \beta_\pm = (\lambda \pm i\kappa_b)^{1/2} \quad . \tag{24}$$

The subscript plus (minus) indicates that the function is analytic on, and to the right (left) of, the inversion path in the λ-plane. It is also convenient to introduce the auxiliary function $D(\lambda)$ defined by

$$D(\lambda) \equiv d(\lambda)/\kappa(\lambda^2 + \gamma^2) \quad , \tag{25}$$

where $\kappa = 2(\kappa_b^2 - \kappa_a^2)$. This function has neither zeros nor poles in the cut λ-plane, and $D(\lambda) \to 1$ as $|\lambda| \to \infty$. A product factorization of D into sectionally analytic functions D_+ and D_- has been presented many times in the literature, for example in [2], and it is merely noted here that such a factorization can be accomplished.

Relation (21a) may then be written

$$\frac{2\kappa_b^2}{i\mu\kappa} \frac{\beta_+ R_1(\lambda)}{D_+(\lambda)} = [-A_1 + (\lambda + i\gamma) L_1(\lambda)] \frac{(\lambda - i\gamma) D_-(\lambda)}{\beta_-} \quad . \tag{26}$$

Applying the usual analytic continuation argument, each side of (26) represents one and the same entire function, say $E(\lambda)$. From the conditions that the displacement must be continuous and that the strain energy density must be integrable at the tip of the crack, it can be shown [3] that $E(\lambda) = o(\lambda^{1/2})$ as $|\lambda| \to \infty$. The extended Liouville theorem then implies that $E(\lambda) = E_o$, a constant, whose value can be determined by setting $\lambda = -i\gamma$ in the right side of (26). The final result of solving (21a) is

$$\overline{\Delta u}_1(\lambda) = \frac{A_1(2i\gamma) \beta_-(\lambda) D_-(-i\gamma)}{(\lambda^2 + \gamma^2) \beta_-(-i\gamma) D_-(\lambda)} \quad . \tag{27}$$

Similarly, (21b) yields

$$\overline{\Delta u}_3(\lambda) = \frac{A_3(2i\gamma) \alpha_-(\lambda) D_-(-i\gamma)}{(\lambda^2 + \gamma^2) \alpha_-(-i\gamma) D_-(\lambda)} \quad . \tag{28}$$

The solution of the Wiener-Hopf equations is thus complete and the transforms may be inverted.

It is worth noting at this point that the function $\bar{u}_i$ is not analytic in a strip of the λ-plane, even when the material is taken to be dissipative. The reason for this is that the displacement u_i represents both incoming (incident) and outgoing (diffracted) waves. This does not mean that the inverse of $\bar{u}_i$ cannot be obtained, however. Since $\bar{u}_i$ is known explicitly as a function of λ , it can be separated into parts representing the incident and the diffracted waves, and these parts can be inverted separately. The desired reflected surface waves are elements of the total diffracted field. The separation of $\bar{u}_i$ is effected simply by substituting (23) into (15). The transform of the displacement of the diffracted waves is then given by

$$\bar{u}_i^d(\lambda, z) = M_{ij}(\lambda, z) L_j(\lambda) \quad . \tag{29}$$

Attention will be limited to the z component of the surface displacement on $z = 0^\pm$. This is sufficient for the discussion of surface waves because a single component of surface displacement completely characterizes a Rayleigh surface wave, if the propagation direction is specified. The inversion integral for $u_3^d(x, 0^{\pm})$ is

$$u_3^d(x, 0^{\pm}) = \frac{1}{2\pi i} \int_{-i\infty}^{i\infty} M_{3j}(\lambda, 0^{\pm})\, L_j(\lambda)\, e^{\lambda x}\, d\lambda \quad , \tag{30}$$

where the integration path is shown in Figure 2. The integrand of (30) has branch points at $\lambda = \pm i\kappa_a$, $\pm i\kappa_b$ and it has a simple pole at $\lambda = i\gamma$. For $x > 0$ the path of integration may be closed by an arc at infinity in the left half-plane. By applying Cauchy's theorem u_3^d may be written as a branch line integral, which represents a contribution due to body waves for $x > 0$. Similar considerations for $x < 0$ show that u_3^d may be written as the sum of the residue of the pole at $\lambda = i\gamma$ plus a branch line integral, the latter again representing a contribution due to body waves. The residue, on the other hand, represents the reflected surface waves, which are the waves of primary interest here. Denoting the residue by $u_3^s(x, 0^{\pm})$, the reflected surface waves are given by

$$u_3^s(x, 0^{\pm}) = \frac{1}{2}\frac{D_-(-i\gamma)}{D_-(i\gamma)} \left\{ \pm A_3 \frac{\alpha_-(i\gamma)}{\alpha_-(-i\gamma)} - i\Gamma A_1 \frac{\beta_-(i\gamma)}{\beta_-(-i\gamma)} \right\} e^{i(\pi + \gamma x)} \tag{31}$$

All quantities in (31) have been defined except the ratio $D_-(-i\gamma)/D_-(i\gamma)$, which is given in [3] as

$$\frac{D_-(-i\gamma)}{D_-(i\gamma)} = \exp\left\{\frac{2\gamma}{\pi}\int_{\kappa_a}^{\kappa_b} \operatorname{atan}\left(\frac{4\xi^2(\xi^2 - \kappa_a^2)^{1/2}(\kappa_b^2 - \xi^2)^{1/2}}{(2\xi^2 - \kappa_b^2)^2}\right)\frac{d\xi}{\xi^2 - \gamma^2}\right\} \tag{32}$$

Thus it can be seen that, even though the inverse transforms for the complete wave motion are far too complicated to allow evaluation, the surface wave contribution may be extracted in a relatively simple form.

DISCUSSION

The analytical result (31) which represents the reflected surface waves has been numerically evaluated for the case when Poisson's ratio is 0.25. The wavenumbers for this case are related by $\kappa_b^2 = 3\,\kappa_a^2$ and $\gamma^2 = 3.549\,\kappa_a^2$. To isolate the amplitude and phase factors of the reflected waves, equation (31) is written in the form

$$u_3^s(x, 0^{\pm}) = B^{\pm}\, e^{i(\gamma x + \psi^{\pm})} \tag{33}$$

where $B^{\pm}$ and $\psi^{\pm}$ are real numbers, the former being non-negative. In the format of (33) the numerical results are presented in the Table for the three cases.

For case (*i*), which is the case of a single incident wave of unit amplitude on $z = 0^+$, the reflected wave on $z = 0^+$ has the comparatively small amplitude 0.085. The amplitude of the reflected wave on $z = 0^-$ is 0.185, which is more than twice as large. For case (*ii*), which is the case when surface waves whose z-components of displacement are symmetric with respect to $z = 0$ are incident, the reflected waves also have the same symmetry. If each incident wave has unit amplitude, the amplitude of each reflected wave is 0.100. The last case, in which the z-components of displacement of the incident surface waves is anti-symmetric with respect to $z = 0$, is essentially a special case of the problem considered in [3]. For incident waves of unit amplitude, the reflected waves are antisymmetric and each has an amplitude 0.270, which is significantly larger than the former case.

It is clear from the Table that case (*i*) is a linear combination of cases (*ii*) and (*iii*), for example, $[0.100\ e^{-i\pi/2} + 0.270\ e^{i\pi/2}]/2 = 0.085\ e^{i\pi/2}$. The factor 1/2 appears on the left side because all amplitudes were normalized with respect to the amplitude of the z-component of the incident wave on $z = 0^+$. Also, the results of cases (*ii*) and (*iii*) can be built up from case (*i*) and its reflections in the x, z-plane and the x, y-plane. For example, to obtain B^+ and ψ^+ for case (*iii*) the result of case (*i*) is added to the result of reflecting case (*i*) in the x, z and x, y-planes, that is, $0.085\ e^{i\pi/2} - 0.185\ e^{-i\pi/2} = 0.270\ e^{i\pi/2}$. The minus sign appears on the left side because only one reflection is made in a plane containing the z-axis, so that the sign of the z-component of displacement changes.

The z-component of displacement adequately describes the reflected surface waves being considered. It is not, however, the only useful parameter in determining the degree of mode conversion from surface waves to body waves upon reflection at the crack tip. It would appear that an energy parameter would also be useful for this purpose. A convenient energy parameter is the square of the amplitude of the z-displacement. The reason that the energy parameter may be derived from a single component of displacement is that the <u>ratio</u> of amplitudes of the components of the displacement for a surface wave <u>is a fixed</u> quantity for a given material. The value of this energy parameter is also included in the Table. These values are also normalized since the energy of a wave of unit amplitude is one. Thus, only about 4.1 percent of the incident surface wave energy is reflected as surface waves in (*i*), about 1.0 percent in (*ii*), and about 7.3 percent in (*iii*).

REFERENCES

[1] A. W. Maue: Die Beugung elastischer Wellen an der Halbebene, Zeitschrift für angewandte Mathematik und Mechanik, Vol. 33, 1953, pp. 1-10.

[2] R. W. Fredricks, and L. Knopoff: The Reflection of Rayleigh Waves by a High Impedance Obstacle on a Half-Space, Geophysics, Vol. 25, 1960, pp. 1195-1202.

[3] L. B. Freund: Guided Surface Waves on an Isotropic Elastic Half-Space, Brown University Report ARPA E72, May, 1970, to appear in Journal of Applied Mechanics.

[4] B. Noble: The Wiener-Hopf Technique, Pergamon, New York, 1958.

[5] G. C. Sih, and J. F. Loeber: Wave Propagation in an Elastic Solid with a Line of Discontinuity or Finite Crack, Quarterly of Applied Mathematics, Vol. 27, 1969, pp. 193-213.

[6] A. T. de Hoop: Representation Theorems for the Displacement in an Elastic Solid and their Application to Elastodynamic Diffraction Theory, D. Sc. Thesis, Technische Hogeschool, Delft, 1958.

[7] L. B. Freund and J. D. Achenbach: Waves in a Semi-Infinite Plate in Smooth Contact with a Harmonically Disturbed Half-Space, International Journal of Solids and Structures, Vol. 4, 1968, pp. 605-621.

[8] G. N. Watson: Theory of Bessel Functions, Cambridge University Press, Second Edition, 1966, p. 416.

TABLE

	(i)	*(ii)*	*(iii)*
B^+	0.085	0.100	0.270
B^-	0.185	0.100	0.270
ψ^+	$\pi/2$	$-\pi/2$	$\pi/2$
ψ^-	$-\pi/2$	$-\pi/2$	$-\pi/2$
E^+	0.0072	0.0100	0.0729
E^-	0.0342	0.0100	0.0729

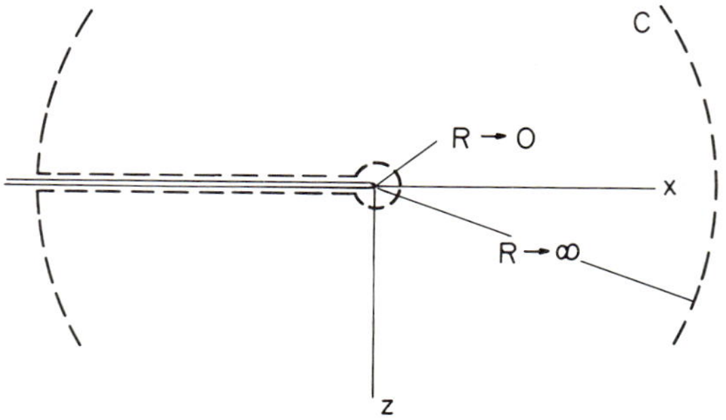

Fig. 1. The physical x, z-plane, showing the curve C of the representation theorem, equation (9).

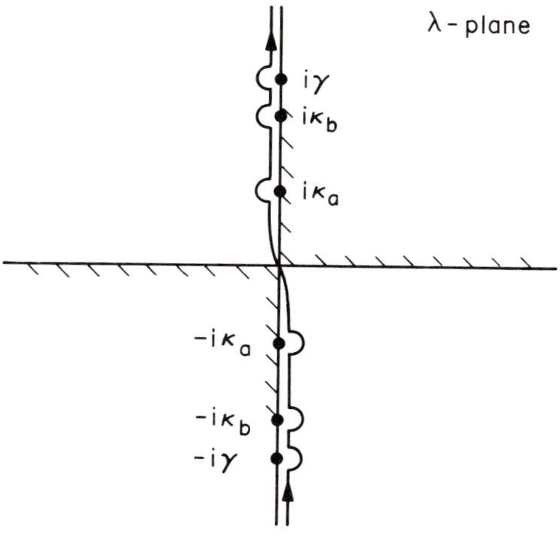

Fig. 2. The complex λ-plane, showing the inversion path for the bilateral Laplace transform.

Developments in Mechanics, Vol. 6. Proceedings of the 12th Midwestern Mechanics Conference: 34

Rayleigh Wave Diffraction in an Elastic Quarter-space with a Rigid Vertical Boundary

L. W. SCHMERR, JR.
IOWA STATE UNIVERSITY

S. A. THAU
ILLINOIS INSTITUTE OF TECHNOLOGY

ABSTRACT

A dynamic elastic quarter-space problem, in which the horizontal surface is struck by a normal line load and the vertical boundary is bonded to a rigid wall, is considered. Integral transforms are applied on space and time, leading to a set of coupled, singular integral equations which are solved iteratively. Both analytical and numerical results in the first iteration are obtained for the stresses on the rigid wall due to the incoming Rayleigh wave and for the surface displacements of the reflected Rayleigh wave.

INTRODUCTION

This paper considers an elastic quarter-space which is subjected to a normal line load of arbitrary time dependence on its horizontal surface and whose vertical surface is clamped to a rigid wall (Fig. 1). This "rigid-clamped" (RC) quarter-space is useful as a model of an abrupt change in topography of the earth's surface, and it can be used to study the interaction of seismic waves near the earth's surface with a buried structure such as the wall of a building foundation.

The first study of wave propagation in an elastic quarter-space was presented by Lapwood [1]* in 1961. The horizontal boundary was subjected to a normal line load while the vertical boundary was free of tractions. By applying spatial Laplace transforms and a Fourier transform on time to the governing equations, Lapwood obtained two coupled pairs of singular integral equations for the displacements along the boundaries. He solved these equations by iteration and obtained results in the first approximation for the Rayleigh wave that is transmitted around the corner.

*Numbers in brackets refer to the references at the end of the paper.

In 1966 Viswanathan [2] extended Lapwood's work by treating the problem of two dissimilar quarter-spaces welded together along a common vertical interface and with a concentrated line load acting on one of the horizontal boundaries. Viswanathan used Fourier integral representations for the displacement potentials and was led to a complicated set of coupled, singular integral equations for certain unknown functions. Following Lapwood's iteration procedure, Viswanathan obtained approximate solutions to these equations and discussed in detail the various waves in the first approximation. However, Viswanathan did not calculate expressions for the stresses nor did he present numerical results for the various waveforms except for the Stonely interface waves. In addition, Viswanathan's expressions for the reflected Rayleigh wave contain integrals whose integrands have poles along their paths of integration, preventing one from making numerical calculations. Furthermore, results for the limiting case in which one of the quarter-spaces becomes infinitely rigid (RC quarter-space problem) cannot be deduced from Viswanathan's solution because certain improper integrals in his formulas become divergent. In this paper, the aforementioned difficulties in Viswanathan's solution are resolved for the RC quarter-space problem and we obtain and evaluate numerically approximate results for both the stresses on the rigid wall due to the incoming Rayleigh wave and for the surface displacements of the reflected Rayleigh wave.

In addition to these two investigations [1], [2], solutions are available in the special case of the "rigid-smooth" (RS) quarter-space problem where the vertical boundary is pressed against an ideally smooth rigid wall. Then an exact solution can be obtained by the method of images, a technique used by a number of authors for a variety of RS quarter-space problems [3]-[6]. Finally, it should be mentioned that Alterman and Rotenberg [7] used a finite-difference scheme to study waves in a quarter-space with a buried line source. They presented results for the displacements both within and along the free boundaries of the quarter-space and discussed the various wave contributions.

GOVERNING EQUATIONS

Consider a homogeneous, isotropic elastic quarter-space which occupies the region $x > 0$, $y > 0$, $-\infty < z < \infty$ in a Cartesian coordinate system (Fig. 1). The quarter-space, initially at rest, is subjected at time $t = 0$ to a concentrated line load of arbitrary time dependence on the surface $y = 0$ at a distance, a, from the corner, while the vertical boundary is rigidly fixed. We assume that plane strain conditions hold. Then, introducing the Laplace transform

$$\bar{f}(x,y,s) = \int_0^\infty f(x,y,t)e^{-st}dt,$$

and applying it to the governing equations of elastodynamics [8], we obtain the transformed equations of motion as

$$\bar{\sigma}_{\alpha\beta,\beta} = s_2^2\bar{u}_\alpha \quad (\alpha = 1,2) \tag{1}$$

and the transformed stress-displacement gradient relations

$$\bar{\sigma}_{\alpha\beta} = (\kappa^2 - 2)\bar{u}_{\gamma,\gamma}\delta_{\alpha\beta} + \bar{u}_{\alpha,\beta} + \bar{u}_{\beta,\alpha} \quad (\alpha,\beta = 1,2). \tag{2}$$

Here $\sigma_{\alpha\beta} = \tau_{\alpha\beta}/\mu$ $(\alpha,\beta = 1,2)$, where $\tau_{\alpha\beta}$ is the stress tensor, u_α is the displacement of the medium in the x_α-direction, and $\delta_{\alpha\beta}$ is the Kronecker delta. Throughout this paper we will take $u_1 = u$, $u_2 = v$, and, as is customary, $x_1 = x$, $x_2 = y$, $x_3 = z$. The symbols s_m, v_m, s_r, v_r and κ are defined by $s_m = s/c_m$, $v_m = 1/c_m$ $(m = 1,2)$, $s_r = s/c_r$, and $\kappa = c_1/c_2$, where c_1, c_2, c_r are the wavespeeds for compressional, shear, and Rayleigh waves, respectively.

The transformed boundary conditions for the RC quarter-space problem similarly are

$$\bar{\sigma}_{\alpha 2}(x,0,s) = -\delta_{\alpha 2}\bar{P}(s)\delta(x-a) \tag{3}$$

$$\bar{u}_{\alpha}(0,y,s) = 0 \qquad (\alpha = 1,2) \tag{4}$$

where $\delta(\cdot)$ is the Dirac delta function, denoting the line load of Fig. 1, and $\bar{P}(s)$ is the Laplace transform of the loading function, $P(t)$.

DERIVATION OF INTEGRAL EQUATIONS

The transformed governing equations of the RC quarter-space problem are equations (1)-(4). Their solution presents considerable difficulties. In this section we apply spatial Fourier sine and cosine transforms to these equations and demonstrate how this procedure leads to a set of coupled, singular integral equations for unknown boundary displacements and stresses in terms of which the complete solution to the RC quarter-space problem can be expressed.

The sine and cosine transforms in x of a function $f(x,y,t)$ are defined as

$$\begin{Bmatrix} f^s(\xi,y,t) \\ f^c(\xi,y,t) \end{Bmatrix} = \int_0^\infty f(x,y,t) \begin{Bmatrix} \sin \xi x \\ \cos \xi x \end{Bmatrix} dx$$

whereas the sine and cosine transforms on y are denoted, through similar relations, by $f^s(x,\eta,t)$ and $f^c(x,\eta,t)$, respectively. Consistent with the above notation, we shall also write double sine transforms in x and y as $f^{ss}(\xi,\eta,t)$, double cosine transforms as $f^{cc}(\xi,\eta,t)$, etc.

If we take (cos, cos) transforms on (x,y), respectively, for Eq. (1) with $\alpha = 1$ and, to be consistent, (sin, sin) transforms on (x,y) for $\alpha = 2$ and then apply the appropriate sine and cosine transforms to Eq. (2), Eq. (3) for $\alpha = 1$, and Eq. (4) for $\alpha = 2$, we obtain two equations for $\bar{u}^{cc}$ and $\bar{v}^{ss}$ which can be solved simultaneously to yield

$$\bar{u}^{cc}(\xi,\eta,s) = \frac{-(\xi^2 + \kappa^2\eta^2 + s_2^2)\bar{\sigma}_{11}^c(0,\eta,s)}{\Delta} \\ \frac{-\xi[(\kappa^2-2)\xi^2 - \kappa^2\eta^2 + (\kappa^2-2)s_2^2]\bar{v}^s(\xi,0,s)}{\Delta} \tag{5}$$

and

$$\bar{v}^{ss}(\xi,\eta,s) = \frac{-(\kappa^2-1)\xi\eta\bar{\sigma}_{11}^c(0,\eta,s)}{\Delta} \\ \frac{+\eta[(3\kappa^2-2)\xi^2 + \kappa^2\eta^2 + \kappa^2 s_2^2]\bar{v}^s(\xi,0,s)}{\Delta} \tag{6}$$

where $\Delta = \kappa^2(\xi^2 + \eta^2 + s_1^2)(\xi^2 + \eta^2 + s_2^2)$. Inverting Eq. (5) on ξ and setting $x = 0$, we obtain, with the replacement $\xi \to \lambda$,

$$\bar{u}^c(0,\eta,s) = \frac{-\bar{\sigma}_{11}^c(0,\eta,s)}{\pi\kappa^2}\int_{-\infty}^{\infty} \frac{(\lambda^2 + \kappa^2\eta^2 + s_2^2)\,d\lambda}{(\lambda^2+\eta^2+s_1^2)(\lambda^2+\eta^2+s_2^2)}$$

$$- \frac{2}{\pi\kappa^2} \int_0^\infty \frac{\lambda[(\kappa^2 - 2)\lambda^2 - \kappa^2\eta^2 + (\kappa^2 - 2)s_2^2]\overline{v}^s(\lambda,0,s)\,d\lambda}{(\lambda^2 + \eta^2 + s_1^2)(\lambda^2 + \eta^2 + s_2^2)} \tag{7}$$

The first integral in Eq. (7) can be evaluated exactly by residues and when Eq. (4) with $\alpha = 1$ is applied, we find

$$\overline{\sigma}_{11}^c(0,\eta,s) = \frac{2s_1^2\sigma_2\sqrt{\eta^2 + s_2^2}}{\pi L(\eta)} \int_0^\infty \lambda\overline{v}^s(\lambda,0,s) \frac{[(\kappa^2 - 2)\lambda^2 - \kappa^2\eta^2 + (\kappa^2 - 2)s_2^2]d\lambda}{(\lambda^2 + \eta^2 + s_1^2)(\lambda^2 + \eta^2 + s_2^2)} \tag{8}$$

where $L(\eta) = \eta^2 - \sigma_1\sigma_2\sqrt{\eta^2 + s_1^2}\sqrt{\eta^2 + s_2^2}$

$\sigma_m = \text{sgn}[\text{Re}\sqrt{\eta^2 + s_m^2}]$ $(m = 1,2)$

Here Re stands for "real part of" and the symbol "sgn" in the σ_m factors stands for "sign of." The σ_m factors arise from integration of the first term in Eq. (7) because in later inversions on η and s, in which the Rayleigh wave contributions alone are extracted, it will be necessary to consider these quantities as being complex. Then the σ_m factors (which are simply unity when the transform variables are real) guarantee the convergence of the inversion integrals [9]. To apply the only remaining boundary condition, Eq. (3) with $\alpha = 2$, we first place Eqs. (5) and (6) into the transformed expression for $\overline{\sigma}_{22}$ obtained from Eq. (2), invert this equation on η, and set $y = 0$, giving finally

$$\overline{v}^s(\xi,0,s) = \frac{\overline{P}(s)\beta_1\sqrt{\xi^2 + s_1^2}\,s_2^2 \sin \xi a}{R(\xi)} + \frac{2\xi s_1^2\beta_1\sqrt{\xi^2 + s_1^2}}{\pi R(\xi)} \int_0^\infty \frac{\overline{\sigma}_{11}^c(0,\lambda,s)[(\kappa^2 - 2)\xi^2 - \kappa^2\lambda^2 + (\kappa^2 - 2)s_2^2]d\lambda}{(\xi^2 + \lambda^2 + s_1^2)(\xi^2 + \lambda^2 + s_2^2)} \tag{9}$$

where $R(\xi) = (2\xi^2 + s_2^2)^2 - 4\xi^2\beta_1\beta_2\sqrt{\xi^2 + s_1^2}\sqrt{\xi^2 + s_2^2}$

and $\beta_m = \text{sgn}[\text{Re}\sqrt{\xi^2 + s_m^2}]$ $(m = 1,2)$

Once solution to the two coupled singular integral equations (8) and (9) is known, through Eqs. (5) and (6) we also have complete solution to the RC quarter-space problem. By making different choices of spatial Fourier sine and cosine transforms, other sets of integral equations could also be derived in exactly the same fashion. For example, taking (cos, sin) transforms instead of (sin, cos) transforms respectively throughout the preceding analysis would give the pair of equations

$$\overline{\sigma}_{12}^c(0,\eta,s) = \frac{\overline{P}(s)}{L(\eta)}\left[\eta^2 e^{-\sigma_1 a\sqrt{\eta^2+s_1^2}} - \sigma_1\sigma_2\sqrt{\eta^2 + s_1^2}\sqrt{\eta^2 + s_2^2}\, e^{-\sigma_2 a\sqrt{\eta^2+s_2^2}}\right] + \frac{2s_1^2\sigma_1\sqrt{\eta^2 + s_1^2}}{\pi L(\eta)} \int_0^\infty \frac{\lambda\overline{u}^s(\lambda,0,s)[\kappa^2\lambda^2 - (\kappa^2 - 2)\eta^2 + s_2^2]d\lambda}{(\lambda^2 + \eta^2 + s_1^2)(\lambda^2 + \eta^2 + s_2^2)} \tag{10}$$

$$\bar{u}^s(\xi,0,s) = \frac{\xi \cos \xi a \, \bar{P}(s)}{R(\xi)} [(2\xi^2 + s_2^2) - 2\beta_1\beta_2\sqrt{\xi^2 + s_1^2}\sqrt{\xi^2 + s_2^2}]$$
$$+ \frac{2\xi s_1^2\beta_2\sqrt{\xi^2 + s_2^2}}{\pi R(\xi)} \int_0^\infty \frac{\bar{\sigma}_{12}^c(0,\lambda,s)[\kappa^2\xi^2 - (\kappa^2 - 2)\lambda^2 + s_2^2]d\lambda}{(\xi^2 + \lambda^2 + s_1^2)(\xi^2 + \lambda^2 + s_2^2)} \tag{11}$$

Both the set of equations (8) and (9) and the set (10) and (11) will be used in the next section to obtain approximate solutions to the RC quarter-space problem.

ITERATIVE SOLUTION

In order to solve the RC problem we must first solve one of the above pairs of integral equations. At the present there is no known way of doing this exactly so we will instead apply the method of successive approximations as done in [1] and [2]. To illustrate this iteration scheme we first consider the set of equations (8) and (9).

NORMAL STRESS AND VERTICAL DISPLACEMENT

ZEROTH AND FIRST ITERATION

To obtain the "zeroth order" approximation for Eqs. (8) and (9) we delete their integral parts entirely, giving

$$\bar{\sigma}_o^c = 0 \tag{12a}$$

$$\bar{v}_o^s = \frac{\bar{P}(s)s_2^2\beta_1\sqrt{\xi^2 + s_1^2}\sin \xi a}{R(\xi)} \tag{12b}$$

where $\bar{v}_n^s$ and $\bar{\sigma}_n^c$ are the nth iterative solutions for $\bar{v}^s(\xi,0,s)$ and $\bar{\sigma}_{11}^c(0,\eta,s)$, respectively. It can be shown [9] that the terms in Eqs. (12a) and (12b) give the solution to a half-space $y > 0$, $-\infty < x < \infty$, with a load $P(t)$ at $x = a$ and $-P(t)$ at $x = -a$. This "image" problem satisfies the RC quarter-space boundary condition $v(0,y,t) = 0$.

To obtain the first approximation for the normal stress and vertical displacement, Eqs. (12a) and (12b) are placed back into Eqs. (8) and (9), giving

$$\bar{v}_1^s = \bar{v}_0^s \tag{13a}$$

$$\bar{\sigma}_1^c = \frac{2s_1^2\sigma_2\sqrt{\eta^2 + s_2^2}}{\pi L(\eta)} \int_0^\infty \frac{[(\kappa^2 - 2)\lambda^2 - \kappa^2\eta^2 + (\kappa^2 - 2)s_2^2]\lambda\bar{v}_0^s d\lambda}{(\lambda^2 + \eta^2 + s_1^2)(\lambda^2 + \eta^2 + s_2^2)} \tag{13b}$$

Even at this early stage the integration in Eq. (13b) is difficult to perform. Thus, we cannot carry out the iteration procedure exactly beyond the "zeroth order" approximation. However, for a large, the predominant incident wave is the Rayleigh wave and we need consider only the normal wall stress it produces. As shown in [9], the effects of incident compressional and shear waves are at least of $O(a^{-1/2})$ when a is large, while the incident Rayleigh wave is theoretically unattenuated with its distance of propagation in a two-dimensional problem [10]. Therefore, for a large, only the residue at the pole $\lambda = is_r$ need be considered [9], a contribution which becomes the exact result for $\bar{\sigma}_1^c$ as $a \to \infty$. Evaluating this Rayleigh wave contribution and inverting the result on η, we find

$$\bar{\sigma}_1(0,y,s) = i\beta s^3 A_1 \bar{P}(s) e^{-s_r a} \int_{-\infty}^{\infty} \frac{\sigma_2\sqrt{\eta^2 + s_2^2}\,[(\kappa^2 - 2)(s_2^2 - s_r^2) - \kappa^2\eta^2]e^{i\eta y}\,d\eta}{L(\eta)(\eta^2 + s_1^2 - s_r^2)(\eta^2 + s_2^2 - s_r^2)} \tag{14}$$

where $A_1 = -\dfrac{2v_1^2 v_2^2 v_r \sqrt{v_r^2 - v_1^2}}{iR'(iv_r)}$ is a real constant,

$R'(iv_r) = [dR(\lambda)/d\lambda]_{\lambda=iv_r}$, and $\beta = -\,\mathrm{sgn}[\mathrm{Im}(s)]$

Here Im stands for "imaginary part of." In evaluating this integral and subsequent integrals we usually can assume s is real and positive. However, in Eq. (14) this would imply that the poles at $\eta = \pm\sqrt{s_r^2 - s_m^2}$ (m = 1,2) are on the real axis. To interpret them, therefore, we consider giving s a nonzero imaginary part as if s were lying along the Bromwich inversion path in a complex s-plane. If $\mathrm{Im}(s) \gtrless 0$, then the poles at $\eta = \pm\sqrt{s_r^2 - s_m^2}$, respectively, are located in the upper half of the η-plane and we must take their residues. The poles and branch contributions then can be shown to give

$$\bar{\sigma}_1(0,y,s) = \sum_{m=1}^{2} s\bar{P}(s)e^{-s_r a}\left[C_{Im}e^{-i\beta s\sqrt{v_r^2 - v_m^2}\,y} + \frac{2i\beta A_1}{\pi}\int_{v_m}^{\infty} G_{Im}(\gamma)e^{-\gamma s y}\,d\gamma\right] \tag{15}$$

where the constants C_{Im} and functions $G_{Im}(\gamma)$ are defined in the Appendix. In treating the problem of two welded quarter-spaces, Viswanathan [2] found expressions similar to those in Eq. (14). However, he then set $\eta = \gamma s$ in the integrand and factored an appropriate power of s outside the integral. But this leaves real singularities on the path of γ-integration without explanation of how the integrals should be interpreted. Also, since Eq. (14) is used in Eq. (9) to obtain the v-displacement of the reflected Rayleigh wave, as will be shown shortly, singular integrals would appear in that result as well. Such difficulties, however, are avoided when the singularities are interpreted with s complex.

The first two terms in Eq. (15) can be inverted directly on a Bromwich path while the last two terms can be inverted by inspection [9]. Thus the final result for the real normal stress on the clamped wall due to the Rayleigh wave may be written in the first iteration as

$$\sigma_1(0,y,t) = \sum_{m=1}^{2}\left[\frac{C_{Im}}{\pi}\int_{-\infty}^{t} \frac{\dot{P}(t-\tau)\sqrt{v_r^2 - v_m^2}\,y\,d\tau}{(\tau - av_r)^2 + (v_r^2 - v_m^2)y^2} + \frac{2A_1}{\pi y}\int_{v_m y}^{\infty} \dot{\Pi}(t - av_r - \tau)\,G_{Im}\!\left(\frac{\tau}{y}\right)d\tau\right] \tag{16}$$

where $\dot{\Pi}(t) = 1/\pi \int_{-\infty}^{\infty} \dot{P}(\xi)d\xi/(t - \xi)$ is the Hilbert transform of $\dot{P}(\xi) = dP(\xi)/d\xi$.

The first two terms in Eq. (16) represent the normal wall stress due to reflection of the incident Rayleigh wave taking place along the wall while the last two terms are the effects of diffracted body waves.

SECOND ITERATION

The iteration procedure can be continued to find the surface displacement $\bar{v}^s(\xi,0,s)$ in the second approximation by placing Eq. (15) back into Eq. (9). First, however, it is convenient to write Eq. (9) in terms of $\bar{\sigma}_{11}(0,y,s)$ instead of $\bar{\sigma}^c_{11}(0,\lambda,s)$. This can be done by using the Fourier cosine transform relation between $\bar{\sigma}_{11}^{c}(0,\lambda,s)$ and $\bar{\sigma}_{11}(0,y,s)$ in Eq. (9) and then performing the λ-integration first. When Eq. (15) is placed in the ensuing expression and the y-integration is performed, we get

$$\bar{v}_2^{s}(\xi,0,s) = \frac{\bar{P}(s)\beta_1\sqrt{\xi^2+s_1^2}\, s_2^2 \sin \xi a}{R(\xi)} + \sum_{m=1}^{2} \frac{s\bar{P}(s)e^{-s_r a}}{R(\xi)}$$

$$\left[\frac{\xi(2\xi^2+s_2^2)C_{Im}}{\beta_1\sqrt{\xi^2+s_1^2}+i\beta s\sqrt{v_r^2-v_m^2}} - \frac{2\beta_1\beta_2\xi\sqrt{\xi^2+s_1^2}\sqrt{\xi^2+s_2^2}\, C_{Im}}{\beta_2\sqrt{\xi^2+s_2^2}+i\beta s\sqrt{v_r^2-v_m^2}}\right.$$

$$+ i\beta A_1\xi(2\xi^2+s_2^2)\int_{v_m}^{\infty}\frac{G_{Im}(\gamma)\,d\gamma}{\gamma s+\beta_1\sqrt{\xi^2+s_1^2}} - 2i\beta A_1\beta_1\beta_2\xi\sqrt{\xi^2+s_1^2}\sqrt{\xi^2+s_2^2}$$

$$\left.\int_{v_m}^{\infty}\frac{G_{Im}(\gamma)\,d\gamma}{\gamma s+\beta_2\sqrt{\xi^2+s_2^2}}\right] \qquad (17)$$

To do both the ξ and γ integrations exactly is prohibitively difficult, but we can estimate the form of the reflected Rayleigh wave due to the incident Rayleigh wave in this approximation by taking the "Rayleigh" pole contribution in the ξ-plane. The result can then be inverted on s directly to give

$$v_2(x,0,t) = B_o P[t-(x+a)v_r] + B_1 II[t-(x+a)v_r] \qquad (18)$$

where B_o and B_1 are two constants given by lengthy expressions [9] involving integrals of $G_{Im}(\gamma)$. Whereas v can be shown to depend only on II(t) in the incident Rayleigh wave [10], we see that the reflected wave in the RC quarter-space problem is a linear combination of P(t) and II(t). Lapwood [1], in determining the Rayleigh wave which is transmitted around the corner of a quarter-space with stress-free boundaries, found a similar result for that wave in his first iteration. In the related problem of two welded quarter-spaces, Viswanathan [2] also found that the surface v-displacement depended on both P(t) and II(t), but the coefficients in his expressions, as mentioned earlier, contain integrals which are not properly defined.

Although the incident body waves in the first approximation are negligibly small for a large, this is not true for the body waves produced by interaction of the Rayleigh wave with the wall. These body waves are contained in the last two integral terms in Eq. (16). Therefore, we must substitute Eq. (17) in its entirety into Eq. (8) to obtain the next approximation. However, in practical terms this can only be done in a formal manner because the resulting integrals are too complicated to evaluate.

SHEAR STRESS AND HORIZONTAL DISPLACEMENT

Having illustrated the method of successive approximations for Eqs. (8) and (9), we will now apply it to Eqs. (10) and (11). Since the analysis follows closely the procedure just outlined, we shall only discuss the details briefly.

ZEROTH AND FIRST ITERATION

As before, we begin by neglecting the integrals in Eqs. (10) and (11), giving

$$\bar{\tau}_o^{c} = \frac{\bar{P}(s)}{L(\eta)}\left[\eta^2 e^{-\sigma_1 a\sqrt{\eta^2+s_1^2}} - \sigma_1\sigma_2\sqrt{\eta^2+s_1^2}\sqrt{\eta^2+s_2^2}\, e^{-\sigma_2 a\sqrt{\eta^2+s_2^2}}\right] \qquad (19a)$$

$$\bar{u}^s_o = \frac{\xi \cos(\xi a)\bar{P}(s)}{R(\xi)} [(2\xi^2 + s_2^2) - 2\beta_1\beta_2\sqrt{\xi^2 + s_1^2}\sqrt{\xi^2 + s_2^2}] \tag{19b}$$

where $\bar{\tau}^c_n$ and $\bar{u}^s_n$ are the nth approximations for $\bar{\sigma}^c_{12}(0,\eta,s)$ and $\bar{u}^s(\xi,0,s)$, respectively.

We now derive the first approximation by placing Eqs. (19a) and (19b) back into Eqs. (10) and (11). As mentioned previously, the Rayleigh wave is the most important part of the incident waves when a is large. The major contribution to the wall shear stress can be extracted by evaluating the residue at the "Rayleigh" pole as before and the result then inverted on η to yield

$$\bar{\tau}_1(0,y,s) = \frac{s^3 A_2 \bar{P}(s) e^{-s_r a}}{\pi} \int_{-\infty}^{\infty} \frac{\sigma_1\sqrt{\eta^2 + s_1^2}\,[s_2^2 - \kappa^2 s_r^2 - (\kappa^2 - 2)\eta^2]e^{i\eta y}d\eta}{L(\eta)(\eta^2 + s_1^2 - s_r^2)(\eta^2 + s_2^2 - s_r^2)} \tag{20}$$

where $A_2 = 2v_r^2 v_1^2[(v_2^2 - 2v_r^2) + 2\sqrt{v_r^2 - v_1^2}\sqrt{v_r^2 - v_2^2}]/iR'(iv_r)$ is a real constant. This integral can also be performed by contour integration as before and the result inverted on s, giving finally

$$\tau_1(0,y,t) = \sum_{m=1}^{2}\left[\frac{C_{2m}}{\pi}\int_{-\infty}^{t} \frac{\dot{P}(t - \tau)(\tau - av_r)d\tau}{(\tau - av_r)^2 + (v_r^2 - v_m^2)y^2} + \frac{2A_2 H(t - av_r - v_m y)}{\pi y}\int_{v_m y}^{t-av_r} \dot{P}(t - av_r - \tau)G_{2m}\left(\frac{\tau}{y}\right)d\tau\right] \tag{21}$$

where the unit step function $H(t) = 1$ for $t > 0$ and is zero for $t < 0$, and the quantities C_{2m} and $G_{2m}(\gamma)$ are defined in the Appendix. Notice that the integrands of the last two terms in Eq. (21) contain $\dot{P}(t)$ in contrast to the normal stress result, Eq. (16), where the factor $\dot{II}(t)$ appeared in the body wave terms instead.

SECOND ITERATION

To obtain the horizontal surface displacement of the reflected Rayleigh wave, we proceed as before and first rewrite Eq. (11) in terms of $\bar{\sigma}_{12}(0,y,s)$, place the first iteration solution for $\bar{\sigma}_{12}(0,y,s)$ into this result and perform the remaining y-integration. Taking the Rayleigh wave contribution in the ξ-plane and inverting on s, we finally find

$$u_2(x,0,t) = D_o P[t - (x + a)v_r] + D_1 II[t - (x + a)v_r] \tag{22}$$

where D_o and D_1 are constants given by lengthy expressions [9]. As seen from Eq. (22) the u-displacement of the reflected Rayleigh wave is also given in terms of P(t) and II(t), while in the incident Rayleigh wave, u depends only on P(t) [10].

This completes the derivation of the iteration solutions. The success of this method depended entirely on our choice of the pairs of integral equations, (8), (9) and (10), (11), since other choices of sine and cosine transforms would lead to integral equations for which the method of successive approximations would fail. In fact the iteration scheme in these cases would predict wall stresses having nonintegrable singularities near the corner of the quarter-space, violating a necessary requirement that the strain energy remain finite in every neighborhood of the corner. Furthermore, the iteration scheme would give a meaningless result for the reflected Rayleigh wave. Explicit proof of both of these statements is given in [9]. This same type of failure occurs if instead we had used spatial half-range Fourier transforms or if we had tried to solve the problem using Fourier integral representations for the displacement potentials, as done by Viswanathan [2].

DISCUSSION OF RESULTS

CORNER BEHAVIOR

The expressions for stresses on the clamped wall in the first iteration were given by Eqs. (16) and (21). We will now examine the behavior of these stresses near the corner of the RC quarter-space, i.e., as $y \to 0$. It is convenient to start from the transform expressions, Eqs. (14) and (20) for $\bar{\sigma}_1(0,y,s)$ and $\bar{\tau}_1(0,y,s)$, respectively. If the function $L(\eta)$ in these two equations is rationalized and the integrands of the resulting expressions then expanded in a sum of partial functions, we obtain

$$\bar{\sigma}_1(0,y,s) = \frac{i\beta s A_1 \bar{P}(s) e^{-s_r a}(\kappa^2 - 2)}{\pi(v_1^2 + v_2^2)} \left[\sum_{m=1}^{2} \int_{-\infty}^{\infty} \frac{e^{i\eta y} d\eta}{\sigma_m \sqrt{\eta^2 + s_m^2}} + \begin{matrix}\text{additional}\\ \text{terms}\end{matrix} \right] \tag{23a}$$

$$\bar{\tau}_1(0,y,s) = \frac{s A_2 \bar{P}(s) e^{-s_r a}(\kappa^2 - 2)}{\pi(v_1^2 + v_2^2)} \left[\sum_{m=1}^{2} \int_{-\infty}^{\infty} \frac{e^{i\eta y} d\eta}{\sigma_m \sqrt{\eta^2 + s_m^2}} + \begin{matrix}\text{additional}\\ \text{terms}\end{matrix} \right] \tag{23b}$$

where the additional terms can be shown to be bounded in y as $y \to 0$ and can be neglected in comparison with the first two terms which are singular at $y = 0$. With s taken as a fixed, positive number, these first terms of Eqs. (23a) and (23b) represent modified Bessel functions of zero order, $K_0(s_m y)$ $(m = 1,2)$, that behave like $-\ell n(y)$ for y small. Thus, as $y \to 0$ we can invert Eqs. (23a) and (23b) on s and obtain

$$\sigma_1(0,y,t) \simeq -\frac{4A_1(\kappa^2 - 2)}{\pi(v_1^2 + v_2^2)} \dot{\Pi}(t - av_r)\ell n(y) \tag{24a}$$

$$\tau_1(0,y,t) \simeq -\frac{4A_2(\kappa^2 - 2)}{\pi(v_1^2 + v_2^2)} \dot{P}(t - av_r)\ell n(y) \tag{24b}$$

Equations (24a) and (24b) show that the stresses on the clamped wall due to the incident Rayleigh wave are singular near the corner of $O(\ell n\ y)$ as $y \to 0$. This type of singularity is permissible since it does not violate the requirement of finite strain energy in any neighborhood of the corner. These results are in contrast to the exact solution to the RS quarter-space problem of Ref. [3], where the wall shear stress is identically zero and the normal stress at the corner is proportional to $\dot{P}(t - av_r)$. Note, however, that the above expressions were derived from the transformed solutions by considering s as a fixed number. We do not expect, therefore, that Eqs. (24a) and (24b) remain valid either as $(t - av_r) \to 0$ or as $(t - av_r) \to \infty$, but that they are meaningful for "moderate" times.

STRESSES ON THE WALL

For the case of step loading $P(t) = PH(t)/\mu$ and for Poisson's ratio ν equal to 1/4, the approximate expressions for the normal and shear stresses on the RC wall due to the incoming Rayleigh wave, Eqs. (16) and (21), can be integrated exactly. The results are given in Figs. 2 and 3, where the stresses are plotted, for various fixed dimensionless times $\tau = t/av_r$, versus the nondimensional depth along the wall $\bar{y} = y/a$.

Figures 2 and 3 show that the Rayleigh wave produces normal and shear stresses for $\tau < 1$, where $\tau = 1$ corresponds to $t = av_r$, i.e., the time at which the major portion of the incident Rayleigh wave arrives at the clamped wall. This is because the Rayleigh wave, when isolated from the body waves, has no distinct front [10]. As $\tau \to 1$, the shear stress in Fig. 3 grows larger and becomes more concentrated near $\bar{y} = 0$ until finally at $\tau = 1$, $\tau_1(0,y,t)$ is zero everywhere except $\bar{y} = 0$ where it is infinite. In contrast, the normal stress at $\tau = 1$ is given by $\sigma_1 = -0.2712/\overline{y}$ instead of Eq. (24a). This curve is not shown in Fig. 2, but is of the same type as obtained at $\tau = 1$ in the RS quarter-space problem of [3]. The sharp "peaks" in Figs. 2 and 3 represent the wavefronts of diffracted compressional and shear waves produced by the incident Rayleigh wave. For the idealized step function loading we have chosen, these diffracted body waves have infinite slopes at their wavefronts, but this behavior can be eliminated by taking a smoother loading profile.

Figure 3 shows that the shear stress τ_1 has finite values at the corner $\overline{y} = 0$ when $P(t) = PH(t)/\mu$. This result is also due to the very special behavior of the step function. For more realistic loadings τ_1 will have the logarithmic singularity at the corner as derived in Eq. (24b). In fact, this equation shows why the singular behavior vanishes for $P(t) = PH(t)/\mu$. For then, $\dot{P}(t - av_r) = P\delta(t - av_r)/\mu$ which is identically zero except at $t = av_r$, i.e., $\tau = 1$.

REFLECTED RAYLEIGH WAVE DISPLACEMENTS

Numerical results for the surface displacements of the reflected Rayleigh wave can be obtained from Eqs. (18) and (22). These displacements depend on the loading P(t) and its Hilbert transform, $\Pi(t)$. Here, we choose $\nu = 1/4$ and $P(t) = Pt_o/(t^2 + t_o^2)$ where t_o and P are constants. This function, used by Lamb in his original classic paper on the dynamic half-space problem [11], approximates the profile of a smooth pulse and gives $\Pi(t) = Pt/(t^2 + t_o^2)$. In this case the displacements are shown in Fig. 4, where u_2 and v_2 are plotted versus the nondimensional variable $\overline{t} = [t - (x + a)v_r]/t_o$. It is interesting to compare these curves with those for the reflected Rayleigh wave displacements contained in u_o and v_o, which are merely images of the incident Rayleigh wave profiles. These latter curves are almost completely opposite those for u_2 and v_2 and thus, it appears the RC problem is indeed different from a "simple" image problem. The diffracted body waves, which arise in order to satisfy the boundary conditions at the clamped wall, are very significant in changing the shape of the reflected Rayleigh wave from that of the incident waveform.

REFERENCES

[1]. Lapwood, E. R., "The Transmission of a Rayleigh Pulse Round a Corner," Geophysical Journal of the Royal Astronomical Society, Vol. 4, 1961, pp. 174-196.

[2]. Viswanathan, K., "Wave Propagation in Welded Quarter-Spaces," Geophysical Journal of the Royal Astronomical Society, Vol. 11, 1966, pp. 293-322.

[3]. Schmerr, L. W., and Thau, S. A., "Dynamic Stresses along a Rigid-Smooth Boundary of a Quarter-Space Due to Incident Rayleigh Waves," ERI Preprint 82700, Iowa State University, August 1970. (To appear in the Journal of Applied Mechanics.)

[4]. Dally, J. W., "A Dynamic Photoelastic Study of a Doubly Loaded Half-Plane," Proceedings of the Tenth Midwestern Mechanics Conference, Fort Collins, Colorado, August 1967, pp. 649-664.

[5]. Thau, S. A., "Dynamic Reactions along a Rigid-Smooth Wall in an Elastic Half-Space with a Moving Boundary Load," International Journal of Solids and Structures, Vol. 4, January 1968, pp. 1-13.

[6]. Wright, T. W., "Impact on an Elastic Quarter-Space," Journal of the Acoustical Society of America, Vol. 45, No. 4, April 1969, pp. 935-944.

[7]. Alterman, Z. S., and Rotenberg, A., "Seismic Waves in a Quarter-Plane," Bulletin of the Seismological Society of America, Vol. 59, No. 1, February 1969, pp. 347-368.

[8]. Sokolnikoff, I. S., Mathematical Theory of Elasticity, McGraw-Hill, New York, Second Edition, 1956.

[9]. Schmerr, L. W., "Elastic Wave Propagation and Scattering in Quarter-Spaces," PhD Thesis, Illinois Institute of Technology, January 1970.

[10]. Thau, S. A., and Dally, J. W., "Subsurface Characteristics of the Rayleigh Wave," International Journal of Engineering Science, Vol. 7, January 1969, pp. 37-52.

[11]. Lamb, H., "On the Propagation of Tremors over the Surface of an Elastic Solid," Philosophical Transactions of the Royal Society of London, Series A, Vol. 203, 1904, pp. 1-42.

APPENDIX

$$C_{mm} = \frac{\sqrt{v_r^2 + v_n^2 - v_m^2}\,(v_2^2 - 2v_r^2)A_m}{v_1^2\sqrt{v_r^2 - v_m^2}\,(v_r^2 - v_m^2 - v_r\sqrt{v_r^2 + v_n^2 - v_m^2})} \qquad \left\{\begin{matrix} m, n = 1, 2 \\ m \neq n \end{matrix}\right\}$$

$$C_{mn} = \frac{2v_r\sqrt{v_r^2 - v_n^2}\,A_m}{v_1^2(v_r^2 - v_n^2 - v_r\sqrt{v_r^2 + v_m^2 - v_n^2})} \qquad \left\{\begin{matrix} m, n = 1, 2 \\ m \neq n \end{matrix}\right\}$$

$$F_{11}(\gamma) = \frac{[(\kappa^2 - 2)(v_2^2 - v_r^2) + \kappa^2\gamma^2]\sqrt{\gamma^2 - v_2^2}}{[\gamma^2(v_1^2 + v_2^2) - v_1^2v_2^2][\gamma^2 - v_1^2 + v_r^2][\gamma^2 - v_2^2 + v_r^2]}$$

$$F_{22}(\gamma) = \frac{[(v_2^2 - \kappa^2 v_r^2) + (\kappa^2 - 2)\gamma^2]\sqrt{\gamma^2 - v_1^2}}{[\gamma^2(v_1^2 + v_2^2) - v_1^2v_2^2][\gamma^2 - v_1^2 + v_r^2][\gamma^2 - v_2^2 + v_r^2]}$$

$$G_{mn}(\gamma) = \sqrt{\gamma^2 - \delta^m_n v_1^2}\sqrt{\gamma^2 - \delta^m_n v_2^2}\,F_{mm} \qquad \delta^m_n = \left\{\begin{matrix} 1 & m = n \\ 0 & m \neq n \end{matrix}\right\}$$

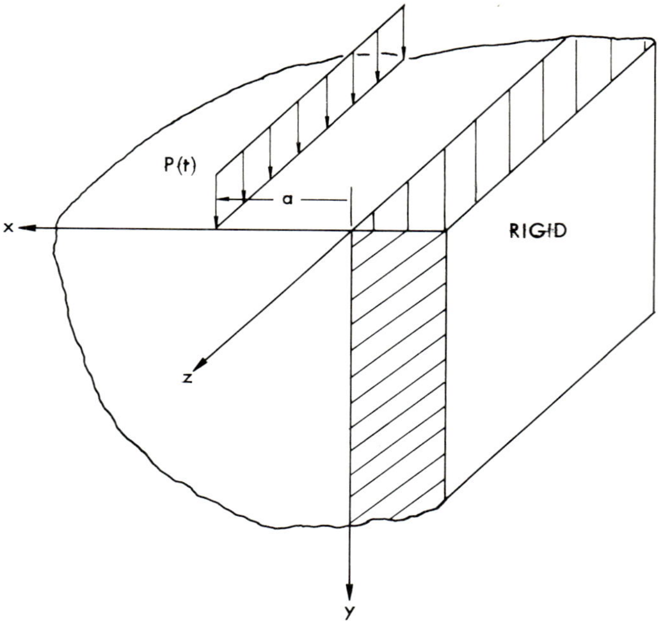

Fig. 1 RC Quarter-Space Geometry

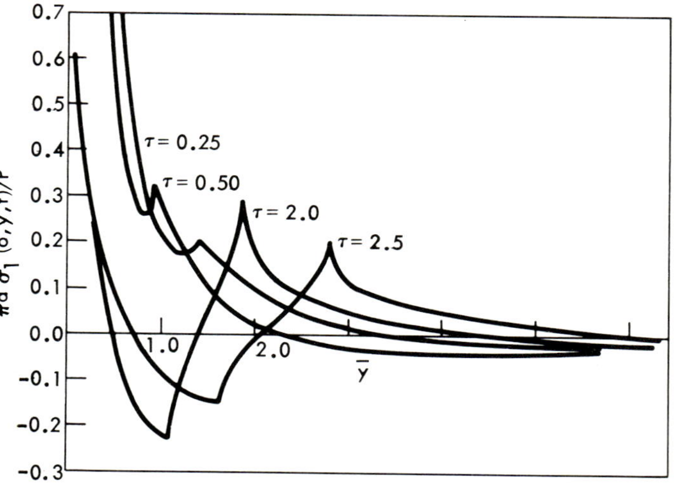

Fig. 2 Dynamic Normal Stress on RC Wall

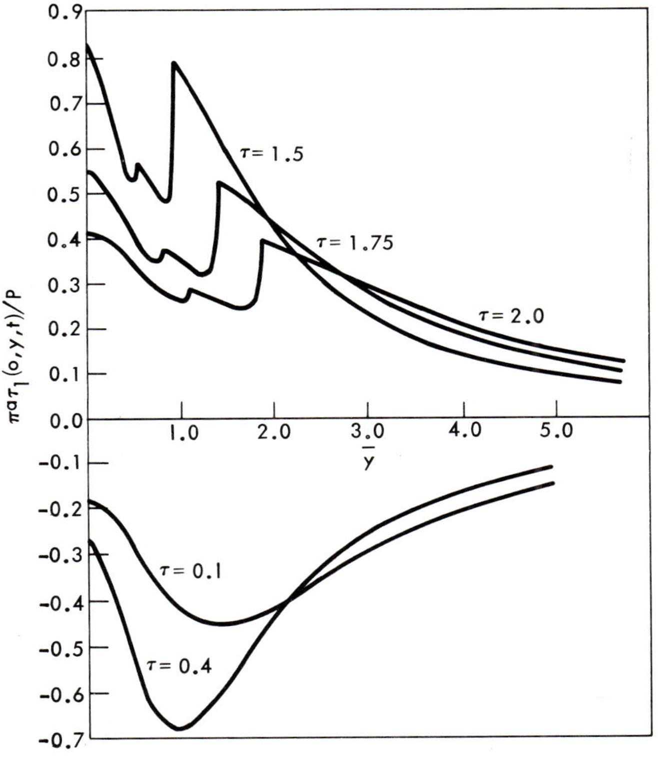

Fig. 3 Dynamic Shear Stress on RC Wall

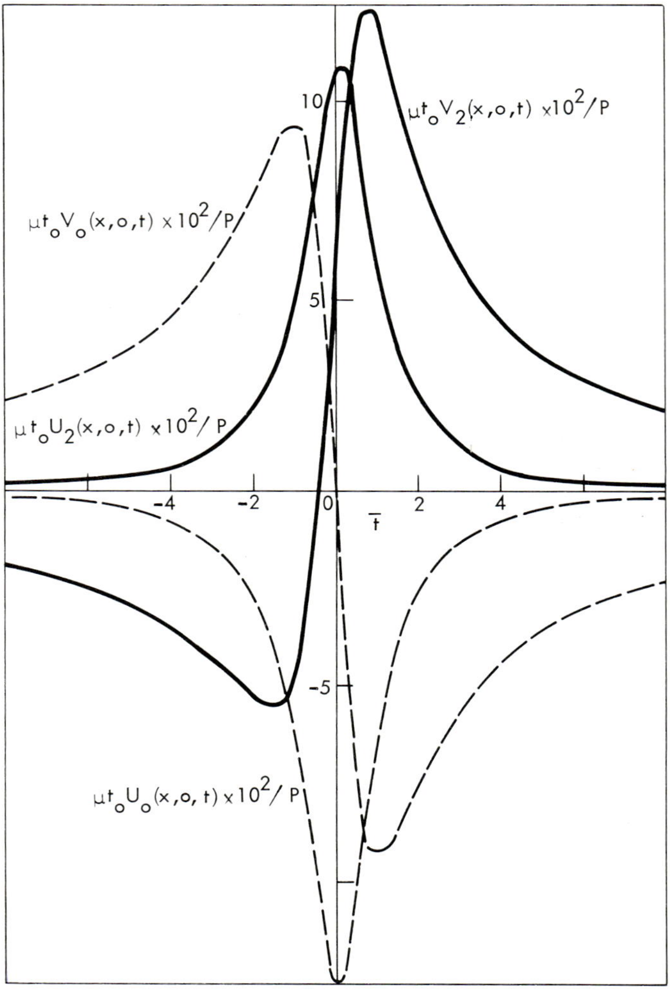

Fig. 4 Reflected Rayleigh Wave Surface Displacements

Developments in Mechanics, Vol. 6. Proceedings of the 12th Midwestern Mechanics Conference: 35

Steady State Wave Propagation in Fiber Reinforced Elastic Materials

S. E. MARTIN, A. BEDFORD and M. STERN
THE UNIVERSITY OF TEXAS AT AUSTIN

ABSTRACT

A theory of composite materials, in which each constituent is modeled as a continuum which undergoes an individual motion but interacts mechanically with the other constituent, is presented in a form proposed for application to a restricted class of fiber reinforced materials. Estimates of the constitutive constants occuring in the theory are provided in terms of the individual constituent properties and the composite geometry.

Some characteristics of the proposed theory are illustrated by considering the propagation of steady state, plane waves in an arbitrary direction in an elastic material containing uniformly distributed, straight, parallel elastic fibers. For a typical set of material parameters, graphs are presented of phase velocity as a function of wave number for various propagation directions relative to the fiber direction. Additional graphs are presented which provide insight into interesting aspects of the behavior of the propagation modes.

INTRODUCTION

It has been shown both analytically [1,2] and experimentally [3] that dynamical processes in composite materials exhibit dispersional effects as a result of the macroscopic composite structure. Because of the complex geometry of even the simplest composite materials, for example, regular plane laminated materials, only a limited class of particularly simple problems can be solved using the classical approach of formulating a boundary value problem in the linear theory of elasticity [4]. For composite materials with more complicated geometries, such as fiber reinforced materials, the classical approach would be formidably complex.

In [5], a new approach to the dynamics of composite materials was introduced in which each constituent of the composite was modeled as a continuum which could undergo an individual motion. Dispersional effects in the composite were then introduced through mechanical coupling between the continua. A one dimensional theory of this type was postulated in [5] and was shown to adequately predict the dispersion of compressional waves propagating in the direction of the layering in a laminated elastic composite.

The concept introduced in [5] was formally developed into a thermomechanical theory for interacting elastic continua in [6]. A set of linearized constitutive equations were obtained for elastic composite materials which exhibit transversely isotropic material symmetry, i.e., laminated materials or materials reinforced by locally unidirectional fibers.

The contribution of this paper is to propose particular forms of the equations obtained in [6] for application to a restricted class of fiber reinforced materials. We consider an isotropic elastic material reinforced by uniformly distributed, thin elastic fibers. It is assumed that the fibers occupy only a small fraction of the total composite volume, and that the fibers are strong in tension but have negligible strength in bending and shear. An example of an actual composite material conforming to these assumptions would be a plastic reinforced by fiberglass filaments. These assumptions make possible a substantial simplification of the general constitutive equations obtained in [6]. In order to make the resulting simplified equations accessible for application to particular composites, we provide estimates for the constitutive constants appearing in the theory in terms of the constituent properties and the fiber geometry.

Finally, we exhibit some characteristics of the obtained equations by deriving the dispersion equation for steady state plane waves propagating in an arbitrary direction relative to the fiber direction in a material containing straight, parallel fibers. Graphs of phase velocity as a function of wave number are presented for each of four resulting modes of propagation. Several interesting and unanticipated features of the results are discussed.

MOMENTUM AND CONSTITUTIVE EQUATIONS

In this section, we present the linear mechanical equations for dynamic processes in composite materials developed in [6]. The theory is founded on the viewpoint that each composite constituent can be regarded as an individual continuum, with interactions with the other constituents being considered as external supplies. The small displacement balance of linear momentum equation for each constituent thus has the form (in Cartesian tensor notation)

$$\rho_{(\xi)} \frac{\partial^2 u_{(\xi)k}}{\partial t^2} = t_{(\xi)jk,j} + \rho_{(\xi)} p_{(\xi)k} + \rho_{(\xi)} f_{(\xi)k} \qquad (1)$$

where ξ denotes the constituent, $\rho_{(\xi)}$ is the constituent mass density per unit composite volume, $u_{(\xi)k}$ is the individual constituent displacement vector, $t_{(\xi)jk}$ is the constituent partial stress tensor, $p_{(\xi)k}$ is the body force density induced in the ξth constituent by its interaction with the other constituent, and $f_{(\xi)k}$ is the constituent external body force density.

The linearized constitutive equations developed in [6] for $t_{(\xi)jk}$ and $p_{(\xi)k}$ for the case of transversely isotropic elastic composites in which the interaction body force terms $p_{(\xi)k}$ depend upon the relative displacement between the constituents, have the form, for two constituents denoted by subscripts m and f,

$$t_{(\xi)jk} = A_{(\xi)}\delta_{jk}u_{(\xi)n,n} + B_{(\xi)}(u_{(\xi)j,k} + u_{(\xi)k,j})$$
$$+ C_{(\xi)}(\delta_{jk}d_m d_n u_{(\xi)m,n} + d_j d_k u_{(\xi)n,n})$$
$$+ D_{(\xi)}[d_k d_m(u_{(\xi)j,m} + u_{(\xi)m,j}) + d_j d_m(u_{(\xi)k,m} + u_{(\xi)m,k}]$$
$$+ E_{(\xi)}d_j d_k d_m d_n u_{(\xi)m,n} \qquad (\xi = m,f) \qquad (2)$$

$$\rho_{(m)}p_{(m)k} = -\rho_{(f)}p_{(f)k} = \nu(u_{(f)k} - u_{(m)k})$$
$$+ \sigma d_k d_j(u_{(f)j} - u_{(m)j}) \qquad (3)$$

where δ_{jk} is the Kronecker delta, d_k is the local direction of the symmetry axis of the material, and $A_{(\xi)}$ through $E_{(\xi)}$, ν and σ are constitutive constants which must be evaluated for a given composite material. Eq. 2 has the form of the usual stress-displacement relation for a single transversely isotropic elastic material.

CONSTITUTIVE CONSTANTS AND DISPLACEMENT EQUATIONS

We now propose simplified forms of Eq. 2 for application to the particular class of fiber reinforced materials consisting of an elastic matrix containing uniformly distributed, continuous, locally unidirectional fibers in which the fibers occupy a small fraction of the total composite volume. Where possible, we also provide estimates for the constitutive constants in the proposed equations in terms of the properties of the composite constituents and the fiber geometry.

Let the material comprising the matrix be elastic, homogeneous and isotropic. Since the fibers occupy a small fraction of the total volume, we assume in our model that the relation between the stress in the matrix and the deformation of the matrix is isotropic. Therefore, for the matrix, denoted by subscript ξ = m, Eq. 2 reduces to

$$t_{(m)jk} = A_{(m)}\delta_{jk}u_{(m)n,n} + B_{(m)}(u_{(m)j,k} + u_{(m)k,j}) \qquad (4)$$

In addition, we approximate the constants $A_{(m)}$ and $B_{(m)}$ by the usual Lame constants λ and μ of the matrix material, so that the constitutive equation for the stress in the matrix material in terms of the individual deformation of the constituent has the same form that it would in the absence of the fibers.

$$t_{(m)jk} = \lambda\delta_{jk}u_{(m)n,n} + \mu(u_{(m)j,k} + u_{(m)k,j}) \qquad (5)$$

We assume that the fibers have negligible bending and shear resistance and thus support only axial forces in the local fiber direction. Thus, denoting the fiber continuum by subscript ξ = f, it is easy to show that Eq. 2 must reduce to

$$t_{(f)jk} = \eta d_j d_k d_m d_n u_{(f)m,n}$$

where the scalar constant $E_{(f)}$ has been replaced by η. The constant η can be related to the fiber geometry and properties by a simple derivation presented in Appendix 1.

Using Eq.s 3, 5 and 6 in Eq. 1, the displacement equations for the case of straight, parallel fibers, with $d_k = \delta_{k1}$ so that the fiber direction is parallel to the x_1 coordinate, become

$$\begin{aligned}\rho_{(m)} \frac{\partial^2 u_{(m)k}}{\partial t^2} = {} & (\lambda+\mu)\, u_{(m)j,jk} + \mu u_{(m)k,jj} \\ & + \nu(u_{(f)k} - u_{(m)k}) \\ & + \sigma\delta_{k1}(u_{(f)1} - u_{(m)1}) \\ & + \rho_{(m)}\, f_{(m)k} \end{aligned} \tag{7}$$

$$\begin{aligned}\rho_{(f)} \frac{\partial^2 u_{(f)k}}{\partial t^2} = {} & \eta\delta_{k1}\, u_{(f)1,11} \\ & - \nu(u_{(f)k} - u_{(m)k}) \\ & - \sigma\delta_{k1}\,(u_{(f)1} - u_{(m)1}) \\ & + \rho_{(f)}\, f_{(f)k} \end{aligned} \tag{8}$$

The constant $\nu+\sigma$, which appears in Eq.s 7 and 8 when $k = 1$, is evaluated in terms of the composite properties in Appendix 2 using a procedure introduced in [5]. The same procedure, when used to evaluate the constant ν appearing in Eq.s 7 and 8 when $k = 2$ or 3, leads to an elasticity problem whose solution cannot be approximated in closed form. In our numerical results, we have therefore assumed for nominal cases that $\sigma=0$, and have also included data for various values of σ to illustrate the sensitivity to this parameter.

STEADY STATE WAVE PROPAGATION

In order to exhibit some characteristics of the proposed theory, we consider the propagation of steady state, plane waves at an arbitrary direction relative to the fiber direction. The propagation direction is assumed to lie in the x_1-x_2 plane and only motion in the x_1-x_2 plane is considered. We therefore seek solutions of Eq.s 7 and 8 of the form

$$u_{(m)j} = A_j e^{ik(\alpha x_1 + \beta x_2 - ct)} \tag{9}$$

$$u_{(f)j} = B_j e^{ik(\alpha x_1 + \beta x_2 - ct)} \tag{10}$$

Substitution of Eq.s 9 and 10 into Eq.s 7 and 8 leads to a set of homogeneous equations in A_j, B_j, resulting in a characteristic equation for the phase velocity c of the form

$$J_8 c^8 + J_6 c^6 + J_4 c^4 + J_2 c^2 + J_0 = 0 \tag{11}$$

where the coefficients J_i are

$$J_8 = \rho^2_{(m)}\rho^2_{(f)}k^8 \tag{12}$$

$$J_6 = \rho_{(m)}\rho_{(f)}k^4 \Big\{\rho_{(m)}k^2[-\eta\alpha^2k^2 - (\sigma+\nu)] + \rho_{(f)}k^2[-(\lambda+\mu)k^2\beta^2 - \mu k^2 - \nu] - \rho_{(f)}k^2[(\lambda+\mu)k^2\alpha^2 + \mu k^2 + (\sigma+\nu)] - \rho_{(m)}k^2\nu\Big\} \tag{13}$$

$$\begin{aligned} J_4 = {} & -\nu^2\rho_{(m)}\rho_{(f)}k^4 + \rho_{(m)}\rho_{(f)}k^4\ [(\lambda+\mu)k^2\beta^2 + \mu k^2 + \nu] \\ & [\eta\alpha^2k^2 + (\sigma+\nu)] + [-\rho_{(f)}k^2\ ((\lambda+\mu)k^2\alpha^2 + \mu k^2 + (\sigma+\nu)) \\ & - \rho_{(m)}k^2\nu]\ [\rho_{(m)}k^2\ (-\eta\alpha^2k^2 - (\sigma+\nu)) + \rho_{(f)}k^2\ (-(\lambda+\mu)k^2\beta^2 \\ & - \mu k^2 - \nu)] + [(\lambda+\mu)k^2\alpha^2 + \mu k^2 + (\sigma+\nu)]\nu\rho_{(m)}\rho_{(f)}\ k^4 \\ & - (\sigma+\nu)^2\rho_{(m)}\rho_{(f)}\ k^4 - \rho_{(f)}k^4\ [(\lambda+\mu)k^2\alpha\beta]^2 \end{aligned} \tag{14}$$

$$\begin{aligned} J_2 = {} & \Big\{\rho_{(m)}k^2\nu^2[\eta\alpha^2k^2 + (\sigma+\nu)] + \rho_{(f)}k^2\nu^2[(\lambda+\mu)k^2\alpha^2 + \mu k^2 \\ & + (\sigma+\nu)]\Big\} + \Big\{[-\rho_{(f)}k^2((\lambda+\mu)k^2\alpha^2 + \mu k^2 + (\sigma+\nu)) - \rho_{(m)}k^2\nu] \\ & [(\lambda+\mu)k^2\beta^2 + \mu k^2 + \nu]\ [\eta\alpha^2k^2 + (\sigma+\nu)] + \nu[(\lambda+\mu)k^2\alpha^2 + \mu k^2 \\ & + (\sigma+\nu)]\ [\rho_{(m)}k^2\ (-\eta\alpha^2k^2 - (\sigma+\nu)) + \rho_{(f)}k^2\ (-(\lambda+\mu)k^2\beta^2 + \mu k^2 \\ & - \nu)]\Big\} + \Big\{-(\sigma+\nu)^2\rho_{(f)}k^2[-(\lambda+\mu)k^2\beta^2 - \mu k^2 - \nu] + (\sigma+\nu)^2\rho_{(m)}\nu k^2\Big\} \\ & + \Big\{-\rho_{(f)}k^2\ [(\lambda+\mu)k^2\alpha\beta]^2\ [-\eta\alpha^2k^2 - (\sigma+\nu) - \nu]\Big\} \end{aligned} \tag{15}$$

$$\begin{aligned} J_0 = {} & \nu^2(\sigma+\nu)^2 + \nu[(\lambda+\mu)k^2\alpha^2 + \mu k^2 + (\sigma+\nu)] + [(\lambda+\mu)k^2\beta^2 + \mu k^2] \\ & [\eta\alpha^2k^2 + (\sigma+\nu)] - \nu(\sigma+\nu)^2\ [(\lambda+\mu)k^2\beta^2 + \mu k^2 + \nu] \\ & - \nu[(\lambda+\mu)k^2\alpha\beta]^2\ [\eta\alpha^2k^2 + (\sigma+\nu)] \end{aligned} \tag{16}$$

The solution of Eq. 11 for $k = 0$, the limiting solution for long wavelength, is

$$c^2\Big|_{k=0} = c_o^2 = \frac{1}{2(\rho_{(m)}+\rho_{(f)})} \Big\{(\lambda + 3\mu + \eta\alpha^2) \pm[(\lambda + 3\mu + \eta\alpha^2)^2$$

$$-4((\lambda+\mu)\mu + (\lambda+\mu)\eta\alpha^2\beta^2 + \mu^2 + \mu\eta\alpha^2)]^{\frac{1}{2}} \tag{17}$$

For the propagation in the direction of the fibers, $\alpha = 1$, $\beta = 0$, and Eq. 17 gives the two solutions

$$c_{oc}^2 = \frac{(\lambda+2\mu+\eta)}{(\rho_{(m)}+\rho_{(f)})} \tag{18}$$

$$c_{os}^2 = \frac{\mu}{(\rho_{(m)}+\rho_{(f)})} \tag{19}$$

Eq.s 18 and 19 clearly represent compressional and shear propagation velocities, respectively. $\lambda+2\mu+\eta$ is the combined compression modulus of the matrix and fibers and $\rho_{(m)} + \rho_{(f)}$ is the composite density. The fibers affect the shear velocity only through their density, since they have no shear resistance.

For propagation normal to the fibers, $\alpha = 0$, $\beta = 1$, and Eq. 17 gives

$$c_{oc}^2 = \frac{(\lambda + 2\mu)}{(\rho_{(m)}+\rho_{(f)})} \tag{20}$$

$$c_{os}^2 = \frac{\mu}{(\rho_{(m)}+\rho_{(f)})} \tag{21}$$

again corresponding to compressional and shear velocities. In this case, no stretch motion of the fibers occurs, the fibers affecting the velocities only through their density.

Numerical solutions of Eq. 11 have been obtained for a set of material parameters corresponding to a glass fiber reinforced phenolic resin [3]:

$\rho_m = 0.00013$ lb-sec^2/in^4 $\qquad$ $Y = 12.4 \times 10^6$ lb/in^2

$\rho_f = 0.00026$ lb-sec^2/in^4 $\qquad$ $\mu_f = 10.2 \times 10^6$ lb/in^2

$\lambda = 0.86 \times 10^6$ lb/in^2 $\qquad$ $s = 0.1$ in

$\mu = 0.37 \times 10^6$ lb/in^2 $\qquad$ $\delta = 0.005$ in

where ρ_m and ρ_f are the matrix and fiber material mass densities, Y is the individual fiber Young's modulus, μ_f is the fiber material shear modulus, s is the spacing between fibers (assumed uniformly distributed in a hexagonal array), and δ is the fiber radius.

In Fig. 1, the phase velocity squared in in^2/sec^2 is plotted as a function of wave number in 1/in. for plane waves propagating in a direction parallel to the fibers. Four curves, corresponding to four distinct propagation modes, are presented. Two of the modes, denoted as curves 1 and 2 in Fig. 1, approach finite propagation velocities for small wave number. These will be called primary modes, in keeping with terminology introduced in [5]. The remaining modes, which approach infinite propagation velocities for small wave number, will be called secondary modes.

The asymptotic values approached by the various modes at low and high wave number are shown in Fig. 1. The values approached by the primary modes for low wave number agree with Eq.s 18 and 19. At high wave number, primary mode 1, which is the compressional mode at low wave number, approaches the compressional propagation velocity in the matrix material, $c^2 = (\lambda + 2\mu)/\rho_{(m)}$. It was shown in [5] that at high wave number this mode approaches the lower of the two constituent compressional propagation velocities, in this case that associated with the matrix. Primary mode 2, which is the shear mode at low wave number, approaches at high wave number the lower of the two constituent shear propagation velocities. In this case, where the shear stiffness of the fibers has been neglected, the shear propagation velocity in the fibers is zero.

Conversely, the secondary modes 3 and 4 approach at high wave number the higher of the constituent compressional and shear propagation velocities. Mode 3 approaches the compressional velocity in the fibers, $c^2 = \eta/\rho_{(f)}$, and mode 4 approaches the shear velocity in the matrix, $c^2 = \mu/\rho_{(m)}$.

In Fig.s 2, 3 and 4, the propagation velocities of the four modes are plotted as functions of wave number for plane waves propagating at 5°, 30° and 60° to the fiber direction. A most interesting phenomenon is found to occur. Mode 1, which is the compressional mode at low wave number for propagation in the direction of the fibers, no longer approaches a compressional velocity at high wave number, but approaches the shear velocity $c^2 = \mu/\rho_{(m)}$. Similarly, mode 4, which approached the shear velocity at high wave number for the zero degree case, now approaches the compressional velocity $c^2 = (\lambda + 2\mu)/\rho_{(m)}$. It appears that a wave which is compressional in nature at low wave number becomes a shear wave at high wave number. That this is indeed what occurs will be shown in what follows.

In Fig. 5, the propagation velocities of the four modes are presented for plane waves propagating at 90° to the fiber direction. The asymptotic values approached by the primary modes at low wave number agree with Eq.s 20 and 21. The interpretation of the high wave number limits is identical to that for propagation parallel to the fibers; the primary modes approach the lower constituent compression and shear velocities while the secondary modes approach the higher compression and shear velocities. For propagation normal to the fibers, the compression and shear resistance of the fibers are zero, so that the lower compression and shear velocities each equal zero, while the higher velocities are those associated with the matrix.

In order to examine further the unanticipated behaviour of modes 1 and 4 observed in Fig.s 2-4, we have also determined the amplitude ratios $u_{(m)2}/u_{(m)1} = A_2/A_1$ and $u_{(f)2}/u_{(f)1} = B_2/B_1$ as functions of wave number. These ratios can then be used to determine the angles $\theta_{(m)} = \tan^{-1}(A_2/A_1)$ and $\theta_{(f)} = \tan^{-1}(B_2/B_1)$, which give the angles between the x_1 axis and the direction of particle motion in each constituent. In Fig.s 6 and 7, $\theta_{(m)}$ and $\theta_{(f)}$ are plotted as functions of wave number for propagation of plane waves at 5° to the fiber direction. Compressional modes are those for which particle motion is approximately parallel to the propagation direction, while shear modes involve particle motion approximately normal to the propagation direction. The "approximate" stipulation is necessary due to the anisotropy introduced by the fibers.

In Fig.s 6 and 7, it is clear that modes 1 and 4 correspond at low wave number to compression and shear modes, respectively. Then, a transition region in wave number is reached where the particle motion directions in modes 1 and 4 shift 90° and they become, respectively, shear and compression modes. This interesting behaviour does not occur in modes 2 and 3.

It can be seen in Fig.s 1-5 that this transition wave number region is associated with the wave number region in which rapid changes in phase velocity of the modes occur. The importance of this transition wave number region or transition frequency region in the behaviour of composites has previously been suggested in [2] and [5].

In Fig.s 1-7, it has been assumed that $\sigma = 0$, so that the coupling force between matrix and fibers for relative motions parallel to the fibers equals the coupling force for the same relative motions normal to the fibers. The sensitivity of the results to this assumption is illustrated in Fig. 8, in which phase velocity as a function of wave number is plotted for propagation at 60° to the fiber direction for $\sigma = \nu$, $\sigma = 0$ and $\sigma = -\frac{1}{2}\nu$ while holding $\sigma + \nu$ constant. The results correspond to a normal coupling force of one-half, one, and two times the parallel coupling force, respectively, for an equal relative displacement. It is clear from Fig. 8 that, even for these extreme variations in σ, the effect on the results is not sufficient to change the conclusions drawn from Fig.s 1-7. In particular, the asymptotic values of the phase velocities are unchanged.

ACKNOWLEDGEMENTS

This research was supported in part by National Science Foundation Grant No. GK-4712.

BIBLIOGRAPHY

[1] C. T. Sun, J. D. Achenbach, and G. Herrmann: Time Harmonic Waves in a Stratified Medium Propagating in the Direction of the Layering. J. of Appl. Mech., Vol. 35, p. 408-411 (1968).

[2] M. Stern, A. Bedford, and C. H. Yew: Wave Propagation in Viscoelastic Laminates. J. of Appl. Mech., (in press).

[3] J. S. Whittier and J. C. Peck: Experiments on Dispersive Pulse Propagation in Laminated Composites and Comparison with Theory. J. of Appl. Mech., Vol. 36, p. 485-490 (1969).

[4] L. M. Brekhovskikh: WAVES IN LAYERED MEDIA, Academic Press, New York (1960).

[5] A. Bedford and M. Stern: Toward a Diffusing Continuum Theory of Composite Materials. Sixth U.S. National Congress of Applied Mechanics, Harvard, J. of Appl. Mech., (in press).

[6] A. Bedford and M. Stern: A Multi-Continuum Theory for Composite Elastic Materials. Acta Mechanica (in press).

APPENDIX 1. EVALUATION OF η

Consider a set of continuous fibers distributed such that their direction can be specified by a continuous vector field d_j and their number density, the number of fibers per unit area measured normal to the local fiber direction, by a continuous scalar field F. It is easy to show that d_j and F are not independent, but are related by $(d_jF)_{,j} = 0$.

The number of fibers crossing an area element ds having normal n_j will be Fd_jn_jds. If t is the mean tensile force in the fibers crossing ds, the stress vector T_k acting on ds can be written as

$$T_k = tFd_jn_jd_k \tag{22}$$

If each fiber has cross sectional area S_f and Young's modulus Y, t can be written as

$$t = S_fY\varepsilon \tag{23}$$

where ε is the mean tensile strain in the fibers crossing ds. Modeling the fibers as a continuum, ε can be identified with the linear strain of the continuum in the fiber direction, given by the usual transformation as

$$\varepsilon = d_md_nu_{(f)m,n} \tag{24}$$

where $u_{(f)m,n}$ is the displacement gradient of the fiber continuum.

With Eq.s 23 and 24, Eq. 22 becomes

$$T_k = S_fYFd_jd_kd_md_nu_{(f)m,n}n_j \tag{25}$$

Defining the stress tensor in the fiber continuum $t_{(f)jk}$ by

$$T_k = t_{(f)jk}n_j \tag{26}$$

and comparing Eq.s 25 and 26, we obtain

$$t_{(f)jk} = S_fYFd_jd_kd_md_nu_{(f)m,n} \tag{27}$$

Comparing this with Eq. 6, the constant η is found to be

$$\eta = S_fYF \tag{28}$$

APPENDIX 2. EVALUATION OF $\nu + \sigma$

Let the fibers be straight, parallel and uniformly distributed throughout the matrix in a hexagonal array, as in Fig. 9. Let the fibers be subjected to a constant, uniformly distributed body force density in a direction parallel to the fibers, and let an oppositely directed body force density be applied to the matrix material, so that the two force systems are in equilibrium. By solving this static problem, both by elasticity theory and by using Eq.s 7 and 8, and requiring the two solutions to correspond, the constant $\nu + \sigma$ can be evaluated.

An approximate elasticity solution can be obtained by replacing the hexagonal problem by a cylindrical problem and considering a single fiber with radius δ contained in a cylinder of matrix material with radius h. This equivalent cylinder is illustrated in Fig. 9. The ratio of matrix mass to fiber mass will be preserved if h is related to the fiber spacing s by $h^2 = (\sqrt{3}/2\pi)s^2$.

Let $f_{(f)1}$ be the body force density acting on the fiber, and let u_{f1} be the axial displacement in the fiber. The boundary condition on u_{f1} at the fiber center is $\partial u_{f1}/\partial r\big|_{r=0} = 0$. The solution of this simple radially symmetric elasticity problem is easily shown to be

$$u_{f1} = (\rho_f f_{(f)1}/4\mu_f)(\delta^2 - r^2) \tag{29}$$

where ρ_f and μ_f are the density and shear modulus of the fiber material and r is the radial coordinate normal to the fiber. The average displacement in the fiber is

$$\overline{u}_{f1} = \frac{\int_A u_{f1}\, dA}{\int_A dA} = \rho_f\, f_{(f)1}\delta^2/8\mu_f \tag{30}$$

Let u_{m1} and $f_{(m)1}$ be the axial displacement component and body force density in the matrix. Due to symmetry, the boundary condition on u_{m1} at the matrix cylinder boundary is $\partial u_{m1}/\partial r\big|_{r=h} = 0$. The resulting solution for u_{m1} is

$$u_{m1} = (\rho_m f_{(m)1}/2\mu)\ (-r^2/2 + h^2 \log r + \delta^2/2 - h^2 \log \delta) \tag{31}$$

where ρ_m and μ are the mass density and shear modulus of the matrix material, and the corresponding average displacement is

$$\overline{u}_{m1} = (\rho_m f_{(m)1} h^2/\mu)\ \frac{\frac{1}{2}\log(h/\delta) + \frac{1}{2}\delta^2/h^2 - \frac{1}{8}\delta^4/h^4 - \frac{3}{8}}{(1 - \delta^2/h^2)} \tag{32}$$

The condition for the body force distributions to be in equilibrium is

$$\rho_m f_{(m)1}(\pi h^2 - \pi\delta^2) + \rho_f f_{(f)1}\pi\delta^2 = 0 \tag{33}$$

Now, solving the same problem using the continuum theory proposed in this paper, we note that the solution is independent of the axial coordinate and, in the continuum theory, must be independent of directions normal to the fibers as well. Eq. 7 for this static problem therefore becomes

$$(\nu + \sigma)\ (u_{(f)1} - u_{(m)1}) + \rho_{(m)} f_{(m)1} = 0 \tag{34}$$

Assuming that the constituent displacements $u_{(f)1}$ and $u_{(m)1}$ correspond to the average displacements $\overline{u}_{f1}$ and $\overline{u}_{m1}$ from the elasticity solution, we equate the relative displacements

$$u_{(f)1} - u_{(m)1} = \bar{u}_{f1} - \bar{u}_{m1} \qquad (35)$$

and using Eq.s 30, 32, 33, 34, and 35, the constant $\nu + \sigma$ is found to be

$$\nu + \delta = \frac{(\mu/h^2)(1-\delta^2/h^2)^2}{\left[\frac{1}{2}\log(h/\delta) + \frac{1}{2}\delta^2/h^2 - \frac{1}{8}\delta^4/h^4 - \frac{3}{8}\right] + \frac{1}{8}(\mu/\mu_f)(1-\delta^2/h^2)^2} \qquad (36)$$

The corresponding result for a laminated material with alternating matrix layers of thickness 2h and reinforcing layers of thickness 2δ is easily shown to be

$$\nu + \sigma = \frac{3\mu/h(h+\delta)}{1 + \frac{\mu\delta}{\mu_f h}} \qquad (37)$$

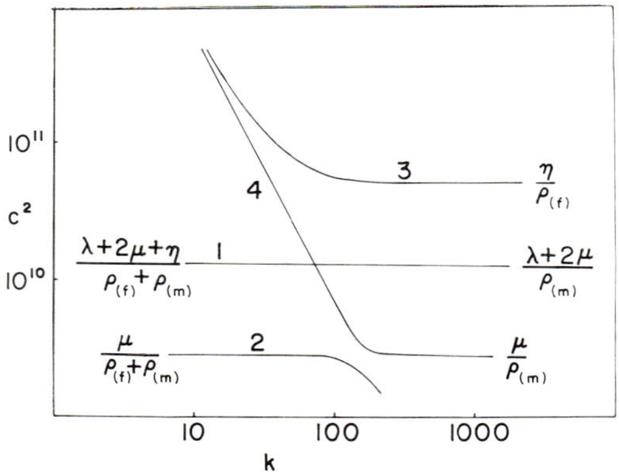

Fig. 1 Phase Velocity vs. Wave Number for Propagation in the Fiber Direction

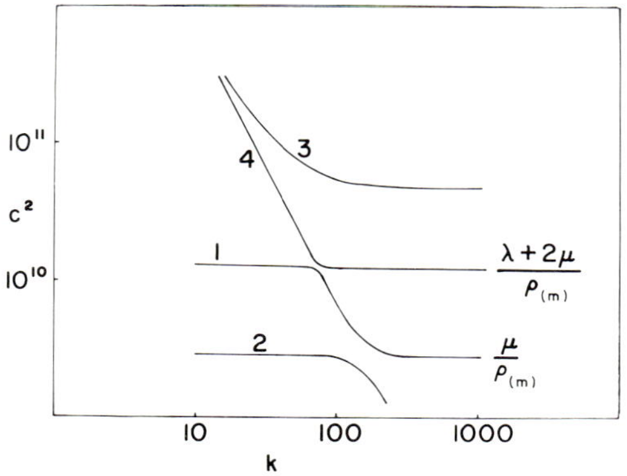

Fig. 2 Phase Velocity vs. Wave Number for Propagation at 5° to the Fiber Direction

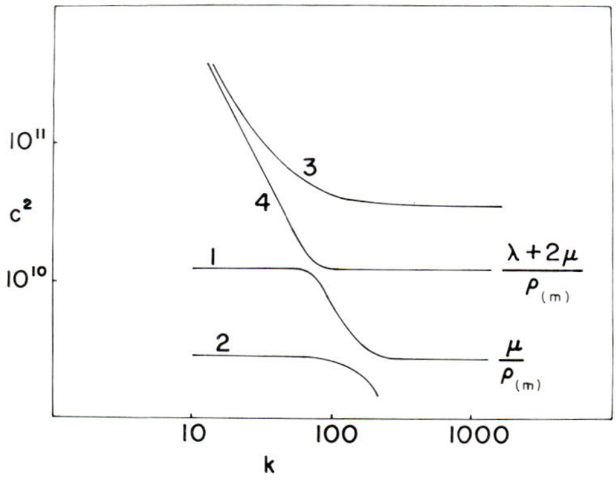

Fig. 3 Phase Velocity vs. Wave Number for Propagation at 30° to the Fiber Direction

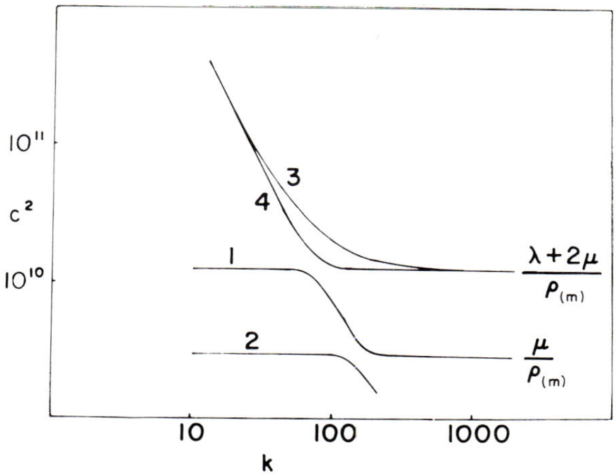

Fig. 4 Phase Velocity vs. Wave Number for Propagation at 60° to the Fiber Direction

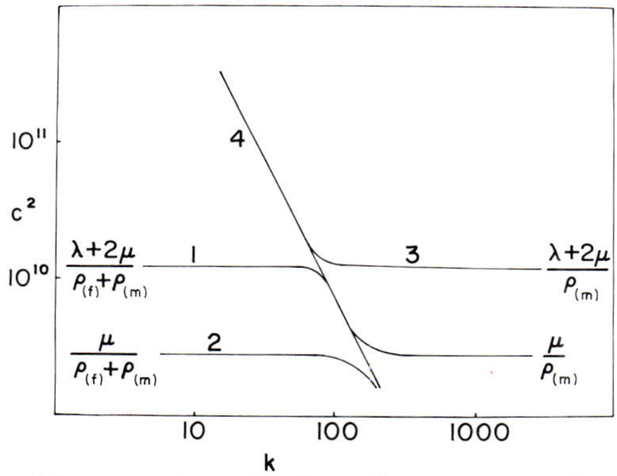

Fig. 5 Phase Velocity vs. Wave Number for Propagation Normal to the Fibers

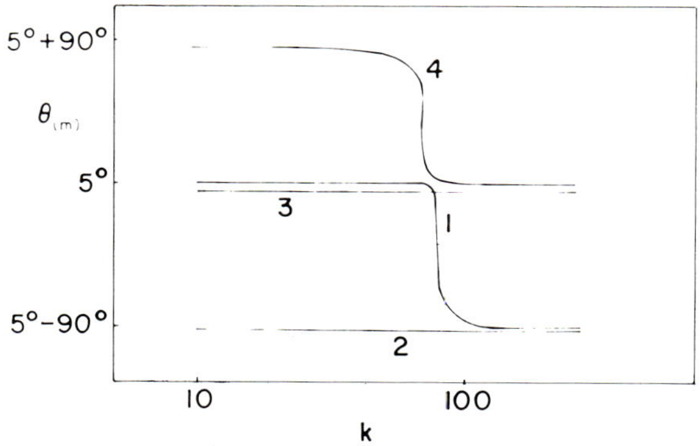

Fig. 6 Direction of Matrix Particle Motion Relative to the x_1-Axis vs. Wave Number for Propagation at 5° to the Fiber Direction

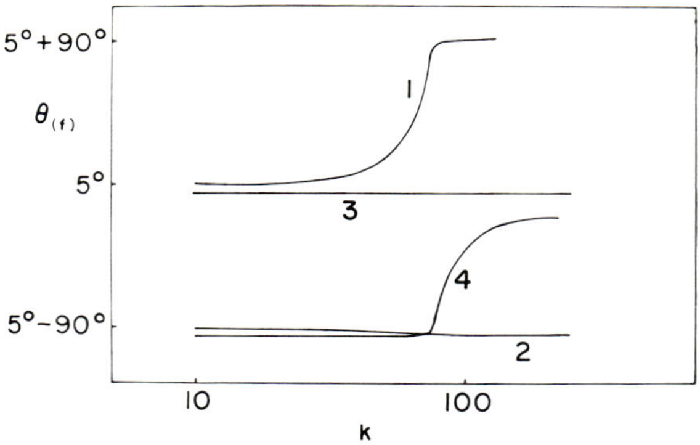

Fig. 7 Direction of Fiber Particle Motion Relative to the x_1-Axis vs Wave Number for Propagation at 5° to the Fiber Direction

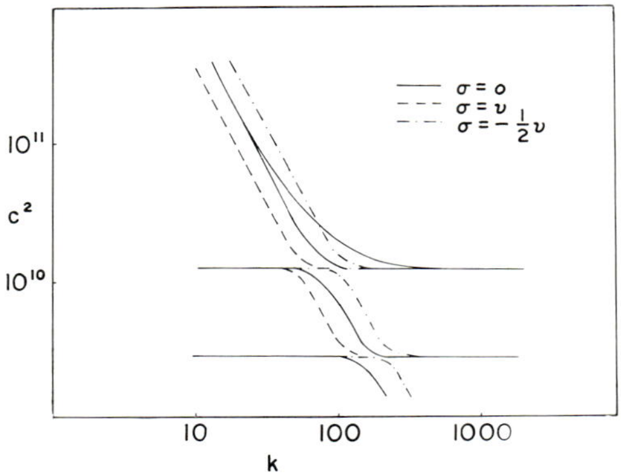

Fig. 8 Phase Velocity vs. Wave Number for Various Values of Normal Coupling Force Constant for Propagation at 60° to the Fiber Direction

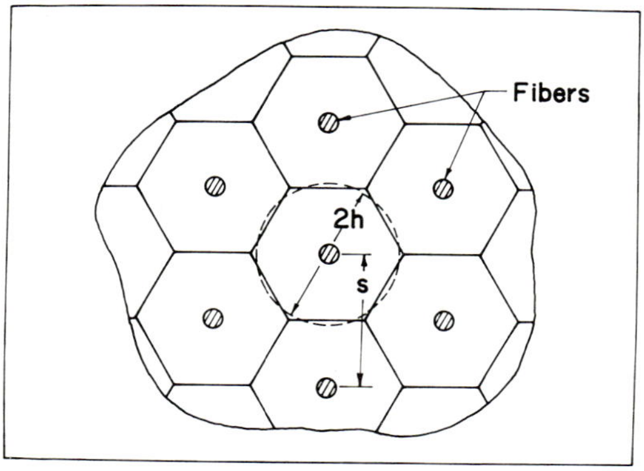

Fig. 9 Uniformly Distributed Fibers in a Hexagonal Array

STABILITY AND BUCKLING

Developments in Mechanics, Vol. 6. Proceedings of the 12th Midwestern Mechanics Conference: 36

Tensile Instability in Orthotropic Sheets Loaded Biaxially at an Arbitrary Orientation

C. W. BERT and J. J. SHAH*
UNIVERSITY OF OKLAHOMA

ABSTRACT

Using Hill's orthotropic plasticity theory, an analysis was developed to predict the nominal stress at which plastic tensile instability occurs in a flat, orthotropic sheet loaded by a biaxial stress system in the plane of the sheet at an arbitrary angle with the major material-symmetry axis. The cases of localized and diffuse necking are both considered, and applied to the case of a glass-fabric reinforced-plastic laminate loaded uniaxially at 45 degrees.

NOMENCLATURE

a_{pi}, a_{qj}	direction cosines
A_η	area associated with load $P_{\eta j}$
A_1, B_1, C_1	quantities defined by Eqs. (19)
C^o	coefficient defined by Eq. (12)
C_{ijkh}	anisotropic plastic stiffness coefficient in tensor notation
f	yield function
K	strength coefficient in Eq. (19)
n	strain-hardening exponent in Eq. (19)
$P_{\eta j}$	load on area A_η
R	anisotropic parameter used in Ref. 5

*Presently associated with Brass Division, Olin Corporation, East Alton, Illinois.

t	monotonically increasing deformation parameter
V	volume
x,y,z	material-symmetry axes of the material
Y_{ij}	yield stress
Z	critical subtangent to the $\bar{\sigma}$ vs. $\bar{\epsilon}$ curve
α	σ_{22}/σ_{11}
$\delta_{\eta i}$	Kronecker delta
ϵ_{ij}	components of the true strain tensor
$\bar{\epsilon}$	effective true strain
σ_{ij}	true stress components
$\bar{\sigma}$	effective true stress
λ	incremental plastic work per unit volume
φ	angle between maximum principal stress and major material-symmetry axis
θ	angle between direction of neck and maximum principal stress

Subscripts:

d	diffuse necking
h,i,j,k, m,n,r,s	summation notation taking values x,y
ℓ	localized necking
p,q	summation notation taking values 1,2
x,y	material-symmetry axes
1,2	applied-load-system axes

Superscripts:

A	actual mode of deformation
o	initial yield condition

INTRODUCTION

Plastic tensile instability (P.T.I.) is exemplified most simply by the necking of a ductile-material specimen in a uniaxial tensile test. In the case of flat sheets, P.T.I. is manifested by a thinning of the sheet thickness in the form of a trough-shaped neck. If the neck forms in a sheet along a direction perpendicular to the maximum principal stress, diffuse necking, first investigated by Swift [1], is taking place. However, in certain instances it has been observed that necking occurred along a direction at an acute angle with the maximum principal stress. This type of P.T.I., called localized necking, was first analyzed by Hill [2], who derived the conditions for determining the inclination of the neck.

As early as 1950, Hill [3] had developed a theory of orthotropic plasticity. Recently, in connection with certain metallic alloys which exhibit plastic orthotropy, there has been interest in P.T.I. of an orthotropic sheet (Keeler and Backofen [4], Moore and Wallace [5], Hillier [6]).

In all of the previous analyses known to the authors, it has been assumed that the material-symmetry axes and the principal-stress axes coincide. The aim of the present work is to study theoretically the plastic tensile instability of orthotropic materials subjected to plane stress when the material-symmetry axes and principal-stress axes do not coincide. Both the localized and diffuse necking criteria are considered. The analyses are applied to flat sheets in biaxial tension and to a numerical example of a glass-fabric-reinforced plastic laminate in uniaxial tension. The results for the example are compared with experimental results.

HYPOTHESES

The analyses presented are based on the following hypotheses:

1. Elastic strains are sufficiently small compared to the plastic strains that only the latter are considered.
2. The incremental or flow theory of plasticity is used rather than the deformation theory.
3. The plastic-range anisotropy is orthotropic and it is not altered by plastic deformation; in other words, proportional stressing is assumed.
4. Imcompressibility (constancy of volume) in the plastic range is also assumed. It was shown by Bert, Mills, and Hyler [7] that this hypothesis has negligible effect on the stress at which plastic tensile instability occurs in isotropic materials.
5. The loading path and the material hardening are such that the ratios of yield stresses remain invariant during deformation.
6. The hydrostatic components of stress do not influence yielding.

STRESS SYSTEMS

Considering a thin sheet element of an orthotropic material, the material-symmetry axes x,y are chosen as the cartesian axes of reference, in the plane of the sheet. It is assumed that directions 1 and 2 are arbitrary orthogonal directions making an angle φ measured positive counter-clockwise.

The transformation equations for transforming stress components from the 1,2 axes to the x,y axes are

$$\sigma_{ij} = \sigma_{pq} a_{pi} a_{qj} \quad ; \; i,j = x,y \; ; \; p,q = 1,2 \tag{1}$$

where a_{pi} and a_{qj} are the direction cosines, σ_{ij} and σ_{pq} are the total stress components with respect to x,y and 1,2 axes, respectively, and repeated indexes denote summation.

YIELD CRITERION AND ASSOCIATED FLOW RULE

The orthotropic yield criterion and associated generalized plastic strain increment suggested by Hill [3] are used. However, following the general approach of Hillier [6], tensor notation is used here.* According to Hill's plasticity theory, the yield criterion for plane stress ($\sigma_{zz} = \sigma_{zx} = 0$) can be written as:

$$2f(\sigma_{ij}) = C_{ijkh}\sigma_{ij}\sigma_{kh} = 1; \; i,j,k,h = x,y \tag{2}$$

where C_{ijkh} are material parameters.

*It should be noted that Hillier expressed the yield criterion in terms of the deviatoric components of the stresses rather than the total stresses.

For the most general anisotropy, there would be a 9 x 9 array if (2) would be expanded. However, due to the symmetry conditions for shear stress ($\sigma_{ij}=\sigma_{ji}$) and shear strains ($\epsilon_{ij}=\epsilon_{ji}$), this array would be reduced to a 6 x 6 array. Still further assuming that the material is orthotropic, i.e., has three mutually orthogonal planes of material symmetry, the array reduces as shown:

$$\begin{matrix} C_{xxxx} & C_{xxyy} & C_{xxzz} & 0 & 0 & 0 \\ & C_{yyyy} & C_{yyzz} & 0 & 0 & 0 \\ & & C_{zzzz} & 0 & 0 & 0 \\ & & & C_{yzyz} & 0 & 0 \\ & \text{(Symmetric)} & & & C_{zxzx} & 0 \\ & & & & & C_{xyxy} \end{matrix} \quad (3)$$

It is seen that for anisotropy of the orthotropic type, there are only twelve non-vanishing coefficients of the original 9 x 9 array. Of these twelve, symmetry about the main diagonal of the array of coefficients requires that

$$C_{jjii} = C_{iijj} \quad (4)$$

Thus, there are only nine independent coefficients. Of these nine, it can be shown that six can be related to the six parameters used by Hill [3].

The remaining three coefficients in the above array can be shown to be as follows (see Shah [9], Appendix A):

$$\begin{aligned} C_{xxyy} &= -\ 1/2\ (C_{xxxx} + C_{yyyy} - C_{zzzz}) \\ C_{yyzz} &= -\ 1/2\ (C_{yyyy} + C_{zzzz} - C_{xxxx}) \\ C_{zzxx} &= -\ 1/2\ (C_{zzzz} + C_{xxxx} - C_{yyyy}) \end{aligned} \quad (5)$$

Hill [3] has defined $d\epsilon_{mn}$ by the following associated flow rule:

$$d\epsilon_{mn} = d\lambda\ [\partial f(\sigma_{mn})/\partial\sigma_{mn}] \quad (6)$$

Thus, for the plane stress condition,

$$\begin{aligned} &d\epsilon_{xx} = d\lambda\ (C_{xxxx}\sigma_{xx} + C_{xxyy}\sigma_{yy}),\ d\epsilon_{yy} = d\lambda\ (C_{yyyy}\sigma_{yy} + C_{xxyy}\sigma_{xx}) \\ &d\epsilon_{zz} = d\lambda\ (C_{zzxx}\sigma_{xx} + C_{yyzz}\sigma_{yy}),\ d\epsilon_{xy} = d\lambda\ (C_{xyxy}\sigma_{xy}) \end{aligned} \quad (7)$$

where $d\lambda$ is a coefficient of physical significance explained later. Equations (7) can be rewritten as:

$$\begin{aligned} &d\epsilon_{xx}/(C_{xxxx}\sigma_{xx} + C_{xxyy}\sigma_{yy}) = d\epsilon_{yy}/(C_{yyyy}\sigma_{yy} + C_{xxyy}\sigma_{xx}) = \\ &d\epsilon_{zz}/(C_{zzxx}\sigma_{xx} + C_{yyzz}\sigma_{yy}) = d\epsilon_{xy}/(C_{xyxy}\sigma_{xy}) = d\lambda \end{aligned} \quad (8)$$

Following Hill [3], the effective yield stress $\bar{\sigma}$ and effective plastic strain increment $d\bar{\epsilon}$ are chosen to satisfy the relation

$$\sigma_{ij}d\epsilon_{ij} = \bar{\sigma}\ d\bar{\epsilon} = d\lambda \quad (9)$$

where $d\lambda$ is the increment of plastic work per unit volume.

It is assumed that ratios of the yield stresses remain invariant during deformation, and hence

$$Y_{ij} = t\, Y^{o}_{ij} \quad ; \quad C_{ijkh} = t^{-2}\, C^{o}_{ijkh} \tag{10}$$

where t is some parameter of deformation, and superscript o denotes the initial yield condition.

The effective stress and effective strain can then be written as:

$$\bar{\sigma} = (C^{o})^{-1/2} \, (C^{o}_{ijkh}\sigma_{ij}\sigma_{kh})^{1/2} \tag{11}$$

where

$$C^{o} = (1/3)\,(C^{o}_{xxxx} + C^{o}_{yyyy} + C^{o}_{zzzz}) \tag{12}$$

and

$$d\bar{\epsilon} = (2/3)^{1/2}\,(-C^{o}_{yyzz} - C^{o}_{zzxx} - C^{o}_{xxyy})^{1/2}$$

$$\Bigg[-C^{o}_{yyzz}\left(\frac{-C^{o}_{zzxx}\, d\epsilon_{yy} + C^{o}_{xxyy}\, d\epsilon_{zz}}{C^{o}_{yyzz}\, C^{o}_{zzxx} + C^{o}_{zzxx}\, C^{o}_{xxyy} + C^{o}_{xxyy}\, C^{o}_{yyzz}}\right)^{2}$$

$$-C^{o}_{zzxx}\left(\frac{-C^{o}_{xxyy}\, d\epsilon_{zz} + C^{o}_{yyzz}\, d\epsilon_{xx}}{C^{o}_{yyzz}\, C^{o}_{zzxx} + C^{o}_{zzxx}\, C^{o}_{xxyy} + C^{o}_{xxyy}\, C^{o}_{yyzz}}\right)^{2}$$

$$-C^{o}_{xxyy}\left(\frac{-C^{o}_{yyzz}\, d\epsilon_{xx} + C^{o}_{xxyy}\, d\epsilon_{yy}}{C^{o}_{yyzz}\, C^{o}_{zzxx} + C^{o}_{zzxx}\, C^{o}_{xxyy} + C^{o}_{xxyy}\, C^{o}_{yyzz}}\right)^{2}$$

$$+ \frac{(d\epsilon_{yz})^{2}}{C^{o}_{yzyz}} + \frac{(d\epsilon_{zx})^{2}}{C^{o}_{zxzx}} + \frac{(d\epsilon_{xy})^{2}}{C^{o}_{xyxy}} \Bigg]^{1/2} \tag{13}$$

CRITICAL SUBTANGENT FOR DIFFUSE NECKING

Following Hillier [6], the principle of maximum rate of plastic work requires that for a volume V of plastic material,

$$\int [(d\sigma^{A}_{ij} - d\sigma_{ij})\, d\epsilon_{ij}]\, dV \geq 0 \tag{14}$$

where here the superscript A refers to the actual mode of deformation.

If the first time derivatives are assumed to be continuous, differentiation of (14) yields

$$\int [\, (d\sigma^{A}_{ij} - d\sigma_{ij})\, d\epsilon_{ij}]\, dV \geq 0 \tag{15}$$

It is postulated that for instability to occur, inequality (15) should be violated for materials subjected to a homogeneous state of stress and strain.

It is convenient to write inequality (15) as follows:

$$Z^{-1} = (\bar{\sigma})^{-1}\, (d\bar{\sigma}/d\bar{\epsilon}) \geq (1/\bar{\sigma})\, (d\bar{\sigma}^{A}/d\bar{\epsilon}^{A}) \tag{16}$$

where Z is the critical subtangent to the curve of effective true stress versus effective true strain.

For purposes of calculation, the critical subtangent for diffuse necking may be calculated from the following formula

$$(Z_d)^{-1} = (C^o\bar{\sigma})^{-1} C^o_{ijrs} C^o_{mnkh} \sigma_{rs}\sigma_{kh}[A^{-1}_\eta (dP_{\eta j}/d\epsilon_{mn}) - A^{-1}_\eta \sigma_{\eta j} (dA_\eta/d\epsilon_{mn})]\delta_{\eta i} \tag{17}$$

where $\delta_{\eta i}$ is the Kronecker delta, and there is no sum over the Greek suffix η.

This expression is the same as that of Hillier [6] except that in the present work the total components of stresses are considered, whereas Hillier had considered the deviatoric components of stresses.

Considering a flat sheet subjected to biaxial tension, i.e. to the stress system σ_{11}, σ_{22} on the principal-stress axes (see Fig. 1) or to the stress system σ_{xx}, σ_{yy}, σ_{xy} on the material-symmetry axes, the critical subtangent for diffuse necking is obtained as follows (Shah [9], Appendix B):

$$\begin{aligned}
&(Z_d)^{-1}\,[(C^o_{xxxx} + C^o_{yyyy} + C^o_{zzzz})/3]^{1/2}\,[C^o_{xxxx}\{(1+\alpha\tan^2\varphi)/(\alpha+\tan^2\varphi)\}^2 \\
&+ 2\,C^o_{xxyy}(1+\alpha\tan^2\varphi)/(\alpha+\tan^2\varphi) + C^o_{yyyy} + (4\,C^o_{xyxy})\{(1-\alpha)(\tan\varphi)/(\alpha+\tan^2\varphi)\}]^{-3/2} \\
&= [\{(C^o_{xxxx})(1+\alpha\tan^2\varphi)/(\alpha+\tan^2\varphi) + C^o_{xxyy}\}^2\{(1+\alpha\tan^2\varphi)/(\alpha+\tan^2\varphi)\} \\
&\quad \{(d\epsilon_{11}/d\epsilon_{22})\tan^2\varphi + 1 + (d\epsilon_{12}/d\epsilon_{22})\tan\varphi\}\{\tan^2\varphi + (d\epsilon_{11}/d\epsilon_{22}) \\
&- (d\epsilon_{12}/d\epsilon_{22})\tan\varphi\}^{-1} + \{C^o_{yyyy} + C^o_{xxyy}(1+\alpha\tan^2\varphi)/(\alpha+\tan^2\varphi)\}^2 \\
&+ 4\,C^o_{xyxy}\{\tan\varphi\,(1-\alpha)/(\alpha+\tan^2\varphi)\}^2\{C^o_{yyyy} + C^o_{xxyy}(1+\alpha\tan^2\varphi)/(\alpha+\tan^2\varphi)\}]
\end{aligned} \tag{18}$$

If $\varphi = 0$ and $C^o_{yyzz}/C^o_{xxyy} = C^o_{zzxx}/C^o_{xxyy} = R$, (18) reduces to the form that Moore and Wallace [5] obtained for the case when the principal-stress axes are lined up with the material-symmetry axes. This serves as a check on (18). It is of interest also to note that Chiang and Kobayashi [8] obtained the same expression as Moore and Wallace by using the plane strain theory and an effective stress (2/3) (2+R)/(1+R) times the effective stress that Moore and Wallace used.

If the work-hardening characteristics of the material are assumed to follow the Hollomon power-law form

$$\bar{\sigma} = K\,\bar{\epsilon}^{\,n} \tag{19}$$

where K and n are empirical constants, it can be shown that the critical effective strain becomes proportional to the magnitude of the critical subtangent (Shah [9], Appendix B).

CRITICAL SUBTANGENT FOR LOCALIZED NECKING

Theortically, the line of the neck should coincide with a characteristic in view of the property of characteristics as curves along which small disturbances propagate (Hill [3]). The slopes dy/dx of the characteristics satisfy the equation

$$d\epsilon_{xx}\ (dx)^2 + 4\ d\epsilon_{xy}(dxdy) + d\epsilon_{yy}(dy)^2 = 0 \tag{20}$$

where $d\epsilon_{xy}$ is the tensor shear strain.

Substituting the values of $d\epsilon_{xx}$, $d\epsilon_{yy}$, $d\epsilon_{xy}$ from Eq. (8), and the values of σ_{xx}, σ_{yy}, σ_{xy} from Eq. (2), Eq. (20) can be written as:

$$[(C^o_{xxxx})(\sigma_{11}\cos^2\varphi + \sigma_{22}\sin^2\varphi - 2\sigma_{12}\sin\varphi\cos\varphi) + C^o_{xxyy}(\sigma_{11}\sin^2\varphi + \sigma_{22}\cos^2\varphi$$

$$-2\sigma_{12}\sin\varphi\cos\varphi)] + [C^o_{yyyy}(\sigma_{11}\sin^2\varphi + \sigma_{22}\cos^2\varphi - 2\sigma_{12}\sin\varphi\cos\varphi)$$

$$+ C^o_{xxyy}(\sigma_{11}\cos^2\varphi + \sigma_{22}\sin^2\varphi - 2\sigma_{12}\sin\varphi\cos\varphi)]\ (dy/dx)^2$$

$$+ 4\ C^o_{xyxy}[\sigma_{11}\sin\varphi\cos\varphi - \sigma_{22}\sin\varphi\cos\varphi + \sigma_{12}(\cos^2\varphi - \sin^2\varphi)](dy/dx) = 0 \tag{21}$$

From Fig. 2,

$$dy/dx = \tan(\theta + \varphi) \tag{22}$$

Then (21) can be rewritten as

$$A_1\ \tan^2\theta + 2B_1\ \tan\theta + C_1 = 0 \tag{23}$$

where

$$A_1 = -\sigma_{11}\Big[-C^o_{xxyy} + \left\{8C^o_{xyxy} + C^o_{yyzz} + C^o_{zzxx} + 4C^o_{xxyy}\right\}\sin^2\varphi\cos^2\varphi$$

$$-\alpha\left\{-C^o_{xxyy}\cos 2\varphi - C^o_{zzxx}\sin^2\varphi - C^o_{yyzz}\cos^2\varphi + 4C^o_{xxyy}\sin^2\varphi\cos^2\varphi\right\}\Big]$$

$$B_1 = \tfrac{1}{2}\sin 2\varphi\ [(-C^o_{yyzz} - 2C_{xxyy} - 4C^o_{xyxy})\sin^2\varphi + (4C^o_{xyxy} + C^o_{zzxx} + C^o_{xxyy})\cos^2\varphi$$

$$+ \alpha\ (-2C^o_{xxyy}\cos^2\varphi - 4C^o_{xyxy}\cos^2\varphi + C^o_{yyzz}\cos^2\varphi + C^o_{zzxx}\sin^2\varphi)]\ \sigma_{11}$$

$$C_1 = [(-C^o_{yyzz}\sin^2\varphi - C^o_{zzxx}\cos^2\varphi) + \alpha(-C^o_{yyzz}\cos^2\varphi - C^o_{zzxx}\sin^2\varphi) - A_1]\sigma_{11}$$

Equation (23) suggests that there are two possible necking directions corresponding to the two roots of the quadratic equation.

Using relation (18), the critical subtangent can be determined in this case of replacing σ_{11}, σ_{22} by σ_{33}, σ_{44} respectively and the angle φ by $(\theta+\varphi)$.

Hence,

$$(Z_\ell)^{-1}\ [(C^o_{xxxx} + C^o_{yyyy} + C^o_{zzzz})/3]^{\frac{1}{2}}\ [C^o_{xxxx}\left\{(1+\alpha\tan^2(\theta+\varphi))/(\alpha+\tan^2(\theta+\varphi))\right\}^2$$

$$+ 2C^o_{xxyy}\left\{1+\alpha\tan^2(\theta+\varphi)\right\}\left\{\alpha+\tan^2(\theta+\varphi)\right\}^{-1} + C^o_{yyyy} + 4C^o_{xyxy}\left\{(1-\alpha)\tan(\theta+\varphi)\right\}^2$$

$$\left\{\alpha + \tan^2(\theta+\varphi)\right\}^{-2}]^{3/2} = [C^o_{xxxx}\left\{1+\alpha\tan^2(\theta+\varphi)\right\}\left\{\alpha+\tan^2(\theta+\varphi)\right\}^{-1} + C^o_{xxyy}]^2$$

$$\left\{1+\alpha\tan^2(\theta+\varphi)\right\}\left\{\alpha+\tan^2(\theta+\varphi)\right\}^{-1}\left\{(d\epsilon_{11}/d\epsilon_{22})\tan^2(\theta+\varphi) + 1 + (d\epsilon_{12}/d\epsilon_{22})\tan(\theta+\varphi)\right\}$$

$$\left\{(d\varepsilon_{11}/d\varepsilon_{22}) + \tan^2(\theta+\varphi) - (d\varepsilon_{12}/d\varepsilon_{22}) \tan(\theta + \varphi)\right\}^{-1} + [C^o_{yyyy} + C^o_{xxyy}\left\{1+\alpha\tan^2(\theta+\varphi)\right\}$$

$$\left\{\alpha+\tan^2(\theta+\varphi)\right\}^{-1}]^2 + 4C^o_{xyxy}\left\{(1-\alpha)\tan(\theta+\varphi)\right\}^2\left\{\alpha+\tan^2(\theta+\varphi)\right\}^{-2}[C^o_{yyyy}+C^o_{xxyy}$$

$$\left\{1+\alpha\tan^2(\theta+\varphi)\right\}\left\{\alpha + \tan^2(\theta + \varphi)\right\}^{-1}] \qquad (24)$$

where

$$d\varepsilon_{11} = (d\lambda/2)(\cos^2\varphi)\{-C^o_{zzxx}\sigma_{11} - 2C^o_{xxyy}(\sigma_{11}-\sigma_{22}) + (\tan^2\varphi)\sigma_{11}(4C^o_{xyxy}+4C^o_{xxyy}) + \sigma_{22}$$

$$(-2C^o_{zzxx} - 2C^o_{yyzz} - 4C^o_{xyxy} - 4C^o_{xxyy}) + (\tan^4\varphi)(-C^o_{yyzz})\sigma_{11} - 2C^o_{xxyy}(\sigma_{11}-\sigma_{22})\}$$

$$d\varepsilon_{22} = (d\lambda/2)(\cos^4\varphi)\{[-2C^o_{xxyy}(\sigma_{22}-\sigma_{11}) - C^o_{yyzz}\sigma_{22}] + (\tan^4\varphi)[-C^o_{zzxx}\sigma_{22}$$

$$-2C^o_{xxyy}(\sigma_{22}-\sigma_{11})] + \tan^2\varphi[\sigma_{11}(-2C^o_{yyzz} - 2C^o_{zzxx} - 4C^o_{xyxy} - 4C^o_{xxyy}) + (4C^o_{xyxy}+4C^o_{xxyy})\sigma_{22}]\}$$

$$d\varepsilon_{12} = (d\lambda/2)\sin^2\varphi\cos^4\varphi\{2C_{xyxy}\sigma_{11} + (4C^o_{yyzz}-2C^o_{xyxy})\sigma_{22} + (4\tan^2\varphi)$$

$$C^o_{xxyy}(-\sigma_{11}+\sigma_{22})\tan^2\varphi[\sigma_{11}(2C^o_{xyxy} + 4C^o_{yyzz}+4C^o_{xxyy}) + (-4C^o_{xxyy}-2C^o_{xyxy})\sigma_{22}]$$

$$+ \sec^2\varphi(-2C^o_{zzxx}\sigma_{11} + 2C^o_{yyzz}\sigma_{22}) + \tan^2\varphi\sec^2\varphi(-2C^o_{yyzz}\sigma_{11} - 2C^o_{zzxx}\sigma_{22})\}$$

APPLICATION

Fig. 3 shows the reciprocal of the critical subtangent plotted as a function of biaxial stress ratio for a typical orthotropic material (glass-cloth-reinforced plastic laminate) loaded at 45° to the axis of major material symmetry and for an isotropic material (Keeler and Backofen [4]). It is seen that in both cases, failure must have occurred due to the diffuse necking criterion, as the value for the critical subtangent is higher for both in diffuse necking and the true stress at which plastic tensile instability occurs is inversely proportional to the critical subtangent. However, in the plot for orthotropic material, the region of localized necking cannot be separated as in the isotropic material. This is a significant difference between the two cases.

REFERENCES

[1]. Swift, H.W., "Plastic Instability Under Plane Stress," Journal of the Mechanics and Physics of Solids, Vol. 1, No. 1, 1952, pp. 1-18.

[2]. Hill, R., "On Discontinuous Plastic States with Special Reference to Localized Necking in Thin Sheets, " Journal of the Mechanics and Physics of Solids, Vol. 1, No. 1, 1952, pp. 19-30.

[3]. Hill, R., The Mathematical Theory of Plasticity, Clarendon Press, Oxford, 1950.

[4]. Keeler, S.P. and Backofen, W.A., "Plastic Instability and Fracture in Sheets Stretched Over Rigid Punches," Transactions of the American Society for Metals, Vol. 56, No. 1, 1963, pp. 25-48.

[5]. Moore, G.G. and Wallace, J.F., "The Effect of Anisotropy on Instability in Sheet-Metal Forming," Journal of the Institute of Metals, Vol. 93, No. 1, 1964, pp. 33-38.

[6]. Hillier, M.J., "On the Tensile Instability of Orthotropic Plastic Material", International Journal of Mechanical Sciences, Vol. 7, June 1965, pp. 441-445.

[7]. Bert, C.W., Mills, E.J., and Hyler, W.S., "Effect of Variation in Poisson's Ratio on Plastic Tensile Instability", Journal of Basic Engineering, Transactions of the American Society of Mechanical Engineers, Vol. 89D, No. 1, Mar. 1967, pp. 35-39.

[8]. Chiang, D.C. and Kobayashi, S., "The Effect of Anisotropy and Work-Hardening Characteristics on the Stress and Strain Distribution in Deep Drawing", Journal of Engineering for Industry, Transactions of the American Society of Mechanical Engineers, Vol. 88B, 1966, pp. 443-448.

[9]. Shah, J.J., "Plastic Tensile Instability in Orthotropic Sheets Subject to Biaxial Loading at an Arbitrary Orientation", master's thesis, University of Oklahoma, 1967.

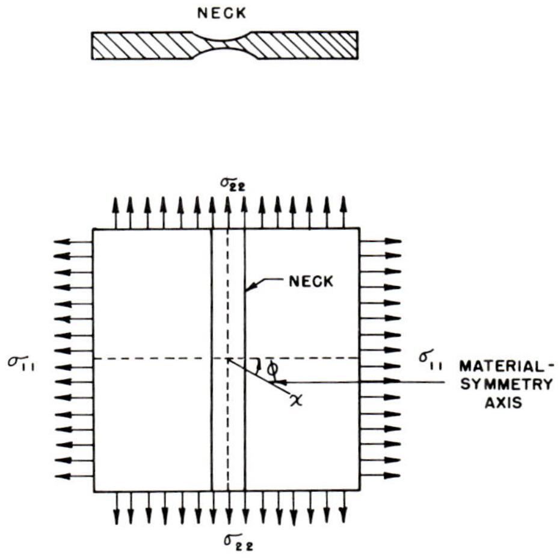

Fig. 1. Schematic diagram of diffuse necking, for the case when σ_{11} is the maximum principal stress.

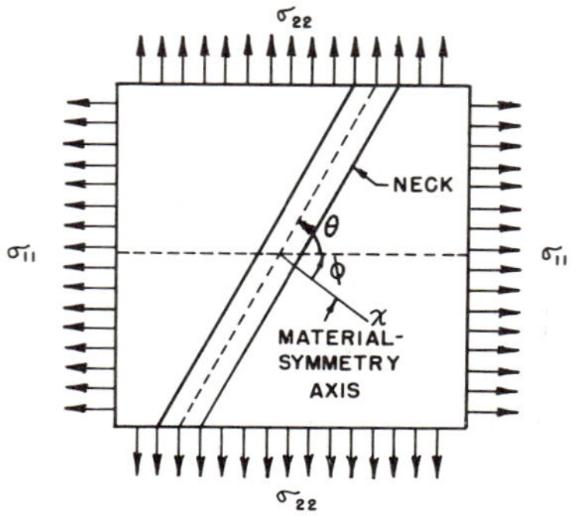

Fig. 2. Schematic diagram of localized necking, showing angular relationships among direction of maximum principal stress (σ_{11}), the major material-symmetry axis (x), and the neck orientation.

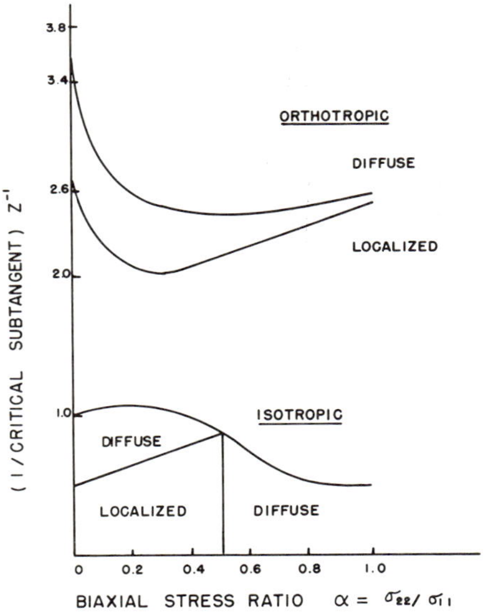

Fig. 3. Critical subtangent for P.T.I. versus biaxial stress ratio for an isotropic and an orthotropic material.

Developments in Mechanics, Vol. 6. Proceedings of the 12th Midwestern Mechanics Conference: 37

Axisymmetric Buckling of Axially Compressed Heated Cylindrical Shells

ATLE GJELSVIK
COLUMBIA UNIVERSITY

ABSTRACT

It is shown that the buckling load of axially loaded geometrically perfect cylindrical shells can be significantly reduced by partial yielding caused by temperature gradients through the wall. The buckling is of the Shanley type. The load on the shell, and the difference between the outside and inside temperature takes the place of the load in the classical Shanley formulation, and it is these quantities which have to change to insure plastic flow as the shell buckles.

INTRODUCTION

The combination of thermal gradients and loads on a structural element may cause partial yielding of the element . If the structure is free to expand it is known from limit analysis that partial yielding will not affect the collapse load of elastic - perfectly plastic structures. For compression elements however, the situation is very different. Bodner [1] and Tao and Gjelsvik [2] have shown that the stability of columns can be substantially reduced by the presence of transverse thermal gradients. Tao and Gjelsvik [2] showed that the buckling process of the column was of the type first discussed by Shanley [3].

In the present paper the buckling of an axially loaded cylindrical shell with a thermal gradient across the wall is analyzed. The outside and inside temperatures are considered spacially constant but slowly varying with time. The temperature gradient across the wall can then be considered linear. The applied load in the shell is also varying with time.

The analysis is restricted to an elastic-perfectly plastic material with temperature independent properties. These restrictions are however not essential. The analysis of the more general problem with work hardening material, temperature dependent properties and non linear temperature distributions can be performed in the same way as the present analysis, but with considerable increase in the complexity of the mathematical manipulations.

The analysis is restricted to axisymmetric deformations. This is felt to be reasonable since the buckling of heated structural elements is essentially plastic and since the plastic buckling of cylindrical shells as shown by Batterman and Lee [4] is usually axisymmetric.

The rate method introduced by Onat [5] and subsequently used by other writers in the study of stability problems is used in the analysis.

The presentation has three main parts. In the first part the governing rate equations for the problem is set down. In the second the stress distribution and deformations prior to buckling are obtained. In the third the buckling and some aspects of the post buckling are investigated.

A closed form solution is obtained for the case with only one plastic zone in the cross section. For a practical problem this is the most important case. For additional plastic zones a step by step numerical procedure is required to obtain the solution. The governing rate equations are given in a sufficiently general form to cover all these cases.

Governing Equations

When written in a convenient non-dimensional form the rate equations of equilibrium are:

$$\frac{d^2\mathring{m}_z}{dz^2} + n_z \frac{d^2 v_r}{dz^2} - \frac{1}{\lambda}\dot{n}_\theta = 0 , \qquad \frac{d\dot{n}_z}{dz} = 0 \tag{1}$$

The non-dimensional quantities are:

$$\dot{m}_z = \frac{\dot{M}_Z}{\sigma_o h^2} , \quad \dot{n}_z = \frac{\dot{N}_Z}{\sigma_o h} , \quad \mathring{n}_\theta = \frac{\dot{N}_\theta}{\sigma_o h} , \quad v_r = \frac{V_r}{h} , \quad z = \frac{Z}{R} ,$$

$$\lambda = \frac{h}{R}, \quad n_z = \frac{N_Z}{\sigma_o h} \tag{2}$$

in which $\dot{M}_Z$ = the rate of change of the meridional bending moment, N_Z = the membrane force in the meridional direction, $\dot{N}_Z$ and $\dot{N}_\theta$ = the rate of change of the meridional and circumferential membrane forces, V_r is the outward radial velocity, Z = axial coordinate, R = radius, h = wall thickness and σ_o = the yield stress. The sign convention and notation used are shown in Fig. 1.

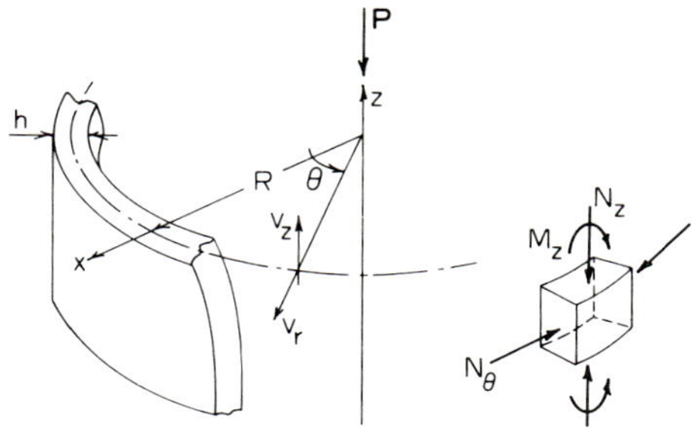

Fig. 1 Notation and sign convention

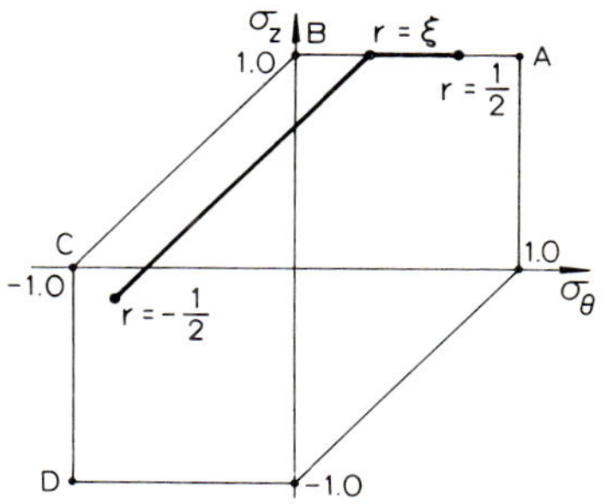

Fig. 2. The yield locus.

When written is a non-dimensional form the strain rates and strain rate velocity relations are:

$$\dot{\epsilon}_z = \dot{e}_z + r\dot{k}_z \ , \quad \dot{\epsilon}_\theta = \dot{e}_\theta \ , \quad \dot{e}_\theta = -\lambda v_r$$

$$\dot{e}_z = -\lambda \frac{dv_z}{dz} \ , \quad \dot{k}_z = \lambda^2 \frac{d^2 v_r}{dz^2} \tag{3}$$

The non-dimensional quantities are:

$$r = \frac{x}{h} \ , \quad \dot{k}_z = h\dot{K}_Z \ , \quad v_z = \frac{V_Z}{h} \tag{4}$$

in which x = a coordinate measured radially outward from the middle surface, $\dot{K}_Z$ = the rate of change of curvature in the meridional direction and V_Z = the velocity in the meridional direction. These rate equations are the same as those used by Batterman [6] in a series of paper on the plastic buckling of cylindrical shells.

The temperature rate distribution $\dot{T}$ in the column wall is when written in a non-dimensional form:

$$\alpha e\mathring{T} = \dot{\tau} + (1 - \nu)\ \dot{t}r \tag{5}$$

The non-dimensional quantities are:

$$e = \frac{E}{\sigma_o} \ , \quad \dot{\tau} = \frac{\alpha E\ (\dot{T}_o + \dot{T}_i)}{2\sigma_o} \ , \quad t = \frac{\alpha E\ (\dot{T}_o - \dot{T}_i)}{(1-\nu)\sigma_o} \tag{6}$$

in which E = Young's modulus, $\dot{T}_o$ and $\dot{T}_i$ = the outside and inside rate of change of temperature, α = the coefficient of linear thermal expansion and ν = Poisson's ratio.

The yield locus is shown in fig. 2. The stress rates required in the analysis are the fully elastic ones and the plastic ones corresponding to the sides AB, BC, and CD including the corners A, B, C, and D of the yield locus. These stress rate-strain rate equations and the criterions for plastic flow are summerized in Table I. The stresses are non-dimensionalized by dividing by σ_o. Compressive stresses are taken positive.

TABLE I MATERIAL RATE EQUATIONS

Position on the Yield Locus	Strain Rate Criterion for Loading	Stress Rate Equations
CORNER A		
$\sigma_z=1$, $\sigma_\theta=1$	$\dot\epsilon_z+\alpha\dot T\ge 0$, $\dot\epsilon_\theta+\dot T\ge 0$	$\dot\sigma_z=0$, $\dot\sigma_\theta=0$
SIDE AB		
$\sigma_z=1$ $0<\sigma_\theta<1$	$\dot\epsilon_z+\nu\dot\epsilon_\theta+\alpha(1+\nu)\dot T\ge 0$	$\dot\sigma_z=0$, $\dot\sigma_\theta=e(\dot\epsilon_z+\alpha\dot T)$
CORNER B		
$\sigma_z=1$, $\sigma_\theta=0$	$\dot\epsilon_\theta+\alpha\dot T\le 0$, $\dot\epsilon_\theta+\dot\epsilon_z+2\alpha\dot T\ge 0$	$\dot\sigma_z=0$, $\dot\sigma_\theta=0$
SIDE BC		
$\sigma_z-\sigma_\theta=1$, $0<\sigma_z<1$	$\dot\epsilon_z-\dot\epsilon_\theta \ge 0$	$\dot\sigma_z=\dot\sigma_\theta=\frac{e}{2(1-\nu)}(\epsilon_z+\epsilon_\theta+2\alpha\dot T)$
CORNER C		
$\sigma_z=0$, $\sigma_\theta=-1$	$\dot\epsilon_z+\alpha\dot T\ge 0$, $\dot\epsilon_\theta+\dot\epsilon_z+2\alpha\dot T\le 0$	$\dot\sigma_z = 0$, $\dot\sigma_\theta = 0$
SIDE CD		
$\sigma_\theta=-1$, $-1<\sigma_z<0$	$\dot\epsilon_\theta+\nu\dot\epsilon_z+\alpha(1+\nu)\dot T\le 0$	$\dot\sigma_z=e(\dot\epsilon_z+\alpha\dot T)$ $\dot\sigma_\theta = 0$
CORNER D		
$\sigma_\theta=-1$, $\sigma_z=-1$	$\dot\epsilon_z+\alpha\dot T\le 0$, $\dot\epsilon_\theta+\alpha\dot T\le 0$	$\dot\sigma_z = 0$, $\dot\sigma_\theta = 0$
ELASTIC	NONE	$\dot\sigma_z=\frac{e}{(1-\nu^2)}[\dot\epsilon_z+\nu\dot\epsilon_\theta+\alpha(1+\nu)\dot T]$
		$\dot\sigma_\theta=\frac{e}{(1-\nu^2)}[\dot\epsilon_\theta+\nu\dot\epsilon_z+\alpha(1+\nu)\dot T]$

Since the stresses are continuous across the wall of the shell, the stress resultant rates are given by:

$$\dot n_z = \int_{-\frac{1}{2}}^{\frac{1}{2}} \dot\sigma_z dr , \quad \dot n_\theta = \int_{-\frac{1}{2}}^{\frac{1}{2}} \dot\sigma_\theta dr , \quad \dot m_z = \int_{-\frac{1}{2}}^{\frac{1}{2}} \dot\sigma_z r dr \tag{7}$$

To conform with the elastic and plastic buckling analysis of unheated cylindrical shells the end of the shell are constrained from rotating but are free to expand. This insures a pre-buckling stress state independent of z. The only kinematic boundary conditions are then:

$$\frac{dv_r}{dz} = 0 \text{ at } z = 0 \quad , \quad z = \frac{L}{R} \tag{8}$$

in which L = length of the shell. For equilibrium at the ends

$$\dot p = \dot n_z = \frac{\dot P}{2\pi R h \sigma_o} \tag{9}$$

in which $\dot P$ is the load rate acting on the shell. The bending moment rate $\dot m_z$ at the ends is also balanced by the end constraint.

For convenience the vector (p,t) will be referred to as the load vector and $(\dot p, \dot t)$ as the load rate vector.

RATE EQUATIONS FOR THE PRE-BUCKLING STRESS STATE

As the temperature and load are increased from zero the stress distribution can go through an elastic and a sequence of elastic-plastic stress states before the shell buckles. The governing equations take a simple form for the pre-buckling stress state, since in this case the stress and strain rates are function of r only. $\dot{k}_z$ is then identically zero. The solution is obtained from the equations in the previous section and can conveniently be written in the form:

$$\dot{e}_z = -\frac{1}{e}\{\dot{\tau} - a^{(i)}\dot{t} - b^{(i)}\dot{p}\}, \quad \dot{e}_\theta = -\frac{1}{e}\{\dot{\tau} - \alpha^{(i)}\dot{t} - \beta^{(i)}\dot{p}\} \quad (10)$$

for the strain rates. The stress rates and the conditions for plastic flow are obtained by substituting in Table I. The superscripts refer to a particular stress state. The correspondence between the superscript and the stress state is summarized in Table II. The stress state corresponding to case 2 is also shown in fig. 2.

TABLE II

Position of Elastic-Plastic and Plastic-Plastic Interfaces

SUPER-SCRIPT	POSITION ON THE YIELD SURFACE								Values of the Functions in Appendices Obtained by Assigning These Values									
	D	DC	ELASTIC	C	CB	B	BA	A	ξ'	ξ	ζ	ζ'	ξ'	ϕ'	ϕ	θ	θ'	ζ'
1			$-\frac{1}{2}\le r\le\frac{1}{2}$						$\frac{1}{2}$	$\frac{1}{2}$	$-\frac{1}{2}$	$-\frac{1}{2}$						
2			$-\frac{1}{2}\le r\le\xi$				$\xi<r\le\frac{1}{2}$		$\frac{1}{2}$		$-\frac{1}{2}$	$-\frac{1}{2}$						
3		$-\frac{1}{2}\le r<\zeta$	$\zeta\le r<\zeta$				$\xi<r\le\frac{1}{2}$		$\frac{1}{2}$			$-\frac{1}{2}$						
4					$-\frac{1}{2}\le r\le\phi'$	$\phi<r\le\phi'$	$\phi'<r\le\frac{1}{2}$						$\frac{1}{2}$			$-\frac{1}{2}$	$-\frac{1}{2}$	$-\frac{1}{2}$
5		$-\frac{1}{2}\le r\le\theta'$		$\theta'<r\le\theta$	$\theta<r\le\phi$	$\phi<r\le\phi$	$\phi'<r\le\frac{1}{2}$						$\frac{1}{2}$					$-\frac{1}{2}$
6		$-\frac{1}{2}\le r\le\zeta$	$\zeta<r\le\xi$				$\xi<r\le\xi'$	$\xi'<r\le\frac{1}{2}$				$-\frac{1}{2}$						
7		$-\frac{1}{2}\le r\le\theta'$		$\theta'\le r\le\theta$	$\theta<r\le\phi$	$\phi<r\le\phi$	$\phi'<r\le\xi'$	$\xi'<r\le\frac{1}{2}$										$-\frac{1}{2}$
8	$-\frac{1}{2}\le r\le\xi'$	$\xi'<r\le\xi$	$\xi<r\le\xi$				$\xi<r\le\xi'$	$\xi'<r\le\frac{1}{2}$										
9	$-\frac{1}{2}\le r\le\xi'$	$\xi'<r\le\theta'$		$\theta<r\le\theta$	$\theta<r\le\phi'$	$\phi<r\le\phi'$	$\phi'<r\le\xi'$	$\xi'<r\le\frac{1}{2}$										
10					$-\frac{1}{2}\le r\le\phi$	$\phi<r\le\phi'$	$\phi'<r\le\xi'$	$\xi<r\le\frac{1}{2}$								$-\frac{1}{2}$	$-\frac{1}{2}$	$-\frac{1}{2}$
11						$-\frac{1}{2}\le r\le\frac{1}{2}$							$\frac{1}{2}$	$\frac{1}{2}$	$\frac{1}{2}$	$-\frac{1}{2}$	$-\frac{1}{2}$	$-\frac{1}{2}$
12	$-\frac{1}{2}\le r\le\zeta'$		$\zeta'<r\le\zeta'$					$\xi'<r\le\frac{1}{2}$		ξ'	ζ'							

The quantities $a^{(i)}$, $b^{(i)}$, $\alpha^{(i)}$ and $\beta^{(i)}$ are functions of the positions of the elastic-plastic and plastic-plastic (corresponding to the corners in the yield surface) interfaces. The actual values of these functions are given in Appendix I. For the cases 1 and 2, the functions take the simple form:

$$a^{(1)} = 0,\ \alpha^{(1)} = 0,\ b^{(1)} = 1,\ \beta^{(1)} = -\nu,\ a^{(2)} = \frac{(1-\nu^2)}{2}\left(\frac{1}{2} - \xi\right),$$

$$\alpha^{(2)} = 0,\ b^{(2)} = 1 + (1-\nu^2)\,\frac{(\frac{1}{2} - \xi)}{(\frac{1}{2} + \xi)},\quad \beta^{(2)} = -\nu \quad (11)$$

The rate equations for the motion of the elastic-plastic and plastic-plastic interfaces through the cross section can be obtained from the condition of continuity of the stress σ_z and σ_θ. These equations and summarized in Appendix II. For case 2, the equation for the single elastic-plastic interface takes the simple form:

$$\dot{\xi} = -\frac{\dot{t}}{2t}(\xi + \frac{1}{2}) - \frac{\dot{p}}{t(\xi+\frac{1}{2})} \tag{12}$$

PRE-BUCKLING STRESS STATE

The pre-buckling stress and strain state can be obtained by integrating the rate equations given in the previous section. The states corresponding to the cases 1 and 2 can be obtained in a closed form. The result is given below. The remaining states can only be obtained through a step by step numerical integration. This integration has to be carried out in the appropriate sequence since the last step of one case forms the initial conditions for the next. The form of the rate equations given are ideally suited for programming on a digital computer.

Positive and negative values of t (positive t corresponds to the outside of the cylinder being the warmest) give identical solutions except the positive direction of r is reversed. Only the solution for r positive outwards is therefore given below.

Before yielding begins the state is given by case 1. The strains are:

$$e_z = -\frac{1}{e}\left\{\tau - p\right\}, \quad e_\theta = -\frac{1}{e}\left\{\tau - \nu p\right\} \tag{13}$$

and the stresses in the complete cross section $-\frac{1}{2} \leq r \leq \frac{1}{2}$ are:

$$\sigma_z = p + tr, \quad \sigma_\theta = tr \tag{14}$$

For positive values of p which is the only case of interest in a buckling analysis, yielding occurs at $r = \frac{1}{2}$ (outside of the wall when t is positive) when the load vector satisfies:

$$t + 2p = 2 \tag{15}$$

This is the line ab in Fig. 3. This figure shows the regions corresponding to cases 1 and 2, and the approximation positions of the regions corresponding to the neighboring cases in the (p,t) plane. Beyond this line the stress state changes to case 2. The strains are:

$$e_z = -\frac{1}{e}\left\{\tau - p - (1-\nu^2)\left[\frac{t}{2} - \sqrt{2t(1-p)} + 1-p\right]\right\}, \quad e_\theta = -\frac{1}{e}\left\{\tau + \nu p\right\} \tag{16}$$

The position of the elastic plastic interface is:

$$\xi = -\frac{1}{2} + \sqrt{\frac{2(1-p)}{t}} \tag{17}$$

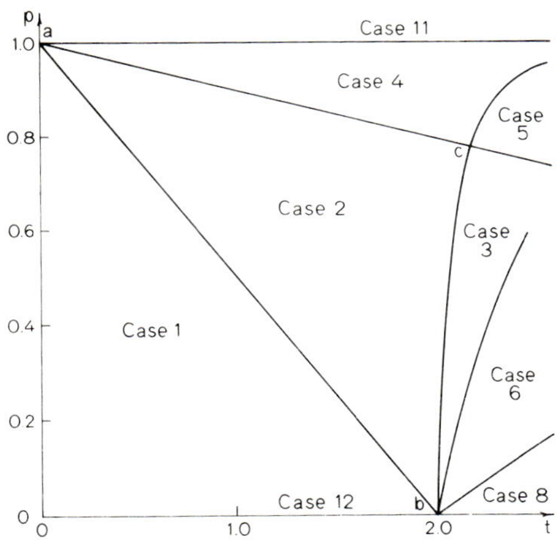

Fig. 3
Regions in the load plane (p,t)

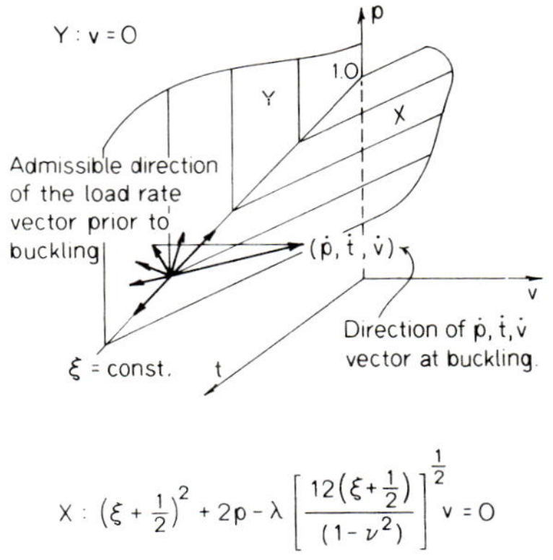

Fig. 4
Direction of the load rate vector

The stresses in the elastic region $-\frac{1}{2} \le r < \xi$ are:

$$\sigma_z = 1 + t \left\{ r + \frac{1}{2} - \sqrt{\frac{2(1-p)}{t}} \right\} \tag{18}$$

$$\sigma_\theta = \nu(1-p) + t \left\{ r + \nu \left[\frac{1}{2} - \sqrt{\frac{2(1-p)}{t}} \right] \right\}$$

and in the plastic region (on the side AB of the yield locus):

$$\sigma_z = 1, \quad \sigma_\theta = \nu(1-p) + (1+\nu)tr \tag{19}$$

For $p < 0.7722$ an additional plastic zone develops in the cross section at $r = -\frac{1}{2}$ when:

$$\nu\sqrt{2t(1-p)} + \frac{(1-\nu)}{2} t + \nu p - (1+\nu) = 0 \tag{20}$$

This is the line bc in fig. 3. Beyond this line the stress state changes to case 3.

For $p \ge 0.7722$ the complete elastic zone becomes plastic with stresses lying along the side BC of the yield locus when:

$$\sqrt{2t(1-p)} + \frac{\nu}{(1-\nu)} (1-p) - \frac{t}{2} = 0 \tag{21}$$

This is the line ac in fig. 3. Beyond this line the stress state changes to case 4.

The criterion for continued flow is:

$$(\xi + \frac{1}{2})^2 \dot{t} + 2\dot{p} \ge 0 \tag{22}$$

This means that the loading vector $(\dot{p},\dot{t})$ must be outward from the constant ξ lines in the (p,t) plane. Outward in this sense means towards decreasing value of ξ, that is increasing extent of the plastic region. This is illustrated in fig. 4. An acceptable loading path in the load plane (p,t) is one where this is satisfied at each point.

BUCKLING

A shell with the appropriate geometry can buckle starting from any of the stress states discussed in the last section. Furthermore, as shown in [2] for each stress state a series of buckling mechanisms exist, each one with an associated buckling load and temperature. This is illustrated schmatically in fig. 5 which shows the family of buckling curves in the load plane (p,t). Each curve or mechanism corresponds to a different degree of unloading of the plastic part of the cross section. The lowest curve corresponds to complete loading, that is continued flow in all the plastic material. It is this curve which will be determined.

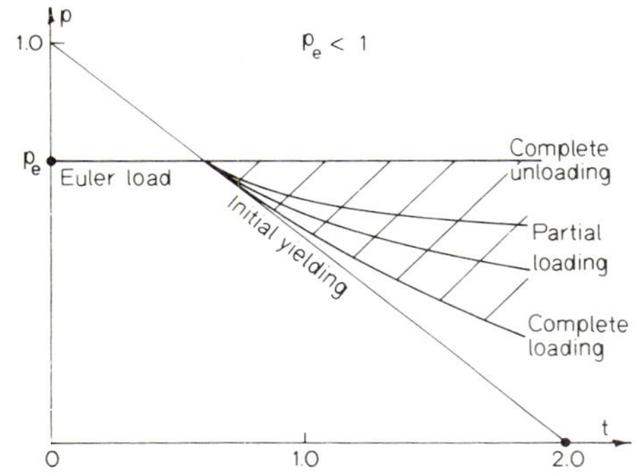

Fig. 5 Family of buckling curves.

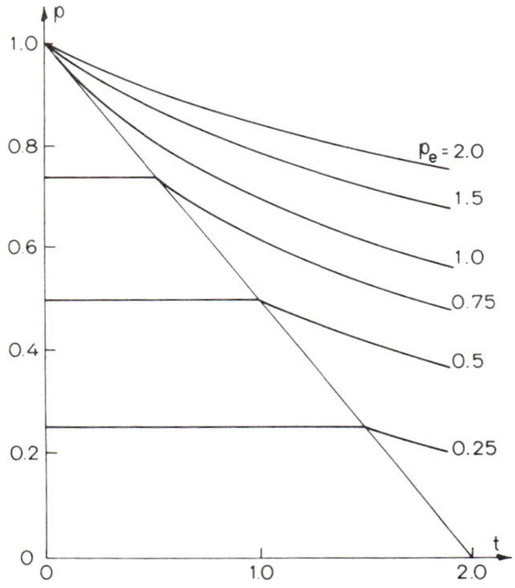

Fig. 6 Buckling load (p,t)

If, therefore, plastic flow is postulated in the plastic part of the cross section the governing equations can be reduced to an equation for the radial velocity v_r at the point of buckling.

$$\lambda^2 A^{(i)} \frac{d^4 v_r}{dz^4} - \left[\lambda B^{(i)} - pC^{(i)}\right] \frac{d^2 v_r}{dz^2} + D^{(i)} v_r =$$

$$\frac{1}{\lambda} \left\{F^{(i)}\dot{p} + G^{(i)} \dot{\tau} + H^{(i)}\dot{t}\right\} \qquad (23)$$

The value of the coefficients in the equation depend on the pre-buckling stress state and are in general functions of the positions of the interfaces. Their values are shown in Appendix III. The superscript refers to a particular stress state in the same way as before. For case 1 and 2 these functions take the simple values:

$$A^{(1)} = \frac{e^2}{12(1-\nu^2)^2}, \quad B^{(1)} = 0, \quad C^{(1)} = \frac{e}{(1-\nu^2)}, \quad D^{(1)} = \frac{e^2}{(1-\nu^2)},$$

$$F^{(1)} = \frac{\nu e}{(1-\nu^2)}, \quad G^{(1)} = \frac{e}{(1-\nu^2)}, \quad H^{(1)} = 0, \quad A^{(2)} = \frac{e^2}{12(1-\nu^2)^2}(\xi+\tfrac{1}{2})^4,$$

$$B^{(2)} = 0, \quad C^{(2)} = \frac{e}{(1-\nu^2)}(\xi+\tfrac{1}{2}), \quad D^{(2)} = \frac{e^2}{(1-\nu^2)}(\xi+\tfrac{1}{2}),$$

$$F^{(2)} = \frac{\nu e}{(1-\nu^2)}(\xi+\tfrac{1}{2}), \quad G^{(2)} = \frac{e}{(1-\nu^2)}(\xi+\tfrac{1}{2}), \quad H^{(2)} = 0 \qquad (24)$$

The solution to this equation satisfying the boundary conditions is:

$$V_r = \frac{1}{\lambda e}\left\{\dot{\tau} - \alpha^{(i)}\dot{t} - \beta^{(i)}\dot{p}\right\} + \dot{V}\cos\frac{m\pi R}{L}z \qquad (25)$$

where $\dot{V}$ is the buckling velocity amplitude and m is an integer. The buckling load is given by

$$p = \frac{\lambda}{C^{(i)}}\left\{B^{(i)} + \lambda A^{(i)}\left[\left(\frac{m\pi R}{L}\right)^2 + \frac{\dfrac{D^{(i)}}{\lambda^2 A^{(i)}}}{\left(\dfrac{m\pi R}{L}\right)^2}\right]\right\} \qquad (26)$$

m has to be determined so as to minimize p. For most cases of practical interest m is a large number (m is in general larger when a temperature gradient is present than with no gradient). When this is the case m can be treated as a continuous variable. The minimum value of p occurs when

$$\left(\frac{m\pi R}{L}\right)^4 = \frac{D^{(i)}}{\lambda^2 A^{(i)}} \qquad (27)$$

The corresponding buckling load is given by:

$$p = \frac{\lambda}{C^{(i)}}\left\{B^{(i)} + 2\left[A^{(i)}D^{(i)}\right]^{\frac{1}{2}}\right\} \qquad (28)$$

At this point it is helpful to recall that $A^{(i)}$, $B^{(i)}$, $C^{(i)}$ and $D^{(i)}$ are functions of the position of the interface, and that these positions in turn are functions of the load vector. To obtain the buckling criterion as a function of the load vector these positions must be eliminated. Except for case 1 and 2 this can only be done numerically.

For case 1 the usual elastic buckling load is obtained:

$$p = p_e = \frac{\lambda e}{\sqrt{3(1-\nu^2)}} \qquad (29)$$

For case 2 the buckling load vector is given by:

$$p^4 t^3 - 8(1-p)^3 p_e^4 = 0 \tag{30}$$

Some typical critical values are shown in fig. 6. This shows that the effect of the temperature is largest for $p_e = 1$.

The strain rate field associated with the solution (25) must also satisfy the criterions for continued flow of the material in the plastic parts of the cross section. Except for cases 1 and 2 this can only be done numerically, but it will in general lead to restrictions on the values of $\dot{p}$, $\dot{t}$ and $\dot{V}$.

For case 2 the criterion reduces to:

$$(\xi + \frac{1}{2})^2 \dot{t} + 2\dot{p} - \lambda \left\{ \frac{12(\xi + \frac{1}{2})}{(1-\nu^2)} \right\} \dot{V} \geq 0 \tag{31}$$

when $\dot{V}$ is positive. Since there is no preference for positive or negative value of $\dot{V}$, the discussion is limited to positive values. The meaning of this inequality is best understood from fig. 4. This figure replaces the two dimensional load deflection curves used in most plastic buckling analysis [See for example [2] and [6]]. The load rate displacement vector $(\dot{p}, \dot{t}, \dot{V})$ must be somewhere between the two planes X and Y.

The largest value of $\dot{V}$ occurs when the equality sign in (31) is satisfied. The load displacement vector then lies in the plane X in fig. 4. When this is the case unloading of the plastic material is just on the point of starting at one side of the plastic region. The most critical loading direction is when the load displacement vector lies in the plane X and is at right angles to the ξ = const. lines. This occurs when:

$$\frac{\dot{p}}{\dot{t}} = \frac{2}{(\xi + \frac{1}{2})^2} \tag{32}$$

which corresponds to the maximum rate of growth of the plastic region.

As the values of p and t are increased above their critical values, a point will be reached where equilibrium can not be maintained except by decreasing the value of p. The corresponding load vector can be defined as the ultimate load on the shell. The value of the ultimate load will require a step by step nonlinear numerical analysis of the type performed in [2]. The buckling load will however in general be a lower bound on the ultimate load. Furthermore, from analogy with plastic buckling of cylindrical shells [6] it can be expected that the buckling load and ultimate load are fairly close.

CONCLUSIONS

It is shown that the theoretical buckling load of axially loaded cylindrical shells can be significantly reduced by partial yielding caused by temperature gradients through the wall. The buckling is of the Shanley type. The load vector (p,t) takes the place of the load in the classical Shanley formulations, and it is this load vector which has to change to insure continued plastic flow as the shell buckles. The ultimate load the shell can carry will depend on the particular path followed by the load vector in the load plane (p,t) after the shell has buckled. From analogy with plastic buckling of cylindrical shells it can be expected that the initial buckling load and ultimate buckling load are fairly close.

REFERENCES

[1] Bodner, S.R., "Buckling of an elastoplastic column subjected to an axial load and a transverse temperature gradient", International Journal of the Mechanical Sciences, Great Britian, Vol. 4, 1962, pp 425-437.

[2] Tao, L.C. and Gjelsvik, A., "Stability of a heated elastic plastic column", Journal of the Engineering Mechanics Division, ASCE, Vol. 95, No. EM5, Oct. 1969, pp 1169-1188.

[3] Shanley, F.R., "Inelastic column theory", Journal of the Aeronautical Sciences, Vol. 14, 1947, pp261-267.

[4] Batterman, S.C., and Lee, L.H.N., "Effect of Modes on Plastic Buckling of Compressed Cylindrical Shells", American Institute of Aeronautics and Astronautics Journal, Vol. 4, Dec. 1966, pp 2255.

[5] Onat, E.T., "The influence of geometry changes on the load deformation behavior of plastic solids," Proceedings, Second Symposium on Naval Structural Mechanics, Office of Naval Research, 1960, pp 225-238.

[6] Batterman, S.C., "Tangent modulus theory for cylindrical shells Buckling under increasing load", Int. J. Solids Structures, 1967, Vol 3, pp 501-512.

APPENDIX I VALUES OF THE FUNCTIONS APPEARING IN THE STRAIN RATES

$$a^{(i)} = \frac{-\left(D^{(i)}\bar{C}^{(i)} - C^{(i)}\bar{D}^{(i)}\right)}{2\left(B^{(i)}\bar{C}^{(i)} - \bar{B}^{(i)}C^{(i)}\right)}, \qquad b^{(i)} = \frac{\bar{C}^{(i)}}{\left(B^{(i)}\bar{C}^{(i)} - \bar{B}^{(i)}C^{(i)}\right)}$$

$$\alpha^{(i)} = \frac{-\left(B^{(i)}\bar{D}^{(i)} - \bar{B}^{(i)}D^{(i)}\right)}{2\left(B^{(i)}\bar{C}^{(i)} - \bar{B}^{(i)}C^{(i)}\right)}, \qquad \beta^{(i)} = \frac{-\bar{B}^{(i)}}{\left(B^{(i)}\bar{C}^{(i)} - \bar{B}^{(i)}C^{(i)}\right)} \tag{33}$$

For the stress states with an elastic region (Cases 1, 2, 3, 6, 8, 12)

$$B^{(i)} = (\zeta - \zeta') + \frac{1}{(1-\nu^2)}(\xi - \zeta), \quad C^{(i)} = \frac{\nu}{(1-\nu^2)}(\xi - \zeta),$$

$$D^{(i)} = (\xi^2 - \zeta^2) + (1-\nu)(\zeta^2 - \zeta'^2), \quad \bar{B}^{(i)} = \frac{\nu}{(1-\nu^2)}(\xi - \zeta), \tag{34}$$

$$\bar{C}^{(i)} = (\xi' - \xi) + \frac{1}{(1-\nu^2)}(\xi - \zeta), \quad \bar{D}^{(i)} = (\xi^2 - \zeta^2) + (1-\nu)(\xi'^2 - \xi^2)$$

For the stress states without an elastic region (Cases 4, 5, 7, 9, 10, 11)

$$B^{(i)} = (\theta' - \zeta') + \frac{1}{2(1-\nu)}(\Phi - \theta), \qquad C^{(i)} = \frac{1}{2(1-\nu)}(\Phi - \theta)$$

$$D^{(i)} = (\Phi^2 - \theta^2) + (1-\nu)(\theta'^2 - \zeta'^2), \quad \bar{B}^{(i)} = \frac{1}{2(1-\nu)}(\Phi - \theta) \tag{35}$$

$$\bar{C}^{(i)} = (\xi' - \Phi') + \frac{1}{2(1-\nu)}(\Phi - \theta), \quad \bar{D}^{(i)} = (\Phi^2 - \theta^2) + (1-\nu)(\xi'^2 - \Phi'^2)$$

APPENDIX II EQUATIONS FOR THE MOTION OF THE INTERFACES

$$\dot{\xi} = \frac{-\dot{t}}{t}\left\{\xi + \frac{1}{(1-\nu^2)}(a^{(i)} + \nu\alpha^{(i)})\right\} - \frac{1}{(1-\nu^2)}(b^{(i)} + \nu\beta^{(i)})\frac{\dot{p}}{t}$$

$$\dot{\xi}' = \frac{-\dot{t}}{t}\left\{\xi' + \frac{1}{(1-\nu)}\alpha^{(i)}\right\} - \frac{1}{(1-\nu)}\beta^{(i)}\frac{\dot{p}}{t}$$

$$\dot{\zeta} = \frac{-\dot{t}}{t}\left\{\zeta + \frac{1}{(1-\nu^2)}(\alpha^{(i)} + \nu a^{(i)})\right\} - \frac{1}{(1-\nu^2)}(\beta^{(i)} + \nu b^{(i)})\frac{\dot{p}}{t}$$

$$\dot{\zeta}' = \frac{-\dot{t}}{t}\left\{\zeta' + \frac{1}{(1-\nu)}a^{(i)}\right\} - \frac{1}{(1-\nu)}b^{(i)}\frac{\dot{p}}{t} \tag{36}$$

$$\dot{\Phi} = \frac{-\dot{t}}{t}\left\{\Phi + \frac{1}{2(1-\nu)}(\alpha^{(i)} + a^{(i)})\right\} - \frac{1}{2(1-\nu)}(b^{(i)} + \beta^{(i)})\frac{\dot{p}}{t}$$

$$\dot{\Phi}' = \frac{-\dot{t}}{t}\left\{\Phi' + \frac{1}{(1-\nu)}\alpha^{(i)}\right\} - \frac{1}{(1-\nu)}\beta^{(i)}\frac{\dot{p}}{t}$$

$$\dot{\theta} = \frac{-\dot{t}}{t}\left\{\theta + \frac{1}{2(1-\nu)}(\alpha^{(i)} + a^{(i)})\right\} - \frac{1}{2(1-\nu)}(b^{(i)} + \beta^{(i)})\frac{\dot{p}}{t}$$

$$\dot{\theta}' = \frac{-\dot{t}}{t}\left\{\theta' + \frac{1}{(1-\nu)}a^{(i)}\right\} - \frac{1}{(1-\nu)}b^{(i)}\frac{\dot{p}}{t}$$

APPENDIX III VALUES OF THE FUNCTIONS IN THE BUCKLING EQUATION

$$A^{(i)} = a^{(i)}\bar{c}^{(i)} - \bar{a}^{(i)}c^{(i)}$$

$$B^{(i)} = (a^{(i)}\bar{b}^{(i)} - \bar{a}^{(i)}b^{(i)}) + (a^{(i)}\bar{\bar{c}}^{(i)} - \bar{\bar{a}}^{(i)}c^{(i)})$$

$$C^{(i)} = a^{(i)}, \quad D^{(i)} = a^{(i)}\bar{\bar{b}}^{(i)} - \bar{\bar{a}}^{(i)}b^{(i)}, \quad F^{(i)} = \bar{\bar{a}}^{(i)} \tag{37}$$

$$G^{(i)} = a^{(i)}\bar{\bar{f}}^{(i)} - \bar{\bar{a}}^{(i)} f^{(i)}, \quad H^{(i)} = a^{(i)}\bar{\bar{g}}^{(i)} - \bar{\bar{a}}^{(i)} g^{(i)}$$

Where for the stress states with an elastic region (Cases 1, 2, 3, 6, 8, 12)

$$a^{(i)} = e\left[(\zeta-\zeta') + \frac{1}{(1-\nu^2)}(\xi - \zeta)\right], \quad b^{(i)} = \frac{\nu e}{(1-\nu^2)}(\xi - \zeta)$$

$$c^{(i)} = \frac{e}{2}\left[(\zeta^2-\zeta'^2) + \frac{1}{(1-\nu^2)}(\xi^2-\zeta^2)\right]$$

$$f^{(i)} = (\zeta-\zeta') + \frac{1}{(1-\nu)}(\xi-\zeta), \quad g^{(i)} = \frac{1}{2}\left[(\xi^2-\zeta^2) + (1-\nu)(\zeta^2-\zeta'^2)\right]$$

$$\bar{a}^{(i)} = \frac{e}{2}\left[(\zeta^2-\zeta'^2) + \frac{1}{(1-\nu^2)}(\xi^2-\zeta^2)\right], \quad \bar{b}^{(i)} = \frac{\nu e}{2(1-\nu^2)}\left[\xi^2-\zeta^2\right]$$

$$\bar{c}^{(i)} = \frac{e}{3}\left[(\zeta^3-\zeta'^3)+ \frac{1}{(1-\nu^2)}(\xi^3-\zeta^3)\right], \quad \bar{f}^{(i)} = \frac{1}{2}\left[(\zeta^2-\zeta'^2)+\frac{1}{(1-\nu)}(\xi^2-\zeta^2)\right]$$

$$\bar{g}^{(i)} = \frac{1}{3}\left[(\xi^3-\zeta^3) + (1-\nu)(\zeta^3-\zeta'^3)\right], \quad \bar{\bar{a}}^{(i)} = \frac{\nu e}{(1-\nu^2)}(\xi-\zeta) \qquad (38)$$

$$\bar{\bar{b}}^{(i)} = e\left[(\xi'-\xi) + \frac{1}{(1-\nu^2)}(\xi-\zeta)\right], \quad \bar{\bar{c}}^{(i)} = \frac{\nu e}{2(1-\nu^2)}(\xi^2-\zeta^2)$$

$$\bar{\bar{f}}^{(i)} = \left[(\xi'-\xi) + \frac{1}{(1-\nu)}(\xi-\zeta)\right], \quad \bar{\bar{g}}^{(i)} = \frac{1}{2}\left[(\xi^2-\zeta^2) + (1-\nu)(\xi'^2-\xi^2)\right]$$

For the stress state without an elastic region (Cases 4, 5, 7, 9, 10, 11)

$$a^{(i)} = e\left[(\theta'-\zeta') + \frac{1}{2(1-\nu)}(\Phi-\theta)\right], \quad b^{(i)} = \frac{e}{2(1-\nu)}(\Phi-\theta)$$

$$c^{(i)} = \frac{e}{2}\left[\theta'^2-\zeta'^2) + \frac{1}{2(1-\nu)}(\Phi^2-\theta^2)\right], \quad f^{(i)} = (\theta'-\zeta') + \frac{1}{(1-\nu)}(\Phi-\theta)$$

$$g^{(i)} = \frac{1}{2}\left[(\Phi^2-\theta^2) + (1-\nu)(\theta'^2-\zeta'^2)\right],$$

$$\bar{a}^{(i)} = \frac{e}{2}\left[(\theta'^2-\zeta'^2) + \frac{1}{2(1-\nu)}(\Phi^2-\theta^2)\right], \quad \bar{b}^{(i)} = \frac{e}{4(1-\nu)}(\Phi^2-\theta^2)$$

$$\bar{c}^{(i)} = \frac{e}{3}\left[\theta'^3-\zeta'^3) + \frac{1}{(1-\nu)}(\Phi^3-\theta^3)\right], \qquad (39)$$

$$\bar{f}^{(i)} = \frac{1}{2}\left[(\theta'^2-\zeta'^2) + \frac{1}{(1-\nu)}(\Phi^2-\theta^2)\right],$$

$$\bar{g}^{(i)} = \frac{1}{3}\left[(\Phi^3-\theta^3) + (1-\nu)(\theta'^3-\zeta'^3)\right], \quad \bar{\bar{a}}^{(i)} = \frac{e}{2(1-\nu)}(\Phi-\theta)$$

$$\bar{\bar{b}}^{(i)} = e\left[(\xi'-\Phi') + \frac{1}{(1-\nu^2)}(\Phi-\theta)\right], \quad \bar{\bar{c}}^{(i)} = \frac{e}{4(1-\nu)}(\Phi^2-\theta^2)$$

$$\bar{\bar{f}}^{(i)} = \left[(\xi'-\Phi') + \frac{1}{(1-\nu)}(\Phi-\theta)\right], \quad \bar{\bar{g}}^{(i)} = \frac{1}{2}\left[(\Phi^2-\theta^2) + (1-\nu)(\xi'^2-\Phi'^2)\right]$$

Developments in Mechanics, Vol. 6. Proceedings of the 12th Midwestern Mechanics Conference: 38

Buckling and Post-buckling Behavior of Shallow Spherical Sandwich Shells under Axisymmetrical Loads

NURI AKKAS and NELSON R. BAULD, JR.
CLEMSON UNIVERSITY

ABSTRACT

The elastic buckling and initial post-buckling behavior of clamped shallow spherical sandwich shells under axisymmetrical loads is investigated numerically for face sheets of the same material and thickness. An analysis is made of the dependence of the buckling and initial post-buckling behavior on the area of the region over which the load is uniformly distributed. It is found that as the area of the loaded region decreases, the buckling behavior changes from the asymmetrical type to axisymmetrical snap-through, and then, for a significant range of radius of the loaded region, the clamped spherical sandwich cap does not buckle either axisymmetrically or asymmetrically.

The buckling behavior of the clamped shallow spherical sandwich cap is studied also for a concentrated load at the apex. Again the face sheets are assumed to be made of the same material and have equal thicknesses. It is found, for the range of material and geometrical parameters considered, that buckling does not occur.

INTRODUCTION

The purpose of this paper is to present a numerical analysis of the buckling and initial post-buckling behavior of clamped shallow spherical sandwich shells (Figure 1) with face sheets made from the same materials and whose thicknesses are equal. The shell is subjected to a pressure that is distributed uniformly over a region surrounding its apex. The core material is assumed to be an isotropic homogeneous continuum capable of transmitting transverse shearing forces only. Consequently, the rigid core theory of sandwich construction is used. The theoretical treatment employed in this study is essentially a perturbation technique proposed by Koiter [1] and which was transcribed into a form suitable for application to the clamped shallow spherical homogeneous shell by Fitch [2].

The buckling and initial post-buckling behaviors of clamped shallow spherical homogeneous shells under various types of loads have been studied extensively in recent years [2-6]. It has been suggested, in the literature, that the clamped shallow spherical homogeneous shell can be regarded as a reasonable approximation of the substructure which supports the ablation material of the heat shield on space vehicles. Since this substructure is likely to be a sandwich construction it is clear that the buckling and initial post-buckling analyses of clamped shallow spherical sandwich shells have significant practical implications.

The buckling and initial post-buckling analysis presented in this paper is based on a set of nonlinear differential equations derived in an earlier investigation by the present authors [7].

THEORY

Basic Equations

The behavior of thin clamped shallow spherical sandwich shells undergoing moderately large deflections can be described by the nondimensional equations

$$\nabla^4 w = \Gamma + \nabla^2 f - 2\left(\frac{1}{x}\dot{f}' - \frac{1}{x^2}\dot{f}\right)\left(\frac{1}{x}\dot{w}' - \frac{1}{x^2}\dot{w}\right) + \left(\frac{1}{x^2}\ddot{w} + \frac{1}{x}w'\right)f'' + \left(\frac{1}{x}f' + \frac{1}{x^2}\ddot{f}\right)w'' - \eta\left\{\frac{1}{x}\left[(x\bar{\alpha})' + \dot{\bar{\beta}}\right] - \nabla^2 w\right\} \quad (1)$$

$$\nabla^4 f = -\nabla^2 w + \left(\frac{1}{x}\dot{w}' - \frac{1}{x^2}\dot{w}\right)^2 - \left(\frac{1}{x}w' + \frac{1}{x^2}\ddot{w}\right)w'' \quad (2)$$

$$\frac{1-\nu}{2}\nabla^2\bar{\alpha} + \frac{1+\nu}{2}\left(\bar{\alpha}'' + \frac{1}{x}\bar{\alpha}' + \frac{1}{x}\dot{\bar{\beta}}'\right) - \frac{1}{x^2}\bar{\alpha} - \frac{3-\nu}{2x^2}\dot{\bar{\beta}} - \Lambda(\bar{\alpha} - w') = 0 \quad (3)$$

$$\frac{1-\nu}{2}\nabla^2\bar{\beta} + \frac{1+\nu}{2}\left(\frac{1}{x}\dot{\bar{\alpha}}' + \frac{1}{x^2}\ddot{\bar{\beta}}\right) + \frac{3-\nu}{2x^2}\dot{\bar{\alpha}} - \frac{1-\nu}{2x^2}\bar{\beta} - \Lambda\left(\bar{\beta} - \frac{1}{x}\dot{w}\right) = 0 \quad (4)$$

with boundary conditions at the clamped edge $(x = \lambda)$

$$w = 0 \quad (5)$$

$$w' = 0 \quad (6)$$

$$f'' - \frac{\nu}{\lambda}f' - \frac{\nu}{\lambda^2}\ddot{f} = 0 \quad (7)$$

$$\lambda\left(f'' - \frac{\nu}{x}f' - \frac{\nu}{x^2}\ddot{f}\right)' - \frac{1}{\lambda}f' - \frac{1}{\lambda^2}\ddot{f} + \nu f'' + 2(1+\nu)\left(\frac{1}{x}\ddot{f}\right)' = 0 \quad (8)$$

$$\bar{\alpha} = 0 \quad (9)$$

$$\bar{\beta} = 0 \quad (10)$$

The nondimensional quantities appearing in Eq. (1) - (10) are related to the corresponding physical quantities through the relations

$$x = \frac{\lambda}{a}r, \quad w = \frac{\lambda^2}{2H}W, \quad \lambda = 2[3(1-\nu^2)]^{1/4}\left(\frac{H}{t}\right)^{1/2},$$
$$f = \frac{\lambda^4}{8EH^2t}F, \quad \bar{\alpha} = \frac{\lambda a}{2H}\alpha, \quad \bar{\beta} = \frac{\lambda a}{2H}\beta, \quad (11)$$
$$\eta = \frac{1}{8}\left(\frac{G_c}{E}\right)\left(\frac{a}{H}\right)^2\frac{\left(1 + \frac{c}{t}\right)^2}{\left(\frac{c}{t}\right)}\lambda^2, \quad \Lambda = \frac{1}{24}\left(\frac{G_c}{E}\right)\left(\frac{a}{H}\right)^2\frac{\lambda^2}{\left(\frac{c}{t}\right)}$$

In the foregoing expressions a and H are, respectively, the base plane radius and apex rise referred to the reference surface of the composite cap. The face sheet thicknesses are assumed to be equal and are denoted by t. The thickness and transverse shearing modulus associated with the core material are denoted by c and G_c, respectively, while E and ν signify the modulus of elasticity and Poisson ratio for the face sheets. The quantities w, f, $\bar{\alpha}$, $\bar{\beta}$ are the nondimensional transverse deflection, Airy stress function and shear angles, respectively; and W, F, α, β are the corresponding physical quantities for the sandwich shell. The notation employed in this paper to indicate differentiation is as follows:

$$(\ )' \equiv \frac{\partial}{\partial x} (\ ) \ , \ (\dot{\ }) \equiv \frac{\partial}{\partial \phi} (\ ) \ ,$$

$$\nabla^2 (\ ) \equiv (\ )'' + \frac{1}{x} (\ )' + \frac{1}{x^2} (\ddot{\ }),$$

$$\nabla^4 (\ ) \equiv \nabla^2 \nabla^2 (\ ) \ .$$

Finally, Γ represents the nondimensional axisymmetrical load and is defined by the relations

$$\Gamma = \begin{cases} p \dfrac{\delta(x)}{x} \quad \text{with} \quad p = \dfrac{PR}{4\pi D} \quad \text{(concentrated load)} \\ 4 \ ph(x_2 - x) \quad \text{with} \quad p = \dfrac{a^4 \lambda^2 q}{64 EtH^3} \quad \text{(centrally distributed pressure)} \end{cases} \tag{12}$$

In Eq. (12), P and q denote the magnitude of the concentrated load and the density of the centrally distributed pressure, respectively. The nondimensional form of r_2 (see Figure 1) is denoted by x_2, R is the radius of curvature of the shell, and

$$D = Et^3/[12 \ (1-\nu^2)] \tag{13}$$

The Dirac delta is defined by the relations

$$\delta(x) = \begin{cases} 0 \ , \ x \neq 0 \\ \infty \ , \ x = 0 \end{cases}$$

and the Heaviside function is defined by

$$h(x_2 - x) = \begin{cases} 1 & x \leq x_2 \\ 0 & x > x_2 \end{cases}$$

Axisymmetrical Behavior

The boundary value problem governing the axisymmetrical deformations of the clamped shallow spherical sandwich shell under the axisymmetrical loads described by Eq. (12) reduces to

$$(x\Theta')' - \frac{1}{x} \Theta + x\Phi - \eta x(\alpha_0 + \Theta) = -\Gamma^* + \Theta\Phi \tag{14}$$

$$(x\Phi')' - \frac{1}{x} \Phi - x\Theta = - \frac{1}{2} \Theta^2 \tag{15}$$

$$\nabla^2 \alpha_0 - \frac{1}{x^2} \alpha_0 - \Lambda(\alpha_0 + \Theta) = 0 \tag{16}$$

$$\Theta(\lambda) = 0 \tag{17}$$

$$\lambda\Phi'(\lambda) - \nu\Phi(\lambda) = 0 \quad (18)$$

$$\alpha_0 (\lambda) = 0 \quad (19)$$

$$\lim_{x\to 0} \{\Theta, \Phi, \alpha_0\} = 0 \quad (20)$$

where $\Theta = -w'$, $\Phi = f'$, and α_0 refers to the axisymmetrical behavior.

The conditions (20) follow from symmetry and boundedness of the membrane forces at the apex. Finally, the nondimensional load parameter Γ^* appearing in Eq. (14) is given by

$$\Gamma^* = \begin{cases} p, & \text{(Concentrated load)} \\ 2px_2^2\, y(x), & \text{(Centrally distributed pressure)} \end{cases} \quad (21)$$

where

$$y(x) = \begin{cases} \frac{x^2}{x_2{}^2} & x \leq x_2 \\ 1 & x > x_2 \end{cases} \quad (22)$$

Average Deflection Parameter

As a suitable nondimensional measure of the deflection of the cap under centrally distributed pressures, the ratio of the average vertical displacement of the region $r \leq r_2$ to the face sheet thickness t is selected. Accordingly,

$$\frac{\overline{W}}{t} \equiv \frac{1}{\pi r_2{}^2 t} \int_0^{2\pi} d\phi \int_0^{r_2} rWdr \quad (23)$$

When the deflections are axisymmetrical, Eq. (23) reduces to

$$\frac{\overline{W}}{t} \equiv \frac{1}{[12\ (1-\nu^2)]^{\frac{1}{2}}} \int_0^{\lambda} y(x)\ \Theta dx \quad (24)$$

Expression (24) can also be used as a measure of the deflection of the cap under a concentrated load at the apex if y = 1. In this case the average deflection parameter is to be interpreted as the ratio of the vertical deflection of the apex to the thickness of a face sheet.

Defining S_0 as the rate of change of the load parameter p with respect to the gross deflection parameter $\overline{W}/t$ at p_c for the axisymmetrical path there results

$$\frac{1}{S_0} = \frac{d}{dp}\left(\frac{\overline{W}}{t}\right) = \frac{1}{[12\ (1-\nu^2)]^{\frac{1}{2}}} \int_0^{\lambda} y \left(\frac{\partial\Theta}{\partial p}\right)_c dx \quad (25)$$

where $(\partial\Theta/\partial p)_c$ is to be evaluated at the bifurcation point.

Buckling Behavior

In order to determine if asymmetrical buckling precedes axisymmetrical buckling a solution of Eq. (1) - (10) is sought in the form

$$\begin{Bmatrix} w \\ f \\ \bar{\alpha} \\ \bar{\beta} \end{Bmatrix} = \begin{Bmatrix} {}_x\int^{\lambda} \Theta dx \\ {}_0\int^{x} \Phi dx \\ \alpha_0 \\ \beta_0 \end{Bmatrix} + \xi \begin{Bmatrix} w_1(x,\phi) \\ f_1(x,\phi) \\ \alpha_1(x,\phi) \\ \beta_1(x,\phi) \end{Bmatrix} \tag{26}$$

where ξ is an infinitesimal scalar parameter. The function $\beta_0(x) \equiv 0$ by symmetry considerations. With no loss in generality, $(w_1, f_1, \alpha_1, \beta_1)$ can be taken in the form

$$\begin{aligned} w_1 \ (x,\phi) &= w_{1n}(x) \cos n\phi \\ f_1 \ (x,\phi) &= f_{1n}(x) \cos n\phi \\ \alpha_1 \ (x,\phi) &= \alpha_{1n}(x) \cos n\phi \\ \beta_1 \ (x,\phi) &= \beta_{1n}(x) \sin n\phi \end{aligned} \tag{27}$$

Substituting Eq. (26) into Eq. (1) - (10), making use of the axisymmetrical Eq. (14) - (20), and linearizing with respect to ξ gives

$$\begin{aligned} L_n^2 \ (w_{1n}) = L_n(f_{1n}) - (\frac{1}{x} f_{1n}' - \frac{n^2}{x^2} f_{1n}) \ \Theta' + \frac{1}{x} w_{1n}'' \Phi - \frac{1}{x} f_{1n}'' \Theta \\ + (\frac{1}{x} w_{1n}' - \frac{n^2}{x^2} w_{1n}) \ \Phi' - \eta[\alpha_{1n}' + \frac{1}{x} \alpha_{1n} + \frac{n}{x} \beta_{1n} - L_n(w_{1n})] \end{aligned} \tag{28}$$

$$L_n^2 \ (f_{1n}) = -L_n \ (w_{1n}) + (\frac{1}{x} w_{1n}' - \frac{n^2}{x^2} w_{1n}) \ \Theta' + \frac{1}{x} w_{1n}'' \Theta \tag{29}$$

$$\begin{aligned} \frac{1-\nu}{2} L_n \ (\alpha_{1n}) + \frac{1+\nu}{2} \ (\alpha_{1n}'' + \frac{1}{x} \alpha_{1n}' + \frac{n}{x} \beta_{1n}') - \frac{1}{x^2}\alpha_{1n} - \frac{(3-\nu)n}{2x^2} \beta_{1n} \\ - \Lambda \ (\alpha_{1n} - w_{1n}') = 0 \end{aligned} \tag{30}$$

$$\begin{aligned} \frac{1-\nu}{2} L_n \ (\beta_{1n}) + \frac{1+\nu}{2} \ (- \frac{n}{x} \alpha_{1n}' - \frac{n^2}{x^2} \beta_{1n}) - \frac{(3-\nu)n}{2x^2} \alpha_{1n} - \frac{1-\nu}{2x^2} \beta_{1n} \\ - \Lambda \ (\beta_{1n} + \frac{n}{x} w_{1n}) = 0 \end{aligned} \tag{31}$$

and boundary conditions at the clamped edge $(x = \lambda)$

$$w_{1n} = 0 \tag{32}$$

$$w_{1n}' = 0 \tag{33}$$

$$f_{1n}'' - \frac{\nu}{\lambda} f_{1n}' + \frac{\nu n^2}{\lambda^2} f_{1n} = 0 \tag{34}$$

$$\lambda f'''_{1n} - \frac{1}{\lambda}[(1-\nu) + n^2 (2+\nu)] f'_{1n} + \frac{3n^2}{\lambda^2} f_{1n} = 0 \tag{35}$$

$$\alpha_{1n} = 0 \tag{36}$$

$$\beta_{1n} = 0 \tag{37}$$

The differential operators appearing in these relations are defined by

$$L_n(\ ) \equiv (\ )'' + \frac{1}{x}(\ )' - \frac{n^2}{x^2}(\ ) \text{ and } L_n^2(\ ) \equiv L_n L_n(\ )$$

To the system of Eq. (28) - (37) there are added the conditions of regularity at the apex

$$\lim_{x\to 0}\{w_{1n}, xw''_{1n}, f_{1n}, xf''_{1n}, \alpha_{1n}, \beta_{1n}\} = 0 \tag{38}$$

The critical bifurcation pressure p_c is the lowest value of p for which Eq. (28)-(38) possess a non-trivial solution for any integer n.

Initial Post-Buckling Behavior

In order to investigate the initial post-buckling behavior of the clamped shallow spherical sandwich shell under either centrally distributed pressures or a concentrated load at the apex the nondimensional transverse deflection, Airy stress function, and transverse shear angles are assumed to have the forms

$$\begin{aligned} w &= \int_x^\lambda \Theta dx + \xi w_1 + \xi^2 w_2 (x,\phi) + \ldots \\ f &= \int_0^x \Phi dx + \xi f_1 + \xi^2 f_2 (x,\phi) + \ldots \\ \bar{\alpha} &= \alpha_0 + \xi\alpha_1 + \xi^2\alpha_2 (x,\phi) + \ldots. \\ \bar{\beta} &= \beta_0 + \xi\beta_1 + \xi^2\beta_2 (x,\phi) + \ldots. \end{aligned} \tag{39}$$

for a slightly buckled configuration. Here, ξ is an infinitesimal scalar parameter. The function β_0 (x) $\equiv$ 0 by symmetry considerations. Substituting the relations (39) into the general nonlinear Eq. (1) - (10), observing that at the bifurcation point both the systems of Eq. (14) - (20) and (28) - (38) hold, and letting $\xi \to 0$ yields the system of equations

$$\begin{aligned} &\nabla^4 w_2 - \nabla^2 f_2 + (\frac{1}{x} f'_2 + \frac{1}{x^2}\ddot{f}_2)\,\Theta'_c - \frac{1}{x} w''_2 \Phi_c + \frac{1}{x} f''_2 \Theta_c \\ &- (\frac{1}{x} w'_2 + \frac{1}{x^2}\ddot{w}_2)\Phi'_c + \eta(\alpha'_2 + \frac{1}{x}\alpha_2 + \frac{1}{x}\dot{\beta}_2 - \nabla^2 w_2) = (\frac{1}{x} f'_1 + \frac{1}{x^2}\ddot{f}_1)\, w''_1 \\ &+ (\frac{1}{x} w'_1 + \frac{1}{x^2}\ddot{w}_1) f''_1 - 2\,(\frac{1}{x^2}\dot{f}_1 - \frac{1}{x}\dot{f}'_1)\,(\frac{1}{x^2}\dot{w}_1 - \frac{1}{x}\dot{w}'_1) \end{aligned} \tag{40}$$

$$\nabla^4 f_2 + \nabla^2 w_2 - (\frac{1}{x^2}\ddot{w}_2 + \frac{1}{x}w_2') \Theta_c' - \frac{1}{x}w_2'' \Theta_c = (\frac{1}{x}\dot{w}_1' - \frac{1}{x^2}\dot{w}_1)^2$$

$$- (\frac{1}{x^2}\ddot{w}_1 + \frac{1}{x}w_1') w_1'' \qquad (41)$$

$$\frac{1-\nu}{2}\nabla^2\alpha_2 + \frac{1+\nu}{2}(\alpha_2'' + \frac{1}{x}\alpha_2' + \frac{1}{x}\dot{\beta}_2') - \frac{1}{x^2}\alpha_2 - \frac{3-\nu}{2x^2}\dot{\beta}_2 - \Lambda(\alpha_2 - w_2') = 0 \qquad (42)$$

$$\frac{1-\nu}{2}\nabla^2\beta_2 + \frac{1+\nu}{2}(\frac{1}{x}\dot{\alpha}_2' + \frac{1}{x^2}\ddot{\beta}_2) + \frac{3-\nu}{2x^2}\dot{\alpha}_2 - \frac{1-\nu}{2x^2}\beta_2 - \Lambda(\beta_2 - \frac{1}{x}\dot{w}_2) = 0 \qquad (43)$$

with boundary conditions (5) - (10) on w_2, f_2, α_2, and β_2.

It can be shown by direct substitution in Eq. (40) - (43) and the associated boundary conditions at $x = \lambda$ and the conditions of regularity at $x = 0$ that the functions (w_2, f_2, α_2, β_2) must be of the form

$$\begin{aligned} w_2(x,\phi) &= -\int^x \zeta dx + \rho(x)\cos 2n\phi \\ f_2(x,\phi) &= \int^x \Psi dx + \chi(x)\cos 2n\phi \\ \alpha_2(x,\phi) &= \gamma(x) + \sigma(x)\cos 2n\phi \\ \beta_2(x,\phi) &= \omega(x) + \tau(x)\sin 2n\phi \end{aligned} \qquad (44)$$

where $\zeta(x)$, $\Psi(x)$, and $\gamma(x)$ must satisfy the boundary value problem

$$(x\zeta')' - (\frac{1}{x} + \Phi_c)\zeta + (x - \Theta_c)\Psi - \eta x(\gamma + \zeta) = g_1(x) \qquad (45)$$

$$(x\Psi')' - \frac{1}{x}\Psi - (x - \Theta_c)\zeta = g_2(x) \qquad (46)$$

$$L_1(\gamma) - \Lambda(\gamma + \zeta) = 0 \qquad (47)$$

$$\zeta(\lambda) = 0, \qquad (48)$$

$$\lambda\Psi'(\lambda) - \nu\Psi(\lambda) = 0 \qquad (49)$$

$$\gamma(\lambda) = 0, \qquad (50)$$

$$\lim_{x\to 0}\{\zeta, \Psi, \gamma\} = 0 \qquad (51)$$

and the functions $\rho(x)$, $\chi(x)$, $\sigma(x)$, and $\tau(x)$ must satisfy the boundary value problem

$$L_{2n}^2(\rho) - L_{2n}(\chi) + (\frac{1}{x}\chi' - \frac{4n^2}{x^2}\chi)\Theta_c' - (\frac{1}{x}\rho' - \frac{4n^2}{x^2}\rho)\Phi_c'$$

$$- \frac{1}{x}\rho''\Phi_c + \frac{1}{x}\chi''\Theta_c + \eta[\frac{1}{x}(x\sigma)' + \frac{2n}{x}\tau - L_{2n}(\rho)] = h_1(x) \qquad (52)$$

$$L_{2n}^2(\chi) + L_{2n}(\rho) - (\frac{1}{x}\rho' - \frac{4n^2}{x^2}\rho)\Theta_c' - \frac{1}{x}\rho''\Theta_c = h_2(x) \qquad (53)$$

$$\frac{1-\nu}{2} L_{2n}(\sigma) + \frac{1+\nu}{2}(\sigma'' + \frac{1}{x}\sigma' + \frac{2n}{x}\tau') - \frac{1}{x^2}\sigma - \frac{(3-\nu)n}{x^2}\tau - \Lambda(\sigma-\rho') = 0 \quad (54)$$

$$\frac{1-\nu}{2} L_{2n}(\tau) + \frac{1+\nu}{2}(-\frac{2n}{x}\sigma' - \frac{4n^2}{x^2}\tau) - \frac{(3-\nu)n}{x^2}\sigma - \frac{1-\nu}{2x^2}\tau - \Lambda(\tau + \frac{2n}{x}\rho) = 0 \quad (55)$$

$$\rho(\lambda) = 0 \quad (56)$$

$$\rho'(\lambda) = 0 \quad (57)$$

$$\chi''(\lambda) - \frac{\nu}{\lambda}\chi'(\lambda) + \frac{4n^2\nu}{\lambda^2}\chi(\lambda) = 0 \quad (58)$$

$$\lambda\chi'''(\lambda) - \frac{1}{\lambda}[(1-\nu) + 4(2+\nu)n^2]\chi'(\lambda) + \frac{12n^2}{\lambda^2}\chi(\lambda) = 0 \quad (59)$$

$$\sigma(\lambda) = 0 \quad (60)$$

$$\tau(\lambda) = 0 \quad (61)$$

$$\lim_{x\to 0}\{\rho, \chi, x\rho'', x\chi'', \sigma, \tau\} = 0 \quad (62)$$

The boundary value problem for the function $\omega(x)$ is

$$\frac{1-\nu}{2} L_1(\omega) - \Lambda\omega = 0 \quad (63)$$

$$\omega(\lambda) = 0 \quad (64)$$

$$\lim_{x\to 0}\{\omega\} = 0 \quad (65)$$

and yields $\omega \equiv 0$.

The functions appearing on the right-hand sides of Eq. (45), (46), (52), and (53) are defined by the relations

$$g_1(x) = \frac{1}{2}\{n^2 (\frac{w_{1n}f_{1n}}{x})' - f_{1n}' w_{1n}'\} \quad (66)$$

$$g_2(x) = \frac{1}{4}\{n^2 (\frac{w_{1n}^2}{x})' - (w_{1n}')^2\} \quad (67)$$

$$h_1(x) = \frac{1}{2}\{(\frac{1}{x}f_{1n}' - \frac{n^2}{x^2}f_{1n}) w_{1n}'' + (\frac{1}{x}w_{1n}' - \frac{n^2}{x^2}w_{1n}) f_{1n}'' +$$

$$2n^2 (-\frac{1}{x^2}f_{1n} + \frac{1}{x}f_{1n}') (-\frac{1}{x^2}w_{1n} + \frac{1}{x}w_{1n}')\} \quad (68)$$

$$h_2(x) = -\frac{1}{2}\{(\frac{1}{x}w_{1n}' - \frac{n^2}{x^2}w_{1n}) w_{1n}'' + n^2 (\frac{1}{x^2}w_{1n} - \frac{1}{x}w_{1n}')^2\} \quad (69)$$

These formulas are precisely the same as those for the homogeneous spherical cap [2,4,5,6].

It is shown in reference [2] that the expansions (39), when they are assumed asymptotically valid for $\xi \ll 1$, imply a relationship of the form

$$\frac{p}{p_c} = 1 + a\xi + b\xi^2 + \ldots \quad (70)$$

It can be shown that the coefficient $a \equiv 0$ for the spherical sandwich cap and that

$$b = -\frac{1}{p_c} \frac{\int_0^\lambda [(\zeta g_1 - \Psi g_2) - \frac{1}{2} x(\rho h_1 - \chi h_2)]dx}{\int_0^\lambda [g_1 (\frac{\partial \Theta}{\partial p})_c - g_2 (\frac{\partial \Phi}{\partial p})_c]dx} \quad (71)$$

By using Eqs. (45) and (46) for g_1 and g_2 and the equations obtained by differentiating the axisymmetrical equilibrium Eq. (14) - (20) with respect to the load parameter p at p_c, the integral in the denominator of Eq. (71) can be written in the more convenient form

$$- \int_0^\lambda \zeta dx, \quad \text{(Concentrated load)}$$

$$- 2x_2^2 \int_0^\lambda y(x)\zeta dx, \quad \text{(Centrally distributed load)} \quad (72)$$

The sign of the coefficient b determines whether the load initially increases or decreases subsequent to asymmetrical buckling. The significance of the coefficient b is connected with the notions of imperfection-sensitive and imperfection-insensitive structures. It has been shown in references [2] and [4] that structures containing geometric imperfections (imperfect structures) are imperfection-sensitive, in the sense that the buckling load for the imperfect structure should be expected to be less than that for the corresponding perfect structure, whenever the load for the perfect structure initially decreases at the bifurcation point. A structure is said to be imperfection-insensitive, in the sense that load-deflection curve for the imperfect structure exhibits a much milder growth of displacement as the load reaches and exceeds the classical buckling load of the corresponding perfect structure, whenever the load for the perfect structure increases subsequent to asymmetrical buckling. Accordingly, a structure is said to be imperfection-sensitive or imperfection-insensitive according to whether b is negative or positive, respectively. In calculating b it has been assumed that the buckling mode has been normalized so that the maximum value of W_1 (= $2H/\lambda^2 w_1$) is equal to the face sheet thickness t.

Let S represent the initial slope of the load-deflection curve corresponding to the bifurcation path. It is not difficult to show that

$$S = \frac{S_0}{1+\delta} \quad (73)$$

where

$$\delta = \frac{S_0}{bp_c[12(1-\nu^2)]^{\frac{1}{2}}} \int_0^\lambda y\zeta dx \quad (74)$$

The function $\zeta(x)$ is defined by the boundary value problem (45) - (51), b is the initial post-buckling coefficient defined by Eq. (71), and p_c is the value of the loading parameter p at the bifurcation point.

Finally, as a measure of the relative stiffness the following parameter is used

$$\vartheta = \frac{2}{\pi} \arctan \frac{1}{\delta} \quad (75)$$

The parameters S_0, S, and ϑ were defined by Fitch and Budiansky [4] and the average deflection parameter defined here is analogous to the one defined by those authors.

NUMERICAL PROCEDURE AND RESULTS

The numerical procedures employed in this study have been described in detail by Fitch [2]. A brief account of our application of these procedures is presented here to provide a clearer understanding of the numerical results to be presented in the sequel.

The system of nonlinear differential Eq. (14) - (20) governing the axisymmetrical buckling of the sandwich cap was solved by Newton's method, which, essentially, replaces these equations with a system of linear correctional equations. This latter system of equations was in turn approximated by a system of finite difference equations using central difference formulas for the derivatives. The linear correctional equations lead, via an iterative process, to the solution of the axisymmetrical problem. The numerical procedure employed was to solve the linearized finite difference analogue of the axisymmetrical Eq. (14) - (20) for a value of p for which the shell response was linear and to use this solution as an initial estimate of the solution for a slightly higher value of p. The finite difference analogue of the linear correctional equations was solved then by an iterative process for the correction functions associated with these initial estimates of the solution. The correction functions were considered to be sufficiently accurate whenever their absolute values at each station were less than 0.01% of the absolute values of the current estimate of the solution at the corresponding station. Having obtained the solution for a given value of p it was used as an initial estimate for a slightly higher value of p and the process was continued until convergence did not occur.

Axisymmetrical cap-snapping was assumed to occur when a local maximum on the axisymmetrical load-deflection curve was reached. The occurence of a local maximum was signified by the failure of the iterative process to converge. In locating the local maximum numerically it was assumed that it had been exceeded if convergence of the system of linear correctional equations had not occurred after fourteen iterations for a fixed value of the loading parameter p. The preceding value of p for which convergence had occurred was then incremented in steps of 1/10 of the original increment until convergence, in the sense described, failed to occur. This latter process was repeated in steps of 1/100 of the original increment until convergence failed to occur. The immediately preceding value of p was taken as the axisymmetrical buckling load. This process gives the critical axisymmetrical buckling load to two places after the decimal point.

The linear eigenvalue problem, Eq. (28) - (38), describing the asymmetrical buckling of the shallow sandwich cap was replaced with a finite difference analogue using central difference formulas. Potters' algorithm [8], with the modifications suggested by Blum and Fulton [9] which eliminate sign changes associated with the singularities of the buckling determinant, was used in determining the asymmetrical buckling load. The value of the modified buckling determinant was plotted against the corresponding value of the load parameter p for several values of the circumferential wave number n. Whenever a change in sign of the modified buckling determinant occurred the preceding value of p was incremented in steps of 1/10 of the original increment and the process was repeated until the sign change was detected again. The asymmetrical buckling load was taken as the value of p corresponding to a linear interpolation between the value of p immediately preceding the sign change for the refined calculations and the value of p associated with the sign change.

The initial post-buckling problem, Eq. (45) - (51) and (52) - (62), was replaced by a finite difference analogue using central difference formulas. Potters' algorithm was used to generate the solution. The integrals appearing in the numerator and denominator of the initial post-buckling coefficient b (equation 71) and in the formula for the relative stiffness parameter (equation 75) were evaluated by Simpon's rule.

The preceding calculations were carried out for a mesh size of 0.5 on the IBM 360-50 computer using double precision throughout. A mesh size of 0.25 was used at two points so as to ascertain the accuracy with which the buckling loads and associated mode forms were given by the 0.5 mesh. The buckling loads associated with the 0.5 mesh were found to be less than 1% larger than those associated with the 0.25 mesh. Moreover, the initial post-buckling coefficient (b) and the relative stiffness parameter (ϑ) differed by approximately 2% and 4%, respectively. These differences were not considered to be significant since the important information extracted from the plots did not change. Consequently, the 0.5 mesh was used throughout the study.

In the present work attention was focused initially on a clamped shallow spherical sandwich shell with equal face sheets for which the pertinent geometrical and material properties were $\lambda = 30$, $(G_c/E)(a/H)^2 = 0.8$, $c/t = 10$. A study was made of the dependence of the buckling and initial post-buckling behavior on the size of the area over which a uniform pressure was distributed. The results of this study are shown in Figure 2.

The top curve in Figure 2 exhibits the critical load as a function of the nondimensional parameter x_2 that measures the extent of the loaded area. The parameter x_2 varies from zero to λ as the loading changes from the equivalent of a concentrated force applied at the apex to a uniform pressure over the entire surface of the shell.

The middle portion of Figure 2 gives the initial post-buckling coefficient b associated with asymmetrical buckling. Finally, the bottom portion of Figure 2 shows the relative stiffness parameter ϑ.

The buckling and initial post-buckling behavior shown in Figure 2 can be described as follows. For $x_2 > 22$ asymmetrical buckling precedes axisymmetrical snap-through buckling and $\bar{b}$ is negative, signifying that the asymmetrical buckling is accompanied by a decrease in load carrying capacity. Since the relative stiffness parameter $\vartheta < -0.5$ in the same interval it follows that a decrease in deflection accompanies the decrease in load carrying capacity; i.e., the initial slope of the bifurcation branch of the load-deflection curve is backward. For $10 \leq x_2 \leq 22$ a local maximum on the axisymmetrical load-deflection curve appears before the lowest bifurcation load. It was observed, for $x_2 < 10$, that neither axisymmetrical nor asymmetrical buckling occurred; instead, the deflection of the cap was characterized by a gradual transition from one equilibrium configuration to another.

Figure 3 shows the axisymmetrical load-deflection curves for several values of x_2. The ordinate of this figure is the average vertical nondimensional axisymmetrical displacement defined by the relation

$$\bar{\rho} = \frac{2}{\lambda^2} \int_0^{\lambda} xw\,dx \tag{76}$$

In making a computer search for possible asymmetrical buckling behavior for $x_2 < 10$, the graphs of the modified buckling determinant vs loading parameter p were considered for the range of wave numbers $n = 1, 2, \ldots, 6$. For example, for $x_2 = 6$, although the load parameter p was increased to 50 in increments of 2.5 the modified buckling determinant did not change sign for any value of n considered. Moreover, there was no evidence that a sign change would likely occur for higher values of n.

The buckling and initial post-buckling behavior of clamped shallow spherical sandwich shells under a concentrated load applied at the apex is studied as a limiting case of the centrally distributed pressure. This behavior was studied for two different sets of material and geometrical parameters: $(a/H)^2(G_c/E) = 0.8$, $c/t = 10$; and $(a/H)^2(G_c/E) = 0.05$, $c/t = 5$. Poisson's ratio was taken as $\nu = 1/3$ in each case. A computer search for possible asymmetrical buckling was made for several values of λ ($\lambda = 30, 35, 40, 45$ for the first set of parameters and $\lambda = 10, 12, 15, 25, 30, 40, 50, 60$ for the second set). The wave number n was given only the values 3, 4, 5 since it is felt that the homogeneous and sandwich spherical caps under a concentrated load at the apex would have the same critical wave numbers. Although the load parameter p was increased up to 2000 in increments of 50, neither axisymmetrical nor asymmetrical buckling was detected: instead the deflection of the cap was characterized by a gradual transition from one equilibrium configuration to another. Several characteristic plots of the pressure vs. axisymmetrical deflection parameter $\bar{\rho}$ are shown in Figure 4 for both sets of parameters.

As a result of these studies, it can be concluded, at least for the range of λ and the material properties considered in this investigation, that the clamped shallow spherical sandwich shell does not buckle under a concentrated load at the apex. This conclusion is in contrast to that observed for the clamped shallow spherical homogeneous shell under a concentrated load at the apex [2]. It is felt that asymmetrical buckling could occur of other sets of material and geometrical parameters.

Finally, the buckling mode shape associated with the asymmetrical buckling behavior is shown in Figure 5 as a function of the nondimensional radial coordinate for $\lambda = 30$, $x_2 = 23$, $(a/H)^2(G_c/E) = 0.8$, and $c/t = 10$. In plotting the buckling mode shape, it has been assumed that the maximum value of $W_1 (\equiv 2H/\lambda^2\ w_1)$ is equal to the face thickness t.

ACKNOWLEDGEMENT

The financial support for this investigation was provided by the Department of Engineering Mechanics at Clemson University. The authors are indebted to Mrs. Linda Lay for typing the manuscript.

BIBLIOGRAPHY

[1] Koiter, W. T., "Elastic Stability and Post-Buckling Behavior," Proc. Symp. Nonlinear Problems, Ed. R. E. Langer, p. 257, Univ. of Wisc. Press (1963).

[2] Fitch, J. R., "The Buckling and Post-Buckling Behavior of Spherical Caps under Concentrated Load, "International Journal of Solids and Structures, Vol. 4, 1968, p. 421.

[3] Huang, N. C., "Unsymmetrical Buckling of Thin Shallow Spherical Shells," Journal of Applied Mechanics, Vol. 31, No. 3, 1964, pp. 447-457.

[4] Fitch, J. R. and Budiansky, B., "Buckling and Post-Buckling Behavior of Spherical Caps under Axisymmetric Load," American Institute of Aeronautics and Astronautics Journal, Vol. 8, No. 4, 1970, pp. 686-693.

[5] Akkas, N. and Bauld, N. R., Jr., "Buckling and Post-Buckling Behavior of Clamped Shallow Spherical Shells under Axisymmetric Ring Loads," To be published in Journal of Applied Mechanics, Paper No. 71-APM-9.

[6] Akkas, N. and Bauld, N. R., Jr., "Buckling and Initial Post-Buckling Behavior of Clamped Shallow Spherical Shells under Axisymmetric Band Type Loads," XIV South American Sessions on Structural Engineering and IV Symposium on Pan American Structures, Buenos Aires, Argentina, 1970.

[7] Akkas, N. and Bauld, N. R., Jr., "Buckling and Initial Post-Buckling Behavior of Clamped Shallow Spherical Sandwich Shells," To be published in International Journal of Solids and Structures.

[8] Potters, M. L.,"A Matrix Method for the Solution of a Second Order Difference Equation in Two Variables," Rept. MR19, Math. Centrum, Amsterdam, 1955.

[9] Blum, R. E. and Fulton, R. E., "A Modification of Potters' Method for Solving Eigenvalue Problems Involving Tridiagonal Matrices," American Institute of Aeronautics and Astronautics Journal, Vol. 4, No. 12, 1966, p. 2231.

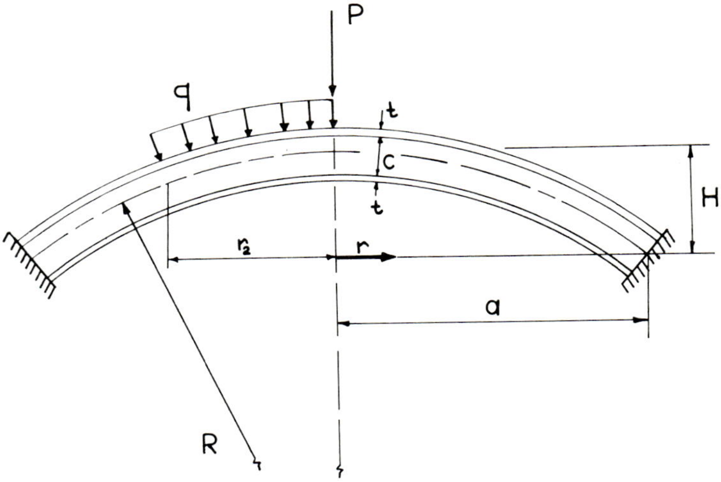

Fig. 1. Geometry of a clamped sandwich spherical cap.

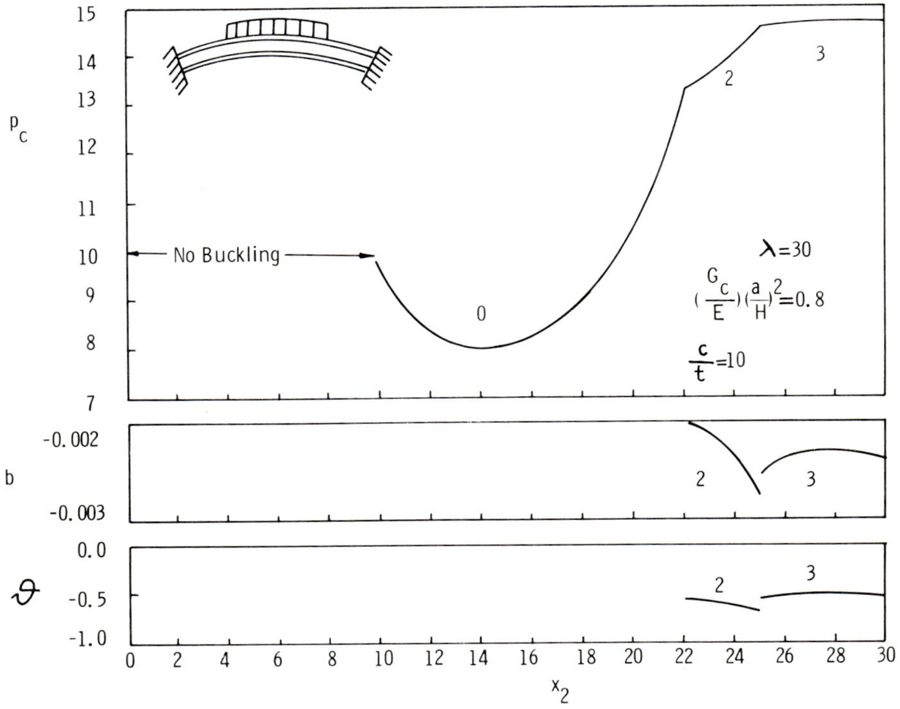

Fig. 2. Buckling and initial post-buckling behavior of clamped shallow sandwich spherical shells under centrally distributed pressure.

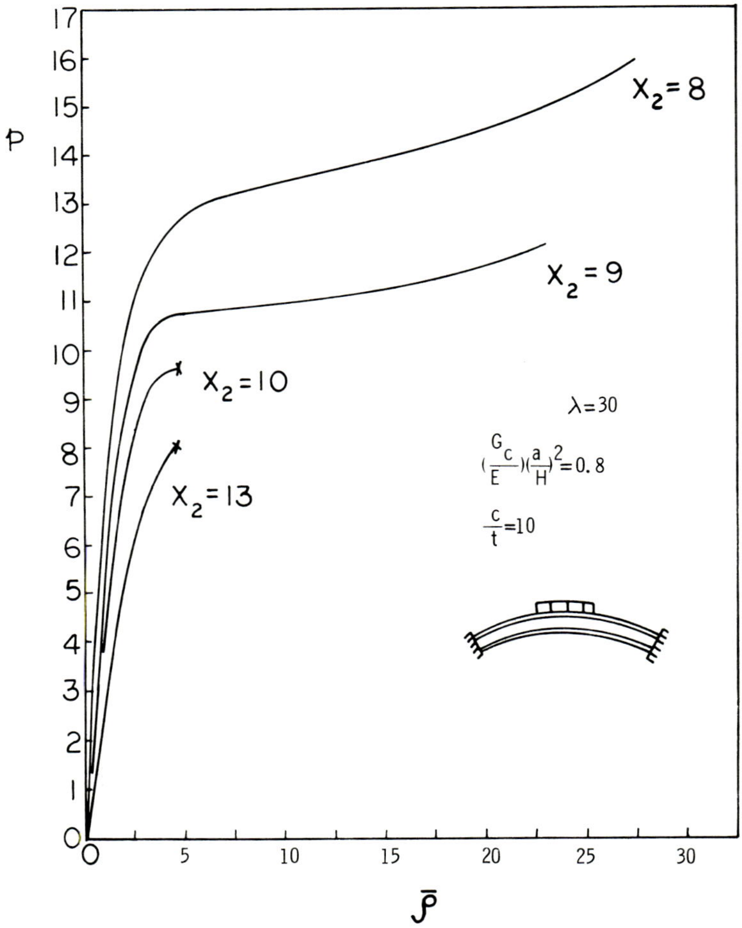

Fig. 3. Nondimensional axisymmetrical load-deflection curves for clamped shallow sandwich spherical shells under centrally distributed pressure.

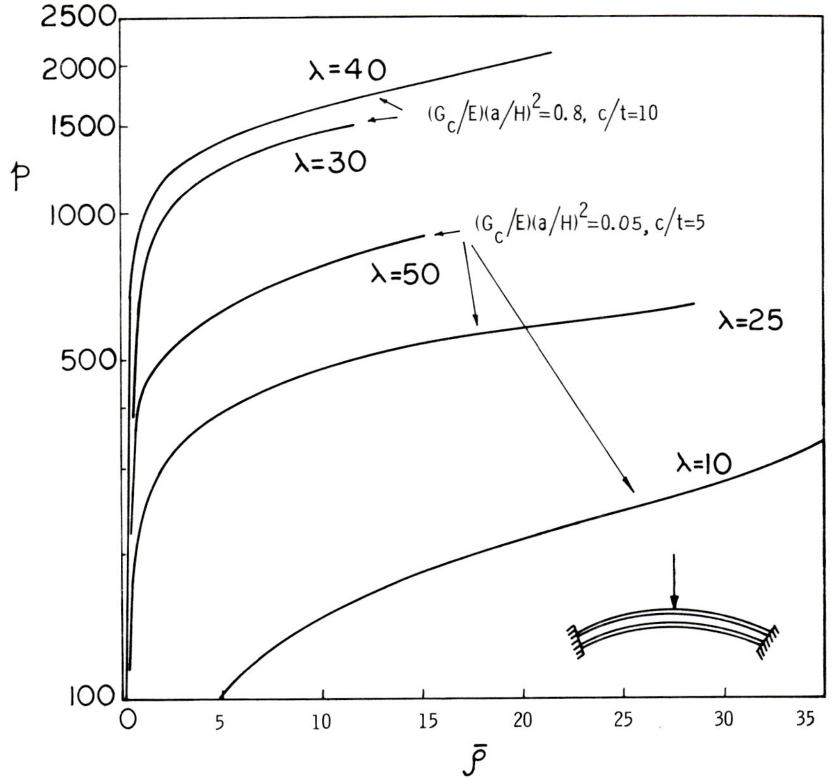

Fig. 4. Nondimensional axisymmetrical load-deflection curves for clamped shallow sandwich spherical shells under concentrated load at the apex.

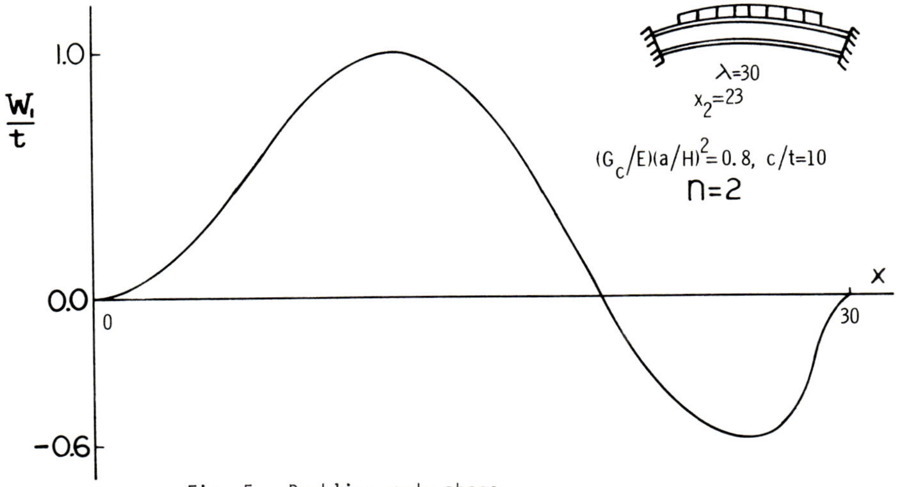

Fig. 5. Buckling mode shape.

Developments in Mechanics, Vol. 6. Proceedings of the 12th Midwestern Mechanics Conference: 39

Stability of Linear Dynamic Systems under Stochastic Parametric Excitations

JOHN A. LEPORE and ROBERT A. STOLTZ
UNIVERSITY OF PENNSYLVANIA

ABSTRACT

This paper deals with a Lyapunov-type analysis of the dynamic stability of linear systems subjected to stochastic parametric excitations. The concept of stochastic convergence and a theorem for mean-square global stability are discussed. The class of systems under consideration is that for which the dynamic system response may be represented by a damped Hill-type equation with stochastic coefficients. A general mean-square stability criterion is obtained for this class of systems, and explicit results are demonstrated in the special case of a rectangular flat plate with simultaneous stochastic in-plane loadings along its length and width. Sufficient conditions are obtained to ensure mean-square global stability in terms of statistical properties of the excitation and response process, and the physical parameters of the system. In the case of stochastically uncorrelated excitation and response, these conditions indicate that excitation variance may increase unboundedly as system damping is increased. This result, for the mean-square stability mode, is unavailable in the literature.

I. INTRODUCTION

Many important problems in the field of dynamic stability can be analyzed in terms of a Hill-type equation, which in general can be taken in the form

$$\frac{d^2g}{dt^2} + 2\alpha \frac{dg}{dt} + \omega^2 \{ 1 - 2\beta \varphi(t) \} g = 0 \tag{1}$$

where α, β, and ω are constants, and $\varphi(t)$ is a time-varying coefficient which may be either deterministic or stochastic. Considerable research has been done on the dynamic stability of systems for which the parameter $\varphi(t)$ in Eq. 1 is deterministic, and, in particular, periodic. Crimi [1] has recently developed a formulation for the dynamic stability of certain systems which can be represented by a set of coupled Hill-type equations with periodic coefficients. An example of such a system is the case of a spinning satellite in elliptic orbit, where the gravity forces affecting satellite attitude are periodic functions of time. A large group of dynamic systems whose stability can be determined from a Hill-type equation is the class of parametrically-excited structural systems. V.V. Bolotin [2] provides a comprehensive treatment of these cases under the restriction of purely deterministic excitations.

The intent of this paper is to obtain a stochastic mean-square stability criterion for systems whose dynamic response can be represented by a Hill-type equation with stochastic coefficients. The difficulty in such an investigation is that, notwithstanding the linear properties of the system, the response displays a nonlinear dependence on the stochastic coefficients. Consequently, the direct method of Lyapunov is used herein to obtain sufficient conditions guaranteeing the mean-square stability of the system. This approach has an advantage over more conventional techniques in that it enables one to deal with the question of the stability of the stochastic system utilizing the governing form of the differential equation without explicit knowledge of its solution. A general mean-square stability criterion will first be established for any system governed by Eq. 1. More detailed explicit results will then be determined for the particular case of an elastic flat plate subjected to stochastic in-plane loadings.

II. STOCHASTIC STABILITY

As with systems which are governed by differential equations with deterministic coefficients, the concept of stability considered here is still essentially a matter of convergence of solutions of the system. But now it is necessary to deal with convergence in a stochastic sense since one is dealing with limits involving random variables. The sequence of random variables x_n converges to x in mean-square if $E\{|x_n - x|^2\} \to 0$ as $n \to \infty$. Of primary concern is the stability which corresponds to convergence in mean-square of the system described by $\dot{\underline{x}} = \underline{f}\{\underline{x}, h(t), t\}$, where $\underline{x}$ is a vector describing the state of the system; $\underline{f}\{x, h(t), t\}$ is a continuous vector function of the stochastic variable $h(t)$, satisfying a Lipschitz condition, and such that $\underline{f}\{\underline{0}, h(t), t\} = 0$ for all t. The null solution $\underline{x}(t) = 0$ is the equilibrium solution to be investigated. Hence, the following definition can be stated:

Definition: An equilibrium state $\underline{x}(t) = 0$ is globally stable in mean-square if, for every $\epsilon > 0$, there exists a real number $\delta(\epsilon) > 0$ such that for any finite initial state, $\|\underline{x}^2(t_0)\| < \delta(\epsilon)$ implies the relationship $E\{\|\underline{x}^2(t)\|\} < \epsilon$, and further that $\lim_{t \to \infty} E\{\|\underline{x}^2(t)\|\} \to 0$.

Using this definition of stability, the behavior of a stochastic system can be determined if a Lyapunov function can be found which satisfies the requirements of the following theorem, proofs of which are given, essentially, by Bertram and Sarachik [3], and will not be repeated here:

Theorem: If there exists a Lyapunov function $V(\underline{x}, t)$ defined over the entire state space, with the following properties: (1) $V(\underline{0}, t) = 0$; (2) $V(\underline{x}, t)$ is continuous in mean-square in both variables $\underline{x}$ and t, and the first partial derivatives of $V(\underline{x}, t)$ in these variables exist; (3) $V(\underline{x}, t)$ is a positive-definite function, i.e., $V(\underline{x}, t) \geq a\,\underline{x}^2(t)$ for some $a > 0$; (4) $\lim V(\underline{x}, t) \to \infty$ for $\|\underline{x}(t)\| \to \infty$, then the equilibrium solution $\underline{x}(t) = 0$ is globally stable in the mean-square sense if $V(x, t)$ is a decrescent function whose expected value of its derivative dV/dt is negative definite over the entire nonzero state space of $\underline{x}(t)$; i.e., $E\{\dot{V}(\underline{x},t)\} < 0, \underline{x}(t) \neq 0$, for all $\|\underline{x}(t_0)\|$.

III. STABILITY ANALYSES IN THE GENERAL CASE

Introducing a new independent variable, $\tau = \omega t$, Eq. 1 becomes

$$g'' + 2\Omega g' + \{1 - 2\beta\varphi(t)\} g = 0 \tag{2}$$

where

$$(\ )' = \frac{d}{d\tau}(\ ) \tag{3}$$

$$\Omega = \frac{\alpha}{\omega} \tag{4}$$

Reducing Eq. (2) to a pair of simultaneous first-order equations, one obtains

$$\underline{Z}' = [\underline{A} + \psi(t)\underline{B}]\underline{Z} \tag{5}$$

where

$$\underline{A} = \begin{bmatrix} 0 & 1 \\ -1 & -2\Omega \end{bmatrix} \tag{6}$$

$$\underline{B} = \begin{bmatrix} 0 & 0 \\ 1 & 0 \end{bmatrix} \tag{7}$$

$$\psi(t) = 2\beta\varphi(t) \tag{8}$$

$$\underline{Z} = \begin{bmatrix} Z_1 \\ Z_2 \end{bmatrix} = \begin{bmatrix} g \\ g' \end{bmatrix} \tag{9}$$

It is assumed that the stochastic process $\{\varphi(t) : t \in (0,\infty)\}$, and consequently $\{\psi(t) : t \in (0,\infty)\}$ satisfies the following conditions: (1) the process is continuous in the interval $0 < t < \infty$ with probability one; and (2) the process is strictly stationary.

In developing the Lyapunov function $V(\underline{Z},t)$ for the system described by Eq. 5, the following quadratic form, defined over the entire state space, is selected as a possible V-function:

$$V(\underline{Z},t) = \underline{Z}^t \, \underline{Y} \, \underline{Z} \tag{10}$$

where Y is a real, symmetric, positive-definite matrix, and $\underline{Z}$ and $\underline{Z}^t$ are the state vector and its transpose, respectively.

In this form, the Lyapunov function approaches zero as the state vector approaches zero, for all time, i.e., it is decrescent. Hence, all the conditions of the preceding theorem are satisfied, save that of the behavior of the mean value of the time derivative of $V(\underline{Z},t)$, which will now be considered.

The time derivative of Eq. 10 is written as

$$\dot{V}(\underline{Z},t) = \omega\{(\underline{Z}^t)' \, \underline{Y} \, \underline{Z} + \underline{Z}^t \, \underline{Y} \, \underline{Z}'\} \tag{11}$$

Evaluating Eq. 11 along the trajectory of the system given by Eq. 5 yields

$$\dot{V}(\underline{Z},t) = \omega \{ \underline{Z}^t [\underline{A}^t \underline{Y} + \underline{Y}\underline{A}] \underline{Z} + \underline{Z}^t [\underline{B}^t \underline{Y} + \underline{Y}\underline{B}] \underline{Z} \; \psi(t) \} \tag{12}$$

It is assumed that Eq. 10 represents the Lyapunov function of the constant coefficient portion of the system described by Eq. 5, i.e.,

$$\underline{Z}' = \underline{A}\,\underline{Z} \tag{13}$$

Since all eigenvalues $\lambda_A = -\Omega + (\Omega^2 - 1)^{1/2}$ of the matrix $\underline{A}$ have negative real parts, this system is asymptotically stable, and consequently

$$\dot{V}_D(\underline{Z},t) = \omega \underline{Z}^t \{ \underline{A}^t \underline{Y} + \underline{Y}\underline{A} \} \underline{Z} < 0 \tag{14}$$

where the subscript D designates the deterministic partial system. The negative-definiteness of $\dot{V}$ will be assured if the matrix sum in the brackets is negative-definite. Hence,

$$\underline{A}^t \underline{Y} + \underline{Y}\underline{A} = -\underline{C} \tag{15}$$

where $\underline{C}$ is a positive-definite matrix, which, in this case, is assumed to be the unit matrix $\underline{I}$. It has been shown by Alimov [4] that the V-function obtained in the foregoing manner is admissable as a Lyapunov function. The three unknown elements of the matrix $\underline{Y}$ can now be determined by the linear simultaneous equations resulting from Eq. 15, with $\underline{C} = \underline{I}$:

$$\underline{Y} = \begin{bmatrix} \Omega + \frac{1}{2\Omega} & \frac{1}{2} \\ \frac{1}{2} & \frac{1}{2\Omega} \end{bmatrix} \tag{16}$$

Having the V-function completely defined, Eq. 12 is written as

$$\dot{V} = \omega \{ -\underline{Z}^t \underline{Z} + \underline{Z}^t \underline{H}\,\underline{Z}\, \psi(t) \} \tag{17}$$

where

$$\underline{H} = \underline{B}^t \underline{Y} + \underline{Y}\underline{B} \tag{18}$$

and

$$H = \begin{bmatrix} 1 & \frac{1}{2\Omega} \\ \frac{1}{2\Omega} & 0 \end{bmatrix} \tag{19}$$

Performing the indicated matrix operations, Eq. 17 becomes

$$\dot{V} = \omega \{ -Z_1^2 - Z_2^2 + \psi(t) Z_1^2 + \frac{\psi(t)}{\Omega} Z_1 Z_2 \} \tag{20}$$

or

$$\dot{V} = \omega \{ -Z_1^2 + \psi(t) Z_1^2 \} + \omega \{ -Z_2^2 + \frac{\psi(t)}{\Omega} Z_1 Z_2 \} \tag{21}$$

Since a sufficient condition for stability is being sought, Eq. 21 may be simplified, eliminating its explicit dependence on the Z_2 variable, by noting that

$$\{ \frac{\psi(t)}{2\Omega} Z_1 - Z_2 \}^2 \geq 0 \tag{22}$$

After expansion, Eq. 22 provides the condition that

$$\{ -Z_2^2 + \frac{\psi(t)}{\Omega} Z_1 Z_2 \} \leq \frac{\psi^2(t)}{4\Omega^2} Z_1^2 \tag{23}$$

Using Eq. 23 in Eq. 21 yields

$$\dot{V} \leq \omega \{ - Z_1^2 + \psi(t) Z_1^2 + \frac{\psi^2(t)}{4\Omega^2} Z_1^2 \} \tag{24}$$

$\dot{V}$ has, equivalently, been maximized with respect to the Z_2 variable, so that the ultimate stability condition will be sufficient over the entire domain of Z_2.

The theorem for global stability in mean-square will be completely satisfied if the mean value of $\dot{V}$ is negative-definite for nonzero values of the state vector. This requirement is ensured if

$$E\{\psi^2 Z_1^2\} + 4\Omega^2 E\{\psi Z_1^2\} - 4\Omega^2 E\{Z_1^2\} < 0 \tag{25}$$

Hence, Eq. 25 represents a general sufficient condition guaranteeing the global mean-square stability of any stochastic linear system whose dynamic response is characterized by Eq. 1. However, in its present form, Eq. 25 cannot be explicitly applied in a convenient manner. Further simplification of Eq. 25 can be attained when a particular dynamic system is specified. This will be demonstrated in the following section.

IV. STABILITY ANALYSIS OF FLAT PLATE

Consider a simply-supported, thin, rectangular flat plate of length a and width b, subjected to simultaneous, uniformly-distributed, stochastic in-plane loadings, as shown in Fig. 1. The two excitations p(t) and q(t) are considered to be stochastically independent, and to have positive-semidefinite mean values, i.e.,

$$E\{p(t)\},\ E\{q(t)\} \geq 0 \tag{26}$$

The condition imposed by Eq. 26 is not particularly restrictive, since, in the great majority of physical cases, this condition is satisfied. Furthermore, both p(t) and q(t) are assumed to be strictly stationary and continuous in the interval $0 < t < \infty$ with probability one.

Utilizing those assumptions customarily made in the study of the dynamics of linear continuous structures, and, since the plate excitations are spatially uniform, the governing partial differential equation is well known to be

$$\nabla^2 \nabla^2 w + \frac{1}{D} \{ p(t) \frac{\partial^2 w}{\partial x^2} + q(t) \frac{\partial^2 w}{\partial y^2} + m \frac{\partial^2 w}{\partial t^2} + c \frac{\partial w}{\partial t} \} = 0 \tag{27}$$

where

w = transverse deflection

D = flexural rigidity

m = mass per unit area

c = viscous damping coefficient

$$\nabla^2 = \frac{\partial^2}{\partial x^2} + \frac{\partial^2}{\partial y^2}$$

The solution of Eq. 27 is assumed to be of the form

$$w(x,y,t) = \sum_{i=1}^{\infty} \sum_{k=1}^{\infty} g_{ik}(t)\, G_{ik}(x,y) \tag{28}$$

where $g_{ik}(t)$ are unknown time-dependent functions, and $G_{ik}(x,y)$ are the spatial-dependent normal modes, forming a complete set and satisfying the simply-supported boundary conditions. For the flat plate,

$$G_{ik}(x,y) = \sin \frac{i\pi x}{a} \sin \frac{k\pi y}{b} \tag{29}$$

Substituting Eq. 28 in Eq. 27, and simplifying, one obtains,

$$\frac{d^2 g_{ik}}{dt^2} + \frac{c}{m} \frac{dg_{ik}}{dt} + \omega_{ik}^2 \left[1 - \frac{p(t)}{p_*} - \frac{q(t)}{q_*} \right] g_{ik} = 0 \tag{30}$$

where

$$\omega_{ik} = \pi^2 \sqrt{\frac{D}{m}} \left[\frac{i^2}{a^2} + \frac{k^2}{b^2} \right] \tag{31}$$

$$p_* = \frac{D\pi^2 i^2}{a^2} \left[1 + \left(\frac{ka}{ib}\right)^2 \right]^2 \tag{32}$$

$$q_* = \frac{D\pi^2 k^2}{b^2} \left[1 + \left(\frac{ib}{ka}\right)^2 \right]^2 \tag{33}$$

It should be noted that the ω_{ik} represent the free vibration frequencies of the unloaded plate; p_* and q_* represent the static critical load values of $p(t)$ and $q(t)$, respectively, each in the absence of the other.

Since it is understood that Eq. 30 governs each mode, the subscripts i,k, may be dropped for convenience. Introducing a new independent variable $\tau = \omega t$, Eq. 30 becomes

$$g'' + 2\Omega g' + \left[1 - (P(t) + Q(t)) \right] g = 0 \tag{34}$$

where

$$\Omega = \frac{c}{2 m \omega} \tag{35}$$

$$P(t) = p(t)/p_* \tag{36}$$

$$Q(t) = q(t)/q_* \tag{37}$$

Since the flat plate problem has been reduced to a damped, stochastic Hill-type equation, its mean-square stability criterion is given directly by Eq. 25, after noting that

$$\psi(t) = P(t) + Q(t) \tag{38}$$

Hence, for mean-square global stability:

$$E\{(P+Q)^2 Z_1^2\} + 4\Omega^2 E\{(P+Q)Z_1^2\} - 4\Omega^2 E\{Z_1^2\} < 0 \tag{39}$$

Since the squared amplitude response process Z_1^2 depends on both excitations P and Q, a multiple regression between Z_1^2 and P and Q is introduced. Consequently, the following linear mean-square regression plane is defined when P and Q are stochastically independent excitations:

$$Z^2 = E[Z_1^2] + \rho_{PZ_1^2} \frac{\sigma_{Z_1^2}}{\sigma_P} \{P - E[P]\} + \rho_{QZ_1^2} \frac{\sigma_{Z_1^2}}{\sigma_Q} \{Q - E[Q]\} \tag{40}$$

Equation 40 provides the best linear estimate of Z_1^2 in terms of P and Q; $\rho_{PZ_1^2}$ and $\rho_{QZ_1^2}$ represent the total correlation coefficients between P and Z_1^2, and Q and Z_1^2, respectively.

Multiplying Eq. 40 by (P + Q), applying the expectation operation to both sides, and simplifying, one obtains

$$\begin{aligned} E\{(P+Q) Z_1^2\} &= E\{P\}E\{Z_1^2\} + E\{Q\}E\{Z_1^2\} \\ &+ \rho_{PZ_1^2} \sigma_{Z_1^2} \sigma_P + \rho_{QZ_1^2} \sigma_{Z_1^2} \sigma_Q \end{aligned} \tag{41}$$

Similarly, multiplying Eq. 40 by $(P + Q)^2$, applying the expectation operator, and simplifying, yields

$$\begin{aligned} E\{(P+Q)^2 Z_1^2\} &= E\{P^2\}E\{Z_1^2\} + E\{Q^2\}E\{Z_1^2\} + 2E\{P\}E\{Q\}E\{Z_1^2\} \\ &+ 2\rho_{PZ_1^2} \sigma_{Z_1^2} \sigma_P \{E[P] + E[Q] + 1/2\, \sigma_P \gamma_P\} \\ &+ 2\rho_{QZ_1^2} \sigma_{Z_1^2} \sigma_Q \{E[P] + E[Q] + 1/2\, \sigma_Q \gamma_Q\} \end{aligned} \tag{42}$$

where γ_P and γ_Q are the coefficients of skewness of the P process and Q process, respectively.

Substituting Eq. 41 and Eq. 42 into Eq. 39, and letting $\upsilon = \sigma_{Z_1^2}/E\{Z_1^2\}$, one obtains after simplification:

$$\begin{aligned} &E^2\{P\} + E^2\{Q\} + 2E\{P\}E\{Q\} + 4\Omega^2 E\{P\} \\ &+ 4\Omega^2 E\{Q\} + \sigma_P^2 + \sigma_Q^2 \\ &+ 2\rho_{PZ_1^2} \upsilon\sigma_P \{E[P] + E[Q] + 1/2\, \sigma_P \gamma_P + 2\Omega^2\} \\ &+ 2\rho_{QZ_1^2} \upsilon\sigma_Q \{E[P] + E[Q] + 1/2\, \sigma_Q \gamma_P + 2\Omega^2\} \\ &- 4\Omega^2 < 0 \end{aligned} \tag{43}$$

Before Eq. 43 can be explicitly applied, the parameter υ, the coefficient of variation of the squared amplitude response process, must be investigated. The response amplitude process Z_1, will, in general, be a nonstationary, nonsymmetric process, attaining both positive and negative values. It is assumed that the probability density function of Z_1 is both continuous and finite at all points in the domain of Z_1, and that finite upper and lower bounds on Z_1 exist, i.e.,

$$-\infty < -a \leq Z_1 \leq b < \infty \tag{44}$$

Consider the associated process $|Z_1|$. In view of Eq. 44, the $|Z_1|$ process is also bounded, i.e.,

$$0 \leq Z_1 \leq \delta < \infty \tag{45}$$

Furthermore, the probability density function of $|Z_1|$ is continuous and finite at all points in the domain of $|Z_1|$.

The normalized beta density function is introduced, given by

$$f_\beta(p/r,s) = \frac{1}{B(r,s-r)} p^{r-1}(1-p)^{s-r-1} \tag{46}$$

where

$$0 \leq p \leq 1 \tag{47}$$

$$s > r > 0 \tag{48}$$

and $B(r,s-r)$ is the complete beta function, given by

$$B(r,s-r) = \int_0^1 p^{r-1}(1-p)^{s-r-1} dp \tag{49}$$

$$= \frac{\Gamma(r)\,\Gamma(s-r)}{\Gamma(s)} \tag{50}$$

The beta density function defines an infinite set of probability distributions dependent on the selection of the parameters r and s. It is particularly useful since its specializations represent a large number of different distributions having practical significance. A particular subset of the complete ensemble of possible beta distributions will be utilized to represent the potential probability distributions of the $|Z_1|$ process. A subset, rather than the entire family, is employed because of the requirement that the density function of $|Z_1|$ be finite at all points in its domain. From Eq. 46, it can be seen that if

$$r < 1: f_\beta(p/r,s) \to \infty \text{ as } p \to 0 \tag{51}$$

$$s < r+1: f_\beta(p/r,s) \to \infty \text{ as } p \to 1 \tag{52}$$

Consequently, the subset utilized will be the family of beta distributions for which

$$s-1 \geq r \geq 1 \tag{53}$$

Since,

$$|Z_1| \leq \delta \tag{54}$$

the normalized beta variable is transformed, such that

$$|Z_1| = \delta p \qquad (0 \leq p \leq 1) \tag{55}$$

Hence,

$$\upsilon^2 = \sigma_{Z_1^2}{}^2 / E^2\{Z_1^2\}$$

$$= \frac{E\{Z_1^4\}}{E^2\{Z_1^2\}} - 1$$

$$= \frac{E\{p^4\}}{E^2\{p^2\}} - 1 \tag{56}$$

The moments of the beta distribution are given in terms of r and s:

$$E\{p^j\} = \frac{(r + j - 1)\,(r + j - 2)\ldots r}{(s + j - 1)\,(s + j - 2)\ldots s} \tag{57}$$

In view of Eq. 53 and Eq. 57, it is easily shown that

$$0 < \upsilon < \sqrt{5} \tag{58}$$

Since it is not known which explicit beta distribution should be used to represent the $|Z_1|$ process, and, indeed, since the probability distribution of the $|Z_1|$ process can be expected to change as the excitation processes are varied, the final stability condition should be a sufficient condition for all cases in which the $|Z_1|$ process can be represented by some member of the previously discussed subset of beta distributions. Since the system correlation coefficients can well be assumed to be positive-semidefinite, and since the excitation mean values are positive-semidefinite, the upper-bound value of the parameter υ, i.e. 5, will be used in Eq. 43 giving

$$\begin{aligned} &E^2\{P\} + E^2\{Q\} + 2E\{P\}E\{Q\} + 4\Omega^2 E\{P\} + 4\Omega^2 E\{Q\} + \sigma_P^2 + \sigma_Q^2 \\ &+ 2\sqrt{5}\,\rho_{QZ_1^2}\,\sigma_Q\{E[P] + E[Q] + 1/2\,\sigma_Q\gamma_Q + 2\Omega^2\} \\ &+ 2\sqrt{5}\,\rho_{PZ_1^2}\,\sigma_P\{E[P] + E[Q] + 1/2\,\sigma_P\gamma_P + 2\Omega^2\} \\ &- 4\Omega^2 < 0 \end{aligned} \tag{59}$$

Equation 59 provides a sufficient condition guaranteeing the global mean-square stability of the rectangular flat plate excited by stochastically independent, in-plane loadings. The criterion can be applied in a straightforward algebraic manner, defining regions of mean-square stability in terms of the excitation mean values, variances, and coefficients of skewness, the total linear correlations between the excitation processes and the response, as well as the physical parameters of the system. An additional simplification of Eq. 59 is available when either or both excitation processes possess symmetric or negatively-skewed probability distributions. In these cases, γ_P and/or γ_Q can be omitted in Eq. 59. This simplification occurs frequently since many excitation processes of physical significance are symmetrically distributed. However, γ_P and/or γ_Q must be retained for excitations having strongly asymmetric probability distributions with positive skewness; e.g., the exponential distribution.

V. DISCUSSION OF RESULTS

Upon examination of Eq. 59, it is easily seen that both plate excitations P(t) and Q(t) enter the stability criterion in a completely parallel manner, since each has been normalized with respect to its corresponding static critical value, as indicated by Eq. 36 and Eq. 37. Therefore, identical stability bounds are obtained when either process acts alone. For simplicity, the primary emphasis of this discussion will be placed on the physically useful case of symmetric excitation processes. In view of this, and allowing $Q(t) \to 0$ in Eq. 59, the stability condition under single, symmetric excitation is found to be

$$E^2\{P\} + \sigma_P^2 + 4\Omega^2 E\{P\} \\ + 2\sqrt{5}\,\rho\sigma_P\{E[P] + 2\Omega^2\} - 4\Omega^2 < 0 \qquad (60)$$

where the subscripts on ρ have been dropped, since only a single correlation is involved.

The fundamental behavior of the mean-square stability criterion is most clearly demonstrated in the case of single excitation; hence, this situation will be investigated first, and will be followed by results obtained under combined excitations.

Figure 2 indicates the regions of mean-square global stability in terms of the mean value and standard deviation of the excitation, for several values of the correlation coefficient ρ. The stability regions are those areas enclosed by a particular curve and the coordinate axes. It is readily apparent that, for any value of ρ, as the excitation variance is reduced, the corresponding allowable mean value may increase. Moreover, as the correlation between the excitation and the response diminishes, the stability region is enlarged. This would be expected since the correlation may be thought of as an indication of the influence of excitation dispersion on the response process. The variation of the stability regions with damping is shown in Figure 3, for a fixed value of ρ. As damping is increased, the overall size of the stability region is enlarged, in terms of both the mean value and variance of the excitation. However, the growth of the mean value with damping is bounded. Indeed, in the deterministic case ($\sigma_P = 0$), P(t) approaches one as damping becomes infinite. On the other hand, the increase of variance with damping is quite a different situation, as shown in Figure 4. For any specified value of damping, the allowable excitation variance increases with decreasing correlation. The growth of allowable variance with increasing damping is, however, bounded, except in the case of an uncorrelated system, in which instance the excitation variance may increase unboundedly with damping. From a physical point of view, this result would be expected, but it has not been obtained in previous analyses of mean-square stability. Figure 4 is based on an excitation mean value of zero; however, similar results are obtained when the mean value is positive and nonzero. Moreover, the unbounded increase of variance with damping for the uncorrelated system holds true for any general excitation process, regardless of symmetry.

Figures 2, 3, and 4 have been based on excitation processes whose probability distributions are either symmetric or negatively skewed. If the excitation process has an exponential probability distribution, for example, the term involving coefficient of skewness in the stability criterion must be retained. Consider a positive excitation process whose lower bound is zero. Figure 5 compares the stability regions for such a case, depending on whether the process has a uniform or an exponential probability distribution. In either case, the moments of the distribution are interrelated, and cannot be varied independently. It can be seen that the region of guaranteed mean-square stability is larger when the process has a uniform distribution. This is essentially due to the fact that the exponential distribution, because of its "tail" extending to infinity, allows small, but finite, probabilities of large fluctuations of the excitation process. Equivalently, for any given excitation mean value, the corresponding variance associated with the exponential distribution is greater than that of the uniform distribution.

At this point, the present results will be compared with those previously obtained for mean-square stability by Lepore and Shah [5], [6]. The stability conditions formerly obtained are dependent, for application, on the particular form of the excitation distribution. Thus, in Figure 6, the former results are depicted for the cases of Gaussian and uniform excitation distributions having zero mean values. The present result is shown for any symmetric excitation with zero mean value. It is seen that the previous result is sensitive to the nature of the excitation, as well as to its mean value and variance. More significantly, although Figure 6 is based on an uncorrelated system ($\rho = 0$), the former conditions display a distinctly asymptotic growth in variance with increased damping. Similar curves are obtained in the case of nonzero excitation mean value.

Stability bounds of other authors are shown in Figure 7. All four authors treated the concept of almost-sure global stability. Both Kozin [7] and Ariaratnam [8] apply the Bellman-Gronwall Lemma to the integral solution of Eq. 2 in order to obtain conditions for almost-sure stability. The bounds of Kozin are considerably smaller than the others, and he shows an unusual decrease near $\Omega = 1$. These characteristics are a consequence of Kozin's choice of an upper bound on the magnitude of the impulse response function and are not characteristic of the system under consideration. Ariaratnam shows that the bounds can be extended, and the sharp decrease eliminated, by introducing three separate ranges of the damping parameter Ω. Caughey and Gray [9] use a Lyapunov-type approach to obtain sufficient conditions guaranteeing the almost-sure stability of Eq. 2. Infante and Plaut [10], likewise, employ a Lyapunov technique to investigate the stability of a column subjected to a stochastic axial excitation. The excitation is restricted to be an ergodic process with a symmetric probability distribution. The results obtained by Infante and Plaut are the sharpest of those available for almost-sure stability; and, indeed, are the only almost-sure criteria having the characteristic of unbounded excitation variance growth with increasing damping. Since the mean-square stability mode is a stronger type of stability than the almost-sure mode, the criteria obtained for mean-square stability are more restrictive than those of the latter; and, indeed, mean-square stability implies almost-sure stability. In the case of an uncorrelated system, the results of this paper are numerically identical to those of Infante and Plaut. However, the present result, for an uncorrelated system, is not limited to ergodic excitation processes having symmetric probability distributions, but is applicable with any stationary, continuous excitation process.

The behavior of Eq. 59, in the presence of both plate excitations, will now be discussed. Unlike the results obtained under single excitation, the regions of mean-square stability will now depend simultaneously on the moments of both excitation processes. Again, for simplicity, the results considered will be limited to cases of symmetric excitations.

Figure 8 depicts the mean-square stability regions under combined excitations, where both correlation coefficients have identical values, and one process has a zero mean value. The regions of mean-square stability are those volumes enclosed by the three coordinate axes and a particular set of three boundary curves. Because both correlation coefficients are identical, the stability region is symmetric with respect to σ_P and σ_Q. Furthermore, the boundary curves in the $E\{P\} - \sigma_P$ plane and the $E\{P\} - \sigma_Q$ plane are identical to the boundary curves obtained in the case of single excitation with the same correlation coefficient (Figure 2). In the case of an uncorrelated system, the boundary curve in the $\sigma_P - \sigma_Q$ plane is merely a quadrant of a circle with radius 2Ω. This is not true, however, for a correlated system. Figure 9 displays the case of unequal correlations. In this instance, the stability region is not symmetric with respect to σ_P and σ_Q. Figure 9 also depicts the shift in the stability region when the correlation coefficients are directly reversed. Allowable excitation variances as a function of damping are shown in Figure 10, for excitations with zero mean values. Again, each boundary curve is identical to that of the single excitation case and, indeed, the solid region of mean-square stability grows unboundedly with damping. Figure 11 is similar to Figure 10, with the exception that the correlations are unequal.

Thus, it is seen that in the case of combined excitations which are stochastically independent of one another, the projections of the stability regions onto the appropriate coordinate planes yield the same boundary curves as obtained under the corresponding single excitation.

VI. CONCLUSIONS

In this paper, sufficient conditions have been developed to ensure the mean-square stability of linear systems whose dynamic response can be described by a Hill-type differential equation. In particular, an explicit stability criterion has been determined for a thin, flat plate subjected to simultaneous, stochastic in-plane loadings. The stability of the flat plate is expressed in terms of its physical parameters, the excitation mean values, variances, and coefficients of skewness, and the correlations between the response process and the excitations.

The result obtained for the mean-square stability of an uncorrelated system provides an unbounded increase in excitation variance as damping grows. A result of this type was formerly unavailable for the mean-square mode of stability.

It should be noted that the approach utilized in this paper leads to the determination of sufficient bounds on the excitation measures for incipient instability. That is, an increase in the statistical characteristics of the excitations somewhat beyond the indicated critical values does not necessarily mean that the plate is physically unable to resist the specified loadings. At that point, however, the assumptions of small deflections and the resulting governing differential equation are no longer valid. One must then investigate the nonlinear characteristics of the system, and this becomes a natural extension of the method used herein.

VII. ACKNOWLEDGMENT

This work was supported by the National Science Foundation under Grant No. NSF-GK-4634.

VIII. NOTATION

$\underline{A}$	matrix of constant coefficient system
$\underline{B}$	excitation matrix
$G(x,y)$	normal modes of flat plate
$\underline{H}$	transformed excitation matrix
$P(t)$, $Q(t)$	normalized plate excitations
V	Lyapunov function
$\underline{Y}$	Lyapunov matrix
$\underline{Z}$	state vector
$g(t)$	time-dependent normal coordinates
$p(t)$, $q(t)$	plate excitations
p_*, q_*	static critical excitation values
Ω	normalized damping parameter
γ	coefficient of skewness
λ_A	eigenvalues of constant coefficient system
ν	coefficient of variation

ρ	correlation coefficient
σ	standard deviation
τ	nondimensionalized time
φ(t)	excitation parameter of Hill-type equation
ψ(t)	normalized excitation parameter

IX. BIBLIOGRAPHY

[1] P. Crimi: Stability of Dynamic Systems with Periodically Varying Parameters. AIAA J., Vol. 8, No. 10, pp. 1760-64 (1970).

[2] V.V. Bolotin: The Dynamic Stability of Elastic Systems. Holden-Day, San Francisco (1964).

[3] J.E. Bertram and P.E. Sarachik: Stability of Circuits with Randomly Time Varying Parameters. Proc. Int'l. Symp. on Circuit and Information Theory, Inst. of Radio Engineers, N.Y., CT-6, pp. 260-270 (1959).

[4] M.S. Alimov: On the Construction of Lyapunov Functions for Systems of Linear Differential Equations with Constant Coefficients. Siberian Math. J., Vol. 2, pp. 3-6 (1961).

[5] J.A. Lepore and H.C. Shah: Dynamic Stability of Circular Plates under Stochastic Excitations. J. of Spacecraft and Rockets, Vol. 7, No. 5, pp. 582-587 (1970).

[6] J.A. Lepore and H.C. Shah: Dynamic Stability of Axially Loaded Columns Subjected to Stochastic Excitations. AIAA J., Vol. 6, No. 8, pp. 1515-1521 (1968).

[7] F. Kozin: On Almost Sure Stability of Linear Systems with Random Coefficients. J. of Math. and Physics, Vol. 42, pp. 59-67 (1963).

[8] S.T. Ariaratnam: Dynamic Stability of a Column under Random Loading. Univ. of Waterloo, Ont. Canada (1967).

[9] T.K. Caughey and A.H. Gray: On the Almost Sure Stability of Linear Dynamic Systems with Stochastic Coefficients. Trans. of the A.S.M.E., Ser., E., J. of Appl. Mech., pp. 365-372 (1965).

[10] E.F. Infante and R.H. Plaut: Stability of a Column Subjected to a Time-Dependent Axial Load. AIAA J., Vol. 7, No. 4, pp. 766-768 (1969).

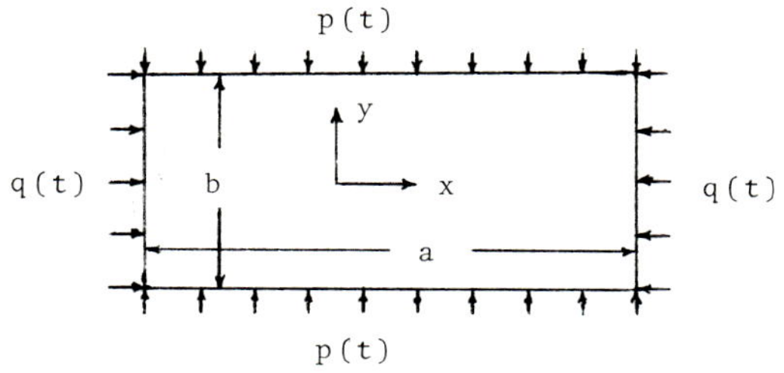

Fig. 1: Thin Rectangular Flat Plate

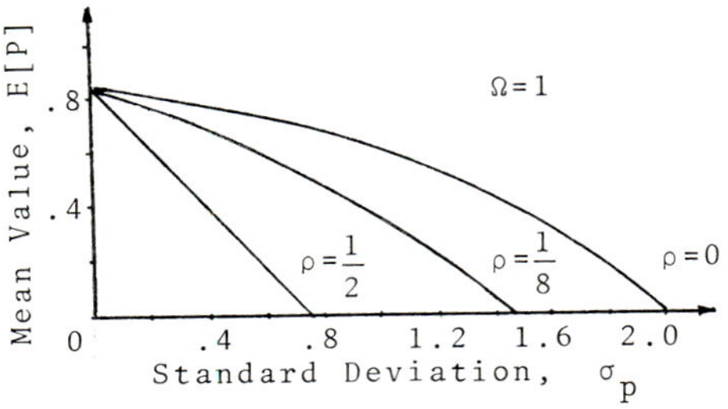

Fig. 2: Regions of Global Mean-Square Stability

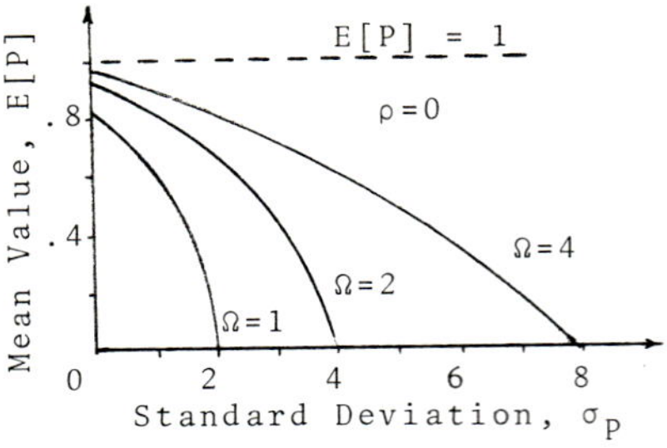

Fig. 3: Variation of Stability Regions with Damping

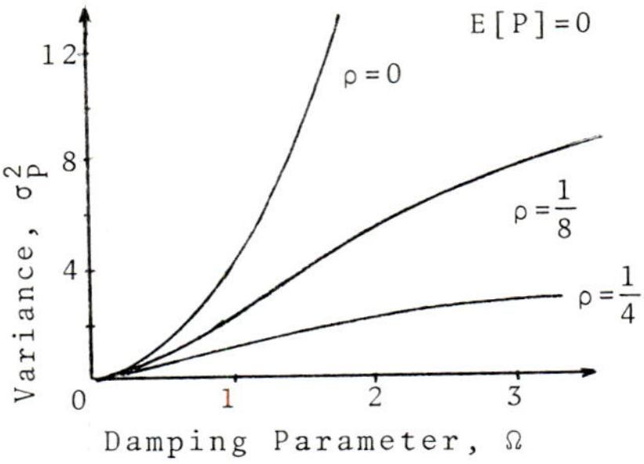

Fig. 4: Excitation Variance vs. Damping

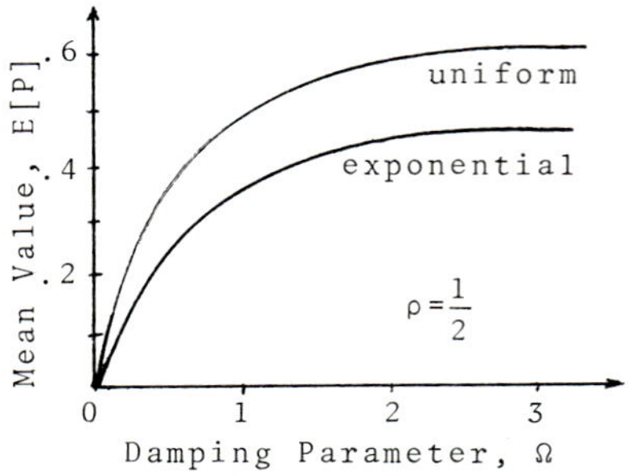

Fig. 5: Comparison of Stability Bounds for Uniform and Exponential Excitation Distributions

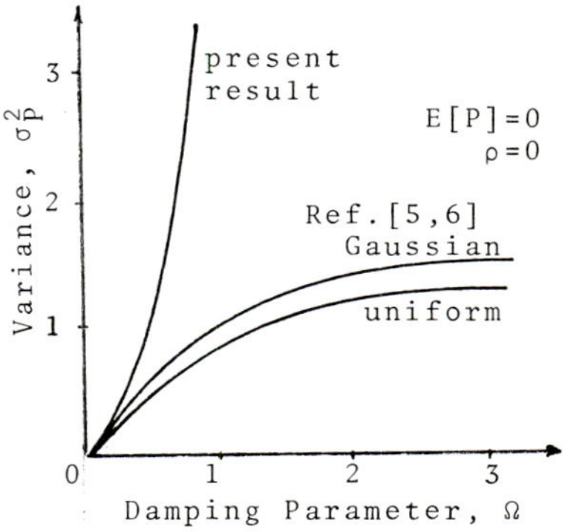

Fig. 6: Excitation Variance vs. Damping - Comparison of Present and Former Results

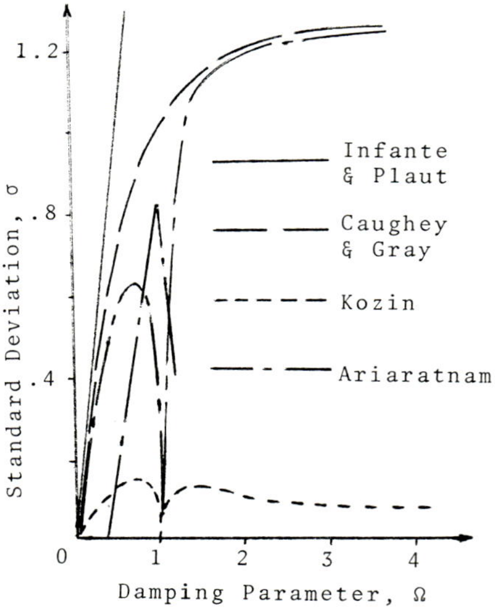

Fig. 7: Almost-Sure Stability Bounds for Gaussian Processes

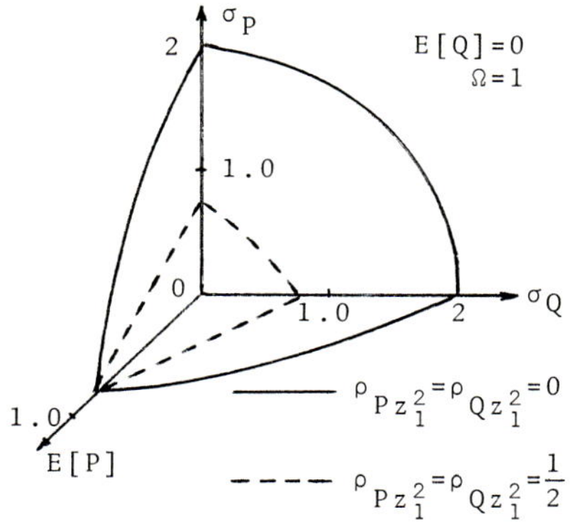

Fig. 8: Regions of Mean-Square Stability Under Combined Excitations;Equal Correlations

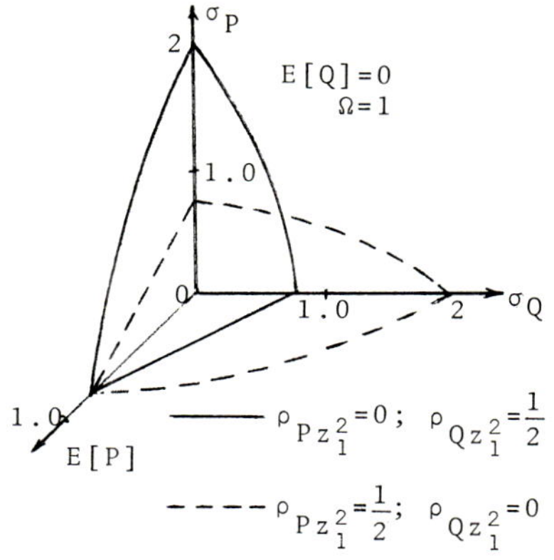

Fig. 9: Regions of Mean-Square Stability Under Combined Excitations; Unequal Correlations

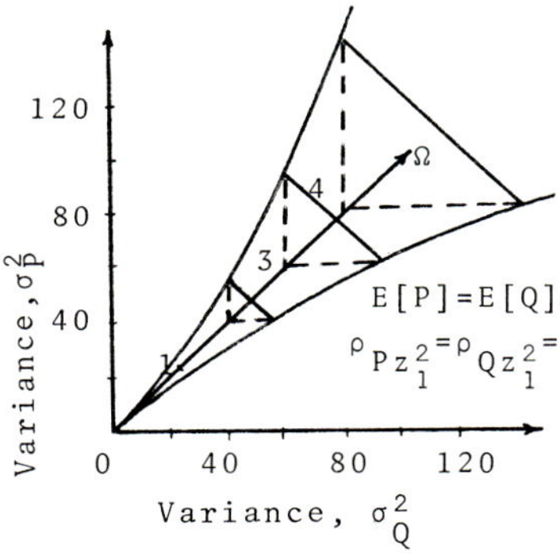

Fig. 10: Excitation Variances vs. Damping; Equal Correlations

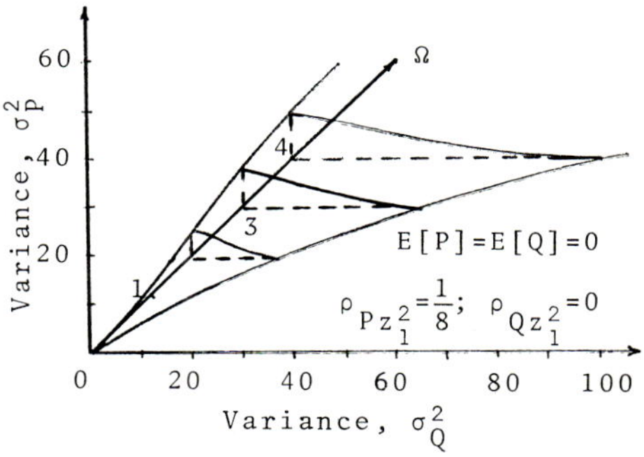

Fig. 11: Excitation Variances vs. Damping; Unequal Correlations

Developments in Mechanics, Vol. 6. Proceedings of the 12th Midwestern Mechanics Conference: 40

Micropolar Medium as a Model for Buckling of Grid Frameworks

ZDENĚK P. BAŽANT
NORTHWESTERN UNIVERSITY

ABSTRACT

Attention is focused on large rectangular frameworks of constant mesh size and constant properties of members in each direction. The framework is considered to be under initial axial loads. A continuous approximation for the expression of potential energy is formulated and, postulating an equivalent micropolar continuum under initial stress, differential equations of equilibrium in terms of displacements and rotations are derived. Expressions for stresses, couple stresses and constitutive relations are also presented.

INTRODUCTION

Although the methods for the analysis of buckling of frames are theoretically well-known, buckling of truly large grid frameworks, such as high-rise buildings, is practically intractable with the classical methods since the size of problem overtaxes the capacity of computers presently available. In practice, the assessment of stability is restricted, as a rule, to local behavior of columns within the frame, and very slender buildings are either avoided, in order to insure that investigation of the overall loss of stability is not important, or very rigid bracings (or stiffening walls) are provided, which secures stability even without the framework. Nevertheless, even in such structures the response to horizontal forces is affected by the initial loads in the columns. In the future development toward higher, lighter and slenderer structures, it can be expected that even the overall stability modes with axial extensions of columns will become an important consideration in design.

Experience with the exact solutions of the overall behavior of large frames indicates that the displacements and rotations of joints usually vary relatively smoothly from floor to floor and bay to bay. This suggests that a certain continuum approximation can be used as one method of overcoming the difficulties.

The development of various theories of structured continua has been accomplished only recently [1, 2, 3]. Their common characteristic feature is the existence of couple stresses and asymmetric shear stresses. As will be shown later, an appropriate approximation to a grid framework is Eringen's micropolar medium [3], characterized by the dependence of the elastic potential on the gradient of microrotation and the difference between micro- and macrorotation, in addition to the dependence on the symmetric part of the displacement gradient as in classical elasticity.

The possibility of applying these theories as approximations to frameworks and lattices has been mentioned in many papers. Some very general discussions were made, e.g., by Woźniak [4]. First specific treatment was presented by Banks and Sokolowski [5]. In their paper, however, the special case of a Cosserat continuum, in which the micro- and macrorotations are equal [1], was assumed. This model is, however, inadequate because the microrotation, which corresponds to the rotation of joints in a framework, and the macrorotation, which characterizes the rotation of a line connecting two adjacent joints, are in general unequal. A further significant contribution was made by Askar and Cakmak [6] who considered a rectangular gridwork with diagonals and correctly arrived at a micropolar medium. However, their model is also not fully consistent because certain important terms in the expression of elastic potential, namely those which contain second derivatives of microrotation but can be transformed on integration by parts to terms with first derivatives only, have been neglected. Buckling and deformations of frameworks under initial stress probably have not yet been treated in this light. The intent of the present paper is to formulate a consistent continuum analogy for such problems.

POTENTIAL ENERGIES OF GRIDWORK AND CONTINUUM

Consider a member of a planar framework (Figure 1) which is initially straight and in equilibrium under a large axial force, P^0. Assume that small end moments M_a, M_b, shear force T and axial force P is superposed at the ends of member,

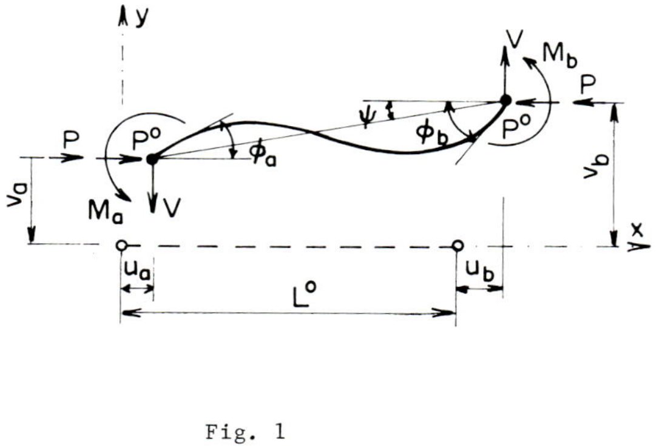

Fig. 1
Incremental forces and deformations of a member of the framework

which thus undergo small rotations φ_a, φ_b, lateral displacements v_a, v_b and longitudinal displacements u_a, u_b. As is well-known [7], the following relationship then applies:

$$\begin{Bmatrix} M_a \\ M_b \\ P \end{Bmatrix} = \begin{bmatrix} ks, & ksc, & 0 \\ ksc, & ks, & 0 \\ 0, & 0, & E' \end{bmatrix} \begin{Bmatrix} \psi - \varphi_a \\ \psi - \varphi_b \\ u_a - u_b \end{Bmatrix} \qquad (1)$$

where $\psi = (v_a - v_b)/L$; L = initial length of member; $\psi - \varphi_a$ and $\psi - \varphi_b$ = rotations relative to ab; k = EI/L; E' = EA/L; I and A = inertia moment and area of the cross-section; E = Young's modulus. Coefficients s and c are functions of P^0, called stability functions. The expressions and tables for these functions are available in the literature [7]. For a zero axial force, s = 4, c = 1/2.

The expression for the incremental strain energy U_1 of a single member is

$$U_1 = \frac{1}{2}\left[M_a(\varphi_a - \psi) + M_b(\varphi_b - \psi) + P(u_b - u_a)\right] - P^0(L\psi^2/2) \tag{2}$$

plus a linear term $P^0(u_b - u_a)$ which need not be considered because it governs only the initial equilibrium. The value $(L\psi^2/2)$ represents, with an error $O(\psi^4)$, the axial extension of the member due to small incremental lateral displacements v_a, v_b. If the expressions for M_a and M_b, and P according to Equation (1) are substituted, Equation (2) may be brought, after rearrangements, to the form:

$$U_1 = \frac{1}{2}E'(u_b - u_a)^2 + \frac{1}{2}ks(\varphi_b - \varphi_a)^2 + ks'(\psi - \varphi_a)(\psi - \varphi_b) - P^0\frac{1}{2}L\psi^2 \tag{3}$$

where

$$s' = s(1 + c) \tag{3a}$$

Consider now a plane rectangular grid framework with members parallel to Cartesian axes x and y (Figure 2). Assume that the properties in each direction are uniform, including the value of the axial forces. Quantities related to the directions x and y will be distinguished by subscripts x and y. Subscripts x or y preceded by a comma will denote partial derivatives, e.g., $v_{,x} = \partial v/\partial x$,

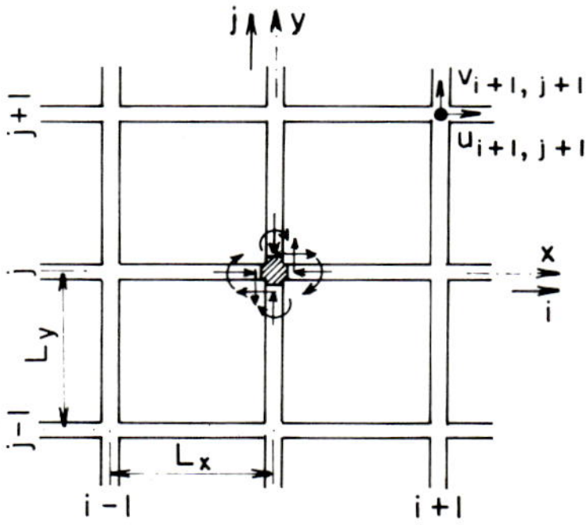

Fig. 2

Notation for the members of framework and forces acting on a joint

$\varphi_{,xx} = \partial^2\varphi/\partial x^2$. The individual joints will be referred to by subscripts i and j expressing the number of the vertical or the horizontal row of members (Figure 2). The displacements of joint (i,j) in the x- and y- directions will be denoted as $u_{i,j}$, $v_{i,j}$, and its rotation as $\varphi_{i,j}$. Here a comma between the subscripts does not refer to a derivative.

The transition from a discrete to a continuous system may be achieved by defining (sufficiently smooth) continuous functions u, v, φ and f_x, f_y, m of the variables x, y, such that their values in points (x_i, y_j) are sufficiently close to the values $u_{i,j}$, $v_{i,j}$, $\varphi_{i,j}$, and $\bar{X}_{i,j}$, $\bar{Y}_{i,j}$, $\bar{M}_{i,j}$, respectively. The latter three values represent prescribed incremental loads and moments applied in the joint and f_x, f_y, m are the equivalent incremental distributed loads and moments per unit area of the gridwork.

The smoothing operation, by which the continuous approximation of gridwork may be obtained, consists in introducing the continuous functions u, v, φ into the expression for potential energy and neglecting higher order derivatives in the Taylor series expansions of u, v, φ. This is, of course, justified only if the change of u, v, φ from joint to joint is sufficiently small.

The incremental strain energy U_x contained in a pair of horizontal members between the joints (i - j, j) and (i + j, j) is a sum of two expressions of form (2). Expanding the values of u, v, and φ in joints (i - 1, j) and (i + 1, j) in Taylor series about the point (i, j) yields the following continuum approximation:

$$U_x = L_x^2 E'_x u_{,x}^2 + L_x^2 k_x s_x \varphi_{,x}^2 + L_x^2 k_x s'_x \varphi\varphi_{,xx} + 2k_x s'_x (v_{,x} - \varphi)^2 - P_x^0 L_x v_{,x}^2 \tag{4}$$

In this expression, the terms with higher than first derivatives of u, v, and φ have in general been dropped. An exception must be made, however, with the term $\varphi\varphi_{,xx}$ because integration by parts in the expression for energy of the whole structure converts this term into a term with first order derivatives. (This point has been overlooked in Reference [6].) It deserves mention that without the term $\varphi\varphi_{,xx}$ an agreement with the continuum approximation derived from the equilibrium equations of a joint could not be reached. The legitimacy of dropping the terms with other combinations of higher derivatives, with regard to integration by parts, can be easily verified.

The incremental strain energy U_y stored in a pair of vertical members meeting in the joint (i, j) can be expressed in a similar manner. The strain energy corresponding to the area $L_x L_y$ of the frame is $(U_x + U_y)/2$.

The incremental potential energy of the whole structure, $\mathcal{U}$, approximately equals

$$\int_{(x)}\int_{(y)} (U_x + U_y - f_x u - f_y v - m\varphi) \frac{dx\ dy}{2L_x L_y} \tag{5}$$

minus the work of the loads applied at the boundary of frame. Integrating the terms involving the products $\varphi\varphi_{,xx}$ and $\varphi\varphi_{,yy}$ by parts (or applying the Green's theorem), the integral (5) takes on the form:

$$\int_{(x)}\int_{(y)} U\ dx\ dy - \int_{(x)}\int_{(y)} (f_x u + f_y v + m\varphi) \frac{dx\ dy}{2L_x L_y} \tag{6}$$

where

$$U = \Big[L_x^2 E'_x u_{,x}^2 + L_y^2 E'_y v_{,y}^2 - L_x^2 k_x s_x c_x \varphi_{,x}^2 - L_y^2 k_y s_y c_y \varphi_{,y}^2 + 2k_x s'_x (v_{,x} - \varphi)^2 + 2k_y s'_y (u_{,y} + \varphi)^2 - P_x^0 L_x v_{,x}^2 - P_y^0 L_y u_{,y}^2\Big]/(2L_x L_y) \tag{7}$$

plus a certain contour integral of terms involving products $\varphi\varphi_{,x}$ and $\varphi\varphi_{,y}$. Expression U can be regarded as the specific incremental elastic potential of the continuum approximating the framework.

Inspecting Equation (7) to determine the mutually independent variables of which U is a function, the special case of our continuum for $P_x^0 = P_y^0 = 0$ is found to represent the micropolar medium as defined by Eringen [3]. This also shows that the classical Cosserat's medium [1] is insufficient, while theories more general than micropolar medium are unnecessarily complex [2, 3].

DIFFERENTIAL EQUILIBRIUM EQUATIONS

The first variation of the incremental potential $\mathcal{U}$ may be written in the form:

$$\delta\mathcal{U} = \int_{(x)}\int_{(y)}\Big[L_x^2 E'_x u_{,x}\,\delta u_{,x} + L_y^2 E'_y v_{,y}\,\delta v_{,y} - L_x^2 k_x s_x c_x \varphi_{,x}\,\delta\varphi_{,x}$$

$$- L_y^2 k_y s_y c_y \varphi_{,y}\,\delta\varphi_{,y} + 2k_x s'_x(\varphi - v_{,x})(\delta\varphi - \delta v_{,x}) + 2k_y s'_y(\varphi + u_{,y})(\delta\varphi + \delta u_{,y})$$

$$- P_x^0 L_x v_{,x}\,\delta v_{,x} - P_y^0 L_y u_{,y}\,\delta u_{,y} - f_x\delta u - f_y\delta v - m\delta\varphi\Big]\frac{dx\,dy}{L_x L_y} \tag{8}$$

plus a certain contour integral expressing the work of prescribed boundary loads. If in Equation (8) the terms containing derivatives of the variations are integrated by parts (or if Green's theorem is applied), the condition that $\delta\mathcal{U} = 0$ for any δu, δv and $\delta\varphi$ results in the following differential equations:

$$L_x^2 E'_x u_{,xx} + k_y s''_y u_{,yy} + 2k_y s'_y \varphi_{,y} + f_x L_x L_y = 0 \tag{9a}$$

$$L_y^2 E'_y v_{,yy} + k_x s''_x v_{,xx} - 2k_x s'_x \varphi_{,x} + f_y L_x L_y = 0 \tag{9b}$$

$$2k_x s'_x(\varphi - v_{,x}) + 2k_y s'_y(\varphi + u_{,y}) + L_x^2 k_x s_x c_x \varphi_{,xx} + L_y^2 k_y s_y c_y \varphi_{,yy} - mL_x L_y = 0 \tag{9c}$$

where s''_x and s''_y are defined as follows (omitting subscript x or y):

$$s'' = 2s' - \pi^2 P^0/P_E\,,\quad P_E = EI\,\pi^2/L^2 \tag{9d}$$

Equations (9a)-(9c) represent the differential equations of equilibrium in terms of displacements and rotations for the continuous medium approximating the framework.

CONSTITUTIVE RELATIONS FOR MICROPOLAR CONTINUUM

The components of stress, σ_{xx}, σ_{xy}, σ_{yx}, σ_{yy}, and couple stress, m_{xz}, m_{yz} for a micropolar medium in plane stress may be defined with the help of the specific potential energy as is indicated in the following relations,

$$\left.\begin{aligned}
\sigma_{xx}^0 + \sigma_{xx} &= \partial U/\partial u_{,x} = \sigma_{xx}^0 + E'_x u_{,x} L_x/L_y\\
\sigma_{yy}^0 + \sigma_{yy} &= \partial U/\partial v_{,y} = \sigma_{yy}^0 + E'_y v_{,y} L_y/L_x\\
\sigma_{xy} &= \partial U/\partial(v_{,x} - \varphi) = (k_x s''_x v_{,x} - 2k_x s'_x\varphi)/(L_x L_y)\\
\sigma_{yx} &= \partial U/\partial(u_{,y} + \varphi) = (k_y s''_y u_{,y} + 2k_y s'_y\varphi)/(L_x L_y)\\
m_{xz} &= \partial U/\partial\varphi_{,x} = -\,k_x s_x c_x \varphi_{,x} L_x/L_y\\
m_{yz} &= \partial U/\partial\varphi_{,y} = -\,k_y s_y c_y \varphi_{,y} L_y/L_x
\end{aligned}\right\} \tag{10}$$

in which also the expressions obtained after substitution of Equation (7) are introduced. The values σ_{xx}^0, σ_{yy}^0 represent initial stresses in the micropolar medium, $\sigma_{xx}^0 = -P_x^0/L_y$, $\sigma_{yy}^0 = -P_y^0/L_x$. (The reason for their appearance in the stress definitions is that the work of the initial stress on the incremental displacement u is $\overset{0}{\sigma}_{xx}u_{,x}$.) Equations (10) have the significance of stress-strain relationships of the micropolar medium.

It is of interest to investigate the relationship of the above stresses to the internal forces in members of the frame. To this end, let us consider their

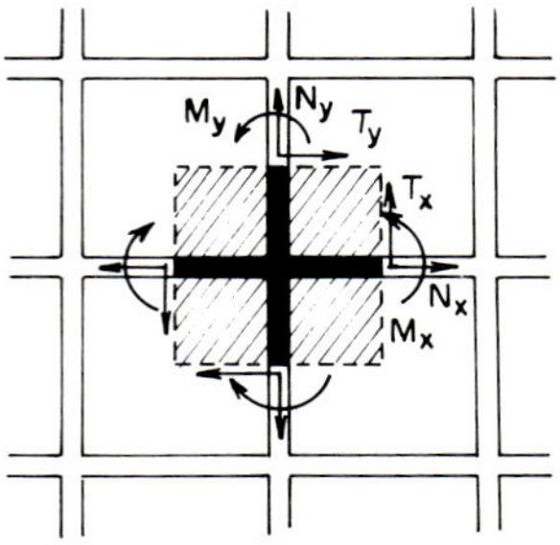

Fig. 3

Internal forces at the midspans and their analogy with the stresses and couple stresses acting on an element of a micropolar continuum

values at the midspan (Figure 3). According to the equilibrium conditions of the member shown in Figure 1,

$$T = (M_a + M_b)/L - \overset{0}{P}\psi \tag{10a}$$

Then, using Equation (1) and considering the equilibrium of the half-length of the member in Figure 1, the internal forces in the midspan can be obtained as follows:

$$\left.\begin{aligned} N &= E'(u_b - u_a) = - P, \quad T = k\left[s''(v_b - v_a)/L - s'(\varphi_a + \varphi_b)\right]/L \quad , \\ M &= (M_b - N_a)/2 = \frac{1}{2} ks(1 - c)(\varphi_b - \varphi_a) \end{aligned}\right\} \tag{11}$$

where N is the axial force, T is the shear force and M is the bending moment at the midspan which is taken, by definition, about the point located on the straight line connecting the ends of member in the deformed position. Notice that these values characterize the end moments as well;

$$M_a = - M -(T + \overset{0}{P}\psi)L/2 \ , \quad M_b = M -(T + \overset{0}{P}\psi)L/2 \tag{11a}$$

Expanding u_a, u_b, v_a, v_b and φ_a, φ_b in Taylor series and dropping all terms containing higher than first order derivatives, the following expressions are obtained:

$$\left.\begin{aligned} N_x &= L_x E'_x u_{,x}, & N_y &= L_y E'_y v_{,y} \ , \\ T_x &= k_x s''_x v_{,x}/L_x - 2k_x s'_x \varphi/L_x, & T_y &= k_y s''_y u_{,y}/L_y + 2k_y s'_y \varphi/L_y, \\ M_x &= \frac{1}{2} L_x k_x s_x (1 - c_x)\, \varphi_{,x}, & M_y &= \frac{1}{2} L_y k_y s_y (1 - c_y)\, \varphi_{,y} \end{aligned}\right\} \tag{12}$$

These expressions may be regarded as the continuous counterparts of the internal forces (11) at the midspan. Values of all functions in these expressions ought to be evaluated for the midspan.

Comparing expressions (12) and (10), it follows that

$$\left.\begin{aligned} N_x &= L_y\sigma_{xx}, & N_y &= L_x\sigma_{yy}, \\ T_x &= L_y\sigma_{yy}, & T_y &= L_x\sigma_{yx}, \\ N_x &= -L_y m_{xz}(1-c_x)/(2c_x), & M_y &= -L_x m_{yz}(1-c_y)/(2c_y) \end{aligned}\right\} \qquad (13)$$

For a medium without initial stress, the latter of these relationships reduces to

$$M_x = -\frac{1}{2}L_y m_{xz}, \qquad M_y = -\frac{1}{2}L_x m_{yz} \qquad (14)$$

It is interesting to note that, in Equation (14), M_x is not equal to the resultant of the couple stresses m_{xz} over length element L_y in the micropolar medium but rather equals minus one-half of it. (The formulation in Reference [5] implies incorrectly that $M_x = L_y m_{xz}$.) With varying initial stress, the ratio m_{xz}/M_x changes. The reason for the lack of any simple, intuitive correspondence between m_{xz} and M_x lies obviously in the fact that M_x varies along the member. By contrast, T and P are constant within each member and N_x or T_x do represent the resultants of stresses σ_{xx} or σ_{yy} over the length element L_y.

Expressions for stresses, Equations (10), and their relations to internal forces in framework, Equation (13), allow to formulate the boundary conditions of micropolar bodies approximating grid grameworks. The boundary conditions can, of course, be also deduced from the first variation of the full expression for potential energy.

CONCLUDING REMARKS

The equations presented above fully define the analogy between a grid framework and a micropolar medium under initial stress.

The equations of equilibrium could have been, alternatively, also derived by determining the continuum approximation to the equations of equilibrium of a joint in the framework (Figure 2). It has been verified that such a procedure does indeed yield the same results. For the correct expression of couple stresses, however, the potential energy approach is inevitable.

Application to practical problems is left to a subsequent paper.

REFERENCES

[1] W. T. Koiter: "Couple Stresses in the Theory of Elasticity," Koninklijke Akademie van Wettenschappen, Proc., Ser. B (Physical Sciences), Vol. 67, pp. 17-44, Amsterdam 1964.

[2] R. D. Mindlin: "Microstructure in Linear Elasticity," Archives for Rational Mech. and Analysis, Vol. 17, pp. 51-78 (1966).

[3] C. E. Eringen: "Linear Theory of Micropolar Elasticity," Journal of Mathematics and Mechanics, Vol. 15, pp. 51-78 (1966).

[4] C. Woźniak: "Load-Carrying Structures of the Dense Lattice Type. The Plane Problem," Archivum Mechaniki Stossowanej, Vol. 18, No. 5, pp. 581-597 (1966);

see also "Bending and Stability Problems with Lattice Structures," No. 6, pp. 781-796.

[5] C. B. Banks, and U. Sokolowski: "On Certain Two-Dimensional Applications of the Couple-Stress Theory," Int. J. of Solids and Structures, Vol. 4, pp. 15-29 (1968).

[6] A. Askar and A. S. Cakmak: "A Structural Model of a Micropolar Continuum," Int. J. of Engineering Sciences, Vol. 6, pp. 583-589 (1968).

[7] M. Z. Horne and W. Merchant: The Stability of Frames, Pergamon Press, New York, 1965.

Developments in Mechanics, Vol. 6. Proceedings of the 12th Midwestern Mechanics Conference: 41

On the Buckling of Pin-connected Beam Gridworks

K. K. HU and P. G. KIRMSER
KANSAS STATE UNIVERSITY

ABSTRACT

It is shown in this paper that the buckling loads and mode shapes for pin-connected beam gridworks loaded by in plane axial loads are characterized by the eigenvalues and eigenmatrices of the Sylvester's matrix operator equation

$$\alpha X + X\beta = \lambda(\gamma_1 X\delta_1 + \gamma_2 X\delta_2).$$

An iterative method for finding the smallest eigenvalue mnd its corresponding eigenmatrix is presented, and used to solve a sample problem of a fairly general nature.

INTRODUCTION

Consider a pin-connected beam gridwork subjected to lateral loads at the joints, as shown in Fig. 1. All the beams in each given direction are alike and parallel to each other, but not necessarily of constant cross-section or uniformly spaced. The boundary conditions, although the same for each beam which ends on a given side of the gridwork, may be different on different sides.

These weak regularity properties are sufficient to insure that the force deflection relationships for such gridworks have a simple and elegant form.

To find this form, let the relationship between forces and deformations for the jth beam in the x direction be given by

$$\underset{(n\times n)}{(\tilde{B})}\ \underset{(n\times 1)}{\{d\}_j} = \underset{(n\times 1)}{\{f\}_j}, \tag{1}$$

where $\tilde{B}$ is the stiffness matrix ; $\{d\}_j$ the column matrix of deflections d_{jk}, and $\{f\}_j$ the corresponding matrix of forces f_{jk}, both measured at stations on the jth beam.

For all of the beams in the x direction this relationship is

$$(\tilde{B})(\{d\}_1 \dots \{d\}_n) = (\{f\}_1 \dots \{f\}_n), \tag{2}$$

or $$\underset{(n\times n)}{\tilde{B}} \quad \underset{(n\times m)}{D} = \underset{(n\times m)}{F}, \tag{3}$$

where the element d_{ij} of the deflection matrix D represents the deflection at station i on beam j in the x direction, and f_{ij} the corresponding force.

The equation for the kth beam in the y direction which corresponds to equation (1) is

$$\underset{(m\times m)}{(A)} \quad \underset{(m\times 1)}{\{\delta\}}_k = \underset{(m\times 1)}{\{\Phi\}}_k, \tag{4}$$

and that which corresponds to equation (3),

$$\underset{(m\times m)}{A} \quad \underset{(m\times n)}{\Delta} = \underset{(m\times n)}{\Phi} \quad . \tag{5}$$

In this equation the element δ_{ij} of Δ is the deflection at station j on the ith beam in the y direction, and ϕ_{ij} the corresponding force.

Now, if an external force system $\underset{m\times n}{(G)}$ is applied at the pin-connected joints, statical equilibrium requires that

$$\underset{(m\times n)}{\Phi} = \underset{(m\times n)}{G} - \underset{(m\times n)}{\tilde{F}}, \tag{6}$$

which on substitution of equations (3) and (5) becomes

$$A\Delta + \tilde{D}B = G. \tag{7}$$

Because the two systems of beams deflect together

$$\Delta = \tilde{D}. \tag{8}$$

Then, letting $\Delta = X$, the unknown deflections caused by the external forces G, equation (7) becomes

$$AX + XB = G, \tag{9}$$

an equation which was studied first by Sylvester, according to C. C. MacDuffee [1].

This equation was used to describe pin-connected beam gridworks by J. Szabo [2] and E. C. Ma [3]; and later generalized to include polar and rigidly connected gridworks by K. K. Hu [4].

Gridworks whose beams in a given direction are not alike, but which have stiffnesses which are single multiples of each other, are described by

$$AXW_1 + W_2XB = G, \tag{10}$$

where W_1 and W_2 are suitable weighting matrices.

Rigidly connected gridworks lead to systems of such equations.

The mathematical nature of equation (9) has been studied by J. J. Sylvester [5], D. E. Rutherford [6], W. G. Bickley and J. McNamee [7], E. C. Ma [8], R. A. Smith [9], and A. Jameson [10]; and operators of the form of the left hand side of equations (9) and 10) by M. Rosenbloom [11], and M. Rosenbloom and G. Lumer [12].

Sylvester's equations are of general interest because they are the finite analogs of linear partial differential equations in two dimensions.

THE BUCKLING EQUATIONS

A different form of Sylvester's equation arises for gridwork buckling problems.

Consider the pin-connected beam gridwork loaded by a scalar multiple of in plane self-equilibrating systems of loads P and Q applied axially to the beams as shown in Fig. 3.

The kth beam in the x direction is shown in Fig. 4.

The forces f_{kj} on the beam are internal forces of the gridwork, and are caused by the pin-connections to the beams in the y direction. These, of course, are unknown.

The bending moment at every section of this beam column is

$$M(x) = -\lambda P_k z_k(x) + \sum_{i=1}^{n} f_{ki}(x-x_i)\ H(x-x_i), \tag{11}$$

where

$$H(x-x_i) = \begin{cases} 1 & x>x_i \\ 0 & x\leq x_i \end{cases} \text{ is the Heaviside function,}$$

and $z_k(x)$ the unknown deflection curve.

From this, the equivalent lateral loading

$$w(x) = M''(x) = -\lambda P_k z_k''(x) + \sum_{i=1}^{n} f_{ki}\delta(x-x_i), \tag{12}$$

in which $\delta(x-x_i)$ is the Dirac delta function and the primes denote derivatives, is easily found by differentiation.

The term $-\lambda P_k z_k''(x)$ represents the equivalent lateral loading caused by the axial load P_k.

As suggested by S. P. Timoshenko [13], this distributed equivalent lateral load can be approximated by concentrated loads placed at the stations of the beam which correspond to mesh points of the gridwork, and the loads themselves obtained from suitable quadrature formulas applied to $P_k y_k''(x)$.

Let the unknown deflection curve be approximated by

$$z_k(x) \simeq \sum_{i=1}^{n} z_{ki}\Psi_i(x), \tag{13}$$

where the $\Psi_i(x)$ are functions which satisfy the boundary conditions for the beams in the x direction, and z_{ki} are the deflections at the mesh points.

Then equation (12) becomes approximately

$$w(x) \simeq -\lambda P_k \sum_{i=1}^{n} z_{ki}\Psi_i''(x) + \sum_{i=1}^{n} f_{ki}\delta(x-x_i) \tag{14}$$

A quadrature formula for this loading can be found to replace it by an equivalent loading of forces and couples applied at stations which correspond to mesh points only.

Consider Fig. 5, which represents a beam fixed at each mesh point.

The reactions F_{kj}, M_{kj} are the forces and moments required to constrain the beam. The negative of these are the equivalent loadings to be found.

Now

$$\underset{(n\times 1)}{\{F\}} = \underset{(n\times 1)}{\{f\}} - \lambda P_k \underset{(m\times n)}{(T_1)} \underset{(n\times 1)}{\{z\}_k} \tag{15}$$

and

$$\underset{(n\times 1)}{\{M\}} = \lambda P_k \underset{(n\times n)}{(T_2)} \underset{(n\times 1)}{\{z\}_k}, \tag{16}$$

where T_1 and T_2 are matrices which transform deflections to forces and couples, respectively.

Using these in the deflection force form of the equations for all of the beams in the x direction yields

$$\underset{(n\times n)}{A_1}\underset{(n\times m)}{Z}\underset{(m\times m)}{W_1} + \underset{(n\times n)}{A_2}\underset{(n\times m)}{\theta}\underset{(m\times m)}{W_1} = \underset{(n\times m)}{f} - \lambda\underset{(n\times n)}{T_1}\underset{(n\times m)}{Z}\underset{(m\times m)}{P} \tag{17}$$

$$\underset{(n\times n)}{A_3}\underset{(n\times m)}{Z}\underset{(m\times m)}{W_1} + \underset{(n\times n)}{A_4}\underset{(n\times m)}{\theta}\underset{(m\times m)}{W_1} = \lambda\underset{(n\times n)}{T_2}\underset{(n\times m)}{Z}\underset{(m\times m)}{P} \tag{18}$$

in which

A_1 is a stiffness matrix which relates deflection to lateral force,
A_2 is a stiffness matrix which relates rotation to lateral force,
W_1 is a weighting matrix for the case where the beams in the x direction have stiffnesses which are multiples of each other,
f is the matrix of internal forces provided by the reactions of the beams in the y direction,
T_1 is the matrix which relates deflections produced by axial forces (P) to those produced by the equivalent lateral forces,
P is the matrix of in plane axial forces on the x beams,
A_3 is the stiffness matrix which relates deflections to moments,
A_4 is the stiffness matrix which relates rotations to moments,
T_2 is the matrix which relates deflections produced by axial forces (P) to those produced by the equivalent lateral forces,
Z is the matrix of lateral deflections of the mesh points,
θ is the matrix of rotations of the x beams at the mesh points,
and λ is a scalar multiplier which controls the intensity of the applied loads (P) and (Q).

Equation (18) can be solved for θ, and this solution substituted into equation (17) to obtain

$$(A_1 - A_2A_4^{-1}A_3)ZW_1 = f - \lambda(T_1ZP - A_2A_4^{-1}T_2ZP). \qquad (19)$$

Repeating this for the beams in the y direction leads to

$$W_2Z(B_1 - B_2B_4^{-1}B_3) = - f - \lambda(QZR_1 - QZR_2B_4^{-1}B_2), \qquad (20)$$

where

B_1 is the stiffness matrix which relates deflection to lateral force,
B_2 is the stiffness matrix which relates rotation to lateral force,
B_3 is the stiffness matrix which relates deflections to moments,
B_4 is the stiffness matrix which relates rotations to moments,
W_2 is a weighting matrix for the case where the stiffnesses of the y beams are multiples of each other,
R_1 is the matrix which relates deflections produced by the axial forces (Q) to those produced by equivalent lateral forces,
R_2 is the matrix which relates deflections produced by the axial forces (Q) to rotations produced by equivalent moments applied to mesh points,
Q is the matrix of axial forces on the y beams,
and Z and λ are the deflections and multiplier previously discussed.

Adding equations (19) and (20) yields

$$(A_1-A_2A_4^{-1}A_3)ZW_1 + W_2Z(B_1-B_2B_4^{-1}B_3) = - \lambda\left((T_1-A_2A_4^{-1}T_2)ZP + QZ(R_1-R_2B_4^{-1}B_2)\right). \qquad (21)$$

Equation (21) is a Sylvester's equation eigenmatrix-eigenvalue problem. The eigenvalues λ_i give the scalar multiples of the axial loadings (P) and (Q) which cause buckling of the beam gridwork, and the corresponding normalized eigenmatrices Z_i, the buckled mode shapes.

As for the case of simple columns, the lowest eigenvalue is of paramount importance for practical buckling problems.

SOLUTION FOR THE LOWEST MODE

Equation (21) is easily transformed to the form

$$\alpha X + X\beta = \lambda(\gamma_1X\delta_1 + \gamma_2X\delta_2), \qquad (22)$$

where

$$\alpha = W_2^{-1}(A_1-A_2A_4^{-1}A_3)$$

$$\beta = (B_1-B_2B_4^{-1}B_3)W_1^{-1}$$

$$\gamma_1 = -W_2^{-1}(T_1-A_2A_4^{-1}T_2)$$

$$\delta_1 = PW_1^{-1}$$

$$\gamma_2 = W_2^{-1}Q$$

and $$\delta_2 = -(R_1-R_2B_4^{-1}B_2)W_1^{-1}.$$

This is the linear operator eigenvalue problem

$$\tau_1(X) = \lambda\tau_2(X), \tag{23}$$

where $$\tau_1(X) = \alpha X + X\beta ,$$

and $$\tau_2(X) = \gamma_1 X\delta_1 + \gamma_2 X\delta_2.$$

Although linear operator eigenvalue problems have been studied extensively, explicit solutions of the generalized eigenmatrix-eigenvalue problem of Sylvester's type are unknown. They are probably not obtainable in closed form.

However, the smallest eigenvalue, which is the important one for gridwork buckling problems, can be found by iteration using known methods [8], [9], [10] for solving the matrix operator equation

$$\tau_1(X) = Y. \tag{24}$$

The procedure is simple, and follows.

Guess X_n and compute $\tau_2(X_n)$. Then solve

$$\tau_1(X^*_{n+1}) = \tau_2(X_n). \tag{25}$$

to get $$X^*_{n+1} = \tau_1^{-1}\tau_2(X_n). \tag{26}$$

Normalize X^*_{n+1} by multiplying by the constant λ_{n+1} which is required to make its largest component 1 (i.e. let

$$\lambda_{n+1}X^*_{n+1} = X_{n+1} \quad), \tag{27}$$

and repeat the process using X_{n+1} as a new guess.

If the process converges, it converges to a solution, for in the limit

$$\frac{1}{\lambda}X = \tau_1^{-1}\tau_2(X), \tag{28}$$

which yields $$\tau_1(X) = \lambda\tau_2(X), \tag{29}$$

the operator equation which was to be solved.

It can be shown that if the eigenvalues are distinct, the process converges to yield the smallest eigenvalue and its corresponding eigenmatrix.

AN EXAMPLE

As an example, the critical value of λ for buckling was found for the pin-connected gridwork shown in Fig. 6.

The various coefficient matrices which occur in equation (22) have the values

$$\alpha = \begin{pmatrix} 3729.56731 & -3590.14423 & 1568.50962 & -418.26923 & 104.56731 \\ -3590.14423 & 5298.07692 & -4008.41346 & 1673.07692 & -418.26923 \\ 1568.50962 & -4008.41346 & 5402.64423 & -4008.41346 & 1568.50962 \\ -418.26923 & 1673.07692 & -4008.41346 & 5298.07692 & -3590.14423 \\ 104.56731 & -418.26923 & 1568.50962 & -3590.14423 & 3729.56731 \end{pmatrix},$$

$$\beta = \begin{pmatrix} 19426.79558 & -22030.38674 & 15621.54696 & -4205.80110 & 1201.65746 \\ -44060.77348 & 59682.32044 & -57479.28177 & 25234.80663 & -7209.94475 \\ 31242.09392 & -57479.28177 & 84917.12707 & -64689.22652 & 28839.77901 \\ 8411.60220 & 25234.80663 & -64689.22652 & 88522.09945 & -71899.17127 \\ 2403.31492 & -7209.94475 & 28839.77901 & -71899.17127 & 113756.90608 \end{pmatrix},$$

$$\gamma_1 = \begin{bmatrix} 0.12346509 & -0.06762958 & 0.00213018 & 0.00257041 & -0.00133491 \\ -0.06762959 & 0.12559526 & -0.06505917 & 0.00079527 & 0.00257041 \\ 0.00213018 & -0.06505917 & 0.12426035 & -0.06505917 & 0.00213018 \\ 0.00257041 & 0.00079527 & -0.06505917 & 0.12559526 & -0.06762959 \\ -0.00133491 & 0.00257041 & 0.00213018 & -0.06762958 & 0.12346509 \end{bmatrix},$$

and

$$\delta_1 = \begin{bmatrix} 4.5 & 0.0 & 0.0 & 0.0 & 0.0 \\ 0.0 & 3.0 & 0.0 & 0.0 & 0.0 \\ 0.0 & 0.0 & 1.0 & 0.0 & 0.0 \\ 0.0 & 0.0 & 0.0 & -1.0 & 0.0 \\ 0.0 & 0.0 & 0.0 & 0.0 & -3.0 \end{bmatrix}.$$

The weighting matrices used in the calculation of α and β are

$$W_1 = \begin{bmatrix} 0.5 & 0.0 & 0.0 & 0.0 & 0.0 \\ 0.0 & 1.0 & 0.0 & 0.0 & 0.0 \\ 0.0 & 0.0 & 1.0 & 0.0 & 0.0 \\ 0.0 & 0.0 & 0.0 & 1.0 & 0.0 \\ 0.0 & 0.0 & 0.0 & 0.0 & 1.0 \end{bmatrix}$$

and $W_2 = [I]$. W_1 is not an identity because the first x beam of the gridwork has only half the stiffness of the others.

The loading matrices are

$$P = \begin{bmatrix} 2.25 & 0.0 & 0.0 & 0.0 & 0.0 \\ 0.0 & 3.0 & 0.0 & 0.0 & 0.0 \\ 0.0 & 0.0 & 1.0 & 0.0 & 0.0 \\ 0.0 & 0.0 & 0.0 & -1.0 & 0.0 \\ 0.0 & 0.0 & 0.0 & 0.0 & -3.0 \end{bmatrix}$$

and $Q = [0]$ (there are no axial forces in the y direction).

It is not necessary to calculate γ_2 and δ_2 because Q is zero in this example.

The IBM 360/50 program used to find the smallest eigenvalue and the corresponding eigenmatrix of equation (22) using the algorithm of equations (25)-(27) used 9 iterations to converge to 6 decimal places. The execution time for the program was about 15 seconds.

The smallest eigenvalue was found to be

$$\lambda_1 = 3379.20 \text{ lbs.},$$

and the eigenmatrix X_1, which is the lowest buckling mode shape matrix of the loaded gridwork,

$$X_1 = \begin{bmatrix} 0.50027324 & 0.86618316 & 1.00000000 & 0.86618316 & 0.50027324 \\ 0.35626692 & 0.61688112 & 0.71220253 & 0.61688112 & 0.35626692 \\ 0.22081071 & 0.38236517 & 0.44146520 & 0.38236517 & 0.22081071 \\ 0.10690479 & 0.18513603 & 0.21376022 & 0.18513603 & 0.10690479 \\ 0.02884139 & 0.04995096 & 0.05767621 & 0.04995096 & 0.02884139 \end{bmatrix}$$

The form of the buckled gridwork is shown in Fig. 7.

CONCLUDING REMARKS

The Sylvester's equation form for beam gridwork problems as discussed in this paper is of a general and convenient form.

The stiffness properties and loadings of the beams of the gridwork are contained in separate matrices instead of being distributed among the elements of large, sparse matrices, as occurs in the usual formulation of such problems. The matrices used here are of the smallest size possible for the given problem.

The methods presented are capable of, and are being generalized to include gridworks with fixed connections between the beams.

They apply to other eigenvalue problems as well, for Sylvester's eigenvalue-eigenmatrix equation is the finite analog of similar partial differential equations.

REFERENCES

[1] MacDuffee, C. C., The Theory of Matrices, Chelsea Publishing Co., New York, 1946.

[2] Szabo, J., Calculation of Bridge Grillages (in German), Stahlbau 27, 6, 141-147 (June 1958).

[3] Ma, E. C., A Matrix Analysis of Beam Gridworks, a Ph. D. Dissertation, Department of Applied Mechanics, Kansas State University, Manhattan, Kansas, 1962.

[4] Hu, K. K., Matrix Analysis of Elastic Gridworks, an M. S. Thesis, Department of Applied Mechanics, Kansas State University, Manhattan, Kansas, 1966.

[5] Sylvester, J. J., C. R. Acad. Sci. Paris 99, 117-118, 409-412, 432-436, 527-529 (1884).

[6] Rutherford, D. E., Koningl. Ned. Akad. Wetenschap., Proc., 35A, 54-59 (1932).

[7] Bickley, W. G., and J. McNamee, Matrix and other direct methods for the solution of systems of linear difference equations, Phil. Trans. Roy. Soc. London (A) 252, 1005, 69-131, (January 1960).

[8] Ma, E. C., A finite series solution of the matrix equation AX - XB = C, SIAM Journal on Applied Mathematics 14, 490-495 (1966).

[9] Smith, R. A., Matrix equation XA + BX = C, SIAM J. Appl. Math 16, 1, 198-201 (1968).

[10] Jameson, A., Solution of the equation AX + XB = C by inversion of an M×M or N×N matrix, SIAM J. Appl. Math. 16, 5, 1020-1023 (1968).

[11] Rosenbloom, M., On the Operator Equation BX - XA = Q, Duke Math. J., 23, 263-269 (1956).

[12] Rosenbloom, M., and G. Lumer, Linear Operator Equations, Proc. Am. Math. Soc. 10, 32-41 (1959).

[13] Timoshenko, S. P. and J. M. Gere, Theory of Elastic Stability, Second Edition, McGraw-Hill Co., New York, 1961.

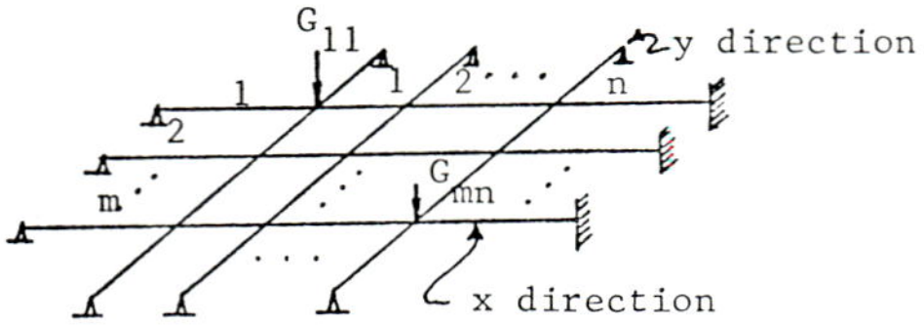

Fig. 1 A laterally loaded pin-connected gridwork

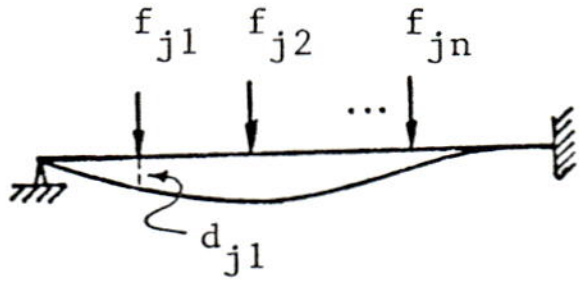

Fig. 2 The jth beam in the x direction

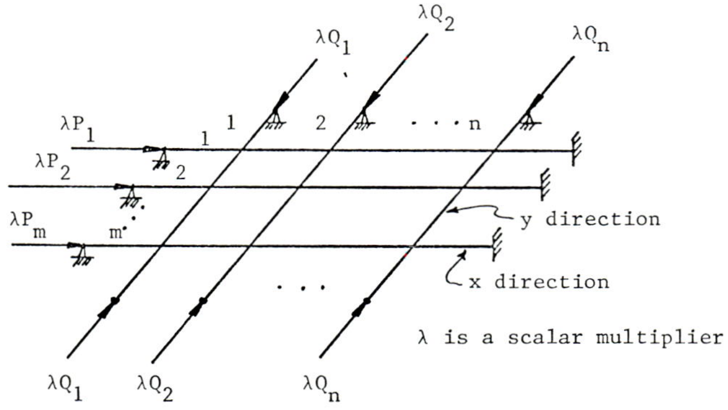

Fig. 3 The loaded gridwork

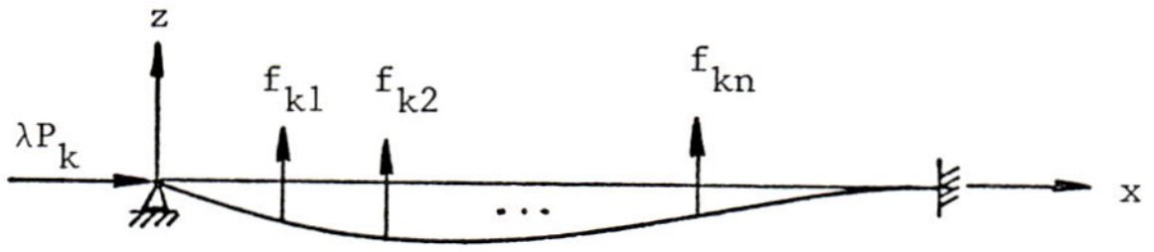

Fig. 4 The kth beam as buckled

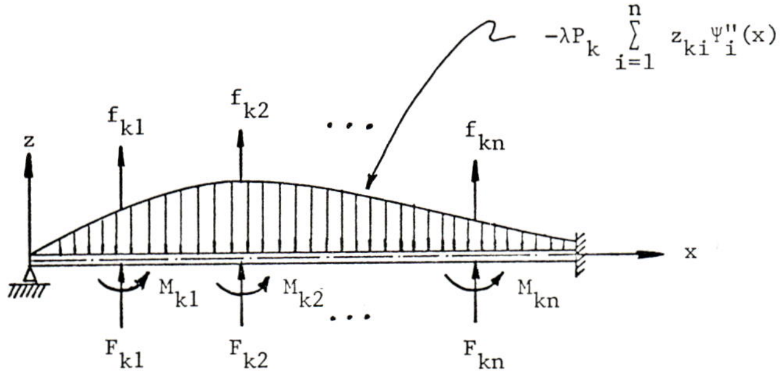

Fig. 5 Loading diagrams for equivalent loadings at mesh points

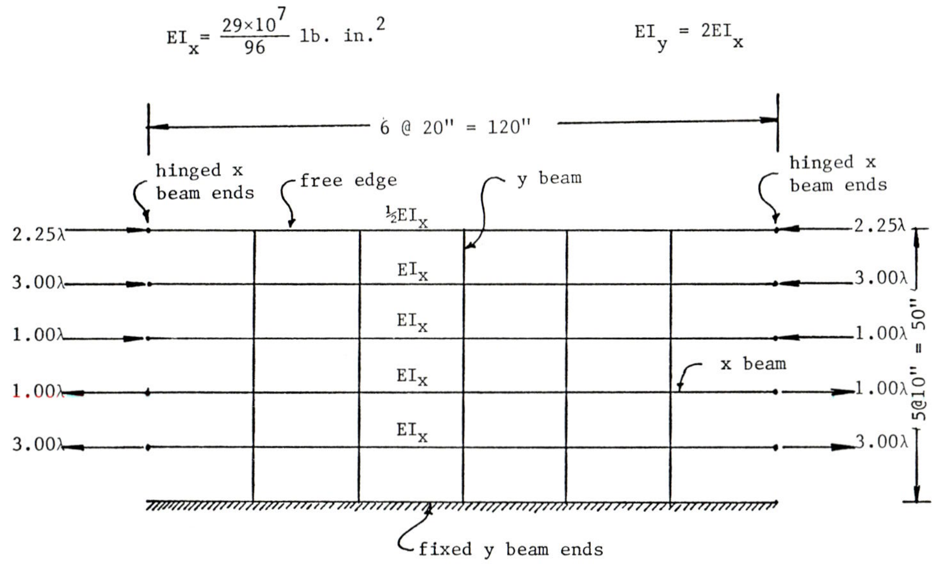

Fig. 6 The sample pin-connected gridwork

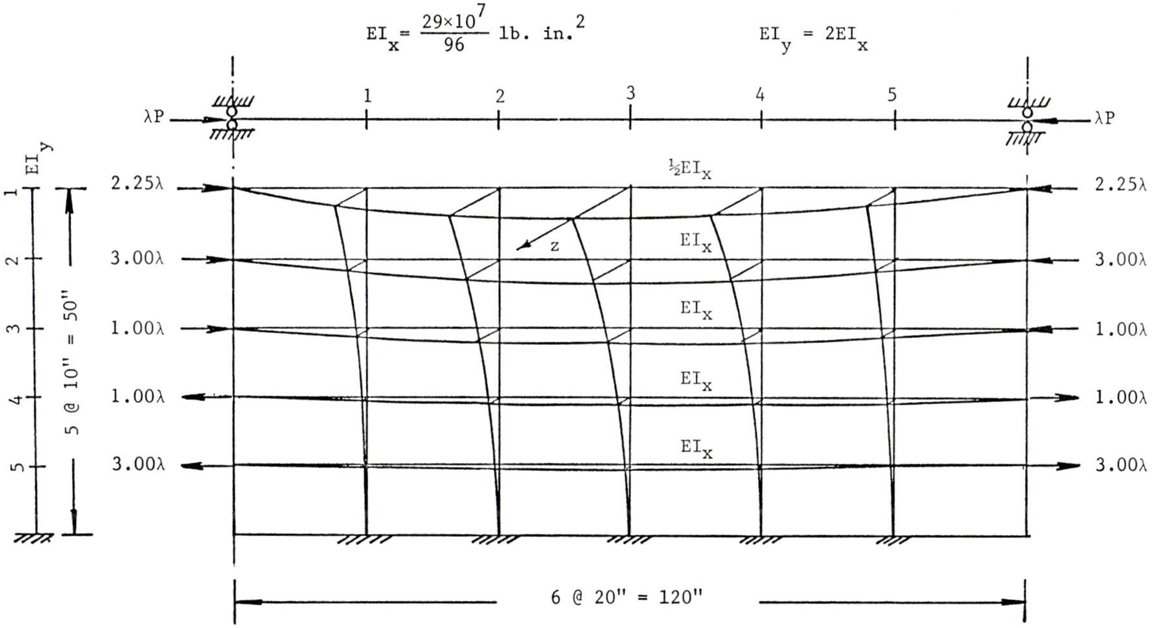

Fig. 7 The buckling mode shape for the smallest eigenvalue

Developments in Mechanics, Vol. 6. Proceedings of the 12th Midwestern Mechanics Conference: 42

Stability Analysis of a Beam Element in a Planar Mechanism

JAMES R. TOBIAS
HONEYWELL, INC.

DARRELL A. FROHRIB
UNIVERSITY OF MINNESOTA

ABSTRACT

The effect of deformations of a slender beam-like element on the stability of small motions of a constrained motion system are derived using Eringen's form of the nonlinear bending-extensibility equations. As the system under consideration contains a small mass ratio between the beam (coupler link) and the output rigid crank, this ratio is used as a perturbation parameter to establish the sensitivity of extensional instability to beam mass. Another perturbation is used to study the coupling between axial extensibility and bending motions which introduce other instabilities into the system. The solution demonstrates that axial vibrational instability develops for certain ranges of the input frequency, and bending instability occurs either when the axial load becomes sufficiently large to cause column-type buckling, or due to parametric resonance.

INTRODUCTION

The effects of distributed elasticity and inertia on the dynamical response of systems with geometric constraint, such as mechanisms, are poorly understood, whereas the literature on rigid element systems is extensive. Typically, the coupling between axial and bending deflections of beam-like mechanism elements is neglected. Meyer zur Capellen (1)* and Neubauer, et al (2) have considered a flexible coupler between rigid rotary elements. In (1), the terms in the equations of motion which include the dynamic bending stability were neglected. These terms are retained in (2), but parametric resonance is disregarded and bending instability is regarded as Euler column buckling.

*Numbers in brackets refer to Bibliography at end of paper.

The objective of this paper is to describe all of the first order small amplitude deflections of a system with a flexible coupler link when driven by a rigid rotary member. The general nonlinear bending equations derived by Eringen (3) are used to describe the motion of a slender coupler link in a planar four bar mechanism. The source of excitation is small oscillation of the input crank which excites planar motion of an elastic coupler and a rigid output link. The solution of the equations of motion shows that axial motions do develop which have an important effect on the bending oscillations of the coupler.

EQUATIONS OF MOTION

The four bar linkage with a flexible homogeneous beam-like coupler (Figure 1) will be analyzed in this paper.

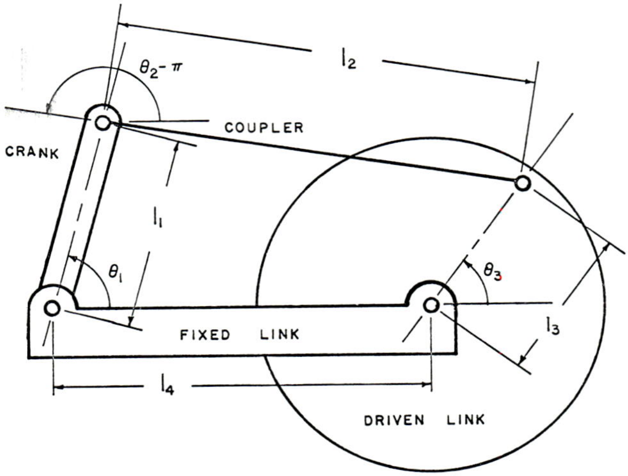

Fig. 1 Planar four bar linkage with an undistorted coupler

The system is contrained in the following way:

1) All motion and deflections are planar
2) The input and output cranks are rigid
3) All revolute (pinned) joints are smooth
4) Gravitational forces are neglected

The coupler is assumed to be a slender beam with isotropic linearly elastic bending stiffness and with distributed inertia small compared to the output link. Rotary inertia is neglected, as the study regards only low mode oscillations.

As the characteristics of motion about a static configuration are sought, the equations are studied as a set of linearized perturbations about a general stationary position. A coordinate system is selected based on the time-varying positions of the system links acting as rigid elements. Writing the angular position of the links as: $\Theta_i = \Theta_{io} + \bar{\Theta}_i(t)$, i = 1,2,3, perturbation of the rigid body constrain equations relates the $\bar{\Theta}_i$'s :

$$\bar{\Theta}_2(t) = \left(\frac{\ell_1}{\ell_2}\right) \frac{\sin(\Theta_{10} - \Theta_{30})}{\sin(\Theta_{30} - \Theta_{20})} \bar{\Theta}_1(t) \tag{1a}$$

$$\bar{\Theta}_3(t) = \left(\frac{\ell_1}{\ell_3}\right) \frac{\sin(\Theta_{10} - \Theta_{20})}{\sin(\Theta_{30} - \Theta_{20})} \bar{\Theta}_1(t) \tag{b}$$

The motion coordinates and forces on an element of the coupler distorted from the rigid body position are described in Figure 2.

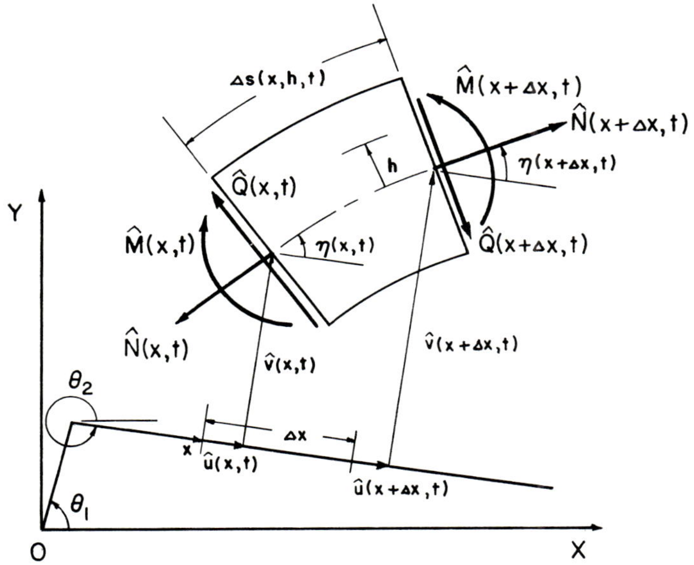

Fig. 2 Free body diagram of small beam element

By (3) the strain-displacement equations are:

$$\varepsilon = \left.\frac{\Delta s - \Delta x}{\Delta x}\right|_{h=0} \tag{2}$$

$$\hat{v}_{,x}(x,t) = [1 + \varepsilon(x,t)] \sin \eta(x,t) \tag{3a}$$

$$1 + \hat{u}_{,x}(x,t) = [1 + \varepsilon(x,t)] \cos \eta(x,t) \tag{b}$$

and the load-deflection equations are

$$\hat{N}(x,t) = EA\,\varepsilon(x,t) \tag{4a}$$

$$\hat{M}(x,t) = EI\,\eta_{,x}(x,t) \tag{b}$$

Using the motion coordinates of Figure 2, the equations of dynamics can be written as:

$$\left[\hat{N}(x,t)\cos\eta(x,t) + \hat{Q}(x,t)\sin\eta(x,t) \right]_0^{\ell_2} = \int_0^{\ell_2} \rho A \left[\hat{a}_x(x,t) + \hat{U}_{,tt}(x,t) \right] dx \tag{5a}$$

$$\left[\hat{N}(x,t)\sin\eta(x,t) - \hat{Q}(x,t)\cos\eta(x,t) \right]_0^{\ell_2} = \int_0^{\ell_2} \rho A \left[\hat{a}_y(x,t) + V_{,tt}(x,t) \right] dx \tag{b}$$

$$\Big[\hat{M}(x,t) + \{ \hat{N}(x,t)\sin\eta(x,t) - \hat{Q}(x,t)\cos\eta(x,t) \} \cdot \{ x + \hat{u}(x,t) \} - \{ \hat{N}(x,t)\cos\eta(x,t) + \hat{Q}(x,t)\sin\eta(x,t) \} \cdot \{ \hat{v}(x,t) \} \Big]_0^{\ell_2} = \int_0^{\ell_2} \rho A \left[\hat{a}_y(x,t) + \hat{V}_{,tt}(x,t) \right] \cdot \left[x + \hat{u}(x,t) \right] dx - \int_0^{\ell_2} \rho A \left[\hat{a}_x(x,t) + \hat{U}_{,tt} \right] \left[\hat{v}(x,t) \right] dx \tag{c}$$

where the rigid body accelerations are:

$$\hat{a}_x(x,t) = \ell_1 \ddot{\theta}_1 \sin(\theta_2 - \theta_1) - \ell_1 \dot{\theta}_1^2 \cos(\theta_2 - \theta_1) - x\,\dot{\theta}_2^2 \tag{6a}$$

$$\hat{a}_y(x,t) = \ell_1 \ddot{\theta}_1 \cos(\theta_2 - \theta_1) + \ell_1 \dot{\theta}_1^2 \sin(\theta_2 - \theta_1) + x\,\ddot{\theta}_2 \tag{b}$$

and the additional accelerations due to beam distortion are:

$$\hat{U}_{,tt}(x,t) = u_{,tt}(x,t) - 2\,\hat{v}_{,t}(x,t)\,\dot{\theta}_2(t) - \hat{v}(x,t)\,\ddot{\theta}_2(t) - \hat{u}(x,t)\,\dot{\theta}_2^2(t) \tag{7a}$$

$$\hat{V}_{,tt}(x,t) = \hat{v}_{,tt}(x,t) + 2\,\hat{u}_{,t}(x,t)\,\dot{\theta}_2(t) + \hat{u}(x,t)\,\ddot{\theta}_2(t) - \hat{v}(x,t)\,\dot{\theta}_2^2(t) \tag{b}$$

Further, the differential equations for the beam internal forces are found by performing a limit analysis as $\Delta x \rightarrow 0$ on the element:

$$[\hat{N}(x,t)\cos\eta(x,t) + \hat{Q}(x,t)\sin\eta(x,t)]_{,x} = \rho A[\hat{a}_x(x,t) + \hat{U}_{,tt}(x,t)] \qquad (8a)$$

$$[\hat{N}(x,t)\sin\eta(x,t) - \hat{Q}(x,t)\cos\eta(x,t)]_{,x} = \rho A[\hat{a}_y(x,t) + V_{,tt}(x,t)] \qquad (b)$$

$$\hat{M}_{,x}(x,t) - \hat{Q}(x,t)[1 + \varepsilon(x,t)] = 0 \qquad (c)$$

BOUNDARY CONDITIONS

As the left end of the coupler is pinned to the crank (Figures 1 and 2) the displacement and moment boundary conditions are zero:

$$\hat{v}(0,t) = 0 \qquad (9a)$$

$$\hat{u}(0,t) = 0 \qquad (b)$$

$$\hat{M}(0,t) = 0 \qquad (c)$$

and by action-reaction at the flexible link end x = 0:

$$\hat{N}(0,t)\cos[\theta_2(t) + \eta(0,t)] + \hat{Q}(0,t)\sin[\theta_2(t)+\eta(0,t)] = -\hat{N}_1(t)\cos\theta_1(t) - \hat{Q}_1(t)\sin\theta_1(t) \qquad (10a)$$

$$\hat{N}(0,t)\sin[\theta_2(t) + \eta(0,t)] - \hat{Q}(0,t)\cos[\theta_2(t)+\eta(0,t)] = -\hat{N}_1(t)\sin\theta_1(t) + \hat{Q}(t)\cos\theta_1(t) \qquad (b)$$

At $x = l_2$, the revolute joint requires:

Displacement matching:

$$\hat{u}(l_2,t)\cos\theta_2(t) - \hat{v}(l_2,t)\sin\theta_2(t) = l_3\{\cos[\theta_3(t) + \eta_3(t)] - \cos\theta_3(t)\} \qquad (11a)$$

$$\hat{u}(l_2,t)\sin\theta_2(t) + \hat{v}(l_2,t)\cos\theta_2(t) = l_3\{\sin[\theta_3(t) + \eta_3(t)] - \sin\theta_3(t)\} \qquad (b)$$

Moment:

$$\hat{M}(l_2,t) = 0 \qquad (c)$$

Action-reaction:

$$\hat{N}(\ell_2,t)\cos[\theta_2(t)+\eta_2(\ell_2,t)] + \hat{Q}(\ell_2,t)\sin[\theta_2(t)+\eta(\ell_2,t)] = -\hat{N}_3(t)\cos[\theta_3(t)+\eta_3(t)] - \hat{Q}_3(t)\sin[\theta_3(t)+\eta_3(t)] \quad (13a)$$

$$\hat{N}(\ell_2,t)\sin[\theta_2(t)+\eta_2(\ell_2,t)] - \hat{Q}(\ell_2,t)\cos[\theta_2(t)+\eta(\ell_2,t)] = -\hat{N}_3(t)\sin[\theta_3(t)+\eta_3(t)] + \hat{Q}_3(t)\cos[\theta_3(t)+\eta_3(t)] \quad (b)$$

A torque equation relating external torque to the motion of the rigid output link supplies:

$$\ell_3\hat{Q}_3(t) = I_3\left[\ddot{\theta}_3(t)+\ddot{\eta}_3(t)\right] \quad (14)$$

SOLUTION

The governing equations were nondimensionalized in the following manner:

$$z = x/\ell_2, \quad u(z,t) = \frac{\hat{u}(x/\ell_2,t)}{\ell_2}, \quad v(z,t) = \hat{v}/\ell_2$$

$$\alpha = \ell_3/\ell_2, \quad \nu = \ell_1/\ell_2, \quad \rho = \ell_4/\ell_2, \quad \gamma = r/\ell_2,$$

$$\mu = \frac{\rho A \ell_2 \ell_3^2}{I_3}, \quad \Omega^2 = \frac{EA\ell_3^2}{\ell_3 I_3}$$

$$a_x = \hat{a}_x/\ell_2, \quad a_y = \hat{a}_y/\ell_2, \quad N = \frac{\hat{N}}{EA}, \quad M = \frac{\hat{M}\ell_2}{EI}$$

$$Q = \frac{\hat{Q}\ell_2^2}{EI}, \quad \ddot{U} = \frac{\hat{U}_{,tt}}{\ell_2}, \quad \ddot{V} = \frac{\hat{V}_{,tt}}{\ell_2}$$

$$Q_i(t) = \frac{\hat{Q}_i}{EA}, \quad N_i = \frac{\hat{N}_i}{EA}$$

Since the inertia of the driven link is large compared to the total mass of the coupler, the parameter

$$\mu = \frac{\rho A \ell_2 \ell_3^2}{I_3}$$

is small. Expanding the unknown quantities in the governing equations in a power series in μ:

$$V(z,t) = \sum_{n=0}^{\infty} \mu^n V_n(z,t)$$

grouping terms of like coefficients of μ^n, n = 1, 2, 3, ... , an expanded set of equations of the system was generated. The zeroth equations describe the system response with an inertialess, flexible coupler. The mechanism then contains a single mass and a time-varying spring constant. The elastic constant changes because of the changing position of the links as the crank executes small angular motion.

To solve the zeroth equations, a coordinate transformation is used to rewrite the equations of motion in terms of a coordinate set identified by the end points of the coupler link shown in Figure 3.

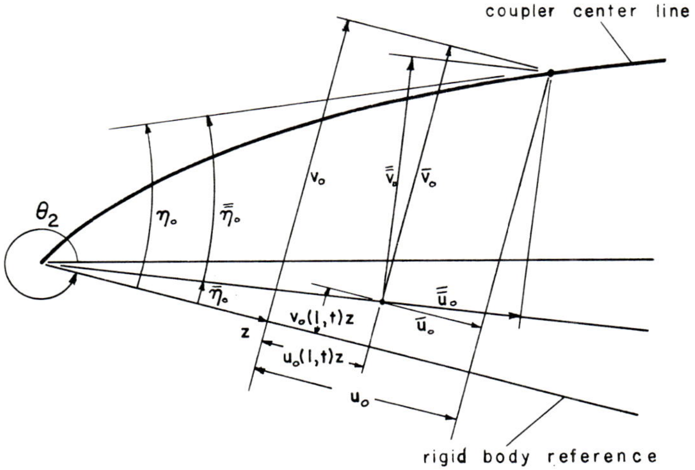

Fig. 3 Coordinate transformation used in zeroth approximation

The new nondimensionalized equations and boundary conditions are:

$$\gamma^2 \bar{\bar{V}}_{0,zzz} - \frac{\varepsilon_0(t)}{1+\varepsilon_0(t)} \bar{\bar{V}}_{0,z} = 0 \tag{15a}$$

$$\bar{\bar{V}}_0(0,t) = 0 \tag{b}$$

$$\bar{\bar{V}}_0(1,t) = 0 \tag{c}$$

$$\bar{\bar{V}}_{0,zz}(0,t) = 0 \tag{d}$$

$$\bar{\bar{V}}_{0,zz}(1,t) = 0 \tag{e}$$

Equation 15a indicates that bending will occur if

$$\frac{-\varepsilon_0}{1+\varepsilon_0} = \gamma^2\pi^2 \tag{16}$$

which is equivalent to the Euler axial compressive force

$$P = \pi^2/\ell^2 \, EI \tag{17}$$

if ε_0 is neglected as small with respect to unity.

As bending will not occur below this critical load, the extensional vibration equations can be written in terms of the angle of the driven link by setting $\eta_o = 0$:

$$\ddot{\eta}_{30} + \Omega^2 \sin^2(\theta_3 - \theta_2)\,\eta_{30} = -\ddot{\theta}_3 \tag{18}$$

The forcing term of Equation 18 is related to the input crank angle

$$\bar{\theta}_1(t) = \bar{\theta}_1 \cos \omega t$$

where $\bar{\theta}_1$ is small. Neglecting nonlinear terms, Equation 18 becomes the forced Mathieu equation

$$\ddot{\eta}_{30} + \bar{\Omega}^2(1 - 2q\cos\omega t)\eta_{30} = \omega^2 f \cos \omega t \tag{19}$$

where:

$$\bar{\Omega}^2 = \Omega^2 \sin^2(\theta_{30} - \theta_{20}) = \frac{EA\,l_3^2}{l_2 I_3} \sin^2(\theta_{30} - \theta_{20})$$

$$q = \frac{\nu\cos(\theta_{30} - \theta_{20})}{\sin^2(\theta_{30} - \theta_{20})}\left[\sin(\theta_{10} - \theta_{30}) - \frac{1}{\alpha}\sin(\theta_{10} - \theta_{20})\right]\bar{\theta}_1$$

$$f = \frac{\nu \sin(\theta_{10} - \theta_{20})}{\alpha \sin(\theta_{30} - \theta_{20})}\,\bar{\theta}_1$$

The stability regions of the homogenous form of Equation 19 have been given by Bolotin (4). The frequencies at which instabilities occur are

$$\omega = \frac{2\bar{\Omega}}{K}, \quad K = 1, 2, \ldots\ldots$$

and are shown in Figure 4.

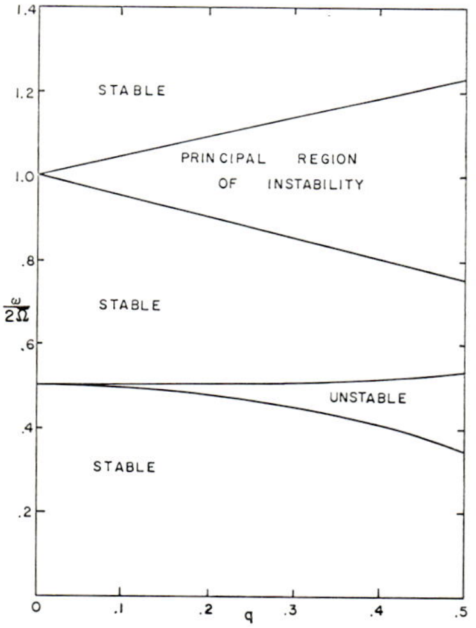

Fig. 4 Regions of axial stability

$\bar{\Omega}$ is the natural frequency of a system consisting of an extensional member with spring constant EA/l_2 and a flywheel with polar moment of inertia I_3. Furthermore, there is no transverse vibration, but buckling can occur if the axial compressive load is large. This solution indicates that axial vibration of the coupler dominates the stability problem.

The motion of the first order approximation is similar to that of the zeroth. Bending instability occurs only when the axial load is greater than the Euler critical load. Axial instabilities occur near the same excitation frequencies. The difference is that bending does occur, and that the axial stability boundaries are changed slightly, due to the inertia loading of the coupler.

To investigate bending instabilities, other than the column effect, it is necessary to perturb the base solution. This will yield a bending equation that can be studied for dynamic stability.

Bolton (4) and Kobrinskiy (5) have shown that the principal region of instability (k = 1) does not change significantly as a small amount of damping is introduced, although the other regions become smaller with damping.

The base solutions

$$V_\mu = \sum_{n=0}^{\infty} \mu^n V_n \,, \quad u_\mu = \sum_{n=0}^{\infty} \mu^n u_n$$

have not introduced the effect of bending due to the distributed inertia of the beam-like coupler. To study this effect, a perturbation about the base solution will be made by expressing the deflections u and V as $u = u_\mu + \tilde{u}$ and $V = V_\mu + \tilde{V}$ and substituting them into the equations of motion.

Another coordinate transformation is used as illustrated in Figure 5 which references the additional perturbation deflections to the distortions introduced earlier from axial extensibility:

$$\tilde{\tilde{u}}(z,t) = \tilde{u}(z,t)\cos\eta_\mu(z,t) + \tilde{V}(z,t)\sin\eta_\mu(z,t) \quad (20a)$$

$$\tilde{\tilde{V}}(z,t) = -\tilde{u}(z,t)\sin\eta_\mu(z,t) + \tilde{V}(z,t)\cos\eta_\mu(z,t) \quad (b)$$

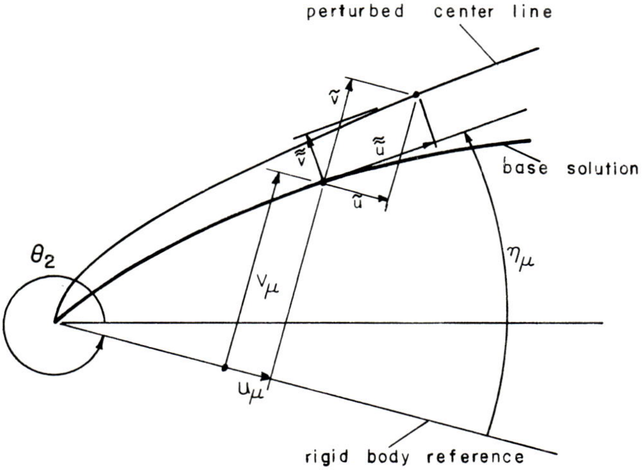

Fig. 5 Perturbed equation coordinates

Neglecting additional axial extension in this new small amplitude bending problem, the deflection equation becomes

$$\gamma^2 \tilde{\tilde{V}}_{,zzzz} - \varepsilon_\mu \tilde{\tilde{V}}_{,zz} - \varepsilon_{\mu,z} \tilde{\tilde{V}}_{,z} + \frac{\mu}{\Omega^2}\left[\tilde{\tilde{V}}_{,tt} - \tilde{\tilde{V}} (\dot{\theta}_2 + \eta_{\mu,t})^2 \right] = 0 \tag{21}$$

with the boundary conditions

$$\tilde{\tilde{V}}(0, t) = 0 \tag{22a}$$

$$\tilde{\tilde{V}}(1, t) = 0 \tag{b}$$

$$\tilde{\tilde{V}}_{,zz}(0, t) = 0 \tag{c}$$

$$\tilde{\tilde{V}}_{,zz}(1, t) = 0 \tag{d}$$

This kind of problem has been studied in detail Bolotin (4), Stoker (6), et al. Using Galerkin's Method, the variable $\tilde{\tilde{V}}$ is represented by the series

$$\tilde{\tilde{V}} = \sum_{j=1}^{K} \phi_j(z)\, p_j(t)$$

where the $\phi_j(z)$ are the eigenfunctions satisfying appropriate boundary conditions. As the eigenfunctions are $\phi_j = \sin j\pi z$ in this case, the equation governing the pj's becomes

$$\ddot{p} + \frac{\gamma^2 \Omega^2 \pi^4}{\mu}\left[1 - 2B\left(\frac{\omega^2/\bar{\Omega}^2}{1 - \omega^2/\bar{\Omega}^2}\right)\cos\omega t\right] p = 0$$

where

$$B = \frac{1}{2}\,\frac{\nu\,\bar{\theta}_1}{\gamma^2 \pi^2}\,\sin(\theta_{10} - \theta_{20}) \tag{23}$$

Bolotin's first approximation for the boundaries of the principal region of instability yields equation

$$1 \pm B\left(\frac{\omega^2/\bar{\Omega}^2}{1 - \omega^2/\bar{\Omega}^2}\right) - \frac{\mu}{4\pi\,\gamma^2}\;\omega^2/\bar{\Omega}^2 = 0 \tag{24}$$

Defining:

$$\hat{\Omega}^2 = \frac{\pi^4 \gamma^2 \Omega^2}{\mu} = \frac{\pi^4 E I}{\ell_2^4 \rho A} \quad \text{and} \quad \beta^2 = \left(\frac{2\hat{\Omega}}{\bar{\Omega}}\right)^2 \quad \text{where } \hat{\Omega}$$

is the natural frequency of the first bending mode, Equation 29 becomes

$$\left\{1 - \beta^2\left(\frac{\omega}{2\hat{\Omega}}\right)^2\right\}\left\{1 - \left(\frac{\omega}{2\hat{\Omega}}\right)^2\right\} = \mp \beta^2 B \left(\frac{\omega}{2\hat{\Omega}}\right)^2 \tag{25}$$

Roots of this equation are:

$$\left(\frac{\omega}{2\hat{\Omega}}\right)^2 = \frac{\beta^2(1 \pm B) + 1 \pm \sqrt{[\beta^2(1 \pm B) + 1]^2 - 4\beta^2}}{2\beta^2} \tag{26}$$

For small excitation amplitude ($B \ll 1$) and for $\beta^2 \neq 1$ the instability regions are:

$$\left(\frac{\omega}{2\hat{\Omega}}\right)^2 = 1 \pm \frac{\beta^2}{\beta^2-1} B = 1 \pm \frac{B}{1-\left(\frac{\bar{\Omega}}{2\hat{\Omega}}\right)^2} \qquad (27a)$$

$$\left(\frac{\omega}{2\hat{\Omega}}\right)^2 = \frac{1}{\beta^2} \mp \frac{1}{\beta^2-1} B = \left(\frac{\bar{\Omega}}{2\hat{\Omega}}\right)^2 \mp \frac{\left(\frac{\bar{\Omega}}{2\hat{\Omega}}\right)^2 B}{1-\left(\frac{\bar{\Omega}}{2\hat{\Omega}}\right)^2} \qquad (b)$$

Hence, there are two principal regions of stability. For $\beta^2 = 1$ the principal regions of instability merge to

$$\left(\frac{\omega}{2\hat{\Omega}}\right)^2 = 1 + \frac{B}{2} \pm \frac{1}{2}\sqrt{B^2+4B} \qquad (28)$$

Likewise, the regions merge for large excitation. The value of B for which this occurs is

$$B^* = |B| = \left(1 - \frac{\bar{\Omega}}{2\hat{\Omega}}\right)^2$$

Figure 6 shows the two regions of instability for small positive B.

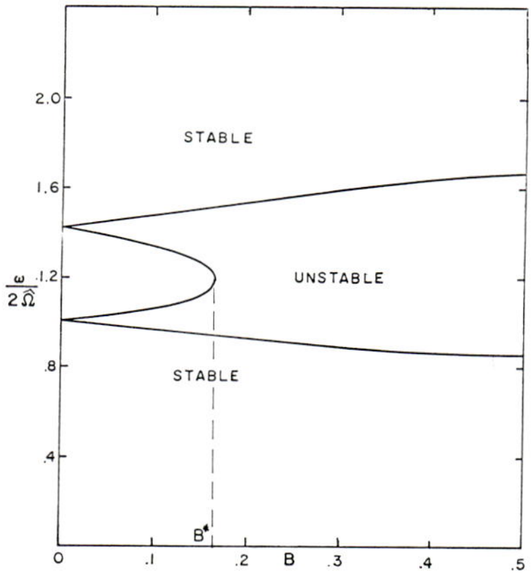

Fig. 6 Principal region of bending instability for $\left(\frac{2\hat{\Omega}}{\bar{\Omega}}\right)^2 = .5$

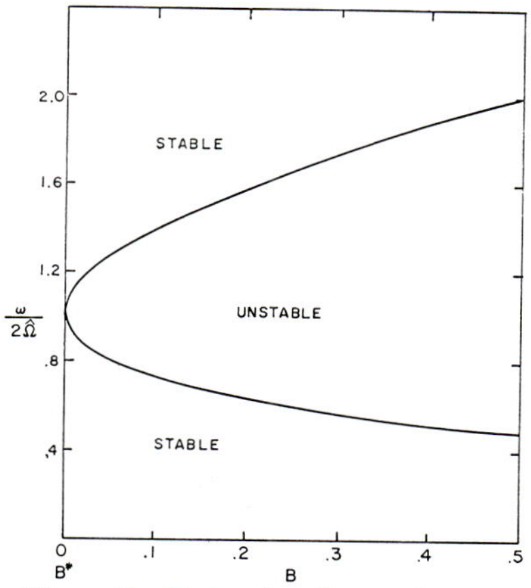

Fig. 7 Principal region of bending instability for $\left(\frac{2\hat{\Omega}}{\bar{\Omega}}\right)^2 = 1$

As the excitation grows to B*, the regions merge. If $\beta = 1$, then $B^* = 0$ and there is only one instability region for all B as shown by Figure 7.

CONCLUSIONS

Based on Eringen's nonlinear bending equations (3), this analysis shows that both axial and bending vibrations are important in regarding the stability of a constrained beam-like coupler. Axial instability can occur, and bending instability also develops when the axial load becomes sufficiently large to cause column-type buckling, or due to parametric resonance.

Axial instabilities occur near excitation frequencies of

$$\frac{2\bar{\Omega}}{K}, \quad K = 1, 2, 3, \ldots \quad \text{where} \quad \bar{\Omega} = \frac{EA\,\ell_3^2}{\ell_2 I_3} \sin^2(\theta_{30} - \theta_{20})$$

is the natural frequency of an inertialess extensible bar pinned with lever arm length $l_3 \sin (\theta_{30} - \theta_{20})$ to a rotary output with polar moment of inertia I_3.

Column buckling occurs when the axial load exceeds the Euler critical load:

$$\frac{\pi^2}{\ell^2} EI .$$

Furthermore, large bending amplitudes can occur due to parametric resonance as well as large axial loads. The principal region for the parametric type of instability occurs near two distinct excitation frequencies (Figures 6 and 7).

One frequency is twice the bending natural frequency while the other is the axial resonant frequency $\bar{\Omega}$. As bending stability is substantially influenced by axial vibration, caution must be exercised in using the classical bending equation which assumes axial inextensibility.

ACKNOWLEDGEMENT

The authors gratefully acknowledge the National Science Foundation traineeship support given Dr. Tobias during the course of his doctoral study in the area reported in this paper.

BIBLIOGRAPHY

1. Meyer zur Capellen, W., "Biegungsschwingungen in der Koppel einer Kurbelschwinge", Osterrichisches Ingenieur-Archiv, 1962, p. 341-348.

2. Neubauer, A. H., Jr., Cohen, R., and Hall, A. S., Jr., "An Analytical Study of the Dynamics of an Elastic Linkage", Journal of Engineering for Industry, August, 1966, p. 311.

3. Eringen, A. C., "On the Non-linear Vibration of Elastic Bars", Quarterly of Applied Mathematics, 9, 1952, p. 361-368.

4. Bolotin, V. V., The Dynamic Stability of Elastic Systems, Holden-Day, Inc., San Francisco, London Amsterdam, 1964.

5. Kobrinskiy, A. Ye., Mechanisms with Elastic Couplings - Dynamics and Stability, Nauka Press, Moscow, 1964, NASA Technical Translation, NASA TT F-534, June, 1969.

6. Lubkin, S. and Stoker, J. J., "Stability of Columns and Strings Under Periodically Varying Forces", Quarterly of Applied Mathematics, Vol. 1, 1943, p. 215-237.

NOMENCLATURE

A	=	cross-sectional area of flexible link
$\hat{a}_x$	=	axial acceleration of the undeformed center line
a_x	=	$\hat{a}_x/l_2$
$\hat{a}_y$	=	transverse acceleration of the undeformed center line
a_y	=	$\hat{a}_y/l_2$
B	=	excitation parameter in the Mathieu Equation
E	=	Young's modulus

f = forcing term in axial vibration equation
I = cross-sectional moment of inertia
l_i = length of links
$\hat{M}$ = bending moment
M = $\hat{M}l_2/EI$, nondimensional bending moment
$\hat{N}$ = resultant axial force on a cross section of the coupler
$\hat{N}_i$ = axial force on ith rigid link
N_i = $\hat{N}_i/EA$ nondimensional axial force
p = time dependent coefficient
$\hat{Q}$ = resultant shear force on cross section of the flexible link
Q = $\hat{Q}l_2^2/EI$, nondimensional shear force
$\hat{Q}_i$ = shear load at end of ith rigid link
Q_i = $\hat{Q}_i/EA$ nondimensional shear force
q = excitation parameter in Mathieu Equation for axial
r = radius of gyration of cross-sectional area
s = arc length of distorted link
t = time
$\hat{U}_{,tt}$ = relative axial acceleration of distorted link
$\hat{u}$ = relative axial coordinate
u = $\hat{u}/l_2$, nondimensional axial coordinate
$\tilde{u}$ = axial perturbation displacement
$\tilde{\tilde{u}}$ = tangential perturbation displacement
$\hat{V}_{,tt}$ = relative transverse acceleration of distorted link
$\hat{v}$ = relative transverse coordinate
v = $\hat{v}/l_2$, nondimensional transverse coordinate
$\tilde{v}$ = transverse perturbation displacement
$\tilde{\tilde{v}}$ = normal perturbation displacement
v_0 = relative transverse zeroth solution displacement
z = x/l_2, nondimensional axial coordinate
α = l_3/l_2, length ratio
β = $2\hat{\Omega}/\Omega$, frequency ratio
γ = r/l_2, slenderness ratio
ε = axial strain
ε_0 = axial strain of zeroth solution
φ = l_4/l_2, length ratio
η = relative rotation of distorted coupler
η_0 = rotation of distorted coupler in zeroth solution
η_{30} = relative rotation of driven link zeroth solution
θ = amplitude of small motion of input crank
θ_i = angle of ith rigid link

θ_{io} = initial angle of ith rigid link

μ = $\rho A l_2 l_3^2 / I_3$, mass ratio

ν = l_1/l_2, length ratio

ρ = density of coupler per unit length

ϕ_i = eigenfunction of bending beam

Ω = $[E A l_3 / I_3 l_2]^{1/2}$

$\bar{\Omega}$ = $\Omega \sin(\theta_{30} - \theta_{20})$, the axial natural vibrational frequency

$\hat{\Omega}$ = $\pi^2 / l_2^2 \, [EI/\rho A]^{1/2}$ first natural frequency of a pinned-pinned beam

ω = frequency of forcing function

ELASTICITY

Developments in Mechanics, Vol. 6. Proceedings of the 12th Midwestern Mechanics Conference: 43

The Generation of Electromagnetic Radiation by Elastic Waves

F. C. MOON
PRINCETON UNIVERSITY

ABSTRACT

The generation of electromagnetic fieldsby both torsion and compressional stress wave pulses in an elastic conducting rod is examined. The analysis is based on a theory for an elastic plasma. For low frequencies, the local electric field outside the rod is found to be proportional to the local acceleration in the rod. Results agree with known experimental observations. A non-contacting strain sensor is proposed.

INTRODUCTION

In this paper the generation of electromagnetic fields by a stress wave pulse in an elastic conductor is examined. This phenomenon raises the possibility of detecting stress waves by a non contacting electronic sensor. The analysis is based on a theory for an elastic plasma posited by the author [1]. This theory, which is similar to that for gaseous plasmas, treats the free electrons in the metal as a compressible gas in a porous, elastic, ion lattice continuum.

The traditional view of metals in continuum mechanics is that of a neutral elastic solid in which currents flow only when the solid is placed in an external field. However, in solid state theory a metal is viewed as an elastic plasma [2] in which free electron-lattice interactions can generate internal fields, currents, and space charge distributions. While many of these phenomena occur at very high frequencies, some effects have been observed at frequencies low enough to be of interest to continuum mechanicians. One of these is the electron inertia effect. In an experiment reported by J. D. Kennedy and C. W. Curtis [3], the angular acceleration of a copper ring by a torsional stress pulse was shown to produce measurable electromagnetic fields.

In this paper the generation of electromagnetic fields by both torsion and compressional stress waves in a rod is shown to be possible. The associated electric field outside the rod is calculated. For low frequencies the local electric field is shown to be proportional to the local acceleration in the rod.

EQUATIONS FOR AN ELASTIC PLASMA

The linearized equations of motion governing the elastic displacement, electron density, and electromagnetic fields are listed below, [1]. Thermoelectric and thermoelastic effects have been neglected. The following notation is used:

ρ^e, ρ^s ... electron and ion mass densities respectively
ν ... free electron gas dilatation
$\underset{\sim}{w}$ ... electron velocity vector
$\underset{\sim}{u}$ ... lattice displacement field
$\underset{\sim}{E}, \underset{\sim}{B}$ ... electric and magnetic fields respectively
q_0 ... free electron density
σ ... electric conductivity
s_1, s_2, s_3, s_4, μ ... elastic constitutive constants

In this theory the net current is given by

$$\underset{\sim}{J} = -q_0(\underset{\sim}{w} - \dot{\underset{\sim}{u}})$$

(dots indicate time differentiation). The net charge density becomes

$$Q = -q_0(\nu + \nabla \cdot \underset{\sim}{u}).$$

In a linearized theory, with no external fields, nonlinear terms in the body force such as $Q\underset{\sim}{E} + \underset{\sim}{J} \times \underset{\sim}{B}$ are dropped. The linearized equations then take the form:

Continuity

$$\frac{\partial \nu}{\partial t} + \nabla \cdot \underset{\sim}{w} = 0 \tag{1}$$

Momentum

$$\rho^s \ddot{\underset{\sim}{u}} - (s_4 + \mu)\nabla(\nabla \cdot \underset{\sim}{u}) - \mu\nabla^2 \underset{\sim}{u} - s_1 \nabla \nu = q_0 \underset{\sim}{E} + \frac{q_0^2}{\sigma}(\underset{\sim}{w} - \dot{\underset{\sim}{u}}) \tag{2}$$

$$\rho^e \dot{\underset{\sim}{w}} + s_2 \nabla \nu + s_3 \nabla(\nabla \cdot \underset{\sim}{u}) = -q_0 \underset{\sim}{E} - \frac{q_0^2}{\sigma}(\underset{\sim}{w} - \dot{\underset{\sim}{u}}) \tag{3}$$

E-M Field Equations

$$\nabla \times \underset{\sim}{E} + \dot{\underset{\sim}{B}} = 0 \quad , \quad \nabla \cdot \underset{\sim}{B} = 0 \tag{4),(5}$$

$$\nabla \times \underset{\sim}{B} - \mu_0 \varepsilon_0 \dot{\underset{\sim}{E}} = -\mu_0 q_0 (\underset{\sim}{w} - \dot{\underset{\sim}{u}}) \tag{6}$$

$$\nabla \cdot \underset{\sim}{E} = -q_0(\nu + \nabla \cdot \underset{\sim}{u})/\varepsilon_0 \tag{7}$$

(μ_0 is the magnetic permeability in a vacuum and ε_0 is the electric permittivity of a vacuum.)
To recover the classical theory of elasticity one sets $\underset{\sim}{J} = 0$, $Q = 0$ or

$$\dot{\underset{\sim}{u}} = \underset{\sim}{w} \qquad \nu + \nabla \cdot \underset{\sim}{u} = 0 \tag{8}$$

The latter condition can be called the condition of electronic equilibrium. These conditions satisfy continuity (1), while the sum of the momentum equations yields

$$(\rho^s + \rho^e)\ddot{\underset{\sim}{u}} - (s_2 - s_1 + s_4 - s_3 + \mu)\nabla(\nabla \cdot \underset{\sim}{u}) - \mu\nabla^2 \underset{\sim}{u} = 0$$

This equation is identical to the governing equations of elastodynamics provided that we identify μ as the elastic shear modulus and the Lame constant λ with

$$\lambda = s_2 - s_1 + s_4 - s_3 \tag{9}$$

Thus except for three of the constants s_i, which can be determined from solid state theory, all the remaining constants are known and familiar to mechanicians. The classical wave speeds for shear and dilatational waves are given by

$$c_s = [\mu/(\rho^s + \rho^e)]^{1/2} \qquad c_d = [(\lambda + 2\mu)/(\rho^s + \rho^e)]^{1/2} \tag{10}$$

Note that the conditions (8) for electronic equilibrium will not, in general, be sufficient for a solution to the set of equations (1) - (7). However one can show that for harmonic waves in an elastic plasma, as the frequency approaches zero, the wave speeds approach those of (10). Thus electronic equilibrium is achieved asymptotically as frequency goes to zero. Further it should be noted that the absence of motion alone will not guarantee electronic equilibrium. In fact, there exist static solutions to the equations (1) - (7) where $Q \neq 0$. Nevertheless, in this paper only wave solutions will be sought and at low frequencies such that the expressions (10) for the elastic waves are valid.

ONE DIMENSIONAL WAVE PROPAGATION

In this section one dimensional wave solutions in an infinite medium are sought. In light of the remarks made in the preceding paragraph, we will assume that the frequencies are low enough such that the lattice waves are non-dispersive and are given by the classical solutions

$$u = \underset{\sim}{f}(x_1 - ct) \tag{11}$$

where c is given by one of the expressions (10). The damping due to induced eddy currents is neglected in (11), and we seek the quantities ν, $\underset{\sim}{w}$, $\underset{\sim}{E}$, $\underset{\sim}{B}$ where $\underset{\sim}{u}$ is assumed to be given by (11).

SHEAR WAVES

Suppose then that $u_1 = u_3 = 0$ and

$$\frac{\partial}{\partial x_2} = 0 \ , \ \frac{\partial}{\partial x_3} = 0$$

The plasma equations become

$$\begin{aligned} \dot{\nu} &= 0 \\ \rho^e \dot{w}_2 &= -q_0 E_2 - \frac{q_0^2}{\sigma}(w_2 - \dot{u}_2) \\ \frac{\partial^2 E_2}{\partial x_1^2} - \mu_0 \varepsilon_0 \ddot{E}_2 &= -\mu_0 q_0 (\dot{w}_2 - \ddot{u}_2) \end{aligned} \tag{12}$$

$$\begin{gathered} \dot{B}_1 = 0, \ \dot{B}_2 + \frac{\partial E_3}{\partial x} = 0, \ \frac{\partial B_2}{\partial x_1} - \mu_0 \varepsilon_0 \dot{E}_3 = 0 \\ \frac{\partial E_1}{\partial x_1} = -q_0 \nu \end{gathered} \tag{13}$$

The equations (13) do not involve any inhomogeneous terms in u_2, thus we may set the fields

$$B_1 = 0, \ B_2 = E_3 = 0, \ E_1 = \nu = 0$$

Suppose then that u_2 is a harmonic wave

$$u_2 = u_0 e^{i\omega(t - x_1/c_s)}$$

Then writing

$$E_2 = E_0 e^{i\omega(t - x_1/c_s)}$$

$$w_2 = w_0 e^{i\omega(t - x_1/c_s)}$$

the equations (12) require

$$\begin{bmatrix} q_0 & \left(\frac{q_0^2}{\sigma} + i\omega\rho^e\right) \\ \omega^2\left(\mu_0\varepsilon_0 - \frac{1}{c_s^2}\right) & i\omega\mu_0 q_0 \end{bmatrix} \begin{bmatrix} E_0 \\ w_0 \end{bmatrix} = u_0 \begin{bmatrix} \frac{q_0^2}{\sigma} i\omega \\ -\mu_0 q_0 \omega^2 \end{bmatrix}$$

$$E_0 = \frac{\rho^e \omega^2}{q_0} u_0 \left[1 + i\omega\left(\mu_0\varepsilon_0 - \frac{1}{c_s^2}\right)\left(\frac{1}{\mu_0\sigma} + \frac{i\omega\rho^e}{\mu_0 q_0^2}\right)\right]^{-1} \tag{14}$$

For normal elastic conductors

$$\mu_0\varepsilon_0 << 1/c_s^2$$

$$1/\sigma >> \omega\rho^e/q^2$$

$$E_0 \simeq \rho^e \frac{\omega^2}{q_0} u_0 \quad \sigma c_s^2\mu_0 [\sigma c_s^2\mu_0 - i\omega]^{-1} \tag{15}$$

For normal conductors $\sigma c_s^2\mu_0 \sim 10^6$ 1/sec. Thus at ultrasonic frequencies the damping term $\omega/\sigma c_s\mu_0$ becomes important. If one considers this expression as the Fourier transform of the response, then using a convolution integral one obtains the response for a stress pulse;

$$E_2(x.t) = -\frac{\rho^e}{q_0}\sigma c_s^2\mu_0 \int_{x/c_s}^{t+x/c_s} \ddot{u}_2(\tau - x/c_s) e^{-\sigma c_s^2\mu_0(t-t_1)} dt_1 \tag{16}$$

where we assume

$$u(t - x/c_s) = 0 \quad , \quad t < x/c_s$$

For very low frequencies we may write

$$E_2(x,t) = -\frac{\rho^e}{q_0}\ddot{u}_2(t - x/c_s).$$

This expression is similar to that assumed in [3] for which experiments checked very closely the theory. The integral (16) accounts for the damping effect which was omitted in the analysis in [3]. The factor ρ^e/q_0 is the universal constant of electron mass to charge ratio. It should be noted that this approximation is tantamount to assuming that the net current is zero or $\dot{u}_2 = w_2$.

DILATATIONAL WAVES

Consider next the case where

$$u_2 = 0,\ w_2 = 0,\ u_3 = 0,\ w_3 = 0$$

$$\frac{\partial}{\partial x_2} = \frac{\partial}{\partial x_3} = 0.$$

These conditions along with the equations (4) - (6) imply that $E_2 = 0$, $B_3 = 0$. The plasma equations become

$$\rho^e\dot{w}_1 + \sigma_2\frac{\partial\nu}{\partial x_1} + \sigma_3\frac{\partial^2 u_1}{\partial x_1^2} = -q_0 E_1 - \frac{q_0^2}{\sigma}(w_1 - \dot{u}_1)$$

$$\dot{\nu} + \frac{\partial w_1}{\partial x_1} = 0 \tag{18}$$

$$\frac{\partial E_1}{\partial x} = -\frac{q_0}{\varepsilon_0}\left(\nu + \frac{\partial u_1}{\partial x_1}\right),\quad \frac{\partial E_1}{\partial t} = \frac{q_0}{\varepsilon_0}(w_1 - \dot{u}_1)$$

The number of unknowns may be reduced if the plasma displacement is used.

$$w_1 = \dot{\eta}_1$$

from which continuity implies

$$\nu = -\frac{\partial \eta_1}{\partial x_1}$$

The last two equations (18) are thus equivalent and are satisfied by the expression

$$E_1 = \frac{q_0}{\varepsilon_0}(\eta_1 - u_1)$$

The electric field is then found to be governed by the equation

$$s_2 \frac{\partial^2 E_1}{\partial x_1^2} + \ddot{E}_1 + \frac{q_0^2}{\sigma \rho^e}\dot{E}_1 + \frac{q_0^2}{\varepsilon_0 \rho^e}E_1 = -\frac{q_0}{\varepsilon_0}\left[u_1 - \frac{(s_2+s_3)}{\rho^e}\frac{\partial^2 u_1}{\partial x_1^2}\right] \tag{19}$$

When the electron gas compressibility s_2 is neglected, the homogeneous solutions for E exhibit the familiar damped plasma oscillations. The term Ω_ρ defined by

$$\Omega_\rho^2 = q_0^2/\varepsilon_0\rho^e$$

is called the plasma frequency at infinite conductivity; for metals $\Omega_\rho \sim 10^{15}$ 1/sec. Thus at low frequencies $\Omega_\rho^2 E_1$ dominates $\ddot{E}_1$. Similarly the order of the damping coefficient of $\dot{E}_1$ is $q_0^2/\sigma\rho^e \sim 10^{13}$ 1/sec for normal conductors and is well below the order of the term $\Omega_\rho^2 E$ for suboptical frequencies. Thus we may use the approximation

$$E_1 = -\frac{\rho^e}{q_0}\left[u_1 - \frac{(s_2+s_3)}{\rho^e}u_1''\right] \tag{20}$$

STRESS PULSE INDUCED FIELDS IN A ROD

In this section we apply the results of the analysis for plane waves in an infinite space to waves in a cylindrical rod. Our object is to determine the magnitude of the induced fields outside the rod and determine their configuration in order to explore the feasibility of measuring them. Hence we are not very concerned about the stress wave itself and will assume that it is non dispersive and given by elementary rod theory. The induced fields inside the rod will be assumed to obey the most elementary approximation neglecting the electron gas compressibility effects as in (20),

$$\underset{\sim}{E} = -\frac{\rho^e}{q_0}\ddot{\underset{\sim}{u}}(z - ct) = -c^2\frac{\rho^e}{q_0}u''(z - ct)$$

where c is either the shear or torsional wave speed or the compressional wave speed in a rod ($c_0 = \sqrt{Y/\rho}$, Y is Young's modulus). For compressional waves we have

$$\underset{\sim}{u} = u_0(z - c_d t)\underset{\sim}{\varepsilon}_z$$

and for torsional waves

$$\underset{\sim}{u} = r u_0(z - c_s t)\underset{\sim}{\varepsilon}_\theta$$

($\underset{\sim}{\varepsilon}_r, \underset{\sim}{\varepsilon}_\theta, \underset{\sim}{\varepsilon}_z$ are orthogonal unit vectors with $\underset{\sim}{\varepsilon}_z$ along the rod axis.) In both cases the Fourier spectrum of the pulse allows one to write

$$u_0'' = \int_{-\infty}^{\infty} A(k)e^{ik(z-ct)}dk$$

Outside the rod the electromagnetic properties are assumed to be those of a vacuum, and the fields $\underset{\sim}{E}$, $\underset{\sim}{B}$ satisfy Maxwell's equations

$$\nabla \times \underset{\sim}{E} + \dot{\underset{\sim}{B}} = 0 \qquad \nabla \cdot \underset{\sim}{B} = 0,\ \nabla \cdot \underset{\sim}{E} = 0$$

$$\nabla \times \underset{\sim}{B} + \mu_0 \varepsilon_0 \dot{\underset{\sim}{E}} = 0$$

At frequencies of interest in solid mechanics one can drop the displacement current term $\dot{\underset{\sim}{E}}$ in the last equation. The electric field then satisfies the equation

$$\nabla^2 \underset{\sim}{E} = 0\ ,\ \nabla \cdot \underset{\sim}{E} = 0$$

There are then two classes of problems for the case $\partial/\partial\theta = 0$.

Torsional waves: $E_z = E_r = 0,\ B_\theta = 0$

$$\nabla^2 E_\theta - E_\theta/r^2 = 0 \tag{21}$$

$$E_\theta = -\frac{\rho^e}{q_0}\ddot{u}_0 \quad \text{on } r = a$$

Compressional waves: $E_\theta = 0,\ B_r = B_z = 0$

$$\nabla^2 E_r - E_r/r^2 = 0$$

$$\nabla^2 E_z = 0 \tag{22}$$

$$E_z = -\frac{\rho^e}{q_0}\ddot{u}_0 \quad \text{on } r = a$$

$$\frac{1}{r}\frac{\partial}{\partial r}(rE_r) = -\frac{\partial E_z}{\partial z} \quad \text{on } r = a$$

For axial variation proportional to e^{ikz} one can show [4] that electric field solutions have the forms

$$E_z = A_1 K_0(kr)e^{ikz}$$

$$E_r = A_2 K_1(kr)e^{ikz}$$

$$E_\theta = A_3 K_1(kr)e^{ikz}$$

where $k > 0$.

K_0, K_1 are hyperbolic Bessel functions which both have the asymptotic behavior.

$$K_0,\ K_1 \underset{kr\to\infty}{\longrightarrow} \left[\frac{\pi}{2kr}\right]^{1/2} e^{-kr}$$

For the non dispersive approximation used here $ka << 1$. At distances $r \sim a$

$$K_0 \sim -\ell n(kr)$$

$$K_1 \sim 1/kr.$$

For a dilatational wave of wavelength $\lambda = 10a$

$$E_z(r=3a)/E_z(r=a) = 0.19.$$

For a torsional wave of wavelength $\lambda = 10a$

$$E_\theta(r=3a)/E_\theta(r=a) = 0.14$$

This rapid fall off with radial distance from the rod means that sensors must be close to the surface to detect these signals.

An actual stress pulse contains many frequencies, and the shape of the induced electric field signal will vary with radial distance r. For example, imagine a symmetric stress pulse for which we can write

$$u_0' = \int_0^\infty S(k) \cos k(z-ct)dk$$

Thus

$$u_0'' = -\int_0^\infty S(k)\, k \sin k(z-ct)dk$$

For a torsional pulse the induced field becomes

$$E_\theta = \frac{\rho^e}{q_0} c_s^2 \int_0^\infty \frac{K_1(kr)}{K_1(ka)} kaS(k)\sin k(z-c_s t)dk$$

or for a dilatational pulse

$$E_z = \frac{\rho^e}{q_0} c_0^2 \int_0^\infty \frac{K_0(kr)}{K_0(ka)} kS(k)\sin k(z-c_d t)dk$$

NON-CONTACTING STRAIN SENSORS

As has been shown, a dilatational, (compressional) wave will induce an axial electric field and a torsional wave will induce a circumferential electric field vector. If we denote the strain by S, the electric field is proportional in both cases to the strain gradient

$$E = -\frac{\rho^e c^2}{q_0}\frac{\partial S(z-ct)}{\partial z}$$

For dilatational waves, the voltage along the rod is given by

$$V = -\int^z E_z dz$$

or

$$V = \frac{\rho^e c^2 S}{q_0}$$

The mass to charge ratio for an electron is $(5.67)10^{-12}$(mks units). For aluminum

$$C_d = 6\ 10^3 \text{ meters/sec.}$$

Thus, the voltage difference along the rod for dilatational waves is

$$V \simeq 2\ 10^{-4} S \text{ volts}$$

For a maximum strain of $S \sim 10^{-4}$, the voltage is of the order of 20 nanovolts. To measure the field, a dipole sensor will detect an axial electric field, as shown in the Figure 1.

For the torsional case, a multiturn coil around the rod will produce a voltage

$$V = E_\theta N\pi d$$

where N is the number of turns and d the coil diameter. This is the method used by Kennedy and Curtis [3] in their experiment. By increasing N, they were able to obtain a voltage of .22 10^{-3} volts with a 3000 turn coil.

A word of caution should be noted in attempting these measurements: the presence of the earth's magnetic field should be accounted for or else nulled out with, for example, a pair of Helmholtz coils, [3].

REFERENCES

1. Moon, F. C., "A Theory for High Current Elastic Conductors," Recent Advances in Engineering Science, V. 5. Gordon & Breach (1970), 87-103.

2. Steele, M. C. and Vural, B., Wave Interactions in Solid State Plasmas. New York: McGraw-Hill Book Co. (1969).

3. Kennedy, J. D. and Curtis, C. W., "Transient Electron-Inertia Field Produced by a Strain Pulse," J. Acoustical Soc. Amer., 41(2), 328-335, Feb. 1967.

4. Abramowitz, M. and Stegun, I. A., editors, Handbook of Mathematical Functions. Nat. Bureau of Standards, U. S. Dept. of Commerce, 374.

ACKNOWLEDGEMENT

This work has been supported in part by the National Science Foundation

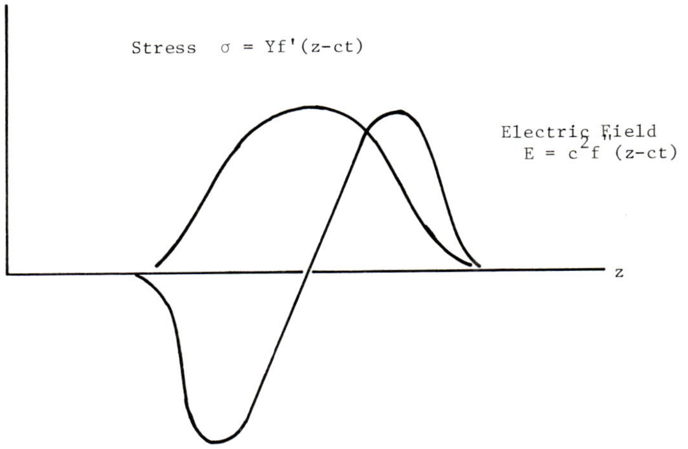

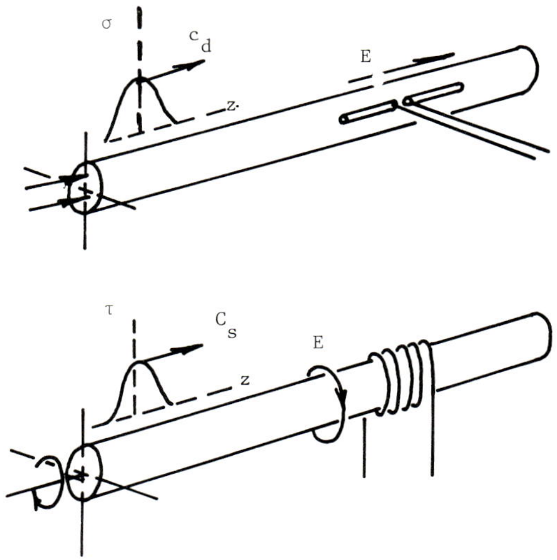

a) Dilatational Wave

Dipole Sensor

b) Torsional Wave Multiturn Coil Sensor

Figure 1

Developments in Mechanics, Vol. 6. Proceedings of the 12th Midwestern Mechanics Conference: 44

Velocity and Stress Distributions in the Earth's Mantle Due to Secular Variations of the Geomagnetic Field

MICHAEL A. CHARLES and PHILLIP R. SMITH
NEW MEXICO STATE UNIVERSITY

ABSTRACT

The possibility that the Earth's mantle is moving under the action of stresses applied at the core-mantle boundary by the motion of the fluid outer-core is studied analytically. First, Maxwell's equations are used to determine as a function of the secular variations of the geomagnetic field, a velocity distribution of the fluid of the outer-core near the core-mantle interface. The stresses associated with the fluid motion in the outer-core are transmitted to the mantle, and the propagation of these stresses through the mantle, taken as a highly viscous fluid, is determined by the solution of the Navier-Stokes equations, using the linear transform method. The displacement velocities and stresses, calculated near the Earth's surface, are in good agreement with observational data on continental drift and the location of the main earthquake zones.

A new point of view for the cause of the motion of the Earth's mantle is thus introduced, which does not involve convective heat transfer. Analytical support is given to the theory which closely relates continental drift and seismic shocks to the secular variations of the geomagnetic field.

INTRODUCTION

It is generally admitted that motions occur on a large scale in the Earth's mantle, which deforms itself and, therefore, the Earth's crust at the rate of a few centimeters per year [1]. Deep seismic shocks occurring in the mantle are attributed to violent readjustments to compressions and dilatations created by these motions. The deformations of the crust explain the continental drift, and, on a lesser scale, the formation of the main features of the Earth's surface. Thus far, the sole plausible mechanism proposed to explain these motions has been convective currents in the mantle [2,3]. However, some doubt is cast upon this hypothesis by observations that the heat flow from beneath the oceans and from beneath the continents is essentially the same [4]. This perhaps indicates that another mechanism may coexist with or even dominate the heat transfer mechanism. We investigate here the possibility that the Earth's mantle is moving under the action of stresses applied at the core-mantle boundary by the motion of the fluid of the outer-core. The velocity of this fluid must be of the order of several kilometers per year if the geomagnetic field is to be produced in the outer-core, as proposed by the Dynamo theory [5,6]. It would seem that, at the core-mantle boundary, such velocities should be able to create large stresses which will propagate through the mantle and set it in motion.

Initially, the fluid velocity of the outer-core near the core-mantle interface is deduced from Maxwell's equations as a function of the secular variations of the geomagnetic field. The mantle is then assumed to be a highly viscous fluid for which the Navier-Stokes equations are valid. Adequate boundary conditions are deduced from the equality of stresses in the radial direction at the core-mantle interface. Velocity and stress distributions are then obtained throughout the mantle by solving the Navier-Stokes equations using the method of linear operators. Finally, the velocities and stresses in the mantle near the crust, calculated by this method, are compared to observational data on continental drift and the location of the major seismic shocks.

DETERMINATION OF THE FLUID VELOCITY IN THE OUTER-CORE NEAR THE CORE-MANTLE BOUNDARY

The magnetohydrodynamic phenomena in the outer-core, taken as an incompressible fluid, can be described by the following equations:

$$\nabla \cdot \underset{\sim}{H_1} = 0, \tag{1}$$

$$\nabla^2 \underset{\sim}{H_1} = \sigma\mu\left[\frac{\partial \underset{\sim}{H_1}}{\partial t} - \nabla \times (\underset{\sim}{V} \times \underset{\sim}{H_1})\right], \tag{2}$$

$$\nabla \cdot \underset{\sim}{V} = 0, \tag{3}$$

and

$$\text{Equation of Motion.} \tag{4}$$

The equation of motion, under the assumptions we will make, is not explicitly needed.

Since we are interested in the part of the outer-core near the core-mantle interface, the following assumption can be made. The magnetic field $\underset{\sim}{H}$ throughout the mantle, taken as an insulator, is given by

$$\underset{\sim}{H} = -\nabla A \tag{5}$$

with

$$A = a \sum_{n=1}^{\infty} \left(\frac{a}{r}\right)^{n+1} \sum_{m=0}^{n} \left[g_n^m(t) \cos m\phi + h_n^m(t) \sin m\phi\right] P_{nm}(\theta) \tag{6}$$

and

$$P_{nm}(\theta) = \left[\frac{2(n-m)!}{(n+m)}\right]^{1/2} P_n^m(\theta) \tag{7}$$

In the transition layer which corresponds to the core-mantle interface, we assume that H_1 can be approximated by H. Then, from (5) and (6),

$$\nabla^2 \underset{\sim}{H_1} \simeq \nabla^2 H = 0 \tag{8}$$

and the fluid velocity in the outer-core, at the core-mantle interface, is determined by

$$\frac{\partial \underset{\sim}{H}}{\partial t} = \nabla x (\underset{\sim}{V} x \underset{\sim}{H}) \tag{9}$$

where H is given by (5) and (6) and is considered known.

Letting

$$\underset{\sim}{V} = V_r i + V_\theta \underset{\sim}{j} + V_\phi \underset{\sim}{k}, \tag{10}$$

$$\underset{\sim}{H} = H_r i + H_\theta \underset{\sim}{j} + H_\phi \underset{\sim}{k} \tag{11}$$

we seek a solution to (9) by assuming the following expansions:

$$V_\theta H_\phi - V_\phi H_\theta = \sum_{n=1}^{\infty} \frac{a^{n+2}}{r^{n+1}} \sum_{m=0}^{n} [A_n^m(t) \cos m\phi + \sin m\phi] \, F_n^m(\theta), \tag{12}$$

$$V_\phi H_r - V_r H_\phi = \sum_{n=1}^{\infty} \frac{a^{n+2}}{r^{n+1}} \sum_{n=0}^{m} [C_n^m(t) \cos m\phi + \sin m\phi] \, G_n^m(\theta), \tag{13}$$

and

$$V_r H_\theta - V_\theta H_r = \sum_{n=1}^{\infty} \frac{a^{n+2}}{r^{n+1}} \sum_{m=0}^{n} [E_n^m(t) \cos m\phi + \sin m\phi] \, H_n^m(\theta), \tag{14}$$

in which $A_n^m(t)$, $C_n^m(t)$, $E_n^m(t)$, $F_n^m(\theta)$, $G_n^m(\theta)$, and $H_n^m(\theta)$ are to be determined.

If we equate term by term, the coefficients of cos mϕ and sin mϕ in (9) and (12) through (14), then we obtain

$$V_\theta H_\phi - V_\phi H_\theta = \sum_{n=1}^{\infty} \frac{a^{n+2}}{r^{n+1}} \sum_{m=1}^{n} \frac{1}{m} [\frac{dh_n^m}{dt} \cos m\phi - \frac{dg_n^m}{dt} \sin m\phi]$$

$$[\frac{n}{dh_n^m/dt} \sin \theta H_n^m(\theta) + \sin \theta \frac{dP_{nm}(\theta)}{d}] \tag{15}$$

$$V_\phi H_r - V_r H_\phi = \sum_{n=1}^{\infty} \frac{a^{n+2}}{r^{n+1}} \sum_{m=1}^{n} \frac{1}{m} [\frac{dh_n^m}{dt} \cos m\phi - \frac{dg_n^m}{dt} \sin m\phi]$$

$$[(n + 1) \sin \theta P_{nm}(\theta) - \frac{1}{dh_n^m/dt} \frac{d}{d\theta} (\sin \theta H_n^m(\theta))] \tag{16}$$

$$V_r H_\theta - V_\theta H_r = \sum_{n=1}^{\infty} \frac{a^{n+2}}{r^{n+1}} \{- \frac{1}{n} \frac{dg_n^{\,o}}{dt} \frac{dP_{no}(\theta)}{d\theta} + \sum_{m=1}^{n} [\frac{dg_n^{\,m}}{dt} \cos m\phi$$

$$+ \frac{dh_n^{\,m}}{dt} \sin m\phi] \frac{H_n^{\,m}(\theta)}{dh_n^{\,m}/dt}\}, \tag{17}$$

where H_r, H_θ, and H_ϕ are known and given by (5) and (6), $H_n^{\,m}(\theta)$ are arbitrary, and $dg_n^{\,m}/dt$ and $dh_n^{\,m}/dt$ are derivatives of the Gauss coefficients which represent the secular variations of the geomagnetic field and are tabulated in References [10] and [11].

Equations (15), (16), and (17), in which V_r, V_θ, and V_ϕ are the unknowns, are not independent since

$$(\underset{\sim}{V} \times \underset{\sim}{H}) \cdot \underset{\sim}{H} = 0. \tag{18}$$

An additional equation is needed, which is the continuity equation, (3). Theoretically, we can now determine V_r, V_θ, and V_ϕ from (3), (15), (16), and (17). Practically, since it is impossible to know if the ratios

$$\frac{dg_n^{\,m}}{dt} / \frac{dh_n^{\,m}}{dt}$$

are bounded as n and m increase, the arbitrary functions $H_n^{\,m}(\theta)$ are taken identical to zero such that the series (15), (16), and (17) surely converge:

$$H_n^{\,m}(\theta) = 0 \text{ for all } (n,m). \tag{19}$$

If the components of the geomagnetic field appearing on the left-hand side of (15), (16), and (17) are approximated by the components of a centered dipole, aligned with the Earth's axis, i.e.

$$H_r = 2(\frac{a}{r})^3 g_1^{\,0} \cos\theta, \tag{20}$$

$$H_\theta = (\frac{a}{r})^3 g_1^{\,0} \sin\theta, \tag{21}$$

and

$$H_\phi = 0, \tag{22}$$

then we can solve for the velocity components. The expression obtained for V_ϕ is

$$V_\phi = \frac{1}{2g_1^{\,0}} \sum_{n=1}^{\infty} \frac{a^{n-1}}{r^{n-2}} \sum_{m=1}^{n} (\frac{n+1}{m}) [\frac{dh_n^{\,m}}{dt} \cos m\phi - \frac{dg_n^{\,m}}{dt} \sin m\phi]$$

$$\tan\theta P_{nm}(\theta). \tag{23}$$

The components V_θ and V_r can also be obtained, but we assume them to be much smaller than V_ϕ. Thus, the approximate velocity in the outer-core near the core-mantle boundary is:

$$\underset{\sim}{V} = V_\phi \underset{\sim}{k} \tag{24}$$

with V_ϕ given by (23) as a function of the secular variations of the geomagnetic field.

DERIVATION OF THE NAVIER-STOKES EQUATIONS IN THE MANTLE

The equations which describe the motions in the mantle, taken as a highly viscous, incompressible, and insulating fluid, are

$$\nabla \cdot \underset{\sim}{W} = 0 \tag{25}$$

$$\frac{\partial \underset{\sim}{W}}{\partial t} + (\underset{\sim}{W} \cdot \nabla) \cdot \underset{\sim}{W} + 2\underset{\sim}{\omega} \times \underset{\sim}{W} = \underset{\sim}{\Omega} - \frac{\nabla p}{\rho} + \frac{\mu_f}{\rho} \nabla^2 \underset{\sim}{W}. \tag{26}$$

Because the mantle is set in motion by the stresses created by the motion of the outer-core, which we assumed has a dominate motion in the longitudinal direction, we assume that the velocity in the mantle can be approximated by

$$\underset{\sim}{W} = W_\phi \underset{\sim}{k} \tag{27}$$

where W_r and W_θ will be neglected compared to W_ϕ, with, however, (25) satisfied. The components of Ω, the summation of the gravitational and centrifugal forces, are

$$R = -g + \omega^2 r \sin^2\theta,$$

$$\Theta = \omega^2 r \sin\theta \cos\theta,$$

and

$$\Phi = 0. \tag{28}$$

The pressure gradient is approximated by the hydrostatic equation:

$$\frac{1}{\rho}\frac{dp}{dr} = -g. \tag{29}$$

Now, the Navier-Stokes equations become, in component form,

$$-\frac{W_\phi^{\,2}}{r} - 2\omega W_\phi \sin\theta = \omega^2 r \sin^2\theta - \frac{\mu_f}{\rho}\left[\frac{2}{r^2 \sin\theta}\frac{\partial W_\phi}{\partial \phi}\right], \tag{30}$$

$$-\frac{W_\phi^{\,2}}{r}\frac{\cos\theta}{\sin\theta} - 2\omega W_\phi \cos\theta = \omega^2 r \sin\theta\cos\theta - \frac{\mu_f}{\rho}\left[\frac{2}{r^2}\frac{\cos\theta}{\sin^2\theta}\frac{\partial W_\phi}{\partial\phi}\right], \tag{31}$$

$$\frac{\partial W_\phi}{\partial t} + \frac{W_\phi}{r\sin\theta}\frac{\partial W_\phi}{\partial \phi} = \frac{\mu_f}{\rho}\left[\nabla^2 W_\phi - \frac{W_\phi}{r^2 \sin^2\theta}\right]. \tag{32}$$

Equations (30) and (31) are identical. If we take the derivative of (30) with respect to ϕ, then we get

$$\frac{W_\phi}{r \sin\theta} \frac{\partial W_\phi}{\partial \phi} = -\omega \frac{\partial W_\phi}{\partial \phi} + \frac{\mu_f}{\rho} \left[\frac{2}{r^2 \sin^2\theta} \frac{\partial^2 W_\phi}{\partial \phi^2}\right], \tag{33}$$

and thus, after expanding $\nabla^2 W_\phi$, (32) becomes

$$\frac{\partial W_\phi}{\partial t} - \omega \frac{\partial W_\phi}{\partial \phi} = \frac{\mu_f}{\rho} \left[\frac{1}{r^2} \frac{\partial}{\partial r} \left(r^2 \frac{\partial W_\phi}{\partial r}\right) + \frac{1}{r^2 \sin\theta} \left(\sin\theta \frac{\partial W_\phi}{\partial \theta}\right) - \frac{W_\phi}{r^2 \sin^2\theta}\right]. \tag{34}$$

But, the minimum value of the kinematic viscosity μ_f/ρ is 10^{20} cm^2/sec. Since the maximum value of r will be r = a = 6370 km., the right-hand side of (34) will be at least of the order of 250 $sec.^{-1}$ Meanwhile, the Earth's angular velocity is

$$\omega = 7 \times 10^{-5} \text{ sec.}^{-1} \tag{35}$$

and it is therefore possible to neglect the term $\omega\, \partial W_\phi/\partial\phi$ in (34) compared to the rest of the terms of the equation. The velocity in the mantle is then the solution of

$$\frac{\partial W_\phi}{\partial t} = \frac{\mu_f}{\rho} \left[\frac{1}{r^2} \frac{\partial}{\partial r} \left(r^2 \frac{\partial W_\phi}{\partial r}\right) + \frac{1}{r^2 \sin\theta} \frac{\partial}{\partial \theta} \left(\sin\theta \frac{\partial W_\phi}{\partial \theta}\right) - \frac{W_\phi}{r^2 \sin^2\theta}\right] \tag{36}$$

subject to two boundary conditions in r, two in θ, and an initial condition.

First, at the inner boundary, for r = b (see Figure 2), the components of the stress tensors in the core and in the mantle must be the same in the r-direction. From (24) and (27), we find that the only non-zero components are in (ϕr), and thus we get only one condition

$$\frac{\partial}{\partial r} \left(\frac{W_\phi}{r}\right)_{r=b} = \frac{\mu c}{\mu_f} \frac{\partial}{\partial r} \left(\frac{V_\phi}{r}\right)_{r=b} = B(\theta,\phi,t) \tag{37}$$

where μc is the viscosity of the outer-core and $B(\theta,\phi,t)$, derived from (23), is

$$B(\theta,\phi,t) = -\frac{1}{2bg_1{}^0} \frac{\mu c}{\mu_f} \sum_{n=1} \left(\frac{a}{b}\right)^{n-1} \sum_{m=1}^{n} \frac{(n^2-1)}{m} \left[\frac{dh_n{}^m}{dt} \cos m\phi\right.$$

$$\left. - \frac{dg_n{}^m}{dt} \sin m\phi\right] \tan\theta P_{nm}(\theta). \tag{38}$$

At the Mohorovicic discontinuity, for r ≃ a, the components of the stress tensor are assumed to be zero in the r-direction, and, here again, we obtain only one condition,

$$\frac{\partial}{\partial r} \left(\frac{W_\phi}{r}\right)_{r=a} = 0. \tag{39}$$

Since the motion is in the ϕ-direction only, by symmetry we take

$$W_\phi\ (r,0,\phi,t) = 0, \tag{40}$$

and

$$W_\phi\ (r,\pi,\phi,t) = 0. \tag{41}$$

Further, we assume that there was no initial velocity, i.e.

$$W_\phi\ (r,\theta,\phi,0) = 0, \tag{42}$$

although, if one could be defined, it would add no complication to the solution of (36).

DISPLACEMENT VELOCITY AND STRESS DISTRIBUTION IN THE MANTLE

A solution to (36), subject to conditions (37), (39), (40), (41), and (42), is sought by using the method of linear operators. First, (37) and (39) are rendered homogeneous by the following transformation:

$$W_\phi\ (r,\theta,\phi,t) = X_\phi\ (r,\theta,\phi,t) + \frac{ab^3}{2(a-b)}\ [\frac{2}{a} - \frac{1}{r}]\ B(\theta,\phi,t) \tag{43}$$

where $B(\theta,\phi,t)$ is given by (38). $X_\phi\ (r,\theta,\phi,t)$ is then expanded in series of, first, Legendre polynomials $P_\ell^{\,1}(\theta)$, ℓ an integer > 0, and then of the eigenfunctions $u_\ell^{\,k}(r)$, written in Bessel functions of half order as follows:

$$u_\ell^{\,k}(r) = r^{-1/2}\ \{J_{(\ell+1/2)}(\gamma_\ell^{\,k} r)\ [(\ell - 1)\ J_{-(\ell+1/2)}(\gamma_\ell^{\,k} a)$$

$$+ \gamma_\ell^{\,k} a\ J_{-(\ell+3/2)}(\gamma_\ell^{\,k} a)]$$

$$- J_{-(\ell+1/2)}(\gamma_\ell^{\,k} r)[-(\ell + 2)\ J_{(\ell+1/2)}(\gamma_\ell^{\,k} a) + \gamma_\ell^{\,k} a\ J_{(\ell-1/2)}(\gamma_\ell^{\,k} a)]\}. \tag{44}$$

The eigenvalues $\gamma_\ell^{\,k}$ are obtained from the zeros of the eigenvalue equation

$$[-(\ell + 2)\ J_{(\ell+1/2)}(\gamma_\ell^{\,k} a) + \gamma_\ell^{\,k} a\ J_{(\ell-1/2)}(\gamma_\ell^{\,k} a)][(\ell - 1)\ J_{-(\ell+1/2)}(\gamma_\ell^{\,k} b)$$

$$+ \gamma_\ell^{\,k} b\ J_{-(\ell+3/2)}(\gamma_\ell^{\,k} b)] - [-(\ell + 2)\ J_{(\ell+1/2)}(\gamma_\ell^{\,k} b)$$

$$+ \gamma_\ell^{\,k} b\ J_{(\ell-1/2)}(\gamma_\ell^{\,k} b)][(\ell - 1)\ J_{-(\ell+1/2)}(\gamma_\ell^{\,k} a) + \gamma_\ell^{\,k} a\ J_{-(\ell+3/2)}(\gamma_\ell^{\,k} a)]$$

$$= 0. \tag{45}$$

The integer k ranges from zero to infinity for $\ell = 1$, but is strictly positive for $\ell > 1$.

The general solution of (36) is thus,

$$W_\phi(r,\theta,\phi,t) = \frac{ab^3}{2(a-b)} \{(\frac{2}{a} - \frac{1}{r}) B(\theta,\phi,t)$$

$$- \frac{15}{4} \frac{rb}{(a^5 - b^5)} P_1{}^1(\theta)[\frac{2\mu_f}{\rho} (1 - \frac{b}{a}) \int_0^t C(1,\phi,\tau)\, d\tau + \frac{b^2}{6} (\frac{a^3}{b^3} + 2 - \frac{3b}{a}) C(1,\phi,t)]$$

$$- \sum_{\ell=1}^{\infty} \frac{(2\ell+1)}{2} \frac{(\ell-1)!}{(\ell+1)!} P_\ell{}^1(\theta) \sum_{k=1}^{\infty} \frac{u_\ell{}^k(r)}{\langle u_\ell{}^k, u_\ell{}^k\rangle} \int_b^a (\frac{2r^2}{a} - r)[\frac{\mu_f}{\rho} (\frac{\ell(\ell+1)}{r^2}$$

$$- (\gamma_\ell{}^k)^2)\, C(\ell,\phi,t) * e^{-\frac{\mu_f}{\rho}(\gamma_\ell{}^k)^2 t} + C(\ell,\phi,t)]\, u_\ell{}^k(r)\, dr\} \qquad (46)$$

in which (*) denotes the convolution, $B(\theta,\phi,t)$ is given by (38), for any value of ℓ

$$C(\ell,\phi,t) = \int_0^\pi B(\theta,\phi,t) \sin\theta P_\ell{}^1(\theta)\, d\theta, \qquad (47)$$

and the inner product is

$$\langle u_\ell{}^k, u_\ell{}^k\rangle = \int_b^a r^2 [u_\ell{}^k(r)]^2\, dr. \qquad (48)$$

We have already pointed out that $\mu_f/(\rho r^2)$ was at least of the order of 250 sec.$^{-1}$, which will also be the minimum value for $\mu_f(\gamma_\ell{}^k)^2/\rho$, since $\gamma_\ell{}^k$ is of the order of α/r, where α is some constant greater than one, for any set (ℓ,k). This fact allows (46) to be greatly simplified. First, since $C(\ell,\phi,t)$ is related to the secular variations, over years, it can be considered constant for time of the order of a second, for which only, the exponential term is not negligible. The convolution term of (46) thus becomes

$$C(\ell,\phi,t) * e^{-\frac{\mu_f}{\rho}(\gamma_\ell{}^k)^2 t} = \frac{C(\ell,\phi,t)}{\frac{\mu_f}{\rho}(\gamma_\ell{}^k)^2} (1 - e^{-\frac{\mu_f}{\rho}(\gamma_\ell{}^k)^2 t}), \qquad (49)$$

and so, as t becomes much larger than a second, the final integral on the right-hand side of (46) becomes

$$\int_b^a (\frac{2r^2}{a} - r) \frac{\ell(\ell+1)}{(\gamma_\ell{}^k)^2 r^2} C(\ell,\phi,t)\, u_\ell{}^k(r)\, dr. \qquad (50)$$

The remaining term in (46) with μ_f/ρ as a coefficient is of the order of

$$\frac{\mu_f}{\rho} \frac{1}{b} \simeq 10^{20}\ \text{cm}^2/\text{sec.}\ \frac{1}{(4 \times 10^8)^3 \text{cm}^3} = \frac{10^{-4}}{64}\ \text{cm}^{-1}\ \text{sec.}^{-1}$$

Every other term in (46) is of the order of

$$\frac{1}{b}\ \text{year}^{-1} \simeq \frac{1}{4 \times 10^8}\ \text{cm}^{-1}\ \text{year}^{-1} \simeq \frac{10^{-15}}{12}\ \text{cm}^{-1}\ \text{sec.}^{-1}$$

Hence, only the term with μ_f/ρ as a coefficient will be retained in (46). Upon integrating and replacing $B(\theta,\phi,t)$ by its value from (38), the final expression for the velocity in the mantle is

$$W_\phi(r,\theta,\phi,t) = \frac{15}{8}\frac{\mu c}{\rho b}\frac{\sin\theta}{g_1^{\ 0}}\frac{r/b}{(a/b)^5 - 1}\sum_{n=1}^{\infty}(n^2 - 1)(\frac{a}{b})^{n-1}$$

$$\sum_{m=1}^{n}\frac{1}{m}[2\frac{(n-m)!}{(n+m)!}]^{1/2}\{[h_n^{\ m}(t) - h_n^{\ m}(0)]\cos m\phi - [g_n^{\ m}(t) - g_n^{\ m}(0)]\sin m\phi\}$$

$$\int_0^\pi \frac{\sin^3\theta}{\cos\theta} P_n^{\ m}(\theta)\ d\theta. \qquad (51)$$

The stress tensor is reduced to one component

$$t_{\phi\phi} = -\frac{2\mu_f}{r\sin\theta}\frac{\partial W_\phi}{\partial\phi}$$

and is a function of only ϕ and t. If $\partial W_\phi/\partial\phi$ is determined from (51), then the stress can be expressed as

$$t_{\phi\phi}(\phi,t) = \frac{15}{4g_1^{\ 0}}\frac{\mu c}{b^2}\frac{\mu_f}{\rho}\frac{1}{(a/b)^5 - 1}\sum_{n=1}^{\infty}(n^2 - 1)(\frac{a}{b})^{n-1}$$

$$\sum_{m=1}^{n}[2\frac{(n-m)!}{(n+m)!}]^{1/2}\{[g_n^{\ m}(t) - g_n^{\ m}(0)]\cos m\phi$$

$$+ [h_n^{\ m}(t) - h_n^{\ m}(0)]\sin m\phi\}\int_0^\pi \frac{\sin^3\theta}{\cos\theta} P_n^{\ m}(\theta)\ d\theta. \qquad (52)$$

RESULTS AND DISCUSSION

The numerical results to be discussed here were obtained from the derived equations for the velocities and the stress assuming that the mean kinematic viscosity and density of the mantle were

$$\frac{\mu_f}{\rho} = 10^{24}\ \text{cm}^2\ \text{sec.}^{-1},$$

and

$$\rho = 5.0\ \text{g.cm.}^{-3},$$

and that the mean kinematic viscosity and density of the outer-core were

$$\frac{\mu c}{\rho c} = \frac{1}{6} \times 10^3 \text{ cm}^2 \text{ sec.}^{-1},$$

and $$\rho c = 10.0 \text{ g.cm.}^{-3}.$$

As a starting point of the theory, we derived a velocity distribution V_ϕ given by (23) for the outer-core, near the core-mantle interface. The calculated magnitude of V_ϕ as a function of the longitude ϕ and the colatitude θ is presented in Figure 3. Note that the velocity attains a value of several kilometers per year and experiences two sign reversals, indicating that at least two cells exist in the flow pattern. These facts agree with the hypotheses of the dynamo theory.

The Calculated Velocity in the Mantle and the Continental Drift

Equation (51) shows that the mantle rotates about the Earth's axis with an angular velocity which is a function of the longitude and time only. Figure 4 gives W_ϕ at the crust-mantle boundary for different years and as a function of ϕ. The initial year, corresponding to time t = 0 in (51), was taken as far back as possible, i.e. 1885, so that the effect of the unknown initial velocity in the mantle would not be too important to the final results for the years 1945, 1955, and 1958. The magnitude of W_ϕ is of the order of a few centimeters per year, which agrees with results obtained by other investigators [7]. The fact that the velocity goes to zero at ϕ = 95°W and ϕ = 115°E requires, by continuity considerations, that the flow turn either radially inward (outward), or north (south) at these longitudes. The numerous fracture zones, located along the western coast of the American continent, which are oriented in an east-west direction, lead us to believe that the flow in the mantle at longitude 95°W must turn north in the northern hemisphere. A possible flow pattern for the cells in the mantle is shown in Figure 5. Note that four cells are pictured and symmetry with respect to the equator is assumed, although this need not be the case. These cells need have no radial component.

If we now assume that the mantle, by friction, slowly deforms the heterogeneous crust, it is possible to compare the above results with the continental drift and some of the seismic features of the Earth's surface. The results of Figure 4, valid for the last 25 years, and drawn on a world map in Figure 6, show that the actual motion of the continents agrees with the pattern often given for the continental drift [1]. It appears clearly that the continents are moving from longitude 115°E towards longitude 95°W with velocities that are small near 115°E, become large in the vicinity of 0° and 180°, and finally diminishes once more near 90°W. This would explain why there is a relative motion between the American continents and Europe and Africa, and why such features, with north-south orientation, as the Red Sea, the North Sea, and the Atlantic Ocean have formed. The seemingly independent nature of the Pacific Ocean is also explained, since it is bracketed by two longitudes of zero velocity.

Near the meridians 115°E and 95°W, where the velocity must begin to turn 90° north or south, the heterogeneous crust is subject to huge strains and stresses. At these meridians is where we find the main shallow earthquakes as shown in Figure 7. Also, at these meridians are located the major seismic faults, all directed north-south, perpendicular to the drift. For example, the San Andreas, the Atacama, the Philippine, and the Alpine faults are shown in Figure 8 [8].

The Calculated Stresses in the Mantle and Major Seismic Shocks

Equation (52) shows that stresses in the mantle are functions of the longitude and time only. The magnitude of these stresses, shown in Figure 9 near the crust-mantle boundary, are of the order of 10^9 g./cm./sec.^{2}, large enough to create earthquakes by compression or dilatation of the mantle. In Figure 7, on a world map showing the position of the main shallow and deep focus earthquakes, which have occurred during the last 50 years [9], two zones are drawn. Within these two zones, the stresses are greater than 60% of the maximum stress. These zones are centered on the meridians 115°E and 95°W; and, as can be seen in Figure 7, these zones appear to coincide with the locations of a great majority of both shallow earthquakes and deep seismic shocks.

CONCLUSION

The theory advanced here, that the mantle is set in motion by stresses applied at the core-mantle interface, does not explain all the main features of the mantle or the Earth's crust. This is probably a consequence of the many restrictive assumptions, such as considering the motion to be longitudinal only, taking the material properties to be constant, neglecting heat transfer, etc.; which were necessary to achieve a solution to the equations of motion. However, because of the close agreement between the theoretically predicted continental drift and the measured drift, and the coincidence of the theoretical zones of maximum stress with the location of the major earthquake zones, we feel that this theory offers a reasonable first approximation to the pattern of motion within the mantle.

Furthermore, an analytical expression is given by this theory which directly relates the continental drift and seismic shocks to the secular variation of the geomagnetic field.

NOMENCLATURE

$A(r,\theta,\phi,t)$	Geomagnetic potential of internal origin.
$A_n^m(t)$	Dimensionless coefficient.
a	Radius of the Earth.
$B(\theta,\phi,t)$	Velocity gradient, $\frac{\partial}{\partial r}\left(\frac{W_\phi}{r}\right)_{r=b}$
b	Inner radius of the mantle.
$C(\ell,\phi,t)$	Transform of $B(\theta,\phi,t)$
$C_n^m(t)$	Dimensionless coefficient.
$E_n^m(t)$	Dimensionless coefficient.
$F_n^m(t)$	Dimensionless coefficient.
$G_n^m(t)$	Dimensionless coefficient.
g	Acceleration of gravity.
$g_n^m(t)$	Gauss coefficient (tabulated in references [10] and [11]).
$\tilde{H}$	Magnetic field.
H_r, H_θ, H_ϕ	The r,θ,ϕ components of $\tilde{H}$, respectively.
$H_n^m(\theta)$	Arbitrary function of θ.
$h_n^m(t)$	Gauss coefficient (tabulated in references [10] and [11]).
$\tilde{i}, \tilde{j}, \tilde{k}$	The unit vectors in the r,θ,ϕ directions, respectively.
J_V	Bessel function of fractional order V.
k, l, m, n	Integers
$p(r)$	Pressure distribution in the Earth.
$P_n^m(\theta)$	Legendre polynomial of order n,m.
$P_{nm}(\theta)$	Schmidt polynomial of order n,m.
r,θ,ϕ	Geographical coordinates.
t	Time
$t_{\phi\phi}$	Stress component $\phi\phi$ in the mantle.
$u_\ell^k(r)$	Eigenfunction.
$\tilde{V}$	Velocity vector of the fluid in the core.
V_r, V_θ, V_ϕ	Components of V in the r,θ,ϕ directions, respectively.
$\tilde{W}$	Velocity vector of the fluid in the mantle.
W_r, W_θ, W_ϕ	Components of W in the r,θ,ϕ directions, respectively.
$X_\phi(r,\theta,\phi,t)$	Variable of substitution used to render boundary conditions of W_ϕ homogeneous.

μ Permeability of the outer core.
μ_c Dynamic viscosity of the outer core.
μ_f Dynamic viscosity of the mantle.
ρ Density of the mantle.
ρ Density of the outer core.
σ Conductivity of the outer core.
$\underset{\sim}{\Omega}$ Body forces, including centrifugal force.
R,Θ,Φ Components of $\underset{\sim}{\Omega}$ in the r,θ,ϕ directions, respectively.
$\underset{\sim}{\omega}$ Angular velocity of the Earth.

REFERENCES

[1] Symposium on Continental Drift, Phil. Trans. Roy. Soc. London, A, 258, (1965).
[2] H. Jeffreys: The Earth, 4th ed., Cambridge University Press, (1959).
[3] S. K. Runcorn: Towards a Theory of Continental Drift, Nature, London, 193, 311, (1962).
[4] E. C. Bullard: Symposium on Continental Drift, Concluding Remarks, Phil. Trans. Roy. Soc. London, A, 258, 322, (1965).
[5] W. M. Elsasser: The Earth's Interior and Geomagnetism, Rev. Mod. Phys., 22, 1, (1950).
[6] E. C. Bullard and H. Gellman: Homogeneous Dynamos and Terrestrial Magnetism, Phil. Trans. Roy. Soc. London, A, 247, 213, (1954).
[7] E. Orowan: Convection in a Non-Newtonian Mantle, Continental Drift and Mountain Building, Phil. Trans. Roy. Soc. London, A, 258, 284, (1965).
[8] C. R. Allen: Transcurrent Faults in Continental Areas, Phil. Trans. Roy. Soc. London, A, 258, 82, (1965).
[9] B. Gutenberg and C. F. Richter: Seismicity of the Earth and Associated Phenomena, Hafner Publishing Company, New York and London, 15, 98, (1965).
[10] E. H. Vestine: On Variations of the Geomagnetic Field, Fluid Motions, and the Rate of the Earth's Rotation, J. Geophys. Res. 58, 127, (1953).
[11] T. Nagata: Benedum Earth Magnetism Symposium, University of Pittsburgh Press, 41, (1962).

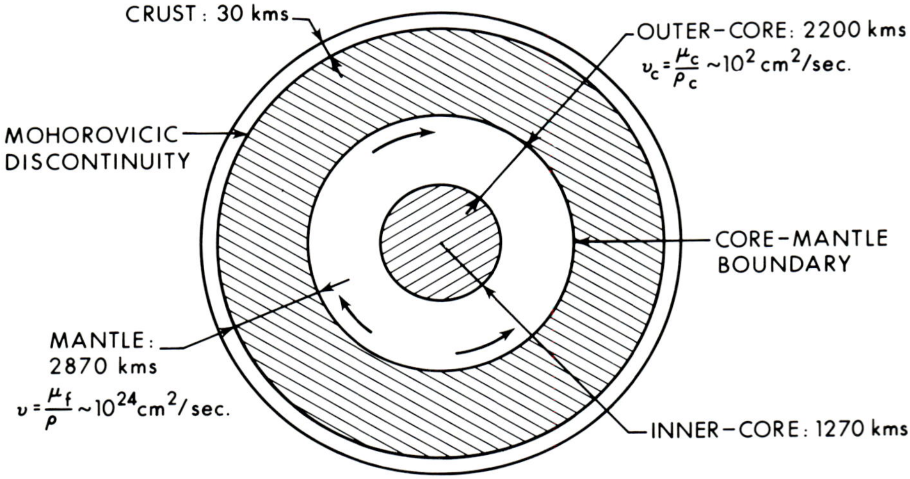

Fig. 1 Main Divisions of the Earth's Interior

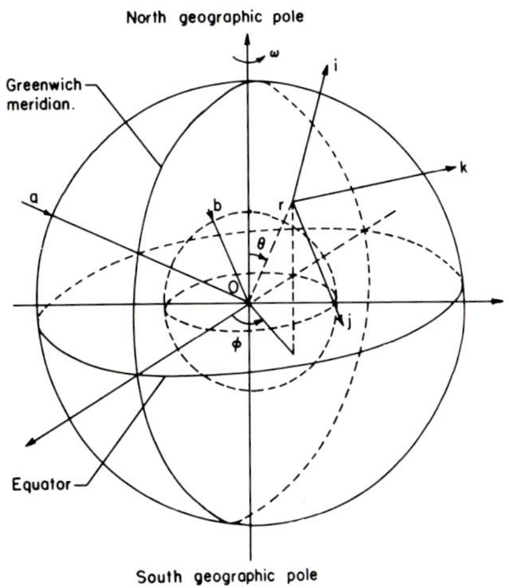

Fig. 2 Model for the Earth's Mantle in Geographical Coordinates

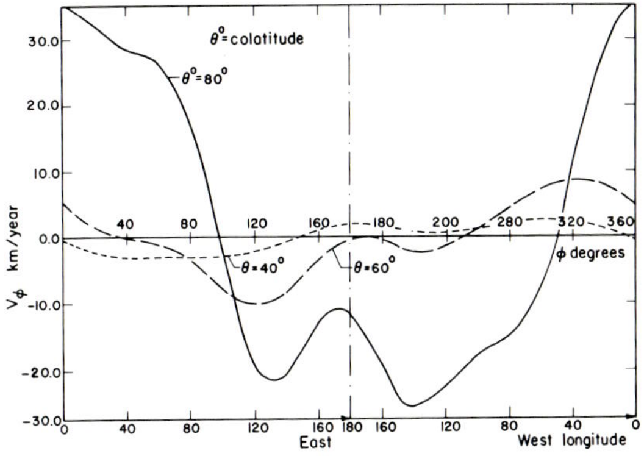

Fig. 3 Velocity Distribution at the Core-Mantle Boundary

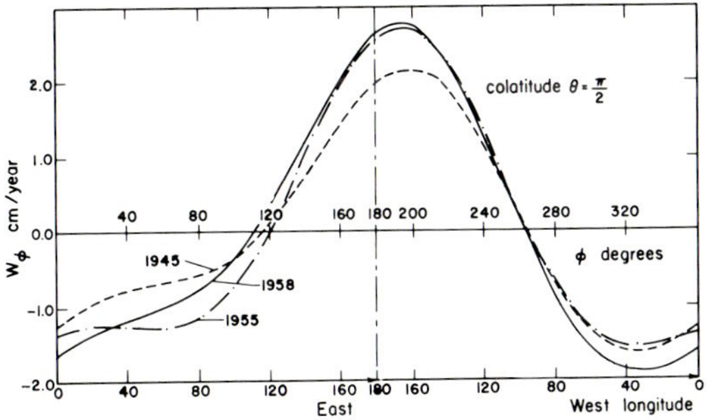

Fig. 4 Velocity Distribution of the Crust-Mantle Boundary

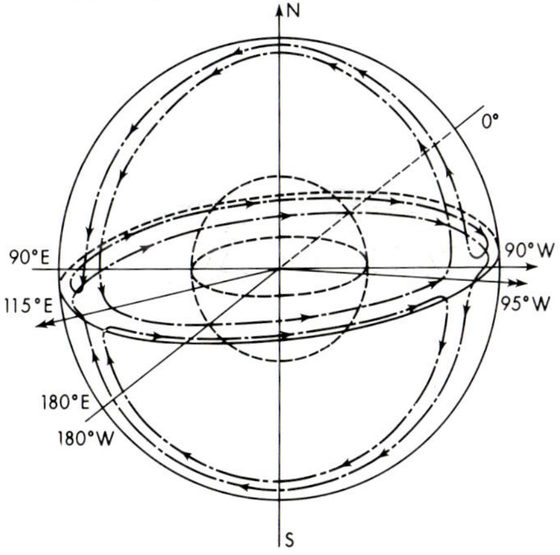

Fig. 5 Possible Flow Pattern in the Mantle

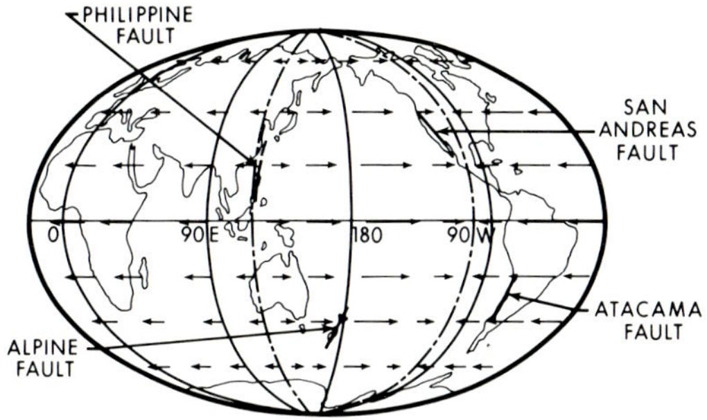

Fig. 6 Velocity Distribution at the Earth's Surface, Showing the Continental Drift

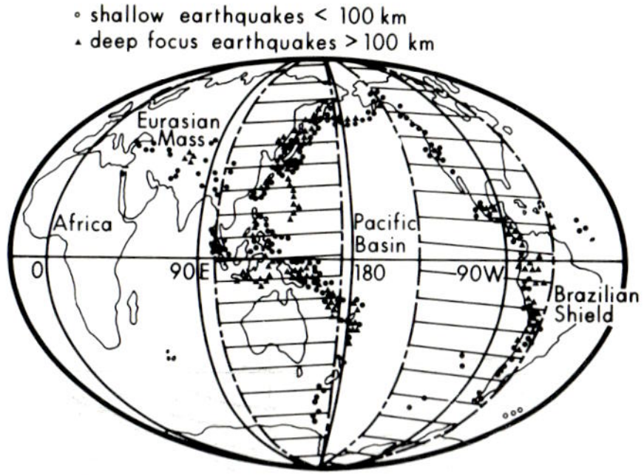

Fig. 7 World Map of Earthquakes and Zones of Maximum Stresses

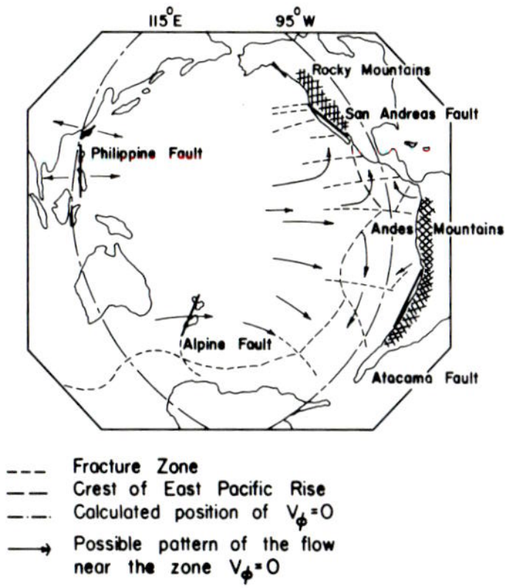

Fig. 8 Major Transcurrent Faults of the Circum Pacific Rim

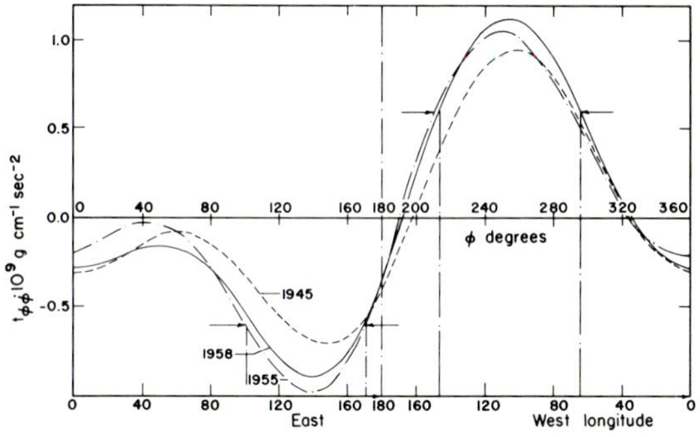

Fig. 9 Stress Distribution in the Mantle

Developments in Mechanics, Vol. 6. Proceedings of the 12th Midwestern Mechanics Conference: 45

Non-Hertzian Contact of an Elastic Sphere Indenting an Elastic Spherical Cavity

M. A. HUSSAIN and S. L. PU
BENET RESEARCH & ENGINEERING LABORATORIES
U. S. ARMY WATERVLIET ARSENAL

ABSTRACT

The indentation of a smooth elastic sphere upon the infinite medium exterior to the sphere is considered. In the absence of a bond between the spherical indentor and the spherical cavity there appear a region of separation and a region of contact. The region where the sphere remains in contact with the external body is a spherical cap. Our primary objective is to determine the colatitude η of the boundary circle of this spherical cap in contact.

By means of Boussinesq potentials of the linear three-dimensional theory of elasticity the mixed boundary value problem is reduced to dual series involving spherical harmonics. The dual series are replaced by an equivalent Fredholm integral equation with unknown η at the upper limit of integration. This integral equation is solved by a variational technique which leads to a transcendental equation for the approximate determination of η.

INTRODUCTION

In this paper we consider the problem of an infinite medium indented by a smooth spherical inclusion to which a concentrated force has been applied. In the absence of a bond at the interface the region of contact between the sphere and the outer body appears to be a spherical cap, Fig.1. The primary objective of this paper is to determine the region of contact.

The problem is essentially non-Hertzian in the sense that the area of contact cannot be considered small. In the classical Hertz problem of two spherical bodies in contact there are two basic results. Firstly, the radius of the circular region of contact is proportional to $P^{1/3}$ [1] where P is the load applied between two bodies in contact. Secondly, the relative approach due to the application of P, i.e., the relative normal displacement of any two points, one in each of the two bodies, which are far away from the contact surface, is proportional to $P^{2/3}$. We propose to investigate these two statements for the present non-Hertzian problem and to compare the results.

The spherical indentation problem is a mixed boundary value problem in which the region of contact is not known a priori. The displacements over the contact region, which is not small as mentioned before, cannot be determined solely from the geometric considerations as done in the Hertz problem or in Reference [2].

We use Boussinesq potentials in spherical coordinates. The mixed conditions lead to dual series involving Legendre functions which are further reduced to a Fredholm integral equation of the first kind for the contact stress. In the integral equation not only the contact stress but also the range of integration (corresponding to the region of contact) is unknown. An approximate solution is obtained by applying a variational method. The reader is referred to [3] for details of the method and its applications.

An approximate solution of the problem for an elastic sphere indenting an elastic cavity is given in [2]. However, the solution is not applicable to the case in which the radii of the sphere and the cavity are equal or nearly equal by virtue of the geometric considerations used in obtaining the displacements, Equation (4) of [2].

For the analogous two-dimensional indentation problem, an approximate solution is given by Sheremetev [4] while the exact solution is obtained in [5].

STATEMENT OF THE PROBLEM

Consider a smooth sphere inside an infinite, homogeneous and isotropic body. The elastic constants of the sphere are G' and ν', those of the external medium are G and ν. We assume that there no bonds and no friction exist between the sphere and the outer body. The sphere is perfectly fitted in the spherical cavity in a state free from stresses and strains. The region of contact becomes a spherical cap as soon as a concentrated force F_z is applied at the center of the sphere in the direction of the positive z-axis, Fig.1. With reference to the spherical coordinates (r, θ, γ), the problem is independent of γ and the boundary surface of the sphere is $r = a$. Our objective is to determine the value of θ, say η, for the point A and the relation between the coefficient $(\delta/2G')$, defined later, and the force F_z. The approximate normal stress distribution in the contact surface will also be obtained.

Using the usual notation, the boundary conditions at the interface $r = a$ are as follows:

$$\tau_{r\theta}(a,\theta) = \tau'_{r\theta}(a,\theta) = 0, \quad 0 \leq \theta \leq \pi \tag{1}$$

$$\sigma_r(a,\theta) = \sigma'_r(a,\theta) = 0, \quad \eta \leq \theta \leq \pi \tag{2}$$

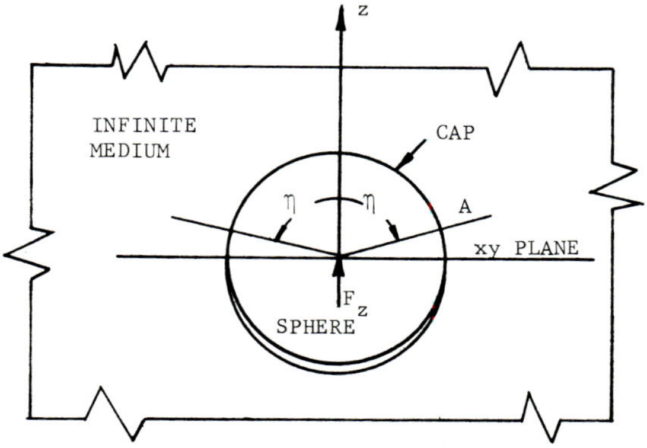

Fig.1 A Sphere Indenting a Spherical Cavity

$$u_r(a,\theta) = u_r'(a,\theta), \quad 0 \le \theta \le \eta \tag{3}$$

$$\sigma_r(a,\theta) = \sigma_r'(a,\theta) \le 0, \quad 0 \le \theta \le \eta \tag{4}$$

Unprimed and primed quantities refer to $r \ge a$ and $r \le a$, respectively.

REDUCTION TO DUAL SERIES

The general axisymmetric solutions of the equations of elasticity in spherical coordinates are given in terms of Boussinesq potentials in [6]. Using matrix notation, we define

$$[S'] = \begin{bmatrix} 2G'u_r' \\ \sigma_r' \\ \tau_{r\theta}' \end{bmatrix}, \qquad [S] = \begin{bmatrix} 2Gu_r \\ \sigma_r \\ \tau_{r\theta} \end{bmatrix}$$

They can be expressed by

$$[S'] = \delta[A_{-2}] - \frac{F_z}{8\pi(1-\nu')}[B_o'] + \sum_{2,3,\ldots}^{\infty} g_n[A_{-n-1}] + \sum_{0,1,\ldots}^{\infty} h_n[B'_{-n-2}]$$

$$[S] = \frac{-F_z}{8\pi(1-\nu)}[B_o] + \sum_{0,1,\ldots}^{\infty} e_n[A_n] + \sum_{2,3,\ldots}^{\infty} f_n[B_{n-1}] \tag{5}$$

where column matrices $[A_n]$, $[B_n]$, $n = \pm$ integer, are given by

$$[A_n] = \begin{bmatrix} -(n+1)r^{-(n+2)} P_n(\cos\theta) \\ (n+1)(n+2)r^{-(n+3)} P_n(\cos\theta) \\ (n+2)r^{-(n+3)}\sin\theta\, P_n'(\cos\theta) \end{bmatrix} \tag{6a}$$

$$[B_n] = \begin{bmatrix} -(n+1)(n+4-4\nu)r^{-(n+1)} P_n(\cos\theta) \\ (n+1)[(n+1)(n+4) - 2\nu]r^{-(n+2)} P_{n+1}(\cos\theta) \\ (n^2+2n-1+2\nu)r^{-(n+2)}\sin\theta\, P_{n+1}'(\cos\theta) \end{bmatrix} \tag{6b}$$

Here P_n denotes the Legendre polynomial of degree n and $P_n'(z) = dP_n(z)/dz$. Solutions $[B_n']$ are obtained from $[B_n]$ by substituting ν' for ν. We note that solution $[A_{-2}]$ corresponds to the rigid body displacements and $(\delta/2G')$ is the relative rigid displacement of the center of the sphere with respect to the center of the cavity in the positive z-axis. Solution $[B_o]$ represents a concentrated force of magnitude $8\pi(1-\nu)$ acting at $r = o$ in an infinite body in the direction of the negative z-axis. Similarly, $[B_o']$ is the solution for an infinite body of Poisson's ratio ν'. Superposition coefficients e_n, f_n, g_n, h_n, and δ are to be determined from (1) through (4).

Substituting (6) into (5), the boundary conditions (1) yield

$$
\left.\begin{aligned}
h_1 &= -\frac{(1-2\nu')}{16\,\pi(1-\nu'^2)a^3}\,F_z \\
e_1 &= -\frac{(1-2\nu)a^2}{24\,\pi(1-\nu)}\,F_z \\
g_n &= \frac{(n^2+2n-1+2\nu')}{n-1}\,a^2 h_n\,, \qquad n \geq 2 \\
f_n &= \frac{-(n+2)}{(n^2-2+2\nu)a^2}\,e_n\,, \qquad n \geq 2
\end{aligned}\right\} \tag{7}
$$

Using (7), the radial components of the stress and displacement for [S′] and [S] at the boundary $r = a$ become

$$
\sigma'_r(a,\theta) = 2(1+\nu')h_o - \frac{3F_z}{4\pi a^2}\cos\theta + \sum_{2,3,\ldots}^{\infty} 2(n^2+n+1+2n\nu'+\nu')a^n h_n P_n \tag{8}
$$

$$
2G'u'_r(a,\theta) = \delta\cos\theta + 2ah_o(1-2\nu') + \frac{3(1+2\nu'-4\nu'^2)F_z}{8\pi a(1-\nu'^2)}\cos\theta + \sum_{2,3,\ldots}^{\infty}\frac{2(2n^2-2n^2\nu'+n\nu'-1+2\nu')}{n-1}\,a^{n+1}h_n P_n \tag{9}
$$

$$
\sigma_r(a,\theta) = \frac{2e_o}{a^3} - \frac{3F_z}{4\pi a^2}\cos\theta - \sum_{2,3,\ldots}^{\infty}\frac{2(n+2)(n^2+n+1-2n\nu-\nu)}{(n^2-2+2\nu)a^{n+3}}\,e_n P_n \tag{10}
$$

$$
2Gu_r(a,\theta) = -\frac{e_o}{a^2} + \frac{(7-8\nu)F_z}{12\pi(1-\nu)a}\cos\theta + \sum_{2,3,\ldots}^{\infty}\frac{2(2n^2-2n^2\nu+4n-5n\nu+1-\nu)}{(n^2-2+2\nu)a^{n+2}}\,e_n P_n \tag{11}
$$

To meet the boundary conditions given by (2), we first define new coefficients F_n as follows

$$
\left.\begin{aligned}
&\frac{2e_o}{a^3} = 2(1+\nu')h_o = F_o\,, \qquad -\frac{F_z}{4\pi a^2} = F_1 \\
&2(n^2+n+1+2n\nu'+\nu')a^n h_n = \frac{-2(n+2)(n^2+n+1-2n\nu-\nu)}{(n^2-2+2\nu)a^{n+3}}\,e_n \\
&\qquad = (2n+1)F_n \qquad \text{for } n = 2, 3, 4 \ldots
\end{aligned}\right\} \tag{12}
$$

so that both (8) and (10) are reduced to

$$
\sigma_r(a,\theta) = \sigma'_r(a,\theta) = \sum_{0,1,2,\ldots}^{\infty}(2n+1)F_n P_n \tag{13}
$$

Equations (9) and (11) become, with the relations of (12),

$$\frac{2G'u_r'(a,\theta)}{a} = \frac{\delta}{a}\cos\theta + \frac{(1-2\nu')}{(1+\nu')}F_o -$$

$$- \frac{3(1+2\nu'-4\nu'^2)}{2(1-\nu'^2)}F_1\cos\theta + \sum_{2,3,\ldots}^{\infty} \gamma_n F_n P_n \tag{14}$$

$$\frac{2Gu_r(a,\theta)}{a} = -\frac{F_o}{2} - \frac{7-8\nu}{3(1-\nu)}F_1\cos\theta + \sum_{2,3,\ldots}^{\infty} \beta_n F_n P_n \tag{15}$$

where γ_n and β_n are abbreviations denoting

$$\gamma_n = \frac{(2n+1)(2n^2 - 2n^2\nu' + n\nu' - 1 + 2\nu')}{(n-1)(n^2 + n + 1 + 2n\nu' + \nu')} \tag{16}$$

$$\beta_n = -\frac{(2n+1)(2n^2 - 2n^2\nu + 4n - 5n\nu + 1 - \nu)}{(n+2)(n^2 + n + 1 - 2n\nu - \nu)} \tag{17}$$

Applying boundary conditions (2) and (3), we obtain the dual series

$$\sum_{0,1,2,\ldots}^{\infty} (2n+1)F_n P_n(\cos\theta) = 0, \qquad \eta \le \theta \le \pi \tag{18}$$

$$\sum_{0,1,2,\ldots}^{\infty} \psi_n F_n P_n(\cos\theta) = -\frac{\alpha\delta\cos\theta}{a}, \qquad 0 \le \theta \le \eta \tag{19}$$

where

$$\alpha = \frac{G}{G'} \tag{20}$$

$$\left.\begin{aligned} \psi_o &= \frac{1}{2} + \frac{\alpha(1-2\nu')}{1+\nu'} \\ \psi_1 &= \frac{7-8\nu}{3(1-\nu)} - \frac{3\alpha(1+2\nu'-4\nu'^2)}{2(1-\nu'^2)} \\ \psi_n &= -\beta_n + \alpha\gamma_n, \qquad n = 2, 3, \ldots \end{aligned}\right\} \tag{21}$$

INTEGRAL EQUATION AND AN APPROXIMATE SOLUTION

Let $\sigma(\theta)$ be the unknown radial stress $\sigma_r(a,\theta)$ at the interface $r = a$. It is obvious that $\sigma(\theta)$ is a continuous function over the entire region $0 \le \theta \le \pi$ and is even in θ. Using Fourier-Legendre analysis, $\sigma(\theta)$ is represented by

$$\sigma(\theta) = \sum_{0,1,2,\ldots}^{\infty} (2n+1)F_n P_n(\cos\theta), \qquad 0 \le \theta \le \pi \tag{22}$$

with $\sigma(\theta)$ vanishing over the region $\eta \le \theta\ \pi$. The coefficients F_n are then given by

$$F_n = \frac{1}{2} \int_0^{\eta} \sigma(t) P_n(\cos t) \sin t \, dt \tag{23}$$

Putting (23) into (19) and interchanging the order of integration and summation, the set of two infinite series (18) and (19) is reduced to a single integral equation for σ

$$\int_0^{\eta} \sigma(t) K(\theta,t) \sin t \, dt = R(\theta), \qquad 0 \le \theta \le \eta \tag{24}$$

where η is unknown, the kernel $K(\theta,t)$ and the right-hand member $R(\theta)$ are known functions given by

$$K(\theta,t) = \sum_{0,1,2,\ldots}^{\infty} \psi_n P_n(\cos\theta) P_n(\cos t) \tag{25}$$

$$R(\theta) = -\frac{2\alpha\delta}{a} \cos\theta \tag{26}$$

Since no closed form solution could be obtained for the Fredholm integral equation (24), we followed an approximate method [3] which has been established in particular for such types of integral equations. To apply the variational method, the kernel has to be symmetric, i.e., $K(t,\theta) = K(\theta,t)$ and the kernel may have at most a logarithmic singularity at $\theta = t$. Equation (25) shows that the kernel is symmetric. The principal part of the kernel can be shown to have a logarithmic singularity at $\theta = t$ by employing the technique introduced in [7] in which a perturbation scheme was established. Hence the variational method is applicable to our problem.

According to the method we must choose a trial function for $\sigma(\theta)$ and construct the functional

$$I(\sigma) = \int_0^{\eta} \sigma(t) \sin t \left\{ \int_0^{\eta} \sigma(\theta) \sin\theta K(\theta,t)\, d\theta - 2R(t) \right\} dt \tag{27}$$

If the function $\sigma(\theta)$ is chosen in the form

$$\sigma(\theta) = B(\cos\theta - \cos\eta)^{-\frac{1}{2}} + A(\cos\theta - \cos\eta)^{\frac{1}{2}}, \quad 0 \le \theta \le \eta \tag{28}$$

where A and B are unknown coefficients their values depend upon the value of η. Substituting (28) into (27) we obtain

$$I = A^2 I_1 + 2ABI_2 + B^2 I_3 - 2AI_4 - 2BI_5 \tag{29}$$

where

$$\begin{matrix} I_1 \\ I_2 \end{matrix} = \int_0^{\eta}\int_0^{\eta} K(\theta,t) \sin\theta \sin t (\cos\theta - \cos\eta)^{\frac{1}{2}} (\cos t - \cos\eta)^{\pm\frac{1}{2}} \, d\theta dt \tag{30}$$

$$\begin{matrix} I_4 \\ I_5 \end{matrix} = \int_0^{\eta} R(t) \sin t (\cos t - \cos\eta)^{\pm\frac{1}{2}} \, dt \tag{31}$$

the positive signs go with I_1, I_4 while the negative signs go with I_2 and I_5. Based on the variational principle the functional is stationary when the trial function varies around the exact σ and the approximate value of η varies around the exact η, provided $B(\eta) = 0$. Hence the values of A and B are formally determined by

$$\frac{\partial I}{\partial A} = \frac{\partial I}{\partial B} = 0 \tag{32}$$

Equations (32) give us

$$A = \frac{I_4 I_3 - I_2 I_5}{I_1 I_3 - I_2^2}, \quad B = \frac{I_1 I_5 - I_2 I_4}{I_1 I_3 - I_2^2} \tag{33}$$

From the boundary conditions (4) there must exist a value of η such that $B(\eta) = 0$ and then the singular term in (28) is dropped out. This gives the equation for the determination of η

$$I_1 I_5 - I_2 I_4 = 0 \tag{34}$$

The expression for A is reduced from (33) to

$$A = I_4 / I_1 \tag{35}$$

which becomes known after η is found from (34). The approximate contact radial stress $\sigma_r(a,\theta)$ is given by (28) with $B = 0$ and A obtained in (35).

Now following the steps outlined in the preceding paragraph, we are required to carry out the integrals (30) and (31). With the results, which are derived in the appendix of this paper,

$$\begin{aligned} I_n^1(\cos\eta) &= \int_0^\eta (\cos\theta - \cos\eta)^{\frac{1}{2}} \sin\theta \, P_n(\cos\theta)\, d\theta \\ &= \frac{\sqrt{2}}{2n+1}\left\{\frac{\sin\left(n - \frac{1}{2}\right)\eta}{2n-1} - \frac{\sin\left(n + \frac{3}{2}\right)\eta}{2n+3}\right\} \end{aligned} \tag{36}$$

and

$$I_n^{-1}(\cos\eta) = 2\sqrt{2}\,\frac{\sin\left(n + \frac{1}{2}\right)\eta}{2n+1} \tag{37}$$

integrals I_1, I_2, I_4, and I_5 become, with the necessary change of order of integration and summation,

$$\begin{aligned} I_1 &= \sum_{0,1,\ldots}^{\infty} \psi_n [I_n^1(\cos\eta)]^2 \\ I_2 &= \sum_{0,1,\ldots}^{\infty} \psi_n [I_n^{-1}(\cos\eta)][I_n^1(\cos\eta)] \\ I_4 &= -\frac{2\alpha\delta}{a} I_1^1(\cos\eta) \\ I_5 &= -\frac{2\alpha\delta}{a} I_1^{-1}(\cos\eta) \end{aligned} \tag{38}$$

where I_1^1 and I_1^{-1} are given respectively by (36) and (37) with $n = 1$. Substituting (36) and (37) into (38) then (38) into (34) the equation for η becomes

$$\sum_{0,1,\ldots}^{\infty} \frac{\psi_n}{(2n+1)^2} \left(\frac{\sin\left(n-\frac{1}{2}\right)\eta}{2n-1} - \frac{\sin\left(n+\frac{3}{2}\right)\eta}{2n+3} \right) \cdot$$

$$\cdot \left[\sin\left(n+\frac{1}{2}\right)\eta \left(\sin\frac{\eta}{2} - \frac{1}{5}\sin\frac{5\eta}{2} \right) - \sin\frac{3}{2}\eta \left(\frac{\sin\left(n-\frac{1}{2}\right)\eta}{2n-1} - \frac{\sin\left(n+\frac{3}{2}\right)\eta}{2n+3} \right) \right] = 0 \tag{39}$$

This equation is of course independent of δ.

RELATION BETWEEN THE CONCENTRATED FORCE F_z AND THE COEFFICIENT $\delta/(2G')$

The ratio $\delta/(2G')$ is the coefficient of rigid body motion or "relative approach" of the inclusion. In the limit, when the inclusion becomes rigid, the force applied to the outer body is due to a rigid body motion of the inclusion only, i.e., $\lim_{G'\to\infty} \delta/(2G')$ is finite. In general there is a definite relation between the force F_z and this rigid body motion of the inclusion. This relation can be obtained as follows.

From the fact that the applied force F_z must be balanced by the net traction on any spherical surface of arbitrary radius centered at the point of application of F_z, we have

$$F_z = -2\pi \int_0^{\pi} (\sigma_r \cos\theta - \tau_{r\theta}\sin\theta) r^2 \sin\theta \, d\theta, \qquad r \geq a \tag{40}$$

and a similar relation for $r \leq a$. At the interface $r = a$, the shearing stress vanishes and the above equation is reduced to

$$F_z = -2\pi a^2 \int_0^{\eta} \sigma(\theta) \cos\theta \sin\theta \, d\theta \tag{41}$$

From equation (28) with $B = 0$, the approximate contact stress is expressed by

$$\sigma(\theta) = A(\cos\theta - \cos\eta)^{\frac{1}{2}}, \qquad 0 \leq \theta \leq \eta \tag{42}$$

where A is, from (35) and (38)

$$A = -\frac{\sqrt{2}\alpha\delta}{3a} \frac{\sin\frac{\eta}{2} - \frac{1}{5}\sin\frac{5}{2}\eta}{\sum_{0,1,\ldots}^{\infty} \psi_n (2n+1)^{-2} \left\{ \frac{\sin\left(n-\frac{1}{2}\right)\eta}{2n-1} - \frac{\sin\left(n+\frac{3}{2}\right)\eta}{2n+3} \right\}^2} \tag{43}$$

Substituting (43) and (42) into (41) and carrying out the integrals with the help of the formulae derived in the appendix, we obtain the following relation

$$\frac{F_z}{Ga(\delta/G')} = \frac{4\pi}{9} \frac{\left(\sin\frac{\eta}{2} - \frac{1}{5}\sin\frac{5}{2}\eta\right)^2}{\sum_{0,1,\ldots}^{\infty} \psi_n (2n+1)^{-2} \left\{ \frac{\sin\left(n-\frac{1}{2}\right)\eta}{2n-1} - \frac{\sin\left(n+\frac{3}{2}\right)\eta}{2n+3} \right\}^2} \tag{44}$$

NUMERICAL RESULTS AND CONCLUSIONS

The quantities of interest in this problem are the angle η of the contact cap, the contact stress σ_r at the interface r = a and the relation between the force F_z and the coefficient $(\delta/2G')$. Among them η is the most fundamental. It depends only upon the material constants but not upon the size of the sphere nor upon the magnitude of the force or the relative approach $(\delta/2G')$. Hence we concentrated our efforts in the computation of η for various values of α, ν, and ν'. To limit the use of computer time it is further assumed that $\nu = \nu'$. The single real nontrivial root of equation (39) has been obtained for $\nu = \nu' = 0$, 0.1, ..., 0.5 and for α from 0 to 10. The results are shown in (a) and (b) of Figure 2. In Figure 2(a) the angle η is plotted versus the ratio α for $\nu = \nu' = 0$, 0.1, 0.2, and 0.3 while Figure 2(b) gives the similar graph for $\nu = \nu' = 0.3$ and 0.5 in different scales. For large values of α, i.e., the indenting sphere is rather soft in comparison with the material being indented, the variation in η depends greatly upon the Poisson ratios. The angle η increases from $\eta \approx 80^\circ$ for $\nu = \nu' = 0$ to $\eta \approx 104^\circ$ for $\nu = \nu' = 0.5$ for $\alpha = 10$. The dependence on ν becomes weak for small values of α. In the case of a rigid sphere, i.e., $\alpha = 0$, the graph of η versus $\nu(= \nu')$ is shown in Figure 3. The angle decreases from $\eta \approx 82.2^\circ$ for $\nu = \nu' = 0$ to $\eta \approx 77.5^\circ$ for $\nu = \nu' = 0.5$. Note that η is in general less than 90°. It becomes greater than 90° only when the sphere is softer $(\alpha > 1)$ and the materials are nearly incompressible.

Contrary to the classical Hertz results, it may be concluded: i) that the region of contact is independent of the force F_z applied and ii) that the relation between the relative approach and F_z is linear. This is in agreement with the results of the analogous two-dimensional problem given in [5].

REFERENCES

[1] Timoshenko, S. and J.N. Goodier: Theory of Elasticity. 2nd Edition, McGraw-Hill, p.375 (1951).

[2] Goodman, L.E. and L.M. Keer: The Contact Stress Problem for an Elastic Sphere Indentry an Elastic Cavity. International Journal of Solids and Structures, Vol. 1, p.407-415 (1965).

[3] Noble, B. and M. Hussain: A Variational Method for Inclusion and Indentation Problems. Journal of the Institute of Mathematics and Its Application, Vol. 5, p.194-205 (1969).

[4] Sheremetev, M.P.: The Solution of the Equations of Certain Contact Problems of the Theory of Elasticity (Equation of the Prandtle Type). Problems of Continuum Mechanics, SIAM, p.419-439 (1961).

[5] Noble, B. and M. Hussain: Exact Solution of Certain Dual Series for Indentation and Inclusion Problems. International Journal of Engineering Science, Vol. 7, p.1149-1161 (1969).

[6] Sternberg, E., R.A. Eubanks, and M.A. Sadowski: On the Axisymmetric Problem of Elasticity Theory for a Region Bounded by Two Concentric Spheres. Proc. 1st U.S. National Congress of Applied Mechanics, p.209-215 (1951).

[7] Noble, B. and M. Hussain: Angle of Contact for Smooth Elastic Inclusions. Developments in Mechanics, Vol.4, Proceedings of the Tenth Midwestern Mechanics Conference, p.459-476 (1967).

[8] Magnus, W. and F. Oberhettinger: Formulas and Theorems for the Special Functions of Mathematical Physics, Chelsa Publishing Co., New York, N.Y., p.66 and p.4 (1949).

APPENDIX

Let us denote

$$I_n^m(\cos \eta) = \int_0^{\eta} (\cos \theta - \cos \eta)^{\frac{m}{2}} P_n(\cos \theta) \sin \theta \, d\theta \qquad (45)$$

The integral representation of Legendre functions [8] is given by

$$P_\nu^\mu(\cos \theta) = \frac{\sqrt{2}}{\sqrt{\pi}} \frac{(\sin \theta)^\mu}{\Gamma\left(-\mu + \frac{1}{2}\right)} \int_0^{\theta} \frac{\cos\left(\nu + \frac{1}{2}\right) \varphi d\varphi}{\left(\cos \varphi - \cos \theta\right)^{\mu + \frac{1}{2}}} \qquad (46)$$

Substituting (46) with $\nu = n$ and $\mu = 0$ into (45) and interchanging the order of integration, we have

$$I_n^m(\cos \eta) = \frac{\sqrt{2}}{\pi} \int_0^{\eta} \cos\left(n + \frac{1}{2}\right)\varphi \left\{ \int_\varphi^{\eta} \frac{(\cos \theta - \cos \eta)^{m/2}}{(\cos \varphi - \cos \theta)^{\frac{1}{2}}} \sin \theta \, d\theta \right\} d\varphi \qquad (47)$$

Introducing a new variable $t = (\cos \varphi - \cos \theta)/(\cos \varphi - \cos \eta)$, we obtain

$$\int_\varphi^{\eta} \frac{(\cos \theta - \cos \eta)^{m/2}}{(\cos \varphi - \cos \theta)^{\frac{1}{2}}} \sin \theta \, d\theta = \frac{\Gamma\left(\frac{1}{2}\right)\Gamma\left(\frac{m}{2} + 1\right)}{\Gamma\left(\frac{m + 3}{2}\right)} (\cos \varphi - \cos \eta)^{\frac{m+1}{2}} \qquad (48)$$

in which $\Gamma(x)$ is the Gamma function. Substituting (48) into (47) gives

$$I_n^m(\cos \eta) = \frac{\sqrt{2}}{\sqrt{\pi}} \frac{m\Gamma\left(\frac{m}{2}\right)}{2\Gamma\left(\frac{m + 3}{2}\right)} \int_0^{\eta} (\cos \varphi - \cos \eta)^{\frac{m+1}{2}} \cos\left(n + \frac{1}{2}\right) \varphi \, d\varphi \qquad (49)$$

Using the expression (46), equation (49) becomes

$$I_n^m(\cos \eta) = \frac{m}{2} \Gamma\left(\frac{m}{2}\right) (\sin \eta)^{\frac{m+2}{2}} P_n^{-\frac{m+2}{2}} (\cos \eta) \qquad (50)$$

For $m = \pm 1$ we immediately obtain equations (36) and (37) which were given previously in the text.

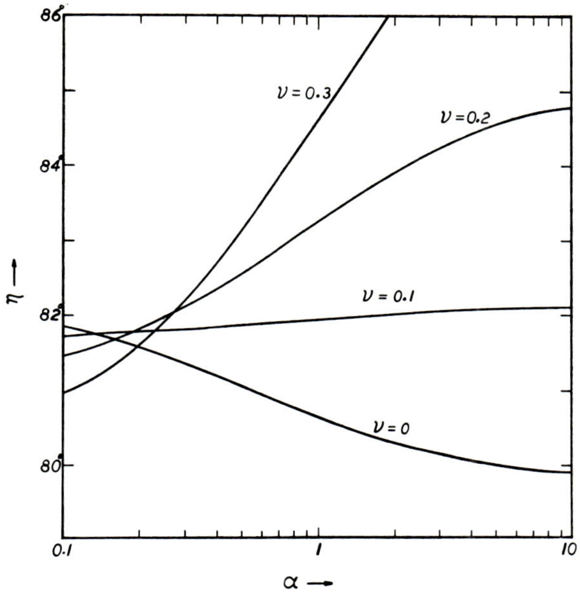

Fig.2a The Colatitude η of the Spherical Cap in Contact vs. the Ratio α of Elastic Constants for $\nu = \nu' = 0$, 0.1, 0.2, and 0.3

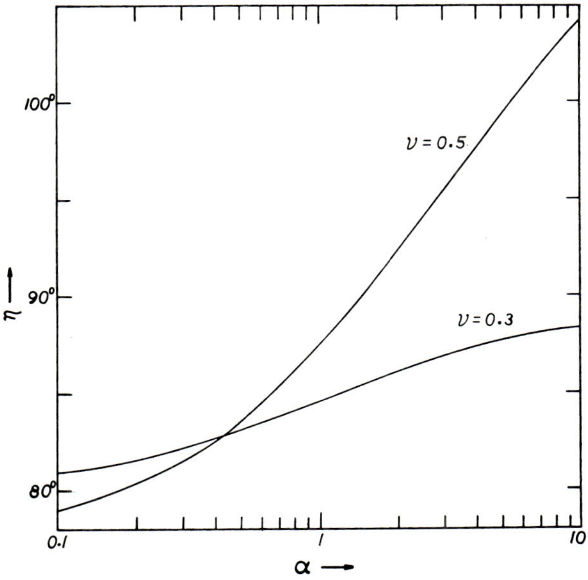

Fig.2b The Colatitude η of the Spherical Cap in Contact vs. the Ratio α of Elastic Constants for $\nu = \nu' = 0.3$ and 0.5

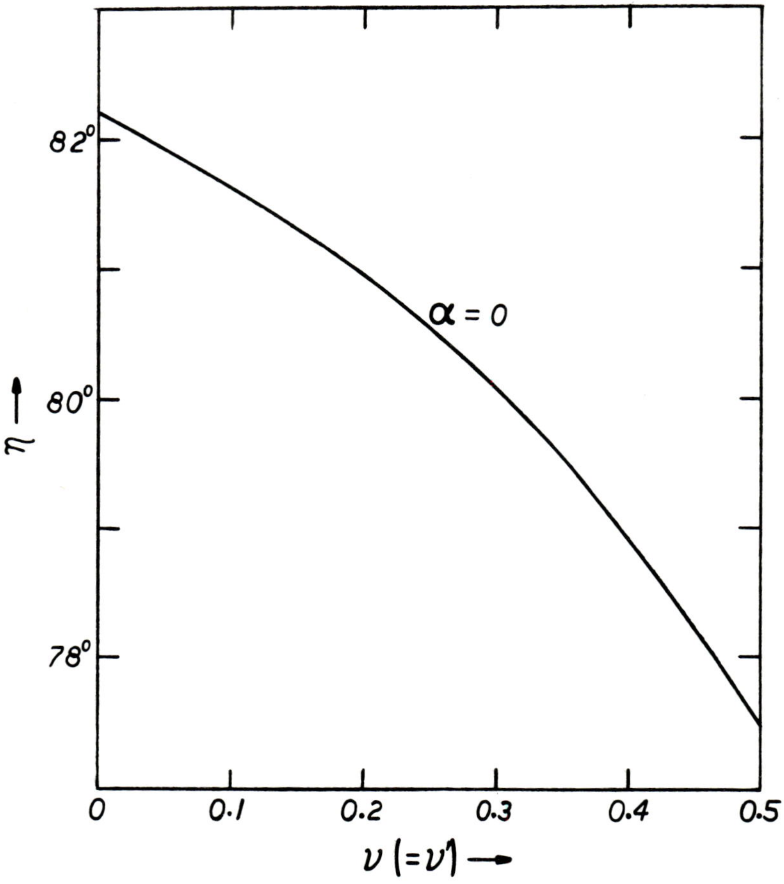

Fig.3 The Colatitude η of the Spherical Cap vs. the Poisson Ratio for the Case of a Rigid Inclusion

Developments in Mechanics, Vol. 6. Proceedings of the 12th Midwestern Mechanics Conference: 46

Analysis of Unbonded Contact Problems Through Application of an Optimization Technique

K. E. YOON
BABCOCK AND WILCOX COMPANY
KWAN RIM
UNIVERSITY OF IOWA

ABSTRACT

In this paper a new approach to the solution of certain contact stress problems is developed by utilizing optimization technique as an analytical tool. It is shown that an unbonded contact problem may be posed as an optimization problem, and that the contact stress distribution as well as the area of contact can be determined through a simple application of the linear programming technique. Of a half dozen contact problems solved by the proposed method, two illustrative examples with known approximate solutions are presented herein for the purpose of making a comparison. By means of this approach it will be possible to conduct an analytical investigation of a large class of unbonded contact problems arising from various structural systems.

INTRODUCTION

This paper is concerned with the symmetrical unbonded contact problems between two elastic bodies. In literature, most of the investigations on contact problems have been carried out on rigid-elastic contact problems, which are called rigid punch problems. In a few elastic-elastic contact problems for which approximate solutions have been obtained, it has been a standard approach to formulate them into integral equation problems. Recently Noble and Hussain [1]* employed a variational method to solve such integral equation problems approximately, followed by Weitsman [2] who solved an unbonded contact problem of an infinite elastic plate pressed against an infinite elastic half space by a concentrated force. Also Pu, Hussain and Anderson [3] solved by using the same technique the problem of lifting of a plate from a foundation due to an axisymmetric pressure.

*Numbers in brackets designate References at the end of this paper.

Another interesting type of elastic-elastic contact problems are the elastic inclusion problems, which are becoming increasingly important in the study of composite materials. Stippes et al [4] solved a two-dimensional smooth circular inclusion problem exactly, and later Noble and Hussain [5] solved the same problem approximately by using dual series relations. An elliptic inclusion problem was investigated by Hussain, Pu and Sadowsky [6], again by using dual series approach and the variational method employed by Noble and Hussain [1].

However, in more general problems, this classical approach quickly becomes mathematically intractable. Therefore, an attempt is made in this paper to develop an approximate but general method of solving a large class of elastic-elastic contact problems by utilizing an optimization technique.

FORMULATION OF PROBLEM

CHARACTERISTICS OF SYMMETRIC UNBONDED CONTACT PROBLEMS

Consider a typical elastic-elastic contact problem; i.e., a problem of contact between an infinite plate and an elastic half space shown in Fig. 1. When two elastic bodies are in contact with each other, the common interface of contact is called the contact area, and the stress acting on the contact area the contact stress. If one assumes that the contact is smooth, then there is no shear stress and only the normal stress acts on the contact area. Furthermore, the unbondedness condition requires that this normal contact stress must be compressive. Thus the fundamental characteristics of an unbonded contact problem are that the contact stress is always compressive and that it acts normal to the contact area.

In general, an unbonded contact problem may be regarded as a set of two boundary value problems, associated with the two bodies in contact with each other, subject to certain conditions of contact. On the common boundary of contact, each boundary value problem must satisfy two boundary conditions simultaneously: namely, the boundary displacement of one body $w_1(r)$ must match that of the other body $w_2(r)$, and the boundary stress must also be the same. (See Fig. 1.) Moreover, the size of the contact area is not known a priori.

In terms of the symbols illustrated in Fig. 1, the contact stress $q(r)$ should satisfy the following condition

$$q(r) \leq 0, \quad 0 \leq r \leq \underline{a} \tag{1}$$

where $\underline{a}$ designates the radius of the contact area. Other conditions of contact may now be written as

$$w_1(r) - w_2(r) = 0, \quad 0 \leq r \leq \underline{a} \tag{2}$$

and $q_1(r) = q_2(r) = q(r), \quad 0 \leq r \leq \underline{a}$

The static equilibrium condition also requires

$$\int_A q(r)\, dA = -P \tag{3}$$

For the simple cases, the boundary displacements $w_i(r)$ are available in an explicit form; however, for the complicated cases, they are given in an operator form as

$$K_i\,[w_i(r)] = Q_i(r, q(r), \underline{a}), \quad 0 \leq r \leq \underline{a}, \ i = 1,2 \tag{4}$$

where K_i are linear differential operators, and Q_i are some functionals of position, the contact stress and the size of the contact area.

POSING THE CONTACT PROBLEM AS AN OPTIMIZATION PROBLEM

The solution of any unbonded contact problem consists of both the contact stress and the size of contact area. If this problem were to be treated as a set of simultaneous boundary value problems of elasticity, one would find that the methods of analysis well established in classical elasticity would not be of much use, because 1) the radius of contact area is not known a priori and 2) the contact stress is subjected to an inequality constraint.

However, such a problem may be readily posed as an optimization problem as described below.

$$\left.\begin{array}{l}
\text{Find } q(r),\ 0 \le r \le \underline{a},\ \varepsilon \text{ and } \underline{a} \text{ which} \\
\text{minimize } \varepsilon \\
\text{subject to the conditions} \\
\qquad \varepsilon = \max[\,|w_1(r) - w_2(r)|\,] \\
\qquad q(r) \le 0, \quad 0 \le r \le \underline{a} \\
\qquad \varepsilon \ge 0 \\
\qquad \int_A q(r)\, dA = -P \\
\text{and} \\
\qquad K_i[w_i(r)] = Q_i(r, q(r), \underline{a}), \quad i = 1,2
\end{array}\right\} \qquad (5)$$

In terms of the terminology used in the optimal design theory [7], one may call the contact stress $q(r)$ a design variable, r the independent variable, and $w_i(r)$, $i = 1,2$, the state variables. In this case ε is the objective or the cost function which is to be minimized and $\underline{a}$ may be considered as one of the design variables.

DISCRETE OPTIMIZATION PROBLEM

In order to obtain an approximate solution to that type of problem defined by Eq. 5, one may discretize the problem by dividing the contact area into n equal intervals and by assuming that the contact stress is piece-wise constant over each interval. If one assumes that the contact stress is symmetric with respect to the center line as it is often the case, then it is necessary to consider only one half of the contact region. And the boundary points along a radius of contact region may be numbered as shown in Fig. 2. The displacement of the point j in the z-direction is denoted by w_{k_j}, where the first subscript k (k = 1,2) designates the upper or the lower body respectively. The discretized contact stress on the j-th interval - between the point (j-1) and the point j - is denoted by q_j. Also the matching of the two boundary displacements along the contact region may be carried out at those n points by initially letting two center points be matched, i.e.,

$$w_{1_0} = w_{2_0}$$

Defining a new design variable $[u] = [u_1, u_2, \ldots, u_n]^T$ such that

$$[u] = -[q]$$

the discrete optimization problem may now be phrased as:

Find u_j, $j = 1,2, . . . ,n$, ε and $\underline{a}$ which

minimize ε

subject to the conditions,

$$\left.\begin{array}{l} w_{1_i} - w_{2_i} + \varepsilon \geq 0, \quad i = 1,2, . . . ,n \\ -w_{1_i} + w_{2_i} + \varepsilon \geq 0, \quad i = 1,2, . . . ,n \\ u_j \geq 0, \; j = 1,2, . . . ,n \\ \varepsilon \geq 0, \\ \sum_{j=1}^{n} u_j A_j = P \\ \text{and} \\ [w_k] = [R_k(r,q_1, . . . ,q_n,\underline{a})], \quad k = 1,2 \end{array}\right\} \quad (6)$$

where $\Delta = \underline{a}/n$, $r_j = j\Delta$, $[w_k] = [w_{k_1}, . . . ,w_{k_n}]^T$, w_{k_i} is the relative displacement of the point i in the z-direction of the upper (if k = 1) or the lower (if k = 2) body and R_k are some functions of design variables and the contact radius. A_j designate area between r_j and r_{j-1}. This optimization problem is almost a linear programming problem, provided that the contact radius $\underline{a}$ is assumed on the outset of each iteration and that $[w_k]$ are linear functions of $[u]$.

ANALYSIS

Once the problem of contact stress had been reduced to a discrete optimization problem of nonlinear programming, it was natural to seek the possibility of solving it through a successive application of the linear programming technique, especially in view of its known effectiveness [8]. Indeed, as it turned out, the discrete optimization problem as defined by Eq. 6 could be solved rather directly as a sequence of linear programming problem. Furthermore, the technique having been ideally suited for computer programming, it was not surprising to find that reasonably accurate solutions were obtainable without undue expenditure of labor for a number of contact problems investigated.

It may be pointed out here that crucial to the application of an optimization technique is the availability of sufficiently accurate elasticity solutions (analytical or otherwise) for both of the contacting bodies. More specifically, in the application of the linear programming technique, one should be able to express accurately the boundary displacement of each contacting body as a linear combination of discrete contact stresses.

Six contact problems have been solved by the authors through an application of the proposed method. The first one is the unbonded contact problem of an infinite elastic beam pressed against the edge of an infinite elastic half plate by a concentrated force. The second problem is that of an infinite elastic plate pressed against an infinite elastic half space by a concentrated force. The third problem is the same as the second except that the plate is finite. The fourth problem is the two-dimensional smooth circular inclusion problem. The fifth is the problem of "loose" inclusion in which there is play between a cylinder and a cylindrical hole in a thick plate. The last one is the contact problem between a circular disk and a ring surrounding it. Because of the space limitation, only two examples are presented in this paper. For the second and the fourth problem there are known analytical solutions; an approximate solution for the second by Weitsman, and an exact solution for the fourth by Stippes, et al. Therefore, these two problems are chosen to serve the purpose of illustrating the effectiveness of the proposed method through comparison of results. Due to close similarity, the third problem is also incorporated into the presentation of the first example.

EXAMPLE 1-A

CONTACT BETWEEN AN INFINITE PLATE AND AN INFINITE HALF SPACE - Consider an infinite plate pressed against an elastic half space by a concentrated load P, as shown in Fig. 1. Assume that:

a) The weight of the plate is negligible.

b) The transverse deformation of the infinite plate is also negligible so that the deflection of the middle surface in the z-direction is equivalent to that of the contacting boundary of the infinite plate.

c) There is no friction between the two contacting bodies.

d) Materials for both the plate and the infinite half space are homogeneous and isotropic, and obey Hooke's law.

This is an axisymmetric problem both in geometry and in loading, hence the contact stress distribution should also be axisymmetric. Discretization of this problem may be carried out by dividing a contact radius into n equal intervals, and by assuming that the contact stress is piece-wise constant over each annular region. Therefore, the discretized contact stresses act on n concentric annular regions. Starting from the center point, all boundary points are numbered along a radius, as shown in Fig. 2.

An influence coefficients t_{ij} is defined as the relative deflection at point i in the z-direction with respect to the origin, due to the j-th annular load of a unit intensity. Then the boundary displacement vector of the plate, w_1, and that of the half space, w_2, are given in matrix equation as

$$\begin{aligned} [w_1] &= [T_p]\,[u] \\ [w_2] &= [T_s]\,[u] \end{aligned} \tag{7}$$

where

$$[T] = \begin{bmatrix} t_{11} \cdot \cdot \cdot \cdot \cdot t_{1n} \\ \cdot \\ \cdot \\ \cdot \\ t_{n1} \cdot \cdot \cdot \cdot \cdot t_{nn} \end{bmatrix}$$

and the subscripts p and s designate the plate and the half space respectively [9 and 10]. Thus, following Eq. 6, this problem may be posed as:

$$\left.\begin{array}{l} \text{Determine } u_j,\ j = 1,2,\ldots,n,\ \varepsilon \text{ and } \underline{a} \text{ which} \\ \text{minimize } \varepsilon \\ \text{subject to the following conditions,} \\ \quad w_{1_i} - w_{2_i} + \varepsilon \geq 0, \quad i = 1,2,\ldots,n \\ \quad -w_{1_i} + w_{2_i} + \varepsilon \geq 0, \quad i = 1,2,\ldots,n \\ \quad u_j \geq 0, \quad j = 1,2,\ldots,n \\ \quad \varepsilon \geq 0 \end{array}\right\} \tag{8}$$

and

$$\pi \cdot \sum_{j=1}^{n} (r_j^2 - r_{j-1}^2)\, u_j = P$$

The last equation is from the static equilibrium condition, where $r_i = i\Delta = i\underline{a}/n$. Rewriting this condition as

$$u_n = \frac{P}{\pi C} - \frac{1}{C} \sum_{j=1}^{n-1} (r_j^2 - r_{j-1}^2)\, u_j \qquad (9)$$

where $C = r_n^2 - r_{n-1}^2 = 2\Delta\underline{a} - \Delta^2$, u_n can be eliminated from Eqs. 7 and 8.

INTRODUCTION OF END MOMENT AS A DESIGN VARIABLE - According to the formulation given by Eq. 8, it fits exactly into a standard form of a linear programming problem, if the value of a is assumed a priori and if u_n is eliminated by using Eq. 9. Therefore, for certain cases, guessing the correct value for a can become a real problem in actual computation. If one happens to choose a value which is far off from the correct one, then he has to go through a large number of trial and error iterations, wasting computer time. One way to circumvent this difficulty is to apply a fictitious uniform radial moment, m ($m \geq 0$), to the plate along the circumference of assumed contact area. If the assumed value of a is smaller than the correct radius, the end moment m will initially take on a large value in enforcing the plate to have a close contact with the half space, and then it will decrease and approach zero as a approaches its correct value. Therefore, starting with a sufficiently small value for a, the linear programming problem is solved iteratively, increasing the value of a each time until the value of the end moment m becomes sufficiently small. With the introduction of m as an additional design variable, Eqs. 7 become

$$\left.\begin{aligned} \begin{bmatrix} w_1 \end{bmatrix} &= \begin{bmatrix} T_p \end{bmatrix} \begin{bmatrix} u \end{bmatrix} + m \begin{bmatrix} v \end{bmatrix} = \left[\begin{array}{c|c} t_{p_{ij}} & \begin{matrix} v_1 \\ \cdot \\ \cdot \\ \cdot \\ v_n \end{matrix} \end{array}\right] \begin{bmatrix} u_1 \\ \cdot \\ \cdot \\ u_n \\ m \end{bmatrix} \\ \begin{bmatrix} w_2 \end{bmatrix} &= \left[\begin{array}{c|c} t_{s_{ij}} & \begin{matrix} 0 \\ \cdot \\ \cdot \\ \cdot \\ 0 \end{matrix} \end{array}\right] \begin{bmatrix} u_1 \\ \cdot \\ \cdot \\ u_n \\ m \end{bmatrix} \end{aligned}\right\} \qquad (10)$$

where v_j is the relative displacement of the point j due to the end moment m, with respect to the origin. Defining $\eta_{ij} = t_{p_{ij}} - t_{s_{ij}}$ and eliminating u_n using Eq. 9, one obtains the following equation:

$$\begin{bmatrix} w_1 \end{bmatrix} - \begin{bmatrix} w_2 \end{bmatrix} = \left[\begin{array}{c|c} \lambda_{ij} - \frac{1}{C}(r_j^2 - r_{j-1}^2)\, \lambda_{in} & \begin{matrix} v_1 \\ \cdot \\ \cdot \\ \cdot \\ v_n \end{matrix} \end{array}\right] \begin{bmatrix} u_1 \\ \cdot \\ \cdot \\ u_{n-1} \\ m \end{bmatrix} + \frac{P}{\pi C} \begin{bmatrix} \lambda_{in} \end{bmatrix} \qquad (11)$$

where $i = 1,2, \ldots, n-1$ and $j = 1,2, \ldots n$.

The optimization problem defined by Eq. 8 may now be reduced exactly into a linear programming problem as shown below:

For an assumed value of a,

determine u_j, $j = 1,2, \ldots, n-1$, m and ε which

minimize ε

subject to the following conditions,

$$\left.\begin{aligned}
&\left[\lambda_{ij} - \frac{1}{C}(r_j^2 - r_{j-1}^2)\lambda_{in} \;\vdots\; \begin{matrix} v_1 \\ \cdot \\ \cdot \\ \cdot \\ v_n \end{matrix} \;\vdots\; \begin{matrix} 1 \\ \cdot \\ \cdot \\ \cdot \\ 1 \end{matrix}\right]\begin{bmatrix} u_1 \\ \cdot \\ \cdot \\ u_{n-1} \\ m \\ \varepsilon \end{bmatrix} + \frac{P}{\pi C}\begin{bmatrix} \\ \lambda_{in} \\ \\ \end{bmatrix} \geq 0, \\
&\left[-\lambda_{ij} + \frac{1}{C}(r_j^2 - r_{j-1}^2)\lambda_{in} \;\vdots\; \begin{matrix} -v_1 \\ \cdot \\ \cdot \\ \cdot \\ -v_n \end{matrix} \;\vdots\; \begin{matrix} 1 \\ \cdot \\ \cdot \\ \cdot \\ 1 \end{matrix}\right]\begin{bmatrix} u_1 \\ \cdot \\ \cdot \\ u_{n-1} \\ m \\ \varepsilon \end{bmatrix} + \frac{P}{\pi C}\begin{bmatrix} \\ -\lambda_{in} \\ \\ \end{bmatrix} \geq 0, \\
&\text{and} \\
&\qquad\qquad \begin{bmatrix} u_1 \\ \cdot \\ \cdot \\ u_{n-1} \\ m \\ \varepsilon \end{bmatrix} \geq 0.
\end{aligned}\right\} \quad (12)$$

NUMERICAL RESULTS - In actual calculation, the magnitude of the contact radius $\underline{a}$ is gradually increased by augmenting the number of divisions, n, in each iteration, while keeping the inverval size Δ fixed, until the end moment becomes negligibly small.

The following input data are used in the computation:

Young's Modulus of Elasticity for both - $E = 10^6$ psi.

Poisson's ratio for both - - - - - - - - $\nu = 0.3$

Flexual rigidity of the plate - - - - - $D = 10^4$ in.-lb.

Total load applied to the plate - - - - $P = 500$ lbs.

Interval size - - - - - - - - - - - - - $\Delta = 0.025$ in.

When n is 30, the end moment is 0.4082 in-lbs/in; and the minimum value of ε is negligible, in the order of 10^{-6}, compared with the magnitude of deflection of the plate, as shown in Table I. The corresponding contact stress distribution is plotted in Fig. 3 by a solid line. When n is increased to 31, the end moment vanishes; but the minimum value of ε increases by approximately 100 times. Thus the correct value for $\underline{a} = n\Delta$ apparently lies in between 30Δ and 31Δ. Contact stress distributions for these two cases are very close to each other.

Recently Weitsman [2] solved the same problem, by formulating it as an integral equation problem and solving the equation approximately by means of a variational method. He gave the following formula

$$\underline{a} = 2.926 \cdot \sqrt[3]{\frac{(1-\nu)D}{G}}$$

as a good approximation for the radius of the contact region. According to the formula, the radius of contact for this particular example is given by

$$\underline{a} = 30.8\Delta$$

This is an excellent agreement with the result of the present analysis, particularly in view of the fact that both methods are approximate and that the influence coefficients used for the plate is not exact. It can be also observed that the magnitude of the total load contributes linearly to both boundary displacements and hence to the error terms; therefore, the contact radius is independent of the magnitude of the total load, which was also pointed out by Weitsman [2].

Instead of an infinite plate, the problem of a circular plate with a finite radius may also be handled by the method without any difficulty. Such a problem should have more practical significance. Also certain types of variable thickness plates may be treated by this method as long as accurate elasticity solutions are available for such plates.

EXAMPLE 1-B

CONTACT BETWEEN A FINITE CIRCULAR PLATE AND AN INFINITE HALF SPACE - This problem is identical with the preceding one except that the radius of the plate is finite (taken as one inch). Influence coefficients for the plate are newly calculated, but all other formulations are exactly the same as the preceding problem. Therefore, only the final results are presented; the contact stress is plotted in Fig. 3 by a dotted line. The magnitude of the maximum error, ε, is again shown to be extremely small as in the preceding example. The end moment m is reduced to about 0.01 in-lbs/in. There wasn't any difficulty in solving this problem by using the optimization technique. However, it would have been very difficult to handle it by other conventional methods of analysis, such as, the integral equation method.

EXAMPLE 2

SMOOTH INCLUSION PROBLEM - An infinite homogeneous isotropic elastic thick plate with a circular cylindrical cavity is subject to a uniform tension, with a circular cylinder of the same size inserted into the cavity. Assume that the thick plate and the cylinder are of the same material. The present method of analysis does not require this assumption; however, it is introduced for the purpose of comparing the result with an existing solution [4]. See Fig. 4. Since smoothness implies the absence of friction on the contacting surfaces, the contact stress will have a compressive radial component only. Furthermore, at the contacting boundary, the radial displacement of the infinite thick plate must match with that of the circular cylinder, and the contact stress distributions for both bodies should be the same. Again the angle of contact is not known a priori.

This problem is slightly different from the one defined by Eq. 6; in this problem the contact takes place on two opposite arcs simultaneously and there is no additional constraint on the design variables u_i. Due to the symmetry of the problem, only the first quadrant will be considered hereafter.

Let $u_r(\theta)$ be the radial displacement positive outward from the origin and α be the angle of contact. Then the following condition must hold:

$$u_r(\theta)_{\text{plate}} - u_r(\theta)_{\text{cyl.}} = 0, \quad 0 \leq \theta \leq \alpha$$

and the unbondedness condition is given by

$$q(\theta) \leq 0, \quad 0 \leq \theta \leq \alpha$$

where $q(\theta)$ is the contact stress.

This problem may also be discretized by dividing the contact angle α into n equal angles and by assuming that the contact stress is piece-wise constant over each interval. For the convenience of computation, the boundary points are numbered as shown in Fig. 5, starting with 1 at the center point of the contact region.

Let

$$w_{1_i} = u_r\,[(i-1)\,\delta]_{\text{plate}}$$

$$w_{2_i} = u_r\,[(i-1)\,\delta]_{cylinder}$$

where $\delta = \alpha/n$; and, let u_j be the negative value of the contact stress on the j-th interval (that is the interval between the point j and the point j + 1). Then, due to the linear property of this problem, the radial displacement of the thick plate can be divided into two parts as

$$u_r(\theta)_{plate} = u_r(\theta)_{\substack{\text{due to contact}\\ \text{stress}}} + u_r(\theta)_{\substack{\text{due to a uniform}\\ \text{tension (S)}}}$$

Accordingly, define an influence coefficient t_{ij} as the radial displacement of the point i due to a uniform edge load of a unit intensity on the j-th interval, and v_i as the radial displacement of the point i due to a unit tension in the x-direction at infinity. Then w_1 and w_2 have the following relations in matrix form [11],

$$\left.\begin{aligned} w_{1_i} &= t_{p_{ij}} \cdot u_j + Sv_i \\ w_{2_i} &= t_{c_{ij}} \cdot u_j \end{aligned}\right\} \quad (13)$$

where i = 1,2, . . . ,n + 1 and j = 1,2, . . . ,n.

This problem is slightly different from the preceding examples in that it does not have an additional constraint on the design variable [u], but it has one more point to be matched on the contact region. Now the problem may be posed as the following linear programming problem:

$$\left.\begin{array}{l} \text{For an assumed value of } \underline{a}, \\ \text{determine } u_j,\ j = 1,2, \ldots ,n \text{ and } \varepsilon \text{ which} \\ \text{Minimize } \varepsilon \\ \text{subject to the following conditions,} \\ \quad w_{1_i} - w_{2_i} + \varepsilon \geq 0, \quad i = 1,2, \ldots ,n+1 \\ \quad -w_{1_i} + w_{2_i} + \varepsilon \geq 0, \quad i = 1,2, \ldots ,n+1 \\ \qquad\qquad u_j \geq 0, \quad j = 1,2, \ldots ,n \\ \qquad\qquad \varepsilon \geq 0. \end{array}\right\} \quad (14)$$

This particular example did not make use of such extra design variable as the end moment, which was used in the previous example problems for the purpose of alleviating the difficulty in correctly assessing the right dimension of the contact area. Here, in searching for the correct value of the contact angle, $\alpha = n\delta$, the angular interval size δ was kept constant, while the number of intervals n was varied. The angle α is determined by varying n in such a way that ε becomes negligibly small. As it turns out, whenever n comes very close to the correct value, the minimum value of ε becomes very small; however, if it deviates from the correct value, the minimum value of ε rapidly increases.

The following input data are used in the calculation:

Young's Modulus of Elasticity for the thick

plate and the cylinder - - - - - - - $E = 10^6$ psi.

Poisson's ratio for the thick plate and

the cylinder - - - - - - - - - - - - - ν = 0.3

Uniform tension for the thick plate S = 100 psi.

Radius of the cylinder - - - - - - - - R = 1.0 in.

Angular interval size - - - - - - - - δ = 0.5 degrees.

The numerical results based on the preceding input data are obtained, and the contact stress distribution is plotted in Fig. 6. Also the analytical solution by Stippes et al [4] is plotted on the same figure, which shows a very close agreement between the two results. The maximum contact stress obtained by the present method is 0.6091 times the uniform tension applied at infinity and the contact angle is 19.5 degrees, compared with Stippes' results of 0.609 and 19.62 degrees respectively. When the angular interval size is increased from 0.5 degrees to 1.0 degree, the resulting contact stress oscillates about the curve for the exact solution, indicating that the interval size is too crude. Also the value of ε becomes too large to be acceptable. The method presented herein is readily applicable not only to the problems with different materials for the plate and the cylinder, but also to the problems of 'loose' or 'tight' inclusions, where the size of the cylinder is either slightly smaller or larger than that of the hole in the thick plate.

CONCLUDING REMARKS

The numerical results obtained in the two example problems indicate that the method presented herein is quite simple and straightforward, yet yields reasonably accurate solutions to the unbonded contact problems. The magnitude of the maximum error for every example problem is indeed negligible, in the order of 10^{-6}, compared to the actual boundary deflections of the contacting bodies. Considering that only a seven digit accuracy was maintained in the numerical calculations, the condition of matching for the boundary displacements was very closely satisfied, as reflected by the smallness of ε.

The results also indicate that, for the two examples of smooth contact considered in this paper, the size of the contact radius or angle is independent of the magnitude of the applied load. The same conclusion was drawn by Weitsman as a result of his investigation of the first example problem. This conclusion implies that such structural elements as stress pads, bearing plates, etc. may possess unique dimensions for their optimal designs, independent of the magnitude of the applied load. And these optimal dimensions can be readily found through an application of the method presented in this paper.

Recent investigators have succeeded in reducing a few elastic-elastic contact problems into integral equation problems. This approach is conceptually elegant, but its applicability is severely limited to a very few problems with simple geometry and loading. Furthermore, even for simple problems, only approximate solutions have been obtained. In more general problems, this classical approach quickly becomes mathematically intractable.

Evidently, there is a strong need in engineering for the development of a general but straightforward method of attacking these elastic-elastic contact problems. This paper is directed toward fulfilling some of this need.

Among several alternatives studied by the authors, the use of optimization technique as an analytical tool seems to be most promising, even though such an approach has not been commonly proposed in the analysis of elasticity problems. Especially, in view of the known effectiveness of linear programming techniques, the possibility of reducing a contact problem into a linear programming problem should be sought. Indeed, the authors have shown that many of the unbonded contact problems can be reduced to problems of simple linear programming and hence reasonably accurate solutions can be obtained without undue expenditure of labor. A similar approach - the use of optimization technique as an analytical tool - is likely to provide a new avenue of solution for many of the more complicated contact problems [12].

ACKNOWLEDGEMENTS

This work was partly supported by the Project Themis, University of Iowa, sponsored by the Department of Defense under DoD Contract No. DAAF03-69-C-0014.

REFERENCES

[1] Noble, B. and Hussain, M. A., "A Variational Method for Inclusion and Indentation Problems," Mathematics Research Center Report, No. 812, U. S. Army, University of Wisconsin, Dec., 1967.

[2] Weitsman, Y., "On the Unbonded Contact Between Plates and an Elastic Half Space," Journal of Applied Mechanics, Vol. 36, E, 2, 1969, pp. 198-202.

[3] Pu, S. L., Hussain, M. A. and Anderson, G., "Lifting of a Plate from the Foundation Due to Axisymmetric Pressure," Development in Mechanics, Vol. 5, (Proceedings of the 11th Midwestern Mechanics Conference), 1969, pp. 577-590.

[4] Stippes, M., Wilson Jr., H. B., and Krull, F. N., "A Contact Stress Problem for a Smooth Disk in an Infinite Plate," Proceedings of the Fourth U. S. National Congress of Applied Mechanics, Vol. 2, June 18-21, 1962, pp. 799-806.

[5] Noble, B. and Hussain, M. A., "Angle of Contact for Smooth Elastic Inclusions," Developments in Mechanics, Vol. 4, (Proceedings of the Tenth Midwestern Mechanics Conference), 1967, pp. 459-476.

[6] Hussain, M. A., Pu, S. L. and Sadowsky, M. A., "Cavitation at the Ends of an Elliptic Inclusion Inside a Plate Under Tension," Journal of Applied Mechanics, Vol. 35, No. 3, Trans. ASME, Vol. 90, Series E, 1968, pp. 505-509.

[7] Haug, E. J. "Theory of Optimization," unpublished Lecture Notes, University of Iowa, Iowa City, Iowa, 1969.

[8] Wilde, D. J., and Beightler, C. S., Foundation of Optimization, Prentice-Hall, Inc., Englewood Cliffs, New Jersey, 1967.

[9] Timoshenko, S. and Goodier, J. N., Theory of Elasticity, Second Edition, McGraw-Hill Book Co., Inc., N. Y., 1951, pp. 95-96 and 367-368.

[10] Timoshenko, S. and Woinowsky-Krieger, S., Theory of Plates and Shells, Second Edition, McGraw-Hill Book Co., Inc., N. Y., 1951, pp. 51-72.

[11] Muskhelishvili, N. I., Some Basic Problems of the Mathematical Theory of Elasticity, translated by Radok, J. R. M., P. Noordhoff Ltd., Groningen, Holland, 1953.

[12] Yoon, K. E., "Analysis of Contact Problems by means of an Optimization Technique," Project Themis Report, No. 24, University of Iowa, Iowa City, Iowa, August, 1970.

NOMENCLATURE

$q(r)$, q_j	contact stress, discretized contact stress, psi.
$\underline{a}$	radius of contact area, in.
$w_k(r)$, w_{k_i}	boundary displacement in z-direction, boundary displacement of point i, in.
K_k	differential operator on w_k
Q_k	functional of position, contact stress and radius of contact area
P	total load applied to the upper body or uniform tension in case of inclusion problem, lb.
ε	maximum of absolute value of the difference of boundary displacements, in.
Δ	interval size, radius of contact area divided by number of division, in.
n	number of division
u_j	negative value of q_j
A_j	area between r_j and r_{j-1}, in^2.
t_{ij}, T	influence coefficients, its matrix
v_i	influence coefficients due to a unit end moment unless otherwise specified
E	Young's Modulus of Elasticity, psi.
G	Modulus of rigidity, psi.
ν	Poisson's ratio
m	end moment, in-lb.
λ_{ij}	difference between two influence coefficients
α	angle of contact, degree
δ	angular interval size, degree
R	radius of a circular boundary in inclusion problems, in.
D	flexural rigidity of a plate, in-lb.

Table I Difference Between Deflections of Plate and Half Space

Pt.	Contact Stress (psi.)	Boundary Deflections (in. x 10^{-6}) Plate	Boundary Deflections (in. x 10^{-6}) Half Space	Difference in Deflections (in. x 10^{-11})
1	1,373.538	4.3259	4.3259	0.134
5	991.701	58.6082	58.6082	1.927
9	704.042	133.7987	133.7987	3.791
13	491.016	210.7851	210.7850	7.549
17	336.098	282.2336	282.2335	11.232
21	224.078	345.6537	345.6535	13.978
25	140.516	400.9401	400.9400	16.203
29	66.588	449.1455	449.1453	19.698

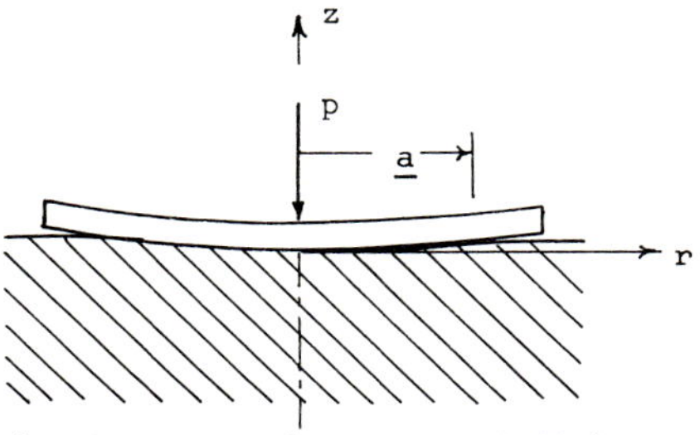

Fig. 1 Contact between an Infinite Plate and Infinite Half Space

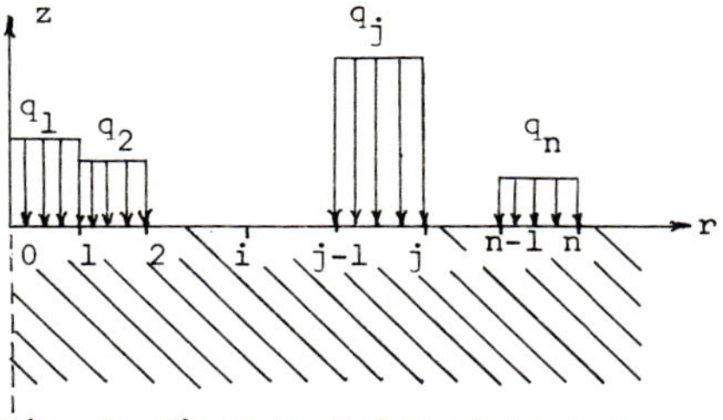

Fig. 2 Discrete Model of Contact Problem

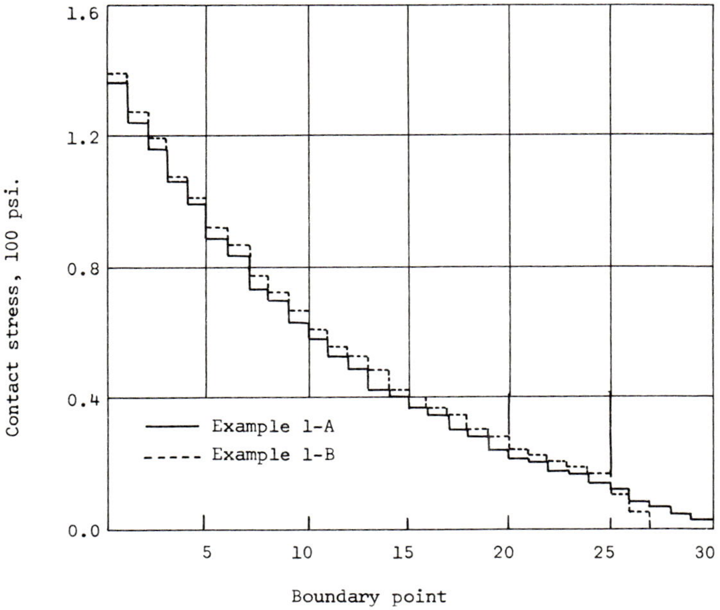

Fig. 3 Contact Stress Distribution

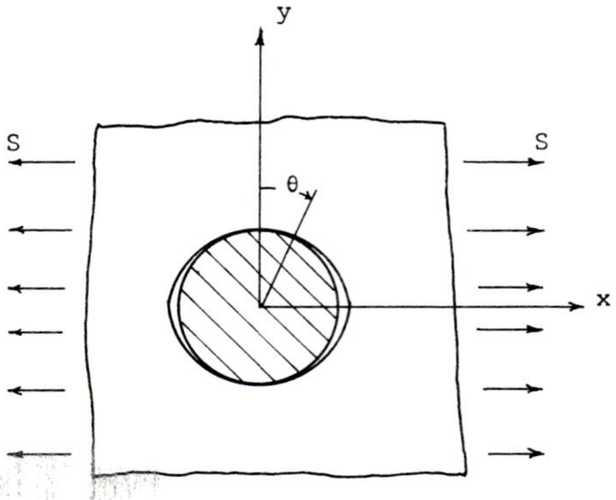

Fig. 4 Smooth Circular Inclusion

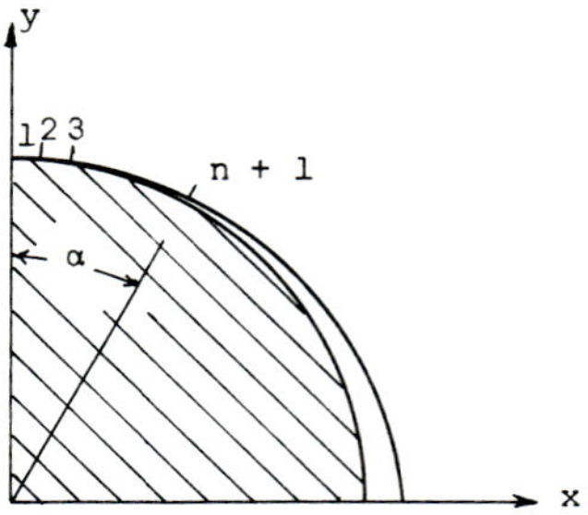

Fig. 5. Boundary Points along the Contact Boundary

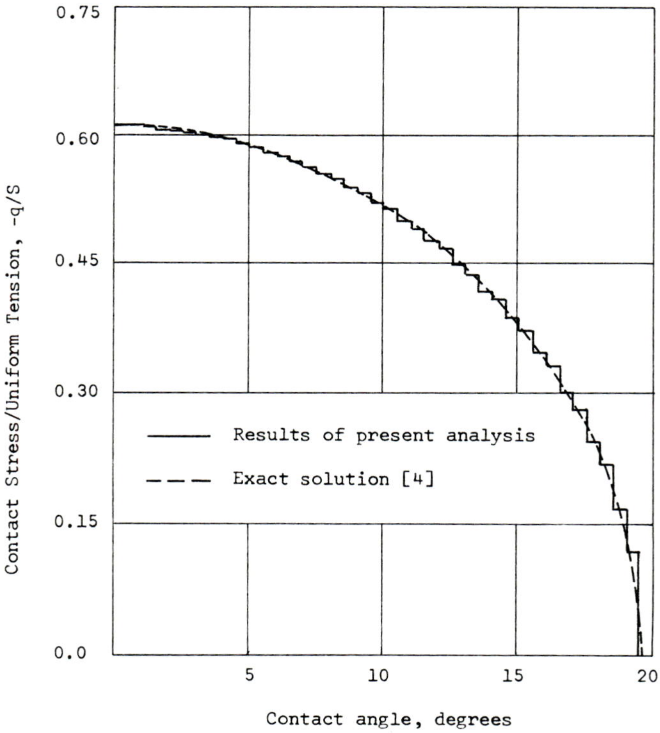

Fig. 6 Contact Stress vs. Contact Angle

VIBRATIONS

Developments in Mechanics, Vol. 6. Proceedings of the 12th Midwestern Mechanics Conference: 47

Elastodynamics of Complex Structural Systems

KENNETH J. SACZALSKI
CLARKSON COLLEGE OF TECHNOLOGY

T. C. HUANG
UNIVERSITY OF WISCONSIN

ABSTRACT

A unified formulation of hybrid elastodynamic equations is developed and used to describe the deterministic and nondeterministic response characteristics of coupled spatial vibratory systems, consisting of continuous elements, point masses and rigid bodies. Dynamic influence coefficients and the Lagrangian-multiplier method have been used in obtaining the energy expressions. The Lagrangian L formed by potential, kinetic and constraint energies provides the basis for the derivation of the equilibrium and compatibility equations of the system.

Exclusion of the explicit time dependence of the elastodynamic equations and operating in the frequency domain, results in a set of transfer functions applicable to both the deterministic and non-deterministic areas. In both cases the results are obtained by using transform techniques. As an example the response of a system due to steady state, transient and random excitations is given.

INTRODUCTION

A vibration analysis technique has been developed recently for spatial vibratory frameworks [1]* based on the continuous element method with the use of dynamic influence coefficients. It is desirable to extend this approach dealing with the deterministic and non-deterministic response characteristics of the more complex structure systems.

The two basic methods used to generate the governing differential equations are referred to as the matrix "force" (flexibility) method and the matrix "displacement" (stiffness) method [2, 3]. The force method yields a set of compatibility equations. The displacement method yields a set of equilibrium equations.

---

* Numbers in brackets designate references at the end of paper.

Equilibrium equations are used in this paper, to denote equations of dynamic equilibrium or equations of motion. Each method has distinct advantages and it seems reasonable to assume that it would be convenient to combine the salient features of both methods.

In this paper a novel hybrid technique has been developed for determining the response characteristics of complex systems excited by steady state sinusoidal, transient and random loadings. The loadings may be in terms of forces or displacements. The resulting set of equilibrium and compatibility equations are designated as the hybrid elastodynamic equations of the complex configuration.

The potential, kinetic and constraint energies of the structural elements, in the form of the Lagrangian L, is the basis for developing the hybrid elastodynamic equations of equilibrium and compatibility [4]. The auxiliary conditions which imply restrictions on the free variations of the time dependent generalized coordinates $q(\xi_j,t)$, are accounted for by including the unknown constraint loads $\lambda_i(\xi_j,t)$, commonly referred to as Lagrangian multipliers, in the elastodynamic equations. The constraint loads are required for nodes constrained to the ground, the plates and shells, and the rigid bodies.

SYSTEM DEFINITION

A given spatial vibratory system is idealized into a number of connected, structural elements of continuous mass distribution, point masses and rigid bodies. For example, straight prismatic elastic members, rings and shear panel elements as well as springs and viscous dampers are considered as structural elements. Associated with each element is a set of geometrical and physical properties. In this paper member elements are considered as prismatic and shear panel elements are considered triangular. Rigid bodies are characterized by their mass and inertia tensors. Several elements may be connected at an intersection, or at one or more locations to any one rigid body, or to the ground.

The intersections of the elements are defined as nodal points or nodal lines. Each nodal point, or any position along a nodal line, has six possible degrees of freedom (three displacements and three rotations). The location of a nodal position j, is specified by spatial coordinates ξ_j, which are expressed in an orthogonal, right-handed, inertial (fixed) coordinate system.

In addition to the inertial system, there is also a local coordinate system attached to each element, in the undeformed state. Transformation of forces and displacements, from the inertial coordinates to the local coordinates is accomplished by using matrices of direction cosines between the two systems. Allowing $\bar{e}_1$, $\bar{e}_2$, and $\bar{e}_3$ to represent unit vectors of the local coordinate system, and $\bar{i}$, $\bar{j}$, $\bar{k}$ the unit vectors of the inertial coordinates, yields the transformation matrix [e] as

$$[e] = \begin{bmatrix} \bar{e}_1\cdot\bar{i} & \bar{e}_1\cdot\bar{j} & \bar{e}_1\cdot\bar{k} \\ \bar{e}_2\cdot\bar{i} & \bar{e}_2\cdot\bar{j} & \bar{e}_2\cdot\bar{k} \\ \bar{e}_3\cdot\bar{i} & \bar{e}_3\cdot\bar{j} & \bar{e}_3\cdot\bar{k} \end{bmatrix} \tag{1}$$

For small deflections the unknown response at a nodal position $P(\xi_A)$ may be represented in inertial coordinates by a three dimensional displacement vector $\bar{q}_A(\delta)$ and a three dimensional rotational vector $\bar{q}_A(\theta)$. These displacements and rotations may be expressed in terms of the local coordinates by

$$\begin{Bmatrix} \bar{\eta} \\ \bar{\psi} \end{Bmatrix}_A = \left[\begin{array}{c|c} e & \text{NULL} \\ \hline \text{NULL} & e \end{array}\right]_A \begin{Bmatrix} \bar{q}(\delta) \\ \bar{q}(\theta) \end{Bmatrix}_A \tag{2}$$

where $\overline{\eta}$ and $\overline{\psi}$ represent the unknown displacement and rotation vectors, respectively, in the local system. Likewise forces $\overline{V}$ and moments $\overline{M}$ in the inertial coordinates may be expressed in terms of local coordinates by

$$\left\{\begin{matrix}\overline{\lambda}(\delta)\\ \overline{\lambda}(\theta)\end{matrix}\right\}_A = \left[\begin{array}{c|c} e & \text{NULL} \\ \hline \text{NULL} & e \end{array}\right]_A \left\{\begin{matrix}\overline{V}\\ \overline{M}\end{matrix}\right\}_A \tag{3}$$

POTENTIAL ENERGY OF DISTRIBUTED MASS ELEMENTS

For a straight, prismatic member element as shown in Fig. 1, having a continuous mass distribution along its length, it is convenient to describe the physical contributions to the elastodynamic equations, by taking the scalar product of the average nodal loads $\overline{\lambda}(\xi)$ at each end, times their respective local response vectors $\overline{\eta}$, $\overline{\psi}$. The scalar product yields the potential energy of the continuous mass element, as

$$U_c = \frac{1}{2}\{\overline{\lambda}(\xi)\}^T \cdot \{\overline{\eta},\overline{\psi}\} \tag{4}$$

The four differential equations, from modified theory, which describe the longitudinal, torsional, and flexural motion of the element [1] are

$$AE\frac{\partial^2\eta_1(t)}{\partial s^2} - \rho A\frac{\partial^2\eta_1(t)}{\partial t^2} = 0 \tag{5}$$

$$GJ\frac{\partial^2\psi_1(t)}{\partial s^2} - \rho I_P\frac{\partial^2\psi_1(t)}{\partial t^2} = 0 \tag{6}$$

$$EI_3\frac{\partial^4\eta_2(t)}{\partial s^4} + \rho A\frac{\partial^2\eta_2(t)}{\partial t^2} - \rho I_3\left(1+\frac{E}{\mu G}\right)\frac{\partial^4\eta_2(t)}{\partial s^2\partial t^2} + \frac{\rho^2 I_3}{\mu G}\frac{\partial^4\eta_2(t)}{\partial t^4} = 0 \tag{7}$$

$$EI_2\frac{\partial^4\eta_3(t)}{\partial s^4} + \rho A\frac{\partial^2\eta_3(t)}{\partial t^2} - \rho I_2\left(1+\frac{E}{\mu G}\right)\frac{\partial^4\eta_3(t)}{\partial s^2\partial t^2} + \frac{\rho^2 I_2}{\mu G}\frac{\partial^4\eta_3(t)}{\partial t^4} = 0 \tag{8}$$

The time dependent solutions to Eqs. (5) through (8) may be expressed in the normal mode form by

$$\overline{\eta}_i(t) = \sum_{n=1}^{12}\eta^*_{in}C_n\delta_n(t); \qquad i = 1,2,3$$
$$\psi_i(t) = \sum_{n=1}^{12}\psi^*_{in}C_n\theta_n(t); \qquad i = 1,2,3 \tag{9}$$

where the C_n are the unknown nodal amplitudes of $\delta_n(t)$ and $\theta_n(t)$, and η^*_{in}, ψ^*_{in} are the shape functions of the normal modes. There are twelve C_n coefficients corresponding to the total number of degrees of freedom for both ends of the element. The solution to Eqs. (5) through (8) in the form of Eqs. (9) are well known and may be expressed by twelve scalar equations representing the local coordinate response, at nodes A and B [1]. These equations may be expressed in matrix form by the relationship:

$$\{\overline{\eta}_A,\overline{\psi}_A,\overline{\eta}_B,\overline{\psi}_B\} = [G_{ij}]\{C_1,C_2,\ldots,C_{12}\} \tag{10}$$

where the matrix $[G_{ij}]$ is a twelve by twelve non-symmetric matrix, containing terms represented by hyperbolic and trigonometric type functions.

The scalar components of the nodal load vector, in local coordinates, are given by

$$\lambda_x(\delta) = AE \frac{\partial \eta_1}{\partial s} \quad (11) \qquad \lambda_x(\theta) = GJ \frac{\partial \psi_1}{\partial s} \quad (14)$$

$$\lambda_y(\delta) = \mu AG \frac{\partial \eta_2'}{\partial s} \quad (12) \qquad \lambda_y(\theta) = - EI_2 \frac{\partial^2 \eta_3''}{\partial s^2} \quad (15)$$

$$\lambda_z(\delta) = \mu AG \frac{\partial \eta_3'}{\partial s} \quad (13) \qquad \lambda_z(\theta) = EI_3 \frac{\partial^2 \eta_2''}{\partial s^2} \quad (16)$$

where η' and η'' denotes shearing components and bending components, respectively, of η. Taking the necessary derivatives of the vectors in Eqs. (9) and substituting into Eqs. (11) through (16) yields a column of twelve loads which are expressed in matrix form by

$$\{\overline{\lambda}_A(\delta), \overline{\lambda}_A(\theta), \overline{\lambda}_B(\delta), \overline{\lambda}_B(\theta)\} = [B_{ij}] \{C_1, C_2, \ldots, C_{12}\} \quad (17)$$

The $[B_{ij}]$ matrix is a twelve by twelve non-symmetric matrix, similar to $[G_{ij}]$ matrix and also containing terms represented by hyperbolic and trigonometric type functions.

Solving for the C vector from Eq. (10), and substituting into Eq. (17), provides the loads as functions of the displacements and rotations in local coordinates

$$\{\overline{\lambda}_A(\delta), \overline{\lambda}_A(\theta), \overline{\lambda}_B(\delta), \overline{\lambda}_B(\theta)\} = [B_{ij}] [G_{ij}]^{-1} \{\overline{\eta}_A, \overline{\psi}_A, \overline{\eta}_B, \overline{\psi}_B\} \quad (18)$$

The matrix formed by $[B_{ij}] [G_{ij}]^{-1}$ is defined as the symmetric, twelve by twelve, matrix of dynamic stiffness coefficients for the continuous mass element. Transforming $\overline{\eta}$, $\overline{\psi}$ from the local coordinates to $q(\delta)$, $q(\theta)$ in the inertial coordinates, and then substituting Eq. (18) into Eq. (4), provides the desired energy expression.

The energy of a curved, prismatic, member element, shown in Fig. 2, with a continuous mass distribution along the length [5], is obtained by a procedure similar to the one used for the straight member element of a continuous mass distribution described in the preceding section.

Assuming a harmonic solution, so as to remove the time dependence, the curvilinear equations of motion for local in-plane deformation in terms of the shape functions [5], are

$$\frac{EA}{R^2}\left(\frac{d\eta_2^*}{d\phi} - \frac{d^2\eta_1^*}{d\phi^2}\right) - \frac{EI_3}{R^4}\left(\frac{d^3\eta_2^*}{d\phi^3} + \frac{d^2\eta_1^*}{d\phi^2}\right) - \rho\omega^2\left(A\eta_1^* + \frac{I_3}{R^2}\left(\eta_1^* + \frac{d\eta_2^*}{d\phi}\right)\right) = 0$$

$$\frac{EA}{R^2}\left(\eta_2^* - \frac{d\eta_1^*}{d\phi}\right) + \frac{EI_3}{R^4}\left(\frac{d^4\eta_2^*}{d\phi^4} + \frac{d^3\eta_1^*}{d\phi^3}\right) - \rho\omega^2\left(A\eta_2^* - \frac{I_3}{R^2}\left(\frac{d\eta_1^*}{d\phi} + \frac{d^2\eta_2^*}{d\phi^2}\right)\right) = 0 \quad (19)$$

For deformations normal to the plane of the element, the equations of motion are

$$\frac{EI_2}{R^4}\left(\frac{d^4\eta_3^*}{d\phi^4} - R\frac{d^2\psi_1^*}{d\phi^2}\right) - \frac{Gk}{R^3}\left(\frac{d^2\psi_1^*}{d\phi^2} + \frac{d^2\eta_3^*}{Rd\phi^2}\right) - \rho\omega^2\left(A\eta_3^* - \frac{I_2}{R^2}\frac{d^2\eta_3^*}{d\phi^2}\right) = 0$$

(20)

$$\frac{EI_2}{R^3}\left(R\psi_1^* - \frac{d^2\eta_3^*}{d\phi^2}\right) - \frac{Gk}{R^3}\left(\frac{d^2\eta_3^*}{d\phi^2} + R\frac{d^2\psi_1^*}{d\phi^2}\right) - \rho\omega^2 J\psi_1^* = 0$$

The removal of time dependence for η and ψ is indicated by the star superscripts. Since the time dependence has been removed, the normal mode shape function for each mode n, of the twelve modes, may be expressed as

$$\eta_{in}^* = \tilde{\eta}_{in} e^{\nu_n \phi}, \quad i = 1,2,3 \; ; \qquad \psi_{in}^* = \tilde{\psi}_{in} e^{\nu_n \phi}, \quad i = 1,2,3 \qquad (21)$$

Substituting η_{1n}^*, η_{2n}^* and their derivatives into Eqs. (19), yields six roots for ν_n (n=1,2,...,6) and six mode shapes which relate η_{1n}^* to η_{2n}^* for each mode. Likewise substitution of η_{3n}^* and ψ_{1n}^* into Eqs. (20), yields six additional roots of ν_n (n=7,8,...,12) which relate η_{3n}^* to ψ_{1n}^*. The shape functions for ψ_{2n}^* and ψ_{3n}^* may be expressed in terms of η_{1n}^*, η_{2n}^*, and η_{3n}^* for each of the n modes by

$$\psi_2^* = -\frac{1}{R}\frac{d\eta_3^*}{d\phi}, \qquad \psi_3^* = \frac{1}{R}\left(\eta_1^* + \frac{d\eta_2^*}{d\phi}\right) \qquad (22)$$

Having determined the shape functions η^* and ψ^* for each mode, the time dependent solutions for the differential equations of motion may now be expressed in the same form as Eq. (9), where the C_n are the unknown modal amplitudes and $\delta_n(t)$, and $\theta_n(t)$, are assumed harmonic. The twelve scalar quantities of responses at nodes A and B, in the local coordinates may be expressed in a matrix relationship similar to Eq. (10) by

$$\{\overline{\eta}_A, \overline{\psi}_A, \overline{\eta}_B, \overline{\psi}_B\} = [G_{ij}]\{C_1, C_2, \ldots, C_{12}\} \qquad (23)$$

where $[G_{ij}]$ matrix is a twelve by twelve, non-symmetric, matrix which contains terms represented by exponential type functions of Eqs. (21).

The scalar components of the load vector, in terms of local curvilinear coordinates response, are given by

$$\lambda_x(\delta) = -\frac{AE}{R}\left(\frac{\partial\eta_1}{\partial\phi} - \eta_2\right) \qquad (24)$$

$$\lambda_y(\delta) = \frac{EI_3}{R^3}\left(\frac{\partial^3\eta_2}{\partial\phi^3} + \frac{\partial^2\eta_1}{\partial\phi^2}\right) - f_1 \qquad (25)$$

$$\lambda_z(\delta) = -\frac{Gk}{R^2}\left(\frac{\partial\psi_1}{\partial\phi} + \frac{1}{R}\frac{\partial\eta_3}{\partial\phi}\right) - \frac{EI_2}{R^3}\left(R\frac{\partial\psi_1}{\partial\phi} - \frac{\partial^3\eta_3}{\partial\phi^3}\right) + f_2 \qquad (26)$$

$$\lambda_x(\theta) = -\frac{Gk}{R}\left(\frac{\partial\psi_1}{\partial\phi} + \frac{1}{R}\frac{\partial\eta_3}{\partial\phi}\right) \qquad (27)$$

$$\lambda_y(\theta) = \frac{EI_2}{R^3}\left(\frac{\partial^3\eta_3}{\partial\phi^3} + \psi_1\right) \qquad (28)$$

$$\lambda_z(\theta) = -\frac{EI_3}{R^2}\left(\frac{\partial^2\eta_2}{\partial\phi^2} + \frac{\partial\eta_1}{\partial\phi}\right) \qquad (29)$$

where f_1 and f_2 are distributed loads through which rotatory inertias are included. Taking the necessary derivatives of the η, ψ vectors and substituting into Eqs. (24) through (29) yields a column of twelve loads, six for each end, which are expressed in matrix form by

$$\{\overline{\lambda}_A(\delta), \overline{\lambda}_A(\theta), \overline{\lambda}_B(\delta), \overline{\lambda}_B(\theta)\} = [B_{ij}]\{C_1, C_2, \ldots, C_{12}\} \qquad (30)$$

From this point on, the procedure is identical to that of the previous section except that in this case $[B_{ij}]\ [G_{ij}]^{-1}$ represents the matrix of dynamic stiffness coefficients for the curved element and the transformation matrix differs for each end node of the curved elements.

POTENTIAL ENERGY OF SHEAR PANEL ELEMENTS

Solving for the in-plane strains ε_{11}, ε_{22}, and ε_{12} of the panel and then transforming to the inertial coordinates of the system is the basis for deriving the potential energy U_p of a triangular shear panel. A similar approach is used if the strain field of a quadrilateral shear panel is desired; however, since the qradrilateral panel may be idealized by two triangular panels, it has been excluded from this work. Let the geometry of the triangular surface be defined by the positions of the vertices at P_A, P_B, and P_C in local coordinates as shown in Fig. 3. The plane strain is expressed [6] as

$$\varepsilon_{11} = [P_{C2}(\eta_{B1}-\eta_{A1})]\ /2\Omega \tag{31}$$

$$\varepsilon_{22} = [P_{C1}(\eta_{A2}-\eta_{B2}) + P_{B1}(\eta_{C2}-\eta_{A2})]\ /2\Omega \tag{32}$$

$$\varepsilon_{12} = [P_{C1}(\eta_{A1}-\eta_{B1}) + P_{B1}(\eta_{C1}-\eta_{A1}) + P_{C2}(\eta_{B2}-\eta_{A2})]\ /2\Omega \tag{33}$$

where Ω is the area of the panel. Substituting Eqs. (31), (32), and (33) into the following expression,

$$U_p = \left[\frac{Eh}{1-\mu^2}(\varepsilon_{11}^2 + 2\mu\varepsilon_{11}\varepsilon_{12} + \varepsilon_{22}^2) + \frac{Eh}{2(1+\mu)}\varepsilon_{12}^2\right]\frac{\Omega}{2} \tag{34}$$

and employing the transformation of Eq. (3), provides the desired form of potential energy for triangular shear panels.

KINETIC ENERGY OF RIGID BODIES

A rigid body is defined by the inertial coordinates of its center of mass (m), its mass and inertial tensor, and the node coordinates of structural elements connected to it. The spacing of all nodes, and positions, on a rigid body will remain invarient with respect to each other.

Assuming the system behaves in a linear manner for small oscillations, the kinetic energy T_m for a rigid body of arbitrary shape as shown in Fig. 4 may be expressed in matrix form [7, 8] as

$$T_m = \frac{1}{2}\left\{\dot{\bar{q}}(\delta), \dot{\bar{q}}(\delta)\right\}_m^T \cdot [M_m] \cdot \left\{\dot{\bar{q}}(\delta), \dot{\bar{q}}(\theta)\right\}_m \tag{35}$$

where

$$M_m = \left[\begin{array}{ccc|ccc} M & & & & & \\ & M & & & \text{NULL} & \\ & & M & & & \\ \hline & & & I_{xx} & -I_{xy} & -I_{xz} \\ & \text{NULL} & & -I_{yx} & I_{yy} & -I_{yz} \\ & & & -I_{zx} & -I_{zy} & I_{zz} \end{array}\right] \tag{36}$$

represents the mass-inertia matrix of the rigid body m, and $\dot{\bar{q}}$ implies time rate of change of $\overline{q}$. If a rigid body is idealized as a point mass, with zero inertia tensor, then the kinetic energy may simply be expressed, in terms of the translational velocity of the node to which the mass is attached, by

$$T_A = \frac{M_A}{2} \left\{ \dot{\overline{q}}_A(\delta) \right\}^T \cdot \left\{ \dot{\overline{q}}_A(\delta) \right\} \tag{37}$$

CONSTRAINT ENERGY

In addition to the potential and kinetic energy of the structural elements, there also exists a Lagrangian form of constraint energy [9] due to the internal loads acting at each node.

Elastic compliance at an unconstrained node in space (i.e. a node free to displace and rotate within the bounds of the elastic properties of the structural elements connected to it), requires that the internal loads at the node be in equilibrium. Therefore, constraint energy is not required for the unconstrained node, and the normal six equations of equilibrium associated with displacement and rotation, are sufficient to characterize the node properties. On the other hand, when a node is constrained to the ground, or to rigid bodies, it is essential to include the constraint energy which will yield unknown constraint load terms in the equilibrium equations, as well as in the associated compatibility equations for the node.

The form of constraint energy due to the load at node r on a rigid body as shown in Fig. 4, is given in terms of the local coordinates by

$$U_r = \begin{Bmatrix} \overline{\lambda}(\delta) \\ \overline{\lambda}(\theta) \end{Bmatrix}_r^T \cdot \left\{ [C_r]\,[T_{mr}]^T \begin{Bmatrix} \overline{q}(\delta) \\ \overline{q}(\theta) \end{Bmatrix}_m + [C_r] \begin{Bmatrix} \overline{q}(\delta) \\ \overline{q}(\theta) \end{Bmatrix}_r \right\} \tag{38}$$

where the transfer matrix T_{mr} between the center of mass and node is given as

$$T_{mr} = \left[\begin{array}{ccc|ccc} -1 & & & & & \\ & -1 & & & \text{NULL} & \\ & & -1 & & & \\ \hline 0 & Z & -Y & -1 & & \\ -Z & 0 & X & & -1 & \\ Y & -X & 0 & & & -1 \end{array} \right] \tag{39}$$

and the C constraint matrix is composed of terms similar to the [e] transformation matrix from Eq. (1), such that

$$C = \left[\begin{array}{c|c} e(\delta) & \text{NULL} \\ \hline \text{NULL} & e(\theta) \end{array} \right] \tag{40}$$

Each row of the two 3 x 3 sub-matrices $e(\delta)$ and $e(\theta)$ of Eq. (40), contains the components of the local unit coordinate vector which is, preferably, in the direction of constraint. If a node is fully constrained, in the direction of the inertial coordinates, then the constraint matrix C reduces to an identity matrix I. If a node is only partially constrained, the rows of the 3 x 3 sub-matrices corresponding to the unconstrained local directions become null or zero rows.

The form of constraint energy when a node A is constrained to the ground is given by

$$U_g = \begin{Bmatrix} \overline{\lambda}(\delta) \\ \overline{\lambda}(\theta) \end{Bmatrix}_A^T \cdot \left\{ [C_g] \begin{Bmatrix} \overline{q}(\delta) \\ \overline{q}(\theta) \end{Bmatrix}_A + [I] \begin{Bmatrix} \overline{w}(\delta) \\ \overline{w}(\theta) \end{Bmatrix}_g \right\} \tag{41}$$

where $\bar{w}_g$ represents displacements and rotations of the ground in local coordinates.

FORMULATION OF THE HYBRID ELASTODYNAMIC EQUATIONS

Assembling the total kinetic, potential and constraint energies of the system into the Lagrangian L and then separating Lagrange's equation into two forms, for the q_j and λ_i variables, gives

$$\frac{d}{dt}\left(\frac{\partial L}{\partial \dot{q}_j}\right) - \frac{\partial L}{\partial q_j} = f_j \tag{42}$$

$$\frac{d}{dt}\left(\frac{\partial L}{\partial \dot{\lambda}_i}\right) - \frac{\partial L}{\partial \lambda_i} = 0 \tag{43}$$

Eq. (42) yield equilibrium equations, some of which contain λ terms, and Eq. (43) yields the compatibility equations for node connections on the ground, and the rigid bodies.

For an assumed harmonic solution, the coupled system of equations, derived from Eqs. (42) and (43) may be assembled in the following form

$$(D_{uu}-\omega^2 M_u)q_u + D_{ur}q_r + D_{ug}q_g = f_u \tag{44}$$

$$D_{ur}^T q_u + (D_{rr}-\omega^2 M_r)q_r + D_{rg}q_g + C_r^T\lambda_r = f_r \tag{45}$$

$$D_{ug}^T q_u + D_{rg}^T q_r + D_{gg}q_g + C_g^T\lambda_g = f_g \tag{46}$$

$$C_r q_r + C_r T_{mr}^T q_m = 0 \tag{47}$$

$$C_g q_g = w_g \tag{48}$$

$$T_{mr} C_r^T \lambda_r - \omega^2 M q_m = f_m \tag{49}$$

where subscripts u, r, g, m represent unconstrained nodes, nodes constrained to the rigid bodies, nodes constrained to the ground, and nodes located at the center of mass, respectively.

The symmetric M matrix is an uncoupled diagonal mass-inertia matrix, made up of smaller 6 x 6 sub-matrices for each rigid body. Each sub-matrix is of the general form given by Eq. (36). The symmetric matrix D, consisting of dynamic stiffness coefficients, is obtained by differentiating the potential energy functions for prismatic members and shear panels, and is composed of D_{uu}, D_{rr}, D_{gg}, D_{ur}, D_{ug}, and D_{rg} in Eqs. (44) through (46) where D_{uu} represents coefficients for unconstrained nodes, D_{rr} represents coefficients for nodes constrained to the rigid bodies, D_{gg} represents the ground constraints, and D_{ur}, D_{ug}, D_{rg}, represent the coupling matrices between the previous matrices. The transfer matrix T, is a non-symmetric matrix composed of smaller 6 x 6 non-symmetric transfer matrices (one for each node constrained to a rigid body). Each sub-matrix is of the general form given by Eq. (39). Finally, the C_r and C_g constraint matrices are, non-symmetric matrices made up of smaller 6 x 6 sub-matrices as defined by Eq. (40), and pertain to nodes constrained to the rigid bodies and the ground, respectively.

Equations (44) through (49) may now be assembled in the basic matrix format, shown in Fig. 5, which represents the hybrid elastodynamic equations for any constrained spatial vibratory system. The effects of viscous damping may be included as damping forces which will contribute imaginary elements to the matrix of real elements as shown in Fig. 5. The resulting matrix operations are then performed in the complex domain.

The elastodynamic equations may now be reduced in size by matrix algebra. For the purpose of illustrating the reduction procedure, it will be assumed that the rigid bodies are idealized as point masses lumped at the nodes -- the procedure can then be expanded to include the more general case of arbitrary shaped rigid bodies. The point mass assumption reduces the elastodynamic equations to the form

$$(D_{uu}-\omega^2 m_u)q_u + D_{ug}q_g = f_u \tag{50}$$

$$D_{ug}^T q_u + D_{gg}q_g + C_g^T\lambda_g = f_g \tag{51}$$

$$C_g q_g = w_g \tag{52}$$

where the r and m node equations (and terms) are eliminated, because r and m coincide with the u or g nodes for the case of point masses. The response q_u is then determined separately for the ground and the force excitations, and the results are added.

For the response due to the ground excitation, the applied forces f_u and f_g are removed (set equal to zero) and, referring to Eq. (52), solving for the motion of q_g as a function of the ground excitation vector w_g yields

$$q_g = C_g^T \cdot w_g \tag{53}$$

From Eq. (50), with $f_u = 0$ and q_g known, the solution for q_u is

$$q_{u1} = -(D_{uu}-\omega^2 M_u)^{-1} \cdot D_{ug} \cdot C_g^T \cdot w_g \tag{54}$$

Conversely, by setting the ground motion equal to zero, the response of q_u due to the applied forces f_u is given as

$$q_{u2} = (D_{uu} - \omega^2 M_u)^{-1} \cdot f_u \tag{55}$$

The total response then is given as the sum of Eqs. (54) and (55).

For the more general case of the hybrid elastodynamic equations as shown in Fig. 5, the center of mass responses, q_{m1} and q_{m2}, due to ground excitation w_g, and due to applied forces f_u, f_r and f_m are given as

$$\{q_{m1}\} = [S_{mr}]^{-1} \cdot \{[T_{mr}]\,[\bar{D}_{rg}]\,[C_g^T]\,\{w_g\}\} \tag{56}$$

$$\{q_{m2}\} = [S_{mr}]^{-1} \cdot \{\{f_m\} + [T_{mr}]\,\{f^*\}\} \tag{57}$$

where

$$[S_{mr}] = [[T_{mr}][[D^*_{rr}] - [D_{ru}][D^*_{uu}]^{-1}[D_{ur}]][T_{mr}^T] - \omega_f^2\,[M]] \tag{58}$$

$$[\bar{D}_{rg}] = [[D_{rg}] - [D^*_{uu}]^{-1}[D_{ug}] \tag{59}$$

$$\{f^*\} = \{[D_{ru}][D^*_{uu}]^{-1}\,\{f_u\} - \{f_r\}\} \tag{60}$$

and the D^*_{rr} and D^*_{uu} matrices contain the inertia effects of point masses attached to the r and u nodes, respectively.

RESPONSE DUE TO STEADY STATE AND TRANSIENT EXCITATION

Exclusion of the explicit time dependence of the elastodynamic equations, by operating in the frequency domain, resulted in a set of frequency spectrums for q_u applicable to both the deterministic and non-deterministic areas. In the Fourier integral approach the frequency spectrum $H(\omega)$ of q_u acts as a transmissibility function and, by the nature of its derivation, is generalized to any forcing function which may be obtained by the superposition of a number of harmonic forcing functions.

The excitation spectrum $F(\omega)$, is obtained from the forward transform of the forcing function $f(t)$. Multiplying the total transmissibility for a particular degree of freedom of a given node by the excitation spectrum yields the frequency response of that node as

$$R(\omega) = H(\omega) \cdot F(\omega) \qquad (61)$$

From Eq. (61) the desired time response $r(t)$ may be obtained by the reverse transform of $R(\omega)$ given as

$$r(t) = \frac{1}{2\pi}\int_{-\infty}^{\infty} R(\omega)\, \varepsilon^{i\omega t}\, d\omega \qquad (62)$$

Several simple Fourier transform algorithms exist [10, 11, 12] which may be used to solve for $F(\omega)$ and Eq. (62). The "fast Fourier transform" was used in this paper.

PROBABILISTIC RESPONSE OF RANDOMLY EXCITED SYSTEMS [13,14,15,16]

In describing the response characteristics of linear systems excited by random dynamic loadings, it will be assumed that a given random excitation will have stationary and ergodic properties, and that a Gaussian random process will yield a Gaussian response. The response of such systems are most conveniently described by use of the power spectral density function. Power is an additive quantity and direct integration of the power spectral density for the frequency range of interest, gives information on the total power of the process. This in turn may be used to obtain the mean square response of the structural system due to the random excitations.

The power spectral density $P(\omega)$ of a linear system at any frequency, is equal to the excitation spectral density $S_e(\omega)$, times the square of the absolute value of the transmissibility function $H(\omega)$ at that frequency, and is given by

$$P(\omega) = |H(\omega)|^2\, S_e(\omega) \qquad (63)$$

Integrating over the frequency range of interest yields the mean square displacement response of the structure as

$$r^2(t) = \int_{\omega_1}^{\omega_2} P(\omega)\, d\omega \qquad (64)$$

where ω_1, ω_2 represent the bandwidth limits of the system.

ILLUSTRATIVE EXAMPLE

The system, shown in Fig. 6, contains four prismatic elastic member elements, and two disk shaped rigid bodies, which are connected rigidly to the lower ground and by momentless connections to the upper ground. The inertial coordinates are located at the rigid body node m_1. A summary of the geometrical and physical properties of the system is given in Table 1. The matrix format of the hybrid elastodynamic equations is illustrated in Fig. 7.

Table 1. Structural Properties of Example

a. Node data

Node	x(in.)	y(in.)	z(in.)	Constraint Remarks
r1	0.5	-2.5	0.0	Full constraint to rigid body m_1
r2	0.5	0.0	2.5	Full constraint to rigid body m_1
r3	-0.5	0.0	0.0	Full constraint to rigid body m_1
r4	-3.0	0.0	0.0	Full constraint to rigid body m_2
r5	-4.0	0.0	0.0	Full constraint to rigid body m_2
s1	8.0	-2.5	0.0	Momentless (pin-point)
s2	8.0	0.0	2.5	Momentless (pin-joint)
g1	-6.5	0.0	0.0	Full constraint to ground
m1	0.0	0.0	0.0	None
m2	-3.5	0.0	0.0	None

b. Member (massless) data

Member	A(in.^{2})	E(psi.)	$I_2 = I_3$(in.^{4})	G(psi)	J(in.^{4})
r1 - s1	.0123	$30x10^6$	$.119x10^{-4}$	$12.5x10^6$	$.238x10^{-4}$
r2 - s2	.0123	$10.5x10^6$	$.119x10^{-4}$	$4.2x10^6$	$.238x10^{-4}$
r3 - r4	.0123	$10.5x10^6$	$.119x10^{-4}$	$4.2x10^6$	$.238x10^{-4}$
r5 - g1	.0123	$10.5x10^6$	$.119x10^{-4}$	$4.2x10^6$	$.238x10^{-4}$

c. Rigid body data

Body	Mass(slugs)	I_{xx}	I_{yy}	I_{zz}	Products of Inertia
m_1	.00924	.0512	.0256	.0256	0
m_2	.00924	.0512	.0256	.0256	0

A harmonic (steady state sinusoidal) forcing function of unit amplitude was applied in the x-direction to the lower ground node. The process was repeated for a series of frequencies to determine the transmissibility (output/input) function, or frequency spectrum, H(ω) of the center of mass nodes. Back substitution in the elastodynamic equations yields the spectrum of the remaining node positions.

Figures 8 and 9 illustrate the spectrum of q_m in the x-direction for the bandwidth of 70 to 900 Hertz (cycles per second). The lower mode natural frequencies were detected at 30, 160, 270, 380, and 620 Hz. Beyond this point the system attenuated rapidly.

After examination of the frequency response, a shock impulse of 1/1000 unit amplitude was applied in the x-direction of the lower ground node. The time response was numerically determined by using the "fast Fourier transform" and checked by using Fourier sine-cosine transforms. The transmissibility functions H(ω), were obtained from the steady state frequency response. Figure 10 illustrates the transient response in the x-direction of both center of mass nodes.

In addition to the transient excitation, a white noise random excitation was also applied to the x-direction of the lower ground node. The white noise spectral band width was assumed to extend from zero to 1000 Hz with an excitation spectral density $S_e(\omega)$ of 1/1000 unit amplitude. The total mean square displacement response was calculated by assuming the integral of Eq. (64) to be a series summation and the resulting root mean square displacement amplitude, in the x-direction of the center of mass nodes, was calculated as 2.375 and 2.264 units for nodes m_1 and m_2 respectively.

CONCLUSIONS

The hybrid elastodynamic formulation provides a consistant and effective technique for determining steady state, transient, and probabilistic response characteristics of complex structural configurations with general constraints. Symmetry and the partitioned form of the elastodynamic matrices minimizes the storage and ill conditioning problem of large order, sparse matrices normally associated with such systems. Furthermore, the response to a general class of excitations has been greatly simplified by operating in the frequency domain and employing numerical fast Fourier transform techniques. The hybrid elastodynamic formulation has also been successfully extended to structural configurations coupled to stiffened plate and shell elements, as well as to the dynamic analysis of structural sub-systems. These results will be reported on in future publications.

REFERENCES

1. T. C. Huang and C. L. Lee, "Free Vibrations of Space Framed Structures," Proc. 11th Midwestern Mechanics Conference, vol. 5, 1969, pp. 861-885.
2. E. C. Pestel and F. A. Leckie, Matrix Methods in Elasto-Mechanics, McGraw-Hill, Inc., New York, 1963.
3. J. H. Argyris and S. Kelsey, Energy Theorems and Structural Analysis, Butterworths, London, 1960.
4. K. J. Saczalski, "Energy Techniques and Matrix Methods for Computer Programming of Dynamic Structural Response," ASME Vibrations Conference, Paper No. 67 - VIBR - 56, 1967.
5. L. Beitch, "Vibration Response of General Continuous Structures," ASME Vibrations Conference, Paper No. 67 - VIRB - 44, 1967.
6. J. H. Argyris and S. Kelsey, Modern Fuselage Analysis and the Elastic Aircraft, Butterworths, 1963.
7. E. T. Whittaker, A Treatise on the Analytical Dynamics of Particles and Rigid Bodies, 4th ed., Dover Pub., New York, 1944.
8. Yu Chen, Vibrations: Theoretical Methods, Addison-Wesley Pub., 1966.
9. C. Lanczos, Variational Principles of Mechanics, 3rd ed., University of Toronto Press, 1966.
10. J. W. Cooley, et al, "The Finite Fourier Transform," IEEE Transactions on Audio and Electroacoustics, vol. Au-17, No. 2, June 1969, pp. 77-85.
11. G. D. Bergland, "A Guided Tour of the Fast Fourier Transform," IEEE Spectrum, July 1969, pp. 41-51.
12. J. W. Cooley and J. W. Tukey, "An Algorithm for the Machine Calculation of Complex Fourier Series," Math. Comput., vol. 19, Apr. 1965, pp. 297-301.
13. L. Meirovitch, Analytical Methods in Vibrations, Macmillan Co., New York, 1967.
14. W. C. Hurty and M. F. Rubinstein, Dynamics of Structures, Prentice-Hall, Inc., New Jersey, 1964.
15. C. Bingham, M. D. Godfrey, and J. W. Tukey, "Modern Techniques of Power Spectrum Estimation," IEEE Transactions on Audio and Electroaucoustics, vol. Au-15, no. 2, June 1967, pp. 56-66.
16. Y. K. Lin, Probabilistic Theory of Structural Dynamics, McGraw-Hill, Inc., New York, 1967.

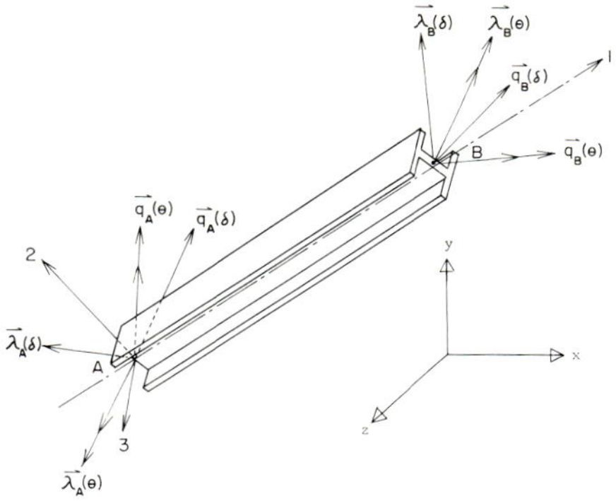

Fig.1 Prismatic Member Element

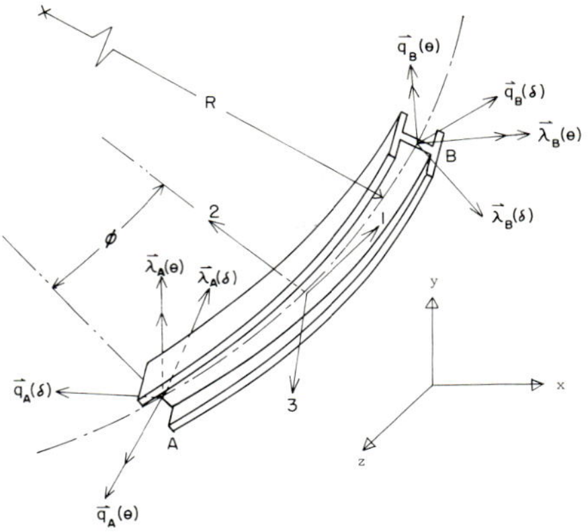

Fig.2 Ring (Curved Member) Element

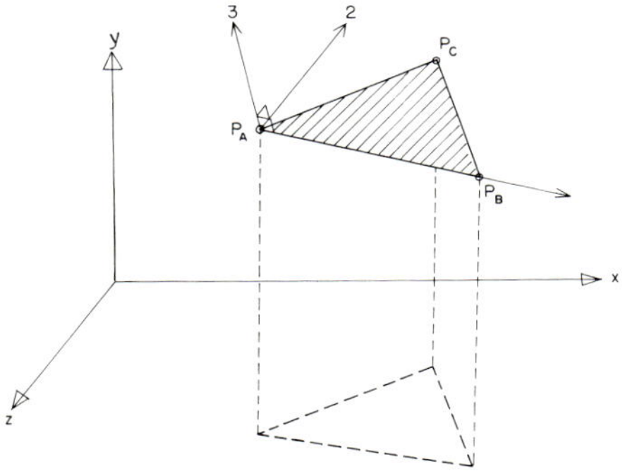

Note : P_A, P_B, P_C are Coordinate Positions Measured in the Local Coordinates (1, 2, 3) of the Panel.

Fig.3 Triangular Shear Panel Element

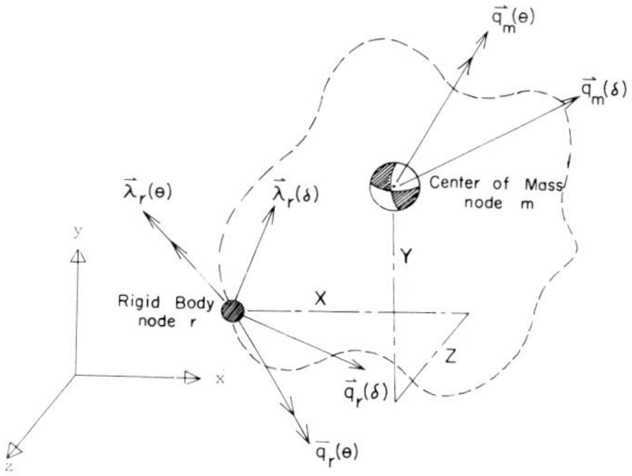

Fig.4 Rigid Body Element

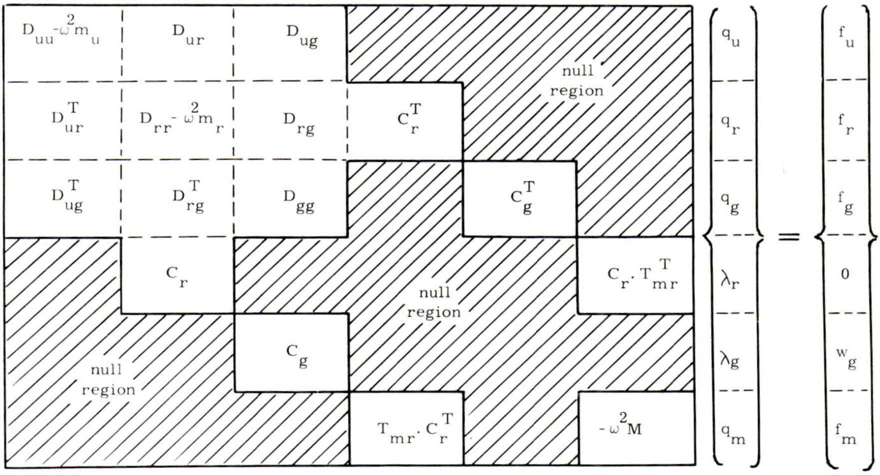

Fig. 5 Matrix Format of the Elastodynamic Equations

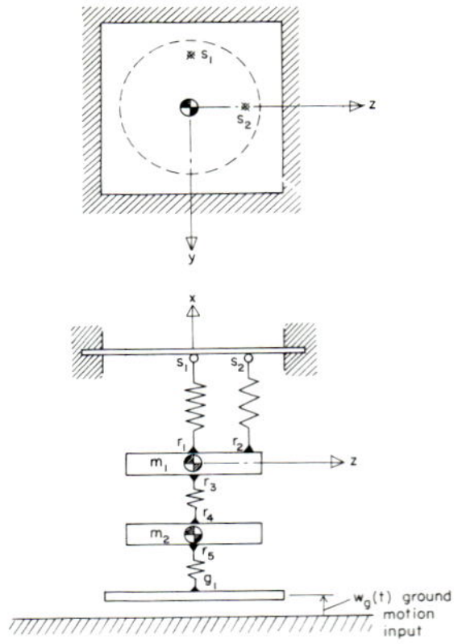

Fig.6   A Coupled System

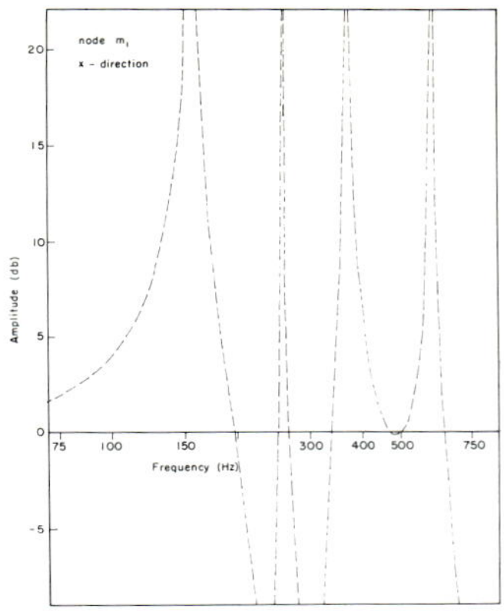

Fig.8   Frequency Spectrum - Node m_1

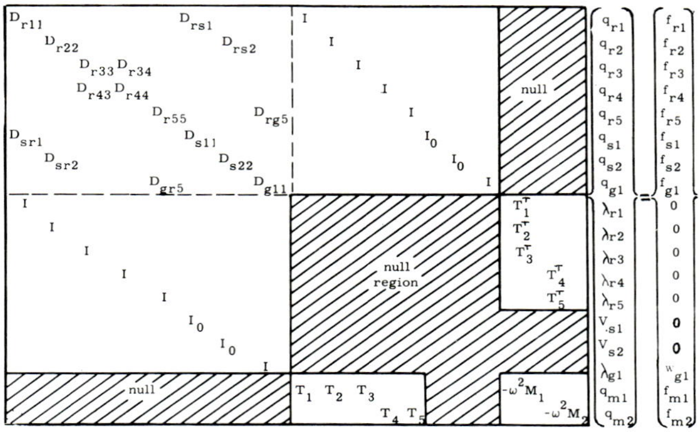

Fig. 7   Matrix Elastodynamic Equations of the Example

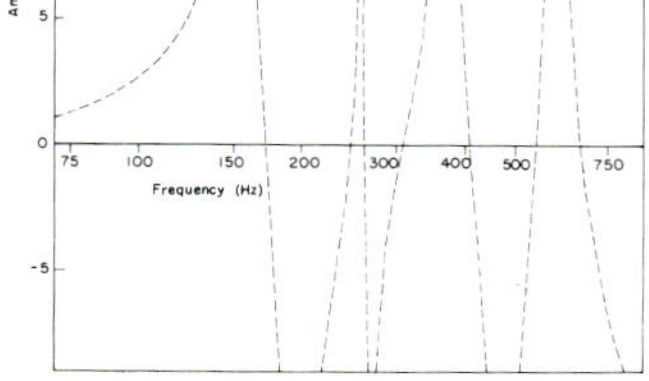

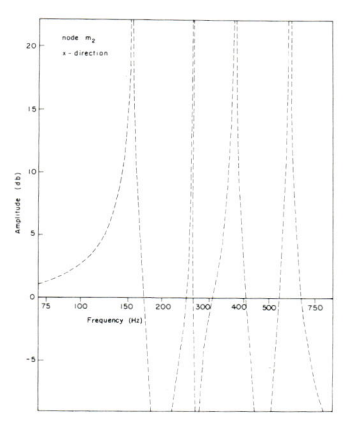

Fig.9   Frequency Spectrum - Node m_2

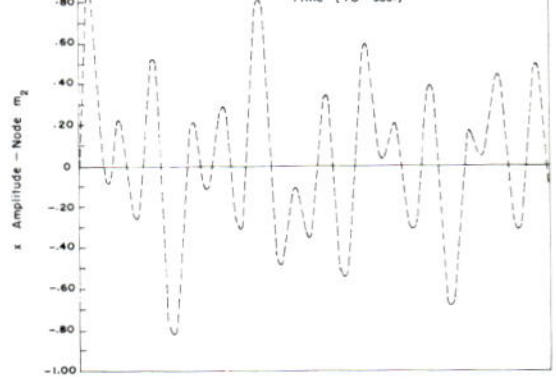

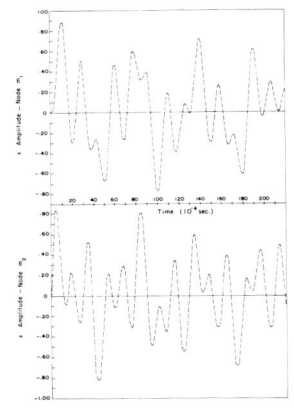

Fig.10   Transient Responses

Developments in Mechanics, Vol. 6. Proceedings of the 12th Midwestern Mechanics Conference: 48

Dynamics of an Automobile in Braking and Acceleration

EDWARD A. SAIBEL and SHANG-LI CHIANG
CARNEGIE-MELLON UNIVERSITY

ABSTRACT

A four wheel automobile is treated as a sprung mass mounted to the wheels by means of springs and dash pots. The non-linear tire characteristics are also introduced.

Transient response of this automobile involving sudden braking or acceleration in rectilinear motion is then obtained numerically by using Euler's self-starting modified method. Pitch angle, angular velocity of pitch, vertical tire forces, load distribution, instantaneous velocity and acceleration of the automobile can be calculated.

Comparison is shown between front wheel, rear wheel, four wheel and cross wheel braking. The effect of a bump or road waviness is also discussed.

INTRODUCTION

Emergency conditions involving heavy braking or acceleration, or sudden manoeuvers are not well described by steady state solutions or linearized systems of forces. To deal completely with vehicle behavior under such conditions transient type solutions must be found. It is rather surprising that the literature is singularly scarce in theoretical approaches to the problem as a whole. Most of the available references give simplified models which ignore the vertical motion of the car or unequal brake action or use a simple two-wheel model.

Saibel, Fox and Bergman [1] in 1961 proposed a model for vehicles in the cornering manoeuver on and off the highway. Saibel and Tsao [5,6] in 1968 extended this model for vehicles under sudden manoeuvers. In 1969 [8] Tsao in his dissertation introduced empirical relations to solve the above problem. Owing to insufficient knowledge of pneumatic tires, empirical equations had to be introduced between tire side forces and the radius of curvature of the path. However, empirical equations limited the application of this model. In doing straight line brake and acceleration analyses the trouble arising from the use of empirical formulas for side forces does not arise.

A model has been developed which includes all important properties such as: location of the center of gravity; moments and products of inertia of sprung mass; wheel mass; spring and damping constants of tires; spring and damping constants of suspension systems; wind effect; car dimensions; tire size; etc.

The analysis leads to five non-linear second order ordinary differential equations which are solved numerically using Euler's self-starting modified method.

The output of the computer program gives the instantaneous velocity of the vehicle, as well as acceleration and distance it travels; all wheel deflections, and deflection rates; all suspension deflections, and deflection rates; vertical tire forces, load distribution; and the point at which it begins to skid for either front wheel drive, rear wheel drive or four wheel drive. It also gives the effect of all physical properties mentioned above to the vehicle response during acceleration or brake processes. For lack of space only a few figures have been shown in this paper. Also, after some simple modifications of the equations of motion, side forces can be calculated if the center of gravity is not on the geometric center line. Since acceleration or braking depends on the manoeuvering of the driver, it is treated as an input in the program.

EQUATIONS OF MOTION

The model we have chosen is that of a sprung mass mounted to the wheels by means of springs and dash pots. Because of the importance of the non-linear properties of the tire we make the springs and dash pots of the suspension system assumed in parallel, connect with the wheel in series with another piar of parallel non-linear tire springs and dash pots as shwon in Fig. 1.

The motion of the sprung mass relative to the ground can be found in terms of the deflections of the springs. These are the Z_{ij}, where i=1,2,3,4, expresses rear right, right front, left front, and rear left, respectively. Road elevation, tire deflection, and spring deflection, Fig. 2, are expressed by j=1,2,3.

If we assume that the motions of the four wheel masses at points 1,2,3,4 are vertical, and assume yaw is unimportant then the roll angle θ and pitch angle φ can be easily expressed as:

$$\theta = \frac{(Z_{43} - Z_{13})}{t} = \frac{(Z_{33} - Z_{23})}{t} \tag{1}$$

$$\varphi = \frac{(Z_{13} - Z_{23})}{\ell} = \frac{(Z_{43} - Z_{33})}{\ell} \tag{2}$$

In order to find the acceleration of the center of gravity, we apply the well known kinematic relationship

$$\ddot{\bar{r}} = \ddot{\bar{R}} + \ddot{\bar{\rho}}_{rel} + 2\,\bar{\omega}x\dot{\bar{\rho}}_{rel} + \dot{\bar{\omega}}x\bar{\rho} + \bar{\omega}x(\bar{\omega}x\bar{\rho}) \tag{3}$$

But because we use three coordinate systems, the acceleration of the center of gravity relative to point O in Fig. 1 will be obtained by first finding the acceleration of the center of gravity relative to point A. Using Eq. 3 we find, Fig. 3.

$$\begin{aligned}\ddot{\bar{r}}_A = {} & \ddot{Z}_{13}\bar{k}' + C\ddot{\varphi}\bar{i}'' - C\ddot{\theta}\bar{j}'' + (b\ddot{\theta} - a\ddot{\varphi})\bar{k}'' \\ & + (b\dot{\theta}\dot{\varphi} - a\dot{\varphi}^2)\bar{i}'' - (b\dot{\theta}^2 - a\dot{\theta}\dot{\varphi})\bar{j}'' \\ & - (C\dot{\theta}^2 + C\dot{\varphi}^2)\bar{k}''\end{aligned} \tag{4}$$

Equation 4 expresses the acceleration of the center of gravity relative to point A. In order to find the acceleration of the center of gravity relative to the inertial frame, we apply Eq. 3 again to the system shown on Fig. 4. In this system, $\ddot{\bar{\rho}}_{rel}$ has the physical meaning of the acceleration of the center of gravity relative to A, which is equal to $\ddot{\bar{r}}_A$, shown on Eq. 4.

Since θ and φ are small quantities, $\dot{\theta}^2, \dot{\varphi}^2$, and $\dot{\theta}\dot{\varphi}$, can be neglected compared with $\ddot{\theta}, \ddot{\varphi}$, and $\ddot{Z}_{13}$ terms. After finding $\bar{\rho}_{rel}$, $\ddot{\bar{R}}$, and transforming the double prime system to the prime system and substituting these quantities into Eq. 3, we have

$$\ddot{\bar{r}} = C\ddot{\varphi}\bar{i}' - C\ddot{\theta}\bar{j}' + (\ddot{Z}_{13} + b\ddot{\theta} - a\ddot{\varphi})\bar{k}' + \ddot{x}_1\bar{i} + \ddot{y}_1\bar{j} \tag{5}$$

In the study of vehicles along a straight line, we know i=i', j=j'. Because the vehicles are constrained to go along a straight line, we will assume vehicles have no roll angle. Thus Eq. 5 reduces to

$$\ddot{\bar{r}} = (\ddot{x}_1 + C\ddot{\varphi})\bar{i}' + (\ddot{Z}_{13} - a\ddot{\varphi})\bar{k}' \tag{6}$$

We shall make use of

$$\bar{G}_G = \dot{\bar{h}}_G \tag{7}$$

where

(1) $\bar{h}_G = h_{xc''}\bar{i}''+h_{yc''}\bar{j}''+h_{zc''}\bar{k}''$

$h_{xc''}=I_x\omega_{x''}-I_{xy}\omega_{y''}-I_{xz}\omega_{z''}$

$h_{yc''}=I_{xy}\omega_{x''}+I_y\omega_{y''}-I_{yz}\omega_{z''}$

$h_{zc''}=-I_{xz}\omega_{x''}-I_{yz}\omega_{y''}+I_z\omega_{z''}$

(2) For convenience, we take axes passing through the center of gravity, parallel to the double prime system, and fixed in the body, (i.e. $x_{c''}y_{c''}z_{c''}$) Because of symmetry, $I_{xy}=I_{yz}=0$.

(3) $\omega_{x''}, \omega_{y''}, \omega_{z''}$ are absolute angular velocities of the double prime system.

(4) "G_G" is the moment of the external impressed forces about the C.G.

The absolute angular velocity of the double prime system is:

$$\omega_{as} = \dot{\theta}\, i'' + \dot{\varphi}\, j'' = \omega_{x''}\, \bar{i}'' + \omega_{y''}\, \bar{j}'' + \omega_{z''}\, \bar{k}'' \tag{8}$$

Substitute Eq. 8 into Eq. 7,

$$\bar{h}_G = I_x\dot{\theta}\bar{i}'' + I_y\dot{\varphi}\bar{j}'' - I_{xz}\dot{\theta}\bar{k}''$$

$$\dot{\bar{h}}_G = I_x\ddot{\theta}\bar{i}'' + I_y\ddot{\varphi}\bar{j}'' - I_{xz}\ddot{\theta}\bar{k}'' + I_x\dot{\theta}\dot{\bar{i}}'' + I_y\dot{\varphi}\dot{\bar{j}}'' - I_{xz}\theta\dot{\bar{k}}''$$

Changing from the double prime to the prime system and neglecting higher order terms,

$$\dot{\bar{h}}_G = I_x\ddot{\theta}\bar{i}' + I_y\ddot{\varphi}\bar{j}' - I_{xz}\ddot{\theta}\bar{k}'$$

For vehicles along a straight line (no roll angle) we have:

$$\dot{\bar{h}}_G = I_y\ddot{\varphi}\bar{j}' \tag{9}$$

In order to write the equations of motion, we use $\bar{G}_G=\dot{\bar{H}}_G$ and $\bar{F}=m\bar{a}$. Let T_i(i=1 to 4) express the effective tractive effort. Since T_i not only accelerates the wheel, but also transfers part of it to accelerate the sprung mass, we have

$$T_{is} = T_i - \frac{W_w}{g}\ddot{x}_i \qquad (10)$$

Couples will be introduced if forces are shifted from one location to another. Thus, couples in the y' direction are:

$$M_{iy'} = -(E_s + Z_{i3} - Z_{i2})T_{is} - (E_w + Z_{i2})T_i \qquad (11)$$

We further define a dynamic wheel load as the instantaneous value of vertical reaction between the tire and the road surface and assume that the engine power is equally divided in the wheels, or that brake forces are equally distributed. In order that the vehicles always go straight, the lateral forces are balanced out. Thus, $T_1=T_4$, $T_2=T_3$, $Z_{12}=Z_{42}$, $Z_{13}=Z_{43}$, $Z_{22}=Z_{32}$, and $Z_{33}=Z_{23}$. We have the equations of motion as follows:

$$2T_1 + 2T_2 - 4\frac{W_w}{g}\ddot{x} - \text{wind} = (\ddot{x} + c\ddot{\varphi})\frac{W_s}{g} \qquad (12)$$

$$2K_s(Z_{12} - Z_{13}) + 2K_s(Z_{22} - Z_{23}) + 2C_s(\dot{Z}_{12} - \dot{Z}_{13}) + 2C_s(\dot{Z}_{22} - \dot{Z}_{23}) = (\ddot{Z}_{13} - a\ddot{\varphi})\frac{W_s}{g} \qquad (13)$$

$$\begin{aligned} &2K_s a(Z_{12} - Z_{13}) + 2C_s a(\dot{Z}_{12} - \dot{Z}_{13}) - 2K_s(\ell - a)(Z_{22} - Z_{23}) \\ &- 2C_s(\ell - a)(\dot{Z}_{22} - \dot{Z}_{23}) + e.\ \text{wind} + (T_{1s} + T_{2s})(a\varphi - C) - T_{2s}\ell\varphi \\ &+ 2\frac{W_w}{g}(2E_s + Z_{13} + Z_{23} - Z_{12} - Z_{22})\ddot{x} - 2(E_s + E_w + Z_{13})T_1 \\ &- 2(E_s + E_w + Z_{23})T_2 = I_y\ddot{\varphi} \end{aligned} \qquad (14)$$

$$\varphi = (Z_{13} - Z_{23})/\ell \qquad (15)$$

$$\frac{W_w}{g}\ddot{Z}_{12} + K_t Z_{12}^{1.76} + C_t\dot{Z}_{12} + K_s(Z_{12} - Z_{13}) + C_s(\dot{Z}_{12} - \dot{Z}_{13}) = 0 \qquad (16)$$

$$\frac{W_w}{g}\ddot{Z}_{22} + K_t Z_{22}^{1.76} + C_t\dot{Z}_{22} + K_s(Z_{22} - Z_{23}) + C_s(\dot{Z}_{22} - \dot{Z}_{23}) = 0 \qquad (17)$$

Where Eq. 16 and Eq. 17 are the equations of motion of the wheel, $Z^{1.76}$ expresses the non-linear tire characteristic. The rolling resistance, and wind effect can be expressed by empirical formulas as for example from Radt and Pacejka [2].

Rolling resistance $= F_v(A' + B'V)$

Wind effect $= \frac{1}{2} C_w A_r \rho V^2$

C_w: 0.45, coefficient of wind

A_r: area of automobile in the largest section, ft^2

V : velocity of the car, ft/sec

F_v: vertical load on each tire, lb_f

We thus have six independent equations with six unknowns x, Z_{12}, Z_{13}, Z_{22},

Z_{23}, and φ. We drop $Z_{13}+Z_{23}-Z_{12}-Z_{22}$ in Eq. 14 because (1) $2E_s \gg Z_{13}+Z_{23}-Z_{12}-Z_{22}$ (2) All other moments in the y' direction $\gg M_{iy'}$

SOLVING TECHNIQUES

After some re-arrangement of Eq. 12 ~ Eq. 17 we have

$$\ddot{x} = C_1(T_1+T_2)-C_2\dot{x}^2 \tag{18}$$

$$\ddot{\varphi} = C_3Z_{12}+[C_4+C_5(T_1+T_2)]Z_{13}+C_6Z_{22}$$
$$+ C_7\dot{Z}_{12}+C_8\dot{Z}_{13}+C_9Z_{22}+[C_{10}T_1+C_{11}]\varphi$$
$$+ C_{12}\dot{\varphi}+C_{13}(T_1+T_2)+C_{14}\dot{x}^2 \tag{19}$$

$$\ddot{Z}_{12} = C_{16}Z_{12}+C_{17}Z_{12}^{1.76}+C_{18}Z_{13}+C_{19}\dot{Z}_{12}+C_{20}Z_{13} \tag{20}$$

$$\ddot{Z}_{22} = C_{21}Z_{13}+C_{22}Z_{22}+C_{23}Z_{22}^{1.76}+C_{24}\dot{Z}_{13}C_{25}\dot{Z}_{22}$$
$$+ C_{26}\varphi+C_{27}\dot{\varphi} \tag{21}$$

$$\ddot{Z}_{13} = C_{28}Z_{12}+[C_{29}+C_{30}(T_1+T_2)]Z_{13}+C_{31}Z_{22}$$
$$+ C_{32}\dot{Z}_{12}+C_{33}\dot{Z}_{13}+C_{34}\dot{Z}_{22}+[C_{35}+C_{36}T_1]\varphi$$
$$+ C_{37}\dot{\varphi}+C_{38}(T_1+T_2)+C_{39}\dot{x}^2 \tag{22}$$

$$Z_{23} = Z_{13}-\ell\varphi \tag{23}$$

where $C_1 \sim C_{39}$ are combinations of car properties.

The initial conditions supply us with values of Z_{13}, Z_{22}, Z_{12}, φ, x, and $\dot{Z}_{13}$, $\dot{Z}_{22}$, $\dot{Z}_{12}$, $\dot{\varphi}$, $\dot{x}$ for t = 0, and we can determine values of $\ddot{Z}_{13}$, $\ddot{Z}_{22}$, $\ddot{Z}_{12}$, $\ddot{\varphi}$, and $\ddot{x}$ when t = 0 by substituting appropriate values into the equations of motion.

We next obtain predicted values of $(\dot{Z}_{13})_1$, $(\dot{Z}_{22})_1$, $(\dot{Z}_{12})_1$, $\dot{\varphi}_1$, $\dot{x}_1$, and $(Z_{13})_1$, $(Z_{22})_1$, $(Z_{12})_1$, φ_1, x_1. Using the predictor equation

$$P(\dot{Z}_{12})_{i+1} = (\dot{Z}_{12})_i + (\ddot{Z}_{12})_i\ \Delta t$$
$$P(Z_{12})_{i+1} = (Z_{12})_i + (\dot{Z}_{12})_i\ \Delta t \tag{24}$$

We thus substitute $p(Z_{12})_1$, $p(Z_{12})_1$ etc, into the equations of motion to obtain approximate values of $(\ddot{Z}_{13})_1$, $(\ddot{Z}_{22})_1$, $(\ddot{Z}_{12})_1$, $\ddot{\varphi}_1$, $\ddot{x}_1$. Then, using the corrected values of $(Z_{13})_1$, and $(\dot{Z}_{13})_1$, etc. are obtained. These values designated as $C(\dot{Z}_{13})_1$ and $C(Z_{13})_1$, etc. are equal to

$$C(\dot{Z}_{12})_{i+1} = (\dot{Z}_{12})_i + [\frac{(\ddot{Z}_{12})_i + P(\ddot{Z}_{12})_{i+1}}{2}] \Delta t$$

$$C(Z_{12})_{i+1} = (Z_{12})_i + [\frac{(\dot{Z}_{12})_i + P(\dot{Z}_{12})_{i+1}}{2}] \Delta t \quad \text{etc.} \qquad (25)$$

The corrected values of $(\dot{Z}_{12})_{i+1}$, $(\dot{Z}_{13})_{i+1}$, $(\dot{Z}_{22})_{i+1}$, $\dot{\varphi}_{i+1}$, $\dot{x}_{i+1}$, and $(Z_{12})_{i+1}$, $(Z_{13})_{i+1}$, $(Z_{22})_{i+1}$, φ_{i+1}, x_{i+1}, etc. are next substituted into the equations of motion to obtain improved values of $(\ddot{Z}_{12})_1$, $(\ddot{Z}_{13})_1$, $(\ddot{Z}_{22})_1$, $\ddot{\varphi}_1$, and $\ddot{x}_1$ which are, in turn, used in the above predictor and corrector equation to obtain still better values of $(\dot{Z}_{12})_1$, $(\dot{Z}_{13})_1$, $(\dot{Z}_{22})_1$, $\dot{\varphi}_1$, $\dot{x}_1$, $(Z_{12})_1$, $(Z_{13})_1$, $(Z_{22})_1$, φ_1, x_1, etc.

The process continues until successive values of $(\dot{Z}_{12})_1$, $(Z_{12})_1$, etc. vary by less than 0.001% of the previous value. At this point we go forward to the next time interval.

SOME RESULTS

Braking of a Vehicle - Figure 5 shows the comparison between front, rear, four wheel, and cross wheel braking. In this case, we take $T_1 = 600\ lb_f$ on each rear wheel or front wheel brake, $T_2 = 300\ lb_f$ on each wheel for four wheel brakes and $T = 600\ lb_f$ for either right front, left reat or right rear, left front cross braking. All physical properties are shown in appendix except $a = (5.5/9.9)\ell$. The initial velocity of the vehicle before braking was taken as 60 MPH. Figure 6 shows the change in normal force when the K_s are different.

Road Waviness and Effect of Obstacles - In the case of road waviness, following the same procedures as before, the kinematic relationship becomes

$$\ddot{\bar{r}} = (\ddot{x}_1 + c\ddot{\varphi})\ \bar{i}' + (\ddot{Z}_{13} - a\ddot{\varphi})\bar{k}' \qquad (6)'$$

Eq. 9 remaining unchanged and Eq. 11 becomes

$$M_{iy'} = -(E_s + Z_{i3} - Z_{i2})T_{is} - (E_w + Z_{i2} - Z_{i1})T_i \qquad (11)'$$

The tire equations change to

$$\frac{W_w}{g}\ddot{Z}_{12} + K_t(Z_{12} - Z_{11})^{1.76} + C_t(\dot{Z}_{12} - \dot{Z}_{11})$$
$$+ K_s(Z_{12} - Z_{13}) + C_s(\dot{Z}_{12} - \dot{Z}_{13}) = 0 \qquad (16)'$$

$$\frac{W_w}{g}\ddot{Z}_{22} + K_t(Z_{22} - Z_{21})^{1.76} + C_t(\dot{Z}_{22} - \dot{Z}_{21})$$
$$+ K_s(Z_{22} - Z_{23}) + C_s(\dot{Z}_{22} - \dot{Z}_{23}) + 0 \qquad (17)'$$

After some rearrangements, the equations reduce to the type shown in Eq. 18 to Eq. 23. Using Euler's improved method for a special case, a vehicle hitting a bump, from Fig. 7 we let

$$x + \ell < L \quad , \quad Z_{21} = 0$$

$$L \le x + \ell \le B + L \quad , \quad Z_{21} = A_m \text{Sin} \frac{\pi[(x+\ell)-L]}{B}$$

$$x + \ell > B + L \quad , \quad Z_{21} = 0$$

also

$$x < L \quad , \quad Z_{11} = 0$$

$$L \le x \le B + L \quad , \quad Z_{11} = A_m \text{Sin} \frac{\pi(x-L)}{B}$$

$$x > B + L \quad , \quad Z_{11} = 0$$

We use the same properties described in Appendix with $a = (5.5/9.9)\ell$. Fig. 8 shows a vehicle starting at 60 MPH. Using front wheel brakes, travelling on a sinusodial road of amplitude $A_m \text{Sin} \frac{\pi X}{B}$ with $A_m = 0.05'$ $B = 7.5'$, 2.44 sec after applying the brake which is 200' from the point of initial braking. Fig. 9 shows what happens to the vehicle when it hits a bump, other conditions being the same as above. Comparison can be seen clearly between sinusodial road (or with a bump) to those without waviness.

In reference [7] Michael has shown an experimental curve displaying the ratio of maximum tire force to static load vs. ratio of bump height to width of wheel. It is interesting to note that using our equations of motion an excellent check is obtained for $\frac{h}{w}$ equal to 0.085 and 0.35 as shown in Fig. 10.

DISCUSSIONS AND CONCLUSIONS

At the present time, most of the available references artifically separate driver reaction from vehicle motion and road factors. However, it is obvious that there are important interactions between these. Man-vehicle-highway systems have been recognized as a new direction in automotive vehicle research but no analyses are available [10].

An attempt has been made in this paper to investigate the transient response of an automobile so that when the driver suddenly locks wheels the effect of road pattern, non-linear characteristics of tires, springs, dampers, and braking characteristics can be found. Since the magnitude of the braking force is much higher than the rolling resistance and wind force, they will have a minor effect on the analytical solution.

Curves shown in Figs. 5, 6 and 9 present the results of some solutions and offer explanations about several important questions [10] such as finding the ideal response times and damping of the vehicle motion which must be considered in studying the interaction of the driver, road, and vehicle.

ACKNOWLEDGEMENT

This work was done under a contract with the National Bureau of Standards. The authors are however solely responsible for the contents. Thanks are due Dr. F. Cecil Brenner, Chief, Tire Systems Section for his help and encouragement.

LIST OF SYMBOLS

A	:	Wheel 1 and ground contact point; origin of X'Y'Z' frame.
A_m	:	Amplitude of the road roughness, in.
Area	:	Largest car sectional area, ft^2.
A',B'	:	Empirical numbers A'=12.5/1000, B'=0.15/1000 ft/sec.

a, b, c : Location of the sprung mass C.G. in X"Y"Z" - frame measured from point 1.

B : One half of the wave length of road rougheness, ft.

C_s : Shock absorber damping constant, lb_f-sec/ft.

C_t : Tire damping constant, lb_f-sec/ft.

e : Location of the wind load below the centroid of the sprung mass, ft.

E_s : Equilibrium length of the steel spring when the car is at rest, ft.

E_w : Equilibrium height of the wheel when the car is at rest, ft.

f : Coefficient of friction.

F_i : Lateral force acting on each tire by the ground in the Y' - axis direction, i=1,2,3,4.

F_{is} : Lateral force acting on the sprung mass in the Y' - axis direction, lb_f, i=1,2,3,4.

F_{iv} : Total vertical load (static load + dynamic load) on i^{th} wheel, lb_f, i=1,2,3,4.

g : Gravitational acceleration 32.2, ft/sec^2.

H : Moment of momentum, lb_f-ft-sec.

I : Moment of inertial of the sprung mass, lb_f-ft-sec^2.

K_s : Steel spring constant, lb_f/ft.

K_t : Tire spring constant (appropriate British units).

ℓ : Wheel base (distance between wheels 1 and 2), ft.

$M_{iX'}$: Moment about the sprung mass C.G. in X' - direction by the lateral forces on the 4 tires, lb-ft, i=1,2,3,4.

$M_{iY'}$: Moment about the sprung mass C.G. in Y' - direction by the longitudinal forces on the 4 tires, lb-ft, i=1,2,3,4.

O : Origin of the inertia frame X Y Z.

T_i : Longitudinal force on each tire in X' - direction, lb_f, i=1,2,3,4.

T_{is} : Longitudinal force on the sprung mass in X' - direction, lb_f, i = 1,2,3,4.

t : Tread, i.e. distance between wheel 1 and 4, ft.

V : Velocity of the car, ft/sec.

W_s : Sprung mass, lb.

W_w : Wheel weight including linkage bar, lb.

X Y Z : Inertia frame origin at O.

X'Y'Z' : Frame origin at A, moving with the car, x' axis passing through right front wheel, and y' axis passing through left rear wheel.

X"Y"Z" : Frame origin at 1, fixed in the sprung mass.

$X''_C Y''_C Z''_C$: Frame origin at C.G. (the centroid of the sprung mass) $X''_C X''$, $Y''_C Y''$, $Z''_C Z''$.

X_i, Y_i : Location of the ith wheel in inertial frame, ft.

$\dot{X}_i, \dot{Y}_i$: Velocity of the ith wheel in inertia frame, ft/sec.

$\ddot{X}_i, \ddot{Y}_i$: Acceleration of the ith wheel in inertia frame, ft/sec^2.

X'_i, y'_i : Location of the ith wheel in terms of i' j' k', ft.

$\dot{X}'_i, \dot{Y}'_i$: Velocity of the ith wheel in terms of i' j' k', ft/sec.

$\ddot{X}'_i. \ddot{Y}'_i$: Acceleration of the ith wheel in terms of i' j' k', ft/sec^2.

Z_{i2} : Tire deflections (functions of time) from their individual equilibrium positions, i.e. measured in the inertia frame, ft.

Z_{i3} : Spring and shock absorber deflections (function of time) from their individual equilibrium positions, i.e. measured in the inertia frame, ft.

θ : Roll angle, vector in the X' - direction, rad.

φ : Pitch angle, vector in the Y' - direction, rad.

ρ : $\rho = ai'' + bj'' + ck''$.

Subscript 1 : Rear right wheel.
Subscript 2 : Front right wheel.
Subscript 3 : Front left wheel.
Subscript 4 : Rear left wheel.

APPENDIX

Car Data

ℓ = wheel base = 119/12 ft

a,b,c = location of C.G. of the sprung mass with respect to x''_c, y''_c, z''_c

= 1/2 ℓ, 1/2t, 0

t = tread = 62/12 ft.

W_s = sprung mass = 4,500 lb

W_w = wheel mass = 40 lb

$I_{x''_c}$ = moment of inertia about C.G. of the sprung mass about X''_c

= 540 lb-ft-sec^2

$I_{y''_c}$ = moment of inertia about Y_c'' through C.G. of the sprung mass

= 3,350 lb-ft-sec^2

$I_{x''_c z''_c}$ = 260 lb-ft-sec^2

Tire Data

K_t = tire spring constant = 19,00 ($F = K_t Z^{1.76}$)

C_t = tire damping constant = 48 ~ 3,200 lb_f-sec/ft (We take C_t = 50)

Estimated Data

E_s = equilibrium distance of steel spring = 0.4'

C_s = shock absorber damping constant = 150 lb_f-sec/ft

K_s = steel spring constant = 108 lb_f/in

BIBLIOGRAPHICAL REFERENCES

[1] Bergman, W., Fox, E. A., and Saibel, E., "Dynamics of an Automobile in a Cornering Manoeuver on and off the Highway," Proc, 1st Int. Cont. Soil Vehicle Systems, Torino, Italy, June 1961, pp. 275-303.

[2] Radt, H. S. and Pacejka, I. H. B., "Analysis of the Steady-State Turning Behavior of an Automobile," Proc, of the Symposium on Control of Vehicle During Braking and Cornering. The Institution of Mechanical Engineers, London, England, 1963, pp. 66-79.

[3] Massaichi Kondo, Fundamentals in Vehicle Dynamics, (In Japanese), Tokyo, 1965.

[4] Bergman, W., "Theoretical Prediction of the Effect of Traction on Cornering Force," SAE Trans., Vol. 67, 1961, p. 614.

[5] Saibel, E. A. and Tsao, M. C., "The Skidding of Vehicles, A Dynamic Analysis," Report 1, National Bureau of Standards CST-430, Washington, D.C., November 1968.

[6] Saibel, E. and Tsao, M. C., "Further Investigations in Vehicle Dynamics," SAE paper no. 700173, January 1970.

[7] Hadekel, R., "The Mechanical Characteristics of Pneumatic Tires," NASA N-48883, NASA Scientific and Technical Information Facility, Fig. 126.

[8] Tsao, M. C., "The Skidding of Vehicles, A Dynamic Analysis," Ph.D. Thesis, Carnegie-Mellon University, June 1969.
[9] Cox, H. L., Aughtie, F., and Brown, A. F., "The Computation of Wheel Impact Forces," Proc. Inst. Automobile Engineers, Vol. 36, (1941-42), p. 45.
[10] Ellis, J. R., Vehicle Dynamics, Lodon, Business Books, 1969, pp. 238-240.

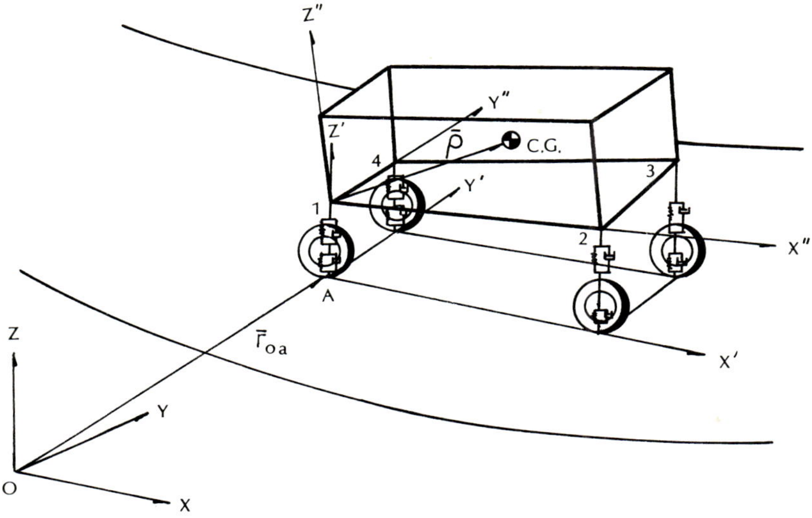

Fig. 1 Coordinate System and Automobile

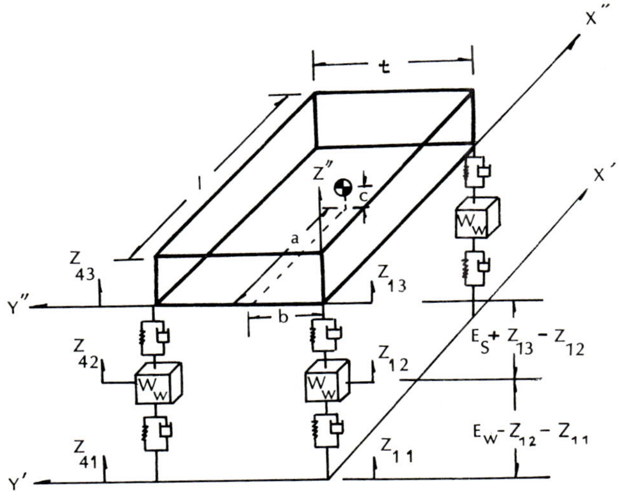

Fig. 2 The Model

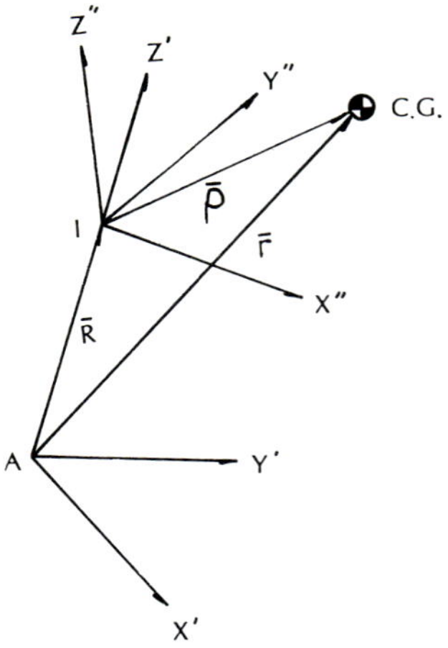

Fig. 3

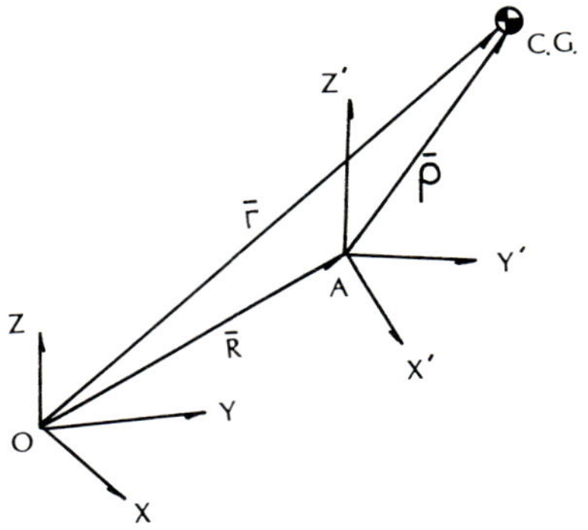

Fig. 4

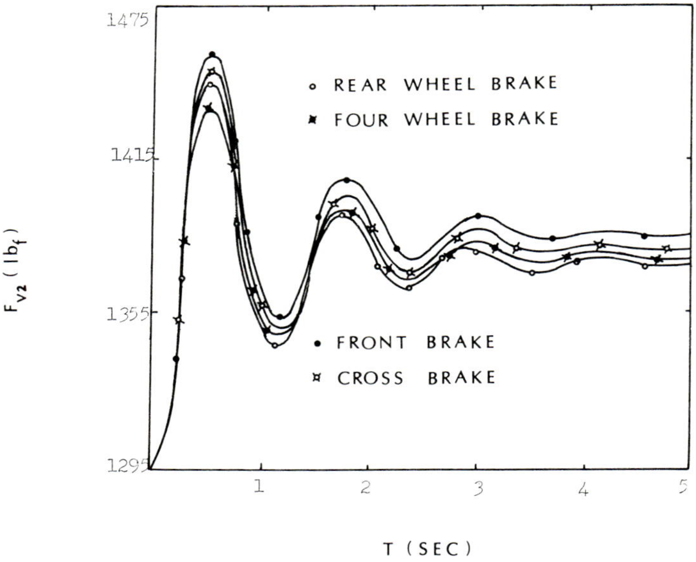

Fig. 5 Vertical Force vs. Time

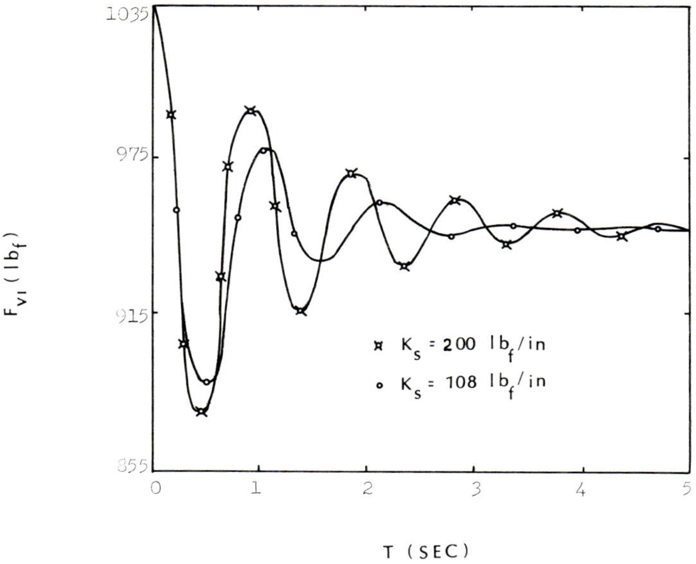

Fig. 6 Vertical Force under Wheel 1 vs. Time, Four Wheel Brake

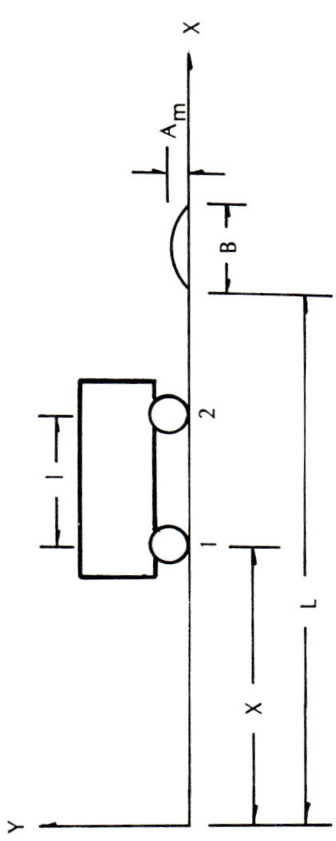

Fig. 7 Bump in Road

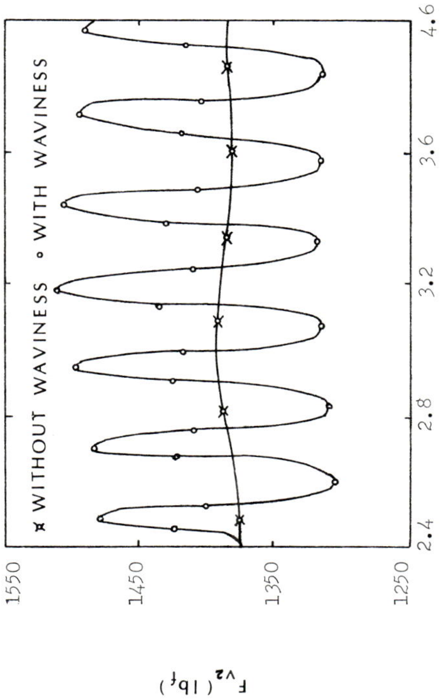

Fig. 8 Normal Force vs. Time

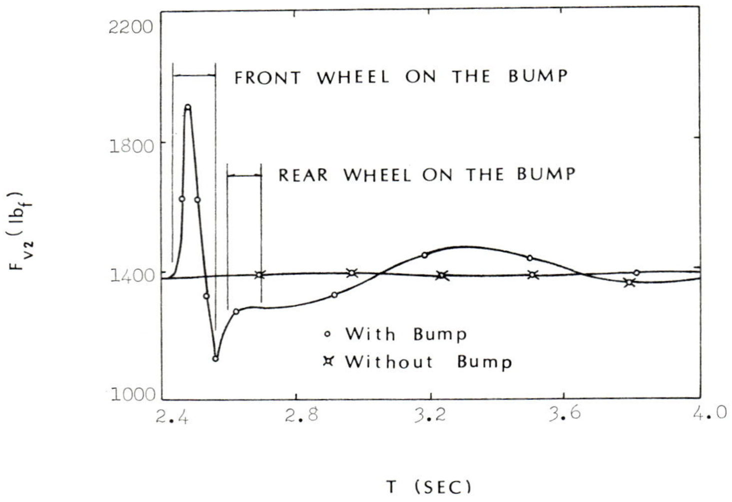

Fig. 9 Normal Force on Front Wheels vs. Time when Vehicle Hits a Bump

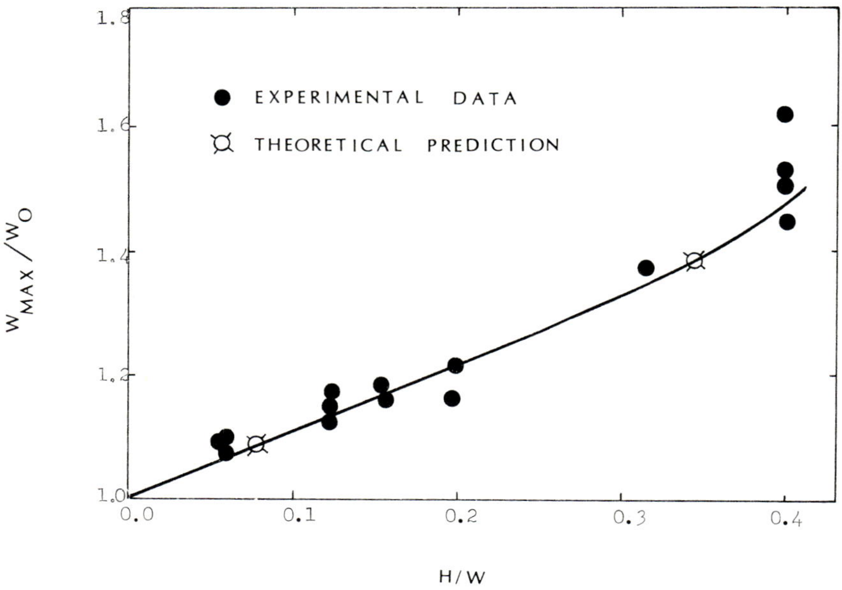

Fig. 10 Comparison with Experimental Results

Developments in Mechanics, Vol. 6. Proceedings of the 12th Midwestern Mechanics Conference: 49

Frictional Effects on the Vibrations of Missiles

H. S. WALKER and CAROL RUBIN
KANSAS STATE UNIVERSITY

ABSTRACT

The natural frequencies and mode shapes of a missile with an idealized air friction force are analyzed. The missile is treated as a homogeneous, elastic, tapered, free-free beam with a constant end thrust. The idealized friction force is a uniformly distributed axial load acting tangent to the instantaneous neutral axis. Simple beam theory, neglecting damping, is used. Using Frobenius' method, a series solution to the equation of motion for a beam of parabolic profile is obtained. The natural frequencies for the first four bending modes of the vibrating beam are found as a function of the thrust and the friction force. The results show that the natural frequencies decrease as the frictional force increases. Thrust versus frequency curves are plotted for cases with and without a control system. Although the results are presented for the beam of parabolic profile only, some other types of tapered beams can be handled with equal facility.

INTRODUCTION

The problem of lateral vibrations in missiles and beams has been treated by several investigators. Sharp [1] discusses a flat free-free beam of non-constant cross-sectional area. Huang and Walker [2] explore the problem of a missile-type body of revolution including a variable end thrust with a directional control system. The lateral vibrations of a beam of constant cross-section and a linearly varying axial load are treated by Fauconneau and Laird [3], and Sun [4]. The axial load is considered to be applied in the direction of an end thrust.

Gorshkov and Shklyarchuk [5] deal with a body of revolution with an end thrust and a directional control system. A friction force is introduced here with a constant end thrust opposing the motion of the missile.

It is the purpose of the present study to introduce an additional variable representing skin friction drag and, using the Huang-Walker paper as a basis for comparison, investigate the effects on the vibration properties. A uniformly distributed axial load is assumed to act tangent to the instantaneous neutral axis of the beam. The simplifying assumptions made for the form of the drag force allow for the derivation and solution of the governing differential equations, and at the same time provide a basis for discussion of the effects of a distributed motion-opposing force on the missile.

The governing differential equation of motion is derived using simple beam theory; and the characteristic equation for the special case of a beam with a parabolic surface of revolution is obtained by means of Frobenius' method.

DERIVATION OF THE EQUATION OF MOTION

Simple beam theory is used to obtain the governing differential equation. The effects of longitudinal vibrations are neglected and the motion is assumed to be planar. Figure 1 shows a differential beam element with the relevant forces and couples acting. The plane of motion is the w,z-plane, as shown, where w is the distance from the z-axis to the beam axis and z is the distance along the beam. The angle α indicates the angle that the vibrating beam axis makes with the equilibrium axis at any given point. A constant axial load representing the drag force is denoted by q and is acting in the direction α. D'Alembert forces are assumed to be acting as shown in Fig. 1, where u is the longitudinal displacement of a material element. It is assumed that u is independent of z, that is, that the rocket material is so stiff that vibrations due to bending occur long before longitudinal vibrations. Therefore, u will be considered dependent only upon t.

D'Alembert's principle yields the following equations of motion:

$$\rho A(z) \frac{\partial^2 u(t)}{\partial t^2} = \frac{\partial}{\partial z} [P(z,t)] - q \tag{1}$$

$$\rho A(z) \frac{\partial^2 w(z,t)}{\partial t^2} = - \frac{\partial}{\partial z} [V(z,t)] + q \frac{\partial w}{\partial z} \tag{2}$$

and

$$P(z,t) \frac{\partial}{\partial z} [w(z,t)] - V(z,t) + \frac{\partial}{\partial z} [M(z,t)] = 0 \tag{3}$$

where a small angle assumption has been made for α.

The cross sectional area and the moment of inertia of the beam at a point z are denoted, respectively, as

$$A(z) = A_o (\frac{z}{\ell})^n$$

and

$$I(z) = I_o (\frac{z}{\ell})^m \tag{4}$$

where A_o and I_o are the respective cross sectional area and moment of inertia of the beam at $z = \ell$. The constants m and n are real numbers and depend upon the choice of rocket shape; for a constant cross sectional area, $m = n = 0$, and a conical cross section has $m = n = 1$.

The following boundary conditions are compatible with the physical system and are assumed to be:

$$P(z,t)\Big|_{z=0} = 0 \quad \text{and} \quad P(z,t)\Big|_{z=\ell} = T_c$$

where T_c is the thrust exerted on the rocket at $z = \ell$. Substituting the boundary conditions into Eq. (1) yields

$$P(z,t) = (T_c - q\ell) (\frac{z}{\ell})^{n+1} + qz \tag{5}$$

Using Eqs. (2), (4) and (5), Eq. (3) becomes

$$\rho A_o (\frac{z}{\ell})^n \frac{\partial^2 w}{\partial t} - q \frac{\partial w}{\partial z} + \frac{\partial}{\partial z} \{[T_c - q\ell] (\frac{z}{\ell})^{n+1} \frac{\partial w}{\partial z} + qz \frac{\partial w}{\partial z}\}$$
$$+ \frac{\partial^2}{\partial z^2} \{[EI_o (\frac{z}{\ell})^m] \frac{\partial^2 w}{\partial z^2}\} = 0 \tag{6}$$

The variables may be non-dimensionalized by taking $\xi = \frac{z}{\ell}$,

$\tau = t\sqrt{\frac{EI_o}{\rho A_o \ell^4}}$, $\bar{T} = \frac{\ell^2 T_c}{EI_o}$, and $\bar{Q} = \frac{\ell^2 q}{EI_o}$. The differential equation of motion now becomes

$$\xi^n \frac{\partial^2 w}{\partial \tau^2} + \frac{\partial^2}{\partial \xi^2}[\xi^m \frac{\partial^2 w}{\partial \xi^2}] + \bar{T}\frac{\partial}{\partial \xi}[\xi^{n+1}\frac{\partial w}{\partial \xi}] - \bar{Q}\frac{\partial}{\partial \xi}[\xi^{n+1}\frac{\partial w}{\partial \xi}] + \bar{Q}\xi\frac{\partial^2 w}{\partial \xi^2} = 0 \tag{7}$$

THE EIGENVALUE PROBLEM

The eigenvalue problem is obtained by assuming

$$w(\xi,\tau) = \psi(\xi)F(\tau) \tag{8}$$

and separating the variables, using λ^2 as the constant of separation. The following equations result:

$$F'' + \lambda^2 F = 0$$

and

$$\frac{d^2}{d\xi^2}[\xi^m \frac{d^2\psi}{d\xi^2}] + (\bar{T}-\bar{Q})\frac{d}{d\xi}[\xi^{n+1}\frac{d\psi}{d\xi}] + \bar{Q}\xi\frac{d^2\psi}{d\xi^2} - \xi^n\lambda^2\psi = 0 \tag{9}$$

The boundary conditions for a free-free tapered beam with end thrust $\bar{T}$ are such that at $z = 0$ and at $z = \ell$ the moment and shear are equal to zero. A simple control system is incorporated at the end of the beam, so that the gimbal angle of the thrust is proportional to the angular rotation at the end of the flexible beam. Thus, in terms of the non-dimensional parameters

$$\text{at} \quad \xi=0: \quad \xi^m \frac{d^2\psi}{d\xi^2} = 0 \quad \text{and} \quad \frac{d}{d\xi}[\xi^m \frac{d^2\psi}{d\xi^2}] = 0$$

$$\text{and at} \quad \xi=1: \quad \frac{d^2\psi}{d\xi^2} = 0 \quad \text{and} \quad \frac{d^3\psi}{d\xi^3} = -\bar{T}k[\frac{d\psi}{d\xi}]_{\xi=1} \tag{10}$$

The parameter k indicates the angle of thrust and ranges from zero to unity. When $k = 0$, there is no shear due to $\bar{T}$, that is, $\bar{T}$ acts along the beam axis; and when $k = 1$, $\bar{T}$ acts at a slope of $\frac{d\psi}{d\xi}$ with respect to the beam axis.

The force $\bar{Q}$ has no effect on either moment or shear since it is acting in an axial direction.

The complete formulation of the eigenvalue problem in terms of the variables ξ, $\bar{T}$, and $\bar{Q}$ consists of Eqs. (9) and (10). Setting T equal to the entire differential operator, that is,

$$T = \frac{d^2}{d\xi^2}[\xi^m \frac{d^2}{d\xi^2}] + (\bar{T}-\bar{Q})\frac{d}{d\xi}[\xi^{n+1}\frac{d}{d\xi}] + \bar{Q}\xi\frac{d^2}{d\xi^2}$$

Eq. (9) becomes:

$$(T-\xi^n\lambda^2)\psi = 0 \tag{11}$$

Orthogonal eigenfunctions and positive eigenvalues cannot be assumed, since it can be shown that the differential operator T is not self adjoint.

A TAPERED BEAM OF PARABOLIC PROFILE

The special case of a beam with a parabolic surface of revolution is now considered. The generating line has an equation of the form

$$x(z) = \sqrt{\frac{A_o}{\pi}}\,(\frac{z}{\ell})^{\frac{1}{2}} \tag{12}$$

The beam cross sectional area and moment of inertia are represented respectively as

$$A = A_o(\frac{z}{\ell}) \quad \text{and} \quad I = I_o(\frac{z}{\ell})^2 \tag{13}$$

Thus, the constants used in the previous calculations are $n = 1$ and $m = 2$.

Now the governing differential equation, Eq. (9) becomes

$$\xi^4 \frac{d^4\psi}{d\xi^4} + 4\xi^3 \frac{d^3\psi}{d\xi^3} + [2\xi^2 + \bar{Q}\xi^3 + (\bar{T}-\bar{Q})\xi^4]\frac{d^2\psi}{d\xi^2} + 2(\bar{T}-\bar{Q})\xi^3 \frac{d\psi}{d\xi} - \xi^3\lambda^2\psi = 0 \quad (14)$$

The boundary conditions are

$$\text{at} \quad \xi=1 \quad ; \quad \frac{d^2\psi}{d\xi^2} = 0 \quad \text{and} \quad \frac{d^3\psi}{d\xi^3} = -\bar{T}k\left[\frac{d\psi}{d\xi}\right]_{\xi=1} \quad (15)$$

It is noted that the boundary conditions at $\xi = 0$ are satisfied identically.

Frobenius' method [6] is used to solve the above boundary value problem. A solution of the form

$$\psi(\xi) = \sum_{r=0}^{\infty} a_r \, \xi^{r+s} \quad (16)$$

is assumed, where s and the a_r's must be determined so that Eq. (14) is satisfied. The following notation is used

$$(s)_n = s(s-1)(s-2)\ldots(s-(n-1))$$

and

$$(s+r)_n = (s+r)(s+r-1)\ldots(s+r-(n-1))$$

Substituting Eq. (16) into Eq. (14) yields

$$\sum_{r=0}^{\infty} a_r(r+s)_4 \, \xi^{r+s} + 4\sum_{r=0}^{\infty} a_r(r+s)_3 \, \xi^{r+s} + \sum_{r=0}^{\infty} a_r(r+s)_2 \, 2\xi^{r+s}$$

$$+ \sum_{r=0}^{\infty} a_r \, \bar{Q}(r+s)_2 \, \xi^{r+s+1} + \sum_{r=0}^{\infty} a_r(\bar{T}-\bar{Q})(r+s)_2\xi^{r+s+2}$$

$$+ \sum_{r=0}^{\infty} 2a_r(\bar{T}-\bar{Q})(r+s)_1 \, \xi^{r+s+2} - \sum_{r=0}^{\infty} a_r\lambda^2\xi^{r+s+3} = 0 \quad (17)$$

Collecting like powers of ξ and equating the coefficient of each power of ξ to zero, a set of equations is obtained, as

$$a_o[(s)_4 + 4(s)_3 + 2(s)_2] = 0 \quad (18a)$$

$$a_1[(s+1)_4 + 4(s+1)_3 + 2(s+1)_2] + a_o[\bar{Q}(s)_2] = 0 \quad (18b)$$

$$a_2[(s+2)_4 + 4(s+2)_3 + 2(s+2)_2] + a_1[\bar{Q}(s+1)_2]$$
$$+ a_o(\bar{T}-\bar{Q})[(s)_2 + 2(s)_1] = 0 \quad (18c)$$

$$\vdots$$

$$a_j[(s+j)_4 + 4(s+j)_3 + 2(s+j)_2] + a_{j-1}[\bar{Q}(s+j-1)_2]$$
$$+ a_{j-2}(\bar{T}-\bar{Q})[(s+j-2)_2 + 2(s+j-2)_1] - a_{j-3}\lambda^2 = 0$$

where $j \geq 3$.

For an arbitrary a_o, Eq. (18a) is a fourth order polynomial in s. Solution of this equation yields four roots,

$$s_o = 0, \quad s_1 = 1, \quad s_2 = 0, \quad \text{and} \quad s_3 = 1 \quad (19)$$

An expression for a_1 is obtained from Eq. (18b)

$$a_1 = \frac{-\bar{Q}(s)_2 \, a_o}{(s+1)_4 + 4(s+1)_3 + 2(s+1)_2} \quad (20a)$$

and Eq. (18c) becomes

$$a_2 = \frac{-a_o(\bar{T}-\bar{Q})[(s)_2 + 2(s)_1] - a_1[\bar{Q}(s+1)_2]}{(s+2)_4 + 4(s+2)_3 + 2(s+2)_2} \quad (20)$$

Rearranging terms in Eq. (18d) gives the general expression for $j \geq 3$,

$$a_j = \frac{a_{j-3}\lambda^2 - (s+j-1)_2[a_{j-2}(\bar{T}-\bar{Q}) + a_{j-1}\bar{Q}]}{[(s+j)_2]^2} \tag{21}$$

It is now possible to solve for the coefficients a_r of the solutions to the fourth order differential equation. Equation (16) represents the form of the solution. From Eq. (19), $s_o = 0$, therefore $\psi_o \Big|_{s=0} = \sum_{r=0}^{\infty} a_r \xi^r$.

Since a_o is assumed arbitrary, let $a_o = 1$; then $a_1 = 0$, and $a_2 = 0$. The general term $(j \geq 3)$ is

$$a_j = \frac{a_{j-3}\lambda^2 - (j-1)_2[a_{j-2}(\bar{T}-\bar{Q}) + a_{j-1}\bar{Q}]}{[(j)_2]^2}$$

For $s_1 = 1$ we have $\psi_1 = \xi \sum_{r=0}^{\infty} a_r \xi^r$. If a_o is again assumed equal to unity, then $a_1 = 0$, $a_2 = \frac{-2(\bar{T}-\bar{Q})}{(3!)^2}$, for $j \geq 3$,

$$a_j = \frac{a_{j-3}\lambda^2 - (j)_2[a_{j-2}(\bar{T}-\bar{Q}) + a_{j-1}\bar{Q}]}{[(j+1)_2]^2}$$

The remaining two solutions [7] are:

$$\psi_2 \Big|_{s=0} = \psi_o \ln \xi + \sum_{r=0}^{\infty} b_r \xi^r$$

and

$$\psi_3 \Big|_{s=1} = \psi_1 \ln \xi + \xi \sum_{r=0}^{\infty} b_r \xi^r \tag{22}$$

The complete solution will be

$$\psi(\xi) = \sum_{i=0}^{3} A_i \psi_i \tag{23}$$

In order to obtain a finite solution at $\xi = 0$, we must set $A_2 = A_3 = 0$. Therefore, the complete solution is

$$\psi(\xi) = A_o \psi_o + A_1 \psi_1 \tag{24}$$

Imposing the boundary conditions from Eq. (15) on Eq. (24) yields the characteristic equation

$$\left[\frac{d^2\psi_o}{d\xi^2}\frac{d^3\psi_1}{d\xi^3} - \frac{d^2\psi_1}{d\xi^2}\frac{d^3\psi_o}{d\xi^3}\right]_{\xi=1} + \bar{T}k\left[\frac{d^2\psi_o}{d\xi^2}\frac{d\psi_1}{d\xi} - \frac{d^2\psi_1}{d\xi^2}\frac{d\psi_o}{d\xi}\right]_{\xi=1} = 0 \tag{25}$$

The variables in this equation are ψ_o, ψ_1, $\bar{T}$, $\bar{Q}$, k, and λ^2. ψ_o and ψ_1 may be evaluated at $\xi = 1$ by computations of the coefficients a_r; and $\bar{T}$, $\bar{Q}$ and k are given for a specific system. Therefore, the characteristic equation may be solved for the eigenvalues λ^2 where λ is the natural frequency of the lateral vibrations.

The mode shapes of the vibrating beam for particular values of $\bar{T}$ and $\bar{Q}$ can be expressed as

$$\psi = A_o\left[\psi_o - \frac{d^2\psi_o/d\xi^2}{d^2\psi_1/d\xi^2}\Bigg|_{\xi=1} \psi_1\right] \tag{26}$$

Solutions of the characteristic equation were obtained on the IBM 360/50. The value k = 0 represents the case where there is no control system. Solutions were obtained, in addition, for the cases k = 0.5 and k = 1.0. For all cases, a friction force of the type assumed here lowers the natural frequencies of the beam. For relatively high $\bar{Q}$ the first two bending modes become indistinguishable from the zero frequency mode. This would indicate that high friction causes such severe damping of the first two modes as to make them indistinguishable from the zero rotational mode of the free-free beam. The natural frequencies for the third and fourth modes decrease considerably as may be seen in Figs. 2 and 3.

An interesting result of the addition of frictional drag may be seen in Fig. 3. The friction force causes the constant frequency curves to come together on the real side of the thrust axis. Since two different mode shapes cannot exist simultaneously, an instability occurs at the point where the lines join. This phenomena also exists for a beam of uniform circular cross section [2]. From this it would seem that frictional drag might tend to decrease the effects of the parabolic shape of the beam. Further analysis would be necessary to determine if there were any numerical correlation between the case of the prismatic beam without friction and the case under consideration.

The effect of the assumed friction force on the mode shape* is shown in Fig. 4 for the third mode and a constant thrust of $\bar{T}$ = 100.0. The results are shown for two different thrust angles.

CONCLUDING REMARKS

The preceding study is based on linear beam theory with zero damping and small vibrations. The results verify the previous work of Huang and Walker for the case with $\bar{Q}$ = 0. The findings indicate that air friction on missiles can lower natural frequencies; this result is important and should bear further investigation since air friction may lower the natural frequencies enough to cause dangerous resonance conditions to occur at low altitude flight. A further study using a more realistic friction force which would be dependent upon the skin friction and pressure drag could provide more information for missile flight calculations.

Experimental evidence would further establish confidence in the results. If a fixed model were used with a fluid flowing past, it would be necessary to secure the model so that the free-free boundary conditions were maintained; the model would have to be fastened at nodal points obtained in the above study.

A static analysis of the system would be in order, so that the critical buckling loads might be determined. For high $\bar{Q}$, buckling must be considered. Timoshenko and Gere [8] discuss the problem of a fixed-free end uniform beam with a distributed, compressive, axial load; a similar analysis would be necessary to determine buckling loads in the present study.

BIBLIOGRAPHY

[1] Sharp, G. R., "An Investigation of the Natural and Forced Lateral Vibration of a Free-Free Beam (Missile)," The American Society of Mechanical Engineers, Vibrations Conference, March, 1967.

[2] Huang, C. L., and Walker, H. S., "Vibration and Stability of Rockets," American Astronautical Society, Advances in Astronautical Sciences, 24, 1967, 5-8.

[3] Fauconneau, G., and Laird, W. M., "The Eigenvalue Problem for Beams and Rectangular Plates with Linearly Varying In-Plane and Axial Load," National Aeronautics and Space Administration, CR-549, 1966.

---

*A damped mode shape has been defined here as one which has the same general shape as the undamped mode shape, or a different shape which may be clearly shown to have progressed from the undamped shape (such as for high damping in Fig. 4). The question of whether these can be considered mode shapes at all is an interesting one since they do represent solutions of the characteristic equation, and yet they do not look like what is generally considered to be a mode shape; however, for the purposes of the present study the above definition has been used.

[4] Sun, S. M., "The Eigenvalue Problem for Natural Frequency of Uniform Beam with Linearly Varying Axial Load," Master's Report, Kansas State University, 1967.

[5] Gorshkov, A. G., and Shklyarchuk, F. N., "Stability of an Elastic Body of Revolution in a Stream of Gas in the Presence of the 'Tracking' Force," In Russian, Inzhenernyi Zhurnal Mekhanika Tverdogo Tela, No. 5, Sept.-Oct., 1966, 151-156.

[6] Ince, E. I., Ordinary Differential Equations, Dover Publications, New York, 1956.

[7] Wayland, H., Differential Equations Applied in Science and Engineering, D. Van Nostrand Company, Inc., Princeton, 1959.

[8] Timoshenko, S. P., and Gere, J. M., Theory of Elastic Stability, McGraw-Hill Book Company, Inc., New York, 1961.

[9] James, M. L., Smith, G. M., and Wolford, J. C., Applied Numerical Methods for Digital Computation with FORTRAN, International Textbook Company, Scranton, 1967.

[10] Tong, K. N., Theory of Mechanical Vibration, John Wiley & Sons, Inc., New York, 1963.

[11] Gregory, M., Elastic Instability: Analysis of Buckling Modes and Loads of Framed Structures, E. & F. N. Spon Limited, London, 1967.

LIST OF SYMBOLS

$A(z)$	Cross sectional area of beam
A_o	Cross sectional area of beam at $z = \ell$
E	Modulus of elasticity (constant)
$I(z)$	Moment of inertia of beam cross section
I_o	Moment of inertia of beam at $z = \ell$
k	Directional control parameter
ℓ	Length of beam
m,n	Real numbers
$P(z,t)$	Axial force distribution in beam
$\bar{Q}$	$\frac{\ell^3 q}{EI_o}$
q	Constant axial load representing friction force
$(S)_n$	Descending factorial
T	Differential operator
T_c	Thrust exerted on beam at $z = \ell$
$\bar{T}$	$\frac{\ell^2 T_c}{EI_o}$
t	Time
$u(t)$	Longitudinal displacement of beam
$V(z,t)$	Transverse shearing force in beam
w	Lateral displacement of beam
z	Longitudinal coordinate of a differential beam element

α Angle between vibrating beam axis and equilibrium beam axis

ξ z/ℓ

σ Volume density of beam (constant)

τ $t\sqrt{\dfrac{EI_o}{\sigma A_o \ell^4}}$

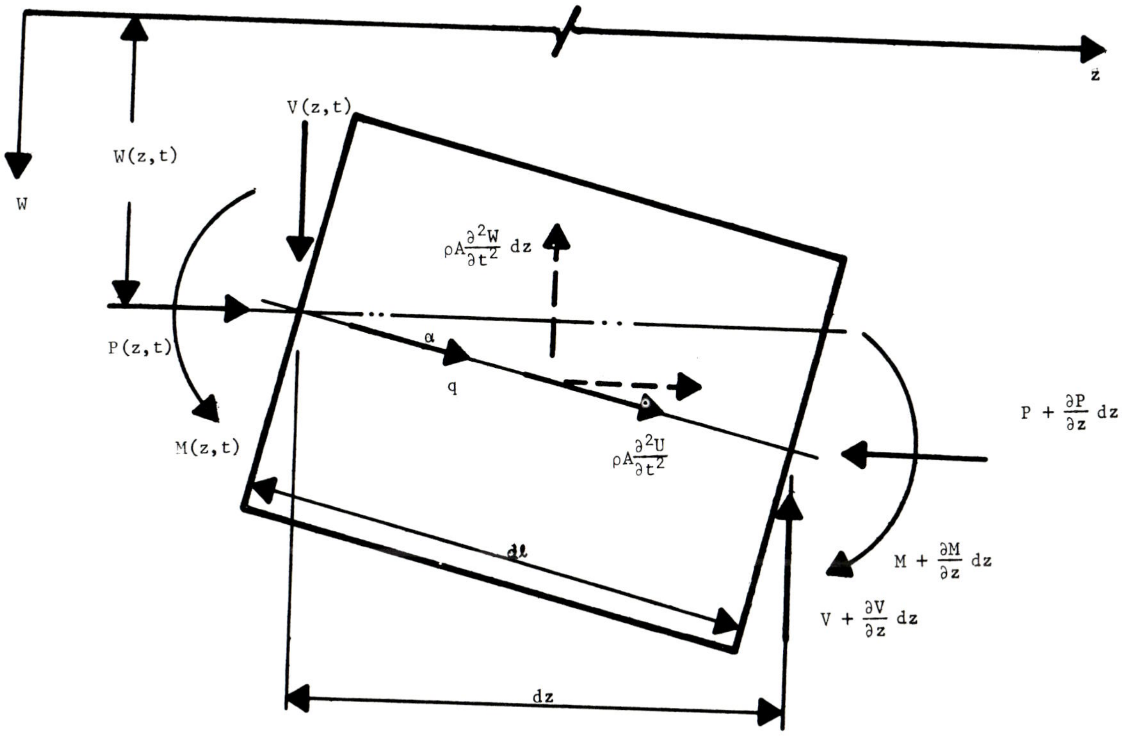

Fig. 1. Differential Element of the Tapered Beam

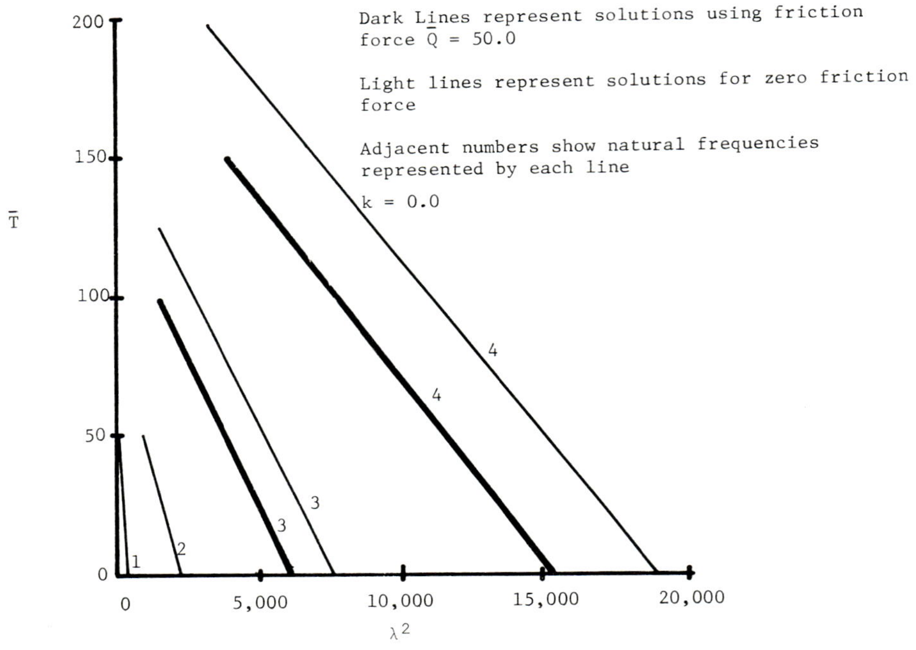

Fig. 2. Thrust vs. λ^2

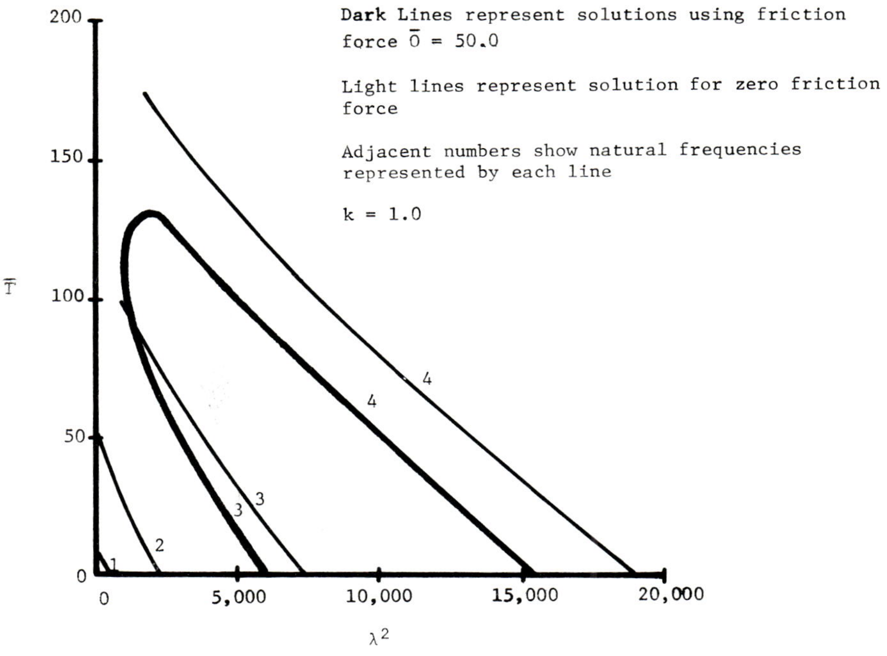

Fig. 3. Thrust vs. λ^2

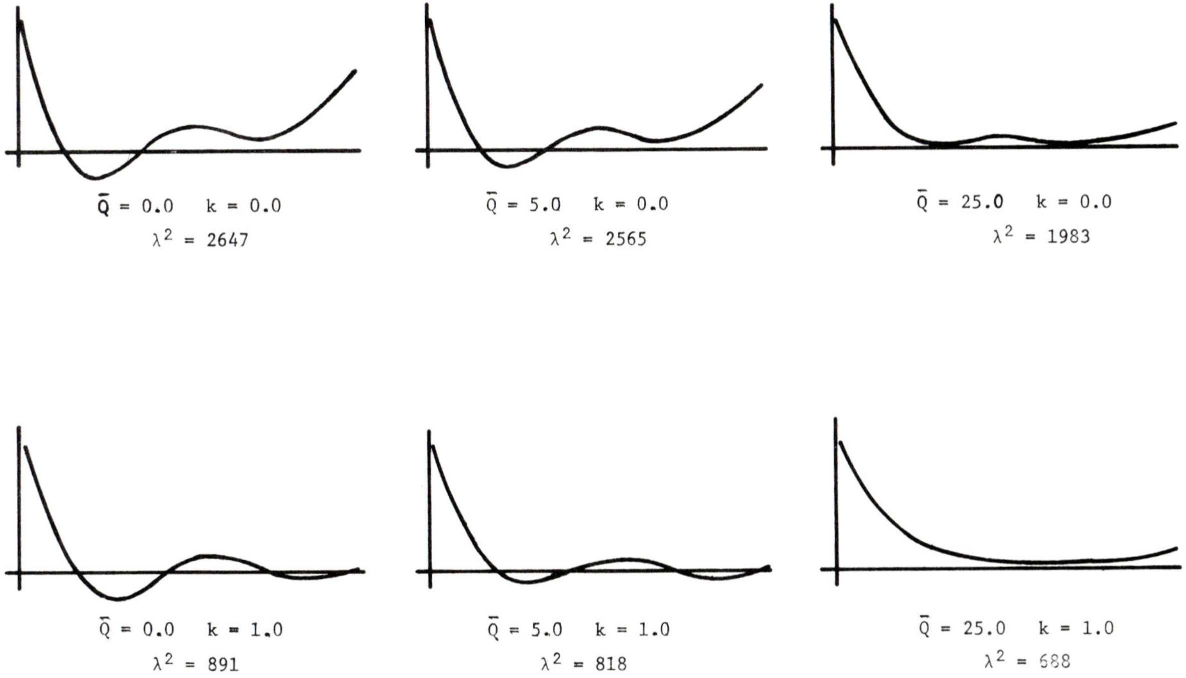

Fig. 4. The Effects of Friction on the Third Mode $\bar{T}$ = 100.0

Developments in Mechanics, Vol. 6. Proceedings of the 12th Midwestern Mechanics Conference: 50

Two-parameter Oscillator Solutions

ROBERT J. MULHOLLAND
OKLAHOMA STATE UNIVERSITY
PIERRE M. HONNELL and KENNETH J. BORGWALD
WASHINGTON UNIVERSITY

ABSTRACT

Amplitude estimates and bounds are derived for a nonlinear two-parameter oscillator. The differential equation considered is characterized by an arctangent nonlinearity and arises as a generalization of van der Pol's equation. Of particular interest is the disclosure of a one-parameter independent lower bound for the radius of the limit cycle solution. The computation of this lower bound requires the solution of a transcendental equation for which representations are obtained by an iterative technique and the method of series reversion. Through the use of computer generated data, upper and lower amplitude bounds are determined for a restricted parameter range.

INTRODUCTION

This paper deals with the analysis of the nonlinear oscillations of a generalization of the two-parameter van der Pol differential equation,

$$\ddot{x} + \mu(\varepsilon x^2 - 1)\dot{x} + x = 0, \tag{1}$$

where μ and ε are known scalar parameters and $\dot{x} = dx/dt$. The van der Pol equation has many important applications in the field of nonlinear oscillations, but in particular the nonlinear damping term in Eq. 1 originally arose as an approximation to the transconductance of a triode electron vacuum tube [1]. Prompted by measurements of the characteristics of modern electronic amplifiers Scott [2] first generalized Eq. 1 to give

$$\ddot{x} + \alpha\left(1 - \frac{\beta}{1+x^2}\right)\dot{x} + x = 0, \tag{2}$$

where α and β are scalar parameters. It should be noted that Eq. 2 becomes Eq. 1 by expanding $1/(1 + x^2)$ in a power series, ignoring powers of x greater than two, and setting $\varepsilon = \beta/\beta-1$ and $\mu = \alpha(\beta-1)$. The validity of this equivalence is clear for x small.

Equation 2 arises from the oscillator illustrated in Fig. 1, for which the following equations apply:

$$L\, di/d\tau = v \tag{3}$$

$$C\, dv/d\tau = i_o - Gv - i\ . \tag{4}$$

The amplifier characteristic is modeled by

$$i_o = A \arctan Bv\ , \tag{5}$$

where A and B are experimental constants chosen to represent the physical properties of amplifier gain and saturation. Eliminating v from Eqs. 3, 4, and 5 and setting

$$t = \tau/\sqrt{LC}\ , \quad \alpha = G/\sqrt{LC}\ , \quad \beta = AB/G\ ,$$

$$\text{and} \quad x = B\sqrt{L/C}\ di/dt \tag{7}$$

gives Eq. 2.

In order to demonstrate rigorously that Eq. 2 possesses a limit cycle, several theorems from the theory of nonlinear oscillations could be employed in a straight forward manner; however, that this is true is easily placed in evidence by a heuristic argument. Since $\alpha = 0$ reduces Eq. 2 to the linear oscillator equation, oscillatory solutions are assumed to exist. For $\alpha > 0$ and $\beta > 1$, the damping coefficient takes negative values for x near zero. This causes the envelope of the solution $x(t)$ to increase. But as x takes larger values the damping coefficient becomes positive, thus limiting an indefinite increase in solution amplitude. This leads to the establishment of an oscillatory equilibrium, which is the presumed limit cycle.

The amplifier of Fig. 1 is of course an analogue device, a nonideal operational amplifier, however, it is interesting to note that Murata [3], [4] in his studies of digital integrated circuits has recently found that NOR logic gates have input/output characteristics not unlike that described by Eq. 5. Murata has arrived independently of Scott at Eq. 2, and has described the asymptotic behavior of the oscillations using perturbation methods. Thus, as with van der Pol's equation, Eq. 2 appears to have wide ranging physical applications. By analogy and the similarity of available physical devices, applications to the analysis of fluidic oscillators are apparent. Bereft of physics, Eq. 2 has the nonlinear feedback control system interpretation given in Fig.2.

The purpose of this paper is to present an analysis of the oscillations of differential equations containing nonlinear arctangent functional characteristics. The analysis focuses specifically upon an analytic result presented by Mulholland [5] regarding a lower bound on the amplitude of oscillation. The accuracy and relevancy of this amplitude bound is discussed with the aid of computer solutions, and a comparison is made with asymptotic results.

ASYMPTOTIC RESULTS

The averaging method of Krylov and Bogoliubov [6] can be employed to obtain an approximation for the approach of the solutions of Eq. 2 to the limit cycle. In the steady-state, the averaging method gives a linear approximation to the limit cycle amplitude and a first-order parameter estimate to the frequency of oscillation. Thus, for small parameter α values, the averaging method gives an accurate perturbation of the linear ($\alpha=0$) periodic solution of Eq. 2.

Consider again Eq. 2, but written as

$$\ddot{x} + x = \alpha\,(1 - \beta + x^2)(1 + x^2)^{-1}\,\dot{x}\,. \tag{8}$$

Equation 8 is obtained from Eq. 2 by multiplying through the damping coefficient by $(1 + x^2)^{-1}$, which is certainly a valid operation over the field of real numbers. The form of the solution sought is

$$x = a\cos(t + \phi) \tag{9}$$

with

$$\dot{x} = -a\sin(t + \phi) \tag{10}$$

where $a = a(t)$ and $\phi = \phi(t)$ are assumed to be slowly varying functions of time for which representations are to be determined. The left side of Eq. 8 thus becomes

$$\ddot{x} + x = -\,\dot{\phi}\,a\cos(t + \phi) - \dot{a}\sin(t + \phi). \tag{11}$$

If the right side of Eq. 8 is expanded in a Fourier series, and consistent with the approximation of Eq. 11 only the linear, or fundamental frequency, terms are retained, then

$$\dot{\phi}\,a\cos(t + \phi) + \dot{a}\sin(t + \phi)$$
$$= \alpha\,(1 - \beta)\,a\sin(t + \phi) + \alpha\,a^3/4\,\sin(t + \phi) \tag{12}$$

results as an approximation to Eq. 8. Since sine and cosine are linearly independent, equating coefficients yields

$$\dot{\phi} = 0 \tag{13a}$$

$$\dot{a} = \alpha\,[a^3/4 - (\beta - 1)\,a]\,, \tag{13b}$$

with solutions

$$\phi(t) = \phi_o \tag{14a}$$

$$a^2(t) = \frac{4\,a_o^2\,(\beta - 1)\,e^{\alpha(\beta - 1)t}}{[a_o^2 - 4\,(\beta - 1)] - e^{\alpha(\beta - 1)t}\,a_o^2} \tag{14b}$$

where $a(0) = a_o$ and $\phi(0) = \phi_o$. For $\mu > 0$, and $\beta > 1$

$$\lim_{t\to\infty} a^2(t) = 4\,(\beta - 1) \tag{15}$$

follows from Eq. 14. This result indicates the possibility of parameter independence with respect to α which is discussed in the sequel. Equation 14b gives only a constant solution from which little can be deduced. Actually this result merely indicates that corrections to the frequency of oscillation $\omega = 1 + \dot{\phi}$ first appear in order α^2 terms, as is clearly shown by the perturbation method [4].

Consider the expression of the solution of Eq. 2 as a series

$$x(t) = x_o(\omega t) + \alpha x_1(\omega t) + \alpha^2 x_2(\omega t) + \dots \qquad (16)$$

where

$$\omega = \omega_o + \alpha\omega_1 + \alpha^2 \omega_2 + \dots \qquad (17)$$

Murata [4] has calculated the first significant terms, giving

$$x(t) = 2\sqrt{\beta - 1}\ \cos\omega t$$
$$+ \alpha/4[(3/4)(\beta-1)^{3/2}(2\beta-1)^{-1}\sin\omega t - (\beta-1)^{3/2}(2\beta-1)^{-1}\sin 3\omega t]$$
$$+ O(\alpha^2) \qquad (18)$$

with

$$\omega = 1 - \alpha^2 \frac{(\beta-1)^3 + (2\beta-1)(\beta-1)^2}{16(2\beta-1)^2(3\beta-2)} + O(\alpha^3). \qquad (19)$$

which compares with the steady-state ($\alpha = 0$) values obtained in Eqs. 14 and 15.

LOWER AMPLITUDE BOUND

Further analysis of the oscillations of Eq. 2 is facilitated by the introduction of Liénard plane coordinates, which results in the following equivalent first order formulation:

$$\dot{x} = -\alpha(x - \beta \arctan x) + y \qquad (20a)$$

$$\dot{y} = -x \qquad (20b)$$

The objective of the following analysis is the derivation of a lower bound on the radius R of the limit cycle L of Eq. 20. The radius of L is a function of the two parameters α and β, and is defined by

$$R(\alpha,\beta) = \inf_L (x^2 + y^2)^{\frac{1}{2}} \qquad (21)$$

where the infinum is taken over all values of (x,y) on L.

If the independent variable t is replaced by $-t$, the solution trajectories in the Liénard plane remain the same in form but with reversed orientation. More importantly, the position and shape of the limit cycle L is invariant under time reflection. Note that replacing α by $-\alpha$ is equivalent to time reflection change of variable.

An investigation of Eq. 20 for $\alpha < 0$ and $\beta > 1$ gives the equilibrium point $x = y = 0$ as asymptotically stable, and the region of stability as that contained within the interior of the limit cycle L. In this region, the properties of an appropriately chosen Liapunov function $V(x,y)$ are known, i.e., V is a positive definite scalar function of the Liénard plane coordinates with a negative definite derivative $\dot{V}$ [7]. In particular, consider

$$V(x,y) = (\tfrac{1}{2})(x^2 + y^2), \qquad (22)$$

hence

$$\dot{V}(x,y) = x\dot{x} + y\dot{y} \tag{23}$$

which by Eq. 20 yields

$$\dot{V}(x,y) = -\alpha x^2 [1 - (\beta/x) \arctan x] . \tag{24}$$

Since $\alpha < 0$, $\dot{V}$ is negative for all $x \neq 0$ satisfying

$$|x| < \beta \, |\arctan x| . \tag{25}$$

For convenience, define the function F by

$$F(x) = x - \beta \arctan x. \tag{26}$$

For $\beta > 1$, there are precisely two non-trivial roots of

$$F(x) = 0 \tag{27}$$

which are denoted by $\pm x_o(\beta)$ where $x_o > 0$. Thus, Eq. 25 holds for $|x| < x_o$.

The roots of Eq. 27 are not easily determined as functions of β, but they are surely bounded away from zero by the roots of the derivative of F,

$$F'(x) = 0 \tag{28}$$

which gives the two local extrema of F in the closed interval $[-x_o, x_o]$. Indeed, a simple calculation gives the roots of Eq. 28 as $\pm x_1(\beta)$, where

$$x_1 = (\beta - 1)^{\frac{1}{2}} . \tag{29}$$

Hence, Eq. 25 holds for $|x| < x_1$ or

$$|x| < (\beta - 1)^{\frac{1}{2}}. \tag{30}$$

It is now clear that the circular region

$$x^2 + y^2 < (\beta - 1)^{\frac{1}{2}} \tag{31}$$

is wholly contained within the region of asymptotic stability. Hence, independent of the parameter α, the limit cycle lies outside the circle with center at the origin and radius equal to $(\beta - 1)^{\frac{1}{2}}$.

The α-independent lower bound on R thus determined,

$$R(\alpha,\beta) \geq (\beta - 1)^{\frac{1}{2}} , \tag{32}$$

appears to be particularly accurate for β close to unity. This point is considered in the next section. It should be noted that Eq. 32 is meaningless for $\beta < 1$, thus correctly implying that in this case no stable limit cycle exists.

COMPUTER SOLUTIONS

It is of interest to compare the actual solutions of Eq. 20 with the asymptotic estimates of Eqs. 14 and 18 and the one-parameter independent bound estimate of Eq. 32. Computer generated solutions of Eq. 20 have been obtained using the Matric Computor. This hybrid computing device, which is still under development, is well suited to tasks of this sort [8], [9]. The mathematical basis for the Matric Computor is the theory of matrices, while the physical basis is electronic network theory. The employment of the computer is based upon a one-to-one correspondence between the mathematical problem to be solved and the equation of physical state which represents the electronic network of the computer. Through the arrangement of the machine into rows and columns of direct-dialing digitized switches, this correspondence provides for the direct entry of matrically formulated problems which contain constant, first-order differential and linear functional coefficients [10], [11].

The following equisolution matrix [11], dialed directly into the Matric Computor, yields on coordinates x_3 and x_4 the solutions of Eq. 20:

$$\begin{bmatrix} 1 & \alpha(1-\beta) & 1 & 0 & 0 & 0 & 1 \\ -1 & d/dt & 0 & 0 & 0 & 0 & 0 \\ 0 & -1 & d/dt & 0 & 0 & 0 & 0 \\ 0 & 0 & -1 & d/dt & 0 & 0 & 0 \\ -1 & -\alpha & -1 & 0 & 1 & 0 & 0 \\ 0 & 0 & -x_3 & 0 & 0 & 1 & 0 \\ 0 & 0 & 0 & 0 & -x_6 & 0 & 1 \end{bmatrix} \begin{bmatrix} x_1 \\ x_2 \\ x_3 \\ x_4 \\ x_5 \\ x_6 \\ x_7 \end{bmatrix} = \begin{bmatrix} 0 \\ 0 \\ 0 \\ 0 \\ 0 \\ 0 \\ 0 \end{bmatrix} \qquad (33)$$

where $x_1 = \ddot{x}$, $x_2 = \dot{x}$, $x_3 = x$, $x_4 = y$, $x_5 = \ddot{x} + \alpha\dot{x} + x$, $x_6 = x^2$ and $x_7 = x^2(\ddot{x} + \alpha\dot{x} + x)$. Equation 33 follows directly from Eq. 2 with the above coordinate equivalences.

For various α and β parameter values the limit cycle solutions of Eq. 33 have been plotted in the Liénard plane. These closed curves will approximate circles with distortion more apparent as either α or β becomes large. The radius $R(\alpha,\beta)$ of each limit cycle then can be determined graphically as shown in Fig. 3 for $\alpha=1$ and $\beta=2$. As illustrated, the radius $R(\alpha,\beta)$ as defined in Eq. 21 is actually equal to the radius of the largest circle which fits wholly within the limit cycle.

Analysis of computer solutions of Eq. 33 indicates that $R(\alpha,\beta)$ is essentially independent of α for at least a parameter range of $0.1 \leq \alpha \leq 2.0$. Practical computing considerations have restricted α to lie in this range, with $\alpha=1$ chosen to be representative for that which follows. Computer data giving the variation of $R(\alpha,\beta)$ with β are produced in Table I along with the lower bound of Eq. 32 and the asymptotic estimate $2(\beta - 1)^{1/2}$ which is order α.

Table I clearly verifies the theoretical prediction that $x_1(\beta)$ provides a lower bound for $R(\alpha,\beta)$, however it should be noted that the asymptotic amplitude provides a closer estimate for the limit cycle radius.

For $\beta \leq 1.2$, the absolute error in approximating $R(\alpha,\beta)$ by $x_1(\beta)$ is nearly constant at 43%, while the asymptotic estimate results in absolute errors which are nearly constant at 13%. As $R(\alpha,\beta)$ passes through unity, somewhere near $\beta=1.3$, the error incurred by the asymptotic approximation appears to pass through zero. For $\beta < 1.3$ $(R < 1)$ the asymptotic estimate provides a rather close upper bound for $R(\alpha,\beta)$, and for $\beta > 1.3$ $(R > 1)$ the estimate becomes a lower bound. This hitherto unknown result exists apparently without analytic verification. Furthermore, from the data of Table I it appears that the perturbation method and that of Krylov and Bogoliubov are particularly valid for $\beta \leq 2$.

Once $x_1(\beta)$ is known, the lower bound on $R(\alpha,\beta)$ can be improved by using iterative scheme to calculate the roots of Eq. 27, thus giving

$$R(\alpha,\beta) \geq x_o(\beta). \qquad (34)$$

Finding the roots of Eq. 27 is equivalent to determining the fixed points of the arctangent mapping, i.e.,

$$x_o = \beta \arctan x_o . \qquad (35)$$

In general, the fixed points of

$$x = f(x) \qquad (36)$$

are sought in terms of the sequence $\{x^n\}$

$$x^{n+1} = f(x^n), \quad n = 1, 2, \ldots \qquad (37)$$

where x^n demptes the n-th iterate.

If the function f maps the closed interval [a,b] into itself and f has a continuous derivative with $|f'(x)| < 1$ on [a,b], then the sequence of Eq. 37 converges to a unique fixed point of Eq. 36 [12]. For $f=\beta$ arctan, the unique fixed point result holds for the interval [a,b], where

$$a = (\beta - 1 + \eta)^{\frac{1}{2}} \qquad (38)$$

for any number $\eta > 0$, and

$$b = \beta\pi/2, \qquad (39)$$

thus giving $x_o(\beta)$ as the limit of the sequence $\{x^n\}$. Clearly,

$$(\beta - 1 + \eta)^{\frac{1}{2}} \leq x_o(\beta) \leq \beta\pi/2, \qquad (40)$$

where the lower bound was established for $\eta=0$ in Eq. 28 and the upper bound is the asymptote to the arctangent curve.

On the interval [a,b], the derivative of f is in absolute value less than one. For consider

$$f'(x) = \beta/(1 + x^2), \qquad (41)$$

and since $x = (\beta - 1 + \eta)^{\frac{1}{2}}$ is the smallest value which x assumes,

$$|f'(x)| < \beta/(\beta + \eta) < 1 \qquad (42)$$

results. Thus, the convergence of the sequence $\{x^n\}$ is guaranteed.

The sequence of Eq. 37 can be easily implemented on a digital computer using $x^o = a$ as a starting value. In practice, it has been found that $\eta=0$ can be chosen to give $x^o = x_1(\beta)$. This result was not unexpected, since η is an arbitrarily small positive number.

Using $x_1(\beta)$ as a starting value, the results of the iterative computation scheme for $x_o(\beta)$ appear in Table I. The iterative calculation was continued until the residue,

$$\Delta_k = |x^k - x^{k-1}| , \qquad (43)$$

was less than 0.001. The value of k at which the iterative process terminates is noted in Table I. As indicated, the process converges rapidly to a close estimate for $R(\alpha,\beta)$ by $x^k(\beta) \simeq x_o(\beta)$ is less than 5%. Furthermore, the estimates $x^k(\beta)$ are monotone, thus retaining the lower bound nature of $x_o(\beta)$.

METHOD OF REVERSION

From Table I it is noted that the radius $R(\alpha,\beta)$ and its lower bound provided by the solution of Eq. 27 are less than unity for $\beta < 1.3$. Thus, under this constraint, the Taylor series

$$\arctan x = x - x^3/3 + x^5/5 - x^7/7 + . . . \quad (44)$$

converges. Equation 27 thus yields

$$1/\beta = 1 - x_o^2/3 + x_o^4/5 - . . . \quad (45)$$

which is rearranged to give

$$\frac{\beta - 1}{1+\beta-1} = x_o^2/3 - x_o^4/5 + . . . \quad (46)$$

The method of series reversion [13], [14] can be applied to Eq. 46 in order to obtain a representation for x_o as a series in β. With reference to previous asymptotic results, a series expansion of the form

$$x_o(\beta) = (\beta - 1)^{\frac{1}{2}} \sum_{k=o}^{\infty} b_k (\beta - 1)^k \quad (47)$$

is sought. Since $x_o(1) = 0$, the expansion of Eq. 47 is made about $\beta=1$. The right side of Eq. 46 is a power series in x_o^2, thus giving rise to the square root factor of Eq. 47. Furthermore,

$$x_o(\beta) = x_1(\beta) \sum_{k=o}^{\infty} b_k (\beta - 1)^k \quad (48)$$

can be written.

If the left side of Eq. 46 is expanded in a geometric series, i.e.,

$$(\beta - 1) - (\beta - 1)^2 + (\beta - 1)^3 - . . .$$

and Eq. 47 is substituted into the right side, then the coefficients b_k are determined by equating like powers in $(\beta - 1)$. In particular,

$$b_o = \sqrt{3} \quad (49a)$$

$$b_1 = 2\sqrt{3}/5 \quad (49b)$$

$$b_2 = -12\sqrt{3}/175 \quad (49c)$$

are easily computed. Therefore, the solution of Eq. 27 is given by

$$x_o(\beta) = \sqrt{3}(\beta - 1)^{\frac{1}{2}} + (2\sqrt{3}/5)(\beta - 1)^{3/2}$$

$$- (12\sqrt{3}/175)(\beta - 1)^{5/2} + 0(7/2) \quad (50)$$

where $0(7/2)$ represents terms that are order $(\beta - 1)^{7/2}$.

Data for the $0(\frac{1}{2})$ estimate for $x_o(\beta)$,

$$x_o(\beta) = \sqrt{3}\, x_1(\beta) + 0(3/2), \quad (51)$$

are compared in Table II with the previously derived bounds $x_1(\beta)$ and $2x_1(\beta)$. Table II clearly establishes $\sqrt{3}\, x_1(\beta)$ as a lower bound for $R(\alpha,\beta)$. The error incurred by approximating $R(\alpha,\beta)$ by $\sqrt{3}\, x_1(\beta)$ is less than 10% for the parameter range $\beta < 1.3$. Thus, it has been shown that

$$\sqrt{3}\ x_1(\beta) \leq R(\alpha,\beta) \leq 2\ x_1(\beta), \tag{52}$$

for $\beta < 1.3$. Either bound of Eq. 52 approximates $R(\alpha,\beta)$ to within 10% absolute error.

CONCLUSION

This paper has been concerned with bounds on the amplitude of oscillation of a two-parameter nonlinear differential equation. In particular, the radius $R(\alpha,\beta)$ of the limit cycle solution has been bounded below by $x_o(\beta)$ independently of α. Since x_o is a solution of a transcendental equation, an explicit representation is impossible. However, x_o has been shown to be bounded below by $x_1(\beta) = (\beta - 1)^{\frac{1}{2}}$. Thus,

$$x_1(\beta) \leq x_o(\beta) \leq R(\alpha,\beta) \tag{53}$$

exists for all parameter values.

Using the method of Krylov and Bogoliubov an asymptotic approximation for the amplitude of oscillation was found to be $2x_1(\beta)$. This estimate along with the lower bounds of Eq. 53 were compared with computer generated solutions for $R(\alpha,\beta)$. It was thereupon found that

$$R(\alpha,\beta) \leq 2x_1(\beta), \tag{54}$$

for $R < 1$, or equivalently $\beta < 1.3$. This result remains a conjecture without apparent mathematical proof. Similarly, α parameter independence is theoretically unclear.

An iterative scheme for the computation of $x_o(\beta)$ using $x_1(\beta)$ as a starting value has been disclosed. The results of such calculations carried out on a digital computer yield approximations for $R(\alpha,\beta)$ with absolute error less than 5%.

The method of series reversion was employed to represent the lower bound $x_o(\beta)$. The first term of this series expansion provides still another lower bound for $R(\alpha,\beta)$, giving

$$\sqrt{3}\ x_1(\beta) \leq R(\alpha,\beta) \leq 2\ x_1(\beta), \tag{55}$$

which is valid under the same constraints imposed upon Eq. 54.

Several possible directions for future investigation of the properties of Eq. 2 are indicated. An upper bound for amplitudes of oscillation greater than unity would be desirable. Such a bound would be necessarily valid for the parameter range $\beta > 1.3$. In another direction, parameter independence of the frequency of oscillation ω has been suggested [5] and actively sought. Equation 19 indicates that for α small

$$\lim_{K\to\infty} \omega = 1 - \alpha^2/64, \tag{56}$$

as shown by Murata [4]. Thus, α independence of ω may result when β is large and α small. But, in theory, this effect may only appear in the limit. In any event, the parameter constraints make a systematic computer study of the oscillations very difficult.

TABLE I

COMPUTER SOLUTIONS

β	$x_1(\beta)$	$2(\beta-1)^{\frac{1}{2}}$	$x^k(\beta)$	k	$R(\alpha\beta)$
1.01	0.100	0.200	0.162	71	0.18
1.02	0.141	0.282	0.222	62	0.25
1.03	0.173	0.346	0.287	53	0.30
1.04	0.200	0.400	0.343	48	0.35
1.05	0.224	0.448	0.386	42	0.40
1.06	0.245	0.490	0.428	38	0.43
1.07	0.265	0.530	0.464	36	0.47
1.08	0.283	0.566	0.501	34	0.51
1.09	0.300	0.600	0.534	32	0.54
1.10	0.316	0.632	0.565	30	0.57
1.20	0.447	0.894	0.831	19	0.84
1.40	0.632	1.264	1.259	12	1.26
1.60	0.774	1.548	1.635	10	1.64
1.80	0.894	1.788	1.989	9	2.00
2.00	1.000	2.000	2.330	8	2.34
3.00	1.414	2.828	3.972	6	4.00
4.00	1.732	3.464	5.572	6	5.58
5.00	2.000	4.000	7.160	6	7.17
10.00	3.000	6.000	15.036	5	15.10

TABLE II

COMPARISON OF AMPLITUDE BOUNDS

β	$x_1(\beta)$	$\sqrt{3}\ x_1(\beta)$	$2x_1(\beta)$	$R(\alpha,\beta)$
1.01	0.100	0.173	0.200	0.18
1.02	0.141	0.244	0.282	0.25
1.03	0.173	0.300	0.346	0.30
1.04	0.200	0.346	0.400	0.35
1.05	0.224	0.388	0.448	0.40
1.06	0.245	0.424	0.490	0.43
1.07	0.265	0.458	0.530	0.47
1.08	0.283	0.489	0.566	0.51
1.09	0.300	0.519	0.600	0.54
1.10	0.316	0.546	0.632	0.57
1.20	0.447	0.773	0.894	0.84

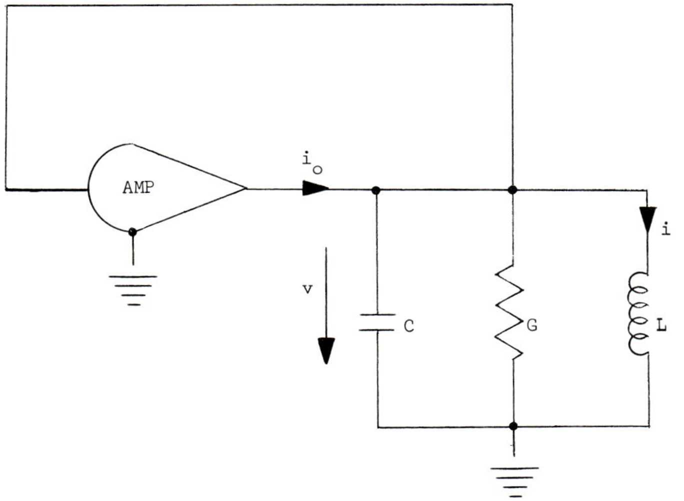

Fig. 1. Nonlinear Arctangent Electronic Oscillator

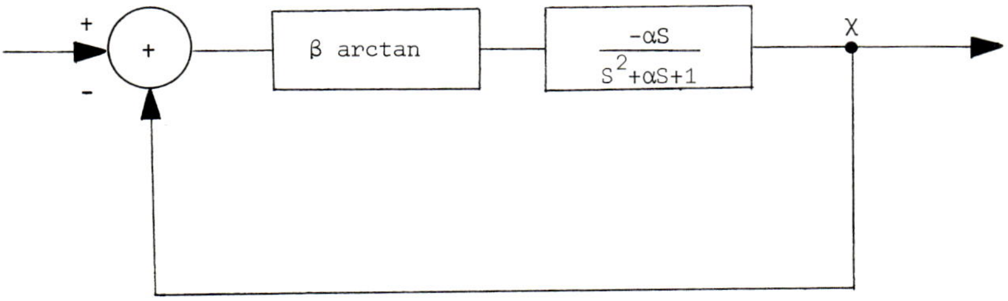

Fig. 2. General Nonlinear Arctangent Oscillator

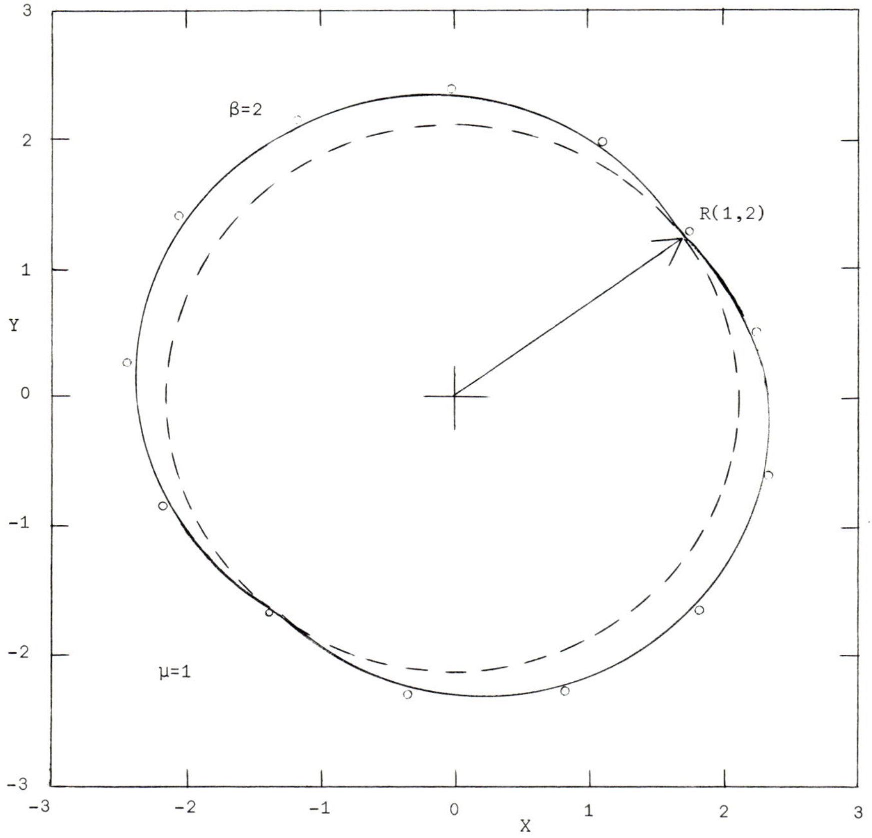

Fig. 3. Example Calculation Showing the Construction Of The Limit Cycle Radius

BIBLIOGRAPHY

[1] van der Pol, B.: On Oscillation Hysteresis in a Triode Generator with Two Degrees of Freedom. Philosophical Magazine, Vol. 6, pp. 700-719 (1922).

[2] Scott, P. R.: Large Amplitude Operation of the Nonlinear Oscillator. Proceedings of the IEEE, Vol. 56, No. 12, pp. 2182-2183 (1968).

[3] Murata, M. and Namekawa, T.: An Analysis of the Oscillator Consisting of Digital Integrated Circuits. Technical Reports of Osaka University, Osaka, Japan, Vol. 20, April (1969).

[4] Murata, M., et. al.: Analysis of an Oscillator Consisting of Digital Integrated Circuits. IEEE Journal of Solid-State Circuits, Vol. SC-5, No. 4, pp. 165-168 (1970).

[5] Mulholland, R. J.: One-Parameter Independent Bound for a Two-Parameter Oscillator. Proceedings of the IEEE, Vol. 57, No. 7, p. 1296 (1969).

[6] Krylov, N. and Bogoliubov, N.: Introduction to Nonlinear Mechanics. Annals of Mathematical Studies No. 11, Princeton University Press, Princeton, New Jersey (1947).

[7] LaSalle, J. and Lefschetz, S.: Stability by Liapunov's Direct Method with Applications. Academic Press, New York, Chapter 2 (1961).

[8] Honnell, P. M.: The Matric Computor. Journal of the Franklin Institute, Vol. 276, pp. 282-296 (1963).

[9] Honnell, P. M.: Kron's Methods and the Theoretical Foundation of the Matric Computor. Journal of the Franklin Institute, Vol. 286, pp. 566-571 (1968).

[10] Honnell, P. M. and Mulholland, R. J.: Matric Formulation of van der Pol's Equation and its Solution on the Matric Computor. Proceedings of the IEEE, Vol. 55, p. 1660 (1967).

[11] Honnell, P. M. and Mulholland, R. J.: Equisolution Matrices for the Matric Computor. Journal of the Franklin Institute, Vol. 288, pp. 261-274 (1969).

[12] Goldstein, A. A.: Constructive Real Analysis. Harper and Row, New York, pp. 4-12 (1967).

[13] Van Orstrand, C. E.: The Reversion of Power Series. Philosophical Magazine, Vol. 19, pp. 366-378 (1910).

[14] Pipes, L. A.: The Reversion Method for Solving Nonlinear Differential Equations. Journal of Applied Physics, Vol. 23, No. 2, pp. 202-207 (1952).

ANALYTICAL METHODS

Developments in Mechanics, Vol. 6. Proceedings of the 12th Midwestern Mechanics Conference: 51

Domain Constriction Method for Determination of Lower Bounds of Eigenvalues

MICHAEL CHI
CATHOLIC UNIVERSITY OF AMERICA
ALFRED MULZET
INTERNATIONAL BUSINESS MACHINE CORP.

ABSTRACT

A method for computing lower bounds of eigenvalues involving complicated regions is presented. It employs a larger but simpler region than the given region, and constrains the former to the latter by the use of Lagrangian undetermined multipliers. Numerical examples are given and results are accurate enough for most engineering applications.

INTRODUCTION

This paper presents a method for determining the lower bounds of eigenvalues involving a region of an arbitrary shape. For complicated shapes, numerical methods, such as the finite-difference and finite-elements methods, can be used but it is necessary to employ very fine grids to occupy the given domain as closely as possible [1]. Rayleigh-Ritz method is the most widely known method to provide upper bounds of eigenvalues [2]. This and its closely related method attributed to Galerkin, require the availability of admissible functions, which, for complicated boundary configuration, may not be easily obtainable. One way to obviate this difficulty is to map the given region conformally onto a simpler one, say a circle [3]. The mapping functions are available for simple configurations but for a general region one must approximate it by a power series [4]. Several methods of obtaining lower bounds of eigenvalues of membranes were cited by Stadter [5]. The most general one was credited to Weinstein, which was extended and reformulated in terms of Hilbert space theory by Aronszajn [6]. This theory has been developed to full generality; many methods including even Rayleigh-Ritz method, may be shown to be related to, or even a special case of it. Others, e.g., B*B method, were developed on the basis of this theory [7].

We remark in passing that, in their methods, the shape and the size of the physical domain are the same as those of the base problem but the boundary conditions are strengthened, or the differential equation is modified. The Domain Constriction Method presented in this paper, on the other hand, selects a base problem whose physical domain is larger than the given one and the former is gradually constrained to approach the latter by a suitable procedure. The dependence of eigenvalues on the domain is well-known and the eigenvalues of the base problem should be pairwise lower than those of the given problem. Since, by the same reason, partial constraints can only raise the eigenvalues, these lower bounds of eigenvalues are improved systematically and progressively. This is of course the same argument as in the general theory and no originality is claimed. The method of constructing the intermediate problems, however, is believed to be new; after a brief outline of the theory it is carefully described in the sequel.

THEORY

The eigenvalue problem may be defined as follows:

"Given a domain L and its boundary ℓ, we seek a constant parameter λ such that a non-zero function v, satisfying a set of prescribed boundary conditions on ℓ, is a solution of the following equation:

$$H v = \lambda v \quad \text{in } L \tag{1}$$

where H is a linear differential operator".

In general, for an arbitrary domain L, the configuration of ℓ may be so complicated that the solution of this problem involves grave mathematical difficulties. We shall restrict our attention to membrane eigenvalue problems. In what follows, we shall refer to this problem as the original problem or L-problem.

Let us choose a domain N which contains the given domain L as a proper sub-domain such that its boundary n is natural to a chosen standard coordinate system, Fig. 1. The membrane eigenvalue problem associated with domain N is easy to solve and its eigenvalues and eigenfunctions are cataloged in standard texts [8]. We shall refer to this problem as the base problem or N-problem.

Let us denote eigenvalues and eigenfunctions of the N-problem by λ_i and ψ_i, i=1,2,....., respectively. It is well-known that the following relation holds:

$$H\psi_i = \lambda_i\psi_i \tag{2}$$

We denote the Hilbert space generated by the orthonormal set ψ_i also by N. The fact that we denote the domain and the space by the same symbol shall not cause any confusion in the present context. Any function u in N has the following representation:

$$u = \sum_{m=1}^{\infty} a_m \psi_m \tag{3}$$

By the energy principle, the solution of the eigenvalue problem is equivalent to the minimization of the following functional [9]:

$$U = (H u, u) - \lambda (u,u) \text{ subjected to } (u,u) = 1 \tag{4}$$

where (,) denotes the inner product of two elements in N. Using Eqs. (2) and (3) and invoking linearity of H and orthonormality of ψ_i, we reduce Eq. (4) to:

$$U = \sum_{m=1}^{\infty} a_m^2 (\lambda_m - \lambda) \tag{5}$$

To solve the L-problem, we must constrain the function u to vanish on ℓ, which is the common boundary of L and B (See Fig. 1). In practice, the constraining condition cannot be satisfied identically. What we shall do is to impose on the function u a sequence of constraints of the following form:

$$F_i (u(x)) = 0 \qquad i=1,2,..... \tag{6}$$

where the F_i (.) are linear. This, on substituting Eq. (3), yields

$$\sum_{m=1}^{\infty} a_m F_i (\psi_m) = 0$$

It will become clear later that the above expression is really dependent only on m, but not the form of the eigenfunctions; thus we can write:

$$\sum_{m=1}^{\infty} a_m I_i(m) = 0 \qquad i=1,2,\ldots.. \tag{7}$$

The details of methods of decomposition of the constraint will be discussed in the next section.

To take account of the constraints in Eq. (6) we employ the Lagrangian undetermined multipliers Θ_j. The functional to be minimized then becomes,

$$V = U - \sum_j \Theta_j F_j \tag{8}$$

Substituting Eq. (5) into the above and using Eq. (7), we have

$$V = \sum_m a_m^2 (\lambda_m - \lambda) - \sum_j \Theta_j \sum_m a_m I_j(m) \tag{9}$$

The necessary condition for the minimum of Eq. (9) to occur is

$$\frac{\partial V}{\partial a_m} = 0 \tag{10}$$

which results in

$$2a_m(\lambda_m - \lambda) - \sum_j \Theta_j I_j(m) = 0$$

or

$$a_m = \frac{\sum_j \Theta_j I_j(m)}{2(\lambda_m - \lambda)} \tag{11}$$

Thus we obtain the relation between a_m and Θ_j. Elimination of a_m between Eqs. (7) and (11) yields

$$\sum_m \sum_j \frac{\Theta_j I_j(m) I_i(m)}{\lambda_m - \lambda} = 0 \qquad i=1,2,\cdots$$

This is a set of linear homogeneous algebraic equations with parameter λ. For nontrivial solution of Θ_j, the coefficient determinant must vanish; or

$$\left| \sum_m \frac{I_i(m) I_j(m)}{\lambda_m - \lambda} \right| = 0, \quad i,j = 1,2,\cdots \tag{12}$$

which is the <u>secular determinant</u> for eigenvalue λ.

Note that the above expression is an infinite determinant. The theory of convergence of this type of determinants is shown in Reference 10.

We note that, in Eq. (12), eigenvalues of the base problem, λ_m, are poles of the secular determinant. This is typical of the constrained problems. Occasionally, however, some of the λ_m will also be the roots of the determinant. In these cases, the original problem and the base problem will have common eigenvalues. These eigenvalues, following Weinstein [6], are called <u>persistent eigenvalues</u>.

In the practical computation, the determinant will be truncated to p by p. This procedure amounts to setting the Lagrangian multipliers of order (p+1) and above to zero. In other words, we approximate the full constraint in the modified problem by partial constraints; following Weinstein, we call it pth intermediate problem ; from the maximum-minimum principle, the eigenvalues of the intermediate problems are expected to bound the eigenvalues of the original problem from below and become progressively better as p increases.

In the evaluation of the infinite series in each term of the secular determinant in Eq. (12), double precision arithmetic must be used to keep the truncation error negligible. It is clear, also from the maximum-minimum principle, that the effect of truncation error tends to increase the eigenvalue approximations. Care must be exercised that, for each intermediate problem, the precision in evaluating the series must be gradually increased until the convergence of eigenvalues is achieved. Otherwise, the truncation error may cause the calculated results to exceed the exact ones and lower bounds cannot be guaranteed. This procedure was used successfully by Budiansky and Hu [11] and shall be discussed again in Example 3.

METHOD OF DECOMPOSITION OF CONSTRAINTS

The theory in the last section depends on the imposition of the boundary conditions on ℓ. To accomplish this we have two alternatives:

(1) - We may require that the function u to vanish at selected points $P(x_i, y_i)$, i=1,2,--along ℓ. Then, it is easily seen that the decomposition is simply $I_i(m) = \psi_m(x_i, y_i)$. As the number of points increases, we expect that, in the limit, the prescribed boundary conditions on ℓ will be satisfied. We shall use this scheme in Examples 1 and 2 in a later section.

This decomposition scheme is relatively simple to use. We remark here that the physical domain N is in effect divided into two parts, L and B, both of which are expected to contribute to the spectrum of eigenvalues. Separation of eigenvalues from these two sources is difficult. In practical application, the domain N is usually only slightly larger than L such that the domain B consists only of crescents. This assures that the first few eigenvalues can be safely attributed to domain L.

If the domain B, which is the complement of L relative to N, is simple we can use one other alternative more effectively, i.e., (2)-we may require that u=0 throughout B. Before we describe the decomposition scheme in a plane region B, we first consider a line interval [a,b].

Let f(t) be a function continuous in [a,b] and let $\{P_n\}$ be a set of normalized orthogonal polynomials which are defined in [a,b] and have a degree (n-1) for all positive integers n. In the geometrical context, f(t) is an element in the space generated by the orthonormal set $\{P_n\}$ and has the following representation:

$$f(t) = \sum_{n=1}^{\infty} (f, P_n) P_n \tag{13}$$

where (,) denotes the inner product and is chosen to be

$$(f,p_n) = \int_a^b p_n(t)\, f(t)\, dt \tag{14}$$

It follows that if $(f,P_n) = 0$, n = 1,2,..., then f(t) vanishes. Since P_n has a degree (n-1), it can be easily verified that Eq. (14) implies

$$\int_a^b t^{n-1} f(t)\, dt = 0 \tag{15}$$

The above theory can easily be generalized to two dimensional problems: Let u(x,y) be a function which is defined and continuous in a set B and let

$$F_i = \iint_B x^{r-1} y^{q-1} u(x,y)\, dxdy \tag{16}$$

If $F_i = 0$, $i=1,2,\ldots$ then $u(x,y)=0$. In Eq. (16) we choose $i=i(r,q) = q+(r+q-1)(r+q-2)/2$ which ensures a one to one correspondence between the pairs (r,q) of natural numbers and the natural numbers i.

In a special case when the domain B and N possesses symmetry in x and y, i.e., $u(x,y)=u(y,x)$, it is sufficient to take r=q and we can take a simple enumeration scheme i=r. Thus Eq.(16) becomes

$$F_i = \iint_B x^{i-1} y^{i-1} u(x,y)\, dx\, dy, \quad i=1,2,\cdots$$

It is obvious that F_i is also linear in this case. Hence, by argument used in deriving Eq. (7) the constraining conditions become

$$\sum_m a_m I_i(m) = 0$$

where
$$I_i(m) = \iint_B x^{r-1} y^{q-1} \psi_m\, dxdy \tag{17}$$

We note that $I_i(m)$ is a function of m only as asserted before. This decomposition scheme is relatively fast in convergence, as illustrated by Example 3.

For symmetrical domains Eq. (17) is specialized to

$$I_i(m) = \iint_B x^{i-1} y^{i-1} \psi_m\, dxdy \tag{18}$$

The detailed justification of the above and the manner how the method fits in the frame work of the general theory are straightforward but too lengthy to be included here; interested readers please refer to reference 12.

EXAMPLES

1. Buckling load of a clamped square plate, Fig. 2.

The governing differential equation is,

$$Aw = \lambda Bw$$

where
$$A = \frac{\partial^4}{\partial x^4} + 2\frac{\partial^4}{\partial x^2 \partial y^2} + \frac{\partial^4}{\partial y^4}$$

$$B = -\frac{\partial^2}{\partial x^2}$$

$$\lambda = \frac{Pa^2}{D}$$

subjected to boundary conditions

$$w(0,y) = w(1,y) = w(x,0) = w(x,1) = 0$$

$$\left.\frac{\partial w}{\partial x}\right|_{x=0,1} = \left.\frac{\partial w}{\partial y}\right|_{y=0,1} = 0$$

The deflection w can be expanded into a Fourier series:

$$w = \sum_m \sum_n a_{mn} \psi_{mn}(x,y)$$

where $\psi_{mn}(x,y)$ is the mnth eigenfunction of the base problem. Following Budiansky and Hu [11], we choose

$\psi_{mn}(x,y) = \cos m\pi x \cos n\pi y$; m,n = even, which satisfies the slope boundary conditions only.

The constraining conditions are

$$w(0,y) = w(1,y) = \sum_{m,n \text{ even}} \sum a_{mn} \cos n\pi y = 0$$

and

$$w(x,0) = w(x,1) = \sum_{m,n \text{ even}} \sum a_{mn} \cos m\pi x = 0$$

Due to symmetry in x,y the above two equations are equivalent and we can use only one expression and omit the other.
The explicit expression of Eq. (10) is as follows:

$$\frac{\partial V}{\partial a_{mn}} = \frac{a_{mn}}{4} \{(m^2 + n^2)^2 \pi^4 - m^2 \pi^2 \lambda\}(1 + \delta_{mo} + \delta_{on}) - \sum_j \Theta_j \cos n\pi y_j = 0$$

where δ_{io} and δ_{oi} = 0 if i = 0; = 1 if i = 0.

Solving for a_{mn} and substituting into the constraining conditions we have,

$$\sum_{m,n \text{ even}} \sum \sum_{j=1}^{p} \Theta_j \frac{\cos n\pi y_i \quad \cos n\pi y_j}{\{(m^2+n^2)^2 - \frac{m^2\lambda}{\pi^2}\} (1 + \delta_{mo} + \delta_{on})} = 0$$

For p=1, i.e., w(0,1/2) = w(1, 1/2) = 0, we obtain $\lambda^{(1)} = 9.71\ \pi^2$.

Comparing with the known exact value $\lambda_1 = 10.065\ \pi^2$, we see that it is a valid lower bound.

Note that this problem is not covered by the general theory inasmuch as the equation does not have the form of Eq. (1); note also that the function spaces associated with N-and L-problems differ not in their physical domains but in the boundary impositions. Hopefully, as the order of the secular determinant increases, the calculated eigenvalue will approach the true eigenvalue of the original problem from below.

2. Two-dimensional problem involving a circular domain

The two-dimensional wave equation

$$\nabla^2 W = k \frac{\partial^2 W}{\partial t^2}$$

governs many physically important problems, such as free vibrations of a membrane, and of transverse magnetic wave guide, axial shear vibrations of a long elastic bar, etc.

For an axisymmetrical circular membrane, it reduces to

$$-\left(\frac{d^2}{dr^2}+\frac{1}{r}\frac{d}{dr}\right) w = \lambda w$$

$$w(1) = 0$$

where $\lambda = k\omega^2 a^2$. In this problem, the eigenvalues can be determined easily and are known to be the square of zeros of Bessel's function of the first kind of zeroth order, namely, 5.7831, 29.3715, etc., [8].

We shall use this problem to demonstrate the validity of the present method. First, we shall construct a base problem. The simplest one is a problem involving a square domain which circumscribes the circle, namely, a square of the size 2 by 2. For the base problem, the eigenvalue problem is

$$-\left(\frac{\partial^2}{\partial x^2}+\frac{\partial^2}{\partial y^2}\right) w = \lambda w \quad \text{in } N$$

$$w = 0 \text{ on } n.$$

Its eigenfunctions are easily determined and, for symmetrical modes, are

$$\psi_{mn} = \cos\frac{m\pi x}{2}\cos\frac{n\pi y}{2}, \qquad m,n = 1,3,5,\cdots$$

We can express w by linear combinations of ψ_{mn} as follows:

$$w = \sum_{m,n \text{ odd}}\sum a_{mn} \cos \beta_m x \cos \beta_n y$$

$$\text{where } \beta_i = \frac{i\pi}{2}$$

The variational principle for the base problem is then,

$$U = \sum_{m,n \text{ odd}}\sum \frac{a_{mn}^2}{4}\{ (\beta_m)^2 + (\beta_n)^2 - \lambda \}$$

The constraining condition is $w(x,y)|_\ell = 0$, where ℓ is described by the equation $x^2 + y^2 - 1 = 0$. Following the decomposition scheme (1) we approximate this constraining condition by $w(x_i, y_i) = 0$, $y_i = \sqrt{1-x_i^2}$, $i = 1,2,\ldots$

Whence, Eq. (10) becomes

$$\sum_{m,n \text{ odd}}\sum a_{mn} \cos \beta_m x_i \cos \beta_n y_i = 0, \quad i=1,2,\cdots,p$$

We apply p Lagrangian multipliers θ_i, $i=1,2\ldots,p$, to constrain the functional U given above. Thus, for the pth intermediate problem, the functional to be minimized becomes

$$V = U - \sum_{j=1}^{p} \Theta_j \sum_{m,n \text{ odd}}\sum a_{mn} \cos \beta_m x_j \cos \beta_n y_j$$

After performing the partial derivative of V with respect to a_{mn}, setting it to zero and eliminating a_{mn} with the help of the constraining conditions we obtain,

$$\left| \sum_{m,n \text{ odd}} \sum \frac{\cos \beta_m x_i \cos \beta_m x_j \cos \beta_n y_i \cos \beta_n y_j}{\beta_m^2 + \beta_n^2 - \lambda} \right| = 0, \quad i,j=1,2,\cdots,p$$

Noting that the domain L, which is a circle in the present case, possesses a large degree of symmetry, we need only consider an octant, $0 \leq r \leq 1$, $0 \leq \theta \leq \pi/4$. There is no definite criterion for selecting the location of constraining points on the circumference of the circle. We elect to choose the points which are as nearly equally spaced as possible; by inserting additional points at half distances between points in each preceding run, we gradually increase the order of the frequency determinant.

We note that each element of the frequency determinant is a double trigonometric series. In the evaluation of the double infinite trigonometric series, the truncation is triggered by a "precision index", PR. First, we sum up on m up to a term whose value is equal or less than $(10)^{-PR}$. The corresponding number of terms included in the evaluation is recorded by a counter. Then we sum up on n and terminate the series similarly by the same precision index. Double precision arithmetic was used throughout.

To locate the roots of the determinant, we evaluate the determinant between two adjacent poles by assigning to λ equal increments. When the determinant changes signs between two adjacent λ, a root is indicated at some intermediate values between these two adjacent λ. Then we subdivide the interval and repeat the process until a desired accuracy is obtained.

Table 1 shows the calculated first eigenvalues for different precision indices and number of points in an octant.

From the table we can see that the eigenvalue for the same precision index increases with the increase in the number of constraining points. For the same number of constraining points the eigenvalue decreases as the precision index increases, but some slowing trend can be observed. This indicates that the series is slowly convergent. Nevertheless, the calculated eigenvalues are consistently lower than the exact value 5.7831 and the results seem to indicate a trend of approaching it from below.

The second eigenvalue is evaluated only for precision index=4.0. See Table 2. Again, the results are consistently lower than the exact value 29.3715. Also, as the number of constraining points increases, the results seem to approach the exact value from below.

The total machine time for a IBM 360-50 used in this example was about 45 minutes. We were unable to calculate the eigenvalues for large p, say 20, since the determinant became ill-conditioned, causing "underflow". We observe that the convergence of scheme of constraining by points is very slow. We shall see in Ex. 3 that the scheme of constraining by area is better.

3 L-Shaped Membrane

L-shaped region is a union of three squares as shown in Fig. 4. We shall determine its eigenvalues by decomposition scheme (2).

The formulation is very similar to Ex. 2 except that it is necessary to put the region N in the first quadrant. Then for the base problem, the differential equation is

$$-\left(\frac{\partial^2 w}{\partial x^2} + \frac{\partial^2 w}{\partial y^2}\right) = \lambda w \quad \text{in } N$$

and boundary condition is $w = 0$ on n.

The eigenfunctions are easily determined to be $\psi_{mn} = \sin \alpha_m x \sin \alpha_n y$

where $\alpha_i = i\pi$, $i = 1,2,3...$ and the variational principle for the base problem is $U = \sum_m \sum_n a_{mn}^2 \{\alpha_m^2 + \alpha_n^2 - \lambda\}$ while the constraining condition is $w(x)\big|_\ell = 0$ where ℓ is described by the following:

$$x = 1/2,\ 0 \leqq y \leqq 1/2 \qquad \text{(a)}$$

$$y = 1/2,\ 0 \leqq x \leqq 1/2 \qquad \text{(b)}$$

Due to complete symmetry about the line x=y, boundary conditions (a) and (b) are identical conditions; we need to use only (a).

By repeated partial integration we obtain

$$I_k(m,n) = J_k(m)\, J_k(n) \quad \text{where}$$

$$J_k(s) = \frac{1}{s\pi}\left[\frac{(1/2)^{k-1}}{(k-1)!} - \frac{1}{s^2\pi^2}\frac{(1/2)^{k-3}}{(k-3)!} + \ldots + f_k(s)\right]$$

$$\text{with } f_k(s) = \begin{cases} (-1)^{k+1/2} \cos s\pi/2 & k \text{ odd} \\ (-1)^{k/2} \sin s\pi/2 & k \text{ even} \end{cases}$$

The secular determinant, Eq.(12), has the following form:

$$\left| \sum_m \sum_n \frac{I_i(m,n)\, I_j(m,n)}{m^2\pi^2 + n^2\pi^2 - \lambda} \right| = 0 \qquad i,j = 1,2,\cdots.$$

Table 3 shows the first three roots of the determinants which are the eigenvalues of different intermediate problems. We observe that the convergence of the eigenvalues of intermediate problems by the scheme of constraining by area is adequately fast for practical applications. Table 4 shows the Rayleigh-Ritz upper bounds of the first eigenvalue for different numbers of terms used in the test function which was suggested by Forsythe and Wasow [1]. The total expenditure of machine time was about 30 minutes for this problem. We note that even the convergence of Rayleigh-Ritz method is slow; we attribute this slow convergence to the presence of a singularity at the re-entrant corner. The "exact" values in Tables 3 and 4 were the average values of bounds obtained by Fox, Henrici and Moler [13], who employed a special solution which effectively removed the singularity at the re-entrant corner and hence obtained extremely close bounds. The curves in Fig. 5 show the variation of eigenvalues vs. precision indeces for different intermediate problems. We recall that the precision index is a measure of truncation error of double series in each element of the secular determinant and that the eigenvalues of pth intermediate problem are the values of λ which make p by p order secular determinant vanish, provided that each element is free from truncation error. Fig. 5 shows that each curve is asymptotic to a definite value as precision index becomes large. This shows that the "asymptotic" eigenvalue of each intermediate problem is obtained with virtually negligible truncation error. It should be emphasized that the "exact" eigenvalues of any intermediate problems are the valid lower bounds of the corresponding eigenvalue of the original problem as established by the general theory. The calculated values corresponding to a finite PR cannot be claimed as the latter unless it can be established numerically that they are undistinquishable from the former within a certain precision.

CONCLUSIONS

The domain constriction method appears to be straight-forward in application. It can deal effectively with a general class of eigenvalue problems involving complicated boundary configurations. Among the advantages, the following are notable:

1. It furnishes guaranteed lower bounds.
2. It provides a systematic procedure to improve the bounds in a progressively better manner.
3. It employs only elementary functions.
4. It is a unified method in the sense that exactly the same procedure is used for all problems of this type.

When this method is used in conjunction with the Rayleigh-Ritz method a reliable error estimate can be obtained. In a variety of problems considered herein, we are able to obtain the eigenvalues within 4 to 6 percent error, which is acceptable for most engineering applications.

On the negative side, the evaluation of double series in each term of the secular determinant and double precision arithmetic used in calculation required considerable expenditure of machine time. The convergence rate is rather slow especially for the scheme of constraining by points. Further development and refinement of the present methods are needed to remove these inadequecies.

REFERENCES

1. G. E. Forsythe and W. R. Wasow, Finite-Difference Methods for Partial Differential Equations, John Wiley & Sons, Inc., New York, 1960.
2. S. G. Mikhlin, Variational Methods in Mathematical Physics, The Macmillan Co., New York, 1964.
3. For an extensive bibliography, see P. A. Laura's discussion of the paper, "The Eigenvalue Problem for Two Dimensional Regions with Irregular Boundaries", Journal of Applied Mechanics, ASME, Ser. E, p. 198, March 1968.
4. L. V. Kantorovich and V. I. Krylov, "Approximate Methods of Higher Analysis", Interscience Publishers, Inc., New York, 1958.
5. J. T. Stadter, "Bounds to Eigenvalues of Rhombical Membranes", SIAM Journal on Applied Mathematics, Vol. 14, No. 2, pp. 324-341, March, 1966.
6. S. H. Gould, Variational Methods for Eigenvalue Problems, University of Toronto Press, 1966.
7. N. W. Bazley and D. W. Fox, "Lower Bounds to Eigenvalues Using Operator Decompositions of the Form B*B, Arch. Rat. Mech. Anal., 10, pp. 352-360, 1962.
8. W. E. Byerly, Fourier's Series, Dover Publications, Inc., New York, 1959.
9. K. Marguerre, "Uber die Anwendung der Energetischen auf Stabitat Probleme", Jahrb. DVL, pp. 252-262, 1938.
10. Frederic Riesz, Les Systemes D'Equations Lineaires a une Infinite D'Inconnues, Gauthier-Villars, Paris, 1913.
11. B. Budiansky and P. C. Hu, "The Lagrangian Multipliers Method of Finding Upper and Lower Limits to Critical Stresses of Clamped Plates", NACA Rep. 848, 1946.

12. M. Chi, "Eigenvalue Problem Involving a Domain of Arbitrary Shape", multilithed informal report No. 13 to NSF under Grant No. GK-2802 to Catholic University, July, 1969.
13. L. Fox, P. Henrici and C. Moler, "Approximations and Bounds for Eigenvalues of Elliptic Operators", SIAM Jour. Numerical Anal., Vol. 4, No. 1, pp. 89-102, 1967.

TABLE 1 First Eigenvalue for Circular Domain

PR	Terms	Order of Determinant, p														
		1	2	3	4	5	6	7	8	9	10	11	12	13	14	15
3.0	79	5.33	5.42	5.55	5.61	5.64	5.67	5.68	5.68	5.68	5.69	5.70	5.75	5.76	5.77	
3.2	125	5.28	5.37	5.49	5.53	5.57	5.62	5.63	5.64	5.64	5.64	5.64	5.65	5.65	5.66	5.67
3.4	203	5.25	5.32	5.45	5.48	5.52	5.58	5.59	5.59	5.59	5.60	5.60	5.60	5.60	5.60	5.60
4.0	797	5.20	5.24	5.35	5.36	5.40	5.44	5.47	5.48	5.48	5.49	5.50	5.51	5.51	5.52	

TABLE 2 Second Eigenvalue for Circular Domain

PR	Terms	Order of Determinant											
		1	2	3	4	5	6	7	8	9	10	11	12
4	797	27.0	27.2	28.0	28.1	28.4	28.8	28.9	28.9	29.0	29.1	29.2	29.3

TABLE 3

Lower Bounds of Eigenvalues by the Present Method

Eigen-value	Order of Intermediate Problem, p									Exact
	3	4	5	6	7	8	9	10	11	
λ_1	34.17	34.99	35.58	36.01	36.33	36.59	36.79	36.95	37.10	38.56
λ_2	51.67	52.94	53.99	54.91	55.56	56.18	56.63	57.05	57.36	60.79
λ_3	75.19	75.58	75.67	75.69	75.70	75.70				78.96

TABLE 4

Upper Bounds of the First Eigenvalue by Rayleigh-Ritz Method

Eigen-value	Order of Intermediate Problem, p									Exact
	1	2	3	4	5	6	7	8	9	10	
λ_1	50.91	41.87	41.52	41.10	41.06	41.00	40.98	40.96	40.95	40.94	38.56

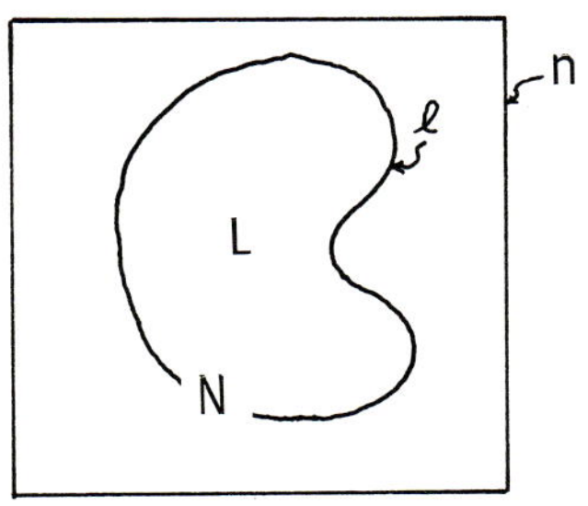

Fig. 1 Domains and Boundaries

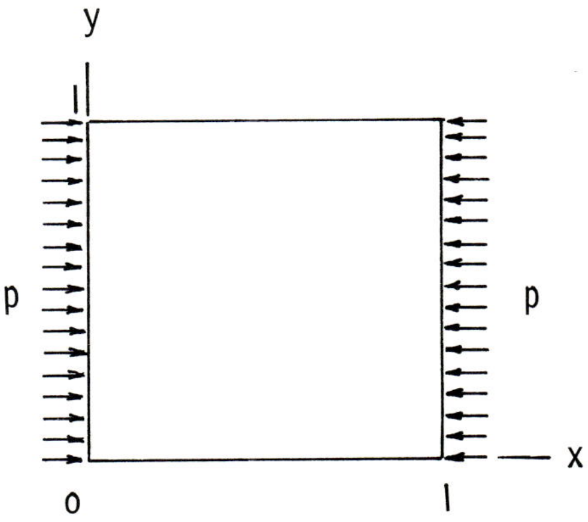

Fig. 2 Buckling of a Clamped Square Plate

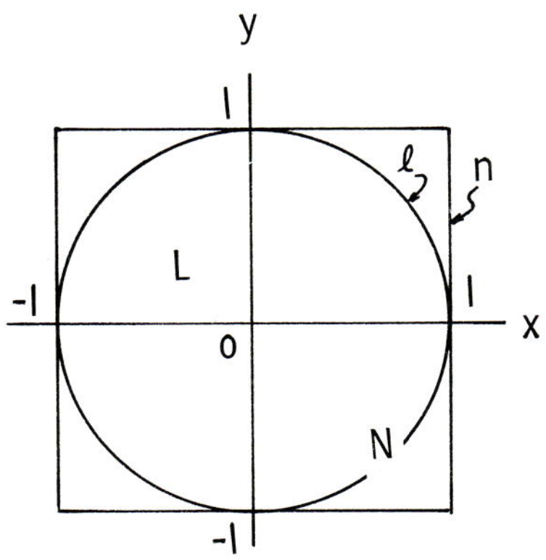

Fig. 3 Circular Domain and Its Base Problem

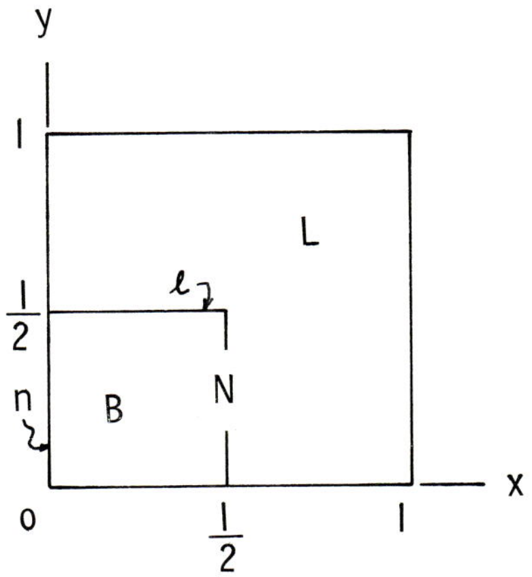

Fig. 4 L-Shaped Domain and Its Base Problem

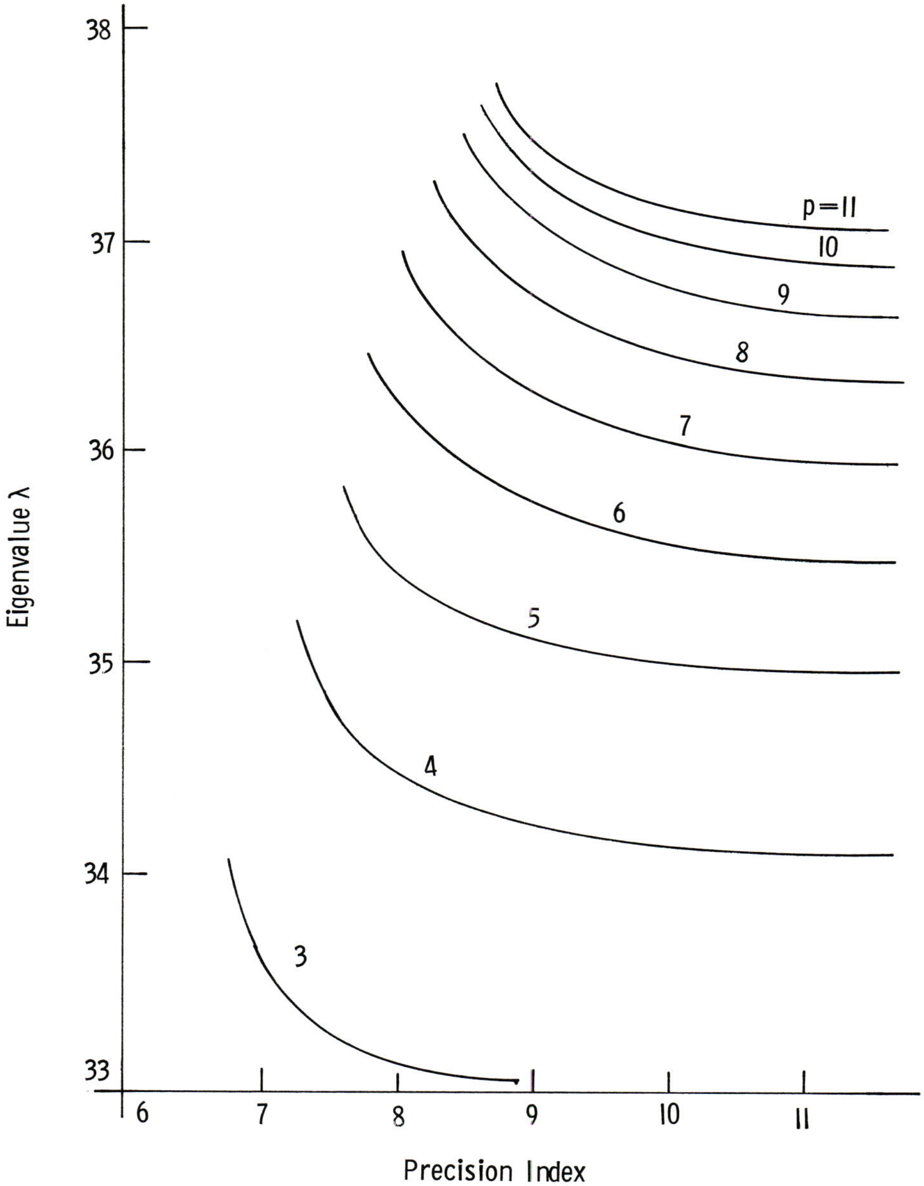

Fig. 5 Convergence of the First Eigenvalue of Different Intermediate Problems (p = 3 to 11) for L-Shaped Domain

Developments in Mechanics, Vol. 6. Proceedings of the 12th Midwestern Mechanics Conference: 52

Extracting Coefficients of Nonlinear Differential Equations from Dynamic Data

G. B. FINDLEY
ADTC, ARMAMENT LABORATORY, EGLIN AIR FORCE BASE
L. D. PARKS
ADTC, MATHEMATICAL LABORATORY, EGLIN AIR FORCE BASE

ABSTRACT

Various methods of extracting coefficients of nonlinear differential equations from data on oscillating systems are investigated. These range from the classical logarithmetic decrement methods to new methods using time derivatives. References to two other methods are given and one example from each of these references is analyzed by methods described in this paper. Numerical results from these two examples indicate the high accuracy obtainable. Advantages of the methods are: differential corrections are not used and hence no initial estimates of the coefficients are required to insure convergence; no segmented fitting of data is required; only a few seconds of computer time is required; linear coefficients are extracted accurately even when more Gaussian noise is introduced into the data than occurs in empirical data. Only single degree of freedom motion is considered but extension to additional degrees of freedom is possible. Nonlinear coefficients are extracted by: fitting of data directly to the equations of motion; using solutions of the quasi-linear differential equations, and using linear points of nonlinear differential equations.

INTRODUCTION

The purpose of this paper is the development of accurate methods for extracting coefficients from data on linear and nonlinear oscillating systems. These coefficients represent physical parameters such as: damping moments, overturning moments, etc., and their accurate values are essential for design purposes and for predicting the motion of the system. We assume the observed physical system, from which data are obtained, is representable by a second order differential equation such as Equation 1.

$$\ddot{\alpha} + f(\alpha, \dot{\alpha})\,\dot{\alpha} + g(\alpha)\,\alpha = 0 \tag{1}$$

The desired coefficients are contained in the f and g functions of this equation. We shall restrict our development to the autonomous single degree of freedom equation, but the methods could be extended to nonautonomous and higher degree of freedom equations.

Various methods of extracting coefficients are investigated which range from the classical logarithmetic decrement method to new methods using time derivatives of the dependent variable α. Several methods of extracting the coefficients of linear terms in Equation 1 are shown to give accurate results even in the presence of a relative large amount of Gaussian noise in the data. The nonlinear coefficients are extracted by: fitting of data directly to the equations of motion; using solutions of the quasi-linear differential equation, and using linear points of nonlinear differential equations. Two examples, one each of damped and undamped nonlinear systems, show the accuracy obtained. Some advantages of the methods are: differential corrections are not used and hence no initial estimates of the coefficients are required to insure convergence; no segmented fitting of data is required; only a few seconds (5 to 20) of computer time is required; coefficients of the linear terms are extracted accurately even when theoretical data contains more Gaussian noise than occurs in empirical data.

METHODS OF ANALYSIS

Assume we have the oscillating amplitude data $\alpha_i(i=1,\cdots,N)$ at times $t_i(i=1,\cdots,N)$ from some physical system, and that Equation 1 is an accurate mathematical model of the system. These amplitude data are to be used to extract coefficients from Equation 1, where the dot notation represents time (or space) derivatives of the dependent variable α.

From these amplitude data it is possible to obtain α_i, $\dot{\alpha}_i,\cdots$, $\alpha_i^{(n)}$, the peak amplitudes α_{pi} at time t_{pi}; the time derivatives at these peak amplitudes; the zero amplitudes α_{0i} at time t_{0i}; the time derivatives at these zero amplitudes; the periods P and their time derivative $\dot{P}$. We assume all these parameters are available from the data. Numerical values for such quantities used in this report were obtained from a computer program utilizing a Truncated Fourier Series (TFS Program) [4] which exactly passes through each data point. This program also calculates certain coefficients of Equation 1.

We consider first the simplest case of damped oscillatory motion obtained when the f and g functions of Equation 1 are constant, thus

$$\ddot{\alpha} + C_1 \dot{\alpha} + C_3 \alpha = 0 \tag{2}$$

In Equation 2, C_1 may be positive or negative but for oscillatory motion $C_3 > C_1^2/4$. With this restriction imposed the solution of Equation 2 is

$$\alpha\ (t) = A_0\ e^{-C_1 t/2}\ \mathrm{Cos}\ (\omega t + \Phi) \tag{3}$$

where

$$\omega = (C_3 - C_1^2/4)^{\frac{1}{2}} = 2\pi\ P^{-1} \tag{4}$$

and A_0 and Φ are constants of integration obtained from values of α and $\dot{\alpha}$ at some epoch time. Since

$$\mathrm{Cos}\ (\omega_{tpi} + \Phi) = \mathrm{Cos}\ (\omega_{tpj} + \Phi) \tag{5}$$

for each $1 \leq i,\ j \leq N$ and $C_1 \neq 0$ we obtain from Equation 3 the classical logarithmetic decrement method for calculating C_1.

$$C_1 = 2\,(t_{pj} - t_{pi})^{-1} \ln (\alpha_{pi}/\alpha_{pj}) \tag{6}$$

To calculate C_3, Equation 4 and Equation 6 are combined to yield

$$C_3 = 4\pi^2 P^{-2} + C_1^{\,2}/4 \tag{7}$$

Equation 6 and Equation 7 have been found to give very accurate values of C_1 and C_3 in the presence of nonlinear effects in the data. It can be shown by taking time derivatives of Equation 3 that in Equation 6 we can set

$$\ln (\alpha_{pi}/\alpha_{pj}) = \ln (\alpha_{pi}^{(n)}/\alpha_{pj}^{(n)}) \tag{8}$$

where (n) denotes the nth time derivative. Sample calculations with Equation 8 substituted into Equation 6 are given in Problem 1. Inspection of Equation 2 reveals that if $\dot{\alpha} = \dot{\alpha}_{pi} = 0$ then

$$C_3 = -\ddot{\alpha}_{pi}/\alpha_{pi} \tag{9}$$

and at zero amplitudes, where $\alpha = \alpha_{0i} = 0$

$$C_1 = -\ddot{\alpha}_{0i}/\dot{\alpha}_{0i} \tag{10}$$

These methods can be used to calculate C_1 and C_3 when both Gaussian noise and nonlinear effects are present in the data. The examples serve as illustrations and give numerical corrections for C_1 and C_3 when correction equations for nonlinear effects, obtained in the Appendix, are added to Equation 6 and Equation 7.

The f and g functions of Equation 1 can be thought of as the damping and static moments respectively. Consider the case when both f and g are nonlinear and defined by the following differential equation

$$\ddot{\alpha} + (C_1 + C_2\,\alpha^2)\,\dot{\alpha} + (C_3 + C_4\,\alpha^2)\,\alpha = 0 \tag{11}$$

An approximate solution of Equation 11 is obtained in the Appendix where expressions for C_3 and C_4 are also given. At zero amplitudes in Equation 11 we again have Equation 10

$$C_1 = -\ddot{\alpha}_{0i}/\dot{\alpha}_{0i} \tag{12}$$

and at peak amplitudes ($\dot{\alpha} = 0$) we have

$$C_3 = -\ddot{\alpha}_{pi}/\alpha_{pi} - C_4\,\alpha_{pi}^2 \tag{13}$$

Equation 43 gives the value of C_4 to use in Equation 13. Equation 13 or Equation 44 can be used to calculate C_3.

EXAMPLES

The autonomous nonlinear differential equation, Equation 1, is useful for describing the motion of a wide variety of physical systems. For an application of the methods described for extracting coefficients we select two examples, both from the field of aerodynamics, which are representable by Equation 11. These two examples provide results which are typical of those obtained from numerous applications.

PROBLEM ONE

Nonlinear Static Moment

$$\ddot{\alpha} + C_1 \dot{\alpha} + C_3\alpha + C_4\alpha^3 = 0 \qquad (14)$$

This example is identical to that analyzed in [1] with the University of Notre Dame WOBBLE program, which uses differential corrections and segmented fitting of data. Relating notations in [1] to notations in this paper we have

$$g(\alpha) = C_3 + C_4\alpha^2 = -C_{m\alpha}\ QSd/I = [150 - 0.5\ \alpha^2]\ QSd/I \qquad (15)$$

$$f(\alpha, \dot{\alpha}) = C_1 = -C_{mq}\ QSd^2/2uI \qquad (16)$$

$C_{m\alpha}$ is the static pitching moment stability coefficient and C_{mq} is the damping moment stability coefficient both having dimensions $(\text{radian})^{-1}$. Q is the dynamic pressure (4.870144 lb/ft), I is the transverse moment of inertia (0.02688 slug-ft^2), d is the diameter of the missile (0.16667 ft), S is the reference area ($\pi d^2/4\ ft^2$), and u is the free stream velocity (64 ft/sec). Determining coefficients C_1 through C_4 from these quantities we have: $C_1 = 0.300255/sec$, $C_3 = 98.8248/sec^2$, $C_4 = -0.329416/sec^2\ rad^2$. These coefficients were substituted into Equation 14, and a fourth-order Runge-Kutta numerical integration computer program was used with initial conditions ($\alpha_0 = 15° = 0.2618$ radian, $\dot{\alpha}_0 = 0$) to generate data in time intervals of 0.01 seconds. The numerically integrated data should be identical to the noise free data used in [1], and was used as input to the TFS program whose outputs are given in Table I, where C_1 was calculated from Equation 6 and C_3 from Equation 7.

Table I
Exact and Calculated Coefficients

COEFFICIENT	EXACT	EXTRACTED	CORRECTED
C_1	0.300255	0.300258	-
C_3	98.8248	98.8210	98.8246
C_4	-0.329416	-0.327000	-

No correction was needed for C_1 since the presence of C_4 had little or no effect on it. A correction to C_3 was necessary due to the presence of C_4 which introduces a third harmonic into the oscillations. The corrected C_3 was calculated from Equation 44, and C_4 was calculated from Equation 43.

It is interesting to note that a slightly more accurate value of C_4 ($C_4 = -0.330376$) was extracted from the data using α_{pi} and $\ddot{\alpha}_{pi}$ in Equation 14 and applying Cramer's rule.

Equation 8 can be used in Equation 6 for calculating C_1. The values of C_1 so calculated, using α_{pi}, $\dot{\alpha}_{pi}$, $\ddot{\alpha}_{pi}$, $\alpha^{(3)}_{pi}$, $\alpha^{(4)}_{pi}$ respectively were: 0.300258; 0.300231; 0.300250; 0.300232; and 0.300355. At peak amplitudes $\dot{\alpha}_{pi} = 0$, therefore

$\dot{\alpha}_{0i}$ was used in obtaining the 0.300231 value. At peak amplitudes $\dot{\alpha}_{pi} = 0$, hence differentiation of Equation 14 gives $C_1 = -\alpha_{pi}^{(3)}/\ddot{\alpha}_{pi}$ which can be used to calculate C_1, or Equation 10 can be used.

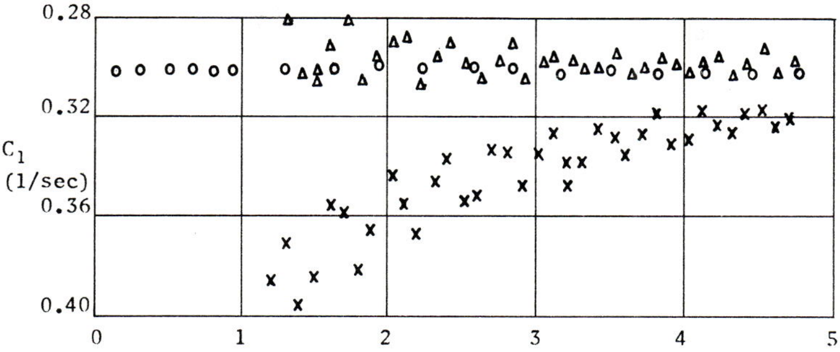

FIGURE 1. C_1 OBTAINED IN REF. [1], + & Δ; OBTAINED FROM EQUATION 6,o

In Figure 1 the (o) values of C_1 were all calculated from Equation 6 except for three values in the zero to one second interval calculated at zero amplitude. All (o) values essentially lie on the line $C_1 = 0.300255$, since the maximum absolute difference is 0.000025. The + values are given in [1] and were calculated by taking overlapping segments of 185 data points. The reader is referred to the cited reference for the equations used to obtain the Δ values. In [1] the + values were used to obtain ω in Equation 3 as a function of time which was then used in segment fitting to remove the errors introduced by the third harmonic. The third harmonic which introduced the scatter when C_1 was calculated by WOBBLE had no detectable influence on C_1 calculated by the TFS program. This is evidently due to the method of obtaining peak and zero values of α in the TFS program.

We now superimpose Gaussian noise on the data and restrict our discussions to extracting C_1 and C_3. Inspection of Equation 9 yields the following equation

$$C_3 = \sum_{i=1}^{N} |\ddot{\alpha}_{pi}| / \sum_{i=1}^{N} |\alpha_{pi}| \tag{17}$$

where positive and negative peaks can be used in the summation. Table II gives results obtained for C_3 from Equation 17 when N = 15.

Table II
C_3 Calculated From Equation 17

3σ NOISE FIGURE (DEGREES)	% NOISE	CALCULATED C_3	% ERROR
0	0	98.754	- 0.072
0.25	2.5	98.251	- 0.581
0.5	5.0	99.054	+ 0.232
1.0	10.0	99.423	+ 0.605
2.5	25.0	116.553	+18.000

In Table II the percent noise is ten times column one because the peak amplitudes ranged from approximately 15° to 5°, and the average peak amplitude was taken to be 10°. In most physical experiments the five percent or less noise range should be the more likely occurrence than the higher values. The large % error corresponding to the 25% noise figure was due to several obviously large values of $\ddot{\alpha}$ in the summation. Table III gives the values of C_3 calculated from the equation $C_3 = (2\pi/P)^2$, where P is the average period for the first eight oscillations. This table shows C_3 can be extracted accurately even from very noisy data. The use of Equation 7 would decrease the % error in all cases.

Table III
C_3 Calculated From Periods

3σ NOISE FIGURE (DEGREES)	% NOISE	CALCULATED C_3	% ERROR
0	0	98.825	0
0.25	2.5	98.679	-0.148
0.5	5.0	98.561	-0.267
1.0	10.0	98.498	-0.331
2.5	25.0	98.423	-0.407

For zero to 10% noise the values of C_3 listed in Tables II & III would be acceptable. The large difference in Tables II and III, for the 25% noise case, simply show the expected result that instantaneous accelerations cannot be obtained as accurately as average periods from very noisy data.

Table IV shows the effect of Gaussian noise on the accuracy of extracting C_1 when Equation 6 was used. The % error is approximately one-half the % noise introduced into the data.

Table IV
C_1 Calculated From Equation 6

3σ NOISE FIGURE (DEGREES)	% NOISE	CALCULATED C_1	% ERROR
0	0	.3003	0
0.1	1.0	.2990	-0.43
0.25	2.5	.2971	-1.07
0.50	5.0	.2942	-2.03
1.00	10.0	.3211	+6.93
2.50	25.0	.3224	+7.36

If C_4 is changed from -0.329416 to -1081.4, while C_1 and C_4 remain unchanged (C_1 = 0.300255, C_3 = 98.8248), this large increase in magnitude of the nonlinear coefficient will cause the logarithmetic decrement method to give inaccurate results. A modified logarithmetic decrement method can be used which will decrease the errors. However, instead of using this modified method, accurate values of C_3 and C_4 can be obtained by using peak amplitudes calculated by the TFS program and applying Cramer's rule. Values of C_1 can then be obtained by using the calculated C_3 and C_4 and applying Cramer's rule to other points than peak amplitudes. Using the above coefficients, with $\alpha_0 = 15°$ and $\dot{\alpha}_0 = 0$, noise free data were generated and C_1, C_3, and C_4 were extracted from the data. Errors in calculated C_1, C_3, and C_4, by Cramer's rule, were respectively less than 0.05%, 0.05%, and 0.2%. The calculated values of C_1 were essentially the same as those given in Figure 1. The presence of a large C_4 is easily detected by the rate of change of the period given by Equation 39. When the percent noise in the data was 0, 2.5, and 5, the percent errors in extracted C_4 were 0.2, 3 and 7 respectively.

PROBLEM TWO

Nonlinear Damping and Static Moment

$$\ddot{\alpha} + C_1 \dot{\alpha} + C_2 \alpha^2 \dot{\alpha} + C_3 \alpha + C_4 \alpha^3 = 0 \tag{18}$$

Equation 18 is identical to Equation 2 of [2] where the dot notations signify space derivatives. The method described in [2] eliminates the need for closed-form solutions by employing numerical solutions to the equations of motion, and it uses a least squares technique employing differential corrections. Values of the four coefficients C_1 through C_4 given in [2] are as follows: $C_1 = -0.00607/\text{ft}$, $C_2 = 0.000105/\text{ft deg}^2$, $C_3 = 0.0208/\text{ft}^2$, $C_4 = -0.0000375/\text{ft}^2 \text{ deg}^2$. These coefficients were obtained from a series of ballistic-range tests on models of the Gemini capsule conducted about six years ago in the Ames Pressurized Ballistic Range. Test results indicated that both the static moment and the damping moment were nonlinear functions of angle of attack α. The distance flown in the tests was approximately 200 ft. Using the above values for C_1 through C_4, initial conditions $\alpha_0 = 4.17$ degree, $\dot{\alpha}_0 = 0$ deg/ft, 200 data points were generated in increments of one foot from the Runge-Kutta integration program. These data points were used as input to the TFS program and results obtained are given in Table V.

Table V
Exact and Calculated Coefficients

COEFFICIENTS	EXACT	EXTRACTED	CORRECTED
C_1	-0.0060700	-0.0060694	-0.0060699
C_2	0.0001050	0.0001057	-
C_3	0.0208000	0.0207996	0.0208000
C_4	-0.0000375	-0.0000377	-

The coefficients in Table V were calculated by the same method used for Table I, Problem 1, except here we obtain a correction for C_1 due to the nonlinear term in the damping moment. Since C_2 was calculated from Equation 47, approximately the same percent error occurs in C_2 and C_4. There are nine linear points (defined in the Appendix) between zero and 200 feet, each of which occur approximately 22.5 degrees after each zero amplitude, and each point will give the same value of C_2 when used in Equation 47.

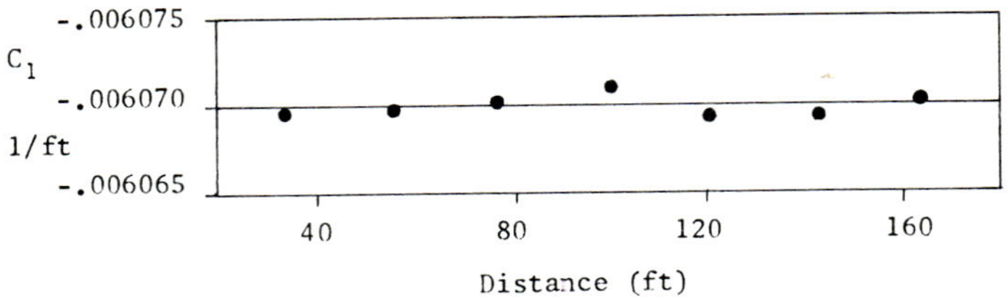

FIGURE 2. C_1 OBTAINED FROM EQUATION 6

In Figure 2, the values of C_1 were all calculated from Equation 6. The results are similar to those obtained in Figure 1. There are no calculations of C_1 available from other methods as was the case for Figure 1.

The result of calculating the coefficients C_1 and C_3 when the data contained Gaussian noise is very similar to the result in Problem 1. Calculated values of C_1 and C_3, obtained from Equations 6 and 7 respectively, are shown in Table VI, where the % noise is based on an average peak amplitude of five degrees. Again, as in Table IV for Problem 1, the % error in calculating C_1 is approximately one-half the % noise in the data. The errors in C_3 are small and are similar to the results obtained in Table III.

Table VI
C_1 and C_3 Calculated From Equations 6 and 7

3σ NOISE FIGURE (DEGREES)	% NOISE	C_1	% ERROR C_1	C_3	% ERROR C_3
0	0	-0.00607	0	0.02080	0
0.1	2	-0.00602	-0.82	0.02080	0
0.25	5	-0.00593	-2.31	0.02081	0.05
0.5	10	-0.00575	-5.27	0.02083	0.14
1.0	20	-0.00681	12.19	0.02085	0.24

CONCLUSIONS

Time derivatives of the dependent variable can be used to extract coefficients of nonlinear differential equations by fitting of data directly to the equations of motion, and accurate coefficients can be extracted not only from noise free data but also from data containing a large amount of Gaussian noise. When these time derivatives are used in conjunction with solutions of the linear and quasi-linear differential equations, accurate coefficients of the nonlinear differential equations are obtained. It has been shown that certain linear points, defined in the Appendix, can be used in extracting the coefficients of nonlinear terms of the differential equation. Other examples, not discussed in this paper, show that coefficients are accurately extracted even when the amplitude of the motion is 60 degrees and nonlinear coefficients are of the same magnitude as linear coefficients. Various methods of extracting coefficients should be evaluated using both noisy and noise free data to determine the precision with which the coefficients can be extracted from empirical data.

ACKNOWLEDGEMENTS

This work was accomplished under the Air Force Office of Aerospace Research, Project 9860. The authors thank Mrs. Jane Colee for her mathematical and programming assistance in this work.

BIBLIOGRAPHICAL REFERENCES

[1] Eikenberry, R. S., "Analysis of the Angular Motion of Missiles", National Bureau of Standards, U.S. Department of Commerce, Springfield, Va., 1970, pp. 81-87.

[2] Chapman, G. T. and Kirk, D. B., "A Method for Extracting Aerodynamic Coefficients from Free-Flight Data", American Institute of Aeronautics and Astronautic Journal, Vol. 8, No. 4, 1970, pp. 753-758.

[3] Minorsky, N., Introduction To Nonlinear Mechanics, J. W. Edwards, Ann Arbor, Michigan, 1947, pp. 233-236.

[4] Colee, P. J., "Fourier Analysis of Discrete Time Series Data", ADTC Mathematical Quarterly, No. 4, May 1969, pp. 11-19.

APPENDIX

DERIVATION OF EQUATIONS

A solution of the nonlinear differential equation

$$\ddot{\alpha} + C_1 \dot{\alpha} + C_2 \alpha^2 \dot{\alpha} + C_3 \alpha + C_4 \alpha^3 = 0 \tag{19}$$

is useful not only for describing α as a function of time but also for extracting coefficients from empirical data. Let Equation 19 be written in the quasi-linear form

$$\ddot{\alpha} + \omega^2 \alpha + \mu f(\alpha, \dot{\alpha}) = 0 \tag{20}$$

where

$$f(\alpha, \dot{\alpha}) = a_1 \dot{\alpha} + a_2 \alpha^2 \dot{\alpha} + a_4 \alpha^3 \tag{21}$$

$$C_1 = \mu a_1, C_2 = \mu a_2, C_3 = \omega^2, C_4 = \mu a_4 \tag{22}$$

and μ is a small parameter.

A solution of Equation 20 can be written as

$$\alpha = A \cos \psi \tag{23}$$

where the amplitude A and the total phase ψ are given by two differential equations of the first order. If one defines two constants λ and K, the equivalent parameters, by the equations

$$\lambda = -\frac{\mu}{\pi A \omega}\int_0^{2\pi} f(A \cos \Phi, -A \omega \sin \Phi) \sin \Phi \, d\Phi \tag{24}$$

$$K = \omega^2 + \frac{\mu}{\pi A}\int_0^{2\pi} f(A \cos \Phi, -A \omega \sin \Phi) \cos \Phi \, d\Phi \tag{25}$$

it can be shown [3] that an "equivalent" linear equation, with coefficients λ and K, approximates the solutions of the quasi-linear equation to an accuracy of order μ^2. The amplitude equation is

$$\dot{A} = -\lambda A/2 \tag{26}$$

and the phase equation is

$$\dot{\psi} = K^{\frac{1}{2}} \tag{27}$$

Insertion of Equation 21 into Equation 24 yields

$$\lambda = \mu/\pi \int_0^{2\pi} [a_1 \operatorname{Sin} \Phi + a_2 A^2 \operatorname{Sin} \Phi \operatorname{Cos}^2 \Phi - a_3 A^2 \omega^{-1} \operatorname{Cos}^3 \Phi] \operatorname{Sin} \Phi \, d\Phi$$

$$= \mu a_1 + \mu A^2 a_2/4 \tag{28}$$

and similarly for K we obtain

$$K = \omega^2 + 3\mu a_4 A^2/4 \tag{29}$$

Substitution of λ into Equation 26 and integration from zero to t we obtain

$$A = A_0 e^{-\mu a_1 t/2} [1 + a_2 A_0^2 (4a_1)^{-1} (1 - e^{-\mu a_1 t})]^{-\frac{1}{2}} \tag{30}$$

where A_0 is the value of A at t = 0. Substitution of Equations 29 and 30 for A and K respectively into Equation 27 and expanding $\dot{\psi}$ in a binomial series and retaining only the first two terms we have

$$\dot{\psi} = \omega + 3\mu a_1 a_4 (2\omega a_2)^{-1} e^{-\mu a_1 t} [1 + 4a_1 C_2^{-1} A_0^{-2} - e^{-\mu a_1 t}]^{-1} \tag{31}$$

On integrating from zero to t we have

$$\psi = \psi_0 + \omega t + \frac{3a_4}{2a_2\omega} \ln [1 + \frac{a_2 A_0^2}{4a_1} (1 - e^{-\mu a_1 t})] \tag{32}$$

where $\psi = \psi_0$ at t = 0. Substitution of Equation 22 into Equations 30 and 31 and A and ψ into Equation 23 gives

$$\alpha = A \operatorname{Cos} \psi$$
$$= A_0 e^{-C_1 t/2} [1 + \frac{C_2 A_0^2}{4C_1} (1 - e^{-C_1 t})]^{-\frac{1}{2}}$$

$$\times \operatorname{Cos} \{\psi_0 + \psi t + \frac{3C_4}{2C_2\omega} \ln [1 + \frac{C_2 A_0^2}{4C_1} (1 - e^{-C_1 t})]\} \tag{33}$$

If $C_2 = 0$, Equation 33 reduces to

$$\alpha = A \operatorname{Cos} \psi = A_0 e^{-C_1 t/2} \operatorname{Cos} \{\psi_0 + \omega t + \frac{3C_4 A_0^2}{8\omega C_1} (1 - e^{-C_1 t})\} \tag{34}$$

Equation 34 is an approximate solution of the differential equation in Problem 1.

Let t_{pi} and t_{pn} be times for peak values of α. Since $\cos\psi(t_{pi}) = \cos\psi(t_{pn})$, we obtain from Equation 34

$$C_1 = \frac{2 \ln (\alpha_{pi}/\alpha_{pn})}{t_{pn} - t_{pi}} \tag{35}$$

We write ψ in the form

$$\psi = \psi_0 + 2\pi t\ P^{-1} \tag{36}$$

where P is the period and is a function of time having the Taylor series

$$P = P_0 + \dot{P}_0 t + \ddot{P}_0 t^2/2 + \ . \ . \ . \tag{37}$$

From Equation 34 and Equation 36 we have, after algebraic simplification, the results

$$\psi = \psi_0 + (\omega + \frac{3}{8\omega} C_4\ A_0{}^2)\ t - \frac{3C_4\ A_0{}^2\ C_1\ t^2}{16\omega} + \ . \ . \ .$$

$$= \psi_0 + 2\pi\ P_0{}^{-1}\ t - 2\pi\ \dot{P}_0\ P_0{}^{-2}\ t^2 + \ . \ . \ . \tag{38}$$

Equating coefficients of the t^2 term we have

$$C_4 = \frac{32\pi\ \omega\ \dot{P}_0}{3\ P_0{}^2\ A_0{}^2\ C_1} = \frac{64\pi^2\ \dot{P}_0}{3\ P_0{}^3\ A_0{}^2\ C_1} \tag{39}$$

where in the last equation we have substituted the approximate value of ω obtained from the coefficients of t in Equation 38. Equation 35 gave C_1 very accurately in Problems 1 and 2, and since P_0 and A_0 can be obtained accurately, the accuracy of C_4 therefore depends mostly on the accuracy of $\dot{P}_0$. Since

$$\alpha_0 = A_0 \cos\psi_0 \tag{40}$$

C_4 can be written in the form

$$C_4 = \frac{64\pi^2 \cos^2\psi_0\ \dot{P}_0}{3\ P_0{}^3\ \alpha_0{}^2\ C_1} \tag{41}$$

From the initial conditions that $\dot{\alpha}$ is zero at $t = 0$, we obtain from Equation 34

$$\cos^2\psi_0 = [1 + \frac{C_1{}^2}{4\omega^2}]^{-1} \approx 1 - \frac{C_1{}^2}{4\omega^2} = 1 - \frac{C_1{}^2}{4C_3} \tag{42}$$

so that Equation 41 can be written as

$$C_4 = \frac{64\pi^2\ (1 - C_1{}^2/4C_3)\ \dot{P}_0}{3\ P_0{}^3\ \alpha_0{}^2\ C_1} \tag{43}$$

Equation 43 was used to calculate C_4 in Problems 1 and 2, and then used in Equation 44 to improve the value of C_3. In most cases the expression $C_1{}^2/4C_3$ can be neglected in Equation 43. Inspection of Equation 38 shows that the equivalent linear form of the differential equation leads to

$$C_3 = 4\pi^2 P_0^{-2} + C_1^2/4 - 3 C_4 A_0^2/4 \tag{44}$$

For $C_2 \neq 0$, since $\cos \psi (t_{pi}) = \cos \psi (t_{pn})$, we obtain from Equation 33 the result

$$\alpha_{pi}/\alpha_{pn} = e^{-C_1(t_{pi} - t_{pn})/2} \left[\frac{1 + C_2 A_0^2 (1 - e^{-C_1 t_{pi}})/4C_1}{1 + C_2 A_0^2 (1 - e^{-C_1 t_{pn}})/4C_1} \right]^{\frac{1}{2}} \tag{45}$$

On expanding Equation 45 in powers of $C_2 A_0^2/4C_1$ and retaining only the first power we have

$$C_1 = \frac{2 \ln (\alpha_{pi}/\alpha_{pn})}{t_{pn} - t_{pi}} - \frac{C_2 A_0^2}{(t_{pn} - t_{pi}) 4 C_1} [e^{-C_1 t_{pn}} - e^{-C_1 t_{pi}}] \tag{46}$$

When C_2 is known, the corrected value of C_1 is obtained from Equation 46 where in the right hand part of this equation C_1 is the uncorrected value of C_1.

There are a number of ways for obtaining C_2. Since methods for obtaining C_1, C_3, and C_4 are available, these coefficients can be used to determine C_2 in Equation 19. Least square fitting of data to Equation 19, where the constants C_1, C_3, C_4 are constrained or not constrained, will also give C_2. Other equations for C_2 can be obtained from the quasi-linear solution.

We introduce another method of obtaining C_2. Since the use of peak and zero amplitudes in Equation 19 leads to accurate extraction of the coefficients, there may be other points on the motion curve of particular usefulness. We therefore define a linear point of a nonlinear differential equation to be a point where the linear part of the equation is zero. An equivalent definition is the point where the nonlinear part of the equation is zero. Applying this definition in Problem 2 we have

$$C_2 = -C_4 \alpha_1/\dot{\alpha}_1 \tag{47}$$

where the subscript 1 is used for linear points. Note that $\alpha = 0$ is a linear point in both Problem 1 and 2, and if $C_4 = 0$ in Equation 19, then $\alpha = 0$ and $\dot{\alpha} = 0$ are linear points of that equation when $C_2 \neq 0$. Nonlinear differential equations may or may not have linear points. Linear points may be of value for extracting the linear coefficients of the equation, but we have only used them for determining nonlinear coefficients when one of the nonlinear coefficients has been determined.

Developments in Mechanics, Vol. 6. Proceedings of the 12th Midwestern Mechanics Conference: 53

Accuracy and Stability for First and Second Order Solutions by Parametric Differentiation

MARVIN C. ALTSTATT and MARTIN C. JISCHKE
UNIVERSITY OF OKLAHOMA

ABSTRACT

Methods of solution by parametric differentiation for the temperature profile and heat flux behind a one-dimensional, hypersonic shock, including thermal radiation, were investigated. Four methods of advance in the parameter--first and second order Taylor series, a Runge-Kutta method and a Milne Predictor-Corrector method--were compared with an exact solution to the problem. All of the methods converged toward the exact solution, and except for the Milne method, the error was always less than that predicted. The Runge-Kutta method provided the most accurate results. Comparison of the methods of advance indicated that, in order to attain a given accuracy, methods for reducing the error that involve a second auxiliary equation are less economical in terms of computer time. Investigation of the stability of the numerical solutions showed regions in which the solutions were unstable and indicated that the stability depended directly on the auxiliary equation and indirectly on the accuracy of the numerical advance.

INTRODUCTION

The purpose of the work reported in this paper was the investigation of the solution by parametric differentiation for the temperature profile and heat flux behind a one-dimensional hypersonic shock, including radiation effects. The problem had been solved previously [3] by this method, but no detailed study of the behavior of the solution had been carried out. As the simplified one-dimensional problem has the characteristics of problems with more complex geometries, and the method of parametric differentiation appears to be of general utility in the solution of these problems, it was felt that such a detailed study would be of value.

---
*Graduate Student, School of Aerospace, Mechanical and Nuclear Engineering.
**Assistant Professor, School of Aerospace, Mechanical and Nuclear Engineering.

The governing equation, a non-linear integro-differential equation, can be solved by iteration after one integration allowing for a check of the solutions obtained by parametric differentiation. These solutions were carried out numerically, by means of first and second order expansions, with optical depth as the parameter. Four solutions by parametric differentiation were obtained: first and second order Taylor expansions, a fourth order Runge-Kutta advance involving first derivations and a Milne Predictor-Corrector method involving second derivatives. It was necessary to expand temperature in a series for small optical depth in order to obtain a solution near zero optical depth.

The stability of the solution with increasing optical depth was also investigated. This was carried out analytically and the results were tested by introducing small disturbances into the computer solutions and comparing with undisturbed solutions.

STATEMENT OF THE PROBLEM

2.1 The Method of Parametric Differentiation

The solution of a problem by parametric differentiation is analogous to the method of solving a steady state problem by casting it as an initial value problem and integrating in time. For parametric differentiation, however, the integration is by steps in a parameter from a known solution at some value of that parameter. In the problem shown here, this procedure results in the stepwise solution of a parabolic type of problem, rather than attempting to integrate an elliptic equation. The advantages are obvious.

Briefly, the method of parametric differentiation is applied as follows [2,3]. Consider an equation of the form

$$L(\Phi_1, \Phi_2, \ldots \Phi_n) = 0 \tag{1}$$

where L is a differential or integro-differential operator and $\Phi = \Phi(\xi, \eta, \lambda)$, with λ the parameter of interest. Assume that a solution is known for some value of the parameter $\lambda = \lambda_o$, and that an auxiliary equation can be found from which the dependence of Φ on λ can be determined. Determining this dependence is usually the most difficult part of the procedure and is generally carried out numerically, as in the present case. The solution can then be perturbed by a small change in λ, $\Delta\lambda$, as long as Φ and its derivatives with respect to λ remain continuous and finite at each step. Each step in λ then produces a new solution for the new value of the parameter.

Questions immediately arise as to the accuracy and region of convergence for the method. The interaction between the error in proceeding by steps in λ, and the error introduced in the numerical solution of the auxiliary equations may produce deviations that are not simply related to the error from either system taken alone. This interaction can also affect the stability of the numerical scheme.

2.2 The Governing Equation

The conservation equations for the one-dimensional, hypersonic, gray radiating shock layer reduce to [3]

$$Bo \frac{dT}{d\xi} = -2 \tau_L T^4 + \tau_L^2 \int_0^1 T^4 E_1(\tau_L |\xi - \zeta|) \, d\zeta \tag{2}$$

with the boundary condition

$$T(0) = 1 \tag{3}$$

Bo is the Boltzmann Number, T is the temperature normalized by its value at the shock and τ_L is the shock layer optical depth defined by

$$\tau_L \equiv \int_0^L \varkappa \, dx \tag{4}$$

$\varkappa$ is the volumetric absorption coefficient. ξ is defined by

$$\xi \equiv \tau / \tau_L \tag{5}$$

where

$$\tau = \int_o^x \varkappa \, dx \tag{6}$$

At $\tau_L = 0$, Eqn. (2) has the solution

$$T(\xi) = 1 \tag{7}$$

It will be seen however, that this solution cannot be used directly as a starting point for the method of parametric differentiation with τ_L as the parameter.

The required auxiliary equation is determined by differentiating Eqn. (2) with respect to τ_L.

$$\begin{aligned} Bo \frac{d\mathcal{J}}{d\xi} + 8 \tau_L T^3 \mathcal{J} - 4 \tau_L^2 \int_o^1 \mathcal{J} T^3 \; E_1(\tau_L | \xi - \zeta |) \, d\zeta \\ = -2T^4 + 2\tau_L \int_o^1 T^4 E_1(\tau_L | \xi - \zeta |) \, d\zeta \\ - \tau_L^2 \int_o^1 T^4 \exp(-\tau_L | \xi - \zeta |) \, d\zeta \end{aligned} \tag{8}$$

where

$$\mathcal{J} = \frac{dT}{d\tau_L} \tag{9}$$

The boundary condition at the shock becomes

$$\mathcal{J}(0) = 0 \tag{10}$$

Equation (8) is a linear in $\mathcal{J}$, and can be solved by straightforward numerical techniques for given values of Bo and τ_L if the temperature profile is known.

To apply the method of parametric differentiation; that is, to find the temperature profile for larger values of the parameter τ_L, a number of techniques may be used. To illustrate a typical technique, consider a simple, first order Taylor expansion,

$$T(\xi;\, \tau_L + \Delta \tau_L) \simeq T(\xi;\, \tau_L) + \Delta \tau_L \mathcal{J}(\xi;\, \tau_L) \tag{11}$$

which allows evaluation of the temperature at $(\tau_L + \Delta\tau_L)$. With the temperature profile known at $\tau_L + \Delta\tau_L$, $(\tau_L + \Delta\tau_L)$ can be computed and the process repeated.

Solution (7) however, cannot be used as a starting point, since Eqn. (8) is singular for $\tau_L = 0$. An expansion for small τ_L must therefore be carried out to provide an initial solution.

2.3 Expansion for Small τ_L

An expansion of T with τ_L small is taken in the form

$$T = 1 + \tau_L T_1 + \tau_L^2 \ln \tau_L T_2 + \tau_L^2 T_3 + \ldots \qquad (12)$$

Substituting Eqn. (12) into Eqn. (2) and setting coefficients of each power of τ_L equal to zero results in the following expression for temperature with τ_L small.

$$T \cong 1 - \frac{2\,\tau_L \xi}{Bo} - \frac{\left(\tau_L^2 \ln \tau_L\right)\xi}{Bo} - \frac{\tau_L^2}{Bo}\left[\frac{8\,\xi^2}{Bo} + \frac{\xi^2}{2}\ln \xi + \frac{(1-\xi^2)}{2}\ln(1-\xi) - \frac{\xi^2 + (1-\xi^2)}{4} - (1-\beta)\xi\right] \qquad (13)$$

This expression was used to compute the temperature profiles as required to initiate the procedure described in Sections 2.1 and 2.2 above.

METHODS OF SOLUTION

3.1 Solutions by Parametric Differentiation

Four methods of advance in τ_L were used: first and second order Taylor Series expansions, a fourth order Runge-Kutta advance using first derivatives and a Milne Predictor-Corrector method involving second derivatives. The methods are described in the following sections.

As was noted in general in Section 2.1, the most difficult part of the method in all cases was the solution of the auxiliary equation, i.e., computing the derivatives of the temperature with respect to the optical depth. This was carried out by writing Eqn. (8), at each of N evenly spaced points across the shock layer, in finite difference form with a simple collocation formula for the integrals, and solving the resulting N equations simultaneously for the value of $\mathcal{J}$ at each point.

The results indicate that all four methods converge toward the exact solution with small step size in optical depth and in the distance across the shock. It will be shown that the accuracy and stability are closely related and are both functions of the step sizes and the Boltzmann Number.

3.1.1 Taylor Expansions

The two Taylor Series advances used were:

$$T(\tau_L + \Delta\tau_L) = T(\tau_L) + \Delta\tau_L \mathcal{J}(\tau_L) + 0(\Delta\tau_L^2) \qquad (14)$$

and

$$T(\tau_L + \Delta\tau_L) = T(\tau_L) + \Delta\tau_L \mathcal{J}(\tau_L) + \frac{\Delta\tau_L^2}{2}\bar{\mathcal{J}}(\tau_L) + 0(\Delta\tau_L^3) \qquad (15)$$

where

$$\bar{\mathcal{J}} = \frac{d^2T}{d\tau_L^2} \qquad (16)$$

Starting from an initial temperature profile for some value of τ_L, which can be computed from Eqn. (13), either Eqn. (14) or Eqn. (15) can be used to proceed stepwise in τ_L. $\mathcal{J}$ is computed as described above. $\dot{\mathcal{J}}$ can then be found by taking the derivative with respect to τ_L of Eqn. (8) and following a similiar numerical procedure.

3.1.2 Runge-Kutta Method

The following expressions were used for this method of advance [2]:

$$T(\tau_L + \Delta\tau_L) = T(\tau_L) + \frac{1}{6} K_1 + \frac{1}{3} K_2 + \frac{1}{3} K_3 + \frac{1}{6} K_4 + O(\Delta\tau_L{}^5) \tag{17}$$

where

$$K_1 = \Delta\tau_L \mathcal{J}(\tau_L, T) \tag{18}$$

$$K_2 = \Delta\tau_L \mathcal{J}(\tau_L + \Delta\tau_L/2,\ T + K_1/2) \tag{19}$$

$$K_3 = \Delta\tau_L \mathcal{J}(\tau_L + \Delta\tau_L/2, T + K_2/2) \tag{20}$$

and

$$K_4 = \Delta\tau_L \mathcal{J}(\tau_L + \Delta\tau_L/2,\ T + K_3) \tag{21}$$

3.1.3 Milne Predictor-Corrector Method

The Milne Predictor-Corrector method was used in the following form [1].

$$\text{P:}\quad T(\tau_L + \Delta\tau_L) = T(\tau_L - 2\Delta\tau_L) + 3[T(\tau_L) - T(\tau_L - \Delta\tau_L)]$$

$$+ \Delta\tau_L{}^2[\dot{\mathcal{J}}(\tau_L) - \dot{\mathcal{J}}(\tau_L - \Delta\tau_L] + O(\Delta\tau_L{}^5) \tag{22}$$

$$\text{C:}\quad T(\tau_L + \Delta\tau_L) = T(\tau_L) + \frac{1}{2}\Delta\tau_L[\mathcal{J}(\tau_L + \Delta\tau_L) + \mathcal{J}(\tau_L)]$$

$$- \frac{1}{12}\Delta\tau_L{}^2[\dot{\mathcal{J}}(\tau_L + \Delta\tau_L) - \dot{\mathcal{J}}(\tau_L)] + O(\Delta\,\tau_L{}^5) \tag{23}$$

Two initial temperature profiles were computed from Eqn. (13) to initiate this advance.

3.2 The Integrated Solution

Equation (2) can be integrated once with the result:

$$T = 1 - \frac{\tau_L}{Bo}\int_0^1 T^4\, E_2(\tau_L\zeta)d\zeta$$

$$- \frac{\tau_L}{Bo}\int_0^1 T^4\, \mathrm{sgn}(\xi - \zeta)\, E_2(\tau_L|\xi - \zeta|)\, d\zeta \tag{24}$$

An approximate solution for the equation above can be obtained by iteration, as in a Neumann Series for linear integral equations.

The region of convergence for the series can be determined by writing, for the Nth iteration

$$T_n(\tau) = 1 - \frac{1}{Bo} \int_0^{\tau_L} \left\{1 - \frac{1}{Bo} \int_0^{\tau_L} \cdots \right.$$

$$\left. \left[1 - \frac{1}{Bo} \int_0^{\tau_L} K(\tau,\eta)\, d\eta\right]^4 \cdots K(\tau,\eta) d\eta \right\}^4 K(\tau,\eta) d\eta \qquad (25)$$

where

$$K(\tau,\eta) = \left[E_2(\eta) + \text{sgn}(\tau - \eta)\, E_2(\tau - \eta)\right] \qquad (26)$$

Integrating the highest order term of Eqn. (25), it can be seen that for

$$Bo < E_3(0) - E_3(\tau_L) \qquad (27)$$

the series does not converge.

It is interesting to note that $\frac{1}{2}(E_3(0) - E_3(\tau_L))$ is the maximum radiative heat flux, so since Bo is the ratio of convective to radiative heat flux, Eqn. (27) can be interpreted as a limitation on the ratio of radiative energy transfer to convective energy transfer.

3.3 Numerical Methods

The auxiliary equation was written to second order accuracy in $\Delta\xi$ by taking

$$\frac{d\mathcal{J}}{d\xi} = \frac{\mathcal{J}_{k+1} - \mathcal{J}_{k-1}}{2\,\Delta\xi} \qquad (28)$$

where $K = 1, 2, \ldots, N$ and N is the number of steps $\Delta\xi$ across the shock layer. To preserve the second order accuracy, it was assumed that

$$\mathcal{J} = a\xi^2 + b\xi + c \qquad (29)$$

near the wall and near the shock.

The integrals were approximated by assuming that for small $\Delta\xi$,

$$\int_{(K-1)\Delta\xi}^{K\Delta\xi} f(T,\mathcal{J})\, g(\xi,\zeta;\tau_L)\, d\zeta \cong f(T,\mathcal{J}) \int_{(K-1)\Delta\xi}^{K\Delta\xi} g(\xi,\zeta;\tau_L) d\zeta \qquad (30)$$

With these approximations the auxiliary equation could be solved as described in Section 3.1.

Computations were carried out on an IBM 360/50 Computer. Time to compile and execute was of the order of two minutes for a typical run. Since an investigation of the various methods required that a large number of solutions be carried out, step sizes in the parameters resulting in only moderate accuracy were used in order to minimize computer time.

ACCURACY

4.1 Comparison of Methods of Advance

Solutions were obtained by means of parametric differentiations for the following ranges of the parameters: $\Delta\tau_L = 0.01 - 0.25$, $\Delta\xi = 0.04 - 0.2$, $Bo = 0.2 - 2.0$ and $\tau_L = 0 - 10$. Most computations were carried out for $Bo - 1.0$,

$\Delta\tau_L = 0.1$ and $\Delta\xi = 0.1$. These values were chosen to produce sufficient accuracy for comparison of the methods of solution while keeping computer time to a minimum.

Figures 1 and 2 show typical temperature profiles and heat flux obtained by parametric differentiation. Comparison of the solutions with each other and with the numerically iterated solutions shows that all of the methods of advance produce profiles exhibiting similar and generally correct behavior. The most noticeable deviations are found in the second order Taylor series and the Milne method.

For the second-order Taylor series, the profile at $\tau_L = 10$ falls far below the interated solution and the other solutions by parametric differentiation. It will be shown in the next section that the advance by second order Taylor series is unstable in that region, which accounts for the large inaccuracy. Also, the heat flux at the wall begins to deviate rapidly from the iterated solution at approximately $\tau_L = 5$.

The temperature for the Milne advance drops sharply near the wall. While a computation of the wall temperature from the iterated solution does indicate a drop in temperature, this is much less than that shown by the Milne advance. This inaccuracy is also reflected in the curves for the heat flux.

Excluding the irregularities noted above, the error varies from a maxium of approximately 25% for the temperature profile at $\tau_L = 10$ for the first order Taylor expansion to less than 3% for the Runge-Kutta advance at $\tau_L = .5$. The heat flux is closer to the iterated solution, with a maximum error for the Runge-Kutta advance of approximately 10%, corresponding to a maximum error in the temperature profile of nearly 20%.

Figures 3 and 4 compare the temperature profiles obtained by parametric differentiation with the iterated solution. The second order Taylor series solution is not included because, for this range of optical depth, it closely parallels the Runge-Kutta advance with a few per cent less accuracy.

At $\tau_L = 0.5$, estimates of the error, assuming that the error in solving the auxiliary equations and the deviation in the method of advance are additive, range from 5% for the Runge-Kutta and Milne methods to 10% for the first order Taylor series. Only the solution by the Milne Predictor-Corrector method shows a greater deviation than that predicted.

The reasons for the unsuitability of the Milne advance are apparent on close examination of Eqns. (22) and (23). In proceeding one step, $\Delta\tau_L$, this advance uses six solutions to the auxiliary equation. Two of these, as used in Eqn (23), are computed from the approximate temperature profile produced by Eqn. (22) and are therefore less accurate. Also, this method depends heavily on the second derivative, which may have errors greater than second order, since it is computed using the first derivative.

However, under other circumstances, the accuracy can be improved through the use of the second derivative. This is shown by the increase in performance of the second order Taylor series over the first order. On the basis of accuracy versus computer time, however, it appears to be more economical to use methods of decreasing the error which do not involve a second auxiliary equation.

4.2 Convergence

Figure 5 shows the convergence of the solutions by parametric differentiation toward the integrated solution for decreasing step size in τ_L. Note that a comparison of these curves does not reflect the relative accuracy of the solutions, since the convergence is indicated at only one point.

It can also be seen that the results obtained by parametric differentiation do not converge exactly to the integrated solution. Decreasing the value of $\Delta\tau_L$

past a certain point, which is different for each method of advance, increases the error produced. This is an effect of the interaction between the error in the solution of the auxiliary equation and the method of advance. In the procedure followed, errors are produced by both the numerical solution of the auxiliary equation and the method of advance in the parameter. Since the auxiliary equation is accurate to second order, decreasing the step size in optical depth, while holding the step size across the shock layer constant eventually results in the error introduced by the method of advance becoming negligible in comparison to the error introduced by the numerical solution of the auxiliary equation. Further reductions in $\Delta\tau_L$ do not improve the accuracy, but only add to the number of steps in optical depth which must be taken to reach a given solution; therefore increasing the error instead of decreasing it. Figure 6 confirms that with a smaller step size across the shock, $\Delta\xi$, the solution by parametric differentiation continues to converge to the integrated solution.

STABILITY

5.1 Analysis

During the course of this investigation of the solution of Eqn. (2) by parametric differentiation, it was noted that for some values of the parameters the solutions were unstable. That is, the temperature profile produced would deviate from the integrated solution, and this deviation rapidly increased with increasing optical depth. In order to predict the occurence of this stability, the analysis shown below was carried out.

A small disturbance, δT, was assumed in the temperature profile. Substituting this disturbance into the finite difference form of the axuiliary equation results in the following expression.

$$\frac{Bo}{2\Delta\xi}\left(\delta\mathcal{J}_{K+1} - \delta\mathcal{J}_{K-1}\right) + 8\,\tau_L T_K^{\,3}\delta\mathcal{J}_K + 24\,\tau_L T_K^{\,3}\mathcal{J}_K\delta T_K$$

$$- 12\,\tau_L \sum_{J=1}^{N} T_J^{\,2}\mathcal{J}_J\delta T_J F_{KJ} - 4\,\tau_L \sum_{J=1}^{N} T_J^{\,3}\delta\mathcal{J}_J F_{KJ}$$

$$= -\,8\,T_K^{\,3}\delta T_K + 8\sum_{J=1}^{N} T_J^{\,3}F_{KJ} - 4\sum_{J=1}^{N} T_J^{\,3}\delta T_J G_{KJ} \tag{31}$$

where F_{KJ} and G_{KJ} are integrals (exchange factors) of the type given in Eqn. (30). δT is taken in the form

$$\delta T = Ae^{i(\lambda\xi + \beta\tau_L)} \tag{32}$$

where β is complex. Then if β_I, the imaginary part of β, is negative, small disturbances will grow with increasing τ_L, and the solution will be unstable.

Substituting Eqn. (32) into Eqn. (31), the resulting expression can be solved for β. If η_K is defined as

$$\eta_K = e^{i\lambda K\Delta\xi} \tag{33}$$

where $\eta_K = \eta_{RK} + \eta_{IK}$, the imaginary part of β_I is found to be:

$$\beta_I = \left(\frac{\lambda Bo\eta_{R2}}{2\Delta\xi} + 8\ \tau_L T_K^{\ 3}\eta_{IK} - 4\ \tau_L \sum_{J=1}^{N} T_J^{\ 3}\eta_{IJ}F_{KJ}\right)$$

$$\left(24\ \tau_L T_K^{\ 2}\mathcal{J}_K\eta_{RK} - 12\ \tau_L \sum_{J=1}^{N} T_J^{\ 2}\mathcal{J}_J\eta_{RJ}F_{KJ}\right.$$

$$\left. + 8\ T_K^{\ 3}\eta_{RK} - 8\sum_{J=1}^{N} T_J^{\ 3}\eta_{RJ}F_{KJ} + 4\sum_{J=1}^{N} T_J^{\ 3}\eta_{RJ}G_{KJ}\right)$$

$$+ \left(8\ \tau_L T_K^{\ 3}\eta_{RK} - 4\ \tau_L \sum_{J=1}^{N} T_J^{\ 3}\eta_{RJ}F_{KJ} - \frac{\lambda Bo}{2\Delta\xi}\eta_{I2}\right)$$

$$\left(24\ \tau_L T_K^{\ 2}\mathcal{J}_K\eta_{IJ} - 12\ \tau_L \sum_{J=1}^{N} T_J^{\ 3}\eta_{IJ}F_{KJ} + 8\ T_K^{\ 3}\eta_{IK}\right.$$

$$\left. - 8\sum_{J=1}^{N} T_J^{\ 3}\eta_{IJ}F_{KJ} + 4\sum_{J=1}^{N} T_J^{\ 3}\eta_{IJ}G_{KJ}\right) \qquad (34)$$

In Eqn. (34) a denominator, which is always positive, is ignored. Only the sign of β_I is of interest.

The procedure leading to Eqn. (34) was also carried out for the auxiliary equation directly, as well as for its finite difference form. The expression resulting from introducing a small disturbance, δT, defined as in Eqn. (37) into equation (8) and solving for β is:

$$\beta = \left(-24\ \tau_L T^2\mathcal{J} - 8\ T^3 + 12\ \tau_L^{\ 2}\int_0^1 T^2\mathcal{J}D(\xi,\zeta)d\zeta\right.$$

$$\left. + 8\ \tau_L\int_0^1 D(\xi,\zeta)T^3 d\zeta - 4\ \tau_L\int_0^1 T^3 E(\xi,\zeta)d\zeta\right) \Big/$$

$$\left(-Bo\lambda + i\left(8\ \tau_L T^3 - 4\ \tau_L^{\ 2}\int_0^1 T^3 D(\xi,\zeta)d\zeta\right)\right) \qquad (35)$$

$D(\xi,\zeta)$ and $E(\xi,\zeta)$ are defined by:

$$D(\xi,\zeta) \equiv e^{i\lambda(\xi-\zeta)}\ E_1(\tau_L|\ \xi - \zeta|\ ) \qquad (36)$$

and

$$E(\xi,\zeta) \equiv e^{i\lambda(\xi-\zeta)}\ \exp(-\ \tau_L|\ \xi - \zeta|\ ) \qquad (37)$$

The real and imaginary parts are indicated by R and I subscripts, as previously.

The imaginary part of Eqn. (35), again neglecting a denominator which is always positive, is

$$\beta_I = \left(\lambda Bo - 4\,\tau_L^2 \int_0^1 D_I T^3\, d\zeta\right)\left(-3\,\tau_L^2 \int_0^1 D_I\, T^2 \mathcal{J}\, d\zeta\right.$$

$$\left. - 2\,\tau_L \int_0^1 D_I T^3 d\zeta + \tau_L \int_0^1 E_I T^3\, d\zeta\right) + \left(8\,\tau_L T^3\right.$$

$$\left. - 4\,\tau_L^2 \int_0^1 D_R T^3\, d\zeta\right)\left(2\, T^3 + 6\,\tau_L T^2 \mathcal{J} + \tau_L \int_0^1 E_R T^3 d\zeta\right.$$

$$\left. - 3\,\tau_L^2 \int_0^1 D_R T^2 \mathcal{J}\, d\zeta - 2\,\tau_L \int_0^1 D_R T^3 d\zeta\right) \qquad (38)$$

5.2 Results

Equations (34) and (38) produced similar results, with the β_I computed from Eqn. (38) generally more positive than that from Eqn. (34), as expected. The results were tested by introducing small disturbances into the computed solutions. Figure 7 demonstrates the behavior of a disturbance in a stable solution. That the damping out of the disturbance is gradual is also predicted by the stability equations, as the values computed for β_I in the stable regions are small.

In the unstable regions, the results from the stability equations are not reliable, since the stability results depend on the temperature profile, and the temperature profile itself is inaccurate and unstable. Figure 8 showing the behavior of a disturbance in an unstable profile, is a rare case. In general, the disturbances introduced into solutions which are unstable cannot be observed, because the solutions become meaningless extremely rapidly.

From Eqn. (34), the stability depends on the auxiliary equation and is a function of Bo, τ_L, $\Delta\xi$ and λ, the wavelength of the disturbance. These parameters, however, do not entirely determine the stability. The temperature profile depends on the step size in optical depth and on the method of advance. The dependence on $\Delta\,\tau_L$ can be seen for one case from Figure 9, where β_I becomes negative for $\Delta\,\tau_L$ approximately equal to .21. As $\Delta\,\tau_L$ increases, the temperature profile becomes sufficiently inaccurate so that the auxiliary equation is unstable.

In Section 3.2, it was shown that the solution by iteration of the integrated equation is unstable for Boltzmann numbers less than approximately one. The solutions by parametric differentiation are also unstable for small Bo, but there is no sharp division in this case; the value of Bo for which the instability occurs depends on the other parameters. For the case shown in Figure 10, β_I is negative for Bo less than approximately 0.2.

In general, the solutions by parametric differentiation become unstable for small Boltzmann number or for increasing error. For example, no stable solutions have been produced for $\Delta\xi$ greater than or equal to 0.2. One would then expect, that since accuracy decreases with the number of steps in optical depth, all the solutions would be unstable for great enough optical depth. This is confirmed by the stability equations, which show that β_I decreases with increasing τ_L. However, only one case, the second order Taylor series, was observed in which a solution that was stable initially became unstable for τ_L less than ten.

CONCLUSIONS

From the analysis of the accuracy and stabilty carried out in the preceding sections, certain general conclusions can be drawn. The error for stable solutions by parametric differentiation can be expected to fall within the limits predicted by summing the errors in solving the auxiliary equation, Eqn. (8), and in integrating the related first-order equation, Eqn. (9). Also the solutions retain their accuracy to a much greater value of the parameter than would be expected from a straightforward estimate of the error.

The key to this accuracy is the stability of the auxiliary equation. If the auxiliary equation has positive stability, errors are damped out, and the range of the parametric differentiation procedure is increased. For neutral stability, the errors would be cumulative, and for the unstable case they would increase.

REFERENCES

[1] Abromowitz, M. and Stegun, I., Handbook of Mathematical Functions with Formulas, Graphs, and Mathematical Tables, National Bureau of Standards, Applied Mathematics Series 55, 1964.

[2] Hildebrand, F., Introduction to Numerical Analysis, McGraw-Hill, 1956.

[3] Jischke, M.C. and Baron, J.R., "Application of the Method of Parametric Differentiation to Radiative Gasdynamics," American Institute of Aeronautics and Astronautics Journal, Vol. 7, No. 7, July 1969, pp 1326-1335.

[4] Rubbert, P.E. and Landahl, M.T., "Solution of a Nonlinear Flow Problem Through Parametric Differentiation," Physics of Fluids, Vol. 10, No. 4, April 1967, pp. 831-835.

[5] Thurston, G.A., "Newton's Method Applied to Problems in Nonlinear Mechanics," Journal of Applied Mechanics, Series E, Vol. 32, June 1965, pp 383-388.

[6] Yoshikawa, K.K. and Chapman, D.R., "Radiative Heat Transfer Behind a Hypersonic Normal Shock Wave," National Aeronautics and Space Administration TN D-1424, September 1962.

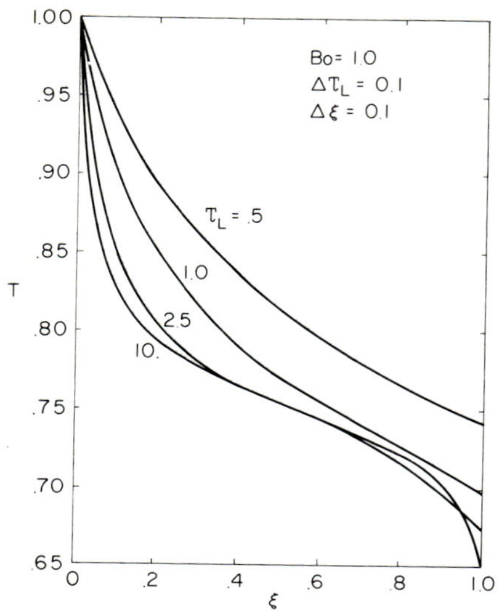

Fig. 1 Shock Layer Temperature Profiles (Runge-Kutta Advance)

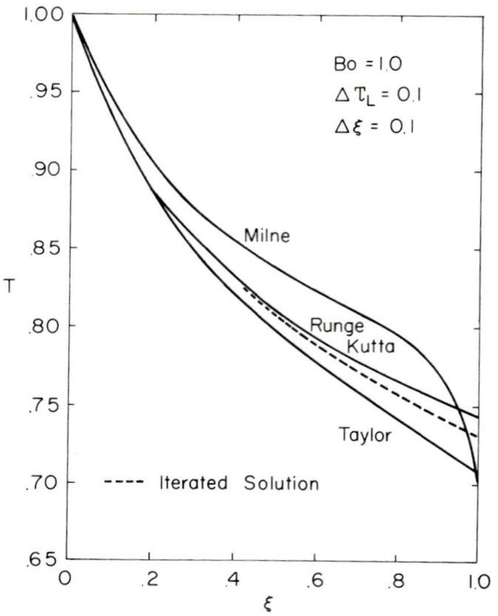

Fig. 3 Comparison of Shock Layer Temperature Profiles at $\tau_L = 0.5$

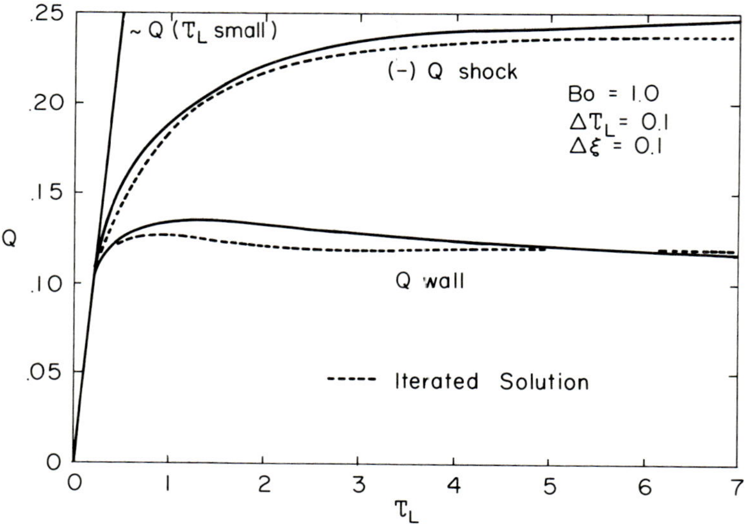

Fig. 2 Heat Transfer (Runge-Kutta Advance)

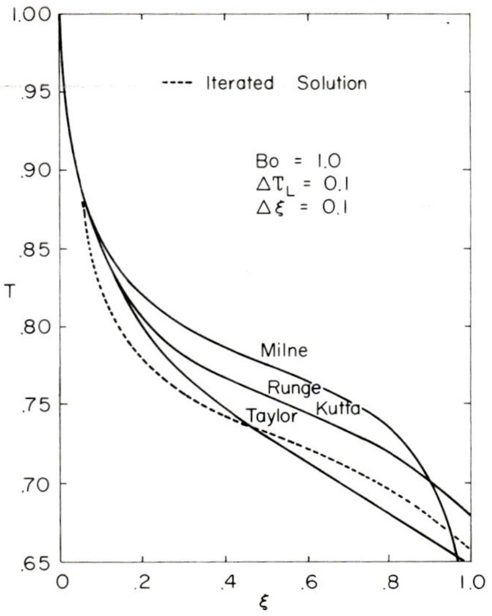

Fig. 4 Comparison of Shock Layer Temperature Profiles at $\tau_L = 2.5$

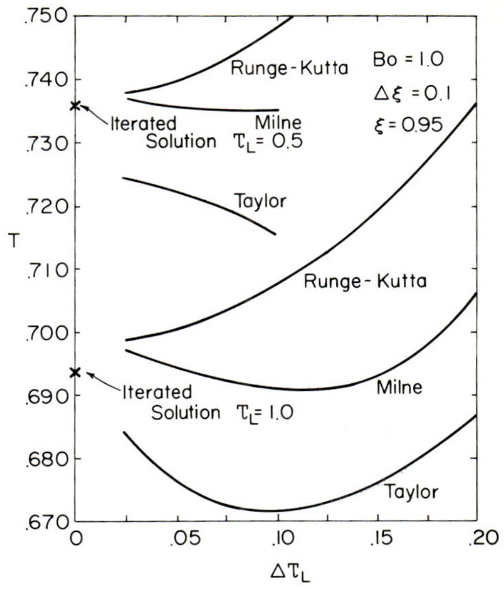

Fig. 5 Convergence of a Solution by Parametric Differentiation

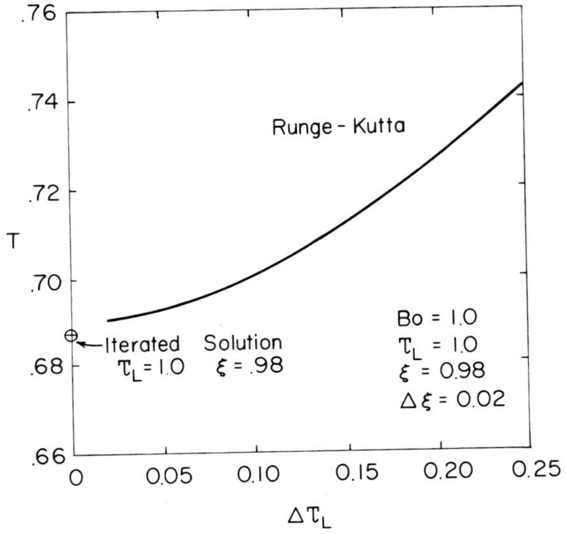

Fig. 6 Convergence of a Solution by Parametric Differentiation

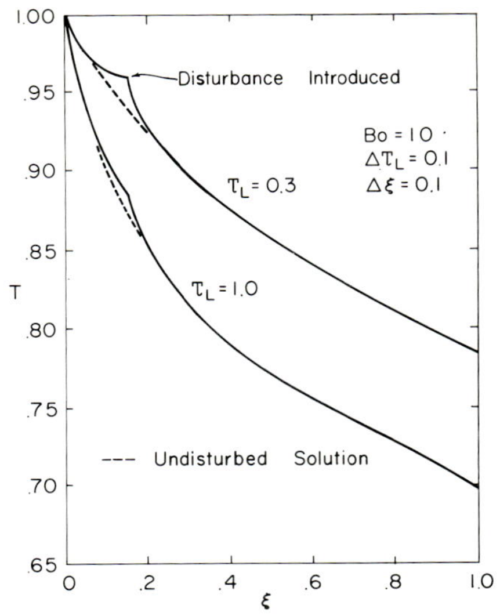

Fig. 7 Propagation of a Small Disturbance in a Stable Temperature Profile (Runge-Kutta Advance)

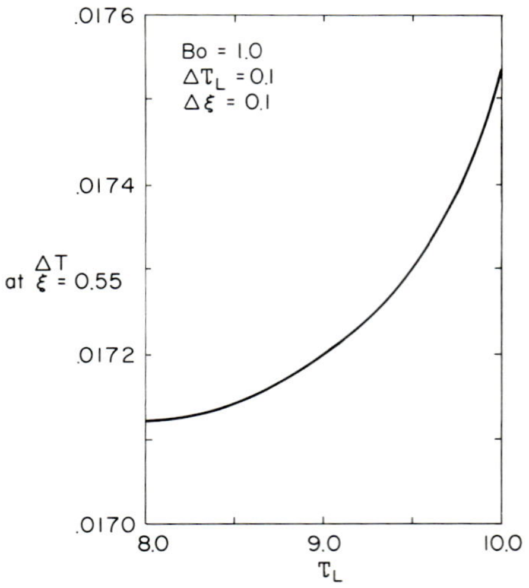

Fig. 8 Behavior of a Small Disturbance in an Unstable Temperature Profile (Second Order Taylor Series Advance)

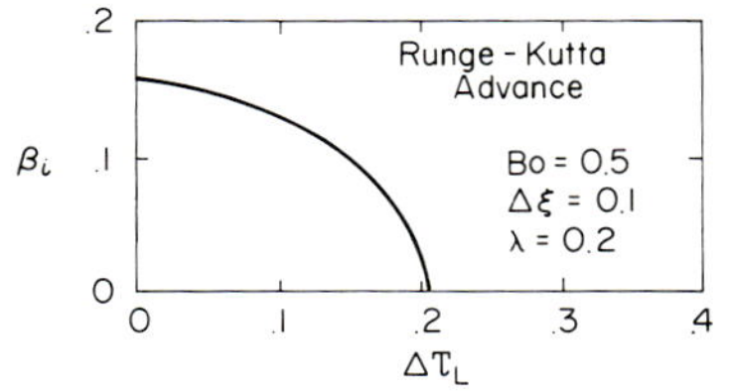

Fig. 9 Stability as a Function of $\Delta\tau_L$

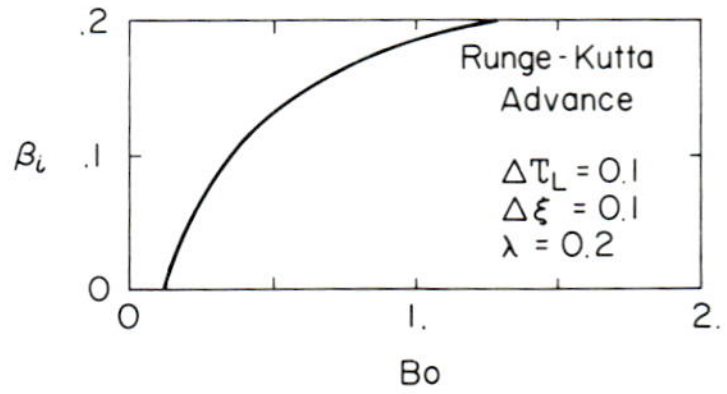

Fig. 10 Stability as a Function of Bo

SHELL THEORY

Developments in Mechanics, Vol. 6. Proceedings of the 12th Midwestern Mechanics Conference: 54

The Contact of Axisymmetric Cylindrical Shells with Smooth Rigid Surfaces

R. A. CHRISTOPHER and F. ESSENBURG
UNIVERSITY OF COLORADO

ABSTRACT

The general problem of an axisymmetric circular cylindrical shell, partially constrained from deflection by smooth rigid surfaces of revolution is considered. The ability of a shell theory which relaxes the Love-Kirchhoff hypothesis to overcome the conceptual difficulties inherent in the classical treatment of such problems is pointed out. The practical advantages of the use of a shear deformation theory and the limitations of design criteria predicated on classical analysis are indicated by way of specific examples.

INTRODUCTION

The present paper is concerned with the behavior of an isotropic axisymmetric circular cylindrical shell a portion of the surface of which is constrained from deflection by the presence of a smooth rigid surface of revolution.

Problems of this type, where the shell deformation is induced by the presence of a temperature field, have been incompletely considered in [1]*. Also of interest to the topic of the present paper are the solutions of the shrink-fit problem given by B. Paul [2] and R. A. Eubanks [3]. I. Malyutin [4] considered the problem of two concentric infinite circular cylindrical shells with the external shell

* The results reported here were obtained in the course of research carried under NSF Grant GK-21 to the University of Colorado.

**Numbers in brackets designate reference at end of paper.

loaded by a compressive ring load. He pointed out the limitation on the validity of his solution in the requirement of either cohesion between the shells or the presence of a sufficiently large compressive axial membrane resultant in the inner shell to insure the maintenance of contact between the shells.

The analysis of shell problems of these general types under the assumptions of a shell theory which invokes the love-Kirchhoff hypothesis involves certain conceptual difficulties (which may also be of practical interest.) The simplest example of such a shell theory is the Theory of Love's First Approximation, hereinafter referred to as classical shell theory. In accordance with the predictions of the classical theory the bending couples and shear stress resultant, as well as the difference between the normal tractions transmitted at the inner and outer surfaces of the shell, are determined solely by the radial deflection together with its axial derivatives and the axial membrane resultant (which is a constant). In a region of the shell which is in contact with a smooth rigid constraining surface the radial deflection is a prescribed function of the axial coordinate and thus the bending couples, shear stress resultant, and the normal traction transmitted to the shell by the constraining surface are determined by the character of the constraining surface and are influenced by the state of stress and deformation existing outside of the region of contact only to the extent that the axial membrane resultant is involved. From a consideration of the order of the system of differential equations governing the state of stress and deformation outside of the region of contact it follows that in addition to the continuity of the axial displacement and axial membrane resultant, only three continuity conditions are admissible at each edge of a region of contact. If the radial deflection, its axial derivative, and the axial bending couple are continuous the shear stress resultant must be discontinuous. This discontinuity can be accounted for physically only by assuming that at each edge of a contact region, a line load is transmitted to the shell from the constraining surface.

In addition it should be noted that the classical treatment may result in a physically unrealistic prediction for the transmitted surface traction in the region of contact. For example, if a circular cylindrical shell is completely constrained from deflection by smooth rigid cylindrical surfaces and edge loaded by an axial bending couple, the deflection is everywhere zero, and the classical theory predicts a zero value for the difference in the normal tractions at the inner and outer surfaces of the shell. (Obviously the classical theory also requires a discontinuity in the bending couple at the edge, which must be accounted for by the assumption that the constraining surfaces transmit a couple to the shell at the edge.)

In view of [1] and [5], it follows that the above indicated physical inconsistencies can be overcome by carrying the analysis within the framework of a linear shell theory which relaxes the Love-Kirchhoff hypothesis. In the present paper a shear deformation shell theory is applied to the general problem of an axisymmetric circular cylindrical shell, loaded in such a manner as to produce contact with a smooth rigid surface of revolution in one or more regions. A general solution which admits continuity of all shell parameters (including the shear stress resultant) at each edge of a contact region, and provides a physically consistent expression for the transmitted surface traction throughout a contact region is presented. Two specific examples involving circular cylindrical constraining surfaces are included in order to illustrate some of the unusual features of shell behavior with such constraints. Numerical illustrations afford a means of comparison of the results of the present treatment with the classical theory prediction.

THE BASIC EQUATIONS

It is convenient to introduce the following dimensionless parameters:

$$n_x = \frac{N_x}{C} \qquad n_s = \frac{N_s}{C} \qquad m_x = \frac{LM_x}{D} \qquad m_s = \frac{LM_s}{D} \qquad v = \frac{L^2V}{D}$$

$$q^+ = \frac{L^3Q^+}{D} \qquad q^- = \frac{L^3Q^-}{D} \qquad u = \frac{U}{L} \qquad w = \frac{W}{L} \qquad x = \frac{X}{L} \tag{1}$$

where

$$C = \frac{E\,h}{(1 - \nu^2)} \qquad\qquad D = \frac{E\,h^3}{12(1 - \nu^2)}$$

In these expressions X is the axial coordinate of the shell; L is some characteristic length; h is the shell thickness; E is Young's modulus and ν is Poisson's ratio; N_x and N_s are respectively the axial and circumferential membrane stress resultants; M_x and M_s are respectively the axial and circumferential bending couples; V is the shear stress resultant; Q^+ and Q^- are the normal surface tractions at the outer and inner surfaces, respectively; U is the axial component of the displacement. A system of equations which suitably characterizes the behavior of an axisymmetric circular cylindrical shell and accounts for the effect of transverse shear deformation is then given by

$$n_x' = 0 \qquad\qquad n_x = u' + \nu\frac{L}{R}w$$

$$v' - 12(\frac{L}{R})(\frac{L}{h})^2 n_s = q^+ - q^- \qquad\qquad n_s = \frac{L}{R}w + \nu u'$$

$$m_x' - v = 0 \qquad\qquad m_x = \beta' \qquad m_s = \nu\beta' \tag{2}$$

$$\beta + w' = \frac{v}{\mu^2}$$

where

$$\mu^2 = 5(1 - \nu)(\frac{L}{h})^2 \qquad\qquad (\ )' = \frac{d}{dx}(\ )$$

and R is the radius of the middle surface of the shell. These equations are essentially those given by Naghdi [6] and, although they include the assumption of shallowness, are suitable for the analysis of this investigation. The natural edge boundary conditions appropriate to the foregoing set of equations (at an edge x = constant) are:

(i) either u specified or n_x specified
(ii) either β specified or m_x specified
(iii) either w specified or v specified (3)

The equations governing the axisymmetric circular cylindrical shell which include the assumptions of the Theory of Love's First Approximation as well as the assumption of shallowness (i.e. classical theory), are readily deduced from the above by letting $1/\mu^2 \to 0$. Thus the classical theory equations are obtained by replacing the last of equations (2) by

$$\beta = -w'$$

and the second of conditions (3) by

(ii) either w' specified or m_x specified.

The equations (2) are readily reduced and are equivalent to

$$n_x = n_x^o$$

$$n_s = (1 - \nu^2) \frac{L}{R} w + \nu n_x^o$$

$$u' = -\nu\frac{L}{R} w + n_x^o$$

$$m_x = \beta'$$

$$v = \beta'' \tag{4}$$

$$\frac{\beta''}{\mu^2} - \beta = w'$$

$$4r^4 w = v' - q^+ + q^- - 4r^4\tilde{n}$$

$$\beta^{iv} - \frac{4r^4}{\mu^2} \beta'' + 4r^4\beta = (q^+ - q^-)'$$

where

$$r^4 = 3(1 - \nu^2) \left(\frac{L}{R}\right)^2 \left(\frac{L}{h}\right)^2 \qquad \tilde{n} = \frac{R}{L} \frac{\nu}{1 - \nu^2} n_x^o$$

and n_x^o is an integration constant.

SOLUTION FOR REGION OF PRESCRIBED SURFACE TRACTION

Let us consider a region of the shell in which the surface traction, $q^+ - q^-$, is a prescribed function of the axial coordinate. If $\frac{h}{R} < \sqrt{\frac{25(1-\nu)}{3(1+\nu)}}$ the last of equations (4) may be written in the form*

$$\left(\frac{d^2}{dx^2} - k^2\right) \left(\frac{d^2}{dx^2} - \bar{k}^2\right)\beta = \frac{d}{dx} (q^+ - q^-) \tag{5}$$

where

$$k = r\left(\sqrt{1 + \frac{r^2}{\mu^2}} + i \sqrt{1 - \frac{r^2}{\mu^2}} \right)$$

the symbol i is the usual imaginary operator and $\bar{k}$ is the complex conjugate of k. The solution of equations (4) may then be written

$$\beta = 2 \operatorname{Re}(A \cosh kx + B \sinh kx) + \beta_p$$

$$m_x = 2 \operatorname{Re}(k A \sinh kx + k B \cosh kx) + \beta_p'$$

$$v = 2 \operatorname{Re}(k^2 A \cosh kx + k^2 B \sinh kx) + \beta_p'' \tag{6}$$

$$4r^4(w + \tilde{n}) = 2 \operatorname{Re}(k^3 A \sinh kx + k^3 B \cosh kx)+\beta_p''' + q^- - q^+$$

---

*It follows that, for all values of ν, a sufficient condition for the roots of the indicial equation associated with the last of equations (4) to be complex is $\frac{h}{R} < \frac{5}{3}$.

$$n_x = n_x^{\circ}$$

$$u = n_x^{\circ} x - \nu \frac{L}{R} \int^x w(\zeta) d\zeta + F$$

where β_p is a particular integral of equation (5), A and B are complex integration constants, and F and n_x° are real integration constants.

The first four of equations (4) may be written in the form

$$\begin{aligned} \beta &= 2 \operatorname{Re}(G e^{kx} + H e^{-kx}) + \beta_p \\ m_x &= 2 \operatorname{Re}(k G e^{kx} - k H e^{-kx}) + \beta_p' \\ v &= 2 \operatorname{Re}(k^2 G e^{kx} + k^2 H e^{-kx}) + \beta_p'' \\ 4r^4(w + \tilde{n}) &= 2 \operatorname{Re}(k^3 G e^{kx} - k^3 H e^{-kx}) + \beta_p''' + q^- - q^+ \end{aligned} \qquad (7)$$

where the complex constants G and H are obviously related to the constants A and B.

It will be noted that the solution of the classical treatment is given by the above with

$$k = r(1 + i) \qquad (8)$$

SOLUTION FOR REGION OF PRESCRIBED SURFACE DISPLACEMENT

In a region in which the deflection w is a prescribed function of the axial coordinate, the surface traction is an unknown of the problem; and β is governed by the sixth of equations (4)

$$\beta'' - \mu^2 = \mu^2 w'$$

Thus the solution is given by

$$\begin{aligned} \beta &= a \cosh \mu x + b \sinh \mu x + \beta_w \\ m_x &= \mu a \sinh \mu x + \mu b \cosh \mu x + \beta_w' \\ v &= \mu^2 a \cosh \mu x + \mu^2 b \sinh \mu x + \beta_w'' \\ q^+ - q^- &= \mu^3 a \sinh \mu x + \mu^3 b \cosh \mu x - 4r^4(w + \tilde{n}) + \beta_w''' \\ n_x &= n_x^{\circ} \\ u &= n_x^{\circ} x - \nu \frac{L}{R} \int^x w(\zeta) d\zeta + c \end{aligned} \qquad (9)$$

where

$$\beta_w = \mu \int^x w'(\zeta) \sinh \mu (x-\zeta) d\zeta$$

and a, b, c, n_x° are integration constants.

The solution of the classic treatment for such a region is given by replacing the first four of Eqs.(9) by

$$\begin{aligned} \beta &= -w' \\ m_x &= -w'' \end{aligned} \qquad (10)$$

$$q^+ - q^- = -w^{iv} - 4r^4(w + \tilde{n})$$

and thus involves only two integration constants.

SOLUTION OF THE CONTACT PROBLEM

Now let us consider a cylindrical shell in the presence of interior and exterior smooth rigid surfaces of revolution, with revolution axis coincident with the axis of the shell. Let the shell be loaded (with surface and edge loadings) in such a manner that in one or more regions the shell is forced into contact with at least one of the rigid surfaces. We shall assume that the geometry of the surfaces of revolution is such that in such regions of contact the shell deflection is sufficiently small that the analysis may be carried within the framework of a linear shell theory. In each such contact region the solution is given by Eqs.(9) (and involves four integration constants) whereas in each region of non-contact the solution is given by Eqs. (6) and involves six integration constants.

If there are N such regions of contact and M regions of non-contact, the problem obviously involves 4N + 6M integration constants. In addition there will be N + M - 1 circular boundaries separating the N regions of contact from the M regions of non-contact. The values of the axial coordinates defining these circular boundaries together with the integration constants constitute 5N + 7M - 1 unknown constants.

We now have three possible cases. If N = M, the unknown constants are 12N - 1 in number. One shell boundary will be in a region of non-contact with three edge boundary conditions and the other edge will be in a contact region with two boundary conditions (since at that edge w is prescribed.) In addition, at each of the 2N - 1 boundaries separating the contact and non-contact regions, the six parameters n_x, v, m_x, u,β and w are regarded as continuous. Thus there are five boundary conditions and 12N - 6 continuity conditions - a total of 12N - 1 conditions - from which the 12N - 1 unknown constants are determined.

Now consider the case where M = N - 1. The number of unknown constants is then 12N - 8. Since both shell boundaries are in contact regions there will be a total of four boundary conditions, which, together with the 6(N+M-1) continuity conditions, gives a total of 12N - 8 conditions.

The remaining possible case is M = N + 1. In this case there will be a total of 12N + 6 conditions. Since both shell boundaries will be in regions of non-contact, there will be six boundary conditions, which, together with the 6(N+M-1) conditions gives a total of 12N + 6 conditions. Thus in all possible cases the number of conditions equals the number of unknown constants and the solution of the problem will satisfy the boundary conditions and admit the continuity of all shell parameters at the boundaries between the regions of contact and non-contact.

In the treatment of problems of this type under the assumptions appropriate to the classical theory, in a contact region the shell parameters and the transmitted surface traction are given by Eqs.(8), (10) and the last pair of Eqs.(9) and thus involve 2 integration constants. The solution for a non-contact region involves six integration constants. The values of the axial coordinate defining the boundaries of the contact regions are also unknowns of the problem. An analysis of the type given above shows that these integration constants and unknown boundaries are determined from the appropriate edge boundary conditions and three continuity conditions at each boundary separating the contact and non-contact regions. It is assumed that n_x, m_x, u, w', and w are continuous at such boundaries and that the discontinuity in v is accounted for by the transmission of a line load by the constraining surface. The validity of such an assumption may be established by deducing the classical theory solution as a limiting case of the shear deformation theory solution, i.e. by letting $\mu^2 \to \infty$.

It should be noted that the number of contact and non-contact regions is determined by the character of the applied surface tractions and edge loadings as well as the geometry of the constraining surface (or surfaces). If the intensity of the applied loading is regarded as increasing monotonically, a contact region will be initiated as a circular line. With further increase in the loading intensity, the line of contact will propagate to a cylindrical region. The validity of any assumption with regard to the existence of a contact region is established by verification that throughout such a region, the transmitted surface traction is a positive quantity (in the sign convention used here.) In the case of the classical treatment it may be possible (as in one of the following examples) for the theory to predict that with an increase in the loading intensity the circular line of initial contact does not propagate to a region but remains a line. In such cases the value of the axial coordinate defining the line of contact changes with an increase in the loading intensity.

EXAMPLES

As an example let us consider an infinite cylindrical shell of radius R in the presence of a smooth rigid circular cylindrical surface of radius R + h/2 + Lδ, where δ is a small dimensionless parameter, subject to a surface load of the form

$$
\begin{aligned}
q^+ &= 0 \\
q^- &= q_o\, e^{-px}, \quad 0 \leq x, \quad 0 < p \qquad (11)\\
&= q_o\, e^{px}, \quad x < 0
\end{aligned}
$$

where q_o and p are constants. In view of the symmetry with respect to the plane x = 0, we will record only the solution for the region $0 \leq x$.

Let q_o^* be the value of q_o (for a prescribed p) for which the shell just touches the rigid constraining surface. Then for $q_o \leq q_o^*$ the solution of the shear deformation theory is given by Eqs.(7) together with the last two of Eqs.(6).

$$
\begin{aligned}
G &= 0 \qquad\qquad F = 0 \\
H &= \frac{p\, q_o}{\left(p^4 - \frac{4r^4}{\mu^2} p^2 + 4r^4\right)} \; \frac{(\bar{k}^2 - p^2)}{(k^2 - \bar{k}^2)} \qquad (12)\\
\beta_p &= \frac{p\, q_o}{\left(p^4 - \frac{4r^4}{\mu^2} p^2 + 4r^4\right)} \; e^{-px}
\end{aligned}
$$

For $q_o = q_o^*$ the shell will contact the constraining surface in a circular line at $x = 0$. For $q_o^* < q_o$ a contact region $x \leq n$ will be propagated and the solution for this region is given by Eqs.(9) with

$$a = 0 \qquad c = 0 \qquad \beta_w = 0$$

$$b = \frac{p\, q_o [2 r^2 - p(k + \bar{k}) + p^2]\, e^{-pn}}{[p^4 - \frac{4r^4}{\mu^2} p^2 + 4r^4][(k\bar{k} + \mu^2)\sinh \mu n + \mu(k + \bar{k})\cosh \mu n]} \tag{13}$$

For $q_o^* < q_o$ the solution for the region $n < x$ is given by Eqs. (7) and the last two of Eqs. (6) with

$$G = 0 \qquad F = 0$$

$$H = \frac{q_o\, e^{(k-p)n}}{(p^4 - \frac{4r^4}{\mu^2} p^2 + 4r^4)} \; \frac{(\bar{k} - p)p[p\bar{k} + \mu^2)\tanh \mu n + \mu(\bar{k} + p)]}{(k - \bar{k})[k\bar{k} + \mu^2)\tanh \mu n + \mu(k + \bar{k})]} \tag{14}$$

$$\beta_p(x) = \frac{p\, q_o}{(p^4 - \frac{4r^4}{\mu^2} p^2 + 4r^4)} \; e^{-px}$$

The transcendental equation governing the contact region boundary n is given by

$$\frac{\delta}{q_o} = \frac{e^{-pn}}{(p^4 - \frac{4r^4}{\mu^2} p^2 + 4r^4)} \; [(1 - \frac{p^2}{\mu^2})$$

$$+ \frac{p}{2r^4} \mathrm{Re}\{ \frac{k^3(p - \bar{k})[p\bar{k} + \mu^2)\tanh \mu n + \mu(p + \bar{k})]}{(k - \bar{k})[2r^2 + \mu^2)\tanh \mu n + \mu(k + \bar{k})]} \}] \tag{15}$$

The expression for q_o^* may be obtained by putting $w/_{x=0} = \delta$ in the solution for $q_o = q_o^*$; it may also be obtained by putting $n = 0$ in Eq.(15) and is given by

$$\frac{q_o^*}{\delta} = [p^4 - \frac{4r^4}{\mu^2} p^2 + 4r^4][(1 - \frac{p^2}{\mu^2}) + 2\, \mathrm{Re}\{\frac{kp(p^2 - \bar{k}^2)}{k^2(k^2 - \bar{k}^2)}\}]^{-1} \tag{16}$$

It may be verified that the expression for the surface traction transmitted to the shell by the constraining surface, i.e. q^+, is positive throughout the contact region $x = n$.

The solution of the classical treatment of this example is given by the following:

For $q_o \leq q_o^*$ the solution is given by Eqs.(7) and (8) together with the last two of Eqs.(6) with

$$G = 0 \qquad F = 0$$

$$H = \frac{p\, q_o}{(p^4 + 4r^4)} \; \frac{(\bar{k}^2 - p^2)}{(k^2 - \bar{k}^2)} \tag{17}$$

$$\beta_p = \frac{p\, q_o}{(p^4 + 4r^4)} e^{-px}$$

For $q_o^* < q_o$ the solution for the region $x < n$ is given by

$$w = \delta \qquad \beta = m_x = v = 0$$
$$q^+ = q_o e^{-px} - 4r^4\delta \tag{18}$$

and for the region $n < x$ the solution is given by Eqs.(7) and (8) together with the last two Eqs.(6) with

$$G = 0 \qquad F = 0$$
$$H = \frac{q_o\, p(\bar{k} - p)\, e^{(k-p)n}}{(p^4 + 4r^4)(k - \bar{k})} \tag{19}$$
$$\beta_p = \frac{q_o p}{(p^4 + 4r^4)} e^{-px}$$

In accordance with the classical theory prediction the transcendental equation governing n is

$$\frac{\delta}{q_o} = \frac{e^{-pn}}{(p^4 + 4r^4)} \left(1 + p\,\frac{(p - 2r)}{2r^2}\right) \tag{20}$$

The value of q_o^* is determined by putting the maximum value of the deflection as indicated by Eqs.(17) equal to δ and is given by

$$\frac{q_o^*}{\delta} = \frac{(p^4 + 4r^4)}{\left[1 + \frac{rp}{4r^4}\,(p^2 - 2r^2)\right]} \tag{21}$$

Let q_o^{**} be the value of q_o at which a region of contact begins to propagate. Thus q_o^{**} is obtained by putting $n = 0$ in Eq.(20) and is given by

$$\frac{q_o^{**}}{\delta} = \frac{(p^4 + 4r^4)}{\left[1 + p\,\frac{(p - 2r)}{2r^2}\right]} \tag{22}$$

Since $q_o^* < q_o^{**}$ it follows that in accordance with the classical theory prediction there is a range of values of q_o, viz., $q_o^* < q_o < q_o^{**}$ for which there will be a circular line of contact at $x = 0$.

As a second example let us consider a finite shell of length 2L in the presence of a smooth rigid surface of radius $R + h/2 + \delta L$, loaded internally with a uniform load $q^- = p_o$, a constant, and $q^+ = 0$. The edge boundary conditions are taken at $x = \pm 1$ as:

$$w = 0,\ m_x = m_o,\ n_x = n_x^o$$

where m_o and n_x^o are prescribed constants.

In view of the symmetry with respect to the plane x = 0 we will consider only the region $0 \leq x \leq 1$.

The point at which contact is initiated will depend on the ratios existing between the constants p_o, m_o, and n_x^o as well as the L/R ratio of the shell. In general, for small L/R ratios the contact region will include the origin. This case was discussed in [1] and will not be considered here. For sufficiently large L/R ratios contact will be initiated at a finite value of x and a contact region $n \leq x \leq n$ will be propagated.

The solution for load values which do not produce contact is perfectly straightforward and will not be recorded. For loading of sufficient magnitude to produce contact the solution for the region $0 \leq x \leq m$ is given by Eqs.(6) with

$$A = 0$$

$$B = \frac{[p_o - 4r^4(\delta + \tilde{n})]\ \bar{k} \sinh \bar{k}m}{k^3\ (k \sinh km \cosh \bar{k}m - \bar{k} \sinh \bar{k}m \cosh km)} \tag{23}$$

$$\beta_p = 0 \qquad\qquad F = 0$$

For the region $m \leq x \leq n$ the solution is given by Eqs.(9) with

$$a = \frac{[p_o - 4r^4(\delta + \tilde{n})]}{4r^4(k \sinh km \cosh \bar{k}m - \bar{k} \sinh \bar{k}m \cosh km)}$$

$$\{ [k^3 \sinh km \cosh \bar{k}m - \bar{k}^3 \sinh \bar{k}m \cosh km]\ \frac{\sinh \mu m}{\mu}$$

$$- [(k^2 - \bar{k}^2)k\bar{k} \sinh km \sinh \bar{k}m]\ \frac{\cosh \mu m}{\mu^2}$$

$$b = \frac{[p_o - 4r^4(\delta + \tilde{n})]}{4r^4(k \sinh km \cosh \bar{k}m - \bar{k} \sinh \bar{k}m \cosh km)}$$

$$\{ [k^3 \sinh km \cosh \bar{k}m - \bar{k}^3 \sinh \bar{k}m \cosh km]\ \frac{\cosh \mu m}{\mu} \tag{24}$$

$$- [(k^2 - \bar{k}^2)\ k\bar{k} \sinh km \sinh \bar{k}m]\ \frac{\sinh \mu m}{\mu^2} \ \}$$

$$\beta_w = 0$$

$$c = -\frac{\nu L}{R} [\ \frac{2}{4r^4}\ \text{Real}\ (k^2\ B \sinh km) + (\ \frac{p_o - 4r^4\tilde{n}}{4r^4}\ )m - m\delta]$$

where B is defined in Eqs.(23).

For the region $n \leq x \leq 1$ the solution is given by Eqs.(6) with

$$A = -\ \frac{[p_o - 4r^4(\delta + \tilde{n})]}{k^2(k^2 - \bar{k}^2)(k \sinh km \cosh \bar{k}m - \bar{k} \sinh \bar{k}m \cosh km)}$$

$$[k\ (\bar{k} \cosh km \sinh \bar{k}m - k \sinh km \cosh \bar{k}m) \sinh km$$

$$+ \sinh km\ (k^2 \sinh km \cosh \bar{k}m - \frac{\bar{k}^3}{k} \cosh km \sinh \bar{k}m)$$

$$- \frac{\bar{k}}{k}\ (k^2 - \bar{k}^2) \sinh km \sinh \bar{k}m \cosh kn\}\ \cosh \mu(n-m)$$

(continued)

$$+ \{ \frac{\bar{k}}{\mu} (k^2 - \bar{k}^2) \sinh^2 km \sinh \bar{k}m$$

$$+ \mu \cosh kn \, (\frac{\bar{k}^3}{k^2} \cosh km \sinh \bar{k}m - k \sinh km \cosh \bar{k}m) \}$$

$$\sinh \mu(n-m) \,]$$

$$B = \frac{[p_o - 4r^4(\delta + \tilde{n})]}{k^2(k^2 - \bar{k}^2)(k \sinh km \cosh \bar{k}m - \bar{k} \sinh \bar{k}m \cosh km)}$$

$$[\frac{\bar{k}}{k} (k^2 - \bar{k}^2)(\sinh km - \sinh \bar{k}m)$$

$$+ k \, (\bar{k} \cosh km \sinh \bar{k}m - k \sinh km \cosh \bar{k}m) \cosh km$$

$$+ \{ \cosh km(k^2 \sinh km \cosh \bar{k}m - \frac{\bar{k}^3}{k} \cosh km \sinh \bar{k}m)$$

$$- \frac{\bar{k}}{k} (k^2 - \bar{k}^2)(\sinh km \sinh \bar{k}m \sinh kn) \} \cosh \mu(n-m)$$

$$+ \{\frac{\bar{k}}{\mu} (k^2 - \bar{k}^2)(\sinh km \cosh km \sinh \bar{k}m)$$

$$+ \mu \sinh kn \, (\frac{\bar{k}^3}{k^2} \cosh km \sinh \bar{k}m - k \sinh km \cosh \bar{k}m) \}$$

$$\sinh \mu(n-m) \,]$$

$$\beta_p = 0$$

$$F = \frac{\nu L}{R} [\frac{2 \text{ Real}}{4r^4} (k^2 A \cosh kn + k^2 B \sinh kn) + (\frac{p_o}{4r^4} - \tilde{n})n - n\delta + c]$$

where A and B are defined immediately above and c is defined in Eqs.(24).

The coupled pair of transcendental equations for the determination of the parameters m and n are given by

$$\frac{m_o}{\delta} = [4r^4 - (\frac{p_o - 4r^4\tilde{n}}{4r^4\delta})]$$

$$\times 2 \text{ Re} \{(k \sinh km \cosh \bar{k}m - \bar{k} \sinh \bar{k}m \cosh km)^{-1}$$

$$[(k \sinh km - \bar{k} \sinh \bar{k}m) \frac{\cosh k}{(k^2 - \bar{k}^2)}$$

$$- (k^3 \sinh km - \bar{k}^3 \sinh \bar{k}m) \frac{\cosh k}{k^2(k^2 - \bar{k}^2)}$$

$$- (k \sinh km \cosh \bar{k}m - \bar{k} \sinh \bar{k}m \cosh km) \frac{\cosh k(1-m)}{(k^2 - \bar{k}^2)}$$

$$+ (k^3 \sinh km \cosh \bar{k}m - \bar{k}^3 \sinh \bar{k}m \cosh km)$$

$$[\frac{\cosh k(1-m)}{k^2(k^2 - \bar{k}^2)} + \frac{\cosh \mu(1-m)}{2(4r^4)}] \qquad (26)$$

(continued)

$$+ [k\bar{k}(k^2 - \bar{k}^2) \sinh km \sinh \bar{k}m]$$

$$\left[\frac{\sinh k(1-m)}{k^3(k^2 - \bar{k}^2)} + \frac{\sinh \mu(1-m)}{2\mu(4r^4)} \right]$$

$$+ (k \sinh km \cosh \bar{k}m - \bar{k} \sinh \bar{k}m \cosh km) \frac{\cosh k(1-n)}{(k^2 - \bar{k}^2)}$$

$$- [(k^3 \sinh km \cosh \bar{k}m - \bar{k}^3 \sinh \bar{k}m \cosh km) \cosh \mu(n-m)$$

$$+ k\bar{k}(k^2 - \bar{k}^2) \sinh km \sinh \bar{k}m \frac{\sinh \mu(n-m)}{\mu}]$$

$$\left[\frac{\cosh k(1-n)}{k^2(k^2 - \bar{k}^2)} + \frac{\cosh \mu(1-n)}{2(4r^4)} \right]$$

$$- [(k^3 \sinh km \cosh \bar{k}m - \bar{k}^3 \sinh \bar{k}m \cosh km)\mu$$

$$\sinh \mu(n-m)$$

$$+ k\bar{k}(k^2 - \bar{k}^2) \sinh km \sinh \bar{k}m \cosh \mu(n-m)]$$

$$\left[\frac{\sinh k(1-n)}{k^3(k^2 - \bar{k}^2)} + \frac{\sinh \mu(1-n)}{2\mu(4r^4)} \right]] \}$$

and

$$\left(\frac{p_o - 4r^4\tilde{n}}{4r^4\delta} \right) = \left[\left(\frac{p_o - 4r^4\tilde{n}}{4r^4\delta} \right) - 4r^4 \right]$$

$$\times 2 \operatorname{Re} \{(k \sinh km \cosh \bar{k}m - \bar{k} \sinh \bar{k}m \cosh km)^{-1}$$

$$[(k \sinh km - \bar{k} \sinh \bar{k}m) \frac{k^2 \cosh k}{(k^2 - \bar{k}^2)}$$

$$- (k^3 \sinh km - \bar{k}^3 \sinh \bar{k}m) \frac{\cosh k}{(k^2 - \bar{k}^2)}$$

$$-(k \sinh km \cosh \bar{k}m - \bar{k} \sinh \bar{k}m \cosh km)k^2 \frac{\cosh k(1-m)}{(k^2 - \bar{k}^2)}$$

$$+(k^3 \sinh km \cosh \bar{k}m - \bar{k}^3 \sinh \bar{k}m \cosh km) \frac{\cosh k(1-m)}{(k^2 - \bar{k}^2)}$$

$$+ [k\bar{k} (k^2 - \bar{k}^2) \sinh km \sinh \bar{k}m] \frac{\sinh k(1-m)}{k(k^2 - \bar{k}^2)}$$

$$+ (k \sinh km \cosh \bar{k}m - \bar{k} \sinh \bar{k}m \cosh km)k^2 \frac{\cosh k(1-n)}{(k^2 - \bar{k}^2)}$$

$$- [(k^3 \sinh km \cosh \bar{k}m - \bar{k}^3 \sinh \bar{k}m \cosh km) \cosh \mu(n-m)$$

$$+k\bar{k}(k^2 - \bar{k}^2) \sinh km \sinh \bar{k}m \frac{\sinh \mu(n-m)}{\mu}] \frac{\cosh k(1-n)}{(k^2 - \bar{k}^2)}$$

$$- [(k^3 \sinh km \cosh \bar{k}m - \bar{k}^3 \sinh \bar{k}m \cosh km)\mu \sinh \mu(n-m)$$

$$+k\bar{k}(k^2 - \bar{k}^2) \sinh km \sinh \bar{k}m \cosh \mu(n-m) \frac{\sinh k(1-n)}{k(k^2 - \bar{k}^2)}]\}$$

The classical theory solution is obtained in a straightforward manner and will not be recorded. It is interesting to note, as will be shown in the specific examples herein below, that for these examples classical theory does not predict a region of contact but rather predicts a line of contact which moves with an increase in loading intensity.

As a specific example let us consider the case where $(p_o - 4r^4\tilde{n})=0$, $m_o > 0$, $\nu = 0.25$, $L/R = 5$, $h/2R = 1/8$. The solution of the transcendental equations is shown in Fig. 1. It will be noted that

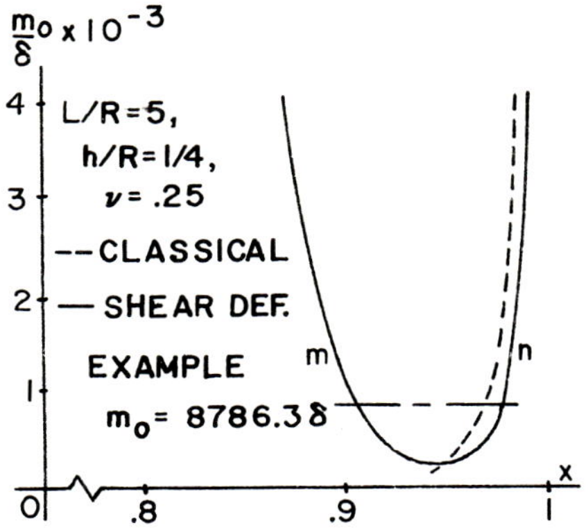

Fig. 1 Contact Region for Finite Shell Loaded by Edge Couple

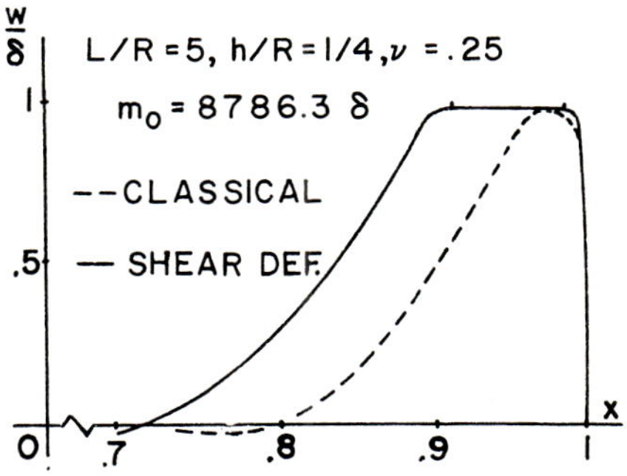

Fig. 2 Deflection Distribution

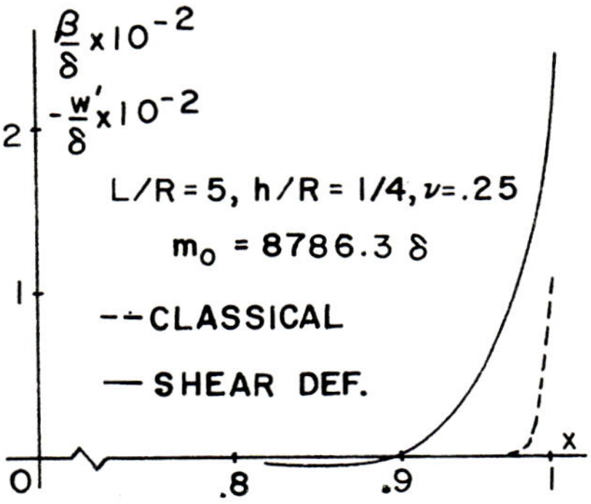

Fig. 3 Rotation Distribution

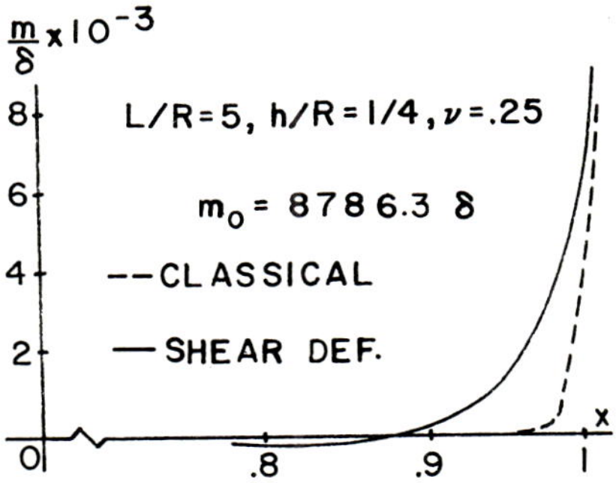

Fig. 4 Axial Bending Couple Distribution

whereas the shear deformation theory predicts a region of contact, the classical theory predicts a line of contact which moves toward the outer edge of the shell as the intensity of m_o increases. It will also be noted that the two theories differ slightly as to the prediction of the value of m_o and position of contact line at which contact is initiated. For the specific value $m_o = 8786.3\delta$, the deflection, rotation, bending couple, shear resultant and surface traction distributions are shown in Fig. 2 through 6. The discrepancies between the predictions of the two theories are clear from these figures. It should be noted that classical theory predicts the transmission of a line load given by the discontinuity in shear shown in Fig. 5.

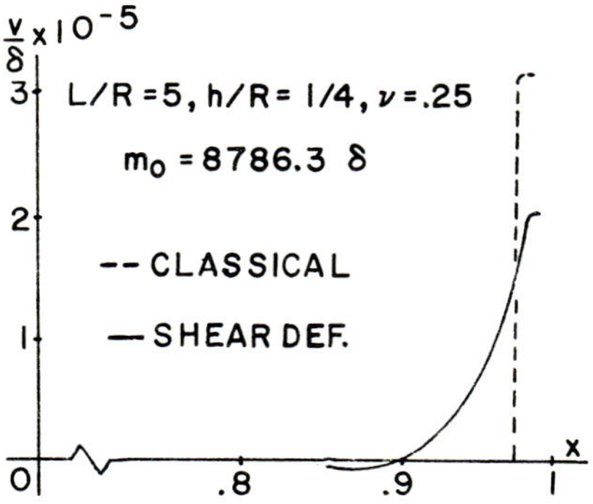

Fig. 5 Shear Resultant Distribution

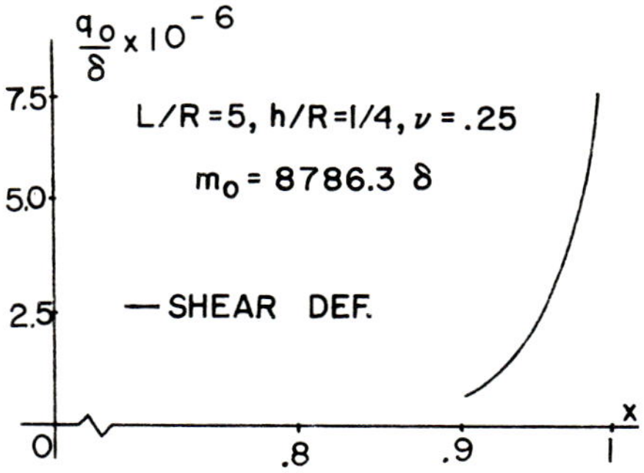

Fig. 6 Traction Transmitted by Constraining Surface to Shell

Conclusion

In the type of problem considered in this paper the use of a shear deformation shell theory removes the conceptual difficulties inherent in the application of the classical theory and provides physically consistent predictions for the parameters of the problem. On the basis of the examples considered it is clear that whereas in some applications the classical theory prediction is adequate for design purposes, in other applications (where predictions of contact regions and transmitted surface tractions are of design interest) the use of a higher order shell theory is essential. Unfortunately there appears to be no a priori method of distinguishing these two classes of application.

Bibliography

[1] F. Essenburg: Thermo-Elastic Cylindrical Shells with Surface Constraints. Non-Classical Shell Problems, Proceedings I.A.S.S. Symposium, Warsaw 1963, pp. 100-115.

[2] B. Paul: Shrink Fits of Moderately Long Bands on Thin Walled Cylinders. Journal of Engineering for Industry, Vol. 84, 1962, pp. 338-342.

[3] R. A. Eubanks: Shrink Fit of Arbitrary Length Sleeves on Thin Cylindrical Shells. Proceedings 8th Midwestern Mechanics Conference, Vol. 2, Part 2, 1963, pp. 84-101.

[4] I.S. Malyutin: On a Contact Problem in Cylindrical Shells. (In Russian), Prikl. Mat. Mekh., Vol. 20, No. 5, 1956, pp.665-666.

[5] F. Essenburg: On Surface Constraints in Plate Problems. Journal of Applied Mechanics, Vol. 29, 1962, pp. 340-344.

[6] P.M. Naghdi: On the Theory of Thin Elastic Shells. Quarterly of Applied Mathematics, Vol. 14, 1957, pp. 369-380.

Developments in Mechanics, Vol. 6. Proceedings of the 12th Midwestern Mechanics Conference: 55

Axisymmetric Deformation Theory of Orthotropic Layered Cylindrical Shells

JAMES TING-SHUN WANG
GEORGIA INSTITUTE OF TECHNOLOGY

ABSTRACT

A linear theory for the axisymmetric deformation of laminated cylindrical shells is presented. The shell is considered to be made up of thin layers of orthotropic homogeneous materials. The effect of shear deformation is considered so that the compatibility of interlaminar stresses and the continuity of deformation at contact surfaces of every two adjacent layers can be maintained where the theory based on the classical Kirchhoff hypotheses fails. It is assumed that the average transverse normal strain is zero, and only one shear function is introduced in the theory. By satisfying the compatibility of stress and the continuity of deformation between interacting surfaces, recurrence relations for the midsurface displacements and interlaminar shear stresses of neighboring layers are established. A single fourth order ordinary differential equation with constant coefficients governing the transverse displacement is finally derived and the solution can easily be obtained. The general solution of a semi-infinite cylindrical shell is presented. A numerical example describing the variation of the interlaminar shear stress for a simple two layered clamped cylindrical shell subjected to uniform internal pressure is also included for illustrative purpose. Such interlaminar shear stress can not otherwise be determined according to the classical theory based upon the Kirchhoff-Love hypotheses.

INTRODUCTION

The demand for increasing structural efficiency of many vehicle designs has resulted in the use of layered composite structural components. The manufacturing techniques and quality control for producing composite materials have advanced rapidly. Flight vehicle structures made of advanced filamentary composites are gradually becoming a reality. However, the existing theories and analyses for such structural components are simple extensions of the classical theories of single layered isotropic or orthotropic plates and shells.

The classical theories for laminated plates and shells were developed according to the Kirchhoff hypotheses. These theories will lead to incompatible or contradicting interlaminar stresses when the laminae are made of different materials. The hypotheses of nondeformable normals will deviate significantly from reality when the differences in stiffnesses of different layers become large. Since the peeling and shearing strengths of the bond material between layers are usually low and the failure in the bond material will severely affect the performance of the structure, it becomes important to estimate accurately the interacting stress levels.

Compared with the analyses of single layered plates and shells of isotropic or orthotropic materials, which have been studied extensively, the analysis of plates and shells of multiple layers, taking into account the interaction of stresses and compatibility of deformations between layers, appears to be a relatively unexplored field. Early in 1924, Shtaerman [1] studied single layered orthotropic shells of revolution axisymmetrically loaded. The basic equations for the theory of orthotropic shells were derived by Mushtari [2], [3] in 1938. These are indicated in a comprehensive review paper by Ambartsumian [4]. The membrane theory of anisotropic shells has been discussed by Ambartsumian [5], Dong [6], and others. The classical theory of laminated plates and shells may be referred to in the work of Dong, Pister, and Taylor [7]. Ashton published a number of papers [8], [9] on the analysis of anisotropic plates according to the classical Kirchhoff hypotheses. Eason [10] studied the free vibration of an anisotropic cylinder; the cylinder was considered to be thick and long and all quantities depend on the radial spatial coordinate only. Das [11] used the Donnell type equations to study a single layered orthotropic cylindrical shell. Mirsky [12] and [13] used the Frobenius series to study the axisymmetric vibration of a single layered orthotropic cylinder. Again the cylinder is considered to be long, and the solution depends on the radial spatial coordinate but is independent of longitudinal coordinates. Kalnins [14], by defining the gross density of the material of all layers, has utilized the method of analysis he developed for single layered rotationally symmetric thin elastic shells to study layered cylindrical shells of isotropic or orthotropic materials as equivalent single layered cylinders.

Ahmed [15] studied the thick shell problem based on more precise theory; however, the investigation was limited to a long cylinder and the problem was reduced to a plane strain problem with cylindrical symmetry. Bert, Baker, and Egle [16] investigated the free vibrations of multilayered anisotropic cylindrical shells based on the Kirchhoff hypotheses. The previous work mentioned above is either based on the assumption of nondeformable normals as a result of the Kirchhoff-Love hypotheses or is applicable only to long cylinders where edge conditions have no effect on the problem. Ambartsumian [4] indicated that the accuracy of the hypothesis of nondeformable normals used for isotropic plates and shells is often quite unacceptable for anisotropic shells even for relatively thin elastic shells. Wang [17] has shown that no solution exists for simply supported anisotropic plates according to the classical laminated plate theory if the twist coupling terms are included in boundary conditions. Whitney [18] investigated the effect of transverse shear deformation on the bending of laminated plates. Pagano [19] obtained exact solutions for the cylindrical bending of composite laminates according to linear elasticity, and some discussion concerning the validity of the classical Kirchhoff hypotheses was presented. By using the Galerkin method, Chamis [20] studied the buckling of anisotropic composite plates with twist coupling terms included. Hsu and Wang [21] studied the vibrations of laminated cylindrical shells consisting of layers of orthotropic laminae by treating each layer separately and then coupling through interactions of interlaminar stresses. The numerical results indicate that for a relatively short cylinder, the differences in frequencies obtained according to the more exact analysis and the analysis based on Kirchhoff hypotheses of nondeformable normals are significant even if the thickness of the shell is small. Detailed discussion of the rotationally symmetric vibrations of layered cylindrical shells may be referred to in the work by Hsu [22].

The work presented in [21] and [22] considers, for axisymmetric case, one unknown shear function for each layer. A system of coupled fourth order governing differential equations is resulted for static problems. The number of equations equal the number of layers of the shell, and the numerical computation for shells having many layers becomes lengthy. The present study considers one shear function for all layers of the shell while continuous deformation and consistent interlaminar stresses at all interfaces are maintained. The assumption though impose additional restriction on the mode of transverse shear stress variation, yet the resulting theory is simplified. A single fourth order governing differential equation is obtained according to the present derivation which is different from that given in [21] and [22].

AXISYMMETRIC DEFORMATION THEORY

Laminated cylindrical shells will be assumed to be made up of thin layers of orthotropic homogeneous materials. The general theory is derived directly from the elasticity equations and a consistent theory is resulted. Since each lamina is assumed to be thin and the transverse normal strain will be small, it is assumed that the average transverse normal strain in each layer is zero. The corresponding average transverse displacement is used in calculating all other related quantities. Such assumption will allow the complete satisfaction of the continuity of deformation and the compatibility of interlaminar stresses. The geometry and some related symbols are shown in Figure 1. For axisymmetric deformation, the strain-displacement relations at a point in a general jth layer are

$$e_{xxj} = \frac{\partial u_j}{\partial x} \qquad (1) \qquad\qquad \gamma_{x\theta j} = 0 \qquad (4)$$

$$e_{\theta\theta j} = \frac{w_j}{R_j} \qquad (2) \qquad\qquad \gamma_{\theta zj} = 0 \qquad (5)$$

$$\bar{e}_{zzj} = \frac{\partial w_j}{\partial \xi_j} = 0 \qquad (3) \qquad\qquad \gamma_{xzj} = \frac{\partial u_j}{\partial \xi_j} + \frac{\partial w_j}{\partial x} \qquad (6)$$

where the subscript j corresponds to the jth layer, $\bar{e}_{zzj}$ and w_j are respectively the average transverse normal strain and displacement.

According to Eq. (3), w_j will be independent of ξ_j. The transverse shear stress in each layer is assumed to vary parabolically as given by the following expression.

$$\tau_{xz}(\xi_j) = \frac{\tau_{xz}^{j} - \tau_{xz}^{j-1}}{h_j}\,\xi_j + \frac{\tau_{xz}^{j} + \tau_{xz}^{j-1}}{2} + (B_{55})_j\varphi(x)\left(\xi_j^2 - \frac{h_j^2}{4}\right) \qquad (7)$$

where superscript j represents the jth contact surface, ξ_j the transverse coordinate measured from the midsurface of each layer along the outward normal, and h_j is the thickness of jth layer. This is an alternative form of Eq. (3) given by Whitney [19] for the study of the bending of laminated plates. The stress strain relations for orthotropic material, where the elasticity axes of symmetry coincide with x and θ coordinates, are

$$\sigma_{xx} = B_{11}\, e_{xx} + B_{12}\, e_{\theta\theta} \qquad (8) \qquad\qquad \tau_{\theta z} = B_{44}\,\gamma_{\theta z} = 0 \qquad (10)$$

$$\sigma_{\theta\theta} = B_{12}\, e_{xx} + B_{22}\, e_{\theta\theta} \qquad (9) \qquad\qquad \tau_{x\theta} = B_{66}\,\gamma_{x\theta} = 0 \qquad (11)$$

$$\tau_{xz} = B_{55}\,\gamma_{xz} \qquad (12)$$

where B_{ij} are nonzero elasticity constants. The transverse normal stress σ_{zz} will be obtained from the following equilibrium equation

$$\frac{1}{R_j+\xi_j} \frac{\partial}{\partial \xi_j} \left[(R_j+\xi_j)\sigma_{zzj} \right] + \frac{\partial \tau_{xzj}}{\partial x} - \frac{\sigma_{\theta\theta j}}{R_j+\xi_j} = 0 \ .$$

Since each layer is considered to be thin, ξ_j is negligible when compared to R_j and the last equation becomes

$$\frac{\partial \sigma_{zzj}}{\partial \xi_j} + \frac{\partial \tau_{xzi}}{\partial x} - \frac{\sigma_{\theta\theta j}}{R_j} = 0 \ . \tag{13}$$

By substituting equation (12) in conjunction with equation (7) into equation (6), and then integrating with respect to ξ_j, one obtains the displacement at any point in a general jth layer,

$$u_j = \bar{u}_j - \xi_j \frac{dw}{dx} + \frac{1}{B_{55j}} \left[\frac{\tau_{xz}^{j} - \tau_{xz}^{j-1}}{2h_j} \xi_j^2 + \frac{\tau_{xz}^{j} + \tau_{xz}^{j-1}}{2} \xi_j \right] + \varphi\left(\frac{\xi_j^3}{3} - \frac{h_j^2}{4} \xi_j \right) \tag{14}$$

where $\bar{u}_j$ is the longitudinal displacement of the midsurface of jth layer.

Substitution of Eq. (14) into Eq. (1) results in the following equation

$$e_{xxj} = \frac{d\bar{u}_j}{dx} - \xi_j \frac{d^2 w}{dx^2} + \frac{1}{B_{55j}} \left[\frac{\xi_j^2}{2h_j} \left(\frac{d\tau_{xz}^{j}}{dx} - \frac{d\tau_{xz}^{j-1}}{dx} \right) + \frac{\xi_j}{2} \left(\frac{d\tau_{xz}^{j}}{dx} + \frac{d\tau_{xz}^{j-1}}{dx} \right) \right] + \frac{d\varphi}{dx} \left(\frac{\xi_j^3}{3} - \frac{h_j}{4} \xi_j^2 \right) \tag{15}$$

To maintain continuity of displacement at all interfaces between every two adjacent layers, the condition

$$u_j \Big|_{\xi_j = \frac{h_j}{2}} = u_{j+1} \Big|_{\xi_{j+1} = -\frac{h_{j+1}}{2}} \tag{16}$$

must be satisfied. By substituting Eq. (14) into Eq. (16), one obtains

$$\bar{u}_{j+1} = \bar{u}_j - \frac{1}{2}(h_j + h_{j+1}) \frac{dw}{dx} + \frac{1}{8} A_{j+1} \tau_{xz}^{j+1} + \frac{3}{8}(A_{j+1} + A_j) \tau_{xz}^{j} + \frac{1}{8} A_j \tau_{xz}^{j-1} + K_j \varphi \tag{17}$$

where

$$A_{j+1} = h_{j+1}(Q_{55})_{j+1} \ , \qquad K_j = -\frac{1}{12}(h_{j+1}^3 + h_j^3)$$

Substitution of Eq. (7) and (9) in conjunction with Eqs. (2) and (15) into Eq. (13) and subsequent integration of the resulting equation with respect to ξ_j yields

$$\sigma_{zz}(\xi_j) = \left(\frac{B_{12}}{R} \right)_j \left(\xi \frac{d\bar{u}}{dx} - \frac{1}{2} \xi \frac{d^2 w}{dx^2} \right)_j + \left(\frac{B_{22}}{R^2} \xi \right)_j w + \left[\frac{B_{12}}{RB_{55}} \left(\frac{\xi^3}{6h} + \frac{\xi^2}{4} \right) - \frac{1}{2} \left(\frac{\xi^2}{h} + \xi \right) \right]_j \frac{d\tau^j}{dx}$$
$$+ \left[-\frac{B_{12}}{RB_{55}} \left(\frac{\xi^3}{6h} - \frac{\xi^2}{4} \right) + \frac{1}{2} \left(\frac{\xi^2}{h} - \xi \right) \right]_j \frac{d\tau^{j-1}}{dx}$$
$$+ \left[\frac{B_{12}}{12R} (\xi^4 - h\xi^3) - B_{55} \left(\frac{\xi^3}{3} - \frac{h^2}{4} \xi \right) \right]_j \frac{d\varphi}{dx} + \sigma_{zzj}^{o} \tag{18a}$$

where $\tau^j = \tau_{xz}^{j}$, and σ_{zzj}^{o} represents the transverse normal stress at the midsurface of the jth layer which can be determined acccrding to the condition

$$\bar{e}_{zzj} = \int_{-h_j/2}^{h_j/2} e_{zzj} d\xi_j = \int_{-h_j/2}^{h_j/2} [B_{33}\sigma_{zz} + B_{31}\sigma_{xx} + B_{32}\sigma_{\theta\theta}]_j \, d\xi_j = 0 \, . \tag{18b}$$

By substituting Eqs. (8), (9) and (18) in conjunction with Eqs. (1), (2) and (7) into Eq. (18b), one obtains

$$\sigma^o_{zzj} = C_{1j} \frac{d^2 w}{dx^2} + C_{2j}\, w + C_{3j} \frac{d\bar{u}_j}{dx} + C_{4j} \frac{d\tau^j}{dx} + C_{5j} \frac{d\tau^{j-1}}{dx} + C_{6j} \frac{d\varphi}{dx}$$

where

$$C_{1j} = \left(\frac{B_{12}h^2}{24R}\right)_j \, , \qquad C_{4j} = \left(\frac{h^2}{24} - \frac{B_{12}h^3}{48RB_{55}} + \frac{1}{24B_{55}} C_3 h^2\right)_j \, ,$$

$$C_{2j} = (B_{32}B_{22} + B_{31}B_{12})_j h_j \, , \qquad C_{5j} = -\left(\frac{B_{12}h^3}{48RB_{55}} + \frac{h^2}{24} + \frac{1}{24B_{55}} C_3 h^2\right)_j \, ,$$

$$C_{3j} = (B_{31}B_{11} + B_{32}B_{12})_j \, , \qquad C_{6j} = -\left(\frac{B_{12}h^4}{960R} + \frac{1}{48} C_3 h^3\right)_j$$

To maintain compatible interacting normal stresses between every two adjacent layers, and to require the satisfaction of normal stress boundary condition on inner or outer surface of a layered cylindrical shell,

$$\sigma_{zz}\Big|_{\xi_j = \frac{h_j}{2}} = \sigma_{zz}\Big|_{\xi_{j+1} = -\frac{h_{j+1}}{2}} \tag{19}$$

$$\sigma_{zz}\Big|_{\xi_N = \frac{h_N}{2}} = p_N$$

The inner surface normal stress boundary condition will be taken care of when the overall equilibrium condition in the transverse direction for a layered shell is satisfied. By using Eq. (18), the condition (19) become

$$\left(\frac{1}{24} Hh^2 + \frac{1}{8} h + C_4\right)_{j+1} \frac{d\tau^{j+1}}{dx} - \left(\frac{1}{24} Hh^2 - \frac{1}{8} h + C_5\right)_j \frac{d\tau^{j-1}}{dx}$$

$$+ \left[\frac{1}{12} (Hh^2)_{j+1} - \frac{1}{12} (Hh^2)_j + \frac{3}{8} (h_{j+1} + h_j) + C_{5(j+1)} - C_{4j}\right] \frac{d\tau^j}{dx}$$

$$+ \left(C_3 - \frac{B_{12}h}{2R}\right)_{j+1} \frac{d\bar{u}_{j+1}}{dx} - \left(C_3 + \frac{B_{12}h}{2R}\right)_j \frac{d\bar{u}_j}{dx}$$

$$= - \left[\left(\frac{B_{12}h}{4R} + C_1\right)_{j+1} + \left(\frac{B_{12}h}{4R} - C_1\right)_j\right] \frac{d^2 w}{dx^2} + \left[\left(\frac{B_{22}h}{2R^2} - C_2\right)_{j+1} + \left(\frac{B_{22}h}{2R^2} + C_2\right)_j\right] w$$

$$- \left[\left(\frac{B_{12}h^4}{64R} - \frac{B_{55}h^3}{12} + C_6\right)_{j+1} + \left(\frac{B_{12}h^4}{192R} - \frac{B_{55}h^3}{12} - C_6\right)_j\right] \frac{d\varphi}{dx} \tag{20a}$$

and

$$\left(\frac{1}{12} Hh^2 - \frac{3}{8} h + C_4\right)_N \frac{d\tau^N}{dx} + \left(\frac{1}{24} Hh^2 - \frac{1}{8} h + C_5\right)_N \frac{d\tau^{N-1}}{dx} + \left(\frac{B_{12}h}{2R} + C_3\right)_N \frac{d\bar{u}_N}{dx}$$

$$= \left(\frac{B_{12}h}{4R} - C_1\right)_N \frac{d^2w}{dx^2} - \left(\frac{B_{22}h}{2R^2} + C_2\right)_N w + \left(\frac{B_{12}h^4}{192R} - \frac{B_{55}}{12} - C_6\right)_N \frac{d\varphi}{dx} + p_N \qquad (20b)$$

where

$$H_j = \left(\frac{B_{12}}{RB_{55}}\right)_j$$

For a shell having N layers, there are a total of (N-1) interacting shear stresses, τ^j_{xz}, and N mid-surface longitudinal displacements $\bar{u}_i$. Eqs. (17) and (20) provide a total of 2N-1 equations. Hence $\frac{d\tau^j}{dx}$ (j = 1,....N-1), and $\frac{d\bar{u}_i}{dx}$ (i = 1,....N) may be expressed in terms of w, $\frac{d^2w}{dx^2}$, $\frac{d\varphi}{dx}$, $\frac{d\tau^o}{dx}$ and $\frac{d\tau^N}{dx}$ by using Eqs. (17) and (20) which may be represented in the following matrix form

$$\left[\begin{array}{c|c} a_{kj} & b_{ki} \\ \hline c_{nj} & d_{ni} \end{array}\right] \left\{\begin{array}{c} \frac{d\tau^j}{dx} \\ \hline \frac{d\bar{u}_i}{dx} \end{array}\right\} = \left[\begin{array}{c} f_{k\ell} \\ \hline f_{n\ell} \end{array}\right] \{X_\ell\} \qquad (21)$$

where

$$\{X_\ell\}^T = \left\{\frac{d^2w}{dx^2} \; w \; \frac{d\varphi}{dx} \; \frac{d\tau^o}{dx} \; \frac{d\tau^N}{dx} \; p_N\right\} .$$

The elements of the coefficient matrices can be easily identified from Eqs. (17) and (20). By solving Eq. (21), one obtains

$$\{Y_m\} = [g_{m\ell}]\{X_\ell\} \qquad (22)$$

where

$$Y_m = \begin{cases} \frac{d\tau^m}{dx} \text{ for } 1 \le m \le N-1 & \text{or } m = j \\ \frac{d\bar{u}_m}{dx} \text{ for } N \le m \le 2N-1 & \text{or } m = N+j-1 \end{cases}$$

and

$$[g_{m\ell}] = \left[\begin{array}{c|c} a_{kj} & b_{ki} \\ \hline c_{nj} & d_{ni} \end{array}\right]^{-1} \left[\begin{array}{c} f_{k\ell} \\ \hline f_{n\ell} \end{array}\right]$$

Substituting Eq. (22) into Eq. (15), one obtains

$$e_{xx}(\xi_j) = k^*_{1j}(\xi_j)\frac{d^2w}{dx^2} + k_{2j}(\xi_j)w + k^{**}_{3j}(\xi_j)\frac{d\varphi}{dx} + k_{4j}(\xi_j)\frac{d\tau^o}{dx} + k_{5j}(\xi_j)\frac{d\tau^N}{dx} + k_{6j}p^N \quad (23)$$

where

$$k_{\ell j}(\xi_j) = g_{r\ell} + \left(\frac{1}{2B_{55}}\right)_j\left[\left(\frac{\xi_j^2}{h_j} + \xi_j\right)g_{j\ell} - \left(\frac{\xi_j^2}{h_j} - \xi_j\right)g_{(j-1)\ell}\right]$$

$$k^*_{1j}(\xi_j) = k_{1j} - \xi_j\ , \qquad k^{**}_{3j}(\xi_j) = k_{3j} + \frac{\xi_j^3}{3} - \frac{h_j^2}{4}\xi_j\ , \qquad r = N + j - 1\ .$$

The stress resultants and stress couples are

$$N_{xx} = \sum_{j=1}^{N}\int_{-h_{j/2}}^{h_{j/2}}\sigma_{xx}(\xi_j)d\xi_j\ , \qquad N_{\theta\theta} = \sum_{j=1}^{N}\int_{-h_{j/2}}^{h_{j/2}}\sigma_{\theta\theta}(\xi_j)d\xi_j$$

$$M_{xx} = \sum_{j=1}^{N}\int_{-h_{j/2}}^{h_{j/2}}(z_j+\xi_j)\sigma_{xx}(\xi_j)d\xi_j\ , \qquad M_{\theta\theta} = \sum_{j=1}^{N}\int_{-h_{j/2}}^{h_{j/2}}(z_j+\xi_j)\sigma_{\theta\theta}(\xi_j)d\xi_j\ .$$

where z_j is the distance measured from the reference surface to the midsurface of jth layer; or $R_j - R_r = z_j$.

Substituting eqs. (8) and (9) in conjunction with eq. (23) into eq. (24), one obtains

$$\begin{bmatrix}\sigma_{xx}\\ \sigma_{\theta\theta}\end{bmatrix} = \begin{bmatrix}B_{11} & B_{12}\\ B_{12} & B_{22}\end{bmatrix}\begin{bmatrix}e_{xx}\\ e_{\theta\theta}\end{bmatrix}$$

$$= \begin{bmatrix}B_{11} & B_{12}\\ B_{21} & B_{22}\end{bmatrix}\begin{bmatrix}k^*_{1j} & k_{2j} & k^{**}_{3j} & k_{4j} & k_{5j} & k_{6j}\\ 0 & \frac{1}{R_j} & 0 & 0 & 0 & 0\end{bmatrix}\begin{Bmatrix}\frac{d^2w}{dx^2}\\ w\\ \frac{d\varphi}{dx}\\ \frac{d\tau^o}{dx}\\ \frac{d\tau^N}{dx}\\ p^N\end{Bmatrix} \quad (25)$$

$$\begin{Bmatrix} N_{xx} \\ N_{\theta\theta} \\ M_{xx} \\ M_{\theta\theta} \end{Bmatrix} = \sum_{j=1}^{N} \int_{-h_{j/2}}^{h_{j/2}} \begin{Bmatrix} \sigma_{xx}(\xi_j) \\ \sigma_{\theta\theta}(\xi_j) \\ (z_j + \xi_j)\, \sigma_{xx}(\xi_j) \\ (z_j + \xi_j)\, \sigma_{xx}(\xi_j) \end{Bmatrix} d\xi_j = [D_{k\ell}]\{X_\ell\} \tag{26}$$

The overall equilibrium equations for a shell element consisting of N layers are

$$\frac{dN_{xx}}{dx} = \tau^N - \tau^o \tag{27}$$

$$\frac{dQ_x}{dx} - \frac{N_{\theta\theta}}{R_r} = p_N - p_0 \tag{28}$$

$$\frac{dM_{xx}}{dx} - Q_x = 0 \tag{29}$$

where R_r is the radius of curvature of the reference surface.

Eqs. (28) and (29) are combined into a single equation,

$$\frac{d^2M_{xx}}{dx^2} - \frac{N_{\theta\theta}}{R_r} = p_0 - p_N \tag{30}$$

substituting Eq. (26) into Eqs. (27) and (30), one obtains

$$D_{11}\frac{d^3w}{dx^3} + D_{12}\frac{dw}{dx} + D_{13}\frac{d^2\varphi}{dx^2} + D_{14}\frac{d^2\tau^o}{dx^2} + D_{15}\frac{d^2\tau^N}{dx^2} + D_{16}\frac{dP_N}{dx} = \tau^N - \tau^o \tag{31}$$

and

$$D_{31}\frac{d^4w}{dx^4} + \left(D_{32} - \frac{D_{21}}{R_r}\right)\frac{d^2w}{dx^2} + D_{33}\frac{d^3\varphi}{dx^3} + D_{34}\frac{d^3\tau^o}{dx^3} + D_{35}\frac{d^3\tau^N}{dx^3} + D_{36}\frac{d^2P^N}{dx^2} - \frac{D_{22}}{R_r}w$$

$$- \frac{D_{23}}{R_r}\frac{d\varphi}{dx} - \frac{D_{24}}{R_r}\frac{d\tau^o}{dx} - \frac{D_{25}}{R_r}\frac{d\tau^N}{dx} - D_{26}\frac{P_N}{R_r} = p_0 - p_N\ . \tag{32}$$

τ^o and τ^N represent tangential loading functions on the inner and outer surfaces of a cylinder which are generally given.

By eliminating φ between Eqs. (31) and (32), one obtains

$$\varphi = -\frac{1}{D_{13}}\left[D_{11}\frac{dw}{dx} + D_{12}\int w\,dx + D_{14}\tau^o + D_{15}\tau^N + D_{16}\int P_N dx + \iint(\tau^o - \tau^N)\,dx\,dx\right] + Fx + G \tag{33}$$

and

$$\frac{d^4w}{dx^4} + b\frac{d^2w}{dx^2} + cw = -\frac{p^o}{D} + \frac{D_{23}}{DR_r}F - \frac{D_{23}}{DR_rD_{13}}\int(\tau^o - \tau^N)\,dx + f_1\frac{d\tau^o}{dx} + f_2\frac{d\tau^N}{dx}$$

$$+ f_3\frac{d^3\tau^o}{dx^3} + f_4\frac{d^3\tau^N}{dx^3} + f_5P_N + f_6\frac{d^2P_N}{dx^2} \tag{34}$$

where

$$b = \left[D_{32} - \frac{D_{33}D_{12}}{D_{13}} - \frac{1}{R_r}\left(D_{21} - \frac{D_{23}D_{11}}{D_{13}}\right)\right] \Big/ D ,$$

$$c = \frac{1}{R_r}\left(\frac{D_{23}D_{12}}{D_{13}} - D_{22}\right) \Big/ D , \qquad D = D_{31} - \frac{D_{11}D_{33}}{D_{13}} ,$$

$$f_1 = \frac{D_{33}}{D_{13}} + \frac{1}{R_r}\left(D_{24} - \frac{D_{23}}{D_{13}} D_{14}\right) , \qquad f_2 = - \frac{D_{33}}{D_{13}} + \frac{1}{R_r}\left(D_{25} - \frac{D_{23}}{D_{13}} D_{15}\right) ,$$

$$f_3 = - D_{34} + \frac{D_{33}}{D_{13}} D_{14} , \qquad f_4 = - D_{35} + \frac{D_{33}}{D_{13}} D_{15} ,$$

$$f_5 = 1 + \frac{1}{R_r}\left(D_{26} - \frac{D_{23}}{D_{13}} D_{16}\right) , \qquad f_6 = - D_{36} + \frac{D_{33}}{D_{13}} D_{16} .$$

SEMI-INFINITE SHELLS

For illustrative purposes, consider a semi-infinite cylindrical shell subjected to uniform edge shear, Q_o, and/or edge moment, M_o. The transverse shear stress at the loaded edge varies in the form of eq. (7). As a result, one requires

$$\varphi\big|_{x=o} = \varphi_o , \tag{35}$$

$$\varphi\big|_{x\to\infty} \text{ must be bounded,} \tag{36}$$

and

$$\tau_{xz}\big|_{x=o} = f(z) . \tag{37}$$

From eq. (33), it is seen that F = 0 in order that condition (36) is satisfied. From Eqs. (27), (31), and (33), it may be noted the constant F represents the longitudinal stress resultant for the present case. By substituting Eq. (35) into Eq. (33), one obtains

$$G = \frac{1}{D_{13}}\left[D_{11}\frac{dw}{dx} + D_{12}\int wdx\right]_{x=o} + \varphi_o$$

The interlaminar shear stresses τ^j are obtained from Eq. (22),

$$\tau^j = \left(g_{j1} - \frac{D_{11}}{D_{13}} g_{j3}\right)\frac{dw}{dx} + \left(g_{j2} - \frac{D_{12}}{D_{13}} g_{j3}\right)\int wdx + g_{j3}G + A_j^* \tag{38}$$

where A_j^* is an integration constant corresponding to jth layer.

The constants A_j^* can be determined immediately by substituting Eq. (38) into Eq. (37) provided that the solution of the transverse displacement, w, is obtained. These constants in general form are

$$A_j^* = f\left(z_j + \frac{h_j}{2}\right) - \left\{g_{j1}\left[\frac{dw}{dx}\right]_{x=o} + g_{j2}\left[\int w dx\right]_{x=o}\right\} - g_{j3}\varphi_o \tag{39}$$

where z_j represents the radial distance measured from the reference surface to the midsurface of the jth layer. Now the governing differential Eq. (34) becomes

$$\frac{d^4w}{dx^4} + b\frac{d^2w}{dx^2} + cw = 0 \ . \tag{40}$$

The boundary conditions are

$$M_{xx}\big|_{x=o} = M_o : \qquad \left[s_1\frac{d^2w}{dx^2} + s_2 w\right]_{x=o} = M_o \tag{41}$$

$$Q_x\big|_{x=o} = -\frac{dM_{xx}}{dx}\bigg|_{x=o} = Q_o : \qquad -\left[s_1\frac{d^3w}{dx^3} + s_2\frac{dw}{dx}\right]_{x=o} = Q_o \tag{42}$$

and w and its derivatives must vanish as x approaches infinity. The quantities s_1 and s_2 in Eqs. (41) and (42) are defined as follows:

$$s_1 = D_{31} - \frac{D_{11}}{D_{13}}D_{33} \ ,$$

$$s_2 = D_{32} - \frac{D_{12}}{D_{13}}D_{33} \ .$$

The general solution of Eq. (40) depends on the type of roots of the characteristic equation as to whether or not they are real and/or imaginary and/or complex conjugate. The type of roots depends on the material constants and the geometry of all layers of the shell. Consider a case, for illustrative purposes, in which the roots $u_k (k = 1,2,3,4)$ are complex conjugate which may be written as

$$\mu_{1,2} = \alpha_1 \pm i\beta \tag{43}$$

and

$$\mu_{3,4} = \alpha^* \pm i\beta^* \ . \tag{44}$$

It can be easily shown that α_1 and α^* have opposite signs. Consider that α_1 is negative and let $\alpha = |\alpha_1|$, the bounded solution of Eq. (40) becomes

$$w = e^{-\alpha x}(A\cos\beta x + B\sin\beta x) \ . \tag{45}$$

By satisfying the boundary conditions (41) and (42), one obtains

$$\begin{aligned} \ell_{11}A + \ell_{12}B &= M_o \\ \ell_{21}A + \ell_{22}B &= -Q_o \end{aligned} \tag{46}$$

where

$$\ell_{11} = s_1(\alpha^2 - \beta^2) + s_2 \ , \qquad \ell_{12} = -2\alpha\beta s_1$$

$$\ell_{21} = s_1(3\alpha\beta^2 - \alpha^3) - \alpha s_2 \ , \qquad \ell_{22} = s_1(3\alpha^2\beta - \beta^3) \ .$$

Solving Eqs. (46), one obtains the constants A and B which are given in the following general form

$$\begin{Bmatrix} A \\ B \end{Bmatrix} = \begin{bmatrix} b_{11} & b_{12} \\ b_{21} & b_{22} \end{bmatrix} \begin{Bmatrix} M_o \\ -Q_o \end{Bmatrix}$$

where the matrix $[b_{ij}]$ is the inverse of the matrix $[\ell_{ij}]$:

The results of this example are summarized below.

$$w = e^{-\alpha x}(A \cos \beta x + B \sin \beta x)$$

$$A = b_{11} M_o - b_{12} Q_o$$

$$B = b_{21} M_o - b_{22} Q_o$$

$$A^*_j = f\left(z_j + \frac{h_j}{2}\right) - \left(g_{j1} + \frac{1}{\alpha^2+\beta^2} g_{j2}\right)\alpha A + \left(g_{j1} - \frac{1}{\alpha^2+\beta^2} g_{j2}\right)\beta B + g_{j3}\varphi_o$$

$$G = -\left(D_{11} + \frac{1}{\alpha^2+\beta^2} D_{12}\right)\frac{\alpha}{D_{13}} A + \left(D_{11} - \frac{1}{\alpha^2+\beta^2} D_{12}\right)\frac{\beta}{D_{13}} B$$

$$\tau^j = -e^{-\alpha x}\left\{\left[\bar{s}_j(\beta A + \alpha B) - s^*_j\left(A - \frac{\alpha}{\beta} B\right)\right]\sin \beta x + \left[s_j(\alpha A - \beta B) + s^*_j\left(\frac{\alpha}{\beta} A + B\right)\right]\cos \beta x\right\} + g_{j3}G + A_j$$

where

$$\bar{s}_j = g_{j1} - \frac{D_{11}}{D_{13}} g_{j3}$$

$$s^*_j = \left(g_{j2} - \frac{D_{12}}{D_{13}} g_{j3}\right)\frac{\beta}{\alpha^2+\beta^2}$$

$$\varphi = \frac{e^{-\alpha x}}{D_{13}}\left\{\left[D_{11}(\beta A+\alpha B) - D_{12}\frac{\beta}{\alpha^2+\beta^2}\left(A - \frac{\alpha}{\beta} B\right)\right]\sin \beta x + \left[D_{11}(\alpha A - \beta B) + D_{12}\frac{\beta}{\alpha^2+\beta^2}\left(\frac{\alpha}{\beta} A + B\right)\right]\cos \beta x\right\} + G$$

NUMERICAL EXAMPLE

As indicated in earlier discussions that the present theory enables one to determine the interlaminar shear stresses where the theory based on the classical Kirchhoff-Love hypotheses fails for such layered construction. A simple case for a cylindrical shell clamped at both ends having two layers of different isotropic materials subjected to uniform internal pressure is considered, and the variation of interlaminar shear stress is presented. For this case, j = 1 corresponds to inner layer, and j = 2 corresponds to outer layer, $B_{11} = B_{22} = E/(1-\nu^2)$,

$B_{12} = -\nu B_{11}$, $B_{55} = G$, E is the modulus of elasticity, ν is the Poisson's ratio, and G is the shear modulus of elasticity. Only one interlaminar shear stress $\tau^1 = \tau$ exists. $h_1 = h_2 = h$, $R_j/h = 25$, $E_2 = 2E_1 = 30 \times 10^6$ psi, and $\nu_1 = \nu_2 = 0.3$ are considered in the example. The characteristic roots of the governing differential eq. (34) are complex conjugates, they are

$$\mu_{1,2,3,4} = \pm \alpha \pm \beta i$$

where $\alpha = 0.149$, and $\beta = 0.344$. For simplicity in computation, half of the shell length $\ell = 18.82$ is chosen so that $\beta\ell = 2\pi$. As a result $\sin \beta\ell = 0$, and $\cos \beta\ell = 1$. The interlaminar shear stress variation is found to be

$$\frac{\tau}{\tau_o} = 0.11465 \cosh\alpha x \sin\beta x - 0.03038 \sinh\alpha x \cos\beta x - 0.04229x$$

where x measured from the midsection of the shell, and τ_o is the interlaminar shear stress at the ends ($x = \pm \ell$). The variation is also plotted in Fig. (2). It is seen that the peak shear stress occurs near but not at the ends, and the stress reduces sharply toward the midsection of the shell.

CONCLUSIONS

A consistent theory for layered cylindrical shells is established. The theory is simpler than the general theory presented in previous work by Hsu and the author. However, the mode of transverse shear stress variation is more restricted. The Kirchhoff hypothesis that the normal to the reference surface of the shell remain normal to it is abandoned in the present theory so that continuous deformation and consistent interlaminar stresses are maintained throughout the shell. Moreover, the present theory enables one to calculate the interlaminar shear stress which can not otherwise be obtained if the Kirchhoff-Love hypothesis is considered.

NOMENCLATURE

B_{ij}	=	elasticity constants
e_{xx}, $e_{\theta\theta}$, e_{zz}	=	lineal strain components
$\gamma_{x\theta}$, $\gamma_{\theta z}$, γ_{xz}	=	shear strain components
h_j	=	thickness of jth layer
M_{xx}, $M_{\theta\theta}$	=	stress couple
M_o, Q_o	=	edge moment and shear
N_{xx}, $N_{\theta\theta}$	=	stress resultants
p^N, τ^N	=	normal and tangential loadings on the exterior surface
p^o, τ^o	=	normal and tangential loadings on the interior surface
Q_x	=	transverse shear stress resultant
R_j	=	radius of curvature of the midsurface of the jth layer shell
R_r	=	radius of curvature of the reference surface
u,w	=	longitudinal and radial displacements

$\bar{u}$	=	jth layer midsurface longitudinal displacement
x, θ	=	longitudinal and circumferential coordinates
σ_{xx}, $\sigma_{\theta\theta}$, σ_{zz}	=	normal stress components
τ_{xz}, $\tau_{x\theta}$, $\tau_{z\theta}$	=	shear stress components
τ_{xz}^{j}	=	interlaminar shear stress at the jth contact surface
τ	=	τ_{xz} for axisymmetric case
ξ_j	=	radial coordinate in the jth layer
z	=	radial coordinate from reference surface
z_j	=	radial distance from reference surface to midsurface of the jth layer

REFERENCES

[1] I. Ia. Shtaerman: On the Theory of Symmetrical Deformation of Anisotropic Elastic Shells. Izv. Kievsk. Polit. i Sel' -khoz, Institute, (1924).

[2] Kh. Mushtari: Some Generalization on the Theory of Thin Shell with Application to the Solution of Stability of Elastic Anisotropic Shells. Izv. Kazanskogo Fiz-Mat. Obshchestva, 11. Kazan, (1938).

[3] Kh. Mushtari: Some Generalization on the Theory of Thin Shells. Prikl. Mat. Mekh. 2, No. 4, p. 439-456, (1939).

[4] S. A. Ambartsumian: Contributions to the Theory of Anisotropic Layered Shells. Applied Mechanics Reviews, Vol. 15, No. 4, p. 245-249, April, (1962).

[5] S. A. Ambartsumian: Theory of Anisotropic Shells. NASA Technical Translation F-118, May, (1964).

[6] S. B. Dong: Membrane Stresses in Laminated Anisotropic Shells. ASCE Journal of Engineering Mechanics, EM3, p. 53-59, June, (1964).

[7] S. B. Dong, K. S. Pister, and R. L. Taylor: On the Theory of Laminated Anisotropic Shells and Plates. J. Aero. Sci., Vol. 28, p. 969, (1962).

[8] J. E. Ashton, and M. E. Waddoups: Analysis of Anisotropic Plates. J. Composite Materials, Vol. 3, p. 148, (1969).

[9] J. E. Ashton: Approximate Solutions for Unsymmetrical Plates. J. Composite Materials, Vol. 3, p. 355, (1969).

[10] G. Eason: On the Vibration of Anisotropic Cylinders and Spheres. Applied Scientific Res. (A), Vol. 12, p. 83-85, (1963).

[11] Y. C. Das: Vibrations of Orthotropic Cylindrical Shells. Applied Scientific Res. (A), Vol. 12, p. 317-327, (1964).

[12] I. Mirsky: Axisymmetric Vibrations of Orthotropic Cylinders. Journal of Acoust. Soc. Amer. 36, 5, p. 2106-2112, November, (1964).

[13] I. Mirsky: Three-dimensional and Shell Theory Analysis for Axisymmetric Vibration of Orthotropic Shells. Journal Acoust. Soc. Amer. 39, 3, p. 549-555, March, (1966).

[14] A. Kalnins: On Free and Forced Vibration of Rotationally Symmetric Layered Shells. Journal of Applied Mechanics, (Brief Note) p. 941-943, Dec. (1965).

[15] N. Ahmed: Axisymmetric Plane-Strain Vibrations of a Thick Layered Orthotropic Cylindrical Shell. Journal of Acoust. Soc. Amer., p. 1509-1516, December (1966).

[16] C. W. Bert: J. L. Baker, and D. M. Egle: Free Vibrations of Multilayer Anisotropic Cylindrical Shells. J. Composite Materials, Vol. 3, p. 480-499, (1969).

[17] J. T. S. Wang: On the Solution of Plates of Composite Materials. J. Composite Materials, Vol. 3, p. 590-592, (1969).

[18] J. M. Whitney: The Effect of Transverse Shear Deformation on the Bending of Laminated Plates. J. Composite Materials, Vol. 3, p. 534-547, (1969).

[19] N. J. Pagano: Exact Solutions of Composite Laminates in Cylindrical Bending. J. Composite Materials, Vol. 3, p. 398-411, (1969).

[20] Christos C. Chamis: Buckling of Anisotropic Plates. J. of the Structural Division, ASCE, p. 2119, October, (1969).

[21] T. M. Hsu, and J. T. S. Wang: A Theory of Laminated Cylindrical Shells Consisting of Layers of Orthotropic Laminae. AIAA Journal, Vol. 8, No. 12, p. 2142-2146, December, (1970).

[22] Teh-Min Hsu: Rotationary Symmetric Vibrations of Orthotropic Layered Cylindrical Shells. Ph.D. Thesis, Georgia Institute of Technology, May, (1969).

ACKNOWLEDGMENT

The author thanks Dr. Teh-Min Hsu of Lockheed-Georgia Company for the fruitful discussions regarding the numerical example.

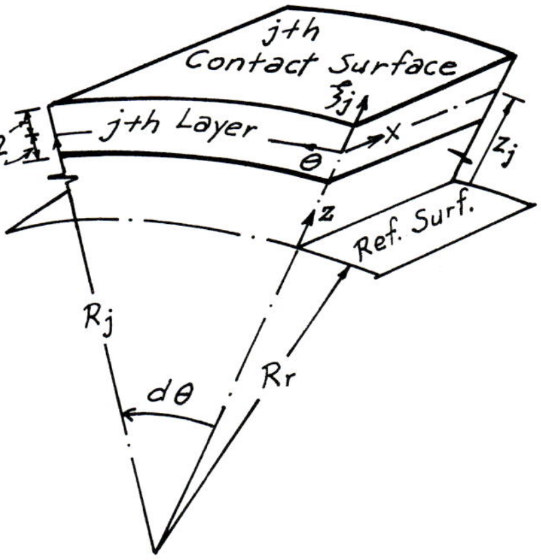

Fig. 1. Geometry

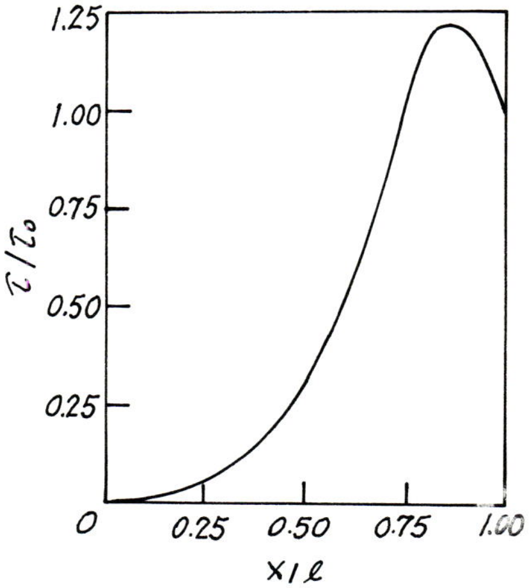

Fig. 2. Shear Stress Variation

Developments in Mechanics, Vol. 6. Proceedings of the 12th Midwestern Mechanics Conference: 56

High-precision Plate and Shell Finite Elements

PHILLIP L. GOULD
WASHINGTON UNIVERSITY
BARNA A. SZABO
WASHINGTON UNIVERSITY
LAWRENCE J. BROMBOLICH
McDONNELL AIRCRAFT CO.
CHUNG-TA TSAI
WASHINGTON UNIVERSITY

ABSTRACT

An examination of the factors influencing the precision of displacement method finite element models for rotational shell and plate bending problems is conducted. Attention is focused on the various methods of determining the coefficient of the polynomials in the assumed interpolation functions with emphasis on the physical meaning and relative influence of the low-order and higher-order terms. For the rotational shell model, the necessary geometrical approximations are also studied.

High-precision displacement functions are proposed for both elements with the objective of achieving convergence through individual element improvement as opposed to the usual method of simply reducing the size of elements and correspondingly increasing the total number of elements. The latter approach, of course, leads to an increased computational effort as well as possible numerical difficulties if the limiting process is carried too far. Along with the refinement in the displacement functions, improved discretizations of the external loading are achieved for both elements.

Examples are presented to demonstrate the striking improvement in the efficiency and accuracy of the analysis which is attainable through the use of high precision elements.

INTRODUCTION

Finite element analysis using the displacement method is usually based on the assumption of a displacement field in which the coefficients of the interpolation functions are directly expressible in terms of a defined set of generalized nodal displacements. A finite element is normally judged acceptable if convergence with respect to progressively reduced element sizes can be demonstrated on two or more standard problems. Experience and subsequert theoretical studies have shown that the low-order terms of the approximating secuence are very important in achieving convergence in this sense. In the finite element displacement method these terms represent rigid body displacement and constant strain components.

The importance of high-order terms can be seen in a convergence process in which the element domains are kept finite ard constant while the order of approximation is progressively increased. Given a complete polynomial or trigonometric sequence, it should then be possible to achieve uniform convergence over each element domain. Some of the theoretical problems associated with this process have not yet been solved. It is not known, for example, how the constant coefficients of the higher-order terms should be treated: should they be assigned to represent additional nodal displacement components or should they be determined from criteria similar to those of the Ritz method.

While questions of this kind may seem to be somewhat removed from the practical aspects of the finite element method, it is clear that they must be answered if finite elements of optimal or near-optimal efficiency are to be identified.

Since very little theoretical guidance is currently available, the writers chose to perform computational experiments employing two common types of finite elements; a rotational shell element and a plate bending element. In both cases the nodal variables were specified in terms of the displacement function and its first derivatives. The treatment of the coefficients of the approximating sequences along with other factors influencing the precision of the approximation are discussed in this paper.

ROTATIONAL SHELL ELEMENT

Displacements

For shells of revolution, the use of Fourier series expansions in the circumferential variable θ reduces the mathematical analysis to a single dimension, thereby eliminating from consideration the problem of specifying complete polynomials for the displacement functions; however, considering the generalized nodal displacements to be the three components of the displacement vector, u, v and w and the two components of the rotation vector β_ϕ and β_θ as shown in Fig. 1, there are ten nodal displacements available to determine the coefficients of the element displacement functions. This constraint is quite restrictive in that an equal order polynomial approximation for each displacement would permit only linear variations for each variable. In most analytical and numerical formulations of the thin shell equations, transverse shearing strains are neglected [1].* With this assumption, β_θ may be expressed directly in terms of v and w, and β_ϕ written as a function of u and $\frac{dw}{ds}$, where s = the nondimensional arc length coordinate as shown in Fig. 2. It is then necessary to construct only three independent displacement functions u(s), v(s) and w(s) in terms of the four nodal displacements, u(0) v(0) w(0) $\beta_\phi(0)$ and u(1) v(1) w(1) $\beta_\phi(1)$, at each end of the element. With eight conditions avai able to evaluate the coefficients for the three displacement functions, an equal order set cannot be specified and several alternatives are evident.

Of the various possibilities, the disp acement functions are usually taken in the form [1]

$$\begin{aligned} u(s) &= c_1 + c_2\, s \\ v(s) &= c_3 + c_4\, s \qquad (1) \\ w(s) &= c_5 + c_6\, s + c_7\, s + c_8\, s^3 \end{aligned}$$

---

* Numbers in brackets refer to items listed in the References.

in which c_1 - c_8 = constant coefficients. The apparent reason for this choice of displacement functions follows from an examination of the strain energy functional from the calculus of variations standpoint as suggested by Stricklin [3]. If the highest order derivative of a particular function appearing in the functional is of order n, the function need be continuous at the end points of the interval only up to the n - 1 derivative in order to establish the necessary conditions for an extremum [4]. Therefore, if the finite element problem is viewed as a piecewise minimization procedure, it is necessary that only the n - 1 derivative of the interpolation polynomial for the function be continuous at the nodes. The strain energy functional in question contains second derivative terms for w and first derivative terms for u and v. Since β_ϕ is also a nodal displacement, $\frac{dw}{ds}$ will be continuous at the nodes and the continuity conditions are satisfied. Thus, it is apparent that the choice of displacement functions given in Eq. 1, including just eight undetermined coefficients corresponding to the eight nodal displacements per element, is admissible and logical although not unique.

It is, however, possible to relax the constraint of selecting a displacement field containing the exact number of coefficients directly expressible in terms of the nodal displacements [2]. When higher-order displacement functions are introduced, the additional coefficients may be evaluated in two ways: (1) by treating each additional coefficient as an added dependent variable in the variational formulation and applying the Principle of Minimum Total Potential Energy to derive an additional equation per coefficient [5,6]; or (2) by defining higher-order derivatives of the basic set of displacements as additional generalized displacements and expressing the coefficients of the higher-order terms as functions of the nodal values of the enlarged set.

It is evident that the first procedure should result in an improved representation of the displacement field in the interior of the element while the second method affects interelement compatibility. Inasmuch as only the first and second derivatives of the displacements enter into the subsequent computations for the stress resultants [2], the efficiency of matching derivatives of a higher order is questionable. For this reason, it appears that the first approach is the more sensible and it has been adopted for this study. In addition, the usual simplification of neglecting transverse shearing strains is not applied since, in this formulation, the inclusion of this effect introduces no additional complications [7].

In view of the foregoing discussion, the displacements are taken in the harmonic form

$$\{D\} = \{u \quad v \quad w \quad \beta_\phi \quad \beta_\theta\}$$

$$= \sum_{j=o}^{\infty} \{u^{(j)}\cos j\theta \;\; v^{(j)}\sin j\theta \;\; w^{(j)}\cos j\theta \;\; \beta_\phi^{(j)}\cos j\theta \;\; \beta_\theta^{(j)}\sin j\theta\} \tag{2}$$

For a typical harmonic $j \geq 1$,*

$$\{D^{(j)}(s,\theta)\} = \lceil\Theta\rfloor\{\overline{D}^{(j)}(s)\} \tag{3a}$$

in which

$$\{\overline{D}^{(j)}(s)\} = \{u^{(j)}(s) \;\; v^{(j)}(s) \;\; w^{(j)}(s) \;\; \beta_\phi^{(j)}(s) \;\; \beta_\theta^{(j)}(s)\} \tag{3b}$$

and

$$\lceil\Theta\rfloor \quad \lceil\cos j\theta \;\; \sin j\theta \;\; \cos j\theta \;\; \cos j\theta \;\; \sin j\theta\rfloor \tag{3c}$$

The symbol ⌈ ⌋ denotes a diagonal matrix.

---

* For harmonic j = 0, all terms corresponding to the displacements v and β_θ are suppressed.

Then each displacement function is represented over a typical element as

$$\overline{d}_i(s) = (1-s)d_i(0) + sd_i(1) + \sum_{m=1}^{k} (s^m - s^{m+1})q_{im} \qquad (i = 1,5) \tag{4}$$

in which

k = total number of higher-order terms in $\overline{d}_i(s)$

$\{\Delta\} = \{d_1(0)\ d_1(1) \ldots d_5(1)\}$ = nodal values of generalized displacement $\overline{d}_i(s)$; also, coefficients of linear terms in the displacement functions.

and

$\{Q\} = \{q_{11}\ q_{1k} \cdots q_{ik}\}$ = coefficients of higher-order terms in the displacement functions.

It should be noted that the higher-order terms vanish at the nodes, s = 0 and s = 1, and that the interpolation polynomials are continuous but the first and higher-order derivatives are not necessarily continuous at these points. With the inclusion of transverse shearing strains in the formulation, each of the displacement terms appear in the strain energy functional only in the first derivative and the aforementioned conditions of continuity are satisfied.

Geometry

The geometrical parameters for a typical shell element are illustrated in Fig. 2 where the middle surface $R = R(Z)$ is shown along with the principal radii of curvature R_ϕ and R_θ and the meridional angle ϕ. The principal geometric terms required for the formulation are R_ϕ, R_θ, $\sin\phi$ and $\cos\phi$. Each of these terms may be written as functions of R, $\frac{dR}{dZ}$ and $\frac{d^2R}{dZ^2}$ [2,7].

The geometrical approximations inherent in the finite element analysis of rotational shells evolve from the selection and combination of comparatively simple surfaces to approximate the actual shell surface in a piecewise sense. The forms which have been adopted range from conical frusta [8] to various curved surfaces [1,9]. It is, however, insufficient and somewhat misleading to focus on the representation of the meridian alone. An examination of the second fundamental form of the differential geometry of curved surfaces reveals that the principal radii of curvature are the most important geometrical parameters entering into the thin shell equations [10]. In order to increase the precision of the finite element proposed in this investigation, Lagrangian interpolation polynomials are used to represent all geometrical terms entering into the strain energy functional. This allows the exact geometrical data to be utilized to the fullest extent and also permits the element stiffness matrix to be computed analytically without resorting to numerical integration [7].

Equilibrium Equations

Substituting the displacement interpolation functions into the applicable strain-displacement relations produces a set of equations relating the strains to $\{\Delta\}$ and $\{Q\}$. Following the introduction of the constitutive matrix to express the strains in terms of the stress resultants, the internal strain energy for an element is

$$U^{(j)} = \frac{1}{2} \lfloor \Delta \mid Q \rfloor [W] \left\{ \frac{\Delta}{Q} \right\} \tag{5}$$

in which

$$[W] = \iint [\tilde{T}]^T [E][\tilde{T}]\, dA = \left[\begin{array}{c|c} S_{11} & S_{12} \\ \hline S_{21} & S_{22} \end{array} \right] \tag{6a}$$

$$dA = LR(s)\ ds\ d\theta \tag{6b}$$

$[E]$ = the constitutive matrix

and

$[T]$ = the compatibility matrix relating the strains and displacements.

Representing the geometrical data in the polynomial form, permits the many integrations contained in Eq. 6(a) to be evaluated by quadrature [7].

After evaluating the potential energy of the external loading and applying the Principle of Minimum Total Potential Energy and static condensation to eliminate $\{Q\}$ [5,6], the equilibrium equations corresponding to the nodal displacements are

$$[\hat{S}]\{\Delta\} = \{\hat{F}\} \tag{7}$$

in which

$$[\hat{S}] = [s_{11} - s_{12}s_{22}^{-1}s_{21}] \tag{8a}$$

$$\{\hat{F}\} = \{\overline{f}_1 + \overline{F} - s_{12}s_{22}^{-1}\overline{f}_2\} \tag{8b}$$

and $\{\overline{f}_1\}$ and $\{\overline{f}_2\}$ = the consistent nodal loading vectors corresponding to $\{\Delta\}$ and $\{Q\}$, respectively $[\hat{S}]$ and $[\hat{F}]$ represent the element stiffness matrix and the generalized element nodal loading vector.

An examination of Eqs. 8(a) and 8(b) reveals that the elements of $[W]$ corresponding to the higher-order terms in the assumed displacement functions, s_{12} and s_{22}, appear in the generalized element loading vector as well as in the element stiffness matrix. Thus, it is apparent that the inclusion of the higher-order terms in the displacement functions permits a more accurate discretization of distributed surface loadings in addition to providing a better representation of the element stiffness matrix. This is especially important when relatively few elements are used in the analysis.

The assembled stiffness matrix for each harmonic is computed using the direct stiffness procedure whereby the stiffness coefficients corresponding to the common coordinates of adjacent elements at each node are algebraically combined. After the specification of the kinematic boundary conditions, the global stiffness matrix results from which the displacements and, subsequently, the stress resultants may be computed [7,12].

At this point, it is of interest to reflect on the method used to derive the high-precision element. The use of the static condensation procedure provided an efficient means of eliminating the coefficients of the higher-order displacement coefficients $\{Q\}$ and subsequently reducing the element stiffness matrix to the basic 10 x 10 size in terms of the nodal displacements. This is an attractive alternative to the more apparent possibility of defining higher-order derivatives of the displacements as additional generalized nodal displacements to be matched between elements. The latter procedure would result in a significant increase in the complexity of the formulation and would also increase the total size of the global stiffness matrix. Also, as noted earlier, it is questionable whether the method of matching higher derivatives would result in a significant improvement in the accuracy of the displacements and stress resultants.

Example

In order to demonstrate the improved accuracy obtained using the high-precision shell element as opposed to less refined models, a hyperboloidal shell of revolution subject to a quasistatic design wind loading condition is treated. The shell dimensions are shown in Fig. 3.

Plots of the meridional force N_ϕ and the meridional moment M_ϕ are presented in Fig. 4 and 5 respectively. The analysis was performed using 20 elements and considering nine harmonics for the description of the wind loading. The results are shown to compare favorably with those given in Reference 13 where 200 less refined elements were used. A more comprehensive study of the applications of the high-precision element is presented in Reference 7.

PLATE FINITE ELEMENT

Several high-precision plate finite elements have been proposed by a number of investigators. The papers of Cowper, Kosko, Lindberg and Olson [14], Butlin and Ford [15], Argyris, Fried and Scharpf [16] are of particular interest since they report on the development of an 18 degrees-of-freedom element which is similar in some respects to the element discussed in this paper.

In attempting to construct a high-precision plate finite element the writers sought to incorporate in the displacement function all available information concerning the solution of the governing differential equation over each element domain. The motivation for this approach was discussed earlier by Szabo and Lee [17] in deriving the stiffness equations for plates from the biharmonic equation using Galerkin's orthogonality criteria.

The interpolation functions were constructed from a set of homogeneous solution terms of the biharmonic equation. This set was obtained from the complete biquintic polynomial sequence by separating linear combinations of the polynomial terms that are homogeneous solutions of the biharmonic equation from those that are particular solutions. The process is illustrated in Table 1.

Table 1.

Separation of the homogeneous and particular solution terms of the biharmonic equation from the complete biquintic polynomial sequence.

HOMOGENEOUS SOLUTION TERMS:	
CONSTANT:	1
LINEAR:	$x \quad y$
QUADRATIC:	$x^2 \quad xy \quad y^2$
CUBIC:	$x^3 \quad x^2y \quad xy^2 \quad y^3$
QUARTIC:	$x^3y \quad xy^3 \quad (x^4-3x^2y^2) \quad (y^4-3x^2y^2)$
QUINTIC:	$(x^5-5x^3y^2) \quad (x^5-5xy^4) \quad (y^5-5x^2y^3) \quad (y^5-5x^4y)$
PARTICULAR SOLUTION TERMS:	
QUARTIC:	$(x^4+3x^2y^2+y^4)$
QUINTIC:	$(x^5+5x^3y^2+5xy^4) \quad (y^5+5x^2y^3+5x^4y)$

Of course, this table can be extended indefinitely to include increasing orders of complete polynomial sequences.

Considering a typical finite element, the coefficients of the particular solution terms depend only upon the distribution of the lateral load over the element. The coefficients of the homogeneous solution terms on the other hand must be determined from the boundary conditions and from suitably defined interelement continuity conditions. If one nodal displacement is defined for each homogeneous solution term in the assumed element displacement function, then as the order of approximation increases so does the number of available nodal displacement components increase. Consequently the boundary conditions and interelement continuity conditions can be satisfied at an increasing number of points. In the limit complete continuity would be established. Given that the governing differential equation is satisfied over each element domain, complete interelement continuity and satisfaction of all boundary conditions would result in the exact solution. This approach, known as the local solution approach [18], establishes a convergence process which does not depend on the reduction of element sizes. It is obvious that the success of this approach depends largely on the definition of the nodal displacement components.

For the purposes of this study an 18 degrees-of-freedom finite plate element was developed from the biquintic polynomial shown in Table 1. Details of the development as well as the results of some computational tests were given in Reference 18.

On performing a number of additional tests it was noted that while a very small number of elements gave accurate solutions, the accuracy did not increase significantly or monotonically when the number of elements was increased. This can be explained, at least heuristically, as being due to two conflicting processes: when the number of elements is increased, the number of nodal displacement components, thus the ability to achieve finer approximation, increases. However, the number of interelement boundaries, the sources of discretization error, also increases. The most efficient use of elements based on the local solution approach appears to be at or near the minimum number required to fit the geometrical configuation of the prototype. This is illustrated for the central deflection of the simply supported square plate in Table 2. The definitions for "P" and "Q" arrangements were given in Reference 14. In the "R" arrangement, the finite element boundaries are parallel with the symmetry axes of the plate.

Table 2a.

Central deflection of the simply-supported square plate under uniform load (entries to be multiplied by qL^4/D)

No. of Elements for 1/4 Plate	Local Solution Approach				Ref. 14		Ref. 16	
	R Arrang.	Error %	Q Arrang.	Error %	Q Arrang.	Error %	P Arrang.	Error %
1	4.0435	0.465	-	-	-	-	-	-
2	4.052045	0.254	4.052045	0.254	4.060937	0.035	4.05748	0.120
4	4.0658	0.084	-	-	-	-	-	-
8	4.074215	0.292	4.085394	0.567	4.062347	0.0	4.06230	0.001
Exact	4.0623527							

Table 2b.

Central deflection of the simply-supported square plate under central point load (entries to be multiplied by PL^2/D)

No. of Elements for 1/4 Plate	Local Solution Approach				Ref. 14		Ref. 16	
	R Arrang.	Error %	Q Arrang.	Error %	Q Arrang.	Error %	P Arrang.	Error %
1	1.126302	2.912	-	-	-	-	-	-
2	1.148046	1.038	1.148046	1.038	1.149279	0.931	1.13847	1.863
4	1.15310	0.602	-	-	-	-	-	-
8	1.15798	0.181	1.1603207	0.020	1.157422	0.229	1.15486	0.450
Exact	1.1600836							

Lack of monotonic convergence is noted for the central deflection of the uniformly loaded square plate. On the other hand monotonic convergence is indicated for the central deflection resulting from a concentrated load applied at the center. In both cases, as in all cases studied, the least number of elements gave satisfactory results in terms of the displacement functions.

Further improvements were gained with the introduction of special elements.

These special elements satisfy the local solution criteria [18] as well as the kinematic boundary conditions corresponding to simply supported, symmetric or fixed boundaries along one or more edges. These elements can be fitted to "general" elements or other "special" elements at their interior sides. Thus the proposed approach is capable of satisfying the governing differential equation everywhere within each element domain and the appropriate kinematic boundary conditions at the external boundaries. Consequently the only sources of error are: (i) pointwise rather than uniform interelement continuity of the displacement function, and (ii) mean rather than uniform satisfaction of the natural boundary conditions.

There are some special cases in which the problem of interelement continuity does not arise. For example, a special element can be constructed to satisfy the kinematic boundary conditions along the sides of a triangular element representing one-eighth of the uniformly loaded square plate. The relationship between this element and a general 18 degrees-of-freedom triangular plate element is illustrated in Fig. 6. The nodal displacement components of the general element are re-defined as indicated in Fig. 6b. Vanishing of the normal slope along the symmetry boundaries and of the displacement along the simply supported boundary is ensured by setting the appropriate nodal variables equal to zero. Three active nodal variables (degrees of freedom) remain as illustrated in Fig. 6c. The errors in terms of these nodal variables are shown in Table 3.

Table 3.

One-eighth of the uniformly loaded,
simply supported square plate.

Solutions in terms of the active nodal displacement components (Fig. 7c)			
Nodal Variable	Exact Value	Computed Value	Error
	(Entries to be multiplied by $\frac{qL^4}{D}$)		per-cent
1 (Displacement)	0.0040625	0.0040640	0.037
2 (Slope)	0.0134758/L	0.0134656/L	0.076
3 (Slope)	0.0098385/L	0.0098521/L	0.138

The gain in accuracy resulting from the use of this element is best illustrated in Fig. 7. The solution in terms of the bending moments along one of the symmetry axes is compared with the exact solution and with the solution obtained with the general element shown in Fig. 6a.

Another interesting problem is the uniformly loaded rectangular plate with fixed, simply supported and free boundaries as shown in Fig. 8a. Since the plate is symmetric, it is advantageous to study just one half of it subdivided into four triangular elements. The displacement function of each element satisfies the governing differential equation and one of the four kinematic boundary conditions. In this example the element displacement functions were constructed from the complete 6th order polynomial sequence which comprises 22 homogeneous solution terms and 6 particular solution terms of the biharmonic equation. The active nodal displacement components are shown in Fig. 8b. The solution obtained through the local solution approach is compared with the series solution, given by Timoshenko and Woinowski-Krieger (19), in Fig. 9. Solution of the same problem, using the mixed method, was reported by Hermann (20,21).

These and other computational tests have shown that the accuracy of solution, particularly in terms of the moment field, largely depends on the definition of nodal displacement components along the interelement boundaries. Research is being continued in this area with the objective of finding that combination of nodal variables which minimizes the discretization error. However, the findings of preliminary studies indicate that the local solution approach already offers an efficient solution technique permitting the use of a minimum number of elements.

CONCLUSION

In this paper, two approaches for deriving high-precision finite elements were discussed. The problems considered were a rotational shell element, and a plate bending element. Both approaches were based on an assumed convergence process in which the element domains are kept finite and constant while the order of approximation is progressively increased. Along with the improvement in the accuracy of the displacement functions, this approach results in a refined discretization of distributed external loading. The examples serve to illustrate that accurate approximations can be obtained with relatively few elements.

ACKNOWLEDGEMENTS

Studies of the local solution approach were supported by the National Science Foundation under Grant No. GK 5258. Computing services were provided by the Washington University Computing Facilities which are supported by NSF Grant G-22296.

REFERENCES

[1] Jones, R. E. and Strome, D. R., "Direct Stiffness Method Analysis of Shells of Revolution Utilizing Curved Elements," AIAA Journal, Vol. 4, No. 9, Sept., 1966, pp. 1519-1525.

[2] Brombolich, L. J. and Gould, P. L., "Finite Element Analysis of Shells of Revolution by Minimization of the Potential Energy Functional," Proceedings of the Conference on Applications of the Finite Element Method in Civil Engineering, Vanderbilt University, Nashville, Tenn., Nov., 1969, pp. 279-307.

[3] Stricklin, J. A., "Geometrically Nonlinear Static and Dynamic Analysis of Shells of Revolution," IUTAM Symposium on High Speed Computing of Elastic Structures, Liege, Belgium, Aug., 1970, p. 11.

[4] Gelfand, I. M. and Fomin, S. V., Calculus of Variations, trans. from Russian by R. A. Silverman, Prentice-Hall, Inc., Englewood Cliffs, N. J., 1963, p. 61-63.

[5] Pian, T. H. H., "Derivation of Element Stiffness Matrices," AIAA Journal, Vol. 2, No. 3, March, 1964, pp. 576-577.

[6] Przemieniecki, J. S., Theory of Matrix Structural Analysis, McGraw-Hill Book Co., New York, 1968.

[7] Brombolich, L. J. and Gould, P. L., "Finite Element Analysis of Shells of Revolution," Research Report No. 16, Structural Division, Dept. of Civil and Environmental Engineering, Washington University, St. Louis, Mo., Feb., 1971.

[8] Meyer, R. R. and Harmon, M. B., "Conical Segment Method for Analyzing Shells of Revolution for Edge Loading," AIAA Journal, Vol. 1, No. 4, April, 1963, pp. 886-891.

[9] Khojastek - Bakht, M. and Popov, E. P., "Analysis of Elastic-Plastic Shells of Revolution," Journal of the Engineering Mechanics Division, ASCE, Vol. 96, No. EM3, Proc. Paper 7354, June, 1970, pp. 327-340.

[10] Kraus, H., Thin Elastic Shells, John Wiley & Sons, Inc., New York, 1967.

[11] Hildebrand, F. B., Introduction to Numerical Analysis, McGraw-Hill Book Co., Inc., New York, 1956.

[12] Navaratna, D. R., "Computation of Stress Resultants in Finite Element Analysis," Tech. Note, AIAA Journal, Vol. 4, No. 11, Nov., 1966, pp. 2058-2060.

[13] Hill, D. W. and Coffin, G. K., "Stresses and Deflections in Cooling Tower Shells Due to Wind Loading," Bulletin of the IASS, No. 35, Sept., 1968, pp. 43-51.

[14] Cowper, G. R., Kosko, E, Lindberg, G. M. and Olson, M. D., "Static and Dynamic Applications of a High-Precision Triangular Plate Bending Element," AIAA Journal, Vol. 7, No. 10, Oct., 1969, pp. 1957-1965.

[15] Butlin, G. A. and Ford, R., "A Compatible Triangular Plate Bending Finite Element," University of Leicester, Engineering Department Report 68-15, Oct., 1968.

[16] Argyris, J. H., Fried, I. and Scharpf, D. W., "The TUBA Family of Plate Elements for the Matrix Displacement Method," The Aeronautical Journal, Royal Aeronautical Society, Vol. 72, 1968, pp. 701-709.

[17] Szabo, B. A. and Lee, G. C., "Stiffness Matrix for Plates by Galerkin's Method," Journal of the Engineering Mechanics Division, ASCE, Vol. 95, No. EM3, Proc. Paper 6608, June, 1969, pp. 571-585.

[18] Szabo, B. A., "The Local Solution Approach in the Finite Element Method," Proceedings, Colloquium of the International Union for Theoretical and Applied Mechanics, University of Liege, Belgium, Aug., 1970.

[19] Timoshenko, S. P. and Woinowsky-Krieger, S., "Theory of Plates and Shells," McGraw-Hill Book Co., Inc., 1959, pp. 208-211.

[20] Hermann, L. R., "A Bending Analysis for Plates," Matrix Methods in Structural Proceedings of the Conference Held at Wright-Patterson Air Force Base, Ohio, 26-28 Oct., 1965, AFFDL-TR-66-80, pp. 577-602.

[21] Hermann, L. R., "Finite Element Bending Analysis for Plates," Journal of the Engineering Mechanics Division, ASCE, Vol. 93, No. EM5, Proc. Paper 5497, Oct., 1967, pp. 13-26.

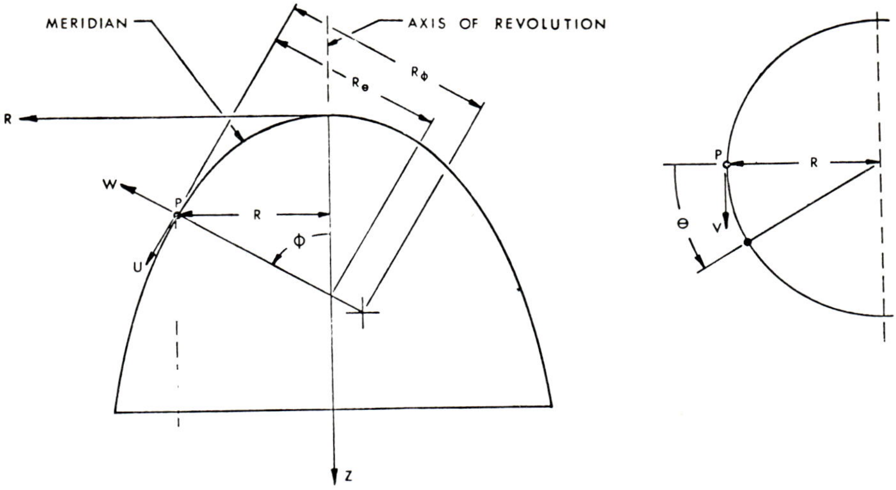

Fig. 1. Shell geometry

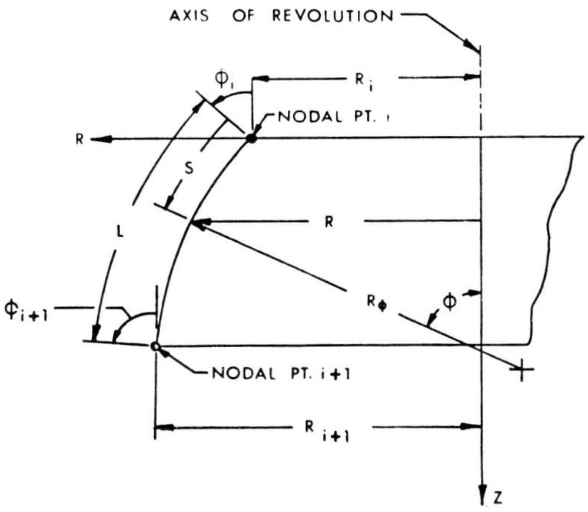

Fig. 2. Finite element geometry

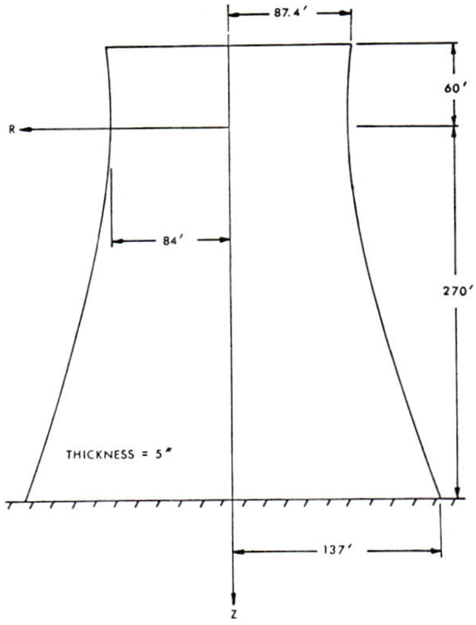

Fig. 3. Hyperboloidal shell of revolution

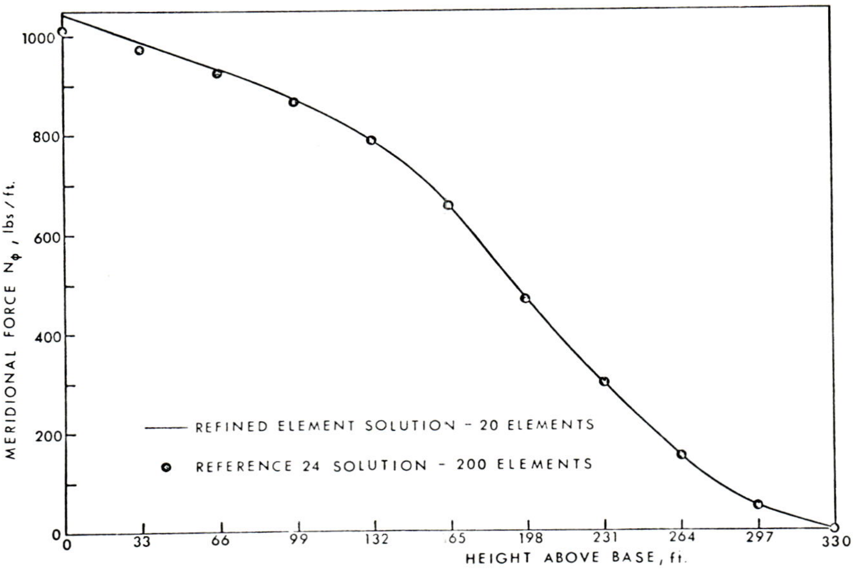

Fig. 4. Meridional stress resultant in shell

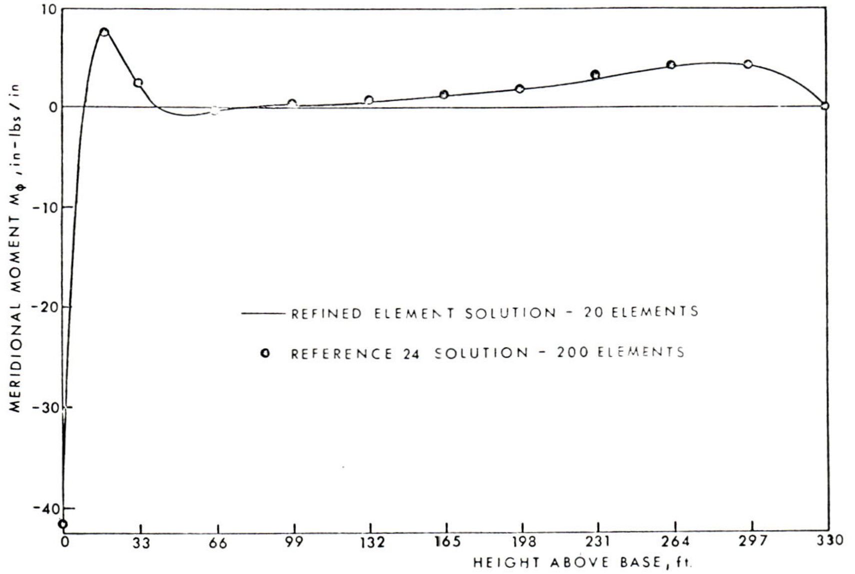

Fig. 5. Meridional moment resultant in shell

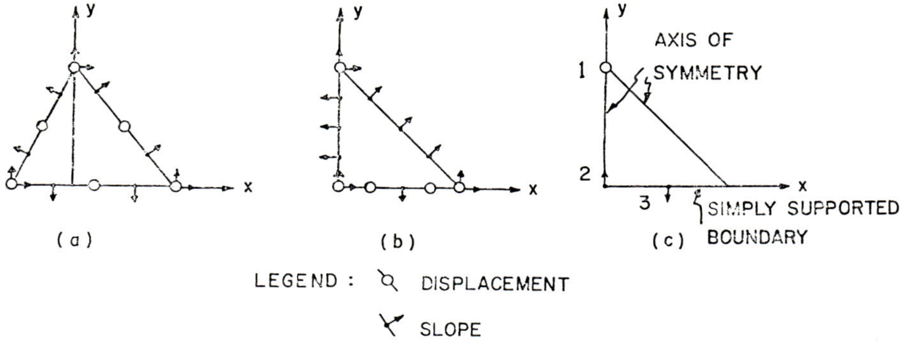

Fig. 6. Definition of nodal variables for 18 degrees-of-freedom finite elements satisfying the local solution criteria.

(a) general element

(b) special element, representing one-eighth of the simply supported square plate

(c) active nodal variables for the special element representing one-eighth of the simply supported square plate.

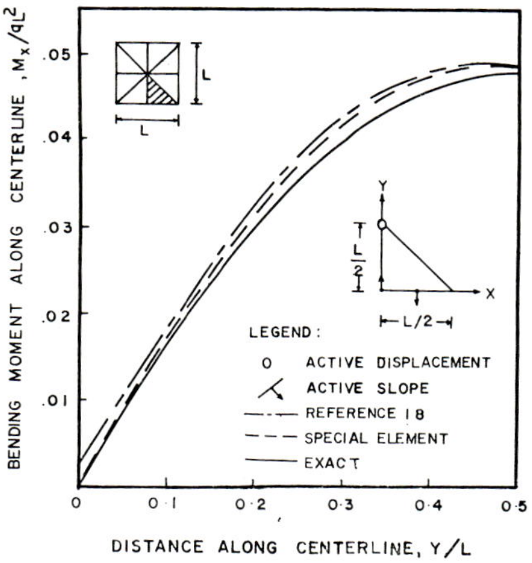

Fig. 7a. One-eighth of the simply-supported square plate under uniform load.

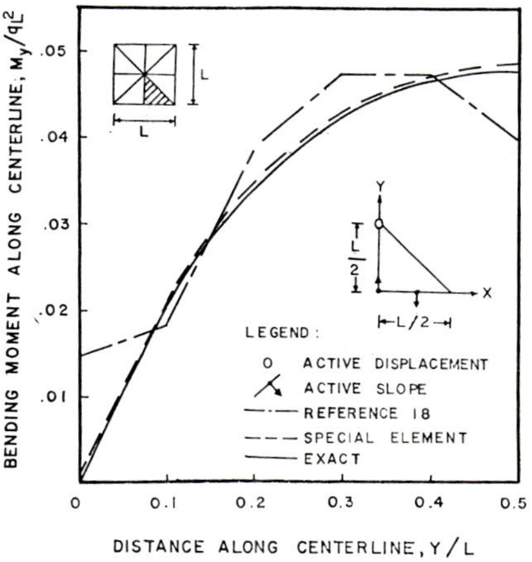

Fig. 7b. One-eighth of the simply-supported square plate under uniform load.

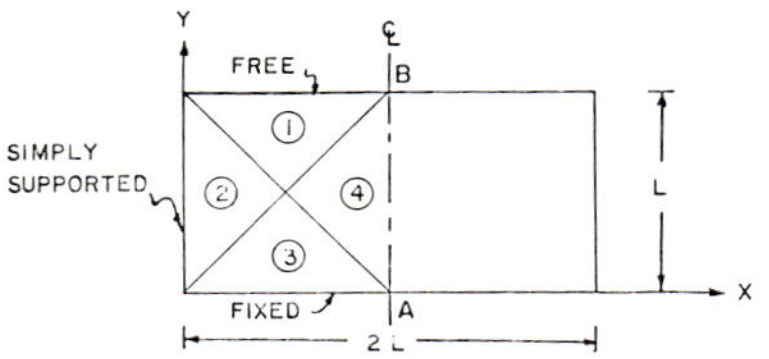

Fig. 8a. Finite element net for the uniformly loaded rectangular plate with two opposite edges simply supported, one edge free, one edge fixed.

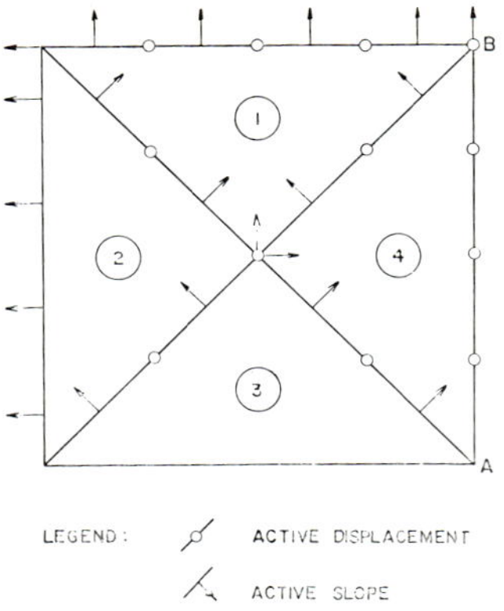

Fig. 8b. Active nodal displacement components.

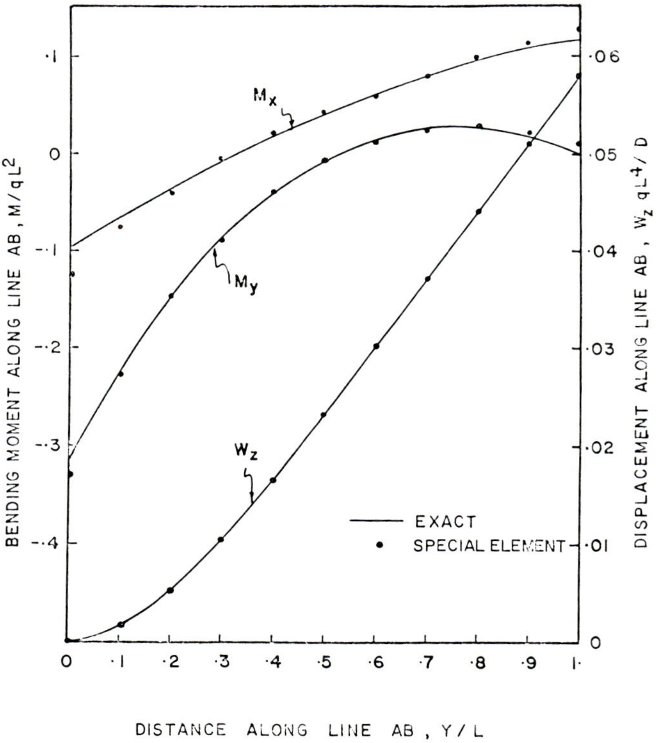

Fig. 9. Solutions for bending moments and displacements along line AB of the problem shown on Fig. 8a.

STRUCTURES

Developments in Mechanics, Vol. 6. Proceedings of the 12th Midwestern Mechanics Conference: 57

Analysis of Elastic Cover Plates*

F. ERDOGAN**
LEHIGH UNIVERSITY

ABSTRACT

The contact problem for a series of cover plates bonded to an elastic half plane is considered. The general problem is reduced to the solution of a Fredholm-type integral equation with a weakly singular kernel and, alternatively, to that of a singular integral equation. The equations for periodically spaced cover plates are obtained from the solution of multiple cover plates through a limiting process. The stress state in the vicinity of the plate ends is examined and it is shown that the stresses have an inverse square root singularity in both elastic as well as inextensible cover plates. For the special case of inextensible cover plates, closed form solutions are given. For one, two and periodically spaced cover plates, the singular integral equations are solved and numerical results are given. The main results of the paper consist of the demonstration of the effect of interaction on the stress intensity factors in the case of more than one cover plate.

INTRODUCTION

In this paper we consider the general contact problem for a series of elastic cover plates bonded to an elastic half space. The special cases of this problem were considered by Bufler [1], and more recently, by Arutiunian [2,3]. The problem of a single cover plate on a half plane and a line inclusion in an infinite plane was studied

---

* This research was supported by the National Science Foundation under Grant GK-11977 and the National Aeronautics and Space Administration under Grant NGR 39-007-011.

** Professor, Department of Mechanical Engineering and Mechanics, Lehigh University, Bethlehem, Pennsylvania 18015.

in detail in [1] under various symmetric and anti-symmetric loading conditions. By applying the technique used by Carafoli [4,5] in connection with the theory of a wing with finite span, Bufler obtained the numerical solution of the integral equation of the problem and determined the distribution of contact stresses between the cover plate and the half plane. However, in [1] the stress singularities, particularly the tensile stresses in the half plane (which are of considerable structural interest) around the singular points were not studied.

In [2] the problem of a single cover plate and in [3] a series of periodically spaced cover plates are considered. In [1-3] as well as in this paper, the cover plates are assumed to be sufficiently "thin" to justify the assumption that the plates are under a state of generalized plane stress with the shear component of the contact stress acting as a body force and the normal component σ_y being negligibly small. The method of solution developed in [2] and [3] is much too cumbersome and complicated to be practical, and the papers do not contain any worked out examples*.

In this paper the general problem for an arbitrary number of cover plates is formulated. This formulation is extended to obtain the basic integral equation for the periodic cover plates under symmetric and anti-symmetric loading conditions. The limiting problem of inextensible cover plates is solved in closed form. The numerical solutions for elastic cover plates are given for one, two, and a periodic array of (identical) plates.

The integral equations for the elastic cover plates are solved by using a relatively simple technique which was outlined in [6]. In addition to its simplicity, the advantage of the method used is that, the quantity which is (from the practical view point) of most immediate interest, namely the interface shear stress, is the unknown function in the integral equations and is readily evaluated. Also the solution directly gives the strength of the stress singularity at the ends of the cover plates, through which the stress state in the half plane around the singular points may easily be evaluated.

FORMULATION OF THE GENERAL PROBLEM

Referring to Figure 1, let a semi-infinite elastic half plane with elastic constants (μ,κ) be reinforced by n cover plates which are bonded to its free surface, y = 0. It is assumed that the dimension of the structure in z direction is either very large and there is complete constraint, in which case, $\varepsilon_z = 0$ and $\kappa = 3-4\nu$, or is small and there is no constraint (in which case $\kappa = (3-\nu)/(1+\nu)$). Thus the problem is one of plane strain or generalized plane stress.

Let the kth cover plate with elastic constants μ_k, κ_k and thickness h_k be located along $a_k < x < b_k$, (y = 0), and be under axial loads T_{1k}, T_{2k} (see Figure 1). Assuming that the normal component of the contact stress is negligible and the shear component is $\bar{\tau}_{xy}(x)$, from

*Largely because of this, some minor errors in [2,3] have gone unnoticed. Even though trivial from the view point of analysis, these errors (particularly that regarding the sign of the parameter λ) change the trend of the numerical results completely (such as having, instead of lower, higher stress concentration around the ends of elastic plates as compared to that of inextensible plates).

the equilibrium of the cover plate we have

$$h_k \sigma_{kx}(x) = \frac{8\mu_k h_k}{1+\kappa_k} \frac{\partial u_k}{\partial x} = T_{1k} + \int_{a_k}^{x} \tau_{xy}^{-}(t)dt \qquad (1)$$

Referring now to [7, Ch. 19], let the Kolosov-Muskhelishvili functions $\phi_1(z)$ and $\Psi_1(z)$ for the half plane S^- be (Figure 1)

$$\phi_1(z) = \phi(z) + C_1$$

$$(2)$$

$$\Psi_1(z) = \Psi(z) + C_2$$

where C_1, C_2 are constant and ϕ and Ψ vanish at infinity. In the problems considered in this paper it will be assumed that at $y = 0$ the normal stress is zero, i.e.,

$$\sigma_y^-(x) = 0, \ (-\infty < x < \infty) \qquad (3)$$

Thus, in terms of the stress components at infinity the constants C_1 and C_2 are given by

$$C_1 = \frac{\sigma_x^\infty}{4}, \ C_2 = -\frac{\sigma_x^\infty}{2} + i\tau_{xy}^\infty \qquad (4)$$

where it is assumed that at $|z| = \infty$ the rotation and σ_y^∞ are zero.

Extending now the definition of $\phi(z)$ into the positive half plane, S^+, as in [7], the stresses and the displacements may be expressed as follows:

$$\sigma_y - \sigma_x + 2i\tau_{xy} = 2[(\bar{z}-z)\phi'(z) - \phi(z) - \bar{\phi}(z) + C_2] \qquad (5)$$

$$\sigma_y - i\tau_{xy} = \phi(z) - \phi(\bar{z}) + (z-\bar{z})\bar{\phi}'(\bar{z}) + 2C_1 + \bar{C}_2 \qquad (6)$$

$$2\mu \frac{\partial}{\partial x}(u+iv) = \kappa\phi(z) + \phi(\bar{z}) + (\bar{z}-z)\bar{\phi}'(\bar{z}) + C_3$$

$$(7)$$

$$C_3 = (\kappa+1)\frac{\sigma_x^\infty}{4} + i\tau_{xy}^\infty$$

From (3) and (6) it follows that

$$\sigma_y^-(x) = 0 = \phi^-(x) + \bar{\phi}^+(x) - \phi^+(x) - \bar{\phi}^-(x) \qquad (8)$$

Since $\text{Im}(\phi) = 0$ at $|z| = \infty$, from (8) we find

$$\bar{\phi}(z) = \phi(z) \qquad (9)$$

in the entire z plane.

Using (9), from (6) and (7) for the relevant quantities on the boundary, $y = 0$, we obtain

$$i(\tau_{xy}^- - \tau_{xy}^\infty) = \phi^+(x) - \phi^-(x) \qquad (10)$$

$$\frac{4\mu}{1+\kappa}\frac{\partial u^-}{\partial x} - \frac{\sigma_x^\infty}{2} = \phi^+ + \phi^- \qquad (11)$$

For n cover plates if we let $(a_k, b_k) = L_k$, $L = \sum_1^n L_k$ and L' the complement of L in $-\infty < x < \infty$, it is seen that (1) is valid for $x \in L$, $\tau_{xy}^{-} = 0$ for $x \in L'$, and since $\sum_{1}^{n} (T_{2k} - T_{1k})$ is finite, τ_{xy}^{∞} is zero. Thus, defining

$$\lambda_k = \frac{1+\kappa_k}{1+\kappa} \frac{\mu}{2\mu_k h_k} \tag{12}$$

from

$$\frac{\partial u^-}{\partial x} = \frac{\partial u_k}{\partial x}, \quad (x \in L) \tag{13}$$

the solution of the Riemann-Hilbert problem (10, 11) may be expressed as [8]

$$\phi(z) = \frac{X(z)}{2\pi i} \sum_{1}^{n} \int_{L_k} \frac{dt}{(t-z)X^+(t)} [\lambda_k (T_{1k} + \int_{a_k}^{t} \tau_{xy}^{-}(s)ds) - \frac{\sigma_x^{\infty}}{2}]$$
$$+ P_{n-1}(z)X(z) \tag{14}$$

where the fundamental function $X(z)$ and its boundary values are given by (see [8])

$$X(z) = \prod_{1}^{n} (z-a_k)^{-1/2}(z-b_k)^{-1/2}$$
$$\tag{15}$$
$$X^+(t) = -X^-(t), \quad t \in L; \quad X^+(t) = X^-(t) = X(t), \quad t \in L'$$

Here the branch of $X(z)$ for which $z^n X(z) \to 1$ as $|z| \to \infty$ will be taken. In (14), $P_{n-1}(z)$ is an arbitrary polynomial of degree n-1.

Part of the integral in (14) may be evaluated as

$$-\frac{\sigma_x^{\infty}}{2} \frac{X(z)}{2\pi i} \int_{L} \frac{dt}{(t-z)X^+(t)} = -\frac{\sigma_x^{\infty}}{4} [1 - F_n(z)X(z)] \tag{16}$$

where the known polynomial of degree n,

$$F_n(z) = z^n + F_{n-1}(z)$$

is the principal part of $1/X(z)$ as $|z| \to \infty$. Thus, since $P_{n-1}(z)$ is an arbitrary polynomial, combining it with $F_{n-1}(z)$, (14) becomes

$$\phi(z) = \frac{\sigma_x^{\infty}}{4} z^n X(z) + P_{n-1}(z)X(z) - \frac{\sigma_x^{\infty}}{4}$$
$$+ \frac{X(z)}{2\pi i} \sum_{1}^{n} \lambda_k \int_{L_k} \frac{dt}{(t-z)X^+(t)} [T_{1k} + \int_{a_k}^{t} \tau_{xy}^{-}(s)ds],$$
$$P_{n-1}(z) = \sum_{0}^{n-1} d_k z^k \tag{17}$$

Noting that $\tau_{xy}^{\infty} = 0$, from (10) and (17) the contact stress may be expressed as

$$\tau_{xy}^{-}(x) = \frac{1}{i} [\phi^+(x) - \phi^-(x)] = \frac{\sigma_x^{\infty}}{2i} x^n X^+(x) + \frac{2}{i} P_{n-1}(x)X^+(x)$$
$$- \frac{X^+(x)}{\pi} \sum_{1}^{n} \lambda_k \int_{L_k} \frac{dt}{(t-x)X^+(t)} [T_{1k} + \int_{a_k}^{t} \tau_{xy}^{-}(s)ds],$$
$$(x \in L) \tag{18}$$

The n arbitrary constants $d_o,---,d_{n-1}$ of the polynomial P_{n-1} are determined from the static equilibrium of the cover plates, which may be expressed as (see Figure 1)

$$\int_{a_k}^{b_k} \tau_{xy}^{-}(t)dt = T_{2k} - T_{1k}, \quad (k = 1,---,n) \tag{19}$$

Thus (18) provides an integral equation for the unknown function $\tau_{xy}^{-}(t)$. As shown in [8], in principle, this equation can be transformed into a Fredholm type equation with a weakly singular kernel.

After determining τ_{xy}^{-} and d_k, other quantities of interest may easily be obtained through quadratures. For example, one such important quantity is σ_x^{-} on the surface of the half plane, which may be expressed, from (5), (6) and (9), as follows

$$\sigma_x^{-}(t) = \sigma_x^{\infty} + 2[\phi^{+}(t) + \phi^{-}(t)], \quad (t\varepsilon L + L') \tag{20}$$

Or, from (17) and (20) we have

$$\sigma_x^{-}(t) = 2\lambda_k[T_{1k} + \int_{a_k}^{t} \tau_{xy}^{-}(s)ds], \quad (t\varepsilon L_k)$$

$$\sigma_x^{-}(t) = [\sigma_x^{\infty}t^n + 4P_{n-1}(t)]X(t) + \frac{2}{\pi i} X(t) \sum_{1}^{n} \lambda_k \int_{a_k}^{b_k} \frac{d\tau}{(\tau-t)X^{+}(\tau)} [T_{1k} + \int_{a_k}^{\tau} \tau_{xy}^{-}(s)ds], \quad (\tau\varepsilon L') \tag{21}$$

From (21) it is seen that under the cover plates, $t\varepsilon L_k$, σ_x^{-} is bounded and around the end points for $t < a_k$ and $t > b_k$, its behavior is

$$\sigma_x^{-} \sim (a_k-t)^{-1/2}, \quad \sigma_x^{-} \sim (t-b_k)^{-1/2}, \quad (t\varepsilon L') \tag{22}$$

Similarly,

$$\begin{aligned} &\tau_{xy}^{-}(t) = 0, \quad t\varepsilon L' \\ &\tau_{xy}^{-} \sim (t-a_k)^{-1/2}, \quad \tau_{xy}^{-} \sim (b_k-t)^{-1/2}, \quad t\varepsilon L \end{aligned} \tag{23}$$

From the view point of numerical analysis the formulation given above is not practical. By defining $\phi(z)$ as

$$\phi(z) = \frac{1}{2\pi i} \int_L \frac{i\tau_{xy}^{-}(t)dt}{t-z} \tag{24}$$

and using (1), (11) and the Plemelj formulae

$$\begin{aligned} &\phi^{+}(x) - \phi^{-}(x) = i\tau_{xy}^{-}(x) \\ &\phi^{+}(x) + \phi^{-}(x) = \frac{1}{\pi i} \int_L \frac{i\tau_{xy}^{-}(t)dt}{t-x} \end{aligned} \tag{25}$$

a slightly different formulation may be obtained as follows:

$$\frac{1}{\pi}\int_L \frac{\tau_{xy}^-(t)dt}{t-x} - \lambda_k \int_{a_k}^{x} \tau_{xy}^-(t)dt = \lambda_k T_{1k} - \frac{\sigma_x^\infty}{2},$$

$$(x\varepsilon L_k,\ L = \sum_1^n L_k) \qquad (26)$$

The index of the singular integral equation (26) is n, and the n arbitrary constants arising from its solution are determined by using the equilibrium conditions (19).

PERIODIC COVER PLATES

First consider (2m+1) identical elastic cover plates with the end points located at

$$a_k = 2kb - a,\ b_k = 2kb + a,\ (k = 0, \mp 1, ---, \mp m) \qquad (27)$$

In this case, the fundamental function X(z) may be expressed as (see, (15))

$$X(z) = [\prod_{-m}^{m} (z-a_k)(z-b_k)]^{-1/2} = [(z+a) \prod_1^m [(z+a^2) - (2kb)^2] \times$$

$$\times (z-a) \prod_1^m [(z-a)^2 - (2kb)^2]]^{-1/2}$$

$$= K[(z+a) \prod_1^m (1 - (\frac{z+a}{2kb})^2)(z-a) \prod_1^m (1 - (\frac{z-a}{2kb})^2)]^{-1/2},$$

$$K = (-1)^m (2b)^{-2m} (m!)^{-2} \qquad (28)$$

Now, going to limit, $m \to \infty$, using the formula [9]

$$\sin\theta = \theta \prod_1^\infty (1 - \frac{\theta^2}{k^2\pi^2}) \qquad (29)$$

and ignoring the multiplicative constant*, we obtain

$$X(z) = [\sin \frac{\pi(z+a)}{2b} \sin \frac{\pi(z-a)}{2b}]^{-1/2}$$

$$= [\sin^2 \frac{\pi z}{2b} - \sin^2 \frac{\pi a}{2b}]^{-1/2} \qquad (30)$$

where the branch which is positive for $z = t$, $a < t < 2b - a$ is taken.

Let h be the thickness and μ_1, κ_1 be the elastic constants of the cover plates. Then defining

$$\lambda = \frac{1+\kappa_1}{1+\kappa} \frac{\mu}{2\mu_1 h_1} \qquad (31)$$

and noting that $\tau_{xy}^-(x) = 0$ for $x\varepsilon L'$, from (1), (13), (10) and (11) we may write

$$\phi^+(x) + \phi^-(x) = \lambda(T_1 + \int_{a_k}^{x} \tau_{xy}^-(s)ds) - \frac{\sigma_x^\infty}{2} = f(x)$$

*The fundamental function X(z), which is the solution of the homogeneous Hilbert problem $\phi^+ + \phi^- = 0$, $t\varepsilon L$; $\phi^+ - \phi^- = 0$, $t\varepsilon L'$, is obviously determinate within an arbitrary multiplicative integral function.

$$x \in L, \quad L = \sum_{-\infty}^{\infty} L_k, \quad L_k = (a_k, b_k)$$

$$\phi^+(x) - \phi^-(x) = -i\tau_{xy}^{\infty}, \quad x \in L' \tag{32}$$

where T_1, T_2 are the loads applied to the ends a_k and b_k, respectively. From the equilibrium of the cover plates and the half plane, it is clear that

$$\int_{a_k}^{b_k} \tau_{xy}^-(t)dt = \int_{-a}^{a} \tau_{xy}^-(t)dt = T_2 - T_1 \tag{33}$$

$$\tau_{xy}^{\infty} = (T_2 - T_1)/2b \tag{34}$$

With the fundamental function X(z) given by (30) the solution of (32) may be expressed as

$$\phi(z) = \frac{X(z)}{2\pi i} \int_L \frac{f(t)dt}{X^+(t)(t-z)} + \frac{X(z)}{2\pi i} \int_{L'} \frac{-i\tau_{xy}^{\infty}}{X^+(t)(t-z)} + P(z)X(z) \tag{35}$$

where P(z) is an integral function. From $\phi \to 0$ as $|z| \to \infty$ it can be shown that $P(z) = 0$.

Using $L = \sum_{-\infty}^{\infty} L_k$, $L_k = (a_k, b_k)$, $a_k = 2kb - a$, $b_k = 2kb + a$, $(k = 0, \mp 1, \ldots)$ and the formulas

$$\frac{1}{\theta} + \sum_{1}^{\infty} \frac{2\theta}{\theta^2 - n^2} = \pi\cot\pi\theta, \quad \sum_{1}^{\infty} \frac{2\theta}{\pi^2\left(\frac{2n-1}{2}\right)^2 - \theta^2} = \tan\theta \tag{36}$$

the infinite series in (35) can be summed and $\phi(z)$ becomes

$$\phi(z) = \frac{X(z)}{4bi} \int_{-a}^{a} \frac{f(t)dt}{X^+(t)\sin\frac{\pi}{2b}(t-z)} + \frac{X(z)}{4bi} \int_{a}^{2b-a} \frac{-i\tau_{xy}^{\infty}dt}{X^+(t)\sin\frac{\pi}{2b}(t-z)} \tag{37}$$

Referring to the definition of f(x) in (32) and evaluating part of the integral, (37) may be written as

$$\begin{aligned}\phi(z) &= \frac{1}{2}\left(\lambda T_1 - \frac{\sigma_x^{\infty}}{2}\right)\left[1 - X(z)\sin\frac{\pi z}{2b}\right] \\ &+ \frac{X(z)}{4bi} \int_{-a}^{a} \frac{\lambda dt}{X^+(t)\sin\frac{\pi}{2b}(t-z)} \int_{-a}^{t} \tau_{xy}^-(s)ds \\ &- \frac{\tau_{xy}^{\infty}X(z)}{4b} \int_{a}^{2b-a} \frac{dt}{X^+(t)\sin\frac{\pi}{2b}(t-z)}\end{aligned} \tag{38}$$

From (10), (20) and (45), after evaluating the definite integral, the stresses at $y = 0$ are found to be

$$\tau_{xy}^{-}(t)\begin{cases} = 0, \quad (a<t<2b-a) \\ = (\sin^2\frac{\pi a}{2b}-\sin^2\frac{\pi t}{2b})^{-1/2}\Big[(\lambda T_1 - \frac{\sigma_x^{\infty}}{2})\sin\frac{\pi t}{2b} \\ \quad + \tau_{xy}^{\infty}\cos\frac{\pi t}{2b} - \frac{\lambda}{2b}\int_{-a}^{a}\frac{(\sin^2\frac{\pi a}{2b}-\sin^2\frac{\pi s}{2b})^{1/2}}{\sin\frac{\pi}{2b}(s-t)}ds \ . \\ \quad \cdot \int_{-a}^{s}\tau_{xy}^{-}(r)dr\Big], \quad (-a<t<a), \quad \tau_{xy}^{\infty} = (T_2-T_1)/2b \end{cases} \qquad (39)$$

$$\sigma_x^{-}(t)\begin{cases} = 2\lambda[T_1 + \int_{-a}^{t}\tau_{xy}^{-}(s)ds], \quad (-a<t<a) \\ = 2\lambda T_1 + (\sin^2\frac{\pi t}{2b}-\sin^2\frac{\pi a}{2b})^{-1/2}\Big[(\sigma_x^{\infty} - 2\lambda T_1)\sin\frac{\pi t}{2b} \\ \quad - 2\tau_{xy}^{\infty}\cos\frac{\pi t}{2b} - \frac{\lambda}{b}\int_{-a}^{a}\frac{(\sin^2\frac{\pi a}{2b}-\sin^2\frac{\pi s}{2b})^{1/2}}{\sin\frac{\pi}{2b}(s-t)}ds \ . \\ \quad \cdot \int_{-a}^{s}\tau_{xy}^{-}(r)dr\Big], \quad (a<t<2b-a) \end{cases} \qquad (40)$$

From (39) and (40) it is not difficult to see that in the immediate vicinity of the plate ends, $x = \mp a$, $y = 0$, the stresses have an integrable singularity shown in (22) and (23). Equation (39) provides an integral equation to determine $\tau_{xy}^{-}(t)$, which may again be reduced to a Fredholm-type equation with a weakly singular kernel. From the view point of numerical analysis, in this problem too, it is advantageous to reformulate the problem in terms of a singular integral equation. For this we note that (26), which is the expression of the boundary condition at $y = 0$, is still valid provided L is taken as $L = \sum_{-\infty}^{\infty} L_k$, $L_k = (2kb-a,\ 2kb+a)$. Taking into account the periodicity of the contact stress, (i.e., $\tau_{xy}^{-}(t) = \tau_{xy}^{-}(t+2kb)$) and using (36), it can easily be shown that

$$\frac{1}{\pi}\int_L \frac{\tau_{xy}^{-}(t)dt}{t-x} = \frac{1}{2b}\int_{-a}^{a}\tau_{xy}^{-}(t)\cot\frac{\pi}{2b}(t-x)dt \qquad (41)$$

Thus from (26) and (41) we obtain

$$\frac{1}{2b}\int_{-a}^{a}\tau_{xy}^{-}(t)\cot\frac{\pi}{2b}(t-x)dt - \lambda\int_{-a}^{x}\tau_{xy}^{-}(t)dt$$
$$= \lambda T_1 - \frac{\sigma_x^{\infty}}{2}, \quad (-a<x<a) \qquad (42)$$

The singular integral equation (42) must be solved under the condition that

$$\int_{-a}^{a}\tau_{xy}^{-}(t)dt = T_2 - T_1 \qquad (43)$$

In passing, we note the following important relations giving the general solution of the singular integral equation with a cotangent kernel:

$$\frac{1}{2b}\int_{-a}^{a} \phi(t)\cot\frac{\pi}{2b}(t-x)dt = h(x) + A_1, \quad (-a<x<a)$$

$$\phi(x) = \left(\sin^2 \frac{\pi a}{2b}-\sin^2 \frac{\pi x}{2b}\right)^{-1/2} \Big[A_1\sin\frac{\pi x}{2b} + A_2\cos\frac{\pi x}{2b}$$

$$- \frac{1}{2b}\int_{-a}^{a} \frac{h(t)dt}{\sin\frac{\pi}{2b}(t-x)} \left(\sin^2 \frac{\pi a}{2b}-\sin^2 \frac{\pi t}{2b}\right)^{1/2}\Big], \quad -a<x<a \tag{44}$$

where A_2 is an arbitrary constant which is determined from

$$\int_{-a}^{a} \phi(x)dx = T \tag{45}$$

T being a specified constant. Equations (44) follow directly from (39) and (42).

SPECIAL CASE: THE INEXTENSIBLE COVER PLATES

In the case of inextensible cover plates $\mu_k = \infty$ and equations (18) and (21) give the solution with $\lambda_k = 0$ $(k = 1,--,n)$. For example, for $n = 1$, $a_1 = -a$, $b_1 = a$, $\sigma_y^\infty = 0 = \tau_{xy}^\infty$, and $T_2 - T_1 = T$ we find

$$\tau_{xy}^-(x) = \begin{cases} \dfrac{1}{(a^2-x^2)^{1/2}}\left(-\dfrac{\sigma_x^\infty}{2} x + \dfrac{T}{\pi}\right), & (-a<x<a) \\ 0, & |x|>a \end{cases}$$

$$\sigma_x^-(x) = \begin{cases} 0, & -a<x<a \\ (x^2-a^2)^{-1/2} \left(\sigma_x^\infty x - \dfrac{2T}{\pi}\right), & |x|>a \end{cases} \tag{46}$$

For $n = 2$, $a_1 = -b$, $b_1 = -a$, $a_2 = a$, $b_2 = b$, $\sigma_y^\infty = 0 = \tau_{xy}^\infty$, $T_{11} = T_{22} = T$, $T_{12} = 0 = T_{21}$, we have (see Figure 2)

$$\tau_{xy}^-(x) = \begin{cases} [(x^2-a^2)(b^2-x^2)]^{-1/2} \left[\dfrac{\sigma_x^\infty}{2} \left(b^2 \dfrac{E(k)}{K(k)} - x^2\right) + \dfrac{bT}{K(k)}\right], & a<|x|<b, \quad k = \left(1-\dfrac{a^2}{b^2}\right)^{1/2} \\ 0, & |x|>b, \ |x|<a \end{cases} \tag{47}$$

$$\sigma_x^-(x) = \begin{cases} 0, & a<|x|<b \\ [(x^2-b^2)(x^2-a^2)]^{-1/2} \left[\sigma_x^\infty \left(x^2-b^2 \dfrac{E(k)}{K(k)}\right) - \dfrac{2bT}{K(k)}\right], & |x|>b, \ |x|<a, \quad k = \left(1-\dfrac{a^2}{b^2}\right)^{1/2} \end{cases}$$

where K(k) and E(k) are complete elliptic integrals of first and second kind, respectively. For the same geometry and the symmetric external loads $T_{11} = 0 = T_{21}$, $T_{12} = T_{22} = T$, we find

$$\tau_{xy}^{-}(x) = \begin{cases} \frac{2}{\pi} Tx[(x^2-a^2)(b^2-x^2)]^{-1/2}, & a<|x|<b \\ 0, & |x|>b, \ |x|<a \end{cases}$$

$$\sigma_x^{-}(x) = \begin{cases} 0, & a<|x|<b \\ -\frac{4}{\pi} Tx[(x^2-a^2)(x^2-b^2)]^{-1/2}, & |x|>b, \ |x|<a \end{cases} \quad (48)$$

Finally, in the case of periodic cover plates $\lambda = 0$ and for $\sigma_y^{\infty} = 0$, $T_2 - T_1 = T$, (39) and (40) give the solution as

$$\tau_{xy}^{-}(x) = \begin{cases} -\left(\sin^2 \frac{\pi a}{2b} - \sin^2 \frac{\pi x}{2b}\right)^{-1/2} \left[\frac{\sigma_x^{\infty}}{2} \sin\frac{\pi x}{2b} - \frac{T}{2b} \cos\frac{\pi x}{2b}\right], & (-a<x<a) \\ 0, & a<|x|<2b-a \end{cases}$$

$$\sigma_x^{-}(x) = \begin{cases} 0, & -a<x<a \\ \left(\sigma_x^{\infty} \sin\frac{\pi x}{2b} - \frac{T}{b} \cos\frac{\pi x}{2b}\right)\left(\sin^2 \frac{\pi x}{2} - \sin^2 \frac{\pi a}{2}\right)^{-1/2}, & a<|x|<2b-a \end{cases} \quad (49)$$

It is not difficult to show that for $a<<b$ (49) reduces to (46), the case of a single cover plate.

From the expressions (46) to (49) it is clearly seen that in the close neighborhood of the plate ends the stresses have the inverse square root singularity as indicated by (22) and (23). For example in the periodic cover plates we have

$$\lim_{t\to a} (t-a)^{1/2} \sigma_x^{-}(t) = \frac{\sqrt{b}}{\sqrt{\pi}} \left(\sigma_x^{\infty}\sqrt{\tan\frac{\pi a}{2b}} - \frac{T}{b}\sqrt{\cot\frac{\pi a}{2b}}\right)$$
$$\lim_{t\to a} (a-t)^{1/2} \tau_{xy}^{-}(t) = -\frac{\sqrt{b}}{\sqrt{\pi}} \left(\frac{\sigma_x^{\infty}}{2}\sqrt{\tan\frac{\pi a}{2b}} - \frac{T}{2b}\sqrt{\cot\frac{\pi a}{2b}}\right) \quad (50)$$

It is interesting to note that as $a\to b$ in periodic cover plates and as $a\to 0$ in two plates, the strength of the stress singularity at $x = a$, $y = 0$ increases indefinitely and the stresses are no longer integrable (i.e., the stresses are proportional to r^{-1}, r being the distance from the singular point). For inextensible cover plates this can easily be seen from (47) and (50). In the general case the proof of the existence of such a singularity may be given by considering the singular behavior of the solution of integral equations (26) and (42). For example for two cover plates the dominant part of (26) which controls the singular behavior of the solution may be expressed as

$$\frac{1}{\pi}\int_a^b \left(\frac{1}{t-x} \mp \frac{1}{t+x}\right) \bar{\tau}_{xy}(t)dt = \begin{cases} 0 \\ -\frac{\sigma_x^\infty}{2}, \quad (a<x<b) \end{cases} \tag{51}$$

where the upper sign in the kernel and the upper right-hand-side represent the case in which $\bar{\tau}_{xy}(t) = \bar{\tau}_{xy}(-t)$, and the lower sign and the lower value represent $\bar{\tau}_{xy}(t) = -\bar{\tau}_{xy}(-t)$. If we now let a = 0 and assume that the function $\bar{\tau}_{xy}(t)$ is of the form

$$\bar{\tau}_{xy}(t) = \frac{f(t)}{t^\alpha(b-t)^\beta}, \quad 0<t<b \tag{52}$$

where f(t) is bounded and $0\leq\alpha\leq 1$, $0\leq\beta\leq 1$, following the procedure of [8, Ch. 4] and using the condition that the integral on the left hand side must be bounded, we obtain the following characteristic equations for the constants α and β:

$$\begin{aligned} &\cot\pi\beta = 0 \\ &\cot\pi\alpha \mp \frac{1}{\sin\pi\alpha} = 0 \end{aligned} \tag{53}$$

In (53) upper and lower signs correspond to that of (51). From (53) it is seen that

$$\begin{aligned} &\beta = 1/2 \\ &\alpha = 0 \text{ for } (-) \text{ sign} \\ &\alpha = 1 \text{ for } (+) \text{ sign} \end{aligned} \tag{54}$$

which clearly verifies the existence of a strong singularity for the anti-symmetric case, and no singularity for the symmetric case at x = a when $a\to 0$.

SOLUTION FOR THE ELASTIC COVER PLATES AND NUMERICAL RESULTS

The numerical solution of the elastic cover plate problems is obtained by applying the method described in [6] to the integral equations (26) and (42). The method, which is extremely simple to apply, is based on the observation that the fundamental function of the singular integral equations (26) and (42) is the weight function of Chebyshev polynomials $T_n(r)$. Thus, after normalization*, by expressing the unknown function $\bar{\tau}_{xy}(r)$ as

$$\bar{\tau}_{xy}(r) = (1-r^2)^{-1/2} \sum_0^\infty A_n T_n(r) \tag{55}$$

the singularity of the integral equation may be removed and the problem may be reduced to an infinite system of linear algebraic equations in the unknown constants A_n. The regularity of the resulting system and the convergence of the computed results have been discussed in [10].

The stress intensity factors at the plate ends are related to the constants A_n through the following simple expression:

*With a change in variable of r = x/a in periodically spaced cover plates, and r = 2x/(b-a) - (b+a)/(b-a) in two cover plates. In the latter case, using symmetry considerations the integral equation is expressed in a<x<b only (see, (51)).

$$k(c) = \lim_{t \to c} \sqrt{2|t-c|}\ \bar{\tau}_{xy}(t) = N \sum_{0}^{\infty} A_n \quad (56)$$

where t = c is the plate end, t is on L and N is a known normalization factor.

For two cover plates and $T_{1k} = 0 = T_{2k}$, (k = 1,2), the results are shown in Figures 2 and 3 and Table I. Figure 2

Table I

Stress Intensity Factors for Two Cover Plates

$\frac{a}{b}$	$\frac{k(a)}{\frac{\sigma_o}{2}\sqrt{b}}$ $\lambda = 0$	$\frac{k(a)}{\frac{\sigma_o}{2}\sqrt{b}}$ $\lambda = 10/6$	$\frac{k(b)}{\frac{\sigma_o}{2}\sqrt{b}}$ $\lambda = 0$	$\frac{k(b)}{\frac{\sigma_o}{2}\sqrt{b}}$ $\lambda = 10/6$
0.0	∞	∞	-1.0	
0.01	1.6685	0.6460	-0.8331	-0.3640
0.05	1.0151	0.4394	-0.7718	-0.3536
0.1	0.8419	0.3864	-0.7287	-0.3450
0.3		0.3232		-0.3168
0.5	0.5088	0.2860	-0.5062	-0.2840
0.8		0.2024		-0.2024
1.0	0	0	0	0

Table II

Stress Intensity Factor $\frac{k(a)}{\frac{\sigma_o}{2}\sqrt{2}}$ for Periodic Cover Plates

$\frac{b}{a}$ \ λ	0	1/3	1/2.1	1/1.5	1/0.9
1.0	∞	∞	∞	∞	∞
1.01		3.421	2.875	2.383	1.728
1.05	3.007	1.905	1.664	1.432	1.098
1.10	2.202	1.516	1.348	1.181	0.9324
1.50	1.289	1.008	0.9286	0.8444	0.7073
2.0	1.129	0.9119	0.8476	0.7782	0.6621
5.0	1.016	0.8415	0.7878	0.7288	0.6280
∞	1.0	0.8334	0.7818	0.7250	0.6272

and the table show the stress intensity factor ratios at x = a, and x = b as functions of a/b for λ = 0 and λ = 10/6. Note that the stress intensity factors in elastic cover plates are considerably smaller than that for the inextensible plates*. Same trend is observed if we compare the contact stresses in the two cases. This is shown in Figure 3.

*λ = 10/6 approximately corresponds to boron-epoxy composite cover plate on aluminum with (h/b) = 0.1.

The results for periodically spaced inextensible and elastic cover plates are shown in Table II and Figure 4. The results for a single cover plate are also given as a limiting case for $b/a \to \infty$. From Figure 4 it is clear that the interaction between the cover plates becomes stronger as b/a decreases. In the case of elastic cover plates, there is very little interaction for $(b/a) > 2$. On the other hand, the ratio of k(a) in periodic cover plates to that in a single plate goes to infinity as $(b/a) \to 1$.

REFERENCES

[1] H. Bufler, "Scheibe mit endlicher, elastischer Versteifung", VDI - Forschungsheft 485 (1961).

[2] N. Kh. Arutiunian, "Contact Problem for a Half Plane with Elastic Reinforcement", J. Appl. Math. Mech. (PMM), Vol. 32, pp. 652-665 (1968).

[3] N. Kh. Arutiunian, "Periodic Contact Problem for a Half-Plane with Elastic Laps", J. Appl. Math. Mech. (PMM), Vol. 33, pp. 791-819 (1969).

[4] E. Carafoli, "Tragflügeltheorie", VEB Verlag Technik, Berlin (1954).

[5] H. Bufler, "Einige strenge Lösungen für den Spannungszustand in ebenen Verbundkörpern", Z. angew.Math. Mech. Vol. 39, pp. 218-236 (1959).

[6] F. Erdogan, "Approximate Solutions of Systems of Singular Integral Equations", SIAM J. Appl. Math. Vol. 17, pp. 1041-1059 (1959).

[7] N. I. Muskhelishvili, "Some Basic Problems of the Mathematical Theory of Elasticity", P. Noordhoff, Groningen, Holland (1953).

[8] N. I. Muskhelishvili, "Singular Integral Equations", P. Noordhoff, Groningen, Holland (1953).

[9] I. S. Gradshteyn and I. M. Ryzhik, "Table of Integrals, Series and Products", Academic Press, New York (1965).

[10] F. Erdogan and G. D. Gupta, "The Problem of Elastic Cover Plate", NASA Report, Lehigh University (1970).

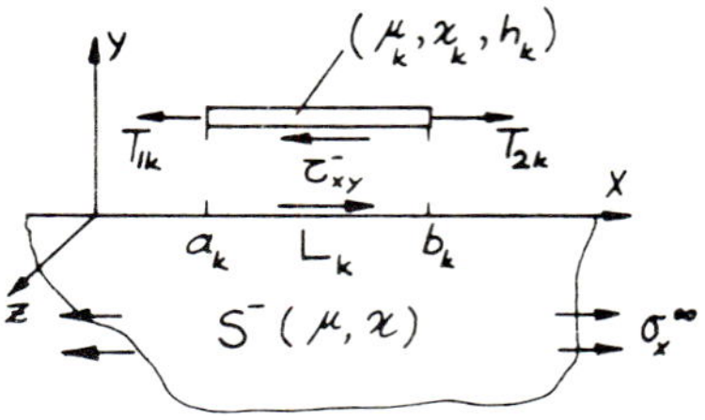

FIGURE 1. GEOMETRY AND NOTATION FOR THE COVER PLATES

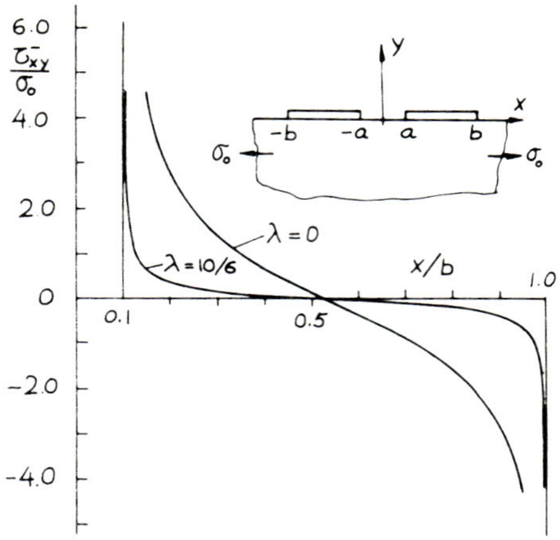

FIGURE 3. DISTRIBUTION OF CONTACT SHEAR FOR TWO COVER PLATES, A/B = 0.1

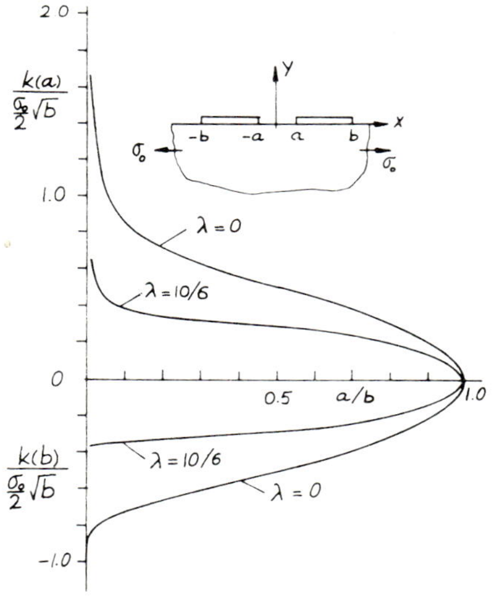

FIGURE 2. STRESS INTENSITY FACTORS FOR TWO COVER PLATES

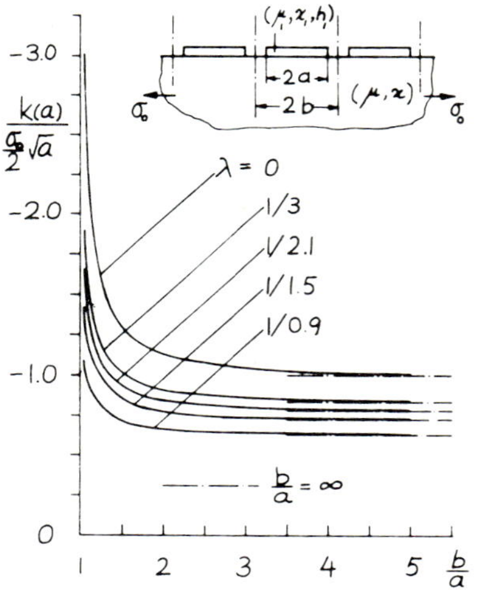

FIGURE 4. STRESS INTENSITY FACTOR FOR PERIODIC COVER PLATES

Developments in Mechanics, Vol. 6. Proceedings of the 12th Midwestern Mechanics Conference: 58

Optimal Design of Trusses under Multiple Constraint Conditions of Mechanics

JASBIR ARORA
UNIVERSITY OF IOWA

EDWARD HAUG, JR.
U.S. ARMY WEAPONS COMMAND

KWAN RIM
UNIVERSITY OF IOWA

ABSTRACT

A steepest-descent method with constraint error compensation is developed for the optimal design of elastic structures. The procedure amalgamates the displacement method of structural analysis, design perturbation relations, and a new form of gradient projection technique to solve a general class of structural design problems. Multiple loading conditions for the structure are treated without difficulty. The structure is required to meet frequency, buckling, stress and displacement requirements simultaneously. Design constraints may be violated by an initial estimate and during the iterative optimization process, but a correction for these violations is made as the analysis progresses. The philosophy of the method is to perturb all governing equations of the problem about an estimated optimum design and retain only the first order terms in them. A step-size constraint is introduced and the optimum design improvement within this constraint is determined explicitly through application of the Kuhn-Tucker necessary conditions. An iterative process is presented, with convergence criterion, for routine solution of a large class of problems. The design method has been developed to be easily applicable to general problems without major computer code re-write. Several problems are solved with the same general code to illustrate the generality and efficiency of the technique.

INTRODUCTION

The posibility of obtaining analytic solutions to optimal structural design problems seems to be remote at this time; therefore, one has to resort to some sort of numerical procedure of obtaining solutions to the problems within some acceptable range of errors. The amount of computation involved is so great that optimal design of commonly used structures has been a formidable task in the past. However, with the advent of high speed digital computers, the subject of optimal design of structures has received a great deal of attention in recent years. It is now possible to handle optimal design of a large class of structures quite effectively. Recent works of Johnson and Brotton [7]*, Venkayya, et. al. [11], Dobbs and Felton [2] and others have greatly renewed an interest in optimal design of trusses.

The complexity of optimal design of trusses depends upon 1) whether the structure is determinate or indeterminate, 2) the type of constraints to be considered, and 3) the type and number of loading conditions for the structure. Optimum design of determinate trusses with only strength constraints is very straightforward. However, in the presence of deflection and frequency constraints, the problem contains nonlinear constraints on member cross-sectional areas. In the case of redundant trusses, a solution for internal forces may be obtained only through a simultaneous consideration of equilibrium and compatability relationships; therefore, overall force distributions are member-size dependent. Thus, one again encounters nonlinear constraints which relate member stresses and the cross-sectional areas. A truss subjected to a moving load may be reduced to the case of multiple critical loading conditions, and again nonlinear programming techniques may be used to solve the problem.

In the present paper linear elastic trusses of fixed geometry, subjected to a number of loading conditions are considered and these are required to satisfy stress, displacement, buckling and frequency constraints. This class of structures is encountered quite frequently in practical situations such as design of industrial buildings, transmission towers, bridges, cooling towers, spacecraft structures and antenas. In all these structures, it is very desirable that the structure should meet stress, deflection, buckling, and frequency requirements simultaneously and be of a minimum weight. In recent years, the optimal design of structures has proceeded on two fronts -- design of structures to meet strength requirements and design to meet frequency requirements. In many cases, optimum distribution of a given amount of material is desired so that the lowest natural frequency of the structure is a maximum. In many other cases, the structure is optimized to satisfy strength and deflection requirements and is then simply checked for resonant frequency. In the present paper, all these considerations are integrated into one formulation and a structure of minimum weight will be sought.

FORMULATION OF PROBLEM

The problem of optimal design of an elastic truss may be stated as follows: the total volume or weight of the truss is to be minimized such that the truss satisfies equations of equilibrium and compatability, stress constraints, displacement constraints, frequency requirements and restrictions on the cross-sectional areas of its members. Let

$$b = [b_1, b_2, \dots, b_m]^T \tag{1}$$

denote a vector whose i^{th} component represents the cross-sectional area of the ith member and superscript T denote matrix transpose. The vector b will be called the design variable vector, since by assigning some value to each component, the truss is completely described. Also let

$$q = [q_1, q_2, \dots, q_n]^T \tag{2}$$

and

$$v = [v_1, v_2, \dots, v_n]^T \tag{3}$$

where q is a vector denoting the nodal displacements of the truss and v is an eigenvector corresponding to a mode of vibration. The vectors q and v will be called the state variable vectors since they describe the state of the truss. These vectors are related to the design variable vector through the governing equations of structural analysis. These equations have the form

*Number in brackets refers to Bibliography.

$$K(b)\ q = Q(b) \tag{4}$$

and

$$K(b)\ v = \zeta\ M(b)\ v \tag{5}$$

where

$$K(b) = [K_{ij}\ (b)]_{nxn} \tag{6}$$

is the stiffness matrix for the structure,

$$M(b) = [M_{ij}\ (b)]_{nxn} \tag{7}$$

is the mass matrix for the structure,

$$Q(b) = [Q_1(b),\ Q_2(b),\ .\ .\ .\ ,\ Q_n(b)]^T \tag{8}$$

is the external load vector and ζ is an eigenvalue related to natural frequency. The objective function for the problem, which is weight of the truss, may be written as

$$J = \sum_{i=1}^{m} \rho_i L_i b_i \tag{9}$$

where ρ_i and L_i are the material density and the length of member i respectively. The objective function is to be minimized subject to following constraints:

$$\zeta \geq \zeta_o \tag{10}$$

$$\sigma_i \leq \sigma_i^c \tag{11}$$

$$\sigma_i \leq \sigma_i^b \tag{12}$$

$$q_j \leq q_j^a \tag{13}$$

and

$$b_i^L \leq b_i \leq b_i^U \tag{14}$$

where ζ_o is a lower limit on natural frequency, σ_i^c and σ_i^b are the critical direct stress and the critical buckling stress for member i respectively, q_j^a is the allowable j^{th} component of displacement and b_i^L and b_i^U are lower and upper bounds on the ith design variable respectively. These constraints may be represented in a compact form:

$$\phi(b,q,\zeta) \leq 0\ , \tag{15}$$

where $\phi(b,q,\zeta)$ is a vector function. With this formulation, many frequency constraints can be included to assure the first few frequencies are outside undesirable bands. The critical buckling stress for the ith member is found from the Euler buckling load P_i^c which is given by $\pi^2 E_i I_i / L_i^2$, where I_i is the moment of inertia of the ith member. It is assumed that the moment of inertia of the cross-section of a number can be written as $I_i = \alpha_i b_i^2$, where α_i is a constant depending upon the cross-sectional geometry of the ith member. The optimal design problem is now reduced to that of finding b, q, v and ζ which satisfy Eqs. 4, 5 and 15 and which minimize J of Eq. 9.

Since analytic methods of finding a solution to the above problem are unwieldy, a numerical procedure will be developed which will give an improved design at each iteration and ultimately converge to an optimum design within some acceptable range of errors. Optimal design problems are generally nonconvex; therefore, a local minimum is obtained. Usually, widely separated starting values are chosen; and, if the method converges to the same point each time, that point is taken as the global minimum. For a detailed development of the method presented, the reader is referred to the previous works [5 & 17].

The philosophy of the method is to make an engineering estimate of the design variable vector $b = b^{(0)}$ which is not required to be feasible and then find changes, δb, in $b^{(0)}$ such that $b^{(1)} = b^{(0)} + \delta b$ is an improved design in some sense. Once an estimate $b = b^{(0)}$ is made, the corresponding value of $q^{(0)}$ can be solved from Eq. 4. Likewise the lowest eigenvalue $\zeta^{(0)}$ and the normalized eigenvector $v^{(0)}$ of Eq. 5 can be obtained for $b = b^{(0)}$. It is shown that δb causes small changes $\delta\zeta$ and δv in ζ and v respectively [5]. The following analysis applies to any iteration of the optimal design procedure; therefore, superscript (0) on the parameters of the arguments of various functions in perturbation equations will be omitted.

The linear expansion of Eq. 4 about nominal values of q and b yields the following relationship between δb and δq:

$$K(b)\,\delta q + \frac{\partial}{\partial b}[K(b)q]\,\delta b = \frac{\partial Q}{\partial b}(b)\,\delta b \tag{16}$$

where the vector calculus notation $\frac{\partial f}{\partial b} = [\frac{\partial f}{\partial b_1}, \ldots, \frac{\partial f}{\partial b_m}]$, etc., is used and q and b are treated as indepencent variables. Likewise, from the first order expansion of Eq. 5 and use of symmetry of K(b) and M(b), one obtains [5],

$$\delta\zeta = \frac{1}{v^T M(b) v}\left[v^T\{\frac{\partial}{\partial b}[K(b)v] - \zeta\frac{\partial}{\partial b}[M(b)v]\}\right]\delta b \tag{17}$$

Similarly, to first order approximation, the change δJ in J is given by

$$\delta J = \ell^{J^T}\,\delta b \tag{18}$$

where

$$\ell^J = \frac{\partial J}{\partial b}^T \tag{19}$$

and the change $\delta\phi$ in ϕ is given by

$$\delta\phi = \frac{\partial\phi}{\partial b}(b, q, \zeta)\,\delta b + \frac{\partial\phi}{\partial q}(b, q, \zeta)\delta q + \frac{\partial\phi}{\partial\zeta}(b, q, \zeta)\delta\zeta \tag{20}$$

where $\phi = [\phi_1(b, q, \zeta), \ldots, \phi_r(b, q, \zeta)]^T$

The validity of all these linear expansions hinges on the fact that δb should be small. To ensure this, it will be required that

$$\delta b^T W \delta b = k^2 \tag{21}$$

where k is a small number and W is a positive definite weighting matrix.

The reduced optimization problem may now be stated as follows: Find δb and δq which satisfy Eqs. 16 & 21, reduce any errors which may exist in Eq. 15 and reduce δJ of Eq. 18 as much as possible. To correct errors and prevent violations in constraint Eq. 15, define

$$\tilde{\phi} = \left[\phi_i(b, q, \zeta), \text{ for each i such that } \phi_i(b, q, \zeta) \geq 0\right] \tag{22}$$

i.e., $\tilde{\phi}$ is a vector of all constraint functions ϕ_i which violate or make Eq. 15 an equality. Violations in the constraints are corrected by demanding $\delta\tilde{\phi} \leq \Delta\tilde{\phi}$, where $\Delta\tilde{\phi}$ is a desired change in the constraint function, or more explicitly,

$$\frac{\partial\tilde{\phi}}{\partial b}(b, q, \zeta)\delta b + \frac{\partial\tilde{\phi}}{\partial q}(b, q, \zeta)\delta q + \frac{\partial\tilde{\phi}}{\partial\zeta}(b, q, \zeta)\delta\zeta \leq \Delta\tilde{\phi} \tag{23}$$

Now in order to eliminate the explicit dependence on δq in the reduced problem, define $\lambda^{\tilde{\phi}}$ to be a solution of

$$K^T(b)\,\lambda^{\tilde{\phi}} = \frac{\partial\tilde{\phi}}{\partial q}^T(b, q, \zeta) \tag{24}$$

where $\lambda^{\tilde\phi}$ is a matrix with as many columns as there are entries in $\tilde\phi$. Taking transpose of Eq. 24, postmultiplying it by δq and using Eq. 16, one obtains

$$\frac{\partial\tilde\phi}{\partial q}(b, q, \zeta)\delta q = \lambda^{\tilde\phi^T} K(b)\delta q = \lambda^{\tilde\phi T}\left[\frac{\partial Q}{\partial b}(b) - \frac{\partial}{\partial b}\{K(b)q\}\right]\delta b$$

Substituting this relation and $\delta\zeta$ from Eq. 17 into Eq. 23, one obtains

$$\delta\tilde\phi = \Lambda^T \delta b \tag{25}$$

in which

$$\Lambda = \frac{\partial\tilde\phi^T}{\partial b}(b, q, \zeta) + \left[\frac{\partial Q}{\partial b}(b) - \frac{\partial}{\partial b}\{K(b)q\}\right]^T \lambda^{\tilde\phi} + \frac{1}{v^T M(b)v}\left[\frac{\partial}{\partial b}\{K(b)v\} - \zeta\frac{\partial}{\partial b}\{M(b)v\}\right]^T v\,\frac{\partial\tilde\phi^T}{\partial\zeta} \tag{26}$$

In terms of this notation, Eq. 23 becomes

$$\Lambda^T\delta b \leq \Delta\tilde\phi \tag{27}$$

It should be noted that inequality (27) is simply a linear approximation of the requested changes in all inequality constraint functions which are violated or are just satisfied as equalities. It is possible that the rows of Λ may be linearly dependent. In such cases they can always be made linearly independent by deleting the linearly dependent constraints.

The reduced problem may be re-summarized as follows: Choose δb so as to minimize δJ of Eq. (18) subject to the constraints Eq. (21) and inequality (27). It may be noted that except for stepsize constraint Eq. 21, the reduced problem is linear. If linear constraints on stepsize had been utilized, a linear programming problem would have resulted. As will be shown below, however, careful application of the Kuhn-Tucker necessary conditions to the above problem yields a closed form solution for δb which provides a convergence criterion. In addition, the necessity for implementing a linear programming code is avoided.

The Kuhn-Tucker conditions of nonlinear programming may now be applied to solve this reduced problem. Accordingly it is necessary that there exists a scalar multiplier $\nu > 0$ and a vector of multipliers μ with the same number of elements as Λ has columns. Furthermore, it is required that $\mu_i \geq 0$ for each i, and $\mu_i(\Lambda_i^T \delta b - \Delta\tilde\phi_i) = 0$ for all i. These multipliers and the optimum δb must satisfy

$$\frac{\partial}{\partial(\delta b)}[H] = 0 \tag{28}$$

where $$H = \ell^{J^T}\delta b + \mu^T(\Lambda^T\delta b - \Delta\tilde\phi) + \nu(\delta b^T W\delta b - k^2)\ .$$

From Eq. 28 one has $\ell^J + \Lambda\mu + 2\nu W\delta b = 0$, or solving for δb,

$$\delta b = -\frac{1}{2\nu}W^{-1}(\ell^J + \Lambda\mu) \tag{29}$$

Substitution of δb from Eq. (29) into Eq. (21) and $\mu_i(\Lambda_i^T\delta b - \Delta\tilde\phi_i) = 0$ yield the proper number of equations for the unknowns μ_i and ν. However, the latter equations are nonlinear and difficult to use. Since only small changes in δb are allowed, one might expect that once a constraint function has reached its allowable limit, or become "tight", that it will stay "tight" for a number of design changes. The approach taken here is to assume this is the case, so that equality in Eq. (27) will hold. A solution is then easily obtained for μ and ν. The validity of the assumption is then checked by examining the algebraic signs of the μ_i. If they are all nonnegative the assumption is correct. If some $\mu_i < 0$ then one concludes that the assumption is incorrect and that the ϕ_i constraint function will become strictly satisfied, $\tilde{\phi}_i < 0$. This function is then removed from the ϕ vector and a recomputation of μ and ν is made, requiring only minimal calculation.

Now, substituting δb from Eq. 29 into inequality 27 and demanding equality, one obtains $-\frac{1}{2\nu}\Lambda^T W^{-1}(\ell^J + \Lambda\mu) = \Delta\tilde{\phi}$, or $B\mu = -2\nu\Delta\tilde{\phi} - \Lambda^T W^{-1}\ell^J$, where $B = \Lambda^T W^{-1}\Lambda$. As noted earlier, rows of Λ can be made linearly independent. Consequently Λ^T has full row rank and matrix B is non-singular. Therefore, from the above expression one obtains

$$\mu = -B^{-1}[2\nu\Delta\tilde{\phi} + \Lambda^T W^{-1}\ell^J] \qquad (30)$$

Substitution of μ from Eq. 30 into Eq. 29 yields

$$\delta b = -\frac{1}{2\nu} W^{-1}[I - \Lambda B^{-1}\Lambda^T W^{-1}]\,\ell^J + W^{-1}\Lambda B^{-1}\Delta\tilde{\phi} \qquad (31)$$

If all the constraints are strictly satisfied then $\delta b = \frac{-1}{2\nu} W^{-1}\ell^J$. The magnitude of $\nu > 0$ may now be determined by substituting Eq. 31 into Eq. 21. However, k must still be chosen in Eq. 21, so it seems logical to choose ν directly. Defining $\eta = 1/2\nu$,

$$\delta b^1 = W^{-1}[I - \Lambda B^{-1}\Lambda^T W^{-1}]\ell^J \qquad (32)$$

and

$$\delta b^2 = W^{-1}\Lambda B^{-1}\Delta\tilde{\phi} \qquad (33)$$

the change δb of Eq. 29 may be written as

$$\delta b = -\eta\delta b^1 + \delta b^2 \qquad (34)$$

where $\eta > 0$ is to be chosen as a step size. This is an arbitrarily small quantity which must be chosen by the designer. It is shown in [5] that δb^1 of Eq. 32 must vanish at the optimum; and, since all the constraints must also be satisfied at the minimum point, it follows from Eq. 34 that δb must also vanish at the optimum. Hence it may be concluded that the value of η should not significantly influence convergence once b is near the optimum. However, it has been observed that a very large step size can cause oscillations about the minimum point and a very small step may slow down the rate of convergence. A scheme of choosing this step size has been tried and has worked well. When all the constraints are satisfied, so that $\Delta\tilde{\phi} \approx 0$, η is chosen so as to yield a few percent reduction in J. One first chooses the desired change $\Delta J < 0$ in J. Then assuming $\Delta\tilde{\phi} = 0$, ΔJ is given by $\Delta J = -\eta\ell^{J^T}\delta b^1$, from which one obtains

$$\eta = -\Delta J/\ell^{J^T}\delta b^1 \qquad (35)$$

The choice of step size η is still an art. An initial reduction of 2 to 10 percent in Eq. 35 has been sought. In problems with only stress constraints a large step size was used and good convergence was obtained. On the other hand, in the presence of other constraints, a smaller step size was required.

STEEPEST-DESCENT COMPUTATIONAL ALGORITHM

The above procedure of successively improving the best available design can be put in a computational algorithm as follows:

Step 1. Obtain the best available engineering estimate of the optimum design variable vector $b^{(j)}$ and solve for $q^{(j)}$ from Eq. 4. Also compute the lowest eigenvalue $\zeta^{(j)}$ and the corresponding eigenvector $v^{(j)}$ from Eq. 5.

Step 2. Check the constraint Eq. 15 and form the constraint function $\tilde{\phi}$ of Eq. 22. Also choose the constraint error correction vector, $\Delta\tilde{\phi}$ of Eq. 27.

Step 3. Compute various matrices such as, $\frac{\partial\tilde{\phi}}{\partial q}$, $\frac{\partial\tilde{\phi}}{\partial b}$, $\frac{\partial\tilde{\phi}}{\partial \zeta}$, $\frac{\partial}{\partial b}$ [K(b)q], etc.

Step 4. Solve for $\lambda^{\tilde{\phi}}$ from Eq. 24.

Step 5. Assemble matrices ℓ^J and Λ of Eqs. 19 & 26 respectively.

Step 6. Choose $\eta = \frac{1}{2\nu} > 0$ and calculate the Lagrange multiplier vector μ from Eq. 30.

Step 7. Check the algebraic sign of each component of μ. If some components of μ are negative, remove the corresponding columns from Λ matrix. Also delete the corresponding elements from the vector $\Delta\tilde{\phi}$ and return to Step 6.

Step 8. Compute δb from Eq. 29 and put

$$b^{(j+1)} = b^{(j)} + \delta b$$

Step 9. Make a check for convergence. If all constraints are satisfied and $||\delta b^1||$ is sufficiently small, terminate the process. Otherwise return to Step 1 with $b^{(j+1)}$ as the best available estimate of the optimum design.

EXAMPLE PROBLEMS AND DISCUSSION

Several trusses were optimized by applying the procedure presented in this paper but due to page limitations results for only two typical structures are presented here. For purposes of comparison first example is taken from [11], and second from [7]. A computer program was written in FORTRAN IV based on the algorithm stated previously. The computations were performed on the University of Iowa IBM 360/65 computer. All structures were optimized with stress, displacement, buckling and frequency constraints. No particular difficulty was encountered in including the frequency and buckling constraints.

Several comments on computational art used in solution of these problems are in order. First, if a feasible design was chosen initially, large steps could be taken until one or more constraints were violated at which time the step size was reduced. Second, it was noted that as the optimum was approached, oscillation occurred. By monitoring the dot product, $\delta b^{(j)^T} \delta b^{(j-1)}$, oscillations were sensed when negative values of the dot product occurred. Thus, step size, η, was divided by two when negative values of the dot product occurred on two successive iterations. Finally, the most effective method of adjusting step size was to monitor successive reductions in cost function after feasibility had occurred. Once insignificant reductions occurred, the step size was reduced to obtain finer convergence.

The Power method used to compute the smallest eigenvalue also worked quite well. At every iteration, the starting value for the eigenvector was taken from the previous iteration which manifested a very rapid rate of convergence. An accuracy of 0.1 percent in each component of the eigenvector was used to compute the new eigenvector. The stiffness matrix for the structure was inverted by the Gauss-Jordan elimination procedure. Another comment that is appropriate here concerns the sign check on the Lagrange multiplier vector μ, called for in Step 7 of the computational algorithm. The algebraic sign of each component of the Lagrange multiplier vector μ was checked at each iteration; and, if some of the components were negative, then the matrix Λ and the vector $\Delta\tilde{\phi}$ were adjusted accordingly. This procedure was particularly useful whenever there were redundant constraint violations. In some cases, the number of constraints violated were more than the number of design variables of the problem, yielding a singular coefficient of μ. In such cases numerical noise yielded a solution such that some of the components of the vector μ were always negative, indicating that the corresponding constraints would be strictly satisfied in the next iteration. The number of constraints with positive components of μ were always less than or equal to the number of design variables of the problem. This procedure of adjusting the constraint set worked very well and it minimized the possibility of divergence of the algorithm.

EXAMPLE 1. TRANSMISSION TOWER

Figure 1 (a) shows the geometry and the dimensions of the Tower. This example is taken from [11]. In this paper, the cross-sectional area of each member is treated as an unknown design variable and the results obtained are given in Table I. The tower was designed first with only stress constraints. The final design weight in this case was 91.13 pounds with a computational time of 38 seconds for 12 iterations. The final design weight reported in [11] was 91.14 pounds with a computational time of 9 seconds for five cycles. The values of final design variables compare quite well with those in [11]. At the final design point all constraints were satisfied within 0.006 percent.

The tower was also designed with stress and displacement constraints and, finally, with all the constraints included. The design weight in the first case was 546.18 pounds with a computational time of 47 seconds for 17 iterations and the maximum constraint violation was 0.00011 percent. The comparable design weight reported in [11] was 555.11 pounds with a computational time of 24 seconds for seven cycles. For a design with all the constraints included, the final weight was 590.32 pounds with a computational time of 129 seconds for 36 iterations and the maximum violation of constraint was 0.028 percent. Figure 1 (b) shows variation of the cost function with respect to the iteration number for the last two cases of this problem. It may be noted that for practical purposes, convergence was obtained in only six iterations.

EXAMPLE 2. 47-BAR PLANE TRUSS

The schematic diagram of the structure with dimensions is shown in Fig. 2 (a). This example is taken from [7] and is optimized for a single loading condition. The design information and the results are shown in Table II. In order to compare the results with [7], the truss is first optimized with stress constraints only. The final design weight was 2,993.37 pounds with a computational time of 115 seconds for 17 iterations. At this point the stress in member 18 was violated by 0.24 percent and all other violations were less than 0.035 percent. Another feasible design occurred at ninth iteration for which the design weight was 2,998.88 pounds and maximum constraint violation was 0.10 percent for stress in seventh member and all other violations were less than 0.016 percent. The final weight reported in [7] was 3,328.5 pounds which is considerably higher than the one reported herein. This may be attributed to the fact that in [7] the members are divided into eight groups so that there are only eight unknown independent design variables, whereas in this paper area of cross-section of each member is treated as an unknown design variable.

The truss was also designed by imposing all the constraints. The starting point, stress limits, upper and lower bounds on the areas are same as those used in [7]. It may be noted that members 21, 22, 23 and 24 had the same upper and lower bounds on areas. The final design weight was 3,771.0 pounds with the computational time of 166 seconds for 24 iterations. The maximum violation of the constraint was 0.27 percent on stress for member 11. Figure 2 (b) shows variation of the cost function with respect to the number of iteration for both the cases. It may be noted that for practical purposes, convergence occurred in approximately six iterations.

CONCLUDING REMARKS

A steepest descent method with constraint error compensation for finding the optimal design of elastic trusses is presented. Although the method is applied to the case of trusses only, it is general enough to apply to frames and other structures. For the general development of the method the reader is referred to [5 and 17]. An application of the method to framed structures is now in progress and the results will be reported in the near future.

The method is relatively automatic in the sense that, for the computer program developed, the input data given is the only pertinent design information required for solution of the problem. All the necessary matrices and their derivatives are automatically generated in the computer. Any person with a reasonable knowledge of FORTRAN language should be able to handle the program without any difficulty.

The method is developed to meet the displacement, strength and frequency requirements on a structure, simultaneously. Hence, it should be very useful in practical applications, particularly in the aerospace and aircraft industries where most of the structures must be designed to meet frequency and strength requirements simultaneously.

Solutions of the example problems presented in this paper were obtained very readily and required no modification of the general computer code developed to implement the computing algorithm presented in this paper. For a detailed discussion of this code the reader is referred to [17]. In addition to the ease of use of this technique, the authors have found that the computing times required were quite competative with more specialized techniques. Refinement of the structural analysis techniques to utilize iterative re-analysis methods are expected to speed up computation significantly. This improvement as well as allowing geometric variables to be altered in design are the subject of continuing work.

ACKNOWLEDGEMENTS

This work was supported by the Rock Island Arsenal, Rock Island, Illinois, under Research Contract No. DAAF01-70-M-A378.

BIBLIOGRAPHY

[1] Corcoran, P.J., "Configurational Optimization of Structures," International Journal of Mechanical Sciences, Vol. 12, No. 5, May, 1970, pp. 459-462.

[2] Dobbs, W.W. and Felton, L.P., "Optimization of Truss Geometry," Journal of the Structural Division, Proceedings of the American Society of Civil Engineers, Vol. 95, No. ST10, October, 1969, pp. 2105-2118.

[3] Felton, L.P., and Hofmeister, L.D., "Optimized Components in Truss Synthesis," AIAA Journal, Vol. 6, No. 12, December, 1968, pp. 2434-2436.

[4] Gellatly, R.A. and Gallaghar, R.H., "A Prodecure for Automated Minimum Weight Structural Design," Aeronautical Quarterly, Vol. 17, No. 3, August, 1966, pp. 216-230 and Vol. 17, No. 4, November, 1966, pp. 332-342.

[5] Haug, E.J., Jr., "Optimal Design of Mechanical Systems," Unpublished class notes on Optimization of Structural Systems, University of Iowa, 1968-69.

[6] Fox, R.L., "Constraint Surface Normals for Structural Synthesis Techniques," AIAA Journal, Vol. 3, No. 8, August, 1965, pp. 1517-1518.

[7] Johnson, D. and Brotton, D.M., "Optimum Elastic Design of Redundant Trusses," Journal of the Structural Division, Proceedings of the American Society of Civil Engineers, Vol. 95, No. ST12, December, 1969, pp. 2589-2610.

[8] Pope, G.G., "The Design of Optimum Structures of Specified Basic Configuration," International Journal of Mechanical Sciences, Vol. 10, No. 4, April, 1968, pp. 251-263.

[9] Sheu, C.Y. and Prager, W., "Recent Developments in Optimal Structural Design," Applied Mechanics Reviews, Vol. 21, No. 10, October, 1968, pp. 985-992.

[10] Sved, G. and Ginos, Z., "Structural Optimization under Multiple Loading," International Journal of Mechanical Sciences, Vol. 10, No. 10, October, 1968.

[11] Venkayya, V.B., Khot, N.S. and Reddy, V.S., "Energy Distribution in an Optimum Structural Design," Technical Report AFFDL-TR-68-158, Wright-Patterson Air Force Base, Ohio 45433, March, 1969.

[12] Young, W.J., Jr. and Christiansen, H.N., "Synthesis of a Space Truss based on Dynamic Criteria," Journal of the Structural Division, Proceedings of the American Society of Civil Engineers, Vol. 92, No. ST6, December, 1966, pp. 425-442.

[13] Schmit, L.A. and Morrow, W.M., "Structural Synthesis with Buckling Constraints," Journal of the Structural Division, Proceedings of the American Society of Civil Engineers, Vol. 89, No. ST2, April, 1963, pp. 107-126.

[14] Schmit, L.A., Jr., and Mallet, R.H., "Structural Synthesis and Design Parameter Hierarchy," Journal of the Structural Division, Proceedings of the American Society of Civil Engineers, Vol. 89, No. ST4, August, 1963, pp. 269-299.

[15] Zarghamee, M.S., "Optimum Frequency Structures," AIAA Journal, Vol. 6, No. 4, April, 1968, pp. 749-50.

[16] Schmit, L.A., "Structural Design by Systematic Synthesis," Proceedings of the Second Conference on Electronic Computation, American Society of Civil Engineers, September, 1968.

[17] Arora, J.S., and Rim, K., "Optimal Design of Trusses Under Multiple Constraint Conditions," Technical Report, No. 3, (DAAF 01-70-M-A378), University of Iowa, Iowa City, October, 1970.

[18] Fox, R.L., and Kapoor, M.P., "Structural Optimization in the Dynamics Response Regime: A Computational Approach," AIAA Journal, Vol. 8, No. 10, October, 1970, pp. 1798-1804.

TABLE I EXAMPLE 1 TRANSMISSION TOWER

Design Information: For each member of the structure, the modulus of elasticity, E_i, the specific weight, ρ_i, the constant, α_i and the stress limits are 10^4 kips/ sq. in., 0.10 lbs/cu. in., 1.0 and ± 40.0 kips/sq. in. respectively. The lower limit on the area of cross-section of each member is 0.10 sq. in. for the case with stress constraints only and 0.01 sq. in. for other cases. There is no upper limit on the member sizes. The resonant frequency for the structure is 173.92 cps and the displacement limits are 0.35 in. on all nodes and in all directions. There are six loading conditions and they are as follows: (all loads are in kips)

Load Cond.	Node	Direction of Load			Load Cond.	Node	Direction of Load		
		x	y	z			x	y	z
1	1	1.0	10.0	-5.0	2	1	0	10.0	-5.0
	2	0	10.0	-5.0		2	-1.0	10.0	-5.0
	3	0.5	0	0		4	-0.5	0	0
	6	0.5	0	0		5	-0.5	0	0
3	1	1.0	-10.0	-5.0	4	1	0	-10.0	-5.0
	2	0	-10.0	-5.0		2	-1.0	-10.0	-5.0
	3	0.5	0	0		4	-0.5	0	0
	6	0.5	0	0		5	-0.5	0	0
5	1	0	20.0	-5.0	6	1	0	-20.0	-5.0
	2	0	-20.0	-5.0		2	0	20.0	-5.0

Output:

	With Stress Constraints only		With Stress and displacement constraints		With all constraints	
Element No.	Starting Values in sq. in.	Final Values in sq. in.	Starting Values in sq. in.	Final Values in sq. in.	Starting Values in sq. in.	Final Values in sq. in.
1	0.200	0.100	1.000	0.010	0.500	0.010
2	0.500	0.376	3.000	2.322	2.500	2.092
3	0.500	0.376	3.000	2.322	2.500	2.075
4	0.500	0.376	3.000	2.322	2.500	2.095
5	0.500	0.376	3.000	2.322	2.500	2.083
6	0.500	0.471	3.000	2.768	2.500	2.357
7	0.500	0.471	3.000	2.768	2.500	2.354
8	0.500	0.471	3.000	2.768	2.500	2.350
9	0.500	0.471	3.000	2.768	2.500	2.335
10	0.200	0.100	1.000	0.010	0.500	0.035
11	0.200	0.100	1.000	0.010	0.500	0.035
12	0.200	0.100	1.000	0.010	0.500	0.087
13	0.200	0.100	1.000	0.010	0.500	0.084
14	0.200	0.100	2.000	0.690	1.500	1.113
15	0.200	0.100	2.000	0.690	1.500	1.113
16	0.200	0.100	2.000	0.690	1.500	1.112
17	0.200	0.100	2.000	0.690	1.500	1.112
18	0.500	0.277	2.000	1.524	2.000	2.056
19	0.500	0.277	2.000	1.524	2.000	2.058
20	0.500	0.277	2.000	1.524	2.000	2.046
21	0.500	0.277	2.000	1.524	2.000	2.058
22	0.500	0.380	3.000	2.733	3.000	2.822
23	0.500	0.380	3.000	2.733	3.000	2.808
24	0.500	0.380	3.000	2.733	3.000	2.803
25	0.500	0.380	3.000	2.733	3.000	2.785
Weight in lbs.	132.37	91.13	772.24	546.18	669.80	590.32

TABLE II EXAMPLE 2 - 47-BAR PLANE TRUSS

Design Information: For each member of the structure, the modulus of elasticity, E_i, the specific weight, ρ_i and the constant, α_i are 3.0 x 10^4 kips/sq. in., 0.284 lbs/cu. in. and 1.0, respectively. The resonant frequency for the structure is 16.0 cps and the displacement limits are, one inch on all nodes and in all directions. There is one loading condition for the truss which is shown on Fig. 2(a).

Output:

Element No.	Lower Area Bound in sq. in.	Upper Area Bound in sq. in.	Tension Stress Limit in Kips per sq. in.	Compression Stress Limit in Kips per sq. in.	Initial Area in sq. in.	Final Area in sq. in.	
						With Stress Constraints Only	With all Constraints
1	3.570	9.620	21.28	14.56	5.690	3.570	7.537
2	3.570	9.620	21.28	14.56	5.690	3.570	5.771
3	3.570	9.620	21.28	14.56	5.690	3.570	3.570
4	3.570	9.620	21.28	14.56	5.690	3.570	4.473
5	3.570	9.620	21.28	14.56	5.690	3.752	6.505
6	3.570	9.620	21.28	14.56	5.690	3.570	6.124
7	3.570	9.620	21.28	14.56	5.690	4.212	7.777
8	3.570	9.620	21.28	14.56	5.690	5.217	9.529
9	1.930	2.940	21.28	15.90	2.210	1.930	1.930
10	1.930	2.940	21.28	15.90	2.210	1.930	1.930
11	1.930	2.940	21.28	15.90	2.210	2.205	2.199
12	1.930	2.940	21.28	15.90	2.210	2.205	2.205
13	1.930	2.940	21.28	15.90	2.210	1.930	2.940
14	1.930	2.940	21.28	15.90	2.210	1.930	2.940
15	1.930	2.940	21.28	15.90	2.210	2.205	2.119
16	1.930	2.940	21.28	15.90	2.210	2.205	2.205
17	1.360	2.190	21.28	15.46	2.100	1.417	2.136
18	1.360	2.190	21.28	15.46	2.100	1.815	1.630
19	1.360	2.190	21.28	15.46	2.100	1.360	1.360
20	1.360	2.190	21.28	15.46	2.100	1.360	1.360
21	0.376	0.376	21.28	3.36	0.376	0.376	0.376
22	0.376	0.376	21.28	3.36	0.376	0.376	0.376
23	0.376	0.376	21.28	3.36	0.376	0.376	0.376
24	0.376	0.376	21.28	3.36	0.376	0.376	0.376
25	1.360	2.190	21.28	12.32	2.100	1.360	1.455
26	1.360	2.190	21.28	12.32	2.100	1.360	1.451
27	1.360	2.190	21.28	12.32	2.100	1.360	2.137
28	1.360	2.190	21.28	12.32	2.100	1.360	1.360
29	1.360	2.190	21.28	12.32	2.100	1.360	1.492
30	1.360	2.190	21.28	12.32	2.100	1.360	1.428
31	2.940	6.040	21.28	17.47	3.850	2.940	3.774
32	2.940	6.040	21.28	17.47	3.850	2.940	2.940
33	2.940	6.040	21.28	17.47	3.850	2.940	2.940
34	2.940	6.040	21.28	17.47	3.850	2.940	5.592
35	2.940	6.040	21.28	17.47	3.850	2.940	3.582
36	2.940	6.040	21.28	17.47	3.850	2.940	2.940
37	0.940	1.320	21.28	4.93	1.200	0.940	0.940
38	0.940	1.320	21.28	4.93	1.200	0.940	0.940
39	0.940	1.320	21.28	4.93	1.200	0.940	0.940
40	2.940	6.040	21.28	10.75	3.500	2.940	2.940
41	2.940	6.040	21.28	10.75	3.500	2.940	2.940
42	2.940	6.040	21.28	10.75	3.500	2.940	2.940
43	2.940	6.040	21.28	10.75	3.500	2.940	2.940
44	2.940	6.040	21.28	10.75	3.500	2.940	2.940
45	2.940	6.040	21.28	10.75	3.500	2.940	2.940
46	2.940	6.040	21.28	10.75	3.500	2.940	2.940
47	2.940	6.040	21.28	10.75	3.500	2.940	2.940
Weight in lbs.					3910.30	2993.37	3771.0

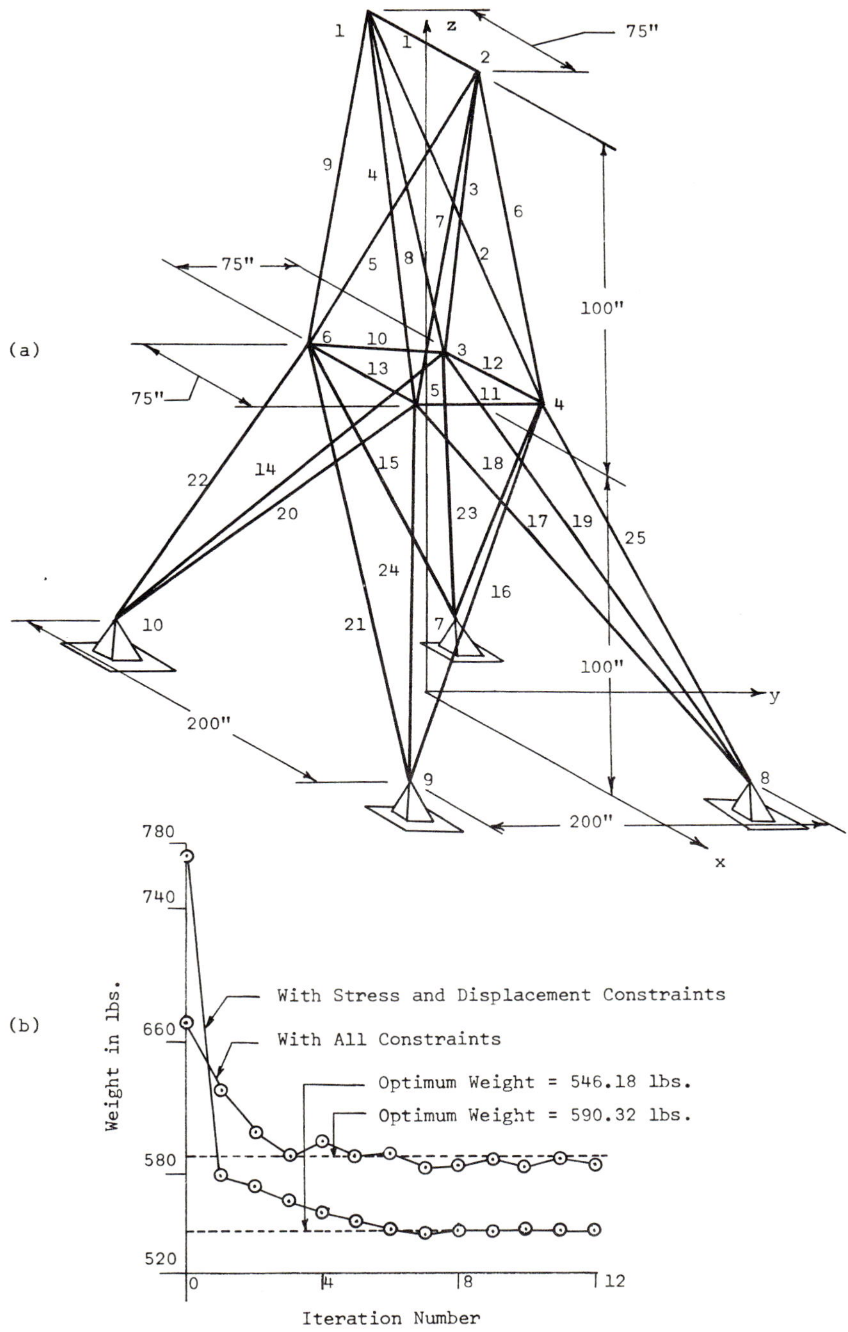

Fig. 1 Transmission Tower

(a)

(b)

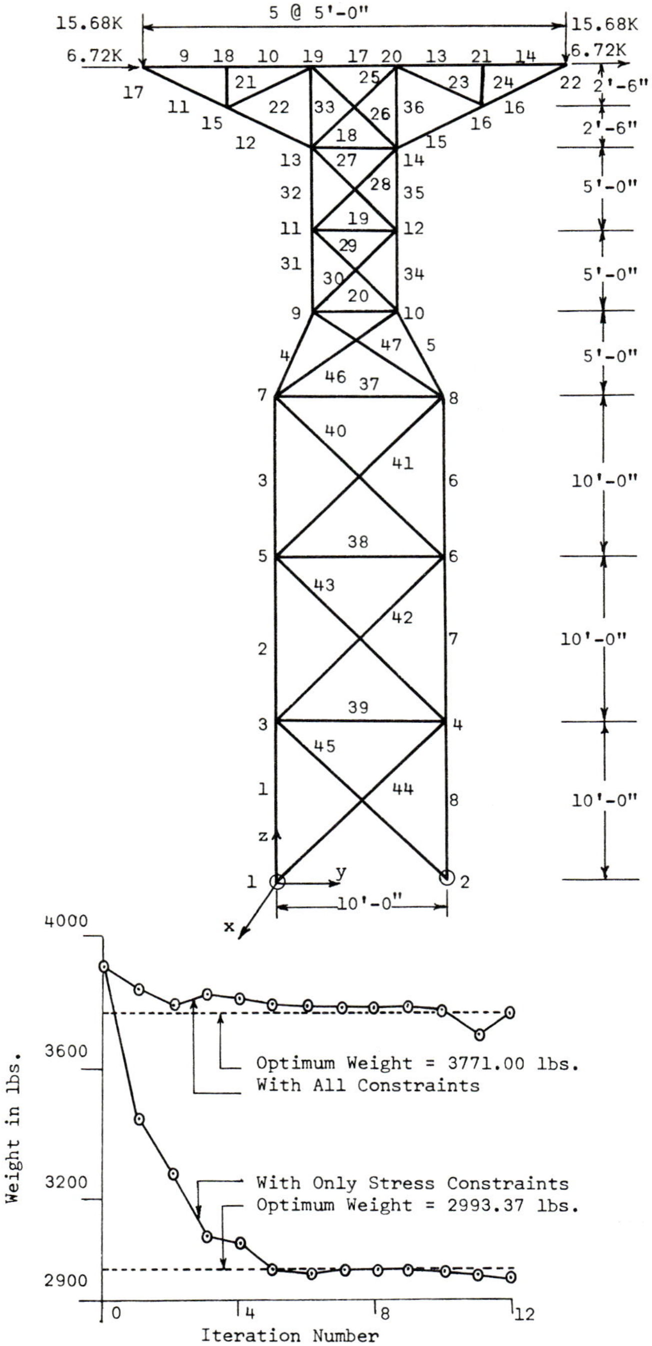

Fig. 2 47-Bar Plane Truss

Developments in Mechanics, Vol. 6. Proceedings of the 12th Midwestern Mechanics Conference: 59

Nonlinear Deflection Analysis of Beams in Space

M. F. MASSOUD
UNIVERSITY OF SHERBROOKE

ABSTRACT

In this paper the formulation of one type of elastica problems is extended to beams of arbitrary space configuration. The shape and loading of the deformed beam are given and the shape of the free beam is required. The geometric curvature, the torsion of the free center line and the twist angle of the cross section are derived using the rotation displacement vector of the center of the cross section. Examples of practical application are presented to explain the use of the derived formulae.

INTRODUCTION Problems in nonlinear deflection of thin beams are, generally, of two types:

(i) the free configuration of the beam is known and the deformed configuration after loading, is thought.
(ii) the deformed configuration of the beam due to given loadings lies upon an assigned curve and the free shape of the beam is required.

The first problem is discussed in references [1],[2],[3] for straight cantilever beams. In reference [3] the same problem is extended for certain space beam configurations acted upon by end loadings. In reference [4] a brief historical introduction is given for work done in the second problem and is followed by a discussion for the solution of coplaner cantilever beams. The aim of this paper is to extend the discussion of the second problem to space beams.

The configuration of the deformed beam is specified by; the curvature $K_d(s)$, the geometric torsion $\tau_d(s)$ of the beam center line and the twist $\Psi_d(s)$ of the beam. The loading of the beam is given in terms of the vector intensities $\underline{f}(s)$, $\underline{c}(s)$, the concentrated load vectors $\underline{F}_i$ and $\underline{C}_i$ (figure 1). Our objective is to determine the free beam geometric parameters $K_f(s)$, $\tau_f(s)$ and $\Psi_f(s)$ in terms of the above known parameters.

The generalization presented in this paper is confined largely to the formulation of the problem with the assumptions that the beam is inextensible, plane cross sections remain plane after deformation and perpendecular to the center line. The change in the shape of the cross section due to the deformation is neglected. The center line of the beam is taken as the locus of the centroids of the cross sections.

EQUATIONS OF EQUILIBRIUM The equations of equilibrium are given by (figure 1),

$$\frac{d\underline{R}(s)}{ds} + \underline{f}(s) = 0 , \tag{1}$$

$$\frac{d\underline{D}(s)}{ds} + \underline{c}(s) + \underline{t}(s) \times \underline{R}(s) = 0 . \tag{2}$$

Integrating equations (1) and (2) from an initial position (s_j) we get,

$$\underline{R}(s) = \underline{R}(s_j) - \int_{s_j}^{s} \underline{f}(\gamma)\, d\gamma \tag{3}$$

$$\underline{D}(s) = \underline{D}(s_j) - \int_{s_j}^{s} \underline{c}(\gamma)\, d\gamma - \int_{s_j}^{s} \underline{t}(\gamma) \times \underline{R}(\gamma)\, d\gamma . \tag{4}$$

Substituting $\underline{t} = d\underline{r}/ds$ (figure 1) in equation (4) and integrating by parts the last term, we obtain another useful expression for $\underline{D}$,

$$\underline{D}(s) = \underline{D}(s_j) + [\underline{r}_j - \underline{r}_s] \times \underline{R}(s_j) - \int_{s_j}^{s} \underline{c}\, d\gamma - \int_{s_j}^{s} [\underline{r}_\gamma - \underline{r}_s] \times \underline{f}\, d\gamma . \tag{5}$$

If at any cross section (s_i) a concentrated force $\underline{F}_i$ and/or a concentrated moment $\underline{C}_i$ act, the equations of equilibrium across this cross section take the forms,

$$\underline{R}(+s_i) - \underline{R}(-s_i) + \underline{F}_i = 0 , \quad \underline{D}(+s_i) - \underline{D}(-s_i) + \underline{C}_i = 0 . \tag{6}$$

DISTORTION - DISPLACEMENT RELATIONS If at each cross section of a beam the triad (n,b,t) coincides with the triad (x,y,t), the beam is called prismatic, otherwise non-prismatic or twisted. The amount of twist in each cross section is given by the angle $\Psi(s)$ between the direction of the axes n and x, (figure 1). As the two triads (n,b,t) and (x,y,t) move along the center line with unit speed, they rotate about their origin with geometric angular velocities $\underline{\omega}$ and $\underline{\Omega}$ respectively (figure 1),

$$\underline{\omega}(s) = K\underline{b} + \tau\underline{t} , \quad \underline{\Omega}(s) = K \sin \Psi \underline{i} + K \cos \Psi \underline{j} + (\tau + \frac{d\Psi}{ds})\underline{t} . \tag{7}$$

Figure (2) shows the free and deformed positions of an arbitrary cross section of the beam. The displacements of this cross section are expressed by two vectors; a linear displacement vector $\underline{u}(s)$ of the centroid of the cross section, and a rotation vector $\underline{\phi}(s)$ around the centroid. The angular distortion vector $\underline{h}(s)$ due to these displacements is,

$$\underline{h}(s) = d\underline{\phi}(s)/ds \ . \tag{8}$$

The rotation vector and the geometric angular velocity are related by the geometric relation [5],

$$\frac{d\underline{\phi}(s)}{ds} = \underline{\Omega}_d(s) - \underline{\dot{\Omega}}_f(s) \ . \tag{9}$$

PARAMETERS OF FREE CONFIGURATION From the thin beam theory, we have the relation [5],

$$\underline{h} = \underline{\underline{\beta}} \cdot \underline{D} \ . \tag{10}$$

From equations (4), (8), (9) and (10) we get,

$$\underline{\Omega}_f = \underline{\Omega}_d - \underline{\underline{\beta}} \cdot [\underline{D}(s_j) + \{(\underline{r}_j - \underline{r}_s) x \underline{R}(s_j) - \int_{s_j}^{s} \underline{c} \, d\gamma - \int_{s_j}^{s} (\underline{r}_\gamma - \underline{r}_s) x \underline{f} \, d\gamma]$$

$$= A\underline{i}_d + B\underline{j}_d + C\underline{t}_d \ . \tag{11*}$$

The following relation can also be obtained from equations (8) to (10),

$$\underline{\phi}(s) = \underline{\phi}(s_j) + \int_{s_j}^{s} [\underline{\underline{\beta}}(\gamma) \cdot \underline{D}(\gamma)] \, d\gamma \ . \tag{12*}$$

The vector $\underline{D}(s)$ and the dyadic $\underline{\underline{\beta}}$ are both known in terms of their components along the principal flexure axes of the deformed configuration (x_d, y_d, t_d). Consequently, the vectors $\underline{\Omega}_f(s)$ and $\underline{\phi}(s)$, from equations (11) and (12), are expressed in terms of their components along the triad (x_d, y_d, t_d) of the deformed beam. The direction cosines of the triad (x_f, y_f, t_f) of the free configuration relative to the triad (x_d, y_d, t_d) can be calculated using three euler rotations, figure (3), where

$$\underline{\phi}(s) = \phi_1 \underline{i}_d + \phi_2 \underline{j}_d + \phi_3 \underline{t}_d = \underline{\beta}(s) + \underline{\gamma}(s) + \underline{\xi}(s) \ , \tag{13}$$

and,

$$\phi_1 = \gamma \sin\xi - \beta \sin\gamma \cos\xi, \ \phi_2 = \gamma \cos\xi + \beta \sin\gamma \cos\xi, \ \phi_3 = \xi + \beta \cos\gamma \ .$$

The values for γ, β and ξ as functions of ϕ_1, ϕ_2 and ϕ_3 are obtained from the solution of the above three equations. One form of solution is given as follows,

$$\phi_3 = \xi + (-\phi_1 \cos\xi + \phi_2 \sin\xi) . \cot(\phi_2 \cos\xi + \phi_1 \sin\xi) \tag{14}$$

$$\gamma = \phi_2 \cos\xi + \phi_1 \sin\xi, \ \beta = \frac{1}{\sin\gamma}(-\phi_1 \cos\xi + \phi_2 \sin\xi) . \tag{15}$$

---

*See appendix

The implicit equation (14) can be solved numerically or graphically for the value of ξ, which is then substituted in equation (15) to give the remaining values of γ and β. Using the above euler angles we can express the geometric angular velocity $\underline{\Omega}_f$, of equation (11), as follows,

$$\Omega_f = A' \underline{i}_f + B' \underline{j}_f + C' \underline{t}_f , \tag{16}$$

where,
$$A' = \alpha_1 \cos\beta \cos\gamma - \alpha_2 \sin\beta + C \cos\beta \sin\gamma$$

$$B' = \alpha_1 \sin\beta \cos\gamma + \alpha_2 \cos\beta + C \sin\beta \sin\gamma$$

$$C' = C \cos\gamma - \alpha_1 \sin\gamma$$

and,
$$\alpha_1 = A \cos\xi - B \sin\xi, \quad \alpha_2 = B \cos\xi + A \sin\xi$$

From equations (7) and (16) we get,

$$K_f(s) = (A'^2 + B'^2)^{\frac{1}{2}}, \quad \Psi_f(s) = \sin^{-1}\left(\frac{A'}{K_f}\right), \quad \tau_f(s) = C' - \frac{d\Psi_f}{ds} \tag{17}$$

If we have concentrated external forces and moments acting on the beam, together with distributed external forces, equations (11) and (12) should be applied piecewise. Starting from position (s_j) where the initial values $\underline{R}(s_j)$ and $\underline{D}(s_j)$ are known we can obtain the parameters $K_f(s)$, $\Psi_f(s)$ and $\tau_f(s)$ valid until the position (s_i) of the next concentrated external force or moment. Applying equation (6) at the position of the concentrated force and/or moment we obtain new initial values $\underline{R}(s_i)$ and $\underline{D}(s_i)$ and so on until the end of the beam.

By specifying the geometric curvature, the geometric torsion and the twist angle of the free configuration the problem is formely solved.

<u>EXAMPLE</u> The deformed shape of the beam is a prismatic cantilever in the form of a part of a cylinderical helix,

$$\underline{r}_d = u \cos\theta \, \underline{i} + u \sin\theta \, \underline{j} + (u \tan\alpha)\, \theta \, \underline{t} .$$

The length of the beam is given by, $s = L$ at $\theta = \theta_L$.

Thus;
$$\Psi_d = 0, \quad \Omega_d = \frac{\cos^2\alpha}{u} \underline{j}_d + \frac{\cos\alpha \sin\alpha}{u} \underline{t}_d . \tag{18}$$

The beam is loaded by a uniform pressure along the n axis hence,

$$\underline{f}(s) = p \, \underline{i}_d, \quad \underline{c}(s) = 0, \quad \underline{R}(s=L) = 0 \text{ and } \underline{D}(s=L) = 0 .$$

We choose a fixed frame of reference $(\overline{X},\overline{Y},Z)$ with Z along the axis of the helix cylinder and the fixed end of the beam ($\theta=0$) with coordinates (u,o,o) hence,

$$\underline{t}(\theta) = -\cos\alpha \sin\theta \, \underline{I} + \cos\alpha \cos\theta \, \underline{J} + \sin\alpha \, \underline{K} ,$$

$$\underline{R}(\theta) = \frac{p\,u}{\cos\alpha} [(\sin\theta - \sin\theta_L)\, \underline{I} - (\cos\theta - \cos\theta_L)\, \underline{J}] ,$$

$$\underline{\underline{\beta}} = \underline{\underline{P}} \cdot \text{diag}\left[\frac{1}{EI_x}\ \frac{1}{EI_y}\ \frac{1}{GI_t}\right] \cdot \underline{\underline{P}}^T ,$$

and,

$$\underline{\underline{P}} = \begin{bmatrix} -\cos\theta & \sin\theta\sin\alpha & -\cos\alpha\sin\theta \\ -\sin\theta & -\cos\theta\sin\alpha & \cos\alpha\cos\theta \\ 0 & \cos\alpha & \sin\alpha \end{bmatrix}$$

Substituting in equation (4), with the initial position at the free end, we get,

$$\underline{\underline{\beta}}\cdot\underline{D} = pu^2\tan\alpha\left[\frac{\sec\alpha}{EI_x}(\theta'\cos\theta'-\sin\theta')\underline{i}_d + \frac{\tan\alpha}{EI_y}\{(1-\cot^2\alpha)(\cos\theta'-1) + \theta'\sin\theta'\}\underline{j}_d + \frac{1}{GI_t}(2-2\cos\theta'-\theta'\sin\theta')\underline{t}_d\right]$$

$$= d_1\underline{i}_d + d_2\underline{j}_d + d_3\underline{t}_d\ , \quad (\theta' = \theta_L - \theta)\ . \tag{19}$$

From equations (11), (18) and (19) we get,

$$\underline{\Omega}_f = d_1\ \underline{i}_d + \left(\frac{\cos^2\alpha}{u} - d_2\right)\underline{j}_d + \left(\frac{\sin\alpha\cos\alpha}{u} - d_3\right)\underline{t}_d\ . \tag{20}$$

With $\underline{\phi}(\theta') = 0$, equations (12) and (19) give,

$$\phi(s) = \phi_1\underline{i}_d + \phi_2\underline{j}_d + \phi_3\underline{t}_d\ ,$$

where,

$$\phi_1 = \frac{pu^3\tan\alpha}{\cos^2\alpha}\Big[\sin\theta\sin\theta_L\Big\{-\frac{3}{4EI_x} + \frac{\sin^2\alpha}{EI_y}\left(\frac{3}{4} - \frac{\cot^2\alpha}{2}\right) + \frac{5\cos^2\alpha}{4GI_t}$$

$$- \theta\sin(\theta_L - \theta)\Big\{\frac{3}{4EI_x} + \frac{\sin^2\alpha}{EI_y}\left(\frac{3}{4} - \frac{\cot^2\alpha}{2}\right) + \frac{5}{4}\frac{\cos^2\alpha}{GI_t}\Big\}$$

$$- \theta\left(\frac{\theta}{4} - \frac{\theta_L}{2}\right)\cos(\theta_L - \theta)\Big\{\frac{1}{EI_x} + \frac{\sin^2\alpha}{EI_y} + \frac{\cos^2\alpha}{GI_t}\Big\}$$

$$+ \frac{\theta}{2}\cos\theta\sin\theta_L\frac{1}{EI_x} - (\cos\theta - 1)(1 - \cot^2\alpha)\frac{\sin^2\alpha}{EI_y}$$

$$+ \frac{\theta_L}{2}\sin\theta\cos\theta_L\left(\frac{1}{EI_x} - \frac{\sin^2\alpha}{EI_y} - \frac{\cos^2\alpha}{GI_t}\right)\Big]$$

$$\phi_2 = \frac{(pu^3)\tan^2\alpha}{\cos\alpha}\Big[\theta\cos(\theta_L - \theta)\Big\{\frac{3}{4EI_x} + \frac{\sin^2\alpha(1-6\cot^2\alpha)}{4EI_y}$$

$$+ \frac{7}{4}\frac{\sin\alpha\cos\alpha}{GI_t}\Big\} + \theta_L\cos(\theta_L - \theta)\Big\{\frac{\cos^2\alpha}{EI_y} + \frac{\cos\alpha\sin\alpha}{GI_t}\Big\}$$

$$- \theta\left(\frac{\theta}{4} - \frac{\theta_L}{2}\right)\sin(\theta_L - \theta)\Big\{\frac{1}{EI_x} + \frac{\sin^2\alpha}{EI_y} + \frac{\cos\alpha\sin\alpha}{GI_t}\Big\}$$

$$+ \sin(\theta_L - \theta)\{-\frac{\cos^2\alpha(2 - \cot^2\alpha)}{EI_y} + \frac{3\cos\alpha\sin\alpha}{GI_t}\}$$

$$+ \sin\theta\cos\theta_L\{-\frac{3}{4EI_x} + \frac{\sin^2\alpha(1 - \cot^2\alpha)}{EI_y} + \frac{\sin^2\alpha}{4EI_y}$$

$$+ \frac{5}{4}\frac{\cos\alpha\sin\alpha}{GI_t}\} + \theta_L\sin\theta\sin\theta_L\{-\frac{1}{2EI_x} + \frac{\sin^2\alpha}{2EI_y}$$

$$+ \frac{\cos\alpha\sin\alpha}{2GI_t}\} + \frac{\theta}{2}\cos\theta\cos\theta_L\frac{1}{EI_x}$$

$$+ \sin\theta\{-\frac{\sin^2\alpha(1 - \cot^2\alpha)}{EI_y} - \frac{2\cos\alpha\sin\alpha}{GI_t}\}$$

$$+ \theta\{-\frac{\cos^2\alpha(1 - \cot^2\alpha)}{EI_y} + \frac{2\cos\alpha\sin\alpha}{GI_t}\}$$

$$+ \frac{\cos^2\alpha(2 - \cot^2\alpha)}{EI_y}\sin\theta_L - \frac{\cos^2\alpha}{EI_y}\theta_L\cos\theta_L$$

$$- \frac{3\cos\alpha\sin\alpha}{GI_t}\sin\theta_L - \frac{\cos\alpha\sin\alpha}{GI_t}\theta_L\cos\theta_L]$$

$$\phi_3 = \frac{(pu^3)\tan\alpha}{\cos\alpha}[\theta\cos(\theta_L - \theta)\{-\frac{3}{4EI_x} - \frac{\sin^2\alpha(1-2\cot^2\alpha)+4\sin^2\alpha}{4EI_y}$$

$$+ \frac{4\sin^2\alpha-3\cos^2\alpha}{4GI_t}\} + \theta_L\cos(\theta_L - \theta)\{\frac{\sin^2\alpha}{EI_y}\frac{\sin^2\alpha}{GI_t}\}$$

$$+ \theta(\frac{\theta}{4} - \frac{\theta_L}{2})\sin(\theta_L - \theta)\{\frac{1}{EI_x} + \frac{\sin^2\alpha}{EI_y} + \frac{\cos^2\alpha}{GI_t}\}$$

$$+ \sin(\theta_L - \theta)\{-\frac{\sin^2\alpha(2 - \cot^2\alpha)}{EI_y} + \frac{3\sin^2\alpha}{GI_t}\}$$

$$+ \sin\theta\cos\theta_L\{\frac{3}{4EI_x} - \frac{\sin^2\alpha(1 - \cot^2\alpha)}{2EI_y} - \frac{\sin^2\alpha}{4EI_y} - \frac{5\cos^2\alpha}{4GI_t}\}$$

$$+ \theta_L\sin\theta\sin\theta_L\{\frac{1}{2EI_x} - \frac{\sin^2\alpha}{2EI_y} - \frac{\cos^2\alpha}{2GI_t}\}$$

$$- \frac{\theta}{2}\cos\theta\cos\theta_L\frac{1}{EI_y}$$

$$+ \sin\theta\{\frac{\sin^2\alpha(1 - \cot^2\alpha)}{EI_y} + \frac{2\cos^2\alpha}{GI_t}\}$$

$$+ \theta\{- \frac{\sin^2\alpha(1 - \cot^2\alpha)}{EI_y} + \frac{2 \sin^2\alpha}{GI_t}\}$$

$$+ \frac{\sin^2\alpha(2 - \cot^2\alpha)}{EI_y} \sin \theta_L - \frac{\sin^2\alpha}{EI_y} \theta_L \cos \theta_L - \frac{3 \sin^2\alpha}{GI_t} \sin \theta_L$$

$$+ \frac{\sin^2\alpha}{GI_t} \theta_L \cos \theta_L].$$

Substituting the above values in equations (14) and (15) we obtain the values of euler angles. The components of $\underline{\Omega_f}$ (equation (20)) and the euler angles are used to derive the components of $\underline{\Omega_f}$ along the principal triad of the free configuration A',B',C' (equation 16). These values are finally used in equation (17) to give the parameters of the free configuration. Figure (4) gives the result of a numerical example, with the following parameters for the deformed configuration;

$u = 3''$, $p = 2$ lb/in, diameter of cross section $= \frac{3}{16}''$, $\theta_L = \pi$

$E = 30 \times 10^6$ lb/in^2, $G = 12 \times 10^6$ lb/in^2, $\alpha = 30°$.

In case $\alpha = 0$, the deformed beam is part of a plane circle and the distributed pressure p is towards the center of the circle, the curvature of the free configuration is,

$$\underline{K_f} = [\frac{1}{u} + \frac{pu^2}{EI_y} \{\cos (\frac{L}{u} - \frac{s}{u}) - 1\}] \underline{j_f} ,$$

and the free beam is prismatic. If in the above case ($\alpha = 0$) the distributed pressure p is parallel to the radius of the circle at one end, the curvature of the free configuration is,

$$\underline{K_f} = [\frac{1}{u} + \frac{pu^2}{EI_y} \{\frac{1}{u} (L-s) \sin (\frac{s}{u}) + \cos (\frac{L}{u}) - \cos (\frac{s}{u})\}] \underline{j_f} ,$$

and the free beam is also prismatic.

A practical application for the last case, is a long beam subjected to wind pressure and the final deformed beam is required to be a part of a circle at a certain wind speed.

<u>CONCLUSION</u> The general formulation of the geometric parameters for the free configuration of a given loaded beam is presented. The calculations of the direction cosines of the principal torsion flexure axes in the free configuration relative to their counterparts in the loaded configuration lend themselves to numerical or graphical evaluation. Geometric interpretation of the free configuration is not easy even for the simple problem presented in the example. The complexity of the solution increases if concentrated forces and moments act upon the beam together with the distributed loading.

<u>APPENDIX</u> Equations (11) and (12) involve integration of vectors expressed in a moving triad. The integration can be performed if the vector can be expressed in a fixed frame of reference $(\overline{X},\overline{Y},Z)$. Let $\underline{P}(s)$ be the rotation matrix between the principal axes at any cross section and the fixed axes. The components of the rotation matrix can be expressed in a vector

form as,

$$\underline{P}(s) = \{\ell_1\ \ell_2\ \ell_3\ m_1\ m_2\ m_3\ n_1\ n_2\ n_3\} .$$

This vector satisfies the ralation,

$$\frac{dP(s)}{ds} = \underline{F}\ \underline{P}(s) , \qquad (a)$$

where $\underline{F}$ is a partitioned diagonal matrix $\underline{F}$ = diag. $\underline{T}$. The square matrix $\underline{T}$ is the rotation operator $(\underline{\Omega}_d\ X)$ represented by the skew symmetrix matrix,

$$\begin{bmatrix} o & -\{\tau + \frac{d\Psi(s)}{ds}\}_d & +K_d \cos \Psi_d \\ \{\tau + \frac{d\Psi(s)}{ds}\}_d & o & -K_d \sin \Psi_d \\ -K_d \cos \Psi_d & +K_d \sin \Psi_d & o \end{bmatrix} \qquad (b)$$

Thus, the coefficients of the matricial first order differential equation (a) are functions of (s). The solution of equation (a) gives the values of the cosine directions of the axes (x_d, y_d, t_d) relative to the fixed frame of reference.

NOMENCLATURE

x,y,t — principal torsion flexure axes at any cross section, with t along the tangent to the center line in the direction of increasing (s); unit vectors $(\underline{i},\underline{j},\underline{t})$.

n,b,t — geometric principal triad at any cross section, with n along the binormal towards the center of curvature, and t along the tangent to the center line and b along the binormal; unit vectors $(\underline{n},\underline{b},\underline{t})$.

$\underline{R}(s)$ — internal force vector acting at the centroid of the cross section.

$\underline{D}(s)$ — internal moment vector acting at the centroid of the cross section.

$\underline{f}(s)$ — vector intensity of external distributed force loading along the center line of the beam.

$\underline{c}(s)$ — vector intensity of external distributed moment loading along the center line of the beam.

$\underline{F}_i$ — external concentrated force at position (s_i).

$\underline{C}_i$ — external concentrated moment at position (s_i).

$\underline{u}(s)$ — linear displacement vector.

$\underline{\phi}(s)$ — rotation displacement vector.

$\underline{r}(s)$ — position vector.

$\underline{h}(s)$ — angular distortion vector.

$\underline{\omega}(s)$ — geometric angular velocity vector of the triad (n,b,t).

$\underline{\Omega}(s)$ — geometric angular velocity vector of the triad (x,y,t).

K(s) — curvature of the center line.

$\tau(s)$ geometric torsion of the center line.

$\Psi(s)$ angle of twist of a cross section.

E,G modulii of elasticity and rigidity.

$\underline{\underline{\beta}}$ the dyadic $\frac{1}{EI_x}\underline{ii} + \frac{1}{EI_y}\underline{jj} + \frac{1}{EI_t}\underline{tt}$.

Subscripts

d pertaining to the deformed configuration.

f pertaining to the free configuration.

REFERENCES

[1] T.P. Mitchell: The Nonlinear Bending of Thin Rods. Journal of Applied Mechanics, March 1959.

[2] K.E. Bisshop and D.C. Drucker: Large Deflection of Cantilever Beams. Quarterly of Applied Mathematics, Vol. III, No. 3, October 1945.

[3] R. Frisch-Fay: Flexible Bars. Butterworths (1962).

[4] C. Truesdall: A New Chapter in the Theory of Elastica. First Midwestern Conference, Solid Mechanics (1953), University of Illinois.

[5] A.E.H. Love: The Mathematical Theory of Elasticity. Dover, fourth edition, p. 396-397.

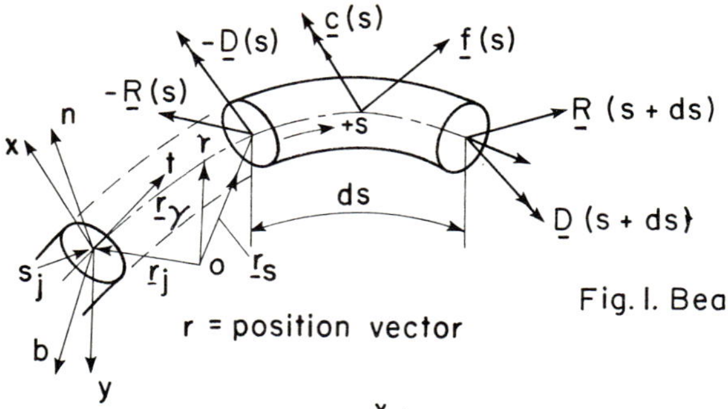

Fig. 1. Beam configuration

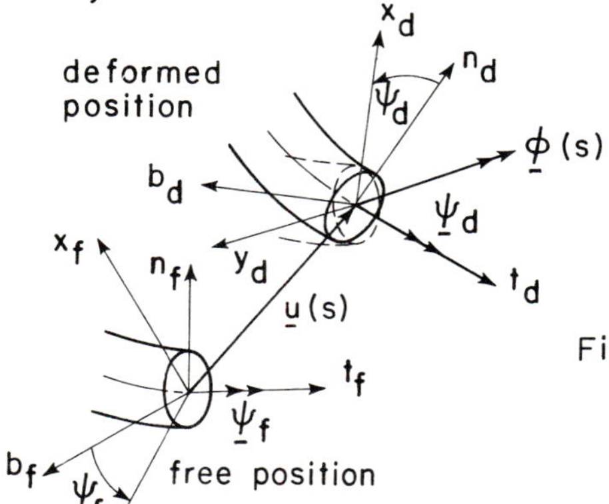

Fig. 2. Displacement vectors

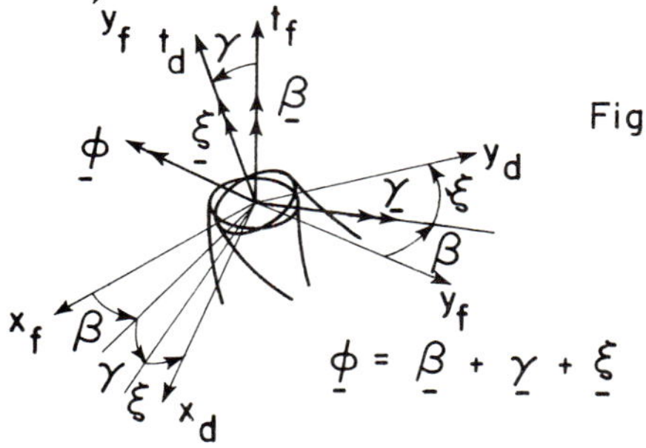

Fig. 3. Euler angles describing the rotation of the cross section

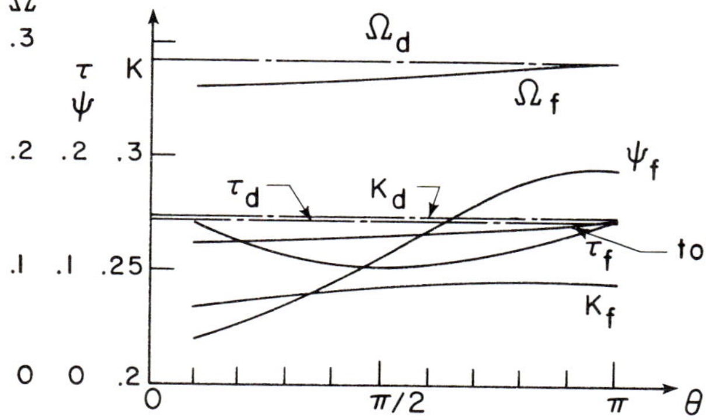

Fig. 4. Results of example

IN MEMORIAL OF

IYADURAI KASIRAJ

MAY 20, 1928 - APRIL 4, 1971

Developments in Mechanics, Vol. 6. Proceedings of the 12th Midwestern Mechanics Conference: 60

Fatigue Damage in Nonlinear Multi-story Frames under Earthquake Load

IYADURAI KASIRAJ
PRAIRIE VIEW A. AND M. COLLEGE

ABSTRACT

A four story shear frame was excited with a random input function from the 1940 El Centro Earthquake NS Component accelerogram. The time history of the relative deflection of each floor with reference to the floor below was determined by assuming an elasto-plastic load deflection hysteresis loop. Another tri-linear deflection-maximum strain developed loop was assumed and the time history of the maximum strain developed at the critical section of each floor was obtained. The reversals of plastic strains were found and the cumulative fatigue damage factor was calculated from each time history of strain for each floor for a particular fundamental frequency of the system. Similar calculations were made for different values of fundamental frequency and for two values of the damping coefficient. The maximum deflection of each floor during the entire excitation was also calculated and a dimensionless maximum deflection ratio was defined.

The comparative study showed that the variation of maximum deflection of a floor did not reflect the true damage that took place in the structure. Though the dimensionless maximum deflection ratio appeared to be proportional to the fatigue damage factor, the calculation of the later may be necessary so as to understand the extent of damage that has taken place during the period of excitation. In general, the structures with higher fundamental frequencies (stiffer structures) were damaged to a greater extent than those with lesser requencies by an excitation similar to the 1940 El Centro Earthquake NS Component. Though the first floor was subjected to a lesser number of plastic reversals compared to the upper floors, its damage was more severe.

This paper presents a procedure for calculating the low-cycle fatigue damage factors in a multi-degree-of-freedom system when it is subjected to a random excitation.

INTRODUCTION

The utilization of the plastic behavior of a material has been investigated extensively in recent years in view of the fact that it absorbs a considerable amount of energy and thus attenuates the dynamic response of structures. Many nonlinear functions have been assumed by different investigators for the load-response relation (1, 2, 3, 4, 5) and the influence of the plastic behavior of the material has been analysed. The plastic design procedures for static loads are already permitted in building codes of many countries. If such a plastic deformation of material is permitted in a structure subjected to a load reversal, like a random vibration, it may be essential to investigate the low-cycle fatigue damage in addition to the maximum response of the structure. In general, there are three distinctly different failure mechanisms in a structure subjected to random vibration (6). One of these is fatigue damage which may be significant if the structure is repeatedly stressed beyond the elastic limit. In such cases, even a few reversals may cause the failure. A numerical procedure to calculate the low-cycle fatigue damage factor produced in a simple structure subjected to earthquake excitation was introduced by Kasiraj and Yao (7) .

The object of this study is to investigate this low-cycle fatigue damage in a multi-degree-of-freedom system. A four story shear building has been analysed for its safety against cumulative fatigue damage when it was excited beyond its elastic limit by the NS Component of the 1940 El Centro Earthquake.

FATIGUE DAMAGE FACTOR

A general hypothesis is describing the cumulative effect of plastic strain on the low-cycle fatigue behavior of metals has been presented (8) . When samples of a material in tension are tested to failure with different precompressions, a graph can be drawn showing the relationship between the failure tension plastic strain, q, and the precompression tension plastic strain ratio, v, as shown in Fig. 1b. The fatigue damage, D_j, resulting from a single reversal of plastic strain from precompression to tension was found to be

$$D_j = \left(\frac{DPP(j)}{q}\right)^{1/S} \quad \ldots\ldots\ldots\ldots\ldots (1)$$

where DPP(j) is the maximum change in plastic strain reached during this single reversal, as shown in Fig. 2, and q is the failure tension plastic strain obtained from Fig. 1b for the particular v ratio in the reversal. For mild steel, an average value of 0.85 can be assumed for the value of q. It was also found that for the jth reversal of load

$$1/S = 1 - 0.86 \left(\frac{DPN(j)}{DPP(j)}\right) \ldots\ldots\ldots\ldots\ldots (\ )$$

where DPN(j) is the precompression plastic strain induced before tension plastic strain is started in the cycle, as shown in Fig. 2. If there are N plastic reversals during the whole duration of response, the total fatigue damage caused is given by

$$DF = \sum_{j=1}^{N} \left(\frac{DPP(j)}{q}\right) \ldots\ldots\ldots\ldots (3)$$

Failure is assumed to occur when DF reaches the value of one. The time history of plastic strain at the critical section of the structure has to be evaluated to calculate the damage factor.

PROBLEM

A multi-story framed structure can be simplified for modal analysis by using the following assumptions: (1) the distributed mass of the structure is lumped at equidistant floor levels (m at each floor level), (2) the shear rigidity at any column level is represented by an elastic spring (K_i at the ith floor), (4) the horixontal deformation of the structure is of the shear type, and there is no rotation of the rigid floor, (5) the damping is such that the dynamic coupling between the various modes do not exist and thus modal analysis can be applied. For a structure such as shown in Fig. 3, the equation of motion can be written as given below: (9) .

$$m\ddot{x}_i + c_i(\dot{x}_i - \dot{x}_{i-1}) - c_{i+1}(\dot{x}_{i+1} - \dot{x}_1) + K_1(x_i - x_{i-1}) - K_{i+1}(x_{i+1} - x_1) = -m\ddot{x}_0 \quad \ldots\ldots (4)$$

$$(i = 1, 2, \ldots\ldots n)$$

where x_1 is a function of time and is the relative displacement of the i-th floor (the i-th mass) from the ground, x_0 is the horizontal distance of the base of the structure with reference to the fixed reference line, as shown in Fig. 3, and the dots indicate differentiation with respect to time. When the base of the structure is vibrated by an earthquake, $\ddot{x}_0$ will represent this earthquake acceleration. The above equation can be modified to give:

$$m\ddot{y}_i + c_i\dot{y}_i - c_{i+1}\dot{y}_{i+1} + K_i y_i - K_{i+1} y_{i+1} = m(\ddot{y}_{i-1} + \ldots + \ddot{y}_1) + \ddot{x}_0) \quad \ldots\ldots (5)$$

$$(i = 1, 2, \ldots\ldots n)$$

where y_i is the relative displacement of the i-th floor with reference to the (i-1) -th floor. That is $y_i = x_{i-1}$. For a four story frame, the following four equations will arise:

$$m\ddot{y}_1 + c_1\dot{y}_1 - c_2\dot{y}_2 + K_1 y_1 - K_2 y_2 = -m\ddot{x}_0 \quad \ldots\ldots (6)$$

$$m\ddot{y}_2 + c_2\dot{y}_2 - c_3\dot{y}_3 + K_2 y_2 - K_3 y_3 = -(\ddot{y}_1 + \ddot{x}_0) \quad \ldots\ldots (7)$$

$$m\ddot{y}_3 + c_3\dot{y}_3 - c_4\dot{y}_4 + K_3 y_3 - K_4 y_4 = -(\ddot{y}_2 + \ddot{y}_1 + \ddot{x}_0 \quad \ldots\ldots (8)$$

$$m\ddot{y}_4 + c_4\dot{y}_4 - K_4 y_4 = -m(\ddot{y}_3 + \ddot{y}_2 + \ddot{y}_1 + \ddot{x}_0) \quad \ldots\ldots (9)$$

There are four relative displacements, namely y_1, y_2, y_3, and y_4, for this four story frame. Each displacement refers to the deflection of one floor relative to the floor below. The time history of these four displacements were evaluated by assuming certain nonlinear load-displacement relationships when the frame was excited by an earthquake. Each of these displacements will create maximum strain at a particular section of the column in each floor. The time history of these maximum strains in each floor were determined by another reasonable nonlinear deflection-maximum strain relationship. The plastic strains were found for each floor and the damage factors were worked out and analysed. In addition, the relative deflections and the number of reversals of each floor have also been analysed.

HYSTERESIS LOOPS

The elasto-plastic load deflection hysteresis loop shown in Fig. 4 was assumed. The load function, $\frac{Q}{m}$, is equal to $\frac{Ky}{m}$ in each floor when y is within the yield deflection, y_y. Beyond this, Q/m is assumed to have a constant value, θg, where θ is a parameter of the system and g is the gratitational acceleration due to gravity. Hence, by changing the parameter θ, the value of yield load, θgm, can be changed. In other words, θ is a parameter which changes the yielding point in the load--deflection curve.

Since steel frames are commonly built with wide flange (WF) shapes, the deflection-maximum strain hysteresis loop shown in Fig. 5 is proposed. The justification for this type of hysteresis loop has been explained in detail by Kasiraj (10.) An average WF section, as listed in the AICE Manual, (11) appears to have a thickness of flange to depth ratio of 0.6 x 10^{-1} and a thickness of web to width ratio of 0.54 x 10^{-1}. There appears to be three slopes in the deflection curve for this WF section. The first slope, K_2, is good when the yield strain is not exceeded in any part of the WF section. The second slope, K_3, is valid when the yield strain is limited to the flange, and the third slope, K_4, is valid when the yield strain extends to the web portion. Within the elastic limit, the slope k_2 of the curve is equal to the deflection at yield, y_y, divided by the yield strain, e_y. When the deflection lies between the absolute values of y_y and $1.08y_y$ the slope k_3 is $1.4k_2$, and the value of k_4 beyond $1.08y_y$ is $8.5k_2$. The above slopes are derived from the deflection calculation of a cantiliver WF section beam when it is loaded beyond the elastic limit for the first time. Though the subsequent curves under load reversals will be different, the same slopes were assumed with the justification that the result will be close to the real value when the elasto-plastic load deflection hysteresis loop shown in Fig. 4 is used with the approximate deflection-maximum strain relation shown in Fig. 5.

NUMERICAL RESULTS

In the four story frame shown in Fig. 3, the stiffness of the columns in the second, third and fourth stories were taken as 0.9, 0.8, and 0.7 of the stiffness of the columns in ground floor. The mass was assumed to be proportional to its stiffness factor. In other words, c_i / K_i was assumed to be constant. Linear viscous damping was assumed even when the frame was stressed beyond the elastic limit.

For this frame, the four natural frequencies of the system were found to be the square root of 0.109 K/m, 0.818 K/m, 1.904 K/m and 2.968 K/m where K is the stiffness coefficient in the ground floor and m is the mass of each floor. The mode shape for these different frequencies are furnished in Fig. 6. The analysis was programmed and the structural responses and fatigue damages in each floor were evaluated numerically. The flow chart of the computer program is shown in Fig. 7. A Runge-Kutta numerical integration procedure was adopted for solving equation 6 through 9. The program was written fixing a permissible error of $1. \times 10^{-5}$. The elasto-plastic force-deflection hysteresis loop shown in Fig. 4 was adopted for the sriffness function in the equations. As the stiffness in each story was different, the yield deflection, y_y, was also different in each story, and hence four different hysteresis loops were used for a particular value of θ in solving the equations. After finding the time history of the response of each floor, the time history of the maximum strain at the critical section of each floor was evaluated by using the tri-linear deflection, maximum strain hysteresis loop shown in Fig. 5. The yield strain, e_y, was assumed as 0.0012, which is a reasonable assumption for mild steel. The plastic strains alone were collected and the low-cycle fatigue damage factor was calculated in each floor using Eq. 3. The above analysis was made for a particular fundamental frequency of the system. The whole analysis was repeated for different fundamental frequencies ranging from 2 to 10 radians per second. The analysis was made for a θ value of 0.15 and for two different values of the damping coefficient in the ground floor; 0.2 times the fundamental frequency divided by the mass, and 0.3 times the fundamental frequency divided by the mass. A damping coefficient of 0.2 times the fundamental frequency divided by the mass will amount to a damping ratio of 0.10 in a single--degree-of-freedom system. As already explained, θ is a parameter which reflects the elastic strength of the columns. (Columns in the first floor for this problem.) The NS Component of the El Centro 1940 earthquake acceleration was taken as the vibrating function at the ground for the system. All possible precautions were taken to check the accuracy of the computer results at various stages.

For purposes of comparison, the absolute values of the maximum relative deflection of each floor, y_{max}, were calculated for systems with different fundamental frequencies. These maximum deflections are plotted fundamental frequencies of the system in Figs. 8 and 9 for two different damping coefficients in the first floor. As expected, the systems with the lower damping coefficient deflect more than the ones with the higher damping coefficient. However, this increase in damping has less influence in higher floors than in the first floor. There is also a tendency of the systems with lower fundamental frequencies to oscillate more than the ones with higher fundamental frequencies. This information gives a misleading feeling of safety for structures with higher fundamental frequencies.

Graphs showing the fatigue damage factors as a function of the fundamental frequencies are furnished in Figs. 10 and 11. These figures show that in general, the first floor is damaged more than the higher floors. This appears to be in agreement with the observation made in Figs. 8 and 9 which also show the first floor deflects more than the upper floors. Figs. 10 and 11 indicate, however, that the structures with higher fundamental frequencies are damaged more than the ones with lower fundamental frequencies. This observation is in contradiction to the one noted in Figs. 8 and 9.

As the deflection of each floor depends on the yield deflection permitted in that floor, a dimensionless maximum deflection ratio, $y_{max.}/y_y$, was obtained by dividing the absolute value of the maximum deflection of a floor by the yield deflection of that floor. These dimensionless response ratios are plotted in Figs. 12 and 13. These graphs appear to be similar to the damage factor graphs presented in Figs. 10 and 11. They also depict the same observation that the structures with higher fundamental frequencies are damaged more than the ones with lower fundamental frequencies by the 1940 El Centro Earthquake. In other words, the structure with lower fundamental frequencies are safer than the ones with higher frequencies when they are hit by a random vibration similar to the 1940 El Centro Earthquake NS Component. It can also be concluded that the dimensionless maximum deflection ratio indirectly reflects the fatigue damage that is being caused in the structure. A question may be raised regarding the necessity of calculating the damage factor which involves an enormous amount of computer time. To analyze this aspect, the ratio of this maximum deflection ratio and the damage factor was plotted against the corresponding fundamental frequency of the system and these graphs are furnished in Figs. 14 and 15. It is seen that though there is a close relationship between the maximum deflection ratio and the damage factor, there is a possibility of underestimating the damage factors in higher floors, and also in the structures with lower frequencies if the maximum deflection ratio alone is studied. Further, the value of the maximum deflection ratio does not give any indication as to the safety of a structure with reference to its ultimate failure.

The following additional observations are also noted. All the floors are found to vibrate with approximately the same number of reversals. However, the upper floors are subjected to a larger number of plastic strain cycles than the first floor, though they are less damaged. The total duration of the earthquake function used in this study was 29.16 seconds. Analysis was also made by considering the record of the first 7.2 seconds duration. It was noticed that almost all the damage caused in the first floor during the entire excitation period of 29.16 seconds was caused during the first 7.2 seconds. However, there was a considerable increase in the damage done to higher floors during the later part of excitation, the greater will be the damage in the upper stories of the structure.

CONCLUSION

When a multi-story structure is stressed beyond the elastic limit by an earthquake excitation, it is possible that it may fail due to low-cycle fatigue damage. The first floor is the one which is damaged most heavily and most of this damage is caused during the early part of the excitation. The upper floors are less damaged, though they are subjected to a larger number of inelastic reversals than the ones induced in the first floor. It takes time for the fatigue damage to be built up in the upper floors and the damage increases if the period of excitation is prolonged. In general, the structures with higher fundamental frequencies (stiffer structures) are damaged to a greater extent than those with lesser frequencies by excitation similar to the 1940 El Centro Earthquake record used in this study. This difference in damage is less visible in the upper stories. The variation of damping coefficient has less effect in the upper stories than in the upper stories than in the first story. Additional strengthening of the first floor may be necessary to prevent a possible fatigue failure.

The study of the maximum relative deflection of each floor may not in itself be sufficient to analyze the safety of the structure. The dimensionless maximum deflection ratio, $y_{max.}/y_y$, appears to be approximately proportional to the fatigue damage factor in the structure. However, the value of this maximum deflection ratio does not reveal any clear meaning as regard to the safety of the structure, and hence it may be necessary to work out the fatigue damage factor values in structures which may undergo inelastic deformations. This paper presents a procedure for calculating the low-cycle fatigue damage factor produced in a multi-degree--freedom-system subjected to a random excitation.

ACKNOWLEDGEMENTS

This investigation was supported in part by the National Aeronautics and Space Administration. The writer wishes to thank Dr. A. I. Thomas, the President of Prairie View A. and M. College, and Mr. A. E. Greaux, the Associate Dean, School of Engineering, for their encouragement and the facilities offered in the college for this research. He also wishes to thank Dr. D. K. McCutchen of the Manned Spacecraft Center, Houston, Texas for his interest in the problem and for making valuable suggestions during the research. The help of Dr. J. M. Henry, Engineering and Applied Science, Mason Laboratory, Yale University, in formulating the computer program is acknowledged. In addition, acknowledgements are due to the staff of the Data Processing centers of Prairie View A. and M. College and of Texas A. and M. University for their cooperation in running the program on the IBM 360/65 Digital Computer.

REFERENCES

1. J. Penzien: Elasto-Plastic Response of Idealized Multi-Story Structures Subjected to a Strong-Motion Earthquake. Proceedings. Second World Conference on Earthquake Engineering, Vol. II., Tokyp and Kyoto, Japan, p. 739-760, (July, 1960).

2. A. S. Veletsos, N. M. Newmark, and C. V. Chelapati: Deformation Spectra for Elasto-Plastic Systems Subjected to Ground Shock and Earthquake Motion. Proceedings, Third World Earthquake Engineering, Vol. II, Auckland and Wellington, New Zealand, pp. 663-682, (January, 1965).

3. W. E. Saul, J. F. Fleming, and S. L. Lee: Dynamic Analysis of Bilinear Inelastic Multiple Storey Shear Buildings. Proceedings, Third World Conference on Earthquake Engineering, Vol. II, Auckland and Wellington, New Zealand, p. 533-551, (January, 1965).

4. S. C. Goel, G. V. Berg: Inelastic Earthquake Response of Tall Steel Frames. Journal of the Structural Division, Proceedings, of the ASCE, Vol. 94, pp. 1907-1934, (August, 1968).

5. H. Y. Yeh, and J. T. P. Yao: Response of Bilinear Structural Systems to Earthquake Loads. Proceedings of the ASME Vibrations Conference, Philadelphia, Pa., (March 30- April 2, 1969).

6. M. Shinozuka: Application of Stochastic Processes to Fatigue, Creep, and Catastrophic Failures. Seminar in the Application of Statistics in Structural Mechanics, Columbia University, (November, 1966).

7. I. Kasiraj, and J T. P. Yao: Fatigue Damage in Seismic Structures. Journal of the Structural Division. Proceedings of the ASCE, Vol. 95, p. 1673-1962, (August, 1969).

8. J. T. P. Yao, and W. H. Munse: Low Cycle Axial Fatigue Behavior of Mild Steel. American Society for Testing and Materials Special Technical Publication No. 338, p. 5-24, (1962).

9. Shinozuka and H. Itahaki: Response of a Multi-Story Frame Structure to Random Excitation. Technical Report No. 29, Institute for the Study of Fatigue and Reliability, Columbia University in the City of New York, (February, 1966).

10. I. Kasiraj: Low Cycle Fatigue Damage in Structures Subjected to Earthquake Excitation. Ph. D. Dissertation Submitted to the University of New Mexico, Albuquerque, (June, 1968).

11. Manual of Steel Construction. 6th Edition, American Institute of Steel Construction, Inc., New York.

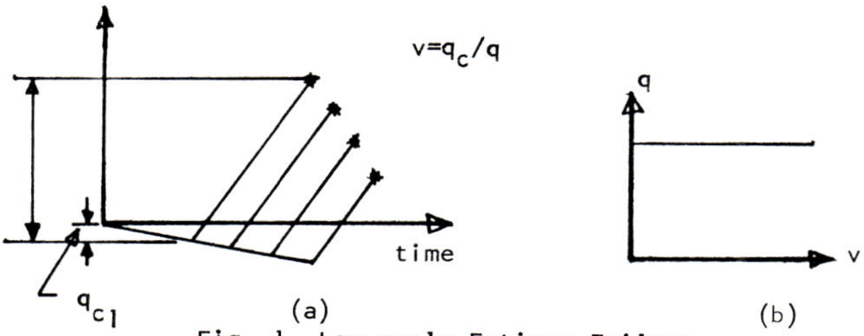

Fig. 1 Low cycle Fatigue Failure

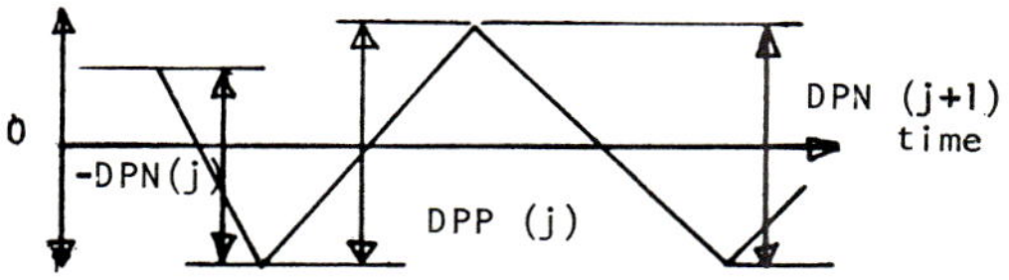

EPN EPP-Positive plastic Strain
EPN= Negative plastic strain

Fig. 2 Fatigue Damage In One Cycle

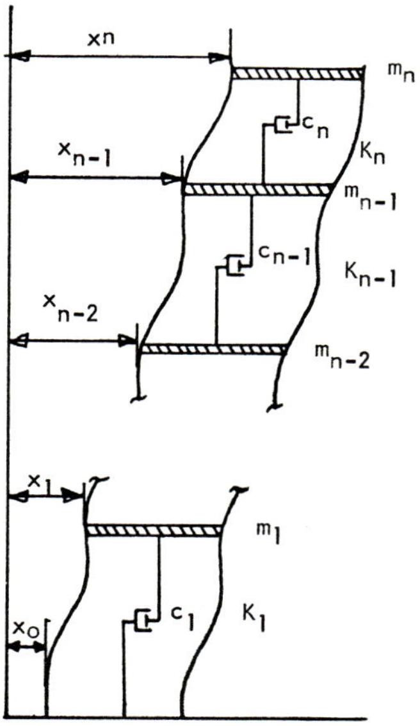

Fig. 3 A Shear Beam Structure

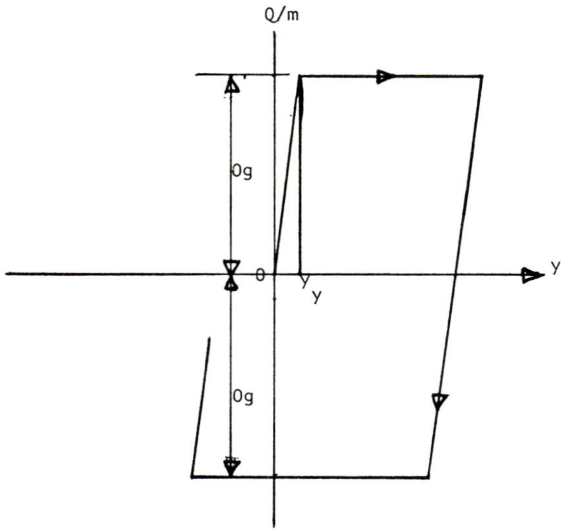

Fig. 4 Elasto-Plastic Force_Deflection Hysteresis Loop

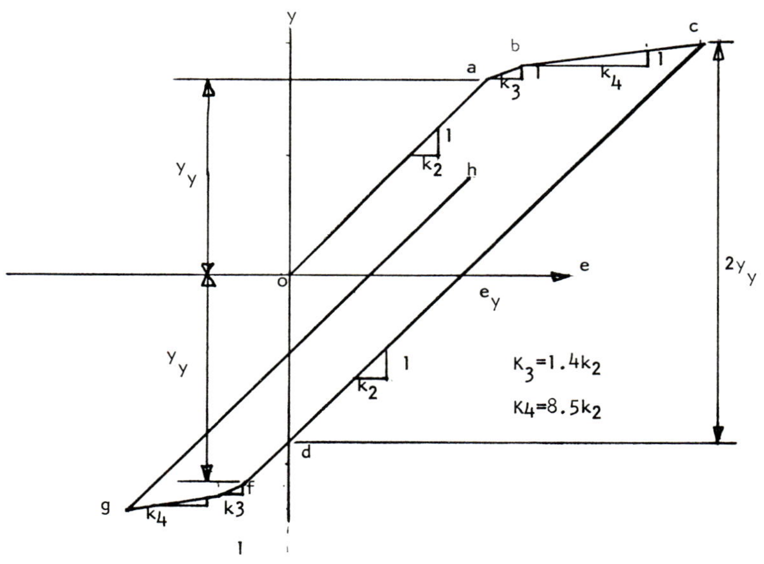

Fig. 5 Deflection-Maximun Strain Hysteresis Loop

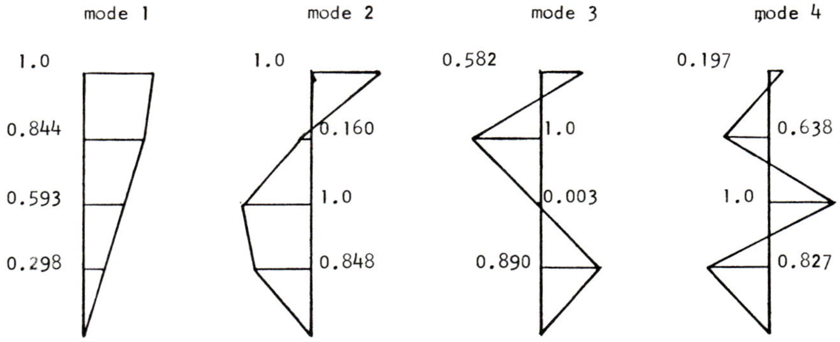

Fig. 6 Mode Shapes

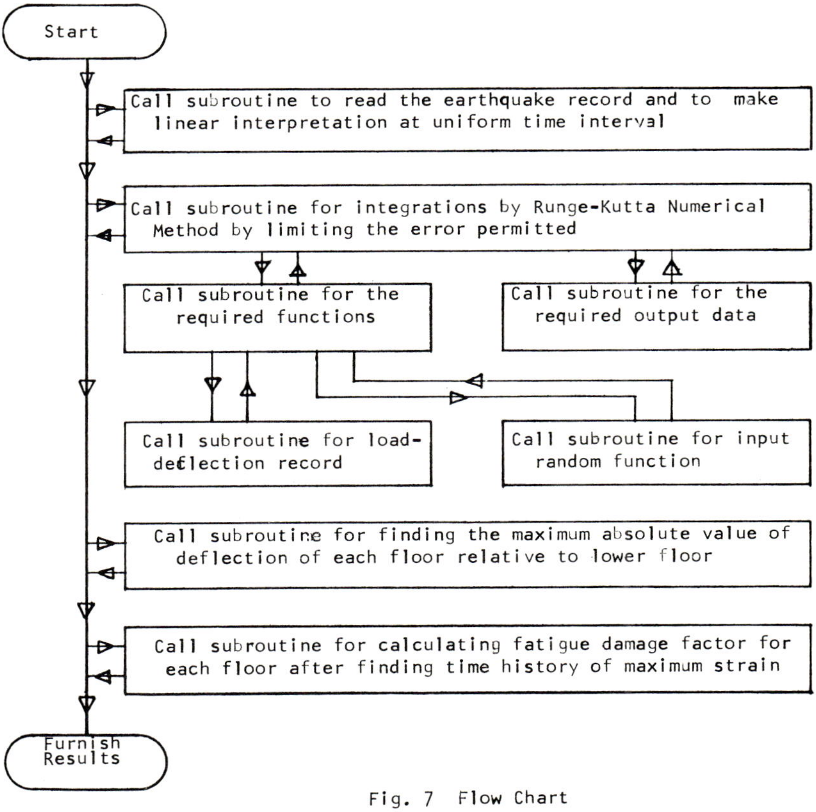

Fig. 7 Flow Chart

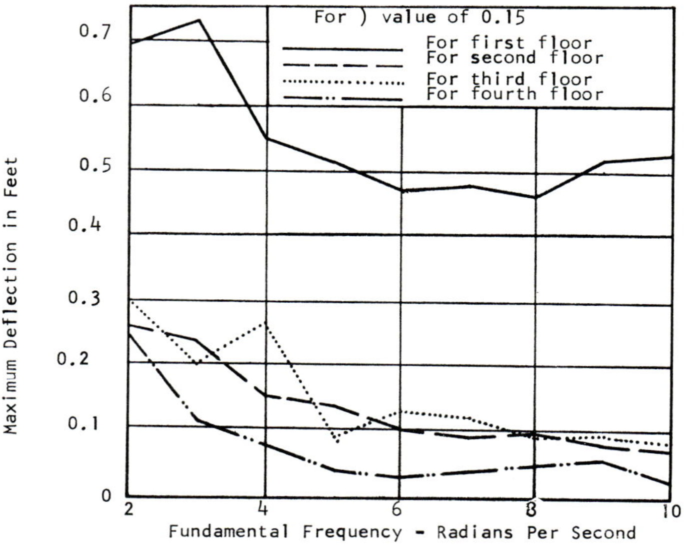

Fig. 8 Maximum Deflection For c = .2p/m

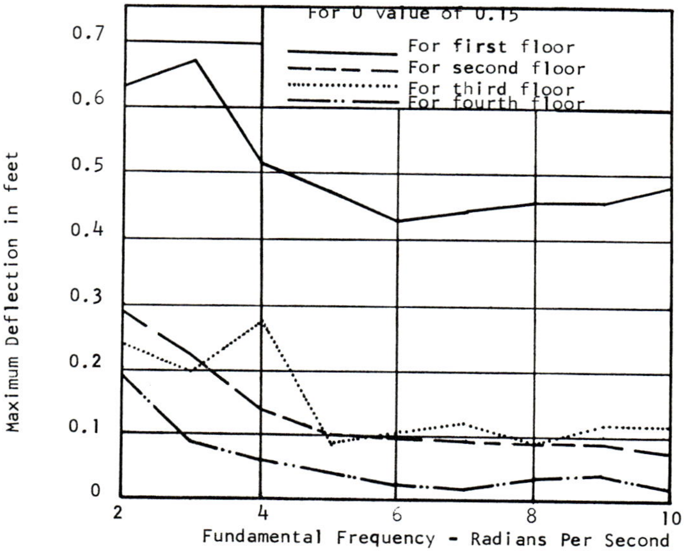

Fig. 9 Maximun Deflection For c = .3p/m

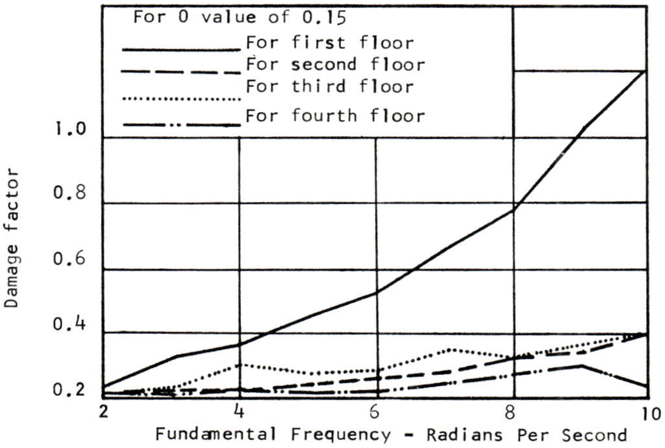

Fig. 10 Damage Factor For c= .2p/m

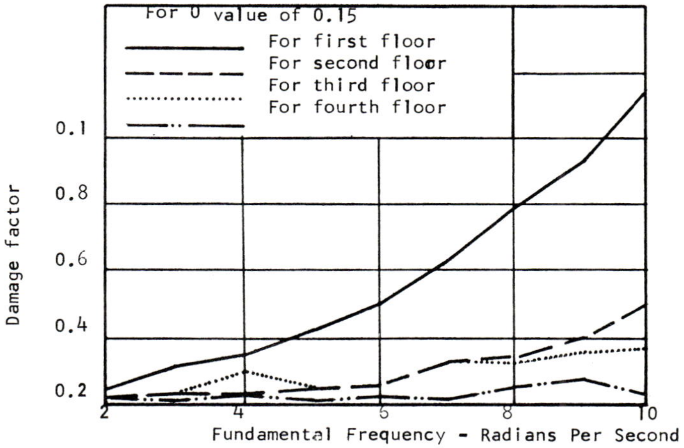

Fig. 11 Damage Factor For c = .3p/m

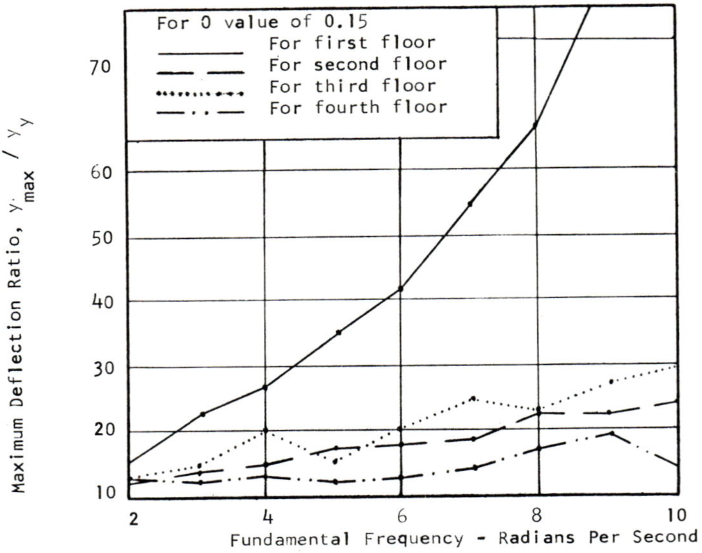

Fig. 12 Maximum Deflection Ratio For c = .2p/m

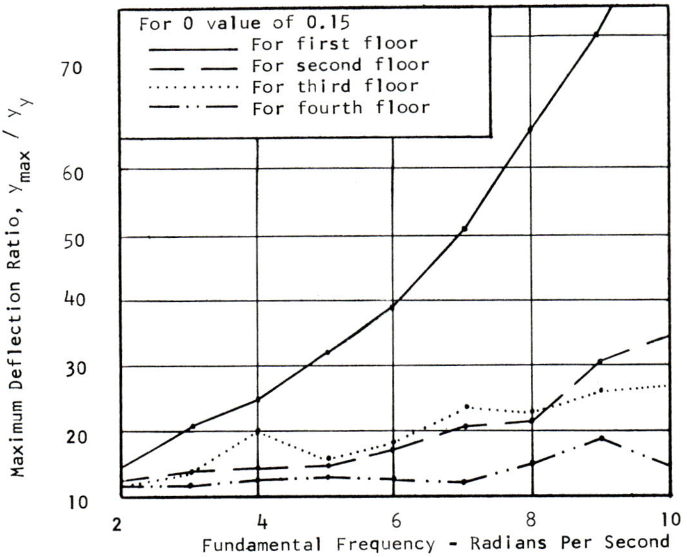

Fig. 13 Maximum Deflection Ratio For c = .3p/m

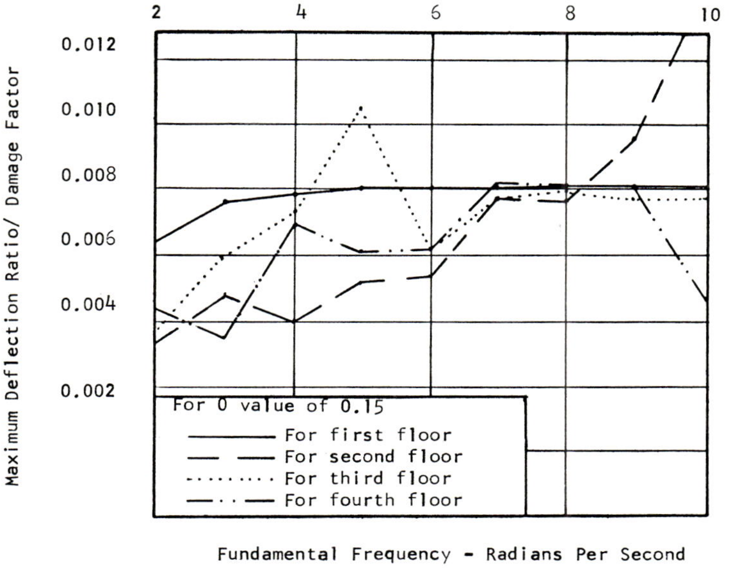

Fig. 14 Maximum Deflection Ratio/DF For c = .2p/m

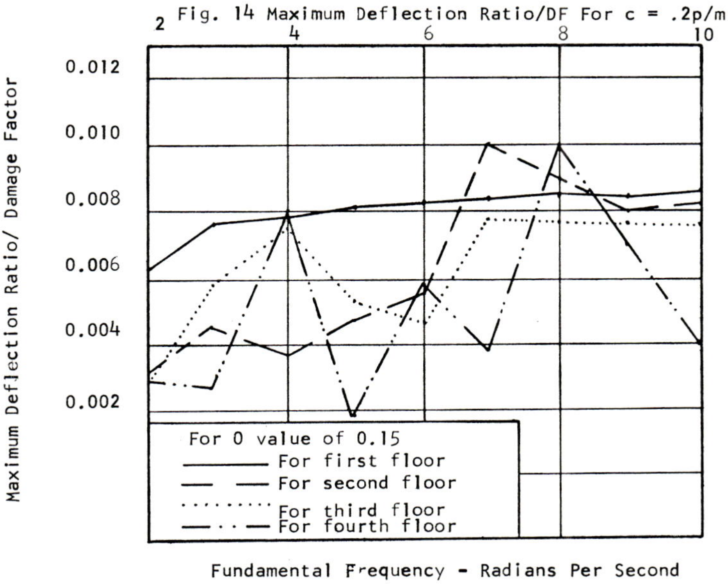

Fig. 15 Maximum Deflection Ratio/DF For c = .3p/m

VIBRATION OF PLATES

Developments in Mechanics, Vol. 6. Proceedings of the 12th Midwestern Mechanics Conference: 61

Transient Response of a Rectangular Plate Subjected to Lateral and Inplane Pressure Pulses

DAVID H. CHENG and LAWRENCE J. KNAPP
THE CITY COLLEGE OF THE CITY UNIVERSITY OF NEW YORK

ABSTRACT

The title problem simulates a window pane exposed to a far-field sonic-boom disturbance. The imposition of lateral and inplane pulses may be simultaneous or separated by a brief time-lag. In addition, there may be an inplane static load. Due to the presence of the inplane pulse load, the equation of motion is of the Mathieu type. An improved procedure in solving Mathieu's equation is presented. The solution is applied to a realistic plate glass and various effects such as the inplane static and dynamic loads, the pulse durations and time-lags are studied.

INTRODUCTION

The effects of sonic boom disturbances on structural elements have been extensively studied in recent years [1]. The disturbance was idealized as an N-Shaped pressure wave moving either parallel or normal to the surface of the structural element [2]. The pressure pulse is therefore laterally applied. It has been found that one of the vulnerable elements is the large plate glass window [3] in a store front. Under a certain unfavorable combination of circumstances, it is possible that a window pane may be subjected to inplane static (due to thermal expansion or other causes) and dynamic (due to oscillation of roof structure) loads in addition to the lateral N-Shaped pressure pulse.

Depending on the direction of propagation of the pressure wave relative to the location of the window in a building, the lateral motion of the window and the roof oscillation, which causes the inplane pulse on the window, do not, in general, start simultaneously. Should the roof oscillate before the pressure wave reaches the window, there will be a time-lag between the inplane and lateral pulses and so on. Since the pressure wave moves at the speed of sound, the interaction with a building in its wake is highly transient. Further, the motion will be tempered by damping which enhances the likelihood that the most severe stress would occur during the brief period of interaction of the lateral and inplane disturbances, or shortly thereafter.

The dynamic stability of a plate subjected to an inplane disturbance which is periodic and in steady state has been extensively studed [4] [5]. Since the primary aim of such studies was to delineate the phenomenon of dynamic stability, they are interested in large amplitude of motion and more importantly the solution behavior at large time. In contrast, the present paper is concerned with the maximum transient response of a plate subjected to an inplane as well as a lateral pulse. The intensity of the pulses is moderate and the support conditions of the plate are such that no dynamic instability would be a problem. Further, the plate response is assumed to involve only small deflections, thus the linear theory is considered valid.

Due to the presence of the inplane dynamic load in the form of a sine pulse, the equation of motion is of the Mathieu type. The homogeneous equation associated with the fundamental mode is solved by a procedure which was first suggested by Floquet [6], then developed by McLachlan [7] and is much improved in the present paper. For higher modes, a perturbation procedure is used to advantage. The nonhomogeneous equation is then solved by the standard method of variation of parameters. With a typical plate and loading condition, the effects of the inplane loads (both static and dynamic), the pulse duration (both inplane and lateral), and the time-delay between the inplane and lateral disturbances are studied in detail.

FORMULATION OF PROBLEM

Consider a simply supported rectangular plate, of sides a and b, and of uniform thickness h, which is subjected to an in-plane as well as a lateral disturbance. The lateral disturbance is characterized by an N-shaped pressure pulse followed with a time delay t_o by the in-plane disturbance characterized by a sine-pulse. The plate and the disturbances are illustrated in Fig. 1.

The equation of motion of the plate is:

$$D\nabla^4 w + [Q_o \sin\frac{2\pi}{\alpha\tau}(t-t_o)\, H(t-t_o)\, H(\alpha\tau+t_o-t) + N_y]\, w_{,yy} + \rho\ddot{w} = P_o(1-t/\tau)H(\tau-t) \quad (1)$$

in which comma represents derivative with respect to space and dot, derivative with respect to time. w is the lateral deflection of the plate, and H is the Heaviside function. Defining:

$$\bar{w} = \frac{w}{h}, \quad \bar{t} = t/\tau_{11}, \quad \bar{t}_o = t_o/\tau_{11}, \quad \bar{\tau} = \tau/\tau_{11} \quad (2)$$

where τ_{11} is the fundamental period of the plate, (1) may be written as:

$$\ddot{\bar{w}} + \tau_{11}^2 \frac{D}{\rho}\nabla^4 \bar{w} + \frac{\tau_{11}^2}{\rho}\Big[Q_o \sin\frac{2\pi}{\alpha\tau}\,\tau_{11}(\bar{t}-\bar{t}_o)\, H(\bar{t}-\bar{t}_o)\, H(\bar{t}-\bar{t}_o)\, H(\alpha\bar{\tau}+\bar{t}_o-\bar{t})$$

$$+ N_y\Big]\, w_{,yy} = \frac{\tau_{11}^2 P_o}{\rho h}(1-2\bar{t}/\bar{\tau})\, H(\bar{\tau}-\bar{t}) \quad (3)$$

The boundary conditions for the plate are:

$$x = 0 \quad \text{and } x = a, \qquad \bar{w} = \bar{w}_{,xx} = 0 \tag{4}$$

$$y = 0 \quad \text{and } y = b, \qquad \bar{w} = \bar{w}_{,yy} = 0 \tag{5}$$

Letting

$$\bar{w}(x,y,\bar{t}) = \sum_{m=1}^{\infty} \sum_{n=1}^{\infty} T_{mn}(\bar{t}) \sin \frac{m\pi x}{a} \sin \frac{n\pi y}{b} \tag{6}$$

and expanding the right side of (3) in double sine series, one gets for any (m,n) the following differential equation:

$$\ddot{T}_{mn} + T_{mn} \left\{ \left[\left(\frac{2\pi\tau_{11}}{\tau_{mn}} \right)^2 - (2n\pi)^2 \frac{N_y}{N_c} \right] - (2n\pi)^2 \frac{Q_o}{N_c} \sin \frac{2\pi}{\alpha\bar{\tau}} (\bar{t} - \bar{t}_o) H(\bar{t} - \bar{t}_o) H(\alpha\bar{\tau} + \bar{t}_o - \bar{t}) \right\}$$

$$= \frac{64 P_o b^2}{mn\pi^2 \sigma_c h^2} (1 - 2\bar{t}/\bar{\tau}) H(\bar{\tau} - \bar{t}) \tag{7}$$

in which the time derivative is taken with respect to $\bar{t}$ and

$$N_c = h\sigma_c \tag{8}$$

$$\sigma_c = \frac{\pi^2 E}{3(1-\nu^2)} \left(\frac{h}{b} \right)^2 \tag{9}$$

using the following transformation,

$$\eta = \frac{\pi}{\alpha\bar{\tau}} (\bar{t} - \bar{t}_o) - \frac{\pi}{4} \tag{10}$$

(7) becomes

$$\frac{d^2 T_{mn}(\eta)}{d\eta^2} + \left\{ A_{mn} - 2 q_n \cos 2\eta \, H\left[\frac{\alpha\bar{\tau}}{\pi}(\eta + \frac{\pi}{4}) \right] H\left[\alpha\bar{\tau} - \frac{\alpha\bar{\tau}}{\pi}(\eta + \frac{\pi}{4}) \right] \right\} \bar{T}_{mn}(\eta)$$

$$= h_{mn}(\eta) \, H(\bar{\tau} - \bar{t}_o - (\eta + \frac{\pi}{4}) \frac{\alpha\bar{\tau}}{\pi}) \tag{11}$$

with

$$A_{mn} = 4(\alpha\bar{\tau})^2 \left[\left(\frac{\tau_{11}}{\tau_{mn}} \right)^2 - \frac{N_y n^2}{N_c} \right] \tag{12}$$

$$q_n = 2 (n\alpha\bar{\tau})^2 \frac{Q_o}{N_c} \tag{13}$$

$$h_{mn} = \frac{64 P_o b^2}{mn\sigma_c h^2} \left(\frac{\alpha\bar{\tau}}{\pi^2} \right)^2 \left(1 - 2\frac{\alpha}{\pi}(\eta + \frac{\pi}{4}) - 2\frac{\bar{t}_o}{\bar{\tau}} \right) \tag{14}$$

Imposing the restriction that no tensile inplane load is transmitted to the plate, the following condition is specified:

$$N_y - Q_o \cos 2\eta \geq 0 \tag{15}$$

Three cases for the time-lag are considered namely, $\bar{t}_o < 0$, $\bar{t}_o = 0$ and $\bar{t}_o > 0$. In the time interval where the inplane pulse load is off, (11) becomes an ordinary differential equation with constant coefficients, and with proper initial conditions, the solution is easily obtainable. In the time interval where there is an inplane pulse but no lateral load, (11) becomes an equation of the Mathieu type. If there is an inplane pulse as well as a lateral load, (11) is an inhomogeneous Mathieu's equation. The procedure is to solve the Mathieu equation supplemented by the particular solution which may be obtained by the standard method of variation of parameters.

SOLUTION TO MATHIEU'S EQUATION

The following equations are the typical ones that require solution:

$$\ddot{T}(\eta) + (A - 2q\cos 2\eta)\, T(\eta) = h(\eta) \tag{16}$$

$$\ddot{T}(\eta) + (A - 2q\cos 2\eta)\, T(\eta) = 0 \tag{17}$$

where $\ddot{T}(\eta)$ represents the second derivative of T with respect to η, A and q are given as specified by (13). The initial conditions may be specified as follows:

$$\text{at} \quad \eta = \eta_i \qquad T(\eta_i) = D_1$$

$$T(\eta_i) = D_2$$

where D_1 and D_2 are prescribed or predetermined.

The solution to (17) may be written as [6] [8]:

$$T = K_1 e^{\mu\eta} \sum_{m=-\infty}^{\infty} C_m e^{m\eta i} + K_2 e^{-\mu\eta} \sum_{m=-\infty}^{\infty} C_m e^{-m\eta i} \tag{18}$$

where K_1 and K_2 will be determined by the initial conditions and μ is a number depending on A and q. Let $\mu = i\beta$, β is a real fraction. Then for m odd and even, (18) takes the following forms respectively:

$$T = K_1 \sum_{r=-\infty}^{\infty} C_{2r+1} \cos(2r+1+\beta)\eta + K_2 \sum_{r=-\infty}^{\infty} C_{2r+1} \sin(2r+1+\beta)\eta \tag{19a}$$

$$T = K_1 \sum_{r=-\infty}^{\infty} C_{2r} \cos(2r+\beta)\eta + K_2 \sum_{r=-\infty}^{\infty} C_{2r} \sin(2r+\beta)\eta \tag{19b}$$

Substituting either term in (19a) into (17) and setting the coefficient of $\cos(2r+1+\beta)\eta$ or $\sin(2r+1+\beta)\eta$ to zero for $r = -\infty$ to ∞, one obtains the recurrence relation:

$$[A - (2r+1+\beta)^2]\, C_{2r+1} - q(C_{2r+3} + C_{2r-1}) = 0 \tag{20a}$$

Similarly, using (19b) one gets:

$$[A-(2r+\beta)^2]C_{2r} - q(C_{2r+2} + C_{2r-2}) = 0 \qquad (20b)$$

Allowing r to take both positive and negative integer values as well as zero, a number of simultaneous equations in the same number of unknown coefficients are obtained by truncating (20) from both ends. The convergence of the series is usually assured by letting the largest $|r| = (\sqrt{A}/2 + N)$ where $N \geq 5$. The value of β may be determined as an eigenvalue from the simultaneous equations. This approach, however, is cumbersome and in most cases (except when A is very small) it is difficult to find β accurately. A more fruitful approach is to find an approximate or trial β and substitute it in the recurrence equations then solve for the coefficients. A method of evaluating the trial β is given in the appendix.

It is noted that an accurate evaluation of the coefficients C's depends on the correctness of the tiral β and the procedure by which they are determined. Once the C's are known, another or a check β may be deduced. If the C's so determined yield a check β which is the same as the trial β, it is an assurance that all the coefficients are correct. Otherwise, the results are in doubt and further improvement is necessary. Now assuming a reasonably accurate trial β has been obtained, the proper procedure is to set aside the equation having the smallest factor in absolute value for C_{2r+1} in (20a) (or for C_{2r} in (20b)). Each coefficient in the remaining equations is then normalized with respect to one of the coefficients, and the equations solved for the normalized coefficients. Finally, the check β is deduced from the equation which was singled out at the start.

The above procedure is different from that used by McLachlan [7] [8] who always sets aside the equation of $r = 0$ in (20) for recovering the check β. When this procedure is followed, Table 1 indicates that the check β is far from confirming the fact that the trial β used and therefore the coefficients C determined are correct. Numerous iterations must be tried and in many cases even this approach is not fruitful.

Table 1. Comparison of Trial and Check β by Existing Method [8]

A	q	Trial β	Check β
4.465	0.744	0.0893	0.0570
17.28	2.88	0.1251	0.1803
38.88	6.48	0.18957	0.12413
38.94	0.028	0.2399	6.2250
41.827	.211	0.4674	0.582
69.12	11.525	0.2535	0.371
84.27	14.045	0.11349	0.0572
108.00	18.00	0.31739	1.4828
125.14	15.68	0.155	-0.681
144.00	18.00	0.952	0.679

It should be pointed out that the trial β listed in Table 1 are obtained by an improved formula supplemented by an iteration procedure as outlined in the Appendix. They are more accurate than those obtained by the existing method [8]. This is confirmed by the fact that by adopting the new procedure outlined above, all the trial β's are confirmed to be exact.

For modes higher than the fundamental, A is usually very large, and much bigger than q, then another procedure, namely perturbation, is more efficient for getting the solution [8]. Rewrite (17) in the following form:

$$\ddot{T} + AT = (2q\cos 2\eta)T \tag{21}$$

First neglect the r.h.s., the solution then consists of $\cos\sqrt{A}\eta$ and $\sin\sqrt{A}\eta$. Now substitute $\cos\sqrt{A}\eta$ for T on the r.h.s. of (21), there results:

$$\ddot{T} + AT = q[\cos(\sqrt{A}+2)\eta + \cos(\sqrt{A}-2)\eta] \tag{22}$$

the particular integral for (22) can then be obtained. Substituting the latter for T in (21) on the r.h.s., another particular integral may be obtained. By repeating this procedure, and using also $\sin\sqrt{A}\eta$, the solution takes the form of an infinite series:

$$T = K_1 \sum_{r=-\infty}^{\infty} C_r \cos(\sqrt{A}-2r)\eta + K_2 \sum_{r=-\infty}^{\infty} C_r \sin(\sqrt{A}-2r)\eta \tag{23}$$

with the recurrence relation as follows:

$$4r(\sqrt{A}-r)C_r - q(C_{r-1}+C_{r+1}) = 0$$

If $r \ll \sqrt{A}$, the recurrence relation may be simplified as:

$$4r\,C_r - \frac{q}{\sqrt{A}}(C_{r-1}+C_{r+1}) = 0 \tag{24}$$

which is identical to the recurrence relation for the J-Bessel function provided that

$$C_r = J_r(q/2\sqrt{A}) \tag{25}$$

When $q/2\sqrt{A} \ll r$, J_r may be represented by the first term of its expansion, or

$$C_r \doteq (q/4\sqrt{A})^r/r! \tag{26}$$

Hence the coefficient decreases very rapidly as r increases. (23) may be expressed as:

$$T = K_1 \sum_{r=-r_o}^{r_o} J_r(q/2\sqrt{A})\cos(\sqrt{A}-2r)\eta + K_2 \sum_{r=-r_o}^{r_o} J_r(q/2\sqrt{A})\sin(\sqrt{A}-2r)\eta \tag{27}$$

where r_o represents the highest r at which the series may be truncated.

Once the solution to (17) becomes available, the solution to (16) can be obtained by the standard method of variation of parameters taking into consideration the proper initial conditions.

DYNAMIC RESPONSE OF A SQUARE PLATE SUBJECTED TO LATERAL AND INPLANE PRESSURE PULSES

The solution is now applied to a square glass plate. In evaluating the time function, the first nine modes are computed and the contribution of higher modes are neglected. Let the dynamic amplification factor (hereafter DAF) be defined as follows:

$$DAF = (\sigma_d + N_y/h + Q_c/h)/\sigma_s \tag{28}$$

in which σ_s is the maximum static stress in the plate under the uniform pressure p_o. The numerator of (28) represents the maximum dynamic stress in the plate. It consists of σ_d which is the dynamic bending stress; Q_c/h, the dynamic stress due to the inplane pulse and N_y/h, the static prestress. Q_c should not be confused with Q_o, which is the peak value whereas Q_c is the magnitude of the inplane pulse at the time the numerator of (28) becomes a maximum.

In the present instance, the DAF is defined merely for convenience. σ_s can be easily computed by the following expression [9]:

$$\sigma_s = 0.271\ p_o (b/h)^2 \tag{29}$$

with the poisson's ratio being assumed as 0.231 [10]. Once σ_s is known, the maximum dynamic stress in the plate can be evaluated with the help of a pre-computed DAF envelope as a function of $\overline{\tau}$ which is the normalized period.

Before constructing the DAF envelopes, a nine mode approximation is compared to a three mode approximation for a typical case and the results are listed in Table 2. The maximum response is obtained by scanning numerical results at small time intervals ($\overline{t}$= 0.01 or 0.02). It is clear that for all practical purposes, the three mode approximation is quite adequate for both deflection and DAF, which reflects an indication of rapid convergence of the proposed solution.

For no time-lag, the envelopes of DAF as functions of $\overline{\tau}$ are given for $Q_o/N_c = 1/4$ and 0 in Fig.2 to illustrate the effect of the dynamic inplane load. Figure 3 shows that the corresponding critical time is advanced by the presence of the inplane load to near the beginning of the lateral disturbance when $\overline{\tau}$ is slightly over one. This confirms the belief that the most severe stress would occur during the interaction of the lateral and inplane pulses with the exception of very flexible plates. It is noted that the effect of the inplane static load is substantial and can be seen in Fig.4. The corresponding critical time follows the same pattern as in the case where inplane static load is absent. The effect of lengthening the period of the inplane dynamic load is seen in Figs. 6 and 7. In general, the longer period gives smaller response.

The effect of time-lag can be seen from Table 3 which is compiled at R = 3 at which the DAF envelope is believed to have stabilized.

It is clear that with proper time-lag, the absolute maximum DAF would be well over 6. A detailed study of time history of lateral deflection reveals that the absolute maximum DAF would take place at the simultaneous occurrence of Q_o (the peak inplane load) and the maximum lateral deflection.

Table 2. Comparison of Maximum Response
Three versus Nine Modes

$p_o = 2, \quad N_y/N_c = 0, \quad Q_o/N_c = 1/4$

$\bar{\tau}$	Three modes		Nine modes	
	$\bar{w}$	DAF	$\bar{w}$	DAF
.60	0.422	2.190	0.424	2.270
.80	0.559	2.821	0.560	2.881
.95	0.581	2.830	0.581	2.869
1.20	0.512	2.565	0.512	2.644
1.40	0.453	2.568	0.454	2.633
1.60	0.438	2.763	0.440	2.831
1.80	0.452	2.882	0.454	2.967
1.97	0.518	2.940	0.519	3.015
2.20	0.504	2.983	0.505	3.073
2.40	0.475	2.999	0.477	3.088
2.60	0.479	3.000	0.480	3.053
2.80	0.482	2.999	0.483	3.091
3.00	0.485	2.999	0.486	3.082

Table 3. Maximum DAF and Corresponding Time t_c

$p_o = 1, \quad Q_o/N_c = 1/4, \quad b/h = 240$

Loading Condition		$\bar{t}_o/\bar{\tau} = -0.25$		$\bar{t}_o/\bar{\tau} = -0.10$		$\bar{t}_o/\bar{\tau} = 0$		$\bar{t}_o/\bar{\tau} = 0.10$		$\bar{t}_o/\bar{\tau} = 0.25$	
		DAF	$\bar{t}_c$	DAF	$\bar{t}_c$	DAF	$\bar{t}_c$	DAF	$\bar{t}_c$	DAF	$\bar{t}_c$
$\alpha=1$	$N_y/N_c = 0$	3.25	.17	3.97	.168	3.50	.18	2.72	.17	2.67	.50
	$N_y/N_c=1/4$	5.13	.18	6.02	.198	5.66	.22	4.64	.17	4.00	.175
$\alpha=2$	$N_y/N_c = 0$	3.85	.165	3.33	.162	2.91	.16	3.10	.748	3.14	.684
	$N_y/N_c=1/4$	6.00	.20	5.41	.18	4.92	.18	5.36	.874	5.92	.85

CONCLUSIONS

It is found that the inplane dynamic load induces substantially higher stress in the plate. The actual stress can be easily evaluated with the help of the DAF envelopes presented here. The highest stress is likely to occur during the interaction of the lateral and inplane pulses with the exception of very flexible plates. The time-delay could cause further increase in the dynamic amplification factor or stress. For a square plate glass window, if the static inplane load and the peak value of the dynamic inplane load are both one-fourth of the static buckling load of the plate, and if the peak value of the lateral pressure is one pound per square foot, the maximum dynamic stress in the plate could reach more than six times that induced statically by the peak lateral pressure uniformly distributed over the plate.

ACKNOWLEDGMENT

The results contained herein are obtained under Grant NGR 33-013-039 which is supported by the Federal Aviation Agency and monitored by the NASA Langley Research Center.

REFERENCES

[1] Cheng, D.H. and Benveniste, J.E., "Dynamic Response of Structural Elements Exposed to Sonic Booms" NASA-CR-1281, March 1969.

[2] Cheng, D.H. and Benveniste, J.E., "Transient Response of Structural Elements to Traveling Pressure Waves of Arbitrary Shape" Int. J. Mech. Sci., 8, 1966, pp 607-618.

[3] Lowery, L. and Andrew, D.K., "Acoustical and Vibrational Studies Relating to an Occurrence of Sonic Boom Induced Damage to a Window in a Store Front," NASA-CR-66170, July 1966.

[4] Bolotin, V.V.: Dynamic Stability of Elastic Systems, Holden-Day, 1964.

[5] Somerset, J.H. and Evan-Iwanowski, R.M., "Influence of Non-linear Inertia on the Parametric Response of Rectangular Plates," Int. J. of Non-linear Mech., Vol.2, 1967.

[6] Floquet, A.M.G., "Sur les equations differentielles lineaires," Ann. de l'Ecole Normale Superieure, 12, 47, 1883.

[7] McLachlan, N.W., "Computation of Solution of Mathieu's Equation," Phil. Mag., 36, 403, 1945.

[8] McLachlan, N.W.: Theory and Application of Mathieu Functions Oxford University Press, 1947.

[9] Timoshenko, S. and Woinowsky-Krieger, S.: Theory of Plates and Shells McGraw-Hill, 1959.

[10] McKinley, R.W., "Response of Glass Windows to Sonic Booms" ASTM Symposium on Effects of Sonic Boom on Buildings, pp. 594-600, November 1964.

Alternatively the solution of (17) may be written after McLachlan [8] as:

$$T = \tilde{K}_1 Ce_d(\eta, q) + \tilde{K}_2 Se_d(\eta, q) \quad \text{(A 1)}$$

in which $\tilde{K}_1$ and $\tilde{K}_2$ are constants to be determined by the initial conditions and

$$Ce_d = \cos d\eta + \sum_{r=1}^{\infty} q^r \tilde{C}_r(\eta, d) \quad \text{(A 2)}$$

$$Se_d = \sin d\eta + \sum_{r=1}^{\infty} q^r \tilde{S}_r(\eta, d) \quad \text{(A 3)}$$

which are known as Mathieu functions of fractional order. The functions $\tilde{C}_r(\eta, d)$ and $\tilde{S}_r(\eta, d)$ are still to be determined. d is a number, its fractional part is represented by β or $0 < \beta < 1$.

It is noted from (17) that A and q must be related so that when q vanishes A reduces to d^2 and the solution degenerates to the first terms of (A 2) and (A 3). Letting

$$A = d^2 + \sum_{r=1}^{\infty} \alpha_r q^r \quad \text{(A 4)}$$

and substituting (A 1) to (A 4) into (17) and collecting coefficients of like powers of q, there results an infinite number of ordinary differential equations in $\tilde{C}_r$ and $\tilde{S}_r$ which can be solved in sequence. $\tilde{C}_r$, $\tilde{S}_r$ and α_r can thus be determined. McLachlan [8] has given the α_r up to α_6. To further improve the value of β, the derivation is extended here to give the term α_8. It can be shown that the α_r with odd indices vanish and those with even indices are:

$$\begin{aligned}
\alpha_2 &= \frac{1}{2(d^2-1)} \\
\alpha_4 &= \frac{5d^2+7}{32(d^2-1)^3(d^2-4)} \\
\alpha_6 &= \frac{9d^4+58d^2+29}{64(d^2-1)^5(d^2-4)(d^2-9)} \\
\alpha_8 &= \frac{1469d^{10}+9144d^8-14035d^6+64228d^4+827565d^2+274748}{64(128)(d^2-1)^7(d^2-4)^3(d^2-9)(d^2-16)}
\end{aligned} \quad \text{(A 5)}$$

It is seen from (A 4) that if $\alpha_2 q^2 << d^2$, the series is rapidly convergent, as a first approximation, one sets $d^2 \doteq A$. Inserting A for d^2 in α_2 of (A 5), then, from (A 4) one gets as a second approximation:

$$d^2 = A - \frac{q^2}{2(A-1)} \quad \text{(A 6)}$$

with higher order terms neglected. Substituting (A 6) for d^2 in α_2, but retaining $d^2 \doteq A$ for the other α_r of (A 5), (A 4) becomes:

$$d^2 \doteq \Big[A - \frac{(A-1)q^2}{2(A-1)^2 - q^2} - \frac{(5A+7)q^4}{32(A-1)^3(A-4)} - \frac{(9A^2+58A+29)q^6}{64(A-1)^5(A-4)(A-9)}$$

$$- \frac{(1469A^5+9144A^4-14035A^3+64228A^2+827565A+274748)q^8}{64(128)(A-1)^7(A-4)^3(A-9)(A-16)}\Big] \qquad (A\ 7)$$

Now β may be computed by the following procedure: substituting (A 7) into (A 5) to evaluate the α's which in turn are substituted into (A 4) to compute a new d. The value of d is substituted into (A 5) to determine the new α's which in turn are substituted into (A 4) to compute another d. This procedure is repeated until no substantial change takes place between two iterations. Once d is obtained, its fractional part, β can be extracted. It was found that β so determined is extremely accurate up to four or five significant figures.

NOTATION

Unless otherwise specified in the text, the following nomenclature will be used throughout:

a,b	Plate lengths
A_{mn} (or A)	Coefficient in Mathieu's equation and defined in Eq.(12)
$\tilde{C}_r$	Cosine function
C_m	Coefficient
Ce_d	Mathieu's function of fractional order d
D	Flexural rigidity of plate, $Eh^3/12(1-\nu^2)$
DAF	Dynamic amplification factor as defined in Eq.(28)
E	Young's modulus
h	Thickness of plate
h_{mn}	As defined in Eq.(14)
H	Heaviside function
J	Bessel's function
N_c	Static buckling load = $h\sigma_c$
N_y	Inplane static load in y-direction
p_o	Peak overpressure of N-shaped pulse, psf
q_n(or q)	Coefficient in Mathieu's equation and defined in Eq.(13)
Q_o	Maximum amplitude of dynamic inplane disturbance
Q_c	Amplitude of dynamic inplane disturbance at the time when the combined dynamic stress is a maximum
$\tilde{S}_r$	Sine function
Se_d	Mathieu's function of fractional order d
t	Time
$\bar{t}$	Dimensionless time ratio t/τ_{11}

t_o Time delay

$\overline{t}_o$ Dimensionless Time delay ratio t_o/τ_{11}

T_{mn} Time-function as in Eq. (6)

w Lateral deflection

$\overline{w}$ Dimensionless lateral deflection w/h

∇^4 Biharmonic operator

α Coefficient defining duration of inplane disturbance relative to lateral disturbance

α_r Coefficients as defined in Eq. (A 4)

σ_c Static buckling stress $= \dfrac{\pi^2 E}{3(1-\nu^2)}\left(\dfrac{h}{b}\right)^2$

σ_d Dynamic stress due to bending

σ_s Maximum static stress in plate due to uniform p_o

η Transformed dimensionless time

τ Duration of N-shaped disturbance

τ_{11} Fundamental period of plate

$\overline{\tau}$ Dimensionless duration of N-shaped disturbance τ/τ_{11}

τ_{mn} Natural period of plate $= \dfrac{2}{\pi}\dfrac{\rho}{D}\left[\left(\dfrac{n}{a}\right)^2 + \left(\dfrac{n}{b}\right)^2\right]^{-1}$

ρ Mass per unit area of plate

ν Poisson's ratio

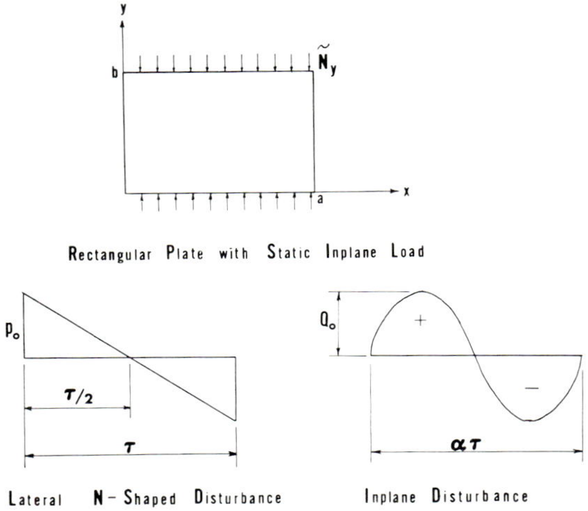

Figure 1

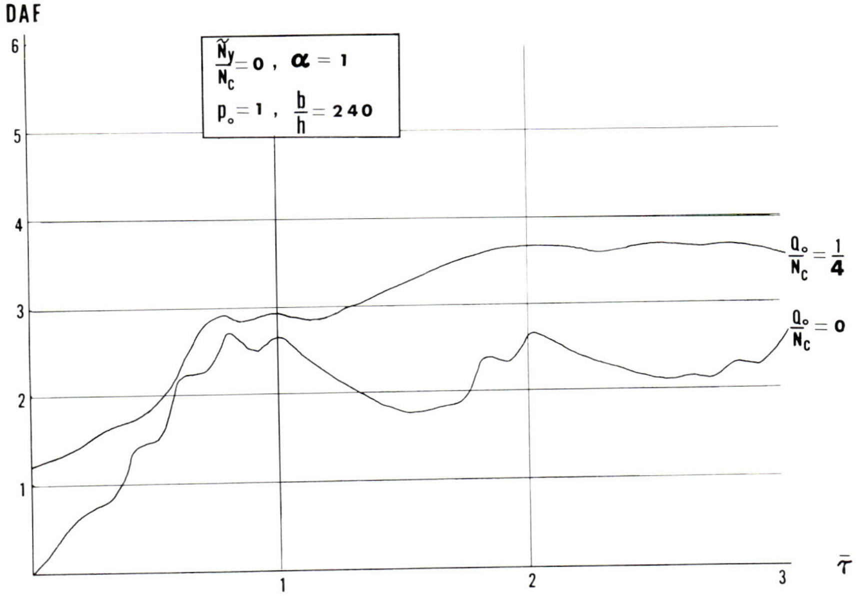

Fig. 2

DAF vs. Period Ratio of a Square Plate

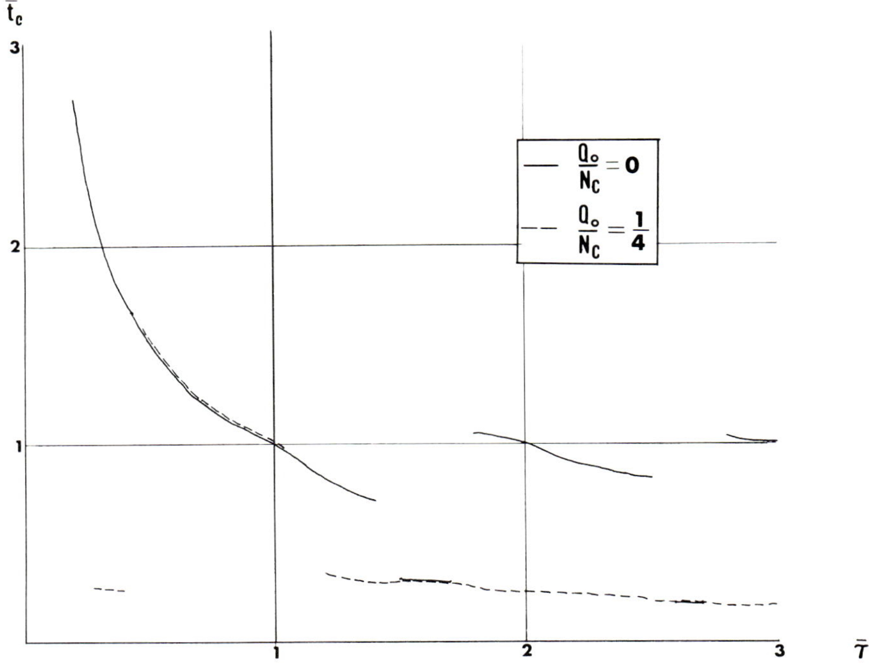

Fig. 3 Critical Time Corresponding to Fig. 2

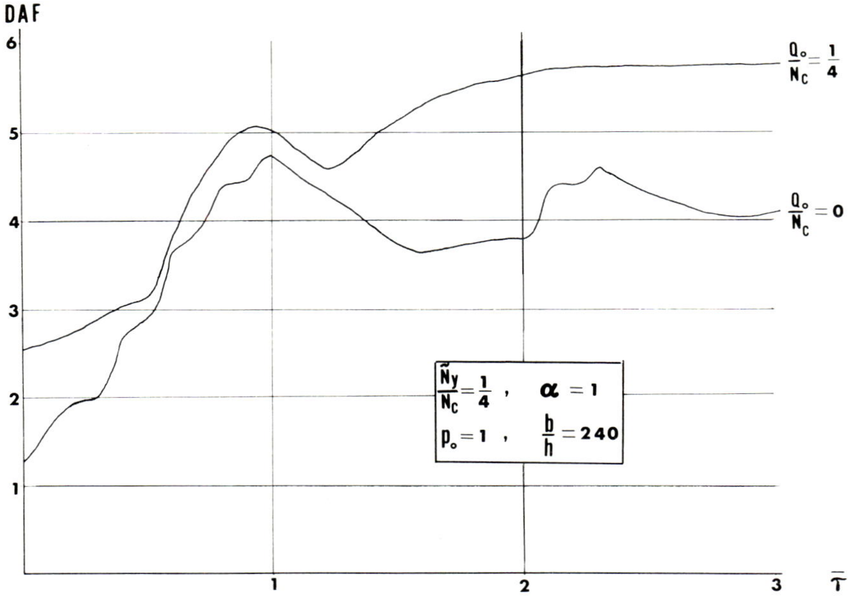

Fig. 4 DAF vs. Period Ratio of a Square Plate

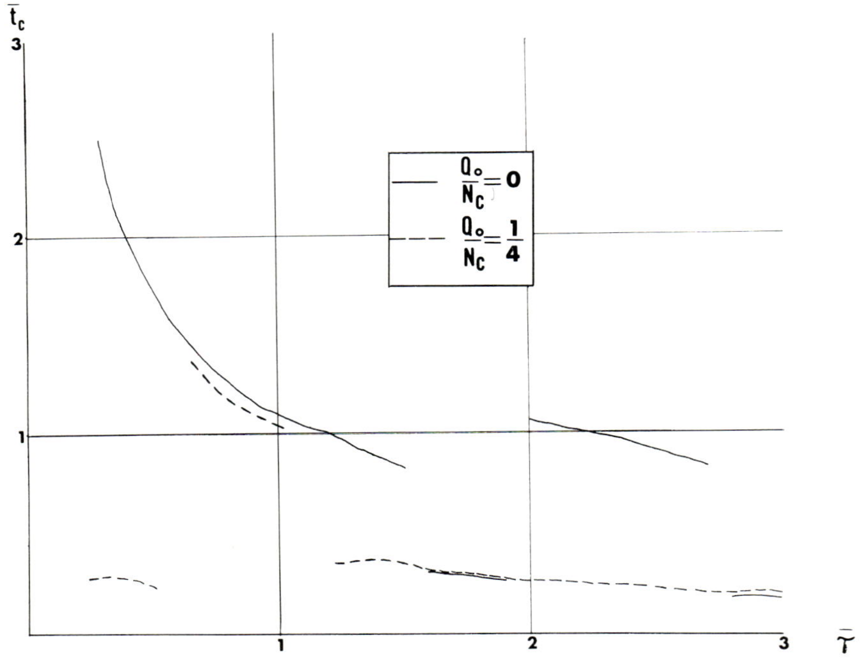

Fig. 5 Critical Time Corresponding to Fig. 4

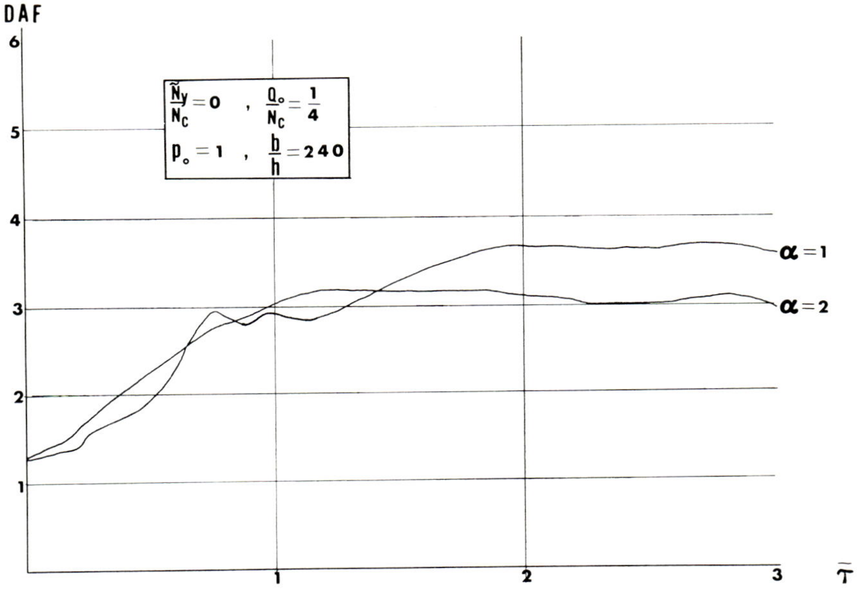

Fig. 6 Effect of Duration of Inplane Dynamic Load — No Static Inplane Load

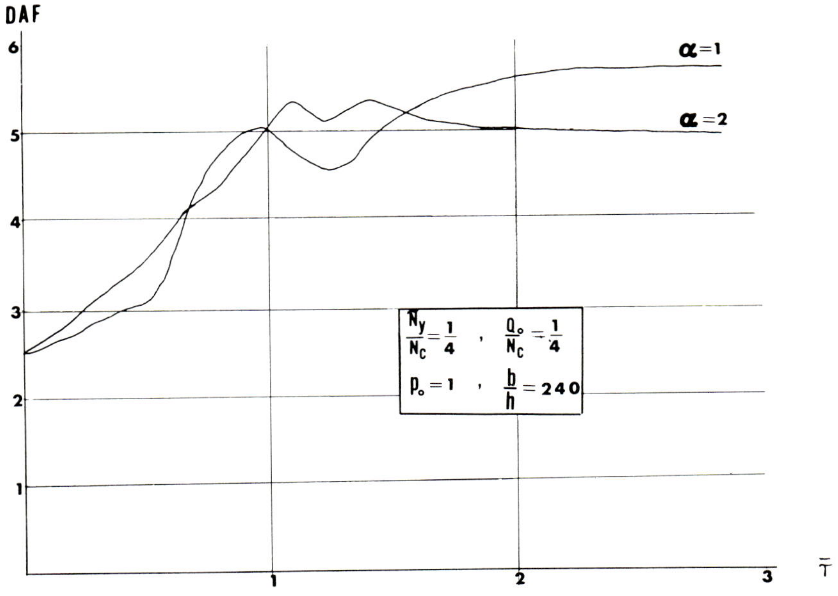

Fig. 7 Effect of Duration of Inplane Dynamic Load—With Static Inplane Load

Developments in Mechanics, Vol. 6. Proceedings of the 12th Midwestern Mechanics Conference: 62

Natural Vibrations of Rectangular Laminated Orthotropic Plates

F. K. W. TSO, S. B. DONG and R. B. NELSON
UNIVERSITY OF CALIFORNIA, LOS ANGELES

ABSTRACT

The vibrations of rectangular laminated orthotropic plates with simple supports are studied using an extended Ritz technique. The analysis is based on the complete theory of elasticity and may be used to study composite plates of a very general type over a large range of frequencies and wave lengths. Examples with known solutions are presented as a check. A three-layer, transversely isotropic plate is investiaged to illustrate the effects of laminated orthotropic structure on the frequencies and the displacement and stress modal patterns.

INTRODUCTION

The interest in and use of composite materials in structural applications have led to substantial research into the behavior of composite structures. An important class of these investigations is concerned with the dynamic behavior of laminated elastic plates.

Most of the work in this area falls into three categories. In the first, the plate is studied using an approximate theory in which statements are made specifying the behavior of the plate through its thickness, the most common of which is the Love-Kirchhoff hypothesis [1,2,3,4]. The validity of these theories, aside from the usual restrictions imposed upon them, is questionable when the material properties differ appreciably from layer to layer and/or when a high degree of orthotropy exists in one or more layers [5].

The second category may be characterized as approximate elasticity solutions [6, 7, 8, 9], in which attempts are made to bridge the gap between the approximate plate and the general theory of elasticity formulations, or to use smoothing and averaging procedures on special laminates with periodic structure through the thickness. These approaches lead to "effective modulus" and "effective stiffness" theories, which are again approximate.

In the final group of investigations, vibrations of (infinite) laminated plates are based on the complete three-dimensional theory of elasticity. In studies of this type, solutions of the field equations for the layers are combined in accordance with the interface traction and displacement continuity and free surface conditions. This leads to a characteristic equation which is solved in order to give natural frequencies. However, for all but the simplest cases, such as two layer or sandwich constructions [10, 11, 12], the algebra involved in both the generation and solution of the frequency equation becomes intractable. To date, much of the work has been confined to the long-wave-length, low-frequency region. Also, the nature of stress and displacement distributions through the thickness has not been well-studied since they cannot be computed until the frequency is known, and then only with substantial effort. As a consequence, the elasticity approach, although quite general, results in such complicated algebraic problems that very little is known at present about the behavior of general laminated plates.

In this paper a method is presented which may be used to determine the frequencies and modal patterns of vibrations of infinite rectangular plates composed of an arbitrary layout of orthotropic laminates with the principal elastic axes coinciding with the coordinate axes. This work is an extension of a previous study in which only plane strain vibrations were considered [13], and thus contains three-dimensional plate behavior. The analysis is based on the complete linear theory of elasticity and is therefore valid for all frequencies and wave lengths. Specifically, a special form of the Ritz technique is used herein, called the "Extended Ritz Technique" by certain authors [14] and commonly known as the finite element method. The main advantage of the method is the circumvention of the formidable algebra associated with the classical method of generating and solving frequency equations. The basic scheme involves expressing the displacement form in the plane of the plate explicitly and solving for the dependence through the thickness using discrete or generalized coordinates. In order to model the behavior with sufficient accuracy, a large number of generalized coordinates is necessary; the resulting algebraic eigenvalue problem is then solved by an efficient solution technique [15], which utilizes a reduced system of generalized coordinates. For this paper, the ten lowest branches of the frequency spectrum for real wave numbers were determined. The modal patterns associated with the frequencies were computed simultaneously from which the stress distributions through the plate thickness were determined. A homogeneous plate and a three-ply isotropic plate which were previously studied [12] are presented as an accuracy check of the present method. Additional examples are presented to illustrate the effect of particular orthotropy in the layers of a laminated plate.

FORMULATION AND SOLUTION METHOD

In this version of the Extended Ritz Technique, the plate is idealized by a system of thin laminas. Each lamina is assigned distinct orthotropic material properties, density and thickness. Here a lamina is not to be confused with a laminate of the plate. It is possible to model a laminate with a number of laminas in order to achieve a better representation of the physical behavior. Let x_i $(i = 1,2,3)$ be a right-handed system of rectangular coordinates, with x_3 normal to the plane of the plate, and t represent time. The displacements along these coordinate directions are denoted by u_i. In the present analysis, an admissible wave form for each lamina is:

$$\begin{aligned} u_1(x_i,t) &= U_1(x_3,t)\cos(\pi x_1/L_1)\sin(\pi x_2/L_2) \\ u_2(x_i,t) &= U_2(x_3,t)\sin(\pi x_1/L_1)\cos(\pi x_2/L_2) \\ u_3(x_i,t) &= U_3(x_3,t)\sin(\pi x_1/L_1)\sin(\pi x_2/L_2) \end{aligned} \tag{1}$$

Although this wave-form is appropriate for a plate with infinite extent in both directions, the conditions satisfied on the four edges ($x_1 = 0,L_1$ and $x_2 = 0,L_2$) of the nodal rectangle (described by $u_3 = 0$) are

$$u_t = 0,\ u_3 = 0 \text{ and } \tau_{nn} = 0$$

where τ_{nn} is the normal stress and u_t is the tangential displacement. Such a system of conditions is equivalent to simple supports on these four edges. The functions $U_i(x_3,t)$ are as yet the undetermined distributions of u_i through the thickness. In this paper, these functions are approximated by quadratic interpolations through the thickness

$$U_i(x_3,t) = U_{ib}(t)\left\{1-3\hat{x}+2\hat{x}^2\right\} + U_{im}(t)\left\{4\hat{x}-4\hat{x}^2\right\} + U_{if}(t)\left\{-\hat{x}+2\hat{x}^2\right\} \quad (i=1,2,3) \tag{3}$$

where the subscripts b, m, f indicate the generalized coordinates at the back, mid and forward nodal planes and $\hat{x}$ is a normalized transverse ordinate defined by:

$$\hat{x} = (x_3-x_{3b})/(x_{3f}-x_{3b}) \quad 0 \leq \hat{x} \leq 1 \tag{4}$$

where x_{3b} and x_{3f} are coordinates of the bounding surfaces of the lamina. Note that a two-point or linear interpolation using coordinates at the back and front nodal surfaces only and deleting the mid-node could have been used from the standpoint of meeting the necessary geometrical continuity requirements. The present three-node interpolation was adopted with the anticipation that fewer laminas would be required for an accurate solution and also with the hope of modeling any high displacement gradients more precisely.

The potential and kinetic energies for a particular lamina are obtained by integrating the strain and kinetic energy densities over the volume consisting of the laminar thickness and rectangular planform defined by one wave length in each of the coordinate directions:

$$V = \frac{1}{2}\int_0^{L_1}\int_0^{L_2}\int_{x_{3b}}^{x_{3f}}\Big\{C_{11}{}_{11}^{2} + C_{22}\epsilon_{22}^2 + C_{33}\epsilon_{33}^2 + 2C_{12}\epsilon_{11}\epsilon_{22} + 2C_{13}\epsilon_{11}\epsilon_{33} + 2C_{23}\epsilon_{22}\epsilon_{33} + 4C_{44}\epsilon_{13}^2 + 4C_{55}\epsilon_{23}^2 + 4C_{66}\epsilon_{12}^2\Big\}dx_3dx_1dx_2 \tag{5}$$

$$T = \frac{1}{2}\int_0^{L_1}\int_0^{L_2}\int_{x_{3b}}^{x_{3f}}\rho\left\{\dot{u}_1^2 + \dot{u}_2^2 + \dot{u}_3^2\right\}dx_3dx_1dx_2 \tag{6}$$

where ϵ_{ij}, C_{ij} and ρ are the cartesian components of the small strain tensor, the elastic moduli and the mass density, respectively, of the lamina. After differentiating the assumed displacement field and substituting the appropriate forms into the potential and kinetic energy expressions, there results

$$V = \frac{1}{2}\{r\}^T[k]\{r\} \tag{7}$$

$$T = \frac{1}{2}\{\dot{r}\}^T[m]\{\dot{r}\} \tag{8}$$

where $\{r\}$ is the ordered set of generalized coordinates defined by:

$$\{r\}^T = \left\{U_{1b}, U_{2b}, U_{3b}, U_{1m}, U_{2m}, U_{3m}, U_{1f}, U_{2f}, U_{3f}\right\} \tag{9}$$

and $[k]$ and $[m]$ are the stiffness and mass matrices of the lamina. Their explicit forms are summarized in the Appendix.

The expression for the potential and kinetic energies of the entire plate is found by summation of all the lamina energies,

$$L = \sum\,[T-V] = \frac{1}{2}\Big[\{\dot{U}\}^T[M]\{\dot{U}\}-\{U\}[K]\{U\}\Big] \tag{10}$$

where $\{U\}$ is an ordered set of generalized coordinates for the entire plate and $[K]$ and $[M]$ are the plate stiffness and mass matrices. Application of Hamilton's Principle on the functional defined by Eq. (10) gives the equations of motion of the plate in discrete coordinates:

$$[K]\{U\} + [M]\{\ddot{U}\} = 0 \tag{11}$$

For simple harmonic motion, the generalized coordinate, $\{U\}$, is taken as

$$\left\{U(x_3,t)\right\} = \left\{U_o(x_3)\right\}e^{i\omega t} \tag{12}$$

Substitution of this form into Eq. (11) gives the algebraic eigenvalue problem

$$\left\{[K] - \omega^2[M]\right\}\{U_o\} = 0 \tag{13}$$

Equation (13) is solved by an efficient direct-iterative eigensolution technique, described in Ref. [15]. The advantage of this solution is a sizeable reduction in computational effort without sacrificing the accurate modeling capability with a large number of generalized coordinates. The essence of the method is a reduction in the rank of the algebraic eigensystem by using a limited number of suitably chosen generalized coordinates, coupled with an iteration technique for obtaining eigenvalues. In this paper, the computer code was written using fifteen reduced generalized coordinates for a given pair of wave lengths, L_1, L_2; iteration was performed until the lowest ten frequencies and their corresponding modal patterns converged. This particular strategy resulted in the computation of eigenvalues and eigenvectors to within very close tolerances in two or three iterations for most cases, and was thus very efficient.

Calculation of stresses is a straightforward step when the modal patterns are known. By evaluating the strains which are in each lamina, the stresses in the lamina can be obtained through the constitutive relations.

$$\begin{aligned}
\tau_{11} &= C_{11}\epsilon_{11} + C_{12}\epsilon_{22} + C_{13}\epsilon_{33} \\
\tau_{22} &= C_{12}\epsilon_{11} + C_{22}\epsilon_{22} + C_{23}\epsilon_{33} \\
\tau_{33} &= C_{13}\epsilon_{11} + C_{23}\epsilon_{22} + C_{33}\epsilon_{33} \\
\tau_{13} &= 2C_{44}\epsilon_{13} \\
\tau_{23} &= 2C_{55}\epsilon_{23} \\
\tau_{12} &= 2C_{66}\epsilon_{12}
\end{aligned} \tag{14}$$

COMPARISON OF ISOTROPIC PLATE RESULTS

Comparisons of two examples on isotropic plates are summarized in Tables I, II, III to assess the accuracy and effectiveness of the present solution method. Both of these cases were solved analytically by Srinivas, Rao and Rao [12] who, in addition, made comparisons with classical and refined plate theories [16,17].

In Tables I and II are comparisons of the frequencies and stresses for selected modes of a homogeneous isotropic plate with Poisson's ratio $\nu=0.3$. Twenty equal-thickness laminas were used to model the plate which corresponds to 41 nodal planes with a total of 123 degrees of freedom. The aspect ratio as well as the thickness of the rectangular plate are characterized by the parameter N. As can be seen, the results for both frequencies and stresses for the various symmetric and antisymmetric thickness modes are in excellent agreement.

The second example is a symmetrically constructed three-ply laminated square isotropic plate. The ratio of thickness of the core to one facing is eight. This plate was again modeled with 20 equal-thickness laminas (41 nodal planes and 123 generalized coordinates). Frequencies were computed for various ratios of relative shear stiffness and relative density. Also in one case, Poisson's ratio of the facing was varied. The present results compared very closely with Srinivas et al. for all cases as shown in Table III.

THREE-LAYER TRANSVERSELY ISOTROPIC PLATE

Since the analysis given here is not restricted by laminate layout or material isotropy, a very large class of problems could be examined. However, for brevity, attention will be given to only one additional example. The intended purpose of this example is to illustrate the effect of substantial differences in material properties from layer to layer on the frequencies of vibration. Consider an unsymmetric three-layer, simply-supported square plate (or vibrations of an infinite plate with equal wave lengths in both directions) with transversely isotropic material properties as listed in Table IV. As can be seen, the core or second layer is considerably less stiff than the facing layers. Also, both the facings and the core are isotropic in the thickness direction, with the one being highly orthotropic. The two relatively thick facing layers have the same properties but have unequal thickness. In this analysis 40 equal-thickness laminas were used to model this three-laminate plate and the corresponding number of degress of freedom is 243 (81 nodal planes).

On Table V, the normalized frequencies are tabulated according to the ratio of total plate thickness to wave length. The normalizing frequency is the 4th thickness mode for H/L = 0.01 (the type of motion for this frequency is a thickness-stretch action). The lowest mode is the usual flexural action, and the second and third modes are longitudinal tension-compression and tension-tension, respectively, in the x_1 and x_2 directions. The 5th and 6th modes are associated with thickness-shear actions. This represents a departure from the homogeneous isotropic plate for the ordering of these modes, as the frequencies associated with thickness-shear motions is lower than that for thickness stretch. Here, because of the particular orthotropy of the plate materials, this is not the case. On Fig. 1 are plotted the modal patterns for the first six modes for H/L = 0.01. On Fig. 2 similar plots are given for the first three modes for a short wave length, H/L = 10.0. Plots of the stresses of the 1st mode for a long wave (H/L = 0.01) and a short wave (H/L = 10.0) are given in Fig. 3. As can be seen from these plots, the laminated construction and the orthotropy lead to an unusual behavior, especially for the short wave lengths. The results for the higher modes (not shown) indicate an increasingly complex behavior.

CONCLUDING REMARKS

An efficient solution method has been presented for the analysis of the three-dimensional behavior of rectangular laminated orthotropic plates. One note of particular interest is the rather drastic departure from the Kirchhoff-Love hypothesis when the degree of orthotropy is high and/or the wave lengths are not sufficiently long. This should provide guidelines as to the range of applicability of the classical laminated anisotropic plate and shell theories [2,3,4].

REFERENCES

[1] K. S. Pister, and S. B. Dong: Elastic Bending of Layered Plates, Journal of Engineering Mechanics Division, ASCE, Vol. 85, No. EM4, pp. 1-10, Oct. 1959.

[2] E. Reissner and Y. Stavsky: Bending and Stretching of Certain Types of Heterogeneous Aeolotropic Elastic Plates, Journal of Applied Mechanics, Vol. 28, No. 3, pp. 402-408, Sept. 1961.

[3] S. A. Ambartsumyan: Theory of Anisotropic Shells, Moscow, 1961.

[4] S. B. Dong, K. S. Pister and R. L. Taylor: On the Theory of Laminated Anisotropic Shells and Plates, Journal of the Aerospace Sciences, Vol. 29, No. 8, pp. 969-975, Aug. 1962.

[5] S. A. Ambartsumyan: Contributions to the Theory of Anisotropic Layered Shells, Applied Mechanics Reviews, Vol. 15, No. 4, pp. 245-249, Apr. 1962.

[6] Y. Y. Yu: A New Theory of Elastic Sandwich Plates - One Dimensional Case, Journal of Applied Mechanics, Vol. 26, No. 3, Sept. 1959.

[7] Y. V. Riznichenko: Seismic Quasi-Anisotropy, Izvestia Akademia Nauk, SSSR, Geofiz. i Geogr. Vol. 19, pp. 518-544, 1949.

[8] G. W. Postma: Wave Propagation in a Stratified Medium, Geophysics, Vol. 20, No. 4, pp. 780-806, Oct. 1955.

[9] C. T. Sun, J. D. Achenbach and G. Herrmann: Continuum Theory for a Laminated Medium, Journal of Applied Mechanics, Vol. 35, No. 3, pp. 467-475, Sept. 1968.

[10] H. Saito and K. Sato: Flexural Wave Propagation and Vibration of Laminated Rods and Beams, Journal of Applied Mechanics, Vol. 84, No. 2, pp. 287-292, June 1962.

[11] J. P. Jones: Wave Propagation in a Two-Layered Medium, Journal of Applied Mechanics, Vol. 31, No. 2, pp. 213-222, June 1964.

[12] S. Srinivas, C. V. Joga Rao and A. K. Rao: An Exact Analysis for Vibration of Simply-Supported Homogeneous and Laminated Thick Rectangular Plates, Journal of Sound and Vibration, Vol. 12, No. 2, pp. 187-199, June 1970.

[13] S. B. Dong and R. B. Nelson: On Natural Vibrations and Waves in Laminated Orthotropic Plates, submitted for publication.

[14] R. E. Nickell and J. L. Sackman: Approximate Solutions in Linear, Coupled Thermoelasticity, Journal of Applied Mechanics, Vol. 35, No. 2, pp. 255-266, June 1968.

[15] S. B. Dong, J. A. Wolf, Jr., and F. E. Peterson: On a Direct Iterative Eigensolution Technique, Internation Journal for Numerical Methods in Engineering (in press-approx. Apr. 1971).

[16] K. S. Pister: Flexural Vibrations of Thin Laminated Plates, Journal of the Acoustical Society of America, Vol. 31, pp. 233, 1959.

[17] R. D. Midlin, A. Schacknow and H. Derisiewicz: Flexural Vibrations of Rectangular Plates, Journal of Applied Mechanics, Vol. 23, No. 3, pp. 430-436, September 1956.

TABLE I

COMPARISON OF FREQUENCIES FOR HOMOGENEOUS ISOTROPIC PLATE

N**	Eigenvalues (λ)* for Homogeneous Isotropic Plates (Poisson's ration ν=0.3)																	
	1-A***		II-A		III-A		IV-A		I-S		II-S		III-S		IV-S		V-S	
	Ref.[12]	Present Results	Ref.[12]	Present Results	Ref.[12]	Present Results	Ref.[12]	Present Results	Ref.[12]	Present Results	Ref.[12]	Present Results	Ref.[12]	Present Results	Ref.[12]	Present Results	Ref.[12]	Present Results
0.0125	0.05893	0.05893	3.1494	3.1611	3.2079	3.2079	9.4232	9.4235	0.35124	0.35124	0.59314	0.59314	5.7980	5.7980	6.2930	6.2930	6.3930	6.3930
0.02	0.09315	0.09315	3.1729	3.1729	3.2465	3.2465	9.4223	9.4225	0.44429	0.44429	0.74983	0.74983	5.7632	5.7632	6.2989	6.2989	6.4461	6.4461
0.05	0.22260	0.22257	3.2192	3.2192	3.3933	3.3932	9.4193	9.4196	0.70248	0.70243	1.1827	1.1826	5.6652	5.6652	6.3223	6.3223	6.6178	6.6178
0.10	0.41714	0.41715	3.2949	3.2949	3.6160	3.6160	9.4166	9.4169	0.99346	0.99348	1.6654	1.6655	5.5688	5.5688	6.3612	6.3612	6.8384	6.8384
0.18	0.68893	0.68904	3.4126	3.4127	3.9310	3.9311	9.4174	9.4176	1.3329	1.3329	2.2171	2.2173	5.4903	5.4903	6.4230	6.4230	7.1199	7.1200
0.20	0.75111	0.75121	3.4414	3.4415	4.0037	4.0038	9.4184	9.4187	1.4050	1.4050	2.3320	2.3322	5.4795	5.4795	6.4383	6.4384	7.1825	7.1825
0.50	1.5158	1.5158	3.8476	3.8476	4.9086	4.9086	9.4698	9.4701	2.2214	2.2214	3.5306	3.5306	5.5554	5.5554	6.6643	6.6643	7.9416	7.9416

* $\lambda = \Omega\sqrt{\rho h^2/G}$ where Ω-angular frequency; ρ-density; h-thickness; and G-shear modulus

** $N = h^2(m^2/L_1^2 + n^2/L_2^2)$ where m and n are mode numbers along x_1 and x_2

*** I-A,........,I-S,...... indicate the first antisymmetric thickness mode,..., the first symmetric thickness mode, etc.

TABLE II

COMPARISON OF STRESSES FOR HOMOGENEOUS ISOTROPIC PLATE

		$\frac{\tau_{11}(z/h)}{\tau_{11}(0)}$		$\frac{\tau_{33}(z/h)}{\tau_{33}(0.4)}$		$\frac{\tau_{12}(z/h)}{\tau_{12}(0)}$		$\frac{\tau_{13}(z/h)}{\tau_{13}(0.5)}$	
Mode	z/h	Ref.[12]	Present Results	Ref.[12]	Present Results	Ref.[12]	Present Results	Ref.[12]	Present Results
I-A	0	1.0000	1.0000	0.0	0.0	1.0000	1.0000	0.0	0.0
	0.1	0.7676	0.7676	1.5336	1.5654	0.7561	0.7561	0.3750	0.3760
	0.2	0.5571	0.5571	2.0238	2.0363	0.5420	0.5420	0.6549	0.6555
	0.3	0.3627	0.3627	1.7578	1.7609	0.3496	0.3496	0.8426	0.8489
	0.4	0.1787	0.1787	1.0000	1.0000	0.1713	0.1713	0.9625	0.9625
	0.5	0.0	0.0	0.0	0.0	0.0	0.0	1.0000	1.0000
II-A	0	1.0000	1.0000	~0.0	~0.0	~0.0	~0.0	0.0	0.0
	0.1	0.9511	0.9511	↓	↓	↓	↓	0.3090	0.3090
	0.2	0.8090	0.8090	↓	↓	↓	↓	0.5878	0.5878
	0.3	0.5878	0.5878	↓	↓	↓	↓	0.8090	0.8090
	0.4	0.3090	0.3090	↓	↓	↓	↓	0.9511	0.9511
	0.5	0.0	0.0	↓	↓	↓	↓	1.0000	1.0000
II-S	0	1.0000	1.0000	0.0	0.0	1.0000	1.0000	0.0	0.0
	0.1	1.0161	1.0161	0.3611	0.3622	1.0438	1.0438	0.7425	0.7514
	0.2	1.0297	1.0296	0.6534	0.6540	1.0797	1.0797	1.0000*	1.0000*
	0.3	1.0400	1.0399	0.8684	0.8686	1.1065	1.1065	0.8813	0.8781
	0.4	1.0464	1.0463	1.0000	1.0000	1.1230	1.1230	0.5057	0.5032
	0.5	1.0486	1.0485	1.0443	1.0442	1.1286	1.1286	0.0	0.0

square plate $h/L_1 = h/L_2 = 0.3$ ($N = 0.18$) and $\nu = 0.3$

* $\tau_{13}(z/h)$ is normalized with $\tau_{13}(0.2)$ as $\tau_{13}(0.5)=0.0$ for the second symmetric thickness mode.

TABLE III

COMPARISON OF FREQUENCIES FOR A SYMMETRIC THREE-PLY LAMINATED ISOTROPIC PLATE

			Eigenvalues $(\lambda)^*$ for Three-Ply Isotropic Plate	
ν_f	ρ_f/ρ_c	G_f/G_c	Ref. [12]	Present Results
0.3	1	1	0.074520	0.074519
0.3	1	2	0.089986	0.089986
0.3	1	5	0.123072	0.123072
0.3	1	15	0.183664	0.183663
0.3	2	15	0.167574	0.167573
0.3	3	15	0.155082	0.155082
0.2	1	1	0.072298	0.072300

$N = h^2(1/L_1^2 + 1/L_2^2)$ $N_f = 0.0002$ $N_c = 0.0028$

$\nu_c = 0.3$

$^*\lambda = \Omega\sqrt{\rho_c h_c^2/G_c}$ where Ω-angular frequency; ρ_c-density of core; h_c-thickness of core; and G_c-shear modulus of core.

TABLE IV

PROPERTIES OF THE THREE-LAYER TRANSVERSELY ISOTROPIC PLATE

LAYER	THICKNESS (in)	MATERIAL PROPERTIES (10^6psi)									DENSITY ρ
		C_{11}	C_{22}	C_{33}	C_{44}	C_{55}	C_{66}	C_{12}	C_{13}	C_{23}	
1	0.2	55.7	55.7	30.9	11.5	11.5	11.5	32.6	26.5	26.5	1.0
2	0.5	3.35	3.35	0.308	1.15	1.15	1.15	1.05	0.132	0.132	1.0
3	0.3	55.7	55.7	30.9	11.5	11.5	11.5	32.6	26.5	26.5	1.0

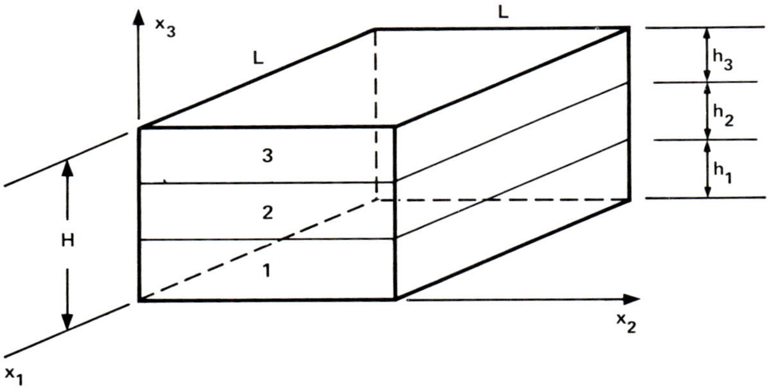

TABLE V

NORMALIZED FREQUENCIES FOR THREE-LAYER TRANSVERSELY ISOTROPIC PLATE

Eigenvalues (Ω) for Three-Ply Laminated Orthotropic Plate

Thickness Mode	$\frac{H}{L}$				
	0.01	0.1	1	5	10
1	0.00158	0.13612	2.79936	11.67493	23.13829
2	0.05830	0.58034	3.11331	11.67973	23.14168
3	0.09854	0.92734	3.77465	12.05680	23.38443
4	1.0*	1.04362	3.95621	12.15296	23.39427
5	1.92074	2.03727	4.33221	12.45416	23.77494
6	1.92292	2.19293	5.55531	12.89858	23.77873
7	2.35316	2.40282	5.79446	13.02873	24.21787
8	3.96025	3.94358	6.24058	13.94925	24.31386
9	4.46094	4.48920	7.23975	14.21500	24.67015
10	4.46168	4.56212	7.72677	15.10654	24.75072

$H = h_1 + h_2 + h_3$

* Normalizing Freq. ω_4 = 1919.6561 rad/sec

$\Omega = \omega/\omega_4$

APPENDIX

The stiffness and mass matrices [k] and [m] are given by the operations:

$$\underset{9\times9}{[k]} = \underset{9\times9}{[G]}^T \underset{9\times9}{[\bar{k}]} \underset{9\times9}{[G]} \tag{A1}$$

$$\underset{9\times9}{[m]} = \underset{9\times9}{[G]}^T \underset{9\times9}{[\bar{m}]} \underset{9\times9}{[G]} \tag{A2}$$

The non-zero components of [G] are:

$$G_{11} = G_{22} = G_{33} = -G_{47} = -G_{58} = -G_{69} = 1$$

$$G_{44} = G_{55} = G_{66} = -G_{74} = -G_{85} = -G_{96} = 4$$

$$G_{77} = G_{88} = G_{99} = G_{71} = G_{82} = G_{93} = 2$$

$$G_{41} = G_{52} = G_{63} = -3 \tag{A3}$$

The non-zero components of the symmetric matrices $[\bar{k}]$ and $[\bar{m}]$ are

$$\bar{k}_{11} = 2\bar{k}_{14} = 3\bar{k}_{17} = \alpha_1$$

$$\bar{k}_{12} = 2\bar{k}_{15} = 3\bar{k}_{18} = 2\bar{k}_{24} = 3\bar{k}_{27} = 3\bar{k}_{45} = 4\bar{k}_{48} = 4\bar{k}_{57} = 5\bar{k}_{78} = \alpha_2$$

$$\bar{k}_{16} = \bar{k}_{19} = -A\xi C_{13}$$

$$\bar{k}_{22} = 2\bar{k}_{25} = 3\bar{k}_{28} = \alpha_3$$

$$\bar{k}_{26} = \bar{k}_{29} = -A\eta C_{23}$$

$$\bar{k}_{33} = 2\bar{k}_{36} = 3\bar{k}_{39} = \alpha_4$$

$$\bar{k}_{34} = \bar{k}_{37} = A\xi C_{44}$$

$$\bar{k}_{35} = \bar{k}_{38} = A\eta C_{55}$$

$$\bar{k}_{44} = \alpha_1/3 + AC_{44}/h$$

$$\bar{k}_{47} = \alpha_1/4 + AC_{44}/h$$

$$\bar{k}_{46} = (C_{44}-C_{13})\xi A/2$$

$$\bar{k}_{49} = (C_{44}-2C_{13})\xi A/3$$

$$\bar{k}_{55} = \alpha_3/3 + AC_{55}/h$$

$$\bar{k}_{58} = \alpha_3/4 + AC_{55}/h$$

$$\bar{k}_{56} = (C_{55}-C_{13})\eta A/2$$

$$\bar{k}_{59} = (C_{55}-2C_{23})\eta A/3$$

$$\bar{k}_{66} = \alpha_4/3 + AC_{33}/h$$

$$\bar{k}_{69} = \alpha_4/3 + AC_{33}/h$$

$$\bar{k}_{67} = (2C_{44}-C_{13})\xi A/3$$

$$\bar{k}_{68} = (2C_{55}-C_{23})\eta A/3 \tag{A4}$$

$$\bar{k}_{77} = \alpha_1/5 + 4AC_{44}/3h$$

$$\bar{k}_{79} = (C_{44}-C_{13})\xi A/2$$

$$\bar{k}_{88} = \alpha_3/5 + 4AC_{55}/3h$$

$$\bar{k}_{89} = (C_{55}-C_{23})\eta A/2$$

$$\bar{k}_{99} = \alpha_4/5 + 4AC_{33}/3h \qquad \text{(A4 Cont.)}$$

$$\bar{m}_{11} = \bar{m}_{22} = \bar{m}_{33} = \rho Ah$$

$$\bar{m}_{14} = \bar{m}_{25} = \bar{m}_{36} = \rho Ah/2$$

$$\bar{m}_{17} = \bar{m}_{28} = \bar{m}_{39} = \bar{m}_{44} = \bar{m}_{55} = \bar{m}_{66} = \rho Ah/3$$

$$\bar{m}_{47} = \bar{m}_{58} = \bar{m}_{69} = \rho Ah/4$$

$$\bar{m}_{77} = \bar{m}_{88} = \bar{m}_{99} = \rho Ah/5 \qquad \text{(A5)}$$

where

$$\alpha_1 = Ah(C_{11}\xi^2 + C_{66}\eta^2)$$

$$\alpha_2 = Ah\xi\eta(C_{12} + C_{66})$$

$$\alpha_3 = Ah(C_{22}\eta^2 + C_{66}\xi^2)$$

$$\alpha_4 = Ah(C_{44}\xi^2 + C_{55}\eta^2) \qquad \text{(A6)}$$

and

$$\xi = \pi/L_1 \;;\quad \eta = \pi/L_2 \;;\quad A = L_1L_2/4 \;;\quad h = x_{3f}-x_{3b} \qquad \text{(A7)}$$

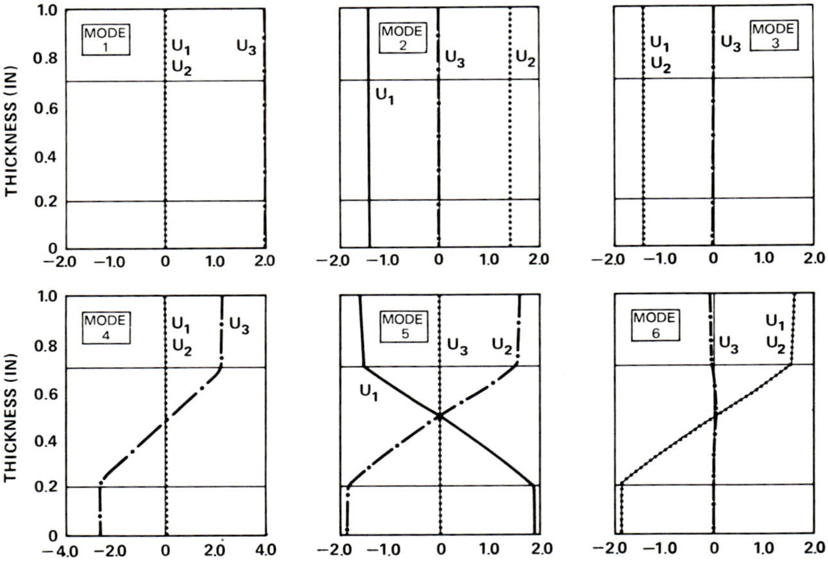

Figure 1. Displacement Modal Patterns for First Six Modes for H/L = 0.01

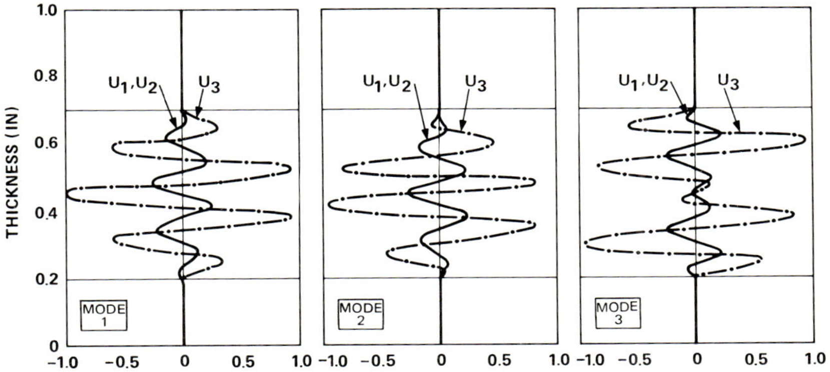

Figure 2. Displacement Modal Patterns for First Three Modes for H/L = 10.0

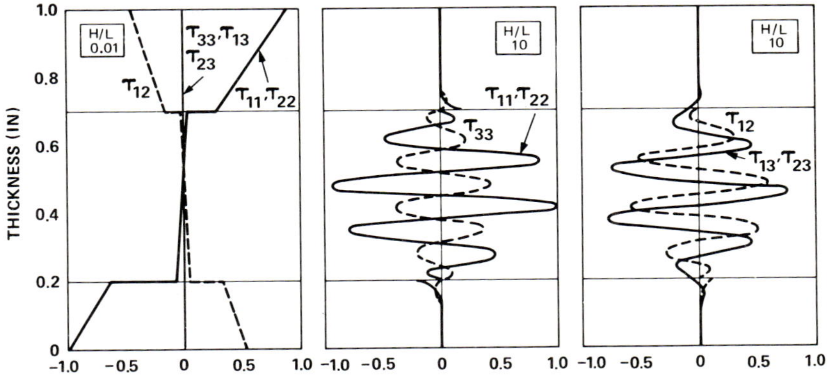

Figure 3. Stress Modal Patterns for Three-Layered Transversely Isotropic Plate for Mode 1

Developments in Mechanics, Vol. 6. Proceedings of the 12th Midwestern Mechanics Conference: 63

On the Nonlinear Vibrations of a Clamped Circular Plate

T. W. LEE
MICHIGAN STATE HIGHWAYS DEPARTMENT
P. T. BLOTTER
UTAH STATE UNIVERSITY
D. H. Y. YEN
MICHIGAN STATE UNIVERSITY

ABSTRACT

Free vibrations of a clamped, circular, elastic plate are considered in this paper. The study is based on a set of fully coupled nonlinear plate equations appropriate for large-amplitude motions of the plate as well as on a simplified single nonlinear plate equation. A systematic perturbation method is used to solve the nonlinear problems. Expressions for the nonlinear mode shapes of vibrations and the amplitude-frequency relationships are obtained. In light of these results, the accuracy of the simplified nonlinear plate theory is then assessed.

INTRODUCTION

We consider here a circular, elastic plate of uniform thickness that is clamped along its edge. We wish to study the free, undamped, axisymmetric vibrations of the plate in which the amplitudes of the transverse motions are large in the sense that they may be of the same order of magnitude as the plate thickness. For such motions of the plate the governing equations are inherently nonlinear, the interaction between the mid-surface forces in the plate and the transverse deflection being a main source of the nonlinearity in the system.

Nonlinear equations governing large-amplitude static deflections of elastic plates were derived by von Kármán as early as 1910 [1]. These so-called von Kármán equations involve the transverse deflection as well as the in-plane displacements and are fully coupled. In 1955, Berger [2] derived a simplified nonlinear plate equation, involving the transverse deflection alone, by neglecting the strain energy due to the second invariant of the mid-surface strains. He then proceeded to solve this equation for static deflections of both circular and rectangular plates under a variety of boundary conditions. Subsequently it was discovered that similar uncoupled nonlinear plate equations can be derived for dynamic problems as well if, in addition to neglecting the strain energy due to the second strain invariant as mentioned above, one also omits the in-plane inertia of the plate. Such simplified plate equations were used in a number of nonlinear vibration studies [3-6], though no rigorous justifications were made of the assumptions that led to the derivation of such plate equations.

The present problem was treated in [3] by Wah, who solved the single nonlinear plate equation using a modified Galerkin method. In the present paper the study will be based on a set of fully coupled nonlinear plate equations involving both the transverse and in-plane displacements. These equations will be derived using Hamilton's principle and without the making of the above mentioned assumptions of neglecting the in-plane inertia or the second strain invariant of the mid-surface strains. The set of nonlinear equations employed here is equivalent to that derived by Herrmann [7] in Cartesian coordinates, which was used by Chu and Herrmann in [8] for studying nonlinear vibrations of rectangular plates.

A perturbation method, recently developed by the authors [9,10], will be used in the present study. Perturbation solutions based on the simplified single plate equation will also be obtained. Expressions for the nonlinear mode shapes of vibrations and the frequency-amplitude relationships will be derived.

These results will be compared and discussed, along with that of Wah [3]. The two sets of solutions here also enable us to assess the validity of the simplified, single nonlinear plate equation. In particular, it is found that the accuracy of the simplified plate equation depends on Poisson's ratio as well as the plate-radius-to-plate-thickness ratio, though the dependence upon the latter ratio is not too sensitive.

EQUATIONS OF MOTION

Consider a circular plate of thickness h and radius a. We choose the polar coordinates (r,θ) for points lying in the mid-surface of the plate, with $r = 0$ being at the center. We denote by $w = w(r,t)$ and $u = u(r,t)$, where t is time, the transverse and radial displacements, respectively, of points in the mid-surface. Under the assumptions mentioned above, the strain energy of the plate due to bending and stretching may be expressed as [11]

$$V = \frac{D}{2}\int_0^{2\pi}\int_0^{a}\{[(\nabla^2 w)^2 + \frac{12}{h^2}e_1^2] - 2(1-\mu)(\frac{12}{h^2}e_2^2 + \frac{1}{r}\frac{\partial w}{\partial r}\frac{\partial^2 w}{\partial r^2})\}rdrd\theta$$

$$= \pi D\int_0^{a}\{[(\nabla^2 w)^2 + \frac{12}{h^2}e_1^2] - 2(1-\mu)(\frac{12}{h^2}e_2^2 + \frac{1}{r}\frac{\partial w}{\partial r}\frac{\partial^2 w}{\partial r^2})\}rdr \tag{1}$$

while the kinetic energy of the plate may be expressed as

$$T = \frac{\rho h}{2}\int_0^{2\pi}\int_0^{a}[(\frac{\partial u}{\partial t})^2 + (\frac{\partial w}{\partial t})^2]\,rdrd\theta = \pi\rho h\int_0^{a}[(\frac{\partial u}{\partial t})^2 + (\frac{\partial w}{\partial t})^2]\,rdr \tag{2}$$

In equations (1) and (2) above $D = Eh^3/12(1-\mu^2)$ is the flexural rigidity of the plate, with E being Young's modulus and μ Poisson's ratio, ρ is the mass density per unit volume, ^{2} here is the operator

$$\nabla^2 = \frac{\partial^2}{\partial r^2} + \frac{1}{r}\frac{\partial}{\partial r} \tag{3}$$

and e_1 and e_2 are, respectively, the first and second strain invariants defined as

$$e_1 = \epsilon_r + \epsilon_\theta = [\frac{\partial u}{\partial r} + \frac{1}{2}(\frac{\partial w}{\partial r})^2] + \frac{u}{r} \tag{4a}$$

$$e_2 = \epsilon_r\epsilon_\theta = [\frac{\partial u}{\partial r} + \frac{1}{2}(\frac{\partial w}{\partial r})^2]\,\frac{u}{r} \tag{4b}$$

with ϵ_r and ϵ_θ being respectively the radial and hoop strains.

In accordance with Hamilton's principle, the motions of the plate must be such as to render the integral

$$I = \int_{t_1}^{t_2}(T - V)\,dt \tag{5}$$

stationary among all admissible displacements $w(r,t)$ and $u(r,t)$, where t_1 and t_2 above are any two fixed instants of time. The Euler-Lagrange equations for this variational problem then follow as

$$\nabla^2\nabla^2 w + \frac{\rho h}{D}\frac{\partial^2 w}{\partial t^2} - \frac{12}{h^2}\left\{\frac{\partial w}{\partial r}\left[\frac{\partial^2 u}{\partial r^2} + \frac{1+\mu}{r}\frac{\partial u}{\partial r} + \frac{1}{2r}\left(\frac{\partial w}{\partial r}\right)^2\right] + \frac{\partial^2 w}{\partial r^2}\left[\frac{\partial u}{\partial r} + \mu\frac{u}{r} + \frac{3}{2}\left(\frac{\partial w}{\partial r}\right)^2\right]\right\} = 0 \tag{6a}$$

$$-\left(\nabla^2 - \frac{1}{r^2}\right)u + \frac{\rho(1-\mu^2)}{E}\frac{\partial^2 u}{\partial t^2} - \frac{\partial w}{\partial r}\frac{\partial^2 w}{\partial r^2} - \frac{1-\mu}{2r}\left(\frac{\partial w}{\partial r}\right)^2 = 0 \tag{6b}$$

Thus we have a set of two coupled nonlinear partial differential equations governing the large-amplitude, free, undamped, axisymmetric vibrations of the circular plate.

The boundary conditions for the clamped plate are

$$w(r,t)\Big|_{r=a} = 0 \tag{7a}$$

$$\frac{\partial w(r,t)}{\partial r}\Big|_{r=a} = 0 \tag{7b}$$

$$u(r,t)\Big|_{r=a} = 0 \tag{7c}$$

In addition, $w(r,t)$, $u(r,t)$ and their r-derivatives must remain finite in $0 \le r \le a$. The initial conditions will be left unspecified at this moment.

We now show how a single, fourth order nonlinear equation for w alone may be derived when simplifying assumptions as those mentioned in the INTRODUCTION are made.

We first neglect the second strain invariant e_2 in equation (1). Using the same Hamilton's principle, we obtain

$$\nabla^2\nabla^2 w + \frac{\rho h}{D}\frac{\partial^2 w}{\partial t^2} - \frac{12}{h^2}\left(e_1\nabla^2 w - r\frac{\partial e_1}{\partial r}\right) = 0 \tag{8a}$$

$$-\frac{\partial e_1}{\partial r} + \frac{\rho(1-\mu^2)}{E}\frac{\partial^2 u}{\partial t^2} = 0 \tag{8b}$$

while the boundary conditions remain the same as in equations (7). We next neglect the in-plane inertia in equation (8b) and obtain

$$\frac{\partial e_1}{\partial r} = 0 \tag{9}$$

Thus the first strain invariant e_1 is independent of r. Upon multiplying both sides of equation (4a) by r and then integrating the resulting equation with respect to r between 0 and a, we obtain

$$e_1 = \frac{2}{a^2}\left[ru\Big|_0^a + \frac{1}{2}\int_0^a r\left(\frac{\partial w}{\partial r}\right)^2 dr\right] \tag{10}$$

The vanishing of u at r = a thus leads to

$$e_1 = \frac{1}{a^2}\int_0^a r\left(\frac{\partial w}{\partial r}\right)^2 dr \tag{11}$$

In view of equations (9) and (11), equation (8a) now becomes uncoupled as

$$\nabla^2\nabla^2 w + \frac{\rho h}{D}\frac{\partial^2 w}{\partial t^2} - \frac{12}{a^2 h^2}\nabla^2 w\int_0^a r\left(\frac{\partial w}{\partial r}\right)^2 dr = 0 \tag{12}$$

which is exactly the same equation of motion used by Wah in [3].

PERTURBATION SOLUTIONS

Let us define the following non-dimensional quantities:

$$\bar{\eta} = w/a, \quad \bar{\gamma} = u/a, \quad \zeta = r/a, \quad \tau = \bar{\omega}\, t, \quad \omega^2 = \rho h a^4 \bar{\omega}^2/D, \quad \lambda = 12a^2/h^2 \tag{13}$$

where ω is a frequency parameter introduced so that the period of vibration may be fixed at 2π in the τ-scale.

In terms of the variables introduced above equations (6a) and (6b) become

$$L_1\bar{\eta} + \omega^2 \ddot{\bar{\eta}} - \lambda[\bar{\eta}'(\bar{\gamma}'' + \frac{1+\mu}{\zeta}\bar{\gamma}' + \frac{1}{2\zeta}\bar{\eta}'^2) + \bar{\eta}''(\bar{\gamma}' + \frac{\mu}{\zeta}\bar{\gamma} + \frac{3}{2}\bar{\eta}'^2)] = 0 \tag{14a}$$

$$L_2\bar{\gamma} + \omega^2 \ddot{\bar{\gamma}} - (\bar{\eta}'\bar{\eta}'' + \frac{1-\mu}{2\zeta}\bar{\eta}'^2) = 0 \tag{14b}$$

with

$$L_1 = \nabla^2\nabla^2 = (\frac{\partial^2}{\partial\zeta^2} + \frac{1}{\zeta}\frac{\partial}{\partial\zeta})^2 \tag{15a}$$

$$L_2 = -(\nabla^2 - \frac{1}{\zeta^2}) = -(\frac{\partial^2}{\partial\zeta^2} + \frac{1}{\zeta}\frac{\partial}{\partial\zeta} - \frac{1}{\zeta^2}) \tag{15b}$$

where the primes above designate partial differentiations with respect to ζ and the dots stand for partial differentiations with respect to τ.

We introduce the perturbation parameter ϵ through the change of variables

$$\bar{\eta} = \epsilon\,\eta \tag{16}$$

where η will be assumed to be of the order unity. So ϵ is just an amplitude parameter of the nondimensional transverse displacement $\bar{\eta}$. As equation (14b) shows that $\bar{\gamma}$ will be of the order $O(\bar{\eta}^2)$, we then write

$$\bar{\gamma} = \epsilon^2\gamma \tag{17}$$

where γ will also be assumed to be of the order unity.

Substitutions of equations (16) and (17) into (14) yield

$$L_1\eta + \omega^2\ddot{\eta} - \epsilon^2\lambda[\eta'(\gamma'' + \frac{1+\mu}{\zeta}\gamma' + \frac{1}{2\zeta}\eta'^2) + \eta''(\gamma' + \frac{\mu}{\zeta}\gamma + \frac{3}{2}\eta'^2)] = 0 \tag{18a}$$

$$L_2\gamma + \omega^2\ddot{\gamma} - (\eta'\eta'' + \frac{1-\mu}{2\zeta}\eta'^2) = 0 \tag{18b}$$

with the boundary conditions

$$\eta(\zeta,\tau)\Big|_{\zeta=1} = 0 \tag{19a}$$

$$\frac{\partial\eta(\zeta,\tau)}{\partial\zeta}\Big|_{\zeta=1} = 0 \tag{19b}$$

$$\gamma(\zeta,\tau)\Big|_{\zeta=1} = 0 \tag{19c}$$

We also pose the following periodicity conditions

$$\eta(\zeta,\tau) = \eta(\zeta,\tau+2\pi) \tag{20a}$$

$$\gamma(\zeta,\tau) = \gamma(\zeta,\tau+2\pi) \tag{20b}$$

and the initial conditions

$$\dot{\eta}(\zeta,0) = 0 \tag{21a}$$

$$\dot{\gamma}(\zeta,0) = 0 \tag{21b}$$

The above implies that the system has zero initial velocities and periodic motions are initiated by releasing the plate from rest in an as yet unspecified configuration.

Following the perturbation method mentioned earlier [9,10], the functions η and γ as well as the parameter ω^2 are expanded as power series in ϵ

$$\eta(\zeta,\tau) = \sum_{i=0}^{\infty} \epsilon^i\,\eta_i(\zeta,\tau) \tag{22}$$

$$\gamma(\zeta,\tau) = \sum_{i=0}^{\infty} \epsilon^i\,\gamma_i(\zeta,\tau) \tag{23}$$

$$\omega^2 = \sum_{i=0}^{\infty} \epsilon^i \omega_i^2 \tag{24}$$

It is clear, either from the appearance of equations (18) or from symmetry considerations, that only even powers of ϵ need be considered in the above expansions. We thus set

$$\eta_i = 0, \quad \gamma_i = 0, \quad \omega_i^2 = 0, \quad \text{for all} \quad i \quad \text{odd} \tag{25}$$

Upon substituting the series in (22) through (24) into equations (18) and collecting like powers of ϵ, the following set of linear equations results:

$$\epsilon^0: \; L_1\eta_0 + \omega_0^2\ddot{\eta}_0 = 0 \tag{26a}$$

$$L_2\gamma_0 + \omega_0^2\ddot{\gamma}_0 = \eta_0'\eta_0'' + \frac{1-\mu}{2\zeta}\eta_0'^2 \tag{26b}$$

$$\epsilon^2: \; L_1\eta_2 + \omega_0^2\ddot{\eta}_2 = -\omega_2^2\ddot{\eta}_0 + \lambda[\eta_0'(\gamma_0'' + \frac{1+\mu}{\zeta}\gamma_0' + \frac{1}{2\zeta}\eta_0'^2)$$

$$+ \eta_0''(\gamma_0' + \frac{\mu}{\zeta}\gamma_0 + \frac{3}{2}\eta_0'^2)] \tag{26c}$$

$$L_2\gamma_2 + \omega_0^2\ddot{\gamma}_2 = -\omega_2^2\ddot{\gamma}_0 + \eta_0'\eta_2'' + \eta_0''\eta_2' + \frac{1-\mu}{\zeta}\eta_0'\eta_2' \tag{26d}$$

$\epsilon^4: \; \cdot \; \cdot \; \cdot \; \cdot \; \cdot$

where the functions η_i and γ_i, $i = 0,2,\ldots$, must individually satisfy the boundary, periodicity and initial conditions as described in equations (19) through (21) and must remain finite in $0\leq\zeta\leq1$.

We now solve the above set of equation in a recursive manner, starting with equation (26a) for η_0. The solutions here are well known in the study of periodic vibrations of a linear clamped circular plate and are of the form

$$\eta_0(\zeta,\tau) = A_{1k}V_k(\zeta) \cos \tau \tag{27}$$

$$\omega_0^2 = \Omega_k^2 \tag{28}$$

where $k = 1,2,\ldots$ and the A_{1k}'s are arbitrary constants. $V_k(\zeta)$ and Ω_k^2 above are, respectively, the k-th eigenfunction and eigenvalue of the following eigenvalue problem for $V(\zeta)$

$$L_1V - \Omega^2V = 0 \tag{29}$$

$$V(1) = 0, \quad V'(1) = 0 \tag{30}$$

with L_1 being given in equation (15a). These eigenfunctions and eigenvalues may be given explicitly as

$$V_k(\zeta) = \frac{\sqrt{2}}{[J_1(\beta_k)+2J_0^2(\beta_k) - \dfrac{J_0^2(\beta_k)}{I_0^2(\beta_k)} I_1^2(\beta_k)]} [J_0(\beta_k\zeta) - \frac{J_0(\beta_k)}{I_0(\beta_k)} I_0(\beta_k\zeta)] \tag{31}$$

$$\Omega_k^2 = \beta_k^4 \tag{32}$$

where β_k is the k-th real positive root of the transcendental equation

$$\frac{J_1(\beta)}{J_0(\beta)} + \frac{I_1(\beta)}{I_0(\beta)} = 0 \tag{33}$$

Here J_ν and I_ν are, respectively, the Bessel and the modified Bessel functions of the first kind, the subscript ν referring to the order of these functions. Numerical values of β_k may be found in [12]. The V_k's also satisfy the orthogonality condition

$$\int_0^1 \zeta V_k(\zeta)V_p(\zeta)d\zeta = \delta_{kp} \tag{34}$$

where δ_{kp} is the Kronecker delta

Let us set k=1 in equations (27) and (28) and consider nonlinear vibrations of the plate in the neighborhood of the first linear mode. The analyses for the cases $k \neq 1$ would proceed similarly. Thus we have

$$\eta_0(\zeta,\tau) = A_{11} V_1(\zeta) \cos \tau \tag{35}$$

$$\omega_0{}^2 = \Omega_1{}^2 = \beta_1{}^4 = (3.1962)^4 \tag{36}$$

as the solution of equation (26a).

In order to solve equation (26b) let us first consider the following eigenvalue problem for $\Gamma(\zeta)$

$$L_2\Gamma - \Lambda^2\Gamma = 0 \tag{37}$$

$$\Gamma(1) = 0 \tag{38}$$

with L_2 being given in equation (15b). The eigenfunctions and eigenvalues are

$$\Gamma_n(\zeta) = \frac{\sqrt{2}}{J_2(j_n)} J_1(j_n\zeta) \tag{39}$$

$$\Lambda_n{}^2 = j_n{}^2 \tag{40}$$

where j_n is the n-th real positive root of $J_1(j) = 0$. It is to be noted that the Γ_n's satisfy the orthogonality condition

$$\int_0^1 \zeta\, \Gamma_n(\zeta)\Gamma_q(\zeta)d\zeta = \delta_{nq} \tag{41}$$

Numerical values of j_n may be found in [13].

With η_0 and $\omega_0{}^2$ now being given by equations (35) and (36), we expand $\gamma_0(\zeta,\tau)$ as

$$\gamma_0(\zeta,\tau) = \sum_{m=0}^{\infty} \sum_{n=1}^{\infty} C_{mn}\, \Gamma_n(\zeta) \cos m\,\tau \tag{42}$$

We substitute equations (35), (36) and (42) into (26b) to obtain

$$\sum_{m=0}^{\infty} \sum_{n=1}^{\infty} (\Lambda_n{}^2 - m^2\Omega_1{}^2) C_{mn}\, \Gamma_n(\zeta) \cos m\,\tau = A_{11}^2 [V_1'(\zeta)V_1''(\zeta) + \frac{1-\mu}{2\zeta} V_1'^2(\zeta)]\cos^2\tau \tag{43}$$

where use has been made of equation (37). We multiply both sides above with $\zeta\Gamma_q(\zeta) \cos p\,\tau$ and integrate the resulting equation with respect to ζ from zero to one and with respect to τ from zero to 2π. Upon using equation (41) and the orthogonality conditions for trigonometric functions, we have

$$C_{0n} = \frac{A_{11}^2}{2\Lambda_n^2} I \tag{44}$$

$$C_{2n} = \frac{A_{11}^2}{2(\Lambda_n^2 - 4\Omega_1^2)} I \tag{45}$$

where n = 1,2,... and

$$I = \int_0^1 [V_1'(\zeta)V_1''(\zeta) + \frac{1-\mu}{2\zeta} V_1'(\zeta)]\zeta\Gamma_n(\zeta)d\zeta \tag{46}$$

All remaining C_{mn} are zero. We thus obtain

$$\gamma_0(\zeta,\tau) = \sum_{n=1}^{\infty} (C_{0n} + C_{2n} \cos 2\,\tau)\Gamma_n(\zeta) \tag{47}$$

where C_{0n} and C_{2n} are to be computed from equations (44) through (46). As no closed-form expression for the integral in equation (46) is available, such computations must be done numerically on a digital computer.

We turn our attention now to equation (26c). We solve (26c) by expanding $\eta_2(\zeta,\tau)$ in the form

$$\eta_2(\zeta,\tau) = \sum_{m=0}^{\infty} \sum_{n=1}^{\infty} A_{mn}^{(2)} V_n(\zeta) \cos m\tau \tag{48}$$

We then substitute equation (48) together with equations (35), (36), and (47) into equation (26c). Making use of equation (29) we obtain

$$\sum_{m=0}^{\infty} \sum_{n=1}^{\infty} (\Omega_n^2 - m^2\Omega_1^2) A_{mn}^{(2)} V_n(\zeta) \cos m\tau = \omega_2^2 A_{11} V_1(\zeta) \cos\tau$$

$$+ \lambda A_{11} V_1'(\zeta) \cos\tau \sum_{q=1}^{\infty} (C_{0q} + C_{2q} \cos 2\tau)[\Gamma_q''(\zeta) + \frac{1+\mu}{\zeta} \Gamma_q'(\zeta)]$$

$$+ \lambda A_{11} V_1''(\zeta) \cos\tau \sum_{q=1}^{\infty} (C_{0q} + C_{2q} \cos 2\tau)[\Gamma_q'(\zeta) + \frac{\mu}{\zeta} \Gamma_q(\zeta)]$$

$$+ \lambda A_{11}^3 \left[\frac{V_1'^3(\zeta)}{2\zeta} + \frac{3V_1''(\zeta)V_1'^2(\zeta)}{2} \right] \cos^3\tau \tag{49}$$

We now multiply both sides above with $\zeta V_s(\zeta)\cos p\tau$ and integrate the resulting equation with respect to ζ from zero to one and with respect to τ from zero to 2π. Again upon using the orthogonality conditions for the trigonometric functions and for the V_n's as given in equation (34), we obtain

$$\omega_2^2 = \lambda\omega_0^2 A_{11}^2 F(\tfrac{a}{h},\mu) \tag{50}$$

with

$$F(\tfrac{a}{h},\mu) = -\frac{1}{\omega_0^2} \int_0^1 \{ \frac{1}{A_{11}^2} \sum_{q=1}^{\infty} (C_{0q} + \tfrac{1}{2} C_{2q}) [\tfrac{\mu}{\zeta} V_1''(\zeta)\Gamma_q(\zeta) + (\tfrac{1+\mu}{\zeta} V_1'(\zeta) + V_1''(\zeta))\Gamma_q'(\zeta)$$

$$+ V_1'(\zeta)\Gamma_q''(\zeta)] + \tfrac{3}{8} [3V_1'^2(\zeta)V_1''(\zeta) + \tfrac{1}{\zeta} V_1'^3(\zeta)]\} \zeta V_1(\zeta) d\zeta \tag{51}$$

and

$$A_{1n}^{(2)} = \frac{\lambda A_{11}}{(\Omega_n^2 - \Omega_1^2)} \int_0^1 \{ \sum_{q=1}^{\infty} (C_{0q} + \tfrac{1}{2} C_{2q}) [\tfrac{\mu}{\zeta} V_1''(\zeta)\Gamma_q(\zeta) + (\tfrac{1+\mu}{\zeta} V_1'(\zeta) + V_1''(\zeta))\Gamma_q'(\zeta)$$

$$+ V_1'(\zeta)\Gamma_q''(\zeta)] + \tfrac{3}{8} A_{11}^2 [3V_1'^2(\zeta)V_1''(\zeta) + \tfrac{1}{\zeta} V_1'^3(\zeta)]\} \zeta V_n(\zeta) d\zeta \tag{52}$$

for $n = 2,3,\ldots$

$$A_{3n}^{(2)} = \frac{\lambda A_{11}}{(\Omega_n^2 - 9\Omega_1^2)} \int_0^1 \{ \sum_{q=1}^{\infty} \tfrac{1}{2} C_{2q} [\tfrac{\mu}{\zeta} V_1''(\zeta)\Gamma_q(\zeta) + (\tfrac{1+\mu}{\zeta} V_1'(\zeta) + V_1''(\zeta))\Gamma_q'(\zeta)$$

$$+ V_1'(\zeta)\Gamma_q''(\zeta)] + \tfrac{1}{8} A_{11}^2 [3V_1'^2(\zeta)V_1''(\zeta) + \tfrac{1}{\zeta} V_1'^3(\zeta)]\} \zeta V_n(\zeta) d\zeta \tag{53}$$

for $n = 1,2,\ldots$. All remaining $A_{mn}^{(2)}$, with the exception of $A_{11}^{(2)}$, are zero. The perturbation scheme does not determine $A_{11}^{(2)}$, which may be determined by requiring the "norm" of $\eta(\zeta,\tau)$,

$$\| \eta \| \equiv \frac{1}{\pi} \int_0^{2\pi} \int_0^1 \zeta \eta^2 (\zeta,\tau) \, d\zeta \, d\tau \tag{54}$$

to be independent of ϵ as $\epsilon \to 0$. This immediately implies

$$\int_0^{2\pi} \int_0^1 \zeta \eta_0(\zeta,\tau) \eta_2(\zeta,\tau) \, d\zeta \, d\tau = 0 \tag{55}$$

from which it follows that

$$A_{11}^{(2)} = 0 \tag{56}$$

We thus have

$$\eta_2(\zeta,\tau) = \sum_{n=2}^{\infty} A_{1n}^{(2)} V_n(\zeta)\cos\tau + \sum_{n=1}^{\infty} A_{3n}^{(2)} V_n(\zeta)\cos 3\tau \tag{57}$$

Again the evaluation of the integrals appearing in equations (51) through (53) must be done on a digital computer.

We terminate our calculations at this order and shall return to examine the results in detail in the next section.

The same perturbation method can be applied to the simplified nonlinear plate equation as given in equation (12). In terms of the same set of nondimensional quantities as defined earlier, we may rewrite equation (12) as

$$L_1\eta - \epsilon^2\lambda(\eta'' + \frac{1}{\zeta}\eta')\int_0^1 \zeta\eta'^2\, d\zeta + \omega^2\ddot{\eta} = 0 \tag{58}$$

where L_1 has been given in equation (15a) and η here satisfies the same boundary, periodicity and initial conditions as given in equations (19a), (19b), (20a) and (21a).

Making similar expansions for η and ω^2 as given in equations (22) and (24) and substituting them into equation (58), we obtain the following set of equations

$$\epsilon^0:\quad L_1\eta_0 + \omega_0^2\ddot{\eta}_0 = 0 \tag{59a}$$

$$\epsilon^2:\quad L_1\eta_2 + \omega_0^2\ddot{\eta}_2 = -\omega_2^2\ddot{\eta}_0 + \lambda(\eta_0'' + \frac{1}{\zeta}\eta_0')\int_0^1 \zeta\eta_0'^2 d\zeta \tag{59b}$$

ϵ^4: ....

As before we set $\eta_0 = A_{11}V_1(\zeta)\cos\tau$ and $\omega_0^2 = \Omega_1^2 = (3.1962)^4$ as in equations (35) and (36) and expand $\eta_2(\zeta,\tau)$ as

$$\eta_2(\zeta,\tau) = \sum_{m=0}^{\infty}\sum_{n=1}^{\infty} A_{mn}^{(2)} V_n(\zeta)\cos m\tau \tag{60}$$

We substitute equations (35), (36) and (60) into (59b) and obtain

$$\sum_{m=0}^{\infty}\sum_{n=1}^{\infty}(\Omega_n^2 - m^2\Omega_1^2)A_{mn}^{(2)} V_n(\zeta)\cos m\tau = \omega_2^2 A_{11}V_1(\zeta)\cos\tau$$
$$+ \lambda(V_1''(\zeta) + \frac{1}{\zeta}V_1'(\zeta))\int_0^1 \zeta V_1'^2(\zeta)d\zeta\cdot\cos^3\tau \tag{61}$$

We now multiply both sides above with $\zeta V_s(\zeta)\cos p\tau$, and perform similar integrations with respect to ζ and τ and obtain finally

$$\omega_2^2 = \frac{3}{4}\lambda\omega_0^2 A_{11}^2 \frac{[J_1^2(\beta_1) + \frac{J_0^2(\beta_1)}{I_0^2(\beta_1)} I_1^2(\beta_1)]^2}{[J_1^2(\beta_1) + 2J_0^2(\beta_1) - \frac{J_0^2(\beta_1)}{I_0^2(\beta_1)} I_1^2(\beta_1)]^2} \tag{62}$$

$$A_{1n}^{(2)} = \frac{3\lambda A_{11}^3}{4(\Omega_n^2 - \Omega_1^2)}\int_0^1 (V_1''(\zeta) + \frac{1}{\zeta}V_1'(\zeta))\zeta V_n(\zeta)d\zeta\int_0^1 \zeta V_1'^2(\zeta)d\zeta \tag{63}$$

for n = 2,3,...

$$A_{3n}^{(2)} = \frac{\lambda A_{11}^3}{4(\Omega_n^2 - 9\Omega_1^2)}\int_0^1 (V_1''(\zeta) + \frac{1}{\zeta}V_1'(\zeta))\zeta V_n(\zeta)d\zeta\int_0^1 \zeta V_1'^2(\zeta)d\zeta \tag{64}$$

for n = 1,2,... . All remaining $A_{mn}^{(2)}$, including $A_{11}^{(2)}$ as before, are zero. We thus have

$$\eta_2(\zeta,\tau) = \sum_{n=2}^{\infty} A_{1n}^{(2)} V_n(\zeta)\cos\tau + \sum_{n=1}^{\infty} A_{3n}^{(2)} V_n(\zeta)\cos 3\tau \tag{65}$$

which is the same as in equation (57), except that now the $A_{mn}^{(2)}$'s are given in equations (63) and (64).

RESULTS AND DISCUSSIONS

Let us now summarize the results that are obtained in the previous section. In terms of the quantities $\bar{\eta}$ and $\bar{\gamma}$ as introduced in equation (13) we have

$$\bar{\eta}(\zeta,\tau) = \epsilon\,\eta_0(\zeta,\tau)+\epsilon^3\eta_2(\zeta,\tau) + 0(\epsilon^5) \quad (66)$$

$$\bar{\gamma}(\zeta,\tau) = \epsilon^2\gamma_0(\zeta,\tau) + 0(\epsilon^4) \quad (67)$$

and

$$\omega^2 = \omega_0{}^2 + \epsilon^2\omega_2{}^2 + 0(\epsilon^4) \quad (68)$$

where η_0 is given by equation (35), $\omega_0{}^2$ by equation (36), γ_0 by equation (47), η_2 by equation (57) and $\omega_2{}^2$ by equations (50) and (51).

As an inspection of the expressions involved above will reveal, the small parameter ϵ appears only through the combination ϵA_{11}. Thus we may set $\epsilon = 1$ and regard A_{11} as being small, or we may simply regard the combination ϵA_{11} as a small parameter.

When the simplified plate equation (58) is used, we have correspondingly

$$\bar{\eta}(\zeta,\tau) = \epsilon\eta_0(\zeta,\tau) + \epsilon^3\eta_2(\zeta,\tau) + 0(\epsilon^5) \quad (69)$$

and

$$\omega^2 = \omega_0{}^2 + \epsilon^2\,\omega_2{}^2 + 0(\epsilon^4) \quad (70)$$

where η_0 and $\omega_0{}^2$ are given, as above, by equations (35) and (36) respectively, but η_2 is now given by equation (65) and $\omega_2{}^2$ by equation (62).

Let us examine in detail the expressions for the nonlinear frequency ω^2 as given in equations (68) and (70). Following equation (68) and recalling that λ is equal to $12(a/h)^2$, we may write

$$\omega^2 = \omega_0{}^2\ [1+12\ F(\tfrac{a}{h},\mu)(\tfrac{a}{h}\,\epsilon\, A_{11})^2] \quad (71)$$

The ratio of the nonlinear period T* to the linear period T is then

$$\frac{T^*}{T} = [1+12\,\frac{F(\frac{a}{h},\mu)}{|V_1|^2_{max}}\,(\tfrac{a}{h}\,\epsilon\,A)^2]^{-1/2} \quad (72)$$

where

$$A = A_{11}|V_1|_{max} \quad (73)$$

Similarly, following equation (70), we have

$$\omega^2 = \omega_0{}^2\ \{1 + \frac{9[J_1{}^2(\beta_1)+\frac{J_0{}^2(\beta_1)}{I_0{}^2(\beta_1)}\,I_1{}^2(\beta_1)]^2}{[J_1{}^2(\beta_1)+2J_0{}^2(\beta_1)-\frac{J_0{}^2(\beta_1)}{I_0{}^2(\beta_1)}\,I_1{}^2(\beta_1)]^2}\,(\tfrac{a}{h}\,\epsilon\,A_{11})^2\} \quad (74)$$

so the ratio of T* to T is now given as

$$\frac{T^*}{T} = \{1 + \frac{9[J_1{}^2(\beta_1)+\frac{J_0{}^2(\beta_1)}{I_0{}^2(\beta_1)}\,I_1{}^2(\beta_1)]^2}{|V_1|^2_{max}[J_1{}^2(\beta_1)+2J_0{}^2(\beta_1)-\frac{J_0{}^2(\beta_1)}{I_0{}^2(\beta_1)}\,I_1{}^2(\beta_1)]^2}\,(\tfrac{a}{h}\,\epsilon\,A)^2\}^{-1/2} \quad (75)$$

We now plot the ratio T*/T against the amplitude parameter $a\epsilon A/h$ using both equation (72) and equation (75). It is observed that the expression $F(a/h,\mu)$

appearing in equation (50) depends on a/h as well as on μ. The integral for $F(a/h,\mu)$, given in equation (51), is now computed numerically on a CDC 6500 digital computer using a Newton-Cotes technique. Four different values of μ: 0.1, 0.2, 0.3 and 0.4, and two different values of a/h: 10 and 100 are used in the computations. The numerical values of $F(a/h,\mu)$ are shown in Table I, where q refers to the number of terms of the infinite series in equation (51) that are used in the computations. It is seen that for an accuracy of four significant figures it is sufficient to take q = 6, i.e. to base the computations on the first six terms of the infinite series. The values of $F(a/h,\mu)$ in Table I also show that the dependence of $F(a/h,\mu)$ on a/h is not sensitive.

In Figures 1 through 4 we plot the curves of T^*/T versus $a\epsilon A/h$, according to equation (72), for μ = 0.1, 0.2, 0.3 and 0.4 respectively. The curves (designated as ③) corresponding to a/h = 10 and a/h = 100 are undistinguishable. In these same figures we also plot the curves of T^*/T versus $a\epsilon A/h$ using equation (75) (designated as ②) as well as those obtained by Wah in [3] (designated as ①). These curves ① and ② do not depend on Poisson's ratio, so they remain the same in Figures 1 through 4.

All these curves show that the nonlinear period of vibration decreases as the amplitude of vibration increases. This is consistent with the fact that the geometric nonlinearity, which is the only kind of nonlinearity considered here, possesses the hard-spring characteristics. The more accurate results obtained using the fully coupled nonlinear plate equations do exhibit the dependence of the solutions upon Poisson's ratio, a fact that can not be accounted for by a simplified single plate equation. These curves show that the numerical errors introduced in the period-amplitude relationship in the free vibrations of a clamped circular plate depend upon μ and this error is minimized for a material with a Poisson's ratio lying between 0.3 and 0.4.

Aside from the fact that the solutions obtained using the fully coupled nonlinear plate equations give more accurate period-amplitude relationships of the vibrations, they also supply information on the in-plane displacement $\overline{\gamma}(\zeta,\tau)$, from which the variations of the in-plane strains and membrane stresses in the radial directions can also be computed. In Tables II and III the two sets of expansion coefficients C_{0n} and C_{2n}, which appear in equations (44), (45) and (47), are computed and presented.

CONCLUSION

Nonlinear perturbation solutions are obtained for the axisymmetric, free vibration of a clamped circular plate. Two sets of plate equations are used - one a more accurate set of fully coupled nonlinear plate equations and one a simplified single plate equation in which the effects of the second strain invariant of the mid-surface strains and the in-plane inertia are neglected. Aside from the fact that solutions obtained using the former set of plate equations give more detailed information on the transverse and in-plane motions of the plate, the results also show that the error introduced when the simplified plate equation is used depends on the plate-radius-to-plate-thickness ratio and on Poisson's ratio.

ACKNOWLEDGMENTS

The computer time provided by the College of Engineering, Michigan State University is gratefully acknowledged. We also wish to thank Mrs. Thelma Liszewski for her superb typing.

BIBLIOGRAPHY

1. von Karman, Th., Encyklopadie der Mathematischen Wissenschaften, Vol. 4, 1910, p. 349.
2. Berger, H.M., "A New Approach to the Analysis of Large Deflections of Plates," Journal of Applied Mechanics, Vol. 22, 1955, pp. 465-472.
3. Wah, T., "Vibration of Circular Plates at Large Amplitudes," Proceedings of the American Society of Civil Engineers, Engineering Mechanics Division, EM5, October 1963, pp. 1-15.
4. Nash, W. A. and Modeer, J.R., "Certain Approximate Analyses of the Nonlinear Behavior of Plates and Shallow Shells," Proceedings of the Symposium on the Theory of Thin Elastic Shells, Delft, the Netherlands, August, 1959, Interscience Publishers, Inc., New York, N.Y., 1960, pp. 331-354.

5. Pal, M. C.,"Large Amplitude Vibration of Circular Plates Subjected to Aerodynamic Heating," International Journal of Solids and Structures, Vol. 6, 1970, pp. 301-313.
6. Wu, C. I. and Vinson, J.R., "On the Nonlinear Oscillations of Plates Composed of Composite Materials," Journal of Composite Materials, Vol. 3, 1969, pp. 548-561.
7. Herrmann, G., "Influence of Large Amplitudes on Flexural Motions of Elastic Plates," National Advisory Committee for Aeronautics Technical Report No. 3578, 1956.
8. Chu, H. and Herrmann, G., "Influence of Large Amplitudes on Free Flexural Vibrations of Rectangular Elastic Plates," Journal of Applied Mechanics, Vol. 23, 1956, pp. 532-540.
9. Blotter, P.T., "Free Periodic Vibrations of Continuous Systems Governed by Nonlinear Partial Differential Equations," Ph.D. Dissertation, Michigan State University, 1968.
10. Lee, T.W., "Free Periodic Vibrations of Continuous Systems Governed by Coupled Nonlinear Partial Differential Equations," Ph.D. Dissertation, Michigan State University, 1969.
11. Timoshenko, S. and Woinowsky-Krieger, S., Theory of Plates and Shells, 2nd Edition, McGraw-Hill, New York, 1959.
12. Volterra, E. and Zachmanoglou, E. C., Dynamics of Vibrations, Merrill Books Inc., Columbus, Ohio 1965.
13. Abramowitz, M. and Stegun, I. A., Handbook of Mathematical Functions, Dover Publications, Inc., New York, 1965.

Table I. Values of $F(\frac{a}{h}, \mu)$

$\frac{a}{h}$ \ $F(\frac{a}{h}, \mu)$ \ q	3	6	9
μ=0.1			
10	0.02908	0.02905	0.02905
100	0.02913	0.02910	0.02910
μ=0.2			
10	0.03031	0.03028	0.03028
100	0.03035	0.03032	0.03032
μ=0.3			
10	0.03129	0.03127	0.03127
100	0.03133	0.03130	0.03130
μ=0.4			
10	0.03204	0.03202	0.03202
100	0.03207	0.03205	0.03205

Table II. Values of C_{0n}/A_{11}^2 (independent of a/h)

μ \ n	1	2	3	4	5	6	7	8	9
0.1	0.16170	-0.23173	-0.02068	-0.00501	-0.00170	-0.00070	-0.00033	-0.00017	-0.00010
0.2	0.12288	-0.22871	-0.02037	-0.00493	-0.00167	-0.00069	-0.00033	-0.00017	-0.00010
0.3	0.08405	-0.22568	-0.02006	-0.00486	-0.00165	-0.00068	-0.00032	-0.00017	-0.00009
0.4	0.04522	-0.22266	-0.01975	-0.00478	-0.00162	-0.00067	-0.00032	-0.00017	-0.00009

Table III. Values of C_{2n}/A_{11}^2

μ	$\frac{a}{h}$ \ n	1	2	3	4	5	6	7	8	9
0.1	10	0.16563	-0.23338	-0.02075	-0.00502	-0.00170	-0.00070	-0.00033	-0.00017	-0.00010
	100	0.16174	-0.23174	-0.02068	-0.00501	-0.00170	-0.00070	-0.00033	-0.00017	-0.00010
0.2	10	0.12586	-0.23033	-0.02044	-0.00494	-0.00167	-0.00069	-0.00033	-0.00017	-0.00010
	100	0.12291	-0.22872	-0.02037	-0.00493	-0.00167	-0.00069	-0.00033	-0.00017	-0.00010
0.3	10	0.08609	-0.22729	-0.02013	-0.00487	-0.00165	-0.00068	-0.00032	-0.00017	-0.00009
	100	0.08407	-0.22570	-0.02006	-0.00486	-0.00165	-0.00068	-0.00032	-0.00017	-0.00009
0.4	10	0.04632	-0.22425	-0.01982	-0.00479	-0.00162	-0.00067	-0.00032	-0.00017	-0.00009
	100	0.04523	-0.22268	-0.01975	-0.00478	-0.00162	-0.00067	-0.00032	-0.00017	-0.00009

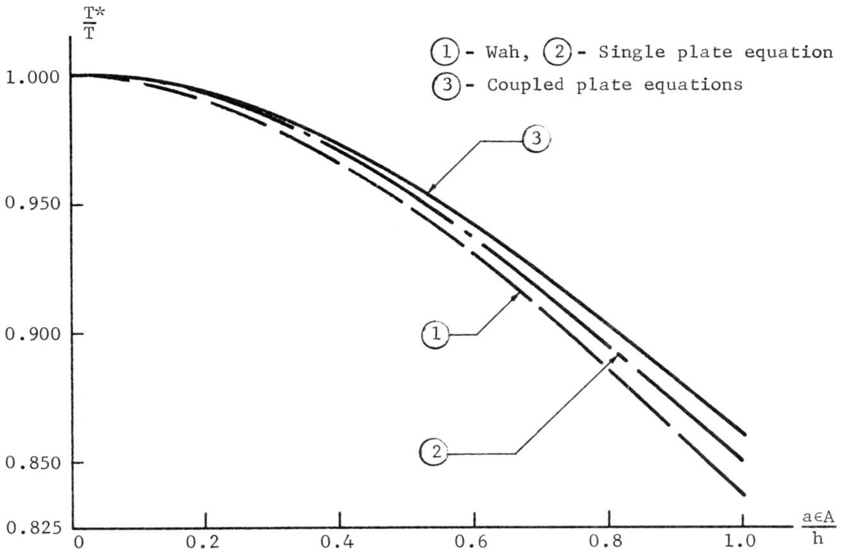

Fig. 1 T*/T versus aεA/h, $\mu = 0.1$

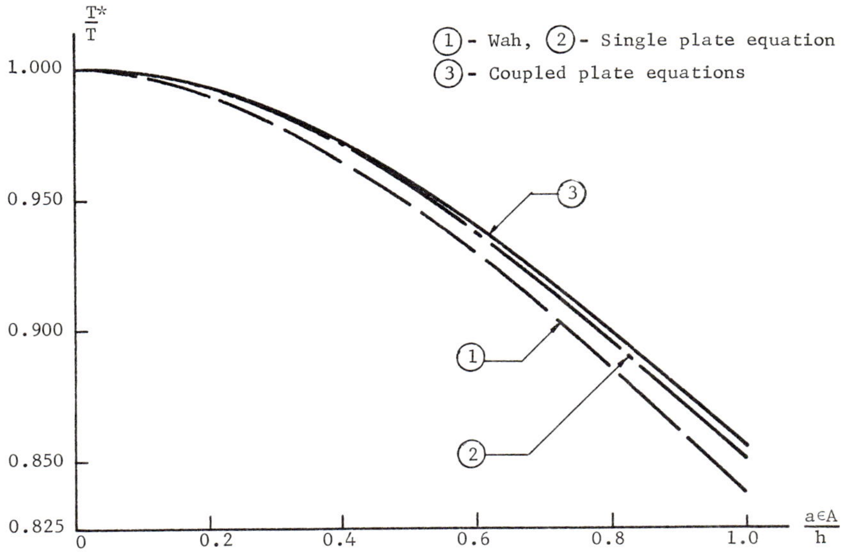

Fig. 2 T*/T versus aεA/h, $\mu = 0.2$

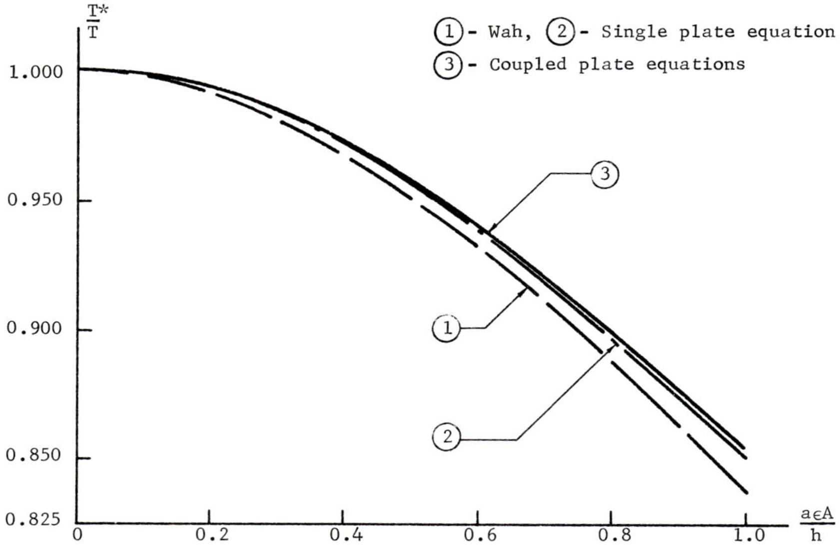

Fig. 3 T*/T versus aεA/h $\mu = 0.3$

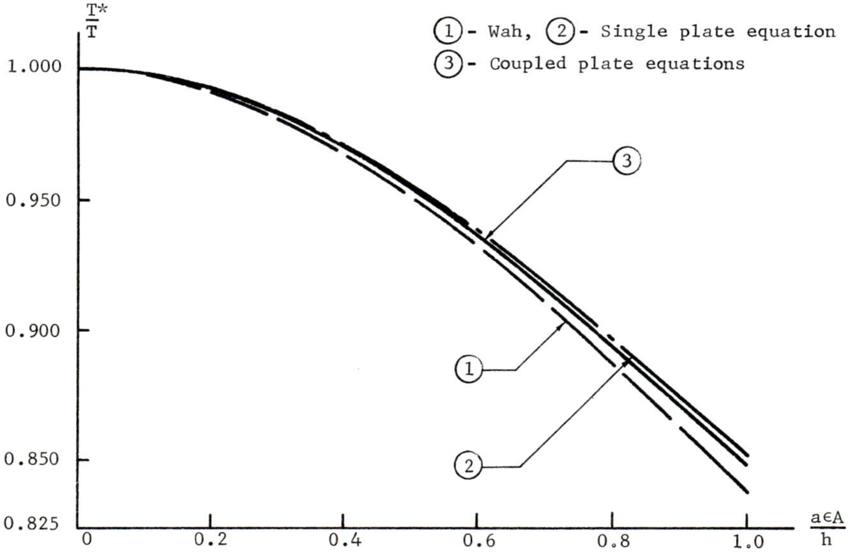

Fig. 4 T*/T versus aεA/h, $\mu = 0.4$

Developments in Mechanics, Vol. 6. Proceedings of the 12th Midwestern Mechanics Conference: 64

Finite Amplitude Oscillations of a Thin Elastic Annulus

B. E. SANDMAN and CHI-LUNG HUANG
KANSAS STATE UNIVERSITY

ABSTRACT

The free and forced nonlinear vibrations of an annular plate with a free inner boundary and a clamped-immovable outer boundary are studied. Steady state harmonic oscillations are assumed and the time variable is eliminated by a Kantorovich averaging method. Thus, the governing equations for the axisymmetric motion of the plate are reduced to a pair of ordinary differential equations which form a nonlinear eigenvalue problem. Approximate solutions are obtained by utilizing the related initial-value problem in conjunction with a numerical method of integration. The results reveal that the transverse shape of vibration, bending and membrane stresses, and frequency are nonlinear functions of amplitude. Experimental resonance data is presented for a direct comparison with the theoretical results.

INTRODUCTION

Modern, light-weight designs often require that thin-plate structures be able to withstand large amplitudes of vibration when they are subjected to severe dynamic loading conditions. If the amplitudes of motion are of the same order of magnitude as the thickness of the plate, then the mathematical description of the motion must be extended from classical linear plate theory to include deformation of the middle plane. In the development of a suitable theory, geometric nonlinearities arise in a coupling of membrane and bending theories for thin plates. Due to the complex nature of the resulting governing equations it appears that the only present means of solution to large amplitude plate vibration problems is by approximate methods of various types.

Approximate solutions are commonly obtained by separating the variables and implementing function space methods (methods which minimize a measure of error on

*This paper is condensed from the first author's doctoral thesis submitted to the Department of Applied Mechanics, Kansas State University.

a function space) for the purpose of eliminating the assumed-space-mode shape function from the governing equations [1,2,3,4,5].

Problems involving the large deflections of annular plates have received limited attention. In this paper the problem of large amplitude vibration of annular plates is formulated in terms of von Karman's [6] dynamic equations. The problem could be formulated in terms of Berger's equations [7], but the neglect of the second strain invariant of the middle plane lacks basic physical justification. Assuming the existence of harmonic vibrations, the time variable is eliminated by employing a Kantorovich averaging method. Thus, the basic governing equations for the problem are reduced to a pair of ordinary differential equations, which form a nonlinear boundary-value problem. A numerical study of these equations is proposed by introducing the related intial-value problem. A few specific cases of free and forced vibration associated with fundamental mode are investigated in detail, and the corresponding results are illustrated. An experimental study of the resonant characteristics of an annular plate is also presented for direct comparison with the theory.

THE GOVERNING EQUATIONS OF MOTION

Consider a thin annular plate located in its initially undeformed configuration by cylindrical coordinates r, θ, and z. The annulus is composed of an elastic, homogeneous and isotropic medium bounded by the planes $z = \pm h/2$ and the cylinders r = b and r = a. The plate is excited in a manner which produces large-amplitude, flexural vibrations.

With appropriate assumptions [9] the non-vanishing components of strain are found to be

$$\varepsilon_r = \frac{\partial u}{\partial r} + \frac{1}{2}\left(\frac{\partial w}{\partial r}\right) - z\frac{\partial^2 w}{\partial r^2} \tag{1a}$$

$$\varepsilon_\theta = \frac{u}{r} - \frac{z}{r}\frac{\partial w}{\partial r} \tag{1b}$$

where u(r,t) and w(r,t) are the radial and transverse components of mid-plane displacement, respectively. In terms of Hooke's laws the radial and circumferential stresses are given by

$$\sigma_r = \frac{E}{1-\nu^2}\,(\varepsilon_r + \nu\varepsilon_\theta) \tag{2a}$$

$$\sigma_\theta = \frac{E}{1-\nu^2}\,(\varepsilon_\theta + \nu\varepsilon_r) \tag{2b}$$

Writing the equilibrium equation for the forces in the radial direction and neglecting the longitudinal inertia, the result is

$$\frac{\partial N_r}{\partial r} + \frac{N_r - N_\theta}{r} = 0 \tag{3a}$$

with

$$N_r = \int_{-h/2}^{+h/2} \sigma_r dz = \frac{Eh}{1-\nu^2}\left[\frac{\partial u}{\partial r} + \frac{1}{2}\left(\frac{\partial w}{\partial r}\right)^2 + \nu\frac{u}{r}\right] \tag{3b}$$

and

$$N_\theta = \int_{-h/2}^{+h/2} \sigma_\theta dz = \frac{Eh}{1-\nu^2}\left[\frac{u}{r} + \nu\frac{\partial u}{\partial r} + \frac{\nu}{2}\left(\frac{\partial w}{\partial r}\right)^2\right] \tag{3c}$$

By eliminating the radial displacement function u(r,t) from equations (3b) and (3c), the compatability equation

$$\frac{\partial}{\partial r}\,(N_\theta + N_r) = -\frac{Eh}{2r}\left(\frac{\partial w}{\partial r}\right)^2 \tag{4}$$

is obtained with the aid of equation (3a). When the stress function $\psi(r,t)$ and the corresponding relations

$$N_r = \frac{\psi}{r} \quad ; \quad N_\theta = \frac{\partial \psi}{\partial r} \tag{5}$$

which satisfy (3a) exactly, are introduced into equation (4), the result is

$$\frac{\partial^2 \psi}{\partial r^2} + \frac{1}{r}\frac{\partial \psi}{\partial r} - \frac{\psi}{r^2} = -\frac{Eh}{2r}\left(\frac{\partial w}{\partial r}\right)^2 \tag{6}$$

The equilibrium equation of moments about a circumferential tangent is

$$\frac{\partial M_r}{\partial r} + \frac{M_r - M_\theta}{r} = V_r \tag{7}$$

where V_r is the shearing force per unit length and

$$M_r = \int_{-h/2}^{+h/2} \sigma_r z dz = -D\left(\frac{\partial^2 w}{\partial r^2} + \frac{\nu}{r}\frac{\partial w}{\partial r}\right) \tag{8a}$$

$$M_\theta = \int_{-h/2}^{+h/2} \sigma_\theta z dz = -D\left(\frac{1}{r}\frac{\partial w}{\partial r} + \nu \frac{\partial^2 w}{\partial r^2}\right) \tag{8b}$$

By applying d'Alembert's principle, dynamic equilibrium of the transverse, z, forces requires that

$$\frac{\partial}{\partial r}(rV_r) = \rho h r \frac{\partial^2 w}{\partial t^2} - rq(r,t) - \frac{\partial}{\partial r}\left(rN_r \frac{\partial w}{\partial r}\right) \tag{9}$$

where ρ is the mass density of the plate and $q(r,t)$ is the lateral loading intensity. Combining equations (7-9), one obtains the equation

$$D\nabla^4 w + \rho h \frac{\partial^2 w}{\partial t^2} - \frac{1}{r}\frac{\partial}{\partial r}\left(\psi \frac{\partial w}{\partial r}\right) = q(r,t) \tag{10}$$

with ψ/r being substituted for N_r. Equations (6) and (10) are dynamic forms of von Karman's equations, which govern the finite-amplitude, axisymmetric vibration of a thin annular plate.

By introducing the dimensionless quantities

$$\chi = w/a \qquad \phi = \frac{1-\nu^2}{Eha}\psi \qquad \xi = r/a$$

$$P = \frac{12(1-\nu^2)}{E}\left(\frac{a}{h}\right)^3 q \qquad \tau = \left(\frac{D}{\rho h a^4}\right)^{1/2} t$$

equations (6) and (10) can be written as

$$\frac{\partial^2 \phi}{\partial \xi^2} + \frac{1}{\xi}\frac{\partial \phi}{\partial \xi} - \frac{\phi}{\xi^2} = -\frac{1-\nu^2}{2\xi}\left(\frac{\partial \chi}{\partial \xi}\right)^2 \tag{11a}$$

$$\nabla^4\chi + \frac{\partial^2\chi}{\partial\tau^2} - 12(\frac{a}{h})^2 \frac{1}{\xi}\frac{\partial}{\partial\xi}(\phi \frac{\partial\chi}{\partial\xi}) = P(\xi,\tau) \tag{11b}$$

APPROXIMATE ANALYSIS

An exact solution of the differential equations (11) is at present unknown. The standard Fourier analysis used in linear vibration problems cannot be applied in an exact sense due to the nonlinear character of the differential equations which causes a coupling of vibration modes. Consequently, the analysis and solution of the problem must be completed in some approximate manner. The following averaging method is proposed in an attempt to find an assumed-time-mode solution [9] of the equations (11).

Kantorovich Averaging Method

The first simplification is imposed by taking the time varying loading intensity to be of the form

$$P(\xi,\tau) = Q(\xi) \sin\omega\tau \tag{12}$$

With the plate being subjected to a sinusoidal loading, it is assumed that the steady state response can be closely approximated by the expressions

$$\chi(\xi,\tau) = G(\xi) \sin\omega\tau \tag{13a}$$

$$\phi(\xi,\tau) = F(\xi) \sin^2\omega\tau \tag{13b}$$

where $G(\xi)$ and $F(\xi)$ are undetermined shape functions of vibration. The assumption (13b) follows from (13a) and the supposition that the resulting membrane stresses should be independent of the up or down position of the plate.

Substituting (13) into the governing equation (11a), one finds

$$\frac{d^2F}{d\xi^2} + \frac{1}{\xi}\frac{dF}{d\xi} - \frac{F}{\xi^2} = - \frac{1-\nu^2}{2\xi} (\frac{dG}{d\xi})^2 \tag{14}$$

Since expressions (12) and (13) cannot satisfy equation (11b) for all τ, the integral

$$I = \int_R^1 [\nabla^4\chi + \frac{\partial^2\chi}{\partial\tau^2} - 12(\frac{a}{h})^2 \frac{1}{\xi}\frac{\partial}{\partial\xi}(\phi \frac{\partial\chi}{\partial\xi}) - P] \, \delta\chi\xi d\xi \tag{15}$$

is employed to obtain a governing equation which closely approximates (11b) within the limits of the assumed form of motion and loading as given in equations (13) and (12), respectively. For any instant of time τ, the above integral is equal to the virtual work of all the tranverse forces as they move through a virtual displacement $\delta\chi = \delta G \sin\omega\tau$. Substituting expressions (12) and (13) into equation (15) and equating the average virtual work over one period oscillation to zero, i.e.,

$$I_A = \int_0^{2\pi/\omega} I \, d\tau = 0, \tag{16}$$

yields

$$\nabla^4 G - \omega^2 G - 9(\frac{a}{h})^2 \frac{1}{\xi}\frac{d}{d\xi}(F \frac{dG}{d\xi}) = Q(\xi) . \tag{17}$$

Thus, the assumed harmonic vibrations become governed by the pair of nonlinear, ordinary differential equations, (14) and (17). For convenience let

$$G(\xi) = Ag(\xi) \text{ and } F(\xi) = A^2 f(\xi) \tag{18}$$

where A is an amplitude parameter, and $g(\xi)$ and $f(\xi)$ are revised shape functions. Substitution of expressions (18) into equations (14) and (17) results in the final forms of governing differential equations

$$\frac{d^2f}{d\xi^2} + \frac{1}{\xi}\frac{df}{d\xi} - \frac{f}{\xi} = - \frac{1-\nu^2}{2\xi}\left(\frac{dg}{d\xi}\right)^2 \qquad (19a)$$

$$\nabla^4 g - \lambda g - 9\,\frac{\alpha}{\xi}\frac{d}{d\xi}\left(f\,\frac{dg}{d\xi}\right) = \frac{Q^*}{\sqrt{\alpha}} \qquad (19b)$$

where $\alpha = (A\frac{a}{h})^2$, $Q^* = (\frac{a}{h})Q$, and $\lambda = \omega^2$.

The boundary conditions for an annual plate with a free inner edge and a clamped immovable outer edge are

$$\left.\begin{aligned} \left(\frac{d^2g}{d\xi^2} + \frac{\nu}{\xi}\frac{dg}{d\xi}\right)_{\xi=R} &= 0 \\ \frac{d}{d\xi}\left(\nabla^2 g\right)_{\xi=R} &= 0 \\ \left(\frac{f}{\xi}\right)_{\xi=R} &= 0 \end{aligned}\right\} \qquad (19c)$$

$$\left.\begin{aligned} (g)_{\xi=1} &= 0 \\ \left(\frac{dg}{d\xi}\right)_{\xi=1} &= 0 \\ \left(\frac{df}{d\xi} - \nu\,\frac{f}{\xi}\right)_{\xi=1} &= 0 \end{aligned}\right\} \qquad (19d)$$

The equations (19) compose a nonlinear two-point boundary-value problem which describes the harmonic response of an annular plate undergoing finite amplitude oscillations. Solving this nonlinear boundary-value problem completes the analysis and reveals the salient characteristics of the plate under investigation.

Initial-Value Method

Due to the nonlinearity in the boundary-value problems posed by equations (19), approximate solutions are proposed through the direct application of numerical integration to the following initial-value method [8].

The governing field equations (19) can be written as the system of six first-order equations

$$\frac{d\overline{Y}}{d\xi} = \overline{H}\,(\xi,\overline{Y};\ \alpha,\lambda,Q^*),\ R<\xi<1, \qquad (20a)$$

where

$$\overline{Y}(\xi) = \begin{Bmatrix} g \\ g' \\ g'' \\ g''' \\ f \\ f' \end{Bmatrix} = \begin{Bmatrix} y_1 \\ y_2 \\ y_3 \\ y_4 \\ y_5 \\ y_6 \end{Bmatrix},\quad (\ )' = \frac{d}{d\xi}$$

and $\overline{H}$ is the appropriately defined $(\overline{6}x1)$ vector function. It is noted that equations (20a) contain two parameters α and λ which may be considered as additional unknowns. In order to obtain a unique relationship between amplitude, $\sqrt{\alpha}$, and frequency, $\sqrt{\lambda}$, an additional restriction, which supplements the boundary conditions, must be placed on $\overline{Y}(\xi)$. This requirement is fulfilled by normalizing a component of $\overline{Y}(R)$. Thus, the boundary conditions are written in the generalized form

$$M\overline{Y}(R) = \begin{Bmatrix} 1 \\ 0 \\ 0 \\ 0 \end{Bmatrix} \tag{20b}$$

and

$$N\overline{Y}(1) = \begin{Bmatrix} 0 \\ 0 \\ 0 \end{Bmatrix} \tag{20c}$$

where

$$M = \begin{pmatrix} 1 & 0 & 0 & 0 & 0 & 0 \\ 0 & \frac{\nu}{R} & 1 & 0 & 0 & 0 \\ 0 & -\frac{1}{R^2} & \frac{1}{R} & 1 & 0 & 0 \\ 0 & 0 & 0 & 0 & \frac{1}{R} & 0 \end{pmatrix}$$

and

$$N = \begin{pmatrix} 1 & 0 & 0 & 0 & 0 & 0 \\ 0 & 1 & 0 & 0 & 0 & 0 \\ 0 & 0 & 0 & 0 & -\nu & 1 \end{pmatrix}$$

The system of equations (20) is studied conveniently by introducing the related initial-value problem

$$\frac{d\overline{Z}}{d\xi} = \overline{H}\ (\xi, \overline{Z};\ \alpha, \lambda, Q^*) \tag{21a}$$

$$\overline{Z}(R) = \overline{\gamma}^*(\eta_1, \eta_2) = \begin{Bmatrix} 1 \\ \eta_1 \\ -\frac{\nu}{R}\ \eta_1 \\ \frac{1+\nu}{R^2}\ \eta_1 \\ 0 \\ \eta_2 \end{Bmatrix}$$

which contains initial values that satisfy the boundary conditions (20b). It is now supposed that for a known function $Q^*(\xi)$, a solution of the initial-value problem (21) is obtained on the closed interval $[R,1]$. This solution is symbolically denoted by

$$\overline{Z} = \overline{Z}\ (\xi; \overline{\eta}, \alpha) \quad ; \quad \overline{\eta} = \begin{Bmatrix} \eta_1 \\ \eta_2 \\ \lambda \end{Bmatrix}$$

The secondary arguments here indicate that the solution depends upon the set of parameters η and α. From this solution and the boundary conditions (20c) the three simultaneous equations

$$N\overline{Z}(1; \overline{\eta}, \alpha) = \overline{0} \tag{22}$$

may be constructed. Thus, it is seen that solving the boundary-value problem (20) is equivalent to obtaining a functional relation $\eta = \overline{\eta}(\alpha)$ such that

$$N\overline{Z}(1; \overline{\eta}(\alpha), \alpha) \equiv \overline{0}$$

Directly,

$$\overline{Y}(\xi; \alpha) = \overline{Z}(\xi; \overline{\eta}(\alpha), \alpha)$$

forms an α-dependent family of solutions to (20).

For a fixed value of α, say $\alpha = \alpha°$, equations (22) reduce to three transcendental equations

$$N\overline{Z}(1;\overline{\eta},\alpha°) = \overline{0} \tag{23}$$

in terms of three unknowns, $\overline{\eta}$. A root, $\overline{\eta}°$, of (23) may be found by a direct application of Newton's method. Starting from an initial guess, η_1, the iterative sequence

$$\overline{\eta}_{k+1} = \overline{\eta}_k + \Delta\overline{\eta}_k \quad ; \quad k = 1,2,3,\ldots \tag{24a}$$

is generated. Retention of first-order terms of a Taylor series expansion of (23) about $\overline{\eta}_k$, provides the linear correction

$$\Delta\overline{\eta}_k = - [N(J_1)_k]^{-1} N\overline{Z}(1;\overline{\eta}_k,\alpha°) \tag{24b}$$

at the kth step of iteration. The (6x3) matrix (J_1) is defined as

$$(J_1) = \left(\frac{\partial\overline{Z}}{\partial\overline{\eta}}\right)_{\xi=1} = \left(\frac{\partial z_i}{\partial\eta_j}\right)_{\xi=1} \quad ; \quad \begin{matrix} i = 1,\ldots,6 \\ j = 1,2,3 \end{matrix} \tag{25}$$

and can be interpreted physically as the change of final values with respect to a change of data, $\overline{\eta}$. The expression for $\Delta\overline{\eta}_k$ in (24b) is the product of the inverse Jacobian matrix and the kth error vector. If the starting guess, $\overline{\eta}_1$, is in a sufficiently small neighborhood of $\overline{\eta}°$, then convergence of the sequence, $\overline{\eta}_k$ to the root, $\overline{\eta}°$, can be expected. Since an explicit solution of the initial-value problem cannot be readily obtained due to the nonlinear character of the vector function $\overline{H}$, the expression for (J_1) in (24) cannot be evaluated directly. Therefore, a method of constructing the (J_1) matrix at any given step of iteration must be devised. Differentiation of the initial-value problem (21) with respect to $\overline{\eta}$ in a formal manner yields the variational problem

$$\frac{d}{d\xi}\left(\frac{\partial\overline{Z}}{\partial\overline{\eta}}\right) = \frac{\partial\overline{H}}{\partial\overline{\eta}} + \left(\frac{\partial\overline{H}}{\partial\overline{Z}}\right)\left(\frac{\partial\overline{Z}}{\partial\overline{\eta}}\right) \tag{26a}$$

$$\left(\frac{\partial\overline{Z}}{\partial\overline{\eta}}\right)_{\xi=R} = \frac{\partial\overline{\gamma}^*}{\partial\overline{\eta}} \tag{26b}$$

which is composed of eighteen first-order equations and a corresponding set of initial values. For a given vector, $\overline{\eta}$, and $\alpha = \alpha°$, this derived problem and the initial-value problem (21) may be integrated simultaneously, at least numerically on the interval [R,1]. Evaluation of the resulting solution to the variational problem at $\xi=1$ provides the (J_1) matrix corresponding to the given values of $\overline{\eta}$ and $\alpha = \alpha°$. Thus, by setting $\overline{\eta} = \overline{\eta}_1$ and integrating equations (21) and (26) from $\xi=R$ to $\xi=1$, the first corrected vector, $\overline{\eta}_2$, can be calculated according to equations (24). Returning with $\overline{\eta} = \overline{\eta}_2$ and repeating the same operations, one obtains the second corrected vector, $\overline{\eta}_3$. Successive repetition of this procedure yields the desired sequence, $\overline{\eta}_k$. In summary, the two initial-value problems (21) and (26) in conjunction with Newton's method provide a convenient mode of investigating the two-point boundary-value problem (20) when the value of α is held fixed.

The analysis of the nonlinear system (20) is completed when the functional relation, $\overline{\eta} = \overline{\eta}(\alpha)$, and accompanying solutions

$$\overline{Y}\ (\xi;\alpha) = \overline{Z}\ (\xi;\overline{\eta}(\alpha),\alpha)$$

are established. The initial-value method may be used to accomplish this in a discrete manner, i.e.,

$$\overline{\eta}^{i} = \overline{\eta}^{i}(\alpha^{i})\ ;\ i = 0,1,2,\ldots,m \tag{27}$$

After having obtained a root, $\overline{\eta}^{\circ}$, which corresponds to $\alpha = \alpha^{\circ}$, the value of α can be perturbed

$$\alpha = \alpha^{\circ} + \Delta\alpha^{\circ} = \alpha^{1}$$

For this value of α iteration is reinstated starting from $\overline{\eta} = \overline{\eta}^{\circ}$. If $\Delta\alpha^{\circ}$ is not exceedingly large, then $\overline{\eta}^{\circ}$ is contained in the new contraction domain of Newton's method, and iteration converges to the root, $\overline{\eta}_{1}$, corresponding to $\alpha = \alpha^{1}$. Successive repetition of this analytic continuation approach leads to the relation given in (27) with

$$\alpha^{i+1} = \alpha^{i} + \Delta\alpha^{i};\ i = 0,1,2,\ldots,m-1$$

NUMERICAL COMPUTATION

The theoretical analysis presented above suggests the employment of a numerical integration technique. Thus, by integrating the initial-value problems (21) and (26) simultaneously with a fourth-order Runge-Kutta-Gill method and performing the successive corrections of $\overline{\eta}$ according to Newton's method, approximate solutions to the governing system (20) can be obtained.

First, the problem is reduced to that of free vibration by letting $Q^* \equiv 0$, and then α is set equal to zero, $\alpha^{\circ} = 0$. In this case the equation (19b), which governs the transverse displacement, becomes linear and has a well-known general solution [10]. This information provides a reasonable starting guess, η_{1}, required by the initial-value method. For $\overline{\eta} = \overline{\eta}_{1}$, the initial-value problems (21) and (26) are integrated numerically on [R,1] with a step-size, $\Delta\mu$. Successive correction and integration are carried out until each of the equations in (23) is satisfied within the range of a prescribed error; this allowable error being consistent with the order, $O(|\Delta\mu|^{4})$, of the Runge-Kutta-Gill method. The corresponding approximate values of $(\eta_{1}^{\circ},\eta_{2}^{\circ},\lambda^{\circ})$, components of $\overline{\eta}^{\circ}$, are stored and the solution is recorded. From this "linear" solution the effects of finite amplitudes can be examined with the continual use of the concept of neighboring solutions.

By gradually incrementing the value of α and restarting the correction and integration procedure from those values of $(\eta_{1},\eta_{2},\lambda)$ obtained in the solution corresponding to the previous value of α, discrete representations of the resonance curve and accompanying solutions are found. This process is terminated momentarily when α reaches a prescribed maximum, α^{m}, and the corresponding values of $(\eta_{1}^{m},\eta_{2}^{m},\lambda^{m})$ are stored. At this stage the steady state response due forced oscillation can be determined by a perturbation technique which takes advantage of the resonant characteristics of the plate. On setting $\alpha = \alpha^{m}$, $(\eta_{1},\eta_{2},\lambda) = (\eta_{1}^{m},\eta_{2}^{m},\lambda^{m})$ and applying a small load, $Q^*(\xi)$, to the plate, the response is easily determined with a slight modification of the above mentioned process. Namely, the type of loading, $Q^*(\xi)$, is held fixed while the value of α is gradually decremented from α^{m} as the integration and correction procedure of the initial-value method yields successive solutions to the forced problem. Since the $\sqrt{\alpha}$ in equation (19b) is considered to be positive, $Q^*(\xi)g(\xi)<0$ implies that the force is 180° out of phase with the response, and $Q^*(\xi)g(\xi)>0$ implies that the force is in phase with the response. Hence, two distinct response curves are obtained for $+ Q^*(\xi)$ and $- Q^*(\xi)$. It is noted that the continuity of solutions with respect to $Q^*(\xi)$ and α allows the utilization of this approach.

Numerical solutions to equations (21) and (26) were obtained with a fourth-order Runge-Kutta-Gill method using a step-size, $\Delta\mu = 1/40$. The suitability of this step-size was examined by conducting several trials with $\Delta\mu = 1/20, 1/40, 1/80$. For each element, α^i, of a sequence which defined discrete values of α, successive corrections of (η_1,η_2,λ) were performed until the final values, $\bar{Z}(1)$, satisfied the norm,

$$\max_{1\leq i\leq 3} \left| \sum_{j=1}^{6} n_{ij}\, z_j(1) \right| \leq 0.1 \times 10^{-5} \tag{28}$$

where $(n_{ij}) = N$ and $z_i(1) = \bar{Z}(1)$. The resulting approximate solutions to equations (20) were recorded in sequential order. The α-dependence of solutions was established by letting $\alpha^{i+1} = \alpha^i + \Delta\alpha^i$, $0<|\Delta\alpha_i|\leq 0.1$, and taking the solution values of (η_1,η_2,λ) corresponding to $\alpha = \alpha^i$ as initial approximations to those solution values corresponding to $\alpha = \alpha^{i+1}$.

Figures 1-6 illustrate the type of results that were found for free vibrations. The resonance curves for radii ratios of R = 0.1 and R = 0.3 are shown in Fig. 1, where W_1 and W_3 are the deflections at the inner edges. Observing the near parallelism of the two curves, the influence of R upon the degree of nonlinearity is apparently small.

The "swelling of the mode" with increasing amplitude (inertial force) is shown in Fig. 2. It is seen that the nonlinear mode shape when $\alpha = 4.0$ differs significantly from the linear ($\alpha = 0.0$) mode shape. Intuitively, the influence of amplitude upon the distribution of bending stress should be of greater significance since the bending stresses are related to derivatives of the transverse shape function, $g(\xi)$.

When time, τ, is equal to $\pi/2\omega$, a maximum excursion occurs and the expressions for the stresses are

$$\frac{\sigma_r^b a^2}{Eh^2} = \frac{\sqrt{\alpha}}{2(1-\nu^2)} \left[\frac{d^2g}{d\xi^2} + \frac{\nu}{\xi}\frac{dg}{d\xi} \right] \tag{29a}$$

$$\frac{\sigma_\theta^b a^2}{Eh^2} = \frac{\sqrt{\alpha}}{2(1-\nu^2)} \left[\frac{1}{\xi}\frac{dg}{d\xi} + \nu \frac{d^2g}{d\xi^2} \right] \tag{29b}$$

$$\frac{\sigma_r^m a^2}{Eh^2} = \frac{\alpha}{1-\nu^2} \left[\frac{f}{\xi} \right] \tag{29c}$$

$$\frac{\sigma_\theta^m a^2}{Eh^2} = \frac{\alpha}{1-\nu^2} \left[\frac{df}{d\xi} \right] \tag{29d}$$

Figures 3-6 show the stress distributions that were established for the free vibration of a free and clamped-immovable annulus with $R = b/a = 0.1$. It is interesting to observe the rapid changes of stress levels in a region around the hole. The extreme rise in the circumferential stresses at the free edge, ξ=R, illustrates the occurrence of a stress concentration phenomenon. It is thus apparent that the circumferential stresses must be included in any failure criteria.

The harmonic response in a neighborhood of the first mode of resonance was investigated by subjecting the annular plate, R = 0.3, to a uniform loading distribution, Q_o^*. The response curves that resulted from solutions with a loading intensity of $Q_o^* = 20$ are presented in Fig. 7, where W_3 is the deflection at the inner edge.

EXPERIMENTAL CONSIDERATIONS

The equipment shown in Fig. 8 was used in determining experimental resonance for the first mode of vibration of an annular plate with a free inside edge and a built-in outside edge.

The plate was forced into motion by four rubber drivers centered around the hole and mounted on a small circular plate which was connected to an electrodynamic exciter. The exciter was powered by a combination audio oscillator and power supply. An electronic counter displayed the driving frequency in digital form. A micrometer probe which extended downward from a plexiglass frame provided a means of measuring the maximum excursions of the plate from its equilibrium position.

For small power inputs to the exciter, preliminary checks on linear resonance and mode shapes were made with a velocity probe and sawdust particles, which were placed on the surface of the plate to indicate nodal lines. Then with the power input held momentarily constant a search for the driving frequency that resulted in a maximum amplitude of response for the first mode of vibration was initiated. The micrometer probe, calibrated in thousandths of an inch, and frequency were adjusted simultaneously in locating the maximum response at the edge of the hole. In particular, the frequency was increased slowly through resonance many times while the probe was readjusted each time until the maximum amplitude and corresponding frequency were found. Readings for this maximum amplitude were taken when the micrometer probe just lacked contact with the plate. The detectable difference between audible contact and no contact was found to be less than one-thousandth of an inch. Repetition of the above procedure for increasing power inputs to the exciter resulted in a set of damped natural frequencies and corresponding amplitudes.

The dimensions and physical properties of the test specimen are:

$$a = 5.045 \text{ in.}$$

$$b = 0.505 \text{ in.}$$

$$h = 0.037 \text{ in.}$$

$$E = 29\text{x}10^6 \text{ lb./in.}^2$$

$$\nu = 0.3$$

$$\rho = 0.00073 \quad \frac{\text{lb.-sec.}^2}{\text{in.}^4}$$

The plate was released and reclamped for each of several experiments in order to ascertain the effectiveness of the clamping fixture. The capabilities of the power supply and exciter along with the choice of flexural stiffness in the specimen limited the maximum attainable amplitudes. Figure 9 compares experimental values to the theoretical resonance curve. The inability to clamp the plate in precisely the same manner for each experiment is the probable cause of the shift in frequencies between the experiments.

DISCUSSION AND CONCLUSION

The theoretical study presented in this paper is based on the assumption of harmonic oscillations. Although the assumptions (13) contradict the inseparability of modes in von Karman's dynamic equations, however, fundamental physical arguments can be made to justify such assumptions. The experimental resonant characteristics that were found for the annular plate are indicative of the suitability of these assumptions.

Previous works, which utilize either von Karman's or Berger's dynamic equations, have approached the problems of large amplitude plate vibration by assuming mode shape functions and applying function space methods to eliminate the space variable. This procedure places limitations upon the number of degrees of freedom that can be maintained in the resulting assumed-space-mode solution. In the present work the time coordinate function is assumed and eliminated by a time averaging method. By eliminating the time variable an infinite number of degrees of freedom in the space coordinate functions are maintained. The employment of numerical integration leads to a discrete approximation of the continuous system. The number of degrees of freedom in the assumed-time-mode solutions and the accuracy of computation are governed by the choice of the step-size, $\Delta\mu$. The numerical solutions, which were obtained in this study, reveal that the mode shape and bending stresses in addition to the membrane stresses are nonlinear functions of the amplitude of vibration. This result is generally not achieved by assumed-space-mode solutions.

The introduction of the concentric hole results in significant magnitudes of membrane stress even at relatively low amplitudes of vibration. This effect is due to a stress concentration factor at the edge of the hole.

Provided that the initial-value method converges for modified forms of the stiffness matrices M and N, generalized boundary conditions can be treated without difficulty.

In carrying out the experiments with the annular plate it became evident that the inclusion of damping and/or acoustical effects would enhance the theory. This, however, may prove to be a difficult task. Also the higher modes and stability of vibration have not been considered and are left open to future investigation.

ACKNOWLEDGEMENT

The authors wish to express their gratitude and appreciation to Dr. Hugh S. Walker and Mr. Wallace M. Johnston whose suggestions and contributions made the experimental work possible.

BIBLIOGRAPHY

1. Chu, Hu-Nan, and Herrmann, G., "Influence of Large Amplitudes on Free Flextural Vibrations of Rectangular Elastic Plate," Journal of Applied Mechanics, Vol. 23, 1959, pp. 532-540.

2. Nowinski, J., "Nonlinear Transverse Vibrations of Circular Elastic Plates Built-in at the Boundary," Proceedings of the 4th U.S. National Congress of Applied Mechanics, 1962, pp. 523-334.

3. Yamaki, N., "Influence of Large Amplitudes on Flexural Vibrations of Elastic Plates," Zeitschrift für angewandte Mathematik and Mechanik, Vol. 41, 1961, pp. 501-510.

4. Wah, T., "Vibration of Circular Plates at Large Amplitudes," Journal of the Engineering Mechanics Division ASCE, Vol. 89, 1963, pp. 1-15.

5. Srinivasan, A. V., "Nonlinear Vibrations of Beams and Plates," International Journal of Non-linear Mechanics, Vol. 1, 1966, pp. 179-191.

6. von Karman, T., "Encyklopädie der Mathematischen Wissenschaften," Vol. IV_4, 1910, p. 349.

7. Berger H. M., "A New Approach to the Analysis of Large Deflections of Plates," Journal of Applied Mechanics, Vol. 22, 1955, pp. 465-472.

8. Keller, H. B., Numerical Methods for Two-Point Boundary-Value Problems, Blaisdell Publishing Co., Waltham, Massachusetts, 1968.

9. Huang, C. L., and Sandman, B. E., "Large Amplitude Vibrations of a Rigidly Clamped Circular Plate," will appear in International Journal of Non-linear Mechanics.

10. Leissa, A. W., Vibration of Plates, National Aeronautics and Space Administration, SP-160, 1969, pp. 7-35.

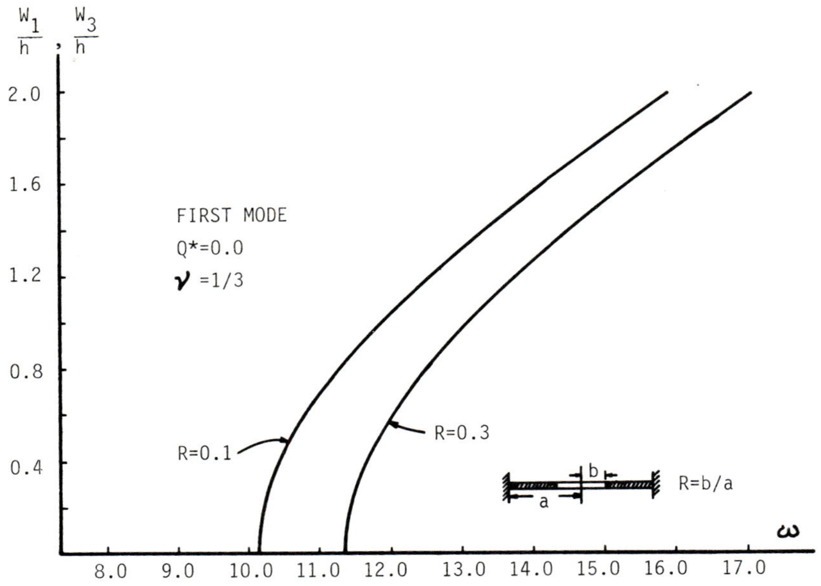

Fig. 1. Harmonic Resonance of an Annular Plate
(Free & Clamped-Immovable)

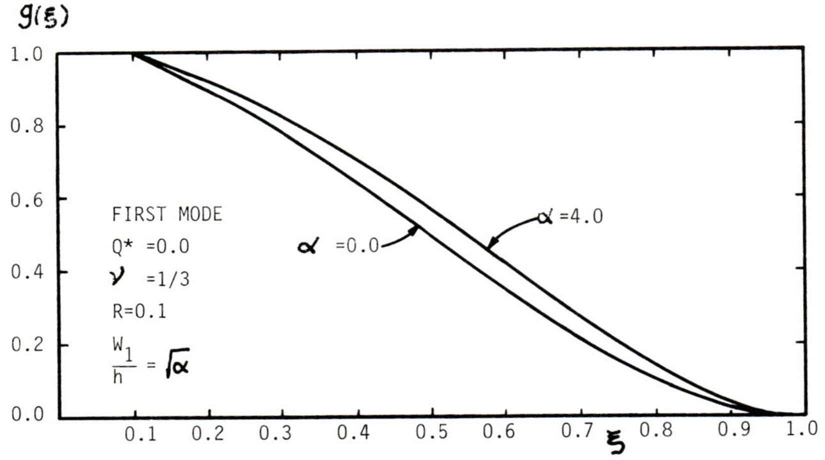

Fig. 2. Shape Function of Vibration for an Annular Plate
(Free & Clamped-Immovable)

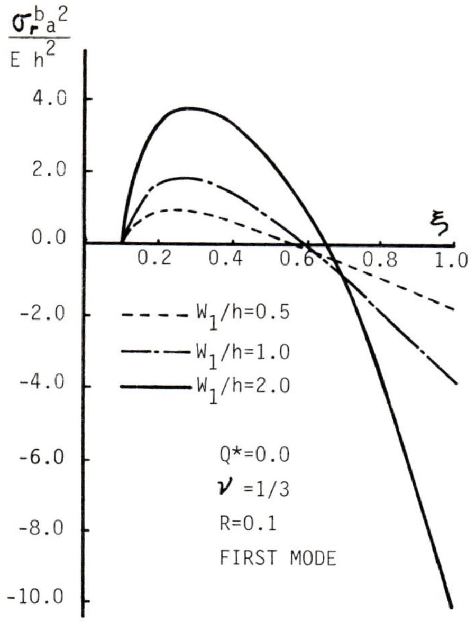

Fig. 3. Radial Bending Stress for an Annular Plate (Free & Clamped-Immovable)

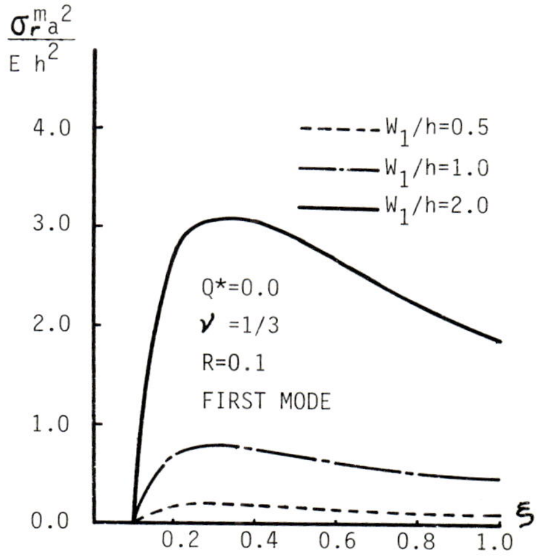

Fig. 4. Radial Membrane Stress for an Annular Plate (Free & Clamped-Immovable)

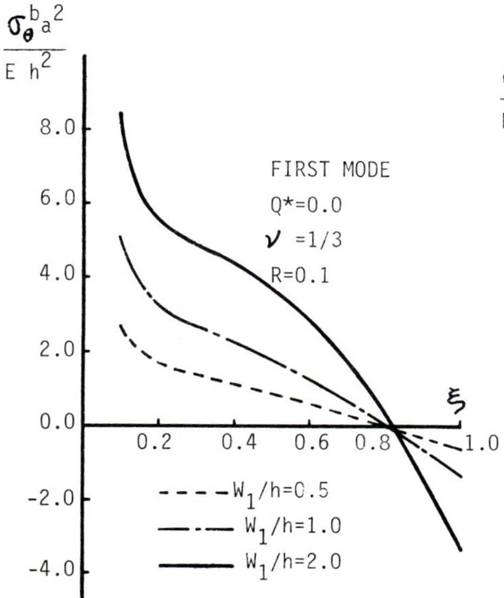

Fig. 5. Circumferential Bending Stress for an Annular Plate (Free & Clamped-Immovable)

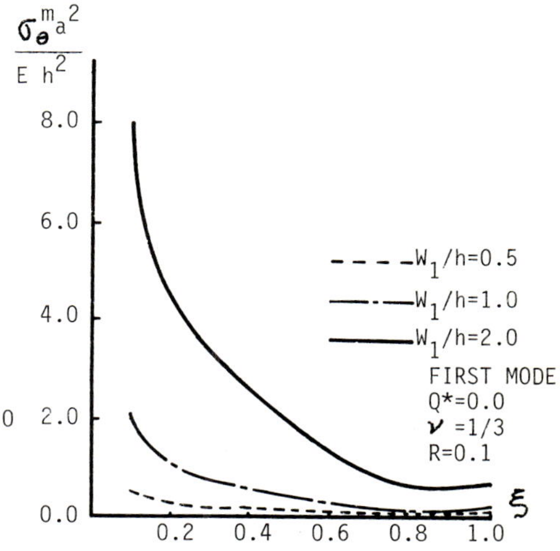

Fig. 6. Circumferential Membrane Stress for an Annular Plate (Free & Clamped-Immovable)

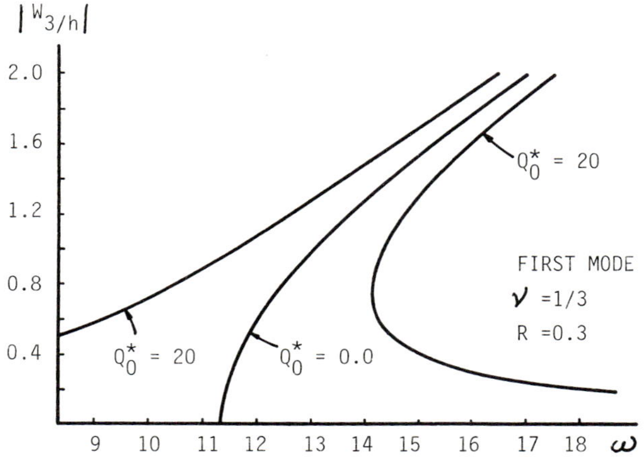

Fig. 7. Harmonic Response of an Annular Plate (Free & Clamped-Immovable)

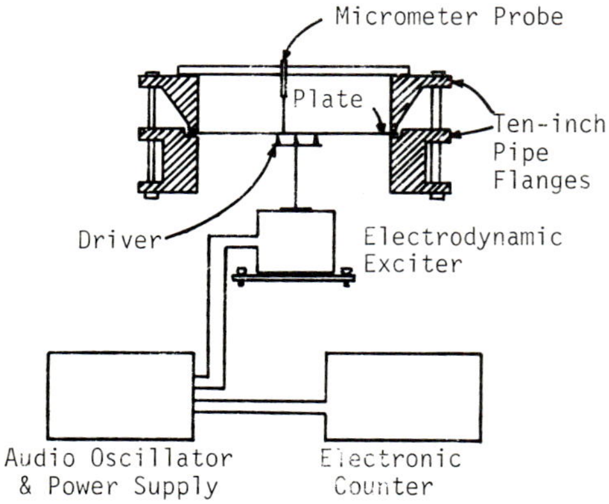

Fig. 8. Schematic Diagram of Experimental Apparatus

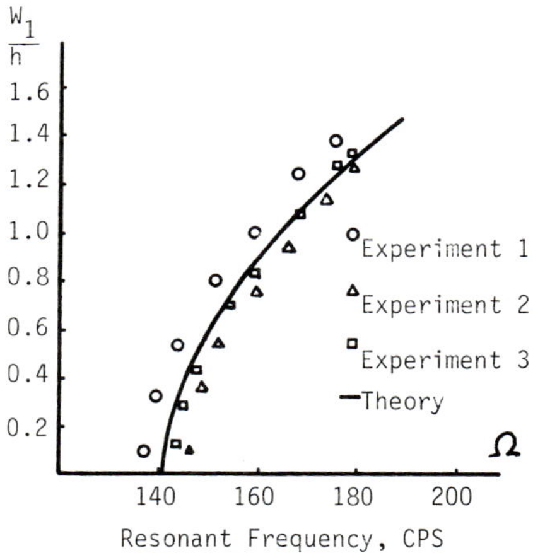

Fig. 9. Comparison of Experimental and Theoretical Resonance

EXPERIMENTAL MECHANICS

Developments in Mechanics, Vol. 6. Proceedings of the 12th Midwestern Mechanics Conference: 65

A Photoelastic Study of Stress Waves in Fiber Reinforced Composites

J. W. DALLY, J. A. LINK and R. PRABHAKARAN
ILLINOIS INSTITUTE OF TECHNOLOGY

ABSTRACT

The methods of dynamic photoelasticity were applied to the study of stress wave propagation in anisotropic materials. Transparent birefringent models were fabricated from glass fiber reinforced plastics and a pulsed ruby laser system was used to record the fringe patterns. Isochromatic fringe patterns in a unidirectionally reinforced model were obtained for several conditions of dynamic loading, including a full-plane with internal loading and a half-plane with edge loading. All the waves observed in homogeneous isotropic materials, namely the P, S, PS and R waves, were identified in the case of the orthotropic models. In addition, a new wave called the PS* wave was observed. It is suggested that the PS* wave was generated by the P wave as it propagated across the glass fibers. The measured velocities of the P wave compared well with the values obtained from the equation for the plate velocity providing the equation was modified to account for the variation in the elastic constants with direction.

I. INTRODUCTION

Wave propagation in anisotropic media is a topic which has received considerable attention beginning with Lord Kelvin's (1) Baltimore Lectures in 1904, Love (2) in 1927 and papers by Musgrave (3) Synge (4) and Buchwald (5) in the 1950's. These papers, highly mathematical in nature, deal with harmonic waves and the establishment of wave surfaces in homogeneous crystalline media with different degrees of anisotropy. From this work it is apparent that three real wave velocities occur regardless of the direction of the propagation of the waves. The displacements associated with these three waves are orthogonal with one wave corresponding to the irrotational dilatation or P wave and the two remaining waves being associated with equivoluminal distortion with the displacement parallel to the wave fronts. Also it was noted by Synge (4) that the Rayleigh wave (for a certain harmonic frequency) either does not exist or that it is propagated in certain discrete directions with corresponding speeds.

In more recent literature, wave propagation studies in anisotropic media have been related to fiber reinforced composites. Achenbach and Herrmann (6) have considered the heterogeneous nature of these materials and have predicted wave velocities along the fiber direction based on an effective stiffness theory. Chou et al (7), using a control volume, derive relations for the dilatational wave speed in the direction of the fibers in a unidirectionally reinforced material.

This paper deals with a photoelastic analysis of wave propagation in fiber reinforced composites. The experimental methods employed are relatively new as orthotropic-birefringent model materials which are adequately transparent for photoelastic analysis are recent developments. Also, the methods for interpretation of the isochromatic and isoclinic data are still in the development stage. Pih and Knight (8) and Sampson (9) have developed techniques to produce the materials and have proposed versions of the stress-optic law for orthotropic materials. Dally and Prabhakaran (10,11,12) have developed simpler methods for producing model materials, predicted fundamental photoelastic constants f_L, f_T and f_{LT} based on the properties of the constituents and have theoretically and experimentally verified a stress-optic law in close correspondence with that advanced by Sampson.

Much work remains to be accomplished in photoorthotropic-elasticity before the method can be applied to general static analyses. However it appears appropriate to initiate work on the dynamic problem in spite of the incompleteness in the development of the experimental method. The full-field visualization of a dynamic fringe pattern will permit identification of the various wave types, estimates of the wave velocities and observation of the effects of the heterogeneous nature of the fiber reinforced model materials.

The experimental procedures involving the fabrication of transparent-birefringent fiber reinforced materials and the dynamic photoelastic recording system are covered in Section II. The photoelastic results are described in Section III and these results are compared with theoretical predictions of wave speeds in Section IV.

II. EXPERIMENTAL PROCEDURE

A. Material Preparation

A photoelastic study of stress-wave propagation in anisotropic materials requires the development of large sheets of model material which are sufficiently transparent for light to be transmitted in a polariscope. Also, the model material must exhibit the required degree of anisotropy and show an adequate degree of photoelastic sensitivity.

The development of transparent anisotropic model materials for photoelasticity is quite recent. Pih and Knight (8) and then Sampson (9) were the first to report on anisotropic materials. These investigators utilized glass roving and epoxy resins and employed an elaborate fabrication process involving filament winding and evacuation of entrapped air, to fabricate transparent materials.

More recently, Dally and Prabhakaran (10) have developed a simpler method of producing both unidirectionally-, and bidirectionally - reinforced model materials by utilizing commercial glass fabrics and a blend of polyester resin specially formulated to match the refractive indices of the two constituents. The glass fiber reinforced laminates produced were larger in size than those obtained before and were sufficiently transparent to give very well defined photoelastic fringe patterns. The reinforcement for the photoelastic model materials was E-glass fibers available in the form of woven fabrics. Two different fabrics were used in fabricating the unidirectionally reinforced laminates. Plies of these glass cloths were cut to size and their ends taped. Glass fabric, style 7500, provided the major share of the reinforcement. Because of its open weave, the filler fibers could be easily removed. A few plies of a second glass fabric, style 1557, were also used in the laminate. The style 1557 cloth having 85 per cent of its glass fibers in the warp direction was used for two reasons. First, the presence of the filler fibers would facilitate machining of holes in the laminate. Secondly, these filler fibers greatly reduce residual birefringence due to shrinkage in curing the laminate. The number of plies of glass fabrics, style 1557 and style 7500, were arranged in the laminate as 1-7-1-7-1.

For the matrix material, Paraplex P444A was selected as the base resin. This polyester resin was blended with 30 per cent styrene monomer (by weight) in order to produce a refractive index of 1.548, which was the same as that of E-glass. The resin system was mixed with 0.5 per cent benzoyl peroxide and 0.5 per cent methylethyl ketone peroxide to accelerate the cure.

The plies of glass fabrics were soaked in the resin bath to obtain complete impregnation. The wet plies were then sandwiched between two sheets of mylar film in a tensioning frame and stretched to obtain parallel alignment. The wet layup was rolled to carefully remove all of the entrapped air and then placed between heavy flat plates to prevent the air from re-entering. The layup was cured in a platen press at 10 psi pressure for 2 hours at 67°C and then for 4 hours at 100°C.

The cured laminates were 12x10 in. in size and 0.1 in. thick. As the transverse fibers in the laminates constituted less than 3 per cent of the total glass content in the longitudinal direction, the material was considered to be unidirectionally reinforced. The volume fraction of the glass was 40 per cent by weight and the resulting product was sufficiently transparent to permit photographic recording of dynamic fringe patterns with exposure times of about 100 nanoseconds.

B. Dynamic Recording Methods

While the composite model materials are sufficiently transparent for static photoelastic applications, they absorb enough light so as to preclude the use of the multiple-spark gap camera (13) for recording the fringe patterns. To circumvent this difficulty, a single-shot recording system was used with a pulsed ruby laser as a light source. This recording system, shown schematically in Fig. 1, is similar to the system described previously by Taylor and Hemann (14). The components in the laser assembly were: the Raytheon LH6 ruby laser head which contains a 3/8 in. diameter by 3 1/4 in. ruby rod and a EGG FX-42 flash-lamp in a highly reflective elliptical cavity; a Crystalab EOM-512D Pockels cell with a quarter wave voltage of 5.7 K volts, a Glan-laser prism and a 99 per cent reflective plane multilayer dielectric mirror.

The laser was operated in the Q-spoiled mode by removing the quarter wave voltage from the Pockels cell with a negative going pulse of extremely short duration from a high voltage pulse amplifier. The pulse of light issued from the laser operated in this mode was sufficiently intense to permit the recording of dynamic fringe patterns with these composite models. The light pulse was also extremely short in duration (about 100 nano seconds) permitting dynamic resolution (15) of a line array of 100 lines/in. propagating at a velocity of 10^5 in./sec.

The beam of light from the laser was expanded with a lens system and passed through two circular polaroids (HNCP-37) to give a light-field polariscope with a nine in. diameter field. Dynamic records were made in an ordinary commercial view camera on Kodak's high-speed infrared film. As the recording system was capable of only a single record during the dynamic event, it was necessary to repeat the event and change the time between the application of the load and the recording of the fringe pattern.

The loads were applied by utilizing small charges of lead azide (200-700 mg.). The time of the initiation of the charges was controlled to ± 2μsec by igniting the lead azide with a bridge wire activated by a 2000 volt pulse from a 20 joule source. A delay generator was employed to fire the ruby laser at a preselected time to ± 1μsec after detonating the explosive. A high speed photo diode was used in conjunction with an oscilliscope to record this event time. By repeating this process, it was possible to generate several time-sequenced records of fringe patterns representing wave propagation in composite models.

III. PHOTOELASTIC RESULTS

A. The Full Plane Model

In the first dynamic test, the load was generated by a 250 mg. charge of lead azide packed into a 3/8 in. diameter hole in the center of a plate 0.1x10x12 in. in size. This model represents a center of dilatation in an entire plane. The record of the isochromatic fringe pattern made 27 μsec. after application of the loading pulse is presented in Fig. 2.

These isochromatic fringes represent the stress waves propagating in an orthotropic full-plane. Three different waves can be clearly identified. The first is the dilatational or P wave which is propagating with the highest velocity. The fringes associated with the P wave are oval in shape because of the dependence of the P wave velocity on direction in an orthotropic medium. Along the L direction (the major axis of the oval), the P wave has developed a significant tail showing a marked oscillation in the stress field behind the front. Along the T direction no oscillations are evident.

The isochromatic fringe patterns due to the shear waves in Fig. 2 are complex because two different waves of this type are evident. The first is the incident shear or S wave which is generated at the source and produces its maximum photoelastic response near the 45 degree diagonals. As the shear stresses cannot exist on the L and T axes, photoelastic response due to the incident S wave front in these regions does not occur.

The second shear wave is connected with the P wave at the T axis. It appears that the P wave generates this trailing PS* wave as it propagates along the T axis, in a manner similar to the generation of a Von-Schmidt wave at the free boundary of a half-plane. The trailing PS* wave appears to be tangent to the incident S wave which is consistent with the concept of the continuous generation of the PS* wave as the P wave propagates along the T axis and across the longitudinally oriented glass fibers.

B. The Half Plane with Edge Loading

Two series of tests were conducted with the half-plane loaded with a 500 mg. lead azide charge at one point on the boundary. In the first series, the longitudinal axis of the orthotropic model was parallel to the boundary. Isochromatic fringe patterns associated with this model were recorded at 18, 25 and 35 μsec after detonating the charge. These patterns presented in Fig. 3 show the expanding wave forms and the development of several distinct wave fronts.

In the first frame at 18 μsec, the fringes associated with the leading P wave are quite evident; however, the PS, S, and Rayleigh waves have not propagated far enough to be clearly distinguished. The fringes in the second frame (t = 25 μsec) have separated far enough to clearly identify the P wave and the plane fronted PS wave generated by the P wave as it moves along the free boundary at grazing incidence. Along the L axis it should be noted that there is clear evidence of a long oscillatory tail following the P wave. The incident shear or S wave appears to show its maximum response along a line making an angle of about 50 degrees with the L axis. The Rayleigh or R wave can be identified although it is still combined to some degree with the tail of the P wave.

The final frame, recorded at 35 μsec, shows the fringes associated with each wave after separation is effectively complete. The P wave has attenuated significantly; however, there is still a very clear indication of its oscillatory tail along the L axis. The planar front of the PS wave is tangent to the incident S wave. Both the surface and subsurface characteristics of the Rayleigh wave can be distinguished although the response is too low to make a detailed comparison between the R wave form in isotropic (16) and orthotropic materials.

Again, it is evident that the P and the S waves appear to be connected near the T axis by the PS* wave. As the P wave propagates across the fibers it appears to generate a new shear wave (PS*) similar to a von-Schmidt wave. This wave was identified previously on the full plane (Fig. 2); however, it is more clearly defined in this half-plane model. The fringe pattern associated with the juncture of the incident S wave, the PS wave and the PS* wave shows the superposition of these three waves.

The second series of tests with the half-plane were made on a model with its longitudinal axis normal to the boundary. Two fringe patterns recorded at times of 18 and 30 μsec in the dynamic event are presented in Fig. 4. The first frame shows the P wave with a large fringe order response along the L axis followed by an oscillatory tail. The response due to the P wave along the T axis is much smaller. The other waves have not separated and cannot be distinguished easily. The second frame taken at 30 μsec shows four different wave forms which include the leading P wave, the von Schmidt PS wave, the incident shear wave and the Rayleigh wave. It is not possible to identify a PS* wave as it should coincide with the PS wave for this model.

IV. ANALYSIS OF PHOTOELASTIC RESULTS

Analysis of the photoelastic results to give the distribution of the dynamic stress fields propagating in orthotropic materials cannot be accomplished. The stress-optic law for photo-orthotropic-elasticity has not as yet been verified in biaxial stress fields. Analysis would be possible along the free boundary of the half plane models where a uniaxial stress field existed and the photoelastic calibration constant was either f_L or f_T. However, in these cases the order of the fringes was too low to give an accurate representation of the boundary stress distribution.

An analysis was conducted to give an estimate of the various wave velocities as a function of the direction of propagation. These estimates of wave velocity were made by computing the velocities of the photoelastic fringe patterns and thus will be slightly lower than the actual velocities of the wave fronts.

Analysis of the fringes associated with the P waves previously shown in Figs. 2, 3 and 4 yielded the distance S from the explosive source to the leading edge of the fringe pattern and the time t required for the fringes to reach these positions. Velocities denoted here as c_{pe} were obtained directly by (s/t). These velocities are shown in Fig. 5 as a function of the position ϕ along the wave front. These results for c_{pe} show a maximum velocity of 160,000 in/sec along the L axis and a minimum of 104,000 in/sec along the T axis.

As the photoelastic fringes representing the P wave propagation are oval in shape, it is apparent that the velocity c_{pe} in general does not represent the velocity in the direction normal to the wave front. To establish this normal velocity c_{pn}, the normal to the fringe pattern was drawn and the component of c_{pe} in this normal direction was computed to give c_{pn} as shown in Fig. 5.

From the velocities, it is evident that the dilatational wave is propagating as if the fiber reinforced model was homogeneous with elastic constants having values between those associated with the glass fibers and the polyester matrix. In order to analytically describe the velocity of the P wave, the familar equation for the velocity in a homogeneous isotropic plate

$$c_p = \sqrt{E/\rho(1 - \nu^2)} \tag{1}$$

was modified. This modification involved the substitution of E_x and ν_{xy} for the elastic constants and the use of an average mass density ρ. The co-ordinate axes x and y are defined on the inset in Fig. 5 and the elastic constants are given by

$$1/E_x = m^4/E_L + m^2n^2 \left(1/G_{LT} - 2\nu_{LT}/E_L\right) + n^4/E_T$$

$$1/\nu_{xy} = \left(E_x/E_L\right) \left[\nu_{LT} - m^2n^2 \left(1 + 2\nu_{LT} + E_L/E_T - E_L/G_{LT}\right)\right] \tag{2}$$

where

$m = \cos\phi$

$n = \sin\phi$

Elastic constants E_L and ν_{LT} were determined with data from a pair of axial and transverse strain gages mounted on a tensile specimen, the fibers oriented along the axis of the specimen. The shear modulus G_{LT} was determined with another pair of axial and transverse strain gages mounted on a tension specimen with the glass fibers oriented at 45 degrees to the axis of the specimen. The modulus E_T was measured with a strain extensometer mounted on a third specimen with its fibers oriented transversely to the axis. Loads were applied statically in an Instron machine. The results obtained for the elastic constants are listed below

$E_L = 4.20 \times 10^6$ psi

$E_T = 1.44 \times 10^6$ psi

$G_{LT} = 0.43 \times 10^6$ psi

$\nu_{LT} = 0.28$

$\rho = 1.64 \times 10^{-4}$ lb.-sec^2/in^4

Utilizing these results in Eqs. 1 and 2 gave the theoretical values of c_p shown in Fig. 5. At the position $\phi = 0$ corresponding to the longitudinal axis, the results from Eq. 1 are slightly larger than the experimental observations. This difference of a few per cent is expected since the fringe front is slightly behind the actual P wave front. At the position $\phi = 90^\circ$ corresponding to the transverse axis, the predicted velocity is about 10 per cent lower than the experimentally observed velocity. This difference may be due to the fact that the dynamic modulus of the matrix, which is important in the velocity of propagation in the T direction, is significantly higher than the static value measured here. These differences in the static and dynamic moduli would influence E_T and account for a portion of the deviation.

The velocity of the incident shear wave c_S was established from the data presented in Figs. 2, 3 and 4 over a range of angular positions ranging from $\phi = 25$ to $\phi = 64$ degrees. These velocities ranged from 73,500 to 78,000 in/sec; however, due to abrupt variations of the shear stress along the wave front and the interaction of the PS and PS* waves with the incident S wave the measurements may be subject to some inaccuracies. Due to these reasons no attempt was made to establish c_S as a function of ϕ.

The velocity of the von-Schmidt or PS wave evident in Figs. 3 and 4 was found to range from 84,000 to 87,000 in/sec. In Fig. 3 where the boundary of the model coincides with the L axis, the normal to the plane front of the PS wave makes an angle of 60 degrees with the L axis. The shear wave velocity in this case was 84,000 $\pm$ 1,000 in/sec. From Fig. 4, it is evident that the normal to the PS wave makes an angle of 56 degrees with the L axis with a value of c_S = 85,5000 $\pm$ 1500 in/sec. While these velocities are modestly higher than those recorded for the incident S waves, they probably are more accurate since the measurement of the angle associated with this von Schmidt wave is not as difficult as positioning the front on the incident S wave.

The Rayleigh waves propagating along the edge of the half-planes illustrated in Fig. 3 and 4 indicate that Rayleigh wave velocity is relatively insensitive to direction of propagation. The velocity along the L direction computed from data obtained in Fig. 3 was 67,000 $\pm$ 2,000 in/sec; however, the velocity in the T direction established from Fig. 4 was 63,000 in/sec. Considering the marked difference in the velocity of the P wave in the L and T directions, this relatively small difference in the R wave velocity is somewhat surprising. Nevertheless, the evidence of the existence of the Rayleigh wave is clear.

V. CONCLUSIONS

This exploratory study has indicated the feasibility of using dynamic photoelasticity to investigate wave propagation in orthotropic materials made from glass fiber reinforced plastics. Several different wave forms have been identified which include the P, S, PS, PS* and the R waves. Of these waves, the PS* wave is the most unusual since it does not occur in homogeneous isotropic materials. It appears that the PS* wave is a type of head wave generated by the P wave as it propagates along the T axis across the glass fibers. Also of interest was the significant oscillatory tail associated with the P wave as it propagates along the L axis in the direction of the glass fibers.

A study of the velocities of the P wave indicated that the glass reinforced composite was acting essentially as a continuum with the velocity c_P depending on direction which could be related closely to the directional dependence of the elastic constants for the composite material. No tendency for wave propagation to occur at velocities associated with the properties of the individual constituents was noted.

Much more work remains to be accomplished on this extremely complex problem. As the methods of photo-orthotropic-elasticity are developed to a more advanced stage it will be possible to study the stress distributions associated with the different wave types and to establish conditions required to generate the PS* wave.

VI. ACKNOWLEDGMENT

This research program was sponsored by the National Science Foundation and the U.S. Army Research Office - Durham. The authors would like to express their appreciation to Drs. M. Gaus and S. Kumar for their support and encouragement.

REFERENCES

(1) Sir W. R. Hamilton: Collected Papers, vol. 1, Cambridge University Press, 1931.

(2) A.E.H. Love: A Treatise on the Mathematical Theory of Elasticity, Cambridge University Press, 4th Ed., 1927.

(3) M.J.P. Musgrave: On the Propagation of Elastic Waves in Aelotropic Media. Proc. of Royal Society of London, Series A, vol. 226, p. 339-355 (1954).

(4) J.L. Synge: Elastic Waves in Anisotropic Media. J. Mathematics and Physics, vol. 35, p. 323-334 (1957).

(5) V. T. Buchwald: Elastic Waves in Anisotropic Media. Proc. of Royal Society of London, Series A, vol. 253, p. 563-580 (1959).

(6) J. D. Achenbach and Herrmann: Dispersion of Free Harmonic Waves in Fiber Reinforced Composites. J. AIAA, vol. 6, No. 10, p. 1832-1836 (1968).

(7) P. C. Chou, A.S.D. Wang and J. L. Rose: Elastic Wave Front Analysis in Unidirectional Composites. Proc. of Army Symposium on Solid Mechanics, Oct. 1970, p. III-93-III-107.

(8) H. Pih and C. E. Knight: Photoelastic Analysis of Anisotropic Fiber Reinforced Composites. J. Comp. Mat., vol. 3, No. 1, p. 94-107 (1969).

(9) R. C. Sampson: A stress-Optic Law for Photoelastic Analysis of Orthotropic Composites. Exp. Mech., vol. 10, No. 5, p. 210-215 (1970).

(10) J.W. Dally and R. Prabhakaran: Photo-Orthotropic-Elasticity-Part I, Photoelastic Constants for Unidirectional Composites. To be published in Experimental Mechanics.

(11) J.W. Dally and R. Prabhakaran: Photo-Orthotropic-Elasticity-Part II, Stress-Optic Law for Orthotropic Composites To be published in Experimental Mechanics.

(12) R. Prabhakaran: Photo-Orthotropic-Elasticity. Ph.D Thesis, Illinois Institute of Technology, 1970.

(13) W. F. Riley and J. W. Dally: Recording Dynamic Fringe Patterns with a Cranz-Schardin Camera. Exp. Mech., vol. 9, No. 8, p. 27N-33N (1969).

(14) C. E. Taylor and J. H. Hemann: An Application of Scattered Light Photoelasticity to Dynamic Stress Analysis. University of Illinois T & A. M Report, No. 294, (1967).

(15) J. W. Dally, A. Henzi and D. Lewis: On the Fidelity of High-Speed Photographic Systems for Dynamic Photoelasticity. Exp. Mech., vol. 9, No. 9, p. 394-400 (1969).

(16) S. A. Thau and J. W. Dally: Subsurface Characteristics of the Rayleigh Wave. Int. J. Eng. Sc., vol. 7, p. 37-52 (1969).

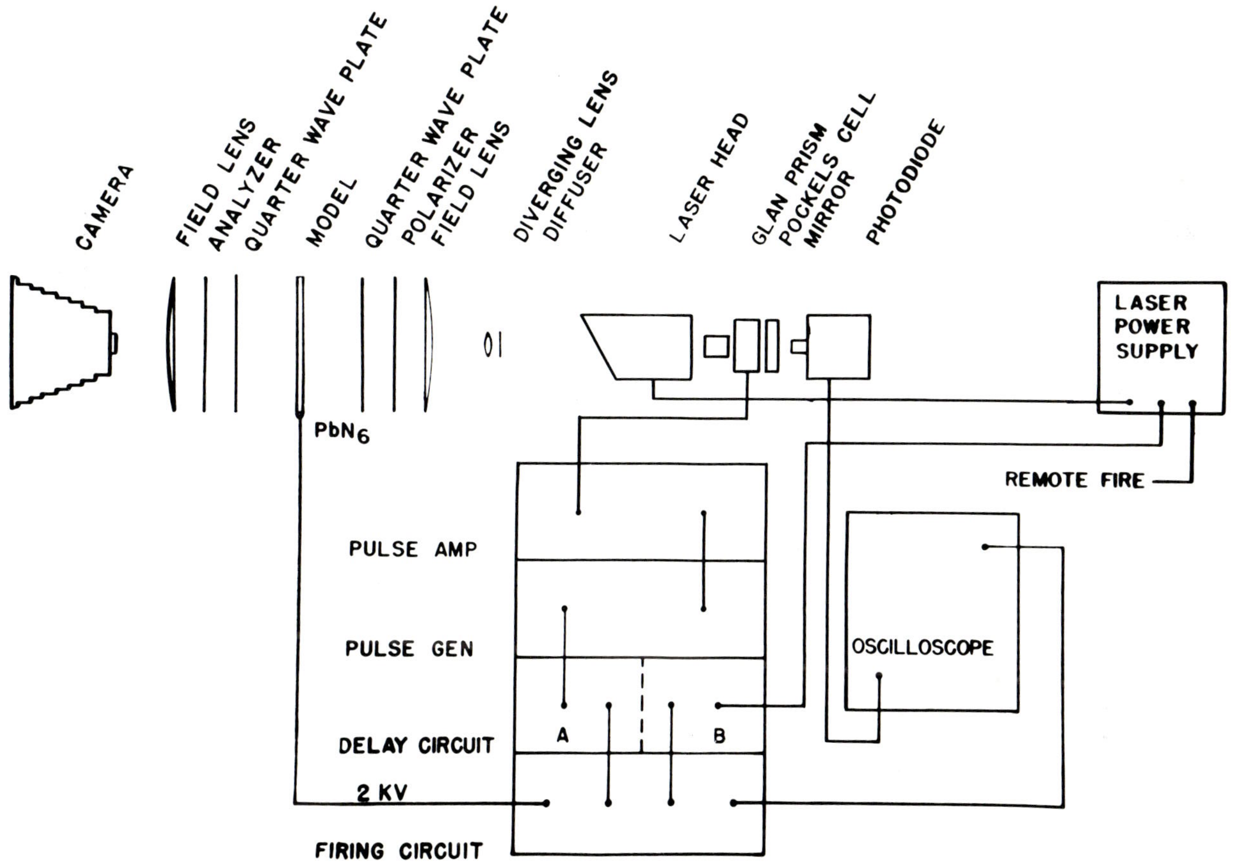

Fig. 1 Block Diagram of the Pulsed Ruby Laser Recording System

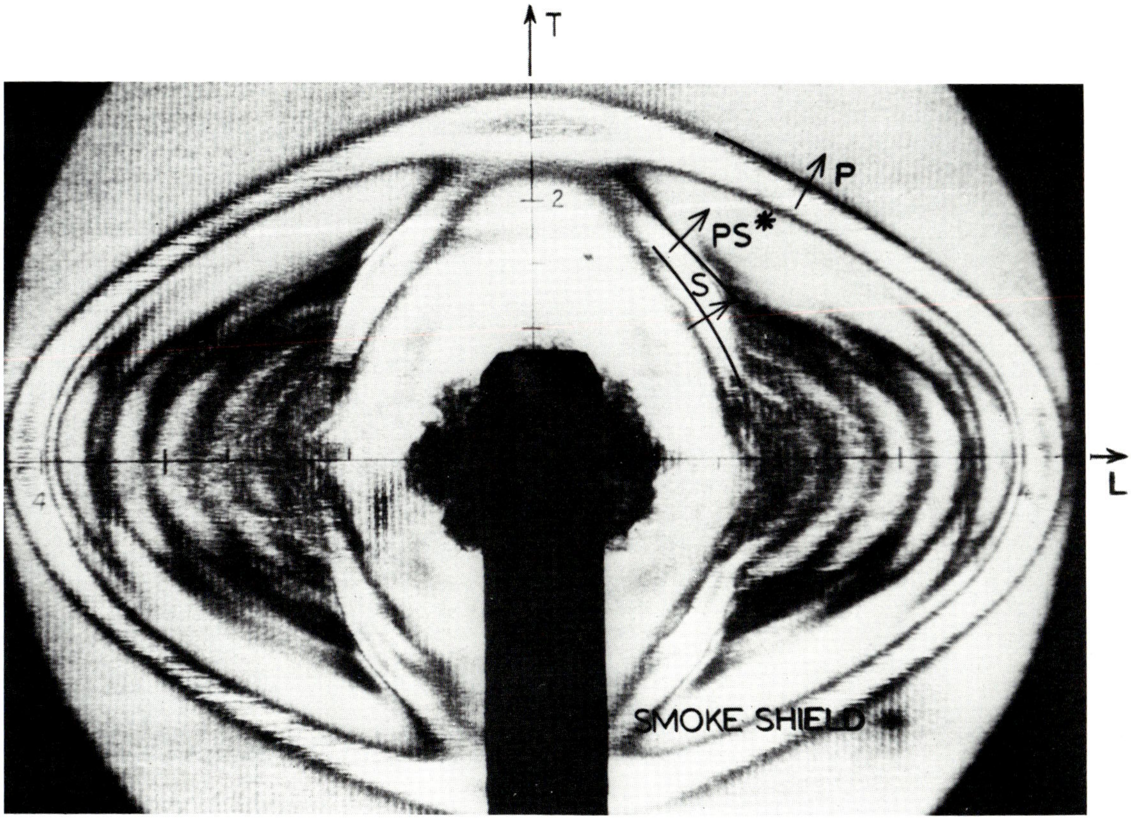

Fig. 2 Isochromatic Fringe Pattern Representing Stress Wave Propagation from a Source of Dilatation in an Orthotropic Full-plane

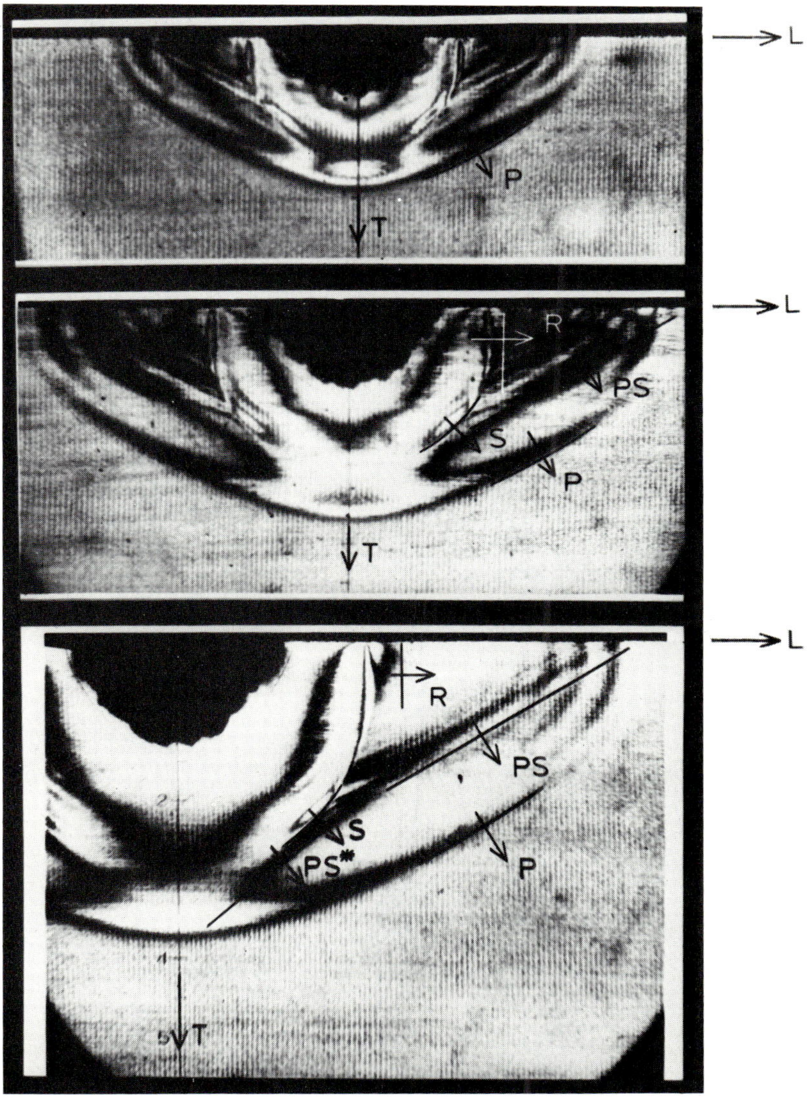

Fig.3 Isochromatic Fringes Associated with Stress Wave Propagation in an Orthotropic Half-plane

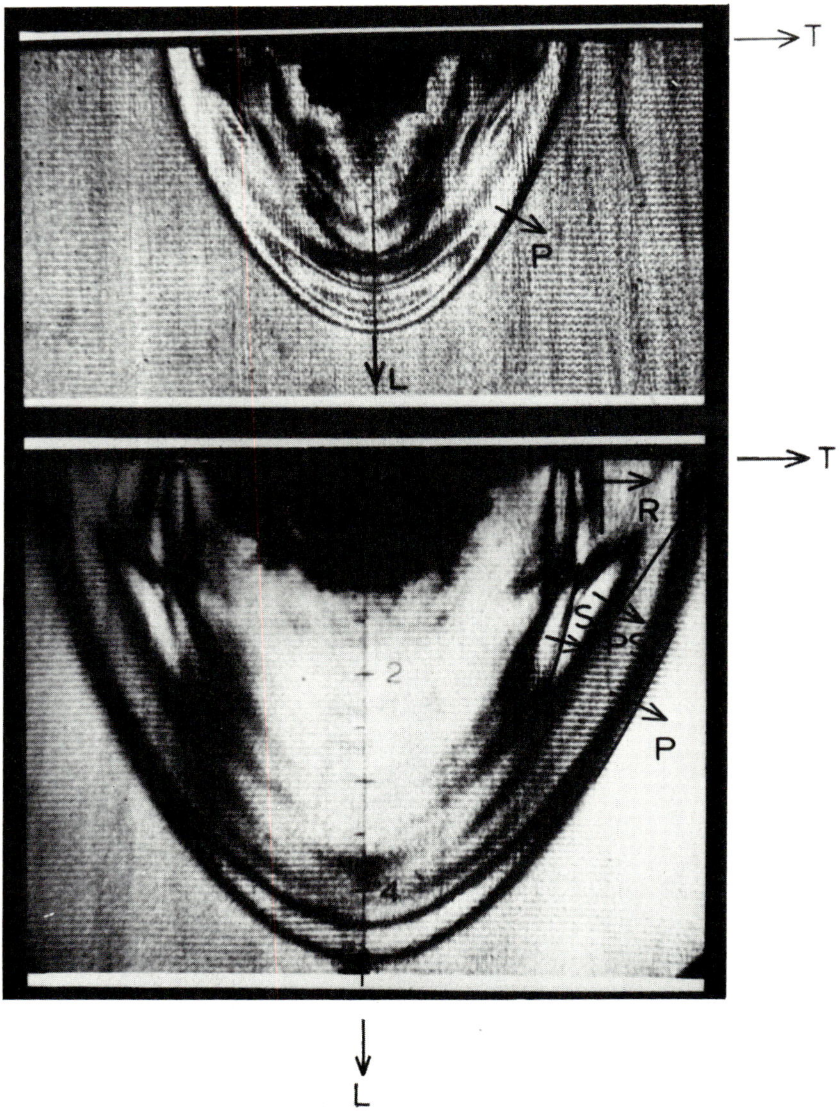

Fig.4 Isochromatic Fringes Associated with Stress Wave Propagation in an Orthotropic Half-plane

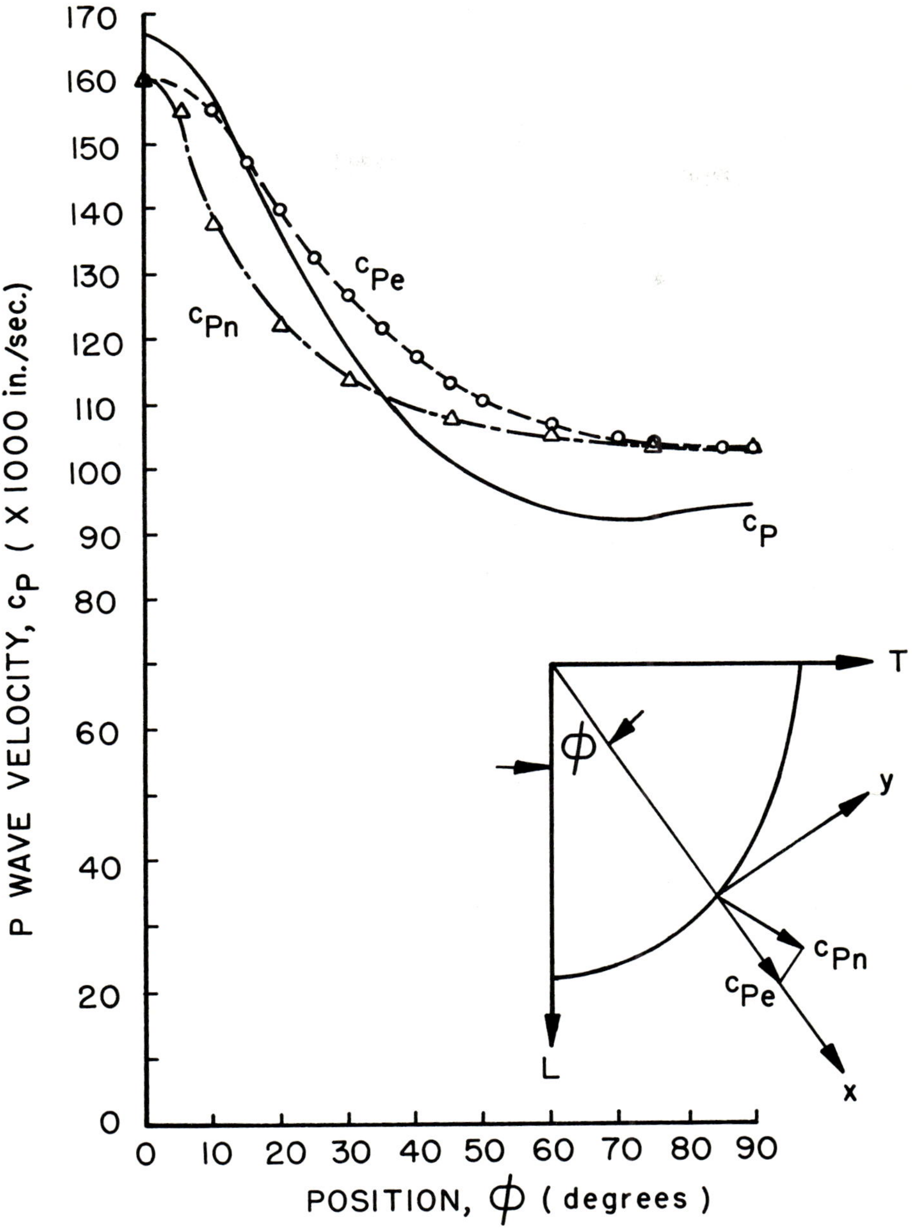

Fig.5 Dilatational Wave Velocity as a Function of Direction of Propagation

Developments in Mechanics, Vol. 6. Proceedings of the 12th Midwestern Mechanics Conference: 66

Applications of Holography to Vibrations of an Axisymmetric Imperfect Conical Shell

T. FURUIKE
NORTH AMERICAN ROCKWELL
V. I. WEINGARTEN
UNIVERSITY OF SOUTHERN CALIFORNIA

ABSTRACT

An experimental and analytical investigation to determine the effects of an axisymmetric imperfection on the free vibrations of conical shells is presented. The time-average holographic interferometry technique is applied to the study of transverse vibrational modes. The images formed during the reconstruction process illustrate the use of this very powerful method in identifying the various vibration modes of conical shells with and without an axisymmetric imperfection. Photographs of the images obtained are presented to clearly illustrate the modal shapes of the conical shells at natural frequencies and the presence of a complex modal coupling pattern.

The finite element method with a higher order Hermitian polynomial displacement field is used to analytically obtain the vibrational characteristics of clamped conical shells with an initial axisymmetric imperfection. The axisymmetric imperfection is described by using many conical elements in that region.

A helium-neon continuous wave laser is used as the source of coherent light. The holography is first applied to perfect conical shells and then extended to a shell containing an initial imperfection. Using the fringe patterns obtained from the reconstructed wavefront, radial deflection shapes of the specimens are plotted. Very good agreement exists when the experimental results were compared with theory. Results indicate that the natural frequencies of the shell increased when an outward axisymmetric imperfection was present.

INTRODUCTION

Recent developments have generated a great resurgence of interest in extensions of Gabor's [1] and [2] wavefront reconstruction process, generally known as holography. A recent review of the historical background on holography is given by Leith and Upatnieks [3] and Pennington [4]. The subject of holography is also discussed in many textbooks as in References [5] - [9]. Many significant contributions on the application of holography occurred around 1964 and 1966 (see References [10] - [14]).

Previous applications of the holographic interferometry technique were limited to feasibility investigations. The first significant application of holographic interferometry to studying and identifying high-frequency transverse vibration modes was limited to beams and plates by Aprahamian and Evensen [15] and [16]. The latter reference, however, also showed the feasibility of obtaining modal patterns for cylindrical shells.

The purpose of this paper is to apply the time-average holographic interferometry technique to the study and identification of the effect of an axisymmetric initial imperfection on the natural frequencies and mode shapes of conical shells.

ANALYTICAL APPROACH

FINITE ELEMENT APPROACH

The shell is assumed to be made of an isotropic homogenous material that obeys Hooke's law. The shell thickness is assumed a constant across any one discrete element and the thickness is also assumed to be small in comparison to the radius of curvature. The Kirchoff Love hyopthesis for a thin shell is assumed to be valid. The Sanders [17] strain-displacement relations for moderate bending were substituted into the strain energy (U) for a thin shell. Only the translational velocities were used in the kinetic energy (T) expression. The finite element representation and the coordinate systems are shown in Fig. 1. The local coordinate displacements are specified as $\tilde{u}$, $\tilde{v}$, and $\tilde{w}$.

The assumed finite element displacement field used, is resolved into Fourier components in the circumferential direction and is expressed in terms of fourth-order Hermitian polynomials in the meridional coordinate. The displacements $\tilde{w}$, $\tilde{u}$, and $\tilde{v}$ shown in Fig. 1 are then expressed as

$$\begin{aligned} \tilde{w} &= (a_1 + a_2 s + a_3 s^2 + a_4 s^3) \cos j\theta \\ \tilde{u} &= (a_5 + a_6 s + a_7 s^2 + a_8 s^3) \cos j\theta \\ \tilde{v} &= (a_9 + a_{10} s + a_{11} s^2 + a_{12} s^3) \sin j\theta \end{aligned} \quad (1)$$

where a_i are constants; j, θ are the circumferential wave number and angle, respectively; s is the meridional distance of each finite element.

The terms containing a_7, a_8, a_{11}, and a_{12} of Eqs. 1 are those terms not normally carried by the constant strain element approximations. The coefficients and the natural frequencies are determined by finding the stationary values of the Hamiltonian subjected to the physical boundary conditions and the compatibility of $\tilde{w}$, $\partial\tilde{w}/\partial s$, $\tilde{u}$, and $\tilde{v}$ at the adjacent elements. Hence the elemental stiffness and consistent mass matrices written in overall system coordinates can be expressed as

$$[k] = [\psi][V^{-1}]^T [A][V^{-1}][\psi]^T \quad (2)$$

and

$$[m] = [\psi][V^{-1}]^T [H][V^{-1}][\psi]^T \quad (3)$$

where $[\psi]$ is the coordinate transformation matrix between overall generalized displacements in terms of the local generalized displacements; $[V]$ relates the local generalized displacements in terms of the coefficients $\{a_i\}$; $[A]$ contains the integrals of the strain energy; and [H] contains the integrals of the kinetic energy. All of the matrices in Eqs. 2 and 3 are given in Reference [18].

The overall stiffness and mass matrices for the complete structure are generated by assuring consistent deformation between the adjacent nth and (n+1)th elements. The finite element representation for the vibration equation to find the natural frequencies ω_j (in radians per second) can be written as

$$\underset{M \times M}{\left[\left[K_R^{\,j} \right] - \omega_j^2 \left[M_R^{\,j} \right] \right]} \underset{M \times 1}{\begin{Bmatrix} q_1 \\ q_2 \\ \vdots \\ q_M \end{Bmatrix}} = \underset{M \times 1}{\begin{Bmatrix} \\ \mathbf{0} \\ \\ \end{Bmatrix}} \tag{4}$$

where $\{q_i\}$ are the generalized displacements and the subscript R denotes the reduced matrices which have accounted for the boundary conditions. The order M of the above equation depends upon the various stiffness [K] and mass [M] matrices generated.

IMPERFECTION REPRESENTATION

The geometric shape w_{im} of the imperfection itself was approximated by

$$w_{im} = \frac{\delta_i}{2}\left(1 - \cos\left(\frac{2\pi s_i}{XL2}\right)\right) \tag{5}$$

where δ_i is the amplitude of the imperfection, XL2 is the wave length of the imperfection for the cosine distribution and s_i is the slant distance measured from the left end of the imperfection (see Fig. 1).

The geometric sketch of the finite element representation of a conical shell with an axisymmetric imperfection is shown in Fig. 1. The imperfection region was idealized into many short conical elements to account for the geometric deviation from the perfect shell.

HOLOGRAPHY

To study the vibration of an object, there are two common holographic interferometry techniques. One is the stored beam technique or the technique to study the live fringe interference at real time.

The second common technique in identifying the mode shapes of a vibrating object is the time-averaging holographic interferometry technique. A beam of coherent light is split into two beams by a beam splitter. One beam is directed to the test specimen and is diffracted by the test specimen. The object wave is now allowed to fall on the photographic plate. The second beam, normally called the reference beam, is reflected via a mirror or mirrors to the same photographic plate. The object and reference waves will form a stable interference pattern when they meet at the photographic recording media. A permanent record of this interference pattern on the photographic emulsion is the hologram.

The reconstruction process consists essentially of illuminating the hologram with the laser beam such that the reconstructed wavefront gives the replica of the original test specimen. The realistic image of the test cone with all the complex vibration patterns can be viewed with the eye and thus can be recorded with a camera.

EXPERIMENTAL PROCEDURE

CONICAL SHELL TEST SPECIMENS

There was a total of three specimens. Two specimens with different thicknesses were made with no prescribed initial imperfection. One specimen, with an axisymmetric initial imperfection was studied to ascertain the effect on the natural frequencies of location, amplitude and wave length of the imperfection. Table I shows the details of the cone geometry and the prescribed imperfection.

The conical specimens were manufactured from a flat sheet of aluminum alloy. A pattern was made to conform to an existing hard plastic mandrel. The test specimens were welded along a meridional seam. They were then spun for truing to circular form. The imperfection was introduced by adding tape around the mandrel and then flowing the soft aluminum alloy material around the tape.

All test specimens were positioned into the base plates by using cerrobend material to simulate the clamped boundary conditions at both ends. The cone surfaces were treated with flat white paint to provide a diffuse optical surface.

EXPERIMENTAL APPARATUS AND TECHNIQUE

A schematic of the test setup for obtaining the hologram is presented in Fig. 2a. A helium-neon continuous wave laser was used to create the beam of coherent light. Two mirror arrangements were used to create the subject beam path to within a quarter of an inch of the reference beam path to the hologram. The mirrors, beam splitter, lens, hologram plateholder, and test specimen were on a heavy shock-isolated granite table. When making measurements of the beam paths, it was necessary to measure the mid-position of the test specimen for the best holographic results.

The holograms were positioned as shown in Fig. 2b to obtain photographs of the images of the vibrational modes as a visual record. Photographs of the various modal patterns are shown in Figs. 3 through 6. The specimens were excited by an electromagnet connected to an amplifier and oscillator. The existence of the natural frequencies of the shell were determined by monitoring the microphone output to an oscilloscope.

Difficulty was encountered in placing the hologram after development exactly in the original position. This made the stored beam technique not too feasible. A solution to this problem is to have the photographic plate developed in position.

INITIAL IMPERFECTION MEASUREMENTS

The initial axisymmetric imperfection was measured by a dial indicator and calibrated scale with a 0.001 vernier reading. The approximation to the actual imperfection was represented by Eq. (5) by letting XL2 equal to 2.10". A comparison between the assumed and measured imperfections is shown in Fig. 7. The imperfection measured actually shows an overshoot of approximately 0.003" at positions B and D of Fig. 7. This, however, is just a little over 3% of the total amplitude of the imperfection and therefore was ignored.

RESULTS AND DISCUSSION

MODAL CHARACTERISTICS IDENTIFICATION

The conical test specimens were marked off in heavy ten degree intervals. These markings on the test shells served to identify the number of circumferential waves. The actual number of circumferential waves j is defined by

$$j = \left(\frac{N-1}{2}\right)\left(\frac{360}{\theta}\right) \tag{6}$$

where N is the total number of black (or white) patterns visible over degrees measured circumferentially. The θ can be summed from the large 10° markings shown in Fig. 3. Equation (6) was used throughout to establish the circumferential wave number for each resonance frequency.

CONICAL TEST SPECIMENS WITH NO INITIAL IMPERFECTIONS

A comparison of the resonance frequency between the finite element solution and the experimental results for Cone 1 is shown in Fig. 8. Figure 8 shows excellent agreement between the experimental and numerical results for all circumferential harmonic waves found by the tests. The boundary conditions used in the finite element program for the small end were $w = \partial w/\partial s = v = 0$ and for the large end $w = \partial w/\partial s = u = v = 0$ where u, v, w are the meridional, circumferential and radial displacements, respectively.

Many mixed modal patterns were obtained even for the shells without imperfections. The primary reason for the twisted patterns for the perfect cone specimens is the existence of a coupling between two or more neighboring modes of nearly equal frequencies. A typical coupled modal pattern is shown in Fig. 3. This problem of distorted modes existed for practically all mode levels. By studying another another shell which was considerably thicker (0.080" compared to 0.016") it was confirmed that the existence of the deformed modal patterns may be attributed to the coupling phenomenon. The very high modes were not obtainable with the use of the existing electromagnet. A typical photograph of the first mode for the perfect 0.080" cone is shown in Fig. 4.

CONICAL SHELL WITH AN INITIAL AXISYMMETRIC IMPERFECTION

A comparison between the finite element solution and the experimentally determined natural frequencies for the conical shell with imperfection (Cone 3) is shown in Fig. 9. This figure shows that the finite element method can indeed account for the initial imperfection in determining the natural frequencies.

Table II shows the comparison between the perfect 0.080" thick conical shell and the 0.080" thick conical shell with the prescribed initial axisymmetric imperfection. From the table one can see that the bump increases the natural frequencies in general. However, the increase is very little. Figure 5 shows that the modal deflection pattern for the imperfect conical shell has a double peak. In Fig. 5 the electromagnet was located in the front and to the right side of the specimen where the pattern is not fully developed. The electromagnet was located in the rear for all other photographs presented.

Figure 6 presents the imperfect cone vibrating at the second mode. From the photograph, it can be seen that the largest deflection is occurring at the small portion of the cone above the imperfection. As the natural frequency being investigated was increased, the influence of the axisymmetric imperfection was less. This observation agrees with the numerical results obtained from the finite element solution for the frequency range investigated.

ANALYSIS OF THE FRINGE PATTERNS

The fringe patterns shown in all of the holograms can be analyzed by using the procedure developed by Powell and Stetson [12]. The occurrence of each fringe pattern is related to the roots of the zero-order Bessel function J_o. This technique was successfully applied to beams and plates by Aprahamian and Evensen [15] and [16]. Each interference is formed when

$$J_o\left(\Omega_i\right) = 0 \tag{7}$$

where

$$\Omega = \frac{2\pi}{\lambda}\left[\cos\theta_1 + \cos\theta_2\right] A_i$$

The λ, θ_1, θ_2 and A_i are defined as follows:

λ is the wave length of the coherent light used to make the hologram.
θ_1 is the angle in degrees between a vector normal to the shell surface and the reflected light to the hologram.
θ_2 is the angle between a vector normal to the shell surface and the incident light.
A_i is the amplitude of vibration for the ith fringe.

Measurements of the position of the beam splitter and hologram relative to the test object and the dimensions of the test cones were made. Values of θ_1 and θ_2 were then determined for each fringe location along a given generator. Detailed dimensions needed for determination of the angles θ_i are given in Table III with the shell geometry in Table I. A helium-neon laser with a wave length of 6328 Angstroms was used in the experiment for the coherent light source.

The fringe patterns of Figs. 4, 5, and 6 were analyzed by using Eq. (7). Since linear theory is used in the numerical results, only the ratio of w/w_{max} is obtained. Thus by using the experimental result for the maximum radial deflection, w_{max}, one can compare the computed answers with the experimental results for the mode shape.

Figure 10a for the 0.080" thick 30° cone shows excellent agreement between deflections obtained from the fringe pattern analysis and the numerical finite element results. Figure 10b shows a slight deviation of the deflection shape, but indicates a relatively good trend between the computer results and the experimental values for the 0.080" thick 30° truncated cone with an initial axisymmetric imperfection (δ_i = 0.082" and XL2 = 2.10") positioned at the middle of the slant length.

Figure 10c shows the excellent agreement between test and computer results for the normal shell surface deflection for the second mode sixth harmonic deflection of the 0.082" thick truncated cone with the axisymmetric initial imperfection.

CONCLUDING REMARKS

An initial axisymmetric imperfection on a conical frustum increases the natural frequencies over that of a similar perfect specimen.

The finite element approach with the higher order Hermitian polynomial displacement field gives very accurate results in determining the natural frequencies of conical shells with and without an axisymmetric imperfection.

The holographic interferometry technique gives very descriptive modal patterns for the natural frequencies of shells. The effect of two near frequencies causing "coupling" can be visualized. The fringe patterns show excellent indications of the magnitude and shape of the lateral deflections on a shell. The effects of a weld and an initial imperfection are all graphically portrayed by the photographs of the virtual images. The holographic interferometry method appears to be faster and more accurate than the conventional microphone one.

ACKNOWLEDGEMENT

The results presented here were obtained at the University of Southern California in the course of research supported by the National Science Foundation. The authors also wish to express their thanks to Dr. R. Aprahamian of TRW Systems for helpful advice on the experimental procedure.

REFERENCES

[1] Gabor, D., "A New Microscopic Principle," Nature, Vol. 161, 1948, pp. 777-778.

[2] Gabor, D., "Microscopy by Reconstructed Wave Fronts," Proceedings of the Royal Society, London, Series A, Vol. 197, 1949, pp. 454-487.

[3] Leith, E. N. and J. Upatnieks, "Recent Advances in Holography," Process in Optics, Vol. VI, North-Holland Publishing Co., Amsterdam, 1967.

[4] Pennington, K. S., "Advances in Holography," Scientific American, Vol. 218, No. 2, Feb. 1968, pp. 40-48.

[5] Mertz, L., Transformations in Optics, John Wiley & Sons, Inc., New York, 1965.

[6] Stroke, G. W., An Introduction to Coherent Optics and Holography, Academic Press, New York, 1966.

[7] Wolf, E., Process in Optics, Vol. VI, North-Holland Publishing Co., Amsterdam, 1967.

[8] DeVelis, J. B. and G. O. Reynolds, Theory and Applications of Holography, Addison-Wesley Publishing Co., Reading, Mass, 1967.

[9] Goodman, J. W., Introduction to Fourier Optics, McGraw-Hill Publishing Co., 1968, pp. 209-268.

[10] Leith, E. N. and J. Upatnieks, "Wavefront Construction with Diffuse Illumination and Three-Dimensional Objects," Journal of the Optical Society of America, Vol. 54, No. 11, Nov. 1964, pp. 1295-1301.

[11] Horman, M. H., "An Application of Wavefront Reconstruction to Interferometry," Applied Optics, Vol. 4, 1965, pp. 333-336.

[12] Powell, R. L. and K. A. Stetson, "Interferometric Vibration Analysis by Wavefront Reconstruction," Journal of the Optical Society of America, Vol. 55, No. 12, Dec. 1965, pp. 1593-1598.

[13] Heflinger, L. O., Wuerker, R. F. and R. E. Brooks, "Holographic Interferometry," Journal of Applied Physics, Vol. 37, No. 2, Feb. 1966, pp. 642-649.

[14] Haines, K. A. and B. P. Hildebrand, "Surface Deformation Measurement Using the Wavefront Reconstruction Technique," Applied Optics, Vol. 5, No. 4, April 1966, pp. 595-602.

[15] Aprahamian, R. and D. A. Evensen, "Applications of Holography to Dynamics: High-Frequency Vibrations of Beams," Journal of Applied Mechanics, Vol. 37, No. 2, June 1970, pp. 287-291.

[16] Aprahamian, R. and D. A. Evensen, "Applications of Holography to Dynamics: High-Frequency Vibrations of Plates," Journal of Applied Mechanics, Vol. 37, No. 4, Dec. 1970, pp. 1083-1090.

[17] Sanders, J. L. Jr., "Nonlinear Theory for Thin Shells," Quarterly of Applied Mathematics, Vol. 21, No. 1, 1963, pp. 21-36.

[18] Furuike, T., "The Effect of An Initial Axisymmetric Imperfection on the Natural Frequencies of Conical Shells," Ph.D. Dissertation, University of Southern California, 1970.

Table I. Aluminum alloy test specimens

	cone 1	cone 2	cone 3
Material	2024	1100	1100
Thickness, t, in.	0.016	0.080	0.080
Diameter (small end), in.	7.00	7.15	7.05
Slant length (not in fixture), in.	9.10	8.90	8.80
Half angle, ϕ, degrees	30.0	30.0	30.0
Amplitude of imperfection, δ_i, in.	0.0	0.0	0.082
Position of imperfection, XL, in.			4.40
Wave length of imperfection, XL2, in.			2.10
Edge clamp, EC, in.			
(a) small end	0.12	0.12	0.12
(b) large end	0.18	0.18	0.18
End plates (aluminum)	Diameter	Thickness	
(a) top, in.	9.0	0.75	
(b) bottom, in.	28.0	0.75	

Table II. Experimental results on the natural frequencies of 0.080" thick 30° truncated cones with and without the initial axisymmetric imperfection

Wave No. j	Natural frequencies, f_j, cycles per sec. First mode Cone 2	First mode Cone 3	Second mode Cone 2	Second mode Cone 3	Third mode Cone 2
4		1396			3888
5	1218	1292			
6	1262	1302	2176	2244	
7	1424	1466			
8	1656	1674			
9	1936	1938	2798	2858	3766
10	2254	2280	3156	3154	
11	2606	2674		3572	
12	2992		4018		
13	3404				

Table III. Detailed dimensions needed for analysis of the fringe patterns

Distance, in. (see Fig. 2)	Test specimens Cone 1	Cone 2	Cone 3
OB to H	32	31 1/2	31
OB to BS	33	33 1/4	32 1/2
H to M_1	7	6 1/2	6 1/2
M_1 to BS	5	4 1/4	4 1/2

Notation: BS = Beamsplitter; M_1 = Mirror 1; H = Hologram; OB = Object (test cone)

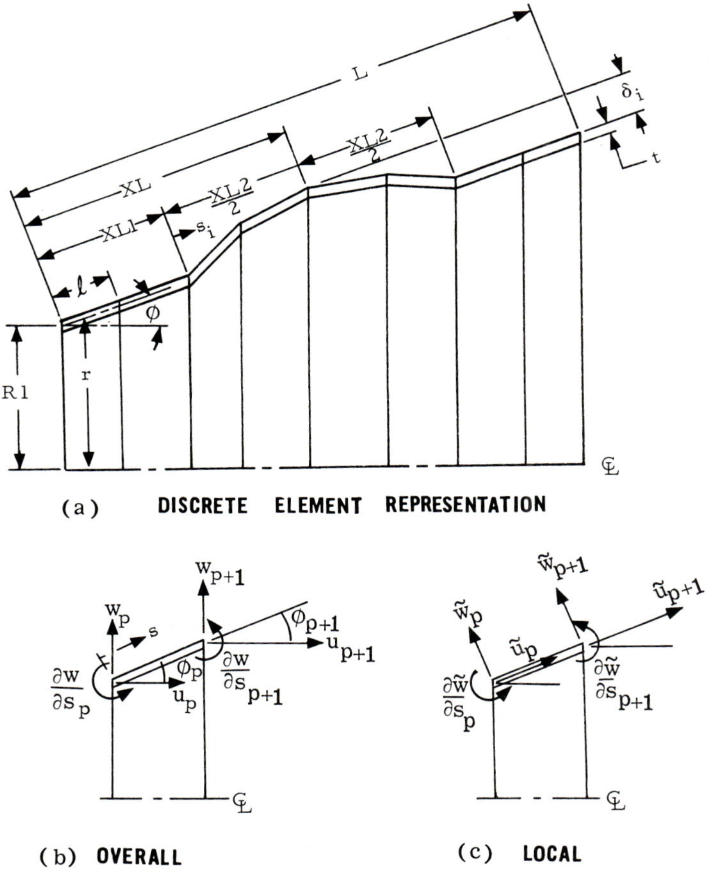

Fig. 1 Finite element representation

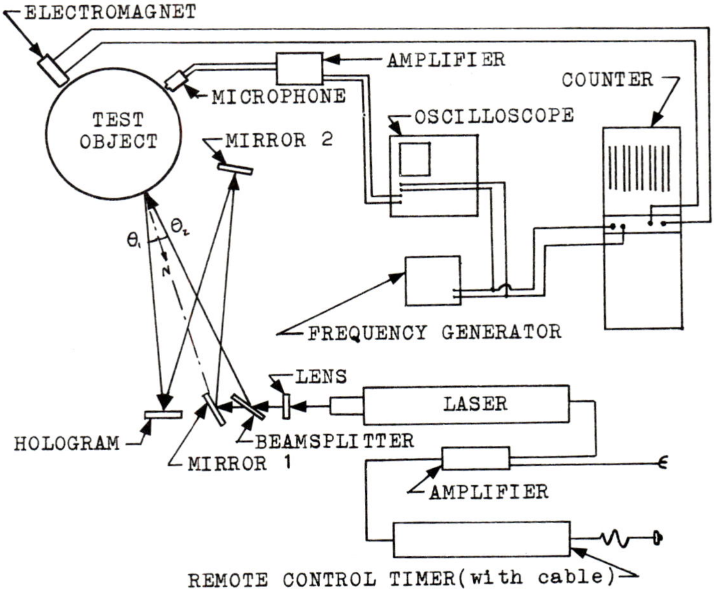

(a) **TEST SETUP TO OBTAIN THE HOLOGRAMS**

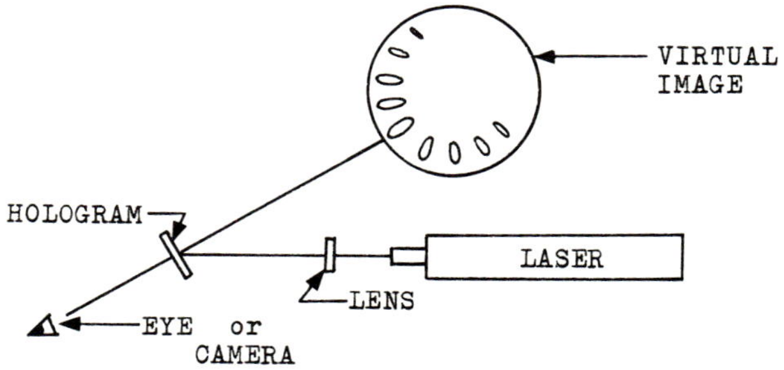

(b) **RECONSTRUCT ON PROCESS TO FORM THE VIRTUAL IMAGE**

Fig. 2 Schematic of test setup

Fig. 3 Photograph of "coupling effect" obtained from the reconstructed image for cone 1

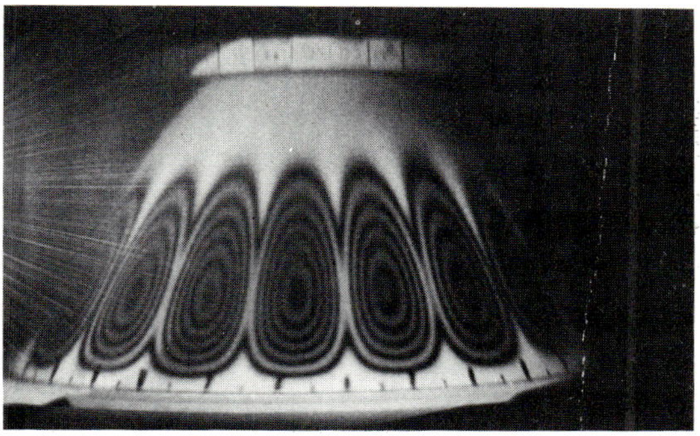

Fig. 4 Photograph of vibration mode (m = 1, j = 11, f = 2606 cps) obtained from the reconstructed image for cone 2

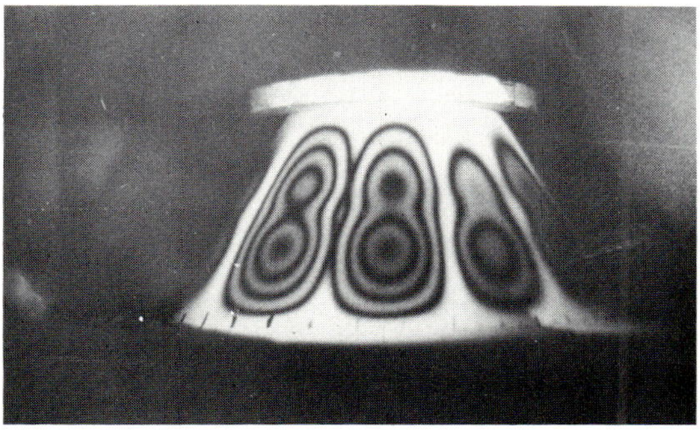

Fig. 5 Photograph of vibration mode (m = 1, j = 6, f = 1302 cps) obtained from the reconstructed image for cone 3

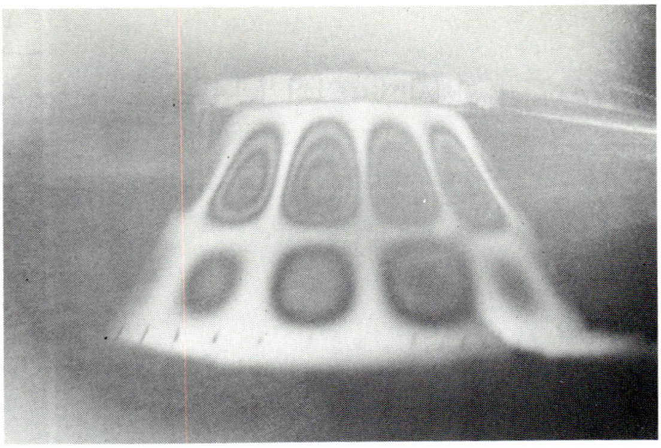

Fig. 6 Photograph of vibration mode (m = 2, j = 7, f = 2032 cps) obtained from the reconstructed image for cone 3

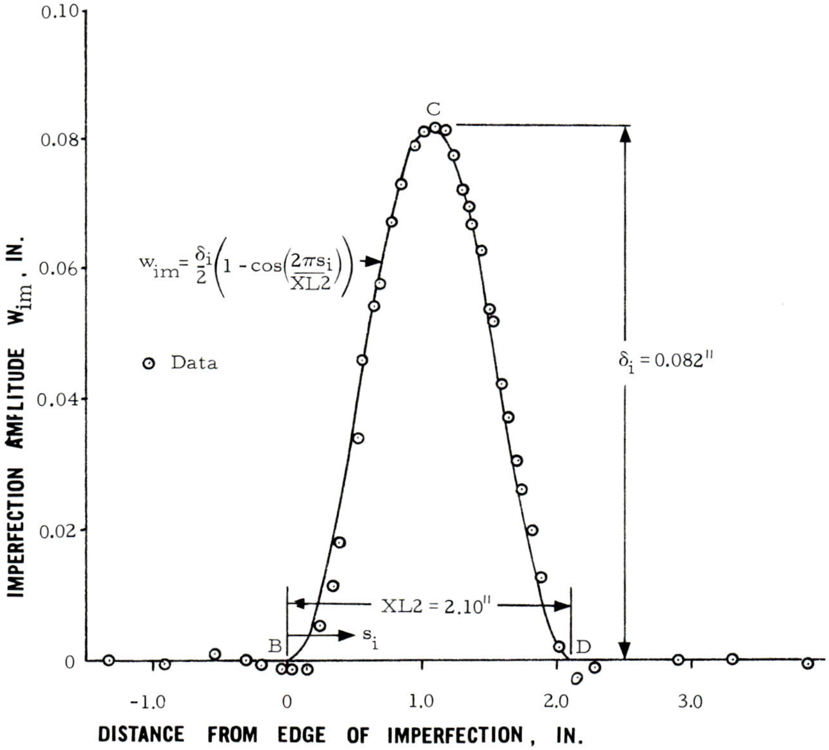

Fig. 7 Comparison between experimentally measured imperfection and assumed imperfection

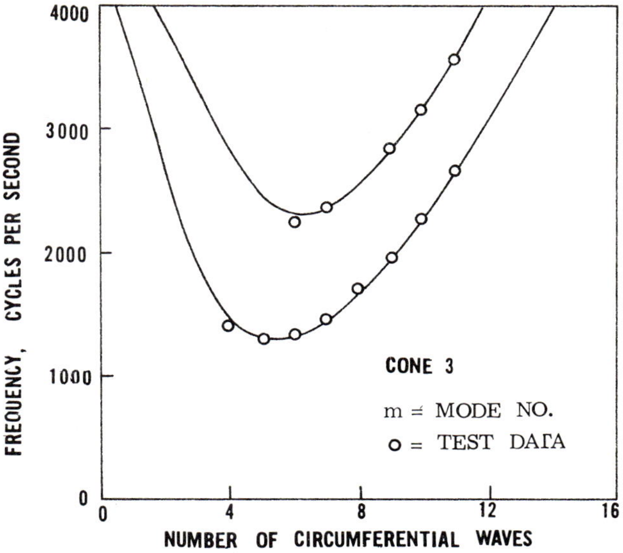

Fig. 8 Comparison between finite element 12 coefficient solution and experimental natural frequencies for cone 1

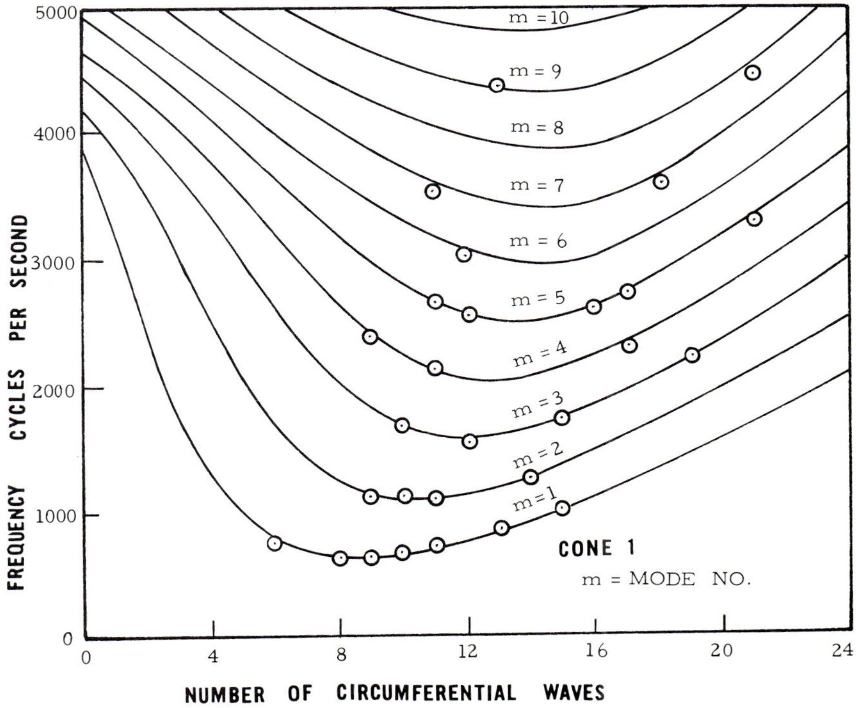

Fig. 9 Comparison between finite element 12 coefficient solution and experimental natural frequencies for cone 3

m = mode no.
j = circumferential wave no.
f = frequency

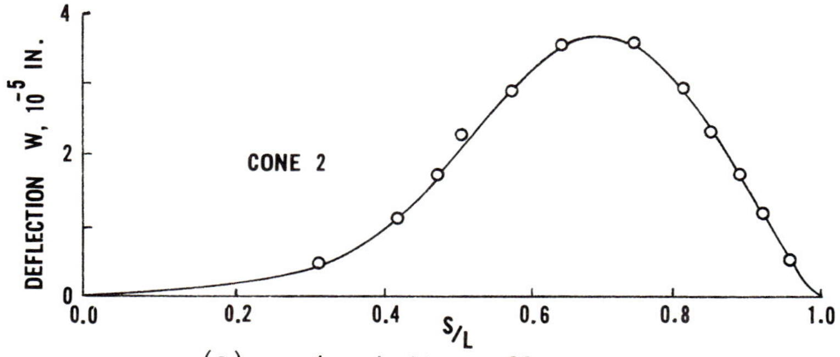

(a) m = 1, j = 11, f = 2606 cps

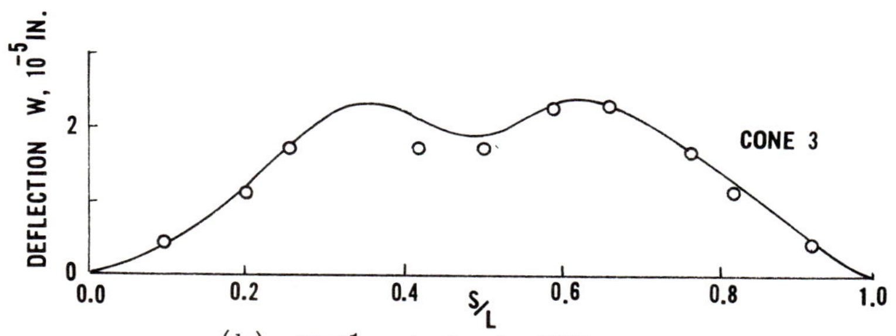

(b) m = 1, j = 6, f = 1302 cps

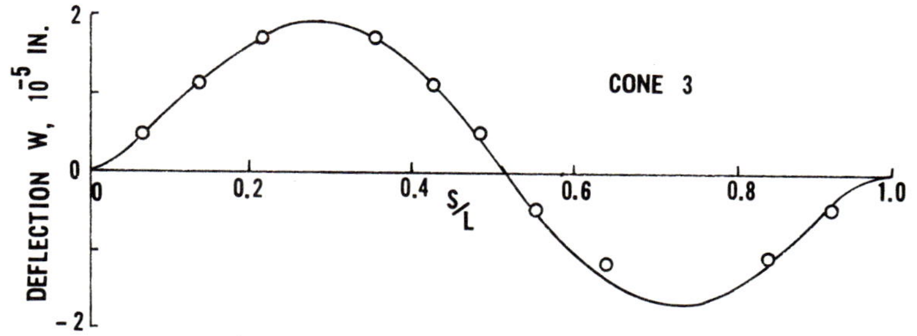

(c) m = 2, j = 6, f = 2244 cps

Fig. 10 Comparison between finite element 12 coefficient solution and experimental mode shapes

Developments in Mechanics, Vol. 6. Proceedings of the 12th Midwestern Mechanics Conference: 67

The Strain-rate Sensitivity of Lead Under Biaxial Stress

JOHN G. WAGNER
UNIVERSITY OF PITTSBURGH

ABSTRACT

Experiments were performed to determine the influence of strain-rate on the biaxial stress-strain behavior of lead. A tension-torsion apparatus is described which is capable of straining short tubular specimens at strain-rates up to 300/second. It is found that the second invariant of the stress deviator, for a given strain-rate, is a good approximation to the effective stress. Moreover the biaxial yield curve for a particular strain-rate lies between the Tresca and von Mises curves.

INTRODUCTION

The study of strain-rate sensitivity in the mechanical behavior of materials is one of increasing importance. To date most of the research conducted in this field has dealt with uniaxial stress-strain behavior. However, even a complete description of uniaxial behavior would only provide information for a rather restricted class of applications. For a more complete formulation of the rate sensitive elastic-plastic behavior of a metal, it is necessary to consider more than one stress and strain component. The current study involves an investigation of the rate sensitivity of the stress-strain behavior of lead under biaxial conditions. Lead was chosen because it requires little energy to deform and because it is highly sensitive to the deformation rate. Strain-rates up to 300/second were achieved through explosive loading.

The paper starts with a brief review of the pertinent research, followed by a description of the present experiments. Results are presented and analyzed in relation to the theory.

Previous Research on Multiaxial Dynamic Plasticity

Perhaps the first attempt to develop a constitutive equation for viscoplastic materials was made by Bingham in 1922. He suggested that, for certain materials in simple shear, the deformation rate is proportional to the difference between the shear stress and the yield stress in shear. This viscoplastic body was generalized by Hohenemser and Prager in 1932 [2] to include multiaxial stress states. In 1963, Perzyna [3], following the example of Malvern's overstress hypothesis [4], introduced a rather general constitutive equation for the elastic-viscoplastic body. In this theory the elastic behavior is presumed time independent and the work hardening viscoplastic response is taken as a function of the excess stress above the static yield stress. An alternative approach to the above is a phenomenological theory of dynamic plasticity based on dislocation studies. Gilman, who is responsible for much of the development of this theory, has given an excellent survy of progress in the field [5]. Applications of the theory to various materials have resulted in quantitative prediction of transient creep curves, stress-strain curves and impact test behavior. References to these investigations are provided in Gilman's review. Recently, Bodner [6] extended the theory to multiaxial stress states by relating it with classical plasticity theory. Others have contributed to the field, particularly in the study of uniaxial behavior. In this respect, a critical survey has been made by Symonds, Ting and Robinson [7]. A brief review and constructive criticism of some of the pertinent literature was given by Lindholm in 1965 [8]. From these reviews, it is evident that few experiments have been conducted involving dynamic multiaxial stress states. However, some of the more important ones are cited below.

In 1949, Kolsky [9] presented an important technique for conducting tests at high rates of strain. The technique has been employed since then in a variety of forms by many researchers. Kolsky himself has presented a review of results in this field [10]. Although most of the early tests involved uniaxial behavior, recently some experiments have been conducted involving more than one component of stress. Gerard and Papirno [11], using shock tube loading of thin circular diaphragms, tested 1100-0 Aluminum and annealed SAE 1010 low carbon steel. The strain rates were on the order of 1/second but only one ratio of biaxial tension ($\sigma_1/\sigma_2 = 1$) could be tested. For the aluminum alloy they found no significant difference between the dynamic and the static response. The results for low carbon steel show increases in the yield strength comparable with those of an equivalent dynamic uniaxial test. Hoge [12] conducted dynamic tests on small tubular specimens under a combination of tension and internal pressure. The ratio of the tensile stresses was roughly 2 with maximum strain-rates on the order of 35/second. The material tested was 6061-T6 Aluminum. His data are consistent with a modified form of Hill's distortion energy criterion [13] for fracture. Lindholm and Yeakley [14] have described a pneumatic tension-torsion machine which can achieve loading rates as high as 1.5×10^7 lb/sec in tension and 4.5×10^5 in-lb/sec in torsion. Preliminary tension-torsion tests were conducted on 1018 mild steel. In 1968, Lindholm [15] showed a good agreement between experimental results and a combination of Perzyna's viscoplastic contitutive equation and the thermal activation dislocation theory. He did this using data for 1100-0 Aluminum obtained over a wide range of strain-rates and temperatures for both uniaxial and biaxial proportional loading. In addition, several tests conducted to demonstrate the effect of sudden changes in the direction of applied load of loading rate show little evidence of the influence of previous loading history on the subsequent mechanical behavior. Under the test conditions described, the expansion of the yield surface by both strain hardening and increasing loading rate appears to be isotropic. More recently, Green, Leasia, Perkins and Maiden [16] have described the design and construction of a dynamic biaxial (internal pressure and tension-compression) testing machine which is capable of shear strain-rates on the order of 10^2/second. Some initial biaxial data have been reported on 6061-T651 Aluminum alloy [17].

Rate-Sensitivity and the Tension-Torsion Test

In investigating the time independent biaxial stress-strain behavior of a metal, a convenient experiment is the tension-torsion test on a thin-walled tubular specimen. With the ratio of applied load to torque constant the principal stress directions remain fixed. Quasi-static stress-strain data have been obtained from experiments of this sort. By appropriate definitions of effective stress (σ_e) and strain (ϵ_e), the data along different radial paths have been correlated so as to yield essentially a single curve. A well known example of such a correlation is that of Osgood's tests on 24-ST Aluminum alloy [18]. Plots of his data on the basis of maximum shear stress versus maximum shear strain show good agreement.

The objective of the present report is to investigate the influence of strain-rate on the biaxial stress-strain behavior of lead. A dynamic tension-torsion test apparatus is devised for this purpose.

STATIC TESTS

The static tests were conducted in a tension-torsion machine in which the tensile load and torque were continuously variable. Large deformations of the specimen were possible. The machine provided a tensile load of up to 300 pounds and a torque of up to 40 inch-pounds. Bending was monitored and never exceeded ± 1% of the resultant strain. Loads were applied through long non-twisting cables in series with load rings; the resulting interaction between tensile load and torque was always less an 0.5%. A continuously variable load was applied at one point and mechanically distributed so as to provide a constant ratio of applied load to torque. Relative elongation and rotation of the specimen was measured by dial gages and recorded photographically. Such an arrangement provided for sensitivities of 100 μ in/in in tension and 200 μ in/in in torsion. With the specimen geometry used, maximum possible strains were 15% in either or both tension or torsion.

Specimens

The static specimens consisted of thin-walled tubular sections with large flanges at each end (Figure 1a). The test sections were right circular cylinders with a nominal gage length of 1.000 inch, an inside diameter of 0.500 inch and a wall thickness of 0.50 inch, thus furnishing a wall thickness to radius ratio of 0.2. Inside and outside diameters were finish machined with very light cuts so as to avoid surface effects. The flanges, ¼ inch thick and 1 inch in outside diameter, afforded a means of gripping the specimen by cementing it in place coaxially with its mating load bars. Specimens were cemented with Eastman 910 in order to facilitate removal while preserving the load bars.

The lead, from which the specimens were machined, was at least 99.99% pure. Heavily cold-working the as-cast material and allowing it to recrystallize produced the necessary fine grain structure. This procedure provided an essentially homogeneous isotropic material with all grain diameters about equal and averaging 0.005 inch (the wall thickness of the specimens was 0.050 inch).

DYNAMIC TESTS

The dynamic tests were performed in an apparatus, described below, and illustrated schematically in Figure 2, which is capable of straining the specimen in either pure tension, pure torsion or a constant ratio of the two. The specimen is tubular and is mounted in the apparatus at its lower end while its top end is attached to a large head-mass. Straining of the specimen is effected by acceleration of the head-mass through the detonation of sheet explosive placed on it. Direct measurements of the load and torque carried by the specimen and of the axial and rotational velocities of the head-mass, are made by means of long slender elastic bars. These bars, properly instrumented with strain gages, are attached to the head-mass and to the lower end of the specimen coaxial with it. The inertia of the head mass is sufficient, when compared with the resistance afforded its motion by the specimen and pressure bar, that it proceeds with a reasonably constant velocity once accelerated by the explosive. The gage length of the specimen is small enough in comparison to the strain pulse so as to insure uniformity of the stress and strain along the length of the test section. Maximum obtainable strain-rates are $\dot{\epsilon}$ = 300/second for tension and $\dot{\gamma}$ = 150/second for torsion, with the ratio of the two strain-rates prescribed by the ratio of the imposed head velocities. Maximum possible strains are essentially limited by the lengths of the pressure bars and range from 10% to 15% depending on the strain-rates. However, mechanical stops are incorporated to provide a check on the maximum value of the strain measured.

Specimens

The dynamic test specimens are thin-walled tubes with keyed flanges at each end to permit gripping (Figure 1b). Test sections are 1.000 inch long, with an inside diameter of 0.500 inch and a 0.050 inch wall thickness (identical to the static specimens). Specimen material is the same as that from which the static specimens are machined. The larger of the two flanged ends is gripped in the moving head mass. This flange is 1 inch long and has a 1 inch outside diameter. Its large size is chosen so as to increase the effective inertia (both axial and rotational) of the head-mass. The smaller flange is 5/8 inch long and has a 7/8 inch outside diameter.

Torque is transmitted from the moving head to the larger flange of the specimen through four 1/8 inch diameter hardened steel keys. The keys are 90° apart. Tensile load is carried in part by a seat in the head-mass and in part by wall friction between the outside diameter of the flange and the head. Expanding inserts are placed within the inside diameter of the flange. They provide the normal force required to generate the necessary wall friction. Similar arrangements to carry torque and tensile load are made at the lower end of the specimen.

Apparatus and Techniques for Dynamic Tests

The specimen is gripped at its lower end by a coupling connected to the lower pressure bar. The upper pressure bar fits down through the inside diameter of the specimen and of the lower pressure bar. This arrangement avoids placing the upper pressure bar in compression and frees the area above the head-mass for the location and initiation of the explosive.

The head-mass is shaped so as to serve several important functions. Its primary purpose is to provide large translational and rotational inertia. The weight of the head is 1.25 pounds, while its rotational inertia is 1.1 pound-inches squared. Both of these values are for the entire head-specimen flange assembly. The head is made in two halves with four lead-filled cavities. The material is cold-rolled steel and the two halves are bolted together after insertion of the specimen flange. The result of such a construction is small overall size, to reduce wave propagation effects within the head, and sufficient toughness to resist large distortions resulting from detonation of the explosive placed on its surface. Each head lasts for approximately three tests before requiring replacement due to excessive geometric changes. The neck on the head is necessary to protect the specimen from crushing under the pressure blast associated with the explosion. The flare on the neck is required to deflect the pressure blast away from the lower pressure bar coupling. An O-ring type seal is installed between the two halves in the wall thickness and flare of the neck to prevent leakage.

The explosive used is Dupont Detasheet C with a thickness of 0.083 inch. It is cut into strips and cemented in shallow grooves machined on the outer edges of the head. The placement of the explosive is symmetric about the specimen axis (Figure 2). The two vertical strips of explosive on either side of the head result in a torque while the two horizontal strips give the tension. Varying the ratio of the explosive charges varies the ratio of the imposed velocities. An explosive leader from the top of one torsion charge to the other is initiated at its mid-point by means of a miniature Dupont detonating cap (Minidet). The torsion charge on each side serves to initiate the tension charge. Each of the two charges (for both tension and torsion) is thus detonated simultaneously and no bending moments are imposed on the specimen. Steel baffle plates are placed between the leader and the head-mass to protect the head-mass from the exploding leader. By means of an independent test, the explosive was found to have a constant detonation speed (2.85×10^5 inches/second) along any finite strip. Consequently, even if the detonating cap is located 1/16 inch from the mid-point of the leader, the primary charges are assured of detonating within 0.44 microseconds of each other. Thus, high detonation speed (higher than the bar wave speed for a structural metal) insures that the charges for torsion and tension detonate simultaneously.

The quantity of explosive one can efficiently utilize has a well defined upper bound. Beyond this limit there is no notable increase in head velocity. Below the limit, however, the head velocity is roughly proportional to the quantity of explosive used. Maximum obtainable average head velocities are 300 inches per second axially and 600 radians per second rotationally.

The electronics and recording instrumentation are quite standard for this type of test. Separate dry cells power each of the strain gage bridges mounted on the pressure bars. The bridge outputs are recorded on two dual-beam oscilloscopes the band-widths for which are from D.C. to 1×10^6 cycles per second. The strain gage bridges are calibrated electrically before each test by means of a precision shunt resistor system.

Sample oscilloscope traces for a tension-torsion test are shown in Figure 3. The outputs, of the strain gage bridges on the upper pressure bar are the top traces in each photograph. The bottom trace in each photograph is the lower pressure bar bridge output corresponding to the same mode as the upper trace, either tension or torsion. The horizontal sweeps for the two oscilloscopes are triggered within 2½ microseconds of each other at the instant the trigger wire is broken by the exploding leader. Sweep times for all four traces are identical. As a result, all time scales are treated as identical and relative to the same zero. The head rotational velocity $\dot{\theta}_1$ is seen to be quite constant during the duration of the test. This is to be anticipated due to the enormous rotational inertia of the head relative to the influence of the specimen and upper pressure bar. The axial head velocity $\dot{x}_1$ is not quite constant, but is still considered reasonably good for an experiment of this nature. The tension and torsion strain gage bridges, on both the upper and lower pressure bars, are located in such a fashion that tensile and torsional waves, are initiated simultaneously on the oscilloscope traces. That is, the ratio of the distances (from the input end) to the tension and torsion bridges equals the ratio of the tensile to torsional wave speeds in the bar material. Consequently, one further observation is the near simultaneity of the $\dot{x}_1$, and $\dot{\theta}_1$, wave fronts ($\dot{x}_1$, commencing 40 microseconds before $\dot{\theta}_1$). This, as explained previously, is attributed to the high detonation speed and arrangement of the explosive used. The 40 microsecond difference is due to the configuration and inhomogeneity of the head-mass. The near simultaneity was shown to be sufficient so as to consider the specimen as having no prestrain in tension prior to the initiation of $\dot{\theta}_1$. In these photographs, the impact with the mechanical stops is not shown. The sudden decrease in is due to the tensile wave interacting with itself after reflection off of the free end of the upper pressure bar. In those tests in which impact does occur, the head displacement and rotation, calculated from integrating the $\dot{x}_1$ and $\dot{\theta}_1$ curves, is always within 3% of the distance to the stops. In Figure 3, the transmitted load trace shows a small compressive pulse leading the tensile wave front. By conducting a series of tests in which no specimen is used, this compressive pulse was shown to be attributable to the shock wave propagating from the explosive and impinging on the lower pressure bar coupling for about 100 microseconds. In other words, the protective baffle arrangement is not completely effective. Since the specimen sees this pulse as a slight elastic tensile pulse it is accounted for in the data reduction.

The relationship in time between the load and torque carried by the specimen, measured at the lower end of the specimen, and the axial and rotational deformation measured at the opposite end is treated in the same manner as that for the split Hopkinson pressure bar experiment. The transit time of a wave in the specimen is small in comparison to the straining pulse length. Due to the resulting numerous reflections within the specimen during the test the stress and strain distribution along the gage length very quickly becomes uniform. In the test shown, for example, the useful pulse length is on the order of 500 microseconds. Typical plastic wave speeds for lead are 3×10^4 inches per second for a tensile wave and 2.1×10^4 inches per second for a torsional wave. There are, therefore, 10 to 20 reflections of both the tensile and torsional plastic waves before the strain-rate decreases. Furthermore, in several of the very early experiments, grids were lightly scribed on the surface of the specimen. Following the test, measurements were made of the resulting permanent strain distribution in both tension and torsion. For both modes of deformation, the deviation from the average permanent strain along the gage length was less than 6%.

The data reduction for the dynamic tests is outlined as follows. The load P_s and the torque T_s carried by the specimen are calculated from the load P_2 and the torque T_2 monitored by the lower pressure bar. By applying equations of motion to the coupling between the specimen and the lower pressure bar one can show that:

$$P_s = P_2 + M_2\dot{P}_2/(\rho C_o A) \quad (1)$$

and

$$T_s = T_2 + I_2\dot{T}_2/(\rho C_1 I_p) \quad (2)$$

where M_2 and I_2 are the mass and moment of inertia respectively of the coupling, ρ is the density, A the cross sectional area and I_2 the polar moment of inertia of the lower pressure bar. C_o is the speed of a tensile wave in a slender rod $[E/\rho]^{1/2}$ and C_1 the torsional wave speed $[G/\rho]^{1/2}$. In equation (1), P_2 is corrected for the small compressive pulse discussed above.

Since the apparatus is not perfectly rigid, motion of the lower end of the specimen must be accounted for in calculating the average strain-rate of the specimen. The upper pressure bar monitors the axial velocity $\dot{x}_1$ and the rotational velocity $\dot{\theta}_1$ of the head-mass. The axial velocity $\dot{x}_2$ and rotational velocity $\dot{\theta}_2$ of the fixed end of the specimen can be determined respectively from the measurements of P_2 and T_2. Consequently the engineering strain-rate of the specimen is:

$$\dot{\epsilon} = (\dot{x}_1 - \dot{x}_2)/L_o \quad (3)$$

and

$$\dot{\gamma} = r_s(\dot{\theta}_1 - \dot{\theta}_2)/L_o \quad (4)$$

where r_s is the mean specimen radius $(r_o + r_i)/2$ and L_o is the gage length. Figure 4 shows the natural strains versus time for the sample oscilloscope traces of Figure 3. The dashed curve for γ_N represents the actual test condition and is due to the 40 microsecond delay of $\dot{\theta}_1$ relative to $\dot{x}_1$. The data reduction assumes that $\dot{\theta}_1$ and $\dot{x}_1$ are initiated simultaneously (solid curves). Several tests were conducted in which the 40 microsecond delay was eliminated so as to establish its effect on results. The stress-strain behavior was not detectably different.

The true stresses and natural strains are calculated as follows:

$$\sigma_T = P_s\,(1 + \epsilon)/[\pi\,(r_o^2 - r_i^2)] \quad (5)$$

$$\tau = r_s T_s/I_p \quad (6)$$

$$\epsilon_N = \ln(1 + \epsilon) \quad (7)$$

$$\gamma_N = \gamma/(1 + \epsilon) \quad (8)$$

where ϵ is the engineering strain $(x_1 - x_2)/L_o$ and P_s, T_s, γ and are obtained from equations (1), (2), (3) and (4). All of the dynamic tests were analyzed in the above manner.

TENSION-TORSION TEST RESULTS

Static Tests

All the tests (both static and dynamic) were conducted at room temperature (76° F). The static tests consisted in probing out along five radial paths in $\sigma-\tau$ space at a constant stress rate.

The stress measure used was the equivalent stress defined as $\sigma_e = \sqrt{3}\, J_2^{1/2}$, J_2 being the second invariant of the stress deviator. In the experiments the stress rate chosen was such that $\dot{\sigma}_e = 8.5$ psi/sec. The data from these tests, when plotted in terms of the equivalent stress σ_e and strain measures $\epsilon_e = 2 I_2^{1/2}/\sqrt{3}$ (I_2 being the second invariant of the strain deviator with $\nu = \frac{1}{2}$), gives the results shown in Figure 5. The true stresses and natural strains are calculated from equations (5), (6), (7) and (8) using the corresponding loads and deflections monitored during the static tests. Each curve represents the average of several tests along the corresponding radial path. The deviation from the average tension or torsion behavior, along any one path, was always less than 1.5% on the stress axis. Average values for the measured Young's modulus and shear modulus were 2.3×10^6 psi and 0.8×10^6 psi, respectively. The static stress-strain behavior in compression, obtained from some independent tests,was found to be virtually identical to that in tension.

The correlation obtained in Figure 5 is reasonably good and comparable to that obtained by Osgood [18] for his 24-ST Aluminum alloy data on an octahedral shear stress plot. Maximum deviation from the mean is roughly one and one-half times as much when the same data are plotted in a maximum shear stress-maximum shear strain space.

More general deformation theories of plasticity may provide for an even better correlation of the data. Prager [19] has shown that the most general correlating variable for an isotropic, plastically incompressible material with identical stress-strain diagrams in pure tension and pure compression is of the form $f(J_2, J_3)$. Drucker [20], using a simple form of the general relationship, established a much better correlation of Osgood's data. However, in the case of the current data, lack of ordering of the curves was almost certainly due to experimental scatter. Consequently, the best one can do to correlate the data is to draw an average curve on the equivalent stress-strain plot.

Since lead is rate sensitive even in the "static" range of testing, several "static" tests were conducted to determine the influence of the rate sensitivity on the above correlation. In these tests the stress rate was stepped from the lowest to the highest limit of the testing machine (3.3 psi/sec to 14.2 psi/sec for tension and 1.9 psi/sec to 7.6 psi/sec for torsion). Only uniaxial tests were conducted. The results show approximately a 3.5% increase in flow stress when the stress rate is increased by a factor of about four. The static tests were conducted at a stress rate about mid-way between the limits of the stepped rate tests. Since the step in stress rate took place mid-way out the strain scales, the resulting change in flow stress is considered to be representative of the rate sensitivity under these conditions. The sensitivity, while real, is insignificant on the equivalent stress-strain plot.

Dynamic Tests

The dynamic tests were conducted by straining at a constant rate along essentially radial paths in strain space. The strain measure used was defined above as the equivalent strain $\epsilon_e = 2I_2^{1/2}/\sqrt{3}$. Two series of tests were conducted, one at an equivalent strain rate of $\dot{\epsilon}_e$ = 75/second and the other at $\dot{\epsilon}_e$ = 280/second. Deviations from these nominal values were in all cases less than 20%. Five radial strain paths were investigated at $\dot{\epsilon}_e$ = 75/second. Due to the limitation of the maximum shear strain-rate obtainable, only two paths were investigated at $\dot{\epsilon}_e$ = 280/second.

Figure 6 shows the data plotted on an equivalent stress-strain plot using the same definitions of σ_e and ϵ_e as in the static tests. Each curve represents the average of several tests along the corresponding radial path. The deviation from the average tension or torsion behavior, along any one path, was always less than 5% on the stress axis. Once again, scatter eliminates the possibility of obtaining a more exact loading function.

In addition to representing the strain hardening behavior of the material, the stress-strain curves provide a means of plotting "small offset" initial yield surfaces. The yield criterion chosen is the first detectable divergence from the elastic loading line. For the dynamic tests the utility of such yield surfaces in a quantitative sense is questionable due to the possible lack of a completely uniform stress and strain state in the early stages of a test. Figure 7 shows the resulting initial yield surfaces obtained. For comparison, the maximum shear yield criterion of Tresca and the J_2 criterion of von Mises have been normalized at the static yield point for pure tension. The initial yield surfaces of lead appear to expand isotropically with increasing strain-rate.

Discussion of Results

In both cases, $\dot{\epsilon}_e$ = 75/second and $\dot{\epsilon}_e$ = 280/second, the tension-torsion data correlate well on the equivalent stress-strain plot. This observation has an important implication. The second invariant of the stress deviator tensor (J_2) is the predominant factor in the plastic deformation of lead at high, as well as static, strain-rates.

We shall compare the static test data, which were obtained along radial paths in stress space, with the dynamic data, obtained along paths in strain space. This is not unreasonable if one remembers that, for an isotropic and incompressible material for which the principal stress directions and ratios have been maintained constant, the axes of principal stress and strain coincide and the ratio of the principal strains remains fixed.

Figure 8 is a plot of equivalent stress against equivalent strain. The three curves represent the average values of the current results for all tests conducted at that particular strain-rate. The intermediate curve has been extrapolated for the purposes of comparison although no data were taken beyond a strain of 5%. The equivalent stress and strain measures are defined such that they reduce to the conventional uniaxial stress and strain values for a state of pure tension or compression. Consequently the results indicate that, for all practical purposes, the curves of Figure 8 may be thought of as representative of tension tests conducted at the strain-rates indicated. Virtually all of the test data accumulated on lead to date have been in investigating uniaxial compression behavior. In general, the flow stress reported in compression is lower than the corresponding value in tension, particularly at the smaller strains. The present data agree reasonably well with the results of Lindholm [21] and with the more recent results of Green et al [22] (Figure 8). The agreement is particularly good in the case of the static behavior. Green's static curve lies slightly above the other two; however, his strain-rate is roughly three times larger. This similarity between the present results and the static compression data of others was perhaps to be expected in view of the excellent agreement obtained between static tension and compression behavior. At the higher strain-rates the flow stresses are undoubtedly higher than those reported separately by Lindholm and Green. At the present time it is difficult to say whether this difference is due to a difference in behavior between tension and compression, to a different grain size or somewhat different material, or whether it is due to specimen geometry.

In addition to the strong influence of strain-rate on the flow stress, the strain hardening characteristics of lead are also of importance. The relative slopes of the curves in Figure 8 give an indication of the influence of strain-rate on work hardening. Thus for strains above 3% the work hardening rate is very similar for the strain rates shown. The implication here is that stress-strain relations of the type

$$\dot{\epsilon} = g[\sigma - \sigma_s(\epsilon)] \tag{9}$$

for which the strain-rate is a simple function of the absolute overstress are possible models of the behavior of lead for large strains. In equation (9) $\sigma_s(\epsilon)$ is the static stress corresponding to the strain ϵ. However, at strains less than 3% the material work hardens more rapidly at higher strain-rates. Under these conditions, functions of the fractional overstress

$$\dot{\epsilon} = f[\sigma/\sigma_s(\epsilon) - 1] \tag{10}$$

are more realistic models than functions of the absolute overstress.

The current investigation provides a relationship between the second invariants of the stress, strain and strain-rate deviators for lead. Figure 9 is a plot of the equivalent stress ($\sigma_e = \sqrt{3}\, J_2^{1/2}$) versus the equivalent strain rate ($\dot{\epsilon}_e = 2\, \dot{I}_2^{1/2}/\sqrt{3}$) on log-log scales at constant levels of the equivalent strain. The curves of constant equivalent strain tend to be linear. (There may be a "kink" in the vicinity of 100/second on the strain-rate scale. If so, there may be a change in the basis dislocation mechanism governing the plastic deformation above and below that rate. A simple mathematical model of the form

$$\sigma_e = A\, \dot{\epsilon}_e^{\,m}\, \epsilon_e^{\,n} \tag{11}$$

will describe the general behavior shown in Figure 9. If one uses for A, m and n the values 5490, 0.047 and 0.255 respectively, the model predicts the solid lines shown. This relationship and the experimental data presented in this paper should be useful when adopting Bodner's [6] approach to constructing a generalized constitutive equation.

CONCLUSIONS

The primary purpose of the research presented is to investigate the strain-rate sensitivity of lead under biaxial stress. Within the limitations of the present experiments, simple uniaxial tests conducted at constant strain-rates are sufficient to describe the strain-rate dependence for biaxial stress states. This was found by plotting the equivalent stress against the corresponding equivalent strain for constant values of the equivalent strain-rate. The second invariants of the stress and strain deviators are, within multiplicative constants, good approximations to the effective stress and strain respectively for high strain-rates as well as for static conditions.

In addition, the present tests provided a number of yield surfaces for lead. The initial yield surface appears to expand isotropically with increasing strain-rate. The static yield surface lies between those of Tresca and von Mises. An increase in the strain rate on the order of 10^5/second will result in an increase in the flow stress of approximately 50%. However, for lead even the commonly referred to static behavior is sensitive to the testing rate. Furthermore, the strain-rate sensitivity is dependent on the strain. The lack of a time independent yield stress or stress-strain curve and the strain dependence of the strain-rate sensitivity, combine to limit the utility of an over-stress type of constitutive theory. The development of constitutive equations through the interaction of classical plasticity theories and dislocation mechanics may be able to overcome these difficulties.

REFERENCES

1 Bingham, E. C., Fluidity and Plasticity, McGraw-Hill Company, New York, 1922, p. 215.

2 Hohenemser, K. and Prager, W., "Uber Die Ansatze Der Mechanik Isotroper Kontinua", Zeitschrift Fur Angewandte Mathematik und Mechanik, Vol. 12, 1932, p. 216.

3 Perzyna, P., "The Constitutive Equations for Work-Hardening and Rate Sensitive Plastic Materials", Proceedings of Vibration Problems, Vol. 4, 1963, p. 281.

4 Malvern, L. E., "Plastic Wave Propagation in a Bar of Material Exhibiting a Strain Rate Effect", Quarterly of Applied Mathematics, Vol. 8, 1951, p. 405.

5 Gilman, J. J., Progress in the Microdynamical Theory of Plasticity, Proceedings of the Fifth U.S. National Congress of Applied Mechanics, 1966, p. 385.

6 Bodner, S. R., "Constitutive Equations for Dynamic Material Behavior", Mechanical Behavior of Materials Under Dynamic Loads, Edited by U. S. Lindholm, Springer-Verlag, New York, 1968, p. 176.

7 Symonds, P. S., Ting, T. C. and Robinson, D. N., Survey of Progress in Plastic Wave Propagation in Solid Bodies, Brown University Technical Report, Contract DA-19-020-ORD-5453(A) June 30, 1967.

8 Lindhold, U. S., "Dynamic Deformation of Metals", Behavior of

Materials Under Dynamic Loading, Edited by N. J. Huffington, Jr. American Society of Mechanical Engineers, New York, 1965, p. 42.

9 Kolsky, H., An Investigation of the Mechanical Properties of Materials at very High Rates of Loading, Proceedings of the Physical Society, Vol. 62B, 1949, p. 676.

10 Kolsky, H., Experimental Studies in Stress Wave Propagation, Proceedings of the Fifth U.S. National Congress of Applied Mechanics, 1966, p.21.

11 Gerard, G. and Papirno, R., "Dynamic Biaxial Stress-Strain Characteristics of Aluminum and Mild Steel", Transactions of the American Society of Mechanical Engineers, Vol. 49, 1957, p. 132.

12 Hoge, K. G., "The Influence of Strain Rate on Mechanical Properties of 6061-T6 Aluminum under Uniaxial and Biaxial Stress States", Experimental Mechanics, Vol. 6, 1966, p. 204.

13 Hill, R., "A Theory on the Yielding and Plastic Flow of Anisotropic Metals", Proceedings of the Royal Society, London, Vol. 193, 1948, p. 281.

14 Lindholm, U. S. and Yeakley, L. M. "A Dynamic Biaxial Testing Machine", Experimental Mechanics, Vol. 7, 1967, p. 1.

15 Lindholm, U. S., "Some Experiments in Dynamic Plasticity under Combined Stress", Mechanical Behavior of Materials Under Dynamic Loading, Edited by U. S. Lindholm, Springer-Verlag, New York, 1968, p. 77.

16 Green, S. J., Leasia, J. D., Perkins, R. D. and Maiden, C. J., "Development of Multiaxial Stress High Strain-Rate Techniques" Materials and Structures Laboratory, Manufacturing Development, General Motors Corporation, Technical Report SAMSO TR 68-71, Vol. III, 1968.

17 Schierloh, F. L., Chaney, R. D. and Green, S. J., "An Application of Acquisition and Reduction of Data of Variable Short-time Tests", Experimental Mechanics, Vol. 10, 1970, p. 23N.

18 Osgood, W. R., "Combined-Stress Tests on 24-ST Aluminum Alloy Tubes," Journal of Applied Mechanics, Transactions of the ASME, Vol. 69, 1947, p. A-147.

19 Prager, W., "Strain Hardening Under Combined Stress", Journal of Applied Physics, Vol. 16, 1945, p. 837.

20 Drucker, D. C., "Relation of Experiments to Mathematical Theories of Plasticity", Journal of Applied Mechanics, Vol. 16, 1949, p. A349.

21 Lindholm, U. S., "Some Experiments with the Split Hopkinson Pressure Bar", Journal of the Mechanics and Physics of Solids, Vol. 12, 1964, p. 317.

22 Green, S. J., Maiden, C. J., Babcock, S. G. and Schierlok, F.L. "The High Strain-Rate Behavior of a Number of Metals," Presented at ASM Session of the Westec Conference, 1969, Los Angeles, California.

FIGURE CAPTIONS

Figure

1. Static (a) and dynamic (b) specimens. The length of each test section is 1 inch.

2. Head mass-specimen assembly of the dynamic tension-torsion apparatus.

3. Sample oscilloscope traces from a dynamic tension-torsion test. The vertical scales for the upper and lower tension traces are $\dot{x}_1$ = 50 in/sec/cm and P_2 = 85 lbs/cm respectively. For the upper and lower torsion traces the vertical scales are $\dot{\theta}_1$ = 225 rad/sec/cm and T_2 = 7 in-lbs/cm respectively. The time base for all four traces is 100 μ sec/cm.

4. Typical curve of natural strain versus time from sample oscilloscope traces.

5. Equivalent stress versus equivalent strain from the static tests.

6. Equivalent stress versus equivalent strain from the dynamic tests.

7. Initial yield surfaces from the static and dynamic tests.

8. Average equivalent stress versus average equivalent strain from the static and dynamic tests.

9. Equivalent stress versus equivalent strain rate.

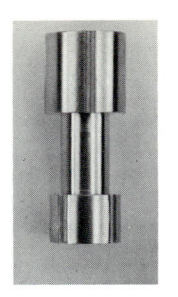

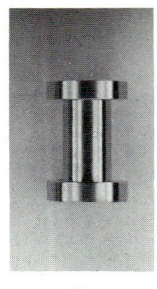

a b

FIGURE 1

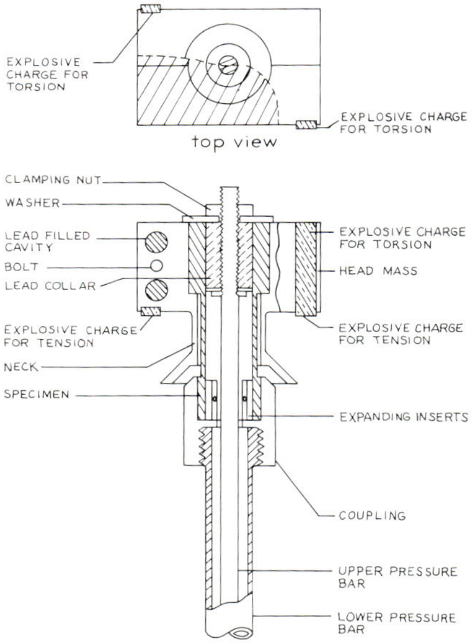

FIGURE 2

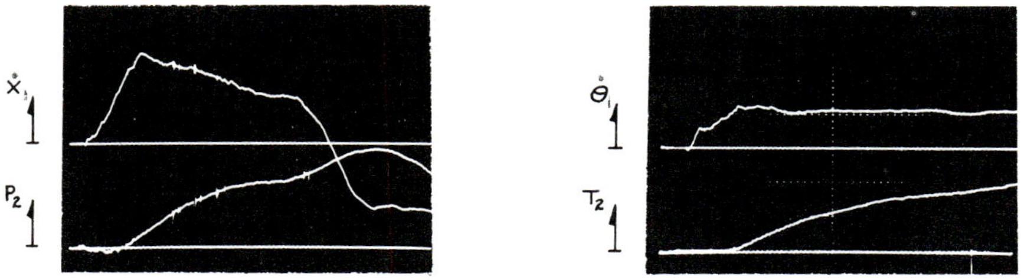

tension torsion

FIGURE 3

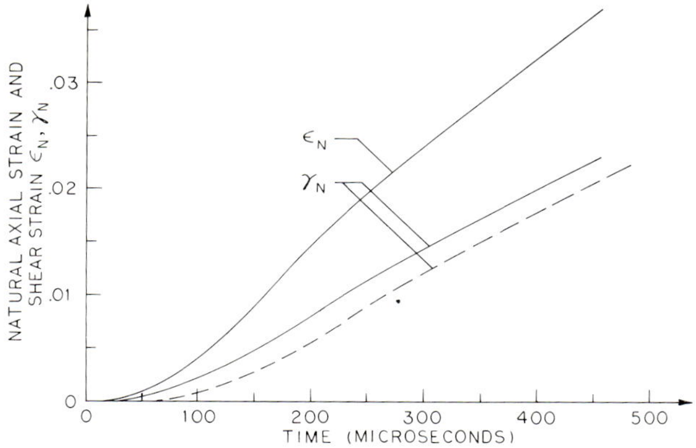

FIGURE 4

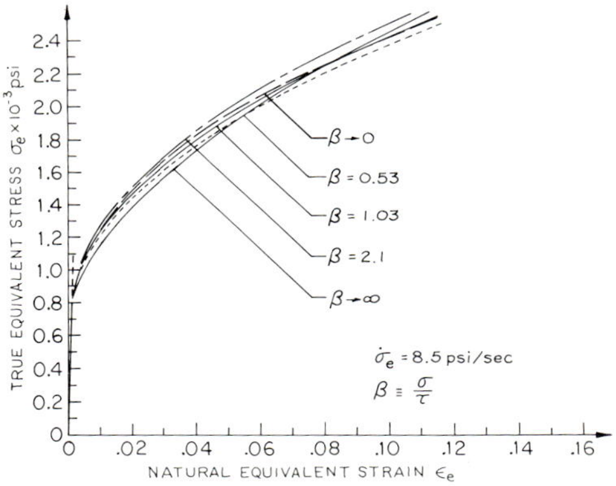

FIGURE 5

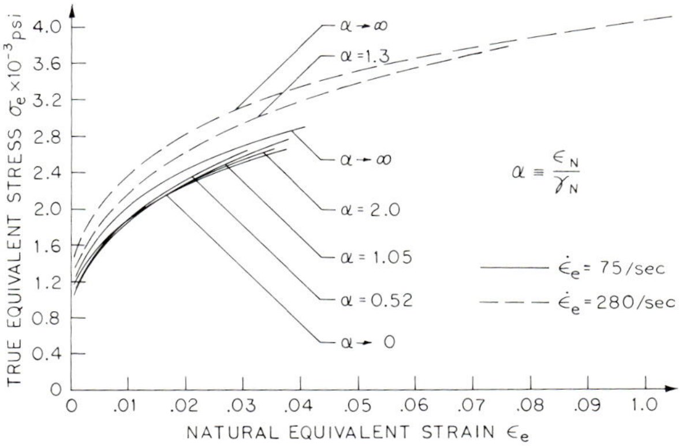

FIGURE 6

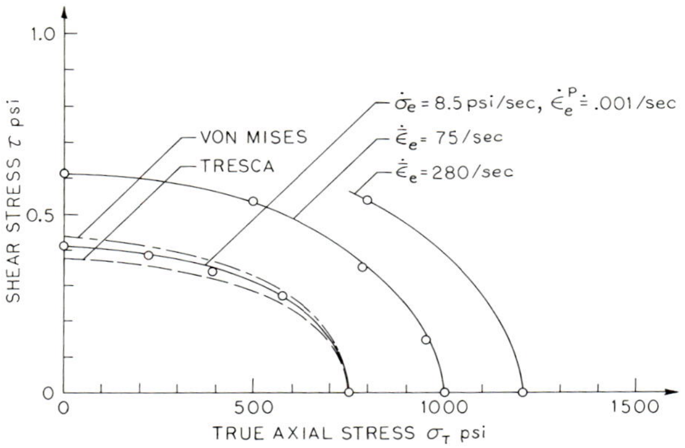

FIGURE 7

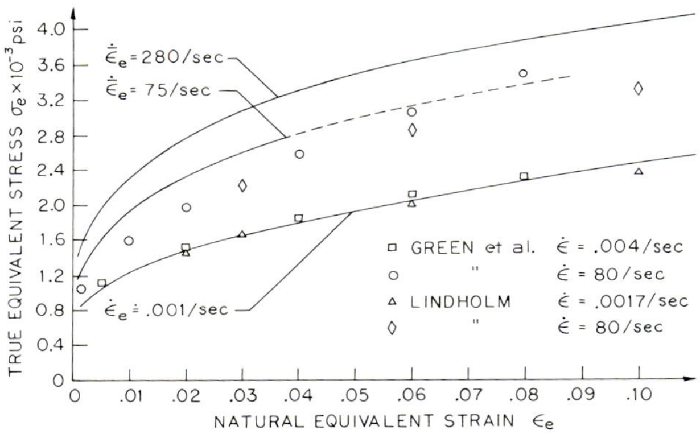

FIGURE 8

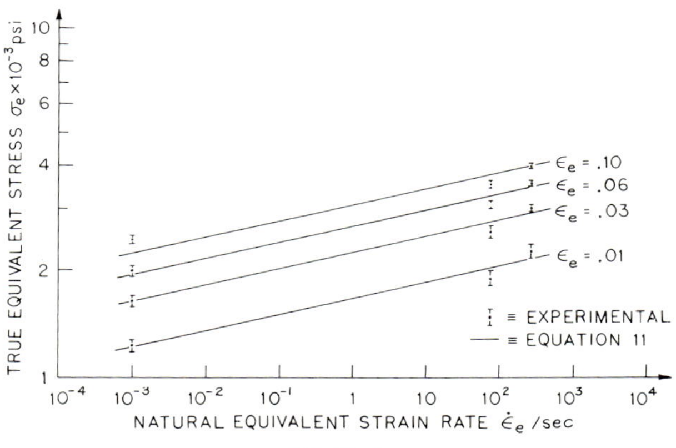

FIGURE 9

VIBRATION OF SHELLS

Developments in Mechanics, Vol. 6. Proceedings of the 12th Midwestern Mechanics Conference: 68

Nonstationary Responses of Cylindrical Shells Near Parametric Resonances

C. L. SUN, YEN YU and R. M. EVAN-IWANOWSKI
SYRACUSE UNIVERSITY

ABSTRACT

A study of nonstationary responses has been conducted for a circular cylindrical shell subjected to a longitudinal (parametric) load employing large deformation shell theory. The checkerboard radial modes have been selected in this study, characterized by the number of longitudinal half waves m and meridional waves n. Shells with initial geometrical imperfections were included in this study. Existence of both parametric and combination additive resonances have been established. The analysis is given for the parametric resonance and passage through resonance (nonstationary responses) for linearly varying frequency of external excitation (sweep).

Stationary responses exhibit hard spring effects. The amount of overhang depends to a great degree on the modes characterized by m and n. Nonstationary responses exhibit beat effect. They were stabilized when the frequency sweep was in the direction away from the instability region. When the sweep was into the instability region, two cases were observed depending on shell characteristics, but primarily on the rate of sweep: (1) nonstationary responses were following stable branches of the stationary responses or (2) they approached their initial stationary value. The effects of initial imperfections selected in this study on stationary and nonstationary regimes of motion are insignificant.

INTRODUCTION. When a cylindrical shell is subjected to an axial load of the form: $P(t) = P_0 + P_1 \cos\nu t$ where P_0, P_1 and are constant, parametric resonances may result, that is, lateral oscillations with half the excitation frequency ν.* This occurs for a certain combination of the shell natural frequencies ω_j and load parameters P_o, P_1 and ν. This type of parametric oscillation is referred to as stationary. Some previously obtained results for stationary parametric oscillations of cylindrical shells are: Wenzke[1] for linear, simply supported shell; Vijayaraghavan and Evan-Iwanowski[2] linear, clamped-free cylindrical shell; experimental work was also conducted confirming analytical calculation; and related problems by Dietz and Evan-Iwanowski[4].

It has been established that shells in general exhibit large prebuckling deformation. This fact has been taken into account in the study resulting in nonlinear equations of motion (field equations). It has been recently established that initial geometrical imperfections play a key role in static stability of shells. In view of this, initial imperfections have been included in the present analysis.

The nonstationary characteristics of the loading are encountered in practice in earthquake tremors with increasing or decreasing frequencies, time varying forces of escaping gases of rockets, water or sound waves impinging on a structure or exciting the inner structure of an ear. The form of nonstationary parametric loads is $P(t) = Po + P_1(t) \cos\theta(t)$ where $P_1(t)$ and $\theta(t) = \nu(t)$ are functions of time.

In the present study, the frequency of external excitation $\nu(t)$ is increased or decreased linearly from a value corresponding to the parametric resonance, and the amplitude of excitation $P_1(t)$ remains constant.

BASIC RELATIONSHIPS. As a starting point in this study, a set of the field equations for a cylindrical shell derived by Dietz[3] was used. These equations include effects of large deformations, stringers and rings reinforcements, rotational inertias and geometrical imperfections. These field equations were adopted for the case of longitudinal (parametric) load. Two checkerboard types for radial displacement were used:

$$w_0(x,y) = f_0 \sin\alpha x \sin\beta y + g_0 \sin2\alpha x \sin2\beta y \tag{1}$$

$$w_1(x,y,t) = f_1(t) \sin\alpha x \sin\beta y + g_1(t) \sin2\alpha x \sin2\beta y \tag{2}$$

where w_o represents initial geometrical imperfections which are in harmony with subsequent oscillatory modes w_1; $\alpha = m\pi/L$; $\beta = n/R$; m is number of half-waves in the longitudinal direction x; and n is number of full waves in the meridianal direction y; L is shell length and R its radius.

The simply supported case was considered, whereby radial displacement and longitudinal moment are vanishing at both ends of the shell, for details, see [5]. Applying Bubnov-Galerkin's method to the spatial-temporal field equations of the shell, we obtain the temporal differential equations of motion for the two modes $f_1(t)$ and $g_1(t)$. The normalized form of these equations is as follows:

$$\ddot{x}_1 + \omega_1^2 x_1 = D_1 x_1^3 + D_2 x_1 x_2^2 + D_3 x_1^2 x_2 + D_4 x_2^3 + D_5 x_1 \cos\theta + D_6 x_2 \cos\theta + D_7 \tag{3}$$

$$\ddot{x}_2 + \omega_2^2 x_2 = E_1 x_1^3 + E_2 x_1 x_2^2 + E_3 x_1^2 x_2 + E_4 x_2^3 + E_5 x_1 \cos\theta + E_6 x_2 \cos\theta + E_7 \tag{4}$$

Equations (3) and (4) represent two coupled nonlinear differential equations for the amplitudes of the two normal modes x_1 and x_2. D's and E's are constant coefficients. Since we will be interested in the isotropic case, the exact form of these coefficients is not presented here (see [5] for details). For the isotropic

* General relationship for the parametric resonance is $\nu/2\omega_j = 1/k$ where ν is excitation frequency, ω_j natural frequency of the j-th mode k = 1, 2, 3,...

case, the coefficients are given in the list of symbols.

ASYMPTOTIC SOLUTIONS AND RESONANCES. Equations (3) and (4) have been solved using the asymptotic method, (see for instance [6]. The first asymptotic approximation has the form:

$$x_1(t) = a_1(t)\cos\psi_1(t) \tag{5}$$

$$x_2(t) = a_2(t)\cos\psi_2(t) \tag{6}$$

where a_1, a_2 and ψ_1, ψ_2 are amplitudes and phase angles of the normal modes, respectively.

After casting the problem in an asymptotic form and performing necessary operations (see [6]), the following relationships resulted from which resonance conditions can be found:

$$\tilde{D}\bar{A}_{11} - 2a_1\omega_1\bar{B}_{11} = \frac{3}{4}a_1^3 D_1 + \frac{1}{2}D_3a_1a_2^2 + \frac{D_5a_1}{2}[\cos(\theta+2\psi_1) + \cos(\theta-2\psi_1)] + \frac{D_6a_2}{2}[\cos(\theta+\psi_1+\psi_2) + \cos(\theta-\psi_1-\psi_2)] \tag{7}$$

$$-(a_1\tilde{D}\bar{B}_{11} + 2\omega_1\bar{A}_{11}) = \frac{D_5a_1}{2}[\sin(\theta+2\psi_1) - \sin(\theta-2\psi_1) + \frac{D_6a_2}{2}[\sin(\theta+\psi_1+\psi_2) - \sin(\theta-\psi_1-\psi_2)] \tag{8}$$

$$\tilde{D}\bar{A}_{12} - 2a_22B_{12} = \frac{1}{2}E_2a_1^2 + \frac{3}{4}E_4a_2^3 + \frac{E_5a_1}{2}[\cos(\theta+\psi_1+\psi_2) + \cos(\theta-\psi_1-\psi_2)] + \frac{E_6a_2}{2}[\cos(\theta+2\psi_2) + \cos(\theta-2\psi_2)] \tag{9}$$

$$-(a_2\tilde{D}\bar{B}_{12} + 2\omega_2A_{12}) = \frac{E_5a_1}{2}[\sin(\theta+\psi_1+\psi_2) - \sin(\theta-\psi_1-\psi_2)] + \frac{E_6a_2}{2}[\sin(\theta+2\psi_2) - \sin(\theta-2\psi_2)] \tag{10}$$

where $\tilde{D} \equiv \nu(t)(\partial/\partial\theta) + \omega_1(\partial/\partial\psi_1) + \omega_2(\partial/\partial\psi_2)$, $\bar{A}_{ij}$, $\bar{B}_{ij}$, $i,j = 1,2$ are functions of time and D's, and E's are constants, and are presented explicitly in [5]. General resonance, i.e., the relationship between excitation frequency ν and natural frequencies ω_1 and ω_2 are:

$$k_o\nu + k_1\omega_1 + k_2\omega_2 = 0 \tag{11}$$

where k_o, k_1 and k_2 are positive or negative integers or zero, (see [7]). For the case on hand, it was found that in the first order asymptotic expansion representation, resonance may occur for the following:

$$\nu = \omega_1 + \omega_2; \qquad (k_o = 1,\ k_1 = -1,\ k_2 = -1), \tag{12}$$

for the combination additive resonance, and

$$\nu = 2\omega_1; \qquad (k_o = 1,\ k_1 = -2,\ k_2 = 0) \tag{13.1}$$

$$\nu = 2\omega_2; \qquad (k_o = 1,\ k_1 = 0,\ k_2 = -2) \tag{13.2}$$

for the parametric resonance.

PARAMETRIC RESONANCE. For the parametric resonance, we consider the case $\nu = 2\omega_1$, and the temporal equation of motion of the f_1 mode reduces to

$$\ddot{f}_1 + B_1 f_1 = B_3 f_1^3 + B_5 f_1 \cos\theta + B_6$$

values of constant coefficients B_1, B_3, B_5 and B_6 for the isotropic case are given in List of Symbols. The amplitude a_1 and phase shift $\phi_1(t)$ of the first mode in the first asymptotic approximation are found from the following two differential equations:

$$\dot{a}_1 = -\frac{a_1 B_5}{2\nu} \sin 2\phi_1 \tag{15}$$

$$\dot{\phi}_1 = \omega_1 - \frac{\nu}{2} - \frac{3a_1^2 B_3}{8\omega_1} - \frac{B_5}{2\nu} \cos 2\phi_1 \tag{16}$$

Stationary Responses*. In order to obtain the stationary responses, we set

$$\dot{a}_1 = 0, \ \dot{\phi}_1 = 0$$

This results in the following relationship between a_1 and ν:

$$a_1 = \left[\frac{8\omega_1\nu(\omega_1 - \frac{\nu}{2}) \pm 4\omega_1 B_5}{3\nu B_3}\right]^{\frac{1}{2}} \tag{17}$$

Nonstationary Responses. Equations (15) and (16) representing nonstationary responses are integrated numerically for the specified values of shell parameters and material constants given in Table I. (These values correspond to the experimental model prepared at the Applied Mechanics Laboratory, Syracuse University). Here h, L, R, and ρ are the shell thickness, length, radius and material density, respectively.

Results. Results of the stationary responses for parametric resonance of the isotropic shell are shown in Figures 1-2 for several longitudinal and meridional modes of radial displacement characterized by the integers m and n. The effects of nonlinearities which were included in the final numerical calculations are represented by the term B_3 in equation (14). It represents the effects of large curvatures due to large deformations, which is a "hard spring" type effect. The right hand overhang depends on the vibratory mode: For a given m (longitudinal half waves) and increasing n (meridional waves), the trend is toward the "harder spring" responses, Figure 1. An opposite trend is observed for fixed n and increasing m, Figure 2. It has to be born in mind that the parametric resonances for various modes occur for different natural frequencies pertaining to a particular mode characterized by numbers m and n, see Table III. The response curves at the end of the paper, however, are plotted for the nondimensional quantities, i.e., ratios of excitation frequency vs. twice the natural frequency.

Figures 3-7 represent plots of stationary and nonstationary responses were obtained starting with some initial stationary value and varying linearly the frequency of external excitation, Table III. The nonstationary responses are characterized by the beat effect, decreased amplitude, if the sweep is away from the instability zone, Figures 5 and 7. When sweep is into the instability zone the following may happen: (1) the nonstationary response may follow the stable stationary branch, Figures 3 and 4; (2) the nonstationary response may approach the initial stationary value, Figures 3 and 6.

Effect of initial geometrical imperfections, which are selected in harmony with the shell spatial oscillatory modes, is shown in Figures 6 and 7. As far as the stationary response is concerned, the solutions obtained for selected values of

* Stability analysis was performed in a usual manner adopting Lyapunov criteria. Using small perturbations for equations (15) and (16), it was found that the upper branches of the responses are stable and the lower, unstable.

imperfections indicates that there is no difference in the parametric response between a perfect shell and a shell with imperfections, see Figures 1 and 6 or 7. In regards to nonstationary responses again show only slight difference. Some slight difference can be noticed in the parametric nonstationary response between perfect and imperfect shells for decreasing ν, Figures 5 and 7.

COMBINATION ADDITIVE RESONANCE. The equations of motion for the two modes $f_1(t) = a_1 \cos\psi_1$ and $g_1(t) = a_2 \cos\psi_2$ indicate the existence of additive combination resonance of the type

$$\nu = \omega_1 + \omega_2; \quad \dot{\theta} = \nu; \quad \dot{\psi}_1 = \omega_1; \quad \dot{\psi}_2 = \omega_2$$

After some calculations, the differential equations for the determination of the amplitudes a_1 and a_2, and the phase shift $\phi = \theta - \psi_1 - \psi_2$ in nonstationary regime has been obtained.

$$\dot{a}_1 = \frac{a_2 D_6}{2} \frac{\sin\phi}{(\nu + \omega_1 - \omega_2)} \tag{18}$$

$$\dot{\phi} = \nu - \omega_1 - \omega_2 + \frac{1}{2\omega_1}\left(\frac{3}{4} a_1^2 D_1 + \frac{1}{2} a_2^2 D_3\right) + \frac{1}{2\omega_2}\left(\frac{1}{2} a_1^2 E_2 + \frac{3}{4} a_2^2 E_4\right) + \left[\frac{a_2 D_6}{2a_1 (\nu + \omega_1 - \omega_2)} + \frac{a_1 E_5}{2a_2 (\nu - \omega_1 + \omega_2)}\right] \cos\phi \tag{19}$$

$$\dot{a}_2 = \frac{a_1 E_5 \sin\phi}{2(\nu - \omega_1 + \omega_2)} \tag{20}$$

Stationary Response. Setting $\dot{a}_1 = \dot{a}_2 = \dot{\phi} = 0$, we obtain the following stationary response for a_1 (similar expressions were obtained for a_2)(with the damping coefficient approaching zero)

$$a_1 = \left\{ \frac{-8\omega_1 \left[\nu - \omega_1 - \omega_2 \pm \left(\frac{D}{4\omega_1\omega_2}\right)^{\frac{1}{2}}\right]}{3D_1 + 3E_4 \frac{\omega_1^2}{\omega_2^2} + 4E_2 \frac{\omega_1}{\omega_2}} \right\}^{\frac{1}{2}} \tag{21}$$

A further simplification was utilized in obtaining the above solution, namely: $D_6 = E_5$ and $D_3 = E_2$.

Nonstationary Response. The excitation frequency, $\nu(t)$, is allowed to vary with time according to relationships specified in Table II. Equations (19) - (20) could be solved numerically in a manner similar to the parametric resonance case.

CONCLUSIONS. The following conclusions can be drawn from the analysis presented in this paper: 1. Stationary parametric responses exhibit typical overhang characteristic of hard spring system with upper branch being stable. No differences have been noted for perfect and imperfect shells studied. 2. The nonstationary responses of perfect shells exhibit beat effect, and amplitudes decrease rapidly for the sweeps proceeding away from the instability zone, i.e., for the decreasing frequency of excitation in the cases analyzed. For the increasing frequency of excitations, i.e., when sweep is directed into the instability zone, the parametric nonstationary responses either follow the stable stationary branch (low sweep) or approach the initial stationary value (high sweep rate). 3. Only slight differences have been observed between the nonstationary parametric responses for perfect and imperfect shells. 4. The existence of the combination additive resonance has been determined. It has been found that combination differential and internal resonances are impossible for the type of analytical model presented in this study.

ACKNOWLEDGMENT. This study was performed on the National Science Foundation Research Participation Program, summer 1970, for which the authors express gratitude. The participants also would like to express their appreciation to the Department of Mechanical and Aerospace Engineering for providing the necessary computer facilities.

BIBLIOGRAPHY

[1] W. Wenzke: Die dynamische Stabilität der axialpulsirend belastend Kreiszylinderschale. Wiss. Z. der techn. Hochschule O.V. Guericke, Magdelburg 7, (1967).

[2] A. Vijayraghavan and R. M. Evan-Iwanowski: Parametric Resonances in Cylindrical Shells. Proceedings CANCAM, Montreal, Canada (1966).

[3] W. K. Dietz: On the Dynamic Stability of Eccentrically Reinforced Circular Cylindrical Shells. TR SURI No. 1620.1245-60, Syracuse, N.Y. (1967).

[4] W. K. Dietz and R. M. Evan-Iwanowski: Dynamic Stability of Eccentrically Reinforced Cylindrical Shells. Proceedings CANCAM, Waterloo, Canada (1968).

[5] E. O. Adams, C. L. Sun, Y. Wu and R. M. Evan-Iwanowski: Nonstationary Responses of Reinforced Cylindrical Shells. SURI TR No. 1690.5006-1, Syracuse, N.Y. (1970).

[6] B. N. Agrawal and R. M. Evan-Iwanowski: Resonances in Nonstationary Nonlinear Mechanical Systems. SURI TR No. 1364-2, Syracuse, N.Y. (February 1970).

[7] R. M. Evan-Iwanowski: Nonstationary Responses in Mechanical Systems. Applied Mechanics Reviews (September 1969).

Table I. Specifications of Shell Parameters

E psi	Po lb	σ Poisson's Ratio	h in	L in	R in	ρ lb-sec²/in⁴
4.38×10^5	20	.4	.021	23.5	6.25	9.5×10^{-5}

Table II. Nonstationary Values of the Excitation Frequencies

Case 1	$\nu = 2\omega_1 + \omega_1 t$
Case 2	$\nu = 2\omega_1 + 0.1\omega_1 t$
Case 3	$\nu = 2.6\omega_1 - 0.1\omega_1 t$

Table III. Natural Frequencies

Modes		Natural Frequencies ω rad/sec	
Longitudinal m	Meridional n	Isotropic perfect	imperfect
1	4	444	441
1	6	352	335
2	4	390	387
3	4	315	312
1	7	466	445

LIST OF SYMBOLS

R - radius
L - length
h - thickness
ρ - material density
E, σ, G - material constants
$\alpha = m\pi/L$
$\beta = n/R$
m = number of longitudinal half-waves
n = number of meridional full waves
$P(t) = P_0 + P_1 \cos\theta(t)$ - total external loading
$\bar{P}_1 = P_1/2\pi R$
$\dot{\theta}(t) = \nu(t)$
$a_i(t)$ - temporal amplitude of the i-th mode
$\psi_i(t)$ = phase angle of the i-th mode
ω_i = natural frequency of the i-th (normalized) mode
ϕ_i, ϕ - phase shift

$$\tilde{D} = \nu(t)\frac{\partial}{\partial\theta} + \omega_1\frac{\partial}{\partial\omega_1} + \omega_2\frac{\partial}{\partial\omega_2}$$

$\bar{A}_{ij}$, $\bar{B}_{ij}$ function to be determined

$$A_{11} = A_{22} = \frac{1}{Eh}$$

$$A_{12} = \frac{1}{2Gh} - \frac{\sigma}{Eh}$$

$$-B_1 = \omega^2 = \frac{-1}{\rho h}\{-\frac{Eh^3}{12(1-\sigma^2)}(\alpha^2+\beta^2)^2 - \frac{\alpha^4}{R^2}\frac{Eh}{(\alpha^2+\beta^2)^2} + \frac{f_o{}^2Eh}{16}(\alpha^4+\beta^4) + \frac{P_o}{2\pi R}(\alpha^2+\sigma\beta^2) + \frac{1}{8}\beta^4\,Eh\,(f_o{}^2+4g_o{}^2)\}$$

$$\bar{m}_1 = \rho h + \frac{1}{12}(\alpha^2+\beta^2)\rho h^3$$

$$B_3 = Eh(\alpha^4+3\beta^4)\,/\,(16\,[\rho h + (\alpha^2+\beta^2)\frac{\rho h^3}{12}])$$

$$B_5 = \bar{P}_1(\alpha^2+\sigma\beta^2)\,/\,[\rho h + (\alpha^2+\beta^2)\frac{\rho h^3}{12}]$$

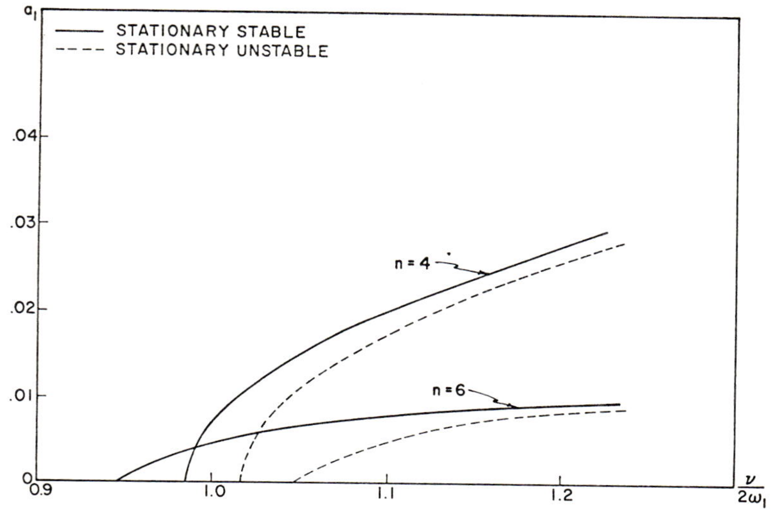

Fig. 1. Parametric stationary response of a perfect isotropic cylindrical shell, m = 1, n = 4 and n = 6.

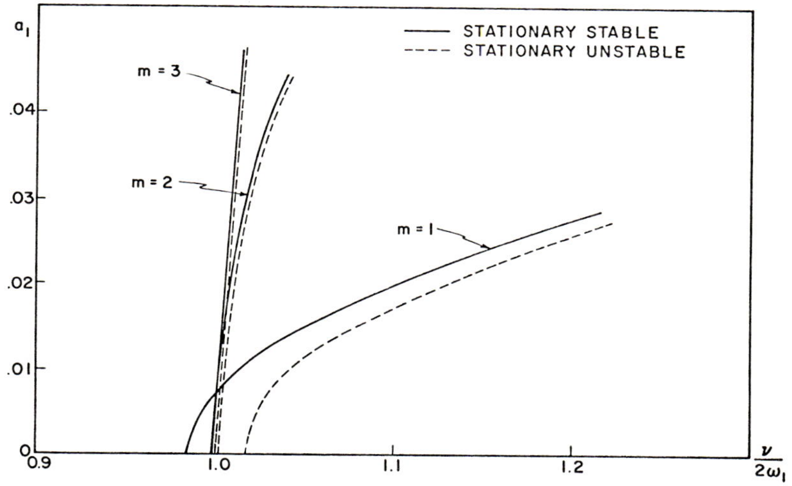

Fig. 2. Parametric stationary response of a perfect isotropic cylindrical shell, m = 1, n = 4; m = 2, n = 4; m = 3; n = 4.

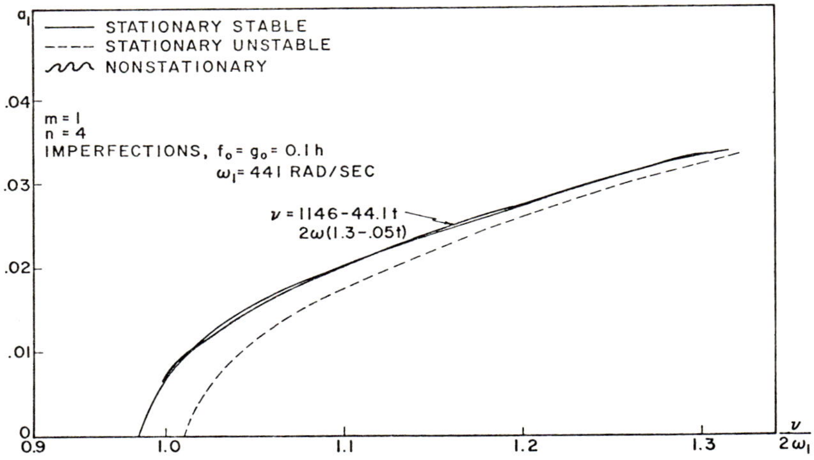

Fig. 7. Parametric stationary and nonstationary (decreasing ν) response of isotropic cylindrical shell with initial geometrical imperfections, $m = 1$, $n = 4$.

Developments in Mechanics, Vol. 6. Proceedings of the 12th Midwestern Mechanics Conference: 69

The Inextensional Vibrations of Paraboloidal Shells of Revolution

DAVID S. MARGOLIAS and VICTOR I. WEINGARTEN
UNIVERSITY OF SOUTHERN CALIFORNIA

ABSTRACT

The inextensional vibrations of paraboloidal shells of revolution are compared to the results obtained using Sanders' shell theory wherein both membrane and bending terms are included. The finite element method is used to solve Sanders' shell equations by employing a curved meridian frustum element. Vibration results of the two theoretical studies are first compared with experimental results and then with each other for both deep and shallow paraboloids.

INTRODUCTION

The analysis of structural shells has become the object of intensive study in the past two to three decades in response to the changing needs of the scientific community. While the application of such structural elements is by no means a new phenomenon, the refinement demanded of the analyses, which define their strength, stiffness and stability characteristics, required a research and development effort to parallel those of the applications for which they were intended. It soon became apparent, however, that only a limited number of such shell problems could, in fact, be solved in a closed, analytical form, and many efforts were redirected to developing approximate methods of solution which would still yield reasonably accurate results.

Because of the availability of the high speed, large memory digital computer, the most attractive approximate solutions to the shell problems were those that numerically approximated the exact or complete theory, since the error in approximation could always be minimized by expanding the size of the numerical solution, i.e., by taking finer increments in the numerical model. The two most widely used numerical solutions were the finite difference method, whereby the shell equations are approximated by a set of simultaneous difference equations, and the finite element method, wherein the total potential and kinetic energies of the shell are approximated as the sum of the energies of the discretized elements into which the continuous shell has been mathematically subdivided. By describing the energy properties of these elements and subsequently mathematically reassembling the continuum, a set of simultaneous equations describing the overall shell is acheived.

While both of these methods have been found to give excellent results for general shell problems, the amount of work necessary to set down the initial formulation, to perform the algebraic steps to bring the problem to its final form, and to program the results for the computer, are considerable indeed. Another approximate method of solution which has been found to yield excellent results for a limited class of shell vibration problems is called the inextensional theory of thin shells. This solution technique compromises the basic shell theory itself and in this way differs from the other methods discussed. The inextensional method admits to no straining of the shell middle surface, so that in the case of the dynamic analysis of such a structure, the assumption is made that the strain energy of the vibrating shell is due entirely to the bending of the middle surface. The vibration problem is then solved using Rayleigh's method of equating the maximum strain and kinetic energies of the shell. One of the limitations of this type of solution is that solutions are obtained only for the first or lowest mode of vibrations at a given circumferential wave number. Since this frequency is usually the one of primary interest, however, then depending on the specific problem being investigated, this may or may not be construed as a serious limitation of the inextensional method. On the other hand, it will be shown that, in the case of the paraboloidal shell for example, a far more serious drawback of this type of solution lies in the fact that the boundary condition for the shell is no longer a variable parameter which can be adjusted to suit the problem, but instead is determined by the analysis itself. As might be expected, since stretching of the shell middle surface is denied in this case, the analysis usually requires that at least one edge of the shell be completely unconstrained and free to deform as the free edge of the shell would. It is also not possible to include the presence of an initial stress field in the shell middle surface, since the presence of such a field implies the existence of the forbidden middle surface strains.

The inextensional method is thus quite restrictive as to the type of shell problem to which it may be applied and is for this reason used primarily as an academic tool rather than as an alternate method of solution. Nevertheless, for those problems which conform to the limitations imposed by the technique, the inextensional theory has been found to yield excellent results with relatively little work. Weingarten [2], for example, has found the inextensional theory to be in excellent agreement with experimental results for thin cylindrical shells, clamped at the one end and free at the other. Watkins and Clary [3] and Weingarten and Gelman [4] have found the same to be true for the case of conical frustum shells, likewise clamped at the one end and free at the other, and Naghdi and Kalnins [5] and Hwang [6] describe similar work for free edge spherical shells.

This paper is not the first to document the inextensional solution for paraboloidal shells. Lin and Lee [7] have presented the solution to such a problem earlier. The purpose of this work is to establish the accuracy of the paraboloid inextensional solution by comparing it to both that of the complete shell theory and to experimentally accumulated results.

COMPLETE THEORY

As was pointed out in the introduction, a closed form, analytic solution for the structural shell problem is possible only in a limited number of cases. The paraboloidal shell of revolution is not one of these select cases, so that solutions utilizing a complete shell theory, i. e., including both membrane and bending distortions, must of necessity involve one of the numerical techniques discussed in the previous section. Accordingly, the finite element method was employed, with the continuous shell being discretized into a number of shell frusta, formed by passing a set of parallel planes through the shell surface and normal to the generating axis or axis of revolution of the paraboloid. The resulting element, shown in Fig. 1, is then a frustum of the original shell, and its meridional geometry exactly duplicates that of the paraboloid.

Employing the finite element method, then, an appropriate set of displacement functions are assumed for the discrete element, the total energy for the shell (strain energy plus kinetic energy) is computed, and an energy principle is used to minimize this total energy with respect to each of the assumed displacement functions. For this reason the finite element method is likened to the Rayleigh-Ritz method, whereby the same procedure is employed without discretizing the continuous body. When the total energy is so minimized with respect to each assumed displacement function (or assumed constants in these displacement functions), a set of simultaneous differential equations are derived, which when banded together into a single matrix equation, succinctly give the familiar eigenvalue equation for the free vibrations problem.

No special assumptions other than those normally made for small deflection thin shell theory are needed or are made. The shell material is assumed to be Hookian, isotropic and perfectly elastic, the shell thickness is assumed to be small in comparison with its principal radii of curvature, the Love-Kirchhoff hypotheses governing normal stresses and normals to the middle surface are assumed to apply, and nonlinear displacement terms in the governing shell equations are assumed to be negligible. Accordingly, the shell strain-displacement relations used in the analysis, which were derived by Sanders [8] are, for the shell of revolution, given by

$$\begin{aligned} \varepsilon_{\theta\theta} &= \frac{1}{r}\left[u\cos\varphi + \frac{\partial v}{\partial\theta} + w\sin\varphi\right] \\ \varepsilon_{ss} &= \left[\frac{\partial u}{\partial s} + w\frac{d\varphi}{ds}\right] \\ \varepsilon_{\theta s} &= \frac{1}{2r}\left[\frac{\partial u}{\partial\theta} - v\cos\varphi + r\frac{\partial v}{\partial s}\right] \end{aligned} \qquad (1)$$

and

$$\begin{aligned} \kappa_{\theta\theta} &= \frac{1}{r^2}\left[ur\cos\varphi\frac{d\varphi}{ds} + \sin\varphi\frac{\partial v}{\partial\theta} - r\cos\varphi\frac{\partial w}{\partial s} - \frac{\partial^2 w}{\partial\theta^2}\right] \\ \kappa_{ss} &= \left[u\frac{d^2\varphi}{ds^2} + \frac{\partial u}{\partial s}\frac{d\varphi}{ds} - \frac{\partial^2 w}{\partial s^2}\right] \\ \kappa_{\theta s} &= \frac{1}{4r^2}\left[\left(3r\frac{d\varphi}{ds} - \sin\varphi\right)\frac{\partial u}{\partial\theta} + \left(r\frac{d\varphi}{ds} - 3\sin\varphi\right)v\cos\varphi\right. \end{aligned} \qquad (2)$$

$$- \left(r \frac{d\varphi}{ds} - 3 \sin \varphi \right) r \frac{\partial v}{\partial s} + 4 \cos\varphi \frac{\partial w}{\partial \theta} - 4r \frac{\partial^2 w}{\partial\theta\partial s} \Bigg]$$

Using the same coordinate system, the strain energy for the shell is given as [1]

$$U = \frac{E}{2(1-\nu^2)} \int_V \left[\left(\varepsilon_{\theta\theta} + \varepsilon_{ss} \right)^2 - 2(1-\nu) \left(\varepsilon_{\theta\theta} \varepsilon_{ss} - \varepsilon_{\theta s}{}^2 \right) \right] r \, d\theta \, ds \, dz$$
$$+ \frac{E}{2(-\nu^2)} \int_V \left[\left(\kappa_{\theta\theta} + \kappa_{ss} \right)^2 - 2(1-\nu) \left(\kappa_{\theta\theta} \kappa_{ss} - \kappa_{\theta s}{}^2 \right) \right] r z^2 \, d\theta \, ds \, dz \tag{3}$$

and the kinetic energy as

$$T = \frac{\rho}{2} \int_V \left[\left(\frac{\partial u}{\partial t} \right)^2 + \left(\frac{\partial v}{\partial t} \right)^2 + \left(\frac{\partial w}{\partial t} \right)^2 \right] r \, d\theta dr \, dz \tag{4}$$

For the frustum element chosen and shown in Fig. 1, four degrees of freedom or generalized displacements are assumed at each node as being most contributive to the total energy value of the shell. In the case at hand, the axisymmetric shell geometry permits the use of rather simple displacement functions to represent these generalized displacements, i.e., it is possible to expand the circumferential (asymmetric) coordinate dependence in an infinite Fourier series. If linear theory is applied, employing the principle of superposition makes it possible to consider each term in the series independently. The displacement functions assumed, therefore, with the meridional coordinate dependence being expressed in the form of Hermitian polynomials, are

$$w_i = \left(a_{1i} + a_{2i} s + a_{3i} s^2 + a_{4i} s^3 \right) e^{j\omega t} \cos n\theta$$
$$u_i = \left(a_{5i} + a_{6i} s \right) e^{j\omega t} \cos n\theta$$
$$v_i = \left(a_{7i} + a_{8i} s \right) e^{j\omega t} \sin n\theta$$

The third order polynomial is used for the normal displacement so that both displacement (w) and rotation ($\beta = \partial w/\partial s - u d\varphi/ds$) continuity are assured at each node, while linear functions are used for the two in-plane displacements, since only displacement continuity is sought for each of these coordinates at each node.

If these displacement functions are used in conjunction with the strain displacement equations, the expressions for the strain and kinetic energies of the shell can be written in matrix form as

$$U = \frac{1}{2} \{a\}^T [\overline{K}] \{a\} e^{2j\omega t}$$
$$K = - \frac{\omega^2}{2} \{a\}^T [\overline{M}] \{a\} e^{2j\omega t}$$

These equations may be altered to a more familiar form by utilizing the equations which relate the coefficients in the assumed displacement functions to the nodal displacements for any given element

$$\{q_i\} = [\Theta_i]^{-1} \{\overline{q}_i\} = [\Psi_i]^{-1} \{a_i\} e^{j\omega t}$$

where

$$\{\overline{q}_i\} = \begin{Bmatrix} u_i \\ v_i \\ w_i \\ \beta_i \\ u_{i+1} \\ v_{i+1} \\ w_{i+1} \\ \beta_{i+1} \end{Bmatrix} \qquad \{a_i\} = \begin{Bmatrix} a_{1i} \\ a_{2i} \\ a_{3i} \\ a_{4i} \\ a_{5i} \\ a_{6i} \\ a_{7i} \\ a_{8i} \end{Bmatrix}$$

$$[\Theta_i]^{-1} = \begin{bmatrix} \sec n\theta & 0 & 0 & 0 & 0 & 0 & 0 & 0 \\ 0 & \csc n\theta & 0 & 0 & 0 & 0 & 0 & 0 \\ 0 & 0 & \sec n\theta & 0 & 0 & 0 & 0 & 0 \\ 0 & 0 & 0 & \sec n\theta & 0 & 0 & 0 & 0 \\ 0 & 0 & 0 & 0 & \sec n\theta & 0 & 0 & 0 \\ 0 & 0 & 0 & 0 & 0 & \csc n\theta & 0 & 0 \\ 0 & 0 & 0 & 0 & 0 & 0 & \sec n\theta & 0 \\ 0 & 0 & 0 & 0 & 0 & 0 & 0 & \sec n\theta \end{bmatrix}$$

$$[\Psi_i]^{-1} = \begin{bmatrix} 0 & 0 & 0 & 0 & 1 & 0 & 0 & 0 \\ 0 & 0 & 0 & 0 & 0 & 0 & 1 & 0 \\ 1 & 0 & 0 & 0 & 0 & 0 & 0 & 0 \\ 0 & 1 & 0 & 0 & -\frac{d\varphi_i}{ds} & 0 & 0 & 0 \\ 0 & 0 & 0 & 0 & 1 & \ell & 0 & 0 \\ 0 & 0 & 0 & 0 & 0 & 0 & 1 & \ell \\ 1 & \ell & \ell^2 & \ell^3 & 0 & 0 & 0 & 0 \\ 0 & 1 & 2\ell & 3\ell^2 & -\frac{d\varphi_{i+1}}{ds} & -\ell\frac{d\varphi_{i+1}}{ds} & 0 & 0 \end{bmatrix}$$

The altered form of the energy expressions is

$$U = \frac{1}{2}\{q\}^T[\Psi]^T[\overline{K}][\Psi]\{q\} = \frac{1}{2}\{q\}^T[K]\{q\}$$

$$T = -\frac{\omega^2}{2}\{q\}^T[\Psi]^T[\overline{M}]\ [\Psi]\{q\} = \frac{1}{2}\{\dot{q}\}^T[M]\{\dot{q}\}$$

since

$$\{\dot{q}\} = j\omega\{q\}$$

for harmonic vibrations.

Applying Lagrange's equation to this conservative system of generalized coordinates gives, in the absence of any externally applied loadings,

$$\left\{\frac{d}{dt}\left(\frac{\partial T}{\partial \dot{q}}\right)\right\} + \left\{\frac{\partial T}{\partial q}\right\} + \left\{\frac{\partial U}{\partial q}\right\} = 0$$

Substituting the differentiated energy expressions into this equation gives the familiar equations of motion for the shell

$$[M]\{\ddot{q}\} + [K]\{q\} = 0$$

which for harmonic vibrations, i.e., $\{\ddot{q}\} = -\omega^2\{q\}$, gives the classical eigenvalue problem

$$[K - \omega^2 M]\{q\} = 0$$

or

$$|K - \omega^2 M| = 0$$

The eigenvalues (frequencies) and eigenvectors (mode shapes) of this problem can be obtained only by use of the digital computer, since there are, for the shell subdivided into N finite elements, approximately 4(N + 1) degrees of freedom for the free edge shell under investigation in this study. The total number of such generalized coordinates must be modified to account for the "effective" boundary conditions at the apex of the closed paraboloid. These conditions are derived from the requirement for finite strains and curvatures at the apex [9], [10] and are given in Table I.

INEXTENSIONAL THEORY

The fundamental assumption made in the inextensional theory, that the middle surface of the shell remains essentially unstrained in the deformed state, gives for Eq. (1)

$$\begin{aligned} \varepsilon_{\theta\theta} &= \frac{1}{r}\left[u\cos\varphi + \frac{\partial v}{\partial \theta} + w\sin\varphi\right] = 0 \\ \varepsilon_{ss} &= \left[\frac{\partial u}{\partial s} + w\frac{d\varphi}{ds}\right] = 0 \\ \varepsilon_{\theta s} &= \frac{1}{2r}\left[\frac{\partial u}{\partial \theta} - v\cos\varphi + r\frac{\partial v}{\partial s}\right] = 0 \end{aligned} \tag{5}$$

The equations to be satisfied by the inextensional displacement functions in terms of the independent variables θ and φ are then

$$u\cos\varphi + \frac{\partial v}{\partial \theta} + w\sin\varphi = 0 \tag{6}$$

$$\frac{\partial u}{\partial \varphi} + w = 0 \tag{7}$$

$$\frac{\partial u}{\partial \theta} - v\cos\varphi + \left(r\frac{d\varphi}{ds}\right)\frac{\partial v}{\partial \varphi} = 0$$

It is observed that only the last of these equations contains a geometry dependent coefficient and hence need be modified to account for different shell of revolution meridional shapes. Accordingly, for the paraboloid of revolution described by the equation $r^2 = Ax$, this equation becomes

$$\frac{\partial u}{\partial \theta} - v\cos\varphi + \sin\varphi\cos^2\varphi\frac{\partial v}{\partial \varphi} = 0 \tag{8}$$

The (unique) solution to the equation set (6) - (8) has been previously determined [7] as

$$\begin{aligned} u &= a_n \sin\varphi \tan^n\varphi\, e^{j\omega t} \cos n\theta \\ v &= a_n \tan^{n+1}\varphi\, e^{j\omega t} \sin n\theta \\ w &= -a_n \tan^n\varphi(\cos\varphi + n\sec\varphi)\, e^{j\omega t} \cos n\theta \end{aligned} \tag{9}$$

Using these displacement functions, the curvatures and twist of the shell middle surface given in Eq. (2) become

$$\kappa_{\theta\theta} = \frac{4n(1-n^2)a_n \sec\varphi \tan^{n-2}\varphi}{A^2} e^{j\omega t}\cos n\theta$$

$$\kappa_{ss} = \frac{4n(n^2-1)a_n \cos\varphi \tan^{n-2}\varphi}{A^2} e^{j\omega t}\cos n\theta$$

$$\kappa_{\theta s} = \frac{4n(1-n^2)a_n \tan^{n-2}\varphi}{A^2} e^{j\omega t}\sin n\theta$$

Substitution of these curvature and twist expressions into the inextensional form of the strain energy given by Eq. (3)

$$U_i = \frac{E}{2(1-\nu^2)}\iiint_V \left[\left(\kappa_{\theta\theta} + \kappa_{ss}\right)^2 - 2(1-\nu)\left(\kappa_{\theta\theta}\kappa_{ss} - \kappa_{\theta s}^2\right)\right] rz^2\, d\theta\, ds\, dz$$

gives, after carrying out the θ and z integrations and performing some algebraic manipulations,

$$U_i = \frac{\pi E}{6(1-\nu^2)}\left[\frac{a_n n(1-n^2)e^{j\omega t}}{A}\right]^2 \int_0^{\varphi_o} [\sec^2\varphi + \cos^2\varphi + 2(1-2\nu)]\, h^3 \sec^3\varphi \tan^{2n-3}\varphi\, d\varphi \tag{10}$$

Substitution of the displacement functions given in Eq. (9) into the kinetic energy expression, which is unchanged from that given in Eq. (4) for the complete theory, similarly gives

$$T_i = \frac{-\pi\rho}{8}[a_n A\omega e^{j\omega t}]^2 \int_0^{\varphi_o} [(1+n^2)\sec^2\varphi + 2n]\, h \sec^3\varphi \tan^{2n+1}\varphi\, d\varphi \tag{11}$$

Using Rayleigh's approach of equating the maximum kinetic and potential (strain) energies gives the final expression for the inextensional natural frequencies

$$f_i = \frac{n(n^2-1)}{\pi A^2}\sqrt{\left[\frac{E}{3\rho(1-\nu^2)}\right]\frac{\int_0^{\varphi_o}[\sec^2\varphi + \cos^2\varphi + 2(1-2\nu)]\, h^3\sec^3\varphi\tan^{2n-3}\varphi\, d\varphi}{\int_0^{\varphi_o}[(1+n^2)\sec^2\varphi + 2n]\, h\sec^3\varphi\tan^{2n+1}\varphi\, d\varphi}} \tag{12}$$

The retention of the thickness of the shell inside the integral allows for the solution of the completely general inextensional case. Since the most practical means of solving the frequency Eq. (12) is by numerical integration in any case, the inclusion of the variable thickness in the problem in fact adds no inconvenience or complication to the solution.

EXPERIMENTAL INVESTIGATION

A limited experimental program was conducted to accumulate data for use in ascertaining the validity of the theoretical results and in particular of the inextensional results. The test specimen used was an 1100-0 aluminum spun paraboloid whose diameter at the free (outer) edge measured 18.0 inches, whose thickness was nominally measured at 0.0375 inches, and whose meridional shape conformed to the paraboloid equation, $r^2 = 6.8936x$. During the experiment the shell was rested on its apex and was stabilized in this orientation with loosely packed wood shavings scattered around the shell. The excitation was provided to the specimen by pulsing an electromagnet positioned in very close proximity to the shell surface. The input to the power supply used to drive the electromagnet was controlled by a frequency oscillator and was monitored with an electronic counter. The response of the shell was measured with a microphone, similarly positioned in close proximity to the paraboloid surface. By centering the paraboloid beneath a spider-like fixture which supported the microphone on a rotary arm, complete circumferential response at any given point along the meridian was measured. The circumferential position of the microphone was indicated by using a potentiometer to measure the relative position of the rorary arm with respect to a given reference on the supporting fixture. With the microphone output (shell response) connected to the ordinate scale and the potentiometer output (microphone circumferential position) connected to the abscissa scale of an x-y plotter, the complete circumferential mode shape was recorded and determined at each frequency for which a resonant rise was detected. These rises in the response spectrum for the shell were found by simultaneously monitoring the microphone output on an oscilloscope. Using this setup, the natural frequencies were determined for the paraboloidal shell for the lowest three vibration modes, despite the fact that only the first mode data is of interest for the inextensional case.

COMPARISON OF RESULTS

Table II shows a comparison of the results of the theoretical and experimental investigations made on the paraboloidal shell, $r^2 = 6.8936x$. The inextensional results were obtained by numerically integrating Eq. (12) on the IBM 360/44 digital computer. These integrations were carried out using the seven point Newton-Cotes integration formula, with the shell divided into 50 segments, or (7) (50) + 1 = 351 integration points. It should be pointed out, however, that the inextensional solutions were found to converge, in fact, for a much smaller number of segments (less than five).

The finite element solutions were similarly obtained using the digital computer, with the larger IBM 360/65 being employed in this case. The program written for this analysis automatically subdivided the shell into 35 frustum elements of equal meridional length. The Cholesky square root method was used to decompose the inertia matrix into an upper and lower triangular form, thereby permitting the ultimate formation of a symmetric dynamical matrix of the form

$$[K'] = [L]^{-1}[K][L^T]^{-1}$$

where

$$[M] = [L][U] = [L][L^T]$$

The resulting eigenvalue problem was then solved by the Given's Method.

As Table II, and the corresponding graphical representations in Fig. 2 and Fig. 3, show, the inextensional theory gives excellent results over the entire range of harmonics investigated for the laboratory specimen. It is apparent, however, that this will not be true for the higher harmonics, since, as both Table II and Fig. 3 demonstrate, the experimental and inextensional results are already beginning the diverge rather rapidly by the thirteenth or fourteenth harmonic.

Upon initial inspection, it appears that the more approximate inextensional theory gives better results than the complete theory, which is represented by the finite element method. Although giving good results, as Fig. 2 shows, it is not until about the fourteenth harmonic that the finite element method gives results which are as good as those of the inextensional theory. It must be emphasized, however, that the finite element solutions, because they are derived using an energy minimization principle, will converge to the exact solutions from values which are greater than the true solutions. Thus, as Fig. 4 shows, the finite element solutions for the experimental shell have failed to converge at the second harmonic using only 35 elements, which is indicative that the solutions shown in Table II and Fig. 2 have, in fact, not converged for any of the harmonics calculated. On the other hand, as was previously indicated, the solutions shown for the inextensional method are all calculated for 50 shell segments and have accordingly reached convergence.

PARAMETRIC ANALYSIS

With the accuracy of the inextensional method established for at least one paraboloidal shell geometry, it is interesting to examine the behavior of this approximate theory for other, more extreme paraboloid shapes. For this investigation, inextensional results are compared to the finite element solutions for a relatively deep (focal length = 0.5) and a relatively shallow (focal length = 4.5) paraboloid, as well as for one which is quite similar (focal length = 1.5) to the experimental shell. The results of these comparisons are given in Table III and in Fig. 5 and Fig. 6. While Fig. 5 shows that the inextensional results continue to give good agreement with the finite element solutions for all paraboloid geometries, an examination of Fig. 6 more clearly shows the differences between the two solutions. Discounting somewhat the rather large discrepancies at the lower harmonics, since it is already established that this is primarily due to the lack of convergence of the finite element solutions, it is observed that the deeper the paraboloid, the better is the agreement between the two solutions. There is, in fact, no real agreement for the paraboloid with the focal length = 4.5, since the difference curve cuts the zero difference axis at a rather significant slope. The curve for the deep shell, on the other hand, is almost constant at about 2 percent for all harmonics greater than the third.

CONCLUSIONS

The foregoing analyses and discussions clearly establish the ease with which the inextensional vibrations of paraboloidal shells can be calculated and the geometries for which such solutions are accurate. Eq. (12), as originally derived by Lin and Lee [7], is found to yield excellent results for the lower harmonics (up to $n \approx 16$) for paraboloids with focal lengths as high as $2.0 \sim 2.5$. For the shallower paraboloids (focal length > 2.5) the range of accuracy is more limited, probably only for the lowest few harmonics. This conclusion is drawn from the supposition that the low harmonic solutions will be in far greater agreement if the finite element solutions are permitted to converge by assuming more than 35 discrete elements.

ACKNOWLEDGMENTS

The authors wish to express their appreciation to Mr. August B. Dontanville for his efforts in performing the experimental portion of this study. The writers also acknowledge the Directors of both the System Simulation Laboratory and the Computer Sciences Laboratory of the University of Southern California for the use of their computer facilities, and the National Science Foundation for its grant which funded the program, of which this study formed an integral part.

BIBLIOGRAPHY

[1] Novozhilov, V. V., The Theory of Thin Shells, P. Noordhoff Ltd., 1959.

[2] Weingarten, V. I., "Free Vibration of Thin Cylindrical Shells," AIAA Journal, Vol. 2, No. 4, April 1964, pp. 717-722.

[3] Watkins, J. D., and Clary, R. R., "Vibrational Characteristics of Thin-Walled Conical Frustum Shells," AIAA Journal, Vol. 2, No. 10, October 1964, pp. 1815-1816.

[4] Weingarten, V. I., and Gelman, A. P., "Free Vibrations of Cantilevered Conical Shells," ASCE Journal of the Engineering Mechanics Division, Vol. 93, No. EM6, December 1967, pp. 127-138.

[5] Naghdi, P. M., and Kalnins, A., "On Vibrations of Elastic Spherical Shells," Journal of Applied Mechanics, March 1962, pp. 65-72.

[6] Hwang, C., "Some Experiments on the Vibration of a Hemispherical Shell," Journal of Applied Mechanics, December 1966, pp. 817-824.

[7] Lin, Y. K., and Lee, F. A., "Vibrations of Thin Paraboloidal Shells of Revolution," Journal of Applied Mechanics, December 1960, pp. 743-744.

[8] Sanders, J. L., Jr., "Nonlinear Theories for Thin Shells," Quarterly of Applied Mathematics, Vol. 21, No. 1, 1963, pp. 21-36.

[9] Budiansky, B., and Radkowski, P. P., "Numerical Analysis of Unsymmetrical Bending of Shells of Revolution," AIAA Journal, Vol. 1, No. 8, August 1963, pp. 1833-1842.

[10] Greenbaum, G. A., "Comments on Numerical Analysis of Unsymmetrical Bending of Shells of Revolution," AIAA Journal, Vol. 2, No. 3, March 1964, pp. 590-591.

[11] Margolias, D. S., "The Effect of External Loading on the Natural Frequencies of Paraboloidal Shells," Ph.D. Thesis, The University of Southern California, 1970.

LIST OF SYMBOLS

A	Coefficient in paraboloidal equation, $r^2 = Ax$
E	Elastic modulus of the shell material
f_i	Inextensional natural frequencies of the shell
h	Thickness of the shell
ℓ	Meridional length of an element
n	Harmonic (circumferential wave) number
r	Radius of the shell
s, θ, z	Orthogonal shell coordinates in the meridional, circumferential and normal directions, respectively
T	Kinetic energy of the shell
T_i	Inextensional kinetic energy of the shell
t	Time
U	Strain energy of the shell
U_i	Inextensional strain energy of the shell
u, v, w	Orthogonal shell displacements in the meridional, circumferential and normal directions, respectively
V	Volume of the shell
β	Rotation of the meridian at the shell middle surface
$\varepsilon_{\theta\theta}$, ε_{ss}, $\varepsilon_{\theta s}$	In-plane strains of the shell middle surface
$\kappa_{\theta\theta}$, κ_{ss}, $\kappa_{\theta s}$	Curvatures and twist of the shell middle surface
ν	Poisson's ratio of the shell material
ρ	Mass density of the shell material
φ	Angle between the shell generating axis and the normal to the middle surface
ω	Natural frequency (radians/second)

Matrices

{a}	Vector of coefficients in the assumed displacement functions
[K]	Stiffness matrix
[M]	Inertia matrix
{q}	Vector of generalized displacemnts
[Ψ]	Matrix relating the element nodal displacements to the coefficients in the assumed displacement functions for the element

Table I

Effective Boundary Conditions
At the Apex of Shells of Revolution

Harmonic	"Effective" Boundary Conditions at the Apex
n=0	$u = v = \beta = 0$
n=1	$u + v = w = 0$
n=2	$u = v = w = \beta = 0$
$n \geq 3$	$u = v = w = 0$

Table II - Comparison of Experimental and Theoretical Results for the Vibrations of a Free Edge Paraboloidal Shell*

Harmonic	Experimental Frequency (cps)	Computed Frequency (cps)		Error In Computed Frequency (percent)	
		Inextensional	35 Finite Elements	Inextensional	35 Finite Elements
2	16	15	22	6.25	37.5
3	40	41		2.50	
4	76	77	83	1.32	9.21
5	121	123		1.65	
6	174	177	184	1.72	5.75
7	236	242	248	2.54	5.08
8	307	315	322	2.61	4.89
9	390	398	404	2.05	3.59
10	477	490	496	2.73	3.98
11	573	592	597	3.32	4.19
12	676	702	707	3.85	4.73
13	784	822		4.85	
14	903	952	953	5.43	5.54
15	1037	1090		5.11	
16	1164	1238	1235	6.36	6.10
18	1412	1562	1552	10.62	9.92

*Equation of Paraboloid is $r^2 = 6.8936x$

Table III - Comparison of Inextension and Complete Shell Theory Results for the Vibrations of Various Free Edge Paraboloidal Shells*

Harmonic Number (n)	Inextensional Theory Results For Natural Frequency (cps)			35 Finite Element Complete Theory Results for Natural Frequency (cps)			Percent Difference** Between Compete Theory and Inextensional Theory Results		
	A=2	A=6	A=18	A=2	A=6	A=18	A=2	A=6	A=18
2	11.56	9.60	11.99	13.45	14.58	27.36	15.11	41.19	78.10
4	60.82	49.07	58.01	61.96	53.36	73.87	1.86	8.38	24.05
6	142.24	112.88	133.15	144.36	117.18	147.89	1.47	3.74	10.49
8	255.46	200.47	236.89	260.47	204.67	248.57	1.94	2.07	4.81
10	400.83	311.70	369.10	409.35	315.65	374.66	2.10	1.26	1.50
12	577.17	446.52	529.73	589.78	450.05	524.77	2.16	0.79	-0.94
14	785.66	604.91	718.75	801.47	607.72	697.60	1.99	0.46	-2.99
16	1025.89	786.85	936.15	1044.88	788.54	892.24	1.83	0.21	-4.80

*The following constants were used for the shells: $E = 10^7$ psi, $\nu = 0.3$, thickness = 0.030 inches, radius = 10.0 inches, and geometry, $r^2 = Ax$

**Based on Percent Error = $\frac{(200)(\text{Finite Element Solution - Inextensional Solution})}{(\text{Finite Element Solution + Inextensional Solution})}$

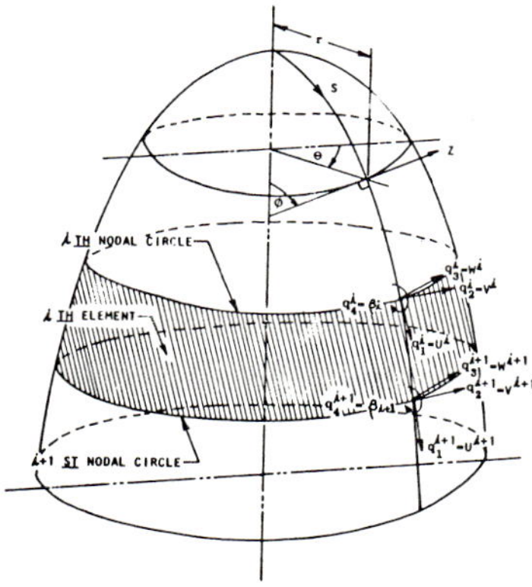

Fig. 1. Finite Element Model, Shell Coordinates and Assumed Nodal Displacements

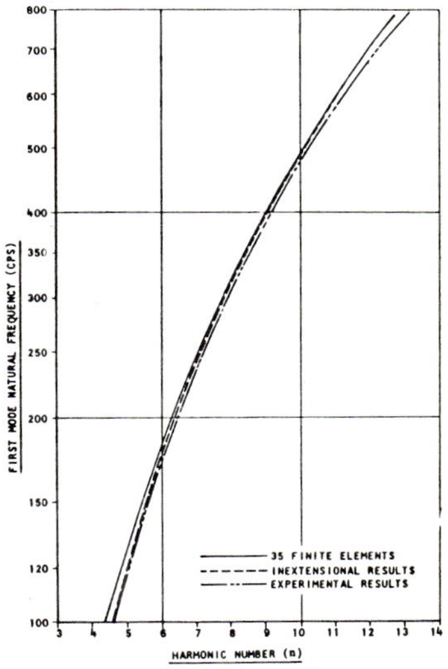

Fig. 2. Comparison of Experimental and Computed Inextensional and Complete Theory Results for the First Mode Natural Frequencies of the Paraboloidal Shell, $r^2 = 6.8936x$.

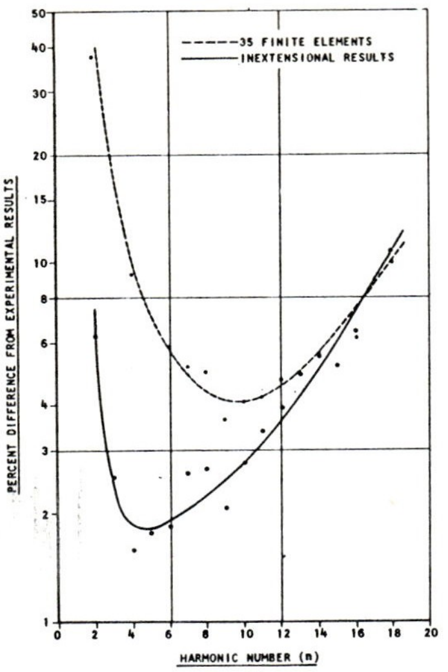

Fig. 3. Percent Difference Between the Theoretical and Experimental Results for the First Mode Natural Frequencies of the Paraboloidal Shell, $r^2 = 6.8936x$.

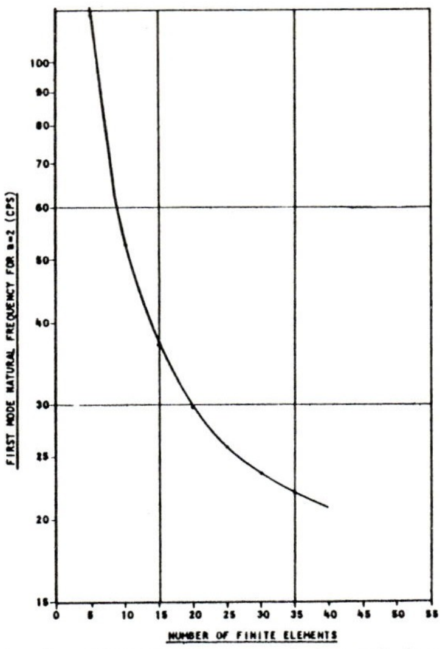

Fig. 4. Convergence of the Finite Element Method for the First Mode Natural Frequencies at the Second Harmonic of the Paraboloidal Shell, $r^2 = 6.8936x$.

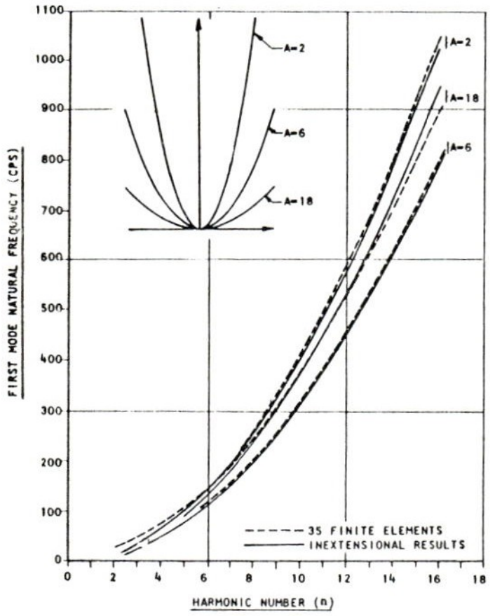

Fig. 5. Comparison of Inextensional and Finite Element Solutions for the First Mode Natural Frequencies of Paraboloidal Shells with Different Focal Lengths.

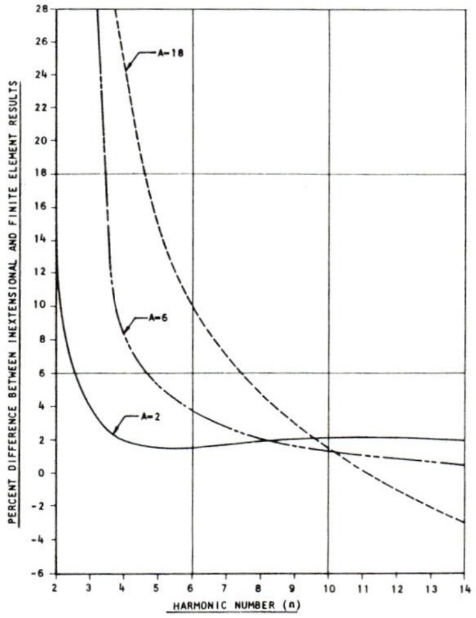

Fig. 6. Percent Difference Between the Inextensional and Finite Element Solutions for the First Mode Natural Frequencies of Paraboloidal Shells with Different Focal Lengths.

Developments in Mechanics, Vol. 6. Proceedings of the 12th Midwestern Mechanics Conference: 70

Impact of Spheres on Elastic Plates of Finite Thickness

Y. M. TSAI and K. DILMANIAN
IOWA STATE UNIVERSITY

ABSTRACT

A general method of calculation is described to determine the stresses and contact time for the impact of spheres on the surface of an elastic plate of finite thickness overlying a rigid foundation. Also described are the stress distribution and contact time for the impact between a rigid sphere and a thin plate.

Experiments are conducted to measure the contact time between steel balls and glass plates of various thicknesses overlying a large steel block. It is shown both theoretically and experimentally that for sufficiently large steel balls the contact time decreases with decreasing plate thickness, and a good agreement between theory and experiment is obtained. The impact velocities required to produce conical fractures are measured to be smaller for thin glass plates than for thick glass plates overlying a hard steel block. This is shown to be a consequence of the fact that for the same impact velocity of a large steel ball the critical tensile stress around the contact circle is larger in a thin plate than in a thick plate.

INTRODUCTION

The measurements of the dynamic and static strength of thick glass plates were carried out by respectively impinging and pressing steel balls onto the surfaces of glass [1]. In the static measurements the load required to produce Hertzian fractures in glass plates was measured and the corresponding fracture stress was calculated as a function of indenter diameters. It was found that the calculated value of the fracture tensile stress is not constant but increases with decreasing radius of the indenter. The relationship between apparent tensile strength and indenter radius is known as Auerbach's Law, and has been discussed in terms of the statistics of the flaw distribution in a glass surface [1]. In the measurements of the dynamic strength in Ref. 1 the impact tests performed were limited to the cases of small indenters. For large indenters whose diameters are compatible with the thickness of glass plates, however, no dynamic measurements were carried out. The difficulty confronting the measurement there appears to be the understanding of local stress distribution around the area of contact between an indenter and a vibrating plate.

The above difficulty is equivalent to that of accounting for the effect of plate thickness on the local stress distribution around the area of contact in static and dynamic loadings. Recently a mathematical study was conducted on the effect of plate thickness on the local stress distribution produced in an elastic plate by pressing spherical indenters on both upper and lower surfaces of the plate [2]. The thickness effect was studied in terms of the nondimensional ratio between contact radius and plate thickness. It was shown that the half-space solution used [1] is an accurate approximation only for the plates, the thicknesses of which are large compared to the radius of the contact area. For thin plates under the pressure of large indenters, however, the maximum tensile stress in the plates may become twice as large as that in a half-space [2]. The above magnification of the tensile stress due to thinness of the plate was demonstrated experimentally by pressing steel balls on the surfaces of glass plates of various thicknesses [3]. In interpreting the experimental data obtained, a general numerical procedure was suggested to determine the fracture stresses of glass specimens. It was shown [3] that the load required to produce fracture in glass plates decreases rapidly as the plate thickness decreases. This was determined to be a consequence of maximum tensile stress being larger in a thin plate than in a thick plate for the same indentation pressure.

The static solutions obtained [2] can be extended to associated dynamic problems of plates. In the Hertz impact theory [4, 5] it was assumed that near the point of contact the stresses and strains may be computed at any instant as though the contact were static. By solving three-dimensional equations of motion, the above assumption was studied as a function of contact time and contact radius [6]. It was shown mathematically that the Hertz theory applies for moderate impact velocities where the contact time is large and contact radius is small. If contact times are 300 μ sec or more and the contact radii are less than 0.045 in., the Hertz stresses in a half-space at the instant of maximum contact radius are shown to differ less than 0.05% from the results obtained by Tsai [6]. In the above dynamic range, therefore, it appears reasonable to apply the Hertz assumption to the associated dynamic problems of plates.

Using the Hertz assumption, the theoretical results obtained by Tsai [2] are extended here to the corresponding dynamic problems. The problem considered is that a plate overlying the surface of a rigid foundation is subjected to the impact of a spherical body. General numerical procedures are suggested to determine contact times and contact stresses for plates of arbitrary thickness.

The theory described here has two apparent parameters, namely, indenter diameter and plate thickness. Measurements associated with the theory are carried out here by impinging steel balls of various diameters onto glass plates of various thicknesses overlying a large steel block. The times of contact and the impact velocities required to produce fractures in glass plates are measured as a function of plate thickness and indenter diameter. The contact times measured are compared to those predicted by the theory developed here. Furthermore, the fracture tensile stresses corresponding to the fracture velocities are also determined on the basis of the present theory. The work described here is three-dimensional in nature. For extremely thin plates, an impact problem similar to that considered here was investigated in Ref. 7. The stress distribution obtained there appears to be a limiting case of the present theory.

THEORETICAL

The problem considered here is the impact of a spherical body on the surface of an elastic plate overlying a rigid foundation. In order to make the problem amenable to a mathematical analysis, the spherical indenter is assumed to be a rigid body. This is also the assumption adapted in the mathematical study of the symmetrical indentations of spherical bodies on the upper and lower surfaces of an elastic plate [2]. The theoretical results obtained there were recently extended to a general method for determining the maximum tensile stress in an elastic plate subjected to arbitrary spherical indentations [3]. The method developed was shown to be useful in analyzing the experimental data obtained in the investigation of fracture in glass plates, and a close agreement between theory and experiment was obtained [3]. Therefore, the assumption mentioned above appears to be reasonable and practical for the current problem.

The theory for the present problem can be extended in cylindrical coordinates (r, θ, z) from an earlier work [2]. The problem considered there is an infinite elastic plate of thickness H which is subjected to a rigid, axisymmetrical indentation on its upper surface z = H, but resting on a smooth, rigid foundation at z = 0. This problem is equivalent to a plate of thickness 2H subjected to the same indentations on both upper and lower surfaces. It is assumed that the shear stress vanishes on the surfaces, and that the normal contact stresses, p(r, t), are formally known functions of the radial distance, r, and the time, t. The function p(r, t) was written as an unknown to be solved in an integral equation [2]. For convenience, some equations in the earlier work [2] are quoted here with proper addition of the time variable, t. These equations are written hereafter following proper reference brackets. In terms of p(r, t), the vertical displacement in the plate surface z = H was written [2] as

$$U_z(r, H, t) = 2C \int_0^\infty \frac{J_0(sr)\sinh^2 sH}{2Hs + \sinh sH} \int_0^{a(t)} p(\lambda, t)\lambda J_0(\lambda s)\,d\lambda\,ds \tag{1}$$

where $C = (1 - \upsilon)/\mu$. μ and υ are, respectively, the shear modulus and Poisson's ratio. $J_0(x)$ is the Bessel function of the first kind of zeroth order, and a(t) is the radius of the contact area between the indenter and the plate. The determination of a(t) will be described later in this paper. When H tends to infinity, the function p(r, t) in Eq. 1 was found to be of the same form as the associated half-space Hertz contact stress [2]. Therefore, the Hertz impact theory applies [5]. The other limiting case of Eq. 1 occurs when H becomes extremely small. If the shape of the vertical displacement produced by the indenter inside the area of contact can be described by g(r, t), Eq. 1 for small H reduces to

$$g(r, t) = \frac{HC}{2} \int_0^\infty sJ_0(sr) \int_0^{a(t)} p(\lambda, t)\lambda J_0(\lambda s)\,d\lambda\,ds = \frac{HC}{2}\, p(r, t)$$

$$\text{for } r \le a(t) \tag{2}$$

Thus the shape of the normal contact stress distribution is the same as that of the indentation for very thin plates. The radial and circumferential stresses in the surfaces of thin plates can also be deduced from the corresponding general expressions obtained in Ref. [2]. After some calculations those expressions for small H reduce to

$$\sigma_{rr} = \begin{cases} \nu \dfrac{1}{r^2} \displaystyle\int_0^r p(\lambda, t)\lambda\,d\lambda & \text{for } r \le a(t) \\[2ex] \nu \dfrac{1}{r^2} \displaystyle\int_0^{a(t)} p(\lambda, t)\lambda\,d\lambda & \text{for } r > a(t) \end{cases} \tag{3}$$

and

$$\sigma_{\theta\theta} = \begin{cases} \nu\left[p(r, t) - \frac{1}{r^2}\int_0^r p(\lambda, t)\lambda d\lambda\right] & \text{for } r \leq a(t) \\ -\frac{\nu}{r^2}\int_0^{a(t)} p(\lambda, t)\lambda d\lambda & \text{for } r > a(t) \end{cases} \tag{4}$$

The general expressions 2 — 4 are valid for any axisymmetrical indenters. If the radius of the contact circle is small compared to the radius R of a rigid spherical indenter, the penetration function can be written [2] as

$$g(r, t) = \frac{r^2}{2R} - \alpha(t) \tag{5}$$

where α is the maximum depth of indentation. If Eq. 5 is used in Eq. 2, p(r, t) is obtained. Since p(r, t) must vanish at r = a, this leads to

$$\alpha = \frac{a^2}{2R} \tag{6}$$

Therefore, the normal stress on the surface of a thin plate is

$$\sigma_{zz} = \begin{cases} p(r, t) = -\frac{a^2}{RHC}\left\{1 - \frac{r^2}{a^2}\right\} & \text{for } r \leq a \\ 0 & \text{for } r > a \end{cases} \tag{7}$$

If Eq. 7 is substituted into Eqs. 3 and 4, the stresses are

$$\sigma_{rr} = \begin{cases} -\frac{\nu a^2}{2RHC}\left\{1 - \frac{r^2}{2a^2}\right\} & \text{for } r \leq a \\ -\frac{\nu a^4}{4RHCr^2} & \text{for } r > a \end{cases} \tag{8}$$

and

$$\sigma_{\theta\theta} = \begin{cases} -\frac{\nu a^2}{2RHC}\left\{1 - \frac{3r^2}{2a^2}\right\} & \text{for } r \leq a \\ +\frac{\nu a^4}{4RHCr^2} & \text{for } r > a \end{cases} \tag{9}$$

The stress distribution inside the contact area expressed in Eqs. 7-9 is the same as that obtained in Ref. [7] from a different approach. Since stresses are assumed to be constant across the thickness of a thin plate, the above results are valid everywhere in a thin plate.

The total applied force from Eq. 7 is equal to

$$P = \int_0^a 2\pi r p(r, t) dr = -\frac{2\pi R}{HC}\alpha^2 \tag{10}$$

If m is the mass of the projectile impinging on a thin elastic plate, the equation of motion can be written as

$$m\ddot{\alpha} = - P \tag{11}$$

where the dot means differentiation with respect to time. If the velocity of approach is V and if the maximum depth of penetration is α_1, the integration of Eq. 11 gives

$$\dot{\alpha}^2 = V^2 - \frac{4\pi R}{3HCm}\alpha^3 \tag{12}$$

and

$$\alpha_1^3 = \frac{3HCm}{4\pi R} V^2 \tag{13}$$

Integrating Eq. 12 over t gives

$$t = \frac{\alpha_1}{V}\int_0^{\alpha/\alpha_1} \frac{dx}{\sqrt{1 - x^3}} \tag{14}$$

The total contact time T is equal to twice the value of t for $\alpha = \alpha_1$. Thus

$$T = \frac{2\alpha_1}{V}\int_0^1 \frac{dx}{\sqrt{1 - x^3}} = \frac{2\alpha_1\sqrt{\pi}\,\Gamma\left(\frac{1}{3}\right)}{V\,3\,\Gamma\left(\frac{5}{6}\right)} = 2.804\left\{\frac{3H(1 - \nu^2)m}{2\pi RE}\right\}^{1/3} V^{-1/3} \tag{15}$$

If Eq. 13 is employed in Eq. 6, the maximum contact radius is determined. The value of a(t) during impact is also obtained from Eq. 6 if $\alpha(t)$ is determined from Eq. 14 and then used in Eq. 6.

Between the above two limiting cases, an elastic plate of finite thickness is of essential interest in the present study. The function p(r, t) was written by Tsai [2] as an unknown in the integral equation

$$p(r, t) = p_H(r, t) + \frac{1}{2\pi H^2}\int_0^a p(\lambda, t)\lambda d\lambda \int_0^\infty Q(u)uJ_0\left(\frac{\lambda u}{2H}\right)du$$

$$\int_r^a \frac{\sin\left(\frac{su}{2H}\right) ds}{\left(s^2 - r^2\right)^{1/2}} \tag{16}$$

where

$$Q(u) = \frac{1 + u - e^{-u}}{u + \sinh u} \tag{17}$$

$p_H(r, t)$ is the associated half-space normal contact stress. For the function g(r, t) described in Eq. 5, the associated normal stress is [2]

$$p_H(r, t) = -4\,\frac{\left(a^2 - r^2\right)^{1/2}}{C\pi R} \tag{18}$$

From the process of successive approximations [2], the solution of Eq. 16 can be written

$$p(r, t) = p_H(o, t)k_z\left(\frac{a}{H}, x\right) \tag{19}$$

where

$$p_H(o, t) = -\frac{4a}{C\pi R} \tag{20}$$

and the nondimensional function of a/H and x = r/a is

$$k_z = (1 - x^2)^{1/2} + \frac{2}{\pi}\left(\frac{a}{2H}\right)^2 \int_o^1 (1 - y^2)^{1/2} ydy \int_o^\infty Q(u)uJ_o\left(\frac{ayu}{2H}\right)du$$
$$\int_x^1 \frac{\sin\left(\frac{a\eta u}{2H}\right)d\eta}{\left(\eta^2 - x^2\right)^{1/2}} + \dots \tag{21}$$

In addition to the above equations, the subsidiary equation for determining the distance of approach g(o, t) = - α(t) was also obtained [2] as follows:

$$g(o, t) + a\int_o^a \frac{g'(m, t)dm}{(a^2 - m^2)^{1/2}} + \frac{C}{2H}\int_o^\infty Q(u)\cos\left(\frac{au}{2H}\right)\int_o^a p(\lambda, t)\lambda J_o\left(\frac{\lambda u}{2H}\right)$$
$$d\lambda du = 0 \tag{22}$$

From Eqs. 5, 19 and 22, α can be written as

$$\alpha = \alpha_H k_\alpha\left(\frac{a}{H}\right) \tag{23}$$

where

$$\alpha_H = \frac{a^2}{R} \tag{24}$$

and the nondimensional function is

$$k_\alpha\left(\frac{a}{H}\right) = 1 - \frac{4}{\pi}\left(\frac{a}{2H}\right)\int_o^\infty Q(u)\cos\left(\frac{au}{2H}\right)\int_o^1 k_z\left(\frac{a}{H}, x\right)xJ_o\left(\frac{aux}{2H}\right)dxdu \tag{25}$$

The first derivative of α from Eq. 23 is

$$\frac{\partial\alpha}{\partial a} = \alpha'_H k_{\alpha'}\left(\frac{a}{H}\right) \tag{26}$$

where

$$\alpha'_H = \frac{2a}{R} \tag{27}$$

and

$$k_{\alpha'} = 1 + \frac{2}{\pi}\left(\frac{a}{2H}\right)^2 \int_0^\infty Q(u)\, \sin\left(\frac{au}{2H}\right)u \int_0^1 k_z\left(\frac{a}{H}, x\right)xJ_o\left(\frac{aux}{2H}\right)dxdu \tag{28}$$

The above derivative will be useful in the following determination of the contact time between a spherical body and plate. During impact, the equation of motion 11 also applies. P is the total force exerted by the plate on the sphere and was written [3] as

$$P = P_H k_p\left(\frac{a}{H}\right) \tag{29}$$

where

$$P_H = -\frac{8a^3}{3CR} \tag{30}$$

and the nondimensional function $k_p\left(\frac{a}{H}\right)$ was calculated and shown graphically by Tsai [3]. If Eq. 11 is multiplied by $\dot{\alpha}$ and if the values of P and α' are substituted from Eqs. 29 and 26, respectively, the integration of Eq. 11 over t gives

$$\frac{m}{2}(\dot{\alpha}^2 - V^2) = -\left\{\int_0^{a(t)} P_H\alpha_H' da + \int_0^{a(t)} P_H\alpha_H'\left[k_p\left(\frac{a}{H}\right)k_{\alpha'}\left(\frac{\alpha}{H}\right) - 1\right]da\right\} \tag{31}$$

where V is the initial velocity of approach of the sphere. If a/H is replaced by y, Eq. 31 can be written as

$$\frac{m}{2}(\dot{\alpha}^2 - V^2) = -\frac{16a^5}{15CR^2}\, k_V^2\left(\frac{a}{H}\right) \tag{32}$$

where the nondimensional function is

$$k_V^2 = 1 + 5\left(\frac{H}{a}\right)^5 \int_0^{a/H} y^4\left[k_p(y)k_{\alpha'}(y) - 1\right] dy \tag{33}$$

When the impact reaches its maximum distance of approach with maximum contact radius a_1, the velocity of the body vanishes, i.e., $\dot{\alpha} = 0$. Under this condition Eq. 32 becomes

$$V = V_H k_V\left(\frac{a_1}{H}\right) \tag{34}$$

where the associated impact velocity is

$$V_H = \left(\frac{32a_1^5}{15CR^2m}\right)^{1/2} \tag{35}$$

For a given impact velocity, Eq. 34 will be used later to determine the maximum contact radius a_1.

To calculate the time of contact, Eq. 32 is written as

$$\dot{\alpha} = V \left[1 - \frac{a^5 k_V^2 \left(\frac{a}{H}\right)}{a_1^5 k_V^2 \left(\frac{a_1}{H}\right)}\right]^{1/2} \tag{36}$$

From Eqs. 36 and 26, the equation relating t and a is

$$t = \int_o^{a(t)} \frac{\left(\frac{2a}{R}\right) k_{\alpha'} \left(\frac{a}{H}\right) da}{V \left[1 - \frac{a^5 k_V^2 \left(\frac{a}{H}\right)}{a_1^5 k_V^2 \left(\frac{a_1}{H}\right)}\right]^{1/2}} \tag{37}$$

Let $x = a/a_1$, then Eq. 37 gives

$$t = \frac{2a_1^2}{RV_H} \int_o^{a/a_1} \frac{x k_{\alpha'} \left(\frac{a_1 x}{H}\right) dx}{\left[k_V^2\left(\frac{a_1}{H}\right) - x^5 k_V^2 \left(\frac{a_1 x}{H}\right)\right]^{1/2}} \tag{38}$$

Eq. 38 can be used to determine the dynamic loading curve and the value of a(t) during impact. When a reaches its maximum a_1, t in Eq. 38 equals half the total contact time T. Thus,

$$T = T_H k_T \left(\frac{a_1}{H}\right) \tag{39}$$

where the associated contact time is

$$T_H = \frac{k a_1^2}{RV_H}, \quad k = 2.94324 \tag{40}$$

and

$$k_T = \frac{4}{k} \int_o^1 \frac{x k_{\alpha'} \left(\frac{a_1 x}{H}\right) dx}{\left[k_V^2\left(\frac{a_1}{H}\right) - x^5 k_V^2 \left(\frac{a_1 x}{H}\right)\right]^{1/2}} \tag{41}$$

If a_1 is determined from Eq. 34, Eq. 39 can be used to determine the contact time T.

All the above complicated nondimensional functions are calculated by using an electronic computer as a function of the nondimensional parameter a/H. Only the curves for $k_V(a/H)$ and $k_T(a/H)$, which are of practical interest here, are shown in Figs. 1 and 2, respectively. The calculation of k_T from Eq. 41 involves singular integrands when x approaches unity. To overcome this difficulty, k_T was calculated by using a common four-point integration method for $x \leq 149/150$. The remaining part of k_T was estimated as

$$\Delta k_T = \frac{4k_{\alpha'}\left(\frac{a_1}{H}\right)}{k k_V\left(\frac{a_1}{H}\right)} \left[I - \int_o^{149/150} \frac{x dx}{(1 - x^5)^{1/2}}\right] \tag{42}$$

where the Beta integral is

$$I = \int_0^1 \frac{x dx}{(1 - x^5)^{1/2}} = 0.73581 \tag{43}$$

The calculated value of k_T (Fig. 2) indicates that the contact time decreases with increasing a/H. In other words, for the same impact velocity and indenter, the contact time decreases with decreasing plate thickness.

The above results can be applied to the problem that a sphere of radius R impinges on the surface of an elastic plate of thickness H overlying a rigid foundation. A trial-and-error method similar to that used by Tsai [3] is to be described here for determining dynamic quantities associated with the problem. For a given V, R, and H, the maximum contact radius a_1 may first be estimated from Eq. 35 by choosing V_H less than V. For this a_1, a_1/H is calculated to determine the value of k_V in Fig. 1. These k_V and V_H values are then substituted in Eq. 34 to see if their product equals the given value V. Usually two or three such trials would give a sufficiently accurate result. Once a_1 has been calculated, T and P can be determined from Eqs. 39 and 29, respectively. Furthermore, the stresses can be calculated as suggested by Tsai [3]. The stress which is of importance in the present work is the following maximum tensile stress along the contact circle

$$\sigma_{rr}(a) = - \frac{1 - 2\nu}{2\pi a^2} P_H k_\sigma \left(\frac{a}{H}\right) \tag{44}$$

The curve for the stress correction factor $k_\sigma(a/H)$ was determined in earlier work [2, 3]. All the results obtained above reduce to the Hertz impact solutions when a/H approaches zero. The procedures suggested above are used to calculate the contact times and critical stresses in the following experiments.

EXPERIMENTAL

The present experiment falls into two parts, namely, the recording of contact times and the measurement of fracture velocities for impinging large steel balls on thin glass plates overlying a large steel block.

The specimen used here consists of a glass plate with 6 x 6 in. surface and 6 x 6 x 4 in. stentor-steel block. The plate is Pittsburgh polished plate glass, and its thickness ranges from 1/8 to 1 in. The steel block was hardened by heat treatment, and one of its largest surfaces was first ground by a machine and then finished with a smooth flat lapping plate. The glass plate was placed on top of the smooth surface of the block with thin oil between the surfaces to insure good contacts [8].

To study dynamic behavior of the specimen in terms of the theoretical results obtained above, the contact time for the impact of steel balls on the free surface of the glass plate was measured on a microsecond counter as a function of plate thickness and ball diameter. The experimental arrangement used is similar to that used by Tsai [9] and Lifshitz and Kolsky [10]. In order to make the free surface of the glass plate electrically conductive with minimum effect on the impact process, an extremely thin silver layer is mounted on the plate surface by a shadow casting process. The glass plate is placed in a vacuum chamber and then exposed to silver vapor until a thin silver layer is formed on its surface. The hard steel ball is suspended by a long fine wire, and both the ball and specimen serve as part of an electrical circuit. The electrical circuit is completed when the ball and the specimens are in contact. The circuit is open, however, when they are not in contact. Therefore, each impact produces a voltage signal which is rectangular in shape for a good electrical contact. The signal produced is monitored by a cathode-ray oscilloscope and fed to the electronic counter. For each measurement of the contact time, five readings very close to each other are recorded from the counter. The mean value of these readings is regarded as the measured value of the contact time. To avoid fracture during the elastic impact, the impact velocity is kept sufficiently low, ranging from 7 to 16 in./sec. The ball diameters used are 1-3/4, 2 and 3 in. The contact times measured are plotted in Fig. 3 for various values of plate thickness, impact velocity, and ball diameter. For a 1-3/4-in. diameter ball, the contact time does not vary appreciably with respect to plate thickness. However, the contact time decreases with decreasing plate thickness for the same impact velocity of a 3-in. diameter ball. If a plate is simply supported and subjected to an impact, the contact time would increase with decreasing plate thickness instead of decrease as observed above.

The variation of contact time with respect to plate thickness can be predicted by the theory developed above. To calculate contact time, Young's modulus $E = 8 \times 10^6$ psi and Poisson's ratio $\nu = 0.25$ are used as given by Fung [11]. The calculated contact times are also shown in Fig. 3 for comparison with the experimental data. The predicted contact time for the 3-in. diameter ball also decreases with decreasing plate thickness, and there is a reasonable agreement between theory and experiment as can be seen in Fig. 3. The agreement suggests that in analysis the steel block may be regarded as a rigid body in comparison to the glass plate. The situation is similar to that shown by Conway _et_. _al_., [7].

The strength of glass plates was measure statically as a function of indenter diameter [1]. The contact diameters involved there were small compared to the plate thickness. The smallness of contact diameter enabled the calculations of contact stresses by using the Hertz half-space solutions. For large indenters which produce contact diameters compatible with the plate thickness, a theory was recently developed to account for the effect of plate thickness on the magnitudes of contact stresses [2]. The magnification of the critical stresses due to the thinness of plates [2] was confirmed experimentally in the static measurements of the strength of thin glass plates [3]. The dynamic strength of glass plates was measured only for small indenters where the Hertz impact theory applies [1]. For large indenters, however, the dynamic measurement was not performed because the local stress distribution could not be determined around the impact area in the glass plates [1]. On the basis of the theory developed above, the dynamic local stress distribution can now be determined, and the theory is used to determine the critical stresses in glass plates for the following experiments.

The specimens used in the tests are similar to those for contact time measurements described above. However, the present procedures for determining the impact velocities required to produce fractures in glass plates are similar to those used in an earlier work [1]. The hard steel ball is suspended by a long fine wire and released from an electromagnet to impinge normally onto the surface of the glass plate. From the arrangements the heights of free fall of the steel balls are measured, and this in turn gives the impact velocities. At sufficiently high impact velocity, conical fractures are produced in glass plates. The shape of the fractures is similar to those observed in Refs. 1, 3 and 8. For each impact velocity, the ball impinges 50 times onto the glass surfaces. The impact velocities are chosen to cover the region from where fractures almost never occur to velocities where fractures almost always occur. Therefore, a distribution curve of the percentage of fractures as a function of impact height is obtained from the measurements for each ball diameter. For each experimental distribution curve obtained, a Gaussian distribution curve which closely fits the experimental curve is determined. The mean value of the Gaussian curve is regarded as the mean impact height corresponding to the fracture velocity. The results obtained are shown in Fig. 4 for various values of indenter diameter, impact heights and plate thickness. It can be seen from Fig. 4 that for the same size of ball the heights of fall required to produce fractures are smaller for the 1/8-in. thick glass plates than the other thick glass plates. This is consistent with the fact that for the same indentation pressure the maximum tensile stress is larger in a thin plate than in a thick plate [2, 3]. For each impact velocity corresponding to each measured mean fracture-height, the contact radius is calculated from

Eq. 34 by the method described above. For each contact radius obtained, the critical stress is calculated from Eqs. 44 and 30. These critical stresses are shown in Fig. 5. It can be seen that the critical stresses are decreasing with increasing indenter diameter. This is consistent with those obtained in static measurements [1, 3].

ACKNOWLEDGMENT

The present research was sponsored by the Engineering Research Institute at Iowa State University, Ames, and the National Science Foundation under Research Initiation Grant GK-4671.

REFERENCES

[1] Y. M. Tsai and H. Kolsky: A Theoretical and Experimental Investigation of the Flaw Distribution on Glass Surfaces. Journal of the Mechanics and Physics of Solids, Vol. 15, 1967, p. 29.

[2] Y. M. Tsai: Stress Distributions in Elastic and Viscoelastic Plates Subjected to Symmetrical Rigid Indentations. Quart. Appl. Math, Vol. 27, 1969, p. 371.

[3] Y. M. Tsai: Thickness Dependence of the Indentation Hardness of Glass Plates. International Journal of Fracture Mechanics, Vol. 5, 1969, p. 157.

[4] H. Hertz: Über die Berührung fester elastischer Körper. Journal für die reine und angewandte Mathematik (Crelle), Vol. 92, 1881, p. 156.

[5] A. E. H. Love: <u>A Treatise of Mathematical Theory of Elasticity</u>. Cambridge University Press, Fourth Edition, 1927, p. 198.

[6] Y. M. Tsai: Dynamic Contact Stresses Produced by the Impact of an Axisymmetrical Projectile on an Elastic Half-Space. Iowa State University Engineering Research Institute Preprint ERI-61800, 1969 (To appear in International Journal of Solids and Structures.)

[7] H. D. Conway, H. C. Lee and R. G. Bayer: The Impact Between a Rigid Sphere and a Thin Layer. J. Appl. Mech., Vol. 159, March 1970.

[8] Y. M. Tsai and H. Kolsky: A Study of the Fractures Produced in Glass Blocks by Impact. Journal of the Mechanics and Physics of Solids, Vol. 15, 1967, p. 263.

[9] Y. M. Tsai : Stress Waves Produced by Impact on the Surface of a Plastic Medium. J. Franklin Inst., Vol. 285, No. 3, 1968, p. 204.

[10] J. M. Lifshitz and H. Kolsky: Some Experiments on An Elastic Rebound. Journal of the Mechanics and Physics of Solids, Vol. 12, 1964, p. 35

[11] Y. C. Fung: <u>Foundations of Solid Mechanics</u>. Prentice-Hall, New Jersey, 1965, p. 131.

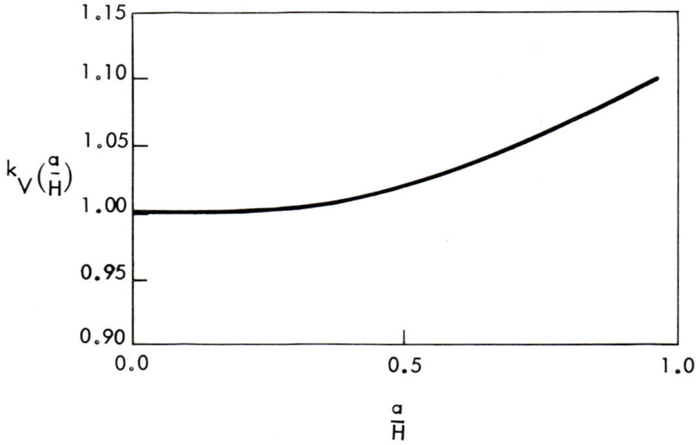

Fig. 1. Velocity Correction Factor $k_V(a/H) = V/V_H$

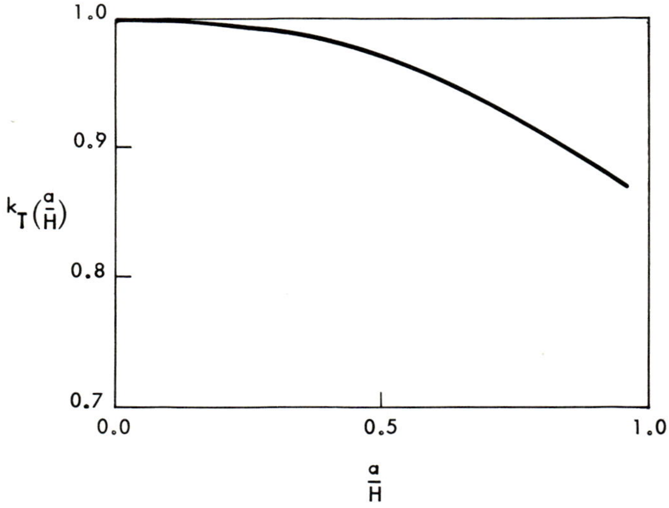

Fig. 2. Contact Time Correction Factor $k_T(a/H) = T/T_H$

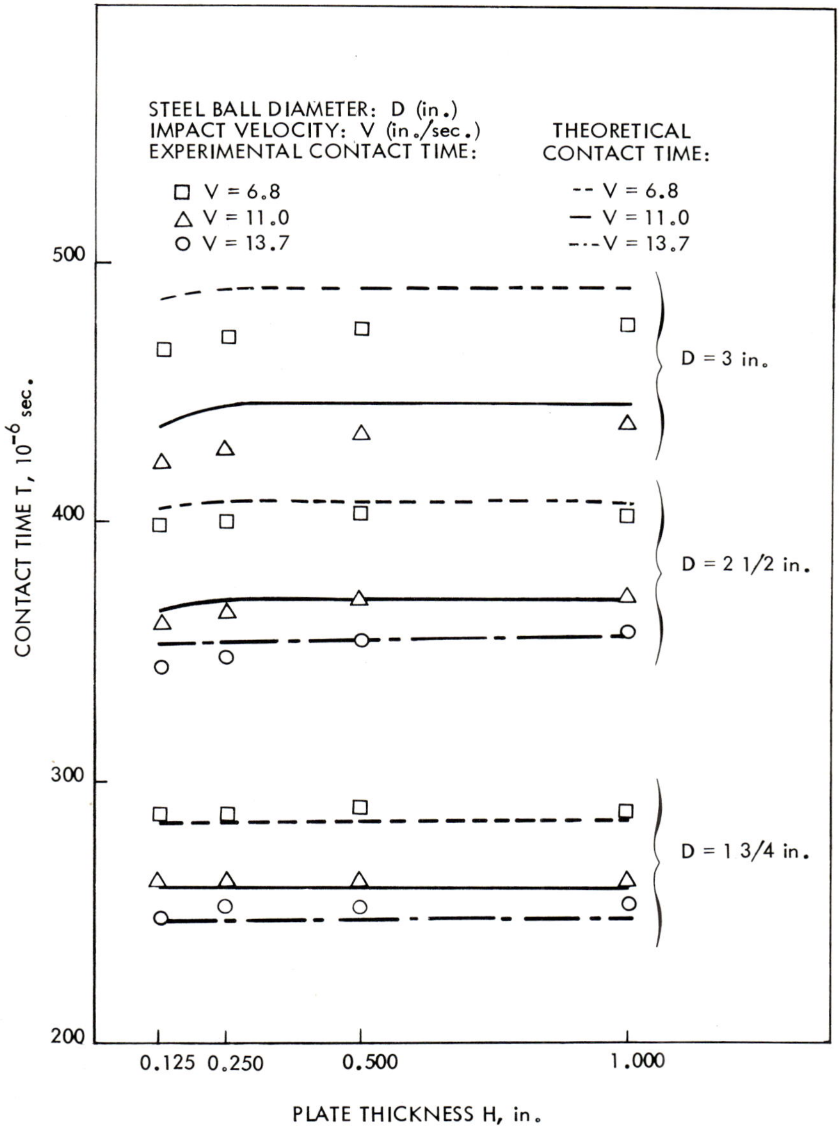

Fig. 3. Contact Time

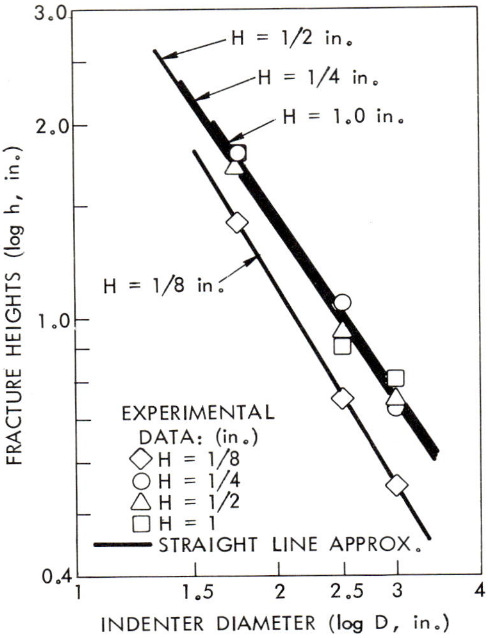

Fig. 4. Fracture Heights

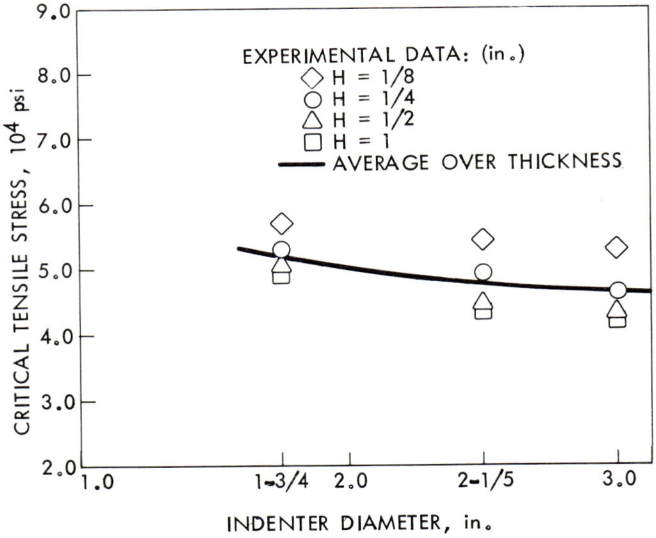

Fig. 5. Fracture Stresses

Authors Index